A TEXTBOOK OF

REFRIGERATION AND AIR CONDITIONING

[For the Students of B.E.; UPSC (Eng
UPSC (Civil Services); Section 'B'
(India) and Diploma Course

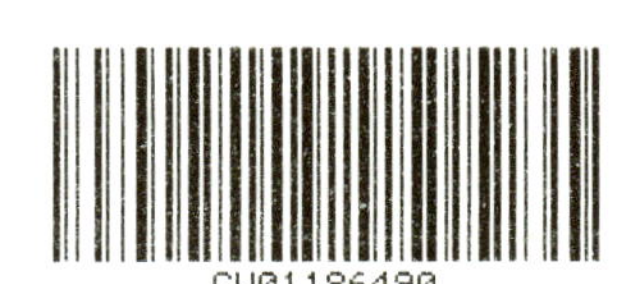

(S.I. UNITS)

R.S. KHURMI
J.K. GUPTA

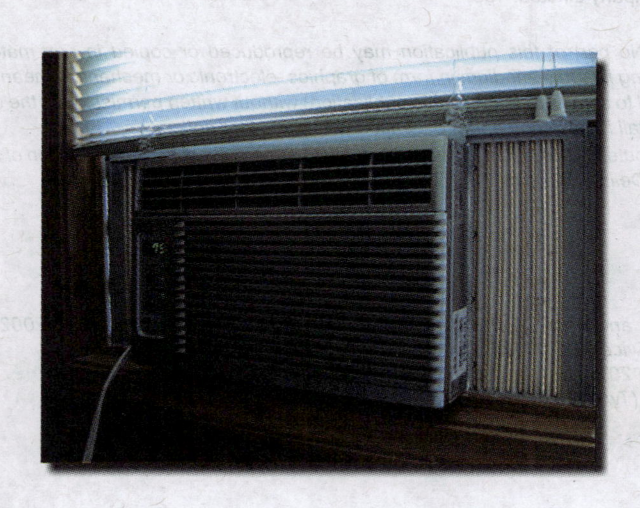

Eurasia Publishing House Private Limited

S Chand And Company Limited

(ISO 9001 Certified Company)

Head Office: Block B-1, House No. D-1, Ground Floor, Mohan Co-operative Industrial Estate, New Delhi – 110 044 | Phone: 011-66672000

Registered Office: A-27, 2nd Floor, Mohan Co-operative Industrial Estate, New Delhi – 110 044 Phone: 011-49731800

www.**schandpublishing.com**; e-mail: **info@schandpublishing.com**

Branches

Chennai	:	Ph: 23632120; chennai@schandpublishing.com
Guwahati	:	Ph: 2738811, 2735640; guwahati@schandpublishing.com
Hyderabad	:	Ph: 40186018; hyderabad@schandpublishing.com
Jalandhar	:	Ph: 4645630; jalandhar@schandpublishing.com
Kolkata	:	Ph: 23357458, 23353914; kolkata@schandpublishing.com
Lucknow	:	Ph: 4003633; lucknow@schandpublishing.com
Mumbai	:	Ph: 25000297; mumbai@schandpublishing.com
Patna	:	Ph: 2260011; patna@schandpublishing.com

First Edition 1987
Subsequent Editions and Reprints 1989, 91, 93, 95, 96, 98 (Twice), 99, 2000, 2001, 2002, 2003, 2004, 2005, 2006, 2007, 2008 (Twice), 2010 (Twice)
Fifth Revised Edition 2011
Reprints 2012, 2013 (Twice), 2014, 2015, 2016 (Twice), 2018 (Twice), 2019, 2020

Reprint 2021

ISBN : 978-81-219-2781-9 **Product Code :** H3RAC62ACRF10ENAE11O

PRINTED IN INDIA

By Vikas Publishing House Private Limited, Plot 20/4, Site-IV, Industrial Area Sahibabad, Ghaziabad – 201 010 and Published by S Chand And Company Limited, A-27, 2nd Floor, Mohan Co-operative Industrial Estate, New Delhi – 110 044.

Preface to the Fifth Edition

We feel satisfied in presenting the new edition of this standard treatise. The favourable and warm reception, which the previous editions of this book have enjoyed all over India and abroad, is a matter of great satisfaction for us.

The present multicolour edition has been thoroughly revised and brought up-to-date. In a chapter of refrigerants, a new article on Substitutes for Chloro-Fluoro - Carbon (CFC) refrigerant has been added, for which we are highly thankful to Mr. Sanjiv Kumar Gupta, B. Tech. (Mechanical Engineering), M.S. (Manufacturing Management). We wish to express our sincere thanks to numerous professors and students, both at home and abroad, for sending their valuable suggestions and recommending the book to their students and friends. We hope, that they will continue to patronise this book in the future also.

We also like to express our sincere thanks to Mr. Navin Joshi, Vice-President (Publishing) for his personal interest taken in printing this book. Our grateful thanks are due to the Editorial Staff of S. Chand & Company Ltd., especially Mr. Shishir Bhatnagar, Manager (Pre-press), Mr. Rupesh Gupta, Subject Editor for their help in conversion of the book into multicolour edition and Mr. Dhan Singh Karki for designing and layouting of this book.

Any errors, omissions and suggestions, for the improvement of this volume brought to our notice, will be thankfully acknowledged and incorporated in the next edition.

R.S. KHURMI
J.K. GUPTA

Contents

Chapter No.		Pages

1. Introduction 1–37

9. Refrigerant Compressors 316 – 357

10. Condensers 358 – 375

11. Evaporators 376 – 397

18. Air Conditioning Systems 549 – 596

19. Cooling Load Estimation 597 – 640

20. Ducts 641 – 698

CHAPTER

1 Introduction

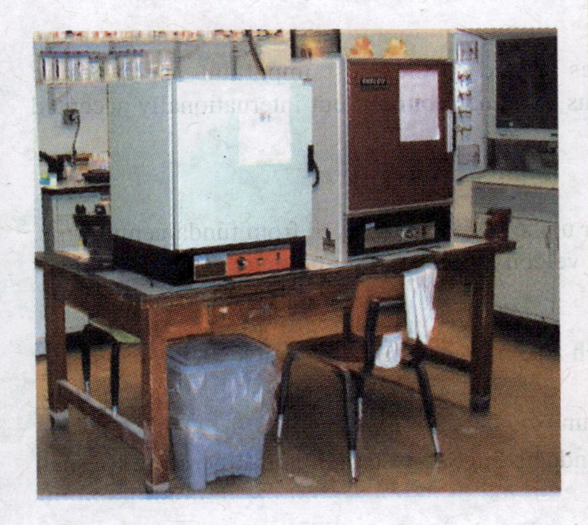

1.1 Definition

The term '*refrigeration*' may be defined as the process of removing heat from a substance under controlled conditions. It also includes the process of reducing and maintaining the temperature of a body below the general temperature of its surroundings. In other words, the refrigeration means a continued extraction of heat from a body whose temperature is already below the temperature of its surroundings.

For example, if some space (say in cold storage) is to be kept at – 2°C (271K), we must continuously extract heat which flows into it due to leakage through the walls and also the heat which is brought into it with the articles stored after the temperature is once reduced to – 2°C (271K). Thus in a refrigerator, heat is virtually being pumped from a lower temperature to a higher temperature. According to *Second Law of Thermodynamics, this process can only be performed with the aid of some external work. It is thus obvious that supply of power (say electric motor) is regularly required to drive a refrigerator. Theoretically, a refrigerator is a reversed heat engine or a heat pump which pumps heat from a cold body and delivers it to a hot body. The substance which works in a heat pump to extract heat from a cold body and to deliver it to a hot body is called a *refrigerant*.

The refrigeration system is known to the man since the middle of nineteenth century. The scientists, of the time, developed a few stray machines to achieve some pleasure. But it paved the way by inviting the attention of scientists for proper studies and research. They were able to build a reasonably reliable machine by the end of nineteenth century for refrigeration jobs. But with the advent of efficient rotary compressors and gas turbines, the science of refrigeration reached the present height. Today it is used for the manufacture of ice and similar products. It is also widely used for the cooling of storage chambers in which perishable foods, drinks and medicines are stored. The refrigeration has also wide applications in submarine ships, aircraft and rockets.

1.2 Fundamental Units

The measurement of physical quantities is one of the most important operations in engineering. Every quantity is measured in terms of some arbitrary, but internationally accepted units, called *fundamental units.*

1.3 Derived Units

Some units are expressed in terms of other units, which are derived from fundamental units, are known as *derived units e.g.* the unit of area, velocity, acceleration, pressure etc.

1.4 System of Units

There are only four systems of units, which are commonly used and universally recognised. These are known as:

1. C.G.S. units, 2. F.P.S. units, 3. M.K.S. units and 4. S.I. units.

Since the present courses of studies are conducted in S.I. system of units, therefore we shall discuss this system of unit only.

1.5 S.I. Units (International System of Units)

The 11th General Conference** of Weights and Measures have recommended a unified and systematically constituted system of fundamental and derived units for international use. This system is now being used in many countries. In India, the standards of Weights and Measures Act, 1956 (vide which we switched over to M.K.S. units) has been revised to recognise all the S.I. units in industry and commerce.

Stopwatch Simple balance

* Refer Art. 1.37.

** It is known as General Conference of Weights and Measures (G.C.W.M.). It is an international organisation, of which most of the advanced and developing countries (including India) are members. The conference has been entrusted with the task of prescribing definitions for various units of weights and measures, which are the very basis of science and technology today.

In this system of units, there are seven fundamental units and two supplementary units, which cover the entire field of science and engineering. These units are shown in Table 1.1.

Table 1.1. Fundamental and supplementary units.

S.No.	Physical quantity	Unit
	Fundamental units	
1.	Length (l)	Metre (m)
2.	Mass (m)	Kilogram (kg)
3.	Time (t)	Second (s)
4.	Temperature (T)	Kelvin (K)
5.	Electric current (I)	Ampere (A)
6.	Luminous intensity (I_v)	Candela (cd)
7.	Amount of substance (n)	Mole (mol)
	Supplementary units	
8.	Plane angle (α, β, θ, ϕ)	Radian (rad)
9.	Solid angle (Ω)	Steradian (sr)

The derived units, which will be commonly used in this book, are given in Table1.2.

Table 1.2. Derived units.

S.No.	Quantity	Symbol	Unit
1.	Linear velocity	V	m/s
2.	Linear acceleration	a	m/s^2
3.	Angular velocity	ω	rad/s
4.	Angular acceleration	α	rad/s^2
5.	Mass density	ρ	kg/m^3
6.	Force, Weight	F, W	N ; 1N = 1kg-m/s^2
7.	Pressure	p	N/m^2
8.	Work, Energy, Enthalpy	W,E,H	J ; 1J = 1N-m
9.	Power	P	W ; 1W = 1J/s
10.	Absolute or dynamic viscosity	μ	N–s/m^2
11.	Kinematic viscosity	ν	m^2/s
12.	Frequency	f	Hz ; 1Hz = 1 cycle/s
13.	Gas constant	R	J/kg K
14.	Thermal conductance	h	W/m^2 K
15.	Thermal conductivity	k	W/m K
16.	Specific heat	c	J/kg K
17.	Molar mass or Molecular mass	M	kg/mol

1.6 Metre

The metre is defined as the length equal to 1 650 763.73 wave lengths in vacuum of the radiation corresponding to the transition between the levels 2 p_{10} and $5d_5$ of the Krypton – 86 atom.

1.7 Kilogram

The kilogram is defined as the mass of the international prototype (standard block of platinum–iridium alloy) of the kilogram, kept at the International Bureau of Weights and Measures at Sevres, near Paris.

1.8 Second

The second is defined as the duration of 9 192 631 770 periods of the radiation corresponding to the transition between the two hyperfine levels of the ground state of the caesium – 133 atom.

1.9 Kelvin

The kelvin is defined as the fraction 1 / 273.16 of the thermodynamic temperature of the triple point of water.

Note: The triple point of water is taken as a fundamental fixed point having a temperature 273.16 K.

1.10 Presentation of Units and their Values

The frequent changes in the present day life are facilitated by an international body known as International Standard Organisation (ISO) which makes recommendations regarding international standard procedures. The implementation of ISO recommendations, in a country, is assisted by its organisation appointed for the purpose. In India, Bureau of Indian Standards (BIS) previously known as Indian Standards Institution (ISI) has been created for this purpose. We have already discussed that the fundamental units in S.I. units for length, mass and time are metre, kilogram and second respectively. But in actual practice, it is not necessary to express all lengths in metres, all masses in kilograms and all times in seconds. We shall, sometimes, use the convenient units, which are multiples or divisions of our basic units in tens. As a typical example, although the metre is the unit of length, yet a small length of one-thousandth of a metre proves to be more convenient unit, especially in the dimensioning of drawings. Such convenient units are formed by using a prefix in front of the basic units to indicate the multiplier. The full list of these prefixes is given in the following table.

Table 1.3. Prefixes used in basic units.

Factor by which the unit is multiplied	*Standard form*	*Prefix*	*Abbreviation*
1 000 000 000 000	10^{12}	tera	T
1 000 000 000	10^{9}	giga	G
1 000 000	10^{6}	mega	M
1 000	10^{3}	kilo	k
100	10^{2}	hecto*	h
10	10^{1}	deca*	da
0.1	10^{-1}	deci*	d
0.01	10^{-2}	centi*	c
0.001	10^{-3}	milli	m
0.000 001	10^{-6}	micro	μ
0.000 000 001	10^{-9}	nano	n
0.000 000 000 001	10^{-12}	pico	p

* These prefixes are generally becoming obsolete probably due to possible confusion. Moreover, it is becoming a conventional practice to use only those powers of ten which conform to 10^{3x}, where x is a positive or negative whole number.

With rapid development of Information Technology, computers are playing a major role in analysis, synthesis and design of machines.

1.11 Rules for S.I. Units

The eleventh General Conference of Weights and Measures recommended only the fundamental and derived units for S.I. system. But it did not elaborate the rules for the usage of the units. Later on many scientists and engineers held a number of meetings for the style and usage of S.I. units. Some of the decisions of the meetings are as follows:

1. For numbers having five or more digits, the digits should be placed in groups of three separated by spaces* (instead of commas) counting both to the left and right to the decimal point.

2. In a four digit number,** the space is not required unless the four digit number is used in a column of numbers with five or more digits.

3. A dash is to be used to separate units that are multiplied together. For example, newton × metre is written as N-m. It should not be confused with mN, which stands for millinewton.

4. Plurals are never used with symbols. For example, metre or metres are written as m.

5. All symbols are written in small letters except the symbols derived from the proper names. For example, N for newton and W for watt.

6. The units with names of scientists should not start with capital letter when written in full. For example, 90 newton and not 90 Newton.

At the time of writing this book, the authors sought the advice of various international authorities, regarding the use of units and their values. Keeping in view the international reputation of the authors, as well as international popularity of their books, it was decided to present units*** and their values as per recommendations of ISO and BIS. It was decided to use :

* In certain countries, comma is still used as the decimal mark.

** In certain countries, a space is used even in a four digit number.

*** In some of the question papers of the universities and other examining bodies, standard values are not used. The authors have tried to avoid such questions in the text of the book. However, at certain places, the questions with sub-standard values have to be included, keeping in view the merits of the question from the reader's angle.

4500	not	4 500	or	4,500
7 589 000	not	7589000	or	7,58,90,00
0.012 55	not	0.01255	or	.01255
30×10^6	not	3,00,00,000	or	3×10^7

The above mentioned figures are meant for numerical values only. Now let us discuss about the units. We know that the fundamental units in S.I. system of units for length, mass and time are metre, kilogram and second respectively. While expressing these quantities, we find it time consuming to write the units such as metres, kilograms and seconds, in full, every time we use them. As a result of this, we find it quite convenient to use some standard abbreviations.

We shall use :

m	for metre or metres
km	for kilometre or kilometres
kg	for kilogram or kilograms
t	for tonne or tonnes
s	for second or seconds
min	for minute or minutes
N-m	for newton × metres (*e.g.* work done)
kN-m	for kilonewton × metres
rev	for revolution or revolutions
rad	for radian or radians

1.12 Mass and Weight

Sometimes much confusion and misunderstanding is created, while using the various systems of units in the measurement of force and mass. This happens, because of the lack of clear understanding of the difference between mass and weight. The following definitions of mass and weight should be clearly understood.

A man whose mass is 60 kg weighs 588.6 N (60 × 9.81 m/s²) on earth, approximately 96 N (60 × 1.6 m/s²) on moon and zero in space. But mass remains the same everywhere.

1. ***Mass.*** It is the amount of matter contained in a given body, and does not vary with the change in its position on the earth's surface. The mass of a body is measured by direct comparison with a standard mass by using a lever balance.

2. ***Weight.*** It is the amount of pull, which the earth exerts upon a given body. Since the pull varies with the distance of the body from the centre of the earth, therefore weight of the body will also vary with its position on the earth's surface (say latitude and elevation). It is thus obvious, that the weight is a *force.*

The earth's pull in metric units, at sea level and 45° latitude, has been adopted as one force unit and named as one kilogram of force. Thus it is a definite amount of force. But, unfortunately, it has the same name as the unit of mass. The weight of a body is measured by the use of a spring balance, which indicates the varying tension in the spring as the body is moved from place to place.

Note: The confusion in the units of mass and weight is eliminated, to a great extent, in S.I. units. In this system, mass is taken in kg and weight in newtons. The relation between the mass (m) and the weight (W) of a body is

$$W = m\,g \quad \text{or} \quad m = W/g$$

where W is in newtons, m is in kg and g is the acceleration due to gravity in m/s^2.

1.13 Force

It is an important factor in the field of Engineering science, which may be defined as an agent which produces or tends to produce, destroy or tends to destroy the motion. According to Newton's Second Law of Motion, the applied force or impressed force is directly proportional to the rate of change of momentum. We know that

Momentum = Mass × Velocity

Let m = Mass of the body,

u = Initial velocity of the body,

v = Final velocity of the body,

a = Constant acceleration, and

t = Time required to change the velocity from u to v.

$\therefore$ Change of momentum $= m\,v - m\,u$

and rate of change of momentum $= \dfrac{mv - mu}{t} = \dfrac{m(v-u)}{t} = ma$ $\quad \ldots \left(\because \dfrac{v-u}{t} = a\right)$

or Force, $F \propto m\,a$ or $F = k\,m\,a$

where k is a constant of proportionality.

For the sake of convenience, the unit of force adopted is such that it produces a unit acceleration to a body of unit mass.

$$\therefore \quad F = m\,a = \text{Mass} \times \text{Acceleration}$$

In S.I. system of units, the unit of force is called newton (briefly written as N). A *newton may be defined as the force while acting upon a mass of one kg produces an acceleration of 1 m/s^2 in the direction of which it acts.* Thus

$$1\text{ N} = 1\text{ kg} \times 1\text{ m/s}^2 = 1\text{ kg-m/s}^2$$

1.14 Absolute and Gravitational Units of Force

We have already discussed that when a body of mass 1 kg is moving with an acceleration of 1 m/s^2, the force acting on the body is 1 newton (briefly written as 1 N). Therefore, when the same body is moving with an acceleration of 9.81 m/s^2, the force acting on the body is 9.81 N. But we denote 1 kg mass attracted towards the earth with an acceleration of 9.81 m/s^2 as 1 kilogram-force (briefly written as kgf) or 1 kilogram-weight (briefly written as kg-wt). It is thus obvious, that

$$1 \text{ kgf} = 1 \text{ kg} \times 9.81 \text{ m/s}^2 = 9.81 \text{ kg-m/s}^2 = 9.81 \text{ N}$$

$$\ldots (\because 1\text{N} = 1 \text{ kg-m/s}^2)$$

The above unit of force *i.e.* kilogram force (kgf) is called *gravitational* or *engineer's unit of force*, whereas newton is the *absolute* or *scientific* or *S.I. unit of force*. It is thus obvious, that the gravitational or engineer's units of force are '*g*' times the unit of force in the absolute or S.I. units.

It will be interesting to know that the *mass of the body in absolute units is numerically equal to the weight of the same body in gravitational units*. For example, consider a body whose mass,

$$m = 100 \text{ kg}$$

Therefore the force, with which the body will be attracted towards the centre of the earth,

$$F = ma = mg = 100 \times 9.81 = 981 \text{ N}$$

Now, as per definition, we know that the weight of a body is the force, by which it is attracted towards the centre of the earth. Therefore weight of the body,

$$W = 981 \text{ N} = 981 / 9.81 = 100 \text{ kgf} \quad \ldots (\because 1\text{kgf} = 9.81 \text{ N})$$

In brief, the weight of a body of mass m kg at a place where gravitational acceleration is 'g' m/s² is $m.g$ newtons.

1.15 Thermodynamic Systems

The *thermodynamic system* (or simply known as *system*) may be broadly defined as a *definite area* or *a space* where some thermodynamic process is taking place. It is a region where our attention is focussed for studying a thermodynamic process. A little observation will show that a thermodynamic system has its boundaries and anything outside the boundaries is called its *surroundings* as shown in Fig. 1.1. These boundaries may be *fixed* like that of a tank enclosing a certain mass of compressed gas, or *movable* like the boundary of a certain volume of liquid in a pipe line.

The thermodynamic systems may be classified into the following three groups :

1. Closed system ; 2. Open system ; and 3. Isolated system.

These systems are discussed, in detail, as follows :

1. *Closed system.* This is a system of fixed mass and identity whose boundaries are determined by the space of the matter (working substance) occupied in it.

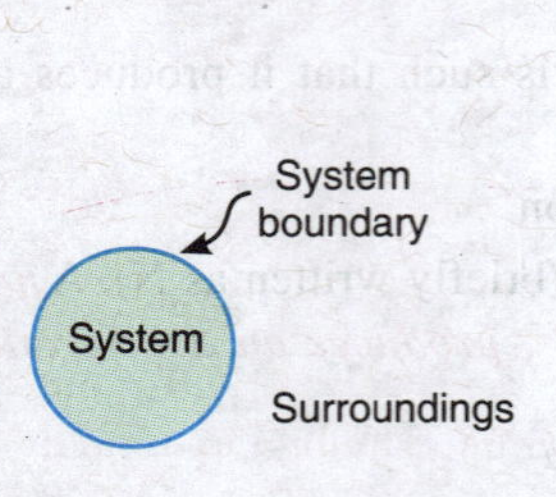

Fig. 1.1. Thermodynamic system.

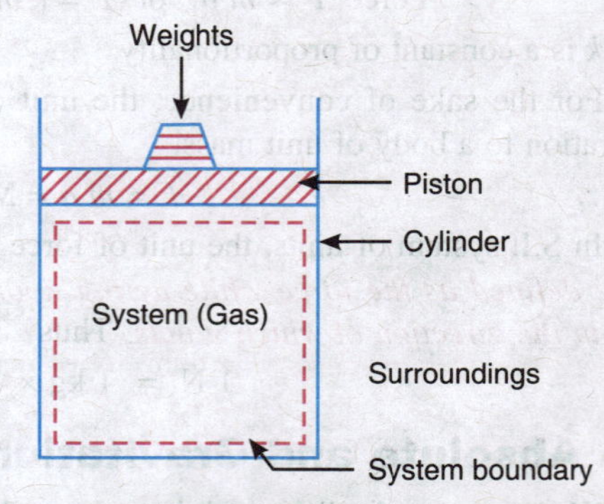

Fig. 1.2. Closed thermodynamic system.

A closed system is shown in Fig. 1.2. The gas in the cylinder is considered as a system. If heat is supplied to the cylinder from some external source, the temperature of the gas will increase and the piston will rise.

As the piston rises, the boundary of the system moves. In other words, the heat and work energy crosses the boundary of the system during this process, but there is no addition or loss of the original mass of the working substance. It is thus obvious, that the mass of the working substance, which comprises the system, is fixed.

Thus, a closed system does not permit any mass transfer across its boundary, but it permits transfer of energy (heat and work).

2. *Open system.* In this system, the mass of the working substance crosses the boundary of the system. Heat and work may also cross the boundary. Fig. 1.3 shows the diagram of an air compressor which illustrates an open system.

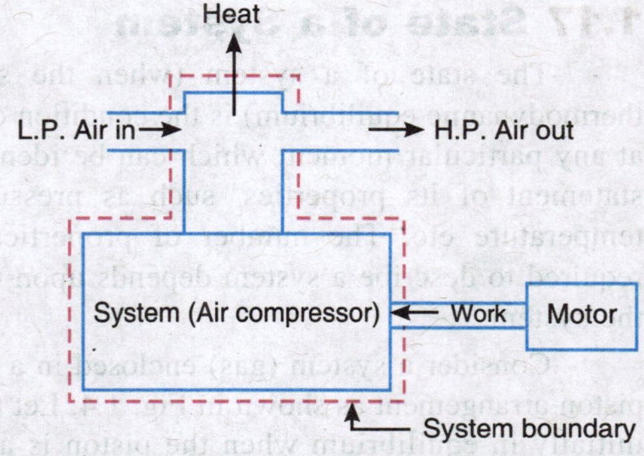

Fig. 1.3. Open thermodynamic system.

The working substance crosses the boundary of the system as the low pressure (L.P.) air enters the compressor and leaves the high pressure (H.P.) air. The work crosses the boundary of the system through the driving shaft and the heat is transferred across the boundary from the cylinder walls.

Thus, an open system permits both mass and energy (heat and work) transfer across the boundaries and the mass within the system may not be constant.

Note: An open system may be referred to as *control volume.* An open system is equivalent in every respect to a control volume, but the term open system is used throughout this text as it specifically implies that the system can have mass and energy crossing the system boundary.

3. *Isolated system.* A system which is completely uninfluenced by the surroundings is called an isolated system. It is a system of fixed mass and no heat or work energy cross its boundary. In other words, an isolated system does not have transfer of either mass or energy (heat or work) with the surroundings. An open system with its surroundings (known as an universe) is an example of an isolated system.

Note: The practical examples of isolated system are rare. The concept of this system is particularly useful in formulating the principles derived from the Second Law of Thermodynamics.

1.16 Properties of a System

The state of a system may be identified or described by certain observable quantities such as volume, temperature, pressure and density etc. All the quantities, which identify the state of a system, are called *properties*.

Note: Thermodynamics deals with those quantities also which are not properties of any system. For example, when there is a flow of energy between a system and its surroundings, the energy transferred is not a property of the system or its surroundings.

The thermodynamic properties of a system may be divided into the following two general classes:

1. *Extensive properties.* A quantity of matter in a given system is divided, notionally into a number of parts. The properties of the system, whose value for the entire system is equal to the sum of their values for the individual parts of the system are called *extensive properties, e.g.* total volume, total mass and total energy of a system are its extensive properties.

2. *Intensive properties.* It may be noticed that the temperature of the system is not equal to the sum of the temperatures of its individual parts. It is also true for pressure and density of the system. Thus properties like temperature, pressure and density are called *intensive properties*.

Note: The ratio of any extensive property of a system to the mass of the system is called an average specific value of that property (also known as intensive property) *e.g.* specific volume of a system (v_s) is the ratio of the total volume (v) of the system to its total mass (m). Mathematically,

$$v_s = v/m$$

The specific volume is an intensive property.

1.17 State of a System

The state of a system (when the system is in thermodynamic equilibrium) is the condition of the system at any particular moment which can be identified by the statement of its properties, such as pressure, volume, temperature etc. The number of properties which are required to describe a system depends upon the nature of the system.

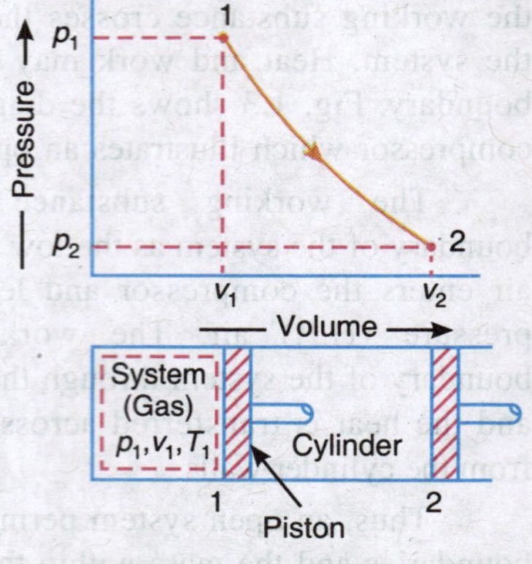

Fig. 1.4. State of a system.

Consider a system (gas) enclosed in a cylinder and piston arrangement as shown in Fig. 1.4. Let the system is initially in equilibrium when the piston is at position 1, represented by its properties p_1, v_1 and T_1. When the system expands, the piston moves towards right and occupies the final position at 2. At this, the system is finally in the equilibrium state represented by the properties p_2, v_2 and T_2. The initial and final states, on the pressure-volume diagram, are shown in Fig. 1.4.

1.18 Temperature

It is an intensive thermodynamic property, which determines the degree of hotness or the level of heat intensity of a body. A body is said to be at a *high temperature* or *hot*, if it shows high level of heat intensity in it. Similarly, a body is said to be at a *low temperature* or *cold*, if it shows a low level of heat intensity.

The temperature of a body is measured with the help of an instrument known as *thermometer* which is in the form of a glass tube containing mercury in its stem. Following are the two commonly used scales for measuring the temperature of a body :

1. Celsius or centigrade scale, and 2. Fahrenheit scale.

Each of these scales is based on two fixed points known as *freezing point of water* under atmospheric pressure or *ice point* and the *boiling point of water* or *steam point*.

1. *Celsius or centigrade scale.* This scale was first used by Celsius in 1742. This scale is mostly used by engineers and scientists. The freezing point of water on this scale is marked as zero, and the boiling point of water as 100. The space between these two points has 100 equal divisions, and each division represents one degree Celsius (briefly written as °C).

2. *Fahrenheit scale.* This scale was first used in 1665. In this scale, the freezing point of water is marked as 32 and the boiling point of water as 212. The space between these two points has 180 equal divisions and each division represents one degree Fahrenheit (briefly written as °F).

Note: The relation between Celsius scale and Fahrenheit scale is given by :

$$\frac{C}{100} = \frac{F-32}{180} \quad \text{or} \quad \frac{C}{5} = \frac{F-32}{9}$$

1.19 Absolute Temperature

As a matter of fact, the zero readings of Celsius and Fahrenheit scales are chosen arbitrarily for the purpose of simplicity. It helps us in our calculations, when changes of temperature in a process are known. But, whenever the value of temperature is used in equations relating to fundamental laws, then the value of temperature, whose reference point is true zero or absolute zero, is used. The temperature, below which the temperature of any substance can not fall, is known as *absolute zero temperature*.

The absolute zero temperature, for all sorts of calculations, is taken as –273°C in case of Celsius scale and – 460°F in case of Fahrenheit scale. The temperatures measured from this zero are called *absolute temperatures*. The absolute temperature in Celsius scale is called degrees Kelvin (briefly written as K)*; such that K = °C + 273. Similarly, absolute temperature in Fahrenheit scale is called degrees Rankine (briefly written as °R); such that °R = °F + 460.

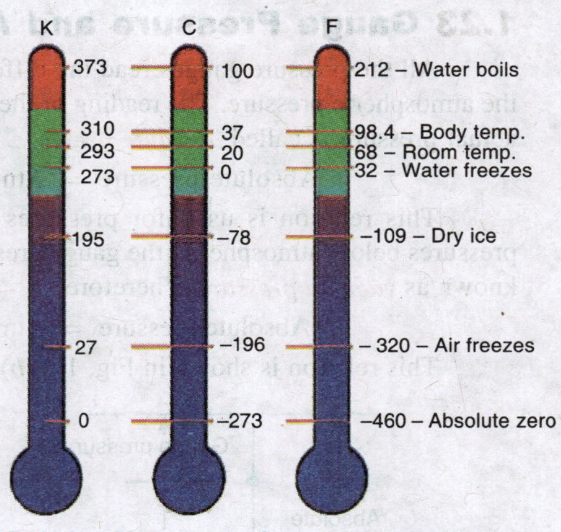

Relation between various temperatures.

1.20 Thermodynamic Equilibrium

A system is said to be in thermodynamic equilibrium, if it satisfies the following three requirements of equilibrium.

1. ***Mechanical equilibrium.*** A system is said to be in mechanical equilibrium, when there is no unbalanced forces acting on any part of the system or the system as a whole. The pressure in the system is same at all points and does not change with time.

2. ***Thermal equilibrium.*** A system is said to be in thermal equilibrium, when there is no temperature difference between the parts of the system or between the system and the surroundings. In other words, the temperature of the system does not change with time and has the same value at all points of the system.

3. ***Chemical equilibrium.*** A system is said to be in chemical equilibrium, when there is no chemical reaction within the system and the chemical composition is same throughout the system, which does not change with time.

1.21 Equality of Temperature

Consider two bodies of the same or different materials, one hot and the other cold. When these bodies are brought in contact, the hot body becomes colder, and the cold body becomes warmer. If these bodies remain in contact for some time, a state reaches when there is no further observable change in the properties of the two bodies. This is a state of thermal equilibrium and at this stage, the two bodies have the equal temperatures. It thus follows that when two bodies are in thermal equilibrium with each other, their temperatures are equal.

1.22 Pressure

The term 'pressure' may be defined as the normal force per unit area. The unit of pressure depends upon the units of force and area.

In S.I. system of units, the practical unit of pressure is N/mm^2, N/m^2, kN/m^2, MN/m^2 etc. But sometimes a bigger unit of pressure (known as bar) is used, such that,

$$1 \text{ bar} = 1 \times 10^5 \text{ N/m}^2 = 0.1 \times 10^6 \text{ N/m}^2 = 0.1 \text{ MN/m}^2$$

Sometimes the pressure is expressed in another unit, called Pa (named after Pascal) and kPa, such that

$$1 \text{ Pa} = 1 \text{ N/m}^2 \quad \text{and} \quad 1 \text{ kPa} = 1 \text{kN/m}^2$$

* In S.I. units, degrees Kelvin is not written as °K but only K.

1.23 Gauge Pressure and Absolute Pressure

All the pressure gauges read the difference between the actual pressure in any system and the atmospheric pressure. The reading of the pressure gauge is known as *gauge pressure*, while the actual pressure is called *absolute pressure*. Mathematically,

Absolute pressure = Atmospheric pressure + Gauge pressure

This relation is used for pressures above atmospheric, as shown in Fig. 1.5 (*a*). For pressures below atmospheric, the gauge pressure will be negative. This negative gauge pressure is known as *vacuum pressure*. Therefore

Absolute pressure = Atmospheric pressure − Vacuum pressure

This relation is shown in Fig. 1.5 (*b*).

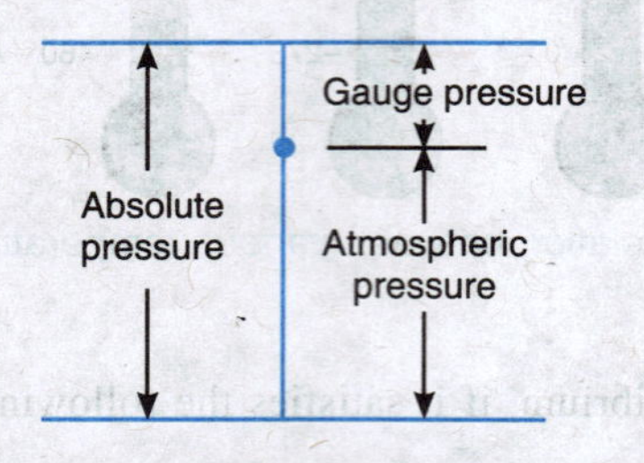

(*a*) Relation between absolute, atmospheric and gauge pressure.

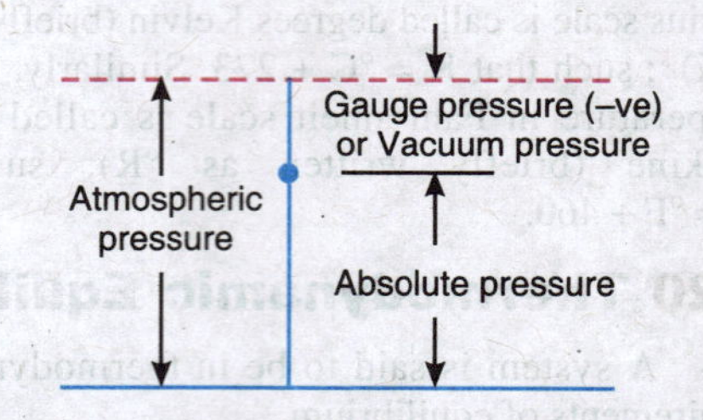

(*b*) Relation between absolute, atmospheric and vacuum pressure.

Fig. 1.5

The standard value of atmospheric pressure is taken as 1.013 bar (or 760 mm of Hg) at sea level.

Note: We know that 1 bar = 10^5 N/m²

∴ Atmospheric pressure = $1.013 \times 10^5 = 1013 \times 10^2$ N/m²

We also know that atmospheric pressure

= 760 mm of Hg

∴ 1 mm of Hg = $1013 \times 10^2 / 760 = 133.3$ N/m²

or 1 N/m² = $760 / 1013 \times 10^2 = 7.5 \times 10^{-3}$ mm of Hg

1.24 Normal Temperature and Pressure (N.T.P.)

The conditions of temperature and pressure at 0°C (273K) temperature and 760 mm of Hg pressure are termed as *normal temperature* and *pressure* (briefly written as N.T.P.).

1.25 Standard Temperature and Pressure (S.T.P.)

The temperature and pressure of any gas, under standard atmospheric conditions, is taken as 15°C (288K) and 760 mm of Hg respectively.

1.26 Energy

The energy is defined as the capacity to do work. In other words, a system is said to possess energy when it is capable of doing work. The energy possessed by a system is of the following two types:

1. Stored energy, and 2. Transit energy (or energy in transition)

The *stored energy* is the energy possessed by a system within its boundaries. The potential energy, kinetic energy and internal energy are the examples of stored energy.

The *transit energy* (or energy in transition) is the energy possessed by a system which is capable of crossing its boundaries. The heat, work and electrical energy are the examples of transit energy.

It may be noted that only the stored energy is a thermodynamic property whereas the transit energy is not a thermodynamic property as it depends upon the path.

1.27 Types of Stored Energy

We have discussed above that the potential energy, kinetic energy and internal energy are the different types of stored energy. These energies are discussed, in detail, as follows:

1. *Potential energy.* It is the energy possessed by a body or a system, for doing work, by virtue of its position above the ground level. For example, a body raised to some height above the ground level possesses potential energy because it can do some work by falling on earth's surface.

Let W = Weight of the body,

m = Mass of the body,

z = Distance through which the body falls, and

g = Acceleration due to gravity = 9.81 m/s^2.

$\therefore$ Potential energy, $PE = W\,z = m\,g\,z$

It may be noted that

(*a*) When W is in newtons and z in metres, then potential energy will be in N-m.

(*b*) When m is in kg and z in metres, then the potential energy will also be in N-m, as discussed below:

We know that potential energy,

$$PE = m\,g\,z = \text{kg} \times \frac{\text{m}}{\text{s}^2} \times \text{m} = \text{N-m} \quad \ldots \left(\because 1\,\text{N} = \frac{1\,\text{kg-m}}{\text{s}^2}\right)$$

2. *Kinetic energy.* It is the energy possessed by a body or a system, for doing work, by virtue of its mass and velocity of motion.

Let m = Mass of the body, and

V = Velocity of the body.

When m is in kg and V is in m/s, then kinetic energy will be in N-m, as discussed below :

We know that kinetic energy,

$$KE = \frac{1}{2} m\,V^2 = \text{kg} \times \frac{\text{m}^2}{\text{s}^2} = \frac{\text{kg-m}}{\text{s}^2} \times \text{m} = \text{N-m} \quad \ldots \left(\because 1\,\text{N} = \frac{1\,\text{kg-m}}{\text{s}^2}\right)$$

3. *Internal energy.* It is the energy possessed by a body or a system due to its molecular arrangement and motion of the molecules. It is usually represented by U.

In the study of thermodynamics, we are mainly concerned with the change in internal energy (dU) which depends upon the change in temperature of the system.

Notes: 1. The total energy of the system (E) is equal to the sum of the above three types of energies. Mathematically,

$$E = PE + KE + U = m\,g\,z + \frac{1}{2} \times m\,V^2 + U$$

Any other form of the energy such as chemical energy, electrical energy etc. is neglected. For unit mass, the above expression is written as

$$e = pe + ke + u = g\,z + \frac{V^2}{2} + u$$

2. When the system is stationary and the effect of gravity is neglected, then $PE = 0$, and $KE = 0$. In such a case,

$$E = U \quad \text{or} \quad e = u$$

1.28 Heat

The heat is defined as the energy transferred, without transfer of mass across the boundary of a system because of a temperature difference between the system and the surroundings. It is usually represented by Q and is expressed in joule (J) or kilo-joule (kJ).

The heat can be transferred in three distinct ways, *i.e.* conduction, convection and radiation. The transfer of heat through solids takes place by *conduction*, while the transfer of heat through fluids is by *convection*. The *radiation* is an electromagnetic wave phenomenon in which energy can be transported through transparent substances and even through a vacuum. These three modes of heat transfer are quite different, but they have one factor in common. All these modes occur across the surface area of a system because of a temperature difference between the system and the surroundings.

The following points are worth noting about heat :

1. The heat is transferred across a boundary from a system at a higher temperature to a system at a lower temperature by virtue of the temperature difference.
2. The heat is a form of transit energy which can be identified only when it crosses the boundary of a system. It exists only during transfer of energy into or out of a system.
3. The heat flowing into a system is considered as *positive* and the heat flowing out of a system is considered as *negative*.

1.29 Sensible Heat

When a substance is heated and the temperature rises as the heat is added, the increase in heat is called *sensible heat*. Similarly, when heat is removed from a substance and the temperature falls, the heat removed (or subtracted) is called sensible heat. It is usually denoted by h_f.

Thus, the sensible heat may be defined as the heat which causes a change in temperature in a substance. For example, the heat absorbed in heating of water upto the boiling temperature is the sensible heat.

1.30 Latent Heat

All pure substances are able to change their state. Solids become liquids and liquids become gas. These changes of state occur at the same temperature and pressure combinations for any given substance. It takes the addition of heat or the removal of heat to produce these changes. The heat which brings about a change of state with no change in temperature is called *latent* (or hidden) *heat*. It is usually denoted by h_{fg}.

The latent heat of ice is 335 kJ/kg. This means that the heat absorbed by 1 kg of ice to change it into water at 0°C and at atmospheric pressure is 335 kJ. This heat is called *latent heat of fusion* (or melting) *of ice*. The water starts vaporising at 100°C (*i.e.* boiling temperature) and changes its state from water to steam (*i.e.* gaseous form). The heat absorbed during this change of state from liquid to gas is called *latent heat of vaporisation* or *condensation*. The latent heat of vaporisation of water at 100°C and at atmospheric pressure is 2257 kJ/kg.

1.31 Specific Heat

The specific heat of a substance may be broadly defined as the amount of heat required to raise the temperature of a unit mass of any substance through one degree. It is generally denoted by c. In S.I. system of units, the unit of specific heat (c) is taken as kJ/kg K. If m kg of a substance of specific heat c is required to raise the temperature from an initial temperature of T_1 to a final temperature of T_2, then heat required,

$$Q = m\,c\,(T_2 - T_1)\text{ kJ}$$

where T_1 and T_2 may be either in °C or K.

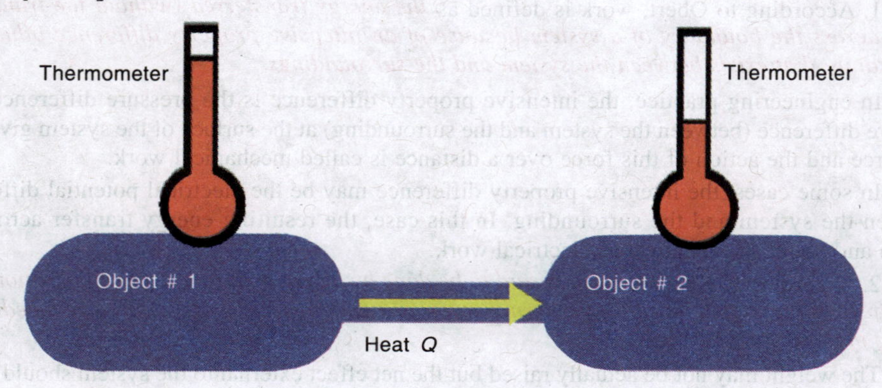

Heat is transferred from warmer object to cooler object.

The average values of specific heats for some commonly used substances are given in following table :

Table 1.4. Values of specific heats for some commonly used substances.

Substance	*Specific heat (kJ/kg K)*	*Substance*	*Specific heat (kJ/ kg K)*
Steel	0.490	Air	1.000
Copper	0.406	Nitrogen	1.010
Glass	0.783	Oxygen	0.925
Mercury	0.138	R-12 (Dichloro-difluoro-methane)	0.892
Brick	0.837	R-22 (Monochloro-difluoro-methane)	1.089
Water	4.187	R-717 (Ammonia)	4.606
Ice	2.110	R-744 (Carbon dioxide)	2.512
Steam	2.094	Salt brine 20%	3.560

1.32 Mechanical Equivalent of Heat

It was established by Joule that heat and mechanical energies are mutually convertible. He established, experimentally, that there is a numerical relation between the unit of heat and the unit of work. This relation is denoted by J (named after Joule) and is known as Joule's equivalent or mechanical equivalent of heat.

Note : In S.I. system of units, the unit of work done is joule or kilo joule (such that 1J = 1 N-m or 1kJ = 1kN-m). The unit of heat is also joule or kilo joule. So we can straightway convert heat units into mechanical units and vice versa.

1.33 Work

In mechanics, work is defined as the product of the force (F) and the distance moved (x) in the direction of the force. Mathematically,

$$\text{Work done} = F \times x$$

The unit of work depends upon the unit of force and the distance moved. In S.I. system of units, the practical unit of work is newton-metre (briefly written as N-m). The work of 1 N-m is known as joule (briefly written as J) such that 1 N-m = 1 J.

In thermodynamics, work may be defined as follows:

1. According to Obert, work is defined as *the energy transferred (without the transfer of mass) across the boundary of a system because of an intensive property difference other than temperature that exists between the system and the surroundings.*

In engineering practice, the intensive property difference is the pressure difference. The pressure difference (between the system and the surrounding) at the surface of the system gives rise to a force and the action of this force over a distance is called mechanical work.

In some cases, the intensive property difference may be the electrical potential difference between the system and the surrounding. In this case, the resulting energy transfer across the system and boundary is known as electrical work.

2. According to Keenan, *work is said to be done by a system during a given operation if the sole effect of the system on things external to the system (surroundings) can be reduced to the raising of a weight.*

The weight may not be actually raised but the net effect external to the system should be the raising of a weight.

For example, consider a system consisting of a storage battery, as shown in Fig. 1.6. The terminals connected to a resistance through a switch constitute external to the system (*i.e.* surroundings). When the switch is closed for a certain period of time, then the current will flow through the battery and the resistance, as a result the resistance becomes warmer. This clearly shows that the system (battery) has interaction with the surroundings. In other words, the energy transfer (electrical energy) has taken place between the system and the surroundings because of potential difference (not the temperature).

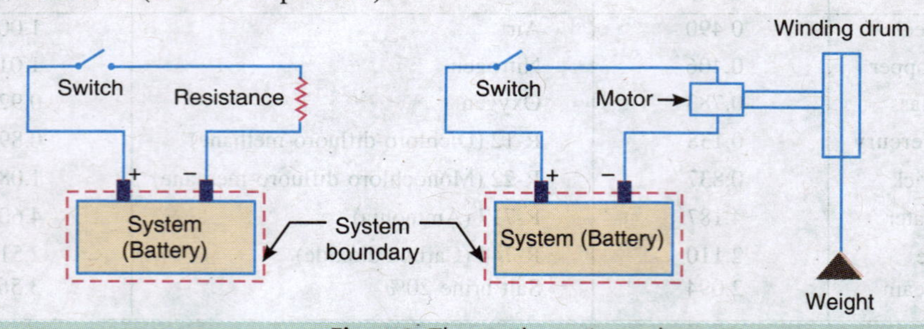

Fig. 1.6. Thermodynamic work.

Now according to the mechanics definition of work, there is no force which moves through a distance. Thus no work is done by the system. However, according to the thermodynamic definition, the work is done by the system because the resistance can be replaced by an ideal motor (100% efficient) driving a winding drum, thereby raising weight. Thus, the sole effect external to the system (surroundings) has been reduced to the raising of a weight. Hence, thermodynamic work is done by the system.

Note : The work done by the system is considered as *positive* work, while the work done on the system is considered as *negative* work.

1.34 Heat and Work - A Path Function

Consider that a system from an initial equilibrium state 1 reaches to a final equilibrium state 2 by two different paths 1-*A*-2 and 1-*B*-2, as shown in Fig. 1.7. The processes are quasi-static.

When the system changes from its initial state 1 to final state 2, the quantity of heat transfer will depend upon the intermediate stages through which the system passes, *i.e.* its path. In other words, heat is a path function. Thus, heat is an inexact differential and is written as δQ. On integrating, for the path 1-*A*-2,

$$\int_1^2 \delta Q = [Q]_1^2 = Q_{1-2} \quad \text{or} \quad {}_1Q_2$$

It may be noted that $\int_1^2 \delta Q \neq Q_2 - Q_1$, because heat is not a point function. Thus, it is meaningless to say 'heat in a system or heat of a system'. The heat can not be interpreted similar to temperature and pressure.*

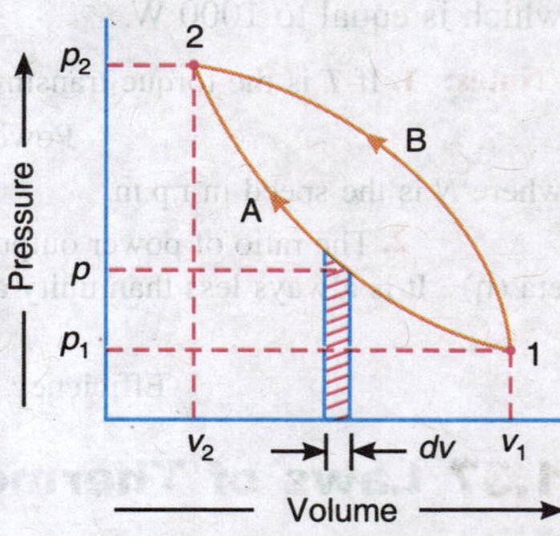

Fig. 1.7. Heat and work–a path function.

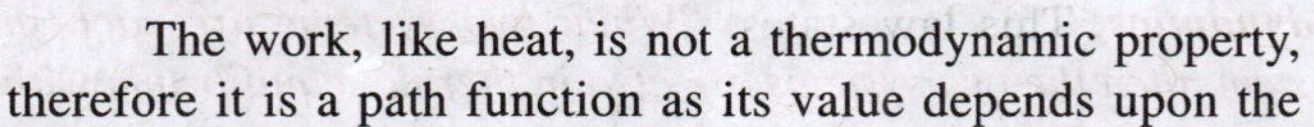

The work, like heat, is not a thermodynamic property, therefore it is a path function as its value depends upon the particular path followed during the process. Since the areas under the curves 1-*A*-2 and 1-*B*-2 are different, therefore work done by these two processes will also be different**. Hence, work is an inexact differential and is written as δW. On integration, for the path 1-*A*-2,

$$\int_1^2 \delta W = [W]_1^2 = W_{1-2} \quad \text{or} \quad {}_1W_2$$

As discussed above, $\int_1^2 \delta W \neq W_2 - W_1$, because work is not a point function. Thus, it is meaningless to say 'work in a system or work of a system'. Since the work can not be interpreted similar to temperature and pressure of the system, therefore it is a path function and it depends upon the process. It is not a point function as the temperature and pressure. The work done in taking the system from state 1 to state 2 will be different for different paths.

1.35 Comparison of Heat and Work

There are many similarities between heat and work. These are

1. The heat and work are both transient phenomena. The systems do not possess heat or work. When a system undergoes a change, heat transfer or work done may occur.
2. The heat and work are boundary phenomena. They are observed at the boundary of the system.
3. The heat and work represent the energy crossing the boundary of the system.
4. The heat and work are path functions and hence they are inexact differentials. They are written as δQ and δW.

1.36 Power

It may be defined as the rate of doing work or work done per unit time. Mathematically,

$$\text{Power} = \frac{\text{Work done}}{\text{Time taken}}$$

* Heat is not a thermodynamic property whereas the temperature and pressure are thermodynamic properties.

** The area under the pressure-volume (pv) diagram represents the work done during the process and is given by $p\ dv$.

In S.I. system of units, the unit of power is watt (briefly written as W) which is equal to 1 J/s or 1 N-m/s. Generally, a bigger unit of power called kilowatt (briefly written as kW) is used which is equal to 1000 W.

Notes: **1.** If T is the torque transmitted in N-m or J and ω is the angular speed in rad/s, then

$$\text{Power, } P = T\,\omega = T \times 2\pi\, N/60 \text{ watt} \qquad \ldots (\because\ \omega = 2\pi N/60)$$

where N is the speed in r.p.m.

2. The ratio of power output to power input is known as *efficiency*. It is denoted by a Greek letter eta (η) . It is always less than unity and is represented as percentage. Mathematically,

$$\text{Efficiency, } \eta = \frac{\text{Power output}}{\text{Power input}}$$

1.37 Laws of Thermodynamics

The following three laws of thermodynamics are important from the subject point of view:

1. *Zeroth Law of Thermodynamics*. This law states, *"When two systems are each in thermal equilibrium with a third system, then the two systems are also in thermal equilibrium with one another."*

This law provides the basis of temperature measurement.

2. *First Law of Thermodynamics*. This law may be stated as follows :

(a) The heat and mechanical work are mutually convertible. According to this law, when a closed system undergoes a thermodynamic cycle, the net heat transfer is equal to the net work transfer. In other words, the cyclic integral of heat transfers is equal to the cyclic integral of work transfers. Mathematically,

$$\oint \delta Q = \oint \delta W$$

where symbol $\oint$ stands for cyclic integral (integral around a complete cycle), and δQ and δW represent infinitesimal elements of heat and work transfers respectively. It may be noted that δQ and δW are expressed in same units.

(b) The energy can neither be created nor destroyed though it can be transformed from one form to another. According to this law, when a system undergoes a change of state (or a thermodynamic process), then both heat transfer and work transfer takes place. The net energy transfer is stored within the system and is known as stored energy or total energy of the system. Mathematically,

$$\delta Q - \delta W = dE$$

The symbol δ is used for a quantity which is inexact differential and symbol d is used for a quantity which is an exact differential. The quantity E is an extensive property and represents the total energy of the system at a particular state.

3. *Second Law of Thermodynamics* : The second law of thermodynamics may be defined in many ways, but the two common statements according to Kelvin-Planck and Clausius are as follows :

According to Kelvin-Planck '*It is impossible to construct an engine working in a cyclic process, whose sole purpose is to convert heat energy from a single thermal reservoir into an equivalent amount of work*'. In other words, no actual heat engine, working on a cyclic process, can convert whole of the heat supplied to it, into mechanical work. It means that there is a degradation of energy in the process of producing mechanical work from the heat supplied. Thus the Kelvin-Planck statement of the second law of thermodynamics, is sometimes known as *law of degradation of energy*.

According to Clausius statement "*It is impossible for a self acting machine, working in a cyclic process, to transfer heat from a body at a lower temperature to a body at a higher temperature without the aid of an external agency*". In other words, heat cannot flow itself from a cold body to a hot body without the help of an external agency (*i.e.* without the expenditure of mechanical work).

1.38 Perfect Gas

A perfect gas (or an ideal gas) may be defined as a state of a substance, whose evaporation from its liquid state is complete,* and strictly obeys all the gas laws under all conditions of temperature and pressure. In actual practice, there is no real or actual gas which strictly obeys the gas laws over the entire range of temperature and pressure. But, the real gases which are ordinarily difficult to liquify, such as oxygen, nitrogen, hydrogen and air, within certain temperature and pressure limits, may be regarded as perfect gases.

1.39 Laws of Perfect Gases

The physical properties of a gas are controlled by the following three variables :

1. Pressure exerted by the gas, 2. Volume occupied by the gas, and 3. Temperature of the gas.

The behaviour of a perfect gas, undergoing any change in the above mentioned variables, is governed by the following laws, which have been established from experimental results.

(a) Boyle's Law. This law was formulated by Robert Boyle in 1662. It states, "*The absolute pressure of a given mass of a perfect gas varies inversely as its volume, when the temperature remains constant.*" Mathematically,

$$p \propto \frac{1}{v} \quad \text{or} \quad p\,v = \text{Constant}$$

The more useful form of the above equation is

$$p_1 v_1 = p_2 v_2 = p_3 v_3 = \ldots = \text{Constant}$$

where suffixes 1, 2 and 3 ... refer to different sets of conditions.

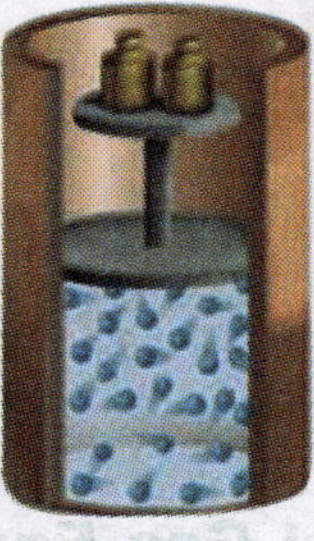

Boyle's law

(b) Charles' Law. This law was formulated by a Frenchman Jacques A.C. Charles in about 1787. It may be stated in the following two different forms :

(*i*) "*The volume of a given mass of a perfect gas varies directly as its absolute temperature, when the absolute pressure remains constant.*" Mathematically,

$$v \propto T \quad \text{or} \quad \frac{v}{T} = \text{Constant}$$

or

$$\frac{v_1}{T_1} = \frac{v_2}{T_2} = \frac{v_3}{T_3} = \ldots = \text{Constant}$$

where suffixes 1, 2 and 3 ... refer to different sets of conditions.

* If its evaporation is partial, the substance is called vapour. A vapour, therefore, contains some particles of liquid in suspension. It is thus obvious that steam, carbon dioxide, sulphur dioxide and ammonia are regarded as vapours. It may be noted that a vapour becomes dry, when it is completely evaporated. If the dry vapour is further heated, the process is called *super heating* and the vapour is called *superheated vapour*. The behaviour of superheated vapour is similar to that of a perfect gas.

(*ii*) "*All perfect gases change in volume by 1 / 273th of its original volume at 0°C for every 1° change in temperature, when the pressure remains constant.*"

Let v_0 = Volume of a given mass of gas at 0°C, and

v_T = Volume of the same mass of gas at t°C.

Then, according to the above statement,

$$v_T = v_0 + \frac{1}{273} \times v_0 \, t = v_0 \left(\frac{273 + t}{273} \right) = v_0 \times \frac{T}{T_0}$$

or
$$\frac{v_T}{T} = \frac{v_0}{T_0}$$

where T = Absolute temperature corresponding to t°C, and

T_0 = Absolute temperature corresponding to 0°C.

A little consideration will show that the volume of a gas goes on decreasing by 1/273th of its original volume for every 1°C decrease in temperature. It is thus obvious, that at a temperature of –273°C, the volume of the gas would become *zero. The temperature at which the volume of a gas becomes zero is called *absolute zero temperature*.

Note: In all calculations of a perfect gas, the pressure and temperature values are expressed in absolute units.

(c) Gay-Lussac Law. This law states, "*The absolute pressure of a given mass of a perfect gas varies directly as its absolute temperature, when the volume remains constant*". Mathematically,

$$p \propto T \quad \text{or} \quad \frac{p}{T} = \text{Constant}$$

or
$$\frac{p_1}{T_1} = \frac{p_2}{T_2} = \frac{p_3}{T_3} = \ldots = \text{Constant}$$

where suffixes 1, 2 and 3 ... refer to different sets of conditions.

1.40 General Gas Equation

In the previous section we have discussed the gas laws which give us the relation between the two variables when the third variable is constant. But in actual practice, all the three variables *i.e.* pressure, volume and temperature, change simultaneously. In order to deal with all practical cases, the Boyle's law and Charles' law are combined together, which give us a *general gas equation*.

According to Boyle's law

$$p \propto \frac{1}{v} \quad \text{or} \quad v \propto \frac{1}{p} \qquad \text{... (Keeping } T \text{ constant)}$$

and according to Charles' law,

$$v \propto T \qquad \text{... (Keeping } p \text{ constant)}$$

It is thus obvious that

$$v \propto \frac{1}{p} \text{ and } T \text{ both} \quad \text{or} \quad v \propto \frac{T}{p}$$

* It is only theoretical. Its exact value is – 273.16° C. But for all practical purposes, this value is taken as – 273° C.

$$\therefore \quad p\,v \propto T \quad \text{or} \quad p\,v = CT$$

where C is a constant, whose value depends upon the mass and properties of the gas concerned. The more useful form of the general gas equation is :

$$\frac{p_1 v_1}{T_1} = \frac{p_2 v_2}{T_2} = \frac{p_3 v_3}{T_3} = \ldots = \text{Constant}$$

where suffixes 1, 2 and 3 .... refer to different sets of conditions.

1.41 Joule's Law

It states, "*The change of internal energy of a perfect gas is directly proportional to the change of temperature*". Mathematically,

$$dE \propto dT \quad \text{or} \quad dE = m\,c\,dT = m\,c\,(T_2 - T_1)$$

where m = Mass of the gas, and

c = A constant of proportionality, known as specific heat.

An important consequence of this law is that if the temperature of a given mass m of a gas changes from T_1 to T_2, then the internal energy will change from E_1 to E_2 and the change in internal energy $(E_2 - E_1)$ will be same irrespective of the manner how the pressure (p) and volume (v) of the gas have changed.

Note: From the Joule's law, we see that whenever a gas expands, without doing any external work and without taking in or giving out heat, its internal energy as well as temperature does not change.

1.42 Characteristic Equation of a Gas

It is a modified form of general gas equation. If the volume (v) in the general gas equation is taken as that of 1 kg of gas (known as its specific volume, and denoted by v_s), then the constant C (in the general gas equation) is represented by another constant R (in the characteristic equation of gas). Thus the general gas equation may be rewritten as :

$$p\,v_s = RT$$

where R is known as *characteristic gas constant* or *simply gas constant.*

For any mass m kg of a gas, the characteristic gas equation becomes :

$$m\,p\,v_s = m\,R\,T$$

or

$$p\,v = m\,R\,T \qquad \ldots (\because m\,v_s = v)$$

Notes: 1. The units of gas constant (R) may be obtained as discussed below :

$$R = \frac{pv}{mT} = \frac{\text{N/m}^2 \times \text{m}^3}{\text{kg} \times \text{K}} = \text{N-m/kg K} = \text{J/kg K} \qquad \ldots (\because 1 \text{ N–m} = 1\text{J})$$

2. The value of gas constant (R) is different for different gases. In S.I. units, its value for atmospheric air is taken 287 J/kg K or 0.287 kJ/kg K.

3. The equation $p\,v = m\,R\,T$ may also be expressed in another form *i.e.*

$$p = \frac{m}{v} R\,T = \rho\,R\,T \qquad \ldots \left(\because \frac{m}{v} = \rho\right)$$

where ρ (rho) is the density of the given gas.

1.43 Specific Heats of a Gas

The specific heat of a substance may be broadly defined as the amount of heat required to raise the temperature of its unit mass through one degree. All the liquids and solids have one specific heat only. But a gas can have any number of specific heats (lying between zero and infinity) depending upon the conditions, under which it is heated. The following two types of specific heats of a gas are important from the subject point of view:

Fig. 1.8. Heat being supplied at constant volume.

1. ***Specific heat at constant volume.*** It is the amount of heat required to raise the temperature of a unit mass of gas through one degree when it is heated at a constant volume. It is generally denoted by c_v.

Consider a gas contained in a container with a fixed lid as shown in Fig. 1.8. Now, if this gas is heated, it will increase the temperature and pressure of the gas in the container. Since the lid of the container is fixed, therefore the volume of the gas remains unchanged.

Let m = Mass of the gas,

T_1 = Initial temperature of the gas, and

T_2 = Final temperature of the gas.

∴ Total heat supplied to the gas at constant volume,

$$Q_{1-2} = \text{Mass} \times \text{Sp. heat at constant volume} \times \text{Rise in temperature}$$

$$= m\, c_v\, (T_2 - T_1)$$

It may be noted that whenever a gas is heated at constant volume, no work is done by the gas*. The whole heat energy is utilised in increasing the temperature and pressure of the gas. In other words, all the amount of heat supplied remains within the body of the gas, and represents the *increase in internal energy of the gas*.

2. ***Specific heat at constant pressure.*** It is the amount of heat required to raise the temperature of a unit mass of gas through one degree, when it is heated at constant pressure. It is generally denoted by c_p.

Consider a gas contained in a container with a movable lid as shown in Fig. 1.9. Now if this gas is heated, it will increase the temperature and pressure of the gas in the container. Since the lid of the container is movable, therefore it will move upwards, in order to counterbalance the tendency for pressure to rise.

Movable

Fig. 1.9. Heat being supplied at constant pressure.

Let m = Mass of the gas,

T_1 = Initial temperature of the gas,

v_1 = Initial volume of the gas, and

T_2, v_2 = Corresponding values for the final condition of the gas.

∴ Total heat supplied to the gas at constant pressure,

$$Q_{1-2} = \text{Mass} \times \text{Sp. heat at constant pressure} \times \text{Rise in temperature}$$

$$= m\, c_p\, (T_2 - T_1)$$

Whenever a gas is heated at a constant pressure, the heat supplied to the gas is utilised for the following two purposes :

* We know that work done by the gas,

$$W = p\, dv = p\,(v_2 - v_1)$$

where p = Pressure of the gas, and

dv = Change in volume = $v_2 - v_1$

When there is no change in volume, then $dv = 0$. Therefore $W = 0$.

1. To raise the temperature of the gas. This heat remains within the body of the gas, and represents the increase in internal energy. Mathematically, increase in internal energy,

$$dU = m\, c_v\, (T_2 - T_1)$$

2. To do some external work during expansion. Mathematically, work done by the gas,

$$W_{1-2} = p\,(v_2 - v_1) = m\, R\, (T_2 - T_1)$$

It is thus obvious, that the specific heat at constant pressure is higher than the specific heat at constant volume.

From above, we may write as

$$Q_{1-2} = dU + W_{1-2} \quad \text{or} \quad Q_{1-2} - W_{1-2} = dU$$

... (First Law of Thermodynamics)

1.44 Enthalpy of a Gas

In Thermodynamics, one of the basic quantities most frequently recurring is the sum of the internal energy (U) and the product of pressure and volume (pv). This sum ($U + pv$) is termed as enthalpy and is written as H. Mathematically,

$$\text{Enthalpy, } H = U + pv$$

Since ($U + pv$) is made up entirely of properties, therefore enthalpy (H) is also a property. For a unit mass, specific enthalpy,

$$h = u + p\, v_s$$

where u = Specific internal energy, and

v_s = Specific volume.

Note: We know that $Q_{1-2} = dU + W_{1-2} = dU + p\, dv$

When gas is heated at constant pressure from an initial condition 1 to a final condition 2, then change in internal energy,

$$dU = U_2 - U_1$$

and work done by the gas, $W_{1-2} = p\, dv = p\,(v_2 - v_1)$

$$\therefore \quad Q_{1-2} = (U_2 - U_1) + p\,(v_2 - v_1)$$

$$= (U_2 + pv_2) - (U_1 + p\, v_1) = H_2 - H_1$$

and for per unit mass, $q_{1-2} = h_2 - h_1$

Thus, for a constant pressure process, the heat supplied to the gas is equal to the change of enthalpy.

1.45 Ratio of Specific Heats

The ratio of two specific heats (*i.e.* c_p / c_v) of a gas is an important constant in the field of Thermodynamics and is represented by a Greek letter gamma (γ). It is also known as *adiabatic index.* Since c_p is always greater than c_v, therefore the value of γ is always greater than unity.

We know that

$$c_p - c_v = R \quad \text{or} \quad c_p = c_v + R$$

Dividing both sides by c_v,

$$\frac{c_p}{c_v} = 1 + \frac{R}{c_v} \quad \text{or} \quad \gamma = 1 + \frac{R}{c_v}$$

The values of c_p, c_v and γ for some common gases are given below :

Table 1.5. Values of c_p and c_v for some common gases.

S.No.	Name of gas	C_p (kJ/kgK)	C_v (kJ/kgK)	$\gamma = \frac{C_p}{C_v}$
1.	Air	1.000	0.720	1.40
2.	Carbon dioxide (CO_2)	0.846	0.657	1.29
3.	Oxygen (O_2)	0.913	0.653	1.39
4.	Nitrogen (N_2)	1.043	0.745	1.40
5.	Ammonia (NH_3)	2.177	1.692	1.29
6.	Carbon monoxide (CO)	1.047	0.749	1.40
7.	Hydrogen (H_2)	14.257	10.133	1.40
8.	Argon (A)	0.523	0.314	1.67
9.	Helium (He)	5.234	3.153	1.66
10.	Methane (CH_4)	2.169	1.650	1.31

1.46 Entropy

The term 'entropy' which literally means transformation, was first introduced by Clausius. It is an important thermodynamic property of a working substance, which increases with the addition of heat, and decreases with its removal. As a matter of fact, it is tedious to define the term entropy. But it is comparatively easy to define change of entropy of a working substance. In a reversible process, over a small range of temperature, the increase or decrease of entropy, when multiplied by the absolute temperature, gives the heat absorbed or rejected by the working substance. Mathematically, heat absorbed by the working substance,

$$\delta Q = T dS \quad \text{or} \quad dS = \frac{\delta Q}{T} \qquad \ldots (i)$$

where T = Absolute temperature, and

dS = Increase in entropy.

Note: The above relation also holds good for heat rejected by the working substance. In that case, dS will be decrease in entropy.

The engineers and scientists use it for providing quick solution to problems dealing with adiabatic expansion. The entropy is usually represented by S.

The adiabatic expansion on the temperature-entropy (T-S) diagram is shown by the curve 1-2 in Fig. 1.10.

The total change in entropy may be obtained by integrating the equation (i) from state 1 to state 2.

$$\therefore \quad \int_1^2 dS = \int_1^2 \frac{\delta Q}{T} \qquad \ldots (ii)$$

The unit of entropy depends upon the unit of heat employed and the absolute temperature.

Therefore, if the heat supplied or rejected is in kJ and the temperature is in K, then the unit of entropy is kJ/K. The entropy may be expressed in so many units of entropy without assigning any dimensional units. Since the entropy is expressed per unit mass of the working substance, it

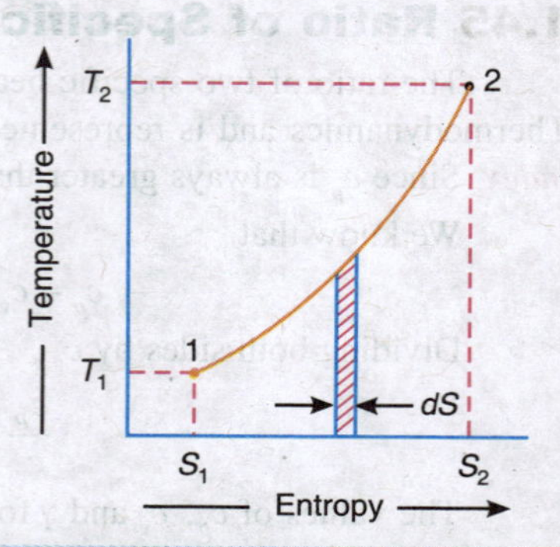

Fig. 1.10. Temperature-entropy diagram.

would be more correct to speak **specific entropy*. The absolute values of entropy cannot be determined, but only the change in entropy may be obtained by using equation (*ii*).

Theoretically, the entropy of a substance is *zero* at *absolute zero temperature*. Hence, in entropy calculations, some convenient datum should be selected from which measurement may be made.

It may be noted that water at 0°C is assumed to have zero entropy, and changes in its entropy are reckoned from this temperature.

Notes: 1. The area under the *T-S* diagram of any thermodynamic process represents the heat absorbed or rejected during that process.

2. Since $\int \frac{\delta Q}{T}$ is same for all reversible paths between states 1 and 2, so we conclude that this quantity is independent of a path and is a function of end states only. In other words, the entropy is a point function and thus it is a property of the system. The entropy may be expressed as a function of other thermodynamic properties of the system, such as the pressure and temperature or pressure and volume.

3. We know that according to First Law of Thermodynamics,

$$\delta Q = dU + \delta W = dU + p\ dv \ \dots\ (\because\ \delta W = p\ dv) \qquad \dots (iii)$$

and

$$\delta Q = T\ dS \qquad \dots (iv)$$

From equations (*iii*) and (*iv*),

$$T\ dS = dU + p\ dv \qquad \dots (v)$$

It is very interesting to note that in equations (*iii*) and (*iv*), δQ and δW are path functions, therefore these equations are true only for reversible processes. But in equation (*v*), *ds*, *dU* and *dv* are point functions as they depend upon the initial and final equilibrium states, therefore equation (*v*) is true for reversible as well as irreversible processes.

4. The entropy remains constant in a reversible process and increases in an irreversible process.

5. The change of entropy is *positive* when heat is absorbed by the gas and there is an increase of entropy.

6. The change of entropy is *negative* when heat is removed from the gas and there is a decrease of entropy.

1.47 Thermodynamic Processes

When a system changes its state from one equilibrium state to another equilibrium state, then the path of successive states through which the system has passed, is known as a *thermodynamic process*. Strictly speaking, no system is in true equilibrium during the process because the properties (such as pressure, volume, temperature etc.) are changing. However, if the process is assumed to take place sufficiently slowly so that the deviation of the properties at the intermediate states is infinitesimally small, then every state passed through by the system will be in equilibrium. Such a process is called *quasi-static* or *reversible process* and it is represented by a continuous curve on the property diagram (*i.e.* pressure-volume diagram) as shown in Fig. 1.11(*a*).

If the process takes place in such a manner that the properties at the intermediate states are not in equilibrium state (except the initial and final state), then the process is said to be *non-equilibrium* or *irreversible process*. This process is represented by the broken lines on the property diagram as shown in Fig. 1.11 (*b*).

* The entropy is an extensive property of the system. The ratio of the extensive property of the system to the mass of the system is the specific value of that property.

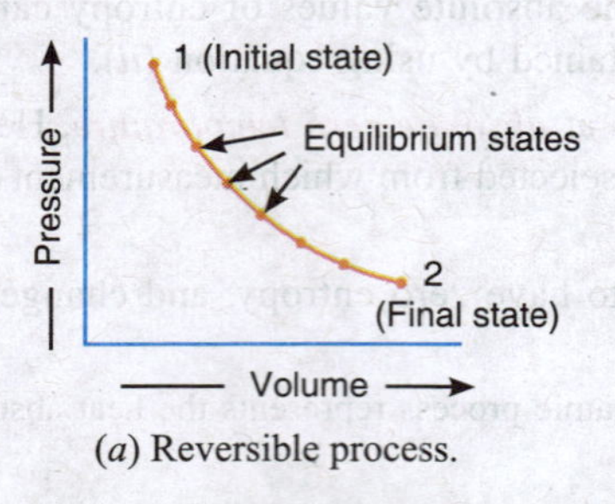

(a) Reversible process.

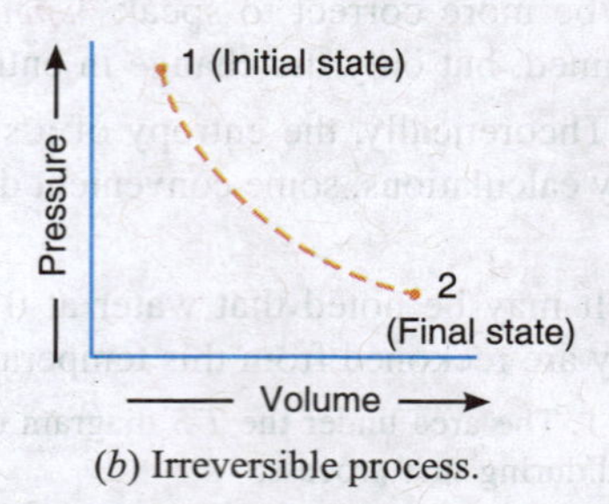

(b) Irreversible process.

Fig. 1.11. Reversible and irreversible process.

All the thermodynamic processes are classified into the following two groups :

1. Non-flow processes, and 2. Flow processes.

The processes occuring in closed systems which do not permit the transfer of mass across their boundaries, are known as *non-flow processes*. It may be noted that in a non-flow process, the energy crosses the system boundary in the form of heat and work, but there is no mass flow into or out of the system.

The processes occuring in open systems which permit the transfer of mass to and from the system, are known as *flow processes*. It may be noted that in a flow process, the mass enters the system and leaves after enhancing energy. The flow processes may be *steady flow* and *non-steady flow processes*.

1.48 Workdone During a Non-flow Process

Consider a system contained in a frictionless piston and cylinder arrangement as shown in Fig. 1.12. As the system expands from its original state 1, it overcomes the external resistance (such as rotation of the flywheel) which opposes the motion of the piston by exerting a force through a distance. The variation of the volume and pressure of the system as it expands to final state 2, is drawn on the pressure-volume diagram (briefly called *p-v* diagram) as shown in Fig. 1.12.

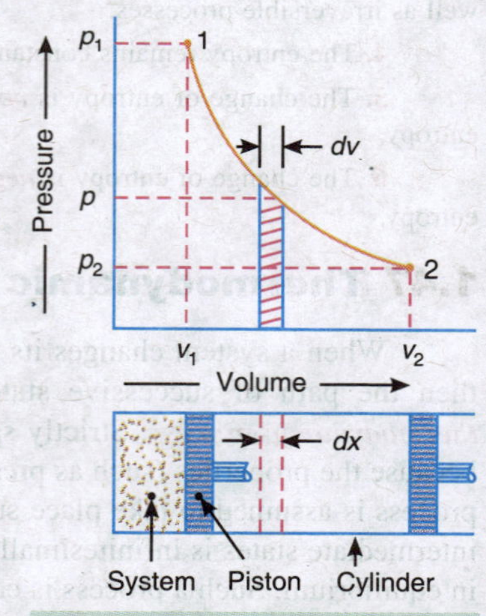

Fig. 1.12. Workdone during a non-flow process.

Let at any small section (shown shaded), the pressure (p) of the system is constant. If A is the cross-sectional area of the piston, then force on the piston ($F = pA$) causes the piston to move through a distance dx. Thus, workdone by the system,

$$\delta W = F\,dx = pA\,dx = p\,dv \qquad \ldots\ (\because\ dv = A\,dx)$$

$\therefore$ Workdone for non-flow process from state 1 to state 2,

$$W_{1-2} = \int_1^2 \delta W = \int_1^2 p\,dv$$

From above, we see that the workdone is given by the area under the *p-v* diagram.

Notes: 1. The workdone by the system is taken as *positive* while the workdone on the system is considered as *negative*.

2. For an irreversible process, $\delta W \neq p\,dv$, because the path of the process is not represented truely on the *p-v* diagram due to its non-equilibrium states in the process.

1.49 Application of First Law of Thermodynamics to a Non-flow Process

We have already discussed in Art. 1.37 that when a system undergoes a change of state or a thermodynamic process, then both the heat transfer and work transfer takes place. The net energy transfer is stored within the system and is known as *stored* or *total energy* of the system. Mathematically,

$$Q_{1-2} - W_{1-2} = dE = E_2 - E_1 \qquad \ldots (i)$$

where Q_{1-2} = Heat transferred or heat supplied to the system during the process *i.e.* from state 1 to state 2,

W_{1-2} = Workdone by the system on the surrounding during the process *i.e.* from state 1 to state 2,

E_2 = *Total energy or stored energy of the system at the end of the process *i.e.* at state 2, and

E_1 = Total energy or stored energy of the system at the start of the process *i.e.* at state 1.

For a non-flow process, the stored energy is the internal energy only. Thus equation (*i*) of the First Law of Thermodynamics, when applied to non-flow process or a static system, may be written as

$$Q_{1-2} - W_{1-2} = dU = U_2 - U_1$$

where dU = Change in internal energy = $U_2 - U_1$

It may be noted that heat and work are not a property of the system, but their difference $(Q_{1-2} - W_{1-2})$ during a process is the numerical equivalent of stored energy. Since the stored energy is a property, therefore $(Q_{1-2} - W_{1-2})$ is also a property.

1.50 Classification of Non-flow Processes

The various non-flow processes which take place in the cycle of a closed system are discussed in the following pages :

1. *Constant volume process or Isochoric process*

We have already discussed that when a gas is heated at a constant volume, its temperature and pressure will increase. Since there is no change in its volume, therefore no work is done by the gas. All the heat supplied to the gas is stored within the gas in the form of internal energy. Now consider *m* kg of a certain gas being heated at constant volume from initial state 1 to a final state 2.

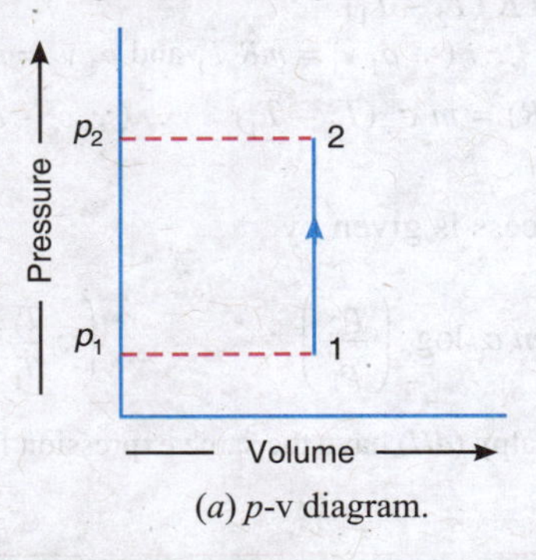

(*a*) *p*-v diagram.

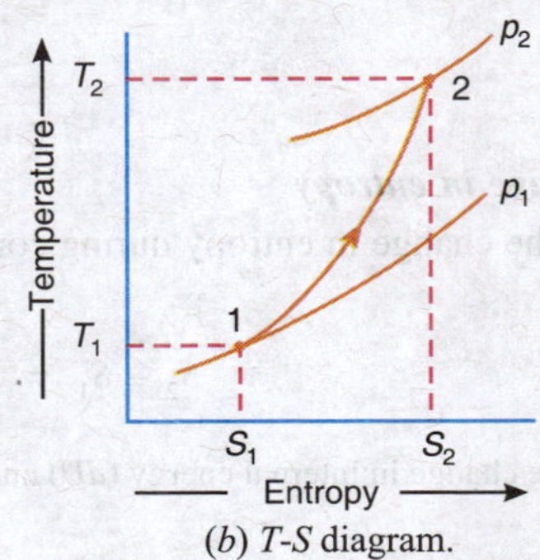

(*b*) *T-S* diagram.

Fig. 1.13. Constant volume process.

* The total energy of a system is sum of potential energy (*PE*), kinetic energy (*KE*) and internal energy (*U*) plus any other form of the energy such as chemical energy, electrical energy etc.

Let p_1, v_1 and T_1 = Pressure, volume and temperature at the initial state 1, and

p_2, v_2 and T_2 = Pressure, volume and temperature at the final state 2.

The process is shown on the pressure-volume* (p-v) diagram and temperature-entropy** (T-S) diagram in Fig. 1.13 (*a*) and (*b*) respectively. It may be noted that the constant volume process is governed by Gay-Lussac law, *i.e.*

$$\frac{p_1}{T_1} = \frac{p_2}{T_2} \quad \text{or} \quad \frac{p}{T} = \text{Constant}$$

The following relations are important for the reversible constant volume process.

(a) Heat supplied or heat transfer

We know that $\delta Q = dU + \delta W$ ... (First Law of Thermodynamics)

On integrating from state 1 to state 2,

$$\int_1^2 \delta Q = \int_1^2 dU + \int_1^2 \delta W$$

or $$Q_{1-2} = (U_2 - U_1) + W_{1-2}$$

Since $W_{1-2} = 0$, therefore heat supplied or heat transfer,

$$Q_{1-2} = U_2 - U_1 = m\, c_v\, (T_2 - T_1)$$

This shows that all the heat supplied to the gas is utilised in increasing the internal energy of the gas.

(b) Change in enthalpy

We know that the change in enthalpy,

$$dH = dU + d\,(pv)$$

On integrating from state 1 to state 2,

$$\int_1^2 dH = \int_1^2 dU + \int_1^2 d\,(pv)$$

or $$H_2 - H_1 = (U_2 - U_1) + (p_2\, v_2 - p_1\, v_1)$$

$$= m\, c_v\, (T_2 - T_1) + m\, R\, (T_2 - T_1)$$

... ($\because p_1\, v_1 = mR\, T_1$ and $p_2\, v_2 = mR\, T_2$)

$$= m\, (T_2 - T_1)\, (c_v + R) = m\, c_p\, (T_2 - T_1)$$ ... ($\because c_p - c_v = R$)

(c) Change in entropy

The change in entropy during constant volume process is given by

$$S_2 - S_1 = m\, c_v\, \log_e \left(\frac{T_2}{T_1}\right) = m\, c_v\, \log_e \left(\frac{p_2}{p_1}\right) \quad \ldots \left(\because \frac{T_2}{T_1} = \frac{p_2}{p_1}\right)$$

Note: The change in internal energy (dU) and the change in enthalpy (dH) have the same expression for each process.

---

* The area below the p-v diagram of any thermodynamic process represents the work done during that process.

** The area below the T-S diagram of a thermodynamic process represents the heat absorbed or rejected during that process.

2. Constant pressure process or Isobaric process

We have already discussed that when a gas is heated at a constant pressure, its temperature and volume will increase. Since there is a change in its volume, therefore the heat supplied to the gas is utilised to increase the internal energy of the gas and for doing some external work.

Now consider m kg of a certain gas being heated at a constant pressure from an initial state 1 to a final state 2.

Let p_1, v_1 and T_1 = Pressure, volume and temperature at the initial state 1, and

p_2, v_2 and T_2 = Pressure, volume and temperature at the final state 2.

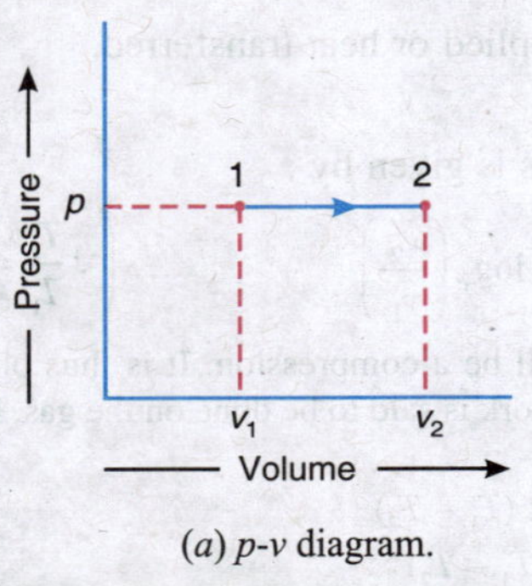

(a) p-v diagram.

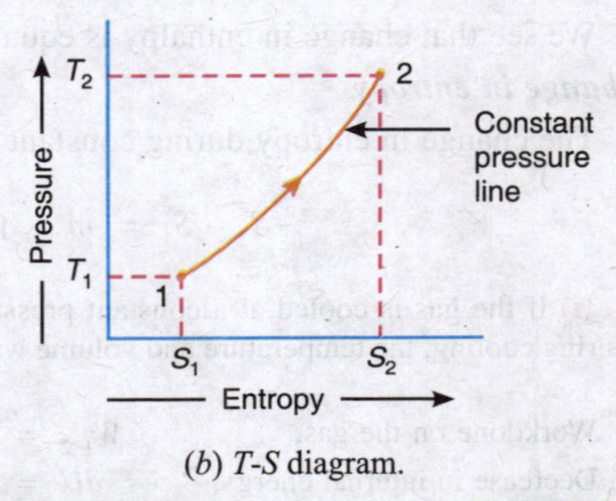

(b) T-S diagram.

Fig. 1.14. Constant pressure process.

The process is shown on the p-v and T-S diagrams in Fig. 1.14 (a) and (b) respectively.

It may be noted that the constant pressure process is governed by Charles' law, *i.e.*

$$\frac{v_1}{T_1} = \frac{v_2}{T_2} \quad \text{or} \quad \frac{v}{T} = \text{Constant}$$

The following relations are important for the reversible constant pressure process.

(a) *Workdone by the gas*

We know that $\delta W = p\, dv$

On integrating from state 1 to state 2,

$$\int_1^2 \delta W = \int_1^2 p\, dv = p \int_1^2 dv$$

or

$$W_{1-2} = p\,(v_2 - v_1) = m\,R\,(T_2 - T_1)$$

$$\ldots (\because p\,v_1 = m\,R\,T_1 \text{ and } p\,v_2 = m\,R\,T_2)$$

(b) *Change in internal energy*

We have discussed above that the change in internal energy (dU) is same for all the processes. Therefore change in internal energy,

$$dU = U_2 - U_1 = m\,c_v\,(T_2 - T_1)$$

(c) *Heat supplied or heat transferred*

We know that $\delta Q = dU + \delta W$

On integrating from state 1 to state 2,

$$\int_1^2 \delta Q = \int_1^2 dU + \int_1^2 \delta W$$

or
$$Q_{1-2} = (U_2 - U_1) + W_{1-2} \quad \ldots (i)$$
$$= m\,c_v\,(T_2 - T_1) + m\,R\,(T_2 - T_1) = m\,(T_2 - T_1)\,(c_v + R)$$
$$= m\,c_p\,(T_2 - T_1) \quad \ldots (\because c_p - c_v = R)$$

The equation (*i*) shows that the heat supplied to the gas is utilised in increasing the internal energy of the gas and for doing some external work.

(*d*) *Change in enthalpy*

We have already discussed in the previous article that the change in enthalpy (dH) is same for all the processes. Therefore, change in enthalpy,

$$dH = H_2 - H_1 = m\,c_p\,(T_2 - T_1)$$

We see that change in enthalpy is equal to the heat supplied or heat transferred.

(*e*) *Change in entropy*

The change in entropy during constant pressure process is given by

$$S_2 - S_1 = m\,c_p\,\log_e\left(\frac{T_2}{T_1}\right) = m\,c_p\,\log_e\left(\frac{v_2}{v_1}\right) \quad \ldots \left(\because \frac{T_2}{T_1} = \frac{v_2}{v_1}\right)$$

Notes: (*i*) If the gas is cooled at a constant pressure,then there will be a compression. It is thus obvious that, during cooling, the temperature and volume will decrease and work is said to be done on the gas. In this case,

Workdone on the gas, $W_{1-2} = p\,(v_1 - v_2) = m\,R\,(T_1 - T_2)$

Decrease in internal energy, $dU = U_1 - U_2 = m\,c_v\,(T_1 - T_2)$

and heat rejected by the gas, $Q_{1-2} = m\,c_p\,(T_1 - T_2)$

(*ii*) During expansion or heating process, work is done by the gas (*i.e.* W_{1-2} is + ve); internal energy of the gas increases (*i.e.* dU is +ve) and heat is supplied to the gas (*i.e.* Q_{1-2} is +*ve*).

(*iii*) During compression or cooling process, work is done on the gas (*i.e.* W_{1-2} is –ve); internal energy of the gas decreases (*i.e.* dU is –ve) and heat is rejected by the gas (*i.e.* Q_{1-2} is –ve).

3. *Constant temperature process or Isothermal process*

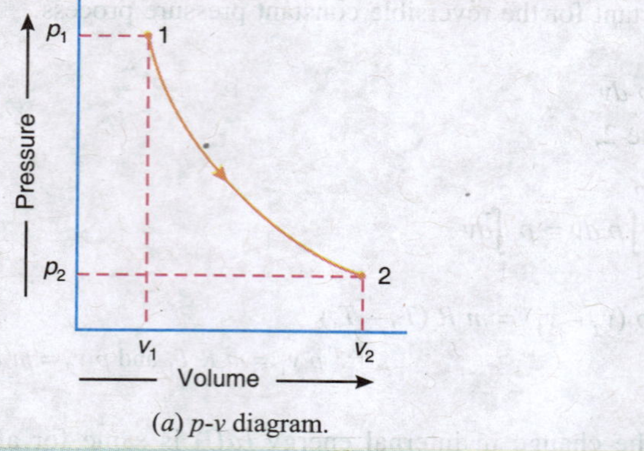

(*a*) *p-v* diagram.

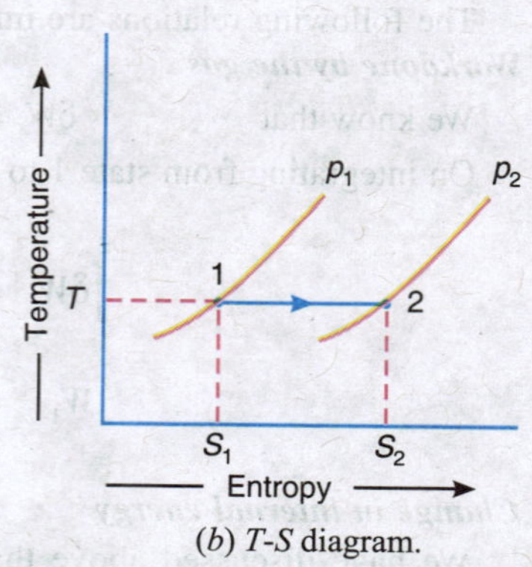

(*b*) *T-S* diagram.

Fig. 1.15. Constant temperature process.

A process, in which the temperature of the working substance remains constant during its expansion or compression, is called constant temperature process or isothermal process. This will happen when the working substance remains in a perfect thermal contact with the surroundings, so that the heat 'sucked in' or 'squeezed out' is compensated exactly for the work done by the gas or on the gas respectively. It is thus obvious that in an isothermal process :

1. there is no change in temperature,
2. there is no change in internal energy, and
3. there is no change in enthalpy.

Now consider m kg of a certain gas being heated at constant temperature from an initial state 1 to final state 2.

Let p_1, v_1 and T_1 = Pressure, volume and temperature at the initial state 1, and

p_2, v_2 and T_2 = Pressure, volume and temperature at the final state 2.

The process is shown on the *p-v* and *T-S* diagrams in Fig. 1.15 (*a*) and (*b*) respectively. It may be noted that the constant temperature process is governed by Boyle's law, *i.e.*

$$p_1 v_1 = p_2 v_2 \quad \text{or} \quad pv = \text{Constant}$$

The following relations are important for the constant temperature process:

(*a*) ***Workdone by the gas***

The workdone by the gas during isothermal expansion is given by

$$W_{1-2} = p_1 v_1 \log_e \left(\frac{v_2}{v_1}\right) = 2.3\, p_1\, v_1 \log r$$

$$= 2.3\, m\, R\, T \log r \qquad \ldots (\because p_1\, v_1 = m\, RT)$$

where r is the expansion ratio or compression ratio if the process is isothermal compression.

Notes: (*i*) Expansion ratio, $r = \dfrac{\text{Volume at the end of expansion}}{\text{Volume at the beginning of expansion}}$

(*ii*) Compression ratio, $r = \dfrac{\text{Volume at the beginning of compression}}{\text{Volume at the end of compression}}$

(*b*) ***Change in internal energy***

We know that change in internal energy,

$$dU = U_2 - U_1 = m\, c_v\, (T_2 - T_1)$$

Since it is a constant temperature process, *i.e.* $T_2 = T_1$, therefore

$$dU = U_2 - U_1 = 0 \quad \text{or} \quad U_1 = U_2$$

(*c*) ***Heat supplied or heat transferred***

We know that heat supplied or heat transferred from state 1 to state 2,

$$Q_{1-2} = dU + W_{1-2} = W_{1-2} \qquad \ldots (\because dU = 0)$$

This shows that all the heat supplied to the gas is equal to the workdone by the gas.

(*d*) ***Change in enthalpy***

We know that change in enthalpy,

$$dH = H_2 - H_1 = m\, c_p\, (T_2 - T_1)$$

Since it is a constant temperature process, *i.e.* $T_2 = T_1$, therefore

$$dH = H_2 - H_1 = 0 \quad \text{or} \quad H_1 = H_2$$

(*e*) ***Change in entropy***

The change in entropy during isothermal process is given by,

$$S_2 - S_1 = m\, R \log_e \left(\frac{v_2}{v_1}\right) = 2.3\, m\, R \log r$$

4. *Reversible adiabatic process or Isentropic process*

A process, in which the working substance neither receives nor gives out heat to its surroundings, during its expansion or compression, is called an **adiabatic process*. This will

* It may be noted that the adiabatic process may be reversible or irreversible. The reversible adiabatic process or frictionless adiabatic process is known as isentropic process (or constant entropy process). But when friction is involved in the process, then the adiabatic process is said to be irreversible, in which case the entropy does not remain constant *i.e.* the entropy increases.

happen when the working substance remains thermally insulated, so that no heat enters or leaves it during the process. It is thus obvious, that in an adiabatic or isentropic (*i.e.* constant entropy) process

1. No heat leaves or enters the gas,
2. The temperature of the gas changes, as the work is done at the cost of internal energy, and
3. The change in internal energy is equal to the work done.

Now consider m kg of a certain gas being heated reversibly and adiabatically from an initial state 1 to a final state 2.

Let p_1, v_1 and T_1 = Pressure, volume and temperature at the initial state 1, and

p_2, v_2 and T_2 = Pressure, volume and temperature at the final state 2.

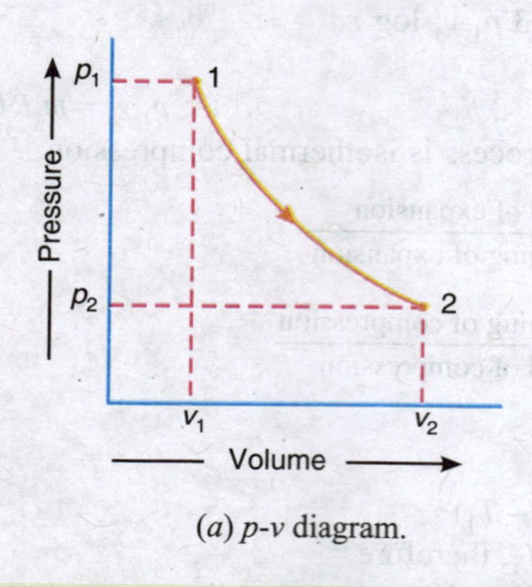

(*a*) *p-v* diagram.

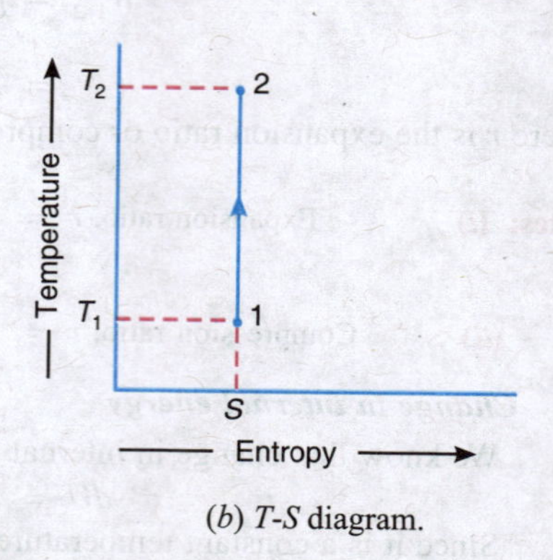

(*b*) *T-S* diagram.

Fig. 1.16. Reversible adiabatic or Isentropic process.

The process is shown on the *p-v* and *T-S* diagrams in Fig 1.16 (*a*) and (*b*) respectively. It may be noted that the reversible adiabatic or isentropic process follows the law pv^{γ} = constant, where γ is called the adiabatic or isentropic index.

The following relations are important for the reversible adiabatic or isentropic process.

(a) Work done during expansion

The work done during reversible adiabatic expansion is given by

$$W_{1-2} = \frac{p_1 v_1 - p_2 v_2}{\gamma - 1} = \frac{m R (T_1 - T_2)}{\gamma - 1} \quad \ldots (\because p_1 v_1 = m\,R\,T_1, \text{ and } p_2 v_2 = m\,R\,T_2)$$

(b) Change in internal energy

We know that change in internal energy,

$$dU = U_2 - U_1 = m\,c_v\,(T_2 - T_1)$$

(c) Heat supplied or heat transferred

We know that heat supplied or heat transferred in case of adiabatic process is zero, therefore

$$Q_{1-2} = 0$$

(d) Change in enthalpy

We know that change in enthalpy,

$$dH = H_2 - H_1 = m\,c_p\,(T_2 - T_1)$$

(e) Change in entropy

Since no heat enters or leaves the system during this process, therefore the change in entropy is zero.

Note: Since the reversible adiabatic or isentropic process follows the law

$$p\,v^{\gamma} = \text{Constant} \quad \text{or} \quad p_1 v_1^{\gamma} = p_2 v_2^{\gamma}$$

$$\therefore \qquad \frac{p_1}{p_2} = \left(\frac{v_2}{v_1}\right)^{\gamma} \qquad \text{... (i)}$$

We know that

$$\frac{p_1 v_1}{T_1} = \frac{p_2 v_2}{T_2} \quad \text{or} \quad \frac{p_1}{p_2} = \frac{T_1}{T_2} \times \frac{v_2}{v_1} \qquad \text{... (ii)}$$

From equations (*i*) and (*ii*),

$$\frac{T_1}{T_2} = \left(\frac{v_2}{v_1}\right)^{\gamma - 1}$$

The equation (*i*) may be written as

$$\frac{v_2}{v_1} = \left(\frac{p_1}{p_2}\right)^{\frac{1}{\gamma}} \qquad \text{... (iii)}$$

and equation (*ii*) may be written as

$$\frac{v_2}{v_1} = \frac{p_1}{p_2} \times \frac{T_2}{T_1} \qquad \text{... (iv)}$$

Now from equations (*iii*) and (*iv*),

$$\frac{T_1}{T_2} = \left(\frac{p_1}{p_2}\right)^{\frac{\gamma - 1}{\gamma}} \quad \text{or} \quad \frac{p_1}{p_2} = \left(\frac{T_1}{T_2}\right)^{\frac{\gamma}{\gamma - 1}}$$

5. *Polytropic process*

The general law for the expansion and compression of gases is given by the relation

$$p\,v^{n} = \text{Constant}$$

where n is called the polytropic index which may have value from zero to infinity, depending upon the manner, in which the expansion or compression has taken place.

The various equations for polytropic process may be expressed by changing the index n for γ in the adiabatic process, *i.e.*

$$\frac{T_1}{T_2} = \left(\frac{v_2}{v_1}\right)^{n - 1} \quad \text{and} \quad \frac{T_1}{T_2} = \left(\frac{p_1}{p_2}\right)^{\frac{n - 1}{n}}$$

Also work done during polytropic process,

$$= \frac{m\,R\,(T_1 - T_2)}{n - 1}$$

and heat absorbed or rejected during polytropic process,

$$= \frac{\gamma - n}{\gamma - 1} \times \text{Work done}$$

1.51 Thermodynamic Cycle

A thermodynamic cycle or a cyclic process consists of a series of thermodynamic operations (processes), which take place in a certain order, and the initial conditions are restored at the end of the process. When the operations or processes of cycle are plotted on *p-v* diagram, they form a closed figure, each operation being represented by its own curve. Since the area under each curve gives the work done to some scale, during each operation, it therefore follows that the net work done during one cycle will be given by the enclosed area of the diagram as shown shaded in Fig. 1.17.

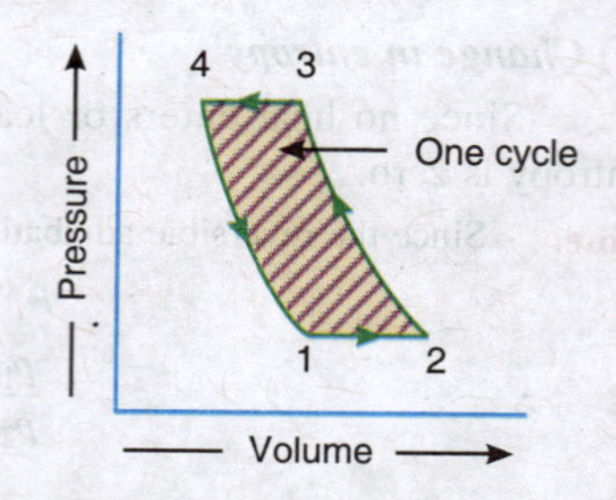

Fig. 1.17. A thermodynamic cycle.

The thermodynamic cycles, in general, may be classified into the following two types :

1. *Reversible cycle.* A thermodynamically reversible cycle consists of reversible processes only. We have already discussed that a reversible process is one which is performed in such a way that at the end of the process, both the system and the surroundings may be restored to their initial states. For example, consider a process in which the system (gas) is expanded from state 1 to state 2 following the path 1-2 as shown in Fig. 1.18. Let during the thermodynamic process 1-2, the workdone by the system is W_{1-2} and the heat absorbed is Q_{1-2}. Now, if by doing the work (W_{1-2}) on the system (*i.e.* by compressing the gas) and extracting heat (Q_{1-2}) from the system, we can bring the system and the surroundings, back from state 2 to state 1 (*i.e.* initial state), following the same path 2-1, then process is said to be a reversible process.

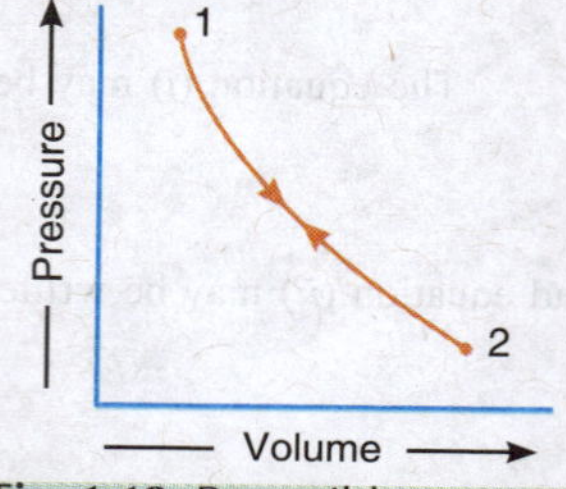

Fig. 1.18. Reversible process.

In a reversible process, there should not be any loss of heat due to friction, radiation or conduction, etc. A cycle will be reversible if all the processes constituting the cycle are reversible. Thus in a reversible cycle, the initial conditions are restored at the end of the cycle.

A little consideration will show that when the operations are performed in the reversed order, the cycle draws heat from the cold body and rejects it to the hot body. This operation requires an external power to drive the mechanism according to second law of thermodynamics. A machine which operates on a reversed cycle is regarded as a "heat pump", such as a refrigerator, because it pumps heat from the cold body to the hot body. Following are the conditions for reversibility of a cycle :

(*a*) The pressure and temperature of the working substance must not differ, appreciably, from those of the surroundings at any stage in the process.

(*b*) All the processes, taking place in the cycle of operation, must be extremely slow.

(*c*) The working parts of the engine must be friction free.

(*d*) There should be no loss of energy during the cycle of operation.

Note: A reversible cycle should not be confused with a mechanically reversible engine. Steam engine cranks may be made to revolve in a reversed direction by mechanically altering the valve settings. But this does not reverse the cycle, on which it works. A two-stroke petrol engine may be made to revolve in reverse direction by altering the timing of ignition. But this also does not reverse the actual cycle.

2. *Irreversible cycle.* We have discussed above that in a reversible process, the heat and work are completely restored back by reversing the process (*i.e.* by compressing the gas). But when the heat and work are not completely restored back by reversing the process, then the process is known as *irreversible process* (also called *natural* or *real process*). In an irreversible process, there is a loss of heat due to friction, radiation or conduction.

In actual practice, most of the processes are irreversible to some degree. The main causes for the irreversibility are :

1. mechanical and fluid friction, 2. unrestricted expansion, and 3. heat transfer with a finite temperature difference. Moreover, friction converts the mechanical work into heat. This heat cannot supply back the same amount of mechanical work, which was consumed for its production. Thus, if there is some friction involved in the process, it becomes irreversible. A cycle will be irreversible if any of the processes, constituting the cycle, is irreversible. Thus in an irreversible cycle, the initial conditions are not restored at the end of the cycle.

1.52 Reversibility and Irreversibility of Thermodynamic Processes

We have already discussed the various thermodynamic processes in Art. 1.50. Now we shall discuss their conditions of reversibility and irreversibility.

1. *Isothermal and adiabatic processes.* It may be noted that a complete process or cycle is only an ideal case. But in actual practice, complete isothermal and adiabatic operations are not achieved. However, they can be approximated. The simple reason for the same is that it is impossible to transfer heat at a constant temperature in case of an isothermal operation. Moreover, it is also impossible to make an absolutely non-conducting cylinder in case of an adiabatic operation. In actual practice, however, an isothermal operation may be approached if the process is so slow that the heat is absorbed or rejected at such a rate that the temperature remains, practically, constant. Similarly, an adiabatic operation may be approached if the process takes place so quickly that no time is given to the heat to enter or leave the gas.

In view of the above, the isothermal and adiabatic processes are taken as reversible processes.

2. *Constant volume, constant pressure and constant pv^n processes.* We know that when the temperature of the hot body, supplying the heat, remains constant during the process, the temperature of the working substance will vary as the operation proceeds. In view of this, the above three operations are irreversible. But, these can be made to approximate to reversibility by manipulating the temperature of the hot body to vary so that at any stage the temperature of the working substance remains constant.

In this way, the constant volume, constant pressure and constant pv^n processes are regarded as reversible processes.

3. *Free expansion and throttling processes.* These processes are irreversible, as there is always a loss of heat due to friction when the working substance passes through an orifice.

1.53 Flow Processes

We have already discussed in Art. 1.47, that the processes occuring in open system which permit the transfer of mass to and from the system, are known as *flow processes*. In a flow process, the mass (working substance) enters the system and leaves after doing the work. The flow process may be classified as:

1. Steady flow process, and 2. Unsteady flow process.

In a steady flow process, the following conditions must be satisfied :

(*a*) The rate of mass flow at inlet and outlet is same, *i.e.* the mass flow rate through the system remains constant.

(*b*) The rate of heat transfer is constant.

(*c*) The rate of work transfer is constant.

(*d*) The state of working substance at any point within the system is same at all times.

(*e*) There is no change in the chemical composition of the system. Thus no chemical energy is involved.

If any one of these conditions are not satisfied, then the process is said to be non-steady flow process. In engineering, we are mainly concerned with steady flow processes.

1.54 Application of First Law of Thermodynamics to a Steady Flow Process

Consider an open system through which the working substance flows at a steady rate, as shown in Fig. 1.19. The working substance enters the system at section 1 and leaves the system at section 2.

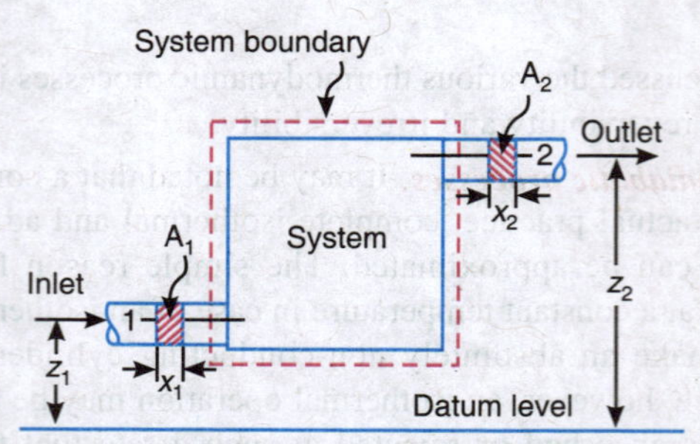

Fig. 1.19. Steady flow process.

Let p_1 = Pressure of the working substance entering the system in N/m^2,

v_{s1} = Specific volume of the working substance entering the system in m^3/kg.

V_1 = Velocity of the working substance entering the system in m/s,

u_1 = Specific internal energy of the working substance entering the system in J/kg,

z_1 = Height above datum level for inlet in metres,

p_2, v_{s2}, V_2, u_2 and z_2 = Corresponding values for the working substance leaving the system.

q_{1-2} = Heat supplied to the system in J/kg, and

w_{1-2} = Work delivered by the system in J/kg.

Consider 1 kg of mass of the working substance.

We know that total energy entering the system per kg of the working substance,

e_1 = Internal energy + *Flow or displacement energy + Kinetic energy + Potential energy + Heat supplied

$$= u_1 + p_1 v_{s1} + \frac{V_1^2}{2} + g\, z_1 + q_{1-2} \text{ (in J/kg)}$$

* The *flow* or *displacement energy* is the energy required to flow or move the working substance against its pressure. It is also known as *flow work.*

For example, let the working substance with pressure p_1 (in N/m^2) flows through area A_1 (in m^2) and moves through a distance x_1 (in metres).

∴ Energy or work required to flow the working substance,

$$FE = \text{Force} \times \text{Distance} = (p_1 A_1)\, x_1 = p_1 v_1 \text{ (in joules)} \quad \ldots (\because v_1 = A_1 . x_1)$$

where v_1 = Volume of the working substance in m^3,

For 1 kg mass of the working substance,

$v_1 = v_{s1}$ = Specific volume of the working substance in m^3/kg.

∴ $FE = p_1 v_{s1}$ (in J/kg)

Similarly, total energy leaving the system per kg of the working substance,

$$e_2 = u_2 + p_2 v_{s2} + \frac{V_2^2}{2} + g\,z_2 + w_{1-2} \text{ (in J/kg)}$$

Assuming no loss of energy during flow, then according to First Law of Thermodynamics (*i.e.* Law of Conservation of Energy), $e_1 = e_2$.

$$\therefore\; u_1 + p_1 v_{s1} + \frac{V_1^2}{2} + g\,z_1 + q_{1-2} = u_2 + p_2 v_{s2} + \frac{V_2^2}{2} + g\,z_2 + w_{1-2}$$

We know that

$u_1 + p_1 v_{s1} = h_1$ = Enthalpy of the working substance entering the system in J/kg, and

$u_2 + p_2 v_{s2} = h_2$ = Enthalpy of the working substance leaving the system in J/kg.

Thus, the above expression may be written as

$$h_1 + \frac{V_1^2}{2} + g\,z_1 + q_{1-2} = h_2 + \frac{V_2^2}{2} + g\,z_2 + w_{1-2} \quad \text{... (i)}$$

or

$$h_1 + ke_1 + pe_1 + q_{1-2} = h_2 + ke_2 + pe_2 + w_{1-2}$$

It may be noted that all the terms in equation (*i*) represent the energy flow per unit mass of the working substance (*i.e.* in J/kg). When the equation (*i*) is multiplied throughout by the mass of the working substance (m) in kg/s, then all the terms will represent the energy flow per unit time (*i.e.* in J/s).

Thus the equation (*i*) may also be written as

$$m\left(h_1 + \frac{V_1^2}{2} + g\,z_1 + q_{1-2}\right) = m\left(h_2 + \frac{V_2^2}{2} + g\,z_2 + w_{1-2}\right) \quad \text{... (ii)}$$

Both the equations (*i*) and (*ii*) are known as *steady flow energy equations*.

Notes: 1. In a steady flow, the mass flow rate (m) of the working substance entering and leaving the system is given by

$$m = \frac{A_1 V_1}{v_{s1}} = \frac{A_2 V_2}{v_{s2}} \text{ (in kg/s)}$$

This equation is known as *equation of continuity.*

2. The steady flow energy equation (*i*), for unit mass flow may be written as

$$q_{1-2} - w_{1-2} = (h_2 - h_1) + \left(\frac{V_2^2}{2} - \frac{V_1^2}{2}\right) + (g\,z_2 - g\,z_1) \quad \text{... (iii)}$$

$$= (h_2 - h_1) + (ke_2 - ke_1) + (pe_2 - pe_1)$$

In differential form, this expression is written as

$$\delta q - \delta w = dh + d\,(ke) - d\,(pe)$$

3. In thermodynamics, the effect of gravity is generally neglected. Therefore equation (*iii*) may be written as

$$q_{1-2} - w_{1-2} = (h_2 - h_1) + \left(\frac{V_2^2}{2} - \frac{V_1^2}{2}\right) \quad \text{... (iv)}$$

If $V_1 = V_2$, then equation (*iv*) reduces to

$$q_{1-2} - w_{1-2} = (h_2 - h_1) \quad \text{... (v)}$$

4. In a non-flow process, the flow or displacement energy at inlet and outlet is zero, *i.e.* $p_1 v_{s1} = 0$ and $p_2 v_{s2} = 0$. Therefore $h_2 = u_2$ and $h_1 = u_1$.

Thus the equation (*v*) may be written as $q_{1-2} - w_{1-2} = u_2 - u_1$; which is same as for non-flow process.

CHAPTER

2

Air Refrigeration Cycles

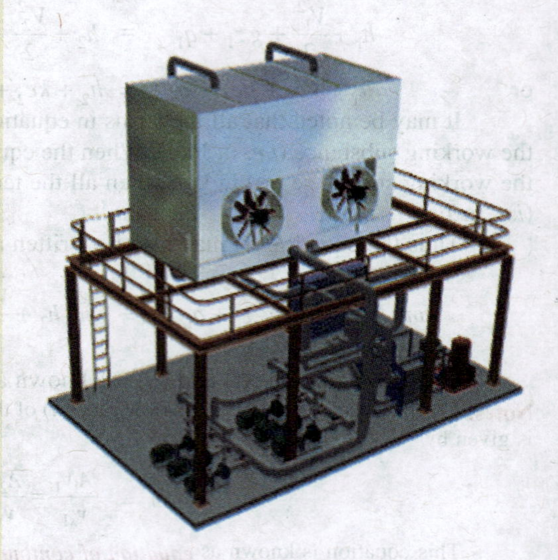

2.1 Introduction

In an air refrigeration cycle, the air is used as a refrigerant. In olden days, air was widely used in commercial applications because of its availability at free of cost. Since air does not change its phase *i.e.* remains gaseous throughout the cycle, therefore the heat carrying capacity per kg of air is very small as compared to vapour absorbing systems. The air-cycle refrigeration systems, as originally designed and installed, are now practically obsolete because of their low coefficient of performance and high power requirements. However, this system continues to be favoured for air refrigeration because of the low weight and volume of the equipment. The basic elements of an air cycle refrigeration system are the compressor, the cooler or heat exchanger, the expander and the refrigerator.

Before discussing the air refrigeration cycles, we should first know about the unit of refrigeration, coefficient of performance of a refrigerator and the difference between the heat engine, a refrigerator and a heat pump.

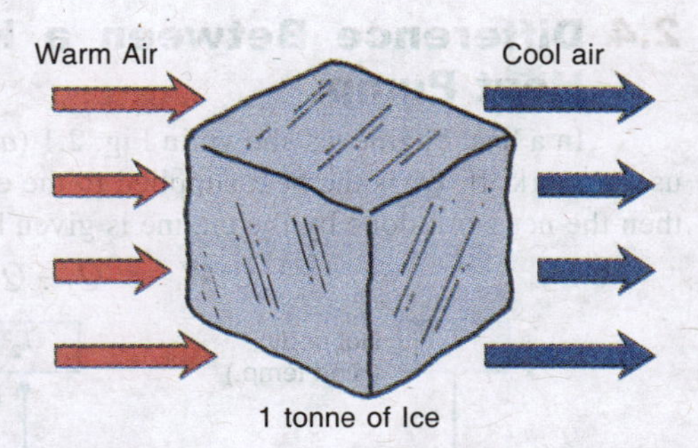

One tonne (1000 kg) of ice requires 335 kJ/kg to melt. When this is accomplished in 24 hours, it is known as a heat transfer rate of 1 tonne of refrigeration (1TR).

2.2 Units of Refrigeration

The practical unit of refrigeration is expressed in terms of 'tonne of refrigeration' (briefly written as TR). A *tonne of refrigeration* is defined as the amount of refrigeration effect produced by the uniform melting of one tonne (1000 kg) of ice from and at 0°C in 24 hours.

Since the latent heat of ice is 335 kJ/kg, therefore one tonne of refrigeration,

$$1\text{TR} = 1000 \times 335 \text{ kJ in 24 hours}$$

$$= \frac{1000 \times 335}{24 \times 60} = 232.6 \text{ kJ/min}$$

In actual practice, one tonne of refrigeration is taken as equivalent to 210 kJ/min or 3.5 kW (*i.e.* 3.5 kJ/s).

2.3 Coefficient of Performance of a Refrigerator

The coefficient of performance (briefly written as C.O.P.) is the ratio of heat extracted in the refrigerator to the work done on the refrigerant. It is also known as theoretical coefficient of performance. Mathematically,

$$\text{Theoretical C.O.P.} = \frac{Q}{W}$$

where Q = Amount of heat extracted in the refrigerator (or the amount of refrigeration produced, or the capacity of a refrigerator), and

W = Amount of work done.

Notes: 1. For per unit mass, $\text{C.O.P.} = \frac{q}{w}$

2. The coefficient of performance is the reciprocal of the efficiency (*i.e.* $1/\eta$) of a heat engine. It is thus obvious, that the value of C.O.P is always greater than unity.

3. The ratio of the actual C.O.P to the theoretical C.O.P. is known as *relative coefficient of performance.* Mathematically,

$$\text{Relative C.O.P.} = \frac{\text{Actual C.O.P.}}{\text{Theoretical C.O.P.}}$$

Example 2.1. *Find the C.O.P. of a refrigeration system if the work input is 80 kJ/kg and refrigeration effect produced is 160 kJ/kg of refrigerant flowing.*

Solution. Given : w = 80 kJ/kg ; q = 160 kJ/kg

We know that C.O.P. of a refrigeration system

$$= \frac{q}{w} = \frac{160}{80} = 2 \textbf{ Ans.}$$

2.4 Difference Between a Heat Engine, Refrigerator and Heat Pump

In a heat engine, as shown in Fig. 2.1 (*a*), the heat supplied to the engine is converted into useful work. If Q_2 is the heat supplied to the engine and Q_1 is the heat rejected from the engine, then the net work done by the engine is given by

$$W_E = Q_2 - Q_1$$

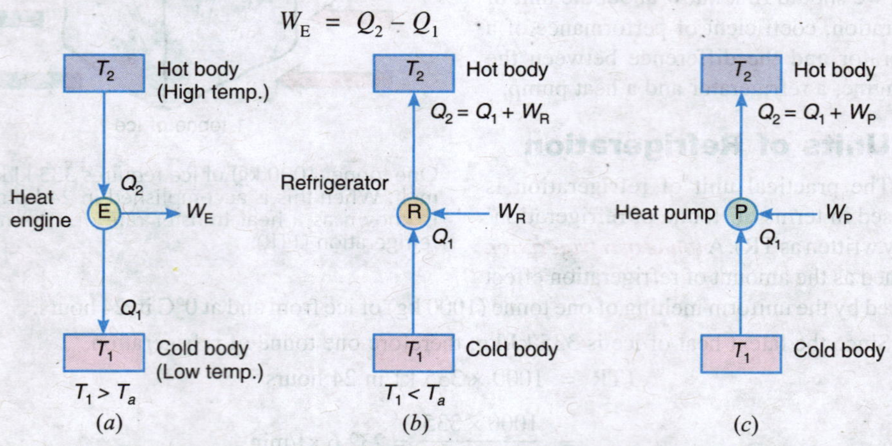

Fig. 2.1. Difference between a heat engine, refrigerator and heat pump.

The performance of a heat engine is expressed by its efficiency. We know that the efficiency or coefficient of performance of an engine,

$$\eta_E \text{ or } (\text{C.O.P.})_E = \frac{\text{Work done}}{\text{Heat supplied}} = \frac{W_E}{Q_2} = \frac{Q_2 - Q_1}{Q_2}$$

A refrigerator as shown in Fig. 2.1 (*b*), is a reversed heat engine which either cool or maintain the temperature of a body (T_1) lower than the atmospheric temperature (T_a). This is done by extracting the heat (Q_1) from a cold body and delivering it to a hot body (Q_2). In doing so, work W_R is required to be done on the system. According to First Law of Thermodynamics,

$$W_R = Q_2 - Q_1$$

The performance of a refrigerator is expressed by the ratio of amount of heat taken from the cold body (Q_1) to the amount of work required to be done on the system (W_R). This ratio is called coefficient of performance. Mathematically, coefficient of performance of a refrigerator,

$$(\text{C.O.P.})_R = \frac{Q_1}{W_R} = \frac{Q_1}{Q_2 - Q_1}$$

Any refrigerating system is a heat pump as shown in Fig. 2.1 (*c*), which extracts heat (Q_1) from a cold body and delivers it to a hot body. Thus there is no difference between the cycle of operations of a heat pump and a refrigerator. The main difference between the two is in their operating temperatures. A refrigerator works between the cold body temperature (T_1) and the atmospheric temperature (T_a) whereas the heat pump operates between the hot body temperature (T_2) and the atmospheric temperature (T_a). A refrigerator used for cooling in summer can be used as a heat pump for heating in winter.

In the similar way, as discussed for refrigerator, we have

$$W_P = Q_2 - Q_1$$

The performance of a heat pump is expressed by the ratio of the amount of heat delivered to the hot body (Q_2) to the amount of work required to be done on the system (W_P). This ratio is called coefficient of performance or energy performance ratio (E.P.R.) of a heat pump. Mathematically, coefficient of performance or energy performance ratio of a heat pump,

$$(\text{C.O.P.})_P \text{ or E.P.R.} = \frac{Q_2}{W_P} = \frac{Q_2}{Q_2 - Q_1}$$

$$= \frac{Q_1}{Q_2 - Q_1} + 1 = (\text{C.O.P.})_R + 1$$

From above we see that the C.O.P. may be less than one or greater than one depending on the type of refrigeration system used. But the C.O.P. of a heat pump is always greater than one.

2.5 Open Air Refrigeration Cycle

In an open air refrigeration cycle, the air is directly led to the space to be cooled (*i.e.* a refrigerator), allowed to circulate through the cooler and then returned to the compressor to start another cycle. Since the air is supplied to the refrigerator at atmospheric pressure, therefore, volume of air handled by the compressor and expander is large. Thus the size of compressor and expander should be large. Another disadvantage of the open cycle system is that the moisture is regularly carried away by the air circulated through the cooled space. This leads to the formation of frost at the end of expansion process and clog the line. Thus in an open cycle system, a drier should be used.

2.6 Closed or Dense Air Refrigeration Cycle

In a closed or dense air refrigeration cycle, the air is passed through the pipes and component parts of the system at all times. The air, in this system, is used for absorbing heat from the other fluid (say brine) and this cooled brine is circulated into the space to be cooled. The air in the closed system does not come in contact directly with the space to be cooled.

The closed air refrigeration cycle has the following thermodynamic advantages :

1. Since it can work at a suction pressure higher than that of atmospheric pressure, therefore the volume of air handled by the compressor and expander are smaller as compared to an open air refrigeration cycle system.
2. The operating pressure ratio can be reduced, which results in higher coefficient of performance.

2.7 Air Refrigerator Working on Reversed Carnot Cycle

In refrigerating systems, the Carnot cycle considered is the reversed Carnot cycle. We know that a heat engine working on Carnot cycle has the highest possible efficiency. Similarly, a refrigerating system working on the reversed Carnot cycle, will have the maximum possible coefficient of performance. We also know that it is not possible to make an engine working on the Carnot cycle. Similarly, it is also not possible to make a refrigerating machine working on the reversed Carnot cycle. However, it is used as the ultimate standard of comparison.

A reversed Carnot cycle, using air as working medium (or refrigerant) is shown on *p-v* and *T-s* diagrams in Fig. 2.2(*a*) and (*b*) respectively. At point 1, let p_1, v_1, T_1 be the pressure,volume and temperature of air respectively.

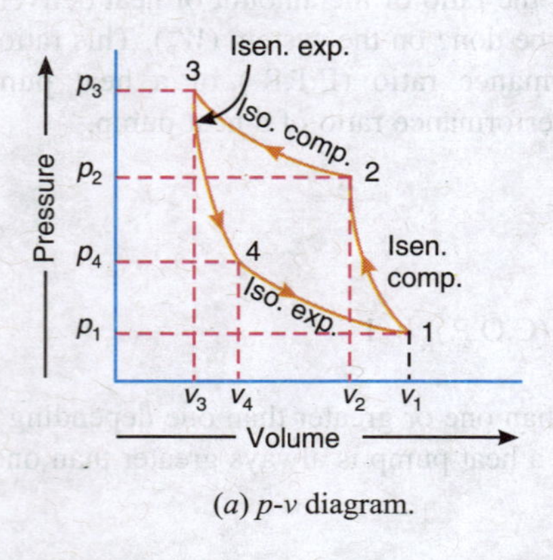

(*a*) *p-v* diagram.

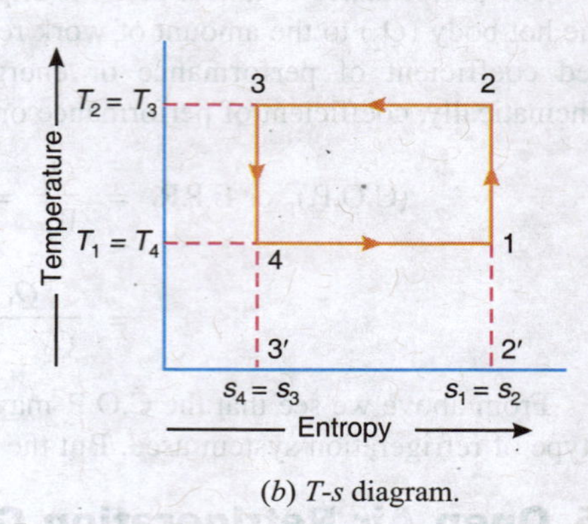

(*b*) *T-s* diagram.

Fig. 2.2. Reversed Carnot cycle.

The four processes of the cycle are as follows :

1. *Isentropic compression process.* The air is compressed isentropically as shown by the curve 1-2 on *p-v* and *T-s* diagrams. During this process, the pressure of air increases from p_1 to p_2, specific volume decreases from v_1 to v_2 and temperature increases from T_1 to T_2. We know that during isentropic compression, no heat is absorbed or rejected by the air.

2. *Isothermal compression process.* The air is now compressed isothermally (*i.e.* at constant temperature, $T_2 = T_3$) as shown by the curve 2-3 on *p-v* and *T-s* diagrams. During this process, the pressure of air increases from p_2 to p_3 and specific volume decreases from v_2 to v_3. We know that the heat rejected by the air during isothermal compression per kg of air,

$$q_R = q_{2-3} = \text{Area } 2\text{–}3\text{–}3'\text{–}2'$$

$$= T_3 (s_2 - s_3) = T_2 (s_2 - s_3)$$

3. *Isentropic expansion process.* The air is now expanded isentropically as shown by the curve 3-4 on *p-v* and *T-s* diagrams. The pressure of air decreases from p_3 to p_4, specific volume increases from v_3 to v_4 and the temperature decreases from T_3 to T_4. We know that during isentropic expansion, no heat is absorbed or rejected by the air.

4. *Isothermal expansion process.* The air is now expanded isothermally (*i.e.* at constant temperature, $T_4 = T_1$) as shown by the curve 4-1 on *p-v* and *T-s* diagrams. The pressure of air decreases from p_4 to p_1, and specific volume increases from v_4 to v_1. We know that the heat absorbed by the air (or heat extracted from the cold body) during isothermal expansion per kg of air,

$$q_A = q_{4-1} = \text{Area } 4\text{–}1\text{–}2'\text{–}3'$$

$$= T_4 (s_1 - s_4) = T_4 (s_2 - s_3) = T_1 (s_2 - s_3)$$

We know that work done during the cycle per kg of air

$$w_R = \text{*Heat rejected} - \text{Heat absorbed} = q_R - q_A = q_{2-3} - q_{4-1}$$

$$= T_2 (s_2 - s_3) - T_1 (s_2 - s_3) = (T_2 - T_1)(s_2 - s_3)$$

∴ Coefficient of performance of the refrigeration system working on reversed Carnot cycle,

$$(\text{C.O.P.})_R = \frac{\text{Heat absorbed}}{\text{Work done}} = \frac{q_A}{q_R - q_A} = \frac{q_{4-1}}{q_{2-3} - q_{4-1}}$$

* In a refrigerating machine, heat rejected is more than heat absorbed.

$$= \frac{T_1(s_2 - s_3)}{(T_2 - T_1)(s_2 - s_3)} = \frac{T_1}{T_2 - T_1}$$

Though the reversed Carnot cycle is the most efficient between the fixed temperature limits, yet no refrigerator has been made using this cycle. This is due to the reason that the isentropic processes of the cycle require high speed while the isothermal processes require an extremely low speed. This variation in speed of air is not practicable.

Note : We have already discussed that C.O.P. of a heat pump,

$$(\text{C.O.P.})_P = (\text{C.O.P.})_R + 1 = \frac{T_1}{T_2 - T_1} + 1 = \frac{T_2}{T_2 - T_1}$$

and C.O.P. or efficiency of a heat engine,

$$(\text{C.O.P.})_E = \frac{w_R}{q_R} = \frac{(T_2 - T_1)(s_2 - s_3)}{T_2(s_2 - s_3)} = \frac{T_2 - T_1}{T_2} = \frac{1}{(\text{C.O.P.})_P}$$

2.8 Temperature Limitations for Reversed Carnot Cycle

We have seen in the previous article that the C.O.P. of the refrigeration system working on reversed Carnot cycle is given by

$$(\text{C.O.P.})_R = \frac{T_1}{T_2 - T_1}$$

where T_1 = Lower temperature, and

T_2 = Higher temperature.

The C.O.P. of the reversed Carnot cycle may be improved by

1. decreasing the higher temperature (*i.e.* temperature of hot body, T_2), or
2. increasing the lower temperature (*i.e.* temperature of cold body, T_1).

This applies to all refrigerating machines, both theoretical and practical. It may be noted that temperatures T_1 and T_2 cannot be varied at will, due to certain functional limitations. It should be kept in mind that the higher temperature (T_2) is the temperature of cooling water or air available for rejection of heat and the lower temperature (T_1) is the temperature to be maintained in the refrigerator. The heat transfer will take place in the right direction only when the higher temperature is more than the temperature of cooling water or air to which heat is to be rejected, while the lower temperature must be less than the temperature of substance to be cooled.

Domestic Air Conditioner.

Thus, if the temperature of cooling water or air (*i.e.* T_2) available for heat rejection is low, the C.O.P. of the Carnot refrigerator will be high. Since T_2 in winter is less than T_2 in summer, therefore, C.O.P. in winter will be higher than C.O.P. in summer. In other words, the Carnot refrigerators work more efficiently in winter than in summer. Similarly, if the lower temperature fixed by the refrigeration application is high, the C.O.P. of the Carnot refrigerator will be high. Thus a Carnot refrigerator used for making ice at 0°C (273 K) will have less C.O.P. than a Carnot refrigerator used for air-conditioned plant in summer at 20°C when the atmospheric temperature is 40°C. In other words, we can say that the Carnot C.O.P. of a domestic refrigerator is less than the Carnot C.O.P. of a domestic air-conditioner.

Example 2.2. *A machine working on a Carnot cycle operates between 305 K and 260 K. Determine the C.O.P. when it is operated as: 1. a refrigerating machine; 2. a heat pump; and 3. a heat engine.*

Solution. Given : $T_2 = 305$ K ; $T_1 = 260$ K

1. C.O.P. of a refrigerating machine

We know that C.O.P. of a refrigerating machine,

$$(\text{C.O.P.})_R = \frac{T_1}{T_2 - T_1} = \frac{260}{305 - 260} = 5.78 \text{ Ans.}$$

2. C.O.P. of a heat pump

*We know that C.O.P. of a heat pump,

$$(\text{C.O.P.})_P = \frac{T_2}{T_2 - T_1} = \frac{305}{305 - 260} = 6.78 \text{ Ans.}$$

3. C.O.P. of a heat engine

**We know that C.O.P. of a heat engine,

$$(\text{C.O.P.})_E = \frac{T_2 - T_1}{T_2} = \frac{305 - 260}{305} = 0.147 \text{ Ans.}$$

Example 2.3. *A Carnot refrigeration cycle absorbs heat at 270 K and rejects it at 300 K.*

1. *Calculate the coefficient of performance of this refrigeration cycle.*
2. *If the cycle is absorbing 1130 kJ/min at 270 K, how many kJ of work is required per second ?*
3. *If the Carnot heat pump operates between the same temperatures as the above refrigeration cycle, what is the coefficient of performance ?*
4. *How many kJ/min will the heat pump deliver at 300 K if it absorbs 1130 kJ/min at 270 K.*

Solution. Given : $T_1 = 270$ K ; $T_2 = 300$ K

1. Coefficient of performance of Carnot refrigeration cycle

We know that coefficient of performance of Carnot refrigeration cycle,

$$(\text{C.O.P.})_R = \frac{T_1}{T_2 - T_1} = \frac{270}{300 - 270} = 9 \text{ Ans.}$$

* We know that C.O.P. of a heat pump, $(\text{C.O.P.})_P = (\text{C.O.P.})_R + 1 = 5.78 + 1 = 6.78$ **Ans.**

** We know that C.O.P. of a heat engine, $(\text{C.O.P.})_E = \frac{1}{(\text{C.O.P.})_P} = \frac{1}{6.78} = 0.147$ **Ans.**

2. *Work required per second*

Let W_R = Work required per second.

Heat absorbed at 270 K (*i.e.* T_1),

Q_1 = 1130 kJ/min = 18.83 kJ/s ...(Given)

We know that $$(\text{C.O.P.})_R = \frac{Q_1}{W_R} \quad \text{or} \quad 9 = \frac{18.83}{W_R}$$

$\therefore$ W_R = 2.1 kJ/s **Ans.**

3. *Coefficient of performance of Carnot heat pump*

We know that coefficient of performance of a Carnot heat pump,

$$(\text{C.O.P.})_P = \frac{T_2}{T_2 - T_1} = \frac{300}{300 - 270} = 10 \text{ Ans.}$$

4. *Heat delivered by heat pump at 300 K*

Let Q_2 = Heat delivered by heat pump at 300 K.

Heat absorbed at 270 K (*i.e.* T_1),

Q_1 = 1130 kJ/min ... (Given)

We know that

$$(\text{C.O.P.})_P = \frac{Q_2}{Q_2 - Q_1} \quad \text{or} \quad 10 = \frac{Q_2}{Q_2 - 1130}$$

$\therefore$ $10Q_2 - 11\,300 = Q_2$ or $Q_2 = 1256$ kJ/min **Ans.**

Example 2.4. *A cold storage is to be maintained at –5°C while the surroundings are at 35°C. The heat leakage from the surroundings into the cold storage is estimated to be 29 kW. The actual C.O.P. of the refrigeration plant is one-third of an ideal plant working between the same temperatures. Find the power required to drive the plant.*

Solution. Given : T_1 = –5°C = –5 + 273 = 268 K;
T_2 = 35°C = 35 + 273 = 308 K ; Q_1 = 29 kW ;

$$(\text{C.O.P.})_{actual} = \frac{1}{3}(\text{C.O.P.})_{ideal}$$

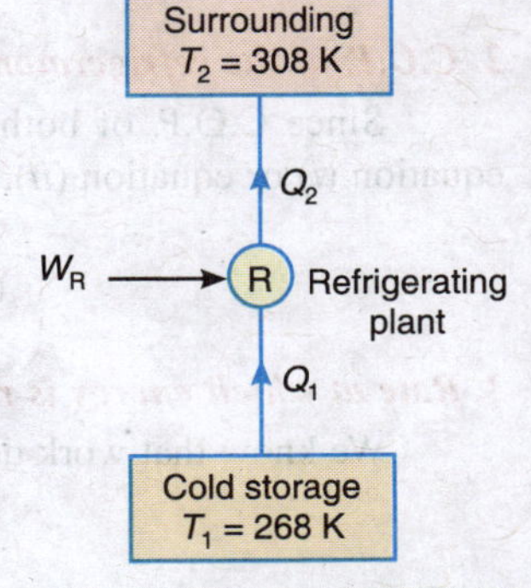

Fig. 2.3

The refrigerating plant operating between the temperatures T_1 and T_2 is shown in Fig. 2.3.

Let W_R = Work or power required to drive the plant.

We know that the coefficient of performance of an ideal refrigeration plant,

$$(\text{C.O.P.})_{ideal} = \frac{T_1}{T_2 - T_1} = \frac{268}{308 - 268} = 6.7$$

$\therefore$ Actual coefficient of performance,

$$(\text{C.O.P.})_{actual} = \frac{1}{3}(\text{C.O.P.})_{ideal} = \frac{1}{3} \times 6.7 = 2.233$$

We also know that $$(\text{C.O.P.})_{actual} = \frac{Q_1}{W_R}$$

$$\therefore \quad W_R = \frac{Q_1}{(\text{C.O.P.})_{actual}} = \frac{29}{2.233} = 12.987 \text{ kW Ans.}$$

Example 2.5. *Two refrigerators A and B operate in series. The refrigerator A absorbs energy at the rate of 1kJ/s from a body at temperature 300 K and rejects energy as heat to a body at temperature T. The refrigerator B absorbs the same quantity of energy which is rejected by the refrigerator A from the body at temperature T, and rejects energy as heat to a body at temperature 1000 K. If both the refrigerators have the same C.O.P., calculate:*

1. The temperature T of the body;

2. The C.O.P. of the refrigerators; and

3. The rate at which energy is rejected as heat to the body at 1000 K.

Solution. Given : $Q_1 = 1$ kJ/s ; $T_1 = 300$ K ; $T_2 = T$; $T_3 = 1000$ K

The arrangement of the refrigerators A and B is shown in Fig. 2.4.

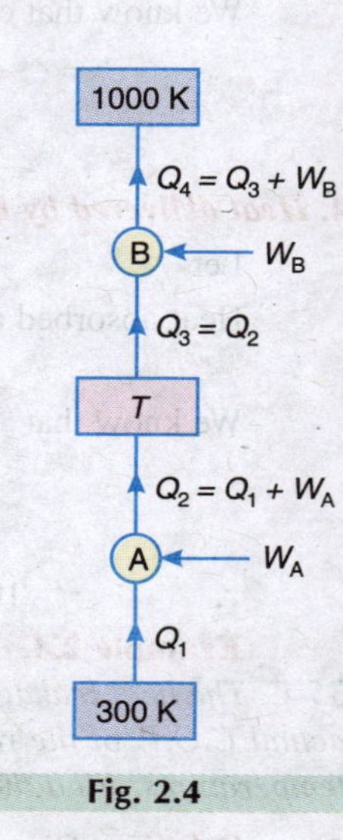

Fig. 2.4

1. *Temperature T of the body*

We know that C.O.P. for refrigerator A,

$$(\text{C.O.P.})_A = \frac{T_1}{T_2 - T_1} = \frac{300}{T - 300} \qquad ...(i)$$

and C.O.P. for refrigerator B,

$$(\text{C.O.P.})_B = \frac{T_2}{T_3 - T_2} = \frac{T}{1000 - T} \qquad ...(ii)$$

Since C.O.P. of both the refrigerators is same, therefore equating equations (*i*) and (*ii*),

$$\frac{300}{T - 300} = \frac{T}{1000 - T}$$

or $\quad 300 \times 1000 - 300\,T = T^2 - 300\,T$

$$\therefore \qquad T = \sqrt{300 \times 1000} = 547.7 \text{ K } \textbf{Ans.}$$

2. *C.O.P. of the refrigerators*

Since C.O.P. of both the refrigerators is same, therefore substituting the value of T in equation (*i*) or equation (*ii*),

$$(\text{C.O.P.})_A = (\text{C.O.P.})_B = \frac{300}{547.7 - 300} = 1.21 \textbf{ Ans.}$$

3. *Rate at which energy is rejected as heat to the body at 1000 K*

We know that work done by refrigerator A,

$$W_A = \frac{Q_1}{(\text{C.O.P.})_A} = \frac{1}{1.21} = 0.826 \text{ kJ/s}$$

and heat rejected by refrigerator A,

$$Q_2 = Q_1 + W_A = 1 + 0.826 = 1.826 \text{ kJ/s}$$

Now workdone by refrigerator B,

$$W_B = \frac{Q_3}{(\text{C.O.P.})_B} = \frac{1.826}{1.21} = 1.51 \text{ kJ/s} \qquad ... (\because Q_3 = Q_2)$$

$\therefore$ Heat rejected to the body at 1000 K,

$$Q_4 = Q_3 + W_B = 1.826 + 1.51 = 3.336 \text{ kJ/s } \textbf{Ans.}$$

Example 2.6. *A refrigerating system operates on the reversed Carnot cycle. The higher temperature of the refrigerant in the system is 35°C and the lower temperature is – 15°C. The capacity is to be 12 tonnes. Determine : 1. C.O.P. ; 2. Heat rejected from the system per hour ; and 3. Power required.*

Solution. Given : $T_2 = 35°C = 35 + 273 = 308$ K ; $T_1 = -15°C = -15 + 273 = 258$ K ; $Q_1 = 12$ TR $= 12 \times 210 = 2520$ kJ/min

The refrigerating system operating on the reversed Carnot cycle is shown in Fig. 2.5.

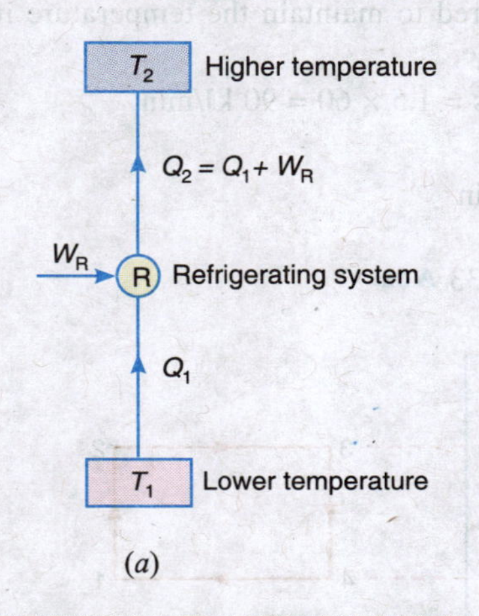

(a)

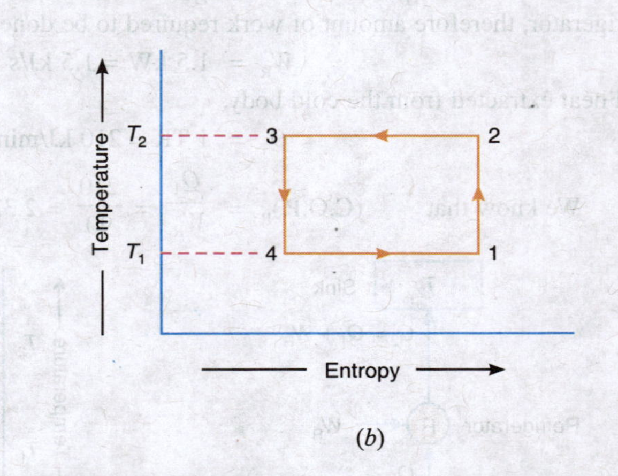

(b)

Fig. 2.5

1. *C.O.P.*

We know that

$$(\text{C.O.P.})_R = \frac{T_1}{T_2 - T_1} = \frac{258}{308 - 258} = 5.16 \textbf{ Ans.}$$

2. *Heat rejected from the system per hour*

Let W_R = Work or power required to drive the system.

We know that $(\text{C.O.P.})_R = \dfrac{Q_1}{W_R}$

$$\therefore \quad W_R = \frac{Q_1}{(\text{C.O.P.})_R} = \frac{2520}{5.16} = 488.37 \text{ kJ / min}$$

and heat rejected from the system,

$$Q_2 = Q_1 + W_R = 2520 + 488.37 = 3008.37 \text{ kJ/min}$$

$$= 3008.37 \times 60 = 180\,502.2 \text{ kJ / h } \textbf{Ans.}$$

3. *Power required*

We know that work or power required,

$$W_R = 488.37 \text{ kJ/ min} = \frac{488.37}{60} = 8.14 \text{ kJ/s or kW } \textbf{Ans.}$$

Example 2.7. *1.5 kW per tonne of refrigeration is required to maintain the temperature of – 40°C in the refrigerator. If the refrigeration cycle works on Carnot cycle, determine the following:*

1. C.O.P. of the cycle ; 2. Temperature of the sink ; 3. Heat rejected to the sink per tonne of refrigeration ; and 4. Heat supplied and E.P.R., if the cycle is used as a heat pump.

Solution. Given : $W_R = 1.5$ kW ; $Q_1 = 1$ TR ; $T_1 = -40°C = -40 + 273 = 233$ K

1. *C.O.P. of the cycle*

The refrigeration cycle working on Carnot cycle is shown in Fig 2.6.

Since 1.5 kW per tonne of refrigeration is required to maintain the temperature in the refrigerator, therefore amount of work required to be done,

$$W_R = 1.5 \text{ kW} = 1.5 \text{ kJ/s} = 1.5 \times 60 = 90 \text{ kJ/min}$$

and heat extracted from the cold body,

$$Q_1 = 1 \text{ TR} = 210 \text{ kJ/min}$$

We know that $$(\text{C.O.P.})_R = \frac{Q_1}{W_R} = \frac{210}{90} = 2.33 \textbf{ Ans.}$$

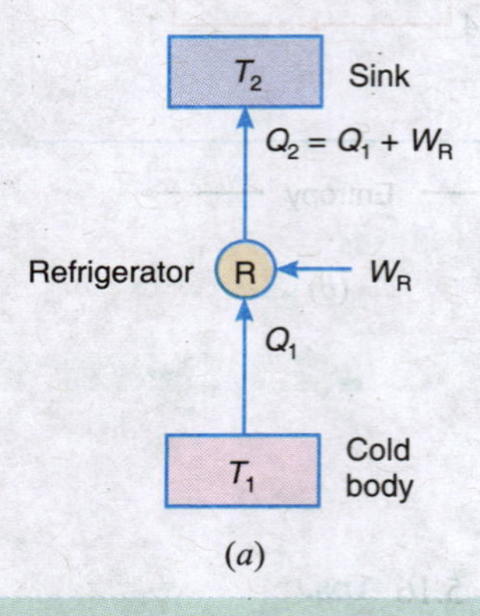

(*a*)

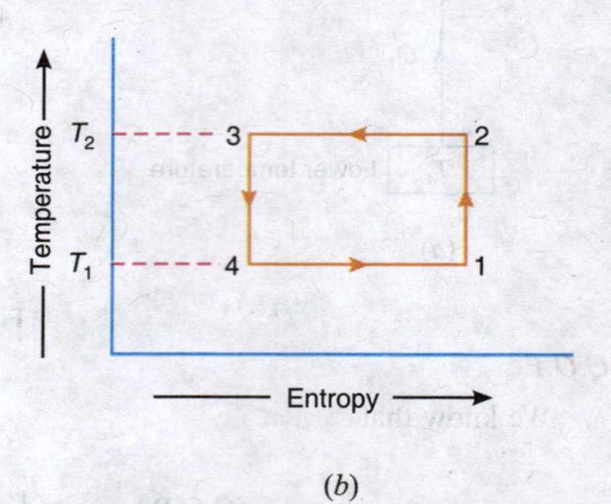

(*b*)

Fig. 2.6

2. *Temperature of the sink*

Let T_2 = Temperature of the sink.

We know that $$(\text{C.O.P.})_R = \frac{T_1}{T_2 - T_1} \quad \text{or} \quad 2.33 = \frac{233}{T_2 - 233}$$

$$\therefore \quad T_2 = \frac{233}{2.33} + 233 = 333 \text{ K} = 333 - 273 = 60°\text{C} \textbf{ Ans.}$$

3. *Heat rejected to the sink per tonne of refrigeration*

We know that heat rejected to the sink,

$$Q_2 = Q_1 + W_R = 210 + 90 = 300 \text{ kJ/min} \textbf{ Ans.}$$

4. *Heat supplied and E.P.R., if the cycle is used as a heat pump*

We know that heat supplied when the cycle is used as a heat pump is

$$Q_2 = 300 \text{ kJ/min} \textbf{ Ans.}$$

and $$\text{E.P.R.} = (\text{C.O.P.})_R + 1 = 2.33 + 1 = 3.33 \textbf{ Ans.}$$

Example 2.8. *The capacity of a refrigerator is 200 TR when working between – 6°C and 25°C. Determine the mass of ice produced per day from water at 25°C. Also find the power required to drive the unit. Assume that the cycle operates on reversed Carnot cycle and latent heat of ice is 335 kJ/kg.*

Solution. Given : $Q = 200$ TR ; $T_1 = -6°C = -6 + 273 = 267$ K ; $T_2 = 25°C = 25 + 273 = 298$ K ; $t_w = 25°C$; $h_{fg(ice)} = 335$ kJ/kg

Mass of ice produced per day

We know that heat extraction capacity of the refrigerator

$= 200 \times 210 = 42\ 000$ kJ/min ... ($\because$ 1TR = 210 kJ/min)

and heat removed from 1 kg of water at 25°C to form ice at 0°C

= Mass × Sp. heat × Rise in temperature + $h_{fg(ice)}$

$= 1 \times 4.187(25 - 0) + 335 = 439.7$ kJ / kg

$\therefore$ Mass of ice produced per min

$$= \frac{42\ 000}{439.7} = 95.52 \text{ kg / min}$$

and mass of ice produced per day $= 95.52 \times 60 \times 24 = 137\ 550$ kg = 137.55 tonnes **Ans.**

Power required to drive the unit

We know that C.O.P. of the reversed Carnot cycle

$$= \frac{T_1}{T_2 - T_1} = \frac{267}{298 - 267} = 8.6$$

Also $$\text{C.O.P.} = \frac{\text{Heat extraction capacity}}{\text{Work done per min}}$$

$$\therefore \quad 8.6 = \frac{42\ 000}{\text{Work done per min}}$$

or Work done per min $= 42\ 000 / 8.6 = 4884$ kJ/min

$\therefore$ Power required to drive the unit

$= 4884 / 60 = 81.4$ kW **Ans.**

Example 2.9. *Five hundred kgs of fruits are supplied to a cold storage at 20°C. The cold storage is maintained at –5°C and the fruits get cooled to the storage temperature in 10 hours. The latent heat of freezing is 105 kJ/kg and specific heat of fruit is 1.256 kJ/kg K. Find the refrigeration capacity of the plant.*

Solution. Given : $m = 500$ kg ; $T_2 = 20°C = 20 + 273 = 293$ K; $T_1 = -5°C = -5 + 273 = 268$ K ; $h_{fg} = 105$ kJ/kg ; $c_F = 1.256$ kJ/kg K

We know that heat removed from the fruits in 10 hours,

$$Q_1 = m\, c_F\, (T_2 - T_1)$$

$$= 500 \times 1.256\ (293 - 268) = 15\ 700 \text{ kJ}$$

and total latent heat of freezing,

$$Q_2 = m \times h_{fg} = 500 \times 105 = 52\ 500 \text{ kJ}$$

$\therefore$ Total heat removed in 10 hours,

$$Q = Q_1 + Q_2 = 15\ 700 + 52\ 500 = 68\ 200 \text{ kJ}$$

and total heat removed in one minute

$= 68\,200/10 \times 60 = 113.7$ kJ/min

$\therefore$ Refrigeration capacity of the plant

$= 113.7/210 = 0.541$ TR **Ans.** ...($\because$ 1TR = 210 kJ/min)

Example 2.10. *A cold storage plant is required to store 20 tonnes of fish. The fish is supplied at a temperature of 30°C. The specific heat of fish above freezing point is 2.93 kJ/kg K. The specific heat of fish below freezing point is 1.26 kJ/kg K. The fish is stored in cold storage which is maintained at – 8°C. The freezing point of fish is – 4°C. The latent heat of fish is 235 kJ/kg. If the plant requires 75 kW to drive it, find :*

1. The capacity of the plant, and 2. Time taken to achieve cooling.

Assume actual C.O.P. of the plant as 0.3 of the Carnot C.O.P.

Solution. Given : $m = 20$ t $= 20\,000$ kg ; $T_2 = 30°C = 30 + 273 = 303$ K ; $c_{AF} = 2.93$ kJ/kg K ; $c_{BF} = 1.26$ kJ/kgK ; $T_1 = -8°C = -8 + 273 = 265$ K ; $T_3 = -4°C = -4 + 273 = 269$ K; $h_{fg(Fish)} = 235$ kJ/kg ; $P = 75$ kW $= 75$ kJ/s

1. Capacity of the plant

We know that Carnot C.O.P.

$$= \frac{T_1}{T_2 - T_1} = \frac{265}{303 - 265} = 6.97$$

$\therefore$ Actual C.O.P. $= 0.3 \times 6.97 = 2.091$

and heat removed by the plant = Actual C.O.P. × Work required

$= 2.091 \times 75 = 156.8$ kJ/s

$= 156.8 \times 60$ kJ/min $= 9408$ kJ/min

$\therefore$ Capacity of the plant

$= 9408 / 210 = 44.8$ TR **Ans.** ... ($\because$ 1TR = 210 kJ/min)

2. Time taken to achieve cooling

We know that heat removed from the fish above freezing point,

$$Q_1 = m \times c_{AF}(T_2 - T_3)$$

$$= 20\,000 \times 2.93\,(303 - 269) = 1.992 \times 10^6 \text{ kJ}$$

Similarly, heat removed from the fish below freezing point,

$$Q_2 = m \times c_{BF}(T_3 - T_1)$$

$$= 20\,000 \times 1.26\,(269 - 265) = 0.101 \times 10^6 \text{ kJ}$$

and total latent heat of fish,

$$Q_3 = m \times h_{fg(Fish)} = 20\,000 \times 235 = 4.7 \times 10^6 \text{ kJ}$$

$\therefore$ Total heat removed by the plant

$$= Q_1 + Q_2 + Q_3$$

$$= 1.992 \times 10^6 + 0.101 \times 10^6 + 4.7 \times 10^6 = 6.793 \times 10^6 \text{ kJ}$$

and time taken to achieve cooling

$$= \frac{\text{Total heat removed by the plant}}{\text{Heat removed by the plant per min}}$$

$$= \frac{6.793 \times 10^6}{9408} = 722 \text{ min} = 12.03 \text{ h}$$ **Ans.**

2.9 Air Refrigerator Working on a Bell-Coleman Cycle (or Reversed Brayton or Joule Cycle)

A Bell-Coleman air refrigeration machine was developed by Bell-Coleman and Light Foot by reversing the Joule's air cycle. It was one of the earliest types of refrigerators used in ships carrying frozen meat. Fig. 2.7 shows a schematic diagram of such a machine which consists of a compressor, a cooler, an expander and a refrigerator.

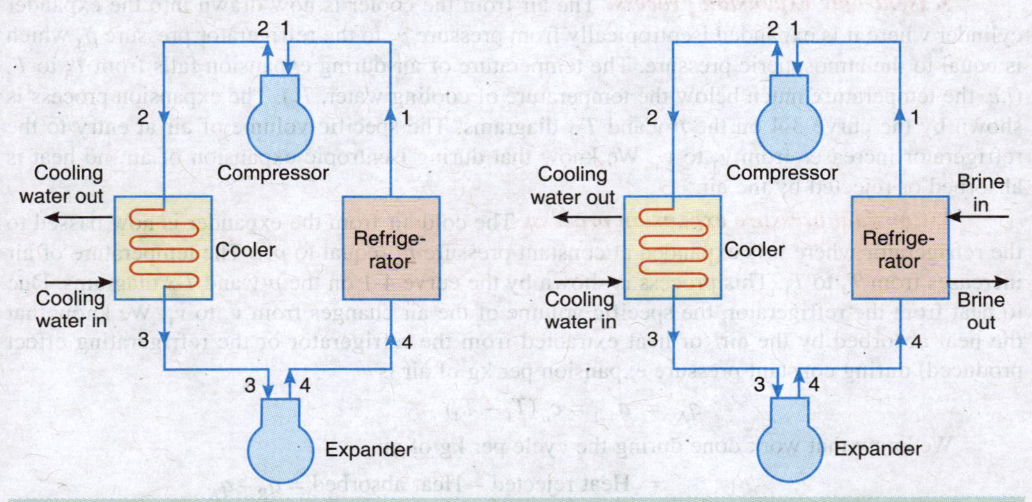

Fig. 2.7. Open cycle air Bell-Coleman Refrigerator. **Fig. 2.8.** Closed cycle or dense air Bell-Coleman Refrigerator.

The Bell-Coleman cycle (also known as reversed Brayton or Joule cycle) is a modification of reversed Carnot cycle. The cycle is shown on *p-v* and *T-s* diagrams in Fig. 2.9 (*a*) and (*b*). At point 1, let p_1, v_1 and T_1 be the pressure, volume and temperature of air respectively. The four processes of the cycle are as follows :

1. *Isentropic compression process.* The cold air from the refrigerator is drawn into the compressor cylinder where it is compressed isentropically in the compressor as shown by the curve 1–2 on *p-v* and *T-s* diagrams. During the compression stroke, both the pressure and temperature increases and the specific volume of air at delivery from compressor reduces from v_1 to v_2. We know that during isentropic compression process, no heat is absorbed or rejected by the air.

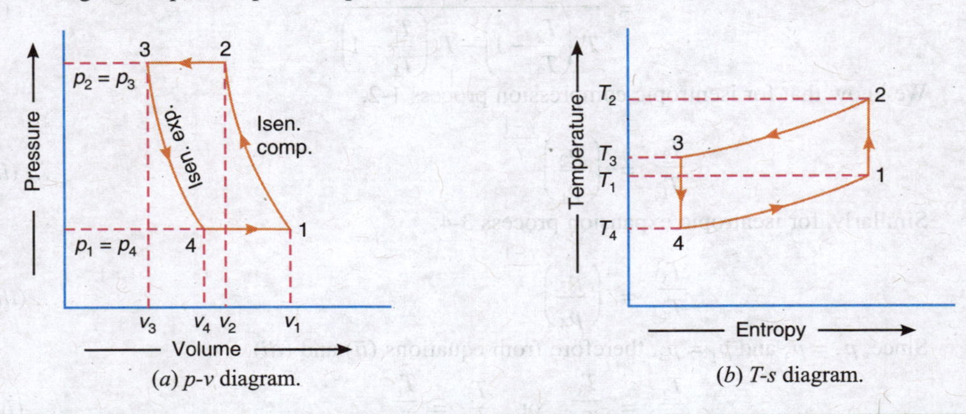

Fig. 2.9. Bell-Coleman cycle.

2. *Constant pressure cooling process*. The warm air from the compressor is now passed into the cooler where it is cooled at constant pressure p_3 (equal to p_2), reducing the temperature from T_2 to T_3 (the temperature of cooling water) as shown by the curve 2-3 on *p-v* and *T-s* diagrams. The specific volume also reduces from v_2 to v_3. We know that heat rejected by the air during constant pressure per kg of air,

$$q_R = Q_{2-3} = c_p (T_2 - T_3)$$

3. *Isentropic expansion process*. The air from the cooler is now drawn into the expander cylinder where it is expanded isentropically from pressure p_3 to the refrigerator pressure p_4 which is equal to the atmospheric pressure. The temperature of air during expansion falls from T_3 to T_4 (*i.e.* the temperature much below the temperature of cooling water, T_3). The expansion process is shown by the curve 3-4 on the *p-v* and *T-s* diagrams. The specific volume of air at entry to the refrigerator increases from v_3 to v_4. We know that during isentropic expansion of air, no heat is absorbed or rejected by the air.

4. *Constant pressure expansion process*. The cold air from the expander is now passed to the refrigerator where it is expanded at constant pressure p_4 (equal to p_1). The temperature of air increases from T_4 to T_1. This process is shown by the curve 4-1 on the *p-v* and *T-s* diagrams. Due to heat from the refrigerator, the specific volume of the air changes from v_4 to v_1. We know that the heat absorbed by the air (or heat extracted from the refrigerator or the refrigerating effect produced) during constant pressure expansion per kg of air is

$$q_A = q_{4-1} = c_p (T_1 - T_4)$$

We know that work done during the cycle per kg of air

$$= \text{Heat rejected} - \text{Heat absorbed} = q_R - q_A$$

$$= c_p (T_2 - T_3) - c_p (T_1 - T_4)$$

∴ Coefficient of performance,

$$\text{C.O.P.} = \frac{\text{Heat absorbed}}{\text{Work done}} = \frac{q_A}{q_R - q_A} = \frac{c_p (T_1 - T_4)}{c_p (T_2 - T_3) - c_p (T_1 - T_4)}$$

$$= \frac{(T_1 - T_4)}{(T_2 - T_3) - (T_1 - T_4)}$$

$$= \frac{T_4\left(\frac{T_1}{T_4} - 1\right)}{T_3\left(\frac{T_2}{T_3} - 1\right) - T_4\left(\frac{T_1}{T_4} - 1\right)} \quad \ldots (i)$$

We know that for isentropic compression process 1-2,

$$\frac{T_2}{T_1} = \left(\frac{p_2}{p_1}\right)^{\frac{\gamma - 1}{\gamma}} \quad \ldots (ii)$$

Similarly, for isentropic expansion process 3-4,

$$\frac{T_3}{T_4} = \left(\frac{p_3}{p_4}\right)^{\frac{\gamma - 1}{\gamma}} \quad \ldots (iii)$$

Since, $p_2 = p_3$ and $p_1 = p_4$, therefore from equations (*ii*) and (*iii*),

$$\frac{T_2}{T_1} = \frac{T_3}{T_4} \quad \text{or} \quad \frac{T_2}{T_3} = \frac{T_1}{T_4} \quad \ldots (iv)$$

Now substituting these values in equation (*i*), we get

$$\text{C.O.P.} = \frac{T_4}{T_3 - T_4} = \frac{1}{\dfrac{T_3}{T_4} - 1}$$

$$= \frac{1}{\left(\dfrac{p_3}{p_4}\right)^{\frac{\gamma-1}{\gamma}} - 1} = \frac{1}{\left(\dfrac{p_2}{p_1}\right)^{\frac{\gamma-1}{\gamma}} - 1} = \frac{1}{(r_p)^{\frac{\gamma-1}{\gamma}} - 1} \quad \ldots (v)$$

where r_p = Compression or Expansion ratio $= \dfrac{p_2}{p_1} = \dfrac{p_3}{p_4}$

Sometimes, the compression and expansion processes take place according to the law pv^n = Constant. In such a case, the C.O.P. is obtained from the fundamentals as discussed below :

We know that work done by the compressor during the process 1-2 per kg of air,

$$w_C = \frac{n}{n-1}(p_2 v_2 - p_1 v_1) = \frac{n}{n-1}(RT_2 - RT_1) \quad \ldots (\because pv = RT)$$

and work done by the expander during the process 3-4 per kg of air,

$$w_E = \frac{n}{n-1}(p_3 v_3 - p_4 v_4) = \frac{n}{n-1}(RT_3 - RT_4)$$

∴ Net work done during the cycle per kg of air,

$$w = w_C - w_E = \frac{n}{n-1} \times R\left[(T_2 - T_1) - (T_3 - T_4)\right]$$

We also know that heat absorbed during constant pressure process 4-1,

$$= c_p (T_1 - T_4)$$

$$\therefore \quad \text{C.O.P.} = \frac{\text{Heat absorbed}}{\text{Work done}} = \frac{q_A}{w} = \frac{c_p(T_1 - T_4)}{\dfrac{n}{n-1} \times R\left[(T_2 - T_1) - (T_3 - T_4)\right]} \quad \ldots (vi)$$

We know that $R = c_p - c_v = c_v (\gamma - 1)$

Substituting the value of R in equation (*vi*),

$$\text{C.O.P.} = \frac{c_p(T_1 - T_4)}{\dfrac{n}{n-1} \times c_v(\gamma - 1)\left[(T_2 - T_1) - (T_3 - T_4)\right]}$$

$$= \frac{\gamma(T_1 - T_4)}{\dfrac{n}{n-1} \times (\gamma - 1)\left[(T_2 - T_1) - (T_3 - T_4)\right]} \quad \ldots \left[\because \frac{c_p}{c_v} = \gamma\right]$$

$$= \frac{T_1 - T_4}{\dfrac{n}{n-1} \times \dfrac{(\gamma - 1)}{\gamma}\left[(T_2 - T_3) - (T_1 - T_4)\right]} \quad \ldots (vii)$$

Notes : 1. In this case, the values of T_2 and T_4 are to be obtained from the following relations:

$$\frac{T_2}{T_1} = \left(\frac{p_2}{p_1}\right)^{\frac{n-1}{n}} \quad \text{and} \quad \frac{T_3}{T_4} = \left(\frac{p_3}{p_4}\right)^{\frac{n-1}{n}}$$

2. For isentropic compression or expansion, $n = \gamma$. Therefore, the equation (*vii*) may be written as

$$\text{C.O.P.} = \frac{T_1 - T_4}{(T_2 - T_3) - (T_1 - T_4)} \quad \text{... (same as before)}$$

3. We have already discussed that the main drawback of the open cycle air refrigerator is freezing of the moisture in the air during expansion stroke which is liable to choke up the valves. Due to this reason, a closed cycle or dense air Bell-Coleman refrigerator as shown in Fig. 2.8 is preferred. In this case, the cold air does not come in direct contact of the refrigerator. The cold air is passed through the pipes and it is used for absorbing heat from the brine and this cooled brine is circulated in the refrigerated space. The term 'dense air system' is derived from the fact that the suction to the compressor is at higher pressure than the open cycle system (which is atmospheric).

Example 2.11. *In a refrigeration plant working on Bell Coleman cycle, air is compressed to 5 bar from 1 bar. Its initial temperature is 10°C. After compression, the air is cooled up to 20°C in a cooler before expanding back to a pressure of 1 bar. Determine the theoretical C.O.P. of the plant and net refrigerating effect. Take* c_p = 1.005 kJ/kg K and c_v = *0.718 kJ/kg K.*

Solution. Given : $p_2 = p_3 = 5$ bar ; $p_1 = p_4 = 1$ bar ; T_1 = 10°C = 10 + 273 = 283 K ; T_3 = 20°C = 20 + 273 = 293 K ; c_p = 1.005 kJ/kg K ; c_v= 0.718 kJ/kg K

The *p-v* and *T-s* diagrams for a refrigeration plant working on Bell-Coleman cycle, is shown in Fig. 2.10 (*a*) and (*b*) respectively.

Let T_2 and T_4 = Temperature of air at the end of compression and expansion respectively.

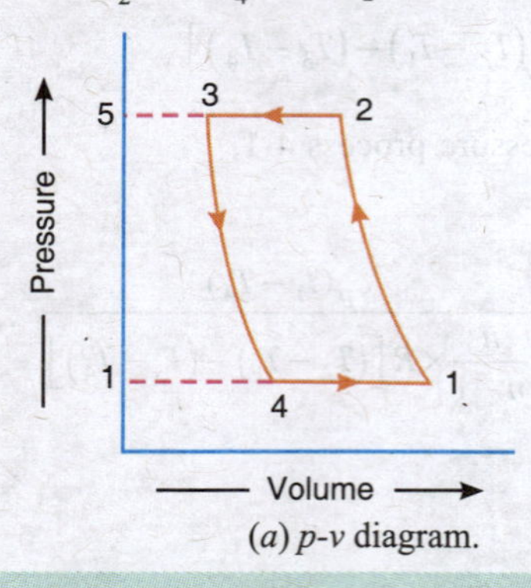

(*a*) *p-v* diagram.

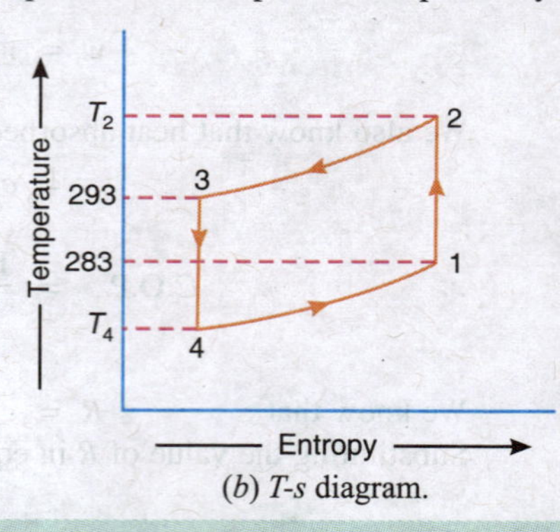

(*b*) *T-s* diagram.

Fig. 2.10

We know that isentropic index for compression and expansion process,

$$\gamma = c_p / c_v = 1.005/0.718 = 1.4$$

For isentropic compression process 1-2,

$$\frac{T_2}{T_1} = \left(\frac{p_2}{p_1}\right)^{\frac{\gamma-1}{\gamma}} = \left(\frac{5}{1}\right)^{\frac{1.4-1}{1.4}} = (5)^{0.286} = 1.584$$

and for isentropic expansion process 3-4

$$\frac{T_3}{T_4} = \left(\frac{p_3}{p_4}\right)^{\frac{\gamma-1}{\gamma}} = \left(\frac{5}{1}\right)^{\frac{1.4-1}{1.4}} = (5)^{0.286} = 1.584$$

$$\therefore \qquad T_4 = T_3 / 1.584 = 293/1.584 = 185 \text{ K}$$

Theoretical C.O.P. of the plant

We know that *theoretical C.O.P. of the plant,

$$= \frac{T_4}{T_3 - T_4} = \frac{185}{293 - 185} = 1.713 \textbf{ Ans.}$$

Net refrigerating effect

We know that net refrigerating effect (*i.e.* heat absorbed during constant pressure process 4–1)

$$= c_p (T_1 - T_4) = 1.005 (283 - 185) = 98.5 \text{ kJ/kg} \textbf{ Ans.}$$

Example 2.12. *A refrigerator working on Bell-Coleman cycle operates between pressure limits of 1.05 bar and 8.5 bar. Air is drawn from the cold chamber at 10°C, compressed and then it is cooled to 30°C before entering the expansion cylinder. The expansion and compression follows the law $pv^{1.3}$ = constant. Determine the theoretical C.O.P. of the system.*

Solution. Given : $p_1 = p_4 = 1.05$ bar ; $p_2 = p_3 = 8.5$ bar ; $T_1 = 10°C = 10 + 273 = 283$ K ; $T_3 = 30°C = 30 + 273 = 303$ K ; $n = 1.3$

The *p-v* and *T-s* diagrams for a refrigerator working on the Bell-Coleman cycle is shown in Fig. 2.11 (*a*) and (*b*) respectively.

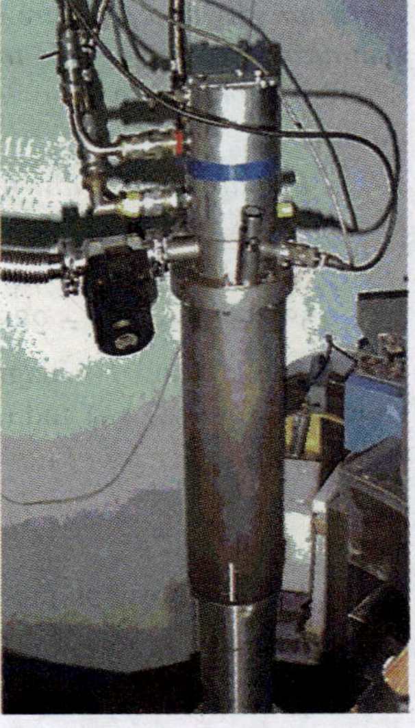

Closed cycle refrigerator.

Let T_2 and T_4 = Temperature of air at the end of compression and expansion respectively.

Since the compression and expansion follows the law $pv^{1.3} = C$, therefore

$$\frac{T_2}{T_1} = \left(\frac{p_2}{p_1}\right)^{\frac{n-1}{n}} = \left(\frac{8.5}{1.05}\right)^{\frac{1.3-1}{1.3}} = (8.1)^{0.231} = 1.62$$

$$\therefore \quad T_2 = T_1 \times 1.62 = 283 \times 1.62 = 458.5 \text{ K}$$

Similarly
$$\frac{T_3}{T_4} = \left(\frac{p_3}{p_4}\right)^{\frac{n-1}{n}} = \left(\frac{8.5}{1.05}\right)^{\frac{1.3-1}{1.3}} = 1.62$$

$$\therefore \quad T_4 = T_3/1.62 = 303/1.62 = 187 \text{ K}$$

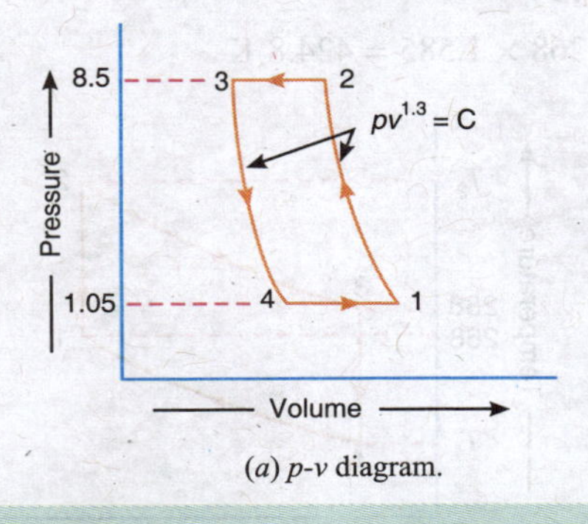

(*a*) *p-v* diagram.

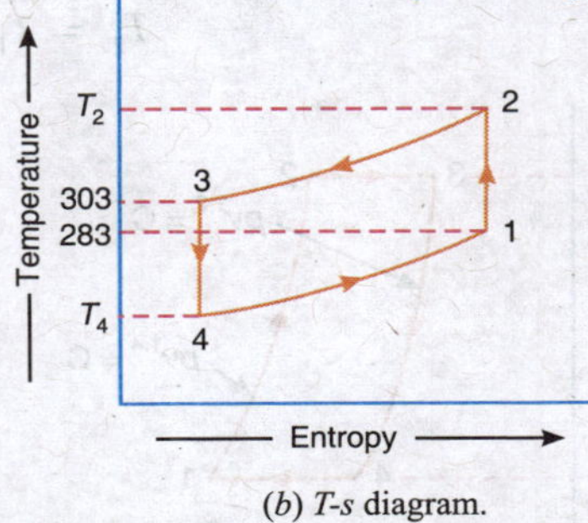

(*b*) *T-s* diagram.

Fig. 2.11

* The theoretical C.O.P. of the plant may also be obtained as follows:
We know that compression or expansion ratio,

$$r_p = \frac{p_2}{p_1} = \frac{p_3}{p_4} = \frac{5}{1} = 5 \quad \text{and} \quad \text{C.O.P.} = \frac{1}{(r_p)^{\frac{\gamma-1}{\gamma}} - 1} = \frac{1}{(5)^{\frac{1.4-1}{1.4}} - 1} = \frac{1}{1.584 - 1} = 1.712 \textbf{ Ans.}$$

We know that theoretical coefficient of performance,

$$\text{C.O.P.} = \frac{T_1 - T_4}{\frac{n}{n-1} \times \frac{(\gamma - 1)}{\gamma}\left[(T_2 - T_3) - (T_1 - T_4)\right]}$$

$$= \frac{(283 - 187)}{\frac{1.3}{1.3-1} \times \frac{(1.4-1)}{1.4}\left[(458.5 - 303) - (283 - 187)\right]} \quad \ldots(\text{Taking } \gamma = 1.4)$$

$$= \frac{96}{1.24 \times 59.5} = 1.3 \textbf{ Ans.}$$

Example 2.13. *The atmospheric air at pressure 1 bar and temperature –5°C is drawn in the cylinder of the compressor of a Bell-Coleman refrigerating machine. It is compressed isentropically to a pressure of 5 bar. In the cooler, the compressed air is cooled to 15°C, pressure remaining the same. It is then expanded to a pressure of 1 bar in an expansion cylinder, from where it is passed to the cold chamber. Find : 1. the work done per kg of air, and 2. C.O.P. of the plant.*

For air assume law for expansion, $pv^{1.2}$ = constant ; law for compression, $pv^{1.4}$ = constant and specific heat of air at constant pressure = 1 kJ/kg K.

Solution. Given : $p_1 = p_4 = 1$ bar ; $T_1 = -5°\text{C} = -5 + 273 = 268$ K ; $p_2 = p_3 = 5$ bar ; $T_3 = 15°\text{C} = 15 + 273 = 288$ K ; $n = 1.2$; $\gamma = 1.4$; $c_p = 1$ kJ/kg K

The *p-v* and *T-s* diagrams for a refrigerating machine working on Bell-Coleman cycle is shown in Fig. 2.12 (*a*) and (*b*) respectively.

1. *Work done per kg of air*

Let T_2 and T_4 = Temperatures at the end of compression and expansion respectively.

The compression process 1-2 is isentropic and follows the law $pv^{1.4}$ = constant.

$$\therefore \quad \frac{T_2}{T_1} = \left(\frac{p_2}{p_1}\right)^{\frac{\gamma-1}{\gamma}} = \left(\frac{5}{1}\right)^{\frac{1.4-1}{1.4}} = (5)^{0.286} = 1.585$$

or $\quad T_2 = T_1 \times 1.585 = 268 \times 1.585 = 424.8$ K

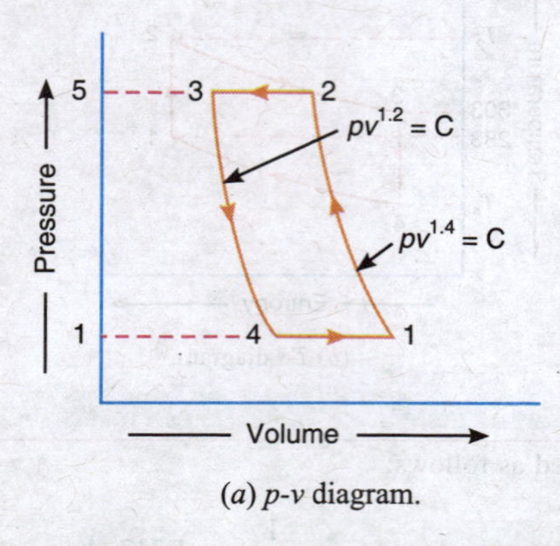

(*a*) *p-v* diagram.

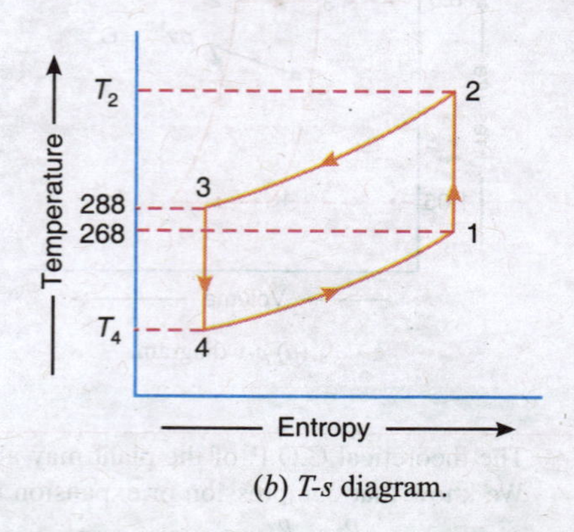

(*b*) *T-s* diagram.

Fig. 2.12

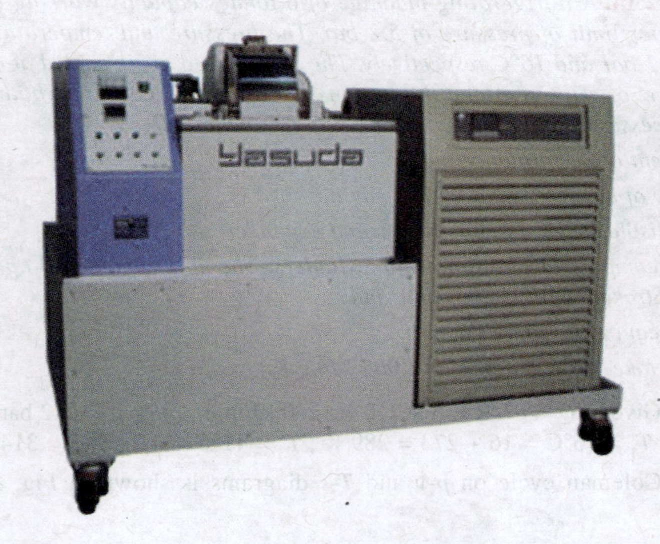

Bell-Coleman Refrigeration Machine

The expansion process 3-4 follows the law $pv^{1.2}$ = constant.

$$\therefore \qquad \frac{T_3}{T_4} = \left(\frac{p_3}{p_4}\right)^{\frac{n-1}{n}} = \left(\frac{5}{1}\right)^{\frac{1.2-1}{1.2}} = (5)^{0.167} = 1.31$$

or $$T_4 = T_3/1.31 = 288/1.31 = 220 \text{ K}$$

We know that workdone by the compressor during the isentropic process 1-2 per kg of air,

$$w_C = w_{1-2} = \frac{\gamma}{\gamma-1} \times R\,(T_2 - T_1)$$

$$= \frac{1.4}{1.4-1} \times 0.287(424.8 - 268) = 159 \text{ kJ/kg}$$

... (Taking R for air = 0.287 kJ/kg K)

and workdone by the expander during the process 3-4 per kg of air,

$$w_E = w_{3-4} = \frac{n}{n-1} \times R\,(T_3 - T_4)$$

$$= \frac{1.2}{1.2-1} \times 0.287(288 - 220) = 118.3 \text{ kJ/kg}$$

∴ Net work done per kg of air,

$$w = w_C - w_E = 159 - 118.3 = 40.7 \text{ kJ/kg}$$ **Ans.**

2. C.O.P. of the plant

We know that heat absorbed during constant pressure process 4-1 per kg of air,

$$q_A = c_p\,(T_1 - T_4) = 1(268 - 220) = 48 \text{ kJ/kg}$$

$$\therefore \qquad \text{C.O.P. of the plant} = \frac{\text{Heat absorbed}}{\text{Work done}} = \frac{q_A}{w} = \frac{48}{40.7} = 1.18$$ **Ans.**

Example 2.14. *A refrigerating machine of 6 tonnes capacity working on Bell-Coleman cycle has an upper limit of pressure of 5.2 bar. The pressure and temperature at the start of compression are 1 bar and 16°C respectively. The compressed air is cooled at constant pressure to a temperature of 41°C, enters the expansion cylinder. Assuming both expansion and compression processes to be isentropic with $\gamma = 1.4$, Calculate :*

1. *Coefficient of performance;*
2. *Quantity of air in circulation per minute;*
3. *Piston displacement of compressor and expander;*
4. *Bore of compressor and expansion cylinders. The unit runs at 240 r.p.m. and is double acting. Stroke length is 200 mm ; and*
5. *Power required to drive the unit.*

For air, take $\gamma = 1.4$, and $c_p = 1.003$ kJ/kg K.

Solution. Given : $Q = 6$ TR $= 6 \times 210 = 1260$ kJ/min ; $p_2 = p_3 = 5.2$ bar ; $p_1 = p_4 = 1$ bar $= 1 \times 10^5$ N/m^2 ; $T_1 = 16°C = 16 + 273 = 289$ K ; $T_3 = 41°C = 41 + 273 = 314$ K ; $\gamma = 1.4$

The Bell-Coleman cycle on *p-v* and *T-s* diagrams is shown in Fig. 2.13 (*a*) and (*b*) respectively.

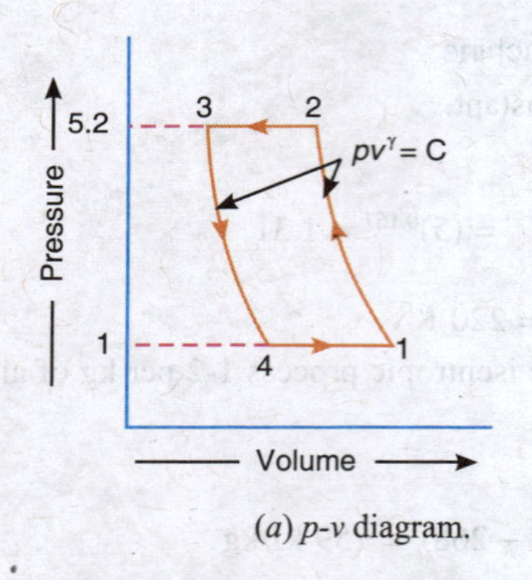

(*a*) *p-v* diagram.

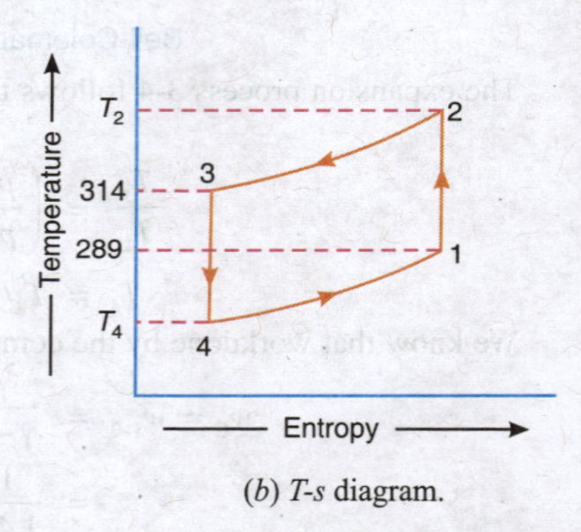

(*b*) *T-s* diagram.

Fig. 2.13

1. *Coefficient of performance*

Let T_2 and T_4 = Temperature at the end of compression and expansion respectively.

The compression and expansion are isentropic (*i.e.* $pv^\gamma = C$) and γ for air = 1.4.

We know that
$$\frac{T_2}{T_1} = \left(\frac{p_2}{p_1}\right)^{\frac{\gamma-1}{\gamma}} = \left(\frac{5.2}{1}\right)^{\frac{1.4-1}{1.4}} = (5.2)^{0.286} = 1.6$$

$$\therefore \quad T_2 = T_1 \times 1.6 = 289 \times 1.6 = 462.4 \text{ K}$$

Similarly
$$\frac{T_3}{T_4} = \left(\frac{p_3}{p_4}\right)^{\frac{\gamma-1}{\gamma}} = \left(\frac{5.2}{1}\right)^{\frac{1.4-1}{1.4}} = (5.2)^{0.286} = 1.6$$

$$\therefore \quad T_4 = T_3 / 1.6 = 314/1.6 = 196.25 \text{ K}$$

We know that coefficient of performance,

$$\text{C.O.P.} = \frac{T_4}{T_3 - T_4} = \frac{196.25}{314 - 196.25} = 1.674 \quad \textbf{Ans.}$$

2. ***Quantity of air in circulation per minute***

Let m_a = Mass of air in circulation in kg per minute.

We know that heat extracted from the refrigerating machine (or refrigerating effect produced) per kg of air

$$= c_p (T_1 - T_4) = 1.003\,(289 - 196.25) = 93 \text{ kJ/kg}$$

and refrigerating capacity of the machine

$$= 6 \text{ TR } = 6 \times 210 = 1260 \text{ kJ/min}$$

∴ Mass of air in circulation,

$$m_a = 1260 / 93 = 13.548 \text{ kg/min} \textbf{ Ans.}$$

3. ***Piston displacement of compressor and expander***

Let v_1 and v_4 = Piston displacement per minute of compressor and expander respectively.

We know that characteristic gas constant,

$$*R_a = c_p\left(\frac{\gamma-1}{\gamma}\right) = 1.003\left(\frac{1.4-1}{1.4}\right) = 0.287 \text{ kJ/kg K} = 287 \text{ J/kg K}$$

We also know that $p_1 v_1 = m_a R_a T_1$

$$\therefore \quad v_1 = \frac{m_a R_a T_1}{p_1} = \frac{13.548 \times 287 \times 289}{1 \times 10^5} = 11.237 \text{ m}^3\text{/min} \textbf{ Ans.}$$

For constant pressure process 4–1,

$$\frac{v_4}{T_4} = \frac{v_1}{T_1}$$

$$\therefore \quad v_4 = v_1 \times \frac{T_4}{T_1} = 11.237 \times \frac{196.25}{289} = 7.63 \text{ m}^3\text{/min} \textbf{ Ans.}$$

4. ***Bore of compressor and expansion cylinders***

Let D and d = Bore of compressor and expansion cylinder in metres, respectively.

N = Speed of the unit = 240 r.p.m. ...(Given)

L = Length of stroke = 200 mm = 0.2 m ...(Given)

We know that piston displacement of compressor cylinder,

$$v_1 = \left[\frac{\pi}{4} \times D^2 \times L \times 2\right] N \quad \text{... (∵ of double acting)}$$

---

* We know that $c_p - c_v = R_a$

Dividing by c_p throughout,

$$1 - \frac{1}{\gamma} = \frac{R_a}{c_p} \quad \text{...}\left(\because \frac{c_v}{c_p} = \frac{1}{\gamma}\right)$$

$$\therefore \quad R_a = c_p\left(\frac{\gamma-1}{\gamma}\right)$$

$$11.237 = \left[\frac{\pi}{4} \times D^2 \times 0.2 \times 2\right] 240 = 75.4\, D^2$$

$\therefore$ $D^2 = 11.237 / 75.4 = 0.149$ or $D = 0.386$ m = 386 mm **Ans.**

Similarly piston displacement of expansion cylinder,

$$v_4 = \left[\frac{\pi}{4} \times d^2 \times 2 \times 2\right] N$$

$$7.63 = \left[\frac{\pi}{4} \times d^2 \times 0.2 \times 2\right] 240 = 75.4\, d^2$$

$\therefore$ $d^2 = 7.63 / 75.4 = 0.1012$ or $d = 0.318$ m = 318 mm **Ans.**

5. Power required to drive the unit

We know that heat absorbed during the constant pressure process 2–3

$$= m_a c_p (T_1 - T_4) = 13.548 \times 1.003\ (289 - 196.25) = 1260 \text{ kJ/min}$$

$$\therefore \quad \text{Workdone per minute} = \frac{\text{Heat absorbed}}{\text{C.O.P.}} = \frac{1260}{1.674} = 752.7 \text{ kJ/min}$$

and power required to drive the unit

$$= 752.7/60 = 12.54 \text{ kJ/s or kW} \quad \textbf{Ans.}$$

Note: The power required to drive the unit may also be calculated from the following relation:

We know that $\text{C.O.P.} = \dfrac{\text{Refrigerating capacity or heat absorbed } (Q)}{\text{Workdone}}$

$$1.674 = \frac{6 \times 210}{\text{Workdone}}$$

$\therefore$ Workdone = 6 × 210/1.674 = 752.7 kJ/min

and power required = 752.7/60 = 12.54 kJ/s or kW **Ans.**

Example 2.15. *An air refrigerator works between the pressure limits of 1 bar and 5 bar. The temperature of the air entering the compressor and expansion cylinder are 10°C and 25°C respectively. The expansion and compression follow the law $pv^{1.3}$ = constant. Find the following :*

1. *The theoretical C.O.P. of the refrigerating cycle ;*
2. *If the load on the refrigerating machine is 10 TR, find the amount of air circulated per minute through the system assuming that the actual C.O.P. is 50% of the theoretical C.O.P.*
3. *The stroke length and piston diameter of single acting compressor if the compressor runs at 300 r.p.m. and the volumetric efficiency is 85 %.*

Take $L / d = 1.5$; $c_p = 1.005$ kJ/kg K ; $c_v = 0.71$ kJ/ kg K.

Solution. Given : $p_1 = p_4 = 1$ bar ; $p_2 = p_3 = 5$ bar ; $T_1 = 10°C = 10 + 273 = 283$ K ; $T_3 = 25°C = 25 + 273 = 298$ K ; $n = 1.3$; $Q = 10$ TR ; Actual C.O.P. = 50% Theoretical C.O.P.; $N = 300$ r.p.m. ; $\eta_v = 85\% = 085$; $L / d = 1.5$; $c_p = 1.005$ kJ/kg K ; $c_v = 0.71$ kJ/kg K

The *p-v* and *T-s* diagrams of the refrigerating cycle are shown in Fig. 2.14 (*a*) and (*b*) respectively.

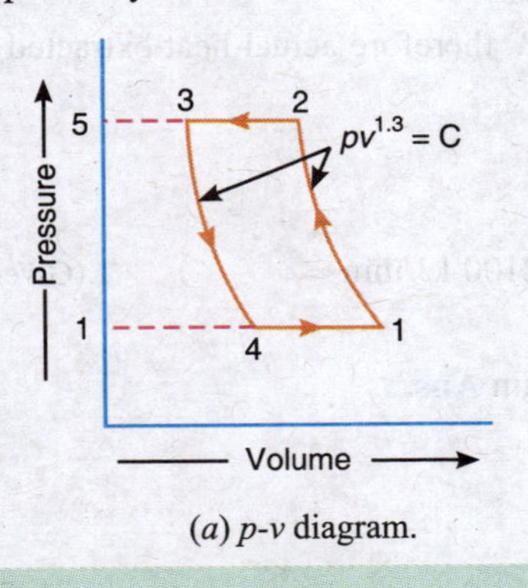

(*a*) *p-v* diagram.

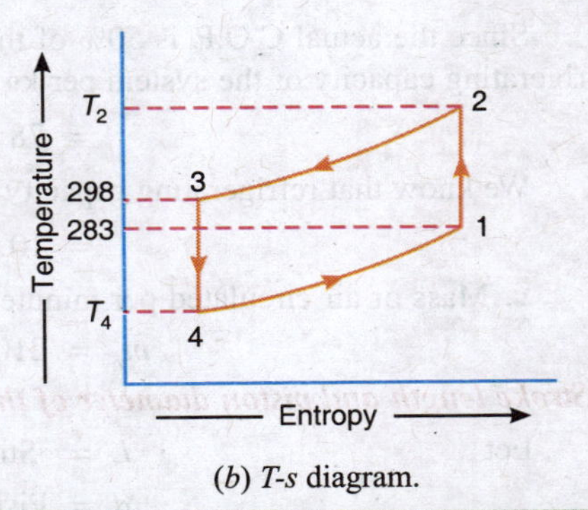

(*b*) *T-s* diagram.

Fig. 2.14

1. *Theoretical C.O.P. of the refrigerating cycle*

Let T_2 and T_4 = Temperature at the end of compression and expansion respectively.

We know that $\dfrac{T_2}{T_1} = \left(\dfrac{p_2}{p_1}\right)^{\frac{n-1}{n}} = \left(\dfrac{5}{1}\right)^{\frac{1.3-1}{1.3}} = (5)^{0.23} = 1.45$

$\therefore \quad T_2 = T_1 \times 1.45 = 283 \times 1.45 = 410.3 \text{ K}$

Similarly $\dfrac{T_3}{T_4} = \left(\dfrac{p_3}{p_4}\right)^{\frac{n-1}{n}} = \left(\dfrac{5}{1}\right)^{\frac{1.3-1}{1.3}} = (5)^{0.23} = 1.45$

$\therefore \quad T_4 = T_3 / 1.45 = 298 / 1.45 = 205.5 \text{ K}$

We know that heat extracted from the refrigerating system per kg of air,

$$q_A = c_p (T_1 - T_4) = 1.005 (283 - 205.5) = 78 \text{ kJ/kg}$$

and characteristic gas constant,

$$R = c_p - c_v = 1.005 - 0.71 = 0.295 \text{ kJ/kg K}$$

Workdone during compression process 1-2 per kg of air,

$$w_C = w_{1-2} = \frac{n}{n-1} \times R\,(T_2 - T_1) = \frac{1.3}{1.3-1} \times 0.295\,(410.3 - 283) \text{ kJ/kg}$$

$$= 162.7 \text{ kJ/kg}$$

and workdone during expansion process 3-4 per kg of air,

$$w_E = w_{3-4} = \frac{n}{n-1} \times R\,(T_3 - T_4) = \frac{1.3}{1.3-1} \times 0.295\,(298 - 205.5) \text{ kJ/kg}$$

$$= 118.2 \text{ kJ/kg}$$

$\therefore$ Net workdone per kg of air supplied,

$$w = w_C - w_E = w_{1-2} - w_{3-4} = 162.7 - 118.2 = 44.5 \text{ kJ/kg}$$

We know that theoretical C.O.P. of the refrigerating cycle

$$= \frac{\text{Heat extracted}}{\text{Workdone}} = \frac{q_A}{w} = \frac{78}{44.5} = 1.75 \text{ **Ans.**}$$

2. *Amount of air circulated per minute*

Let m_a = Mass of air circulated per minute.

Since the actual C.O.P. is 50% of the theoretical C.O.P., therefore actual heat extracted or refrigerating capacity of the system per kg of air

$$= 78 \times 0.5 = 39 \text{ kJ/kg}$$

We know that refrigerating capacity of the system

$$= 10 \text{ TR} = 10 \times 210 = 2100 \text{ kJ/min} \quad \text{...(Given)}$$

∴ Mass of air circulated per minute,

$$m_a = 2100/39 = 53.8 \text{ kg / min} \quad \textbf{Ans.}$$

3. *Stroke length and piston diameter of the compressor*

Let L = Stroke length, and

d = Piston diameter.

Since the mass of air supplied to the compressor at point 1 is m_a = 53.8 kg / min, therefore its volume,

$$v_1 = \frac{m_a R T_1}{p_1} = \frac{53.8 \times 295 \times 283}{1 \times 10^5} = 45 \text{ m}^3\text{/min}$$

... (R is taken in J/kg K and p is in N/m²)

We also know that volume (v_1),

$$45 = \left(\frac{\pi}{4} \times d^2 \times L\right) N \times \eta_v = \left[\frac{\pi}{4} \times d^2 \times 1.5\, d\right] 300 \times 0.85$$

$$= 300\, d^3 \quad \text{... } (\because L/d = 1.5)$$

$$\therefore \quad d^3 = 45/300 = 0.15 \quad \text{or} \quad d = 0.53 \text{ m} = 530 \text{ mm} \quad \textbf{Ans.}$$

and $$L = 1.5\, d = 1.5 \times 530 = 795 \text{ mm} \quad \textbf{Ans.}$$

Example 2.16. *A dense closed cycle refrigeration system working between 4 bar and 16 bar extracts 126 MJ of heat per hour. The air enters the compressor at 5°C and into the expander at 20°C. Assuming the unit runs at 300 r.p.m., find out 1. Power required to run the unit; 2. Bore of compressor ; and 3. Refrigerating capacity in tonnes of ice at 0°C per day. Take the following :*

The compressor and expander are double acting and stroke for compressor and expander is 300 mm. The mechanical efficiency of compressor is 80%. The mechanical efficiency of expander is 85%. Assume the compression and expansion are isentropic.

Solution. Given : $p_1 = p_4$ = 4 bar = 4 × 10⁵ N/m² ; $p_2 = p_3$ = 16 bar = 16 × 10⁵ N/m²; Q = 126 MJ/h = 2100 kJ/min ; T_1 = 5°C = 5 + 273 = 278 K ; T_3 = 20°C = 20 + 273 = 293 K ; N = 300 r.p.m ; L = 300 mm = 0.3 m; η_C = 80% = 0.8 ; η_E = 85% = 0.85

The cycle on *p-v* and *T-s* diagrams is shown in Fig. 2.15 (*a*) and (*b*) respectively.

1. *Power required to run the unit*

Let T_2 and T_4 = Temperatures at the end of compression and expansion respectively.

The compression and expansion are isentropic (*i.e.* $pv^\gamma = C$) and γ for air = 1.4. We know that

$$\frac{T_2}{T_1} = \left(\frac{p_2}{p_1}\right)^{\frac{\gamma-1}{\gamma}} = \left(\frac{16}{4}\right)^{\frac{1.4-1}{1.4}} = (4)^{0.286} = 1.486$$

$\therefore \quad T_2 = T_1 \times 1.486 = 278 \times 1.486 = 413$ K

Similarly $\quad \dfrac{T_3}{T_4} = \left(\dfrac{p_3}{p_4}\right)^{\frac{\gamma-1}{\gamma}} = \left(\dfrac{16}{4}\right)^{\frac{1.4-1}{1.4}} = (4)^{0.286} = 1.486$

$\therefore \quad T_4 = T_3 / 1.486 = 293 / 1.486 = 197$ K

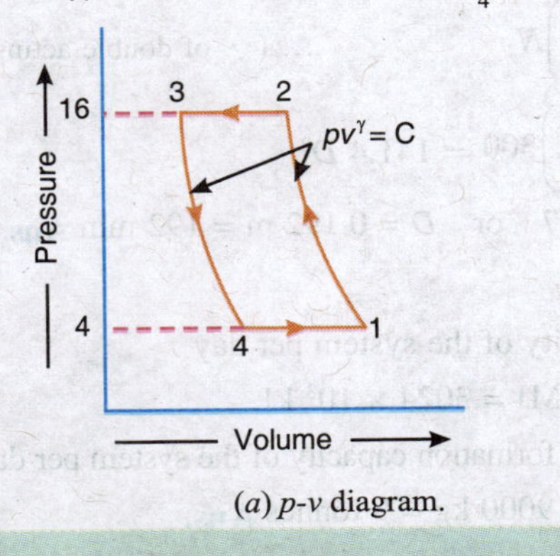

(a) p-v diagram.

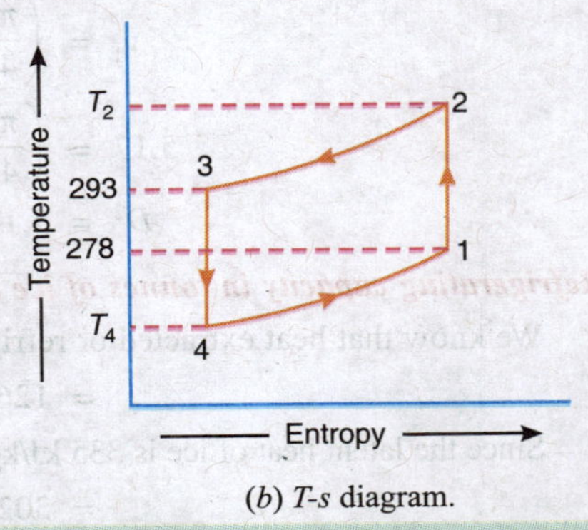

(b) T-s diagram.

Fig. 2.15

We know that heat extracted from the refrigeration system per kg of air,

$$q_A = c_p (T_1 - T_4) = 1(278 - 197) = 81 \text{ kJ/kg}$$

... (Taking c_p for air = 1 kJ/kg K)

$\therefore$ Mass of air circulated,

$$m_a = \frac{\text{Heat extracted / min}}{\text{Heat extracted / kg}} = \frac{2100}{81} = 25.9 \text{ kg / min}$$

We know that workdone during compression process 1-2 per kg of air,

$$w_C = w_{1-2} = \frac{\gamma}{\gamma - 1} \times R\,(T_2 - T_1) \frac{1}{\eta_C}$$

$$= \frac{1.4}{1.4 - 1} \times 0.287(413 - 278) \frac{1}{0.8} = 169.5 \text{ kJ/kg}$$

and workdone during expansion process 3-4 per kg of air,

$$w_E = w_{3-4} = \frac{\gamma}{\gamma - 1} \times R\,(T_3 - T_4) \eta_E$$

$$= \frac{1.4}{1.4 - 1} \times 0.287(293 - 197)0.85 = 82 \text{ kJ/kg}$$

$\therefore$ Net workdone per kg of air supplied in the system,

$$w = w_C - w_E = w_{1-2} - w_{3-4} = 169.5 - 82 = 87.5 \text{ kJ/kg}$$

and power required to run the system

$$= \frac{m_a \times w}{60} = \frac{25.9 \times 87.5}{60} = 37.8 \text{ kW} \quad \textbf{Ans.}$$

2. Bore of compressor

Let $\quad D$ = Bore of compressor in metres.

Since the mass of air supplied to the compressor at point 1 is m_a = 25.9 kg/min, therefore its volume,

$$v_1 = \frac{m_a R T_1}{p_1} = \frac{25.9 \times 287 \times 278}{4 \times 10^5} = 5.17 \text{ m}^3\text{/min}$$

... ($\because$ R for air = 287 J/kg K)

We also know that volume,

$$v_1 = \left(\frac{\pi}{4} \times D^2 \times L \times 2\right) N \quad \text{... (}\because\text{ of double acting)}$$

$$5.17 = \left(\frac{\pi}{4} \times D^2 \times 0.3 \times 2\right) 300 = 141.4\ D^2$$

$\therefore$ D^2 = 5.17/141.4 = 0.037 or D = 0.192 m = 192 mm **Ans.**

3. *Refrigerating capacity in tonnes of ice at 0°C per day*

We know that heat extracted or refrigerating capacity of the system per day

= 126 × 24 = 3024 MJ = 3024 × 10^3 kJ

Since the latent heat of ice is 335 kJ/kg, therefore ice formation capacity of the system per day

= 3024 × 10^3/335 = 9000 kg = 9 tonnes **Ans.**

Example 2.17. *In an open cycle air refrigeration machine, air is drawn from a cold chamber at – 2°C and 1 bar and compressed to 11 bar. It is then cooled at this pressure, to the cooler temperature of 20°C and then expanded in expansion cylinder and returned to the cold room. The compression and expansion are isentropic, and follows the law $pv^{1.4}$ = constant. Sketch the p-v and T-s diagrams of the cycle and for a refrigeration of 15 tonnes, find : 1. theoretical C.O.P; 2. rate of circulation of the air in kg/min ; 3. piston displacement per minute in the compressor and expander ; and 4. theoretical power per tonne of refrigeration.*

Solution. Given : T_1 = –2°C = –2 + 273 = 271 K ; $p_1 = p_4$ = 1 bar = 1 × 10^5 N/m²; $p_2 = p_3$ = 11 bar ; T_3 = 20°C = 20 + 273 = 293 K ; γ = 1.4 ; Q = 15 TR

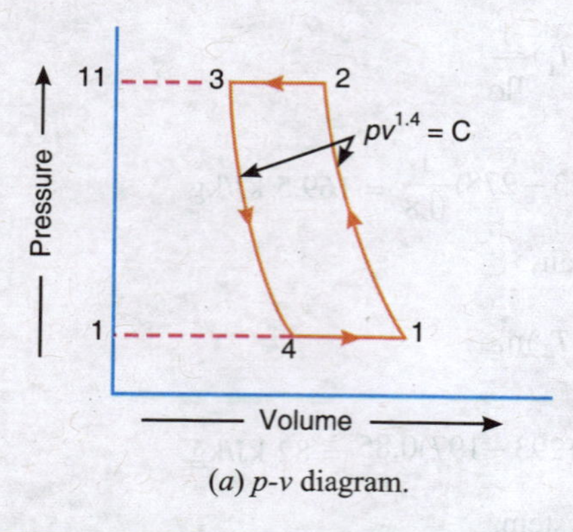

(a) p-v diagram.

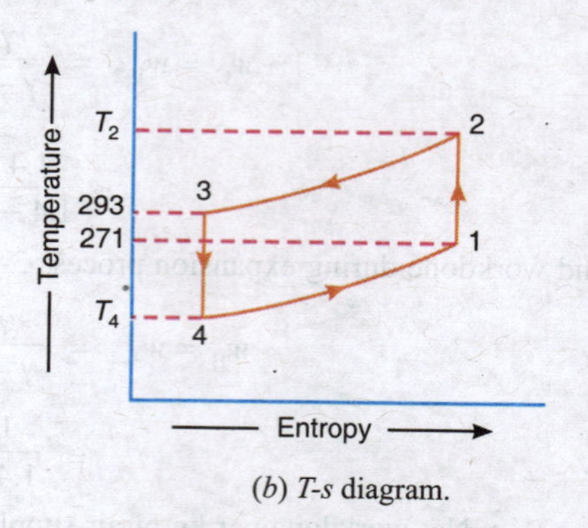

(b) T-s diagram.

Fig. 2.16

The p-v and T-s diagrams of the cycle are shown in Fig. 2.16 (a) and (b) respectively.

1. *Theoretical C.O.P.*

Let T_2 and T_4 = Temperatures at the end of compression and expansion respectively.

We know that

$$\frac{T_2}{T_1} = \left(\frac{p_2}{p_1}\right)^{\frac{\gamma-1}{\gamma}} = \left(\frac{11}{1}\right)^{\frac{1.4-1}{1.4}} = (11)^{0.286} = 1.985$$

$\therefore \quad T_2 = T_1 \times 1.985 = 271 \times 1.985 = 538 \text{ K}$

Similarly $\quad \dfrac{T_3}{T_4} = \left(\dfrac{p_3}{p_4}\right)^{\frac{\gamma-1}{\gamma}} = \left(\dfrac{11}{1}\right)^{\frac{1.4-1}{1.4}} = (11)^{0.286} = 1.985$

$\therefore \quad T_4 = T_3 / 1.985 = 293/1.985 = 147.6 \text{ K}$

We also know that theoretical C.O.P.

$$= \frac{T_1 - T_4}{(T_2 - T_3) - (T_1 - T_4)} = \frac{271 - 147.6}{(538 - 293) - (271 - 147.6)}$$

$$= \frac{123.4}{121.6} = 1.015 \textbf{ Ans.}$$

2. *Rate of circulation of the air in kg/min*

Refrigeration capacity = 15 TR ...(Given)

$\therefore$ Heat extracted/min = 15 × 210 = 3150 kJ/min ... ($\because$ 1 TR = 210 kJ/min)

We know that heat extracted from cold chamber per kg of air,

$$q_A = c_p (T_1 - T_4) = 1(271 - 147.6) = 123.4 \text{ kJ/kg}$$

... ($\because$ c_p for air = 1kJ/kg K)

$\therefore$ Rate of circulation of air,

$$m_a = \frac{\text{Heat extracted / min}}{\text{Heat extracted / kg}} = \frac{3150}{123.4} = 25.5 \text{ kg/min} \textbf{ Ans.}$$

3. *Piston displacement per minute in the compressor and expander*

Let v_1 and v_4 = Piston displacement per minute in the compressor and expander respectively.

We know that $\quad p_1 v_1 = m_a R_a T_1$

$$\therefore \quad v_1 = \frac{m_a R_a T_1}{p_1} = \frac{25.5 \times 287 \times 271}{1 \times 10^5} = 19.8 \text{ m}^3\text{/min} \textbf{ Ans.}$$

...(Taking R_a = 287 J/kg K)

For constant pressure process 4-1,

$$\frac{v_4}{T_4} = \frac{v_1}{T_1}$$

$$\therefore \quad v_4 = v_1 \times \frac{T_4}{T_1} = 19.8 \times \frac{147.6}{271} = 10.8 \text{ m}^3 \textbf{ Ans.}$$

4. *Theoretical power per tonne of refrigeration*

We know that net workdone on the refrigeration machine per minute

$= m_a$ (Heat rejected – Heat extracted)

$= m_a c_p [(T_2 - T_3) - (T_1 - T_4)]$

$= 25.5 \times 1[(538 - 293) - (271 - 147.6)] = 3100$ kJ/min

$\therefore$ Theoretical power of the refrigerating machine

= 3100/60 = 51.67 kW

and theoretical power per tonne of refrigeration

= 51.67/15 = 3.44 kW/TR **Ans.**

Example 2.18. *A dense air machine operates on reversed Brayton cycle and is required for a capacity of 10 TR. The cooler pressure is 4.2 bar and the refrigerator pressure is 1.4 bar. The air is cooled in the cooler at a temperature of 50°C and the temperature of air at inlet to compressor is –20°C. Determine for the ideal cycle : 1. C.O.P.; 2. mass of air circulated per minute ; 3. theoretical piston displacement of compressor ; 4. theoretical piston displacement of expander ; and 5. net power per tonne of refrigeration. Show the cycle on p-v and T-s planes.*

Solution. Given : Q = 10 TR; $p_2 = p_3$ = 4.2 bar ; $p_1 = p_4$ = 1.4 bar = 1.4 × 10^5 N/m^2; T_3 = 50°C = 50 + 273 = 323 K ; T_1 = –20°C = –20 + 273 = 253 K

The cycle on *p-v* and *T-s* planes is shown in Fig. 2.17 (*a*) and (*b*) respectively.

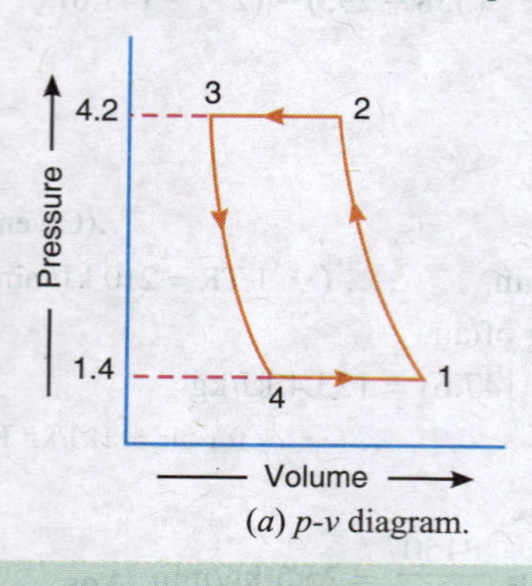

(*a*) *p-v* diagram.

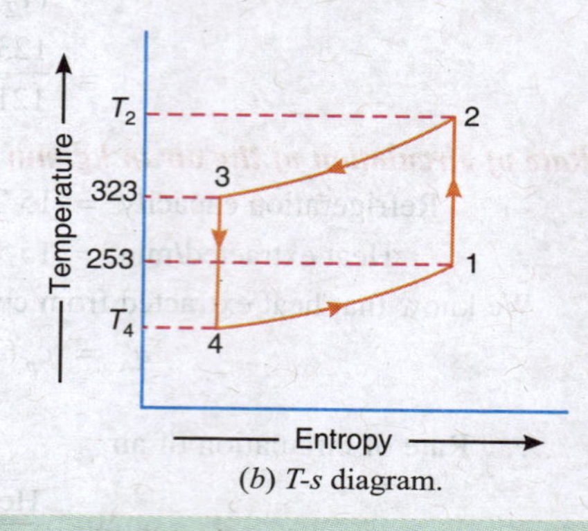

(*b*) *T-s* diagram.

Fig. 2.17

1. *Coefficient of performance (C.O.P.)*

Let T_2 and T_4 = Temperatures at the end of compression and expansion respectively.

Let us assume the compression and expansion to be isentropic and γ for air as 1.4. We know that

$$\frac{T_2}{T_1} = \left(\frac{p_2}{p_1}\right)^{\frac{\gamma-1}{\gamma}} = \left(\frac{4.2}{1.4}\right)^{\frac{1.4-1}{1.4}} = (3)^{0.286} = 1.369$$

$$\therefore \quad T_2 = T_1 \times 1.369 = 253 \times 1.369 = 346 \text{ K}$$

Similarly

$$\frac{T_3}{T_4} = \left(\frac{p_3}{p_4}\right)^{\frac{\gamma-1}{\gamma}} = \left(\frac{4.2}{1.4}\right)^{\frac{1.4-1}{1.4}} = (3)^{0.286} = 1.369$$

$$\therefore \quad T_4 = T_3 / 1.369 = 323 / 1.369 = 236 \text{ K}$$

We know that

$$\text{C.O.P.} = \frac{T_1 - T_4}{(T_2 - T_3) - (T_1 - T_4)} = \frac{253 - 236}{(346 - 323) - (253 - 236)}$$

$$= \frac{17}{6} = 2.83 \text{ **Ans.**}$$

2. *Mass of air circulated per minute*

Since the capacity of the machine is 10 TR, therefore heat extracted per min

$$= 10 \times 210 = 2100 \text{ kJ/min} \qquad \ldots (\because 1 \text{ TR} = 210 \text{ kJ/min})$$

We know that heat extracted from the refrigerator per kg of air

$$= c_p (T_1 - T_4) = 1 \,(253 - 236) = 17 \text{ kJ/kg}$$

∴ Mass of air circulated per minute.

$$m_a = \frac{\text{Heat extracted / min}}{\text{Heat extracted / kg}} = \frac{2100}{17} = 123.5 \text{ kg/min } \textbf{Ans.}$$

3. *Theoretical piston displacement of compressor*

Let v_1 = Theoretical piston displacement of compressor per min.

We know that
$$v_1 = \frac{m_a R_a T_1}{p_1} = \frac{123.5 \times 287 \times 253}{1.4 \times 10^5} = 64 \text{ m}^3 \textbf{ Ans.}$$

4. *Theoretical displacement of expander*

Let v_4 = Theoretical displacement of expander per minute.

We know that for constant pressure process 4-1,

$$\frac{v_4}{T_4} = \frac{v_1}{T_1}$$

or
$$v_4 = \frac{v_1 \times T_4}{T_1} = 64 \times \frac{236}{253} = 60 \text{ m}^3 \textbf{ Ans.}$$

5. *Net power per tonne of refrigeration*

We know that net work done on the refrigerating machine per minute

$$= m_a \text{ (Heat rejected – Heat extracted)}$$
$$= m_a c_p [(T_2 - T_3) - (T_1 - T_4)]$$
$$= 123.5 \times 1 [(346 - 323) - (253 - 236)] = 741 \text{ kJ/min}$$

∴ Net power of the refrigerating machine

$$= 741/60 = 12.35 \text{ kW}$$

and net power per tonne of refrigeration

$$= 12.35/10 = 1.235 \text{ kW/TR } \textbf{Ans.}$$

Example 2.19. *An air refrigeration used for food storage provides 25 TR. The temperature of air entering the compressor is 7°C and the temperature at exit of cooler is 27°C. Find :1. C.O.P. of the cycle; and 2. power per tonne of refrigeration required by the compressor. The quantity of air circulated in the system is 3000 kg / h. The compression and expansion both follows the law $pv^{1.3}$ = constant and take $\gamma = 1.4$; and $c_p = 1$ kJ/kg K for air.*

Solution. Given : $Q = 25$ TR ; $T_1 = 7°C = 7 + 273 = 280$ K ; $T_3 = 27°C = 27 + 273 = 300$ K ; $m_a = 3000$ kg / h = 50 kg / min

The refrigeration cycle on *p-v* and *T-s* diagram is shown in Fig. 2.18 (*a*) and (*b*) respectively.

1. *C.O.P. of the cycle*

Let T_2 and T_4 = Temperature of air at the end of compression and expansion respectively.

Since the capacity of the refrigerator is 25 TR, therefore heat extracted from the refrigerator

$$= 25 \times 210 = 5250 \text{ kJ/min} \quad \text{... (i)}$$

Also the heat extracted from the refrigerator

$$= m_a c_p (T_1 - T_4) = 50 \times 1 (280 - T_4)$$
$$= 50 (280 - T_4) \text{ kJ/min} \quad \text{... (ii)}$$

From equations (*i*) and (*ii*),

$$50\,(280 - T_4) = 5250$$

$$280 - T_4 = 105 \quad \text{or} \quad T_4 = 280 - 105 = 175 \text{ K}$$

We know that
$$\frac{T_3}{T_4} = \left(\frac{p_3}{p_4}\right)^{\frac{n-1}{n}} = \left(\frac{p_2}{p_1}\right)^{\frac{n-1}{n}} \quad \ldots (iii)$$

and
$$\frac{T_2}{T_1} = \left(\frac{p_2}{p_1}\right)^{\frac{n-1}{n}} \quad \ldots (iv)$$

From equations (*iii*) and (*iv*),

$$\frac{T_2}{T_1} = \frac{T_3}{T_4} \quad \text{or} \quad T_2 = \frac{T_1 \times T_3}{T_4} = \frac{280 \times 300}{175} = 480 \text{ K}$$

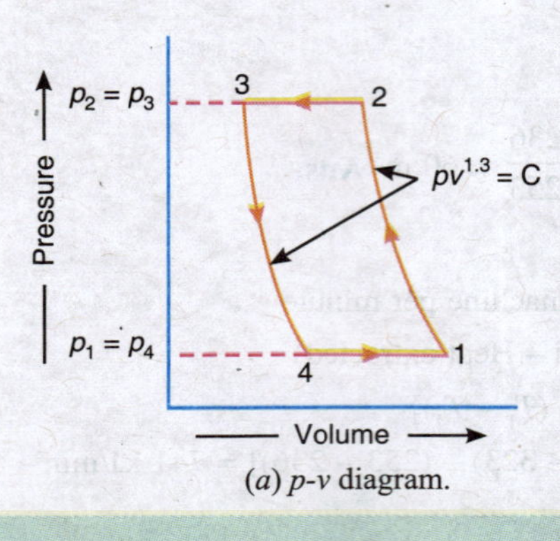

(*a*) *p-v* diagram.

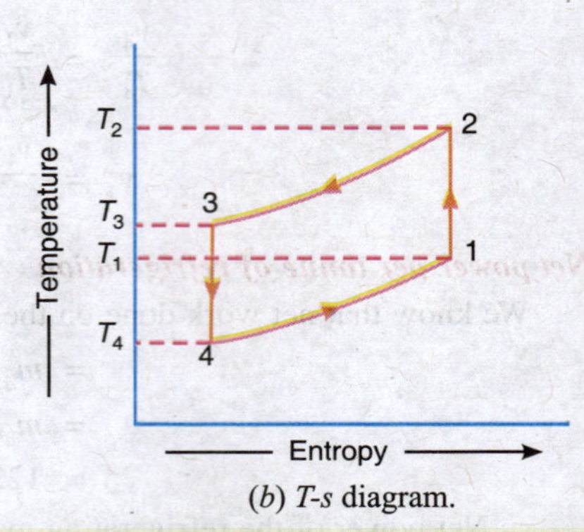

(*b*) *T-s* diagram.

Fig. 2.18

We know that C.O.P. of the cycle

$$= \frac{T_1 - T_4}{\frac{n}{n-1} \times \frac{\gamma - 1}{\gamma}\left[(T_2 - T_3) - (T_1 - T_4)\right]}$$

$$= \frac{280 - 175}{\frac{1.3}{1.3-1} \times \frac{1.4-1}{1.4}\left[(480 - 300) - (280 - 175)\right]} = 1.13 \textbf{ Ans.}$$

2. Power per tonne of refrigeration

We know that heat absorbed during the constant pressure process 4–1

$$= m_a\, c_p\, (T_1 - T_4) = 50 \times 1\, (280 - 175) = 5250 \text{ kJ/min}$$

$$\therefore \quad \text{Work done/min} = \frac{\text{Heat absorbed}}{\text{C.O.P.}} = \frac{5250}{1.13} = 4646 \text{ kJ/min}$$

and power per tonne of refrigeration

$$= \frac{4646}{60 \times 25} = 3.1 \text{ kW/ TR} \textbf{ Ans.}$$

Example 2.20. *A dense air refrigeration system of 10 tonnes capacity works between 4 bar and 16 bar. The air leaves the cold chamber at 0°C and discharges air at 25°C to the expansion cylinder after air cooler. The expansion and compression cylinders are double acting. The mechanical efficiency of compressor and expander are 85% and 80% respectively. The compressor speed is 250 r.p.m. and has a stroke of 250 mm. Determine:*

1. C.O.P. ; 2. Power required; and 3. Bore of compression and expansion cylinders.

Assume isentropic compression and expansion as polytropic with n = 1.25.

Solution. Given : $Q = 10$ TR ; $p_1 = p_4 = 4$ bar ; $p_2 = p_3 = 16$ bar ; $T_1 = 0°C = 273$ K ; $T_3 = 25°C = 25 + 273 = 298$ K ; $\eta_{mc} = 85\% = 0.85$; $\eta_{me} = 80\% = 0.8$; $N = 250$ r.p.m. ; $L = 250$ mm $= 0.25$ m ; $n = 1.25$

The *p-v* and *T-s* diagrams for the cycle are shown in Fig. 2.19 (*a*) and (*b*) respectively.

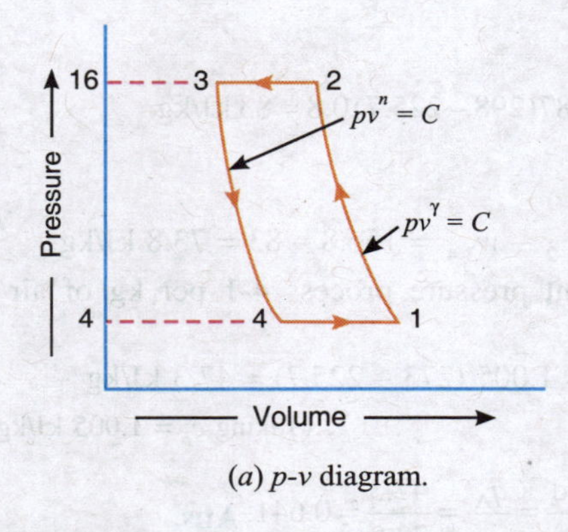

(*a*) *p-v* diagram.

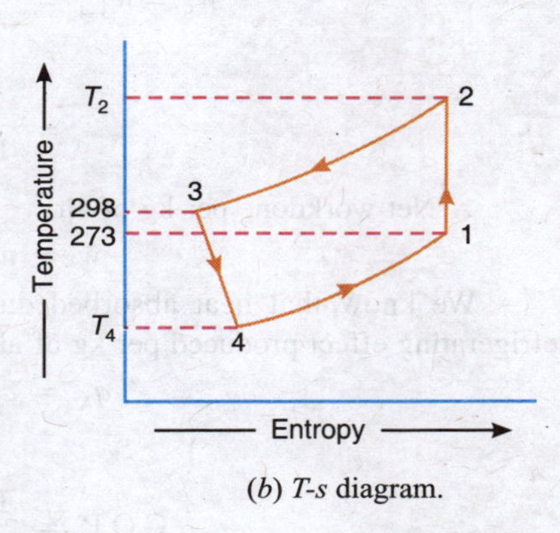

(*b*) *T-s* diagram.

Fig. 2.19

Let T_2 = Temperature of air at the end of isentropic compression in the compressor, and

T_4 = Temperature of air at the end of polytropic expansion in the turbine.

We know that for isentropic compression process 1-2,

$$\frac{T_2}{T_1} = \left(\frac{p_2}{p_1}\right)^{\frac{\gamma-1}{\gamma}} = \left(\frac{16}{4}\right)^{\frac{1.4-1}{1.4}} = (4)^{0.286} = 1.486$$

... [Taking γ for air = 1.4]

$$\therefore \quad T_2 = T_1 \times 1.486 = 273 \times 1.486 = 405.7 \text{ K}$$

Similarly, for polytropic expansion process 3-4,

$$\frac{T_3}{T_4} = \left(\frac{p_3}{p_4}\right)^{\frac{n-1}{n}} = \left(\frac{16}{4}\right)^{\frac{1.25-1}{1.25}} = (4)^{0.2} = 1.32$$

$$\therefore \quad T_4 = T_3 / 1.32 = 298 / 1.32 = 225.7 \text{ K}$$

1. *C.O.P.*

We know that workdone by the compressor during the isentropic process 1-2 per kg of air,

$$w_C = w_{1-2} = \frac{\gamma}{\gamma-1} \times R\,(T_2 - T_1) \times \frac{1}{\eta_{mc}}$$

$$= \frac{1.4}{1.4-1} \times 0.287(405.7-273) \times \frac{1}{0.85} = 156.8 \text{ kJ/kg}$$

... (Taking R for air = 0.287 kJ/kg K)

and workdone by the expander during the process 3-4 per kg of air,

$$w_E = w_{3-4} = \frac{n}{n-1} \times R\,(T_3 - T_4)\eta_{me}$$

$$= \frac{1.25}{1.25-1} \times 0.287(298-225.7)0.8 = 83 \text{ kJ/kg}$$

∴ Net workdone per kg of air,

$$w = w_C - w_E = w_{1\text{-}2} - w_{3\text{-}4} = 156.8 - 83 = 73.8 \text{ kJ/kg}$$

We know that heat absorbed during constant pressure process 4-1 per kg of air (or refrigerating effect produced per kg of air)

$$q_A = c_p\,(T_1 - T_4) = 1.005\,(273 - 225.7) = 47.3 \text{ kJ/kg}$$

...(Taking c_p = 1.005 kJ/kg K)

∴ $$\text{C.O.P.} = \frac{\text{Heat absorbed}}{\text{Workdone}} = \frac{q_A}{w} = \frac{47.3}{73.8} = 0.641$$ **Ans.**

2. *Power required*

Let m_a = Mass of air circulated in kg per minute.

We know that refrigeration capacity of the system,

$$Q = 10 \text{ TR} = 10 \times 210 = 2100 \text{ kJ/min}$$

and refrigerating effect per kg of air

$$= c_p\,(T_1 - T_4) = 1.005\,(273 - 225.7) = 47.3 \text{ kJ/kg}$$

∴ Mass of air circulated,

$$m_a = 2100/47.3 = 44.4 \text{ kg/min}$$

and workdone per minute = $m_a \times w = 44.4 \times 73.8 = 3277$ kJ/min

∴ Power required = 3277/60 = 54.6 kJ/s or kW **Ans.**

3. *Bore of compression and expansion cylinders*

Let D and d = Bore of compression and expansion cylinders respectively, and

v_1 and v_4 = Piston displacement of compressor and expander respectively.

We know that

$$p_1 v_1 = m_a R_a T_1$$

$$\therefore \quad v_1 = \frac{m_a R_a T_1}{p_1} = \frac{44.4 \times 287 \times 273}{4 \times 10^5} = 8.7 \text{ m}^3/\text{min}$$

$$\ldots (\because R_a = 287 \text{ J/kg K})$$

We also know that

$$v_1 = \left[\frac{\pi}{4} \times D^2 \times L \times 2\right] N \qquad \ldots (\because \text{ of double acting})$$

$$8.7 = \left[\frac{\pi}{4} \times D^2 \times 0.25 \times 2\right] 250 = 98.2\, D^2$$

$$\therefore \quad D^2 = 8.7/98.2 = 0.0886 \quad \text{or} \quad D = 0.2976 \text{ m} = 297.6 \text{ mm} \quad \textbf{Ans.}$$

Now for constant pressure process 4-1,

$$\frac{v_4}{T_4} = \frac{v_1}{T_1}$$

$$\therefore \quad v_4 = v_1 \times \frac{T_4}{T_1} = 8.7 \times \frac{225.7}{273} = 7.2 \text{ m}^3/\text{min}$$

We know that

$$v_4 = \left[\frac{\pi}{4} \times d^2 \times L \times 2\right] N$$

$$7.2 = \left[\frac{\pi}{4} \times d^2 \times 0.25 \times 2\right] 250 = 98.2\, d^2$$

$$\therefore \quad d^2 = 7.2 / 98.2 = 0.0733 \quad \text{or} \quad d = 0.271 \text{ m} = 271 \text{ mm} \quad \textbf{Ans.}$$

Example 2.21. *A dense air refrigeration cycle operates between pressures of 4 bar and 16 bar. The air temperature after heat rejection to surroundings is 37°C and air temperature at exit of refrigerator is 7°C. The isentropic efficiencies of turbine and compressor are 0.85 and 0.8 respectively. Determine compressor and turbine work per TR ; C.O.P.; and power per TR. Take* $\gamma = 1.4$ *and* $c_p = 1.005$ *kJ/kg K.*

Solution. Given : $p_1 = p_4 = 4$ bar ; $p_2 = p_3 = 16$ bar ; $T_3 = 37°\text{C} = 37 + 273 = 310$ K ; $T_1 = 7°\text{C} = 7 + 273 = 280$ K ; $\eta_T = 0.85$; $\eta_C = 0.8$; $\gamma = 1.4$; $c_p = 1.005$ kJ/kg K

The *p-v* and *T-s* diagrams for the cycle are shown in Fig. 2.20 (*a*) and (*b*) respectively.

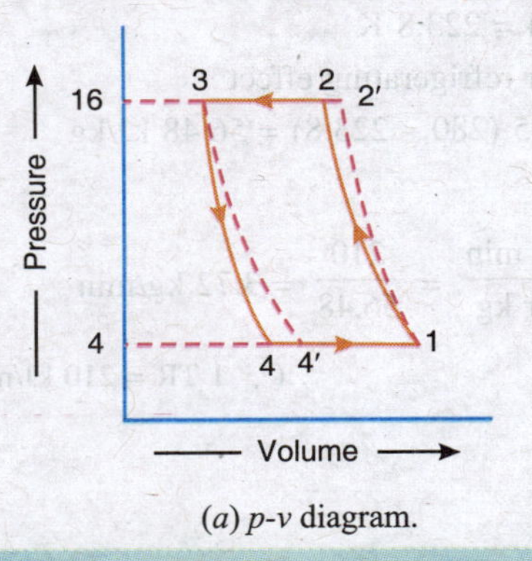

(*a*) *p-v* diagram.

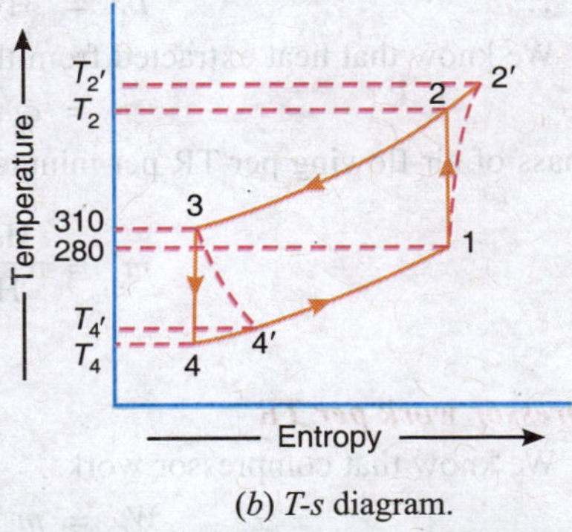

(*b*) *T-s* diagram.

Fig. 2.20

Let T_2 = Temperature of air at the end of isentropic compression in the compressor,

$T_{2'}$ = Actual temperature of air leaving the compressor,

T_4 = Temperature of air at the end of isentropic expansion in the turbine, and

$T_{4'}$ = Actual temperature of air leaving the turbine.

We know that for isentropic compression process 1-2,

$$\frac{T_2}{T_1} = \left(\frac{p_2}{p_1}\right)^{\frac{\gamma-1}{\gamma}} = \left(\frac{16}{4}\right)^{\frac{1.4-1}{1.4}} = (4)^{0.286} = 1.486$$

$$\therefore \quad T_2 = T_1 \times 1.486 = 280 \times 1.486 = 416 \text{ K}$$

Similarly, for isentropic expansion process 3-4,

$$\frac{T_3}{T_4} = \left(\frac{p_3}{p_4}\right)^{\frac{\gamma-1}{\gamma}} = \left(\frac{16}{4}\right)^{\frac{1.4-1}{1.4}} = (4)^{0.286} = 1.486$$

$$\therefore \quad T_4 = T_3 / 1.486 = 310 / 1.486 = 208.6 \text{ K}$$

We know that isentropic efficiency of the compressor,

$$\eta_C = \frac{\text{Isentropic increase in temperature}}{\text{Actual increase in temperature}} = \frac{T_2 - T_1}{T_{2'} - T_1}$$

$$0.8 = \frac{416 - 280}{T_{2'} - 280} = \frac{136}{T_{2'} - 280}$$

$$\therefore \quad T_{2'} = \frac{136}{0.8} + 280 = 450 \text{ K}$$

and isentropic efficiency of the turbine,

$$\eta_T = \frac{\text{Actual increase in temperature}}{\text{Isentropic increase in temperature}} = \frac{T_3 - T_{4'}}{T_3 - T_4}$$

$$0.85 = \frac{310 - T_{4'}}{310 - 208.6} = \frac{310 - T_{4'}}{101.4}$$

$$\therefore \quad T_{4'} = 310 - 0.85 \times 101.4 = 223.8 \text{ K}$$

We know that heat extracted from the refrigerator or refrigerating effect

$$= c_p (T_1 - T_{4'}) = 1.005 (280 - 223.8) = 56.48 \text{ kJ/kg}$$

and mass of air flowing per TR per minute,

$$m_a = \frac{\text{Heat extracted per min}}{\text{Heat extracted per kg}} = \frac{210}{56.48} = 3.72 \text{ kg/min}$$

...($\because$ 1 TR = 210 kJ/min)

Compressor work per TR

We know that compressor work,

$$W_C = m_a \, c_p (T_{2'} - T_1)$$

$$= 3.72 \times 1.005 (450 - 280) = 635.6 \text{ kJ/min}$$ **Ans.**

Turbine work per TR

We know that turbine work,

$$W_T = m_a c_p (T_3 - T_4')$$

$$= 3.72 \times 1.005\ (310 - 223.8) = 322.3 \text{ kJ/min } \textbf{Ans.}$$

C.O.P.

We know that net workdone per TR,

$$W_{net} = W_C - W_T = 635.6 - 322.3 = 313.3 \text{ kJ/min}$$

$$\therefore \quad \text{C.O.P.} = \frac{\text{Heat extracted per min}}{\text{Net workdone}} = \frac{210}{313.3} = 0.67 \textbf{ Ans.}$$

Power per TR

We know that net workdone per TR,

$$W_{net} = 313.3 \text{ kJ/min}$$

$$\therefore \quad \text{Power per TR} = 313.3/60 = 5.22 \text{ kW } \textbf{Ans.}$$

EXERCISES

1. A Carnot cycle machine operates between the temperature limits of 47°C and –30°C. Determine the C.O.P. when it operates as 1. a refrigerating machine ; 2. a heat pump ; and 3. a heat engine. **[Ans. 3:16 ; 4.16 ; 0.24]**

2. A heat pump is used for heating the interior of a house in a cold climate. The ambient temperature is –5°C and the desired interior temperature is 25°C . The compressor of the heat pump is to be driven by a heat engine working between 1000°C and 25°C. Treating both cycles as reversible, calculate the ratio in which the heat pump and the heat engine share the heating load. **[Ans. 7]**

3. A refrigerating plant is required to produce 2.5 tonnes of ice per day at – 4°C from water at 20°C. If the temperature range in the compressor is between 25°C and – 6°C, calculate power required to drive the compressor. Latent heat of ice = 335 kJ/kg and specific heat of ice = 2.1 kJ/kg K. **[Ans. 1.437 kW]**

4. A refrigerator using Carnot cycle requires 1.25 kW per tonne of refrigeration to maintain a temperature of – 30°C. Find : 1. C.O.P. of the Carnot refrigerator; 2. Temperature at which heat is rejected; and 3. Heat rejected per tonne of refrigeration. **[Ans. 2.8 ; 55.4°C ; 284 kJ/min]**

5. Ten tonnes of fish is frozen to –30°C per day. The fish enters the freezing chamber at 30°C and freezing occurs at –3°C. The frozen fish is cooled to –30°C. The specific heats of fresh and frozen fish are 3.77 kJ/kg K and 1.67 kJ/kg K respectively while latent heat of freezing is 251.2 kJ/kg K. Find the tonnage of the plant which runs for 18 hours per day. The evaporator and condensor temperatures are –40°C and 45°C respectively. If the C.O.P. of the plant is 1.8, determine the power consumption of the plant in kW. Also find the refrigerating efficiency of the plant. **[Ans. 18.6 TR ; 36.1 kW ; 65.7%]**

6. A Carnot refrigeration system has working temperature of –30°C and 40°C. What is the maximum C.O.P. possible ? If the actual C.O.P. is 75% of the maximum, calculate the actual refrigerating effect produced per kilowatt hour. **[Ans. 3.47 ; 0.743 TR]**

7. A refrigerator storage is supplied with 30 tonnes of fish at a temperature of 27°C. The fish has to be cooled to – 9°C for preserving it for long period without deterioration. The cooling takes place in 10 hours. The specific heat of fish is 2.93 kJ/kg K above freezing point of fish and 1.26 kJ/kg K below freezing point of fish which is – 3°C. The latent heat of freezing is 232 kJ/kg. What is the capacity of the plant in tonnes of refrigeration for cooling the fish ? What would be the ideal C.O.P. between this temperature range ? If the actual C.O.P. is 40% of the ideal, find the power required to run the cooling plant. **[Ans. 78 TR ; 7.33 ; 93.3 kW]**

8. A refrigerating system working on Bell-Coleman cycle receives air from cold chamber at – 5°C and compresses it from 1 bar to 4.5 bar. The compressed air is then cooled to a temperature of 37°C before it is expanded in the expander. Calculate the C.O.P. of the system when compression and expansion are (*i*) isentropic ; and (*ii*) follow the law $pv^{1.25}$ = constant. **[Ans.** 1.86 ; 1.98]

9. A Bell-Coleman refrigerator works between 4 bar and 1 bar pressure limits. After compression, the cooling water reduces the air temperature to 17°C. What is the lowest temperature produced by the ideal machine ? Compare the coefficient of performance of this machine with that of the ideal Carnot cycle machine working between the same pressure limits, the temperature at the beginning of compression being –13°C. **[Ans.** –78°C ; 2.07, 1.02]

10. An air refrigerator working on Bell-Coleman cycle takes air into the compressor at 1 bar and 268 K. It is compressed in the compressor to 5 bar and cooled to 298 K at the same pressure. It is further expanded in the expander to 1 bar and discharged to take the cooling load. The isentropic efficiencies of the compressor and expander are 85% and 90% respectively. Determine ; 1. refrigeration capacity of the system if the air circulated is 40 kg / min ; 2. power required for the compressor ; and 3. C.O.P. of the system. **[Ans.** 13.14 TR ; 46 kW; 0.812]

11. An air refrigeration system having pressure ratio of 5 takes air at 0°C. It is compressed and then cooled to 19°C at constant pressure. If the efficiency of the compressor is 95% and that of expander is 75%, determine: 1. the refrigeration capacity of the system, if the flow of air is 75 kg/min ; 2. the power of the compressor ; and 3. C.O.P. of the system. Assume compression and expansion processes to be isentropic. Take $\gamma = 1.4$; $c_p = 1$ kJ/kg K ; and $c_v = 0.72$ kJ/kg K. **[Ans.** 31.68 TR; 106.6 kW; 1.71]

12. A 5 tonne refrigerating machine operating on Bell Coleman cycle has an upper limit of pressure of 12 bar. The pressure and temperature at the start of compression are 1 bar and 17°C respectively. The compressed air cooled at constant pressure to a temperature of 40°C enters the expansion cylinder. Assuming both the expansion and compression processes to be isentropic with $\gamma = 1.4$; Determine : 1. C.O.P.; 2. quantity of air in circulation per minute; 3. piston displacement of compressor and expander; 4. bore of compressor and expansion cylinders. The unit runs at 250 r.p.m. and is double acting. Stroke length is 200 mm ; and 5. power required to drive the unit.Take $c_p = 1$ kJ/kg K ; $c_v = 0.71$ kJ/kg K ; $R = 0.287$ kJ/kg K. **[Ans.** 0.952 ; 7.65 kg/min ; 6.37 m^3/min, 3.35 m^3/min ; 284 mm ; 18.4 kW]

13. An air refrigerator used for food storage, provides 50 TR. The temperature of air entering the compressor is 7°C and the temperature before entering into the expander is 27°C. Assuming a 70% mechanical efficiency, find : 1. actual C.O.P; and 2. the power required to run the compressor.
The quantity of air circulated in the system is 100 kg/min. The compression and expansion follow the law $pv^{1.3}$ = constant.
Take $\gamma = 1.4$; $c_p = 1$ kJ/kg K for air. **[Ans.** 1.13 ; 110.6 kW]

14. A dense air refrigerating system operating between pressures of 17.5 bar and 3.5 bar is to produce 10 tonnes of refrigeration. Air leaves the refrigerating coils at –7°C and it leaves the air cooler at 15.5°C. Neglecting losses and clearance, calculate the net work done per minute and the coefficient of performance. For air $c_p = 1.005$ kJ/kg K and $\gamma = 1.4$. **[Ans.** 1237 kJ/min ; 1.7]

15. A dense air refrigeration machine operating on Bell-Coleman cycle operates between 3.4 bar and 17 bar. The temperature of air after the cooler is 15°C and after the refrigerator is 6°C. For a refrigeration capacity of 6 tonnes, find : 1. Temperature after compression and expansion; 2. Air circulation required in the cycle per minute; 3. Work of compressor and expander; 4. Theoretical C.O.P.; and 5. Rate of water circulation required in the cooler in kg/min, if the rise in temperature is limited to 30°C.
[Ans. 169°C, –91.2°C; 12.9 kg/min; 2112 kJ/min; 1377 kJ/min; 1.72; 199.6 kg/min]

QUESTIONS

1. How is the effectiveness of a refrigeration system measured ?
2. Explain the term "tonne of refrigeration".

3. Discuss the advantages of the dense air refrigerating system over an open air refrigeration system.
4. What is the difference between a refrigerator and a heat pump ? Derive an expression for the performance factor for both if they are running on reversed Carnot cycle.
5. Prove that the performance factor of a Bell-Coleman cycle refrigeration system is given by

$$\text{C.O.P.} = \frac{T_2}{T_3 - T_2}$$

where T_2 and T_3 are the temperatures of air at the inlet and discharge of compressor respectively. Explain, with a neat sketch, the working of this cycle.

OBJECTIVE TYPE QUESTIONS

1. The heat removing capacity of one tonne refrigerator is equal to
(*a*) 21 kJ/min (*b*) 210 kJ/min (*c*) 420 kJ/min (*d*) 620 kJ/min
2. One tonne refrigerating machine means that
(*a*) one tonne is the total mass of the machine
(*b*) one tonne of refrigerant is used
(*c*) one tonne of water can be converted into ice
(*d*) one tonne of ice when melts from and at 0°C in 24 hours, the refrigeration effect produced is equivalent to 210 kJ/min
3. The coefficient of performance is always .... one.
(*a*) equal to (*b*) less than (*c*) greater than
4. The ratio of heat extracted in the refrigerator to the workdone on the refrigerant is called
(*a*) coefficient of performance of refrigeration (*b*) coefficient of performance of heat pump
(*c*) relative coefficient of performance (*d*) refrigerating efficiency
5. The relative coefficient of performance is equal to
(*a*) $\frac{\text{Theoretical C.O.P.}}{\text{Actual C.O.P.}}$ (*b*) $\frac{\text{Actual C.O.P.}}{\text{Theoretical C.O.P.}}$
(*c*) Actual C.O.P. × Theoretical C.O.P.
6. In a refrigerating machine, if the lower temperature is fixed, then the C.O.P. of the machine can be increased by
(*a*) increasing the higher temperature (*b*) decreasing the higher temperature
(*c*) operating the machine at a lower speed (*d*) operating the machine at a higher speed
7. If the condenser and evaporator temperatures are 312 K and 273 K respectively, then reversed Carnot C.O.P. is
(*a*) 5 (*b*) 7 (*c*) 9 (*d*) 10
8. The C.O.P. of a reversed Carnot cycle is most strongly depend upon
(*a*) evaporator temperature (*b*) condenser temperature
(*c*) specific heat (*d*) refrigerant
9. The efficiency of Carnot heat engine is 80%. The C.O.P. of a refrigerator operating on the reversed Carnot cycle is equal to
(*a*) 0.25 (*b*) 0.40 (*c*) 0.60 (*d*) 0.80
10. The C.O.P. for a reversed Carnot refrigerator is 4. The ratio of its highest temperature to the lowest temperature will be
(*a*) 1 (*b*) 1.25 (*c*) 1.75 (*d*) 2
11. In a closed or dense air refrigeration cycle, the operating pressure ratio can be reduced, which results in .... coefficient of performance.
(*a*) lower (*b*) higher
12. Air refrigeration cycle is used in
(*a*) commercial refrigerators (*b*) domestic refrigerators
(*c*) air-conditioning (*d*) gas liquefaction

13. In a refrigerating machine, heat rejected is .... heat absorbed.
(*a*) equal to (*b*) less than (*c*) greater than

14. Air refrigerator works on
(*a*) Carnot cycle (*b*) Rankine cycle
(*c*) reversed Carnot cycle (*d*) Bell-Coleman cycle

15. In air-conditioning of aeroplanes, using air as a refrigerant, the cycle used is
(*a*) reversed Carnot cycle (*b*) reversed Joule cycle
(*c*) reversed Brayton cycle (*d*) reversed Otto cycle

ANSWERS

1. (*b*)	**2.** (*d*)	**3.** (*c*)	**4.** (*a*)	**5.** (*b*)
6. (*b*)	**7.** (*b*)	**8.** (*a*)	**9.** (*a*)	**10.** (*b*)
11. (*b*)	**12.** (*d*)	**13.** (*c*)	**14.** (*c*), (*d*)	**15.** (*c*)

CHAPTER

3 Air Refrigeration Systems

1. Introduction.
2. Merits and Demerits of Air Refrigeration System.
3. Methods of Air Refrigeration Systems.
4. Simple Air Cooling System.
5. Simple Air Evaporative Cooling System.
6. Boot-strap Air Cooling System.
7. Boot-strap Air Evaporative Cooling System.
8. Reduced Ambient Air Cooling System.
9. Regenerative Air Cooling System.
10. Comparison of Various Air Cooling Systems for Air-craft.

3.1 Introduction

The advent of high-speed passenger aircraft, jet aircraft and missiles has introduced the need for compact, and simple refrigeration systems, capable of high capacity, with minimum reduction of pay load. When the power requirements, needed to transport the additional weight of the refrigerating system are taken into account, the air cycle systems usually prove to be the most efficient. The cooling demands per unit volume of space, are heavy. An ordinary passenger aircraft requires a cooling system capable of 8 TR capacity and a super constellation requires a cooling system of more than 8 TR capacity. A jet fighter travelling at 950 km/h needs a cooling system capable of 10 to 20 TR capacity. To dissipate the heat load from 10 kW of electronic equipment in a missile or other high speed flight system, approximately 3 TR of cooling capacity are required. The miniaturization of electronic equipment concentrates a heavy cooling load in a small area. It creates difficulty in transferring heat to air at high altitudes. Moreover, low pressure of air further complicates the refrigeration design requirements.

3.2 Merits and Demerits of Air Refrigeration System

Following are the merits and demerits of air refrigeration system:

Merits

1. The air is easily available and there is no cost of the refrigerant.
2. The air is non-toxic and non-inflammable.
3. The leakage of air in small amounts is tolerable.
4. Since the main compressor is employed for the compressed air source, therefore there is no problem of space for extra compressor.
5. The air is light in weight per tonne of refrigeration.
6. The chilled air is directly used for cooling, there by eliminating the cost of separate evaporator.
7. Since the pressure in the whole system is quite low, therefore the piping, ducting etc. are quite simple to design, fabricate and maintain.

Demerits

1. It has low coefficient of performance.
2. The rate of air circulation is relatively large.

3.3 Methods of Air Refrigeration Systems

The various methods of air refrigeration systems used for aircrafts these days are as follows:

1. Simple air cooling system,
2. Simple air evaporative cooling system,
3. Boot strap air cooling system,
4. Boot strap air evaporative cooling system,
5. Reduced ambient air cooling system, and
6. Regenerative air cooling system.

Now we shall discuss all the above mentioned cooling systems, one by one, in the following pages.

3.4 Simple Air Cooling System

A simple air cooling system for aircrafts is shown in Fig. 3.1. The main components of this system are the main compressor driven by a gas turbine, a heat exchanger, a cooling turbine and a cooling air fan. The air required for refrigeration system is bled off from the main compressor. This high pressure and high temperature air is cooled initially in the heat exchanger where ram air is used for cooling. It is further cooled in the cooling turbine by the process of expansion. The work of this turbine is used to drive the cooling fan which draws cooling air through the heat exchanger. This system is good for ground surface cooling and for low flight speeds.

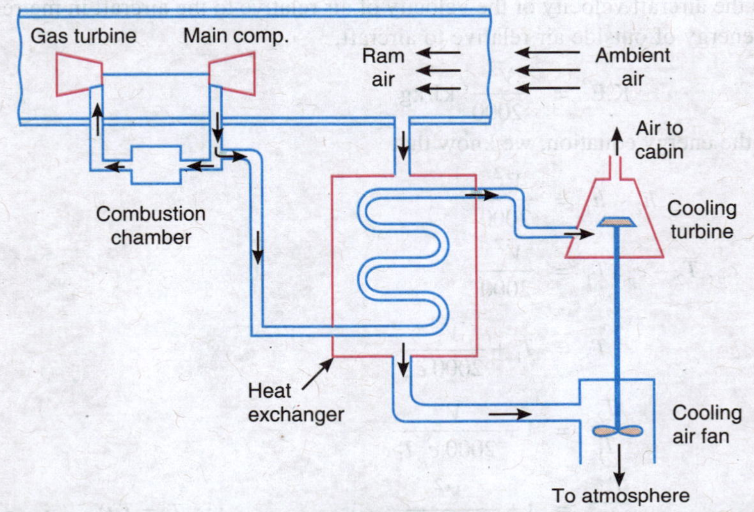

Fig. 3.1. Simple air cooling system.

The *T-s* diagram for a simple air cooling system is shown in Fig. 3.2.

The various processes are discussed below :

1. *Ramming process.* Let the pressure and temperature of ambient air is p_1 and T_1 respectively. The ambient air is rammed isentropically from pressure p_1 and temperature T_1 to the pressure p_2 and temperature T_2. This ideal ramming action is shown by the vertical line 1-2 in Fig. 3.2. In actual practice, because of internal friction due to irreversibilities, the temperature of the rammed air is more than T_2. Thus the actual ramming process is shown by the curve 1-2′ which is adiabatic but not isentropic due to friction. The pressure and temperature of the rammed air is now $p_{2'}$ and $T_{2'}$ respectively. During the ideal or actual ramming process, the total energy or enthalpy remains constant *i.e.* $h_2 = h_{2'}$ and $T_2 = T_{2'}$.

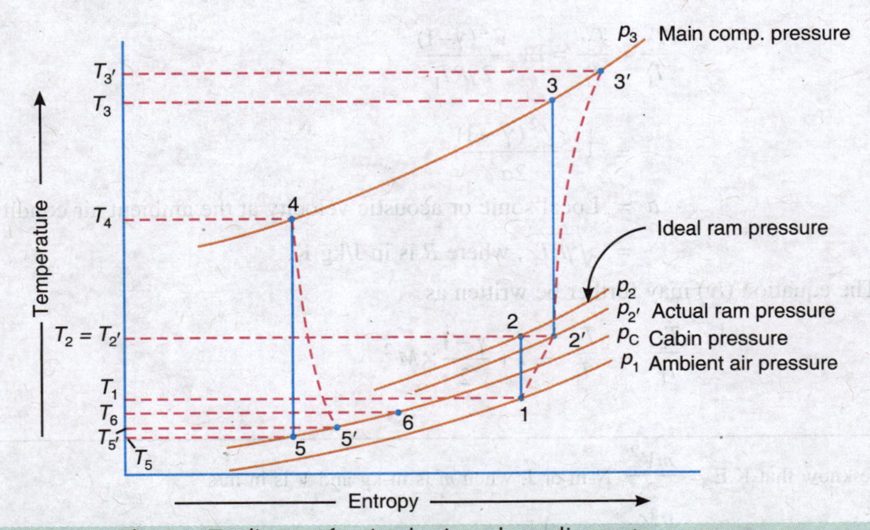

Fig. 3.2. *T-s* diagram for simple air cycle cooling system.

If V is the aircraft velocity or the velocity of air relative to the aircraft in metres per second, then kinetic energy of outside air relative to aircraft,

$$\text{*K.E.} = \frac{V^2}{2000} \text{ kJ/kg} \qquad \ldots (i)$$

From the energy equation, we know that

$$h_2 - h_1 = \frac{V^2}{2000}$$

$$c_p \cdot T_2 - c_p \cdot T_1 = \frac{V^2}{2000}$$

$$\therefore \quad T_2 = T_1 + \frac{V^2}{2000\, c_p}$$

or

$$\frac{T_2}{T_1} = 1 + \frac{V^2}{2000\, c_p T_1}$$

and

$$\frac{T_{2'}}{T_1} = 1 + \frac{V^2}{2000\, c_p T_1} \qquad \ldots (\because T_2 = T_{2'}) \qquad \ldots (ii)$$

We know that, $c_p - c_v = R$

or

$$c_p \left[1 - \frac{c_v}{c_p}\right] = R \quad \text{or} \quad c_p \left[1 - \frac{1}{\gamma}\right] = R \qquad \ldots (\because c_p / c_v = \gamma)$$

$$\therefore \quad c_p = \frac{\gamma R}{\gamma - 1}$$

Substituting the value of c_p in equation (*ii*), we have

$$\frac{T_2}{T_1} = \frac{T_{2'}}{T_1} = 1 + \frac{V^2(\gamma - 1)}{2000\, \gamma R T_1} = 1 + \frac{V^2(\gamma - 1)}{2\gamma(1000 R) T_1} \quad \text{(when } R \text{ is in kJ/kgK)} \ldots (iii)$$

It may be noted that when R is in J/kg K, then equation (*iii*) may be written as follows:

$$\frac{T_2}{T_1} = \frac{T_{2'}}{T_1} = 1 + \frac{V^2(\gamma - 1)}{2\gamma R T_1}$$

$$= 1 + \frac{V^2(\gamma - 1)}{2a^2} \qquad \ldots (iv)$$

where a = Local sonic or acoustic velocity at the ambient air conditions.

$= \sqrt{\gamma R T_1}$, where R is in J/kg K.

The equation (*iv*) may further be written as

$$\frac{T_2}{T_1} = \frac{T_{2'}}{T_1} = 1 + \frac{\gamma - 1}{2} \times M^2$$

---

* We know that K.E. $= \dfrac{mV^2}{2}$ N-m or J, when m is in kg and V is in m/s

For m = 1 kg, K.E. $= \dfrac{V^2}{2}$ J/kg $= \dfrac{V^2}{2000}$ kJ/kg

where M = Mach number of the flight. It is defined as the ratio of air craft velocity (V) to the local sonic velocity (a).

The temperature $T_2 = T_{2'}$ is called the *stagnation temperature* of the ambient air entering the main compressor. The stagnation pressure after isentropic compression (p_2) is given by

$$\frac{p_2}{p_1} = \left(\frac{T_2}{T_1}\right)^{\frac{\gamma}{\gamma-1}}$$

Due to the irreversible compression in the ram, the air reaches point 2′ instead of point 2 at the same stagnation temperature but at a reduced stagnation pressure $p_{2'}$. The pressure $p_{2'}$ may be obtained from the expression of ram efficiency (η_R) which is given as

$$\eta_R = \frac{\text{Actual rise in pressure}}{\text{Isentropic rise in pressure}} = \frac{p_{2'} - p_1}{p_2 - p_1}$$

2. *Compression process.* The isentropic compression of air in the main compressor is represented by the line 2′– 3. In actual practice, because of internal friction, due to irreversiblities, the actual compression is represented by the curve 2′– 3′. The work done during this compression process is given by

$$W_C = m_a\, c_p\, (T_{3'} - T_{2'})$$

where m_a = Mass of air bled from the main compressor for refrigeration purposes.

3. *Cooling process.* The compressed air is cooled by the ram air in the heat exchanger. This process is shown by the curve 3′– 4 in Fig. 3.2. In actual practice, there is a pressure drop in the heat exchanger which is not shown in the figure. The temperature of air decreases from $T_{3'}$ to T_4. The heat rejected in the heat exchanger during the cooling process is given by

$$Q_R = m_a\, c_p\, (T_{3'} - T_4)$$

4. *Expansion process.* The cooled air is now expanded isentropically in the cooling turbine as shown by the curve 4 – 5. In actual practice, because of internal friction due to irreversibilities, the actual expansion in the cooling turbine is shown by the curve 4 – 5′. The work done by the cooling turbine during this expansion process is given by

$$W_T = m_a\, c_p\, (T_4 - T_{5'})$$

The work of this turbine is used to drive the cooling air fan which draws cooling air from the heat exchanger.

5. *Refrigeration process.* The air from the cooling turbine (*i.e.* after expansion) is sent to the cabin and cock pit where it gets heated by the heat of equipment and occupancy. This process is shown by the curve 5′– 6 in Fig. 3.2. The refrigerating effect produced or heat absorbed is given by

$$R_E = m_a\, c_p\, (T_6 - T_{5'})$$

where T_6 = Inside temperature of cabin.

$T_{5'}$ = Exit temperature of cooling turbine.

We know that C.O.P. of the air cycle

$$= \frac{\text{Refrigerating effect produced}}{\text{Work done}}$$

$$= \frac{m_a c_p (T_6 - T_{5'})}{m_a c_p (T_{3'} - T_{2'})} = \frac{T_6 - T_{5'}}{T_{3'} - T_{2'}}$$

If Q tonnes of refrigeration is the cooling load in the cabin, then the air required for the refrigeration purpose,

$$m_a = \frac{210\,Q}{c_p(T_6 - T_{5'})} \text{ kg / min}$$

Power required for the refrigeration system,

$$P = \frac{m_a c_p (T_{3'} - T_{2'})}{60} \text{ kW}$$

and C.O.P. of the refrigerating system

$$= \frac{210\,Q}{m_a c_p (T_{3'} - T_{2'})} = \frac{210\,Q}{P \times 60}$$

Note : 1. We have discussed above that C.O.P. of the air cycle

$$= \frac{m_a c_p (T_6 - T_{5'})}{m_a c_p (T_{3'} - T_{2'})} = \frac{T_6 - T_{5'}}{T_{3'} - T_{2'}}$$

This expression excludes the ram work and the cooling turbine work. In case the ram work and the cooling turbine work is taken into account, then C.O.P. of the air cycle

$$= \frac{m_a c_p (T_6 - T_{5'})}{m_a c_p (T_{3'} - T_{2'}) + m_a c_p (T_{2'} - T_1) - m_a c_p (T_4 - T_{5'})}$$

2. Work or power required for pressurization (*i.e.* to maintain normal atmospheric pressure in air-craft cabin at high altitude),

$$W_{PR} = \frac{m_a c_p T_1}{\eta_c}\left[\left(\frac{p_c}{p_1}\right)^{\frac{\gamma-1}{\gamma}} - 1\right]$$

and work or power required for pressurization (excluding ram work),

$$W_{PR} = \frac{m_a c_p T_{2'}}{\eta_c}\left[\left(\frac{p_c}{p_{2'}}\right)^{\frac{\gamma-1}{\gamma}} - 1\right]$$

where p_c = Cabin pressure ; and η_c = Efficiency of the compressor.

Example 3.1. *An air conditioning system unit of a pressurised aircraft receives its air from the jet engine compressor at a pressure of 1.25 bar. The ambient pressure and temperature are 0.2 bar and 237 K respectively. The air-conditioning unit consists of a free wheeling compressor and turbine mounted on one shaft. The work produced by the turbine is sufficient to drive the compressor. The compressed air is then cooled in the cooler at constant pressure and then expanded in the turbine to the cabin pressure of 1 bar and temperature of 280 K. Calculate the compressor discharge pressure and the cooler exit temperature.*

Solution. Given : p_2 = 1.25 bar ; p_1 = 0.2 bar; T_1 = 237 K ; p_5 = 1 bar ; T_5 = 280 K

The schematic diagram and *T-s* diagram of the given air-conditioning system is shown in Fig. 3.3 (*a*) and (*b*) respectively.

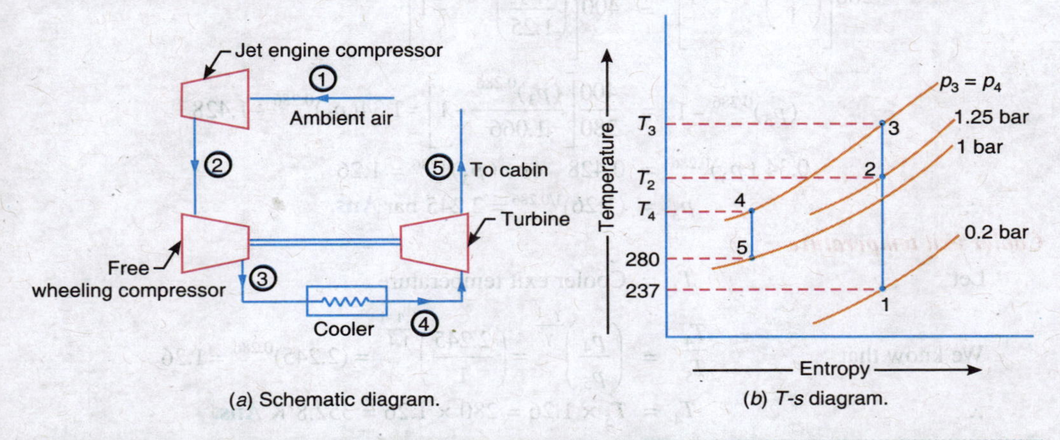

(*a*) Schematic diagram. (*b*) *T-s* diagram.

Fig. 3.3

Compressor discharge pressure

Let p_3 = Compressor discharge pressure in bar,

T_2 = Temperature of air leaving the jet engine compressor or entering the free wheeling compressor, and

T_3 = Temperature of air leaving the free wheeling compressor.

We know that
$$\frac{T_2}{T_1} = \left(\frac{p_2}{p_1}\right)^{\frac{\gamma-1}{\gamma}} = \left(\frac{1.25}{0.2}\right)^{\frac{1.4-1}{1.4}} = (6.25)^{0.286} = 1.69$$

... (Taking γ for air = 1.4)

$$\therefore \quad T_2 = T_1 \times 1.69 = 237 \times 1.69 = 400 \text{ K}$$

Work done by the free wheeling compressor in compressing the air,

$$W_C = c_p(T_3 - T_2)$$

and work done by the turbine is expanding the air,

$$W_T = c_p(T_4 - T_5)$$

Since the work produced by the turbine is sufficient to drive the compressor, therefore $W_T = W_C$, or

$$T_4 - T_5 = T_3 - T_2$$

$$T_5\left[\frac{T_4}{T_5} - 1\right] = T_2\left[\frac{T_3}{T_2} - 1\right]$$

$$T_5\left[\left(\frac{p_4}{p_5}\right)^{\frac{\gamma-1}{\gamma}} - 1\right] = T_2\left[\left(\frac{p_3}{p_2}\right)^{\frac{\gamma-1}{\gamma}} - 1\right]$$

$$280\left[\left(\frac{p_3}{1}\right)^{\frac{1.4-1}{1.4}}-1\right] = 400\left[\left(\frac{p_3}{1.25}\right)^{\frac{1.4-1}{1.4}}-1\right]$$

$$(p_3)^{0.286}-1 = \frac{400}{280}\left[\frac{(p_3)^{0.286}}{1.066}-1\right] = 1.34(p_3)^{0.286}-1.428$$

$$0.34\,(p_3)^{0.286} = 0.428 \quad \text{or} \quad (p_3)^{0.286} = 1.26$$

$$\therefore \quad p_3 = (1.26)^{1/0.286} = 2.245 \text{ bar } \textbf{Ans.}$$

Cooler exit temperature

Let T_4 = Cooler exit temperature.

We know that
$$\frac{T_4}{T_5} = \left(\frac{p_4}{p_5}\right)^{\frac{\gamma-1}{\gamma}} = \left(\frac{2.245}{1}\right)^{\frac{1.4-1}{1.4}} = (2.245)^{0.286} = 1.26$$

$$\therefore \quad T_4 = T_5 \times 1.26 = 280 \times 1.26 = 352.8 \text{ K } \textbf{Ans.}$$

Example 3.2. *A simple air cooled system is used for an aeroplane having a load of 10 tonnes. The atmospheric pressure and temperature are 0.9 bar and 10°C respectively. The pressure increases to 1.013 bar due to ramming. The temperature of the air is reduced by 50°C in the heat exchanger. The pressure in the cabin is 1.01 bar and the temperature of air leaving the cabin is 25°C. Determine : 1 Power required to take the load of cooling in the cabin; and 2. C.O.P. of the system.*

Assume that all the expansions and compressions are isentropic. The pressure of the compressed air is 3.5 bar.

Solution. Given : Q = 10 TR ; p_1 = 0.9 bar ; T_1 = 10°C = 10 + 273 = 283 K ; p_2 = 1.013 bar ; $p_5 = p_6$ = 1.01 bar ; T_6 = 25°C = 25 + 273 = 298 K ; p_3 = 3.5 bar

1. Power required to take the load of cooling in the cabin

First of all, let us find the mass of air (m_a) required for the refrigeration purpose. Since the compressions and expansions are isentropic, therefore the various processes on the *T-s* diagram are as shown in Fig. 3.4.

Let T_2 = Temperature of air at the end of ramming or entering the main compressor,

T_3 = Temperature of air leaving the main compressor after isentropic compression,

T_4 = Temperature of air leaving the heat exchanger, and

T_5 = Temperature of air leaving the cooling turbine.

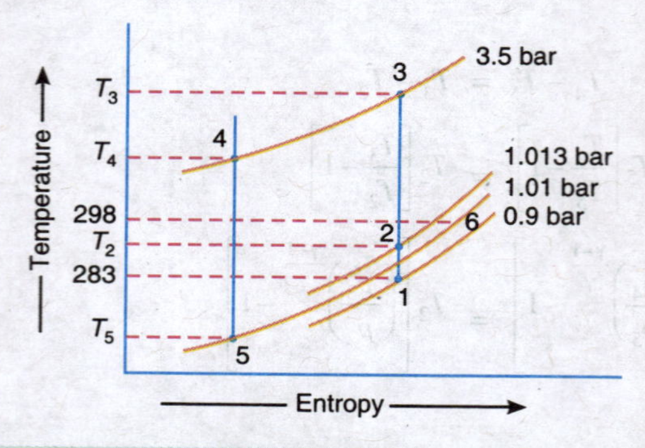

Fig. 3.4

We know that $$\frac{T_2}{T_1} = \left(\frac{p_2}{p_1}\right)^{\frac{\gamma-1}{\gamma}} = \left(\frac{1.013}{0.9}\right)^{\frac{1.4-1}{1.4}} = (1.125)^{0.286} = 1.034$$

$$\therefore \quad T_2 = T_1 \times 1.034 = 283 \times 1.034 = 292.6 \text{ K}$$

Similarly $$\frac{T_3}{T_2} = \left(\frac{p_3}{p_2}\right)^{\frac{\gamma-1}{\gamma}} = \left(\frac{3.5}{1.013}\right)^{\frac{1.4-1}{1.4}} = (3.45)^{0.286} = 1.425$$

$$\therefore \quad T_3 = T_2 \times 1.425 = 292.6 \times 1.425 = 417 \text{ K} = 144°\text{C}$$

Since the temperature of air is reduced by 50°C in the heat exchanger, therefore temperature of air leaving the heat exchanger,

$$T_4 = 144 - 50 = 94°\text{C} = 367 \text{ K}$$

We know that $$\frac{T_5}{T_4} = \left(\frac{p_5}{p_4}\right)^{\frac{\gamma-1}{\gamma}} = \left(\frac{1.01}{3.5}\right)^{\frac{1.4-1}{1.4}} = (0.288)^{0.286} = 0.7$$

$$\therefore \quad T_5 = T_4 \times 0.7 = 367 \times 0.7 = 257 \text{ K}$$

We know that mass of air required for the refrigeration purpose,

$$m_a = \frac{210\,Q}{c_p(T_6 - T_5)} = \frac{210 \times 10}{1\,(298 - 257)} = 51.2 \text{ kg/min}$$

... (Taking c_p for air = 1 kJ/kg K)

∴ Power required to take the load of cooling in the cabin,

$$P = \frac{m_a c_p (T_3 - T_2)}{60} = \frac{51.2 \times 1\,(417 - 292.6)}{60} = 106 \text{ kW}$$ **Ans.**

Rotary engine power generator sets for air-crafts.

2. C.O.P. of the system

We know that C.O.P. of the system

$$= \frac{210\,Q}{P \times 60} = \frac{210 \times 10}{106 \times 60} = 0.33$$ **Ans.**

Example 3.3. *An aircraft refrigeration plant has to handle a cabin load of 30 tonnes. The atmospheric temperature is 17°C. The atmospheric air is compressed to a pressure of 0.95 bar and temperature of 30°C due to ram action. This air is then further compressed in a compressor to 4.75 bar, cooled in a heat exchanger to 67°C, expanded in a turbine to 1 bar pressure and supplied to the cabin. The air leaves the cabin at a temperature of 27°C. The isentropic efficiencies of both compressor and turbine are 0.9. Calculate the mass of air circulated per minute and the C.O.P. For air,* c_p *= 1.004 kJ/kg K and* c_p / c_v *= 1.4*

Solution. Given : Q = 30 TR ; T_1 = 17°C = 17 + 273 = 290 K ; p_2 = 0.95 bar ; T_2 = 30°C = 30 + 273 = 303 K ; $p_3 = p_{3'}$ = 4.75 bar ; T_4 = 67°C = 67 + 273 = 340 K ; $p_5 = p_{5'}$ = 1 bar ; T_6 = 27°C = 27 + 273 = 300 K ; $\eta_C = \eta_T$ = 0.9 ; c_p = 1.004 kJ/kg K ; $c_p / c_v = \gamma$ = 1.4

The *T-s* diagram for the simple air refrigeration cycle with the given conditions is shown in Fig. 3.5.

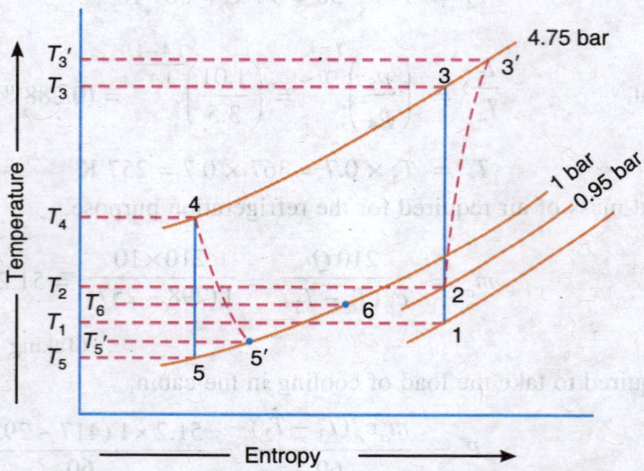

Fig. 3.5

Let T_3 = Temperature of the air after isentropic compression in the compressor,

$T_{3'}$ = Actual temperature of the air leaving the compressor,

T_5 = Temperature of the air leaving the turbine after isentropic expansion, and

$T_{5'}$ = Actual temperature of the air leaving the turbine.

We know that for isentropic compression process 2-3,

$$\frac{T_3}{T_2} = \left(\frac{p_3}{p_2}\right)^{\frac{\gamma-1}{\gamma}} = \left(\frac{4.75}{0.95}\right)^{\frac{1.4-1}{1.4}} = (5)^{0.286} = 1.584$$

$$\therefore \quad T_3 = T_2 \times 1.584 = 303 \times 1.584 = 480 \text{ K}$$

and isentropic efficiency of the compressor,

$$\eta_C = \frac{\text{Isentropic increase in temperature}}{\text{Actual increase in temperature}} = \frac{T_3 - T_2}{T_{3'} - T_2}$$

$$0.9 = \frac{480 - 303}{T_{3'} - 303} = \frac{177}{T_{3'} - 303}$$

$$\therefore \quad T_{3'} - 303 = 177/0.9 = 196.7 \quad \text{or} \quad T_{3'} = 303 + 196.7 = 499.7 \text{ K}$$

Now for the isentropic expansion process 4-5,

$$\frac{T_4}{T_5} = \left(\frac{p_4}{p_5}\right)^{\frac{\gamma-1}{\gamma}} = \left(\frac{4.75}{1}\right)^{\frac{1.4-1}{1.4}} = (4.75)^{0.286} = 1.561$$

$$\therefore \quad T_5 = T_4 / 1.561 = 340 / 1.561 = 217.8 \text{ K}$$

and isentropic efficiency of the turbine,

$$\eta_T = \frac{\text{Actual increase in temperature}}{\text{Isentropic increase in temperature}} = \frac{T_4 - T_{5'}}{T_4 - T_5}$$

$$0.9 = \frac{340 - T_{5'}}{340 - 217.8} = \frac{340 - T_{5'}}{122.2}$$

$$\therefore \quad T_{5'} = 340 - 0.9 \times 122.2 = 230 \text{ K}$$

Mass of air circulated per minute

We know that mass of air circulated per minute,

$$m_a = \frac{210\, Q}{c_p (T_6 - T_{5'})} = \frac{210 \times 30}{1.004\,(300 - 230)} = 89.64 \text{ kg/min}$$ **Ans.**

C.O.P.

We know that $$\text{C.O.P.} = \frac{210\, Q}{m_a c_p (T_{3'} - T_2)} = \frac{210 \times 30}{89.64 \times 1.004\,(499.7 - 303)} = 0.356$$ **Ans.**

Example 3.4. *An aircraft moving with speed of 1000 km/h uses simple gas refrigeration cycle for air-conditioning. The ambient pressure and temperature are 0.35 bar and –10°C respectively. The pressure ratio of compressor is 4.5. The heat exchanger effectiveness is 0.95. The isentropic efficiencies of compressor and expander are 0.8 each. The cabin pressure and temperature are 1.06 bar and 25°C. Determine temperatures and pressures at all points of the cycle. Also find the volume flow rate through compressor inlet and expander outlet for 100 TR. Take c_p = 1.005 kJ/kg K ; R = 0.287 kJ/kg K and c_p / c_v = 1.4 for air.*

Solution. Given : $V = 1000$ km / h = 277.8 m / s ; $p_1 = 0.35$ bar ; $T_1 = -10°C = -10 + 273 = 263$ K ; $p_3 / p_2 = 4.5$; $\eta_E = 0.95$; $\eta_C = \eta_T = 0.8$; $p_5 = p_{5'} = 1.06$ bar ; $T_6 = 25°C = 25 + 273 = 298$ K ; $Q = 100$ TR ; $c_p = 1.005$ kJ/kg K ; $R = 0.287$ kJ / kg K = 287 J / kg K ; $c_p / c_v = \gamma = 1.4$.

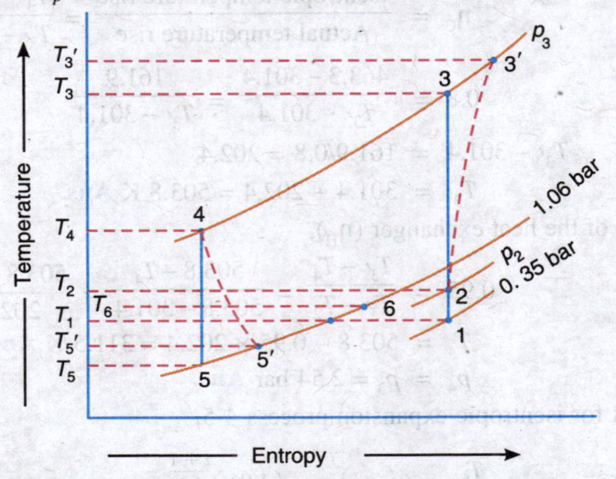

Fig. 3.6

Temperatures and pressures at all points of the cycle

The *T-s* diagram for the simple gas refrigeration cycle with the given conditions is shown in Fig. 3.6.

Let T_2 and p_2 = Stagnation temperature and pressure of the ambient air entering the compressor,

T_3 and p_3 = Temperature and pressure of the air leaving the compressor after isentropic compression,

$T_{3'}$ = Actual temperature of the air leaving the compressor,

T_4 = Temperature of the air leaving the heat exchanger or entering the expander,

p_4 = Pressure of the air leaving the heat exchanger or entering the expander = $p_3 = p_{3'}$,

T_5 = Temperature of the air leaving the expander after isentropic expansion,

$T_{5'}$ = Actual temperature of the air leaving the expander.

We know that

$$T_2 = T_1 + \frac{V^2}{2000\, c_p} = 263 + \frac{(277.8)^2}{2000 \times 1.005}$$

$$= 263 + 38.4 = 301.4 \text{ K } \textbf{Ans.}$$

and

$$\frac{p_2}{p_1} = \left(\frac{T_2}{T_1}\right)^{\frac{\gamma}{\gamma-1}} = \left(\frac{301.4}{263}\right)^{\frac{1.4}{1.4-1}} = (1.146)^{3.5} = 1.611$$

$\therefore$

$$p_2 = p_1 \times 1.611 = 0.35 \times 1.611 = 0.564 \text{ bar } \textbf{Ans.}$$

Since $p_3 / p_2 = 4.5$ (Given), therefore

$$p_3 = p_2 \times 4.5 = 0.564 \times 4.5 = 2.54 \text{ bar } \textbf{Ans.}$$

We know that for isentropic compression process 2-3,

$$\frac{T_3}{T_2} = \left(\frac{p_3}{p_2}\right)^{\frac{\gamma-1}{\gamma}} = (4.5)^{\frac{1.4-1}{1.4}} = (4.5)^{0.286} = 1.537$$

$\therefore$

$$T_3 = T_2 \times 1.537 = 301.4 \times 1.537 = 463.3 \text{ K}$$

We also know that isentropic efficiency of the compressor,

$$\eta_C = \frac{\text{Isentropic temperature rise}}{\text{Actual temperature rise}} = \frac{T_3 - T_2}{T_{3'} - T_2}$$

$$0.8 = \frac{463.3 - 301.4}{T_{3'} - 301.4} = \frac{161.9}{T_{3'} - 301.4}$$

$$T_{3'} - 301.4 = 161.9/0.8 = 202.4$$

$\therefore$

$$T_{3'} = 301.4 + 202.4 = 503.8 \text{ K } \textbf{Ans.}$$

Effectiveness of the heat exchanger (η_H),

$$0.95 = \frac{T_{3'} - T_4}{T_{3'} - T_2} = \frac{503.8 - T_4}{503.8 - 301.4} = \frac{503.8 - T_4}{202.4}$$

$\therefore$

$$T_4 = 503.8 - 0.95 \times 202.4 = 311.5 \text{ K } \textbf{Ans.}$$

and

$$p_4 = p_3 = 2.54 \text{ bar } \textbf{Ans.}$$

We know that for isentropic expansion process 4-5,

$$\frac{T_5}{T_4} = \left(\frac{p_5}{p_4}\right)^{\frac{\gamma-1}{\gamma}} = \left(\frac{1.06}{2.54}\right)^{\frac{1.4-1}{1.4}} = (0.417)^{0.286} = 0.7787$$

$\therefore$

$$T_5 = T_4 \times 0.7787 = 311.5 \times 0.7787 = 243 \text{ K}$$

Now isentropic efficiency of the expander,

$$\eta_E = \frac{\text{Actual temperature rise}}{\text{Isentropic temperature rise}} = \frac{T_4 - T_{5'}}{T_4 - T_5}$$

$$0.8 = \frac{311.5 - T_{5'}}{311.5 - 243} = \frac{311.5 - T_{5'}}{68.5}$$

$$\therefore \quad T_{5'} = 311.5 - 0.8 \times 68.5 = 256.7 \text{ K} \quad \textbf{Ans.}$$

Volume flow rate

Let v_2 = Volume flow rate through the compressor inlet, and

$v_{5'}$ = Volume flow rate through the expander outlet.

We know that mass flow rate of air,

$$m_a = \frac{210\,Q}{c_p(T_6 - T_{5'})} = \frac{210 \times 100}{1.005\,(298 - 256.7)} = 506 \text{ kg/min}$$

and $$p_2 v_2 = m_a R T_2$$

$$\therefore \quad v_2 = \frac{m_a R T_2}{p_2} = \frac{506 \times 287 \times 301.4}{0.564 \times 10^5} = 776 \text{ m}^3\text{/min} \quad \textbf{Ans.}$$

... (R is taken in J/kg K and p_2 is taken in N/m^2)

Similarly $$p_{5'} v_{5'} = m_a R\, T_{5'}$$

$$\therefore \quad v_{5'} = \frac{m_a R T_{5'}}{p_{5'}} = \frac{506 \times 287 \times 256.7}{1.06 \times 10^5} = 351.7 \text{ m}^3\text{/min} \quad \textbf{Ans.}$$

Example 3.5. *The cock pit of a jet plane flying at a speed of 1200 km/h is to be cooled by a simple air cooling system. The cock pit is to be maintained at 25°C and the pressure in the cock pit is 1 bar. The ambient air pressure and temperature are 0.85 bar and 30°C. The other data available is as follows :*

Cock-pit cooling load = 10 TR ; Main compressor pressure ratio = 4 ; Ram efficiency = 90% ; Temperature of air leaving the heat exchanger and entering the cooling turbine = 60°C ; Pressure drop in the heat exchanger = 0.5 bar ; Pressure loss between the cooler turbine and cock pit = 0.2 bar.

Assuming the isentropic efficiencies of main compressor and cooler turbine as 80%, find the quantity of air passed through the cooling turbine and C.O.P. of the system. Take $\gamma = 1.4$ and $c_p = 1$ kJ/kg K.

Solution. Given : $V = 1200$ km / h = 333.3 m / s ; $T_6 = 25°\text{C} = 25 + 273 = 298$ K ; $p_6 = 1$ bar ; $p_1 = 0.85$ bar ; $T_1 = 30°\text{C} = 30 + 273 = 303$ K ; $Q = 10$ TR ; $p_3 / p_{2'} = 4$; $\eta_R = 90\% = 0.9$; $T_4 = 60°\text{C} = 60 + 273 = 333$ K ; $p_4 = (p_{3'} - 0.5)$ bar ; $p_5 = p_{5'} = p_6 + 0.2 = 1 + 0.2 = 1.2$ bar ; $\eta_C = \eta_T = 80\% = 0.8$; $\gamma = 1.4$; $c_p = 1$ kJ/kg K

The *T-s* diagram for the simple air cooling system with the given conditions is shown in Fig. 3.7.

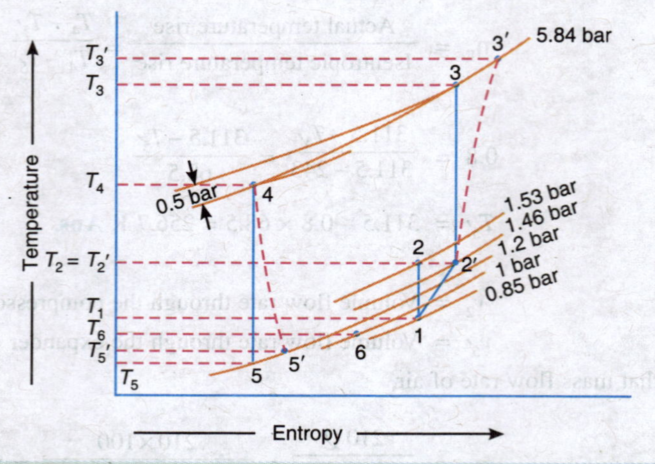

Fig. 3.7

Let $T_{2'}$ = Stagnation temperature of the ambient air entering the main compressor = T_2,

p_2 = Pressure of air after isentropic ramming, and

$p_{2'}$ = Stagnation pressure of air entering the main compressor.

We know that
$$T_2 = T_{2'} = T_1 + \frac{V^2}{2000\, c_p} = 303 + \frac{(333.3)^2}{2000 \times 1}$$
$$= 303 + 55.5 = 358.5 \text{ K}$$

and
$$\frac{p_2}{p_1} = \left(\frac{T_2}{T_1}\right)^{\frac{\gamma}{\gamma-1}} = \left(\frac{358.5}{303}\right)^{\frac{1.4}{1.4-1}} = (1.183)^{3.5} = 1.8$$

$\therefore$
$$p_2 = p_1 \times 1.8 = 0.85 \times 1.8 = 1.53 \text{ bar}$$

We know that ram efficiency,
$$\eta_R = \frac{\text{Actual pressure rise}}{\text{Isentropic pressure rise}} = \frac{p_{2'} - p_1}{p_2 - p_1}$$
$$0.9 = \frac{p_{2'} - 0.85}{1.53 - 0.85} = \frac{p_{2'} - 0.85}{0.68}$$

$\therefore$
$$p_{2'} = 0.9 \times 0.68 + 0.85 = 1.46 \text{ bar}$$

Now for the isentropic process 2′– 3,
$$\frac{T_3}{T_{2'}} = \left(\frac{p_3}{p_{2'}}\right)^{\frac{\gamma-1}{\gamma}} = (4)^{\frac{1.4-1}{1.4}} = (4)^{0.286} = 1.486$$

$\therefore$
$$T_3 = T_{2'} \times 1.486 = 358.5 \times 1.486 = 532.7 \text{ K}$$

and isentropic efficiency of the compressor,
$$\eta_C = \frac{\text{Isentropic temperature rise}}{\text{Actual temperature rise}} = \frac{T_3 - T_{2'}}{T_{3'} - T_{2'}}$$
$$0.8 = \frac{532.7 - 358.5}{T_{3'} - 358.5} = \frac{174.2}{T_{3'} - 358.5}$$

$\therefore$
$$T_{3'} = \frac{174.2}{0.8} + 358.5 = 576 \text{ K}$$

Since the pressure ratio of the main compressor (p_3 / p_2') is 4, therefore pressure of air leaving the main compressor,

$$p_3 = p_3' = 4\,p_2' = 4 \times 1.46 = 5.84 \text{ bar}$$

Pressure drop in the heat exchanger

$$= 0.5 \text{ bar}$$

∴ Pressure of air after passing through the heat exchanger or at entrance to the cooling turbine,

$$p_4 = p_3' - 0.5 = 5.84 - 0.5 = 5.34 \text{ bar}$$

Also there is a pressure loss of 0.2 bar between the cooling turbine and the cock pit. Therefore pressure of air leaving the cooling turbine,

$$p_5 = p_5' = p_6 + 0.2 = 1 + 0.2 = 1.2 \text{ bar}$$

Now for the isentropic process 4-5,

$$\frac{T_4}{T_5} = \left(\frac{p_4}{p_5}\right)^{\frac{\gamma-1}{\gamma}} = \left(\frac{5.34}{1.2}\right)^{\frac{1.4-1}{1.4}} = (4.45)^{0.286} = 1.53$$

$$\therefore \quad T_5 = T_4 / 1.53 = 333 / 1.53 = 217.6 \text{ K}$$

We know that isentropic efficiency of the cooling turbine,

$$\eta_T = \frac{\text{Actual temperature rise}}{\text{Isentropic temperature rise}} = \frac{T_4 - T_5'}{T_4 - T_5}$$

$$0.8 = \frac{333 - T_5'}{333 - 217.6} = \frac{333 - T_5'}{115.4}$$

$$\therefore \quad T_5' = 333 - 0.8 \times 115.4 = 240.7 \text{ K}$$

Quantity of air passed through the cooling turbine

We know that quantity of air passed through the cooling turbine,

$$m_a = \frac{210\,Q}{c_p(T_6 - T_5')} = \frac{210 \times 10}{1(298 - 240.7)} = 36.6 \text{ kg / min} \quad \textbf{Ans.}$$

C.O.P. of the system

We know that C.O.P. of the system

$$= \frac{210\,Q}{m_a c_p (T_3' - T_2')} = \frac{210 \times 10}{36.5 \times 1(576 - 358.5)} = 0.264 \quad \textbf{Ans.}$$

Example 3.6. *In an aeroplane, a simple air refrigeration is used. The main compressor delivers the air at 5 bar and 200°C. The bled air taken from compressor is passed through a heat exchanger, cooled with the help of ram air so that the temperature of air leaving the heat exchanger is 45°C and the pressure is 4.5 bar. The cooling turbine drives the exhaust fan which is used to force the ram air through the heat exchanger. The air leaving the heat exchanger passes through the cooling turbine and then supplied to cabin at 1 bar. The pressure loss between the cooling turbine and cabin is 0.2 bar. If the rate of flow of air through the cooling turbine is 20 kg/min, determine the following :*

1. The temperature of the air leaving the expander ;

2. The power delivered to the ram air which is passed through the heat exchanger ; and

3. The refrigeration load in tonnes when the temperature of the air leaving the cabin is limited to 25°C.

Assume that the isentropic efficiency of the cooling turbine is 75% and no loss of heat from air between the cooling turbine and cabin. Take γ = 1.4 and c_p = 1 kJ/kg K.

Solution. Given : $p_3 = 5$ bar ; $T_3 = 200°C = 200 + 273 = 473$ K ; $T_4 = 45°C = 45 + 273 = 318$ K ; $p_4 = 4.5$ bar ; $p_6 = 1$ bar ; $p_5 = p_{6'} = p_6 + 0.2 = 1 + 0.2 = 1.2$ bar ; $m_a = 20$ kg / min ; $T_6 = 25°C = 25 + 273 = 298$ K ; $\eta_T = 75\% = 0.75$; $\gamma = 1.4$; $c_p = 1$ kJ/kg K

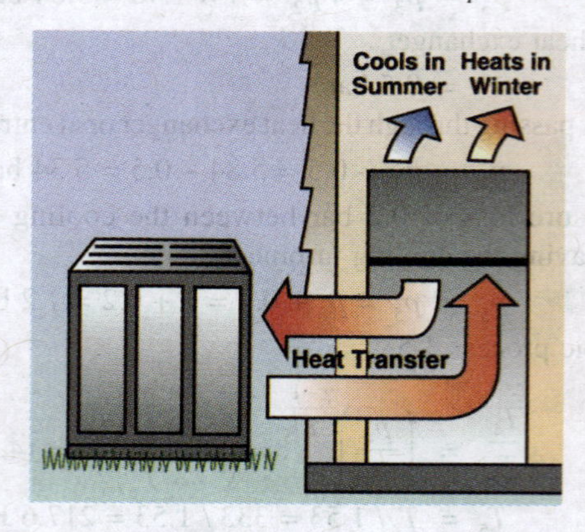

Simple air-refrigeration unit.

The schematic diagram for the simple air refrigeration system is shown in Fig. 3.1. The various processes on the *T-s* diagram are shown in Fig. 3.8. The point 3 represents the air delivered from the compressor to heat exchanger and the point 4 shows the condition of air leaving the heat exchanger. The vertical line 4 – 5 represents the isentropic expansion of air in the cooling turbine and the curve 4 – 5′ shows the actual expansion of air in the cooling turbnine due to internal friction.The line 5′– 6 represents the refrigeration process.

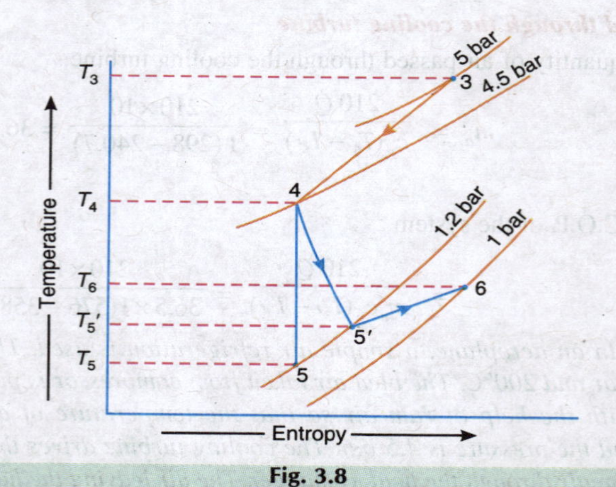

Fig. 3.8

1. *Temperature of air leaving the expander*

Let T_5 = Temperature of air at the end of isentropic expansion in the cooling turbine or expander, and

$T_{5'}$ = Actual temperature of air leaving the cooling turbine or expander.

We know that
$$\frac{T_5}{T_4} = \left(\frac{p_5}{p_4}\right)^{\frac{\gamma-1}{\gamma}} = \left(\frac{1.2}{4.5}\right)^{\frac{1.4-1}{1.4}} = (0.267)^{0.286} = 0.685$$

$$\therefore \quad T_5 = T_4 \times 0.685 = 318 \times 0.685 = 217.8 \text{ K}$$

Isentropic efficiency of the cooling turbine,

$$\eta_T = \frac{\text{Actual temperature rise}}{\text{Isentropic temperature rise}} = \frac{T_4 - T_{5'}}{T_4 - T_5}$$

$$0.75 = \frac{318 - T_{5'}}{318 - 217.8} = \frac{318 - T_{5'}}{100.2}$$

$\therefore \quad T_{5'} = 318 - 0.75 \times 100.2 = 242.85$ K **Ans.**

2. *Power delivered to the ram air which is passed through the heat exchanger*

We know that work delivered to the ram air which is passed through the heat exchanger,

$$= m_a c_p (T_4 - T_{5'}) = 20 \times 1\ (318 - 242.85) = 1503 \text{ kJ/min}$$

$\therefore$ Power delivered $= 1503/60 = 25.05$ kW **Ans.**

3. *Refrigeration load*

We know that the refrigeration load taken from the cabin

$$= m_a c_p (T_6 - T_{5'}) = 20 \times 1\ (298 - 242.85) = 1103 \text{ kJ/min}$$

$$= 1103/210 = 5.25 \text{ TR}$$ **Ans.**

Example 3.7. *The speed of an aircraft flying at an altitude of 8000 metres, where the ambient air is at 0.341 bar pressure and 263 K temperature, is 900 km/h. The compression ratio of the air compressor is 5. The cabin pressure is 1.013 bar and the temperature is 27°C. Determine: 1. The power requirement of the aircraft for pressurization (excluding the ram work); 2. The additional power required for refrigeration; and 3. The refrigerating capacity for simple air craft refrigeration cycle on the basis of 1 kg / s flow of air. Specific heat of air = 1.005 kJ / kg K and γ = 1.4.*

Also find the change in their values, if the following data is to be taken into consideration:

Compressor efficiency = 82%; Expander or turbine efficiency = 77%; Effectiveness of heat exchanger = 80%; and Ram efficiency = 84%.

Solution. Given : *$h = 8000$ m ; $p_1 = 0.341$ bar ; $T_1 = 263$ K ; $V = 900$ km/h $= 250$ m/s ; $p_3 / p_2 = 5$; $p_c = p_5 = p_6 = 1.013$ bar ; $T_6 = 27°C = 27 + 273 = 300$ K ; $m_a = 1$ kg/s ; $c_p = 1.005$ kJ/kg K ; $\gamma = 1.4$

The *T-s* diagram for the simple air refrigeration cycle with the given conditions is shown in Fig. 3.9.

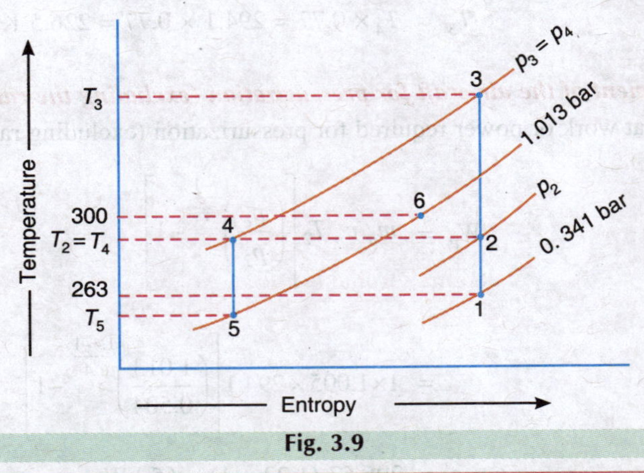

Fig. 3.9

* Superfluous data.

Let T_2 and p_2 = Stagnation temperature and pressure of the ambient air entering the compressor,

T_3 and p_3 = Temperature and pressure of the air leaving the compressor after isentropic compression, and

T_5 = Temperature of the air leaving the expander after isentropic expansion.

We know that $$T_2 = T_1 + \frac{V^2}{2000\, c_p}$$

$$= 263 + \frac{(250)^2}{2000 \times 1.005} = 263 + 31.1 = 294.1 \text{ K}$$

and $$\frac{p_2}{p_1} = \left(\frac{T_2}{T_1}\right)^{\frac{\gamma}{\gamma-1}} = \left(\frac{294.1}{263}\right)^{\frac{1.4}{1.4-1}} = (1.118)^{3.5} = 1.477$$

$\therefore$ $$p_2 = p_1 \times 1.477 = 0.341 \times 1.477 = 0.504 \text{ bar}$$

Since $p_3 / p_2 = 5$ (Given), therefore

$$p_3 = p_2 \times 5 = 0.504 \times 5 = 2.52 \text{ bar}$$

We know that for isentropic compression process 2-3,

$$\frac{T_3}{T_2} = \left(\frac{p_3}{p_2}\right)^{\frac{\gamma-1}{\gamma}} = (5)^{\frac{1.4-1}{1.4}} = (5)^{0.286} = 1.584$$

$\therefore$ $$T_3 = T_2 \times 1.584 = 294.1 \times 1.584 = 465.8 \text{ K}$$

and for isentorpic expansion process 4-5,

$$\frac{T_5}{T_4} = \left(\frac{p_5}{p_4}\right)^{\frac{\gamma-1}{\gamma}} = \left(\frac{1.013}{2.52}\right)^{\frac{1.4-1}{1.4}} = (0.402)^{0.286} = 0.77$$

... ($\because p_4 = p_3 = 2.52$ bar)

$\therefore$ $$T_5 = T_4 \times 0.77 = 294.1 \times 0.77 = 226.5 \text{ K}$$

... (Assuming $T_4 = T_2$)

1. *Power requirement of the air craft for pressurization (excluding the ram work)*

We know that work or power required for pressurization (excluding ram work),

$$W_P = m_a\, c_p\, T_2 \left[\left(\frac{p_c}{p_2}\right)^{\frac{\gamma-1}{\gamma}} - 1\right]$$

$$= 1 \times 1.005 \times 294.1 \left[\left(\frac{1.013}{0.504}\right)^{\frac{1.4-1}{1.4}} - 1\right]$$

$$= 295.57\ (1.22 - 1) = 65 \text{ kW}$$ **Ans.**

2. *Additional power required for refrigeration*

We know that workdone during compression process 2-3,

$$W_C = m_a\, c_p\, (T_3 - T_2)$$
$$= 1 \times 1.005\ (465.8 - 294.1) = 172.56 \text{ kW}$$

and work done during expansion process 4-5,

$$W_T = m_a\, c_p\, (T_4 - T_5)$$
$$= 1 \times 1.005\ (294.1 - 226.5) = 68 \text{ kW}$$

∴ Power required for refrigeration (excluding ram work),

$$W_R = W_C - W_T = 172.56 - 68 = 104.56 \text{ kW}$$

and additional power required for refrigeration

$$= W_R - W_P = 104.56 - 65 = 39.56 \text{ kW} \quad \textbf{Ans.}$$

3. *Refrigerating Capacity*

We know that refrigerating capacity or refrigerating effect produced,

$$R_E = m_a\, c_p\, (T_6 - T_5)$$
$$= 1 \times 1.005\ (300 - 226.5) = 73.87 \text{ kW} \quad \textbf{Ans.}$$

Change in power requirement, additional power required and refrigerating capacity

Given : $\eta_C = 82\% = 0.82$; $\eta_T = 77\% = 0.77$; $\eta_H = 80\% = 0.8$; $\eta_R = 84\% = 0.84$

The *T-s* diagram for the simple air refrigeration cycle with the given conditions is shown in Fig. 3.10.

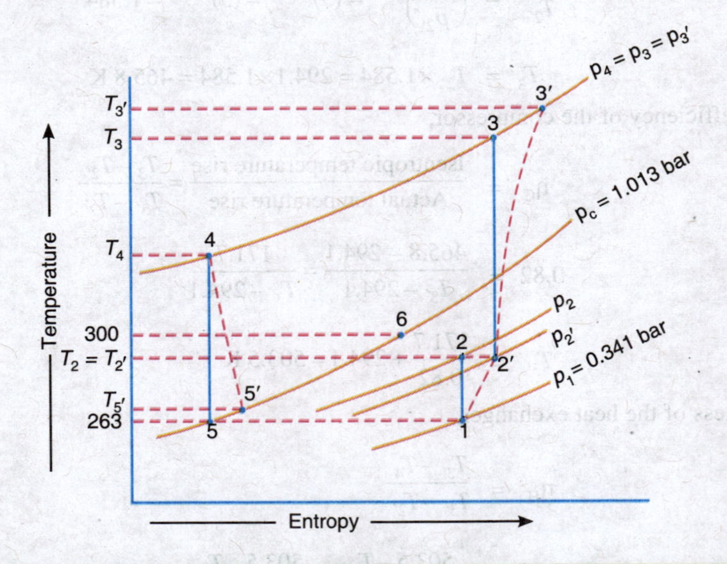

Fig. 3.10

Let $\quad T_{2'}$ = Stagnation temperature of the ambient air entering the compressor = T_2 = 294.1 K (Calculated already),

$p_{2'}$ = Stagnation pressure of the air entering the compressor,

$T_{3'}$ = Actual temperature of the air leaving the compressor,

T_4 = Temperature of the air leaving the heat exchanger or entering the expander or cooling turbine,

$T_{5'}$ = Actual temperature of the air leaving the expander or cooling turbine.

We know that ram efficiency,

$$\eta_R = \frac{\text{Actual pressure rise}}{\text{Isentropic pressure rise}} = \frac{p_{2'} - p_1}{p_2 - p_1}$$

$$0.84 = \frac{p_{2'} - 0.341}{0.504 - 0.341} = \frac{p_{2'} - 0.341}{0.163}$$

... ($\because p_2 = 0.504$ bar)

$$\therefore \quad p_{2'} = 0.84 \times 0.163 + 0.341 = 0.478 \text{ bar}$$

Since $p_3 / p_{2'} = 5$ (Given), therefore

$$p_3 = p_{2'} \times 5 = 0.478 \times 5 = 2.39 \text{ bar}$$

We know that for isentropic compression process 2′-3,

$$\frac{T_3}{T_{2'}} = \left(\frac{p_3}{p_{2'}}\right)^{\frac{\gamma-1}{\gamma}} = (5)^{\frac{1.4-1}{1.4}} = (5)^{0.286} = 1.584$$

$$\therefore \quad T_3 = T_{2'} \times 1.584 = 294.1 \times 1.584 = 465.8 \text{ K}$$

and isentropic efficiency of the compressor,

$$\eta_C = \frac{\text{Isentropic temperature rise}}{\text{Actual temperature rise}} = \frac{T_3 - T_{2'}}{T_{3'} - T_{2'}}$$

$$0.82 = \frac{465.8 - 294.1}{T_{3'} - 294.1} = \frac{171.7}{T_{3'} - 294.1}$$

$$\therefore \quad T_{3'} = \frac{171.7}{0.82} + 294.1 = 503.5 \text{ K}$$

Effectiveness of the heat exchanger,

$$\eta_H = \frac{T_{3'} - T_4}{T_{3'} - T_{2'}}$$

$$0.8 = \frac{503.5 - T_4}{503.5 - 294.1} = \frac{503.5 - T_4}{209.4}$$

$$\therefore \quad T_4 = 503.5 - 0.8 \times 209.4 = 336 \text{ K}$$

We know that for isentropic expansion process 4-5,

$$\frac{T_4}{T_5} = \left(\frac{p_4}{p_5}\right)^{\frac{\gamma-1}{\gamma}} = \left(\frac{2.39}{1.013}\right)^{\frac{1.4-1}{1.4}} = (2.36)^{0.286} = 1.278$$

$$\dots (\because p_4 = p_3 = p_{3'})$$

$$\therefore \quad T_5 = T_4 / 1.278 = 336/1.278 = 263 \text{ K}$$

Now for isentropic efficiency of the expander or turbine,

$$\eta_T = \frac{\text{Actual temperature rise}}{\text{Isentropic temperature rise}} = \frac{T_4 - T_{5'}}{T_4 - T_5}$$

$$0.77 = \frac{336 - T_{5'}}{336 - 263} = \frac{336 - T_{5'}}{73}$$

$$\therefore \quad T_{5'} = 336 - 0.77 \times 73 = 279.8 \text{ K}$$

Power requirement of the air craft for pressurization (excluding ram work)

We know that work or power required for pressurization (excluding ram work),

$$W_P = \frac{m_a\, c_p\, T_{2'}}{\eta_C}\left[\left(\frac{p_c}{p_{2'}}\right)^{\frac{\gamma-1}{\gamma}} - 1\right]$$

$$= \frac{1 \times 1.005 \times 294.1}{0.82}\left[\left(\frac{1.013}{0.478}\right)^{\frac{1.4-1}{1.4}} - 1\right]$$

$$= 360.45\,(1.239 - 1) = 86.15 \text{ kW}$$ **Ans.**

Additional power required for refrigeration

We know that work done during compression process 2′-3′,

$$W_C = m_a c_p (T_{3'} - T_{2'})$$

$$= 1 \times 1.005\,(503.5 - 294.1) = 210 \text{ kW}$$

and work done during expansion process, 4-5′,

$$W_T = m_a c_p (T_4 - T_{5'})$$

$$= 1 \times 1.005\,(336 - 279.8) = 56.5 \text{ kW}$$

∴ Power required for refrigeration (excluding ram work)

$$W_R = W_C - W_T = 210 - 56.5 = 153.5 \text{ kW}$$

and additional power required for refrigeration

$$= W_R - W_P = 153.5 - 86.15 = 67.35 \text{ kW}$$ **Ans.**

Refrigerating Capacity

We know that refrigerating capacity or refrigerating effect produced,

$$R_E = m_a c_p (T_6 - T_{5'})$$

$$= 1 \times 1.005\,(300 - 279.8) = 20.3 \text{ kW}$$ **Ans.**

3.5 Simple Air Evaporative Cooling system

A simple air evaporative cooling system is shown in Fig. 3.11. It is similar to the simple cooling system except that the addition of an evaporator between the heat exchanger and cooling turbine. The evaporator provides an additional cooling effect through evaporation of a refrigerant such as water. At high altitudes, the evaporative cooling may be obtained by using alcohol or ammonia. The water, alcohol and ammonia have different refrigerating effects at different altitudes. At 20 000 metres height, water boils at 40°C, alcohol at 9°C and ammonia at – 70°C.

Air evaporative cooling system

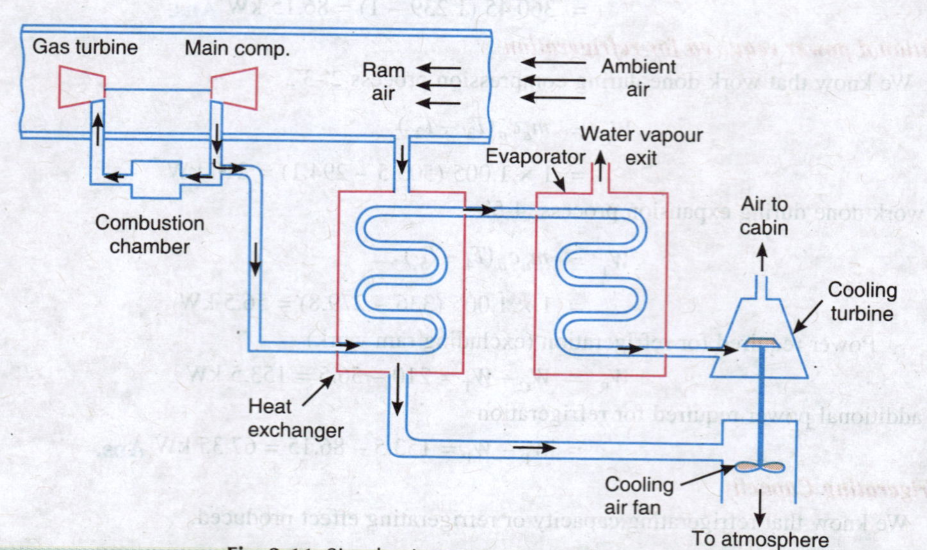

Fig. 3.11. Simple air evaporative cooling system.

The T-s diagram for a simple air cycle evaporative cooling system is shown in Fig. 3.12.

The various processes are same as discussed in the previous article, except that the cooling process in the evaporator as shown by 4 – 4′ in Fig. 3.12.

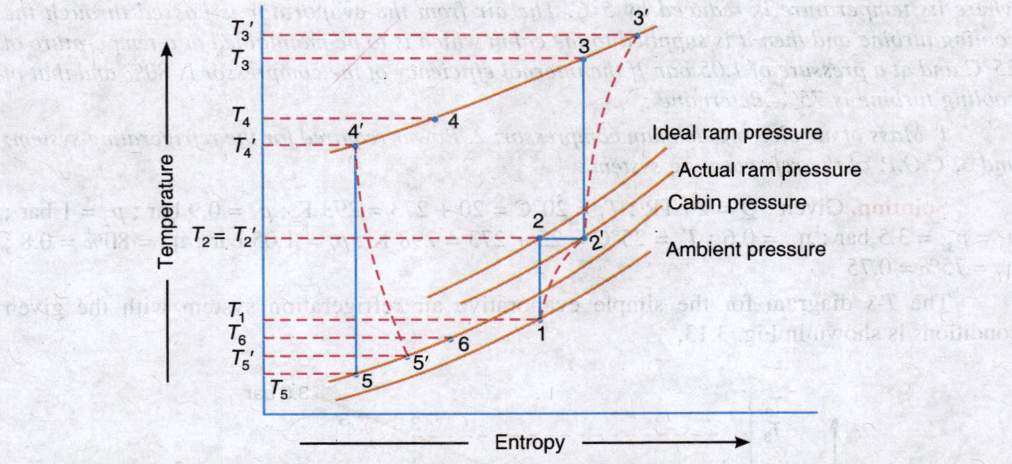

Fig. 3.12. *T-s* diagram for simple evaporative cooling system.

If Q tonnes of refrigeration is the cooling load in the cabin, then the air required for the refrigeration purpose,

$$m_a = \frac{210\,Q}{c_p\,(T_6 - T_{5'})} \text{ kg / min}$$

Power required for the refrigerating system,

$$P = \frac{m_a c_p\,(T_{3'} - T_{2'})}{60} \text{ kW}$$

and C.O.P. of the refrigerating system

$$= \frac{210\,Q}{m_a c_p\,(T_{3'} - T_{2'})} = \frac{210\,Q}{P \times 60}$$

The initial mass of evaporant (m_e) required to be carried for the given flight time is given by

$$m_e = \frac{Q_e\,.\,t}{h_{fg}}$$

where Q_e = Heat to be removed in evaporation in kJ/min,

t = Flight time in minutes, and

h_{fg} = Latent heat of vaporisation of evaporant in kJ/kg.

Notes : 1. In *T-s* diagram as shown in Fig. 3.12, the thick lines show the ideal condition of the process, while the dotted lines show actual conditions of the process.

2. If cooling of 45 minutes duration or less is required, it may be advantageous to use evaporative cooling alone.

Example 3.8. *A simple evaporative air refrigeration system is used for an aeroplane to take 20 tonnes of refrigeration load. The ambient air conditions are 20°C and 0.9 bar. The ambient air is rammed isentropically to a pressure of 1 bar. The air leaving the main compressor at pressure 3.5 bar is first cooled in the heat exchanger having effectiveness of 0.6 and then in the evaporator where its temperature is reduced by 5°C. The air from the evaporator is passed through the cooling turbine and then it is supplied to the cabin which is to be maintained at a temperature of 25°C and at a pressure of 1.05 bar. If the internal efficiency of the compressor is 80% and that of cooling turbine is 75% ,determine :*

1. Mass of air bled off the main compressor; 2. Power required for the refrigerating system; and 3. C.O.P. of the refrigerating system.

Solution. Given : Q = 20 TR ; T_1 = 20°C = 20 + 273 = 293 K ; p_1 = 0.9 bar ; p_2 = 1 bar ; $p_3 = p_{3'}$ = 3.5 bar ; η_H = 0.6 ; T_6 = 25°C = 25 + 273 = 298 K ; p_6 = 1.05 bar ; η_C = 80% = 0.8 ; η_T = 75% = 0.75

The *T-s* diagram for the simple evaporative air refrigeration system with the given conditions is shown in Fig. 3.13.

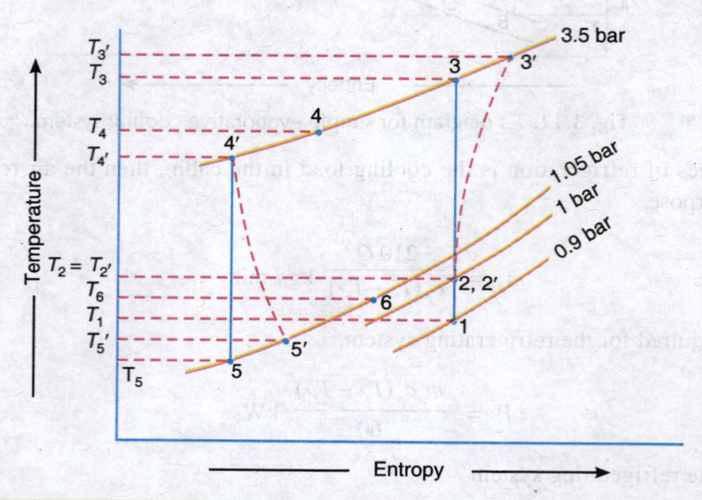

Fig. 3.13

Let T_2 = Temperature of air entering the main compressor,

T_3 = Temperature of air after isentropic compression in the main compressor,

$T_{3'}$ = Actual temperature of air leaving the main compressor, and

T_4 = Temperature of air entering the evaporator.

We know that for an isentropic ramming process 1-2,

$$\frac{T_2}{T_1} = \left(\frac{p_2}{p_1}\right)^{\frac{\gamma-1}{\gamma}} = \left(\frac{1}{0.9}\right)^{\frac{1.4-1}{1.4}} = (1.11)^{0.286} = 1.03$$

... (Taking $\gamma = 1.4$)

$\therefore \quad T_2 = T_1 \times 1.03 = 293 \times 1.03 = 301.8$ K

Now for the isentropic compression process 2-3,

$$\frac{T_3}{T_2} = \left(\frac{p_3}{p_2}\right)^{\frac{\gamma-1}{\gamma}} = \left(\frac{3.5}{1}\right)^{\frac{1.4-1}{1.4}} = (3.5)^{0.286} = 1.43$$

$$\therefore \quad T_3 = T_2 \times 1.43 = 301.8 \times 1.43 = 431.6 \text{ K}$$

We know that efficiency of the compressor,

$$\eta_C = \frac{\text{Isentropic increase in temperature}}{\text{Actual increase in temperature}} = \frac{T_3 - T_2}{T_{3'} - T_2}$$

$$0.8 = \frac{431.6 - 301.8}{T_{3'} - 301.8} = \frac{129.8}{T_{3'} - 301.8}$$

$$\therefore \quad T_{3'} = 301.8 + 129.8/0.8 = 464 \text{ K}$$

Effectiveness of the heat exchanger (η_H),

$$0.6 = \frac{T_{3'} - T_4}{T_{3'} - T_{2'}} = \frac{464 - T_4}{464 - 301.8} = \frac{464 - T_4}{162.2} \quad \dots (\because T_{2'} = T_2)$$

$$\therefore \quad T_4 = 464 - 0.6 \times 162.2 = 366.7 \text{ K} = 93.7°\text{C}$$

Since the temperature of air in the evaporator is reduced by 5°C, therefore the temperature of air leaving the evaporator and entering the cooling turbine,

$$T_{4'} = T_4 - 5 = 93.7 - 5 = 88.7°\text{C} = 361.7 \text{ K}$$

Now for the isentropic expansion process 4′– 5,

$$\frac{T_{4'}}{T_5} = \left(\frac{p_3}{p_6}\right)^{\frac{\gamma-1}{\gamma}} = \left(\frac{3.5}{1.05}\right)^{\frac{1.4-1}{1.4}} = (3.33)^{0.286} = 1.41$$

$$\therefore \quad T_5 = T_{4'} / 1.41 = 361.7/1.41 = 256.5 \text{ K}$$

Efficiency of the cooling turbine,

$$\eta_T = \frac{\text{Actual increase in temperature}}{\text{Isentropic increase in temperature}} = \frac{T_{4'} - T_{5'}}{T_{4'} - T_5}$$

$$0.75 = \frac{361.7 - T_{5'}}{361.7 - 256.5} = \frac{361.7 - T_{5'}}{105.2}$$

$$\therefore \quad T_{5'} = 361.7 - 0.75 \times 105.2 = 282.8 \text{ K}$$

1. *Mass of air bled off the main compressor*

We know that mass of air bled off the main compressor,

$$m_a = \frac{210\,Q}{c_p(T_6 - T_{5'})} = \frac{210 \times 20}{1\,(298 - 282.8)} = 276 \text{ kg / min} \quad \textbf{Ans.}$$

2. *Power required for the refrigerating system*

We know that power required for the refrigerating system,

$$P = \frac{m_a c_p (T_{3'} - T_{2'})}{60} = \frac{276 \times 1\,(464 - 301.8)}{60} = 746 \text{ kW} \quad \textbf{Ans.}$$

3. *C.O.P. of the refrigerating system*

We know that C.O.P. of the refrigerating system

$$= \frac{210\,Q}{P \times 60} = \frac{210 \times 20}{746 \times 60} = 0.094 \quad \textbf{Ans.}$$

3.6 Boot-strap Air Cooling System

A boot-strap air cooling system is shown in Fig. 3.14. This cooling system has two heat exchangers instead of one and a cooling turbine drives a secondary compressor instead of cooling fan. The air bled from the main compressor is first cooled by the ram air in the first heat exchanger. This cooled air, after compression in the secondary compressor, is led to the second heat exchanger where it is again cooled by the ram air before passing to the cooling turbine. This type of cooling system is mostly used in transport type aircraft.

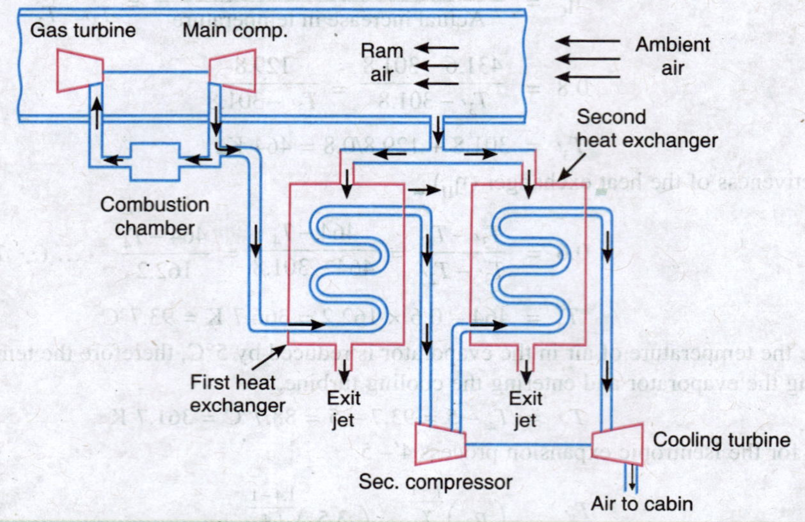

Fig. 3.14. Boot-strap air cooling system.

The *T-s* diagram for a boot-strap air cycle cooling system is shown in Fig. 3.15. The various processes are as follows :

1. The process 1– 2 represents the isentropic ramming of ambient air from pressure p_1 and temperature T_1 to pressure p_2 and temperature T_2. The process 1– 2′ represents the actual ramming process because of internal friction due to irreversibilities.
2. The process 2′– 3 represents the isentropic compression of air in the main compressor and the process 2′– 3′ represents the actual compression of air because of internal friction due to irreversibilities.
3. The process 3′–4 represents the cooling by ram air in the first heat exchanger. The pressure drop in the heat exchanger is neglected.
4. The process 4 – 5 represents the isentropic compression of cooled air, from first heat exchanger, in the secondary compressor. The process 4 – 5′ represents the actual compression process because of internal friction due to irreversibilities.
5. The process 5′– 6 represents the cooling by ram air in the second heat exchanger. The pressure drop in the heat exchanger in neglected.
6. The process 6 – 7 represents the isentropic expansion of cooled air in the cooling turbine upto the cabin pressure. The process 6 – 7′ represents actual expansion of the cooled air in the cooling turbine.
7. The process 7′– 8 represents the heating of air upto the cabin temperature T_8.

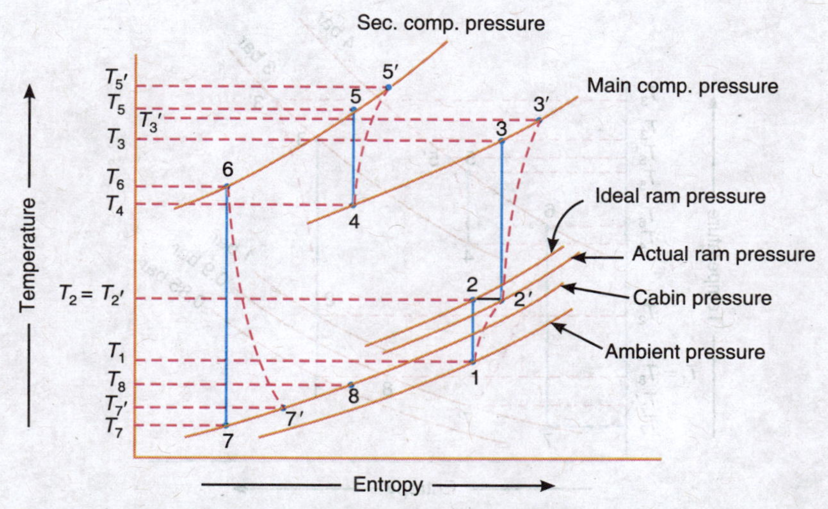

Fig. 3.15. *T-s* diagram for boot strap air cooling system.

If Q tonnes of refrigeration is the cooling load in the cabin, then the quantity of air required for the refrigeration purpose will be

$$m_a = \frac{210\,Q}{c_p(T_8 - T_{7'})} \text{ kg / min}$$

Power required for the refrigerating system,

$$P = \frac{m_a c_p (T_{3'} - T_{2'})}{60} \text{ kW}$$

and C.O.P. of the refrigerating system

$$= \frac{210\,Q}{m_a c_p (T_{3'} - T_{2'})} = \frac{210}{P \times 60}$$

Example 3.9. *A boot-strap cooling system of 10 TR capacity is used in an aeroplane. The ambient air temperature and pressure are 20°C and 0.85 bar respectively. The pressure of air increases from 0.85 bar to 1 bar due to ramming action of air. The pressure of air discharged from the main compressor is 3 bar. The discharge pressure of air from the auxiliary compressor is 4 bar. The isentropic efficiency of each of the compressor is 80%, while that of turbine is 85%. 50% of the enthalpy of air discharged from the main compressor is removed in the first heat exchanger and 30% of the enthalpy of air discharged from the auxiliary compressor is removed in the second heat exchanger using rammed air. Assuming ramming action to be isentropic, the required cabin pressure of 0.9 bar and temperature of the air leaving the cabin not more than 20°C, find : 1. the power required to operate the system; and 2. the C.O.P. of the system. Draw the schematic and temperature -entropy diagram of the system. Take* $\gamma = 1.4$ *and* $c_p = 1$ *kJ/kg K.*

Solution. Given : $Q = 10$ TR ; $T_1 = 20°C = 20 + 273 = 293$ K ; $p_1 = 0.85$ bar ; $p_2 = 1$ bar ; $p_3 = p_{3'} = p_4 = 3$ bar ; $p_5 = p_{5'} = p_6 = 4$ bar ; $\eta_{C1} = \eta_{C2} = 80\% = 0.8$; $\eta_T = 85\% = 0.85$; $p_7 = p_{7'} = p_8 = 0.9$ bar ; $T_8 = 20°C = 20 + 273 = 293$ K ; $\gamma = 1.4$; $c_p = 1$ kJ/kg K

The schematic diagram for a boot-strap cooling system is shown in Fig. 3.14. The temperature- entropy (*T-s*) diagram with the given conditions is shown in Fig. 3.16.

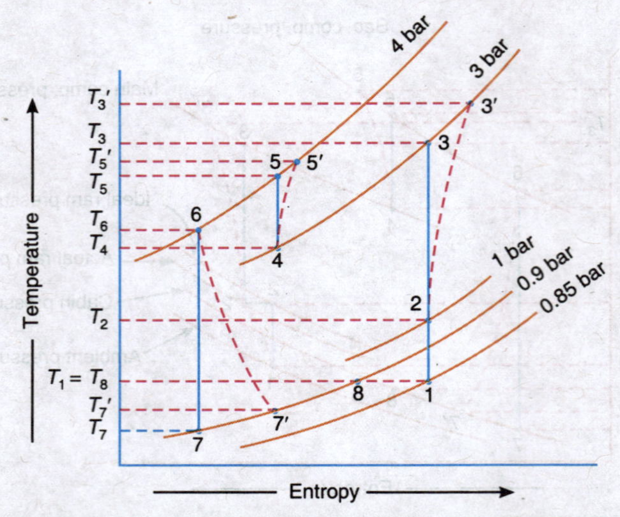

Fig. 3.16

We know that for isentropic ramming process 1-2,

$$\frac{T_2}{T_1} = \left(\frac{p_2}{p_1}\right)^{\frac{\gamma-1}{\gamma}} = \left(\frac{1}{0.85}\right)^{\frac{1.4-1}{1.4}} = (1.176)^{0.286} = 1.047$$

$$\therefore \quad T_2 = T_1 \times 1.047 = 293 \times 1.047 = 306.8 \text{ K} = 33.8°\text{C}$$

Now for isentropic process 2-3,

$$\frac{T_3}{T_2} = \left(\frac{p_3}{p_2}\right)^{\frac{\gamma-1}{\gamma}} = \left(\frac{3}{1}\right)^{\frac{1.4-1}{1.4}} = (3)^{0.286} = 1.37$$

$$\therefore \quad T_3 = T_2 \times 1.37 = 306.8 \times 1.37 = 420.3 \text{ K} = 147.3°\text{C}$$

We know that isentropic efficiency of the compressor,

$$\eta_{C1} = \frac{\text{Isentropic increase in temperature}}{\text{Actual increase in temperature}} = \frac{T_3 - T_2}{T_{3'} - T_2}$$

$$0.8 = \frac{420.3 - 306.8}{T_{3'} - 306.8} = \frac{113.5}{T_{3'} - 306.8}$$

$$\therefore \quad T_{3'} = 306.8 + 113.5/0.8 = 448.7 \text{ K} = 175.7°\text{C}$$

Since 50% of the enthalpy of air discharged from the main compressor is removed in the first heat exchanger (*i.e.* during the process 3′– 4), therefore temperature of air leaving the first heat exchanger,

$$T_4 = 0.5 \times 175.7 = 87.85°\text{C} = 360.85 \text{ K}$$

Now for the isentropic process 4–5,

$$\frac{T_5}{T_4} = \left(\frac{p_5}{p_4}\right)^{\frac{\gamma-1}{\gamma}} = \left(\frac{4}{3}\right)^{\frac{1.4-1}{1.4}} = (1.33)^{0.286} = 1.085$$

$$\therefore \quad T_5 = T_4 \times 1.085 = 360.85 \times 1.085 = 391.5 \text{ K} = 118.5°\text{C}$$

We know that isentropic efficiency of the auxiliary compressor,

$$\eta_{C2} = \frac{T_5 - T_4}{T_{5'} - T_4}$$

$$0.8 = \frac{391.5 - 360.85}{T_{5'} - 360.85} = \frac{30.65}{T_{5'} - 360.85}$$

$$\therefore \quad T_{5'} = 360.85 + 30.65/0.8 = 399.16 \text{ K} = 126.16°\text{C}$$

Since 30% of the enthalpy of air discharged from the auxiliary compressor is removed in the second heat exchanger (*i.e.* during the process 5′– 6), therefore temperature of air leaving the second heat exchanger,

$$T_6 = 0.7 \times 126.16 = 88.3°\text{C} = 361.3 \text{ K}$$

For the isentropic process 6–7,

$$\frac{T_7}{T_6} = \left(\frac{p_7}{p_6}\right)^{\frac{\gamma-1}{\gamma}} = \left(\frac{0.9}{4}\right)^{\frac{1.4-1}{1.4}} = (0.225)^{0.286} = 0.653$$

$$\therefore \quad T_7 = T_6 \times 0.653 = 361.3 \times 0.653 = 236 \text{ K} = -37°\text{C}$$

We know that turbine efficiency,

$$\eta_T = \frac{\text{Actual increase in temperature}}{\text{Isentropic increase in temperature}} = \frac{T_6 - T_{7'}}{T_6 - T_7}$$

$$0.85 = \frac{361.3 - T_{7'}}{361.3 - 236} = \frac{361.3 - T_{7'}}{125.3}$$

$$\therefore \quad T_{7'} = 361.3 - 0.85 \times 125.3 = 254.8 \text{ K} = -18.2°\text{C}$$

1. *Power required to operate the system*

We know that amount of air required for cooling the cabin,

$$m_a = \frac{210\,Q}{c_p(T_8 - T_{7'})} = \frac{210 \times 10}{1\,(293 - 254.8)} = 55 \text{ kg / min}$$

and power required to operate the system,

$$P = \frac{m_a c_p (T_{3'} - T_2)}{60} = \frac{55 \times 1\,(448.7 - 306.8)}{60} = 130 \text{ kW}$$ **Ans.**

2. *C.O.P. of the system*

We know that C.O.P. of the system

$$= \frac{210\,Q}{m_a c_p (T_{3'} - T_2)} = \frac{210 \times 10}{55 \times 1\,(448.7 - 306.8)} = 0.27$$ **Ans.**

3.7 Boot-strap Air Evaporative Cooling System

A boot-strap air cycle evaporative cooling system is shown in Fig. 3.17. It is similar to the boot-strap air cycle cooling system except that the addition of an evaporator between the second heat exchanger and the cooling turbine.

The *T-s* diagram for a boot-strap air evaporative cooling system is shown in Fig 3.18. The various processes of this cycle are same as a simple boot-strap system except the process 5′–6 which represents cooling in the evaporator using any suitable evaporant.

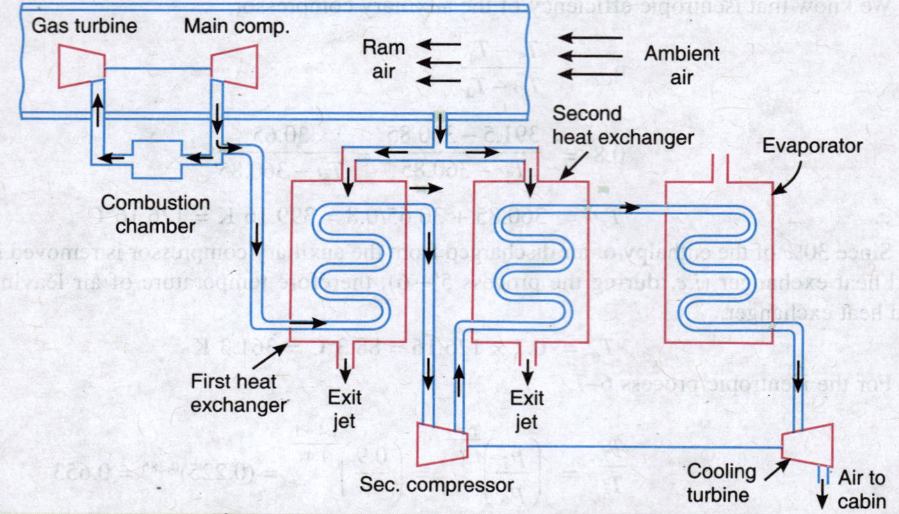

Fig. 3.17. Boot-strap air evaporative cooling system.

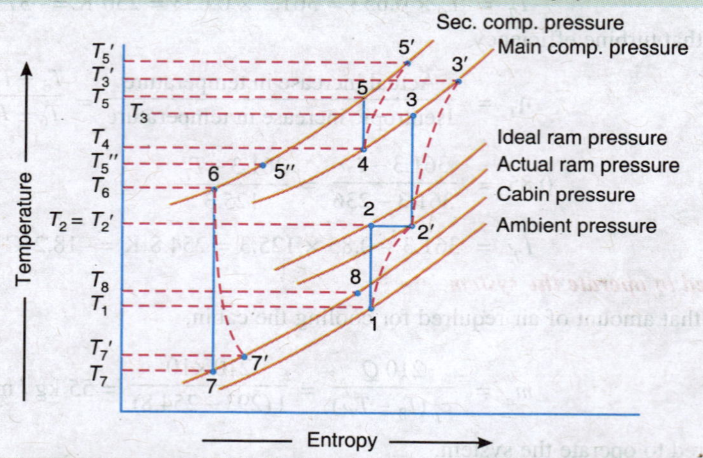

Fig. 3.18. *T-s* diagram for boot-strap air evaporative cooling system.

If Q tonnes of refrigeration is the cooling load in the cabin, then the quantity of air required for the refrigeration purpose will be

$$m_a = \frac{210\,Q}{c_p(T_8 - T_{7'})} \text{ kg / min}$$

Power required for the refrigeration system is given by

$$P = \frac{m_a c_p (T_{3'} - T_{2'})}{60} \text{ kW}$$

and C.O.P. of the refrigerating system

$$= \frac{210\,Q}{m_a c_p (T_{3'} - T_{2'})} = \frac{210\,Q}{P \times 60}$$

Note: Since the temperature of air leaving the cooling turbine in boot-strap evaporative system is lower than the simple boot-strap system, therefore mass of air (m_a) per tonne of refrigeration will be less in boot- strap evaporative system.

Example 3.10. *The following data refer to a boot strap air cycle evaporative refrigeration system used for an aeroplane to take 20 tonnes of refrigeration load :*

Ambient air temperature	= 15°C
Ambient air pressure	= 0.8 bar
Mach number of the flight	= 1.2
Ram efficiency	= 90%
Pressure of air bled off the main compressor	= 4 bar
Pressure of air in the secondary compressor	= 5 bar
Isentropic efficiency of the main compressor	= 90%
Isentropic efficiency of the secondary compressor	= 80%
Isentropic efficiency of the cooling turbine	= 80%
Temperature of air leaving the first heat exchanger	= 170°C
Temperature of air leaving the second heat exchanger	= 155°C
Temperature of air leaving the evaporator	= 100°C
Cabin temperature	= 25°C
Cabin pressure	= 1 bar

Find : 1. Mass of air required to take the cabin load, 2. Power required for the refrigeration system, and 3. C.O.P. of the system.

Solution. Given : Q = 20 TR ; T_1 = 15°C = 15 + 273 = 288 K ; p_1 = 0.8 bar ; M = 1.2 ; η_R = 90% = 0.9 ; $p_3 = p_{3'} = p_4$ = 4 bar ; $p_5 = p_{5'} = p_{5''} = p_6$ = 5 bar ; η_{C1} = 90% = 0.9 ; η_{C2} = 80% = 0.8 ; η_T = 80% = 0.8 ; T_4 = 170°C = 170 + 273 = 443 K; T_5'' = 155°C = 155 + 273 = 428 K ; T_6 = 100°C = 100 + 273 = 373 K ; T_8 = 25°C = 25 + 273 = 298 K ; $p_8 = p_7 = p_7'$ = 1 bar

The *T-s* diagram for the boot-strap air cycle evaporative refrigeration system, with the given conditions, is shown in Fig. 3.19.

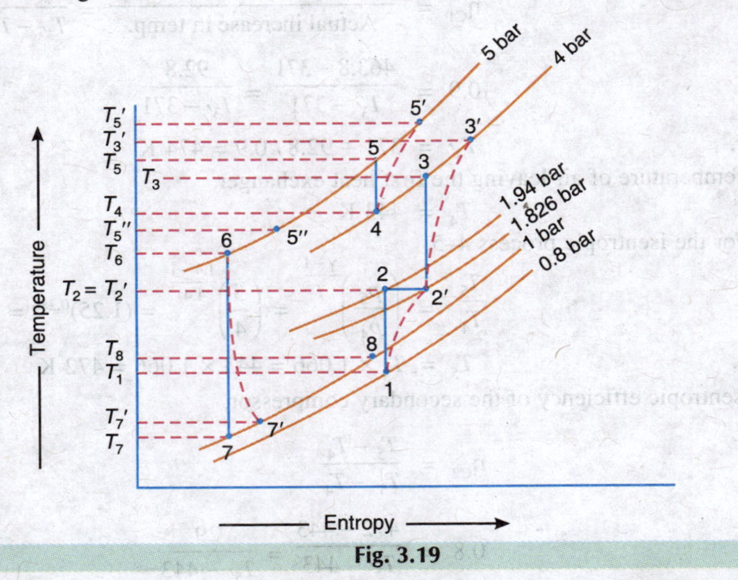

Fig. 3.19

Let $\quad T_{2'}$ = Stagnation temperature of ambient air entering the main compressor,

p_2 = Pressure of air at the end of isentropic ramming, and

$p_{2'}$ = Stagnation pressure of ambient air entering the main compressor.

We know that $$\frac{T_{2'}}{T_1} = 1 + \frac{\gamma - 1}{2} M^2 = 1 + \frac{1.4 - 1}{2}(1.2)^2 = 1.288$$

$\therefore$ $$T_{2'} = T_1 \times 1.288 = 288 \times 1.288 = 371 \text{ K}$$

For isentropic process 1-2,

$$\frac{p_2}{p_1} = \left(\frac{T_2}{T_1}\right)^{\frac{\gamma}{\gamma-1}} = \left(\frac{371}{288}\right)^{\frac{1.4}{1.4-1}} = (1.288)^{3.5} = 2.425$$

$\therefore$ $$p_2 = p_1 \times 2.425 = 0.8 \times 2.425 = 1.94 \text{ bar}$$

We know that ram efficiency,

$$\eta_R = \frac{\text{Actual pressure rise}}{\text{Isentropic pressure rise}} = \frac{p_{2'} - p_1}{p_2 - p_1}$$

$$0.9 = \frac{p_{2'} - 0.8}{1.94 - 0.8} = \frac{p_{2'} - 0.8}{1.14}$$

$\therefore$ $$p_{2'} = 0.9 \times 1.14 + 0.8 = 1.826 \text{ bar}$$

Now for the isentropic process 2′–3,

$$\frac{T_3}{T_{2'}} = \left(\frac{p_3}{p_{2'}}\right)^{\frac{\gamma-1}{\gamma}} = \left(\frac{4}{1.826}\right)^{\frac{1.4-1}{1.4}} = (2.19)^{0.286} = 1.25$$

$\therefore$ $$T_3 = T_{2'} \times 1.25 = 371 \times 1.25 = 463.8 \text{ K}$$

We know that isentropic efficiency of the main compressor,

$$\eta_{C1} = \frac{\text{Isentropic increase in temp.}}{\text{Actual increase in temp.}} = \frac{T_3 - T_{2'}}{T_{3'} - T_{2'}}$$

$$0.9 = \frac{463.8 - 371}{T_{3'} - 371} = \frac{92.8}{T_{3'} - 371}$$

$\therefore$ $$T_{3'} = 371 + 92.8 / 0.9 = 474 \text{ K}$$

Temperature of air leaving the first heat exchanger,

$$T_4 = 443 \text{ K}$$ ... (Given)

For the isentropic process 4–5,

$$\frac{T_5}{T_4} = \left(\frac{p_5}{p_4}\right)^{\frac{\gamma-1}{\gamma}} = \left(\frac{5}{4}\right)^{\frac{1.4-1}{1.4}} = (1.25)^{0.286} = 1.066$$

$\therefore$ $$T_5 = T_4 \times 1.066 = 443 \times 1.066 = 472 \text{ K}$$

Isentropic efficiency of the secondary compressor,

$$\eta_{C2} = \frac{T_5 - T_4}{T_{5'} - T_4}$$

$$0.8 = \frac{472 - 443}{T_{5'} - 443} = \frac{29}{T_{5'} - 443}$$

$\therefore$ $$T_{5'} = 443 + 29/0.8 = 479 \text{ K}$$

Temperature of air leaving the second heat exchanger,

$$T_{5''} = 428 \text{ K}$$ ... (Given)

Temperature of air leaving the evaporator,

$$T_6 = 373 \text{ K} \quad \text{... (Given)}$$

Now for the isentropic process 6–7,

$$\frac{T_6}{T_7} = \left(\frac{p_6}{p_7}\right)^{\frac{\gamma-1}{\gamma}} = \left(\frac{5}{1}\right)^{\frac{1.4-1}{1.4}} = (5)^{0.286} = 1.584$$

$$\therefore \quad T_7 = T_6 / 1.584 = 373/1.584 = 235.5 \text{ K}$$

We know that isentropic efficiency of the cooling turbine,

$$\eta_T = \frac{\text{Actual increase in temp.}}{\text{Isentropic increase in temp.}} = \frac{T_6 - T_{7'}}{T_6 - T_7}$$

$$0.8 = \frac{373 - T_{7'}}{373 - 235.5} = \frac{373 - T_{7'}}{137.5}$$

$$\therefore \quad T_{7'} = 373 - 0.8 \times 137.5 = 263 \text{ K}$$

1. *Mass of air required to take the cabin load*

We know that mass of air required to take the cabin load,

$$m_a = \frac{210\, Q}{c_p (T_8 - T_7)} = \frac{210 \times 20}{1\,(298 - 263)} = 120 \text{ kg / min} \quad \textbf{Ans.}$$

2. *Power required for the refrigeration system*

We know that power required for the refrigeration system,

$$P = \frac{m_a c_p (T_{3'} - T_{2'})}{60} = \frac{120 \times 1\,(474 - 371)}{60} = 206 \text{ kW} \quad \textbf{Ans.}$$

3. *C.O.P. of the system*

We know that C.O.P. of the system $= \dfrac{210\, Q}{P \times 60} = \dfrac{210 \times 20}{206 \times 60} = 0.34$ **Ans.**

3.8 Reduced Ambient Air Cooling System

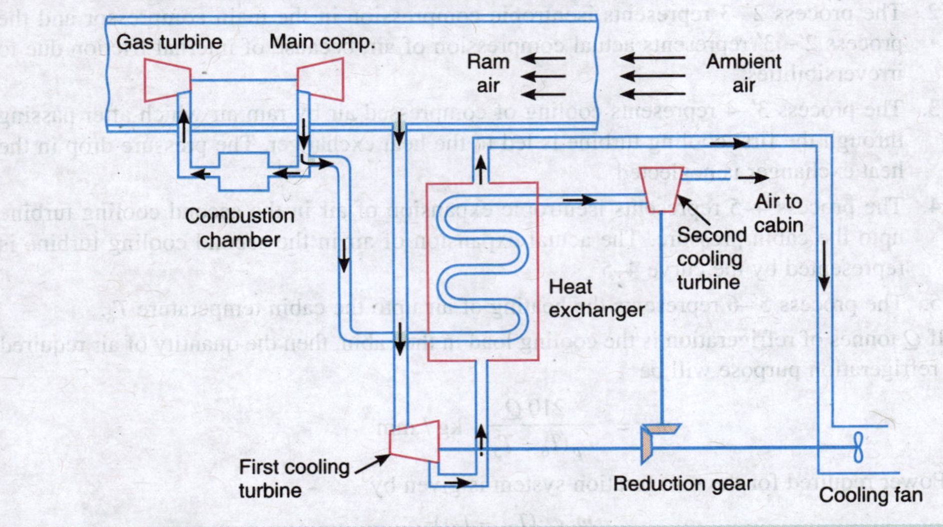

Fig. 3.20. Reduced ambient air cooling system.

The reduced ambient air cooling system is shown in Fig. 3.20. This cooling system includes two cooling turbines and one heat exchanger. The air reduced for the refrigeraion system is bled off from the main compressor. This high pressure and high temperature air is cooled initially in the heat exchanger. The air for cooling is taken from the cooling turbine which lowers the high temperature of rammed air. The cooled air from the heat exchanger is passed through the second cooling turbine from where the air is supplied to the cabin. The work of the cooling turbine is used to drive the cooling fan (through reduction gears) which draws cooling air from the heat exchanger. The reduced ambient air cooling system is used for very high speed (supersonic) aircrafts, when the ram temperature is too high.

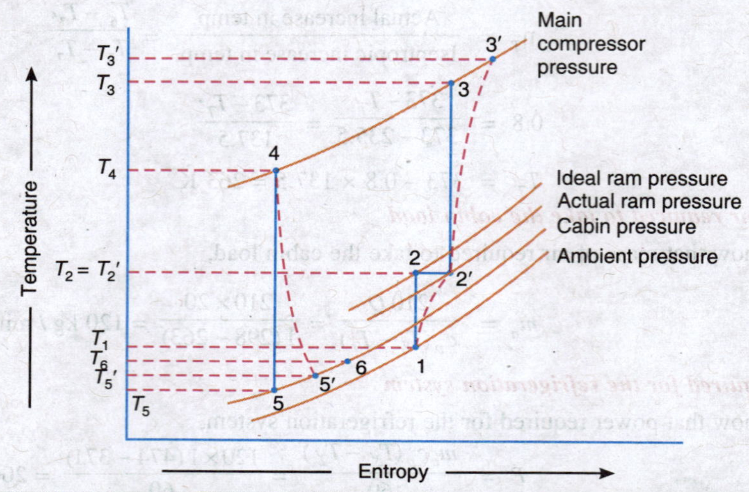

Fig. 3.21. *T-s* diagram for reduced ambient air cycle cooling system.

The *T-s* diagram for the reduced ambient air cycle cooling system is shown in Fig. 3.21. The various processes are as follows :

1. The process 1–2 represents isentropic ramming of air and the process 1–2′ represents actual ramming of air because of internal friction due to irreversibilities.
2. The process 2′–3 represents isentropic compression in the main compressor and the process 2′– 3′ represents actual compression of air, because of internal friction due to irreversibilities.
3. The process 3′–4 represents cooling of compressed air by ram air which after passing through the first cooling turbine is led to the heat exchanger. The pressure drop in the heat exchanger is neglected.
4. The process 4–5 represents isentropic expansion of air in the second cooling turbine upto the cabin pressure. The actual expansion of air in the second cooling turbine is represented by the curve 4–5′.
5. The process 5′–6 represents the heating of air upto the cabin temperature T_6.

If Q tonnes of refrigeration is the cooling load in the cabin, then the quantity of air required for the refrigeration purpose will be

$$m_a = \frac{210\,Q}{c_p(T_6 - T_{5'})} \text{ kg / min}$$

Power required for the refrigeration system is given by

$$P = \frac{m_a c_p (T_{3'} - T_{2'})}{60} \text{ kW}$$

and C.O.P. of the system $= \dfrac{210\,Q}{m_a c_p (T_{3'} - T_{2'})} = \dfrac{210\,Q}{P \times 60}$

Example 3.11. *The reduced ambient air refrigeration system used for an aircraft consists of two cooling turbines, one heat exchanger and one air cooling fan. The speed of aircraft is 1500 km/h. The ambient air conditions are 0.8 bar and 10°C. The ram efficiency may be taken as 90%. The rammed air used for cooling is expanded in the first cooling turbine and leaves it at a pressure of 0.8 bar. The air bled from the main compressor at 6 bar is cooled in the heat exchanger and leaves it at 100°C. The cabin is to be maintained at 20°C and 1 bar. The pressure loss between the second cooling turbine and cabin is 0.1 bar. If the isentropic efficiency for the main compressor and both of the cooling turbines are 85% and 80% respectively, find :*

1. mass flow rate of air supplied to cabin to take a cabin load of 10 tonnes of refrigeration;

2. quantity of air passing through the heat exchanger if the temperature rise of ram air is limited to 80 K;

3. power used to drive the cooling fan; and

4. C.O.P. of the system.

Solution. Given : $V = 1500$ km / h = 417 m / s ; $p_1 = 0.8$ bar ; $T_1 = 10°C = 10 + 273 = 283$ K ; $\eta_R = 90\% = 0.9$; $p_3 = p_4 = 6$ bar ; $T_4 = 100°C = 100 + 273 = 373$ K ; $T_6 = 20°C = 20 + 273 = 293$ K ; $p_6 = 1$ bar ; $\eta_C = 85\% = 0.85$; $\eta_{T1} = \eta_{T2} = 80\% = 0.8$; $Q = 10$ TR

The *T-s* diagram for the reduced ambient air refrigeration system with the given conditions is shown in Fig. 3.22.

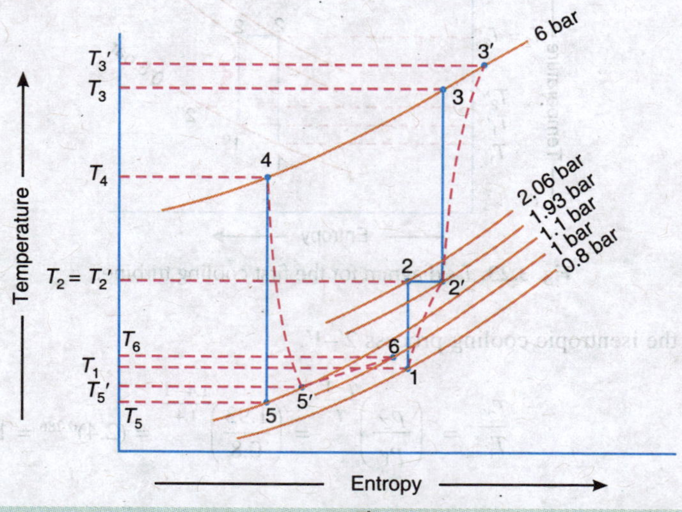

Fig. 3.22

Let $T_{2'}$ = Stagnation temperature of ambient air entering the main compressor,

p_2 = Pressure of air at the end of isentropic ramming, and

$p_{2'}$ = Stagnation pressure of air entering the main compressor.

We know that
$$T_{2'} = T_1 + \frac{V^2}{2000\,c_p} = 283 + \frac{(417)^2}{2000 \times 1} = 370 \text{ K}$$

... (Taking $c_p = 1$ kJ/kg K)

For the isentropic ramming process 1-2,

$$\frac{p_2}{p_1} = \left(\frac{T_2}{T_1}\right)^{\frac{\gamma}{\gamma-1}} = \left(\frac{370}{283}\right)^{\frac{1.4}{1.4-1}} = (1.31)^{3.5} = 2.57$$

$$\therefore \quad p_2 = p_1 \times 2.57 = 0.8 \times 2.57 = 2.06 \text{ bar}$$

We know that ram efficiency,

$$\eta_R = \frac{\text{Actual rise in pressure}}{\text{Isentropic rise in pressure}} = \frac{p_{2'} - p_1}{p_2 - p_1}$$

$$0.9 = \frac{p_{2'} - 0.8}{2.06 - 0.8} = \frac{p_{2'} - 0.8}{1.26}$$

$$\therefore \quad p_{2'} = 0.9 \times 1.26 + 0.8 = 1.93 \text{ bar}$$

The *T-s* diagram for the expansion of ram air in the first cooling turbine is shown in Fig. 3.23. The vertical line 2′–1′ represents the isentropic cooling process and the curve 2′–2″ represents the actual cooling process.

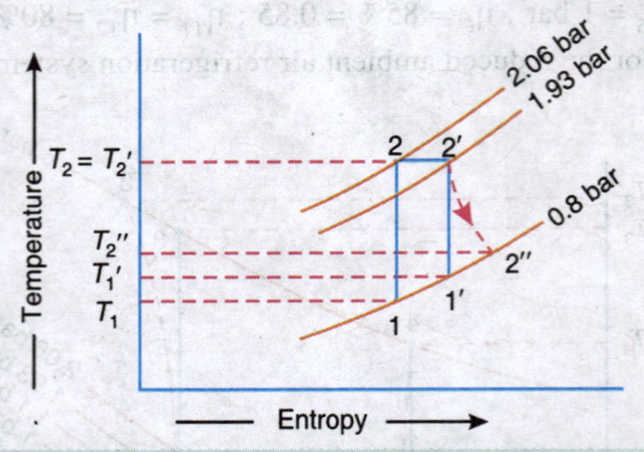

Fig. 3.23. *T-s* diagram for the first cooling turbine.

Now for the isentropic cooling process 2′–1′,

$$\frac{T_{2'}}{T_{1'}} = \left(\frac{p_{2'}}{p_{1'}}\right)^{\frac{\gamma-1}{\gamma}} = \left(\frac{1.93}{0.8}\right)^{\frac{1.4-1}{1.4}} = (2.4)^{0.286} = 1.284$$

... ($\because p_{1'} = p_1$)

$$\therefore \quad T_{1'} = T_{2'}/1.284 = 370/1.284 = 288 \text{ K}$$

Isentropic efficiency of the first cooling turbine,

$$\eta_{T1} = \frac{\text{Actual increase in temp.}}{\text{Isentropic increase in temp.}} = \frac{T_{2'} - T_{2''}}{T_{2'} - T_{1'}}$$

$$0.8 = \frac{370 - T_{2''}}{370 - 288} = \frac{370 - T_{2''}}{82}$$

$$\therefore \quad T_{2''} = 370 - 0.8 \times 82 = 304 \text{ K}$$

For the isentropic compression process 2′–3,

$$\frac{T_3}{T_{2'}} = \left(\frac{p_3}{p_{2'}}\right)^{\frac{\gamma-1}{\gamma}} = \left(\frac{6}{1.93}\right)^{\frac{1.4-1}{1.4}} = (3.11)^{0.286} = 1.38$$

$$\therefore \quad T_3 = T_{2'} \times 1.38 = 370 \times 1.38 = 511 \text{ K}$$

We know that isentropic efficiency of the compressor,

$$\eta_C = \frac{\text{Isentropic increase in temp.}}{\text{Actual increase in temp.}} = \frac{T_3 - T_{2'}}{T_{3'} - T_{2'}}$$

$$0.85 = \frac{511 - 370}{T_{3'} - 370} = \frac{141}{T_{3'} - 370}$$

$$\therefore \quad T_{3'} = 370 + 141/0.85 = 536 \text{ K}$$

Since there is a pressure drop of 0.1 bar between the second cooling turbine and the cabin, therefore pressure of air at exit from the second cooling turbine,

$$p_5 = p_{5'} = p_6 + 0.1 = 1 + 0.1 = 1.1 \text{ bar}$$

Now for the isentropic expansion of air in the second cooling turbine (process 4–5),

$$\frac{T_4}{T_5} = \left(\frac{p_4}{p_5}\right)^{\frac{\gamma-1}{\gamma}} = \left(\frac{6}{1.1}\right)^{\frac{1.4-1}{1.4}} = (5.45)^{0.286} = 1.62$$

$$\therefore \quad T_5 = T_4 / 1.62 = 373 / 1.62 = 230 \text{ K}$$

We know that the isentropic efficiency of the second cooling turbine,

$$\eta_{T2} = \frac{\text{Actual increase in temp.}}{\text{Isentropic increase in temp.}} = \frac{T_4 - T_{5'}}{T_4 - T_5}$$

$$0.8 = \frac{373 - T_{5'}}{373 - 230} = \frac{373 - T_{5'}}{143}$$

$$\therefore \quad T_{5'} = 373 - 0.8 \times 143 = 258.6 \text{ K}$$

1. *Mass flow rate of air supplied to cabin*

We know that mass flow rate of air supplied to cabin,

$$m_a = \frac{210\,Q}{c_p (T_6 - T_{5'})}$$

$$= \frac{210 \times 10}{1\,(293 - 258.6)} = 61 \text{ kg / min} \quad \textbf{Ans.}$$

2. *Quantity of ram air passing through the heat exchanger*

Let $\quad m_R$ = Quantity of ram air passing through the heat exchanger.

The compressed air bled off at temperature $T_{3'}$ = 536 K is cooled in the heat exchanger to a temperature T_4 = 373 K by the ram air from the first cooling turbine at a temperature T_2'' = 304 K. The temperature rise of ram air in the heat exchanger is limited to 80 K. Considering perfect heat transfer in the heat exchanger,

$$m_R \times c_p \times 80 = m_a \times c_p (T_{3'} - T_4)$$

$$m_R \times 1 \times 80 = 61 \times 1\,(536 - 373) = 9943$$

$$\therefore \quad m_R = 9943 / 80 = 124.3 \text{ kg / min} \quad \textbf{Ans.}$$

3. *Power used to drive the cooling fan*

Since the work output of both the cooling turbines is used to drive the cooling fan, therefore work output from the first cooling turbine,

$$W_{T1} = m_R \times c_p (T_{2'} - T_{2''})$$
$$= 124.3 \times 1 (370 - 304) = 8204 \text{ kJ/min}$$

and work output from the second cooling turbine,

$$W_{T2} = m_a \times c_p (T_4 - T_{5'})$$
$$= 61 \times 1 (373 - 258.6) = 6978 \text{ kJ/min}$$

∴ Combined work output from both the cooling turbines,

$$W_T = W_{T1} + W_{T2} = 8204 + 6978 = 15\,182 \text{ kJ/min}$$

and power used to drive the cooling fan

$$= 15\,182 / 60 = 253 \text{ kW}$$ **Ans.**

4. *C.O.P. of the system*

We know that C.O.P. of the system

$$= \frac{210\,Q}{m_a c_p (T_{3'} - T_{2'})} = \frac{210 \times 10}{61 \times 1\,(536 - 370)} = 0.21$$ **Ans.**

Example 3.12. *The reduced ambient system of air refrigeration for cooling an aircraft cabin consists of two cooling turbines, one heat exchanger and one fan. The first cooling turbine is supplied with the ram air at 1.1 bar and 15° C and delivers after expansion to the heat exchanger at 0.9 bar for cooling the air bled off from the main compressor at 3.5 bar. The cooling air from the heat exchanger is sucked by a fan and discharged to the atmosphere. The cooled air from the heat exchanger is expanded upto 1 bar in the second cooling turbine and discharged into air cabin to be cooled. The air from the cabin is exhausted at 22°C. The refrigerating capacity required is 10 tonnes. If the compression index for the main compressor is 1.5 and the expansion index for both the cooling turbines is 1.35, determine :*

1. Mass flow rate of the cabin air ;

2. Cooling capacity of the heat exchanger and flow rate of the ram air when compressed air is to be cooled to 60°C in the heat exchanger and temperature rise in the heat exchanger for the ram air is not to exceed 30 K ;

3. Combined output of both cooling turbines driving the air fan with transmission efficiency of 60% ; and

4. C.O.P. of the refrigerating system considering only power input to the compressor.

Solution. Given : $p_2 = 1.1$ bar ; $T_2 = 15°\text{C} = 15 + 273 = 288$ K ; $p_{2'} = 0.9$ bar ; $p_3 = p_4 = 3.5$ bar ; $p_5 = p_6 = 1$ bar ; $T_6 = 22°\text{C} = 22 + 273 = 295$ K ; $Q = 10$ TR ; $\gamma_1 = 1.5$; $\gamma_2 = 1.35$; $T_4 = 60°\text{C} = 60 + 273 = 333$ K ; $\eta_T = 60\% = 0.6$

The *T-s* diagram for the reduced ambient system of air refrigeration for cooling an aircraft cabin with the given conditions is shown in Fig. 3.24. The process of cooling the ram air in the first cooling turbine is shown by the curve 2–2′.

Let $T_{2'}$ = Temperature of ram air after expansion in the first cooling turbine, and

T_3 = Temperature of air bled off from the main compressor.

We know that for the expansion in the first cooling turbine (process 2–2′),

$$\frac{T_2}{T_{2'}} = \left(\frac{p_2}{p_{2'}}\right)^{\frac{\gamma_2 - 1}{\gamma_2}} = \left(\frac{1.1}{0.9}\right)^{\frac{1.35-1}{1.35}} = (1.22)^{0.26} = 1.053$$

$$\therefore \quad T_{2'} = T_2 / 1.053 = 288 / 1.053 = 273.5 \text{ K}$$

For the process 2–3,

$$\frac{T_3}{T_2} = \left(\frac{p_3}{p_2}\right)^{\frac{\gamma_1 - 1}{\gamma_1}} = \left(\frac{3.5}{1.1}\right)^{\frac{1.5-1}{1.5}} = (3.18)^{0.333} = 1.47$$

$$\therefore \quad T_3 = T_2 \times 1.47 = 288 \times 1.47 = 423.4 \text{ K}$$

and for the process 4–5,

$$\frac{T_4}{T_5} = \left(\frac{p_4}{p_5}\right)^{\frac{\gamma_2 - 1}{\gamma_2}} = \left(\frac{3.5}{1}\right)^{\frac{1.35-1}{1.35}} = (3.5)^{0.26} = 1.385$$

$$\therefore \quad T_5 = T_4 / 1.385 = 333 / 1.385 = 240.4 \text{ K}$$

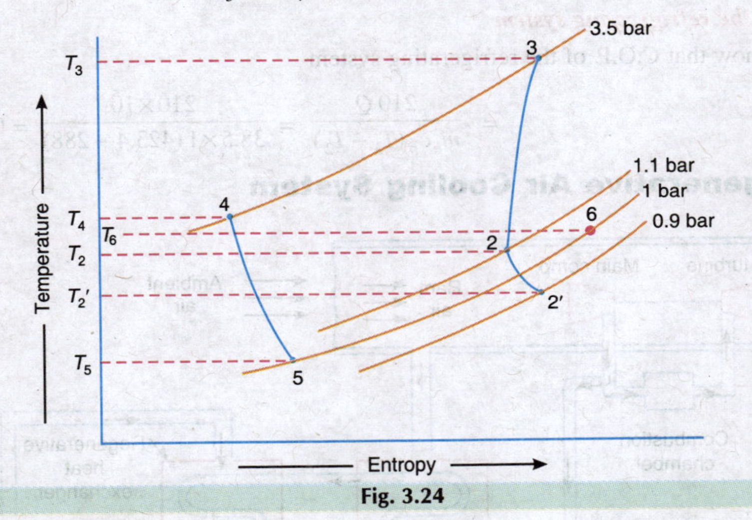

Fig. 3.24

1. *Mass flow rate of cabin air*

We know that mass flow rate of cabin air,

$$m_a = \frac{210\,Q}{c_p\,(T_6 - T_5)} = \frac{210 \times 10}{1\,(295 - 240.4)} = 38.5 \text{ kg / min}$$

... (c_p for air = 1kJ/kg K)

2. *Cooling capacity of the heat exchanger and flow rate of ram air*

We know that cooling capacity of the heat exchanger

$$= m_a\,c_p\,(T_3 - T_4) = 38.5 \times 1\,(423.4 - 333) \text{ kJ / min}$$

$$= 3480 \text{ kJ/min} \quad \textbf{Ans.}$$

In order to find the flow rate of ram air (m_R), equate the enthalpy lost by compressed air to the enthalpy gained by ram air.

We know that enthalpy lost by compressed air

$$= m_a\,c_p\,(T_3 - T_4) = 38.5 \times 1(423.4 - 333) \text{ kJ / min}$$

$$= 3480 \text{ kJ/min} \qquad \text{... (i)}$$

Since it is given that the temperature rise for the ram air is not to exceed 30 K, therefore enthalpy gained by ram air

$$= m_R \times 1 \times 30 = 30\,m_R \text{ kJ/min} \qquad \text{... (ii)}$$

Equating equations (*i*) and (*ii*),

$$m_R = 116 \text{ kg / min} \quad \textbf{Ans.}$$

3. Combined output of both cooling turbines

We know that output of first cooling turbine

$$= m_a c_p (T_2 - T_2') = 116 \times 1\ (288 - 273.5) = 1682 \text{ kJ/min}$$
$$= 1682 / 60 = 28 \text{ kW}$$

and output of second cooling turbine

$$= m_a c_p (T_4 - T_5) = 38.5 \times 1\ (333 - 240.4) = 3565 \text{ kJ/min}$$
$$= 3565 / 60 = 59.4 \text{ kW}$$

Since the transmission efficiency is 60%, therefore combined output of both the cooling turbines

$$= (28 + 59.4)\ 0.6 = 52.44 \text{ kW}$$ **Ans.**

4. C.O.P. of the refrigerating system

We know that C.O.P. of the refrigerating system

$$= \frac{210\ Q}{m_a c_p (T_3 - T_2)} = \frac{210 \times 10}{38.5 \times 1\ (423.4 - 288)} = 0.4$$ **Ans.**

3.9 Regenerative Air Cooling System

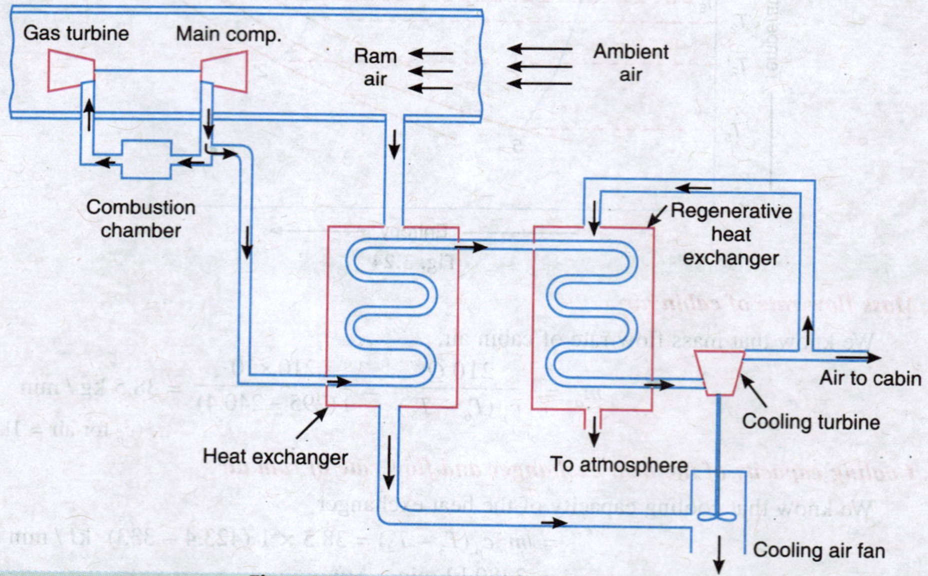

Fig. 3.25. Regenerative air cooling system.

The regenerative air cooling system is shown in Fig. 3.25. It is a modification of a simple air cooling system with the addition of a regenerative heat exchanger. The high pressure and high temperature air from the main compressor is first cooled by the ram air in the heat exchanger. This air is further cooled in the regenerative heat exchanger with a portion of the air bled after expansion in the cooling turbine. This type of cooling system is used for supersonic aircrafts and rockets.

The *T-s* diagram for the regenerative air cooling system is shown in Fig. 3.26. The various processes are as follows :

1. The process 1–2 represents isentropic ramming of air and process 1– 2′ represents actual ramming of air because of internal friction due to irreversibilities.

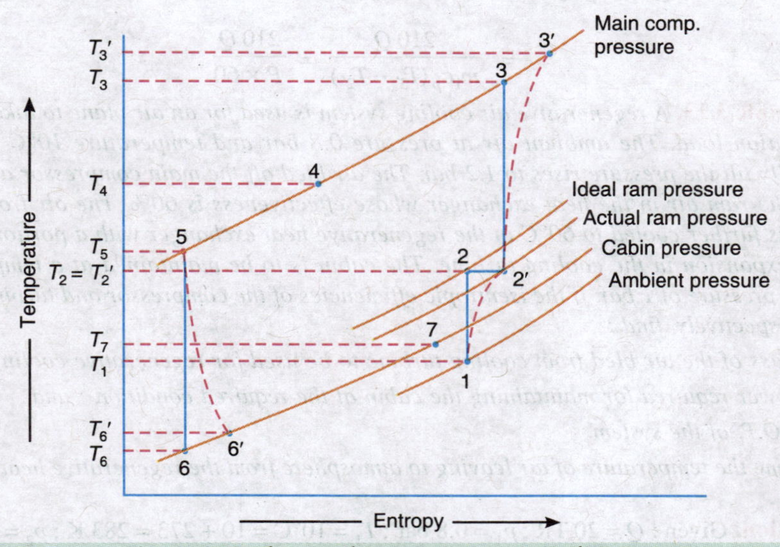

Fig. 3.26. *T-s* diagram for regenerative air cooling system.

2. The process 2′–3 represents isentropic compression of air in the main compressor and the process 2′–3′ represents actual compression of air because of internal friction due to irreversibilities.
3. The process 3′–4 represents cooling of compressed air by ram air in the heat exchanger.
4. The process 4–5 represents cooling of air in the regenerative heat exchanger.
5. The process 5–6 represents isentropic expansion of air in the cooling turbine upto the cabin pressure and the process 5–6′ represents actual expansion of air in the cooling turbine.
6. The process 6′–7 represents heating of air upto the cabin temperature T_7.

If Q tonnes of refrigeration is the cooling load in the cabin, then the quantity of air required for the refrigeration purpose will be

$$m_a = \frac{210\,Q}{c_p(T_7 - T_{6'})} \text{ kg / min}$$

Let m_1 = Total mass of air bled from the main compressor, and

m_2 = Mass of cold air bled from the cooling turbine for regenerative heat exchanger.

For the energy balance of regenerative heat exchanger, we have

$$m_2\, c_p\,(T_8 - T_{6'}) = m_1 c_p\,(T_4 - T_5)$$

$$\therefore \quad m_2 = \frac{m_1(T_4 - T_5)}{(T_8 - T_{6'})}$$

where T_8 = Temperature of air leaving to atmosphere from the regenerative heat exchanger.

Power required for the refrigeration system,

$$P = \frac{m_1 c_p (T_{3'} - T_{2'})}{60} \text{ kW}$$

and C.O.P. of the refrigerating system

$$= \frac{210\,Q}{m_1 c_p (T_{3'} - T_{2'})} = \frac{210\,Q}{P \times 60}$$

Example 3.13. *A regenerative air cooling system is used for an air plane to take 20 tonnes of refrigeration load. The ambient air at pressure 0.8 bar and temperature 10°C is rammed isentropically till the pressure rises to 1.2 bar. The air bled off the main compressor at 4.5 bar is cooled by the ram air in the heat exchanger whose effectiveness is 60%. The air from the heat exchanger is further cooled to 60°C in the regenerative heat exchanger with a portion of the air bled after expansion in the cooling turbine. The cabin is to be maintained at a temperature of 25°C and a pressure of 1 bar. If the isentropic efficiencies of the compressor and turbine are 90% and 80% respectively, find :*

1. Mass of the air bled from cooling turbine to be used for regenerative cooling ;

2. Power required for maintaining the cabin at the required condition ; and

3. C.O.P. of the system.

Assume the temperature of air leaving to atmosphere from the regenerative heat exchanger as 100°C.

Solution. Given : Q = 20 TR ; p_1 = 0.8 bar ; T_1 = 10°C = 10 + 273 = 283 K ; p_2 = 1.2 bar ; $p_3 = p_4 = p_5$ = 4.5 bar ; η_H = 60% = 0.6 ; T_5 = 60°C = 60 + 273 = 333 K ; T_7 = 25°C = 25 + 273 = 298 K; $p_7 = p_6 = p_{6'}$ = 1 bar ; η_C = 90% = 0.9 ; η_T = 80% = 0.8 ; T_8 = 100°C = 100 + 273 = 373 K

The *T-s* diagram for the regenerative air cooling system with the given conditions is shown in Fig. 3.27.

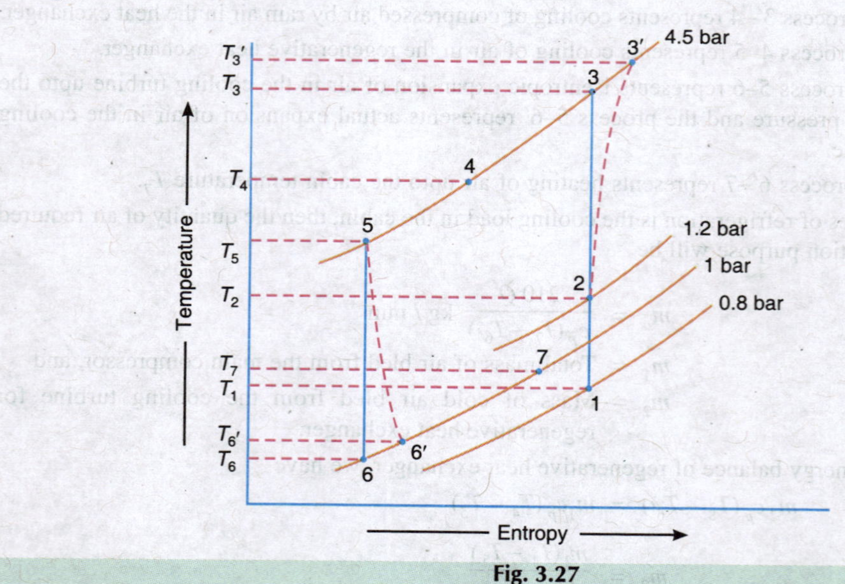

Fig. 3.27

Let T_2 = Temperature of air at the end of ramming and entering to the main compressor,

T_3 = Temperature of air after isentropic compression in the main compressor, and

$T_{3'}$ = Actual temperature of air leaving the main compressor.

We know that for the isentropic ramming of air (process 1–2),

$$\frac{T_2}{T_1} = \left(\frac{p_2}{p_1}\right)^{\frac{\gamma-1}{\gamma}} = \left(\frac{1.2}{0.8}\right)^{\frac{1.4-1}{1.4}} = (1.5)^{0.286} = 1.123$$

$$\therefore \quad T_2 = T_1 \times 1.123 = 283 \times 1.123 = 317.8 \text{ K}$$

and for the isentropic compression process 2–3,

$$\frac{T_3}{T_2} = \left(\frac{p_3}{p_2}\right)^{\frac{\gamma-1}{\gamma}} = \left(\frac{4.5}{1.2}\right)^{\frac{1.4-1}{1.4}} = (3.75)^{0.286} = 1.46$$

$$\therefore \quad T_3 = T_2 \times 1.46 = 317.8 \times 1.46 = 464 \text{ K}$$

Isentropic efficiency of the compressor,

$$\eta_C = \frac{\text{Isentropic increase in temp.}}{\text{Actual increase in temp.}} = \frac{T_3 - T_2}{T_{3'} - T_2}$$

$$0.9 = \frac{464 - 317.8}{T_{3'} - 317.8} = \frac{146.2}{T_{3'} - 317.8}$$

$$\therefore \quad T_{3'} = 317.8 + 146.2 / 0.9 = 480 \text{ K}$$

We know that effectiveness of the heat exchanger (η_H),

$$0.6 = \frac{T_{3'} - T_4}{T_{3'} - T_2} = \frac{480 - T_4}{480 - 317.8} = \frac{480 - T_4}{162.2}$$

$$\therefore \quad T_4 = 480 - 0.6 \times 162.2 = 382.7 \text{ K}$$

Now for the isentropic cooling in the cooling turbine (process 5–6),

$$\frac{T_5}{T_6} = \left(\frac{p_5}{p_6}\right)^{\frac{\gamma-1}{\gamma}} = \left(\frac{4.5}{1}\right)^{\frac{1.4-1}{1.4}} = (4.5)^{0.286} = 1.54$$

$$\therefore \quad T_6 = T_5 / 1.54 = 333 / 1.54 = 216 \text{ K}$$

and isentropic efficiency of the cooling turbine,

$$\eta_T = \frac{\text{Actual increase in temp.}}{\text{Isentropic increase in temp.}} = \frac{T_5 - T_{6'}}{T_5 - T_6}$$

$$0.8 = \frac{333 - T_{6'}}{333 - 216} = \frac{333 - T_{6'}}{117}$$

$$\therefore \quad T_{6'} = 333 - 0.8 \times 117 = 239.4 \text{ K}$$

1. *Mass of air bled from the cooling turbine to be used for regenerative cooling*

Let m_a = Mass of air bled from the cooling turbine to be used for regenerative cooling,

m_1 = Total mass of air bled from the main compressor, and

m_2 = Mass of cold air bled from the cooling turbine for regenerative heat exchanger.

We know that the mass of air supplied to the cabin,

$$m_a = m_1 - m_2$$

$$= \frac{210\,Q}{c_p (T_7 - T_{6'})} = \frac{210 \times 20}{1(298 - 239.4)} = 71.7 \text{ kg / min} \quad \ldots (i)$$

and
$$m_2 = \frac{m_1(T_4 - T_5)}{(T_8 - T_{6'})} = \frac{m_1(382.7 - 333)}{(373 - 239.4)} = 0.372\, m_1 \quad \text{... (ii)}$$

From equation (*i*), we find that

$$m_1 - m_2 = 71.7 \quad \text{or} \quad m_1 - 0.372\, m_1 = 71.7$$

$$\therefore \quad m_1 = \frac{71.7}{1 - 0.372} = 113.4 \text{ kg / min}$$

and
$$m_2 = 0.372\, m_1 = 0.372 \times 113.4 = 42.2 \text{ kg / min} \quad \textbf{Ans.}$$

Note: From equation (*ii*), $m_2 / m_1 = 0.372$. Therefore we can say that the air bled from the cooling turbine for regenerative cooling is 37.2% of the total air bled from the main compressor.

2. Power required for maintaining the cabin at the required condition

We know that the power required for maintaining the cabin at the required condition,

$$P = \frac{m_1 c_p (T_{3'} - T_2)}{60} = \frac{113.4 \times 1\,(480 - 317.8)}{60} = 307 \text{ kW} \quad \textbf{Ans.}$$

3. C.O.P. of the system

We know that C.O.P. of the system

$$= \frac{210\, Q}{m_1 c_p (T_{3'} - T_2)} = \frac{210 \times 20}{113.4 \times 1\,(480 - 317.8)} = 0.23 \quad \textbf{Ans.}$$

3.10 Comparison of Various Air Cooling Systems used for Aircraft

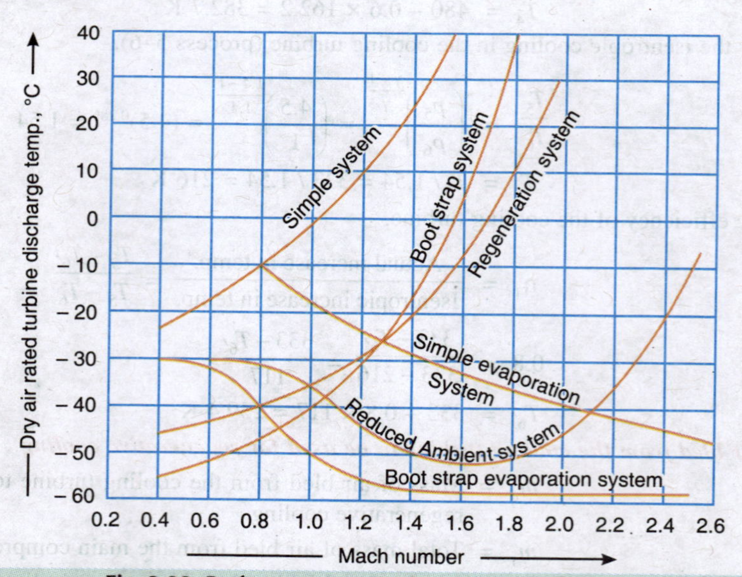

Fig. 3.28. Performance curves for various air cooling systems.

The performance curves for the various air cooling systems used for aircraft are shown in Fig. 3.28. These curves show the dry air rated turbine discharge temperature (DART) against the Mach number. From Fig. 3.28, we see that the simple air cooling system gives maximum cooling effect on the ground surface and decreases as the speed of aircraft increases. The boot strap system, on the other hand, requires the air plane to be in flight so that the ram air can be used for cooling in the heat exchangers. One method of overcoming this drawback of boot strap system is to use part of work derived from turbine to drive a fan which pulls air over the secondary heat exchanger, thus combining the features of a simple and boot strap system. As the speed of aircraft increases, the temperature of ram cooling air rises and the ram air becomes less effective as a coolant in the heat exchanger. In such cases, a suitable evaporant is used with the ram air so that the cabin temperature does not rise. For high speed aircrafts, the boot strap evaporative or regenerative systems are used because they give lower turbine discharge temperature than the simple cooling system. In some cases, aeroplanes carry an auxiliary gas turbine for cabin pressurisation and air conditioning.

From Fig. 3.28, we see that the turbine discharge temperature of the air is variable. Therefore, in order to maintain the constant temperature of supply air to the cabin, it requires some control system.

EXERCISES

1. In an aircraft cooling system, air enters the compressor at 1 bar and 4°C. It is compressed to 3 bar with an isentropic efficiency of 72%. After being cooled to 55°C at constant pressure in a heat exchanger, the air expands in a turbine to 1 bar with an isentropic efficiency of 78%. The low temperature air absorbs a cooling load of 3 tonnes of refrigeration at constant pressure before reentering the compressor which is driven by the turbine. Assuming air to be an ideal gas, determine the C.O.P. of the refrigerator, the driving power required and the air mass flow rate

[**Ans.** 0.247; 42.5 kW ; 0.58 kg/s]

2. A simple air refrigeration system is used for an aircraft to take a load of 20 TR. The ambient pressure and temperature are 0.9 bar and 22°C respectively. The pressure of air is increased to 1 bar due to isentropic ramming action. The air is further compressed in a compressor to 3.5 bar and then cooled in a heat exchanger to 72°C. Finally, the air is passed through the cooling turbine and then it is supplied to the cabin at a pressure of 1.03 bar. The air leaves the cabin at a temperature of 25°C. Assuming the isentropic efficiencies of the compressor and turbine as 80 per cent and 75 percent respectively, find :

1. Power required to take the load in the cooling cabin; and 2. C.O.P. of the system.

Take c_p = 1.005 kJ / kg K ; and γ = 1.4. [**Ans.** 390 kW ; 0.18]

3. The ambient conditions for an air craft cruising at 1000 km/h are 0.35 bar and – 15°C. The cabin temperature is 25°C and turbine exit pressure is 1.06 bar. The pressure ratio of the compressor is 3. Assuming 100% efficiency for ram effect, compressor and turbine and ideal heat exchangers, determine for simple gas refrigeration cycle of 20TR capacity:

1. Temperatures and pressures at all points of the cycle;
2. Mass flow rate and volume flow rates at compressor inlet and turbine outlet; and
3. Work requirement and C.O.P.

Assume $c_p = 1.005 \times 10^3$ J/kg K; R_{air} = 286 J/kg K; and γ = 1.4.

[**Ans.** 296.4 K, 0.569 bar; 405.8 K, 1.707 bar; 296.4 K, 1.707 bar; 258.6 K; 106 kg/min, 158 m³/min, 74 m³/min; 11654.4 kJ/min, 0.36]

4. The cockpit of a jet plane is to be cooled by a simple air refrigeration system. The data available is as follows :

Cock-pit cooling load	= 20 TR
Speed of the plane	= 1000 km / h
Ambient air pressure	= 0.35 bar
Ambient air temperature	= –15°C
Ram efficiency	= 90%
Pressure ratio in the main compressor	= 3
Pressure drop in the heat exchanger	= 0.1 bar
Isentropic efficiencies of main compressor and cooling turbine	= 80%

Temperature of air entering the cooling turbine	= 30°C
Pressure of air leaving the cooling turbine	= 1.06 bar
Pressure in the cock-pit	= 1 bar

If the cock-pit is to be maintained at 25°C, find :

1. Stagnation temperature and pressure of air entering the main compressor ;
2. Mass flow rate of air to cock-pit ;
3. Power required to drive the refrigerating system ; and
4. C.O.P. of system. [**Ans.** 210 kg / min ; 477 kW ; 0.147]

5. A boot-strap air refrigeration system of 20 TR capacity is used for an aeroplane flying at an altitude of 2000 m. The ambient air pressure and temperature are 0.8 bar and 0°C. The ram air pressure and temperature are 1.05 bar and 17°C. The pressure of air after isentropic compression in the main compressor is 4 bar. This air is now cooled to 27°C in another auxiliary heat exchanger and then expanded isentropically upto the cabin pressure of 1.01 bar. If the air leaves the cabin at 25°C and the efficiencies for the main compressor, auxiliary compressor and the cooling turbine are 80%; 75% and 80% respectively; find :1. Power required to operate the system ; and 2. C.O.P. of the system. [**Ans.** 106.2 kW ; 0.66]

6. A boot strap air refrigeration system is used for an aeroplane to take 10 tonnes of refrigeration load. The ambient air conditions are 15°C and 0.9 bar. This air is rammed isentropically to a pressure of 1.1 bar. The pressure of the air bled off the main compressor is 3.5 bar and this is further compressed in secondary compressor to a pressure of 4.5 bar. The isentropic efficiency of both the compressors is 90% and that of cooling turbine is 85%. The effectiveness of both the heat exchangers is 0.6. If the cabin is to be maintained at 25°C and the pressure in the cabin is 1 bar, find : 1. mass of air passing through the cabin ; 2. power used for the refrigeration system ; and 3. C.O.P. of the system. [**Ans.** 55.3 kg / min ; 125 kW ; 0.28]

7. The following data refer to a reduced ambient refrigeration system :

Ambient pressure	= 0.8 bar
Pressure of ram air	= 1.1 bar
Temperature of ram air	= 20°C
Pressure at the end of main compressor	= 3.3 bar
Efficiency of main compressor	= 80%
Heat exchanger effectiveness	= 80%
Pressure at the exit of the auxiliary turbine	= 0.8 bar
Efficiency of auxiliary turbine	= 85%
Temperature of air leaving the cabin	= 25°C
Pressure in the cabin	= 1.013 bar
Flow rate of air through cabin	= 60 kg / min

Find : 1.Capacity of the cooling system required ; 2. Power needed to operate the system ; and 3. C.O.P. of the system. [**Ans.** 19.5 TR ; 136 kW ; 0.504]

8. The following data refers to a reduced ambient air refrigeration system used for an aircraft :

Speed of aircraft	= 1500 km / h
Ambient pressure	= 0.8 bar
Ambient temperature	= 5°C
Ram efficiency	= 100 %
Pressure of cooled air leaving the first cooling turbine	= 0.8 bar
Temperature of cooled air leaving the heat exchanger	= 100°C
Pressure ratio of the main compressor	= 3
Pressure loss between the outlet of second cooling turbine and the cabin	= 0.1 bar
Pressure in the cabin	= 1 bar

Temperature in the cabin = 22°C
Load in the cabin = 10 TR
Isentropic efficiency of compressor = 85%
Isentropic efficiency of both cooling turbines = 80%

Find: 1. mass flow of the air passing through the second cooling turbine ; 2. quantity of ram air passing through the heat exchanger, if the rise in temperature is limited to 80 K ; and 3. C.O.P. of the system. **[Ans.** 54 kg / min ; 79.5 kg / min ; 0.26]

9. A regenerative air refrigeration system for an aeroplane is designed to take a load of 30 TR. The temperature and pressure conditions of the atmosphere are 5°C and 0.85 bar. The pressure of the air is increased from 0.85 bar to 1.2 bar due to ramming action. The pressure of air leaving the main compressor is 4.8 bar. 60% of the total heat of the air leaving the main compressor is removed in the heat exchanger and then it is passed through the cooling turbine. The temperature of the rammed air which is used for cooling purposes in the heat exchanger is reduced to 50°C by mixing the air coming out from the cooling turbine, and in the process, the cooling air from the turbine gets heated to 100°C before it is discharged. The isentropic efficiencies of the compressor and turbine are 90% and 80% respectively. The pressure and temperature required in the cabin are 1 bar and 25°C respectively. Assuming isentropic ramming and mass of cooled air passing through the heat exchanger equal to the mass of cooling air, find : 1. the ratio of by-passed air to ram air used for cooling purposes ; and 2. the power required for maintaining the cabin at required condition. **[Ans.** 41.2% ; 445 kW]

QUESTIONS

1. Explain the working of a simple air cycle cooling system used for aircrafts.
2. List the names of three evaporative coolants that can be used in aircraft refrigeration system at an altitude of 15 000 metres. How will you estimate the amount of the coolant required for a given flight of aircraft ?
3. Describe boot-strap cycle of air refrigeration system, with a schematic diagram and show the cycle on *T-s* diagram.
4. Explain, with a neat sketch, the working principle of boot-strap evaporative type of air refrigeration system. Draw *T-s* diagram for the system.
5. Describe with a diagram the reduced ambient air cooling system.
6. Describe with a sketch a regenerative air cooling system.
7. Compare the various air cooling systems used for aircraft.

OBJECTIVE TYPE QUESTIONS

1. An ordinary passenger aircraft requires a cooling system of capacity
   (*a*) 2 TR (*b*) 4 TR (*c*) 8 TR (*d*) 10 TR
2. A jet fighter travelling at 950 km / h needs a cooling system of capacity
   (*a*) 2 to 4 TR (*b*) 4 to 8 TR (*c*) 8 to 10 TR (*d*) 10 to 20 TR
3. The simple air cooling system is good for ....flight speeds.
   (*a*) low (*b*) high
4. The water, alcohol and ammonia have .... refrigerating effects at different altitudes.
   (*a*) same (*b*) different
5. A boot-strap air cooling system has
   (*a*) one heat exchanger (*b*) two heat exchangers
   (*c*) three heat exchangers (*d*) four heat exchangers
6. The air cooling system mostly used in transport type aircrafts is
   (*a*) simple air cooling system (*b*) simple evaporative air cooling system
   (*c*) boot-strap air cooling system (*d*) all of these
7. In a boot strap air evaporative cooling system, the evaporator is provided
   (*a*) between the combustion chamber and the first heat exchanger

(b) between the first heat exchanger and the secondary compressor
(c) between the secondary compressor and the second heat exchanger
(d) between the second heat exchanger and the cooling turbine

8. The reduced ambient air cooling system has
(a) one cooling turbine and one heat exchanger
(b) one cooling turbine and two heat exchangers
(c) two cooling turbines and one heat exchanger
(d) two cooling turbines and two heat exchangers

9. The reduced ambient air cooling system is used for very................ speed aircrafts.
(a) low (b) high

10. The cooling system used for supersonic aircrafts and rockets is
(a) simple air cooling system (b) boot-strap air cooling system
(c) reduced ambient air cooling system (d) regenerative air cooling system

ANSWERS

1. (c)	2. (d)	3. (a)	4. (b)	5. (b)
6. (c)	7. (d)	8. (c)	9. (b)	10. (d)

CHAPTER

4 Simple Vapour Compression Refrigeration Systems

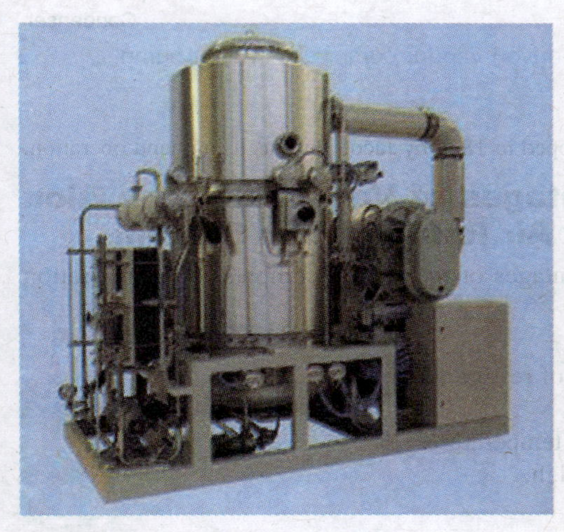

4.1 Introduction

A vapour compression refrigeration system* is an improved type of air refrigeration system in which a suitable working substance, termed as refrigerant, is used. It condenses and evaporates at temperatures and pressures close to the atmospheric conditions. The refrigerants, usually, used for this purpose are ammonia (NH_3), carbon dioxide (CO_2) and sulphur dioxide (SO_2). The refrigerant used, does not leave the system, but is circulated

* Since low pressure vapour refrigerant from the evaporator is changed into high pressure vapour refrigerant in the compressor, therefore it is named as vapour compression refrigeration system.

throughout the system alternately condensing and evaporating. In evaporating, the refrigerant absorbs its latent heat from the brine* (salt water) which is used for circulating it around the cold chamber. While condensing, it gives out its latent heat to the circulating water of the cooler. The vapour compression refrigeration system is, therefore a latent heat pump, as it pumps its latent heat from the brine and delivers it to the cooler.

The vapour compression refrigeration system is now-a-days used for all purpose refrigeration. It is generally used for all industrial purposes from a small domestic refrigerator to a big air conditioning plant.

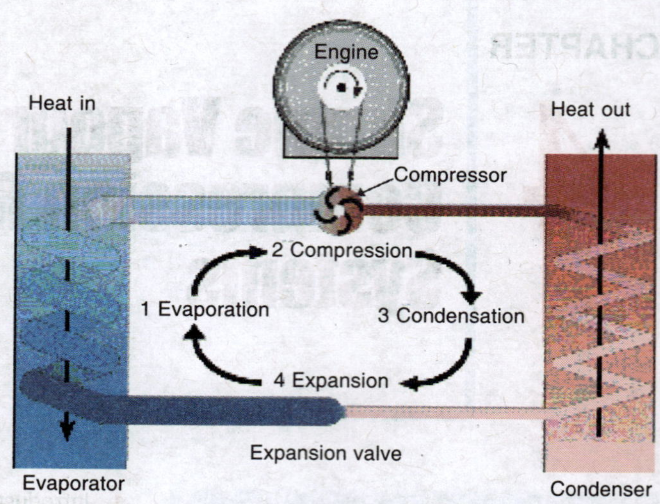

Engine-driven vapour compression heat pump.

Note: The first vapour compression system was developed in 1834 by Jacob Perkins using hand operation.

4.2 Advantages and Disadvantages of Vapour Compression Refrigeration System over Air Refrigeration System

Following are the advantages and disadvantages of the vapour compression refrigeration system over air refrigeration system :

Advantages

1. It has smaller size for the given capacity of refrigeration.
2. It has less running cost.
3. It can be employed over a large range of temperatures.
4. The coefficient of performance is quite high.

Disadvantages

1. The initial cost is high.
2. The prevention of leakage of the refrigerant is the major problem in vapour compression system.

4.3 Mechanism of a Simple Vapour Compression Refrigeration System

Fig. 4.1 shows the schematic diagram of a simple vapour compression refrigeration system. It consists of the following five essential parts :

1. *Compressor.* The low pressure and temperature vapour refrigerant from evaporator is drawn into the compressor through the inlet or suction valve *A*, where it is compressed to a high pressure and temperature. This high pressure and temperature vapour refrigerant is discharged into the condenser through the delivery or discharge valve *B*.

2. *Condenser.* The condenser or cooler consists of coils of pipe in which the high pressure and temperature vapour refrigerant is cooled and condensed. The refrigerant, while passing through the condenser, gives up its latent heat to the surrounding condensing medium which is normally air or water.

* Brine is used as it has a very low freezing temperature.

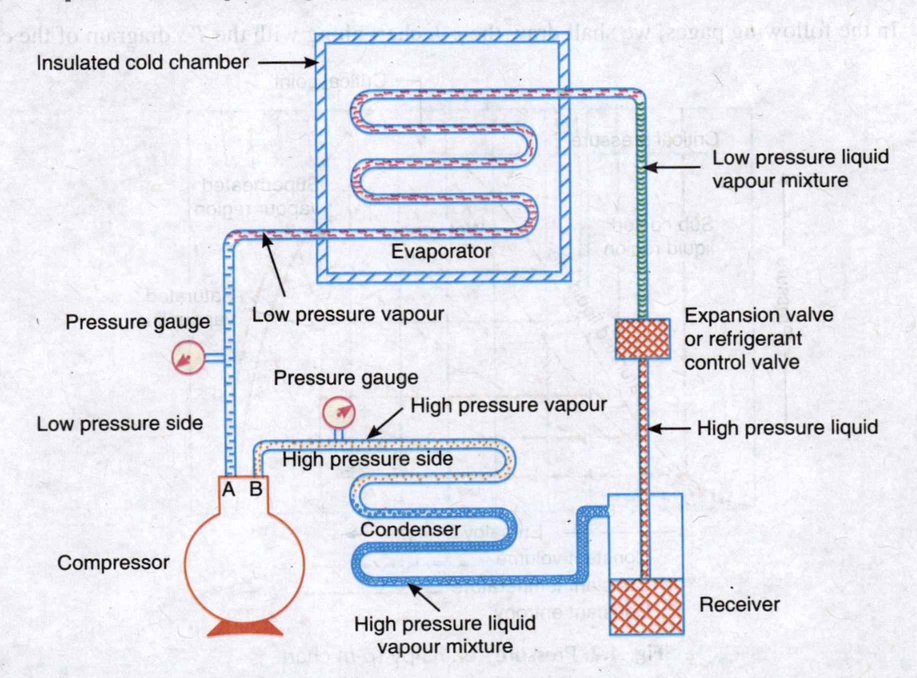

Fig. 4.1. Simple vapour compression refrigeration system.

3. *Receiver.* The condensed liquid refrigerant from the condenser is stored in a vessel known as receiver from where it is supplied to the evaporator through the expansion valve or refrigerant control valve.

4. *Expansion valve.* It is also called throttle valve or refrigerant control valve. The function of the expansion valve is to allow the liquid refrigerant under high pressure and temperature to pass at a controlled rate after reducing its pressure and temperature. Some of the liquid refrigerant evaporates as it passes through the expansion valve, but the greater portion is vaporised in the evaporator at the low pressure and temperature.

5. *Evaporator.* An evaporator consists of coils of pipe in which the liquid-vapour refrigerant at low pressure and temperature is evaporated and changed into vapour refrigerant at low pressure and temperature. In evaporating, the liquid vapour refrigerant absorbs its latent heat of vaporisation from the medium (air, water or brine) which is to be cooled.

Note: In any compression refrigeration system, there are two different pressure conditions. One is called the *high pressure side* and other is known as *low pressure side.* The high pressure side includes the discharge line (*i.e.* piping from delivery valve *B* to the condenser), condenser, receiver and expansion valve. The low pressure side includes the evaporator, piping from the expansion valve to the evaporator and the suction line (*i.e.* piping from the evaporator to the suction valve *A*).

4.4 Pressure-Enthalpy (*p-h*) Chart

The most convenient chart for studying the behaviour of a refrigerant is the *p-h* chart, in which the vertical ordinates represent pressure and horizontal ordinates represent enthalpy (*i.e.* total heat). A typical chart is shown in Fig. 4.2, in which a few important lines of the complete chart are drawn. The saturated liquid line and the saturated vapour line merge into one another at the critical point. A saturated liquid is one which has a temperature equal to the saturation temperature corresponding to its pressure.The space to the left of the saturated liquid line will, therefore, be sub-cooled liquid region. The space between the liquid and the vapour lines is called wet vapour region and to the right of the saturated vapour line is a superheated vapour region.

In the following pages, we shall draw the *p-h* chart along with the *T-s* diagram of the cycle.

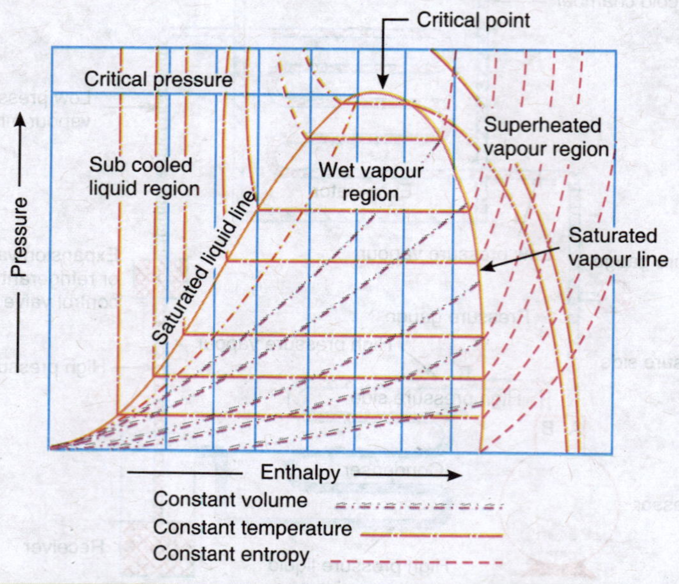

Fig. 4.2. Pressure - enthalpy (*p-h*) chart.

4.5 Types of Vapour Compression Cycles

We have already discussed that vapour compression cycle essentially consists of compression, condensation, throttling and evaporation. Many scientists have focussed their attention to increase the coefficient of performance of the cycle. Though there are many cycles, yet the following are important from the subject point of view :

1. Cycle with dry saturated vapour after compression,
2. Cycle with wet vapour after compression,
3. Cycle with superheated vapour after compression,
4. Cycle with superheated vapour before compression, and
5. Cycle with undercooling or subcooling of refrigerant.

Now we shall discuss all the above mentioned cycles, one by one, in the following pages.

4.6 Theoretical Vapour Compression Cycle with Dry Saturated Vapour after Compression

A vapour compression cycle with dry saturated vapour after compression is shown on *T-s* and *p-h* diagrams in Fig. 4.3 (*a*) and (*b*) respectively. At point 1, let T_1, p_1 and s_1, be the temperature, pressure and entropy of the vapour refrigerant respectively. The four processes of the cycle are as follows :

1. *Compression process.* The vapour refrigerant at low pressure p_1 and temperature T_1 is compressed isentropically to dry saturated vapour as shown by the vertical line 1-2 on *T-s* diagram and by the curve 1-2 on *p-h* diagram. The pressure and temperature rises from p_1 to p_2 and T_1 to T_2 respectively.

The work done during isentropic compression per kg of refrigerant is given by

$$w = h_2 - h_1$$

where h_1 = Enthalpy of vapour refrigerant at temperature T_1, *i.e.* at suction of the compressor, and

h_2 = Enthalpy of the vapour refrigerant at temperature T_2, *i.e.* at discharge of the compressor.

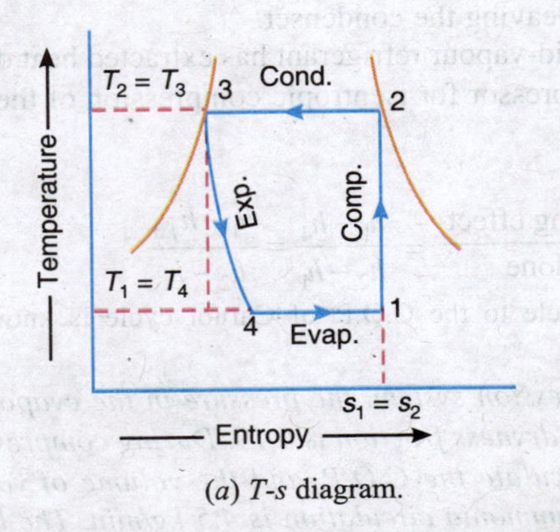

(*a*) *T-s* diagram.

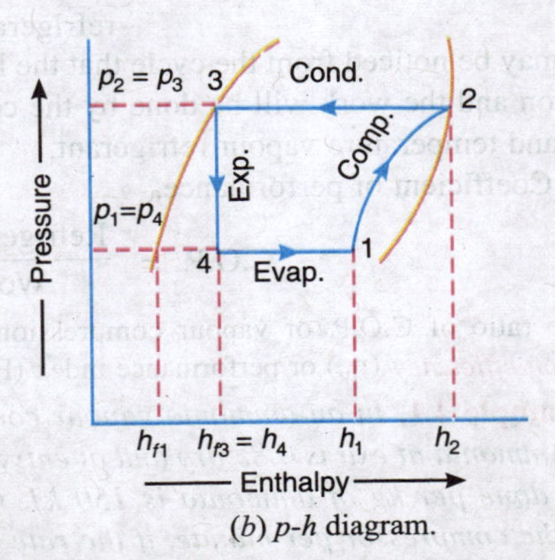

(*b*) *p-h* diagram.

Fig. 4.3. Theoretical vapour compression cycle with dry saturated vapour after compression.

2. Condensing process. The high pressure and temperature vapour refrigerant from the compressor is passed through the condenser where it is completely condensed at constant pressure p_2 and temperature T_2, as shown by the horizontal line 2-3 on *T-s* and *p-h* diagrams. The vapour refrigerant is changed into liquid refrigerant. The refrigerant, while passing through the condenser, gives its latent heat to the surrounding condensing medium.

3. Expansion process. The liquid refrigerant at pressure $p_3 = p_2$ and temperature $T_3 = T_2$ is expanded by *throttling process through the expansion valve to a low pressure $p_4 = p_1$ and temperature $T_4 = T_1$, as shown by the curve 3-4 on *T-s* diagram and by the vertical line 3-4 on *p-h* diagram. We have already discussed that some of the liquid refrigerant evaporates as it passes through the expansion valve, but the greater portion is vaporised in the evaporator. We know that during the throttling process, no heat is absorbed or rejected by the liquid refrigerant.

Notes: (*a*) In case an expansion cylinder is used in place of throttle or expansion valve to expand the liquid refrigerant, then the refrigerant will expand isentropically as shown by dotted vertical line on *T-s* diagram in Fig. 4.3 (*a*). The isentropic expansion reduces the external work being expanded in running the compressor and increases the refrigerating effect. Thus, the net result of using the expansion cylinder is to increase the coefficient of performance.

Since the expansion cylinder system of expanding the liquid refrigerant is quite complicated and involves greater initial cost, therefore its use is not justified for small gain in cooling capacity. Moreover, the flow rate of the refrigerant can be controlled with throttle valve which is not possible in case of expansion cylinder which has a fixed cylinder volume.

(*b*) In modern domestic refrigerators, a capillary (small bore tube) is used in place of an expansion valve.

4. Vaporising process. The liquid-vapour mixture of the refrigerant at pressure $p_4 = p_1$ and temperature $T_4 = T_1$ is evaporated and changed into vapour refrigerant at constant pressure and temperature, as shown by the horizontal line 4-1 on *T-s* and *p-h* diagrams. During evaporation, the liquid-vapour refrigerant absorbs its latent heat of vaporisation from the medium (air, water or brine) which is to be cooled. This heat which is absorbed by the refrigerant is called *refrigerating effect* and it is briefly written as R_E. The process of vaporisation continues upto point 1 which is the starting point and thus the cycle is completed.

* The throttling process is an irreversible process.

We know that the refrigerating effect or the heat absorbed or extracted by the liquid-vapour refrigerant during evaporation per kg of refrigerant is given by

$$R_E = h_1 - h_4 = h_1 - h_{f3} \qquad \ldots (\because h_{f3} = h_4)$$

where h_{f3} = Sensible heat at temperature T_3, *i.e.* enthalpy of liquid refrigerant leaving the condenser.

It may be noticed from the cycle that the liquid-vapour refrigerant has extracted heat during evaporation and the work will be done by the compressor for isentropic compression of the high pressure and temperature vapour refrigerant.

∴ Coefficient of performance,

$$\text{C.O.P.} = \frac{\text{Refrigerating effect}}{\text{Work done}} = \frac{h_1 - h_4}{h_2 - h_1} = \frac{h_1 - h_{f3}}{h_2 - h_1}$$

Note: The ratio of C.O.P. of vapour compression cycle to the C.O.P. of Carnot cycle is known as *refrigeration efficiency* (η_R) or performance index (P.I.).

Example 4.1. *In an ammonia vapour compression system, the pressure in the evaporator is 2 bar. Ammonia at exit is 0.85 dry and at entry its dryness fraction is 0.19. During compression, the work done per kg of ammonia is 150 kJ. Calculate the C.O.P. and the volume of vapour entering the compressor per minute, if the rate of ammonia circulation is 4.5 kg/min. The latent heat and specific volume at 2 bar are 1325 kJ/kg and 0.58 m³/kg respectively.*

Solution. Given : $p_1 = p_4 = 2$ bar ; $x_1 = 0.85$; $x_4 = 0.19$; $w = 150$ kJ/kg ; $m_a = 4.5$ kg/min ; $h_{fg} = 1325$ kJ/kg ; $v_g = 0.58$ m³/kg

C.O.P.

The *T-s* and *p-h* diagrams are shown in Fig. 4.3 (*a*) and (*b*) respectively.

Since the ammonia vapour at entry to the evaporator (*i.e.* at point 4) has dryness fraction (x_4) equal to 0.19, therefore enthalpy at point 4,

$$h_4 = x_4 \times h_{fg} = 0.19 \times 1325 = 251.75 \text{ kJ/kg}$$

Similarly, enthalpy of ammonia vapour at exit *i.e.* at point 1,

$$h_1 = x_1 \times h_{fg} = 0.85 \times 1325 = 1126.25 \text{ kJ/kg}$$

∴ Heat extracted from the evaporator or refrigerating effect,

$$R_E = h_1 - h_4 = 1126.25 - 251.75 = 874.5 \text{ kJ/kg}$$

We know that work done during compression,

$$w = 150 \text{ kJ/kg}$$

$$\therefore \quad \text{C.O.P.} = R_E / w = 874.5 / 150 = 5.83 \textbf{ Ans.}$$

Volume of vapour entering the compressor per minute

We know that volume of vapour entering the compressor per minute

= Mass of refrigerant / min × Specific volume

$= m_a \times v_g = 4.5 \times 0.58 = 2.61$ m³/min **Ans.**

Example 4.2. *The temperature limits of an ammonia refrigerating system are 25°C and –10°C. If the gas is dry at the end of compression, calculate the coefficient of performance of the cycle assuming no undercooling of the liquid ammonia. Use the following table for properties of ammonia :*

Temperature (°C)	*Liquid heat (kJ/kg)*	*Latent heat (kJ/kg)*	*Liquid entropy (kJ/kg K)*
25	298.9	1166.94	1.1242
–10	135.37	1297.68	0.5443

Solution. Given : $T_2 = T_3 = 25°C = 25 + 273 = 298$ K ; $T_1 = T_4 = -10°$ C $= -10 + 273 = 263$ K ; $h_{f3} = h_4 = 298.9$ kJ/kg ; $h_{fg2} = 1166.94$ kJ/kg ; $s_{f2} = 1.1242$ kJ/kg K ; $h_{f1} = 135.37$ kJ/ kg ; $h_{fg1} = 1297.68$ kJ/kg ; $s_{f1} = 0.5443$ kJ/kg K

The *T-s* and *p-h* diagrams are shown in Fig. 4.4 (*a*) and (*b*) respectively.

Let x_1 = Dryness fraction at point 1.

We know that entropy at point 1,

$$s_1 = s_{f1} + \frac{x_1 h_{fg1}}{T_1} = 0.5443 + \frac{x_1 \times 1297.68}{263}$$

$$= 0.5443 + 4.934\, x_1 \quad \ldots (i)$$

Similarly, entropy at point 2,

$$s_2 = s_{f2} + \frac{h_{fg2}}{T_2} = 1.1242 + \frac{1166.94}{298} = 5.04 \quad \ldots (ii)$$

Since the entropy at point 1 is equal to entropy at point 2, therefore equating equations (*i*) and (*ii*),

$$0.5443 + 4.934\, x_1 = 5.04 \quad \text{or} \quad x_1 = 0.91$$

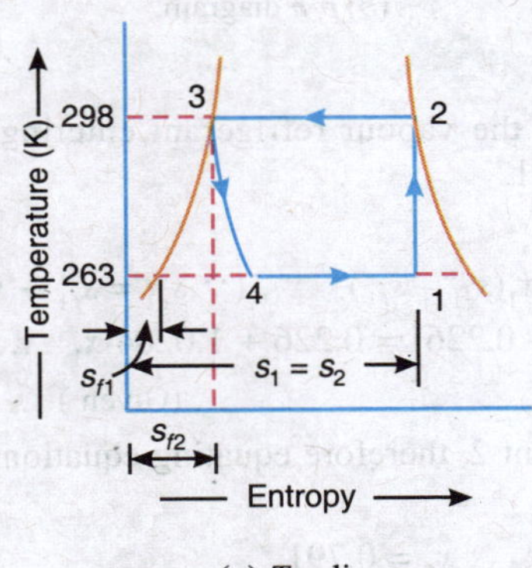

(*a*) *T-s* diagram.

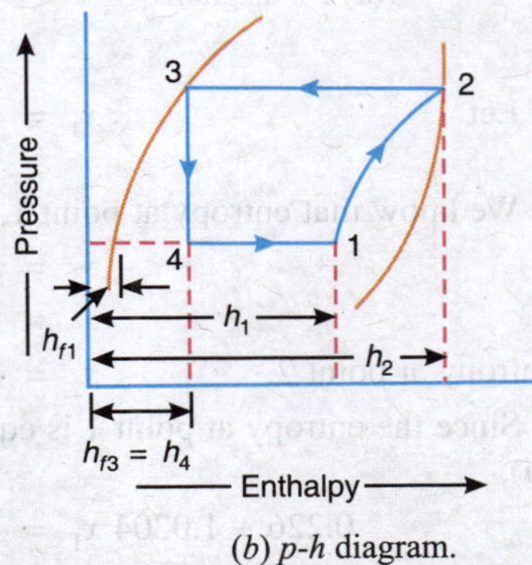

(*b*) *p-h* diagram.

Fig. 4.4

We know that enthalpy at point 1,

$$h_1 = h_{f1} + x_1\, h_{fg1} = 135.37 + 0.91 \times 1297.68 = 1316.26 \text{ kJ/kg}$$

and enthalpy at point 2,

$$h_2 = h_{f2} + h_{fg2} = 298.9 + 1166.94 = 1465.84 \text{ kJ/kg}$$

∴ Coefficient of performance of the cycle

$$= \frac{h_1 - h_{f3}}{h_2 - h_1} = \frac{1316.26 - 298.9}{1465.84 - 1316.26} = 6.8 \textbf{ Ans.}$$

Example 4.3. *A vapour compression refrigerator works between the pressure limits of 60 bar and 25 bar. The working fluid is just dry at the end of compression and there is no under-cooling of the liquid before the expansion valve. Determine : 1. C.O.P. of the cycle ; and 2. Capacity of the refrigerator if the fluid flow is at the rate of 5 kg/min.*

Data :

Pressure (bar)	*Saturation temperature (K)*	*Enthalpy (kJ/kg)*		*Entropy (kJ/kg K)*	
		Liquid	*Vapour*	*Liquid*	*Vapour*
60	*295*	*151.96*	*293.29*	*0.554*	*1.0332*
25	*261*	*56.32*	*322.58*	*0.226*	*1.2464*

Solution. Given : $p_2 = p_3 = 60$ bar ; $p_1 = p_4 = 25$ bar ; $T_2 = T_3 = 295$ K ; $T_1 = T_4 = 261$ K ; $h_{f3} = h_4 = 151.96$ kJ/kg ; $h_{f1} = 56.32$ kJ/kg ; $h_{g2} = h_2 = 293.29$ kJ/kg ; $h_{g1} = 322.58$ kJ/kg ; *$s_{f2} = 0.554$ kJ/kg K ; $s_{f1} = 0.226$ kJ/kg K ; $s_{g2} = s_2 = 1.0332$ kJ/kg K ; $s_{g1} = 1.2464$ kJ/kg K

1. C.O.P. of the cycle

The *T-s* and *p-h* diagrams are shown in Fig. 4.5 (*a*) and (*b*) respectively.

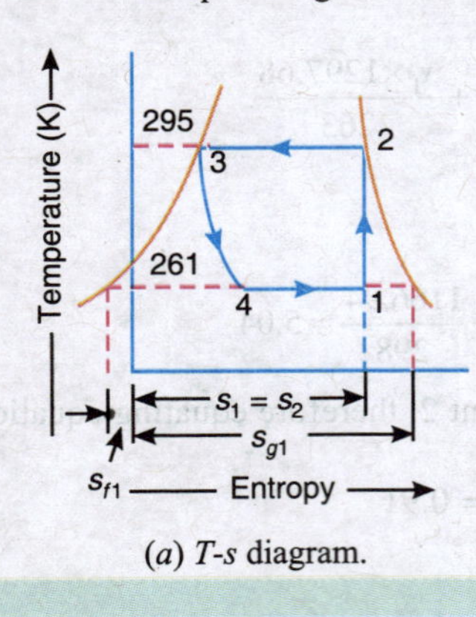

(*a*) *T-s* diagram.

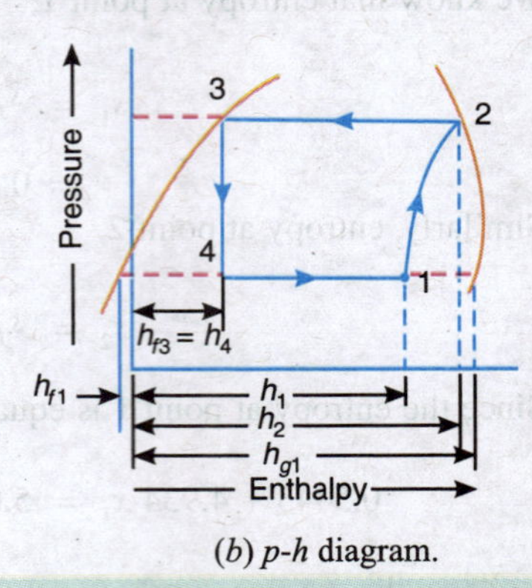

(*b*) *p-h* diagram.

Fig. 4.5

Let x_1 = Dryness fraction of the vapour refrigerant entering the compressor at point 1.

We know that entropy at point 1,

$$s_1 = s_{f1} + x_1\, s_{fg1} = s_{f1} + x_1(s_{g1} - s_{f1}) \quad \ldots (\because s_{g1} = s_{f1} + s_{fg1})$$

$$= 0.226 + x_1\,(1.2464 - 0.226) = 0.226 + 1.0204\, x_1 \quad \ldots (i)$$

and entropy at point 2,

$$s_2 = s_{g2} = 1.0332 \text{ kJ/kg K} \quad \ldots \text{(Given)} \ldots (ii)$$

Since the entropy at point 1 is equal to entropy at point 2, therefore equating equations (*i*) and (*ii*),

$$0.226 + 1.0204\, x_1 = 1.0332 \quad \text{or} \quad x_1 = 0.791$$

We know that enthalpy at point 1,

$$h_1 = h_{f1} + x_1\, h_{fg1} = h_{f1} + x_1\,(h_{g1} - h_{f1}) \ldots (\because h_{g1} = h_{f1} + h_{fg1})$$

$$= 56.32 + 0.791\,(322{,}58 - 56.32) = 266.93 \text{ kJ/kg}$$

∴ C.O.P. of the cycle

$$= \frac{h_1 - h_{f3}}{h_2 - h_1} = \frac{266.93 - 151.96}{293.29 - 266.93} = 4.36 \textbf{ Ans.}$$

2. Capacity of the refrigerator

We know that the heat extracted or refrigerating effect produced per kg of refrigerant

$$= h_1 - h_{f3} = 266.93 - 151.96 = 114.97 \text{ kJ/kg}$$

Since the fluid flow is at the rate of 5 kg / min, therefore total heat extracted

$$= 5 \times 114.97 = 574.85 \text{ kJ/min}$$

∴ Capacity of the refrigerator

$$= \frac{574.85}{210} = 2.74 \text{ TR } \textbf{Ans.} \quad \ldots (\because 1 \text{ TR} = 210 \text{ kJ/min})$$

* Superfluous data

Example 4.4. *28 tonnes of ice from and at 0°C is produced per day in an ammonia refrigerator. The temperature range in the compressor is from 25°C to –15°C. The vapour is dry and saturated at the end of compression and an expansion valve is used. There is no liquid subcooling. Assuming actual C.O.P. of 62% of the theoretical, calculate the power required to drive the compressor. Following properties of ammonia are given:*

Temperature	*Enthalpy (kJ/kg)*		*Entropy (kJ/kg K)*	
0°C	*Liquid*	*Vapour*	*Liquid*	*Vapour*
25	*298.9*	*1465.84*	*1.1242*	*5.0391*
–15	*112.34*	*1426.54*	*0.4572*	*5.5490*

Take latent heat of ice = 335 kJ/kg.

Solution. Given: Ice produced = 28 t/day ; $T_2 = T_3 = 25°C = 25 + 273 = 298 K$; $T_1 = T_4 = -15°C = -15 + 273 = 258$ K ; $h_{f3} = h_4 = 298.9$ kJ/kg ; $h_{f1} = 112.34$ kJ/kg ; $h_{g2} = h_2 = 1465.84$ kJ/kg ; $h_{g1} = 1426.54$ kJ/kg ; *$s_{f2} = 1.1242$ kJ/kg K ; $s_{f1} = 0.4572$ kJ/kg K ; $s_{g2} = s_2 = 5.0391$ kJ/kg K ; $s_{g1} = 5.5490$ kJ/kg K.

The *T-s* and *p-h* diagrams are shown in Fig. 4.6 (*a*) and (*b*) respectively.

First of all, let us find the dryness fraction (x_1) of the vapour refrigerant entering the compressor at point 1.

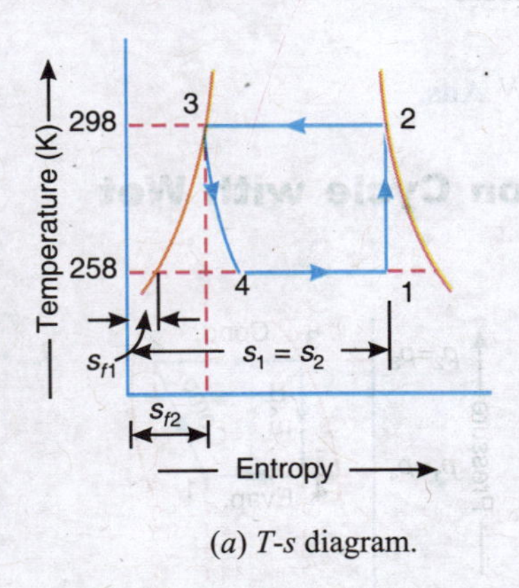

(*a*) *T-s* diagram.

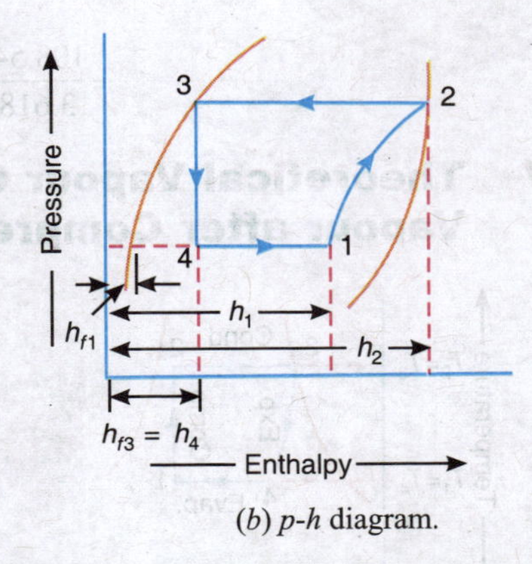

(*b*) *p-h* diagram.

Fig. 4.6

We know that entropy at point 1,

$$s_1 = s_{f1} + x_1\, s_{fg1} = s_{f1} + x_1(s_{g1} - s_{f1}) \qquad \ldots(\because s_{g1} = s_{f1} + s_{fg1})$$

$$= 0.4572 + x_1(5.5490 - 0.4572)$$

$$= 0.4572 + 5.0918\, x_1 \qquad \ldots(i)$$

and entropy at point 2,

$$s_2 = s_{g2} = 5.0391 \text{ kJ/kg K} \qquad \ldots \text{(Given)} \quad \ldots(ii)$$

Since the entropy at point 1 is equal to entropy at point 2, therefore equating equations (*i*) and (*ii*),

$$0.4572 + 5.0918\, x_1 = 5.0391 \quad \text{or} \quad x_1 = 0.9$$

We know that enthalpy at point 1,

$$h_1 = h_{f1} + x_1 h_{fg1} = h_{f1} + x_1(h_{g1} - h_{f1}) \qquad \ldots(\because h_{g1} = h_{f1} + h_{fg1})$$

$$= 112.34 + 0.9\,(1426.54 - 112.34) = 1295.12 \text{ kJ/kg}$$

* Superfluous data.

$$\therefore \quad \text{Theoretical C.O.P.} = \frac{h_1 - h_{f3}}{h_2 - h_1} = \frac{1295.12 - 298.9}{1465.84 - 1295.12} = \frac{996.22}{170.72} = 5.835$$

Since actual C.O.P. is 62% of theoretical C.O.P., therefore

$$\text{Actual C.O.P} = 0.62 \times 5.835 = 3.618$$

We know that ice produced from and at 0°C

$$= 28 \text{ t/day} = \frac{28 \times 1000}{24 \times 3600} = 0.324 \text{ kg / s}$$

Latent heat of ice = 335 kJ/kg ...(Given)

∴ Refrigeration effect produced

$$= 0.324 \times 335 = 108.54 \text{ kJ/s}$$

We know that actual C.O.P.,

$$3.618 = \frac{\text{Refrigeration effect}}{\text{workdone}} = \frac{108.54}{\text{workdone}}$$

∴ Workdone or power required to drive the compressor

$$= \frac{108.54}{3.618} = 30 \text{ kJ/s or kW} \quad \textbf{Ans.}$$

4.7 Theoretical Vapour Compression Cycle with Wet Vapour after Compression

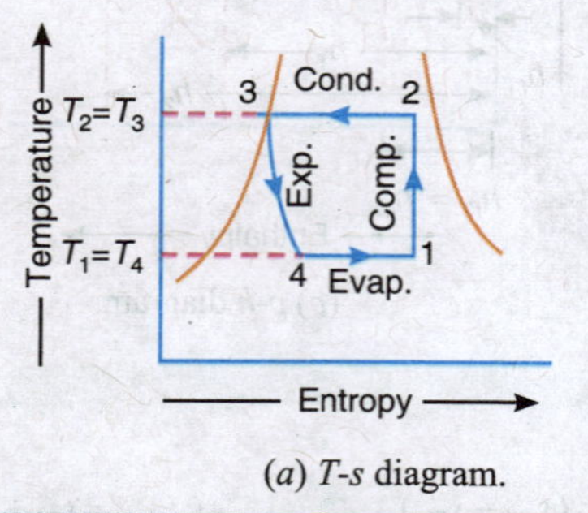

(*a*) *T-s* diagram.

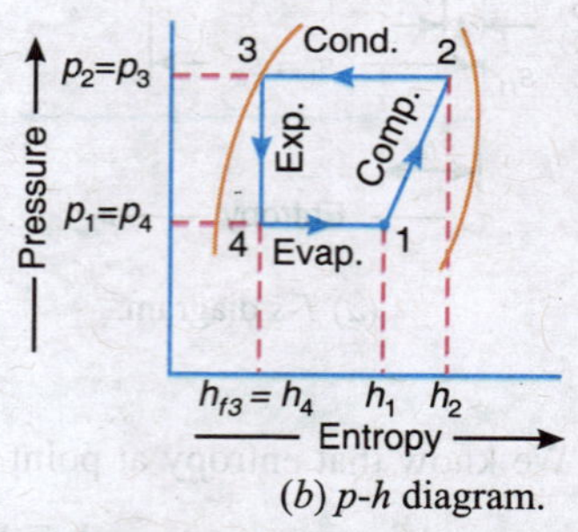

(*b*) *p-h* diagram.

Fig. 4.7. Theoretical vapour compression cycle with wet vapour after compression.

A vapour compression cycle with wet vapour after compression is shown on *T-s* and *p-h* diagrams in Fig. 4.7 (*a*) and (*b*) respectively. In this cycle, the enthalpy at point 2 is found out with the help of dryness fraction at this point. The dryness fraction at points 1 and 2 may be obtained by equating entropies at points 1 and 2.

Now the coefficient of performance may be found out as usual from the relation,

$$\text{C.O.P.} = \frac{\text{Refrigerating effect}}{\text{Work done}} = \frac{h_1 - h_{f3}}{h_2 - h_1}$$

Note: The remaining cycle is same as discussed in the last article.

Example 4.5. *Find the theoretical C.O.P. for a CO_2 machine working between the temperature range of 25°C and – 5°C. The dryness fraction of CO_2 gas during the suction stroke is 0.6. Following properties of CO_2 are given :*

Temperature °C	*Liquid*		*Vapour*		*Latent heat kJ/kg*
	Enthalpy kJ/kg	*Entropy kJ/kg K*	*Enthalpy kJ/kg*	*Entropy kJ/kg K*	
25	164.77	0.5978	282.23	0.9918	117.46
–5	72.57	0.2862	321.33	1.2146	248.76

Solution. Given : $T_2 = T_3 = 25°C = 25 + 273 = 298$ K ; $T_1 = T_4 = -5°C = -5 + 273 = 268$ K ; $x_1 = 0.6$; $h_{f3} = h_{f2} = 164.77$ kJ/kg ; $h_{f1} = h_{f4} = 72.57$ kJ/kg ; $s_{f2} = 0.5978$ kJ/kg K; $s_{f1} = 0.2862$ kJ/kg K ; $h_{2'} = 282.23$ kJ/kg ; $h_{1'} = 321.33$ kJ/kg ; $^*s_{2'} = 0.9918$ kJ/kg K ; $^*s_{1'} = 1.2146$ kJ/kg K ; $h_{fg2} = 117.46$ kJ/kg ; $h_{fg1} = 248.76$ kJ/kg

The *T-s* and *p-h* diagrams are shown in Fig. 4.8 (*a*) and (*b*) respectively.

First of all, let us find the dryness fraction at point 2, *i.e.* x_2. We know that the entropy at point 1,

$$s_1 = s_{f1} + \frac{x_1 h_{fg1}}{T_1} = 0.2862 + \frac{0.6 \times 248.76}{268} = 0.8431 \quad ...(i)$$

Similarly, entropy at point 2,

$$s_2 = s_{f2} + \frac{x_2 h_{fg2}}{T_2} = 0.5978 + \frac{x_2 \times 117.46}{298}$$

$$= 0.5978 + 0.3941\, x_2 \quad ...(ii)$$

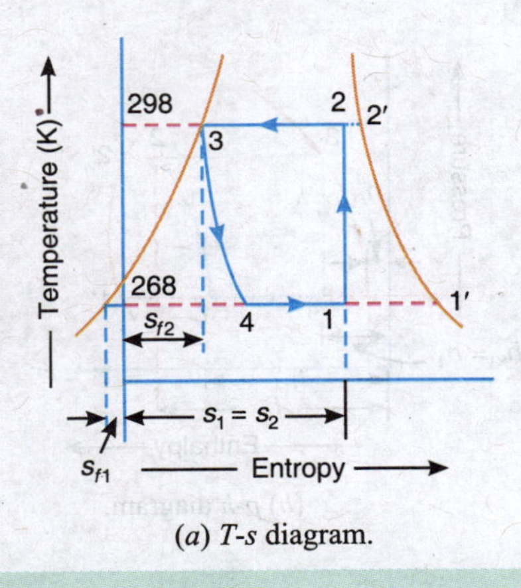

(*a*) *T-s* diagram.

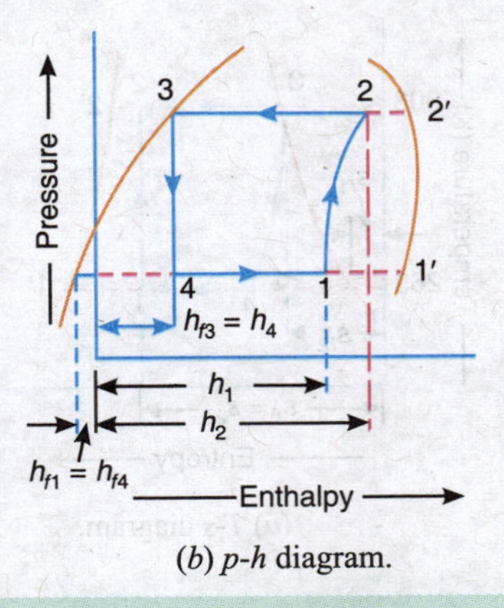

(*b*) *p-h* diagram.

Fig. 4.8

Since the entropy at point 1 (s_1) is equal to entropy at point 2 (s_2), therefore equating equations (*i*) and (*ii*),

$$0.8431 = 0.5978 + 0.3941\, x_2 \quad \text{or} \quad x_2 = 0.622$$

We know that enthalpy at point 1,

$$h_1 = h_{f1} + x_1 h_{fg1} = 72.57 + 0.6 \times 248.76 = 221.83 \text{ kJ/kg}$$

and enthalpy at point 2,

$$h_2 = h_{f2} + x_2 h_{fg2} = 164.77 + 0.622 \times 117.46 = 237.83 \text{ kJ/kg}$$

* Superfluous data

$\therefore$ Theoretical C.O.P. $= \dfrac{h_1 - h_{f3}}{h_2 - h_1} = \dfrac{221.83 - 164.77}{237.83 - 221.83} = \dfrac{57.06}{16} = 3.57$ **Ans.**

Example 4.6. *An ammonia refrigerating machine fitted with an expansion valve works between the temperature limits of – 10°C and 30°C. The vapour is 95% dry at the end of isentropic compression and the fluid leaving the condenser is at 30°C. Assuming actual C.O.P. as 60% of the theoretical, calculate the kilograms of ice produced per kW hour at 0°C from water at 10°C. Latent heat of ice is 335 kJ/kg. Ammonia has the following properties :*

Temperature °C	*Liquid heat (h_f) kJ/kg*	*Latent heat (h_{fg}) kJ/kg*	*Liquid entropy (s_f)*	*Total entropy of dry saturated vapour*
30	*323.08*	*1145.80*	*1.2037*	*4.9842*
–10	*135.37*	*1297.68*	*0.5443*	*5.4770*

Solution. Given : $T_1 = T_4 = -10°C = -10 + 273 = 263$ K ; $T_2 = T_3 = 30°C = 30 + 273 = 303$ K ; $x_2 = 0.95$; $h_{f3} = h_{f2} = 323.08$ kJ/kg ; $h_{f1} = h_{f4} = 135.37$ kJ/kg ; $h_{fg2} = 1145.8$ kJ/kg ; $h_{fg1} = 1297.68$ kJ/kg , $s_{f2} = 1.2037$; $s_{f1} = 0.5443$; *$s_2' = 4.9842$; *$s_1' = 5.4770$

The *T-s* and *p-h* diagrams are shown in Fig. 4.9 (*a*) and (*b*) respectively.

Let x_1 = Dryness fraction at point 1.

We know that entropy at point 1,

$$s_1 = s_{f1} + \frac{x_1 h_{fg1}}{T_1} = 0.5443 + \frac{x_1 \times 1297.68}{263}$$

$$= 0.5443 + 4.934\, x_1 \qquad \ldots (i)$$

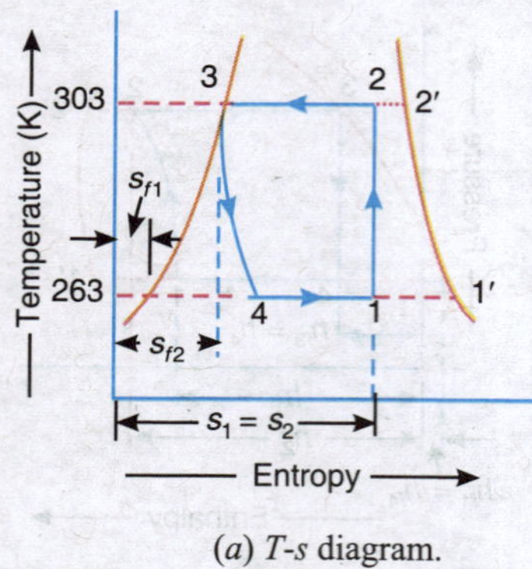

(*a*) *T-s* diagram.

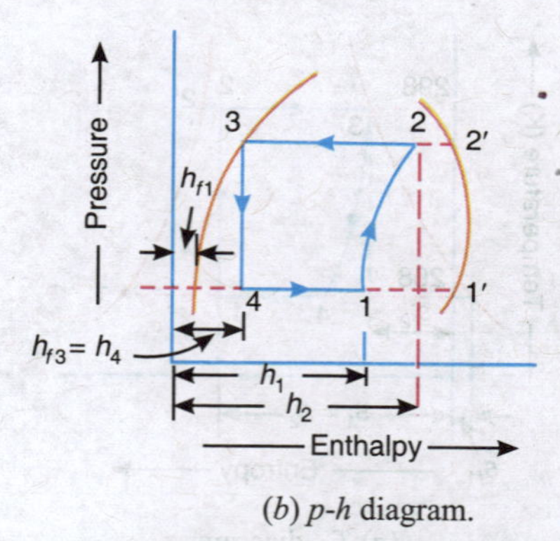

(*b*) *p-h* diagram.

Fig. 4.9

Similarly, entropy at point 2,

$$s_2 = s_{f2} + \frac{x_2 h_{fg2}}{T_2} = 1.2037 + \frac{0.95 \times 1145.8}{303} = 4.796 \qquad \ldots (ii)$$

Since the entropy at point 1 (s_1) is equal to entropy at point 2 (s_2), therefore equating equations (*i*) and (*ii*),

$$0.5443 + 4.934\, x_1 = 4.796 \quad \text{or} \quad x_1 = 0.86$$

* Superfluous data.

∴ Enthalpy at point 1, $h_1 = h_{f1} + x_1 h_{fg1} = 135.37 + 0.86 \times 1297.68 = 1251.4$ kJ/kg

and enthalpy at point 2, $h_2 = h_{f2} + x_2 h_{fg2} = 323.08 + 0.95 \times 1145.8 = 1411.6$ kJ/kg

We know that theoretical C.O.P.

$$= \frac{h_1 - h_{f3}}{h_2 - h_1} = \frac{1251.4 - 323.08}{1411.6 - 1251.4} = 5.8$$

∴ Actual C.O.P. $= 0.6 \times 5.8 = 3.48$

Work to be spent corresponding to 1 kW hour,

$$W = 3600 \text{ kJ}$$

∴ Actual heat extracted or refrigeration effect produced per kW hour

$$= W \times \text{Actual C.O.P.} = 3600 \times 3.48 = 12\ 528 \text{ kJ}$$

We know that heat extracted from 1 kg of water at 10°C for the formation of 1 kg of ice at 0°C

$$= 1 \times 4.187 \times 10 + 335 = 376.87 \text{ kJ}$$

∴ Amount of ice produced

$$= \frac{12528}{376.87} = 33.2 \text{ kg / kW hour}$$ **Ans.**

4.8 Theoretical Vapour Compression Cycle with Superheated Vapour after Compression

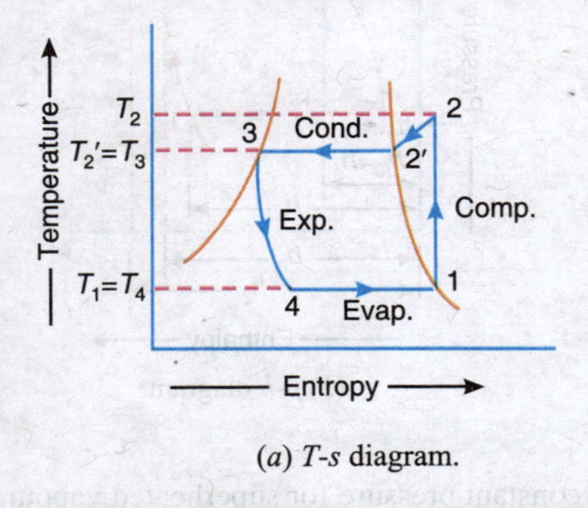

(*a*) *T-s* diagram.

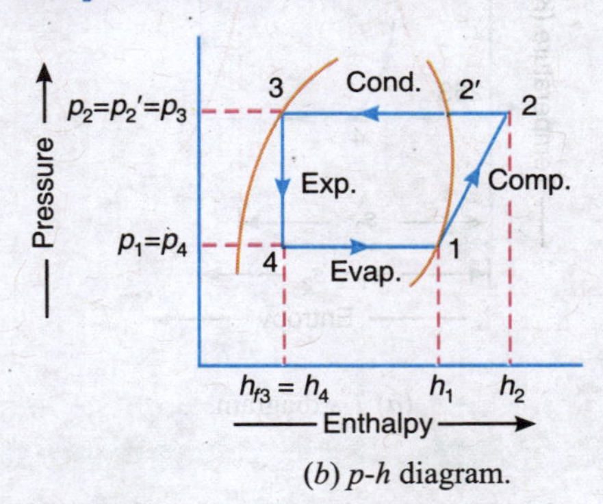

(*b*) *p-h* diagram.

Fig. 4.10. Theoretical vapour compression cycle with superheated vapour after compression.

A vapour compression cycle with superheated vapour after compression is shown on *T-s* and *p-h* diagrams in Fig. 4.10 (*a*) and (*b*) respectively. In this cycle, the enthalpy at point 2 is found out with the help of degree of superheat. The degree of superheat may be found out by equating the entropies at points 1 and 2.

Now the coefficient of performance may be found out as usual from the relation,

$$\text{C.O.P.} = \frac{\text{Refrigerating effect}}{\text{Work done}} = \frac{h_1 - h_{f3}}{h_2 - h_1}$$

A little consideration will show that the superheating increases the refrigerating effect and the amount of work done in the compressor. Since the increase in refrigerating effect is less as compared to the increase in work done, therefore, the net effect of superheating is to have low coefficient of performance.

Note : In this cycle, the cooling of superheated vapour will take place in two stages. Firstly, it will be condensed to dry saturated stage at constant pressure (shown by graph 2-2′) and secondly, it will be condensed at constant temperature (shown by graph 2′-3). The remaining cycle is same as discussed in the last article.

Example 4.7. *A vapour compression refrigerator uses methyl chloride (R-40) and operates between temperature limits of – 10°C and 45°C. At entry to the compressor, the refrigerant is dry saturated and after compression it acquires a temperature of 60°C. Find the C.O.P. of the refrigerator. The relevant properties of methyl chloride are as follows :*

Saturation temperature in °C	*Enthalpy in kJ/kg*		*Entropy in kJ/kg K*	
	Liquid	*Vapour*	*Liquid*	*Vapour*
–10	*45.4*	*460.7*	*0.183*	*1.637*
45	*133.0*	*483.6*	*0.485*	*1.587*

Solution. Given : $T_1 = T_4 = -10°C = -10 + 273 = 263$ K ; $T_{2'} = T_3 = 45°C = 45 + 273 = 318$ K ; $T_2 = 60°C = 60 + 273 = 333$ K ; $^*h_{f1} = 45.4$ kJ/kg ; $h_{f3} = 133$ kJ/kg ; $h_1 = 460.7$ kJ/kg ; $h_{2'} = 483.6$ kJ/kg ; $^*s_{f1} = 0.183$ kJ/kg K ; $^*s_{f3} = 0.485$ kJ/kg K ; $s_1 = s_2 = 1.637$ kJ/kg K; $s_{2'} = 1.587$ kJ/kg K

The *T-s* and *p-h* diagrams are shown in Fig. 4.11 (*a*) and (*b*) respectively.

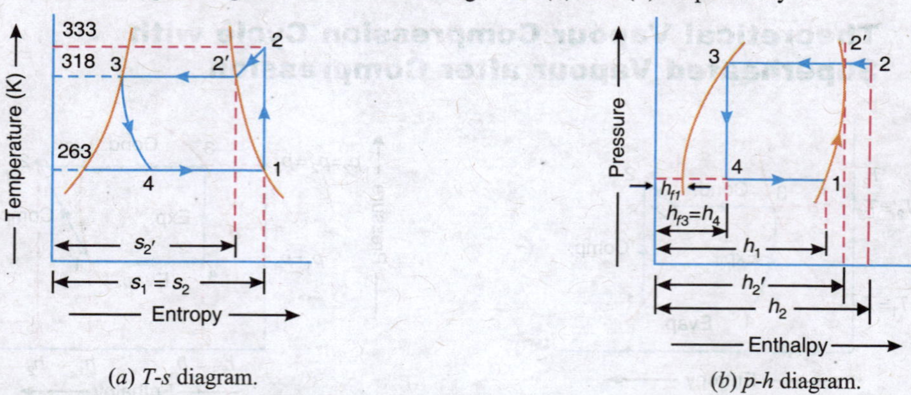

(*a*) *T-s* diagram. (*b*) *p-h* diagram.

Fig. 4.11

Let c_p = Specific heat at constant pressure for superheated vapour.

We know that entropy at point 2,

$$s_2 = s_{2'} + 2.3\, c_p \log\left(\frac{T_2}{T_{2'}}\right)$$

$$1.637 = 1.587 + 2.3\, c_p \log\left(\frac{333}{318}\right)$$

$$= 1.587 + 2.3\, c_p \times 0.02 = 1.587 + 0.046\, c_p$$

$$\therefore \quad c_p = 1.09$$

* Superfluous data.

and enthalpy at point 2, $h_2 = h_{2'} + c_p \times \text{Degree of superheat} = h_{2'} + c_p\,(T_2 - T_{2'})$

$= 483.6 + 1.09\,(333 - 318) = 500$ kJ/kg

∴ C.O.P. of the refrigerator

$$= \frac{h_1 - h_{f3}}{h_2 - h_1} = \frac{460.7 - 133}{500 - 460.7} = 8.34 \text{ Ans.}$$

Example 4.8. *A simple refrigerant 134a (tetrafluroethane) heat pump for space heating, operates between temperature limits of 15°C and 50°C. The heat required to be pumped is 100 MJ/h. Determine : 1. The dryness fraction of refrigerant entering the evaporator; 2. The discharge temperature assuming the specific heat of vapour as 0.996 kJ/kg K; 3. The theoretical piston displacement of the compressor; 4. The theoretical power of the compressor; and 5. The C.O.P.*

The specific volume of refrigerant 134a saturated vapour at 15°C is 0.04185 m³/kg. The other relevant properties of R-134a are given below:

Saturation temperature (°C)	*Pressure (bar)*	*Specific enthalpy (kJ/kg)*		*Specific entropy (kJ/kg K)*	
		Liquid	*Vapour*	*Liquid*	*Vapour*
15	*4.887*	*220.26*	*413.6*	*1.0729*	*1.7439*
50	*13.18*	*271.97*	*430.4*	*1.2410*	*1.7312*

Solution: Given: $T_1 = T_4 = 15°C = 15 + 273 = 288$ K ; $T_{2'} = T_3 = 50°C = 50 + 273 = 323$ K ; $Q = 100$ MJ/h $= 100 \times 10^3$ kJ/h ; $c_p = 0.996$ kJ/kg K ; $v_1 = 0.04185$ m³/kg ; $h_{f1} = 220.26$ kJ/kg ; $h_{f3} = h_4 = 271.97$ kJ/kg ; $h_1 = 413.6$ kJ/kg ; $h_{2'} = 430.4$ kJ/kg ; $s_{f1} = 1.0729$ kJ/kg K ; $s_1 = s_2 = 1.7439$ kJ/kg K ; $s_{f3} = 1.2410$ kJ/kg K ; $s_{2'} = 1.7312$ kJ/kg K

The *T-s* and *p-h* diagrams are shown in Fig. 4.12 (*a*) and (*b*) respectively.

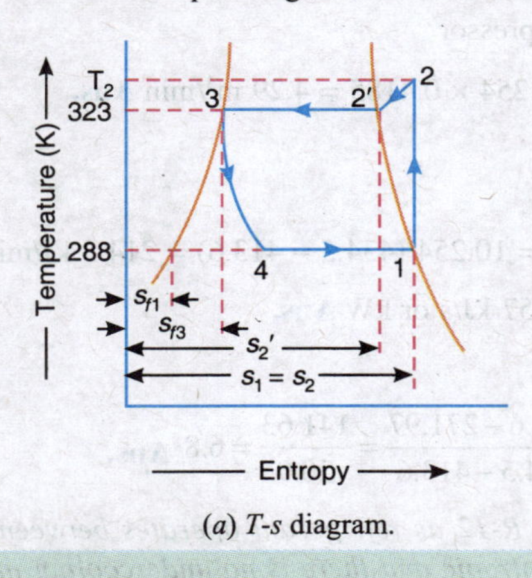

(*a*) *T-s* diagram.

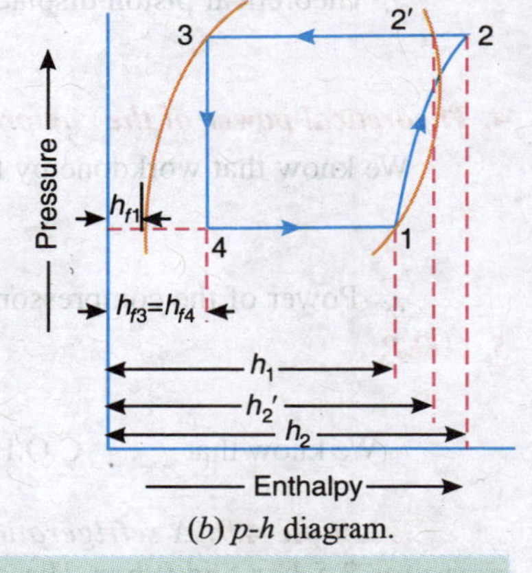

(*b*) *p-h* diagram.

Fig. 4.12

1. Dryness fraction of refrigerant entering the evaporator

We know that dryness fraction of refrigerant entering the evaporator *i.e.* at point 4,

$$x_4 = \frac{h_4 - h_{f1}}{h_1 - h_{f1}} = \frac{271.97 - 220.26}{413.6 - 220.26} = \frac{51.71}{193.34} = 0.2675 \text{ Ans.}$$

2. Discharge temperature

Let T_2 = Discharge temperature.

We know that entropy at discharge *i.e.* at point 2,

$$s_2 = s_{2'} + 2.3\, c_p \log\left(\frac{T_2}{T_{2'}}\right)$$

$$1.7439 = 1.7312 + 2.3 \times 0.996 \log\left(\frac{T_2}{T_{2'}}\right)$$

$$\log\left(\frac{T_2}{T_{2'}}\right) = \frac{1.7439 - 1.7312}{2.3 \times 0.996} = 0.005\,54$$

$$\frac{T_2}{T_{2'}} = 1.0128 \quad \text{...(Taking antilog of 0.005 54)}$$

$$T_2 = T_{2'} \times 1.0128 = 323 \times 1.0128 = 327.13 \text{ K} = 54.13°\text{C}$$ **Ans.**

3. Theoretical piston displacement of the compressor

We know that enthalpy at discharge *i.e.* at point 2,

$$h_2 = h_{2'} + c_p (T_2 - T_{2'})$$

$$= 430.4 + 0.996\,(327.13 - 323) = 434.5 \text{ kJ/kg}$$

and mass flow rate of the refrigerant,

$$m_R = \frac{Q}{h_2 - h_{f3}} = \frac{100 \times 10^3}{434.5 - 271.97} = 615.3 \text{ kg/h} = 10.254 \text{ kg/min}$$

∴ Theoretical piston displacement of the compressor

$$= m_R \times v_1 = 10.254 \times 0.4185 = 4.29 \text{ m}^3/\text{min}$$ **Ans.**

4. Theoretical power of the compressor

We know that workdone by the compressor

$$= m_R (h_2 - h_1) = 10.254\,(434.5 - 413.6) = 214.3 \text{ kJ/min}$$

∴ Power of the compressor = 214.3/60 = 3.57 kJ/s or kW **Ans.**

5. C.O.P.

We know that
$$\text{C.O.P.} = \frac{h_1 - h_{f3}}{h_2 - h_1} = \frac{413.6 - 271.97}{434.5 - 413.6} = \frac{141.63}{20.9} = 6.8$$ **Ans.**

Example 4.9. *A refrigeration machine using R-12 as refrigerant operates between the pressures 2.5 bar and 9 bar. The compression is isentropic and there is no undercooling in the condenser.*

The vapour is in dry saturated condition at the beginning of the compression. Estimate the theoretical coefficient of performance. If the actual coefficient of performance is 0.65 of theoretical value, calculate the net cooling produced per hour. The refrigerant flow is 5 kg per minute. Properties of refrigerant are :

Pressure, bar	Saturation temperature, °C	Enthalpy, kJ/kg		Entropy of saturated vapour, kJ/kg K
		Liquid	Vapour	
9.0	36	70.55	201.8	0.6836
2.5	–7	29.62	184.5	0.7001

Take c_p for superheated vapour at 9 bar as 0.64 kJ/kg K.

Solution. Given : $T_{2'} = T_3 = 36°C = 36 + 273 = 309$ K ; $T_1 = T_4 = -7°C = -7 + 273 = 266$ K ; $(C.O.P.)_{actual} = 0.65\ (C.O.P.)_{th}$; $m = 5$ kg /min ; $h_{f3} = h_4 = 70.55$ kJ / kg ; *$h_{f1} = h_{f4} = 29.62$ kJ/kg ; $h_2' = 201.8$ kJ / kg ; $h_1 = 184.5$ kJ/kg ; $s_2' = 0.6836$ kJ/kg K ; $s_1 = s_2 = 0.7001$ kJ / kg K ; $c_p = 0.64$ kJ / kg K

The *T-s* and *p-h* diagrams are shown in Fig. 4.13 (*a*) and (*b*) respectively.

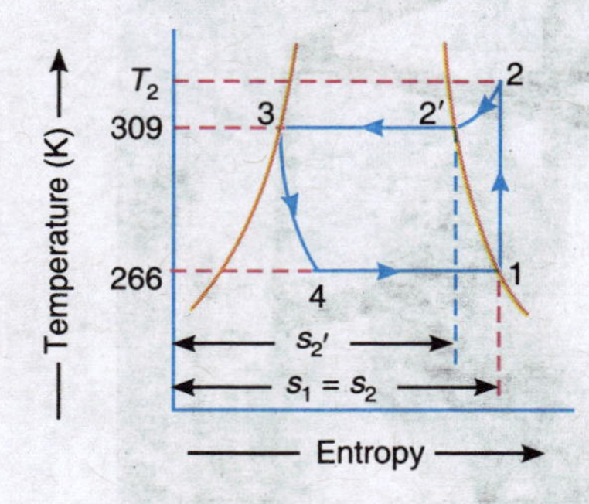

(*a*) *T-s* diagram.

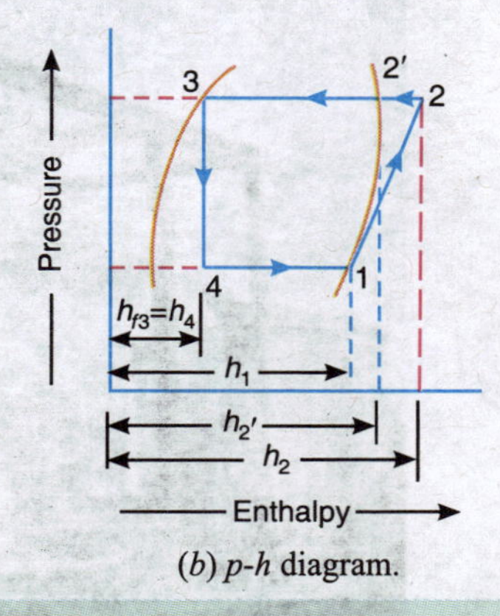

(*b*) *p-h* diagram.

Fig. 4.13

Theoretical coefficient of performance

First of all, let us find the temperature at point 2 (T_2).

We know that entropy at point 2,

$$s_2 = s_{2'} + 2.3\ c_p\ \log\left(\frac{T_2}{T_{2'}}\right)$$

$$0.7001 = 0.6836 + 2.3 \times 0.64\ \log\left(\frac{T_2}{309}\right)$$

$$\log\left(\frac{T_2}{309}\right) = \frac{0.7001 - 0.6836}{2.3 \times 0.64} = 0.0112$$

$$\frac{T_2}{309} = 1.026 \qquad \text{... (Taking antilog of 0.0112)}$$

$$\therefore \quad T_2 = 1.026 \times 309 = 317 \text{ K}$$

We know that enthalpy of superheated vapour at point 2,

$$h_2 = h_{2'} + c_p\ (T_2 - T_{2'})$$

$$= 201.8 + 0.64\ (317 - 309) = 206.92 \text{ kJ/kg}$$

* Superfluous data.

∴ Theoretical coefficient of performance,

$$(\text{C.O.P.})_{th} = \frac{h_1 - h_{f3}}{h_2 - h_1} = \frac{184.5 - 70.55}{206.92 - 184.5} = 5.1 \text{ Ans.}$$

Net cooling produced per hour

We also know that actual C.O.P. of the machine,

$$(\text{C.O.P.})_{actual} = 0.65 \times (\text{C.O.P.})_{th} = 0.65 \times 5.1 = 3.315$$

and actual work done, $w_{actual} = h_2 - h_1 = 206.92 - 184.5 = 22.42$ kJ/kg

Refrigeration machine.

We know that net cooling (or refrigerating effect) produced per kg of refrigerant

$$= w_{actual} \times (\text{C.O.P.})_{actual} = 22.42 \times 3.315 = 74.3 \text{ kJ/kg}$$

∴ Net cooling produced per hour

$$= m \times 74.3 = 5 \times 74.3 = 371.5 \text{ kJ/min}$$

$$= \frac{371.5}{210} = 1.77 \text{ TR Ans.} \qquad \ldots (\because 1 \text{ TR} = 210 \text{ kJ/min})$$

Example 4.10. *A simple saturation cycle using R-12 is designed for taking a load of 10 tonnes. The refrigerator and ambient temperature are –0°C and 30°C respectively. A minimum temperature difference of 5°C is required in the evaporator and condenser for heat transfer. Find : 1. Mass flow rate through the system ; 2. Power required in kW ; 3. C.O.P. ; and 4. Cylinder dimensions assuming L / D = 1.2, for a single cylinder, single acting compressor if it runs at 300 r.p.m. with volumetric efficiency of 90%.*

Solution. Given : $Q = 10\text{TR} = 10 \times 210 = 2100$ kJ/min

Since a minimum temperature difference of 5°C is required in the evaporator and condenser, therefore evaporator temperature would be

$$T_1 = T_4 = 0 - 5 = -5°C = -5 + 273 = 268 \text{ K}$$

and condenser temperature,

$$T_{2'} = T_3 = 30 + 5 = 35°C = 35 + 273 = 308 \text{ K}$$

The *T-s* and *p-h* diagrams are shown in Fig. 4.14 (*a*) and (*b*) respectively.

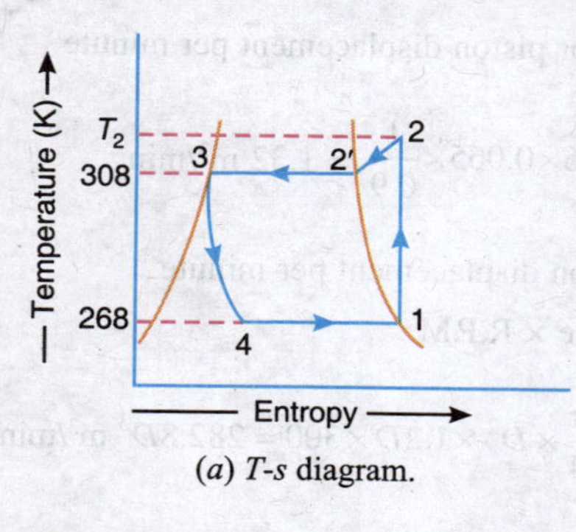

(*a*) *T-s* diagram.

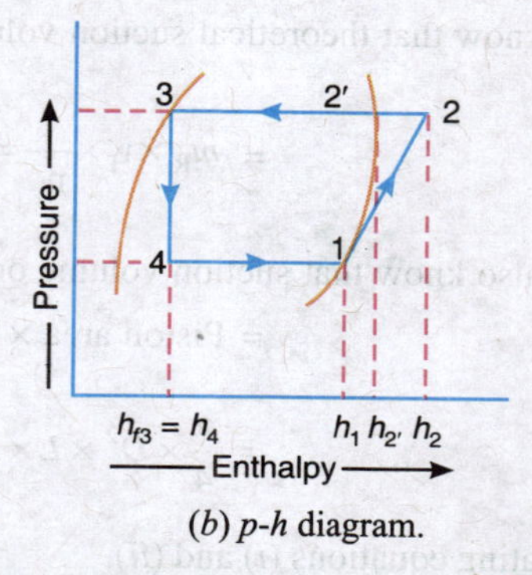

(*b*) *p-h* diagram.

Fig. 4.14

From *p-h* diagram, we find that enthalpy of dry saturated vapour at –5°C (268 K) *i.e.* at point 1,

$$h_1 = 185 \text{ kJ/kg}$$

Enthalpy of superheated vapour at point 2,

$$h_2 = 206 \text{ kJ/kg}$$

Enthalpy of saturated liquid at 35°C (308 K) *i.e.* at point 3,

$$h_{f3} = h_4 = 70 \text{ kJ/kg}$$

and specific volume of dry saturated vapour at –5°C (268 K) *i.e.* at point 1,

$$v_1 = 0.065 \text{ m}^3/\text{kg}$$

1. Mass flow rate through the system

We know that refrigerating effect per kg of the refrigerant

$$= h_1 - h_{f3} = 185 - 70 = 115 \text{ kJ/kg}$$

$$\therefore \quad \text{Mass flow rate, } m_R = \frac{\text{Refrigerating capacity}}{\text{Refrigerating effect}} = \frac{10 \times 210}{115} = 18.26 \text{ kg / min} \textbf{ Ans.}$$

2. Power required

We know that workdone during compression of the refrigerant

$$= m_R (h_2 - h_1) = 18.26 (206 - 185) = 383.46 \text{ kJ/min}$$

$$\therefore \quad \text{Power required} = 383.46/60 = 6.4 \text{ kJ/s or kW} \textbf{ Ans.}$$

3. C.O.P.

We know that $\quad \text{C.O.P.} = \dfrac{h_1 - h_{f3}}{h_2 - h_1} = \dfrac{185 - 70}{206 - 185} = \dfrac{115}{21} = 5.476$ **Ans.**

4. Cylinder dimensions

Let D = Bore of cylinder,

L = Length of stroke = 1.2 D ...(Given)

N = Speed of compressor = 300 r.p.m. ...(Given)

η_v = Volumetric efficiency = 90% = 0.9 ...(Given)

We know that theoretical suction volume or piston displacement per minute

$$= m_R \times v_1 \times \frac{1}{\eta_v} = 18.26 \times 0.065 \times \frac{1}{0.9} = 1.32 \text{ m}^3/\text{min} \quad ...(i)$$

We also know that suction volume or piston displacement per minute

= Piston area × Stroke × R.P.M.

$$= \frac{\pi}{4} \times D^2 \times L \times N = \frac{\pi}{4} \times D^2 \times 1.2D \times 300 = 282.8D^3 \text{ m}^3/\text{min} \quad ...(ii)$$

Equating equations (*i*) and (*ii*),

$D^3 = 1.32 / 282.8 = 4.667 \times 10^{-3}$ or $D = 0.167$m = 167 mm **Ans.**

and $L = 1.2\,D = 1.2 \times 167 = 200.4$ mm **Ans.**

Example 4.11. *A water cooler using R-12 works on the condensing and evaporating temperatures of 26°C and 2°C respectively. The vapour leaves the evaporator saturated and dry. The average output of cold water is 100 kg / h cooled from 26°C to 6°C.*

Allowing 20% of useful heat into water cooler and the volumetric efficiency of the compressor as 80% and mechanical efficiency of the compressor and the electric motor as 85% and 95% respectively, find :1. volumetric displacement of the compressor ; and 2. power of the motor. Data for R-12 is given below :

Temperature °C	*Pressure bar*	*Enthalpy, kJ/kg*		*Entropy, kJ/kg K*		*Specific heat kJ/kg K*		*Specific volume of vapour m³/kg*
		Liquid	*Vapour*	*Liquid*	*Vapour*	*Liquid*	*Vapour*	
26	6.69	60.64	198.10	0.2270	0.6865	0.996	0.674	0.026
2	3.297	37.92	188.39	0.1487	0.6956	1.067	0.620	0.052

Solution. Given : $T_2' = T_3 = 26°C = 26 + 273 = 299$ K ; $T_1 = T_4 = 2°C = 2 + 273 = 275$ K ; $m_w = 100$ kg/h ; $T_{w1} = 26°C = 26 + 273 = 299$ K ; $T_{w2} = 6°C = 6 + 273 = 279$ K ; $\eta_v = 80\% = 0.80$; $\eta_{m1} = 85\% = 0.85$; $\eta_{m2} = 95\% = 0.95$; $h_{f3} = 60.64$ kJ/kg ; $*h_{f1} = 37.92$ kJ/kg ; $h_2' = 198.10$ kJ/kg ; $h_1 = 188.39$ kJ/kg ; $*s_{f3} = 0.2270$ kJ/kg K ; $*s_{f1} = 0.1487$ kJ/kg K ; $*s_2' = 0.6865$ kJ/kg K ; $s_1 = s_2 = 0.6956$ kJ/kg K ; $*c_{p3} = 0.996$ kJ/kg K ; $*c_{p4} = 1.067$ kJ/kg K; $*c_{p2}' = 0.674$ kJ/kg K ; $c_{p1} = 0.620$ kJ/kg K ; $*v_2' = 0.026$ m³ / kg ; $v_1 = 0.052$ m³ / kg

* Superfluous data.

1. *Volumetric displacement of the compressor*

The *T-s* and *p-h* diagrams are shown in Fig. 4.15 (*a*) and (*b*) respectively. Since 20% of the useful heat is lost into water cooler, therefore actual heat extracted from the water cooler,

$$h_E = 1.2\, m_w \times c_w\, (T_{w1} - T_{w2})$$

$$= 1.2 \times 100 \times 4.187\,(299 - 279) = 10\,050 \text{ kJ/h} = 167.5 \text{ kJ/min}$$

... ($\because$ Sp. heat of water, $c_w = 4.187$ kJ/kg K)

We know that heat extracted or the net refrigerating effect per kg of the refrigerant

$$= h_1 - h_{f3} = 188.39 - 60.64 = 127.75 \text{ kJ/kg}$$

$\therefore$ Mass flow of the refrigerant,

$$m_R = \frac{167.5}{127.75} = 1.3 \text{ kg/min}$$

and volumetric displacement of the compressor

$$= \frac{m_R \times v_1}{\eta_v} = \frac{1.3 \times 0.052}{0.80} = 0.085 \text{ m}^3\text{/min} \textbf{ Ans.}$$

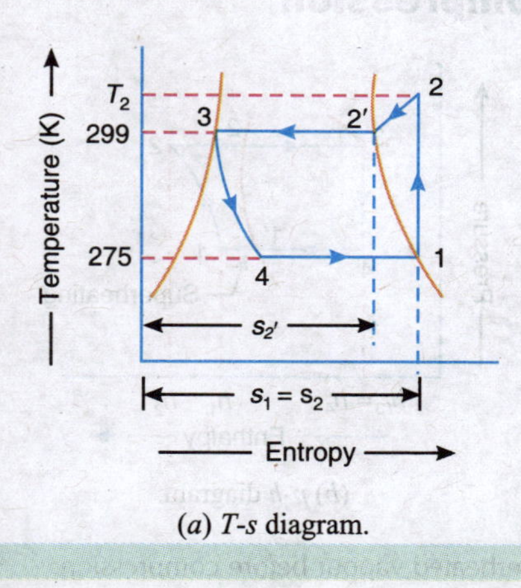

(*a*) *T-s* diagram.

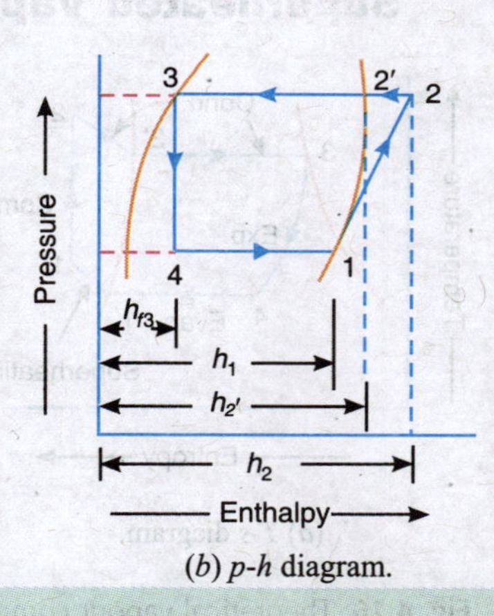

(*b*) *p-h* diagram.

Fig. 4.15

2. *Power of the motor*

First of all, let us find the temperature at point 2 (T_2). We know that entropy at point 2,

$$s_2 = s_{2'} + 2.3\, c_{p2'} \log\left(\frac{T_2}{T_{2'}}\right)$$

$$0.6956 = 0.6865 + 2.3 \times 0.674 \log\left(\frac{T_2}{299}\right)$$

$$\log\left(\frac{T_2}{299}\right) = \frac{0.6956 - 0.6865}{2.3 \times 0.674} = 0.005\,87$$

or $$\frac{T_2}{299} = 1.0136$$ ... (Taking anti-log of 0.005 87)

$$\therefore \quad T_2 = 299 \times 1.0136 = 303 \text{ K}$$

We know that enthalpy at point 2,

$$h_2 = h_{2'} + c_{p2'}(T_2 - T_{2'})$$
$$= 198.10 + 0.674\,(303 - 299) = 200.8 \text{ kJ/kg}$$

We also know that work done by the compressor per kg of the refrigerant

$$= h_2 - h_1 = 200.8 - 188.39 = 12.41 \text{ kJ/kg}$$

and work done per minute $= m_R \times 12.41 = 1.3 \times 12.41 = 16.133$ kJ/min

$$= 0.27 \text{ kJ/s} = 0.27 \text{ kW}$$

∴ Power required for the compressor

$$= \frac{0.27}{\eta_{m1}} = \frac{0.27}{0.85} = 0.317 \text{ kW}$$

and power of the motor $= \dfrac{0.317}{\eta_{m2}} = \dfrac{0.317}{0.95} = 0.334$ kW **Ans.**

4.9 Theoretical Vapour Compression Cycle with Superheated Vapour before Compression

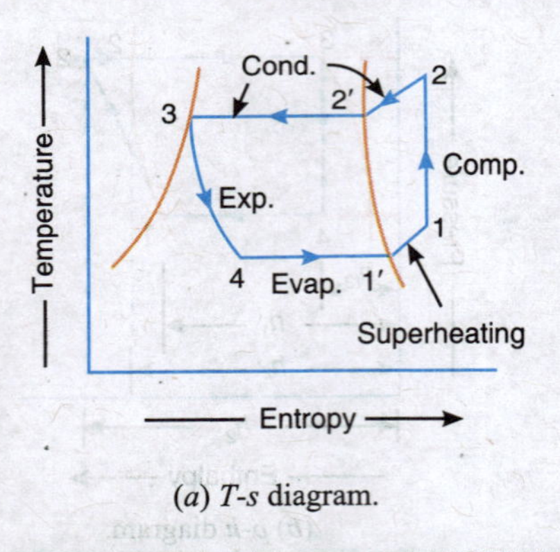

(*a*) *T-s* diagram.

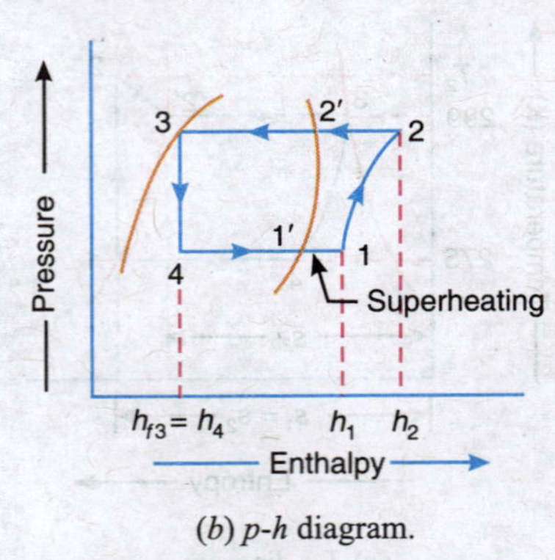

(*b*) *p-h* diagram.

Fig. 4.16. Theoretical vapour compression cycle with superheated vapour before compression.

A vapour compression cycle with superheated vapour before compression is shown on *T-s* and *p-h* diagrams in Fig. 4.16 (*a*) and (*b*) respectively. In this cycle, the evaporation starts at point 4 and continues upto point 1′, when it is dry saturated. The vapour is now superheated before entering the compressor upto the point 1.

The coefficient of performance may be found out as usual from the relation,

$$\text{C.O.P.} = \frac{\text{Refrigerating effect}}{\text{Work done}} = \frac{h_1 - h_{f3}}{h_2 - h_1}$$

Note: In this cycle, the heat is absorbed (or extracted) in two stages. Firstly from point 4 to point 1′ and secondly from point 1′ to point 1. The remaining cycle is same as discussed in the previous article.

Example 4.12. *A vapour compression refrigeration plant works between pressure limits of 5.3 bar and 2.1 bar. The vapour is superheated at the end of compression, its temperature being 37°C. The vapour is superheated by 5°C before entering the compressor.*

If the specific heat of superheated vapour is 0.63 kJ/kg K, find the coefficient of performance of the plant. Use the data given below :

Pressure, bar	*Saturation temperature, °C*	*Liquid heat, kJ/kg*	*Latent heat, kJ/kg*
5.3	15.5	56.15	144.9
2.1	–14.0	25.12	158.7

Solution. Given : p_2 = 5.3 bar ; p_1 = 2.1 bar ; T_2 = 37°C = 37 + 273 = 310 K ; $T_1 - T_1'$ = 5°C ; c_p = 0.63 kJ/kg K ; T_2' = 15.5°C = 15.5 + 273 = 288.5 K ; T_1' = – 14°C = – 14 + 273 = 259 K ; $h_{f3} = h_{f2}'$ = 56.15 kJ/kg ; h_{f1}' = 25.12 kJ/kg ; h_{fg2}' = 144.9 kJ/kg ; h_{fg1}' = 158.7 kJ/kg

The *T-s* and *p-h* diagrams are shown in Fig. 4.17 (*a*) and (*b*) respectively.

We know that enthalpy of vapour at point 1,

$$h_1 = h_1' + c_p (T_1 - T_1') = (h_{f1}' + h_{fg1}') + c_p (T_1 - T_1')$$
$$= (25.12 + 158.7) + 0.63 \times 5 = 186.97 \text{ kJ/kg}$$

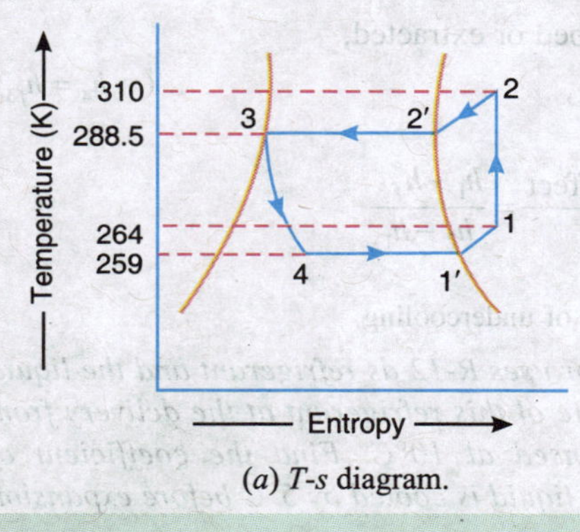

(*a*) *T-s* diagram.

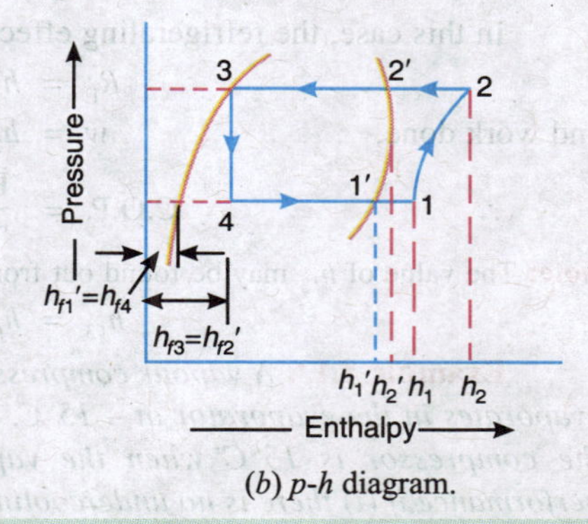

(*b*) *p-h* diagram.

Fig. 4.17

Similarly, enthalpy of vapour at point 2,

$$h_2 = h_2' + c_p (T_2 - T_2') = (h_{f2}' + h_{fg2}') + c_p (T_2 - T_2')$$
$$= (56.15 + 144.9) + 0.63 (310 - 288.5) = 214.6 \text{ kJ/kg}$$

∴ Coefficient of performance of the plant,

$$\text{C.O.P.} = \frac{h_1 - h_{f3}}{h_2 - h_1} = \frac{186.97 - 56.15}{214.6 - 186.97} = \frac{130.82}{27.63} = 4.735 \textbf{ Ans.}$$

4.10 Theoretical Vapour Compression Cycle with Under-cooling or Subcooling of Refrigerant

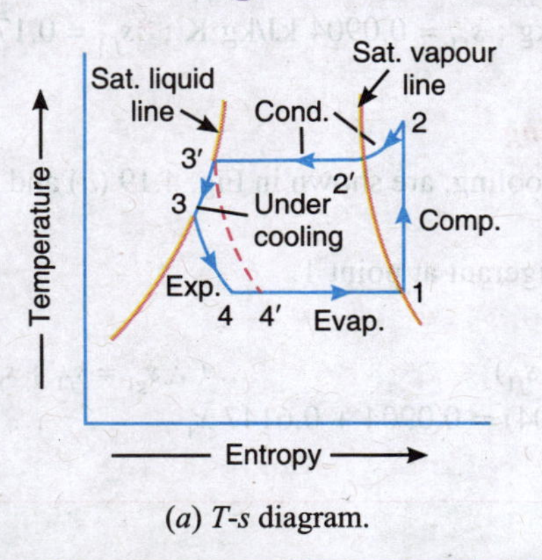

(*a*) *T-s* diagram.

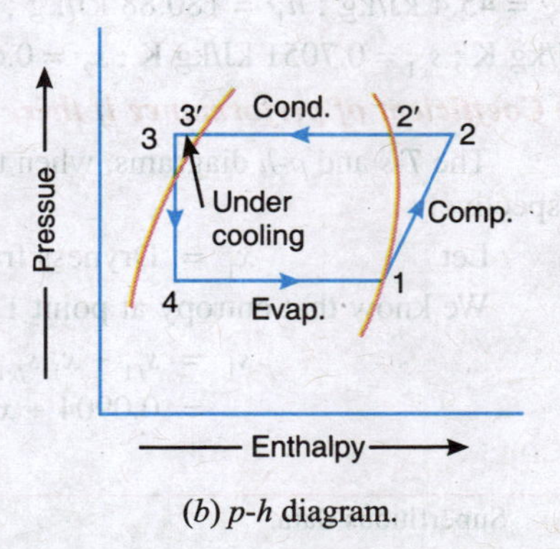

(*b*) *p-h* diagram.

Fig. 4.18. Theoretical vapour compression cycle with undercooling or subcooling of the refrigerant.

Sometimes, the refrigerant, after condensation process 2′-3′, is cooled below the saturation temperature (T_3') before expansion by throttling. Such a process is called *undercooling* or *subcooling* of the refrigerant and is generally done along the liquid line as shown in Fig. 4.18 (*a*) and (*b*). The ultimate effect of the undercooling is to increase the value of coefficient of performance under the same set of conditions.

The process of undercooling is generallly brought about by circulating more quantity of cooling water through the condenser or by using water colder than the main circulating water. Sometimes, this process is also brought about by employing a heat exchanger. In actual practice, the refrigerant is superheated after compression and undercooled before throttling, as shown in Fig. 4.18 (*a*) and (*b*). A little consideration will show, that the refrigerating effect is increased by adopting both the superheating and undercooling process as compared to a cycle without them, which is shown by dotted lines in Fig. 4.18 (*a*).

In this case, the refrigerating effect or heat absorbed or extracted,

$$R_E = h_1 - h_4 = h_1 - h_{f3} \quad \ldots (\because h_4 = h_{f3})$$

and work done,

$$w = h_2 - h_1$$

$$\therefore \quad \text{C.O.P.} = \frac{\text{Refrigerating effect}}{\text{Work done}} = \frac{h_1 - h_{f3}}{h_2 - h_1}$$

Note: The value of h_{f3} may be found out from the relation,

$$h_{f3} = h_{f3}' - c_p \times \text{Degree of undercooling}$$

Example 4.13. *A vapour compression refrigerator uses R-12 as refrigerant and the liquid evaporates in the evaporator at – 15°C. The temperature of this refrigerant at the delivery from the compressor is 15°C when the vapour is condensed at 10°C. Find the coefficient of performance if (i) there is no undercooling , and (ii) the liquid is cooled by 5°C before expansion by throttling.*

Take specific heat at constant pressure for the superheated vapour as 0.64 kJ/kg K and that for liquid as 0.94 kJ/kg K. The other properties of refrigerant are as follows :

Temperature in °C	*Enthalpy in kJ/ kg*		*Specific entropy in kJ / kg K*	
	Liquid	*Vapour*	*Liquid*	*Vapour*
–15	*22.3*	*180.88*	*0.0904*	*0.7051*
+10	*45.4*	*191.76*	*0.1750*	*0.6921*

Solution. Given : $T_1 = T_4 = -15°C = -15 + 273 = 258$ K ; $T_2 = 15°C = 15 + 273 = 288$ K ; $T_2' = 10°C = 10 + 273 = 283$ K ; $c_{pv} = 0.64$ kJ/kg K ; $c_{pl} = 0.94$ kJ/kg K ; $h_{f1} = 22.3$ kJ/kg ; $h_{f3}' = 45.4$ kJ/kg ; $h_1' = 180.88$ kJ/kg ; $h_2' = 191.76$ kJ/kg ; $s_{f1} = 0.0904$ kJ/kg K ; *$s_{f3} = 0.1750$ kJ/kg K ; $s_{g1} = 0.7051$ kJ/kg K ; $s_2' = 0.6921$ kJ/kg K

(i) Coefficient of performance if there is no undercooling

The *T-s* and *p-h* diagrams, when there is no undercooling, are shown in Fig. 4.19 (*a*) and (*b*) respectively.

Let x_1 = Dryness fraction of the refrigerant at point 1.

We know that entropy at point 1,

$$s_1 = s_{f1} + x_1 s_{fg1} = s_{f1} + x_1 (s_{g1} - s_{f1}) \quad \ldots (\because s_{g1} = s_{f1} + s_{fg1})$$

$$= 0.0904 + x_1 (0.7051 - 0.0904) = 0.0904 + 0.6147 x_1 \quad \ldots (i)$$

* Superfluous data.

and entropy at point 2,

$$s_2 = s_{2'} + 2.3\ c_{pv}\ \log\left(\frac{T_2}{T_{2'}}\right)$$

$$= 0.6921 + 2.3 \times 0.64\ \log\left(\frac{288}{283}\right)$$

$$= 0.6921 + 2.3 \times 0.64 \times 0.0077 = 0.7034 \qquad \ldots (ii)$$

Since the entropy at point 1 is equal to entropy at point 2, therefore equating equations (*i*) and (*ii*),

$$0.0904 + 0.6147\ x_1 = 0.7034 \quad \text{or} \quad x_1 = 0.997$$

(*a*) *T-s* diagram.

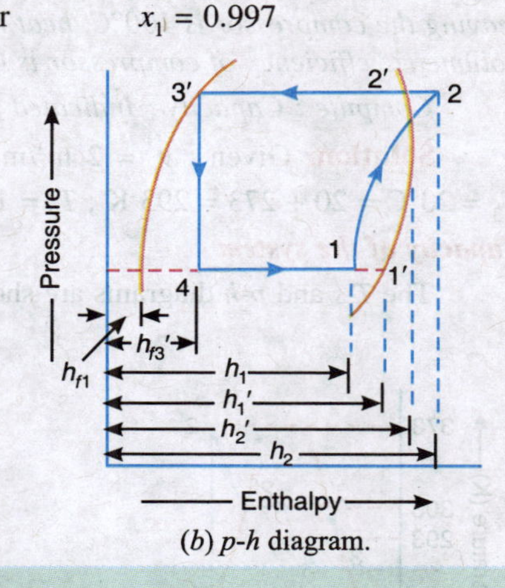

(*b*) *p-h* diagram.

Fig. 4.19

We know that the enthalpy at point 1,

$$h_1 = h_{f1} + x_1\ h_{fg1} = h_{f1} + x_1\ (h_{g1} - h_{f1})$$

$$= 22.3 + 0.997\ (180.88 - 22.3) = 180.4 \text{ kJ/kg}$$

$$\ldots (\because\ h_{g1} = h_1')$$

and enthalpy at point 2,

$$h_2 = h_{2'} + c_{pv}\ (T_2 - T_{2'})$$

$$= 191.76 + 0.64\ (288 - 283) = 194.96 \text{ kJ/kg}$$

$$\therefore \quad \text{C.O.P.} = \frac{h_1 - h_{f3'}}{h_2 - h_1} = \frac{180.4 - 45.4}{194.96 - 180.4} = 9.27 \ \textbf{Ans.}$$

(ii) Coefficient of performance when there is an undercooling of 5°C

The *T-s* and *p-h* diagrams, when there is an undercooling of 5°C, are shown in Fig. 4.20 (*a*) and (*b*) respectively.

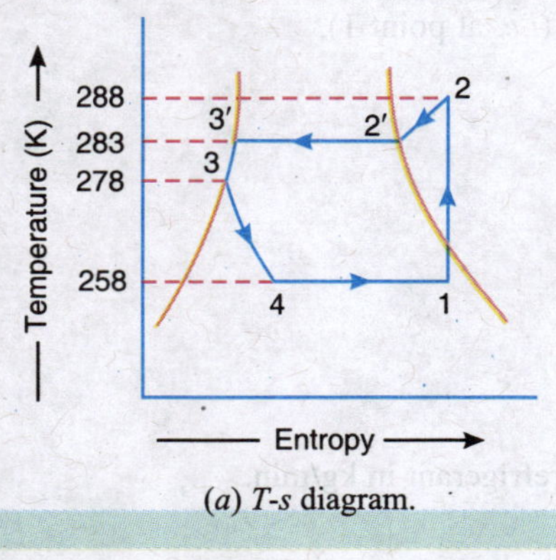

(*a*) *T-s* diagram.

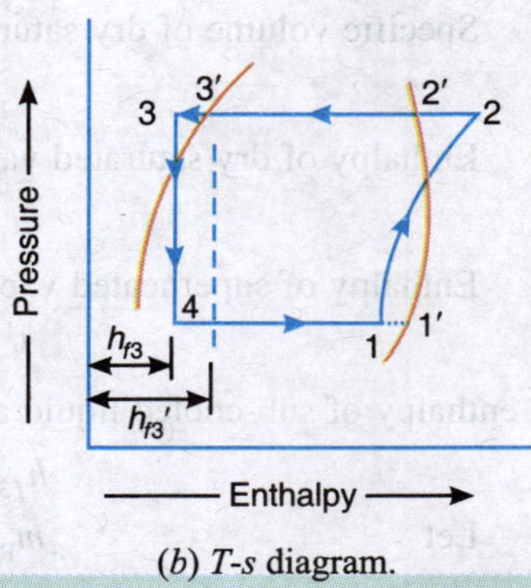

(*b*) *T-s* diagram.

Fig. 4.20

We know that enthalpy of liquid refrigerant at point 3,

$$h_{f3} = h_{f3'} - c_{pl} \times \text{Degree of undercooling}$$
$$= 45.4 - 0.94 \times 5 = 40.7 \text{ kJ/kg}$$

$$\therefore \quad \text{C.O.P.} = \frac{h_1 - h_{f3}}{h_2 - h_1} = \frac{180.4 - 40.7}{194.96 - 180.4} = 9.59 \textbf{ Ans.}$$

Example 4.14. *A simple NH_3 vapour compression system has compressor with piston displacement of 2 m^3/min, a condenser pressure of 12 bar and evaporator pressure of 2.5 bar. The liquid is sub-cooled to 20°C by soldering the liquid line to suction line. The temperature of vapour leaving the compressor is 100°C, heat rejected to compressor cooling water is 5000 kJ/hour, and volumetric efficiency of compressor is 0.8.*

Compute : Capacity ; Indicated power ; and C.O.P. of the system.

Solution. Given : $v_p = 2$ m³/min ; $p_2 = p_{2'} = p_{3'} = p_3 = 12$ bar ; $p_1 = p_4 = 2.5$ bar ; $T_3 = 20°C = 20 + 273 = 293$ K ; $T_2 = 100°C = 100 + 273 = 373$ K ; $\eta_v = 0.8$

Capacity of the system

The *T-s* and *p-h* diagrams are shown in Fig. 4.21 (*a*) and (*b*) respectively.

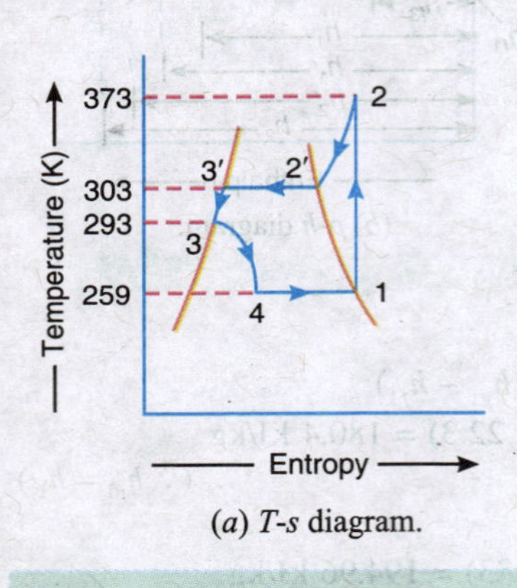

(*a*) *T-s* diagram.

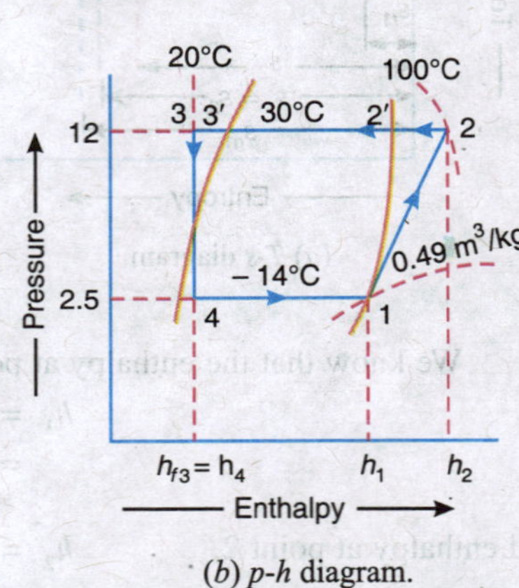

(*b*) *p-h* diagram.

Fig. 4.21

From *p-h* diagram, we find that the evaporating temperature corresponding to 2.5 bar is

$$T_1 = T_4 = -14°C = -14 + 273 = 259 \text{ K}$$

Condensing temperature corresponding to 12 bar is

$$T_{2'} = T_{3'} = 30°C = 30 + 273 = 303 \text{ K}$$

Specific volume of dry saturated vapour at 2.5 bar (*i.e.* at point 1),

$$v_1 = 0.49 \text{ m}^3/\text{kg}$$

Enthalpy of dry saturated vapour at point 1,

$$h_1 = 1428 \text{ kJ/kg}$$

Enthalpy of superheated vapour at point 2,

$$h_2 = 1630 \text{ kJ/kg}$$

and enthalpy of sub-cooled liquid at 20°C at point 3,

$$h_{f3} = h_4 = 270 \text{ kJ/kg}$$

Let $\quad m_R$ = Mass flow of the refrigerant in kg/min.

We know that piston displacement,

$$v_p = \frac{m_R \times v_1}{\eta_v} \quad \text{or} \quad m_R = \frac{v_p \times \eta_v}{v_1} = \frac{2 \times 0.8}{0.49} = 3.265 \text{ kg / min}$$

We know that refrigerating effect per kg of refrigerant

$$= h_1 - h_{f3} = 1428 - 270 = 1158 \text{ kJ/kg}$$

and total refrigerating effect $= m_R (h_1 - h_{f3}) = 3.265 (1428 - 270) = 3781$ kJ/min

∴ Capacity of the system $= 3781 / 210 = 18$ TR **Ans.**

Indicated power of the system

We know that work done during compression of the refrigerant

$$= m_R (h_2 - h_1) = 3.265 (1630 - 1428) = 659.53 \text{ kJ/min}$$

Heat rejected to compressor cooling water

$$= 5000 \text{ kJ/h} = 5000/60 = 83.33 \text{ kJ/min} \quad \text{... (Given)}$$

∴ Total work done by the system

$$= 659.53 + 83.33 = 742.86 \text{ kJ/min}$$

and indicated power of the system

$$= 742.86/60 = 12.38 \text{ kW} \quad \textbf{Ans.}$$

C.O.P. of the system

We know that C.O.P. of the system

$$= \frac{\text{Total refrigerating effect}}{\text{Total work done}} = \frac{3781}{742.86} = 5.1 \quad \textbf{Ans.}$$

Example 4.15. *Saturated ammonia at 2.5 bar enters a 160mm × 150mm (bore × stroke) twin cylinder, single acting compressor whose volumetric efficiency is 79% and speed is 250 r.p.m. The head pressure is 12 bar. The subcooled liquid ammonia at 22°C enters the expansion valve. For a standard refrigeration cycle, find: 1. The ammonia circulated in kg / min.; 2. The refrigeration in TR ; and 3. The C.O.P. of the refrigeration cycle. Refer to the following table for the properties of ammonia:*

Pressure (bar)	*Saturation temperature (°C)*	*Specific volume of vapour (m³/kg)*	*Specific enthalpy (kJ/kg)*		*Specific entropy (kJ/kg K)*	
			Liquid	*Vapour*	*Liquid*	*Vapour*
2.5	–15	0.5098	112.4	1426.58	0.4572	5.5497
12	30	0.1107	323.08	1468.87	1.2037	4.9842

Assume specific heat at constant pressure for liquid ammonia as 4.606 kJ/kg K and for superheated ammonia vapour as 2.763 kJ/kg K.

Solution. Given : $p_1 = p_4 = 2.5$ bar ; $D = 160$ mm $\doteq 0.16$ m ; $L = 150$ mm $= 0.15$ m ; No. of cylinders = 2 ; $\eta_v = 79\% = 0.79$; $N = 250$ r.p.m. ; $p_2 = p_3 = 12$ bar ; $T_3 = 22°C = 22 + 273 = 295$ K ; $T_1 = T_4 = -15°C = -15 + 273 = 258$ K ; $T_{2'} = T_{3'} = 30°C = 30 + 273 = 303$ K ; $v_1 = 0.5098$ m³/kg; * $v_{2'} = 0.1107$ m³/kg * $h_{f1} = 112.4$ kJ/kg ; $h_{f3'} = 323.08$ kJ/kg ; $h_1 = 1426.58$ kJ/kg ; $h_{2'} = 1468.87$ kJ/kg ; $s_{f1} = 0.4572$ kJ/kg K ; $s_{f3} = 1.2037$ kJ/kg K ; $s_1 = s_2 = 5.5497$ kJ/kg K ; $s_{2'} = 4.9842$ kJ/kg K ; $c_{pl} = 4.606$ kJ/kg K ; $c_{pv} = 2.763$ kJ/kg K

* Superfluous data.

The *T-s* and *p-h* diagrams are shown in Fig. 4.22 (*a*) and (*b*) respectively.

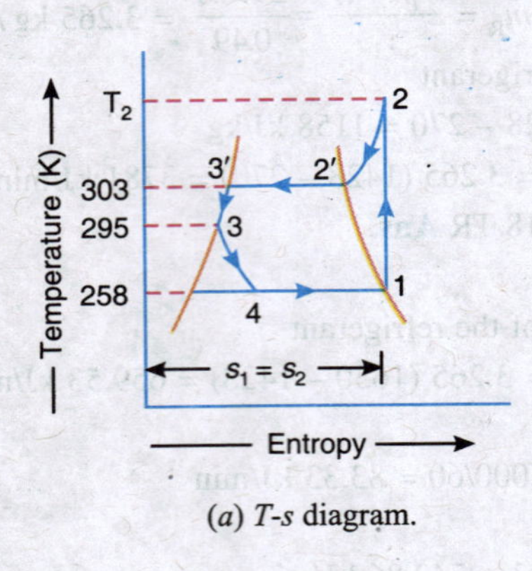

(*a*) *T-s* diagram.

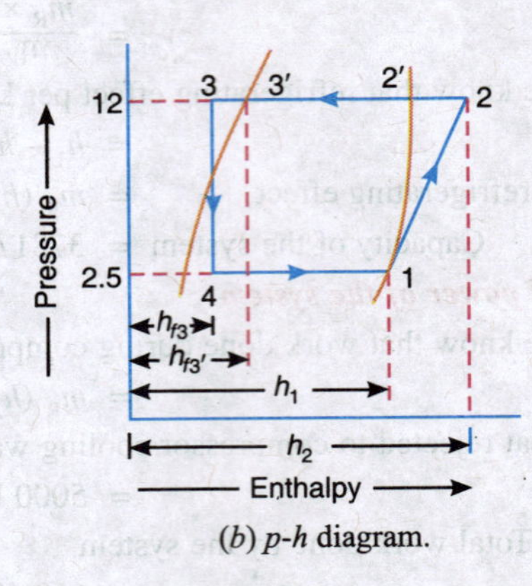

(*b*) *p-h* diagram.

Fig. 4.22

1. Ammonia circulated in kg/min

Let m_R = Mass flow rate of ammonia in kg/min.

We know that suction volume or piston displacement per minute

= Piston area × Stroke × R.P.M × No. of cylinders

$$= \frac{\pi}{4} \times D^2 \times L \times N \times 2$$

$$= \frac{\pi}{4}(0.16)^2 0.15 \times 250 \times 2 = 1.508 \text{ m}^3/\text{min} \quad ...(i)$$

We also know that piston displacement per minute

$$= m_R \times v_1 \times \frac{1}{\eta_v} = m_R \times 0.5098 \times \frac{1}{0.79} = 0.6453\, m_R \quad ...(ii)$$

Equating equations (*i*) and (*ii*),

m_R = 1.508 / 0.6453 = 2.34 kg / min **Ans.**

2. Refrigeration in TR

We know that enthalpy of liquid refrigerant at point 3,

$$h_{f3} = h_{f3'} - c_{pl}(T_{3'} - T_3)$$

= 323.08 – 4.606(303 – 295) = 286.23 kJ/kg

and total refrigeration effect = $m_R(h_1 - h_{f3})$

= 2.34 (1426.58 – 286.23) = 2668.4 kJ/min

∴ Refrigeration or capacity of the system

= 2668.4 / 210 = 12.7 TR **Ans.** ... (∵ 1TR = 210 kJ/min)

3. C.O.P. of the refrigeration cycle

First of all, let us find the temperature of the superheated vapour at point 2 (T_2). We know that entropy at point 2,

$$s_2 = s_{2'} + 2.3\, c_{pv} \log\left(\frac{T_2}{T_{2'}}\right)$$

$$5.5497 = 4.9842 + 2.3 \times 2.763 \log\left(\frac{T_2}{T_{2'}}\right)$$

$$\log\left(\frac{T_2}{T_{2'}}\right) = \frac{5.5497 - 4.9842}{2.3 \times 2.763} = 0.089$$

or $$\frac{T_2}{T_{2'}} = 1.227 \qquad \text{...(Taking antilog of 0.089)}$$

$\therefore$ $$T_2 = T_{2'} \times 1.227 = 303 \times 1.227 = 371.78 \text{ K}$$
$$= 371.78 - 273 = 98.78°\text{C}$$

We know that enthalpy at point 2,

$$h_2 = h_{2'} + c_{pv}(T_2 - T_{2'})$$
$$= 1468.87 + 2.763\,(371.78 - 303) = 1658.9 \text{ kJ/kg}$$

$\therefore$ C.O.P. of the refrigeration cycle

$$= \frac{h_1 - h_{f3}}{h_2 - h_1} = \frac{1426.58 - 286.23}{1658.9 - 1426.58} = \frac{1140.35}{232.32} = 4.91 \textbf{ Ans.}$$

Example 4.16. *A vapour compression refrigerator uses methyl chloride (R-40) and operates between pressure limits of 177.4 kPa and 967.5 kPa. At entry to the compressor, the methyl chloride is dry saturated and after compression has a temperature of 102°C. The compressor has a bore and stroke of 75 mm and runs at 8 rev/s with a volumetric efficiency of 80%. The temperature of the liquid refrigerant as it leaves the condenser is 35°C and its specific heat capacity is 1.624 kJ/kg K. The specific heat capacity of the superheated vapour may be assumed to be constant. Determine : 1. refrigerator C.O.P.; 2. mass flow rate of refrigerant; and 3. cooling water required by the condenser if its temperature rise is limited to 12°C. Specific heat capacity of water = 4.187 kJ/kg K.*

The relevant properties of methyl chloride are as follows :

Sat. temp. °C	*Pressure kPa*	*Specific volume m³/kg*		*Specific enthalpy kJ/kg*		*Specific entropy kJ/kg K*	
		Liquid	*Vapour*	*Liquid*	*Vapour*	*Liquid*	*Vapour*
– 10	*177.4*	*0.00102*	*0.233*	*45.38*	*460.76*	*0.183*	*1.762*
45	*967.5*	*0.00115*	*0.046*	*132.98*	*483.6*	*0.485*	*1.587*

Solution. Given : $p_1 = p_4 = 177.4$ kPa ; $p_2 = p_3 = 967.5$ kPa ; $T_2 = 102°\text{C} = 102 + 273 = 375$ K ; $D = L = 75$ mm $= 0.075$ m ; $N = 8$ r.p.s. $= 480$ r.p.m. ; $\eta_v = 80\% = 0.8$; $T_3 = 35°\text{C} = 35 + 273 = 308$ K ; $c_{pl} = c_{pv} = 1.624$ kJ/kg K ; $c_{pw} = 4.187$ kJ/kg K ; $T_1 = T_4 = -10°\text{C} = -10 + 273 = 263$ K ; $T_2' = T_3' = 45°\text{C} = 45 + 273 = 318$ K ; $v_1 = 0.233$ m³/kg ; $v_2' = 0.046$ m³/kg ; $^*h_{f1} = 45.38$ kJ/kg ; $h_{f3}' = 132.98$ kJ/kg ; $h_1 = 460.76$ kJ/kg ; $h_2' = 483.6$ kJ/kg ; $s_{f1} = 0.183$ kJ/kg K ; $^*s_{f3} = 0.485$ kJ/kg K ; $^*s_1 = s_2 = 1.762$ kJ/kg K ; $s_2' = 1.587$ kJ/kg K

* Superfluous data.

The *T-s* and *p-h* diagrams are shown in Fig. 4.23 (*a*) and (*b*) respectively.

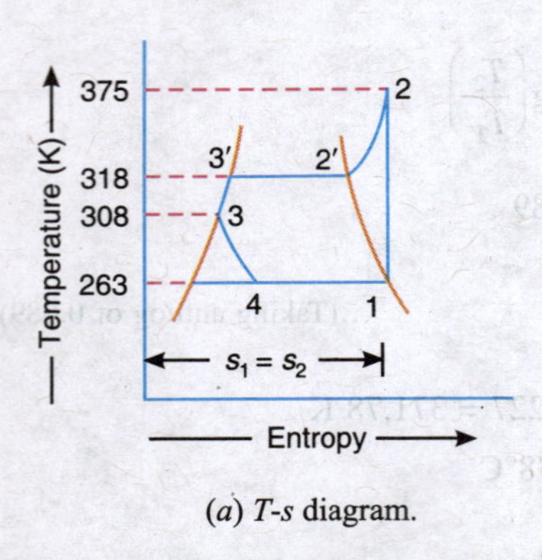

(*a*) *T-s* diagram.

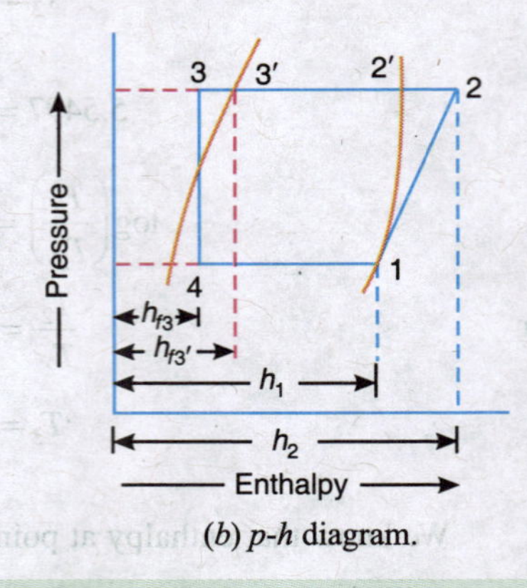

(*b*) *p-h* diagram.

Fig. 4.23

1. *Refrigerator C.O.P.*

We know that enthalpy at point 2,

$$h_2 = h_{2'} + c_{pv}(T_2 - T_2') = 483.6 + 1.624\,(375 - 318) = 576.2 \text{ kJ/kg}$$

and enthalpy of liquid refrigerant at point 3,

$$h_{f3} = h_{f3'} - c_{pl}(T_{3'} - T_3) = 132.98 - 1.624\,(318 - 308) = 116.74 \text{ kJ/kg}$$

We know that refrigerator C.O.P.

$$= \frac{h_1 - h_{f3}}{h_2 - h_1} = \frac{460.76 - 116.74}{576.2 - 460.76} = 2.98 \text{ Ans.}$$

2. *Mass flow rate of refrigerant*

Let m_R = Mass flow rate of refrigerant in kg / min.

We know that suction volume or piston displacement per minute,

$$= \text{Piston area} \times \text{Stroke} \times \text{R.P.M.}$$

$$= \frac{\pi}{4}(0.075)^2\ 0.075 \times 480 = 0.16 \text{ m}^3\text{/min} \qquad \ldots (i)$$

We also know that piston displacement per minute

$$= m_R \times v_1 \times \frac{1}{\eta_v} = m_R \times 0.233 \times \frac{1}{0.8} = 0.29\ m_R \qquad \ldots (ii)$$

Equating equations (*i*) and (*ii*),

$$m_R = 0.16 / 0.29 = 0.55 \text{ kg / min } \textbf{Ans.}$$

3. *Cooling water required by the condenser*

Let m_w = Cooling water required by the condenser in kg / min.

We know that heat given out by the refrigerant in the condenser

$$= m_R\,(h_2 - h_{f3}) = 0.55\,(576.2 - 116.74) \text{ kg / min}$$

$$= 252.7 \text{ kJ/min} \qquad \ldots (iii)$$

and heat taken by water in the condenser

$$= m_w \times c_{pw} \times \text{Rise in temperature}$$

$$= m_w \times 4.187 \times 12 = 50.244\ m_w \quad \dots (iv)$$

Equating equations (*iii*) and (*iv*),

$$m_w = 252.7/50.244 = 5.03 \text{ kg / min} \quad \textbf{Ans.}$$

Example 4.17. *The following data refer to a single stage vapour compression system:*

Refrigerant used (Ozone friendly) R-134a; Condensing temperature = 35°C; Evaporator temperature = –10°C; Compressor R.P.M. = 2800; Clearance volume/Swept volume = 0.03; Swept volume = 269.4 × 10⁻⁶ m³; Expansion index = 1.12; Compression efficiency = 0.8; Condensate subcooling = 5°C.

Find : 1. Capacity of the system in TR ; 2. Power required ; 3. C.O.P.; 4. Heat rejection to condenser ; and 5. Refrigeration efficiency.

The properties of R-134a are given below:

Sat.temp. °C	*Pressure bar*	*Specific volume of vapour, m³/kg*	*Specific enthalpy kJ/kg*		*Specific entropy kJ/kg K*	
			Liquid	*Vapour*	*Liquid*	*Vapour*
–10	*2.014*	*0.0994*	*186.7*	*392.4*	*0.9512*	*1.733*
35	*8.870*	*—*	*249.1*	*417.6*	*1.1680*	*1.715*

Assume isentropic compression and suction vapour as dry saturated. The specific heat of vapour refrigerant may be taken as 1.1 kJ/kg K and for liquid refrigerant as 1.458 kJ/kg K.

Solution. Given : $T_{2'} = T_{3'} = 35°C = 35 + 273 = 308$ K ; $T_1 = T_4 = -10°C = -10 + 273 = 263$ K ; $N = 2800$ r.p.m.; $v_c / v_s = C = 0.03$; $v_s = 269.4 \times 10^{-6}$ m³ ; $n = 1.12$; $\eta_c = 0.8$; $T_3 = 35 - 5 = 30°C = 30 + 273 = 303$ K ; $p_1 = p_4 = 2.014$ bar ; $p_{2'} = p_{3'} = 8.870$ bar ; $v_1 = 0.0994$ m³/kg ; $^*h_{f1} = 186.7$ kJ/kg ; $h_{f3'} = 249.1$ kJ/kg ; $h_1 = 392.4$ kJ/kg ; $h_{2'} = 417.6$ kJ/kg ; $s_{f1} = 0.9512$ kJ/kg K ; $^*s_{f3} = 1.1680$ kJ/kg K ; $s_1 = s_2 = 1.733$ kJ/kg K ; $s_{2'} = 1.715$ kJ/kg K ; $c_{pv} = 1.1$ kJ/kg K ; $c_{pl} = 1.458$ kJ/kg K

The *T-s* and *p-h* diagrams are shown in Fig. 4.24 (*a*) and (*b*) respectively.

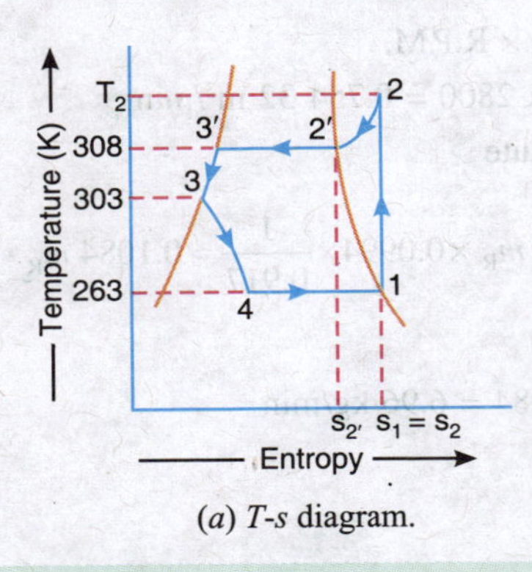

(*a*) *T-s* diagram.

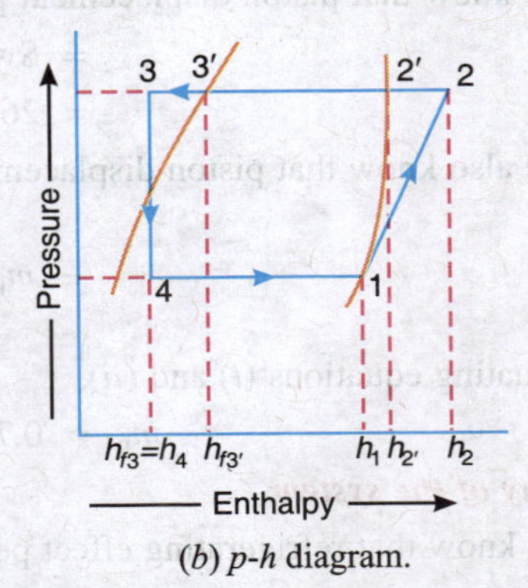

(*b*) *p-h* diagram.

Fig. 4.24

* Superfluous data.

First of all, let us find the temperature of superheated vapour at point 2 (T_2). We know that entropy at point 2,

$$s_2 = s_{2'} + 2.3\, c_{pv} \log\left(\frac{T_2}{T_{2'}}\right)$$

$$1.733 = 1.715 + 2.3 \times 1.1 \log\left(\frac{T_2}{308}\right)$$

$$\log\left(\frac{T_2}{308}\right) = \frac{1.733 - 1.715}{2.3 \times 1.1} = 0.007\,114$$

$$\frac{T_2}{308} = 1.0165 \qquad \text{... (Taking antilog of 0.007 114)}$$

$$\therefore \quad T_2 = 1.0165 \times 308 = 313.08 \text{ K}$$

We know that enthalpy at point 2,

$$h_2 = h_{2'} + c_{pv}(T_2 - T_{2'})$$

$$= 417.6 + 1.1(313.08 - 308) = 423.2 \text{ kJ/kg}$$

and enthalpy of liquid refrigerant at point 3,

$$h_{f3} = h_{f3'} - c_{pl}(T_{3'} - T_3)$$

$$= 249.1 - 1.458\,(308 - 303) = 241.81 \text{ kJ/kg}$$

We know that volumetric efficiency of the compressor,

$$\eta_v = 1 + C - C\left(\frac{p_2}{p_1}\right)^{1/n} = 1 + 0.03 - 0.03\left[\frac{8.87}{2.014}\right]^{1/1.12}$$

$$= 1.03 - 0.113 = 0.917$$

Let m_R = Mass flow rate of the refrigerant in kg / min.

We know that piston displacement per minute

= Swept volume × R.P.M.

$$= 269.4 \times 10^{-6} \times 2800 = 0.754\,32 \text{ m}^3\text{/min} \qquad \text{...}(i)$$

We also know that piston displacement per minute

$$= m_R \times v_1 \times \frac{1}{\eta_v} = m_R \times 0.0994 \times \frac{1}{0.917} = 0.1084\, m_R \qquad \text{...}(ii)$$

Equating equations (*i*) and (*ii*),

$$m_R = 0.75432 / 0.1084 = 6.96 \text{ kg/min}$$

1. Capacity of the system

We know that refrigerating effect per minute

$$= m_R\,(h_1 - h_{f3}) = 6.96\,(392.4 - 241.81) = 1048 \text{ kJ/min}$$

∴ Capacity of the system

$= 1048 / 210 = 4.991$ TR **Ans.** ... ($\because$ 1 TR = 210 kJ/min)

2. Power required

We know that workdone during compression of the refrigerant

$= m_R (h_2 - h_1) = 6.96 (423.2 - 392.4) = 214.47$ kJ/min

∴ Power required $= 214.47/60 = 3.57$ kJ/s or kW **Ans.**

3. C.O.P.

We know that
$$\text{C.O.P.} = \frac{\text{Refrigerating effect}}{\text{Workdone}} = \frac{1048}{214.4} = 4.89 \text{ Ans.}$$

4. Heat rejected to condenser

We know that heat rejected to condenser

$= m_R (h_2 - h_{f3})$

$= 6.96 (423.2 - 241.81) = 1262.47$ kJ/min **Ans.**

Note: The heat rejected to condenser in also equal to refrigerating effect + workdone (1048 + 214.47 = 1262.47 kJ/min).

5. Refrigeration efficiency

We know that C.O.P of the Carnot cycle,
$$(\text{C.O.P.})_{\text{Carnot}} = \frac{T_1}{T_{2'} - T_1} = \frac{263}{308 - 263} = 5.844$$

$$\therefore \text{ Refrigeration efficiency} = \frac{(\text{C.O.P.})_{\text{Cycle}}}{(\text{C.O.P.})_{\text{Carnot}}} = \frac{4.89}{5.844} = 0.8367 \text{ or } 83.67\% \text{ Ans.}$$

Example 4.18. *A commercial refrigerator operates with R-12 between 1.2368 bar and 13.672 bar. The vapour is dry and saturated at the compressor inlet. Assuming isentropic compression, determine the theoretical C.O.P. of the plant. The isentropic discharge temperature is 64.86°C. If the actual C.O.P. of the plant is 80% of the theoretical, calculate the power required to run the compressor to obtain a refrigerating capacity of 1 TR. If the liquid is sub-cooled through 10°C after condensation, calculate the power required. The properties of R-12 are given below :*

Saturation temp. (°C)	*Saturation pressure (bar)*	*Enthalpy (kJ/kg)*		*Entropy (kJ/kg K)*	
		Liquid	*Vapour*	*Liquid*	*Vapour*
–25	1.2368	13.33	176.48	0.0552	0.7126
55	13.672	90.28	207.95	0.3197	0.6774

Properties of superheated R-12

Temperature (°C)	*Pressure (bar)*	*Enthalpy (kJ/kg)*	*Entropy (kJ/kg K)*
64.86	13.672	220.6	0.7126

Assume specific heat of liquid to be 1.055 kJ/kg K.

Solution. Given : $p_1 = p_4 = 1.2368$ bar ; $p_2 = p_3 = 13.672$ bar ; $T_2 = 64.86°C = 64.86 + 273 = 337.86$ K ; $(C.O.P.)_{actual} = 80\%\ (C.O.P)_{th}$; Q = 1TR ; $T_1 = -25°C = -25 + 273 = 248$ K ; $T_{2'} = 55°C = 55 + 273 = 328$ K ; $^*h_{f1} = 13.33$ kJ/kg ; $h_1 = 176.48$ kJ/kg ; $^*s_{f1} = 0.0552$ kJ/kg K ; $^*s_1 = s_2 = 0.7126$ kJ/kg K ; $h_{f3'} = 90.28$ kJ/kg ; $^*h_{2'} = 207.95$ kJ/kg K ; $^*s_{f3} = 0.3197$ kJ/kg K ; $^*s_{2'} = 0.6774$ kJ/kg K ; $h_2 = 220.6$ kJ/kg ; $c_{pl} = 1.055$ kJ/kg K

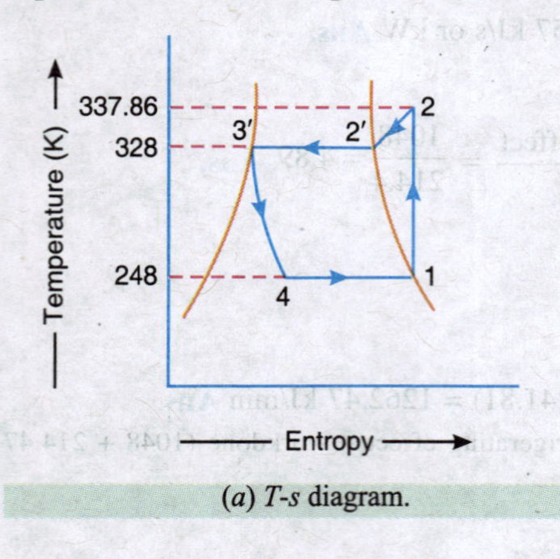

(a) T-s diagram.

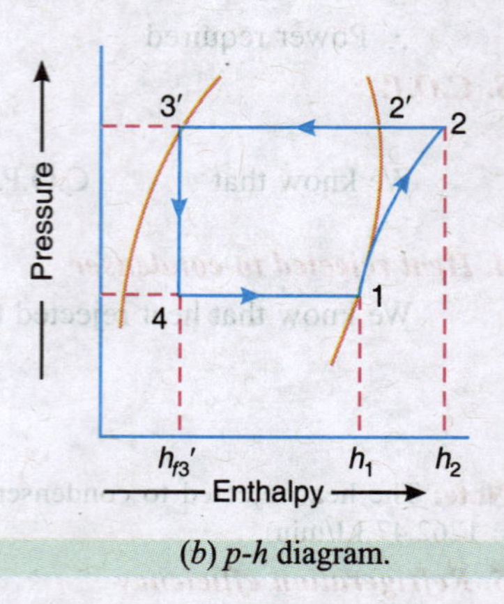

(b) p-h diagram.

The *T-s* and *p-h* diagrams are shown in Fig. 4.25 (*a*) and (*b*) respectively.

Theoretical C.O.P. of the plant

We know that theoretical C.O.P. of the plant

$$= \frac{h_1 - h_{f3'}}{h_2 - h_1} = \frac{176.48 - 90.28}{220.6 - 176.48} = 1.95 \textbf{ Ans.}$$

Power required to run the compressor

Since the actual C.O.P. of the plant is 80% of the theoretical, therefore,

$$(C.O.P.)_{actual} = 0.8 \times 1.95 = 1.56$$

and actual work done by the compressor,

$$w_{actual} = h_2 - h_1 = 220.6 - 176.48 = 44.12 \text{ kJ/kg}$$

∴ Net refrigerating effect produced per kg of refrigerant

$$= w_{actual} \times (C.O.P.)_{actual}$$
$$= 44.12 \times 1.56 = 68.83 \text{ kJ/kg}$$

We know that refrigerating capacity

$$= 1 \text{ TR} = 210 \text{ kJ/min}$$

∴ Mass flow of refrigerant,

$$m_R = 210/68.83 = 3.05 \text{ kg/min}$$

and work done during compression of the refrigerant

$$= m_R (h_2 - h_1) = 3.05\ (220.6 - 176.48) = 134.57 \text{ kJ/min}$$

∴ Power required to run the compressor

$$= 134.57/60 = 2.24 \text{ kW } \textbf{Ans.}$$

* Superfluous data.

Power required if the liquid is subcooled through 10°C

The *T-s* and *p-h* diagrams with subcooling of liquid are shown in Fig. 4.26 (*a*) and (*b*) respectively.

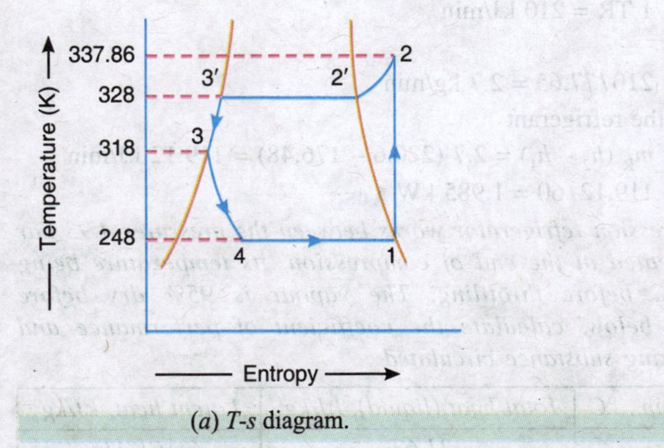

(*a*) *T-s* diagram.

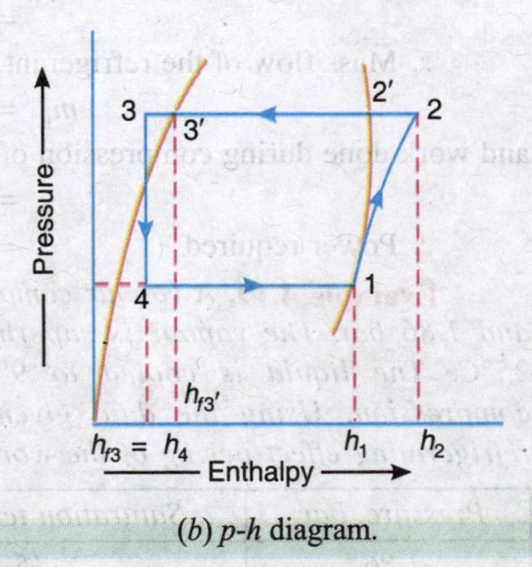

(*b*) *p-h* diagram.

We know that enthalpy of liquid refrigerant at point 3,

$$h_{f3} = h_{f3'} - c_{pl} \times \text{Degree of subcooling}$$
$$= 90.28 - 1.055 \times 10 = 79.73 \text{ kJ/kg}$$

$$\therefore \quad \text{Theoretical C.O.P.} = \frac{h_1 - h_{f3}}{h_2 - h_1} = \frac{176.48 - 79.73}{220.6 - 176.48} = 2.2$$

and $\quad (\text{C.O.P.})_{actual} = 0.8 \times 2.2 = 1.76$

Actual work done by the compressor,

$$w_{actual} = h_2 - h_1 = 220.6 - 176.48 = 44.12 \text{ kJ/kg}$$

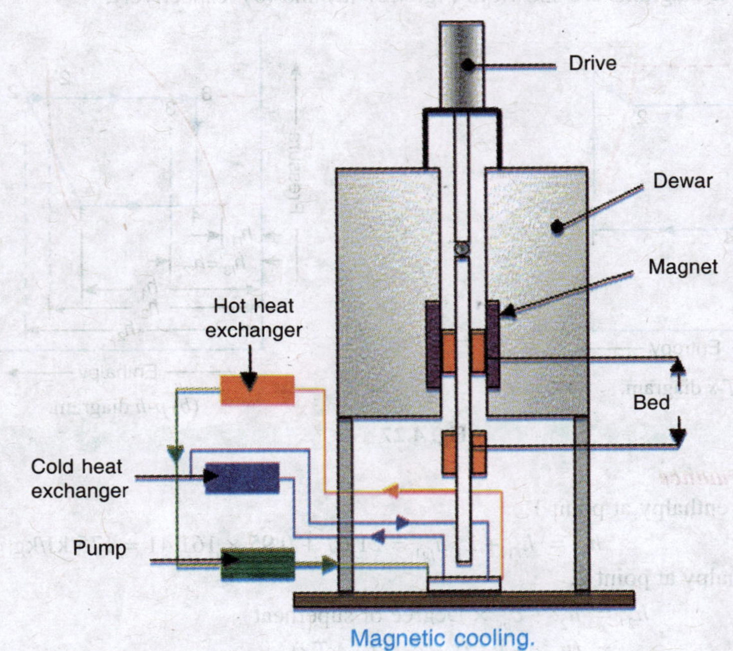

Magnetic cooling.

∴ Net refrigeration effect produced per kg of refrigerant

$$= w_{actual} \times (\text{C.O.P.})_{actual} = 44.12 \times 1.76 = 77.65 \text{ kJ/kg}$$

We know that refrigerating capacity

$$= 1 \text{ TR} = 210 \text{ kJ/min}$$

∴ Mass flow of the refrigerant,

$$m_R = 210/77.65 = 2.7 \text{ kg/min}$$

and work done during compression of the refrigerant

$$= m_R (h_2 - h_1) = 2.7 (220.6 - 176.48) = 119.12 \text{ kJ/min}$$

∴ Power required $= 119.12/60 = 1.985$ kW **Ans.**

Example 4.19. *A vapour compression refrigerator works between the pressures 4.93 bar and 1.86 bar. The vapour is superheated at the end of compression, its temperature being 25°C. The liquid is cooled to 9°C before throttling. The vapour is 95% dry before compression. Using the data given below, calculate the coefficient of performance and refrigerating effect per kg of the working substance circulated :*

Pressure ,bar	*Saturation temp., °C*	*Total heat (liquid), kJ/kg*	*Latent heat, kJ/kg*
1.86	*–15*	*21.67*	*161.41*
4.93	*14.45*	*49.07*	*147.80*

The specific heat at constant pressure for the superheated vapour is 0.645 kJ/kg K and for the liquid is 0.963 kJ/kg K.

Solution. Given : $p_2 = p_3 = 4.93$ bar ; $p_1 = p_4 = 1.86$ bar ; $T_2 = 25°C = 25 + 273 = 298$ K ; $T_3 = 9°C = 9 + 273 = 282$ K ; $x_1 = 95\% = 0.95$; $T_{3'} = T_{2'} = 14.45°C = 14.45 + 273 = 287.45$ K; $T_1 = T_4 = -15°C = -15 + 273 = 258$ K ; $h_{f1} = 21.67$ kJ/kg ; $h_{f3'} = h_{f2'} = 49.07$ kJ/kg ; $h_{fg1} = 161.41$ kJ/kg ; $h_{fg2'} = 147.8$ kJ/kg ; $c_{pv} = 0.645$ kJ/kg K ; $c_{pl} = 0.963$ kJ/kg K

The *T-s* and *p-h* diagrams are shown in Fig. 4.27 (*a*) and (*b*) respectively.

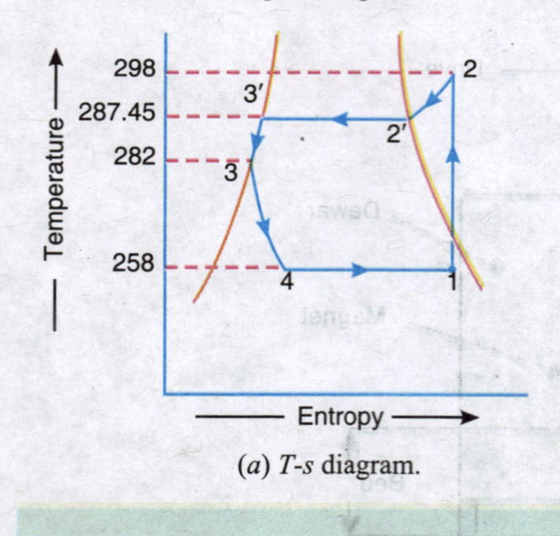

(*a*) *T-s* diagram.

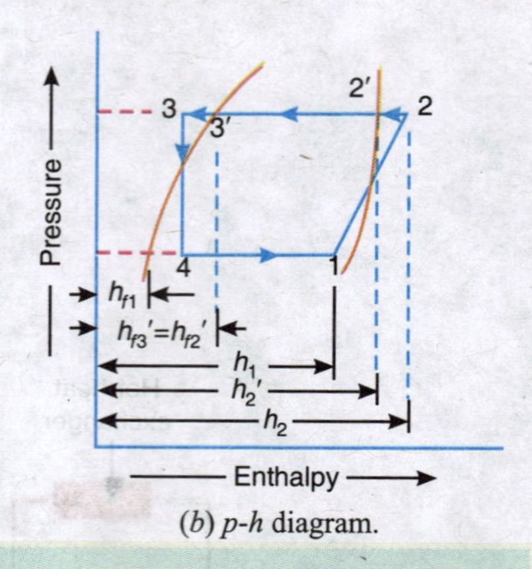

(*b*) *p-h* diagram.

Fig. 4.27

Coefficient of performance

We know that enthalpy at point 1,

$$h_1 = h_{f1} + x_1 h_{fg1} = 21.67 + 0.95 \times 161.41 = 175 \text{ kJ/kg}$$

Similarly, enthalpy at point 2,

$$h_2 = h_{2'} + c_{pv} \times \text{Degree of superheat}$$

$$= (h_{f2'} + h_{fg2'}) + c_{pv} (T_2 - T_{2'})$$

$$= (49.07 + 147.8) + 0.645 (298 - 287.45) = 203.67 \text{ kJ/kg}$$

and enthalpy of liquid refrigerant at point 3,

$$h_{f3} = h_{f3}' - c_{pl} \times \text{Degree of undercooling}$$
$$= h_{f3}' - c_{pl}(T_3' - T_3)$$
$$= 49.07 - 0.963\,(287.45 - 282) = 43.82 \text{ kJ/kg}$$

∴ Coefficient of performance,

$$\text{C.O.P.} = \frac{h_1 - h_{f3}}{h_2 - h_1} = \frac{175 - 43.82}{203.67 - 175} = 4.57 \textbf{ Ans.}$$

Refrigerating effect per kg of the working substance

We know that the refrigerating effect

$$= h_1 - h_{f3} = 175 - 43.82 = 131.18 \text{ kJ/kg} \textbf{ Ans.}$$

Example 4.20. *In a 15 TR ammonia refrigeration plant, the condensing temperature is 25°C and evaporating temperature – 10°C. The refrigerant ammonia is sub-cooled by 5°C before passing through the throttle valve. The vapour leaving the evaporator is 0.97 dry. Find : 1. coefficient of performance ; and 2. power required.*

Use the following properties of ammonia : —

Saturation temperature, °C	*Enthalpy, kJ/kg*		*Entropy, kJ/kg K*		*Specific heat, kJ/kg K*	
	Liquid	*Vapour*	*Liquid*	*Vapour*	*Liquid*	*Vapour*
25	*298.9*	*1465.84*	*1.1242*	*5.0391*	*4.6*	*2.8*
–10	*135.37*	*1433.05*	*0.5443*	*5.4770*	—	—

Solution. Given : $Q = 15$ TR ; $T_2' = T_3' = 25°\text{C} = 25 + 273 = 298$ K ; $T_1 = T_4 = -10°\text{C} = -10 + 273 = 263$ K ; $T_3 = 25 - 5 = 20°\text{C} = 20 + 273 = 293$ K ; $x_1 = 0.97$; $h_{f3}' = 298.9$ kJ/kg ; $h_2' = 1465.84$ kJ/kg ; *$s_{f3}' = 1.1242$ kJ/kg K ; $s_{g2}' = s_2' = 5.0391$ kJ/kg K ; $c_{pl} = 4.6$ kJ/kg K ; $c_{pv} = 2.8$ kJ/kg K ; $h_{f1} = 135.37$ kJ/kg ; $h_1' = 1433.05$ kJ/kg ; $s_{f1} = 0.5443$ kJ/kg ; $s_{g1} = 5.4770$ kJ/kg K

The *T-s* and *p-h* diagrams are shown in Fig. 4.28 (*a*) and (*b*) respectively.

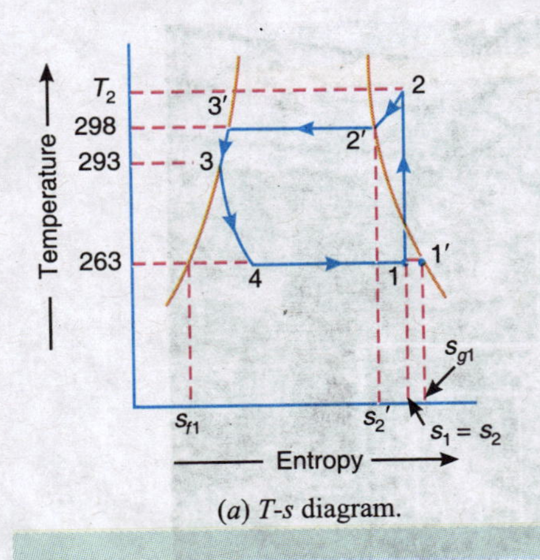

(*a*) *T-s* diagram.

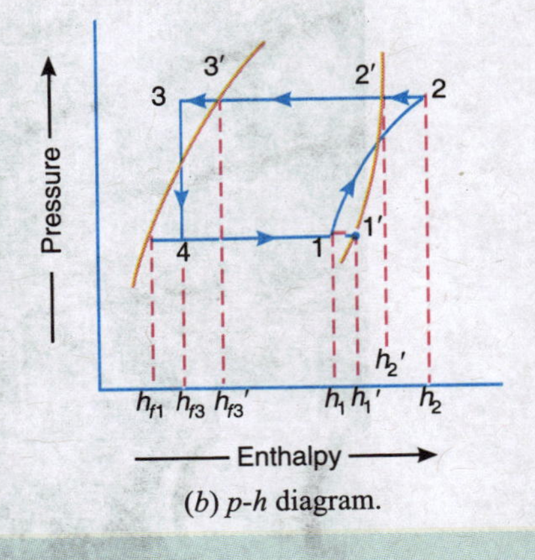

(*b*) *p-h* diagram.

Fig. 4.28

First of all, let us find the temperature of refrigerant at point 2 (T_2).

* Superfluous data.

We know that entropy at point 1,

$$s_1 = s_{f1} + x_1 s_{fg1} = s_{f1} + x_1 (s_{g1} - s_{f1}) \quad \ldots (\because s_{g1} = s_{f1} + s_{fg1})$$

$$= 0.5443 + 0.97 (5.4770 - 0.5443) = 5.329 \text{ kJ/kg K} \quad \ldots(i)$$

and entropy at point 2,

$$s_2 = s_{2'} + 2.3\, c_{pv} \log\left(\frac{T_2}{T_{2'}}\right) = 5.0391 + 2.3 \times 2.8 \log\left(\frac{T_2}{298}\right)$$

$$= 5.0391 + 6.44 \log\left(\frac{T_2}{298}\right) \quad \ldots (ii)$$

Since entropy at point 1 is equal to entropy at point 2, therefore equating equations (*i*) and (*ii*),

$$5.329 = 5.0391 + 6.44 \log\left(\frac{T_2}{298}\right)$$

$$\log\left(\frac{T_2}{298}\right) = \frac{5.329 - 5.0391}{6.44} = 0.045$$

or

$$\frac{T_2}{298} = 1.109 \quad \ldots \text{(Taking antilog of 0.045)}$$

$$\therefore \quad T_2 = 298 \times 1.109 = 330 \text{ K}$$

1. *Coefficient of performance*

We know that enthalpy at point 1,

$$h_1 = h_{f1} + x_1 h_{fg1} = h_{f1} + x_1 (h_{g1} - h_{f1}) \quad \ldots (\because h_{g1} = h_{f1} + h_{fg1})$$

$$= 135.37 + 0.97 (1433.05 - 135.37) = 1394.12 \text{ kJ/kg}$$

$$\ldots (\because h_{g1} = h_{1'})$$

Enthalpy at point 2,

$$h_2 = h_{2'} + c_{pv} \times \text{Degree of superheat} = h_{2'} + c_{pv} (T_2 - T_{2'})$$

$$= 1465.84 + 2.8 (330 - 298) = 1555.44 \text{ kJ/kg}$$

and enthalpy of liquid refrigerant at point 3,

$$h_{f3} = h_{f3'} - c_{pl} \times \text{Degree of undercooling}$$

$$= 298.9 - 4.6 \times 5 = 275.9 \text{ kJ/kg}$$

Ammonia refrigeration plant.

∴ Coefficient of performance,

$$\text{C.O.P.} = \frac{h_1 - h_{f3}}{h_2 - h_1} = \frac{1394.12 - 275.9}{1555.44 - 1394.12} = 6.93 \text{ Ans.}$$

2. *Power required*

We know that the heat extracted or refrigerating effect produced per kg of refrigerant

$$R_E = h_1 - h_{f3} = 1394.12 - 275.9 = 1118.22 \text{ kJ/kg}$$

and refrigerating capacity of the system,

$$Q = 15 \text{ TR} = 15 \times 210 = 3150 \text{ kJ/min} \quad \text{... (Given)}$$

∴ Mass flow of the refrigerant,

$$m_R = \frac{Q}{R_E} = \frac{3150}{1118.22} = 2.81 \text{ kg/min}$$

Work done during compression of the refrigerant

$$= m_R (h_2 - h_1) = 2.81 (1555.44 - 1394.12) = 453.3 \text{ kJ/min}$$

∴ Power required $= 453.3 / 60 = 7.55$ kW **Ans.**

Example 4.21. *A refrigeration system of 10.5 tonnes capacity at an evaporator temperature of –12°C and condenser temperature of 27°C is needed in a food storage locker. The refrigerant ammonia is subcooled by 6°C before entering the expansion valve. The vapour is 0.95 dry as it leaves the evaporator coil. The compression in the compressor is isentropic. Find:*

1. Condition of vapour at outlet of the compressor ; 2. Condition of vapour at entrance to the evaporator ; 3. C.O.P. ; and 4. Power required in kW.

Neglect valve throttling and clearance effect.

Solution. Given : $Q = 10.5$ TR ; $T_1 = T_4 = -12°\text{C} = -12 + 273 = 251\text{K}$; $T_{2'} = T_{3'} = 27°\text{C} = 27 + 273 = 300\text{K}$; $T_3 = 27 - 6 = 21°\text{C} = 21 + 273 = 294\text{K}$; $x_1 = 0.95$

The *T-s* and *p-h* diagrams are shown in Fig. 4.29 (*a*) and (*b*) respectively.

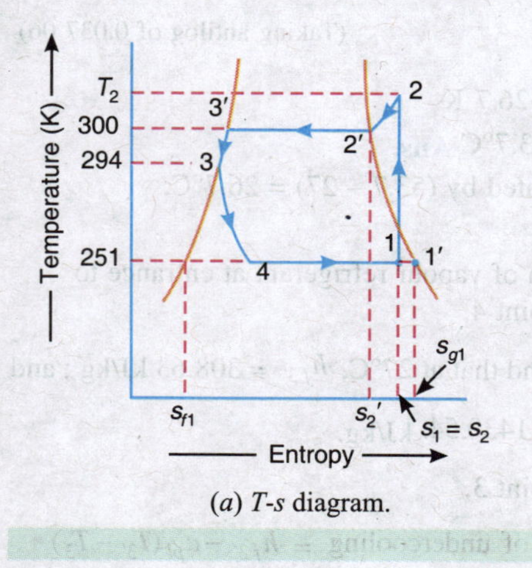

(*a*) *T-s* diagram.

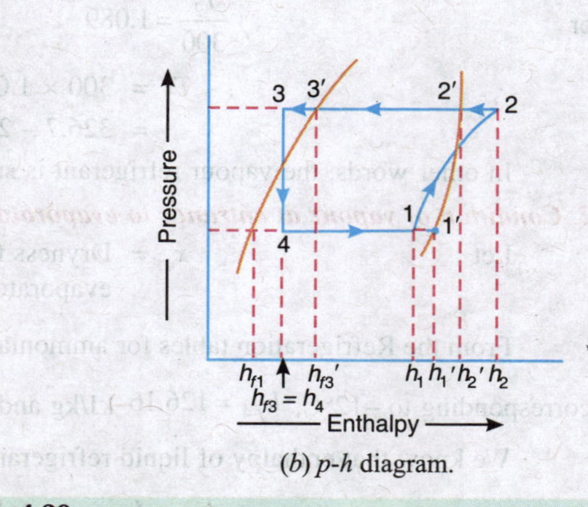

(*b*) *p-h* diagram.

Fig. 4.29

1. Condition of vapour at outlet of the Compressor

Let T_2 = Temperature of the vapour refrigerant at the outlet of the compressor.

From the Refrigeration tables for ammonia, we find that corresponding to 27°C, entropy at point 2′,

$$s_{2'} = s_{g2'} = 5.0170 \text{ kJ/kg K}$$

and corresponding to –12°C, $s_{f1} = 0.5096$ kJ/kg K, and $s_{g1} = 5.5055$ kJ/kg K

We know that entropy at point 1,

$$s_1 = s_{f1} + x_1\, s_{fg1} = s_{f1} + x_1(s_{g1} - s_{f1})$$

$$= 0.5096 + 0.95(5.5055 - 0.5096) = 5.2557 \text{ kJ/kg K} \quad ...(i)$$

Since s_1 (at –12°C) is greater than $s_{2'}$ (at 27°C), therefore condition of vapour at point 2 is superheated. Taking the specific heat at constant pressure for the superheated vapour (c_{pv}) as 2.8 kJ/kg K, we have entropy at point 2,

$$s_2 = s_{2'} + 2.3\, c_{pv} \log\left(\frac{T_2}{T_{2'}}\right) = 5.0170 + 2.3 \times 2.8 \log\left(\frac{T_2}{300}\right)$$

$$= 5.0170 + 6.44 \log\left(\frac{T_2}{300}\right) \quad ...(ii)$$

Since the entropy at point 1 is equal to entropy at point 2, therefore equating equations (*i*) and (*ii*),

$$5.2557 = 5.0170 + 6.44 \log\left(\frac{T_2}{300}\right)$$

$$\log\left(\frac{T_2}{300}\right) = \frac{5.2557 - 5.0170}{6.44} = 0.037\,06$$

or

$$\frac{T_2}{300} = 1.089 \quad \text{(Taking antilog of 0.037 06)}$$

$$\therefore \quad T_2 = 300 \times 1.089 = 326.7 \text{ K}$$

$$= 326.7 - 273 = 53.7°\text{C} \quad \textbf{Ans.}$$

In other words, the vapour refrigerant is superheated by (53.7 – 27) = 26.7°C.

2. Condition of vapour at entrance to evaporator

Let x_4 = Dryness fraction of vapour refrigerant at entrance to evaporator at point 4.

From the Refrigeration tables for ammonia, we find that at 27°C, $h_{f3'} = 308.63$ kJ/kg ; and corresponding to –12°C, $h_{f4} = 126.16$ kJ/kg and $h_{g4} = 1430.54$ kJ/kg.

We know that enthalpy of liquid refrigerant at point 3,

$$h_{f3} = h_{f3'} - c_{pl} \times \text{Degree of undercooling} = h_{f3'} - c_{pl}(T_{3'} - T_3)$$

$$= 308.63 - 4.6(300 - 294) = 281.06 \text{ kJ/kg} \quad ...(iii)$$

...(Taking specific heat of liquid ammonia as 4.6 kJ/kg K)

and enthalpy of vapour refrigerant at point 4,

$$h_4 = h_{f4} + x_4 h_{fg4} = h_{f4} + x_4 (h_{g4} - h_{f4}) \quad ...(\because h_{g4} = h_{f4} + h_{fg4})$$

$$= 126.16 + x_4\,(1430.54 - 126.16)$$

$$= 126.16 + 1304.38\,x_4 \quad ...(iv)$$

Since enthalpy of liquid refrigerant at point 3 (h_{f3}) is equal to enthalpy of vapour refrigerant at point 4 (h_4), therefore equating equations (*iii*) and (*iv*),

$$281.06 = 126.16 + 1304.38 x_4 \quad \text{or} \quad x_4 = 0.1187 \quad \textbf{Ans.}$$

3. C.O.P.

From Refrigeration tables for ammonia, we find that corresponding to –12°C (*i.e.* at point 1),

$$h_{f1} = 126.16 \text{ kJ/kg} \; ; \text{ and } h_{g1} = 1430.54 \text{ kJ/kg}$$

and corresponding to 27°C (*i.e.* at point 2′),

$$h_{2'} = 1467.22 \text{ kJ/kg}$$

We know that enthalpy at point 1,

$$h_1 = h_{f1} + x_1\, h_{fg1} = h_{f1} + x_1 (h_{g1} - h_{f1})$$

$$= 126.16 + 0.95\,(1430.54 - 126.16) = 1365.32 \text{ kJ/kg}$$

and enthalpy at point 2,

$$h_2 = h_{2'} + c_{pv} \times \text{Degree of superheat} = h_{2'} + c_{pv} (T_2 - T_{2'})$$

$$= 1467.22 + 2.8\,(326.7 - 300) = 1541.98 \text{ kJ/kg}$$

We know that

$$\text{C.O.P.} = \frac{h_1 - h_{f3}}{h_2 - h_1} = \frac{1365.32 - 281.06}{1541.98 - 1365.32} = \frac{1084.26}{176.66} = 6.137 \quad \textbf{Ans.}$$

4. Power required

We know that refrigerating effect produced per kg of refrigerant,

$$R_E = h_1 - h_{f3} = 1365.32 - 281.06 = 1084.26 \text{ kJ/kg}$$

and refrigerating capacity of the system,

$$Q = 10.5 \text{ TR} = 10.5 \times 210 = 2205 \text{ kJ/min}$$

∴ Mass flow of the refrigerant,

$$m_R = \frac{Q}{R_E} = \frac{2205}{1084.26} = 2.0336 \text{ kg / min}$$

Workdone during compression of the refrigerant

$$= m_R (h_2 - h_1) = 2.0336\,(1541.98 - 1365.32) = 359.26 \text{ kJ/min}$$

∴ Power required $= 359.26 / 60 = 5.987$ kJ/s or kW **Ans.**

Example 4.22. *The following data refers to a 20 TR ice plant using ammonia as refrigerant :*

The temperature of water entering and leaving the condenser are 20°C and 27°C and temperature of brine in the evaporator is – 15°C.

Before entering the expansion valve, ammonia is cooled to 20°C and the ammonia enters the compressor dry saturated.

Calculate for one tonne of refrigeration the power expended, the amount of cooling water in the condenser and the coefficient of performance of the plant.

Use the properties given in the table below :

Saturation temperature, °C	*Enthalpy, kJ/kg*		*Entropy, kJ/kg K*		*Specific heat, kJ/kg K*	
	Liquid	*Vapour*	*Liquid*	*Vapour*	*Liquid*	*Vapour*
–15	*112.34*	*1426.54*	*0.4572*	*5.5490*	*4.396*	*2.303*
25	*298.90*	*1465.84*	*1.1242*	*5.0391*	*4.606*	*2.805*

Solution. Given : $Q = 20$ TR ; $T_{w2} = 20°C = 20 + 273 = 293$ K ; $T_{w1} = 27°C = 27 + 273 = 300$ K ; $T_{2'} = T_{3'} = 25°C = 25 + 273 = 298$ K ; $T_1 = T_4 = -15°C = -15 + 273 = 258$ K ; $T_3 = 20°C = 20 + 273 = 293$ K ; *$h_{f1} = 112.34$ kJ/kg ; $h_{f3'} = 298.90$ kJ/kg ; $h_1 = 1426.54$ kJ/kg ; $h_{2'} = 1465.84$ kJ/kg ; *$s_{f1} = 0.4572$ kJ/kg K ; *$s_{f3} = 1.1242$ kJ/kg K ; $s_1 = s_2 = 5.5490$ kJ/kg K ; $s_{2'} = 5.0391$ kJ/kg K ; *$c_{pl4} = 4.396$ kJ/kg K ; $c_{pl3} = 4.606$ kJ/kg K ; *$c_{pv1} = 2.303$ kJ/kg K ; $c_{pv2'} = 2.805$ kJ/kg K

Power expended per TR

The *T-s* and *p-h* diagrams are shown in Fig. 4.30 (*a*) and (*b*) respectively. First of all, let us find the temperature of refrigerant at point 2 (T_2). We know that entropy at point 2,

$$s_2 = s_{2'} + 2.3\, c_{pv2'} \log\left(\frac{T_2}{T_{2'}}\right)$$

$$5.5490 = 5.0391 + 2.3 \times 2.805 \log\left(\frac{T_2}{298}\right)$$

$$\log\left(\frac{T_2}{298}\right) = \frac{5.5490 - 5.0391}{2.3 \times 2.805} = 0.079$$

$$\frac{T_2}{298} = 1.2 \quad \text{... (Taking antilog of 0.079)}$$

or $$T_2 = 1.2 \times 298 = 357.6 \text{ K or } 84.6° \text{ C}$$

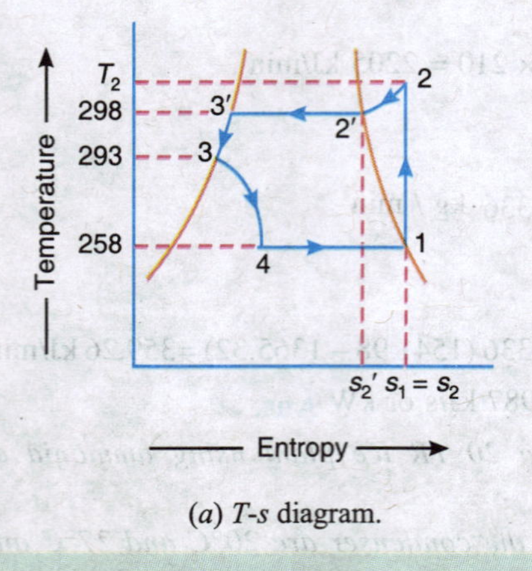

(*a*) *T-s* diagram.

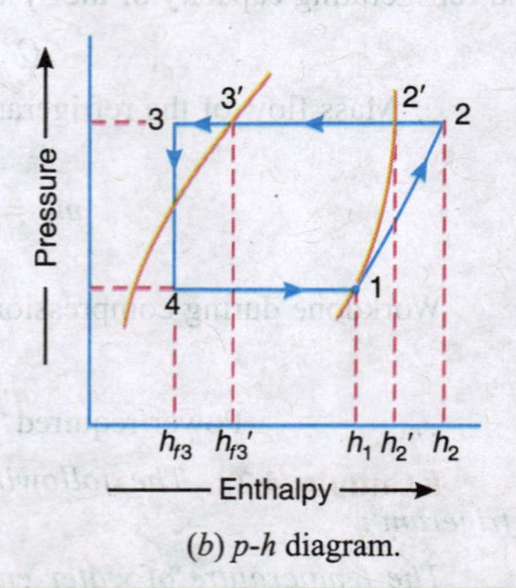

(*b*) *p-h* diagram.

Fig. 4.30

* Superfluous data.

∴ Enthalpy at point 2,

$$h_2 = h_{2'} + c_{pv2'} (T_2 - T_{2'})$$
$$= 1465.84 + 2.805 (357.6 - 298) = 1633.02 \text{ kJ/kg}$$

and enthalpy of liquid refrigerant at point 3,

$$h_{f3} = h_{f3'} - c_{pl3} \times \text{Degree of undercooling}$$
$$= h_{f3'} - c_{pl3} (T_{3'} - T_3)$$
$$= 298.9 - 4.606 (298 - 293) = 275.87 \text{ kJ/kg}$$

We know that heat extracted or refrigerating effect produced per kg of the refrigerant,

$$R_E = h_1 - h_{f3} = 1426.54 - 275.87 = 1150.67 \text{ kJ/kg}$$

and capacity of the ice plant,

$$Q = 20 \text{ TR} = 20 \times 210 = 4200 \text{ kJ/min}$$

∴ Mass flow of the refrigerant,

$$m_R = \frac{Q}{R_E} = \frac{4200}{1150.67} = 3.65 \text{ kg/min}$$

Work done by the compressor per minute

$$= m_R (h_2 - h_1) = 3.65 (1633.02 - 1426.54) = 753.65 \text{ kJ/min}$$

∴ Power expended per TR

$$= \frac{753.65}{60 \times 20} = 0.628 \text{ kW/TR}$$ **Ans.**

Amount of cooling water in the condenser

Let m_w = Amount of cooling water in the condenser.

We know that heat given out by the refrigerant in the condenser

$$= m_R (h_2 - h_{f3})$$
$$= 3.65 (1633.02 - 275.87) = 4953.6 \text{ kJ/min} \quad \ldots (i)$$

Since the specific heat of water, c_w = 4.187 kJ/kg K, therefore heat taken by water in the condenser

$$= m_w \times c_w (T_{w1} - T_{w2})$$
$$= m_w \times 4.187 (300 - 293) = 29.3 \, m_w \text{ kJ/min} \quad \ldots (ii)$$

Since the heat given by the refrigerant in the condenser is equal to the heat taken by water in the condenser, therefore equating equations (*i*) and (*ii*),

$$29.3 \, m_w = 4953.6 \quad \text{or} \quad m_w = 169 \text{ kg/min}$$ **Ans.**

Example 4.23. *A vapour compression refrigeration machine, with Freon-12 as refrigerant, has a capacity of 12 tonne of refrigeration operating between – 28°C and 26°C. The refrigerant is subcooled by 4°C before entering the expansion valve and the vapour is superheated by 5°C before leaving the evaporator. The machine has a six-cylinder single-acting compressor with stroke equal to 1.25 times the bore. It has a clearance of 3% of the stroke volume. Determine : 1. Theoretical power required ; 2. C.O.P., 3. Volumetric efficiency ; and 4. Bore and stroke of cylinder. The speed of compressor is 1000 r.p.m.*

The following properties of Freon-12 may be used :

Sat. temp., °C	Pressure, bar	Sp. volume of vapour, m³/kg	Enthalpy, kJ/kg		Entropy, kJ/kg K	
			Liquid	Vapour	Liquid	Vapour
–28	1.093	0.1475	10.64	175.11	0.0444	0.7153
26	6.697	0.0262	60.67	198.11	0.2271	0.6865

Specific heat of liquid refrigerant = 0.963 kJ/kg K and specific heat of superheated vapour = 0.615 kJ/kg K.

Solution. Given : $Q = 12$ TR ; $T_{1'} = -28°C = -28 + 273 = 245$ K ; $T_{2'} = T_{3'} = 26°C = 26 + 273 = 299$ K ; $T_{3'} - T_3 = 4°C$ or $T_3 = 22°C = 22 + 273 = 295$ K ; $T_1 - T_{1'} = 5°C$ or $T_1 = -23°C = -23 + 273 = 250$ K ; $L = 1.25D$; Clearance volume = 3% Stroke volume ; $N = 1000$ r.p.m. ; $v_{1'} = 0.1475$ m³/kg ; $v_{2'} = 0.0262$ m³/kg ; $h_{f1} = 10.64$ kJ/kg ; $h_{f3'} = 60.67$ kJ/kg ; $h_{1'} = 175.11$ kJ/kg ; $h_{2'} = 198.11$ kJ/kg ; $s_{f1} = 0.0444$ kJ/kg K ; $s_{f3} = 0.2271$ kJ/kg K ; $s_{1'} = 0.7153$ kJ/kg K ; $s_{2'} = 0.6865$ kJ/kg K ; $c_{pl} = 0.963$ kJ/kg K ; $c_{pv} = 0.615$ kJ/kg K

The *T-s* and *p-h* diagrams are shown in Fig. 4.31 (*a*) and (*b*) respectively.

1. Theoretical power required

First of all, let us find the temperature of superheated vapour at point 2 (T_2).

We know that entropy at point 1,

$$s_1 = s_{1'} + 2.3\, c_{pv} \log\left(\frac{T_1}{T_{1'}}\right)$$

$$= 0.7153 + 2.3 \times 0.615 \log\left(\frac{250}{245}\right) = 0.7277 \qquad \dots (i)$$

and entropy at point 2,

$$s_2 = s_{2'} + 2.3\, c_{pv} \log\left(\frac{T_2}{T_{2'}}\right) = 0.6865 + 2.3 \times 0.615 \log\left(\frac{T_2}{299}\right)$$

$$= 0.6865 + 1.4145 \log\left(\frac{T_2}{299}\right) \qquad \dots (ii)$$

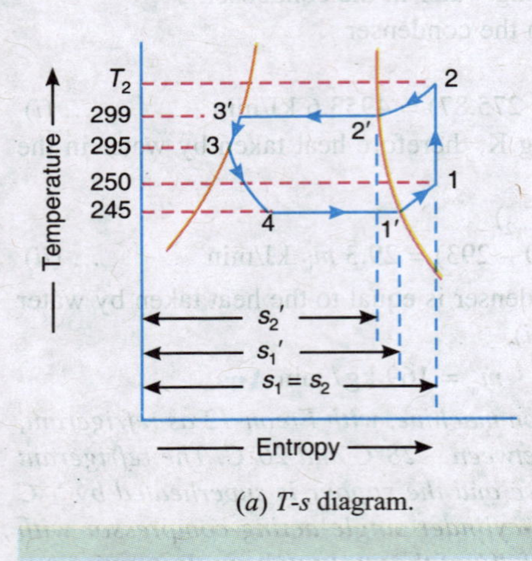

(*a*) *T-s* diagram.

(*b*) *p-h* diagram.

Fig. 4.31

Since the entropy at point 1 is equal to entropy at point 2, therefore equating equations (*i*) and (*ii*),

$$0.7277 = 0.6865 + 1.4145 \log\left(\frac{T_2}{299}\right)$$

$$\log\left(\frac{T_2}{299}\right) = \frac{0.7277 - 0.6865}{1.4145} = 0.0291$$

$$\frac{T_2}{299} = 1.0693 \qquad \dots \text{(Taking antilog of 0.0291)}$$

$$\therefore \quad T_2 = 299 \times 1.0693 = 319.7 \text{ K}$$

We know that enthalpy at point 1,

$$h_1 = h_{1'} + c_{pv}\,(T_1 - T_{1'})$$
$$= 175.11 + 0.615\,(250 - 245) = 178.18 \text{ kJ/kg}$$

Enthalpy at point 2,

$$h_2 = h_{2'} + c_{pv}\,(T_2 - T_{2'})$$
$$= 198.11 + 0.615\,(319.7 - 299) = 210.84 \text{ kJ/kg}$$

and enthalpy of liquid refrigerant at point 3,

$$h_{f3} = h_{f3'} - c_{pl}\,(T_{3'} - T_3)$$
$$= 60.67 - 0.963\,(299 - 295) = 64.52 \text{ kJ/kg}$$

We know that heat extracted or refrigerating effect per kg of the refrigerant,

$$R_E = h_1 - h_{f3} = 178.18 - 64.52 = 113.66 \text{ kJ/kg}$$

and refrigerating capacity of the system,

$$Q = 12 \text{ TR} = 12 \times 210 = 2520 \text{ kJ/min} \quad \text{...(Given)}$$

∴ Mass flow of the refrigerant,

$$m_R = \frac{Q}{R_E} = \frac{2520}{113.66} = 22.17 \text{ kg/min}$$

Work done during compression of the refrigerant

$$= m_R\,(h_2 - h_1) = 22.17\,(210.84 - 178.18) = 724 \text{ kJ/min}$$

∴ Theoretical power required

$$= 724/60 = 12.07 \text{ kW } \textbf{Ans.}$$

2. C.O.P.

We know that C.O.P.

$$= \frac{h_1 - h_{f3}}{h_2 - h_1} = \frac{178.18 - 64.52}{210.84 - 178.18} = 3.48 \textbf{ Ans.}$$

3. Volumetric efficiency

Let v_2 = Specific volume at point 2, and

C = Clearance = 3% = 0.03. ... (Given)

First of all, let us find the specific volume at suction to the compressor, *i.e.* at point 1. Applying Charles' law to points 1 and 1′,

$$\frac{v_1}{T_1} = \frac{v_{1'}}{T_{1'}} \quad \text{or} \quad v_1 = v_{1'} \times \frac{T_1}{T_{1'}} = 0.1475 \times \frac{250}{245} = 0.1505 \text{ m}^3\text{/kg}$$

Again applying Charles' law to points 2 and 2′,

$$\frac{v_2}{T_2} = \frac{v_{2'}}{T_{2'}} \quad \text{or} \quad v_2 = v_{2'} \times \frac{T_2}{T_{2'}} = 0.0262 \times \frac{319.7}{299} = 0.028 \text{ m}^3\text{/kg}$$

We know that volumetric efficiency,

$$\eta_v = 1 + C - C\left(\frac{v_1}{v_2}\right) = 1 + 0.03 - 0.03\left(\frac{0.1505}{0.028}\right)$$
$$= 0.87 \text{ or } 87\% \textbf{ Ans.}$$

4. Bore and stroke of cylinder

Let D = Bore of cylinder,

L = Length of cylinder = 1.25 D, and ... (Given)

N = Speed of compressor = 1000 r.p.m. ... (Given)

We know that theoretical suction volume or piston displacement per minute

$$= m_R \times v_1 \times \frac{1}{\eta_v} = 22.17 \times 0.1505 \times \frac{1}{0.87} = 3.84 \text{ m}^3/\text{min}$$

Since the machine has six cylinder, single acting compressor, therefore, theoretical suction volume or piston displacement per cylinder per minute

$$= \frac{3.84}{6} = 0.64 \text{ m}^3/\text{min} \quad \text{... (iii)}$$

We also know that suction volume or piston displacement per minute

$$= \text{Piston area} \times \text{Stroke} \times \text{R.P.M.}$$

$$= \frac{\pi}{4} \times D^2 \times L \times N = \frac{\pi}{4} \times D^2 \times 1.25\, D \times 1000$$

$$= 982\, D^3 \text{ m}^3/\text{min} \quad \text{... (iv)}$$

Equating equations (*iii*) and (*iv*),

$$D^3 = 0.64 / 982 = 0.000\,652$$

$\therefore$ D = 0.0867 m = 86.7 mm **Ans.**

and L = 1.25 × 86.7 = 108.4 mm **Ans.**

Example 4.24. *A food storage locker requires a refrigeration capacity of 12 TR and works between the evaporating temperature of – 8°C and condensing temperature of 30°C. The refrigerant R-12 is subcooled by 5°C before entry to expansion valve and the vapour is superheated to – 2°C before leaving the evaporator coils. Assuming a two cylinder, single acting compressor operating at 1000 r.p.m. with stroke equal to 1.5 times the bore, determine : 1. coefficient of performance; 2. theoretical power per tonne of refrigeration; and 3. bore and stroke of compressor when (a) there is no clearance; and (b) there is a clearance of 2%.*

Use the following data for R-12 :

Saturation temperature, °C	*Pressure, bar*	*Enthalpy, kJ/kg*		*Entropy, kJ/kg K*		*Specific volume of vapour, m³/kg*
		Liquid	*Vapour*	*Liquid*	*Vapour*	
–8	2.354	28.72	184.07	0.1149	0.7007	0.0790
30	7.451	64.59	199.62	0.2400	0.6853	0.0235

The specific heat of liquid R-12 is 1.235 kJ/kg K, and of vapour R-12 is 0.733 kJ/kg K.

Solution. Given : Q = 12 TR ; T_1' = – 8°C = – 8 + 273 = 265 K ; T_2' = 30°C = 30 + 273 = 303 K ; $T_3' - T_3$ = 5°C ; T_1 = – 2°C = – 2 + 273 = 271 K ; h_{f1} = 28.72 kJ/kg ; h_{f3}' = 64.59 kJ/kg ; h_1' = 184.07 kJ/kg ; h_2' = 199.62 kJ/kg ; *s_{f1} = 0.1149 kJ/kg K ; *s_{f3} = 0.2400 kJ/kg K ; s_1' = 0.7007 kJ/kg K ; s_2' = 0.6853 kJ/kg K ; v_1' = 0.079 m³/kg ; v_2' = 0.0235 m³/kg ; c_{pl} = 1.235 kJ/kg K ; c_{pv} = 0.733 kJ/kg K

The *T-s* and *p-h* diagrams are shown in Fig. 4.32 (*a*) and (*b*) respectively.

1. *Coefficient of performance*

First of all, let us find the temperature of superheated vapour at point 2 (T_2).

* Superfluous data.

We know that entropy at point 1,

$$s_1 = s_{1'} + 2.3\ c_{pv}\ \log\left(\frac{T_1}{T_{1'}}\right)$$

$$= 0.7007 + 2.3 \times 0.733\ \log\left(\frac{271}{265}\right) = 0.7171 \qquad \ldots (i)$$

and entropy at point 2,

$$s_2 = s_{2'} + 2.3\ c_{pv}\ \log\left(\frac{T_2}{T_{2'}}\right) = 0.6853 + 2.3 \times 0.733\ \log\left(\frac{T_2}{303}\right)$$

$$= 0.6853 + 1.686\ \log\left(\frac{T_2}{303}\right) \qquad \ldots (ii)$$

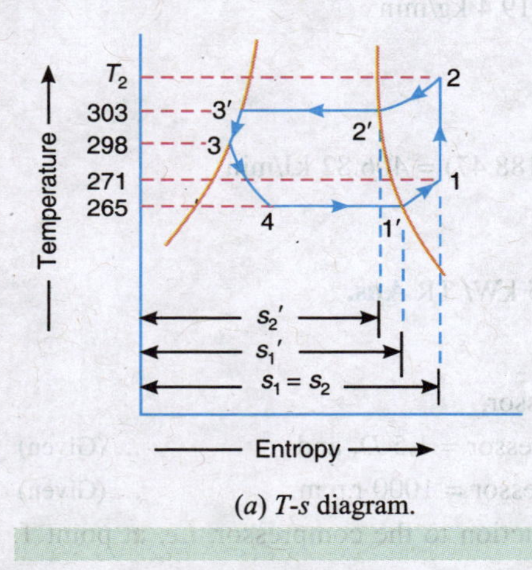

(a) T-s diagram.

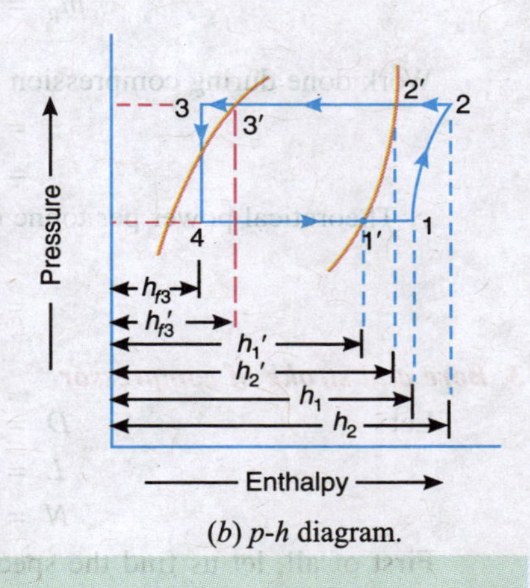

(b) p-h diagram.

Fig. 4.32

Since the entropy at point 1 is equal to entropy at point 2, therefore equating equations (i) and (ii),

$$0.7171 = 0.6853 + 1.686\ \log\left(\frac{T_2}{303}\right)$$

or

$$\log\left(\frac{T_2}{303}\right) = \frac{0.7171 - 0.6853}{1.686} = 0.0188$$

$$\left(\frac{T_2}{303}\right) = 1.0444 \qquad \ldots \text{(Taking antilog of 0.0188)}$$

$$\therefore \quad T_2 = 1.0444 \times 303 = 316.4\ \text{K}$$

We know that enthalpy at point 1,

$$h_1 = h_{1'} + c_{pv}\ (T_1 - T_{1'})$$

$$= 184.07 + 0.733\ (271 - 265) = 188.47\ \text{kJ/kg}$$

Enthalpy at point 2,

$$h_2 = h_{2'} + c_{pv}\ (T_2 - T_{2'})$$

$$= 199.62 + 0.733\ (316.4 - 303) = 209.44\ \text{kJ/kg}$$

and enthalpy of liquid refrigerant at point 3,

$$h_{f3} = h_{f3'} - c_{pl}\ (T_{3'} - T_3)$$

$$= 64.59 - 1.235 \times 5 = 58.42\ \text{kJ/kg}$$

$$\therefore \quad \text{C.O.P.} = \frac{h_1 - h_{f3}}{h_2 - h_1} = \frac{188.47 - 58.42}{209.44 - 188.47} = \frac{130.05}{20.97} = 6.2 \textbf{ Ans.}$$

2. Theoretical power per tonne of refrigeration

We know that the heat extracted or refrigerating effect per kg of the refrigerant,

$$R_E = h_1 - h_{f3} = 188.47 - 58.42 = 130.05 \text{ kJ/kg}$$

and the refrigerating capacity of the system,

$$Q = 12 \text{ TR} = 12 \times 210 = 2520 \text{ kJ/min} \quad \text{... (Given)}$$

∴ Mass flow of the refrigerant,

$$m_R = \frac{Q}{R_E} = \frac{2520}{130.05} = 19.4 \text{ kg/min}$$

Work done during compression of the refrigerant

$$= m_R (h_2 - h_1)$$

$$= 19.4 (209.44 - 188.47) = 406.82 \text{ kJ/min}$$

∴ Theoretical power per tonne of refrigeration

$$= \frac{406.82}{60 \times 12} = 0.565 \text{ kW/ TR } \textbf{Ans.}$$

3. Bore and stroke of compressor

Let D = Bore of compressor,

L = Stroke of compressor = 1.5 D, and ... (Given)

N = Speed of compressor = 1000 r.p.m. ... (Given)

First of all, let us find the specific volume at suction to the compressor, *i.e.* at point 1. Applying Charles' law,

$$\frac{v_1}{T_1} = \frac{v_1'}{T_1'}$$

or

$$v_1 = v_1' \times \frac{T_1}{T_1'} = 0.0790 \times \frac{271}{265} = 0.081 \text{ m}^3\text{/kg}$$

(a) When there is no clearance

We know that theoretical suction volume or piston displacement per minute

$$= m_R \times v_1 = 19.4 \times 0.081 = 1.57 \text{ m}^3\text{/min}$$

and theoretical suction volume or piston displacement per cylinder per minute

$$= 1.57 / 2 = 0.785 \text{ m}^3\text{/min} \quad \text{... (iii)}$$

Also theoretical suction volume or piston displacement per minute

$$= \text{Piston area} \times \text{Stroke} \times \text{R.P.M.}$$

$$= \frac{\pi}{4} \times D^2 \times L \times N = \frac{\pi}{4} \times D^2 \times 1.5 D \times 1000$$

$$= 1178.25\, D^3 \text{ m}^3\text{/min} \quad \text{...(iv)}$$

Equating equations (*iii*) and (*iv*),

$$1178.25\, D^3 = 0.785$$

or

$$D^3 = 0.785 / 1178.25 = 0.000\,666$$

$\therefore \qquad D = 0.087 \text{ m} = 87 \text{ mm}$ **Ans.**

and $\qquad L = 1.5\,D = 1.5 \times 87 = 130.5 \text{ mm}$ **Ans.**

(b) When there is a clearance of 2%

Let $\qquad v_2$ = Specific volume at point 2, and

C = Clearance = 2% = 0.02 ... (Given)

Applying Charles' law to points 2 and 2′,

$$\frac{v_2}{T_2} = \frac{v_2'}{T_2'} \quad \text{or} \quad v_2 = v_2' \times \frac{T_2}{T_2'} = 0.0235 \times \frac{361.4}{303} = 0.0245 \text{ m}^3/\text{kg}$$

We know that volumetric efficiency of the compressor,

$$\eta_v = 1 + C - C\left(\frac{v_1}{v_2}\right) = 1 + 0.02 - 0.02\left(\frac{0.081}{0.0245}\right)$$

$$= 1.02 - 0.066 = 0.954$$

$\therefore$ Piston displacement per cylinder per min

$$= \frac{m_R \times v_1}{2} \times \frac{1}{\eta_v} = \frac{19.4 \times 0.081}{2} \times \frac{1}{0.954} = 0.8236 \text{ m}^3/\text{min} \qquad \text{... } (v)$$

Equating equations (*iv*) and (*v*),

$$1178.25\,D^3 = 0.8236$$

or $\qquad D^3 = 0.8236 / 1178.25 = 0.000\,70$

$\therefore \qquad D = 0.0887 \text{ m} = 88.7 \text{ mm}$ **Ans.**

and $\qquad L = 1.5\,D = 1.5 \times 88.7 = 133 \text{ mm}$ **Ans.**

4.11 Actual Vapour Compression Cycle

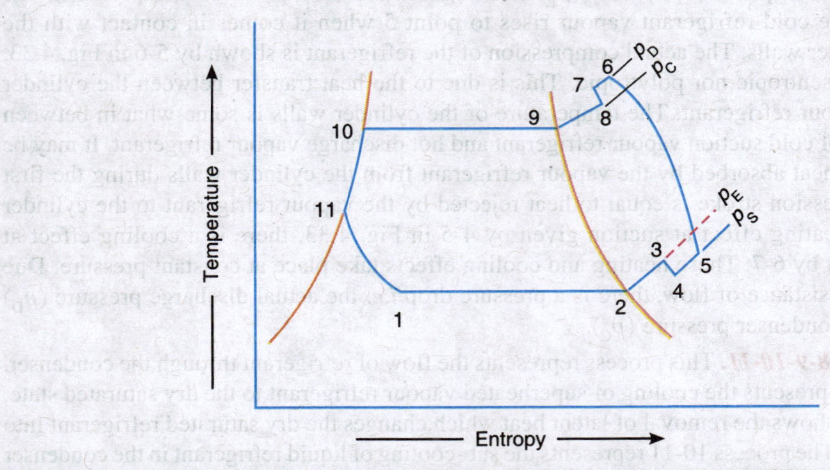

Fig. 4.33. Actual vapour compression cycle.

The actual vapour compression cycle differs from the theoretical vapour compression cycle in many ways, some of which are unavoidable and cause losses. The main deviations between the theoretical cycle and actual cycle are as follows :

1. The vapour refrigerant leaving the evaporator is in superheated state.
2. The compression of refrigerant is neither isentropic nor polytropic.
3. The liquid refrigerant before entering the expansion valve is sub-cooled in the condenser.
4. The pressure drops in the evaporator and condenser.

The actual vapour compression cycle on *T-s* diagram is shown in Fig. 4.33. The various processes are discussed below :

(a) Process 1-2-3. This process shows the flow of refrigerant in the evaporator. The point 1 represents the entry of refrigerant into the evaporator and the point 3 represents the exit of refrigerant from evaporator in a superheated state. The point 3 also represents the entry of refrigerant into the compressor in a superheated condition. The superheating of vapour refrigerant from point 2 to point 3 may be due to :

(*i*) automatic control of expansion valve so that the refrigerant leaves the evaporator as the superheated vapour.

(*ii*) picking up of larger amount of heat from the evaporator through pipes located within the cooled space.

(*iii*) picking up of heat from the suction pipe, *i.e.* the pipe connecting the evaporator delivery and the compressor suction valve.

In the first and second case of superheating the vapour refrigerant, the refrigerating effect as well as the compressor work is increased. The coefficient of performance, as compared to saturation cycle at the same suction pressure may be greater, less or unchanged.

The superheating also causes increase in the required displacement of compressor and load on the compressor and condenser. This is indicated by 2-3 on *T-s* diagram as shown in Fig. 4.33.

(b) Process 3-4-5-6-7-8. This process represents the flow of refrigerant through the compressor. When the refrigerant enters the compressor through the suction valve at point 3, the pressure falls to point 4 due to frictional resistance to flow. Thus the actual suction pressure (p_S) is lower than the evaporator pressure (p_E). During suction and prior to compression, the temperature of the cold refrigerant vapour rises to point 5 when it comes in contact with the compressor cylinder walls. The actual compression of the refrigerant is shown by 5-6 in Fig. 4.33, which is neither isentropic nor polytropic. This is due to the heat transfer between the cylinder walls and the vapour refrigerant. The temperature of the cylinder walls is some-what in between the temperatures of cold suction vapour refrigerant and hot discharge vapour refrigerant. It may be assumed that the heat absorbed by the vapour refrigerant from the cylinder walls during the first part of the compression stroke is equal to heat rejected by the vapour refrigerant to the cylinder walls. Like the heating effect at suction given by 4-5 in Fig. 4.33, there is a cooling effect at discharge as given by 6-7. These heating and cooling effects take place at constant pressure. Due to the frictional resistance of flow, there is a pressure drop *i.e.* the actual discharge pressure (p_D) is more than the condenser pressure (p_C).

(c) Process 8-9-10-11. This process represents the flow of refrigerant through the condenser. The process 8-9 represents the cooling of superheated vapour refrigerant to the dry saturated state. The process 9-10 shows the removal of latent heat which changes the dry saturated refrigerant into liquid refrigerant. The process 10-11 represents the sub-cooling of liquid refrigerant in the condenser before passing through the expansion valve. This is desirable as it increases the refrigerating effect per kg of the refrigerant flow. It also reduces the volume of the refrigerant partially evaporated from the liquid refrigerant while passing through the expansion valve. The increase in refrigerating effect can be obtained by large quantities of circulating cooling water which should be at a temperature much lower than the condensing temperatures.

(d) Process 11-1. This process represents the expansion of subcooled liquid refrigerant by throttling from the condenser pressure to the evaporator pressure.

4.12 Effect of Suction Pressure

We have discussed in the previous article that in actual practice, the suction pressure (or evaporator pressure) decreases due to the frictional resistance of flow of the refrigerant. Let us consider a theoretical vapour compression cycle 1′-2′-3-4′ when the suction pressure decreases from p_S to p_S' as shown on *p-h* diagram in Fig. 4.34.

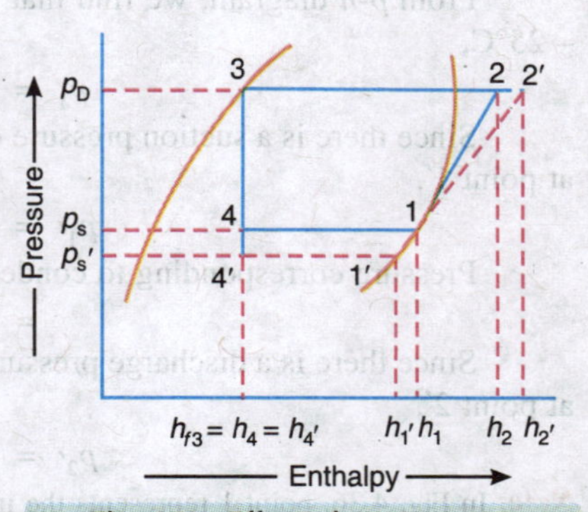

Fig. 4.34. Effect of suction pressure.

It may be noted that the decrease in suction pressure

1. decreases the refrigerating effect from $(h_1 - h_4)$ to $(h_1' - h_4')$, and
2. increases the work required for compression from $(h_2 - h_1)$ to $(h_2' - h_1')$.

Since the C.O.P. of the system is the ratio of refrigeraing effect to the work done, therefore with the decrease in suction pressure, the net effect is to decrease the C.O.P. of the refrigerating system for the same amount of refrigerant flow. Hence with the decrease in suction pressure, the refrigerating capacity of the system decreases and the refrigeration cost increases.

4.13 Effect of Discharge Pressure

We have already discussed that in actual practice, the discharge pressure (or condenser pressure) increases due to frictional resistance of flow of the refrigerant. Let us consider a theoretical vapour compression cycle 1-2′-3′-4′ when the discharge pressure increases from p_D to p_D' as shown on *p-h* diagram in Fig. 4.35. It may be noted that the increase in discharge pressure

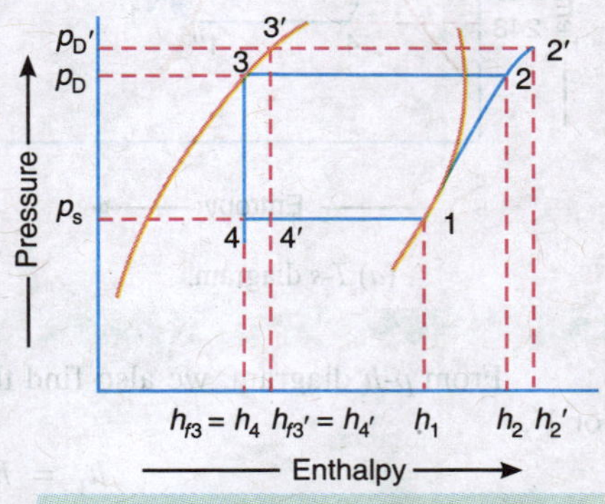

Fig. 4.35. Effect of discharge pressure.

1. decreases the refrigerating effect from $(h_1 - h_4)$ to $(h_1 - h_4')$, and
2. increases the work required for compression from $(h_2 - h_1)$ to $(h_2' - h_1)$.

From above, we see that the effect of increase in discharge pressure is similar to the effect of decrease in suction pressure. But the effect of increase in discharge pressure is not as severe on the refrigerating capacity of the system as that of decrease in suction pressure.

Example 4.25. *A simple ammonia-compression system operates with a capacity of 150 tonnes. The condensation temperature in the condenser is 35°C. The evaporation temperature in brine cooler is – 25°C. The ammonia leaves the evaporator and enters the compressor at – 8°C. Ammonia enters the expansion valve at 30°C. Wire drawing through the compressor valves :*

Suction = 0.118 bar ; Discharge = 0.23 bar ; Compression index = 1.22 ; Volumetric efficiency = 0.75.

Calculate : 1. Power ; 2. Heat transferred to cylinder water jacket ; 3. Piston displacement ; 4. Heat transfer in condenser ; and 5. Coefficient of performance.

Solution. Given : $Q = 150$ TR ; $T_2 = T_2'' = T_3' = 35°C = 35 + 273 = 308$ K ; $T_1'' = T_4 = -25°C = -25 + 273 = 248$ K ; $T_1 = -8°C = -8 + 273 = 265$ K ; $n = 1.22$; $\eta_v = 0.75$

The *T-s* and *p-h* diagrams are shown in Fig. 4.36 (*a*) and (*b*) respectively.

From *p-h* diagram, we find that the pressure corresponding to evaporation temperature of – 25°C,

$$p_1 = p_1'' = p_4 = 1.518 \text{ bar}$$

Since there is a suction pressure drop of 0.118 bar due to *wire drawing, therefore pressure at point 1′,

$$p_1' = 1.518 - 0.118 = 1.4 \text{ bar} = 1.4 \times 10^5 \text{ N/m}^2$$

Pressure corresponding to condensation temperature of 35°C

$$= 13.5 \text{ bar}$$

Since there is a discharge pressure drop of 0.23 bar due to wire drawing, therefore pressure at point 2′,

$$p_2' = 13.5 + 0.23 = 13.73 \text{ bar} = 13.73 \times 10^5 \text{ N/m}^2$$

Note: In Fig. 4.36, point 1 represents the inlet of the suction valve and point 1′ is the outlet of the suction valve. The point 2′ represents the inlet of discharge valve and point 2 is the outlet of the discharge valve.

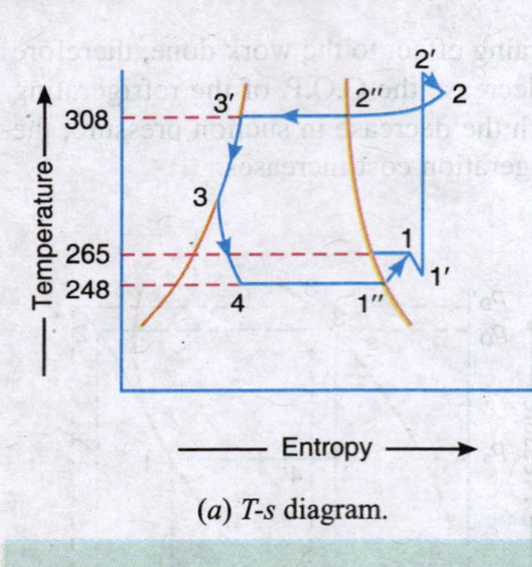

(*a*) *T-s* diagram.

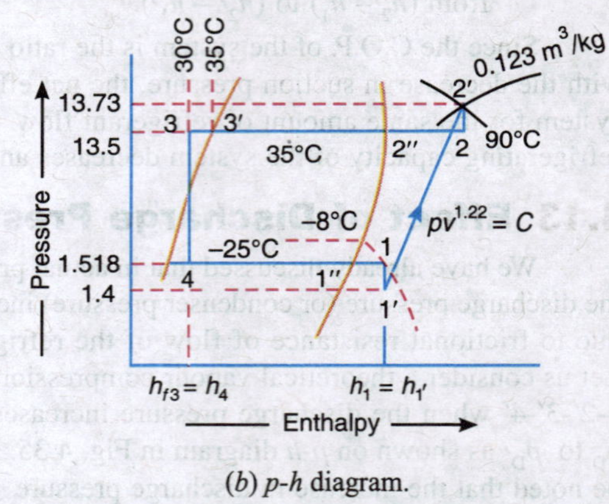

(*b*) *p-h* diagram.

Fig. 4.36

From *p-h* diagram, we also find that enthalpy of superheated ammonia vapours at point 1 or 1′,

$$h_1 = h_1' = 1440 \text{ kJ/kg}$$

Specific volume at point 1′,

$$v_1' = 0.8 \text{ m}^3\text{/kg}$$

Temperature at point 1′,

$$T_1' = -9°C = -9 + 273 = 264 \text{ K}$$

Let v_2' = Specific volume at point 2′.

Since the compression is according to $pv^{1.22} = C$, therefore

$$p_1' (v_1')^n = p_2' (v_2')^n$$

or

$$v_2' = v_1' \left(\frac{p_1'}{p_2'}\right)^{\frac{1}{n}} = 0.8 \left(\frac{1.4}{13.73}\right)^{\frac{1}{1.22}} = 0.123 \text{ m}^3\text{/kg}$$

* Wire drawing is a constant enthalpy process.

Now plot a point 2′ on the *p-h* diagram corresponding to $p_{2'}$ = 13.73 bar and $v_{2'}$ = 0.123 m³/kg. From the *p-h* diagram, we find that

Enthalpy of superheated ammonia vapours at point 2 or 2′,

$$h_2 = h_{2'} = 1620 \text{ kJ/kg}$$

Temperature at point 2′.

$$T_{2'} = 90°\text{C}$$

and enthalpy of liquid ammonia at point 3,

$$h_{f3} = h_4 = 320 \text{ kJ/kg}$$

1. *Power*

We know that refrigerating effect per kg,

$$R_E = h_1 - h_{f3} = 1440 - 320 = 1120 \text{ kJ/kg}$$

and refrigerating capacity $= 150 \text{ TR} = 150 \times 210 = 31\,500 \text{ kJ/min}$ ... (Given)

∴ Mass flow of the refrigerant,

$$m_R = \frac{31\,500}{1120} = 28.12 \text{ kg / min}$$

We know that work done by the compressor per minute

$$= m_R \times \frac{n}{n-1}(p_{2'}v_{2'} - p_{1'}v_{1'})$$

$$= 28.12 \times \frac{1.22}{1.22 - 1}(13.73 \times 10^5 \times 0.123 - 1.4 \times 10^5 \times 0.8)$$

$$= 89 \times 10^5 \text{ J/min} = 8900 \text{ kJ/min}$$

∴ Power = 8900/60 = 148.3 kW **Ans.**

2. *Heat transferred to cylinder water jacket*

We know that actual work done by the compressor

$$= m_R (h_{2'} - h_{1'}) = 28.12 (1620 - 1440) = 5062 \text{ kJ/min}$$

∴ Heat transferred to cylinder water jacket

$$= 8900 - 5062 = 3838 \text{ kJ/min}$$ **Ans.**

3. *Piston displacement*

We know that piston displacement

$$= \frac{m_R \times v_{1'}}{\eta_v} = \frac{28.12 \times 0.8}{0.75} = 30 \text{ m}^3\text{/min}$$ **Ans.**

4. *Heat transfer in condenser*

We know that heat transfer in condenser

$$= m_R (h_2 - h_{f3}) = 28.12 (1620 - 320) = 36\,556 \text{ kJ/min}$$ **Ans.**

5. *Coefficient of performance*

We know that coefficient of performance,

$$\text{C.O.P} = \frac{\text{Refrigerating capacity}}{\text{Work done}} = \frac{31\,500}{8900} = 3.54$$ **Ans.**

Geared turbo - compressor for the compression of vapours.

4.14 Improvements in Simple Saturation Cycle

The simple saturation cycle may be improved by the following methods :

1. By introducing the flash chamber between the expansion valve and the evaporator.
2. By using the accumulator or pre-cooler.
3. By subcooling the liquid refrigerant by the vapour refrigerant.
4. By subcooling the liquid refrigerant leaving the condenser by liquid refrigerant from the expansion valve.

The effect of the above mentioned methods on the simple saturation cycle are discussed, in detail, in the following pages :

4.15 Simple Saturation Cycle with Flash Chamber

We have already discussed that when the high pressure liquid refrigerant from the condenser passes through the expansion valve, some of it evaporates. This partial evaporation of the liquid refrigerant is known as *flash*. It may be noted that the vapour formed during expansion is of no use to the evaporator in producing refrigerating effect as compared to the liquid refrigerant which carries the heat in the form of latent heat. This formed vapour, which is incapable of producing any refrigeration effect, can be by - passed around the evaporator and supplied directly to the suction of the compressor. This is done

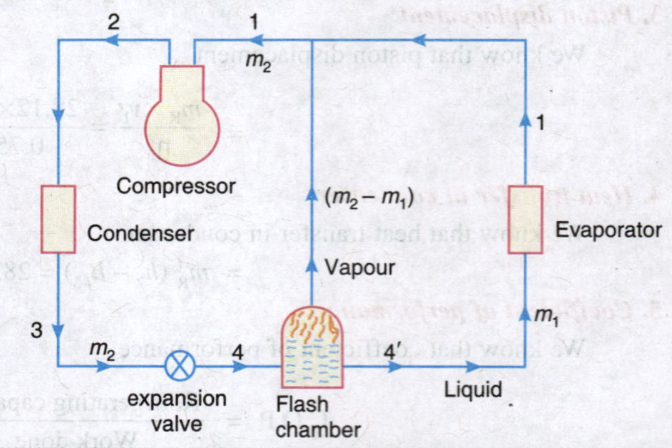

Fig. 4.37. Simple saturation cycle with flash chamber.

by introducing a flash chamber between the expansion valve and the evaporator as shown in Fig. 4.37. The flash chamber is an insulated container and it separates the liquid and vapour due to centrifugal effect. Thus the mass of the refrigerant passing through the evaporator reduces.

Let us consider that a certain amount of refrigerant is circulating through the condenser. This refrigerant after passing through the expansion valve, is supplied to the flash chamber which separates the liquid and vapour refrigerant. The liquid refrigerant from the flash chamber is supplied to the evaporator and the vapour refrigerant flows directly from the flash chamber to the suction of the compressor. The *p-h* diagram of the cycle is shown in Fig. 4.38.

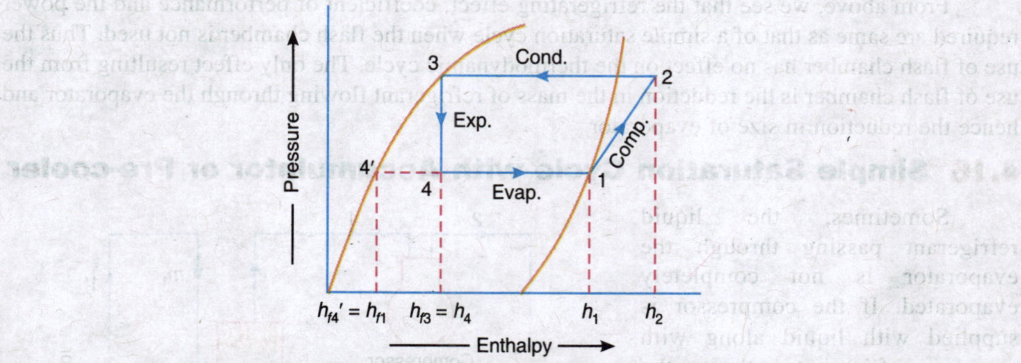

Fig. 4.38. *p-h* diagram of simple saturation cycle with flash chamber.

Let m_1 = Mass of liquid refrigerant supplied to the evaporator, and

m_2 = Mass of refrigerant (liquid and vapour) circulating through the condenser, or leaving the expansion valve.

∴ Mass of vapour refrigerant flowing directly from the flash chamber to the suction of the compressor

$$= m_2 - m_1$$

Now considering the thermal equilibrium of the flash chamber. Since the flash chamber is an insulated vessel, therefore, there is no heat exchange between the flash chamber and the atmosphere. In other words, the heat taken and given out by the flash chamber are same. Mathematically,

Heat taken by the flash chamber

= Heat given out by the flash chamber

or
$$m_2 h_4 = m_1 h_{f4}' + (m_2 - m_1)\, h_1$$

$$m_2\,(h_1 - h_4) = m_1\,(h_1 - h_{f4}')$$

$$\therefore \quad m_1 = m_2\left[\frac{h_1 - h_4}{h_1 - h_{f4'}}\right] = m_2\left[\frac{h_1 - h_{f3}}{h_1 - h_{f4'}}\right] \quad \ldots (\because h_4 = h_{f3}) \ldots (i)$$

We know that the heat extracted or refrigerating effect,

$$R_E = m_1\,(h_1 - h_{f4}')$$

$$= m_2\left(\frac{h_1 - h_{f3}}{h_1 - h_{f4'}}\right)(h_1 - h_{f4}') \quad \ldots \text{[From equation (i)]}$$

$$= m_2\,(h_1 - h_{f3})$$

and workdone in compressor,

$$W = m_2(h_2 - h_1)$$

$$\therefore \quad \text{C.O.P.} = \frac{R_E}{W} = \frac{m_2(h_1 - h_{f3})}{m_2(h_2 - h_1)} = \frac{h_1 - h_{f3}}{h_2 - h_1}$$

and power required to drive the compressor,

$$P = \frac{m_2(h_2 - h_1)}{60} \text{ kW}$$

From above, we see that the refrigerating effect, coefficient of performance and the power required are same as that of a simple saturation cycle when the flash chamber is not used. Thus the use of flash chamber has no effect on the thermodynamic cycle. The only effect resulting from the use of flash chamber is the reduction in the mass of refrigerant flowing through the evaporator and hence the reduction in size of evaporator.

4.16 Simple Saturation Cycle with Accumulator or Pre-cooler

Sometimes, the liquid refrigerant passing through the evaporator is not completely evaporated. If the compressor is supplied with liquid along with vapour refrigerant, then the compressor has to do an additional work of evaporating and raising the temperature of liquid refrigerant. It will also upset the normal working of the compressor which is meant only for compressing the pure vapour refrigerant.

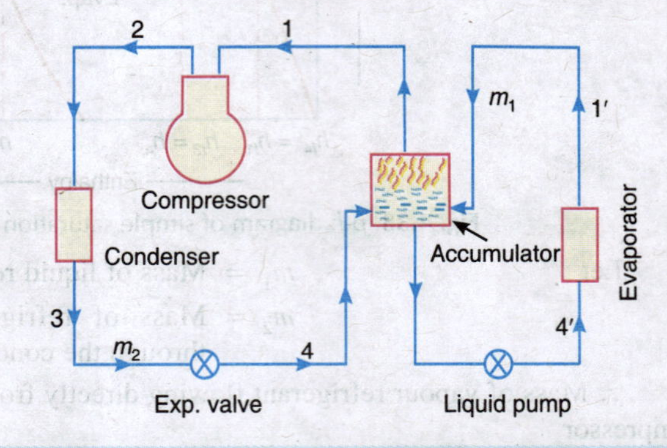

Fig. 4.39. Simple saturation cycle with accumulator or pre-cooler.

In order to avoid this difficulty, an insulated vessel, known as accumulator or pre-cooler, is used in the system, as shown in Fig. 4.39. The accumulator receives the discharge (a mixture of liquid and vapour refrigerant) from the expansion valve and supplies the liquid refrigerant only to the evaporator, as in the case of flash chamber.

The discharge from the evaporator is sent again to the accumulator which helps to keep off the liquid from entering the compressor. Thus the accumulator supplies dry and saturated vapour to the compressor. A liquid pump is provided in the system in order to maintain circulation of the refrigerant in the evaporator.

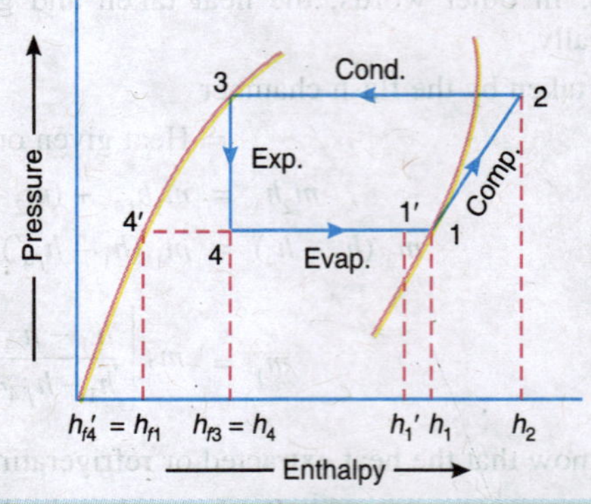

Fig. 4.40. *p-h* diagram of simple saturation cycle with accumulator.

Let $\quad m_1$ = Mass of liquid refrigerant circulating through the evaporator, and

m_2 = Mass of refrigerant flowing in the condenser.

When all the liquid refrigerant does not evaporate in the evaporator, it is represented by point 1′ on *p-h* diagram as shown in Fig. 4.40. Let the mass of refrigerant that leaves the evaporator at point 1′ is same *i.e.* m_1.

Consider the thermal equilibrium of the accumulator. Since the accumulator is an insulated vessel, therefore there is no heat exchange between the accumulator and the atmosphere. In other words, the heat taken in and given out by the accumulator is equal. Mathematically,

Heat taken in by the accumulator

= Heat given out by the accumulator

$$m_2 h_4 + m_1 h_{1'} = m_2 h_1 + m_1 h_{f4'}$$

$$m_1 (h_{1'} - h_{f4'}) = m_2 (h_1 - h_4)$$

$$m_1 = m_2 \left[\frac{h_1 - h_4}{h_{1'} - h_{f4'}}\right] = m_2 \left[\frac{h_1 - h_{f3}}{h_{1'} - h_{f4'}}\right] \quad \ldots (\because h_4 = h_{f3}) \ldots (i)$$

We know that the heat extracted or refrigerating effect,

$$R_E = m_1 (h_{1'} - h_{f4'})$$

$$= m_2 \left[\frac{h_1 - h_{f3}}{h_{1'} - h_{f4'}}\right] (h_{1'} - h_{f4'}) \quad \ldots \text{[From equation } (i)]$$

$$= m_2 (h_1 - h_{f3})$$

and workdone in compressor, $W = m_2 (h_2 - h_1)$

$$\therefore \quad \text{C.O.P.} = \frac{R_E}{W} = \frac{m_2 (h_1 - h_{f3})}{m_2 (h_2 - h_1)} = \frac{h_1 - h_{f3}}{h_2 - h_1}$$

and power required to drive the compressor,

$$P = \frac{m_2 (h_2 - h_1)}{60} \text{ kW}$$

From above, we see that when the accumulator is used in the system, the refrigerating effect, coefficient of performance, and power required is same as the simple saturation cycle. The accumulator is used only to protect the liquid refrigerant to flow into the compressor and thus dry compression is always ensured.

4.17 Simple Saturation Cycle with Sub-cooling of Liquid Refrigerant by Vapour Refrigerant

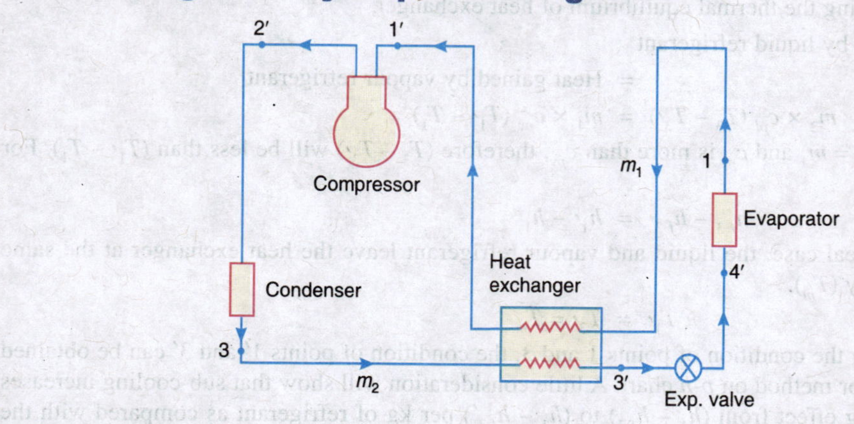

Fig. 4.41. Simple saturation cycle with sub-cooling of liquid refrigerant by vapour refrigerant.

We know that the liquid refrigerant leaving the condenser is at a higher temperature than the vapour refrigerant leaving the evaporator. The liquid refrigerant leaving the condenser can be subcooled by passing it through a heat exchanger which is supplied with saturated vapour from the evaporator as shown in Fig. 4.41. In the heat exchanger, the liquid refrigerant gives heat to the vapour refrigerant. The *p-h* diagram of the cycle is shown in Fig. 4.42.

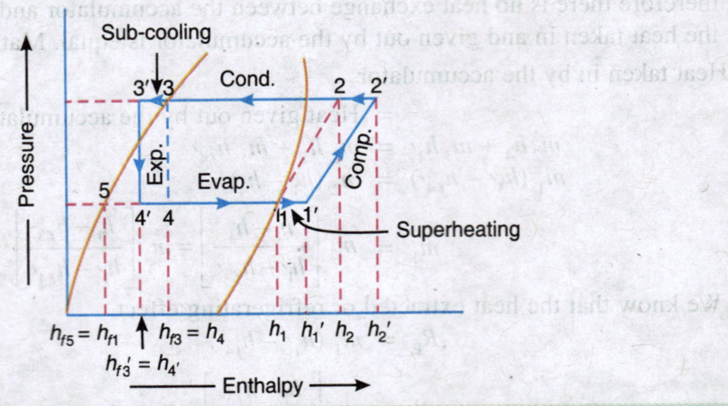

Fig. 4.42. *p-h* diagram of simple saturation cycle with sub-cooling of liquid refrigerant by vapour refrigerant.

Let m_1 = Mass of the vapour refrigerant,

m_2 = Mass of the liquid refrigerant,

T_3 = Temperature of liquid refrigerant entering the heat exchanger,

T_3' = Temperature of liquid refrigerant leaving the heat exchanger,

T_1 = Temperature of vapour refrigerant entering the heat exchanger,

T_1' = Temperature of vapour refrigerant leaving the heat exchanger,

c_{pv} = Specific heat of vapour refrigerant, and

c_{pl} = Specific heat of liquid refrigerant.

Considering the thermal equilibrium of heat exchanger,

Heat lost by liquid refrigerant

= Heat gained by vapour refrigerant

i.e.
$$m_2 \times c_{pl}\,(T_3 - T_3') = m_1 \times c_{pv}\,(T_1' - T_1)$$

Since $m_2 = m_1$ and c_{pl} is more than c_{pv}, therefore $(T_3 - T_3')$ will be less than $(T_1' - T_1)$. For energy balance,

$$h_{f3} - h_{f3}' = h_1' - h_1$$

In the ideal case, the liquid and vapour refrigerant leave the heat exchanger at the same temperature, say (T_m).

$$\therefore \quad T_1' = T_3' = T_m$$

Knowing the condition of points 1 and 3, the condition of points 1′ and 3′ can be obtained by trial and error method on *p-h* chart. A little consideration will show that sub-cooling increases the refrigerating effect from $(h_1 - h_{f3})$ to $(h_1 - h_{f3}')$ per kg of refrigerant as compared with the simple saturation cycle.

If Q tonnes of refrigeration is the load on the evaporator, then the mass of refrigerant (m_R) required to be circulated through the evaporator for sub-cooled cycle is given by

$$m_R = \frac{210\,Q}{h_1 - h_{f3'}} \text{ kg / min}$$

∴ Power required to drive the compressor,

$$P_1 = \frac{m_R(h_{2'} - h_{1'})}{60} \text{ kW}$$

$$= \frac{210\,Q}{60}\left[\frac{h_{2'} - h_{1'}}{h_1 - h_{f3'}}\right] \text{kW}$$

and power required to drive the compressor without heat exchanger (*i.e.* for simple saturation cycle),

$$P_2 = \frac{m_R(h_2 - h_1)}{60} \text{ kW}$$

$$= \frac{210\,Q}{60}\left[\frac{h_2 - h_1}{h_1 - h_{f3}}\right] \text{kW}$$

∴ Excess power required to drive the compressor as compared to simple saturation cycle,

$$P_{excess} = P_1 - P_2$$

$$= \frac{210\,Q}{60}\left[\frac{h_{2'} - h_{1'}}{h_1 - h_{f3'}} - \frac{h_2 - h_1}{h_1 - h_{f3}}\right] \text{kW}$$

From above, we see that sub-cooling the liquid refrigerant by vapour refrigerant, the coefficient of performance of the cycle is reduced. Even with theoretical loss resulting from the above type of sub-cooling, there are many actual installations which adopt this process.

Example 4.26. *For a vapour compression refrigeration system using R-22 as refrigerant, condenser outlet temperature is 40°C and the evaporator inlet temperature is –20°C. In order to avoid flashing of the refrigerant, a liquid-suction vapour heat exchanger is provided where liquid is subcooled to 26°C. The refrigerant leaves the evaporator as saturated vapour. The compression process is isentropic. Find the power requirement and coefficient of performance if capacity of the system is 10 kW at –20°C. Show the cycle on temperature-entropy and pressure-enthalpy diagrams. The specific heat of vapour refrigerant is 1.03 kJ/kg K. The thermodynamic properties are given below:*

Saturation temperature °C	*Pressure bar*	*Specific volume of vapour m³/kg*	*Specific enthalpy, kJ/kg*		*Specific entropy, kJ/kg K*	
			Liquid	*Vapour*	*Liquid*	*Vapour*
–20	*2.458*	*0.093*	*22.21*	*243.25*	*0.0908*	*0.9638*
26	*10.819*	*0.0218*	*79.74*	*260.64*	*0.2935*	*0.9014*
40	*15.489*	*0.0148*	*97.94*	*263.21*	*0.3563*	*0.8822*

Solution. Given : $T_2 = T_3 = 40°C = 40 + 273 = 313$ K ; $T_1 = T_{4'} = -20°C = -20 + 273 = 253$ K ; $T_{3'} = 26°C = 26 + 273 = 299$K ; $Q = 10$ kW = 10 kJ/s ; $c_{pv} = 1.03$ kJ/kg K ; $*v_1 = 0.093$ m³/kg ; $*h_{f1} = 22.21$ kJ/kg ; $h_1 = 243.25$ kJ/kg ; $*s_{f1} = 0.0908$ kJ/kg K ; $s_1 = 0.9638$ kJ/kg K ; $h_{f3'} = h_{4'} = 79.74$ kJ/kg; $* v_2 = 0.0148$ m³/kg ; $h_{f3} = 97.94$ kJ/kg ; $h_2 = 263.21$ kJ/kg ; $*s_{f3} = 0.3563$ kJ/kg K ; $s_2 = 0.8822$ kJ/kg K

The *T-s* and *p-h* diagrams are shown in Fig. 4.43 (*a*) and (*b*) respectively. Considering the thermal equilibrium of the heat exchanger,

Heat lost by liquid refrigerant = Heat gained by vapour refrigerant

i.e. $$h_{f3} - h_{f3'} = h_{1'} - h_1$$

$$97.94 - 79.74 = h_{1'} - 243.25$$

$$\therefore \quad h_{1'} = 97.94 - 79.74 + 243.25 = 261.45 \text{ kJ/kg}$$

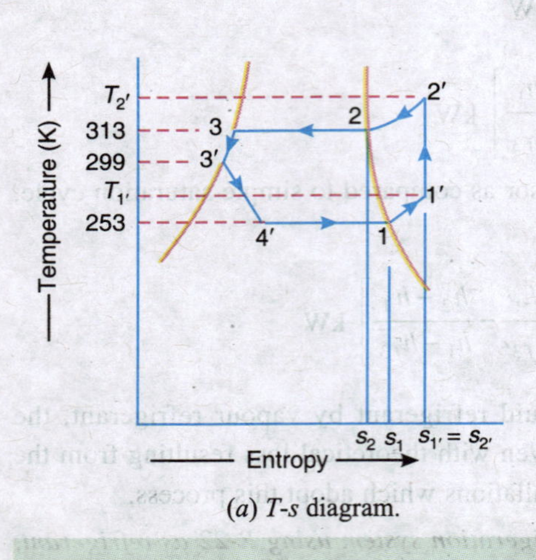

(*a*) *T-s* diagram.

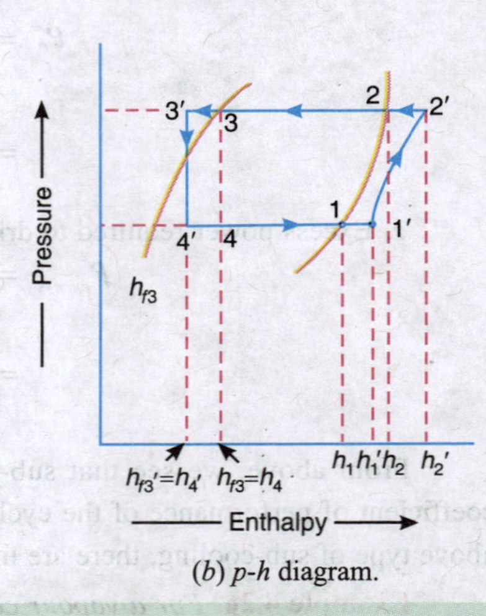

(*b*) *p-h* diagram.

Fig. 4.43

Let $T_{1'}$ = Temperature of vapour refrigerant leaving the heat exchanger.

We know that $$c_{pv}(T_{1'} - T_1) = h_{1'} - h_1$$

$$1.03(T_{1'} - 253) = 261.45 - 243.25 = 18.2$$

$$\therefore \quad T_{1'} = \frac{18.2}{1.03} + 253 = 270.67 \text{K}$$

Now let as find the temperature of superheated vapour at point $2'(T_{2'})$. We know that

* Superfluous data.

Entropy at point 1′,

$$s_{1'} = s_1 + 2.3\, c_{pv} \log\left(\frac{T_{1'}}{T_1}\right)$$

$$= 0.9638 + 2.3 \times 1.03 \log\left(\frac{270.67}{253}\right)$$

$$= 0.9638 + 2.3 \times 1.03 \times 0.0293 = 1.0332 \quad \text{...}(i)$$

and entropy at point 2′, $s_{2'} = s_2 + 2.3 c_{pv} \log\left(\frac{T_{2'}}{T_2}\right) = 0.8822 + 2.3 \times 1.03 \log\left(\frac{T_{2'}}{313}\right)$

$$= 0.8822 + 2.369 \log\left(\frac{T_{2'}}{313}\right) \quad \text{...}(ii)$$

Since the entropy at point 1′ is equal to entropy at point 2′, therefore equating equations (*i*) and (*ii*),

$$1.0332 = 0.8822 + 2.369 \log\left(\frac{T_{2'}}{313}\right)$$

or

$$\log\left(\frac{T_{2'}}{313}\right) = \frac{1.0332 - 0.8822}{2.369} = 0.063\,74$$

$$\frac{T_{2'}}{313} = 1.158 \quad \text{...(Taking antilog of 0.063 74)}$$

$$\therefore \quad T_{2'} = 1.158 \times 313 = 362.45 \text{ K}$$

and enthalpy at point 2′, $h_{2'} = h_2 + c_{pv}(T_{2'} - T_2)$

$$= 263.21 + 1.03\,(362.45 - 313) = 314.14 \text{ kJ/kg}$$

Power requirement

We know that refrigerating effect per kg of the refrigerant,

$$R_E = h_1 - h_{f3'} = 243.25 - 79.74 = 163.51 \text{ kJ/kg}$$

∴ Mass flow of the refrigerant,

$$m_R = \frac{Q}{R_E} = \frac{10}{163.51} = 0.0611 \text{ kg/s}$$

and workdone during compression of the refrigerant

$$= m_R\,(h_{2'} - h_{1'}) = 0.0611(314.14 - 261.45) = 3.22 \text{ kJ/s}$$

∴ Power required = 3.22 kJ/s or kW **Ans.**

2. Coefficient of performance

We know that coefficient of performance,

$$\text{C.O.P.} = \frac{h_1 - h_{f3'}}{h_{2'} - h_{1'}} = \frac{243.25 - 79.74}{314.14 - 261.45} = \frac{163.51}{52.69} = 3.1 \text{ Ans.}$$

4.18 Simple Saturation Cycle with Sub-cooling of Liquid Refrigerant by Liquid Refrigerant

We know that the liquid refrigerant leaving the condenser is at a higher temperature than the liquid refrigerant leaving the expansion valve. The liquid refrigerant leaving the condenser

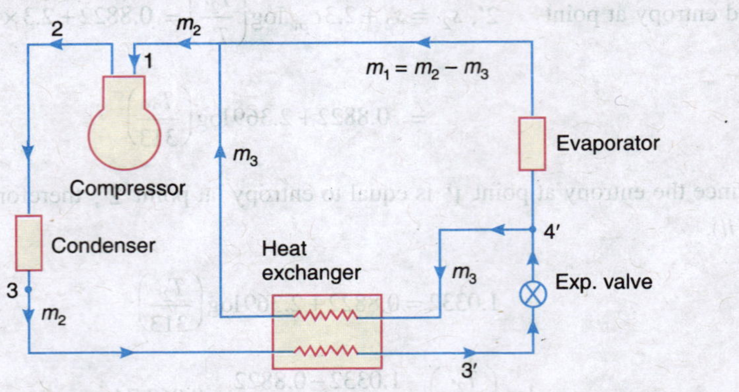

Fig. 4.44. Simple saturation cycle with sub-cooling of liquid refrigerant by liquid refrigerant.

can be sub-cooled by passing it through a heat exchanger which is supplied with liquid refrigerant from the expansion valve, as shown in Fig. 4.44. In the heat exchanger, the liquid refrigerant from the condenser gives heat to the liquid refrigerant from the expansion valve. The *p-h* diagram of the cycle is shown in Fig. 4.45.

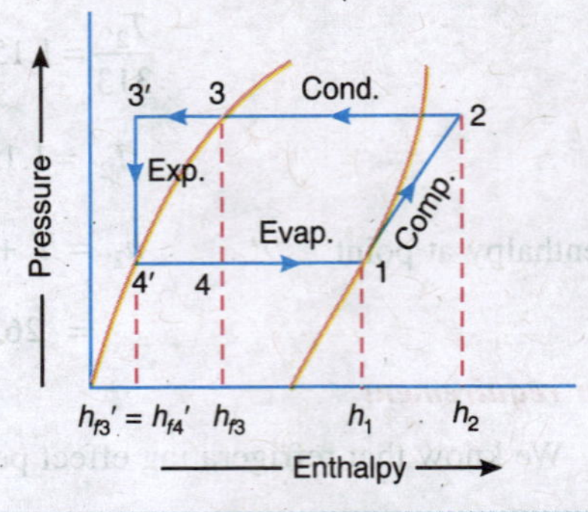

Fig. 4.45. *p-h* diagram of simple saturation cycle with sub-cooling of liquid refrigerant by liquid refrigerant.

Let m_1 = Mass of refrigerant leaving the evaporator,

m_2 = Mass of liquid refrigerant passing through the condenser, and

m_3 = Mass of liquid refrigerant supplied to the heat exchanger, from the expansion valve.

Considering the thermal equilibrium of the heat exchanger,

Heat lost by liquid refrigerant from condenser

= Heat gained by liquid refrigerant from expansion valve

i.e. $$m_2 (h_{f3} - h_{f3}') = m_3 (h_1 - h_{f4}')$$

$$\therefore \quad m_3 = m_2 \left[\frac{h_{f3} - h_{f3'}}{h_1 - h_{f4'}}\right] = m_2 \left[\frac{h_{f3} - h_{f3'}}{h_1 - h_{f3'}}\right]$$

$$\ldots (\because h_{f4'} = h_{f3'}) \quad \ldots (i)$$

We know that refrigerating effect,

$$R_E = m_1(h_1 - h_{f4}') = (m_2 - m_3)(h_1 - h_{f4}') \quad \ldots (\because m_1 = m_2 - m_3)$$

$$= \left[m_2 - m_2 \left(\frac{h_{f3} - h_{f3'}}{h_1 - h_{f4'}} \right) \right] (h_1 - h_{f4}')$$

... [From equation (*i*)]

$$= m_2 (h_1 - h_{f4}') - m_2 (h_{f3} - h_{f3}')$$

$$= m_2 h_1 - m_2 h_{f4}' - m_2 h_{f3} + m_2 h_{f3}'$$

$$= m_2 h_1 - m_2 h_{f3} = m_2 (h_1 - h_{f3}) \quad \ldots (\because m_2 h_{f4}' = m_2 h_{f3}')$$

and work done in compressor,

$$W = m_2 (h_2 - h_1)$$

$$\therefore \quad \text{C.O.P.} = \frac{R_E}{W} = \frac{m_2 (h_1 - h_{f3})}{m_2 (h_2 - h_1)} = \frac{h_1 - h_{f3}}{h_2 - h_1} \quad \ldots (ii)$$

If Q tonnes of refrigeration is the load on the evaporator, then the mass of refrigerant (m_1) required to be circulated through the evaporator is given by

$$m_1 = \frac{210\,Q}{h_1 - h_{f4'}}$$

or

$$m_2 - m_3 = \frac{210\,Q}{h_1 - h_{f4'}}$$

Substituting the value of m_3 from equation (*i*),

$$m_2 - m_2 \left[\frac{h_{f3} - h_{f3'}}{h_1 - h_{f4'}} \right] = \frac{210\,Q}{h_1 - h_{f4'}}$$

$$m_2 \left[1 - \frac{h_{f3} - h_{f3'}}{h_1 - h_{f4'}} \right] = \frac{210\,Q}{h_1 - h_{f4'}}$$

$$m_2 \left[\frac{h_1 - h_{f4'} - h_{f3} + h_{f3'}}{h_1 - h_{f4'}} \right] = \frac{210\,Q}{h_1 - h_{f4'}}$$

$$\therefore \quad m_2 = \frac{210\,Q}{h_1 - h_{f3}} \quad \ldots (\because h_{f4'} = h_{f3'})$$

We know that power required to drive the compressor,

$$P = \frac{m_2 (h_2 - h_1)}{60} \text{ kW}$$

Substituting the value of m_2 in the above equation,

$$P = \frac{210\,Q}{60} \left[\frac{h_2 - h_1}{h_1 - h_{f3}} \right] \text{ kW}$$

Notes: 1. From equation (*i*), we see that the mass of refrigerant required in the heat exchanger is exactly equal to the mass of flash that would form in a simple saturation cycle.

2. Since the C.O.P. of the cycle and power required to drive the compressor is same as that of simple saturation cycle, therefore this arrangement of sub-cooling the liquid refrigerant is of no advantage. In other words, the above method of subcooling is thermodynamically same as the simple saturation cycle.

EXERCISES

1. An ammonia refrigerator works between – 6.7°C and 26.7° C, the vapour being dry at the end of isentropic compression. There is no under-cooling of liquid ammonia and the liquid is expanded through a throttle valve after leaving the condenser. Sketch the cycle on the *T-s* and *p-h* diagram and calculate the refrigeration effect per kg of ammonia and the theoretical coefficient of performance of the unit with the help of the properties given below :

Temperature, °C	*Enthalpy, kJ/kg*		*Entropy, kJ/kg K*	
	Liquid	*Vapour*	*Liquid*	*Vapour*
–6.7	152.18	1437.03	0.6016	5.4308
26.7	307.18	1467.03	1.1515	5.0203

[**Ans.** 1028.3 kJ/kg ; 7.2]

2. An ammonia refrigerator produces 30 tonnes of ice from and at 0°C in 24 hours. The temperature range of the compressor is from 25°C to – 15°C. The vapour is dry saturated at the end of compression and an expansion valve is used. Assume a coefficient of performance to be 60% of the theoretical value. Calculate the power required to drive the compressor. Latent heat of ice = 335 kJ/ kg. Properties of ammonia are :

Temperature, °C	*Enthalpy, kJ/kg*		*Entropy, kJ/kg K*	
	Liquid	*Vapour*	*Liquid*	*Vapour*
25	298.9	1465.84	1.1242	5.0391
–15	112.34	1426.54	0.4572	5.5490

[**Ans.** 33.24 kW]

3. An ammonia refrigerating machine fitted with an expansion valve works between the temperature limits of – 10°C and 30°C. The vapour is 95% dry at the end of isentropic compression and the fluid leaving the condenser is at 30°C. If the actual coefficient of performance is 60% of the theoretical, find the ice produced per kW hour at 0°C from water at 10°C. The latent heat of ice is 335 kJ/kg. The ammonia has the following properties :

Temperature, °C	*Liquid heat, kJ/kg*	*Latent heat, kJ/kg*	*Entropy, kJ/kg K*	
			Liquid	*Vapour*
30	323.08	1145.79	1.2037	4.9842
–10	135.37	1297.68	0.5443	5.4770

[**Ans.** 33.24 kg/kWh]

4. A vapour compression works on a simple saturation cycle with R-12 as the refrigerant which operates between the condenser temperature of 40°C and an evaporator temperature of –5°C. For the modified cycle, the evaporator temperature is changed to –10°C and other operating conditions are the same as the original cycle. Compare the power requirement for both cycles. Both system develops 15 tonnes of refrigeration.

[**Ans.** 10.7 kW, 12.47 kW]

5. A R-12 refrigerating machine works on vapour-compression cycle. The temperature of refrigerant in the evaporator is – 20°C. The vapour is dry saturated when it enters the compressor and leaves it in a superheated condition. The condenser temperature is 30°C. Assuming specific heat at constant pressure for R-12 in the superheated condition as 1.884 kJ/kg K, determine :
   1. condition of vapour at the entrance to the condenser ;
   2. condition of vapour at the entrance to the evaporator ; and
   3. theoretical C.O.P. of the machine.

The properties of R-12 are :

Temperature, °C	*Enthalpy, kJ/kg*		*Entropy, kJ/kg K*	
	Liquid	*Vapour*	*Liquid*	*Vapour*
−20	17.82	178.73	0.0731	0.7087
30	64.59	199.62	0.2400	0.6843

[Ans. 33.8°C ; 29% dry ; 4.07]

6. It is proposed to replace R-12 by ozone friendly R-134a in a refrigeration plant of 10TR capacity with evaporator and condenser temperatures of 0°C and 40°C respectively. Considering standard saturation cycle (evaporator exit and condenser exit as saturated states), compare the mass flow rate, compressor work (in kW), condenser heat rejection (in kW) and C.O.P for the two refrigerants. The saturation properties and vapour specific heats are as follows:

Refrigerant	*Saturation temperature (°C)*	*Enthalpy (kJ/kg)*		*Entropy (kJ/kg K)*		*Specific heat (kJ/kg K)*
		Liquid	*Vapour*	*Liquid*	*Vapour*	
R-134a	0	200.81	398.78	1.0025	1.7261	—
	40	255.73	419.63	1.1884	1.7128	1.068
R-12	0	36.1	187.5	0.0142	0.6966	—
	40	74.6	203.2	0.2718	0.6825	0.776

[Ans. For R-134a : 0.245 kg/s ; 6.13 kW ; 41.13 kW ; 5.7
For R-12 : 0.31 kg/s ; 6.25 kW ; 41.25 kW ; 5.6]

7. A CO_2 refrigerating plant fitted with an expansion valve, works between the pressure limits of 54.81 bar and 20.93 bar. The vapour is compressed isentropically and leaves the compressor cylinder at 32°C. The condensation takes place at 18°C in the condenser and there is no undercooling of the liquid. Determine the theoretical coefficient of performance of the plant. The properties of CO_2 are :—

Pressure, bar	*Saturation temperature, °C*	*Enthalpy, kJ/kg*		*Entropy, kJ/kg K*	
		Liquid	*Vapour*	*Liquid*	*Vapour*
54.81	+18	137.48	302.55	0.5065	1.0738
20.93	−18	43.27	323.06	0.1733	1.2692

[Ans. 4.92]

8. A vapour compression plant using R-12 operates between 35°C condensing temperature and −5°C evaporation temperature with saturated vapour leaving the evaporator. The plant consists of twin cylinder, single acting compressor with 100 mm diameter and 120 mm stroke running at 300 r.p.m. The volumetric efficiency is 85% and the mechanical efficiency is 90%. Assuming isentropic compression, determine : 1. C.O.P. ; 2. Power required ; and 3. Tonnage capacity of the plant.

[Ans. 5.476 ; 2.87 kW ; 4.1 TR]

9. A single stage NH_3 refrigeration system has cooling capacity of 500 kW. The evaporator and condenser temperatures are − 10°C and 30°C respectively. Assuming saturation cycle, determine : 1. mass flow rate of refrigerant ; 2. adiabatic discharge temperature ; 3. compressor work in kW ; 4. condenser heat rejection ; 5. C.O.P. ; and 6. compressor swept volume in m^3/min, if volumetric efficiency is 70%.

The following values may be taken :

h_g (– 10°C) = 1431.6 kJ/kg ; h_f (30°C) = 322.8 kJ/kg ; v_g (–10°C) = 0.4185 m³/kg ;
s_g (– 10°C) = 5.4717 kJ/kg K.

The properties of superheated NH_3 at condenser pressure of 11.66 bar (30°C) are as follows :

At 85°C, h = 1621.8 kJ/kg ; s = 5.5484 kJ/kg K. At 90°C, h = 1634.5 kJ/kg ; s = 5.4838 kJ/kg K.

[Ans. 0.45 kg/s ; 88.3°C ; 89.5 kW ; 590 kW ; 5.585 ; 16.2 m³/min]

10. A refrigeration system of 10TR capacity at an evaporator temperature of –12°C, needs a condenser temperature of +28°C. The refrigerant NH_3 is subcooled by 5°C before entering the expansion valve. The vapour is 0.95 dry when it leaves the evaporator. Using p-h chart for NH_3, find : 1. Conditions of vapour at the outlet of compressor ; 2. Condition of vapour at entrance of evaporator ; 3. C.O.P. ; and 4. Power required. **[Ans.** 60°C ; 0.12 ; 5.7 ; 6.14 kW]

11. The evaporator and condenser temperatures in an NH_3 refrigeration system are – 10°C and 40°C respectively. Determine per TR basis : 1. mass flow rate ; 2. compressor work ; 3. condenser heat rejection; 4. C.O.P. ; and 5. refrigerating efficiency. Use only the properties of NH_3 given below :

Saturation temperature (t) °C	*Pressure (p) bar*	*Specific volume of vapour (v_g) m³/kg*	*Enthalpy, kJ/kg*		*Entropy, kJ/kg K*	
			Liquid	*Vapour*	*Liquid*	*Vapour*
–10	*2.908*	*0.418*	*134.95*	*1431.41*	*0.5435*	*5.4712*
40	*15.55*	*0.0833*	*371.47*	*1472.02*	*1.3574*	*4.8728*

For superheated NH_3 at 15.55 bar, the following values may be taken :

Superheat K	*Specific volume (v) m³/kg*	*Enthalpy (h) kJ/kg*	*Entropy (s) kJ/kg K*
60	*0.108*	*1647.9*	*5.3883*
80	*0.116*	*1700.3*	*5.5253*

[Ans. 0.198 kg/min ; 0.82 kW ; 4.32 kW ; 4.27 ; 81.2%]

12. In a vapour compression refrigeration system using R-12, the evaporator pressure is 1.4 bar and the condenser pressure is 8 bar. The refrigerant leaves the condenser sub-cooled to 30°C. The vapour leaving the evaporator is dry and saturated. The compression process is isentropic. The amount of heat rejected in the condenser is 13.42 MJ/min. Determine : 1. refrigerating effect in kJ/kg; 2. refrigerating load in TR; 3. compressor input in kW ; and 4. C.O.P.
Show the cycle on a *p-h* diagram. **[Ans.** 114 kJ/kg ; 49 TR ; 51.4 kW ; 3.35]

13. A vapour compression refrigerator works between the temperature limits of – 20°C and 25°C. The refrigerant leaves the compressor in dry saturated condition. If the liquid refrigerant is undercooled to 20°C before entering the throttle valve, determine :

1. work required to drive the compressor ;
2. refrigerating effect produced per kg of the refrigerant; and
3. theoretical C.O.P.

Assume specific heat of the refrigerant as 4.8. The properties of the refrigerant are :

Temperature, °C	*Enthalpy, kJ/kg*		*Entropy, kJ/kg K*	
	Liquid	*Vapour*	*Liquid*	*Vapour*
–20	*89.78*	*1420.02*	*0.3684*	*5.6244*
25	*298.90*	*1465.84*	*1.1242*	*5.0391*

[Ans. 189.7 kJ/kg ; 990.2 kJ/kg ; 5.01]

14. A food storage chamber requires a refrigeration system of 12 TR capacity with an evaporator temperature of –8°C and condenser temperature of 30°C. The refrigerant R-12 is subcooled by 5°C before entering the throttle valve, and the vapour is superheated by 6°C before entering the compressor. If the liquid and vapour specific heats are 1.235 and 0.733 kJ/kg K respectively, find : 1. refrigerating effect per kg; 2. mass of refrigerant circulated per minute; and 3. coefficient of performance.

The relevant properties of the refrigerant R-12 are given below :

Saturation temperature, °C	*Enthalpy, kJ/kg*		*Entropy, kJ/kg K*	
	Liquid	*Vapour*	*Liquid*	*Vapour*
–8	*28.70*	*184.06*	*0.1148*	*0.7007*
30	*64.59*	*199.62*	*0.2400*	*0.6853*

[Ans. 130.05 kJ/kg ; 19.4 kg ; 6.2 **]**

15. The following data refer to a single cylinder, single acting compressor of an ammonia refrigeration system :

Bore	= 100 mm
Stroke	= 150 mm
Speed	= 200 r.p.m.
Indicated mean effective pressure	= 3.2 bar
Condenser pressure	= 10 bar
Evaporator pressure	= 3 bar
Temperature of water at entry to condenser	= 55°C
Temperature of water at exit from condenser	= 20°C
Rate of cooling water flowing in the condenser	= 12.5 kg/min
Inlet water temperature	= 12.5°C
Outlet water temperature	= 20.5°C

If the mass of ice produced per hour from water at 15°C is 50 kg and the latent heat of ice is 335 kJ/kg, find ; (*a*) coefficient of performance ; (*b*) mass flow of ammonia per minute ; and (*c*) condition of ammonia entering the compressor.

The relevant properties of ammonia are given below :

Pressure, bar	*Saturation temperature, °C*	*Enthalpy, kJ/kg*		*Specific heat, kJ/kg K*	
		Liquid	*Vapour*	*Liquid*	*Vapour*
12	*31*	*327.9*	*1469.5*	*4.6*	*2.8*
2.9	*–10*	*135.4*	*1433*	—	—

[Ans. 4.46 ; 0.326 kg ; 0.86 **]**

16. A freezer of 20 TR capacity has evaporator and condenser temperatures of – 30°C and 25°C respectively. The refrigerant R-12 is sub-cooled by 4°C before it enters the expansion valve and is superheated by 5°C before leaving the evaporator. The compression is isentropic and the valve throttling and clearance are to be neglected. If a six cylinder, single acting compressor with stroke equal to bore running at 1000 r.p.m. is used, determine (*a*) C.O.P. of the refrigerating system, (*b*) mass of refrigerant to be circulated per min, (*c*) theoretical piston displacement per minute, and (*d*) theoretical bore and stroke of the compressor. The specific heat of liquid R-12 is 1.235 kJ/kg K and of vapour R-12 is 0.733 kJ/kg K.

The properties of R-12 are given below :

Saturation temp., °C	*Pressure, bar*	*Enthalpy, kJ/kg*		*Entropy, kJ/kg K*		*Specific volume, m³/kg*	
		Liquid	*Vapour*	*Liquid*	*Vapour*	*Liquid*	*Vapour*
–30	*1.0044*	*8.86*	*174.20*	*0.0371*	*0.7171*	*0.006 73*	*0.1596*
25	*6.5184*	*59.7*	*197.73*	*0.2239*	*0.6868*	*0.007 64*	*0.0269*

[**Ans.** 3.64 ; 34.12 kg/min ; 5.56 m³/min ; 0.106 m]

17. A vapour compression refrigeration system of 2400 kJ/min capacity works at an evaporator temperature of 263 K and a condenser temperature of 303 K. The refrigerant used is R-12 and is subcooled by 6°C before entering the expansion valve and vapour is superheated by 7°C before leaving the evaporator coil. The compression of refrigerant is reversible adiabatic. The refrigeration compressor is two cylinder, single acting with stroke equal to 1.25 times the bore and runs at 1000 r.p.m. Take liquid specific heat = 1.235 kJ/kg K and vapour specific heat = 0.733 kJ/kgK. Determine : 1. Refrigerating effect per kg ; 2. Mass of refrigerant circulated per minute ; 3. Theoretical piston displacement per minute ; 4. Power required to run the compressor ; 5. Heat removed in through condenser ; and 6. Bore and stroke of the compressor.

The properties of R-12 are given below:

Saturation temp., K	*Pressure bar*	*Specific volume of vapour, m³/kg*	*Enthalpy, kJ/kg*		*Entropy, kJ/kg K*	
			Liquid	*Vapour*	*Liquid*	*Vapour*
263	*2.19*	*0.0767*	*26.9*	*183.2*	*0.1080*	*0.7020*
303	*7.45*	*0.0235*	*64.6*	*199.6*	*0.2399*	*0.6854*

[**Ans.** 131.14 kJ/kg; 18.3 kg/min; 1.44 m³/min; 7 kW; 2820 kJ/min; 90mm, 112.5 mm]

18. A refrigeration plant of 8 TR capacity has its evaporation temperature of – 8°C and condenser temperature of 30°C. The refrigerant is sub-cooled by 5°C before entering into the expansion valve and vapour is superheated by 6°C before leaving the refrigerator. The suction pressure drop is 0.2 bar in the suction valve and discharge pressure drop is 0.1 bar in the discharge valve.

If the refrigerant used is R-12, find out the C.O.P. of the plant and theoretical power required for the compressor. Assume compression is isentropic. Use *p-h* chart for calculation.

QUESTIONS

1. Mention the advantages of vapour compression refrigeration system over air refrigeration system.
2. Describe the mechanism of a simple vapour compression refrigeration system.
3. Explain with reference to *T-s* diagram, the stages involved in vapour compression process of refrigeration. Establish an expression for the coefficient of performance.
4. Sketch the *T-s* and *p-h* diagrams for the vapour compression cycles when the vapour after compression is (*i*) dry saturated, and (*ii*) wet.
5. Why in practice a throttle valve is used in vapour compression refrigerator rather than an expansion cylinder to reduce pressure between the condenser and the evaporator ?
6. Explain the effect of subcooling of condensate with the help of *T-s* and *p-h* diagrams in vapour compression systems.
7. Describe a simple vapour compression refrigeration system without liquid subcooling and with superheated vapour after compression. Show the entire system on *T-s* and *p-h* planes. Why is superheating considered to be good in certain cases ?
8. How does an actual vapour compression cycle differ from that of a theoretical cycle ?

OBJECTIVE TYPE QUESTIONS

1. The coefficient of performance of vapour compression refrigeration system is quite .... as compared to air refrigeration system.
(*a*) low (*b*) high
2. During a refrigeration cycle, heat is rejected by the refrigerant in a
(*a*) compressor (*b*) condenser (*c*) evaporator (*d*) expansion valve
3. In a vapour compression refrigeration system, the condition of refrigerant before entering the compressor is
(*a*) saturated liquid (*b*) wet vapour
(*c*) dry saturated liquid (*d*) superheated vapour
4. The highest temperature during the cycle, in a vapour compression refrigeration system, occurs after
(*a*) compression (*b*) condensation (*c*) expansion (*d*) evaporation
5. In a vapour compression refrigeration system, the lowest temperature during the cycle occurs after
(*a*) compression (*b*) condensation (*c*) expansion (*d*) evaporation
6. In a vapour compression refrigeration system, the effect of superheating the vapour before suction to compression
(*a*) increases the work of compression
(*b*) increases the heat rejection in the condenser
(*c*) may increase or decrease C.O.P. depending upon the refrigerant used
(*d*) all of the above
7. In a domestic vapour compression refrigerator, the refrigerant commonly used is
(*a*) CO_2 (*b*) Ammonia (*c*) Freon - 12 (*d*) all of these
8. The subcooling is a process of cooling the refrigerant in vapour compression refrigeration system
(*a*) before compression (*b*) after compression
(*c*) before throttling (*d*) after throttling
9. In a vapour compression refrigeration system, subcooling the liquid refrigerant is to .... coefficient of performance.
(*a*) increase (*b*) decrease
10. The process of undercooling is generally brought about by
(*a*) circulating more quantity of cooling water through the condenser
(*b*) using water colder than the main circulating water
(*c*) employing a heat exchanger
(*d*) any one of the above

ANSWERS

1. (*b*)	2. (*b*)	3. (*d*)	4. (*a*)	5. (*d*)
6. (*d*)	7. (*c*)	8. (*c*)	9. (*a*)	10. (*d*)

CHAPTER

5

Compound Vapour Compression Refrigeration Systems

1. Introduction.
2. Advantages of Compound (or Multi-stage) Vapour Compression with Intercooler.
3. Types of Compound Vapour Compression with Intercooler.
4. Two Stage Compression with Liquid Intercooler.
5. Two Stage Compression with Water Intercooler and Liquid Sub-cooler.
6. Two Stage Compression with Water Intercooler, Liquid Sub-cooler and Flash Chamber.
7. Two Stage Compression with Water Intercooler, Liquid Sub-cooler and Flash Intercooler.
8. Three Stage Compression with Water Intercoolers.
9. Three Stage Compression with Flash Chambers.
10. Three Stage Compression with Flash Intercoolers.
11. Three Stage Compression with Multiple Expansion Valves and Flash Intercoolers.

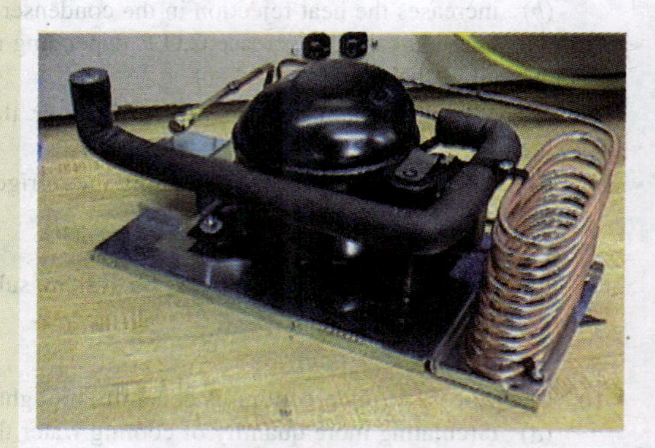

5.1 Introduction

In the previous chapter, we have discussed the simple vapour compression refrigeration system in which the low pressure vapour refrigerant from the evaporator is compressed in a single stage (or a single compressor) and then delivered to a condenser at a high pressure. But sometimes, the vapour refrigerant is required to be delivered at a very high pressure as in the case of low temperature refrigerating systems. In such cases either we should compress the vapour refrigerant by employing a single stage compressor with a very high pressure ratio between the condenser and evaporator or compress it in two or more compressors placed in series. The compression carried out in two or more compressors is called *compound* or *multistage compression.*

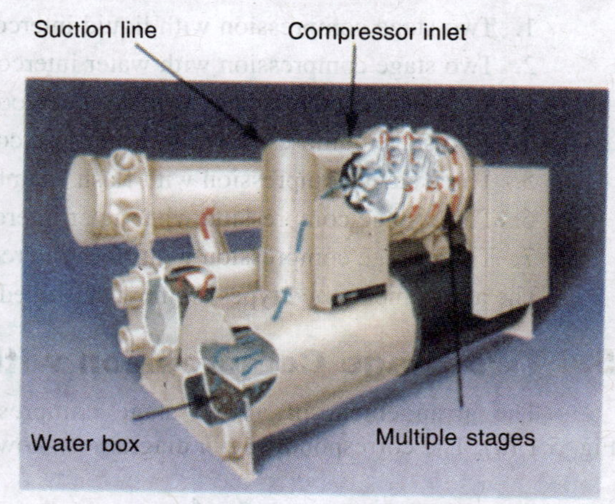

Multiple stages of compression.

In vapour compression refrigeration systems, the major operating cost is the energy input to the system in the form of mechanical work. Thus any method of increasing coefficient of performance is advantageous so long as it does not involve too heavy an increase in other operating expenses, as well as initial plant cost and consequent maintenance.

Since the coefficient of performance of a refrigeration system is the ratio of refrigerating effect to the compression work, therefore the coefficient of performance can be increased either by increasing the refrigerating effect or by decreasing the compression work. A little consideration will show that in a vapour compression system, the compression work is greatly reduced if the refrigerant is compressed very close to the saturated vapour line. This can be achieved by compressing the refrigerant in more stages with intermediate intercooling. But it is economical only where the pressure ratio is considerable as would be the case when very low evaporator temperatures are desired or when high condenser temperature may be required. The compound compression is generally economical in large plants.

The refrigerating effect can be increased by maintaining the condition of the refrigerant in more liquid state at the entrance to the evaporator. This can be achieved by expanding the refrigerant very close to the saturated liquid line. It may be noted that by subcooling the refrigerant and by removing the flashed vapour, as they are during multistage expansion, the expansion can be brought close to the liquid line.

5.2 Advantages of Compound (or Multi-stage) Vapour Compression with Intercooler

Following are the main advantages of compound or multi-stage compression over single stage compression :

1. The work done per kg of refrigerant is reduced in compound compression with intercooler as compared to single stage compression for the same delivery pressure.
2. It improves the volumetric efficiency for the given pressure ratio.
3. The sizes of the two cylinders (*i.e.* high pressure and low pressure) may be adjusted to suit the volume and pressure of the refrigerant.
4. It reduces the leakage loss considerably.
5. It gives more uniform torque, and hence a smaller size flywheel is needed.
6. It provides effective lubrication because of lower temperature range.
7. It reduces the cost of compressor.

5.3 Types of Compound Vapour Compression with Intercooler

In compound compression vapour refrigeration systems, the superheated vapour refrigerant leaving the first stage of compression is cooled by suitable method before being fed to the second stage of compression and so on. Such type of cooling the refrigerant is called *intercooling.* Though there are many types of compound compression with intercoolers, yet the following are important from the subject point of view :

1. Two stage compression with liquid intercooler.
2. Two stage compression with water intercooler.
3. Two stage compression with water intercooler, liquid subcooler and liquid flash chamber.
4. Two stage compression with water intercooler, liquid subcooler and flash intercooler.
5. Three stage compression with flash chambers.
6. Three stage compression with water intercoolers.
7. Three stage compression with flash intercoolers.

The above mentioned types are now discussed, in detail, one by one in the following pages.

5.4 Two Stage Compression with Liquid Intercooler

The arrangement of a two stage compression with liquid intercooler is shown in Fig. 5.1 (*a*). The corresponding *p-h* diagram is shown in Fig. 5.1 (*b*).

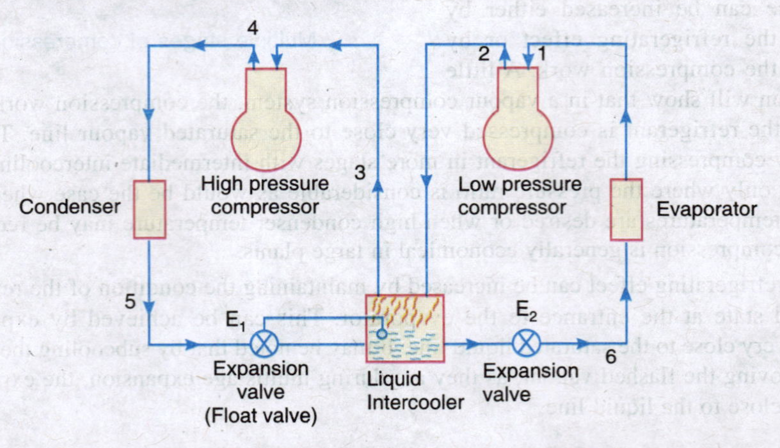

(*a*) Two stage compression with liquid intercooler.

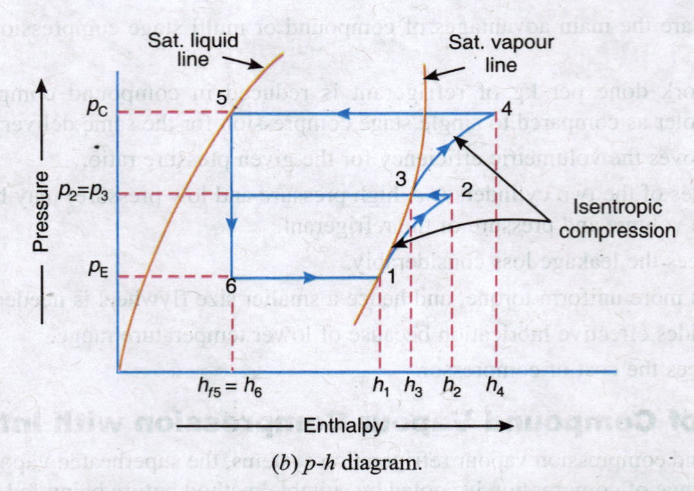

(*b*) *p-h* diagram.

Fig. 5.1

The various points on the *p-h* diagram are plotted as discussed below :

1. First of all, draw a horizontal pressure line representing the evaporator pressure p_E (or suction pressure of low pressure compressor) which intersects the saturated vapour line at point 1. At this point, the saturated vapour is supplied to the low pressure compressor. Let, at point 1, the enthalpy of the saturated vapour is h_1 and entropy s_{v1}.
2. The saturated vapour refrigerant admitted at point 1 is compressed isentropically in the low pressure compressor and delivers the refrigerant in a superheated state. The pressure rises from p_E to p_2. The curve 1-2 represents the isentropic compression in the low pressure compressor. In order to obtain point 2, draw a line from point 1, with entropy equal to s_{v1}, along the constant entropy line intersecting the intermediate pressure line p_2 at point 2. Let enthalpy at this point is h_2.
3. The superheated vapour refrigerant leaving the low pressure compressor at point 2 is cooled (or desuperheated) at constant pressure $p_2 = p_3$ in a liquid intercooler by the liquid refrigerant from the condenser. The refrigerant leaving the liquid intercooler is in saturated vapour state. The line 2-3 represents the cooling or desuperheating process. Let the enthalpy and entropy at point 3 is h_3 and s_{v3} respectively.
4. The dry saturated vapour refrigerant is now supplied to high pressure compressor where it is compressed isentropically from intermediate or intercooler pressure p_2 to condensor pressure p_C. The curve 3-4 represents the isentropic compression in the high pressure compressor. The point 4 on the *p-h* diagram is obtained by drawing a line of entropy equal to s_{v3} along the constant entropy line as shown in Fig. 5.1 (*b*). Let the enthalpy of superheated vapour refrigerant at point 4 is h_4.
5. The superheated vapour refrigerant leaving the high pressure compressor at point 4 is now passed through the condenser at constant pressure p_C as shown by a horizontal line 4-5. The condensing process 4-5 changes the state of refrigerant from superheated vapour to saturated liquid.
6. The high pressure saturated liquid refrigerant from the condenser is passed to the intercooler where some of liquid refrigerant evaporates in desuperheating the superheated vapour refrigerant from the low pressure compressor. In order to make up for the liquid evaporated, *i.e.* to maintain a constant liquid level, an expansion valve E_1 which acts as a float valve, is provided.
7. The liquid refrigerant from the intercooler is first expanded in an expansion valve E_2 and then evaporated in the evaporator to saturated vapour condition, as shown in Fig. 5.1(*b*).

Let m_1 = Mass of refrigerant passing through the evaporator (or low pressure compressor) in kg/min, and

m_2 = Mass of refrigerant passing through the condenser (or high pressure compressor) in kg/min.

The high pressure compressor in a given system will compress the mass of refrigerant from low pressure compressor (m_1) and the mass of liquid evaporated in the liquid intercooler during cooling or desuperheating of superheated vapour refrigerant from low pressure compressor. If m_3 is the mass of liquid evaporated in the intercooler, then

$$m_3 = m_2 - m_1$$

The value of m_2 may be obtained by considering the thermal equilibrium for the liquid intercooler as shown in Fig. 5.2, *i.e.*

Heat taken by the liquid intercooler = Heat given by the liquid intercooler

or $$m_2 h_{f5} + m_1 h_2 = m_1 h_6 + m_2 h_3$$

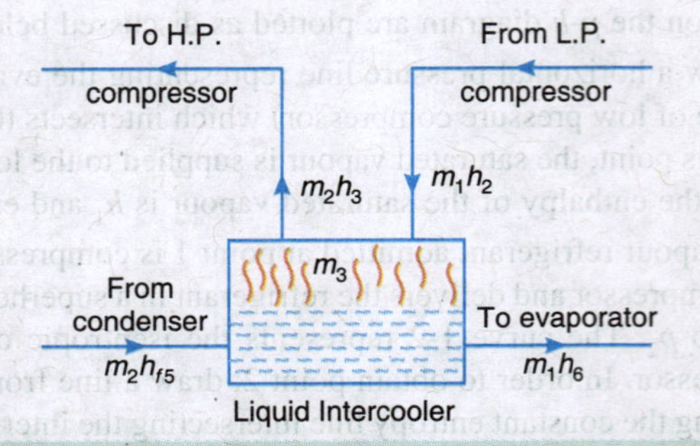

Fig. 5.2. Thermal equilibrium for liquid intercooler.

$$\therefore \quad m_2 = \frac{m_1(h_2 - h_6)}{h_3 - h_{f5}} = \frac{m_1(h_2 - h_{f5})}{h_3 - h_{f5}} \quad \ldots (\because h_6 = h_{f5})$$

and mass of liquid refrigerant evaporated in the intercooler,

$$^*m_3 = m_2 - m_1 = \frac{m_1(h_2 - h_{f5})}{h_3 - h_{f5}} - m_1 = \frac{m_1(h_2 - h_3)}{h_3 - h_{f5}}$$

We know that refrigerating effect,

$$R_E = m_1(h_1 - h_f) = m_1(h_1 - h_{f5}) = 210\ Q \text{ kJ/min}$$

where Q is the load on the evaporator in tonne of refrigeration.

Total workdone in both the compressors,

$$W = m_1(h_2 - h_1) + m_2(h_4 - h_3)$$

∴ Power required to drive the system,

$$P = \frac{m_1(h_2 - h_1) + m_2(h_4 - h_3)}{60} \text{ kW}$$

and C.O.P. of the system
$$= \frac{R_E}{W} = \frac{m_1(h_1 - h_{f5})}{m_1(h_2 - h_1) + m_2(h_4 - h_3)} = \frac{210\ Q}{P \times 60}$$

Notes: 1. In case of ammonia, when liquid refrigerant is used for intercooling, the total power requirement will decrease. It is due to the fact that the mass of liquid evaporated during intercooling is extremely small because of its high latent heat of vaporisation and the constant entropy lines of ammonia become very flat in the superheat region. Thus the intercooling by liquid refrigerant is commonly used in multi-stage ammonia plants, because of less power requirement.

2. In case of refrigerant R-12, when liquid refrigerant is used for intercooling, the total power requirements may actually increase. It is due to the fact that the latent heat of vaporisation is small and the constant entropy lines of R-12 does not change very much with the temperature. Thus in R-12 systems, the saving in work by performing the compression close to the saturated vapour line does not compensate for the increased mass flow rate through the high stage compressor. Therefore, intercooling by liquid refrigerant in R-12 systems, is never employed.

* The value of m_3 may be calculated by using the following heat balance equation :

$$m_3\, h_{f5} + m_1\, h_2 = m_2\, h_3$$

$$m_3\, h_{f5} + m_1\, h_2 = (m_1 + m_3)\, h_3 \quad \ldots (\because m_2 = m_1 + m_3)$$

$$\therefore \quad m_3 = \frac{m_1(h_2 - h_3)}{h_3 - h_{f5}}$$

Example 5.1. *Calculate the power needed to compress 20 kg/min of ammonia from saturated vapour at 1.4 bar to a condensing pressure of 10 bar by two-stage compression with intercooling by liquid refrigerant at 4 bar. Assume saturated liquid to leave the condenser and dry saturated vapours to leave the evaporator. Use the p-h chart.*

Determine, also, the power needed when intercooling is not employed.

Solution. Given : $m_1 = 20$ kg/min ; $p_E = 1.4$ bar ; $p_C = 10$ bar ; $p_2 = p_3 = 4$ bar

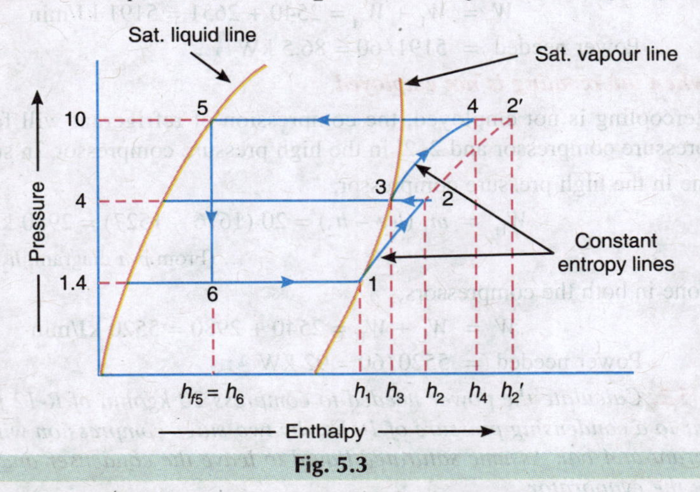

Fig. 5.3

The *p-h* diagram for a two stage compression with intercooling by liquid refrigerant is shown in Fig. 5.3. The various values for ammonia as read from the *p-h* diagram are as follows :

Enthalpy of saturated vapour refrigerant entering the low pressure compressor at point 1,

$$h_1 = 1400 \text{ kJ/kg}$$

Entropy of saturated vapour refrigerant entering the low pressure compressor at point 1,

$$s_1 = 5.75 \text{ kJ/kg K}$$

Enthalpy of superheated vapour refrigerant leaving the low pressure compressor at point 2,

$$h_2 = 1527 \text{ kJ/kg}$$

Enthalpy of saturated vapour refrigerant leaving the intercooler or entering the high pressure compressor at point 3,

$$h_3 = 1428 \text{ kJ/kg}$$

Entropy of saturated vapour refrigerant leaving the intercooler or entering the high pressure compressor at point 3,

$$s_3 = 5.39 \text{ kJ/kg K}$$

Enthalpy of superheated vapour refrigerant leaving the high pressure compressor at point 4,

$$h_4 = 1550 \text{ kJ/kg}$$

Enthalpy of saturated liquid refrigerant passing through the condenser at point 5,

$$h_{f5} = h_6 = 284 \text{ kJ/kg}$$

We know that mass of refrigerant passing through the condenser (or high pressure compressor),

$$m_2 = \frac{m_1(h_2 - h_{f5})}{h_3 - h_{f5}} = \frac{20\,(1527 - 284)}{1428 - 284} = 21.73 \text{ kg/min}$$

Work done in low pressure compressor,

$$W_L = m_1 (h_2 - h_1) = 20 (1527 - 1400) = 2540 \text{ kJ/min}$$

Work done in high pressure compressor,

$$W_H = m_2 (h_4 - h_3) = 21.73 (1550 - 1428) = 2651 \text{ kJ/min}$$

and total work done in both the compressors,

$$W = W_L + W_H = 2540 + 2651 = 5191 \text{ kJ/min}$$

$\therefore$ Power needed $= 5191/60 = 86.5$ kW **Ans.**

Power needed when intercooling is not employed

When intercooling is not employed, the compression of refrigerant will follow the path 1-2 in the low pressure compressor and 2–2′ in the high pressure compressor. In such a case,

Work done in the high pressure compressor,

$$W_H = m_1 (h_{2'} - h_2) = 20 (1676 - 1527) = 2980 \text{ kJ /min}$$

... (From *p-h* diagram, $h_{2'} = 1676$ kJ/kg)

and total workdone in both the compressors,

$$W = W_L + W_H = 2540 + 2980 = 5520 \text{ kJ/min}$$

$\therefore$ Power needed $= 5520/60 = 92$ kW **Ans.**

Example 5.2. *Calculate the power needed to compress 20 kg/min of R-12 from saturated vapour at 1.4 bar to a condensing pressure of 10 bar by two-stage compression with intercooling by liquid refrigerant at 4 bar. Assume saturated liquid to leave the condenser and dry saturated vapours to leave the evaporator.*

Use the p-h chart. Sketch the cycle on a skeleton p-h chart and label the values of enthalpy at salient points.

Solution. Given : $m_1 = 20$ kg/min ; $p_E = 1.4$ bar ; $p_C = 10$ bar ; $p_2 = p_3 = 4$ bar

The *p-h* diagram for a two-stage compression with intercooling by liquid refrigerant is shown in Fig. 5.4. The various values for R-12 as read from the *p-h* diagram are as follows :

Enthalpy of saturated vapour refrigerant entering the low pressure compressor at point 1,

$$h_1 = 178 \text{ kJ/kg}$$

Entropy of saturated vapour refrigerant entering the low pressure compressor at point 1,

$$s_1 = 0.71 \text{ kJ/kg K}$$

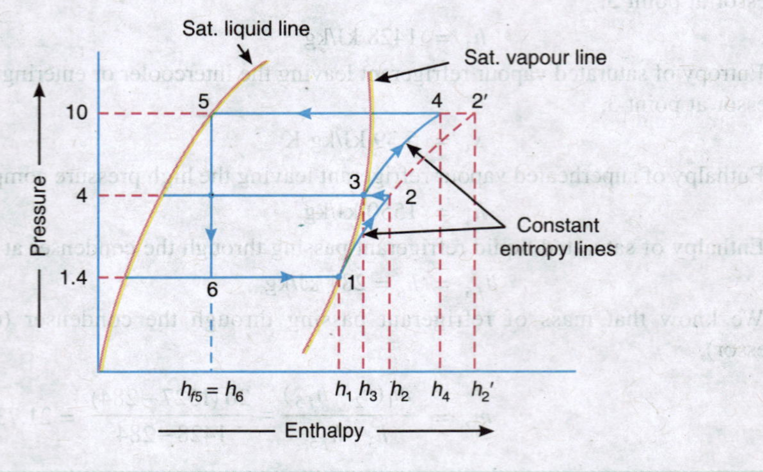

Fig. 5.4

Enthalpy of superheated vapour refrigerant leaving the low pressure compressor at point 2,

$$h_2 = 195 \text{ kJ/kg}$$

Enthalpy of saturated vapour refrigerant leaving the intercooler or entering the high pressure compressor at point 3,

$$h_3 = 191 \text{ kJ/kg}$$

Entropy of saturated vapour refrigerant entering the high pressure compressor at point 3,

$$s_3 = 0.695 \text{ kJ/kg K}$$

Enthalpy of superheated vapour refrigerant leaving the high pressure compressor at point 4,

$$h_4 = 210 \text{ kJ/kg}$$

Enthalpy of saturated liquid refrigerant leaving the condenser at point 5,

$$h_{f5} = h_6 = 77 \text{ kJ/kg}$$

We know that mass of refrigerant passing through the condenser (or high pressure compressor),

$$m_2 = \frac{m_1(h_2 - h_{f5})}{h_3 - h_{f5}} = \frac{20(195-77)}{191-77} = 20.7 \text{ kg/min}$$

Work done in low pressure compressor,

$$W_L = m_1 (h_2 - h_1) = 20 (195 - 178) = 340 \text{ kJ/min}$$

Work done in high pressure compressor,

$$W_H = m_2 (h_4 - h_3) = 20.7 (210 - 191) = 393 \text{ kJ/min}$$

and total work done in both the compressors,

$$W = W_L + W_H = 340 + 393 = 733 \text{ kJ/min}$$

$\therefore$ Power needed $= 733/60 = 12.2$ kW **Ans.**

Note: When intercooling is not employed, the compression of refrigerant will follow the path 1-2 in the low pressure compressor and 2-2′ in the high pressure compressor. In such a case

Work done in high pressure compressor,

$$W_H = m_1 (h_2' - h_2) = 20 (211 - 195) = 320 \text{ kJ/min}$$

... (From *p-h* chart, $h_2' = 211$ kJ/kg)

and total work done in both compressors,

$$W = W_L + W_H = 340 + 320 = 660 \text{ kJ/min}$$

$\therefore$ Power needed $= 660/60 = 11$ kW

From above we see that the power needed is more when intercooling by liquid refrigerant is employed than without intercooling.

5.5 Two Stage Compression with Water Intercooler and Liquid Sub-cooler

The arrangement of a two-stage compression with water intercooler and liquid sub-cooler is shown in Fig. 5.5 (*a*). The corresponding *p-h* diagram is shown in Fig. 5.5 (*b*). The various processes in this system are as follows :

1. The saturated vapour refrigerant at the evaporator pressure p_E is admitted to low pressure compressor at point 1. In this compressor, the refrigerant is compressed isentropically from the evaporator pressure p_E to the water intercooler pressure p_2, as shown by the curve 1-2 in Fig. 5.5 (*b*).
2. The refrigerant leaving the low pressure compressor at point 2 is in superheated state. This superheated vapour refrigerant is now passed through the water intercooler at constant pressure, in order to reduce the degree of superheat. The line 2-3 represents the water intercooling or desuperheating process.

3. The refrigerant leaving the water intercooler at point 3 (which is still in the superheated state) is compressed isentropically in the high pressure compressor to the condenser pressure p_C. The curve 3-4 shows the isentropic compression in high pressure compressor.
4. The discharge from the high pressure compressor is now passed through the condenser which changes the state of refrigerant from superheated vapour to saturated liquid as shown by process 4-5.
5. The temperature of the saturated liquid refrigerant is further reduced by passing it through a liquid sub-cooler as shown by process 5-6.
6. The liquid refrigerant from the sub-cooler is now expanded in an expansion valve (process 6-7) before being sent to the evaporator for evaporation (process 7-1).

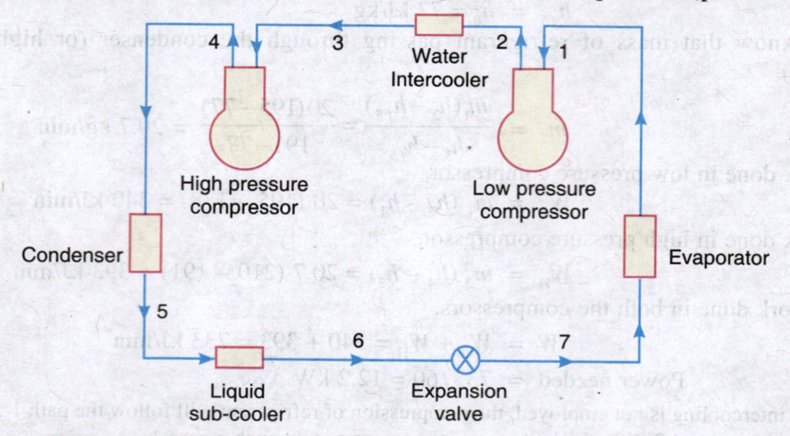

(*a*) Two stage compression with water intercooler and liquid sub-cooler.

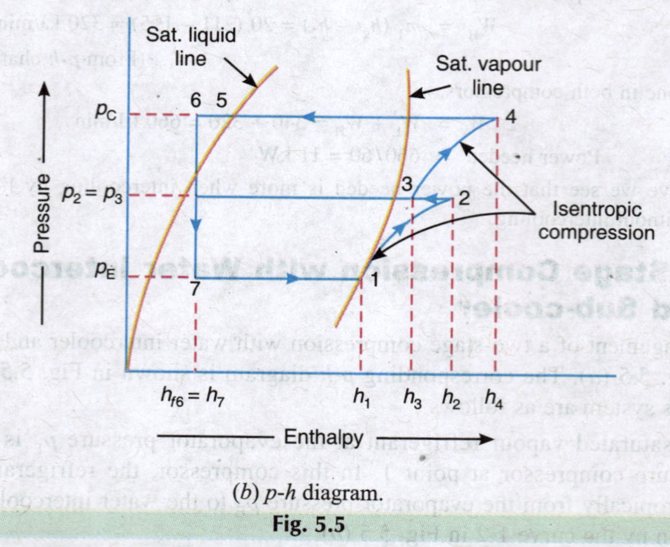

(*b*) *p-h* diagram.

Fig. 5.5

It may be noted that water intercooling reduces the work to be done in high pressure compressor. It also reduces the specific volume of the refrigerant which requires a compressor of less capacity (or stroke volume). The complete desuperheating of the vapour refrigerant is not possible in case of water intercooling. It is due to the fact that temperature of the cooling water used in the water intercooler is not available sufficiently low so as to desuperheat the vapour completely.

Let Q = Load on the evaporator in tonnes of refrigeration.

∴ Mass of refrigerant passing through the evaporator (or passing through the L.P. compressor),

$$m = \frac{210\,Q}{h_1 - h_7} = \frac{210\,Q}{h_1 - h_{f6}} \text{ kg/min} \qquad \ldots (\because h_7 = h_{f6})$$

Since the mass of refrigerant passing through the compressors is same, therefore, total work done in both the compressors,

$$W = \text{Work done in L.P. compressor + Work done in H.P. compressor}$$
$$= m\,(h_2 - h_1) + m\,(h_4 - h_3) = m\,[(h_2 - h_1) + (h_4 - h_3)]$$

∴ Power required to drive the system,

$$P = \frac{m[(h_2 - h_1) + (h_4 - h_3)]}{60} \text{ kW}$$

We know that refrigerating effect,

$$R_E = m\,(h_1 - h_{f6}) = 210\,Q \text{ kJ/min}$$

∴ C.O.P. of the system $= \dfrac{R_E}{W} = \dfrac{m\,(h_1 - h_{f6})}{m\,[(h_2 - h_1) + (h_4 - h_3)]} = \dfrac{210\,Q}{P \times 60}$

Example 5.3. *The following data refer to a two stage compression ammonia refrigerating system with water intercooler.*

Condenser pressure = 14 bar ; Evaporator pressure = 2 bar ; Intercooler pressure = 5 bar ; Load on the evaporator = 2TR.

If the temperature of the de-superheated vapour and sub-cooled liquid refrigerant are limited to 30°C, find (a) the power required to drive the system, and (b) C.O.P. of the system.

Solution. Given : $p_C = 14$ bar ; $p_E = 2$ bar ; $p_2 = p_3 = 5$ bar ; $Q = 10$ TR ; $t_3 = t_6 = 30°C$

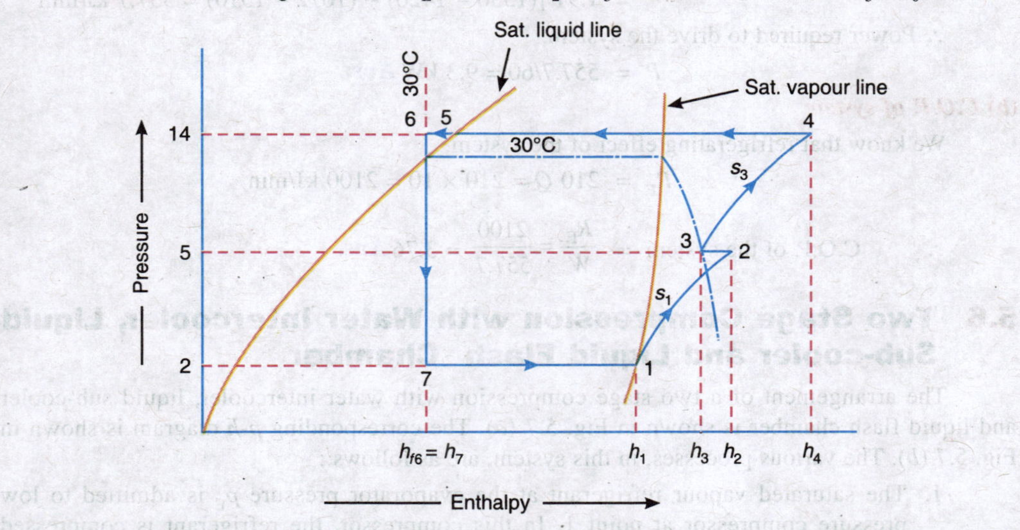

Fig. 5.6

The *p-h* diagram for a two stage compression system with water intercooler is shown in Fig. 5.6. The various values as read from the *p-h* diagram for ammonia are as follows :

Enthalpy of saturated vapour refrigerant entering the low pressure compressor at point 1,

$$h_1 = 1420 \text{ kJ/kg}$$

Entropy of saturated vapour refrigerant at point 1,

$$s_1 = 5.6244 \text{ kJ/kg K}$$

Enthalpy of superheated vapour refrigerant leaving the water intercooler at point 3,

$$h_3 = 1510 \text{ kJ/kg}$$

Entropy of superheated vapour refrigerant at point 3,

$$s_3 = 5.424 \text{ kJ/kg K}$$

Enthalpy of superheated vapour refrigerant leaving the high pressure compressor at point 4,

$$h_4 = 1672 \text{ kJ/kg}$$

Enthalpy of liquid refrigerant leaving the liquid sub-cooler,

$$h_{f6} = h_7 = 323 \text{ kJ/kg}$$

The points 2 and 4 on the *p-h* diagram are obtained in the similar way as discussed in Art. 5.3.

From the *p-h* diagram, we find that enthalpy of superheated vapour refrigerant at point 2,

$$h_2 = 1550 \text{ kJ/kg}$$

(a) Power required to drive the system

We know that mass of refrigerant circulating through the system,

$$m = \frac{210\,Q}{h_1 - h_{f6}} = \frac{210 \times 10}{1420 - 323} = 1.91 \text{ kg/min}$$

Total work done in both the compressors,

$$W = m\,[(h_2 - h_1) + (h_4 - h_3)]$$

$$= 1.91\,[(1550 - 1420) + (1672 - 1510) = 557.7 \text{ kJ/min}$$

∴ Power required to drive the system,

$$P = 557.7/60 = 9.3 \text{ kW}$$ **Ans.**

(b) C.O.P. of system

We know that refrigerating effect of the system,

$$R_E = 210\,Q = 210 \times 10 = 2100 \text{ kJ/min}$$

∴ C.O.P. of the system $= \dfrac{R_E}{W} = \dfrac{2100}{557.7} = 3.76$ **Ans.**

5.6 Two Stage Compression with Water Intercooler, Liquid Sub-cooler and Liquid Flash Chamber

The arrangement of a two stage compression with water intercooler, liquid sub-cooler and liquid flash chamber is shown in Fig. 5.7 (*a*). The corresponding *p-h* diagram is shown in Fig. 5.7 (*b*). The various processes, in this system, are as follows :

1. The saturated vapour refrigerant at the evaporator pressure p_E is admitted to low pressure compressor at point 1. In this compressor, the refrigerant is compressed isentropically from evaporator pressure p_E to water intercooler (or flash chamber) pressure p_F as shown by the curve 1-2 in Fig. 5.7 (*b*).

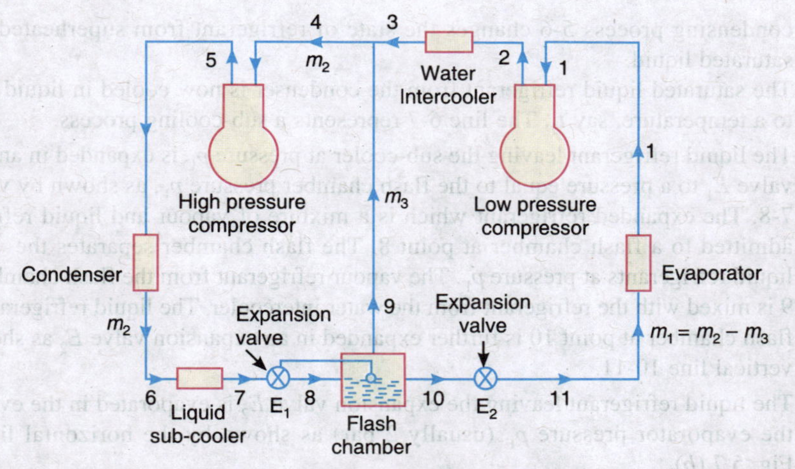

(*a*) Two stage compression with water intercooler, liquid sub-cooler and liquid flash chamber.

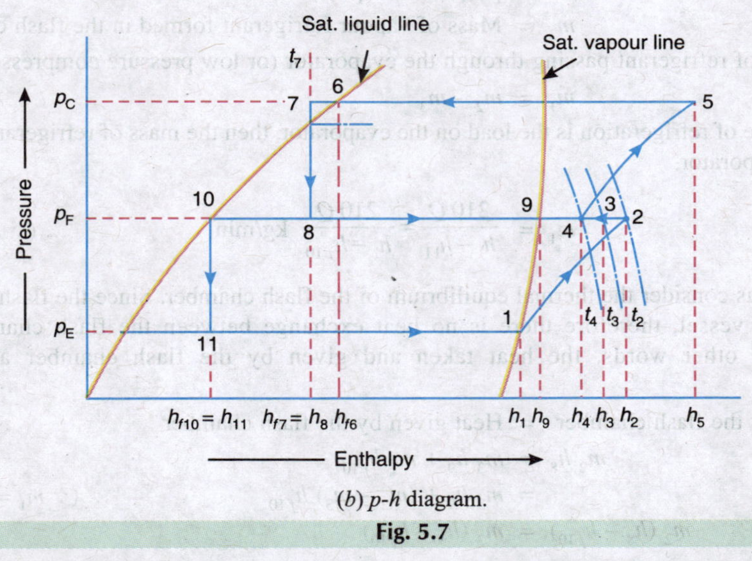

(*b*) *p-h* diagram.

Fig. 5.7

2. The superheated vapour refrigerant leaving the low pressure compressor at point 2 is now passed through the water intercooler at constant pressure p_F in order to reduce the degree of superheat (*i.e.* from temperature t_2 to t_3). The line 2-3 represents the water intercooling or de-superheating process.
3. The superheated vapour refrigerant leaving the water intercooler at point 3 is mixed with the vapour refrigerant supplied by the flash chamber at point 9. The condition of refrigerant after mixing is shown by point 4 which is in superheated state. Let the temperature at this point is t_4.
4. The superheated vapour refrigerant admitted at point 4 to the high pressure compressor is compressed isentropically from the intercooler or flash chamber pressure p_F to condenser pressure p_C as shown by the curve 4-5. The temperature rises from t_4 to t_5.
5. The superheated vapour leaving the high pressure compressor at pressure p_C is passed through a condenser at constant pressure as shown by a horizontal line 5-6. The

condensing process 5-6 changes the state of refrigerant from superheated vapour to saturated liquid.

6. The saturated liquid refrigerant from the condenser is now cooled in liquid sub-cooler to a temperature, say t_7. The line 6-7 represents a sub-cooling process.
7. The liquid refrigerant leaving the sub-cooler at pressure p_C is expanded in an expansion valve E_1 to a pressure equal to the flash chamber pressure p_F, as shown by vertical line 7-8. The expanded refrigerant which is a mixture of vapour and liquid refrigerants is admitted to a flash chamber at point 8. The flash chamber separates the vapour and liquid refrigerants at pressure p_F. The vapour refrigerant from the flash chamber at point 9 is mixed with the refrigerant from the water intercooler. The liquid refrigerant from the flash chamber at point 10 is further expanded in an expansion valve E_2 as shown by the vertical line 10-11.
8. The liquid refrigerant leaving the expansion valve E_2 is evaporated in the evaporator at the evaporator pressure p_E (usually 2 bar) as shown by the horizontal line 11-1 in Fig. 5.7 (*b*).

Let m_2 = Mass of refrigerant passing through the condenser (or high pressure compressor), and

m_3 = Mass of vapour refrigerant formed in the flash chamber.

∴ Mass of refrigerant passing through the evaporator (or low pressure compressor),

$$m_1 = m_2 - m_3$$

If Q tonne of refrigeration is the load on the evaporator, then the mass of refrigerant passing through the evaporator,

$$m_1 = \frac{210\,Q}{h_1 - h_{11}} = \frac{210\,Q}{h_1 - h_{f10}} \text{ kg/min} \qquad \ldots (\because h_{11} = h_{f10})$$

Now let us consider the thermal equilibrium of the flash chamber. Since the flash chamber is an insulated vessel, therefore there is no heat exchange between the flash chamber and atmosphere. In other words, the heat taken and given by the flash chamber are same. Mathematically,

Heat taken by the flash chamber = Heat given by the flash chamber

or
$$m_2\, h_8 = m_3\, h_9 + m_1\, h_{f10}$$
$$= m_3\, h_9 + (m_2 - m_3)\, h_{f10} \qquad \ldots (\because m_1 = m_2 - m_3)$$
$$m_2\,(h_8 - h_{f10}) = m_3\,(h_9 - h_{f10})$$
$$\therefore \quad m_3 = m_2\left(\frac{h_8 - h_{f10}}{h_9 - h_{f10}}\right) = m_2\left(\frac{h_{f7} - h_{f10}}{h_9 - h_{f10}}\right) \qquad \ldots (\because h_8 = h_{f7}) \quad \ldots (i)$$

The vapour refrigerant from the water intercooler (represented by point 3) is mixed with vapour refrigerant m_3 from the flash chamber (represented by point 9) at the same pressure before entering the high pressure compressor. The enthalpy of the mixed refrigerant (represented by point 4) may be calculated by using the equation,

$$m_2\, h_4 = m_3\, h_9 + m_1\, h_3$$
$$= m_3\, h_9 + (m_2 - m_3)\, h_3$$

We know that refrigerating effect of the system,

$$R_E = m_1\,(h_1 - h_{11}) = 210\,Q \text{ kJ/min}$$

Work done in low pressure compressor,

$$W_L = m_1 (h_2 - h_1)$$

Work done in high pressure compressor,

$$W_H = m_2 (h_5 - h_4)$$

Total workdone in both the compressors,

$$W = W_L + W_H = m_1 (h_2 - h_1) + m_2 (h_5 - h_4)$$

∴ Power required to drive the system,

$$P = \frac{m_1(h_2 - h_1) + m_2(h_5 - h_4)}{60} \text{ kW}$$

and C.O.P. of the system

$$= \frac{R_E}{W} = \frac{m_1(h_1 - h_{11})}{m_1(h_2 - h_1) + m_2(h_5 - h_4)} = \frac{210\,Q}{P \times 60}$$

Note: Since the mass of vapour refrigerant m_1 is cooled in the water intercooler from condition 2 to 3, therefore cooling capacity of the intercooler

$$= m_1 (h_2 - h_3)$$

Example 5.4. *A two stage refrigerating system is operating between the pressure limits of 8 bar and 1.4 bar. The working fluid is R-134a. The refrigerant leaves the condener as a saturated liquid and is throttled to a flash chamber operating at 3.2 bar. The part of refrigerant evaporates during the flashing process and this vapour is mixed with the refrigerant leaving the low pressure compressor. The mixture is then compressed to the condensor pressure by the high pressure compressor. The liquid in the flash chamber is throttled to the evaporator pressure and cools the refrigerated space as it vaporises in the evaporator. Assuming the refrigerant leaves the evaporator as a saturated vapour and both compressions are isentropic, determine :*

1. *The fraction of refrigerant that evaporates as it is throttled to the flash chamber ;*
2. *The amount of heat removed from the refrigerated space and the compressor work per unit mass of refrigerant flowing through the condensor ; and*
3. *The coefficient of performance.*

Solution. Given : $p_C = 8$ bar ; $p_E = 1.4$ bar ; $p_F = 3.2$ bar

The arrangement for a two stage refrigerating system with the given conditions is shown in Fig 5.8 (*a*) and the corresponding *p-h* diagram is shown in Fig. 5.8 (*b*). The various values as read from the *p-h* diagram for *R*-134*a* are as follows :

Enthalpy of saturated vapour refrigerant entering the low pressure compressor at point 1,

$$h_1 = 387 \text{ kJ/kgK}$$

Entropy of saturated vapour refrigerant entering the low pressure compressor at point 1,

$$s_1 = 1.7387 \text{ kJ/kgK}$$

Enthalpy of superheated vapour refrigerant leaving the low pressure compressor at point 2,

$$h_2 = 404 \text{ kJ/kg}$$

Enthalpy of saturated vapour refrigerant leaving the flash chamber at point 8,

$$h_8 = 400 \text{ kJ/kg}$$

Enthalpy of saturated liquid refrigerant leaving the condenser at point 5,

$$h_{f5} = h_6 = 244 \text{ kJ/kg}$$

Enthalpy of saturated liquid refrigerant leaving the flash chamber at point 7,

$$h_{f7} = h_9 = 203 \text{ kJ/kg}$$

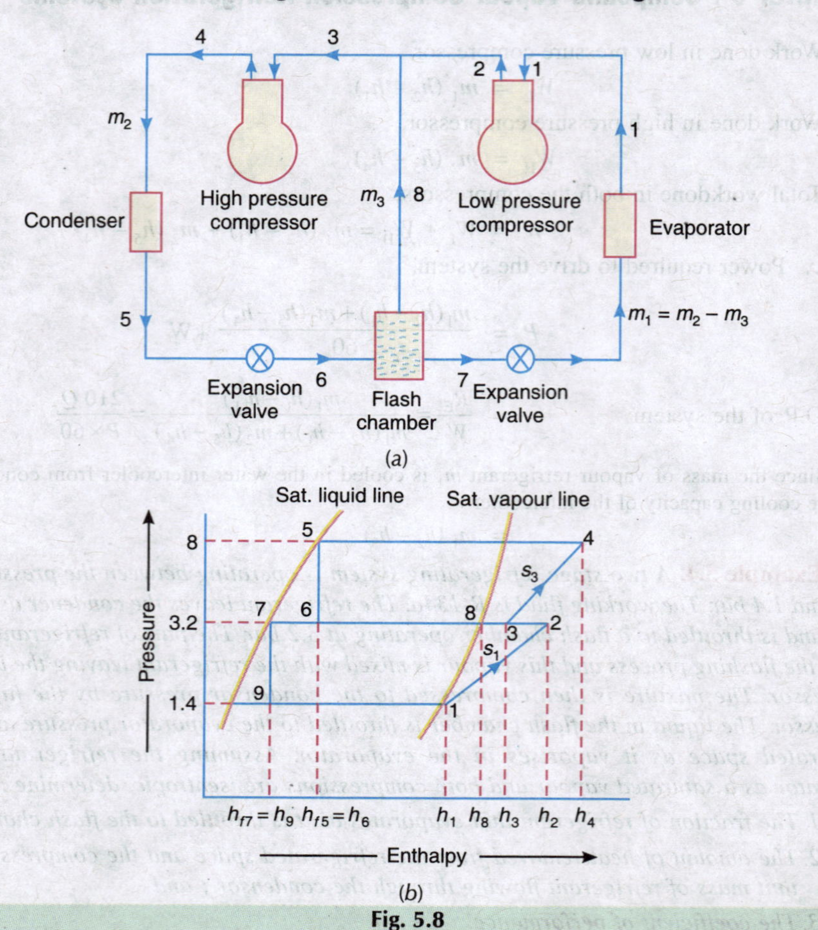

Fig. 5.8

1. *Fraction of refrigerant that evaporates in the flash chamber*

Considering the mass of refrigerant passing through the condenser (or high pressure compressor) be 1 kg, *i.e.* taking

$$m_2 = 1 \text{ kg}$$

Let m_3 = Mass of vapour refrigerant formed in the flash chamber.

$\therefore$ Mass of refrigerant passing through the evaporator (or low pressure compressor),

$$m_1 = m_2 - m_3$$

For the thermal equilibrium of the flash chamber, the heat taken by the flash chamber must be equal to the heat given by the flash chamber. In other words

$$m_2 h_6 = m_3 h_8 + m_1 h_{f7}$$

$$1 \times 244 = m_3 \times 400 + (1 - m_3) 203$$

$$= 400\, m_3 + 203 - 203\, m_3 = 197\, m_3 + 203$$

$$\therefore \quad m_3 = \frac{244 - 203}{197} = 0.208 \text{ kg}$$

It means that fraction of refrigerant that evaporates in the flash chamber is 0.208 of the refrigerant passing through the condenser. In other words, $m_3 / m_2 = 0.208$ **Ans.**

2. Amount of heat removed from the rerigerated space and the compressor work

The vapour refrigerant as represented by point 2 is mixed with vapour refrigerant from the flash chamber as represented by point 8. The enthalpy of the mixed refrigerant entering the high pressure compressor as represented by point 3 is given by

$$m_2 h_3 = m_3 h_8 + m_1 h_2 = m_3 h_8 + (m_2 - m_3) h_2$$

or

$$h_3 = \frac{m_3}{m_2} \times h_8 + h_2 \left(1 - \frac{m_3}{m_2}\right)$$

$$= 0.208 \times 400 + 404\ (1 - 0.208) = 403 \text{ kJ/kg}$$

We see from *p-h* diagram that at point 3 (intersection of pressure 3.2 bar and enthalpy (h_3) of 403 kJ/kg), the entropy is $s_3 = 1.736$ kJ/kg. Now from point 3, draw a line of entropy equal to 1.736 kJ/kg K along the constant entropy line which intersects the condenser pressure (8 bar) line at point 4. From *p-h* diagram, we find that enthalpy of refrigerant leaving the high pressure compressor (or entering the condenser) at point 4 is

$$h_4 = 422 \text{ kJ/kg}$$

∴ Amount of heat removed from the refrigerated space or refrigerated effect,

$$R_E = m_1\ (h_1 - h_9) = (m_2 - m_3)\ (h_1 - h_{f7})$$

$$\ldots (\because m_1 = m_2 - m_3 \text{ ; and } h_9 = h_{f7})$$

$$= (1 - 0.208)\ (387 - 203) = 145.7 \text{ kJ} \textbf{ Ans.}$$

We know that compressor work *i.e.* workdone in both the compressors,

$$W = \text{Workdone is L.P. compressor + Workdone in H.P. compressor}$$

$$= m_1\ (h_2 - h_1) + m_2\ (h_4 - h_3)$$

$$= (1 - 0.208)\ (404 - 387) + 1(422 - 403)$$

$$= 13.46 + 19 = 32.46 \text{ kJ} \textbf{ Ans.}$$

3. Coefficient of performance

We know that coefficient of performance

$$= \frac{R_E}{W} = \frac{145.7}{32.46} = 4.5 \textbf{ Ans.}$$

Example 5.5. *A two stage compression ammonia refrigeration system operates between overall pressure limits of 14 bar and 2 bar. The temperature of the desuperheated vapour and subcooled liquid refrigerant are limited to 30°C. The flash tank separates dry vapour at 5 bar pressure and the liquid refrigerant then expands to 2 bar.*

Estimate the C.O.P. of the machine and power required to drive the compressor, if the mechanical efficiency of the drive is 80% and load on the evaporator is 10 TR.

Solution. Given : $p_C = 14$ bar ; $p_E = 2$ bar ; $p_F = 5$ bar ; $t_3 = t_7 = 30°C$; $\eta_m = 80\% = 0.8$; $Q = 10$ TR

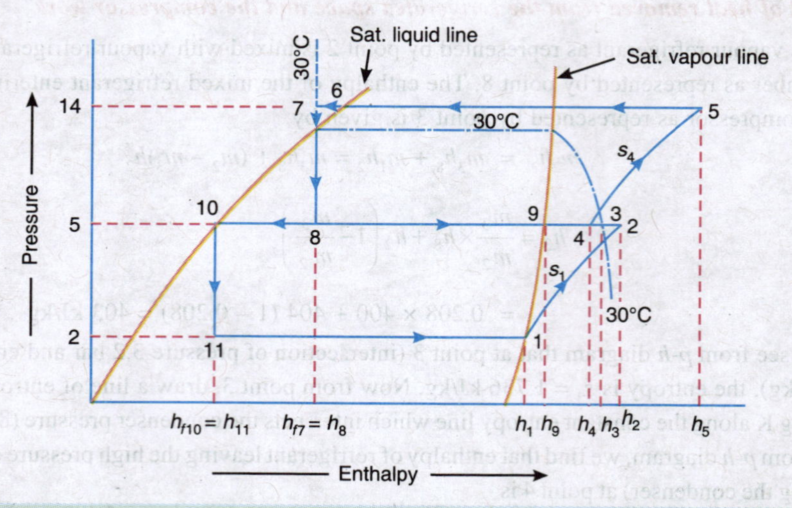

Fig. 5.9

The *p-h* diagram for a two-stage compression system with given conditions is shown in Fig. 5.9. The values as read from *p-h* diagram for ammonia, are as follows :

Enthalpy of saturated vapour refrigerant entering the low pressure compressor at point 1,

$$h_1 = 1420 \text{ kJ/kg}$$

Entropy of saturated vapour refrigerant entering the low pressure compressor at point 1,

$$s_1 = 5.6244 \text{ kJ/kg K}$$

Enthalpy of superheated vapour refrigerant leaving the low pressure compressor at point 2,

$$h_2 = 1550 \text{ kJ/kg}$$

Enthalpy of superheated vapour refrigerant leaving the water intercooler at point 3,

$$h_3 = 1510 \text{ kJ/kg}$$

Enthalpy of saturated vapour refrigerant leaving the flash tank at point 9,

$$h_9 = 1432 \text{ kJ/kg}$$

Enthalpy of liquid refrigerant leaving the subcooler at point 7,

$$h_{f7} = h_8 = 323 \text{ kJ/kg}$$

Enthalpy of saturated liquid refrigerant leaving the second expansion valve at point 10,

$$h_{f10} = h_{11} = 198 \text{ kJ/kg}$$

Let m_2 = Mass of refrigerant passing through the condenser.

We know that mass of the vapour refrigerant formed in the flash tank,

$$m_3 = m_2\left(\frac{h_8 - h_{f10}}{h_9 - h_{f10}}\right) = m_2\left(\frac{323-198}{1432-198}\right) = 0.1\ m_2 \quad \text{... } (i)$$

and mass of refrigerant passing through the evaporator,

$$m_1 = m_2 - m_3 = \frac{210\,Q}{h_1 - h_{f10}} = \frac{210 \times 10}{1420-198} = 1.72 \text{ kg / min} \quad \text{... } (ii)$$

From equations (*i*) and (*ii*),

$$m_2 - 0.1\ m_2 = 1.72 \quad \text{or} \quad m_2 = 1.9 \text{ kg / min}$$

and

$$m_3 = 0.1\ m_2 = 0.1 \times 1.9 = 0.19 \text{ kg / min}$$

The desuperheated vapour refrigerant $(m_2 - m_3)$ as represented by point 3 is mixed with the vapour refrigerant from the flash tank as represented by point 9. The enthalpy of the mixed refrigerant entering the high pressure compressor as represented by point 4 is given by

$$m_2 \, h_4 = m_3 \, h_9 + (m_2 - m_3) \, h_3$$

$$= 0.1 \, m_2 \, h_9 + (m_2 - 0.1 \, m_2) \, h_3 \qquad \text{... [From equation } (i)]$$

or

$$h_4 = 0.1 \times h_9 + 0.9 \times h_3 = 0.1 \times 1432 + 0.9 \times 1510 = 1502 \text{ kJ/kg}$$

We see from *p-h* diagram that at point 4 (intersection of pressure 5 bar and enthalpy 1502 kJ/kg), the entropy is $s_4 = 5.51$ kJ/kg K. Now from point 4, draw a line of entropy equal to 5.51 kJ/kg K along the constant entropy line which intersects the condenser pressure (14 bar) line at point 5. Thus, the point 5 is located. From *p-h* diagram, we find that enthalpy of refrigerant leaving the high pressure compressor at point 5 is

$$h_5 = 1650 \text{ kJ/kg}$$

C.O.P. of the machine

We know that refrigerating effect

$$= m_1 \, (h_1 - h_{f10}) = 210 \, Q = 210 \times 10 = 2100 \text{ kJ/min}$$

Work done in both the compressors

$$= m_1 \, (h_2 - h_1) + m_2 \, (h_5 - h_4)$$

$$= 1.72 \, (1550 - 1420) + 1.9 \, (1650 - 1502)$$

$$= 223.6 + 281.2 = 504.8 \text{ kJ/min}$$

Since the mechanical efficiency of the drive is 80%, therefore actual workdone in both the compressors

$$= 504.8 / 0.8 = 631 \text{ kJ/min}$$

$$\therefore \quad \text{Actual C.O.P.} = \frac{\text{Refrigerating effect}}{\text{Actual work done}} = \frac{2100}{631} = 3.32 \ \textbf{Ans.}$$

Power required to drive the compressors

We know that power required to drive the compressors

$$= \frac{\text{Actual work done}}{60} = \frac{631}{60} = 10.5 \text{ kW } \textbf{Ans.}$$

5.7 Two Stage Compression with Water Intercooler, Liquid Sub-cooler and Flash Intercooler

A two stage compression with water intercooler, liquid sub-cooler and flash intercooler is shown in Fig. 5.10 (*a*). The corresponding *p-h* diagram is shown in Fig. 5.10 (*b*).

We have seen in the previous article that when the vapour refrigerant from the low pressure compressor is passed through the water intercooler, its temperature does not reduce to the saturated vapour line or even very near to it, before admitting it to the high pressure compressor [Refer point 4 of Fig. 5.7 (*b*)]. In fact, with water cooling there may be no saving of work in compression. But the improvement in performance and the reduction in compression work may be achieved by using a flash chamber as an intercooler as well as flash separator, as shown in Fig. 5.10 (*a*). The corresponding *p-h* diagram is shown in Fig. 5.10 (*b*). The various processes, in this system, are as follows :

1. The saturated vapour refrigerant at the evaporator pressure p_E is admitted to the low pressure compressor at point 1. In this compressor, the refrigerant is compressed isentropically from evaporator pressure p_E to the flash intercooler pressure p_F, as shown by the curve 1-2 in Fig. 5.10 (*b*).

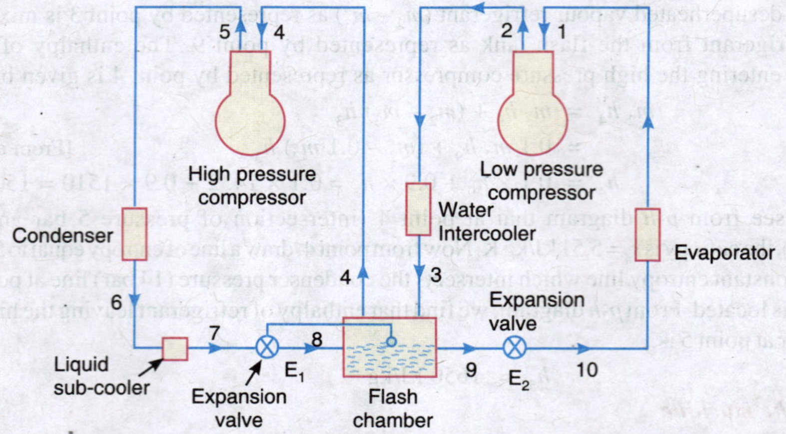

(*a*) Two stage compression with water intercooler, liquid sub-cooler and flash intercooler.

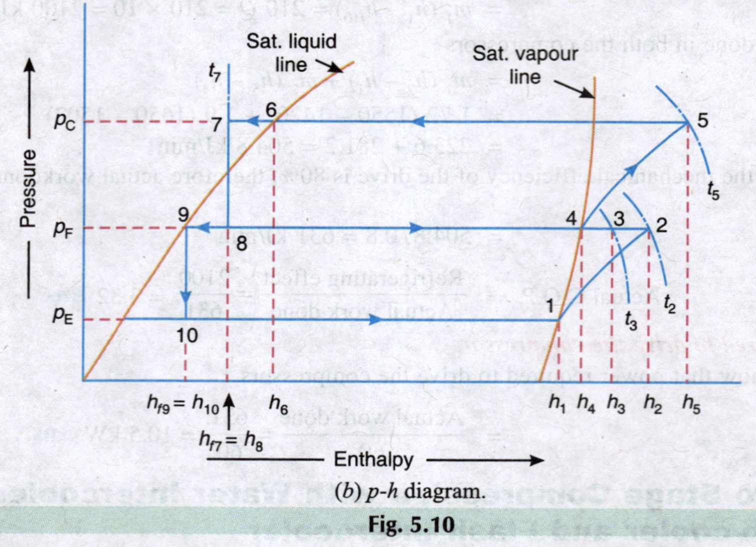

(*b*) *p-h* diagram.

Fig. 5.10

2. The superheated vapour refrigerant leaving the low pressure compressor at point 2 is now passed through the water intercooler at constant pressure p_F, in order to reduce the degree of superheat (*i.e.* from temperature t_2 to t_3). The line 2-3 represents the water intercooling or desuperheating process.
3. The superheated vapour refrigerant leaving the water intercooler at point 3 is passed through a flash intercooler which cools the superheated vapour refrigerant to saturated vapour refrigerant as shown by the line 3-4. The cooling of superheated vapour refrigerant is done by the evaporation of a part of the liquid refrigerant from the flash intercooler placed at point 8.
4. The saturated vapour refrigerant leaving the flash intercooler enters the high pressure compressor at point 4 where it is compressed isentropically from flash intercooler pressure p_F to condenser pressure p_C, as shown by the curve 4-5.
5. The superheated vapour refrigerant leaving the high pressure compressor at pressure p_C is passed through a condenser at constant pressure. The condensing process as shown by line 5-6 changes the state of refrigerant from superheated vapour to saturated liquid.

6. The saturated liquid refrigerant leaving the condenser at point 6 is now cooled at constant pressure p_C in the liquid sub-cooler to a temperature t_7 as shown in Fig. 5.9 (*b*). The line 6-7 shows the sub-cooling process.
7. The liquid refrigerant leaving the sub-cooler at point 7 is expanded in an expansion valve E_1 to a pressure equal to the flash intercooler pressure p_F, as shown by the vertical line 7-8. The expanded refrigerant (which is a mixture of vapour and liquid refrigerant) is admitted to flash intercooler at point 8 which also acts as a flash separator.
8. The liquid refrigerant leaving the flash intercooler at point 9 is passed through the second expansion valve E_2 (process 9-10) and then evaporated in the evaporator as shown by the horizontal line 10-1.

Let m_1 = Mass of the refrigerant passing through the evaporator (or low pressure compressor), and

m_2 = Mass of the refrigerant passing through the condenser (or high pressure compressor).

If Q tonne of refrigeration is the load on the evaporator, then the mass of refrigerant passing through the evaporator is given by,

$$m_1 = \frac{210\,Q}{h_1 - h_{10}} = \frac{210\,Q}{h_1 - h_{f9}} \text{ kg / min} \qquad \ldots (\because h_{10} = h_{f9})$$

Now for the thermal equilibrium of the flash intercooler,

Heat taken by the flash intercooler

= Heat given by the flash intercooler

$$m_2\,h_8 + m_1\,h_3 = m_2\,h_4 + m_1\,h_{f9}$$

$$m_1\,(h_3 - h_{f9}) = m_2\,(h_4 - h_8)$$

Vapour compression refrigeration system.

$$\therefore \quad m_2 = m_1\left(\frac{h_3 - h_{f9}}{h_4 - h_8}\right) = m_1\left(\frac{h_3 - h_{f9}}{h_4 - h_{f7}}\right) \text{ kg/min} \qquad (\because h_8 = h_{f7})$$

We know that refrigerating effect,

$$R_E = m_1 (h_1 - h_{10}) = m_1 (h_1 - h_{f9}) = 210\, Q \text{ kJ/min}$$

and work done in both the compressors,

$$W = \text{Work done in L.P. compressor + Work done in H.P. compressor}$$

$$= m_1 (h_2 - h_1) + m_2 (h_5 - h_4)$$

∴ Power required to drive the system,

$$P = \frac{m_1 (h_2 - h_1) + m_2 (h_5 - h_4)}{60} \text{ kW}$$

and coefficient of performance of the system,

$$\text{C.O.P.} = \frac{R_E}{W} = \frac{m_1 (h_1 - h_{f9})}{m_1 (h_2 - h_1) + m_2 (h_5 - h_4)} = \frac{210\, Q}{P \times 60}$$

Example 5.6. *In a 15 TR ammonia plant, compression is carried out in two stages with water and flash intercooling and water subcooling. The particulars of the plant are as follows:*

Condenser pressure = *12 bar*

Evaporator pressure = *3 bar*

Flash intercooler pressure = *6 bar*

Limiting temperature for intercooling and sub-cooling = *20°C*

Draw the cycle on p-h chart and estimate (a) the coefficient of performance of the plant, (b) the power required for each compressor, and (c) the swept volume for each compressor if the volumetric efficiency of both the compressors is 80%.

Solution. Given : $Q = 15$ TR ; $p_C = 12$ bar ; $p_E = 3$ bar ; $p_F = 6$ bar ; $t_3 = t_7 = 20°$ C ; $\eta_v = 80\% = 0.8$

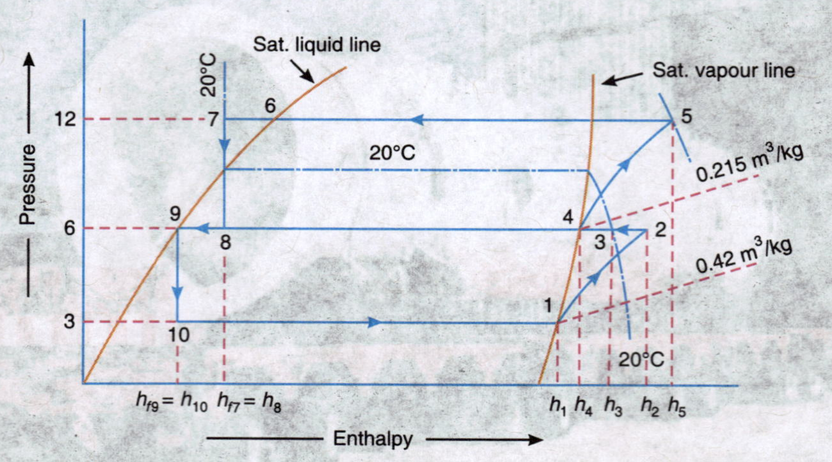

Fig. 5.11

The cycle on the *p-h* chart may be drawn as shown in Fig. 5.11. The various values as read from the *p-h* diagram for ammonia are as follows :

Enthalpy of saturated vapour refrigerant entering the low pressure compressor at point 1,

$$h_1 = 1422 \text{ kJ/kg}$$

Entropy of saturated vapour refrigerant entering the low pressure compressor at point 1,

$$s_1 = 5.49 \text{ kJ/kg K}$$

Specific volume of saturated vapour refrigerant entering the low pressure compressor at point 1,

$$v_1 = 0.42 \text{ m}^3\text{/kg}$$

Enthalpy of superheated vapour refrigerant leaving the low pressure compressor at point 2,

$$h_2 = 1505 \text{ kJ/kg}$$

Enthalpy of superheated vapour refrigerant leaving the water intercooler at point 3,

$$h_3 = 1465 \text{ kJ/kg}$$

Enthalpy of saturated vapour refrigerant leaving the flash intercooler at point 4,

$$h_4 = 1440 \text{ kJ/kg}$$

Entropy of saturated vapour refrigerant at point 4,

$$s_4 = 5.25 \text{ kJ/kg K}$$

Specific volume of saturated vapour refrigerant at point 4,

$$v_4 = 0.215 \text{ m}^3\text{/kg}$$

Enthalpy of superheated vapour refrigerant at point 5,

$$h_5 = 1530 \text{ kJ/kg}$$

Enthalpy of liquid refrigerant leaving the liquid sub-cooler at point 7,

$$h_{f7} = h_8 = 265 \text{ kJ/kg}$$

Enthalpy of saturated liquid refrigerant leaving the flash intercooler at point 9,

$$h_{f9} = h_{10} = 224 \text{ kJ/kg}$$

(a) Coefficient of performance of the plant

We know that mass of refrigerant passing through the evaporator (or low pressure compressor),

$$m_1 = \frac{210\,Q}{h_1 - h_{f9}} = \frac{210 \times 15}{1422 - 224} = 2.63 \text{ kg/min}$$

and mass of refrigerant passing through the condenser (or high pressure compressor),

$$m_2 = m_1 \left(\frac{h_3 - h_{f9}}{h_4 - h_{f7}} \right) = 2.63 \left(\frac{1465 - 224}{1440 - 265} \right) = 2.78 \text{ kg/min}$$

∴ Coefficient of performance of the plant,

$$\text{C.O.P.} = \frac{m_1 (h_1 - h_{f9})}{m_1 (h_2 - h_1) + m_2 (h_5 - h_4)}$$

$$= \frac{2.63\,(1422 - 224)}{2.63\,(1505 - 1422) + 2.78\,(1530 - 1440)} = 6.725 \text{ **Ans.**}$$

(b) Power required for each compressor

We know that work done in low pressure compressor,

$$W_L = m_1 (h_2 - h_1) = 2.63\,(1505 - 1422) = 218.3 \text{ kJ/min}$$

∴ Power required for low pressure compressor,

$$P_L = 218.3/60 = 3.64 \text{ kW} \quad \textbf{Ans.}$$

Similarly, work done in high pressure compressor,

$$W_H = m_2 (h_5 - h_4) = 2.78 (1530 - 1440) = 250.2 \text{ kJ/min}$$

∴ Power required for high presssure compressor,

$$P_H = 250.2/60 = 4.17 \text{ kW} \quad \textbf{Ans.}$$

(c) Swept volume for each compressor

We know that swept volume for low pressure compressor

$$= \frac{m_1 \times v_1}{\eta_v} = \frac{2.63 \times 0.42}{0.8} = 1.46 \text{ m}^3\text{/min} \quad \textbf{Ans.}$$

and swept volume for high pressure compressor

$$= \frac{m_2 \times v_4}{\eta_v} = \frac{2.78 \times 0.215}{0.8} = 0.747 \text{ m}^3\text{/min} \quad \textbf{Ans.}$$

5.8 Three Stage Compression with Water Intercoolers

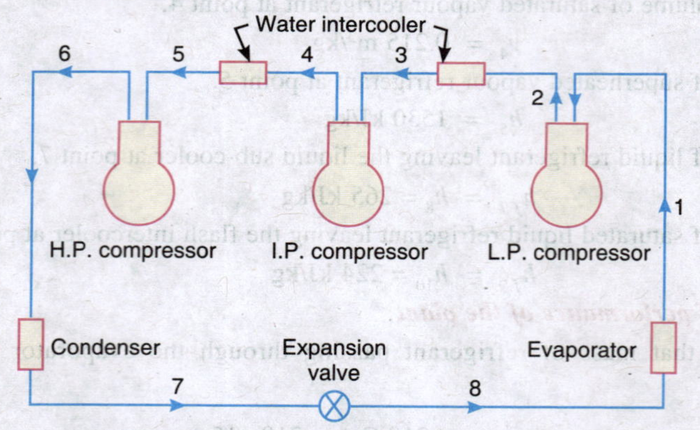

(*a*) Three stage compression with water intercoolers.

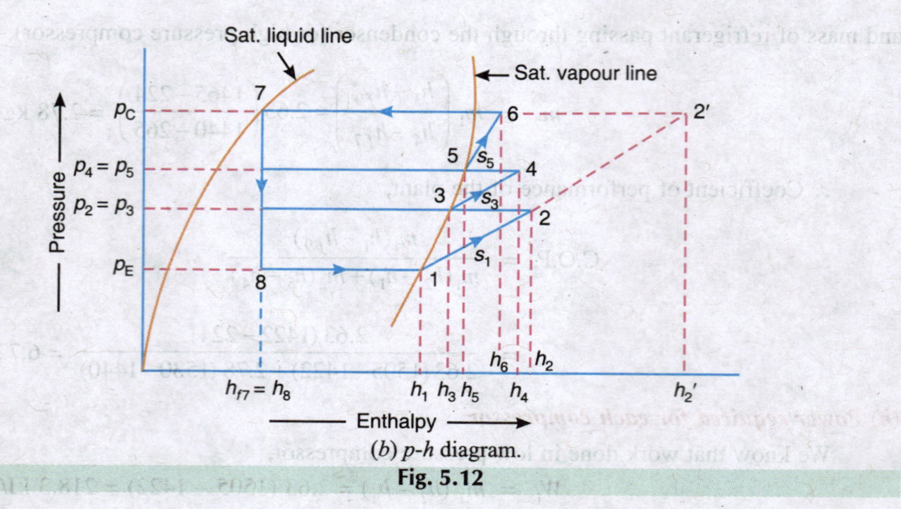

(*b*) *p-h* diagram.

Fig. 5.12

The arrangement of a three stage compression with water intercoolers is shown in Fig. 5.12 (*a*). The corresponding *p-h* diagram is shown in Fig. 5.12 (*b*). The work done in the high pressure compressor may be greatly reduced with such an arrangement. The water intercooling between the stages reduces the degree of superheat of the refrigerant. It also reduces the specific volume of the refrigerant which requires a compressor of less capacity (or stroke volume). We see from *p-h* diagram that the water intercoolers reduce the temperature of superheated vapour to its saturation value after each stage, as shown by points 3 and 5 in Fig. 5.12 (*b*). It may be noted that the complete desuperheating of the vapour is not possible because the temperature of cooling water used in water intercoolers is not available sufficiently low so as to desuperheat the vapour completely.

We know that for a load of Q tonnes of refrigeration on the evaporator, the mass of refrigerant circulating through the evaporator is given by

$$m = \frac{210\,Q}{h_1 - h_8} = \frac{210\,Q}{h_1 - h_{f7}} \qquad \ldots (\because h_8 = h_{f7})$$

Work done in low pressure (L.P.) compressor,

$$W_L = m\,(h_2 - h_1)$$

Work done in intermediate pressure (I.P.) compressor,

$$W_I = m\,(h_4 - h_3)$$

and work done in high pressure (H.P.) compressor,

$$W_H = m\,(h_6 - h_5)$$

∴ Total work done in the three compressors,

$$W = W_L + W_I + W_H$$
$$= m\,[(h_2 - h_1) + (h_4 - h_3) + (h_6 - h_5)]$$

∴ Power required to drive the three compressors,

$$P = \frac{m\,[(h_2 - h_1) + (h_4 - h_3) + (h_6 - h_5)]}{60}\ \text{kW}$$

We know that refrigerating effect,

$$R_E = m\,(h_1 - h_{f7}) = 210\,Q\ \text{kJ/min}$$

∴ Coefficient of performance of the system,

$$\text{C.O.P.} = \frac{R_E}{W} = \frac{m\,(h_1 - h_{f7})}{m\,[(h_2 - h_1) + (h_4 - h_3) + (h_6 - h_5)]}$$

$$= \frac{h_1 - h_{f7}}{(h_2 - h_1) + (h_4 - h_3) + (h_6 - h_5)} = \frac{210\,Q}{P \times 60}$$

Example 5.7. *A vapour compression system with ammonia as the refrigerant works between the pressure limits of 2 bar and 12 bar with three stage compression. The vapours leaving the water intercoolers at pressures 4 bar and 8 bar are in a saturated state. If the load is 10 TR, find the power required to drive the three compressors and compare the C.O.P. of this system with that of a simple saturation cycle working between the same overall pressure limits.*

Solution. Given : $p_E = 2$ bar ; $p_C = 12$ bar ; $p_2 = p_3 = 4$ bar ; $p_4 = p_5 = 8$ bar ; $Q = 10$ TR

The *p-h* diagram for a three stage compression with water intercooling is shown in Fig. 5.13. The various values as read from the *p-h* diagram for ammonia are as follows :

Enthalpy of saturated vapour refrigerant entering the first compressor at point 1,

$$h_1 = 1420\ \text{kJ/kg}$$

Entropy of saturated vapour refrigerant at point 1,

$$s_1 = 5.564\ \text{kJ/kg K}$$

Enthalpy of superheated vapour refrigerant leaving the first compressor at point 2,

$$h_2 = 1515 \text{ kJ/kg}$$

Enthalpy of saturated vapour refrigerant leaving the water intercooler and entering the second compressor at point 3,

$$h_3 = 1442 \text{ kJ/kg}$$

Entropy of saturated vapour refrigerant at point 3,

$$s_3 = 5.367 \text{ kJ/kg K}$$

Enthalpy of superheated vapour refrigerant leaving the second compressor at point 4,

$$h_4 = 1525 \text{ kJ/kg}$$

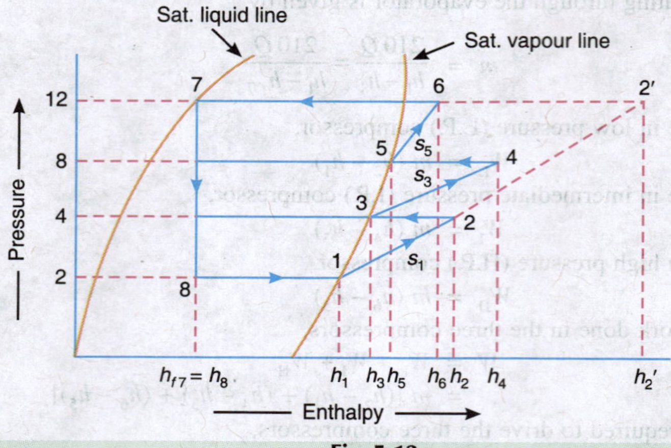

Fig. 5.13

Enthalpy of saturated vapour refrigerant leaving the water intercooler and entering the third compressor at point 5,

$$h_5 = 1461 \text{ kJ/kg}$$

Entropy of saturated vapour refrigerant at point 5,

$$s_5 = 5.1186 \text{ kJ/kg K}$$

Enthalpy of superheated vapour refrigerant leaving the third compressor at point 6,

$$h_6 = 1500 \text{ kJ/kg}$$

Enthalpy of saturated liquid refrigerant leaving the condenser at point 7,

$$h_{f7} = h_8 = 328 \text{ kJ/kg}$$

Power required to drive the three compressors

We know that mass of refrigerant required to be circulated through the evaporator,

$$m = \frac{210\,Q}{h_1 - h_{f7}} = \frac{210 \times 10}{1420 - 328} = 1.92 \text{ kg/min}$$

Work done in the three compressors,

$$W = m\,[(h_2 - h_1) + (h_4 - h_3) + (h_6 - h_5)]$$

$$= 1.92\,[(1515 - 1420) + (1525 - 1442) + (1500 - 1461)] \text{ kJ/min}$$

$$= 416.64 \text{ kJ/min}$$

∴ Power required to drive the three compressors,

$$P = 416.64/60 = 6.94 \text{ kW}$$ **Ans.**

Comparison of C.O.P. of the system with that of simple saturation cycle

We know that refrigerating effect of the system,

$$R_E = 210\,Q = 210 \times 10 = 2100 \text{ kJ/min}$$

$$\therefore \quad \text{C.O.P. of the system} = \frac{R_E}{W} = \frac{2100}{416.64} = 5.04$$

For a simple saturation cycle working between the same pressure limits of 2 bar and 12 bar, the enthalpy of superheated vapour leaving the compressor at point 2′ is

$$h_{2'} = 1670 \text{ kJ/kg}$$

Work done in the compressor for simple saturation cycle,

$$W_1 = m\,(h_{2'} - h_1) = 1.92\,(1670 - 1420) = 480 \text{ kJ/min}$$

and C.O.P. of the simple saturation cycle

$$= \frac{R_E}{W_1} = \frac{2100}{480} = 4.375$$

∴ Percentage increase in C.O.P. of the system as compared to simple saturation cycle

$$= \frac{5.04 - 4.375}{4.375} \times 100 = 15.2\% \quad \textbf{Ans.}$$

5.9 Three Stage Compression with Flash Chambers

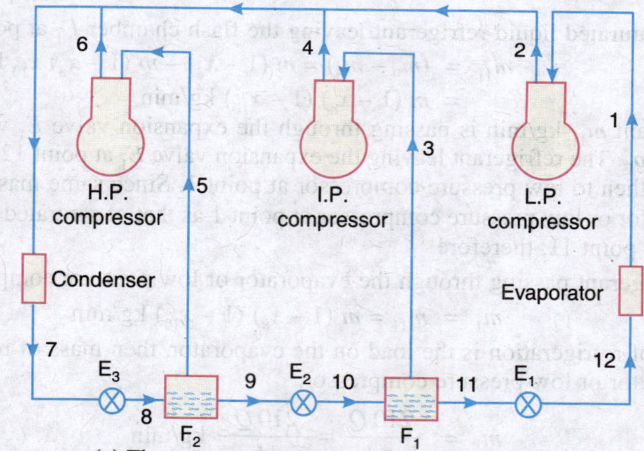

(*a*) Three stage compression with flash chambers.

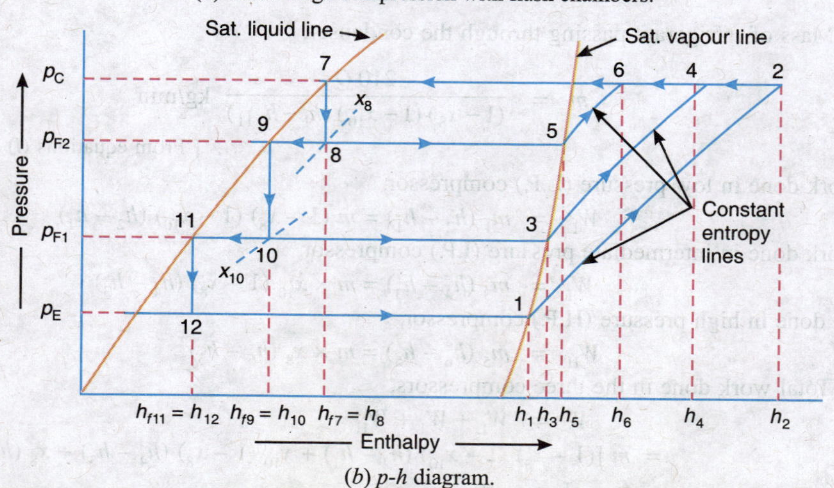

(*b*) *p-h* diagram.

Fig. 5.14

The arrangement of three compressors with multiple expansion valves E_1, E_2 and E_3 and two flash chambers F_1 and F_2 is shown in Fig. 5.14 (*a*). The corresponding *p-h* diagram is shown in Fig. 5.14 (*b*).

Let *m* kg/min be the mass of refrigerant leaving the condenser at point 7. This refrigerant while passing through the expansion valve E_3 reduces its pressure from p_C to p_{F2}. The refrigerant leaving the expansion valve E_3 at point 8 is separated by the flash chamber F_2. If x_8 is the dryness fraction of the refrigerant at point 8, then mass of saturated vapour refrigerant separated at point 8 and delivered to high pressure compressor at point 5 will be

$$m_5 = m \times x_8 \text{ kg/min}$$

∴ Mass of saturated liquid refrigerant leaving the flash chamber F_2 at point 9,

$$m_9 = m - m_5 = m - m \times x_8$$

$$= m(1 - x_8) \text{ kg/min} \quad \ldots (\because m_5 = m \times x_8)$$

This saturated liquid refrigerant (*i.e.* m_9 kg/min) is now passed through the second expansion valve E_2 where its pressure reduces from p_{F2} to p_{F1}. The refrigerant leaving the expansion valve E_2 is separated by the flash chamber F_1. If x_{10} is the dryness fraction of the refrigerant at point 10, then mass of saturated vapour separated at point 10 and delivered to intermediate pressure compressor at point 3 will be

$$m_3 = m_9 \times x_{10} = m(1 - x_8)\, x_{10} \text{ kg/min} \quad \ldots [\because m_9 = m(1 - x_8)]$$

∴ Mass of saturated liquid refrigerant leaving the flash chamber F_1 at point 11,

$$m_{11} = (m_9 - m_3) = m(1 - x_8) - m(1 - x_8)\, x_{10} \text{ kg/min}$$

$$= m(1 - x_8)(1 - x_{10}) \text{ kg/min}$$

This refrigerant m_{11} kg/min is passing through the expansion valve E_1 where its pressure reduces from p_{F1} to p_E. The refrigerant leaving the expansion valve E_1 at point 12 is passed through the evaporator and then to low pressure compressor at point 1. Since same mass of refrigerant is supplied to evaporator or low pressure compressor at point 1 as that of saturated liquid leaving the flash chamber F_1 at point 11, therefore

Mass of refrigerant passing through the evaporator or low pressure compressor,

$$m_1 = m_{11} = m(1 - x_8)(1 - x_{10}) \text{ kg/min} \quad \ldots (i)$$

If *Q* tonnes of refrigeration is the load on the evaporator, then mass of refrigerant passing through the evaporator or low pressure compressor,

$$m_1 = \frac{210\, Q}{h_1 - h_{12}} = \frac{210\, Q}{h_1 - h_{f11}} \text{ kg/min} \quad \ldots (\because h_{12} = h_{f11}) \ldots (ii)$$

∴ Mass of refrigerant passing through the condenser,

$$m = \frac{210\, Q}{(1 - x_8)(1 - x_{10})(h_1 - h_{f11})} \text{ kg/min}$$

... [From equations (*i*) and (*ii*)]

Work done in low pressure (L.P.) compressor,

$$W_L = m_1 (h_2 - h_1) = m(1 - x_8)(1 - x_{10})(h_2 - h_1)$$

Work done in intermediate pressure (I.P.) compressor,

$$W_I = m_3 (h_4 - h_3) = m \times x_{10} (1 - x_8)(h_4 - h_3)$$

and work done in high pressure (H.P.) compressor,

$$W_H = m_5 (h_6 - h_5) = m \times x_8 (h_6 - h_5)$$

∴ Total work done in the three compressors,

$$W = W_L + W_I + W_H$$

$$= m[(1 - x_8)(1 - x_{10})(h_2 - h_1) + x_{10}(1 - x_8)(h_4 - h_3) + x_8 (h_6 - h_5)]$$

and power required to drive the three compressors.

$$P = W/60 \text{ kW}$$

We know that refrigerating effect,

$$R_E = 210\,Q$$

∴ C.O.P. of the system

$$= \frac{R_E}{W} = \frac{210\,Q}{P \times 60}$$

Notes: 1. By using multiple expansion valves in the above arrangement, the refrigerant can be expanded close to the liquid line and by using the flash chambers, the total work done per kg of refrigerant is reduced.

2. Thermodynamically, this arrangement leads to more C.O.P. as compared to simple saturation cycle for the operating pressure range, because of decrease in total compression work without affecting the refrigerating effect produced at the evaporator.

Example 5.8. *An ice plant working on ammonia as refrigerant works between overall pressure limits of 2.5 bar and 15 bar. It is fitted with expansion valve with vapour extraction at 5 bar and 10 bar. The load on the plant is 10 TR. Find the circulation of the refrigerant through the condenser and the power required to drive the three compressors. Use p-h chart.*

Solution. Given : $p_E = 2.5$ bar ; $p_C = 15$ bar ; $p_{F2} = 5$ bar ; $p_{F1} = 10$ bar ; $Q = 10$ TR

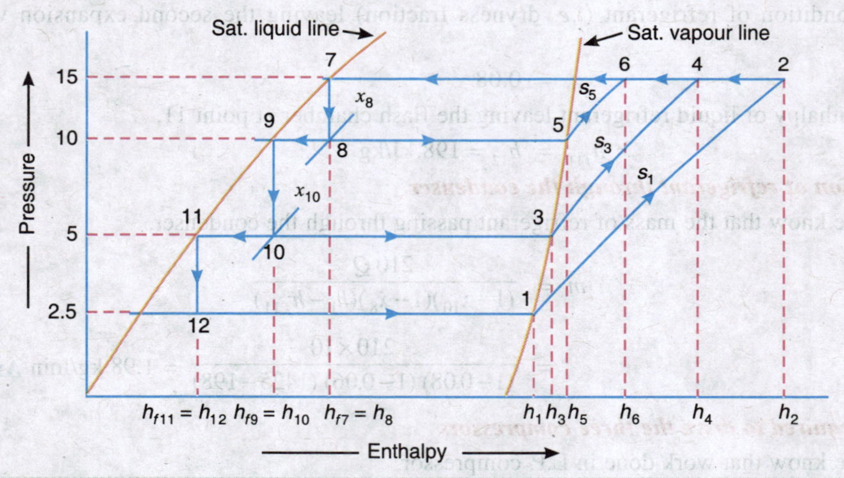

Fig. 5.15

The *p-h* diagram for a three stage compression system with the given conditions is shown in Fig. 5.15. The various values as read from the *p-h* diagram for ammonia are as follows :

Enthalpy of saturated vapour refrigerant entering the low pressure compressor at point 1,

$$h_1 = 1425 \text{ kJ/kg}$$

Entropy of saturated vapour refrigerant at point 1,

$$s_1 = 5.55 \text{ kJ/kg K}$$

Enthalpy of saturated vapour refrigerant entering the intermediate pressure compressor at point 3,

$$h_3 = 1432 \text{ kJ/kg}$$

Entropy of saturated vapour refrigerant at point 3,

$$s_3 = 5.32 \text{ kJ/kg K}$$

Enthalpy of saturated vapour refrigerant entering the high pressure compressor at point 5,

$$h_5 = 1445 \text{ kJ/kg}$$

Entropy of saturated vapour refrigerant at point 5,

$$s_5 = 5.05 \text{ kJ/kg K}$$

Enthalpy of superheated vapour refrigerant leaving the low pressure compressor at point 2,

$$h_2 = 1660 \text{ kJ/kg}$$

Enthalpy of superheated vapour refrigerant leaving the intermediate pressure compressor at point 4,

$$h_4 = 1600 \text{ kJ/kg}$$

Enthalpy of superheated vapour refrigerant leaving the high pressure compressor at point 6,

$$h_6 = 1510 \text{ kJ/kg}$$

Enthalpy of saturated liquid refrigerant leaving the condenser at point 7,

$$h_{f7} = h_8 = 352 \text{ kJ/kg}$$

Condition of refrigerant (*i.e.* dryness fraction) leaving the first expansion valve at point 8,

$$x_8 = 0.06$$

Enthalpy of saturated liquid refrigerant leaving the flash chamber at point 9,

$$h_{f9} = h_{10} = 290 \text{ kJ/kg}$$

Condition of refrigerant (*i.e.* dryness fraction) leaving the second expansion valve at point 10,

$$x_{10} = 0.08$$

Enthalpy of liquid refrigerant leaving the flash chamber at point 11,

$$h_{f11} = h_{12} = 198 \text{ kJ/kg}$$

Circulation of refrigerant through the condenser

We know that the mass of refrigerant passing through the condenser,

$$m = \frac{210\,Q}{(1-x_{10})(1-x_8)(h_1-h_{f11})}$$

$$= \frac{210 \times 10}{(1-0.08)\,(1-0.06)\,(1425-198)} = 1.98 \text{ kg/min} \textbf{ Ans.}$$

Power required to drive the three compressors

We know that work done in L.P. compressor,

$$W_L = m\,(1-x_{10})\,(1-x_8)\,(h_2-h_1)$$

$$= 1.98\,(1-0.08)\,(1-0.06)\,(1660-1425) = 402.4 \text{ kJ/min}$$

Work done in I.P. compressor,

$$W_I = m \times x_{10}\,(1-x_8)\,(h_4-h_3)$$

$$= 1.98 \times 0.08\,(1-0.06)\,(1600-1432) = 25 \text{ kJ/min}$$

Work done in H.P. compressor,

$$W_H = m \times x_8\,(h_6-h_5)$$

$$= 1.98 \times 0.06\,(1510-1445) = 7.72 \text{ kJ/min}$$

and total workdone in the three compressors,

$$W = W_L + W_I + W_H$$

$$= 402.4 + 25 + 7.72 = 435.12 \text{ kJ/min}$$

∴ Power required to drive the three compressors,

$$P = W/60 = 435.12/60 = 7.25 \text{ kW} \textbf{ Ans.}$$

5.10 Three Stage Compression with Flash Intercoolers

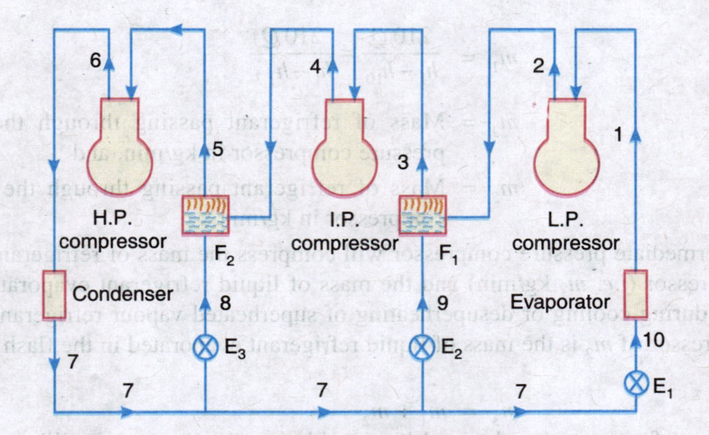

(a) Three stage compression with flash intercoolers.

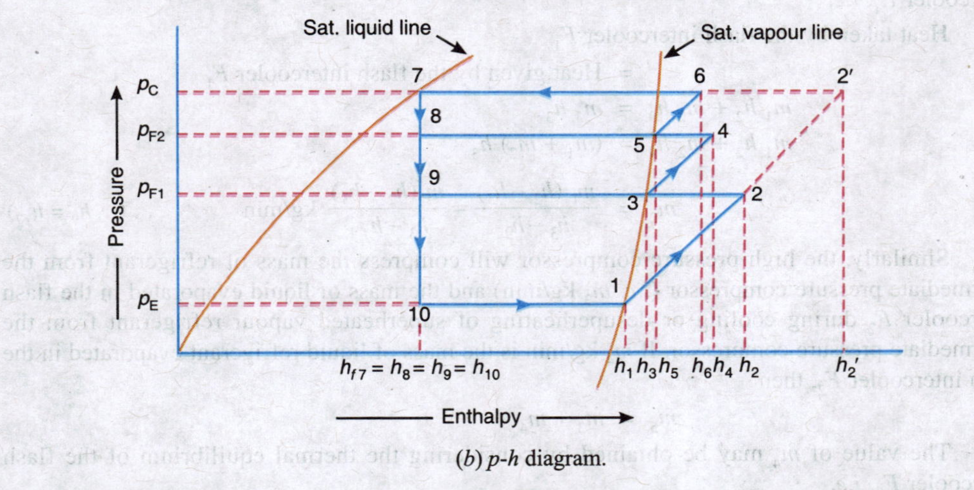

(b) p-h diagram.

Fig. 5.16

We have already discussed that due to non-availability of cooling water at low temperature for intercooling, the superheated vapour at the end of each stage cannot be completely desuperheated. This difficulty can be overcome by adopting the flash intercoolers F_1 and F_2 between the stages as shown in Fig. 5.16 (*a*). The *p-h* diagram for the arrangement is shown in Fig. 5.16 (*b*). The superheated vapour from low pressure compressor (as represented by point 2) is cooled to saturated vapour in the flash intercooler F_1 by the liquid refrigerant from expansion valve E_2 (as represented by point 9). The flash intercooling process for the first stage as represented by 2-3 is carried out at pressure p_{F1}. Similarly, the superheated vapour from intermediate pressure compressor (as represented by point 4) is cooled to saturated vapour in the flash intercooler F_2 by the liquid refrigerant from expansion valve E_3 (as represented by point 8). The flash intercooling process for the second stage as represented by 4-5 is carried out at pressure p_{F2}.

If Q tonnes of refrigeration is the load on the evaporator, then the mass of refrigerant passing through the evaporator or low pressure compressor at point 1,

$$m_1 = \frac{210\,Q}{h_1 - h_{10}} = \frac{210\,Q}{h_1 - h_{f7}} \qquad \ldots (\because h_{10} = h_{f7})$$

Let m_3 = Mass of refrigerant passing through the intermediate pressure compressor in kg/min, and

m_5 = Mass of refrigerant passing through the high pressure compressor in kg/min.

The intermediate pressure compressor will compress the mass of refrigerant from the low pressure compressor (*i.e.* m_1 kg/min) and the mass of liquid refrigerant evaporated in the flash intercooler F_1 during cooling or desuperheating of superheated vapour refrigerant from the low pressure compressor. If m_2 is the mass of liquid refrigerant evaporated in the flash intercooler F_1, then

$$m_3 = m_1 + m_2$$

The value of m_2 may be obtained by considering the thermal equilibrium of the flash intercooler F_1, *i.e.*

Heat taken by the flash intercooler F_1

= Heat given by the flash intercooler F_1

or
$$m_1\,h_2 + m_2\,h_9 = m_3\,h_3$$
$$m_1\,h_2 + m_2\,h_9 = (m_1 + m_2)\,h_3$$

$$\therefore \quad m_2 = \frac{m_1\,(h_2 - h_3)}{h_3 - h_9} = \frac{m_1(h_2 - h_3)}{h_3 - h_{f7}} \text{ kg/min} \qquad \ldots (\because h_9 = h_{f7})$$

Similarly, the high pressure compressor will compress the mass of refrigerant from the intermediate pressure compressor (*i.e.* m_3 kg/min) and the mass of liquid evaporated in the flash intercooler F_2 during cooling or desuperheating of superheated vapour refrigerant from the intermediate pressure compressor. If m_4 kg/min is the mass of liquid refrigerant evaporated in the flash intercooler F_2, then

$$m_5 = m_3 + m_4$$

The value of m_4 may be obtained by considering the thermal equilibrium of the flash intercooler F_2, *i.e.*

Heat taken by the flash intercooler F_2

= Heat given by the flash intercooler F_2

or
$$m_3\,h_4 + m_4\,h_8 = m_5\,h_5$$
$$m_3\,h_4 + m_4\,h_8 = (m_3 + m_4)\,h_5$$

$$\therefore \quad m_4 = \frac{m_3\,(h_4 - h_5)}{h_5 - h_8} = \frac{m_3(h_4 - h_5)}{h_5 - h_{f7}} \text{ kg/min} \qquad \ldots (\because h_8 = h_{f7})$$

We know that work done in L.P. compressor,

$$W_L = m_1\,(h_2 - h_1)$$

Work done in I.P. compressor,

$$W_I = m_3\,(h_4 - h_3)$$

Work done in H.P. compressor,

$$W_H = m_5\,(h_6 - h_5)$$

and total work done in the three compressors,

$$W = W_L + W_I + W_H$$
$$= m_1 (h_2 - h_1) + m_3 (h_4 - h_3) + m_5 (h_6 - h_5)$$

$\therefore$ Power required to drive the three compressors,

$$P = W / 60 \text{ kW}$$

We know that refrigerating effect,

$$R_E = m_1 (h_1 - h_{10}) = 210\, Q \text{ kJ /min}$$

$$\therefore \quad \text{C.O.P. of the system} = \frac{R_E}{W} = \frac{210\, Q}{P \times 60}$$

Hyper form refrigeration system.

Example 5.9. *A three stage ammonia refrigeration system with flash intercooling operates between the overall pressure limits of 2 bar and 12 bar. The flash intercooler pressures are 4 bar and 8 bar. If the load on the evaporator is 10 TR, find the power required to run the system and compare the C.O.P. of the system with that of simple saturation cycle working between the same overall pressure limits.*

Solution. Given : $p_E = 2$ bar ; $p_C = 12$ bar ; $p_{F1} = 4$ bar ; $p_{F2} = 8$ bar ; $Q = 10$ TR

The *p-h* diagram for three stage compression system with flash intercooling is shown in Fig. 5.17. The various values as read from the *p-h* diagram are as follows :

Enthalpy of saturated vapour refrigerant entering the low pressure compressor at point 1,

$$h_1 = 1420 \text{ kJ/kg}$$

Entropy of saturated vapour refrigerant at point 1,

$$s_1 = 5.564 \text{ kJ/kg K}$$

Enthalpy of superheated vapour refrigerant leaving the low pressure compressor at point 2,

$$h_2 = 1515 \text{ kJ/kg}$$

Enthalpy of saturated vapour refrigerant leaving the first flash intercooler or entering the intermediate pressure compressor at point 3,

$$h_3 = 1442 \text{ kJ/kg}$$

Entropy of saturated vapour refrigerant at point 3,

$$s_3 = 5.367 \text{ kJ/kg K}$$

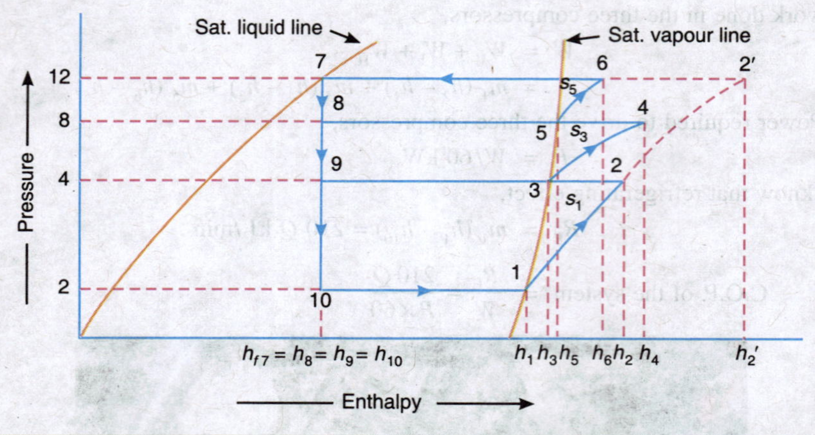

Fig. 5.17

Enthalpy of superheated vapour refrigerant leaving the intermediate pressure compressor at point 4,

$$h_4 = 1525 \text{ kJ/kg}$$

Enthalpy of saturated vapour refrigerant leaving the second flash chamber or entering the high pressure compressor at point 5,

$$h_5 = 1461 \text{ kJ/kg}$$

Entropy of saturated vapour refrigerant at point 5,

$$s_5 = 5.1186 \text{ kJ/kg K}$$

Enthalpy of superheated vapour refrigerant leaving the high pressure compressor or entering the condenser,

$$h_6 = 1500 \text{ kJ/kg}$$

Enthalpy of saturated liquid refrigerant leaving the condenser at point 7,

$$h_{f7} = h_8 = h_9 = h_{10} = 328 \text{ kJ/kg}$$

Power required to run the system

We know that mass of refrigerant passing through the evaporator or L.P. compressor,

$$m_1 = \frac{210\,Q}{h_1 - h_{f7}} = \frac{210 \times 10}{1420 - 328} = 1.92 \text{ kg/min}$$

Mass of liquid refrigerant evaporated in the flash intercooler after the first stage of compression,

$$m_2 = \frac{m_1(h_2 - h_3)}{h_3 - h_{f7}} = \frac{1.92\,(1515 - 1442)}{1442 - 328} = 0.126 \text{ kg/min}$$

∴ Mass of refrigerant passing through I.P. compressor,

$$m_3 = m_1 + m_2 = 1.92 + 0.126 = 2.046 \text{ kg/min}$$

Mass of liquid refrigerant evaporated in the flash intercooler after the second stage of compression,

$$m_4 = \frac{m_3(h_4 - h_5)}{h_5 - h_{f7}} = \frac{2.046\,(1525 - 1461)}{1461 - 328} = 0.116 \text{ kg/min}$$

∴ Mass of refrigerant passing through H.P. compressor,

$$m_5 = m_3 + m_4 = 2.046 + 0.116 = 2.162 \text{ kg/min}$$

We know that work done in L.P. compressor,

$$W_L = m_1 (h_2 - h_1) = 1.92 (1515 - 1420) = 182.4 \text{ kJ/min}$$

Work done in I.P. compressor,

$$W_I = m_3 (h_4 - h_3) = 2.046 (1525 - 1442) = 169.8 \text{ kJ/min}$$

Work done in H.P. compressor,

$$W_H = m_5 (h_6 - h_5) = 2.162 (1500 - 1461) = 84.3 \text{ kJ/min}$$

and total work done in the three compressors,

$$W = W_L + W_I + W_H = 182.4 + 169.8 + 84.3 = 436.5 \text{ kJ/min}$$

∴ Power required to run the system,

$$= 436.5/60 = 7.27 \text{ kW} \quad \textbf{Ans.}$$

Comparison of C.O.P. of the system with that of simple saturation cycle

We know that refrigerating effect of the system,

$$R_E = m_1 (h_1 - h_{10}) = 210\, Q = 210 \times 10 = 2100 \text{ kJ/min}$$

$$\therefore \quad \text{C.O.P. of the system} = \frac{R_E}{W} = \frac{2100}{436.5} = 4.81$$

For a simple saturation cycle working between the same pressure limits of 2 bar and 12 bar, the enthalpy of superheated vapour leaving the compressor at point 2′ is

$$h_{2'} = 1670 \text{ kJ/kg} \quad \text{... (From } p\text{-}h \text{ chart)}$$

Work done in the compressor for simple saturation cycle,

$$W_1 = m_1 (h_{2'} - h_1) = 1.92 (1670 - 1420) = 480 \text{ kJ/min}$$

and C.O.P. of the simple saturation cycle

$$= \frac{R_E}{W_1} = \frac{2100}{480} = 4.375$$

∴ Percentage increase in C.O.P. as compared to simple saturation cycle

$$= \frac{4.81 - 4.375}{4.375} \times 100 = 9.9\% \quad \textbf{Ans.}$$

5.11 Three Stage Compression with Multiple Expansion Valves and Flash Intercoolers

In the previous article of three stage compression with flash intercoolers, a single expansion valve was used along the flow line to evaporator. But in this arrangement, multi-expansion valves are used to increase the coefficient of performance of the system as shown in Fig. 5.18 (*a*).

If Q tonnes of refrigeration is the load on the evaporator, then the mass of refrigerant passing through the evaporator or low pressure compressor,

$$m_1 = \frac{210\, Q}{h_1 - h_{12}} = \frac{210\, Q}{h_1 - h_{f11}} \text{ kg/min} \quad \text{... } (\because h_{12} = h_{f11})$$

From Fig. 5.18 (*a*), we see that the superheated vapour refrigerant discharged from the low pressure compressor is brought into the flash intercooler F_1 where it is desuperheated by the liquid refrigerant received from the expansion valve E_2 at point 10. During the process of desuperheating, some of the liquid refrigerant gets evaporated and supplied to the intermediate pressure compressor.

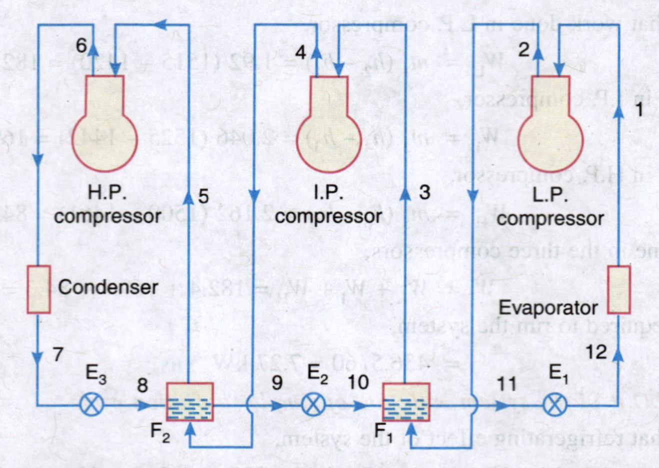

(*a*) Three stage compression with multiple expansion valves and flash intercoolers.

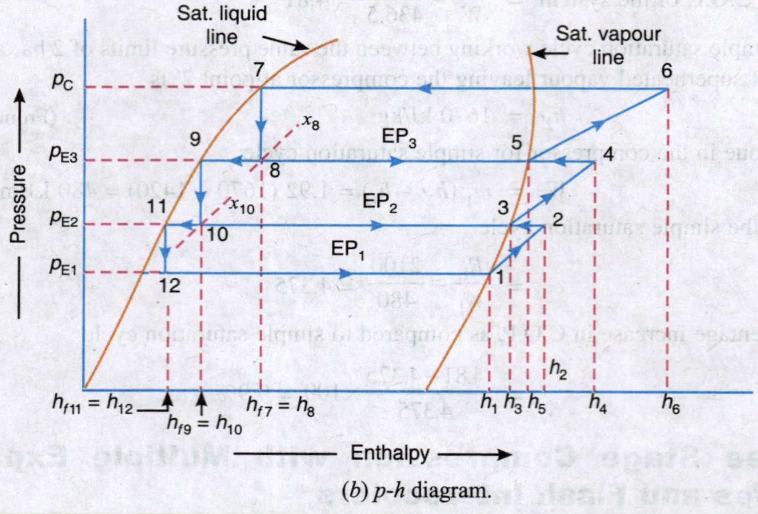

(*b*) *p-h* diagram.

Fig. 5.18

Thus the mass of vapour refrigerant passing through the intermediate pressure compressor (or second stage of compression),

m_2 = Mass of vapour refrigerant from L.P. compressor + Mass of vapour refrigerant or flash resulting from expansion valve E_2 + Mass of vapour refrigerant resulting from evaporation in the flash intercooler F_1

$$= m_1 + m_1\left(\frac{x_{10}}{1-x_{10}}\right) + m_1\left(\frac{h_2 - h_3}{h_3 - h_{10}}\right)$$

$$= m_1\left[1 + \frac{x_{10}}{1-x_{10}} + \frac{h_2 - h_3}{h_3 - h_{10}}\right]$$

where x_{10} is the dryness fraction of refrigerant leaving the expansion valve E_2 at point 10.

Similarly, the mass of vapour refrigerant passing through the high pressure compressor (or third stage of compression),

m_3 = Mass of vapour refrigerant from I.P. compressor + Mass of vapour refrigerant or flash resulting from expansion valve E_3 + Mass of vapour refrigerant resulting from evaporation in the flash intercooler F_2.

$$= m_2 + \frac{m_1 x_8}{(1-x_{10})(1-x_8)} + m_2 \left(\frac{h_4 - h_5}{h_5 - h_8} \right)$$

$$= m_2 \left(1 + \frac{h_4 - h_5}{h_5 - h_8} \right) + \frac{m_1 x_8}{(1-x_{10})(1-x_8)}$$

Replacement refrigeration system.

We know that work done in L.P. compressor,

$$W_L = m_1 (h_2 - h_1)$$

Work done in I.P. compressor,

$$W_I = m_2 (h_4 - h_3)$$

Work done in H.P. compressor,

$$W_H = m_3 (h_6 - h_5)$$

and total work done in three compressors,

$$W = W_L + W_I + W_H$$

$$= m_1 (h_2 - h_1) + m_2 (h_4 - h_3) + m_3 (h_6 - h_5)$$

∴ Power required to drive the compressors,

$$P = W/60 \text{ kW}$$

We know that refrigerating effect,

$$R_E = m_1 (h_1 - h_{f11}) = 210\, Q \text{ kJ/min}$$

∴ C.O.P. of the system $= \dfrac{R_E}{W} = \dfrac{210\, Q}{P \times 60}$

Example 5.10. *The following data refer to a three stage compression with three stage expansion valve and flash intercooling.*

Condenser pressure	= *12 bar*
Evaporator pressure	= *2 bar*
Flash intercooler pressures	= *4 bar and 8 bar*
Load on the evaporator	= *10 TR*

Find the power required to drive the system and compare the C.O.P. of this system with that of simple saturation cycle working between the same overall pressure limits.

Solution. Given : $p_C = 12$ bar ; $p_E = 2$ bar ; $p_{F1} = 4$ bar ; $p_{F2} = 8$ bar ; $Q = 10$ TR

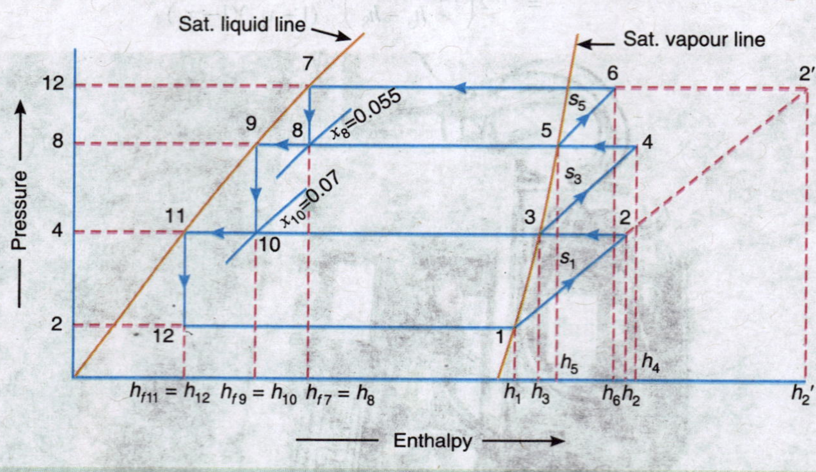

Fig. 5.19

The *p-h* diagram for a three stage compression with three stage expansion valve and flash intercooling is shown in Fig. 5.19. The various values as read from the *p-h* diagram for ammonia are as follows :

Enthalpy of saturated vapour refrigerant entering the low pressure compressor at point 1,

$$h_1 = 1420 \text{ kJ/kg}$$

Entropy of saturated vapour refrigerant at point 1,

$$s_1 = 5.564 \text{ kJ/kg K}$$

Enthalpy of superheated vapour refrigerant leaving the low pressure compressor or entering the first flash intercooler at point 2,

$$h_2 = 1515 \text{ kJ/kg}$$

Enthalpy of saturated vapour refrigerant leaving the first flash intercooler or entering the intermediate pressure compressor at point 3,

$$h_3 = 1442 \text{ kJ/kg}$$

Entropy of saturated vapour refrigerant at point 3,

$$s_3 = 5.367 \text{ kJ/kg K}$$

Enthalpy of superheated vapour refrigerant leaving the intermediate pressure compressor or entering the second flash intercooler at point 4,

$$h_4 = 1525 \text{ kJ/kg}$$

Enthalpy of saturated vapour refrigerant leaving the second flash intercooler or entering the high pressure compressor at point 5,

$$h_5 = 1461 \text{ kJ/kg}$$

Entropy of saturated vapour refrigerant at point 5,

$$s_5 = 5.1186 \text{ kJ/kg K}$$

Enthalpy of superheated vapour refrigerant leaving the high pressure compressor or entering the condenser at point 6,

$$h_6 = 1500 \text{ kJ/kg}$$

Enthalpy of liquid refrigerant leaving the condenser at point 7,

$$h_{f7} = h_8 = 328 \text{ kJ/kg}$$

Condition of refrigerant (dryness fraction) leaving the expansion valve at point 8,

$$x_8 = 0.055$$

Enthalpy of refrigerant leaving the expansion valve at point 10,

$$h_{10} = h_{f9} = 265 \text{ kJ/kg}$$

Condition of refrigerant (dryness fraction) leaving the expansion valve at point 10,

$$x_{10} = 0.07$$

Enthalpy of refrigerant entering the evaporator at point 12,

$$h_{12} = h_{f11} = 171.5 \text{ kJ/kg}$$

Power required to drive the system

We know that mass of refrigerant required to be circulated through the evaporator or L.P. compressor,

$$m_1 = \frac{210\,Q}{h_1 - h_{f11}} = \frac{210 \times 10}{1420 - 171.5} = 1.682 \text{ kg/min}$$

Mass of refrigerant passing through I.P. compressor,

$$m_2 = m_1 \left[1 + \frac{x_{10}}{1 - x_{10}} + \frac{h_2 - h_3}{h_3 - h_{10}}\right]$$

$$= 1.682 \left[1 + \frac{0.07}{1 - 0.07} + \frac{1515 - 1442}{1442 - 265}\right] \text{ kg/min}$$

$$= 1.682\ (1 + 0.075 + 0.062) = 1.91 \text{ kg/min}$$

and mass of refrigerant passing through H.P. compressor,

$$m_3 = m_2 \left[1 + \frac{h_4 - h_5}{h_5 - h_8}\right] + \frac{m_1 \times x_8}{(1 - x_{10})\,(1 - x_8)}$$

$$= 1.91 \left[1 + \frac{1525 - 1461}{1461 - 328}\right] + \frac{1.682 \times 0.055}{(1 - 0.07)\,(1 - 0.055)} \text{ kg/min}$$

$$= 1.91 \times 1.056 + 0.105 = 2.12 \text{ kg/min}$$

We know that total work done in three compressors,

$$W = m_1\,(h_2 - h_1) + m_2\,(h_4 - h_3) + m_3\,(h_6 - h_5)$$

$$= 1.682\ (1515 - 1420) + 1.91\ (1525 - 1442) + 2.12\ (1500 - 1461)$$

$$= 159.8 + 158.5 + 82.7 = 401 \text{ kJ/min}$$

$\therefore$ Power required to drive the system,

$$P = 401/60 = 6.68 \text{ kW} \quad \textbf{Ans.}$$

Comparison of C.O.P. of this system with that of simple saturation cycle

We know that refrigerating effect of the system,

$$R_E = 210\,Q = 210 \times 10 = 2100 \text{ kJ/min}$$

$\therefore$ C.O.P. of the system $= \dfrac{R_E}{W} = \dfrac{2100}{401} = 5.23$

For a simple saturation cycle, working between the same pressure limits of 2 bar and 12 bar, the enthalpy of superheated vapour leaving the single compressor at point 2′ is

$h_{2'} = 1670$ kJ/kg ... (From *p-h* chart)

We know that refrigerating effect for simple saturation cycle,

$R_{E1} = h_1 - h_{f7} = 1420 - 328 = 1092$ kJ/kg

Work done in compressor for simple saturation cycle,

$W_1 = h_{2'} - h_1 = 1670 - 1420 = 250$ kJ/kg

and C.O.P. of the system $= \dfrac{R_{E1}}{W_1} = \dfrac{1092}{250} = 4.37$

$\therefore$ Percentage increase in C.O.P.

$$= \frac{5.23 - 4.37}{4.37} \times 100 = 19.68 \% \text{ Ans.}$$

EXERCISES

1. A two stage compression ammonia refrigerating system with intercooler working between the pressure limits of 1.55 bar and 14 bar, is used to take a load of 50 TR. The intercooler pressure is 4.92 bar. The ammonia is cooled to 32°C in the water intercooler and subcooled as liquid to 30°C. Find (*a*) the rate of ammonia circulation per minute ; (*b*) power required to drive the compressors ; and (*c*) C.O.P. of the system.

2. The following data refer to a two stage compression ammonia refrigeration system with water intercooler, liquid sub-cooler and a liquid flash chamber :

Condenser pressure	= 14 bar
Evaporator pressure	= 1.55 bar
Intercooler pressure	= 4.92 bar
Temperature of ammonia leaving the intercooler	= 32°C
Temperature of ammonia leaving the sub-cooler	= 30°C
Volumetric efficiency of low pressure compressor	= 85%
Volumetric efficiency of high pressure compressor	= 80%

If the load on the evaporator is 50 TR, find (*a*) rate of ammonia circulation per minute ; (*b*) power required to drive the compressors ; (*c*) piston displacement for both low pressure and high pressure compressors in m³/min; (*d*) diameters of the cylinders for the single acting low pressure and high pressure compressors, when the speed of compressors is 300 r.p.m. and the stroke is equal to the diameter of low pressure cylinder ; and (*e*) C.O.P. of the system.

[Ans. 8.62 kg/min ; 53 kW ; 7.9 m^3/min, 3.35 m^3/min ; 322 mm, 210 mm ; 3.32]

3. A two stage compression ammonia refrigeration system with water and flash intercooling and water sub-cooling, operates between overall pressure limits of 13.89 bar and 1.9 bar. The flash intercooler pressure is 4.97 bar. The temperature of refrigerant leaving the water intercooler and the water sub-cooler is limited to 30°C. If the load is 10 TR, find : (*a*) coefficient of performance of the system ; (*b*) power required to drive each compressor ; and (*c*) swept volume for each compressor, assuming the volumetric efficiency for both the compressors as 80%.

[Ans. 3.76 ; 4.14 kW, 5.15 kW ; 1.166 m^3/min, 0.56 m^3/min]

4. A vapour compression system using ammonia works between the pressure limits of 1.9 bar and 12 bar. It is fitted with expansion valves and flash chambers such that the vapours are extracted at 3.98 bar and 8 bar. If the load is 10 TR, find the mass of refrigerant flowing through the condenser. Also determine the power required to drive the three compressors and compare the coefficient of performance of this system with that of a simple saturation cycle working within the same pressure limits.

[Ans. 1.9 kg/min ; 7.36 kW ; 8.9% increase]

5. The following data refer to a 10 TR three stage compression system :

   Condenser pressure = 15 bar

   Evaporator pressure = 2.5 bar

   Intermediate pressures = 5 bar and 10 bar

   Find the power required to drive the system when it is provided with (*a*) water intercooling ; and (*b*) flash intercooling.

   Compare the C.O.P. of the above system with that of simple saturation cycle working between the same overall pressure limits. **[Ans.** (*a*) 8 kW ; 6.96% increase, (*b*) 8.5 kW ; 0.87% increase]

6. If the system given in Exercise 5 above, operates with three stage expansion valve and flash intercooling, find the power required to run the system and compare the C.O.P. of the system with that of simple saturation cycle working within the same overall pressure limits. **[Ans.** 7.6 kW ; 9.9% increase]

QUESTIONS

1. What are the advantages of compound compression with intercooler over single stage compression ?
2. Explain a two stage compression system with liquid intercooler.
3. Describe, with the help of schematic and *p-h* diagrams, the working of a two stage compression system with water intercooler, liquid intercooler and a liquid flash chamber.
4. What is the function of a flash intercooler provided in a compound vapour compression refrigerating system ?
5. Discuss the relative merits and demerits of flash and water intercooling employed with multiple compression.
6. Explain a three stage compression with multiple expansion valves and flash intercooler.

OBJECTIVE TYPE QUESTIONS

1. In a compound vapour compression refrigeration system, the compression of refrigerant is carried out in .... compressor.

   (*a*) a single (*b*) more than one
2. The compound compression with intercooling is economical in .... plants.

   (*a*) small (*b*) large
3. In case of multi-stage ammonia plants, when liquid refrigerant is used for intercooling, the total power requirement will

   (*a*) increase (*b*) decrease
4. In case of multi-stage plants using R-12 as refrigerant, the total power requirements will ...., when liquid refrigerant is used for intercooling.

   (*a*) increase (*b*) decrease
5. When water intercooling is used in a multi-stage compression, it

   (*a*) reduces the work to be done in high pressure compressor

   (*b*) reduces the specific volume of the refrigerant

   (*c*) requires a compressor of less stroke volume

   (*d*) all of the above

ANSWERS

1. (*b*)	2. (*b*)	3. (*b*)	4. (*a*)	5. (*d*)

CHAPTER

6

Multiple Evaporator and Compressor Systems

1. Introduction.
2. Types of Multiple Evaporator and Compressor Systems.
3. Multiple Evaporators at the Same Temperature with Single Compressor and Expansion Valve.
4. Multiple Evaporators at Different Temperatures with Single Compressor, Individual Expansion Valves and Back Pressure Valves.
5. Multiple Evaporators at Different Temperatures with Single Compressor, Multiple Expansion Valves and Back Pressure Valves.
6. Multiple Evaporators at Different Temperatures with Individual Compre-ssors and Individual Expansion Valves.
7. Multiple Evaporators at Different Temperatures with Individual Compre-ssors and Multiple Expansion Valves.
8. Multiple Evaporators at Different Temperatures with Compound Compre-ssion and Individual Expansion Valves.
9. Multiple Evaporators at Different Temperatures with Compound Compre-ssion, Individual Expansion Valves and Flash Intercoolers.
10. Multiple Evaporators at Different Temperatures with Compound Compression, Multiple Expansion Valves and Flash Intercoolers.

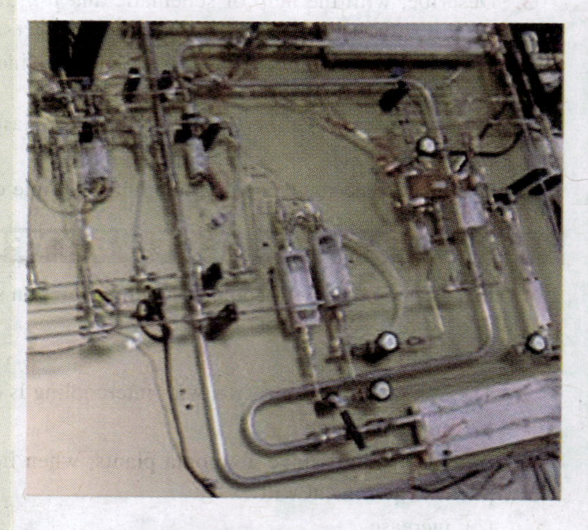

6.1 Introduction

In the previous chapter, we have discussed a single evaporator system in which the entire load is carried by a single evaporator at one temperature. But in many refrigeration installations, different temperatures are required to be maintained at various points in the plant such as in hotels, large restaurants, institutions, industrial plants and food markets where the food products are received in large quantities and stored at different temperatures. For example, the fresh fruits, fresh vegetables, fresh cut meats, frozen products, dairy products, canned goods, bottled goods have all different conditions of temperature and humidity for storage. In such cases, each location is cooled by its own evaporator in order to obtain more satisfactory control of the condition.

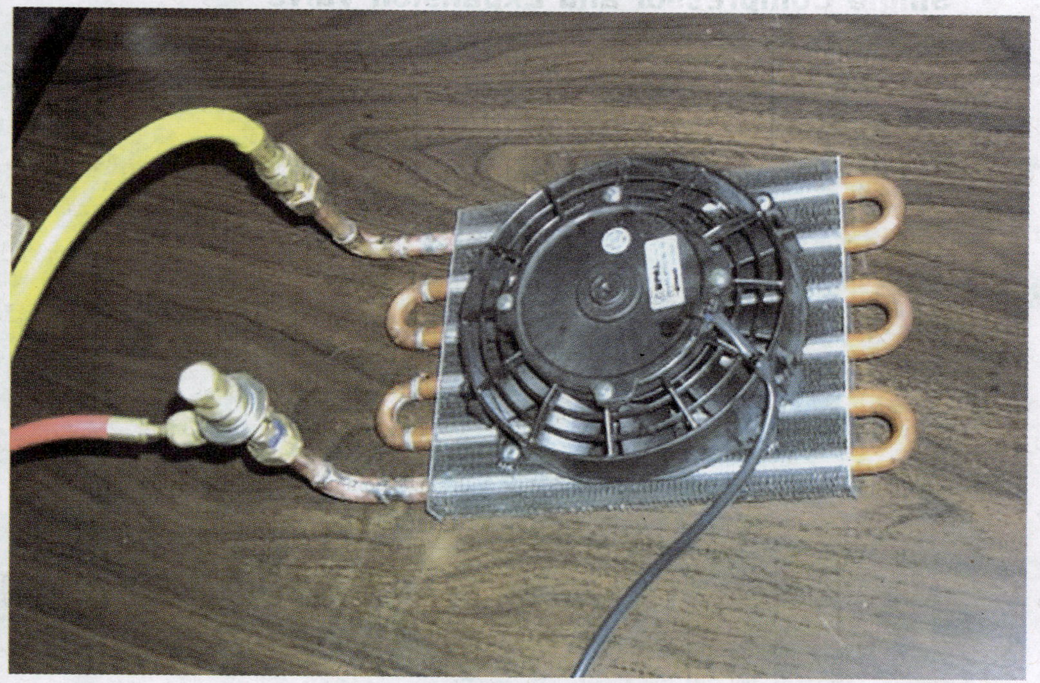

Phase 2 evaporator.

6.2 Types of Multiple Evaporator and Compressor Systems

Following types of multiple evaporator and compressor systems are important from the subject point of view :

1. Multiple evaporators at the same temperature with single compressor and expansion valve.
2. Multiple evaporators at different temperatures with single compressor, individual expansion valves and back pressure valves.
3. Multiple evaporators at different temperatures with single compressor, multiple expansion valves and back pressure valves.
4. Multiple evaporators at different temperatures with individual compressors and individual expansion valves.
5. Multiple evaporators at different temperatures with individual compressors and multiple expansion valves.
6. Multiple evaporators at different temperatures with compound compression and individual expansion valves.
7. Multiple evaporators at different temperatures with compound compression, individual expansion valves and flash intercoolers.
8. Multiple evaporators at different temperatures with compound compression, multiple expansion valves and flash intercoolers.

The above mentioned types of multiple evaporators and compressor systems are discussed in detail, one by one, in the following pages.

6.3 Multiple Evaporators at the Same Temperature with Single Compressor and Expansion Valve

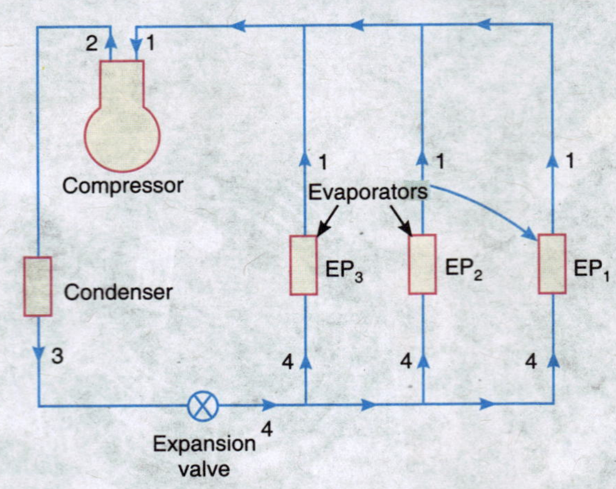

(a) Multiple evaporators at the same temperature with single compressor and expansion valve.

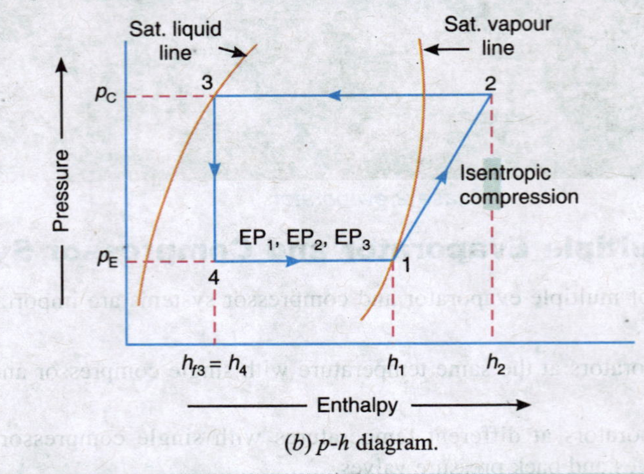

(b) p-h diagram.

Fig. 6.1

This system is generally used when the number of loads such as the food products or other hygroscopic materials kept in different compartments are to be maintained at the same temperature. The arrangement of such a system with three evaporators EP_1, EP_2 and EP_3 is shown in Fig. 6.1 (a). The corresponding p-h diagram is shown in Fig. 6.1 (b).

Let Q_1, Q_2 and Q_3 = Loads on the evaporators EP_1, EP_2 and EP_3 respectively, in tonne of refrigeration.

∴ Mass of refrigerant required to be circulated through the first evaporator EP_1,

$$m_1 = \frac{210\, Q_1}{h_1 - h_4} \text{ kg/min}$$

Similarly, mass of refrigerant required to be circulated through the second evaporator EP_2,

$$m_2 = \frac{210\, Q_2}{h_1 - h_4} \text{ kg/min}$$

and mass of refrigerant required to be circulated through the third evaporator EP_3,

$$m_3 = \frac{210\, Q_3}{h_1 - h_4} \text{ kg/min}$$

∴ Total mass of refrigerant flowing through the evaporators or a compressor,

$$m = m_1 + m_2 + m_3$$

Work done in the compressor,

$$W = (m_1 + m_2 + m_3)\,(h_2 - h_1)$$

∴ Power required to drive the compressor,

$$P = \frac{(m_1 + m_2 + m_3)\,(h_2 - h_1)}{60} \text{ kW}$$

Total refrigerating effect,

$$R_E = (m_1 + m_2 + m_3)\,(h_1 - h_4)$$

∴ Coefficient of performance of the system,

$$\text{C.O.P.} = \frac{R_E}{W} = \frac{(m_1 + m_2 + m_3)\,(h_1 - h_4)}{(m_1 + m_2 + m_3)\,(h_2 - h_1)}$$

$$= \frac{(h_1 - h_4)}{h_2 - h_1} = \frac{h_1 - h_{f3}}{h_2 - h_1} \qquad \ldots (\because h_4 = h_{f3})$$

Example 6.1. *A single compressor using R-12 as refrigerant has three evporators of capacity 10 TR, 20 TR and 30 TR. All the evaporators operate at – 10°C and the vapours leaving the evaporators are dry and saturated. The condenser temperature is 40°C. The liquid refrigerant leaving the condenser is subcooled to 30°C. Assuming isentropic compression, find (a) the mass of refrigerant flowing through each evaporator ; (b) the power required to drive the compressor; and (c) the C.O.P. of the system.*

Solution. Given : Q_1 = 10 TR ; Q_2 = 20 TR ; Q_3 = 30 TR ; t_E = – 10°C ; t_C = 40°C ; t_3 = 30°C

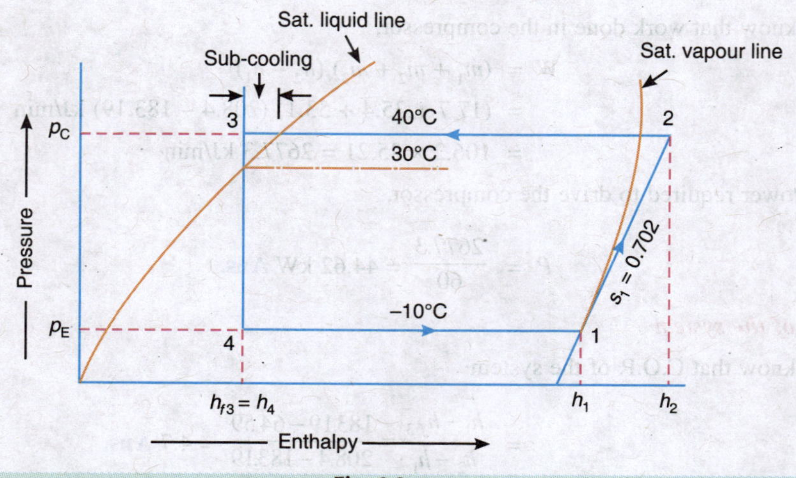

Fig. 6.2

The *p-h* diagram for a single compressor system with three evaporators at the same temperature of – 10°C is shown in Fig. 6.2. The various values as read from the *p-h* diagram for R-12 are as follows :

Enthalpy of saturared vapour refrigerant leaving the evaporators at – 10°C or entering the compressor at point 1,

$$h_1 = 183.19 \text{ kJ/kg}$$

Entropy of saturated vapour refrigerant at point 1,

$$s_1 = 0.702 \text{ kJ/kg K}$$

Enthalpy of superheated vapour refrigerant leaving the compressor at point 2,

$$h_2 = 208.4 \text{ kJ/kg}$$

Enthalpy of sub-cooled liquid refrigerant at 30°C *i.e.* at point 3,

$$h_{f3} = h_4 = 64.59 \text{ kJ/kg}$$

(a) Mass of refrigerant flowing through each evaporator

We know that mass of refrigerant flowing through the first evaporator,

$$m_1 = \frac{210\,Q_1}{h_1 - h_4} = \frac{210 \times 10}{183.19 - 64.59} = 17.7 \text{ kg/min } \textbf{Ans.}$$

Mass of refrigerant flowing through the second evaporator,

$$m_2 = \frac{210\,Q_2}{h_1 - h_4} = \frac{210 \times 20}{183.19 - 64.59} = 35.4 \text{ kg/min } \textbf{Ans.}$$

and mass of refrigerant flowing through the third evaporator,

$$m_3 = \frac{210\,Q_3}{h_1 - h_4} = \frac{210 \times 30}{183.19 - 64.59} = 53.1 \text{ kg/min } \textbf{Ans.}$$

(b) Power required to drive the compressor

We know that work done in the compressor,

$$W = (m_1 + m_2 + m_3)(h_2 - h_1)$$

$$= (17.7 + 35.4 + 53.1)(208.4 - 183.19) \text{ kJ/min}$$

$$= 106.2 \times 25.21 = 2677.3 \text{ kJ/min}$$

∴ Power required to drive the compressor,

$$P = \frac{2677.3}{60} = 44.62 \text{ kW } \textbf{Ans.}$$

(c) C.O.P. of the system

We know that C.O.P. of the system

$$= \frac{h_1 - h_{f3}}{h_2 - h_1} = \frac{183.19 - 64.59}{208.4 - 183.19} = 4.7 \textbf{ Ans.}$$

6.4 Multiple Evaporators at Different Temperatures with Single Compressor, Individual Expansion Valves and Back Pressure Valves

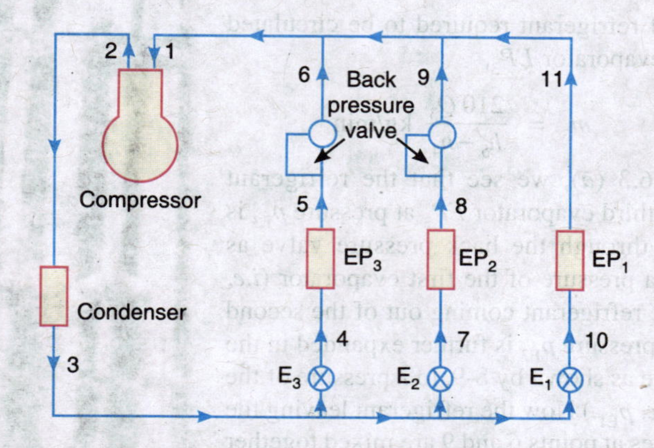

(*a*) Multiple evaporators at different temperatures with single compressor, individual expansion valves and back pressure valves.

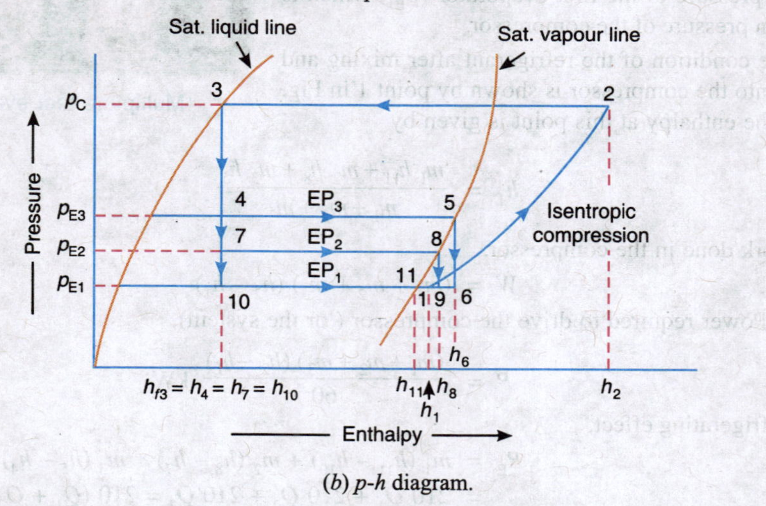

(*b*) *p-h* diagram.

Fig. 6.3

The arrangement, as shown in Fig. 6.3 (*a*), consists of three evaporators EP_1, EP_2 and EP_3 operating at different temperatures with single compressor, three individual expansion valves E_1, E_2 and E_3 and back pressure valves. The corresponding *p-h* diagram is shown in Fig. 6.3 (*b*).

Let Q_1, Q_2 and Q_3 = Loads on the evaporators EP_1, EP_2 and EP_3 respectively, in tonnes of refrigeration.

∴ Mass of refrigerant required to be circulated through the first evaporator EP_1,

$$m_1 = \frac{210\,Q_1}{h_{11} - h_{10}} \text{ kg/min}$$

Similarly, mass of refrigerant required to be circulated through the second evaporator EP_2,

$$m_2 = \frac{210\,Q_2}{h_8 - h_7} \text{ kg/min}$$

and mass of refrigerant required to be circulated through the third evaporator EP_3,

$$m_3 = \frac{210\,Q_3}{h_5 - h_4} \text{ kg/min}$$

Multiple effect evaporator.

From Fig. 6.3 (*a*), we see that the refrigerant coming out of the third evaporator EP_3 at pressure p_{E3} is further expanded through the back pressure valve as shown by 5-6, to a pressure of the first evaporator (*i.e.* p_{E1}). Similarly, the refrigerant coming out of the second evaporator EP_2 at pressure p_{E2} is further expanded in the back pressure valve as shown by 8-9, to a pressure of the first evaporator (*i.e.* p_{E1}.) Now the refrigerant leaving the back pressure valves at points 6 and 9 are mixed together with the refrigerant leaving the first evaporator at point 11, at the pressure of the first evaporator (p_{E1}) which is the suction pressure of the compressor.

The condition of the refrigerant after mixing and entering into the compressor is shown by point 1 in Fig. 6.3 (*b*). The enthalpy at this point is given by

$$h_1 = \frac{m_1\,h_{11} + m_2\,h_8 + m_3\,h_5}{m_1 + m_2 + m_3}$$

Work done in the compressor,

$$W = (m_1 + m_2 + m_3)\,(h_2 - h_1)$$

∴ Power required to drive the compressor (or the system),

$$P = \frac{(m_1 + m_2 + m_3)\,(h_2 - h_1)}{60} \text{ kW}$$

Refrigerating effect,

$$R_E = m_1\,(h_{11} - h_{10}) + m_2\,(h_8 - h_7) + m_3\,(h_5 - h_4)$$
$$= 210\,Q_1 + 210\,Q_2 + 210\,Q_3 = 210\,(Q_1 + Q_2 + Q_3)$$

∴ C.O.P. of the system $= \dfrac{R_E}{W} = \dfrac{210\,(Q_1 + Q_2 + Q_3)}{(m_1 + m_2 + m_3)\,(h_2 - h_1)} = \dfrac{210(Q_1 + Q_2 + Q_3)}{P \times 60}$

Example 6.2. *A single compressor using R-12 as refrigerant has three evaporators of capacity 30 TR, 20 TR and 10 TR. The temperature in the three evaporators is to be maintained at – 10°C, 5°C and 10°C respectively. The condenser pressure is 9.609 bar. The liquid refrigerant leaving the condenser is sub-cooled to 30°C. The vapours leaving the evaporators are dry and saturated. Assuming isentropic compression, find (a) the mass of refrigerant flowing through each evaporator ; (b) the power required to drive the compressor ; and (c) C.O.P. of the system.*

Solution. Given : Q_1 = 30 TR ; Q_2 = 20 TR ; Q_3 = 10 TR ; t_{10} = – 10°C ; t_7 = 5°C ; t_4 = 10°C ; *p_C = 9.609 bar ; t_3 = 30°C

* The saturation temperature corresponding to 9.609 bar is 40°C.

The *p-h* diagram for a single compressor with three evaporators at – 10°C, 5°C and 10°C is shown in Fig. 6.4. The various values as read from the *p-h* diagram for *R*-12 are as follows :

Enthalpy of saturated vapour refrigerant leaving the first evaporator at – 10°C at point 11,

$$h_{11} = 183.19 \text{ kJ/kg}$$

Enthalpy of saturated vapour refrigerant leaving the second evaporator at 5°C at point 8,

$$h_8 = 189.65 \text{ kJ/kg}$$

Enthalpy of saturated vapour refrigerant leaving the third evaporator at 10°C at point 5,

$$h_5 = 191.74 \text{ kJ/kg}$$

Enthalpy of sub-cooled liquid refrigerant at 30°C at point 3,

$$h_{f3} = h_4 = h_7 = h_{10} = 64.59 \text{ kJ/kg}$$

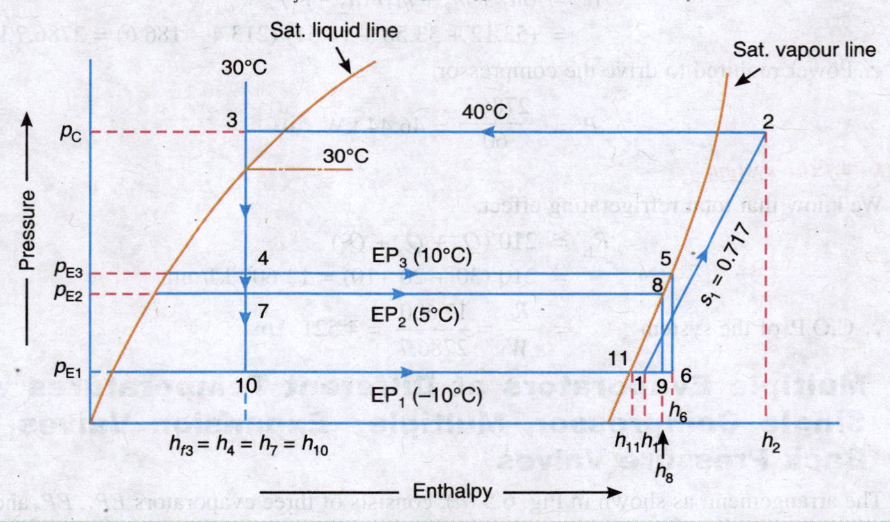

Fig. 6.4

(a) Mass of refrigerant flowing through each evaporator

We know that mass of refrigerant flowing through the first evaporator,

$$m_1 = \frac{210\, Q_1}{h_{11} - h_{10}} = \frac{210 \times 30}{183.19 - 64.59} = 53.12 \text{ kg/min } \textbf{Ans.}$$

Mass of refrigerant flowing through the second evaporator,

$$m_2 = \frac{210\, Q_2}{h_8 - h_7} = \frac{210 \times 20}{189.19 - 64.59} = 33.58 \text{ kg/min } \textbf{Ans.}$$

and mass of refrigerant flowing through the third evaporator,

$$m_3 = \frac{210\, Q_3}{h_5 - h_4} = \frac{210 \times 10}{191.74 - 64.59} = 16.51 \text{ kg/min } \textbf{Ans.}$$

(b) Power required to drive the compressor

The refrigerant coming out of the three evaporators are mixed together before entering into the compressor. The condition of mixed refrigerant entering into the compressor is shown by point 1 on the *p-h* diagram. The enthalpy at point 1 is given by

$$h_1 = \frac{m_1\, h_{11} + m_2\, h_8 + m_3\, h_5}{m_1 + m_2 + m_3}$$

$$= \frac{53.12 \times 183.19 + 33.58 \times 189.65 + 16.51 \times 191.74}{53.12 + 33.58 + 16.51}$$

$$= \frac{9731.05 + 6368.45 + 3165.63}{103.21} = 186.6 \text{ kJ/kg}$$

Mark point 1 on *p-h* diagram such that h_1 = 186.6 kJ/kg. The specific entropy at this point is s_1 = 0.717 kJ/kg K. Now from point 1, draw a constant entropy line intersecting the horizontal condenser pressure line at point 2 as shown in Fig. 6.4. The enthalpy at point 2 as read from *p-h* diagram is

$$h_2 = 213.4 \text{ kJ/kg}$$

We know that work done in the compressor,

$$W = (m_1 + m_2 + m_3)(h_2 - h_1)$$

$$= (53.12 + 33.58 + 16.51)(213.4 - 186.6) = 2786.7 \text{ kJ/min}$$

∴ Power required to drive the compressor,

$$P = \frac{2786.7}{60} = 46.44 \text{ kW} \textbf{ Ans.}$$

(c) C.O.P of the system

We know that total refrigerating effect,

$$R_E = 210(Q_1 + Q_2 + Q_3)$$

$$= 210(30 + 20 + 10) = 12\ 600 \text{ kJ/min}$$

∴ C.O.P. of the system $= \dfrac{R_E}{W} = \dfrac{12\ 600}{2786.7} = 4.521$ **Ans.**

6.5 Multiple Evaporators at Different Temperatures with Single Compressor, Multiple Expansion Valves and Back Pressure Valves

The arrangement, as shown in Fig. 6.5 (*a*), consists of three evaporators EP_1, EP_2 and EP_3 operating at different temperatures with single compressor, multiple expansion valves E_1, E_2 and E_3 and back pressure valves. The corresponding *p-h* diagram is shown in Fig. 6.5 (*b*). In this system, the refrigerant flows from the condenser through expansion valve E_3 where its pressure is reduced from the condenser pressure p_C to the pressure of third evaporator (*i.e.* highest temperature evaporator) EP_3 (*i.e.* p_{E3}). All the vapour formed after leaving the expansion valve E_3 plus enough liquid to take care of the load of evaporator EP_3 passes through this evaporator EP_3. The remaining refrigerant then flows through the expansion valve E_2 where its pressure is reduced from p_{E3} to p_{E2}. Again all the vapour formed after leaving the expansion valve E_2 plus enough liquid to take care of the load of evaporator EP_2 passes through the evaporator EP_2. The remaining liquid now flows through the expansion valve E_1 and supplies it to first evaporator (*i.e.* lowest temperature evaporator) EP_1. The vapour refrigerants coming out of the second and third evaporators EP_2 and EP_3 are further expanded through the back pressure valves to reduce their pressures to p_{E1}, as shown by 9-9′ and 6-6′ respectively. Now the refrigerants leaving the back pressure valves at points 6 and 9 are mixed together with the refrigerant leaving the first evaporator at point 11, at pressure p_{E1} which is the suction pressure of the compressor.

Let Q_1, Q_2 and Q_3 = Loads on the evaporators EP_1, EP_2 and EP_3 respectively in tonnes of refrigeration.

We know that the mass of refrigerant required to be circulated (at point 10) through the first evaporator or the lowest temperature evaporator EP_1,

$$m_{e1} = m_1 = \frac{210\,Q}{h_{11} - h_{10}} \text{ kg/min}$$

Mass of refrigerant required to be circulated (at point 7) through the second evaporator or the intermediate temperature evaporator EP_2,

$$m_2 = \frac{210\,Q_2}{h_9 - h_7} \text{ kg/min}$$

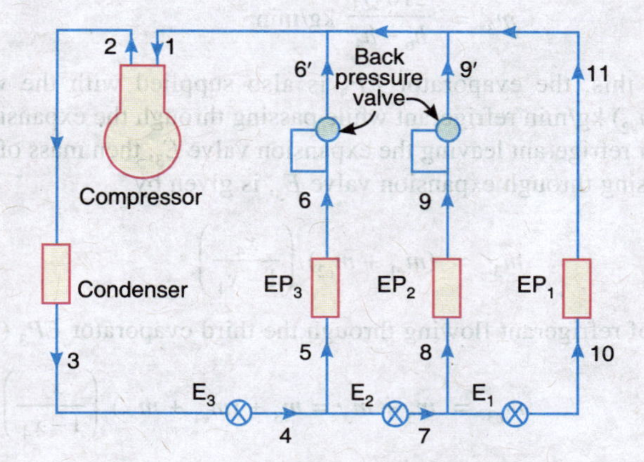

(*a*) Multiple evaporators at different temperatures with single compressor, multiple expansion valves and back pressure valves.

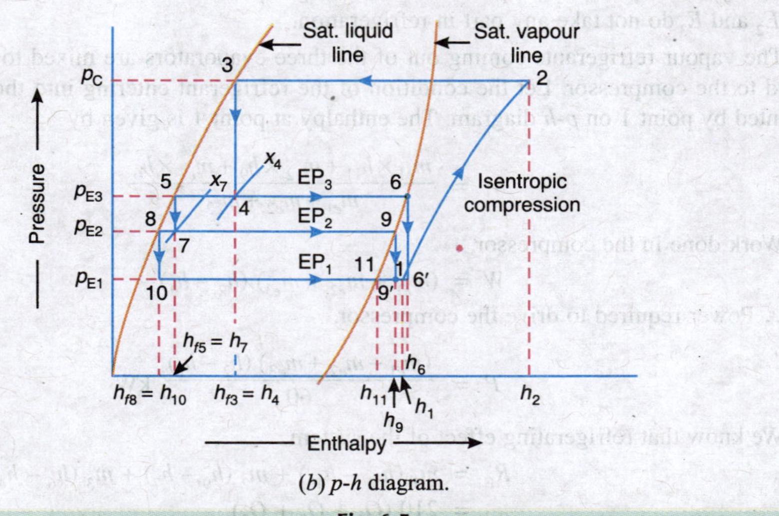

(*b*) *p-h* diagram.

Fig. 6.5

We have discussed above that the second evaporator EP_2 is also supplied with the vapours formed during expansion of m_{e1} kg/min refrigerant while passing through expansion valve E_2. If x_7 is the dryness fraction of the refrigerant leaving the expansion valve E_2, then the mass of vapours formed by m_{e1} while passing through the expansion valve E_2, is given by

$$m_{2'} = m_{e1}\left(\frac{x_7}{1 - x_7}\right)$$

∴ Total mass of refrigerant flowing through the second evaporator EP_2 (*i.e.* at point 8),

$$m_{e2} = m_2 + m_2' = m_2 + m_{e1}\left(\frac{x_7}{1-x_7}\right)$$

Similarly, mass of refrigerant required to be circulated (at point 4) through the third evaporator or the highest temperature evaporator EP_3,

$$m_3 = \frac{210\,Q_3}{h_6 - h_4}\ \text{kg/min}$$

In addition to this, the evaporator EP_3 is also supplied with the vapours formed by expansion of $(m_{e1} + m_{e2})$ kg/min refrigerant while passing through the expansion valve E_3. If x_4 is the dryness fraction of refrigerant leaving the expansion valve E_3, then mass of vapours formed by $(m_{e1} + m_{e2})$ while passing through expansion valve E_3, is given by

$$m_3' = (m_{e1} + m_{e2})\left(\frac{x_4}{1-x_4}\right)$$

∴ Total mass of refrigerant flowing through the third evaporator EP_3 (*i.e.* at point 5),

$$m_{e3} = m_3 + m_3' = m_3 + (m_{e1} + m_{e2})\left(\frac{x_4}{1-x_4}\right)$$

∴ Total mass of refrigerant flowing through the condenser

$$= m_{e1} + m_{e2} + m_{e3}$$

It may be noted that the vapours formed during expansion while passing through expansion valves E_2 and E_3 do not take any part in refrigeration.

The vapour refrigerants coming out of the three evaporators are mixed together and then supplied to the compressor. Let the condition of the refrigerant entering into the compressor is represented by point 1 on *p-h* diagram. The enthalpy at point 1 is given by

$$h_1 = \frac{m_{e1} \times h_{11} + m_{e2} \times h_9 + m_{e3} \times h_6}{m_{e1} + m_{e2} + m_{e3}}$$

Work done in the compressor,

$$W = (m_{e1} + m_{e2} + m_{e3})\,(h_2 - h_1)$$

∴ Power required to drive the compressor,

$$P = \frac{(m_{e1} + m_{e2} + m_{e3})\,(h_2 - h_1)}{60}\ \text{kW}$$

We know that refrigerating effect of the system,

$$R_E = m_1\,(h_{11} - h_{10}) + m_2\,(h_9 - h_7) + m_3\,(h_6 - h_4)$$
$$= 210\,(Q_1 + Q_2 + Q_3)$$

∴ C.O.P. of the system

$$= \frac{R_E}{W} = \frac{210\,(Q_1 + Q_2 + Q_3)}{(m_{e1} + m_{e2} + m_{e3})\,(h_2 - h_1)}$$

$$= \frac{210\,(Q_1 + Q_2 + Q_3)}{P \times 60}$$

Example 6.3. *A single compressor using R-12 as refrigerant has three evaporators of capacity 30 TR, 20 TR and 10 TR. The temperature in the three evaporators is to be maintained at – 10°C, 5°C and 10°C respectively. The system is provided with multiple expansion valves and back pressure valves. The condenser temperature is 40°C. The liquid refrigerant leaving the condenser is sub cooled to 30°C. The vapours leaving the evaporators are dry and saturated. Assuming isentropic compression, find (a) the mass of refrigerant flowing through each evaporator ; (b) the power required to drive the compressor ; and (c) the C.O.P. of the system.*

Solution. Given : Q_1 = 30 TR ; Q_2 = 20 TR ; Q_3 = 10 TR ; t_{10} = – 10°C ; t_8 = 5°C ; t_5 = 10°C

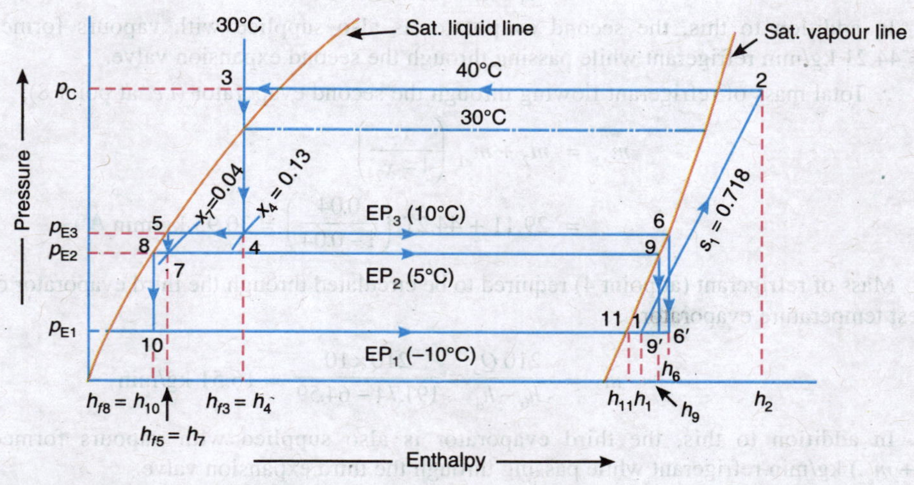

Fig. 6.6

The *p-h* diagram for a single compressor with three evaporators and multiple expansion valves is shown in Fig. 6.6. The various values as read from the *p-h* diagram for *R*-12 are as follows :

Enthalpy of saturated vapour refrigerant leaving the first evaporator at 10°C at point 11,

$$h_{11} = 183.19 \text{ kJ/kg}$$

Enthalpy of saturated vapour refrigerant leaving the second evaporator at 5°C at point 9,

$$h_9 = 189.65 \text{ kJ/kg}$$

Enthalpy of saturated vapour refrigerant leaving the third evaporator at 10°C at point 6,

$$h_6 = 191.74 \text{ kJ/kg}$$

Enthalpy of sub-cooled liquid refrigerant at 30°C at point 3,

$$h_{f3} = h_4 = 64.59 \text{ kJ/kg}$$

Condition of refrigerant (*i.e.* dryness fraction) leaving the expansion valve at point 4,

$$x_4 = 0.13$$

Enthalpy of saturated liquid refrigerant at 10°C at point 5,

$$h_{f5} = h_7 = 45.37 \text{ kJ/kg}$$

Condition of refrigerant (*i.e.* dryness fraction) leaving the expansion valve at point 7,

$$x_7 = 0.04$$

Enthalpy of saturated liquid refrigerant at 5°C at point 8,

$$h_{f8} = h_{10} = 40.69 \text{ kJ/kg}$$

(a) Mass of refrigerant flowing through each evaporator

We know that mass of refrigerant flowing through the first evaporator or the lowest temperature evaporator,

$$m_{e1} = m_1 = \frac{210\,Q_1}{h_{11} - h_{10}} = \frac{210 \times 30}{183.19 - 40.69} = 44.21 \text{ kg/min} \quad \textbf{Ans.}$$

Mass of refrigerant (at point 7) required to be circulated through the second evaporator,

$$m_2 = \frac{210\,Q_2}{h_9 - h_7} = \frac{210 \times 20}{189.65 - 45.37} = 29.11 \text{ kg/min}$$

In addition to this, the second evaporator is also supplied with vapours formed by m_{e1} = 44.21 kg/min refrigerant while passing through the second expansion valve.

∴ Total mass of refrigerant flowing through the second evaporator (*i.e.* at point 8),

$$m_{e2} = m_2 + m_{e1}\left(\frac{x_7}{1 - x_7}\right)$$

$$= 29.11 + 44.21\left(\frac{0.04}{1 - 0.04}\right) = 30.95 \text{ kg/min} \quad \textbf{Ans.}$$

Mass of refrigerant (at point 4) required to be circulated through the third evaporator or the highest temperature evaporator,

$$m_3 = \frac{210\,Q_3}{h_6 - h_4} = \frac{210 \times 10}{191.74 - 64.59} = 16.51 \text{ kg/min}$$

In addition to this, the third evaporator is also supplied with vapours formed by $(m_{e1} + m_{e2})$ kg/min refrigerant while passing through the third expansion valve.

∴ Total mass of refrigerant flowing through the third evaporator (*i.e.* at point 5),

$$m_{e3} = m_3 + (m_{e1} + m_{e2})\left(\frac{x_4}{1 - x_4}\right)$$

$$= 16.51 + (44.21 + 30.95)\left(\frac{0.13}{1 - 0.13}\right) = 27.78 \text{ kg/min} \quad \textbf{Ans.}$$

(b) Power required to drive the compressor

The refrigerant coming out of the three evaporators is mixed together before entering into the compressor. The condition of mixed refrigerant entering into the compressor is shown by point 1 on the *p-h* diagram. The enthalpy at point 1 is given by

$$h_1 = \frac{m_{e1} \times h_{11} + m_{e2} \times h_9 + m_{e3} \times h_6}{m_{e1} + m_{e2} + m_{e3}}$$

$$= \frac{44.21 \times 183.19 + 30.95 \times 189.65 + 27.78 \times 191.74}{44.21 + 30.95 + 27.78}$$

$$= \frac{8098.8 + 5869.7 + 5326.5}{103.12} = 187 \text{ kJ/kg}$$

Mark point 1 on *p-h* diagram such that h_1 = 187 kJ/kg. The specific entropy at this point is s_1 = 0.718 kJ/kg K. Now from point 1, draw a constant entropy line intersecting the horizontal condenser pressure line at point 2, as shown in Fig. 6.6. The enthalpy at point 2 as read from *p-h* diagram is

$$h_2 = 214.2 \text{ kJ/kg}$$

We know that work done in the compressor,

$$W = (m_{e1} + m_{e2} + m_{e3})\,(h_2 - h_1)$$
$$= (44.21 + 30.95 + 27.78)\,(214.2 - 187) = 2800 \text{ kJ/min}$$

∴ Power required to drive the compressor,

$$P = \frac{2800}{60} = 46.7 \text{ kW } \textbf{Ans.}$$

(c) C.O.P. of the system

We know that total refrigerating effect,

$$R_E = 210\,(Q_1 + Q_2 + Q_3)$$
$$= 210\,(30 + 20 + 10) = 12\,600 \text{ kJ/min}$$

∴ C.O.P. of the system $= \dfrac{R_E}{W} = \dfrac{12600}{2800} = 4.5$ **Ans.**

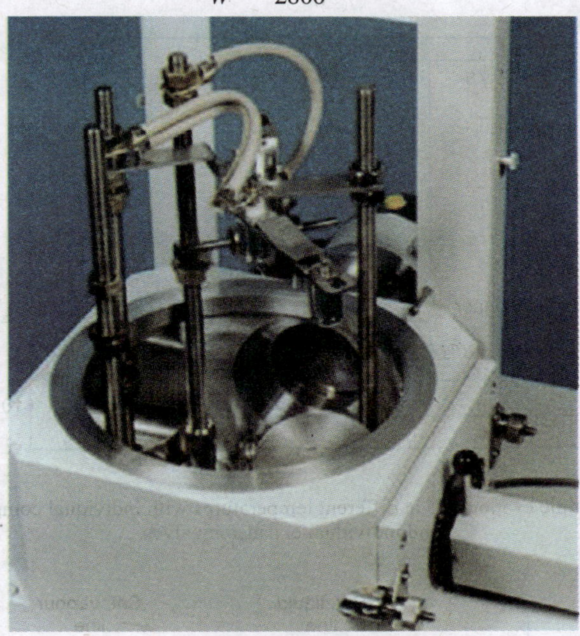

Multiple evaporator.

6.6 Multiple Evaporators at Different Temperatures with Individual Compressors and Individual Expansion Valves

The arrangement, as shown in Fig. 6.7 (*a*), consists of three evaporators EP_1, EP_2 and EP_3 at different temperatures with individual compressors C_1, C_2 and C_3 and individual expansion valves E_1, E_2 and E_3. The corresponding *p-h* diagram is shown in Fig. 6.7 (*b*).

Let Q_1, Q_2 and Q_3 = Loads on the evaporators EP_1, EP_2 and EP_3 respectively in tonnes of refrigeration.

∴ Mass of refrigerant required to be circulated through the first evaporator EP_1,

$$m_1 = \frac{210\,Q_1}{h_1 - h_{10}} \text{ kg/min}$$

Similarly, mass of refrigerant required to be circulated through the second evaporator EP_2,

$$m_2 = \frac{210\, Q_2}{h_3 - h_9} \text{ kg/min}$$

and mass of refrigerant required to be circulated through the third evaporator EP_3,

$$m_3 = \frac{210\, Q_3}{h_5 - h_8} \text{ kg/min}$$

∴ Power required to drive the compressor C_1,

$$P_1 = \frac{m_1 (h_2 - h_1)}{60} \text{ kW}$$

Similarly, power required to drive the compressor C_2,

$$P_2 = \frac{m_2 (h_4 - h_3)}{60} \text{ kW}$$

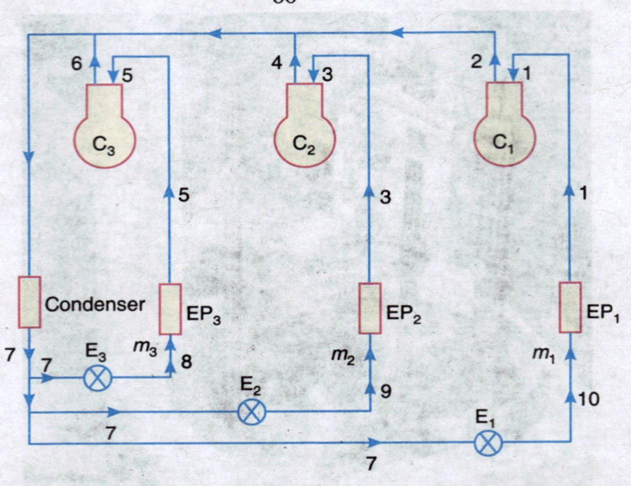

(*a*) Multiple evaporators at different temperatures with individual compressors and individual expansion valves.

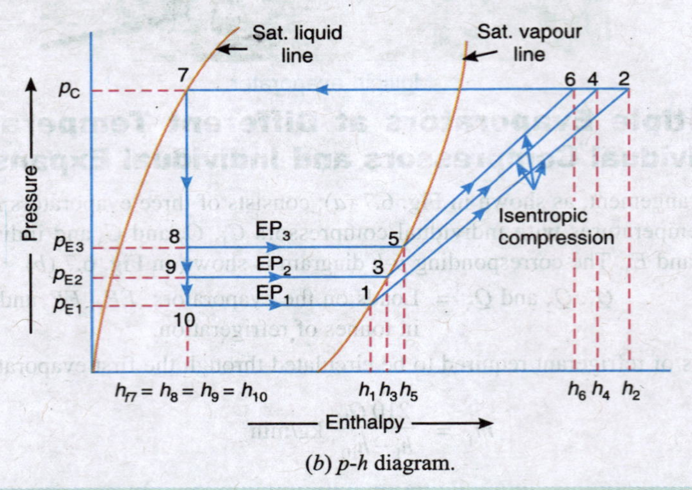

(*b*) *p-h* diagram.

Fig. 6.7

and power required to drive the compressor C_3,

$$P_3 = \frac{m_3 (h_6 - h_5)}{60} \text{ kW}$$

∴ Total power required to drive the system,

$$P = P_1 + P_2 + P_3$$

We know that the refrigerating effect of the system,

$$R_E = 210 (Q_1 + Q_2 + Q_3)$$

and total work done in the three compressors,

$$W = m_1 (h_2 - h_1) + m_2 (h_4 - h_3) + m_3 (h_6 - h_5)$$

∴ C.O.P. of the system

$$= \frac{R_E}{W} = \frac{210(Q_1 + Q_2 + Q_3)}{m_1 (h_2 - h_1) + m_2 (h_4 - h_3) + m_3 (h_6 - h_5)}$$

$$= \frac{210(Q_1 + Q_2 + Q_3)}{P \times 60}$$

Example 6.4. *The refrigeration system using R-12 as refrigerant consists of three evaporators of capacities 20 TR, 30 TR and 10 TR with individual expansion valves and individual compressors. The temperature in the three evaporators is to be maintained at – 10°C, 5°C, and 10°C respectively. The vapours leaving the evaporators are dry and saturated. The condenser temperature is 40°C and the liquid refrigerant leaving the consdenser is subcooled to 30°C. Assuming isentropic compression in each compressor, find (a) the mass of refrigerant flowing through each evaporator ; (b) the power required to drive the system ; and (c) C.O.P. of system.*

Solution. Given : $Q_1 = 20$ TR ; $Q_2 = 30$ TR ; $Q_3 = 10$ TR ; $t_1 = t_{10} = -10°C$; $t_3 = t_9 = 5°C$; $t_5 = t_8 = 10°C$; $t_c = 40°C$; $t_7 = 30°C$

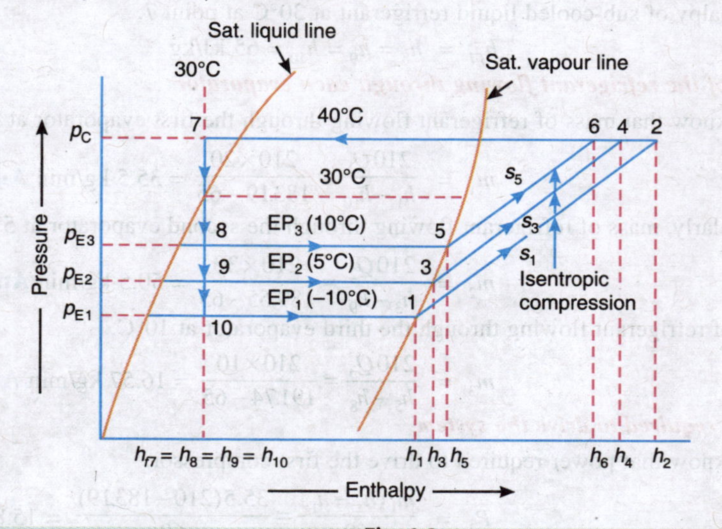

Fig. 6.8

The *p-h* diagram of a refrigeration system consisting of three evaporators with individual expansion valves and individual compressors is shown in Fig. 6.8. The various values as read from the *p-h* diagram for *R*-12 are as follows :

Enthalpy of saturated vapour refrigerant leaving the first evaporator at – 10°C or entering the first compressor at point 1,

$$h_1 = 183.19 \text{ kJ/kg}$$

Entropy of saturated vapour refrigerant at point 1,

$$s_1 = 0.7019 \text{ kJ/kg K}$$

Enthalpy of saturated vapour refrigerant leaving the second evaporator at 5°C or entering the second compressor at point 3,

$$h_3 = 189.65 \text{ kJ/kg}$$

Entropy of saturated vapour refrigerant at point 3,

$$s_3 = 0.6943 \text{ kJ/kg K}$$

Enthalpy of saturated vapour refrigerant leaving the third evaporator at 10°C or entering the third compressor at point 5,

$$h_5 = 191.74 \text{ kJ/kg}$$

Entropy of saturated vapour refrigerant at point 5,

$$s_5 = 0.6921 \text{ kJ/kg K}$$

From points 1, 3 and 5, draw the constant entropy lines of s_1, s_2 and s_3 intersecting the horizontal condenser pressure line (corresponding to 40°C) at points 2, 4 and 6 respectively. From *p-h* diagram, we find that,

Enthalpy of superheated vapour refrigerant leaving the first compressor at point 2,

$$h_2 = 210 \text{ kJ/kg}$$

Enthalpy of superheated vapour refrigerant leaving the second compressor at point 4,

$$h_4 = 208 \text{ kJ/kg}$$

Enthalpy of superheated vapour refrigerant leaving the third compressor at point 6,

$$h_6 = 206 \text{ kJ/kg}$$

Enthalpy of sub-cooled liquid refrigerant at 30°C at point 7,

$$h_{f7} = h_8 = h_9 = h_{10} = 65 \text{ kJ/kg}$$

(a) Mass of the refrigerant flowing through each evaporator

We know that mass of refrigerant flowing through the first evaporator at – 10°C,

$$m_1 = \frac{210\,Q_1}{h_1 - h_{10}} = \frac{210 \times 20}{183.19 - 65} = 35.5 \text{ kg/min } \textbf{Ans.}$$

Similarly, mass of refrigerant flowing through the second evaporator at 5°C,

$$m_2 = \frac{210\,Q_2}{h_3 - h_9} = \frac{210 \times 30}{189.65 - 65} = 50.5 \text{ kg/min } \textbf{Ans.}$$

and mass of refrigerant flowing through the third evaporator at 10°C,

$$m_3 = \frac{210\,Q_3}{h_5 - h_8} = \frac{210 \times 10}{191.74 - 65} = 16.57 \text{ kg/min } \textbf{Ans.}$$

(b) Power required to drive the system

We know that power required to drive the first compressor,

$$P_1 = \frac{m_1 (h_2 - h_1)}{60} = \frac{35.5(210 - 183.19)}{60} = 15.86 \text{ kW}$$

Power required to drive the second compressor,

$$P_2 = \frac{m_2 (h_4 - h_3)}{60} = \frac{50.5(208 - 189.65)}{60} = 15.44 \text{ kW}$$

and power required to drive the third compressor,

$$P_3 = \frac{m_3 (h_6 - h_5)}{60} = \frac{16.57 (206 - 191.74)}{60} = 3.94 \text{ kW}$$

∴ Total power required to drive the system,

$$= P_1 + P_2 + P_3 = 15.86 + 15.44 + 3.94 = 35.24 \text{ kW} \quad \textbf{Ans.}$$

(c) C.O.P. of the system

We know that refrigerating effect of the system,

$$R_E = 210 (Q_1 + Q_2 + Q_3) = 210 (20 + 30 + 10) = 12\,600 \text{ kJ/min}$$

and work done in the three compressors,

$$W = m_1 (h_2 - h_1) + m_2 (h_4 - h_3) + m_3 (h_6 - h_5)$$
$$= 35.5 (210 - 183.19) + 50.5 (208 - 189.65) + 16.57 (206 - 191.74)$$
$$= 951.7 + 926.7 + 236.3 = 2114.7 \text{ kJ/min}$$

$$\therefore \text{ C.O.P. of the system} = \frac{R_E}{W} = \frac{12\,600}{2114.7} = 5.96 \quad \textbf{Ans.}$$

6.7 Multiple Evaporators at Different Temperatures with Individual Compressors and Multiple Expansion Valves

The arrangement, as shown in Fig. 6.9 (*a*), consists of three evaporators EP_1, EP_2 and EP_3 at different temperatures with individual compressors C_1, C_2 and C_3 and multiple expansion valves E_1, E_2 and E_3. The corresponding *p-h* diagram is shown in Fig. 6.9 (*b*). In this system, all the refrigerant from the condenser flows through the expansion valve E_3 where its pressure is reduced from the condenser pressure p_C to the pressure of the third evaporator *i.e.* p_{E3}. All the vapours formed after leaving the expansion valve E_3 plus enough liquid to take care of the load of evaporator EP_3 passes through this evaporator EP_3. The remaining refrigerant then flows through the expansion valve E_2 where its pressure is reduced from p_{E3} to p_{E2}. Again, all the vapour formed after leaving the expansion valve E_2 plus enough liquid to take care of the load of evaporator EP_2 passes through this evaporator EP_2. The remaining liquid now flows through the expansion valve E_1 which supplies it to the first evaporator (lowest temperature evaporator) EP_1.

Multiple expansion valves.

Let Q_1, Q_2 and Q_3 = Loads on the evaporators EP_1, EP_2 and EP_3 respectively in tonnes of refrigeration.

We know that mass of refrigerant required to be circulated (at point 12) through the first evaporator (or lowest temperature evaporator) EP_1,

$$m_{e1} = m_1 = \frac{210\, Q_1}{h_1 - h_{12}} \text{ kg/min}$$

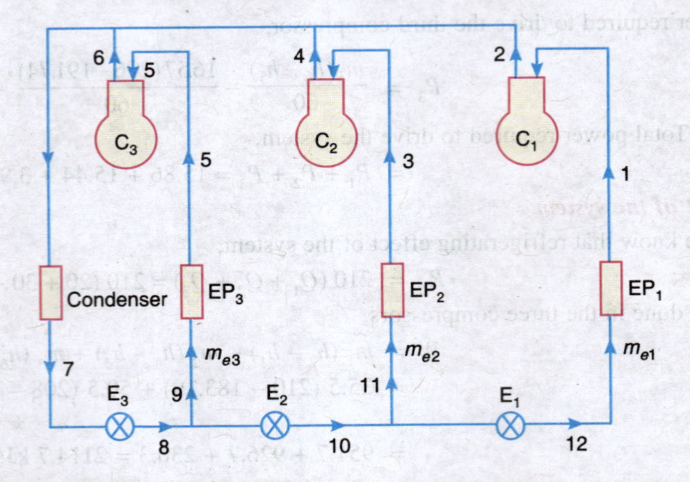

(*a*) Multiple evaporators at different temperatures with individual compressors and multiple expansion valves.

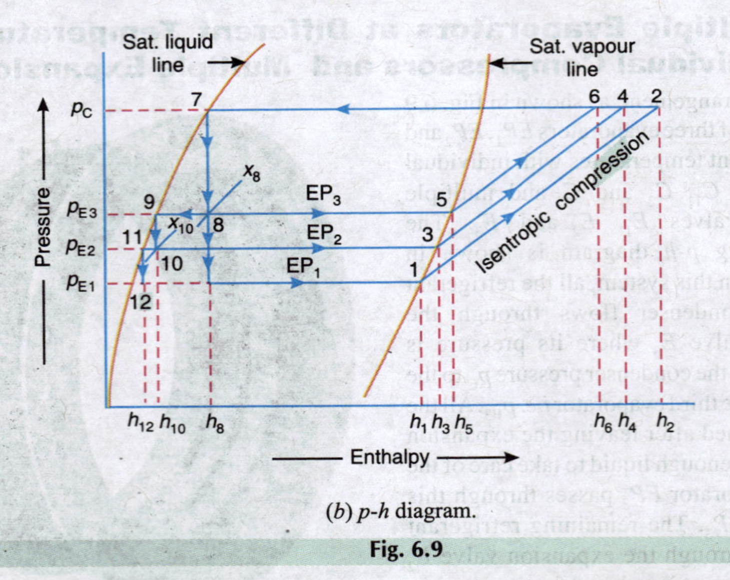

(*b*) *p-h* diagram.

Fig. 6.9

Mass of refrigerant required to be circulated (at point 10) through the second evaporator (or intermediate temperature evaporator) EP_2,

$$m_2 = \frac{210\, Q_2}{h_3 - h_{10}} \text{ kg/min}$$

We have discussed above that the second evaporator EP_2 is also supplied with vapours formed during expansion of m_{e1} kg/min refrigerant while passing through the expansion valve E_2. If x_{10} is the dryness fraction of the refrigerant leaving the expansion valve E_2 at point 10, then the mass of vapours formed by m_{e1} while passing through the expansion valve E_2 is given by

$$m_2' = m_{e1}\left(\frac{x_{10}}{1 - x_{10}}\right)$$

∴ Total mass of refrigerant flowing through the second evaporator EP_2 at point 11,

$$m_{e2} = m_2 + m_2' = m_2 + m_{e1}\left(\frac{x_{10}}{1 - x_{10}}\right)$$

Similarly, mass of refrigerant required to be circulated (at point 8) through the third evaporator (or highest temperature evaporator) EP_3,

$$m_3 = \frac{210\,Q_3}{h_5 - h_8} \text{ kg/min}$$

In addition to this, the evaporator EP_3 is also supplied with vapours formed by expansion of $(m_{e1} + m_{e2})$ kg/min refrigerant while passing through the expansion valve E_3. If x_8 is the dryness fraction of refrigerant leaving the expansion valve E_3, then mass of vapours formed by $(m_{e1} + m_{e2})$ while passing through expansion valve E_3 is given by

$$m_3' = (m_{e1} + m_{e2})\left(\frac{x_8}{1 - x_8}\right)$$

∴ Total mass of refrigerant flowing through the third evaporator EP_3 at point 9,

$$m_{e3} = m_3 + m_3' = m_3 + (m_{e1} + m_{e2})\left(\frac{x_8}{1 - x_8}\right)$$

∴ Total mass of refrigerant flowing through the condenser,

$$m = m_{e1} + m_{e2} + m_{e3}$$

It may be noted that the vapours formed during expansion while passing through expansion valves E_2 and E_3 do not take any part in refrigeration.

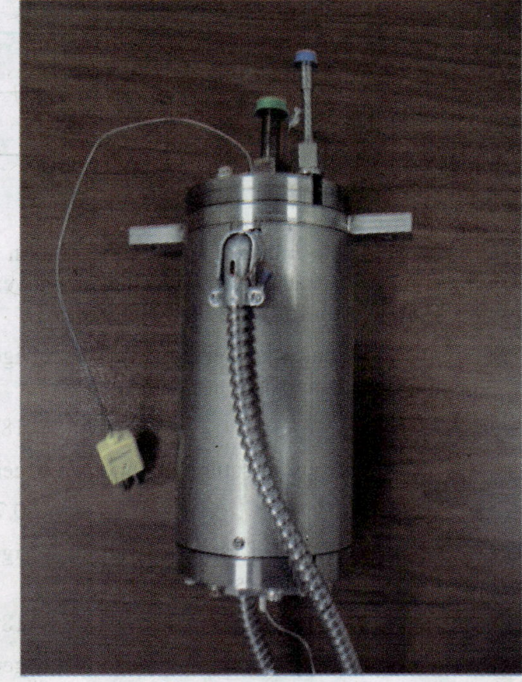

Flash evaporator.

We know that power required to drive the first compressor,

$$P_1 = \frac{m_{e1}(h_2 - h_1)}{60} \text{ kW}$$

Similarly, power required to drive the second compressor,

$$P_2 = \frac{m_{e2}(h_4 - h_3)}{60} \text{ kW}$$

and power required to drive the third compressor,

$$P_3 = \frac{m_{e3}(h_6 - h_5)}{60} \text{ kW}$$

∴ Total power required to drive the system,

$$P = P_1 + P_2 + P_3$$

We know that refrigerating effect of the system,

$$R_E = 210\,(Q_1 + Q_2 + Q_3)$$

and total work done in the three compressors,

$$W = m_{e1}\,(h_2 - h_1) + m_{e2}\,(h_4 - h_3) + m_{e3}\,(h_6 - h_5)$$
$$= P \times 60 \text{ kJ/min}$$

∴ C.O.P. of the system $= \dfrac{R_E}{W} = \dfrac{210\,(Q_1 + Q_2 + Q_3)}{P \times 60}$

Example 6.5. *A refrigeration system using R-12 as refrigerant consists of three evaporators of capacities 20 TR at – 10°C, 30 TR at 5°C and 10 TR at 10°C . The vapours leaving the three evaporators are dry and saturated. The system is provided with individual compressors and multiple expansion valves. The condenser temperature is 40°C and the liquid refrigerant leaving the condenser is subcooled to 30°C. Assuming isentropic compression in each compressor, find (a) the mass of refrigerant flowing through each evaporator ; (b) the power required to drive the system ; and (c) the C.O.P. of the system.*

Solution. Given : Q_1 = 20 TR ; $t_1 = t_{12}$ = – 10°C ; Q_2 = 30 TR ; $t_3 = t_{10} = t_{11}$ = 5°C ; Q_3 = 10 TR ; $t_5 = t_8 = t_9$ = 10°C ; t_c = 40°C ; t_7 = 30°C

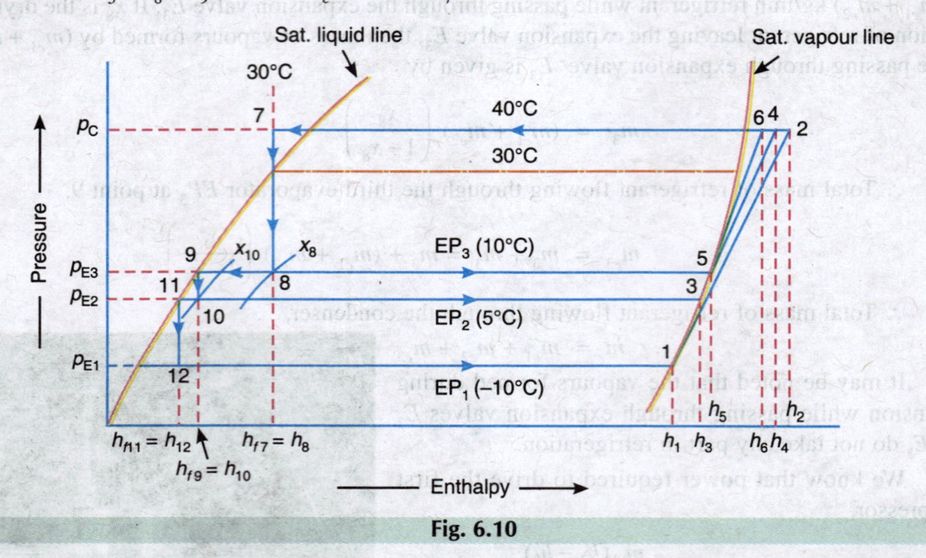

Fig. 6.10

The *p-h* diagram of a refrigeration system consisting of three evaporators with individual compressors and multiple expansion valves is shown in Fig. 6.10. The various values as read from *p-h* diagram for *R*-12 are as follows :

Enthalpy of saturated vapour refrigerant leaving the first evaporator at – 10°C or entering the first compressor at point 1,

$$h_1 = 183.19 \text{ kJ/kg}$$

Entropy of saturated vapour refrigerant at point 1,

$$s_1 = 0.702 \text{ kJ/kg K}$$

Enthalpy of saturated vapour refrigerant leaving the second evaporator at 5°C or entering the second compressor at point 3,

$$h_3 = 189.65 \text{ kJ/kg}$$

Entropy of saturated vapour refrigerant at point 3,

$$s_3 = 0.6943 \text{ kJ/kg K}$$

Enthalpy of saturated vapour refrigerant leaving the third evaporator at 10°C or entering the third compressor at point 5,

$$h_5 = 191.74 \text{ kJ/kg}$$

Entropy of saturated vapour refrigerant at point 5,

$$s_5 = 0.6921 \text{ kJ/kg K}$$

From points 1, 3 and 5, draw constant entropy lines of s_1, s_3 and s_5 intersecting the condenser pressure line (corresponding to 40°C) at points 2, 4 and 6 respectively. From *p-h* diagram, we find that,

Enthalpy of superheated vapour refrigerant leaving the first compressor at point 2,

$$h_2 = 209.3 \text{ kJ/kg}$$

Enthalpy of superheated vapour refrigerant leaving the second compressor at point 4,

$$h_4 = 208.2 \text{ kJ/kg}$$

Enthalpy of superheated vapour refrigerant leaving the third compressor at point 6,

$$h_6 = 206.7 \text{ kJ/kg}$$

Enthalpy of sub-cooled liquid refrigerant at 30°C at point 7,

$$h_{f7} = h_8 = 64.59 \text{ kJ/kg}$$

Condition of refrigerant (dryness fraction) leaving the expansion valve at point 8,

$$x_8 = 0.13$$

Enthalpy of saturated liquid refrigerant at point 9,

$$h_{f9} = h_{10} = 45.37 \text{ kJ/kg}$$

Condition of refrigerant (dryness fraction) leaving the expansion valve at point 10,

$$x_{10} = 0.04$$

Enthalpy of saturated liquid refrigerant at point 11,

$$h_{f11} = h_{12} = 40.69 \text{ kJ/kg}$$

(a) Mass of refrigerant flowing through each evaporator

We know that mass of refrigerant flowing through the first evaporator at – 10°C,

$$m_{e1} = m_1 = \frac{210Q_1}{h_1 - h_{12}} = \frac{210 \times 20}{183.19 - 40.69} = 29.47 \text{ kg/min} \quad \textbf{Ans.}$$

Mass of refrigerant required to be circulated through the second evaporator at 5°C,

$$m_2 = \frac{210Q_2}{h_3 - h_{10}} = \frac{210 \times 30}{189.65 - 45.37} = 43.66 \text{ kg/min}$$

In addition to this, the second evaporator is also supplied with vapours formed during expansion of m_{e1} = 29.47 kg/min refrigerant while passing through the second expansion valve.

∴ Mass of vapour formed by m_{e1} kg/min refrigerant (at point 10),

$$m_2' = m_{e1}\left(\frac{x_{10}}{1 - x_{10}}\right) = 29.47\left(\frac{0.04}{1 - 0.04}\right) = 1.23 \text{ kg/min}$$

∴ Total mass of refrigerant flowing through the second evaporator at 5°C,

$$m_{e2} = m_2 + m_2' = 43.66 + 1.23 = 44.89 \text{ kg/min} \quad \textbf{Ans.}$$

Mass of refrigerant required to be circulated through the third evaporator at 10°C,

$$m_3 = \frac{210\,Q_3}{h_5 - h_8} = \frac{210 \times 10}{191.74 - 64.59} = 16.51 \text{ kg/min}$$

In addition to this, the third evaporator is also supplied with vapour formed during expansion of $(m_{e1} + m_{e2})$ kg/min of refrigerant while passing through the third expansion valve.

∴ Mass of vapours formed by $(m_{e1} + m_{e2})$ kg/min of refrigerant (at point 8),

$$m_3' = (m_{e1} + m_{e2}) \left(\frac{x_8}{1 - x_8} \right)$$

$$= (29.47 + 44.89) \left(\frac{0.13}{1 - 0.13} \right) = 11.11 \text{ kg/min}$$

∴ Total mass of refrigerant flowing through the third evaporator at 10°C,

$$m_{e3} = m_3 + m_3' = 16.51 + 11.11 = 27.62 \text{ kg/min} \quad \textbf{Ans.}$$

(b) Power required to drive the system

We know that power required to drive the first compressor,

$$P_1 = \frac{m_{e1}(h_2 - h_1)}{60} = \frac{29.47\,(209.3 - 183.19)}{60} = 12.82 \text{ kW}$$

Power required to drive the second compressor,

$$P_2 = \frac{m_{e2}(h_4 - h_3)}{60} = \frac{44.89\,(208.2 - 189.65)}{60} = 13.88 \text{ kW}$$

and power required to drive the third compressor,

$$P_3 = \frac{m_{e3}(h_6 - h_5)}{60} = \frac{27.62\,(206.7 - 191.74)}{60} = 6.90 \text{ kW}$$

∴ Total power required to drive the system,

$$P = P_1 + P_2 + P_3$$

$$= 12.82 + 13.88 + 6.90 = 33.6 \text{ kW} \quad \textbf{Ans.}$$

(c) C.O.P. of the system

We know that refrigerating effect of the system,

$$R_E = 210\,(Q_1 + Q_2 + Q_3)$$

$$= 210\,(20 + 30 + 10) = 12\,600 \text{ kJ/min}$$

and total work done in the three compressors,

$$W = P \times 60 = 33.6 \times 60 = 2016 \text{ kJ/min}$$

∴ C.O.P. of the system $= \dfrac{R_E}{W} = \dfrac{12\,600}{2016} = 6.25$ **Ans.**

6.8 Multiple Evaporators at Different Temperatures with Compound Compression and Individual Expansion Valves

The arrangement, as shown in Fig. 6.11(*a*), consists of three evaporators EP_1, EP_2 and EP_3 at different temperatures with compound compression (three stage compression) and individual expansion valves E_1, E_2 and E_3. The corresponding *p-h* diagram is shown in Fig. 6.11 (*b*).

Let Q_1, Q_2 and Q_3 = Loads on the evaporators EP_1, EP_2 and EP_3 respectively in tonnes of refrigeration.

Mass of refrigerant flowing through the first evaporator EP_1 or the first compressor C_1,

$$m_1 = \frac{210\,Q_1}{h_1 - h_{12}} \text{ kg/min}$$

Similarly, mass of refrigerant flowing through the second evaporator EP_2,

$$m_2 = \frac{210\,Q_2}{h_2 - h_{11}} \text{ kg/min}$$

and mass of refrigerant flowing through the third evaporator EP_3,

$$m_3 = \frac{210\, Q_3}{h_6 - h_{10}} \text{ kg/min}$$

From Fig. 6.11 (*a*), we see that the mass of refrigerant (m_1) coming out from the first compressor C_1 is mixed with the mass of refrigerant (m_2) coming out from the second evaporator EP_2, before entering into the second compressor C_2. The condition of the mixed refrigerant entering into the second compressor C_2 is shown by point 4 on the *p-h* diagram. The enthalpy at point 4 is given by

$$(m_1 + m_2)\, h_4 = m_1\, h_2 + m_2\, h_3$$

$$\therefore \quad h_4 = \frac{m_1\, h_2 + m_2\, h_3}{m_1 + m_2}$$

(*a*) Multiple evaporators at different temperatures with compound compression and individual expansion valves.

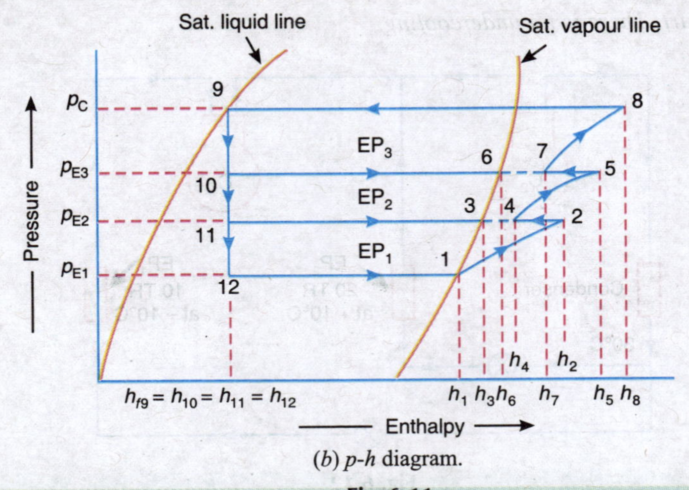

(*b*) *p-h* diagram.

Fig. 6.11

Similarly, the refrigerant coming out from the second compressor C_2 (*i.e.* $m_1 + m_2$) is mixed with the refrigerant coming out from the third evaporator EP_3 (*i.e.* m_3), before entering into the third compressor C_3. The condition of the mixed refrigerant entering into the third compressor C_3 is shown by point 7 on the *p-h* diagram. The enthalpy at point 7 is given by

$$(m_1 + m_2 + m_3)\, h_7 = (m_1 + m_2)\, h_5 + m_3\, h_6$$

$$\therefore \quad h_7 = \frac{(m_1 + m_2)\, h_5 + m_3\, h_6}{m_1 + m_2 + m_3}$$

We know that power required to drive the first compressor C_1,

$$P_1 = \frac{m_1 (h_2 - h_1)}{60} \text{ kW}$$

Similarly, power required to drive the second compressor C_2,

$$P_2 = \frac{(m_1 + m_2)(h_5 - h_4)}{60} \text{ kW}$$

and power required to drive the third compressor C_3,

$$P_3 = \frac{(m_1 + m_2 + m_3)(h_8 - h_7)}{60} \text{ kW}$$

∴ Total power required to drive the system,

$$P = P_1 + P_2 + P_3$$

We know that refrigerating effect of the system,

$$R_E = 210\,(Q_1 + Q_2 + Q_3) \text{ kJ/min}$$

and total work done in the three compressors,

$$W = P \times 60 \text{ kJ/min}$$

$$\therefore \text{ C.O.P. of the system} = \frac{R_E}{W} = \frac{210\,(Q_1 + Q_2 + Q_3)}{P \times 60}$$

Example 6.6. *A compound refrigeration system is used for multi- load purposes, as shown in Fig. 6.12. R-12 is used as refrigerant.*

Find (a) the power required to run the system; and (b) the coefficient of performance of the combined system.

Use p-h chart. There is no undercooling.

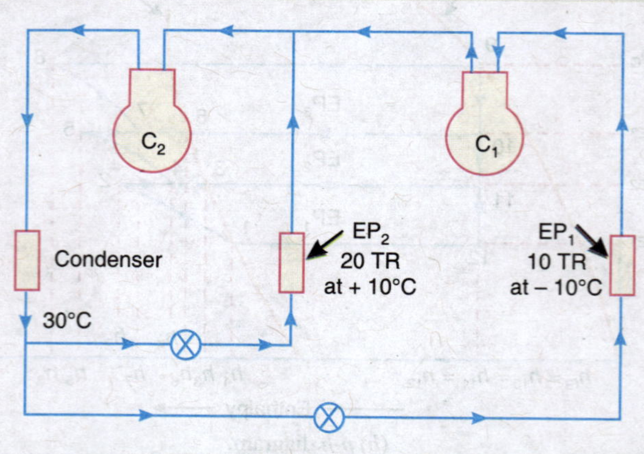

Fig. 6.12

Solution. Given : Q_1 = 10 TR ; $t_1 = t_8 = -$ 10°C ; Q_2 = 20 TR ; $t_3 = t_7$ = 10°C ; t_c = 30° C

The *p-h* diagram of a given compound refrigeration system is shown in Fig. 6.13. The various values, as read from the *p-h* diagram for *R*-12 are as follows :

Enthalpy of saturated vapour refrigerant leaving the first evaporator at –10°C or entering the first compressor at point 1,

$$h_1 = 183.19 \text{ kJ/kg}$$

Entropy of saturated vapour refrigerant at point 1,

$$s_1 = 0.702 \text{ kJ/kg K}$$

Enthalpy of superheated vapour refrigerant leaving the first compressor at point 2,

$$h_2 = 195 \text{ kJ/kg}$$

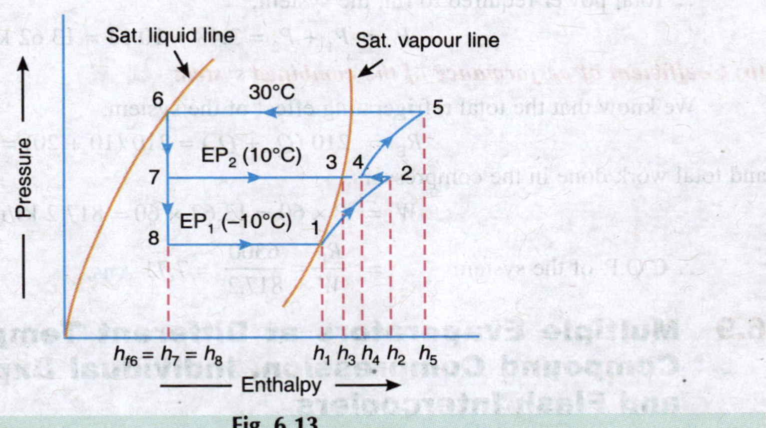

Fig. 6.13

Enthalpy of saturated vapour refrigerant leaving the second evaporator EP_2 at 10°C at point 3,

$$h_3 = 191.74 \text{ kJ/kg}$$

Enthalpy of saturated liquid refrigerant leaving the condenser at 30°C at point 6,

$$h_{f6} = h_7 = h_8 = 64.6 \text{ kJ/kg}$$

We know that the mass of refrigerant passing through the first evaporator or the first compressor,

$$m_1 = \frac{210\,Q_1}{h_1 - h_8} = \frac{210 \times 10}{183.19 - 64.6} = 17.7 \text{ kg/min}$$

and mass of refrigerant passing through the second evaporator,

$$m_2 = \frac{210\,Q_2}{h_3 - h_7} = \frac{210 \times 20}{191.74 - 64.6} = 33 \text{ kg/min}$$

The refrigerant leaving the first compressor at point 2 is mixed with the refrigerant coming out the second evaporator at point 3 before entering into the second compressor. The condition of refrigerant after mixing is shown by point 4 on the *p-h* diagram. The enthalpy at point 4 is given by

$$h_4 = \frac{m_1\,h_2 + m_2\,h_3}{m_1 + m_2} = \frac{17.7 \times 195 + 33 \times 191.74}{17.7 + 33} \text{ kJ/kg}$$

$$= \frac{3451.5 + 6327.4}{50.7} = 193 \text{ kJ/kg}$$

Mark point 4 on *p-h* diagram such that h_4 = 193 kJ/kg. The entropy at this point is 0.7 kJ/kg K. Now from point 4, draw a constant entropy line intersecting the horizontal condenser pressure line at point 5 as shown in the figure. The enthalpy at point 5 as read from the *p-h* diagram is

$$h_5 = 205 \text{ kJ/kg}$$

(a) Power required to run the system

We know that power required to run the first compressor,

$$P_1 = \frac{m_1(h_2 - h_1)}{60} = \frac{17.7(195 - 183.19)}{60} = 3.48 \text{ kW}$$

and power required to run the second compressor,

$$P_2 = \frac{(m_1 + m_2)(h_5 - h_4)}{60} = \frac{(17.7 + 33)(205 - 193)}{60} = 10.14 \text{ kW}$$

∴ Total power required to run the system,

$$P = P_1 + P_2 = 3.48 + 10.14 = 13.62 \text{ kW}$$ **Ans.**

(b) Coefficient of performance of the combined system

We know that the total refrigerating effect of the system,

$$R_E = 210\,(Q_1 + Q_2) = 210\,(10 + 20) = 6300 \text{ kJ/min}$$

and total work done in the compressors,

$$W = P \times 60 = 13.62 \times 60 = 817.2 \text{ kJ/min}$$

∴ C.O.P. of the system

$$= \frac{R_E}{W} = \frac{6300}{817.2} = 7.71$$ **Ans.**

6.9 Multiple Evaporators at Different Temperatures with Compound Compression, Individual Expansion Valves and Flash Intercoolers

The power requirements may be reduced by using compound or multi- stage compression with intercooling. The intercooling may be done economically by allowing some of the refrigerant to by-pass the evaporators and be used in flash type intercoolers.

The arrangement, as shown in Fig. 6.14 (*a*), consists of three evaporators EP_1, EP_2 and EP_3 at different temperatures with compound compression (three-stage compression) and individual expansion valves E_1, E_2 and E_3. The flash intercoolers F_1 and F_2 are provided in the system for cooling the superheated vapours leaving the compressors C_1 and C_2 respectively. The corresponding *p-h* diagram of the system is shown in Fig. 6.14 (*b*)

Let Q_1, Q_2 and Q_3 = Loads on the evaporators EP_1, EP_2 and EP_3 respectively, in tonnes of refrigeration.

∴ Mass of refrigerant required to be circulated through the first evaporator EP_1 or passing through the first compressor C_1,

$$m_{c1} = m_1 = \frac{210\,Q_1}{h_1 - h_{10}} \text{ kg/min}$$

Mass of refrigerant required to be circulated through the second evaporator EP_2,

$$m_2 = \frac{210\,Q_2}{h_3 - h_9} \text{ kg/min}$$

Mass of refrigerant required to be by-passed (at point 9) to the flash intercooler F_1 for desuperheating the superheated vapour refrigerant m_{c1} coming from first compressor C_1 to the dry saturated condition as at point 3 is given by

$$m_2' = \frac{m_{c1}(h_2 - h_3)}{h_3 - h_9}$$

∴ Total mass of refrigerant passing through the second compressor C_2,

$$m_{c2} = m_{c1} + m_2 + m_2'$$

Mass of refrigerant required to be circulated through the third evaporator EP_3,

$$m_3 = \frac{210\, Q_3}{h_5 - h_8} \text{ kg/min}$$

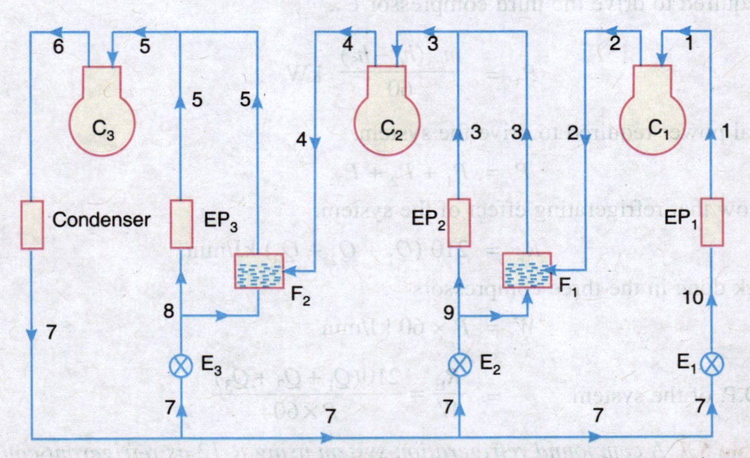

(*a*) Multiple evaporators at different temperatures with compound compression, individual expansion valves and flash intercoolers.

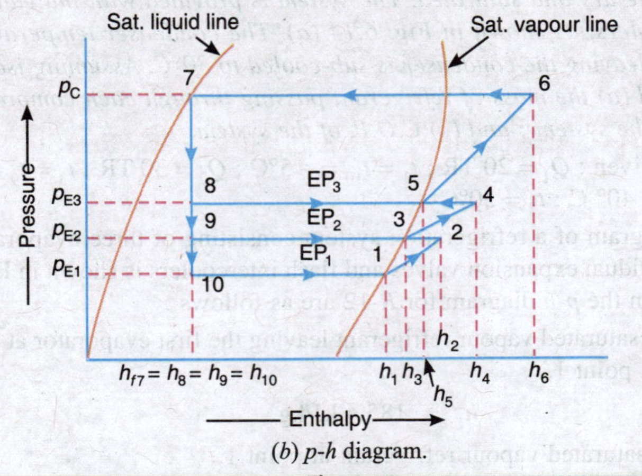

(*b*) *p-h* diagram.

Fig. 6.14

Mass of refrigerant required to be by-passed (at point 8) to the flash intercooler F_2 for desuperheating the superheated vapour refrigerant m_{c2} coming from compressor C_2 to the dry saturated condition as at point 5 is given by

$$m_3' = \frac{m_{c2}(h_4 - h_5)}{h_5 - h_8}$$

∴ Total mass of refrigerant passing through the third compressor C_3,

$$m_{c3} = m_{c2} + m_3 + m_3'$$

We know that power required to drive the first compressor C_1,

$$P_1 = \frac{m_{c1}(h_2 - h_1)}{60} \text{ kW}$$

Similarly, power required to drive the second compressor C_2,

$$P_2 = \frac{m_{c2}(h_4 - h_3)}{60} \text{ kW}$$

and power required to drive the third compressor C_3,

$$P_3 = \frac{m_{c3}(h_6 - h_5)}{60} \text{ kW}$$

∴ Total power required to drive the system,

$$P = P_1 + P_2 + P_3$$

We know that refrigerating effect of the system,

$$R_E = 210\,(Q_1 + Q_2 + Q_3) \text{ kJ/min}$$

and total work done in the three compressors

$$W = P \times 60 \text{ kJ/min}$$

$$\therefore \text{ C.O.P. of the system } = \frac{R_E}{W} = \frac{210(Q_1 + Q_2 + Q_3)}{P \times 60}$$

Example 6.7. *A compound refrigeration system using R-12 as refrigerant consists of three evaporators of capacities 20 TR at – 5°C, 30 TR at 0°C and 10 TR at 5°C. The vapours leaving the evaporators are dry and saturated. The system is provided with individual expansion valves and flash intercoolers, as shown in Fig. 6.14 (a). The condenser temperature is 40°C and the liquid refrigerant leaving the condenser is sub-cooled to 30°C. Assuming isentropic compression at each stage, find (a) the mass of refrigerant passing through each compressor ; (b) the power required to drive the system ; and (c) C.O.P. of the system.*

Solution. Given : Q_1 = 20 TR ; $t_1 = t_{10}$ = – 5°C ; Q_2 = 30 TR ; $t_3 = t_9$ = 0°C ; Q_3 = 10 TR ; $t_5 = t_8$ = 5°C ; t_c = 40° C ; t_7 = 30°C

The *p-h* diagram of a refrigeration system consisting of three evaporators with compound compression, individual expansion valves and flash intercoolers is shown in Fig. 6.15. The various values as read from the *p-h* diagram for *R*-12 are as follows :

Enthalpy of saturated vapour refrigerant leaving the first evaporator at – 5°C or entering the first compressor at point 1,

$$h_1 = 185.4 \text{ kJ/kg}$$

Entropy of saturated vapour refrigerant at point 1,

$$s_1 = 0.6991 \text{ kJ/kg K}$$

Enthalpy of saturated vapour refrigerant leaving the second evaporator at 0°C or entering the second compressor at point 3,

$$h_3 = 187.5 \text{ kJ/kg}$$

Entropy of saturated vapour refrigerant at point 3,

$$s_3 = 0.6965 \text{ kJ/kg K}$$

Enthalpy of saturated vapour refrigerant leaving the third evaporator at 5°C or entering the third compressor at point 5,

$$h_5 = 188.6 \text{ kJ/kg}$$

Entropy of saturated vapour refrigerant at point 5,

$$s_5 = 0.6943 \text{ kJ/kg K}$$

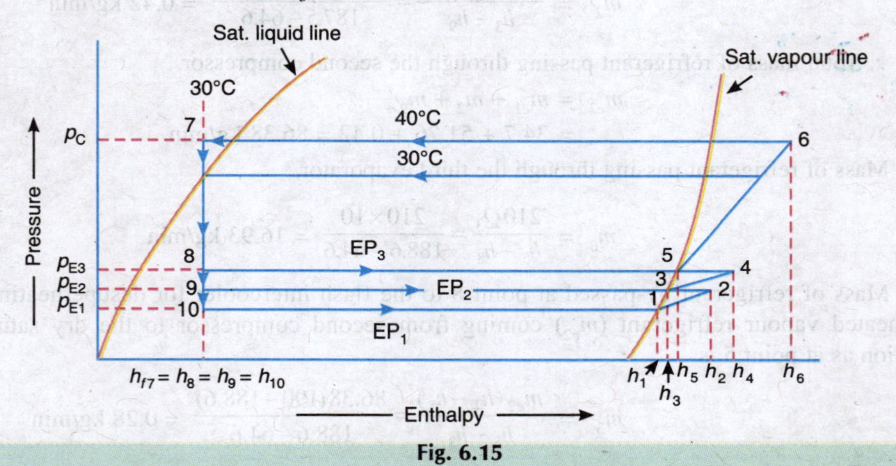

Fig. 6.15

In order to obtain the condition of refrigerant leaving the first compressor (*i.e.* point 2), draw a constant entropy line from point 1 with $s_1 = 0.6991$ intersecting the second evaporator pressure line corresponding to 0°C at point 2. Similarly, the condition of refrigerant leaving the second compressor at point 4 and third compressor at point 6 may be obtained by drawing the constant entropy lines from points 3 and 5 with $s_3 = 0.6965$ and $s_5 = 0.6943$ as shown in Fig. 6.15. Now from *p-h* diagram,

Enthalpy of superheated vapour refrigerant leaving the first compressor at point 2,

$$h_2 = 189 \text{ kJ/kg}$$

Enthalpy of superheated vapour refrigerant leaving the second compressor at point 4,

$$h_4 = 190 \text{ kJ/kg}$$

Enthalpy of superheated vapour refrigerant leaving the third compressor at point 6,

$$h_6 = 206 \text{ kJ/kg}$$

Enthalpy of sub-cooled liquid refrigerant at 30°C at point 7,

$$h_{f7} = h_8 = h_9 = h_{10} = 64.6 \text{ kJ/kg}$$

(a) Mass of refrigerant passing through each compressor

We know that mass of refrigerant passing through the first evaporator or the first compressor,

$$m_{c1} = m_1 = \frac{210\,Q_1}{h_1 - h_{10}} = \frac{210 \times 20}{185.4 - 64.6} = 34.7 \text{ kg/min } \textbf{Ans.}$$

Mass of refrigerant passing through the second evaporator,

$$m_2 = \frac{210 Q_2}{h_3 - h_9} = \frac{210 \times 30}{187.5 - 64.6} = 51.26 \text{ kg/min}$$

Mass of refrigerant required to be by-passed at point 9 to the flash intercooler for desuperheating the superheated refrigerant (m_{c1}) coming from first compressor to the dry saturated condition as at point 3 is given by

$$m_2' = \frac{m_{c1}(h_2 - h_3)}{h_3 - h_9} = \frac{34.7(189 - 187.5)}{187.5 - 64.6} = 0.42 \text{ kg/min}$$

∴ Total mass of refrigerant passing through the second compressor,

$$m_{c2} = m_{c1} + m_2 + m_2'$$
$$= 34.7 + 51.26 + 0.42 = 86.38 \text{ kg/min}$$

Mass of refrigerant passing through the third evaporator,

$$m_3 = \frac{210 Q_3}{h_5 - h_8} = \frac{210 \times 10}{188.6 - 64.6} = 16.93 \text{ kg/min}$$

Mass of refrigerant by-passed at point 8 to the flash intercooler for desuperheating the superheated vapour refrigerant (m_{c2}) coming from second compressor to the dry saturated condition as at point 5 is

$$m_3' = \frac{m_{c2}(h_4 - h_5)}{h_5 - h_8} = \frac{86.38(190 - 188.6)}{188.6 - 64.6} = 0.28 \text{ kg/min}$$

∴ Total mass of refrigerant passing through the third compressor,

$$m_{c3} = m_{c2} + m_3 + m_3'$$
$$= 86.38 + 16.93 + 0.28 = 103.59 \text{ kg/min} \quad \textbf{Ans.}$$

(b) Power required to drive the system

We know that power required to drive the first compressor,

$$P_1 = \frac{m_{c1}(h_2 - h_1)}{60} = \frac{34.7\,(189 - 185.4)}{60} = 2.08 \text{ kW}$$

Power required to drive the second compressor,

$$P_2 = \frac{m_{c2}(h_4 - h_3)}{60} = \frac{86.38\,(190 - 187.5)}{60} = 3.6 \text{ kW}$$

and power required to drive the third compressor,

$$P_3 = \frac{m_{c3}(h_6 - h_5)}{60} = \frac{103.59\,(206 - 188.6)}{60} = 30.04 \text{ kW}$$

∴ Total power required to drive the system

$$P = P_1 + P_2 + P_3$$
$$= 2.08 + 3.6 + 30.04 = 35.72 \text{ kW} \quad \textbf{Ans.}$$

(c) C.O.P. of the system

We know that C.O.P. of the system

$$= \frac{210\,(Q_1 + Q_2 + Q_3)}{P \times 60} = \frac{210\,(20 + 30 + 10)}{35.72 \times 60} = 5.88 \quad \textbf{Ans.}$$

6.10 Multiple Evaporators at Different Temperatures with Compound Compression, Multiple Expansion Valves and Flash Intercoolers

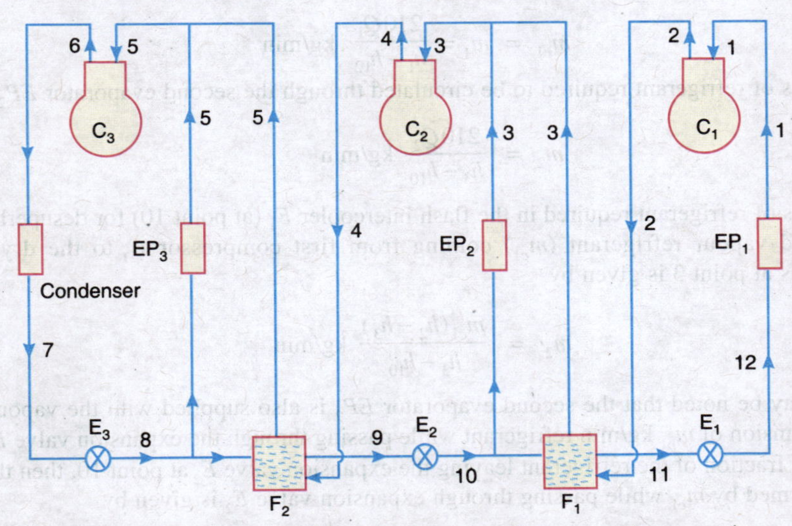

(*a*) Multiple evaporators at different temperatures with compound compression, multiple expansion valves and flash intercoolers.

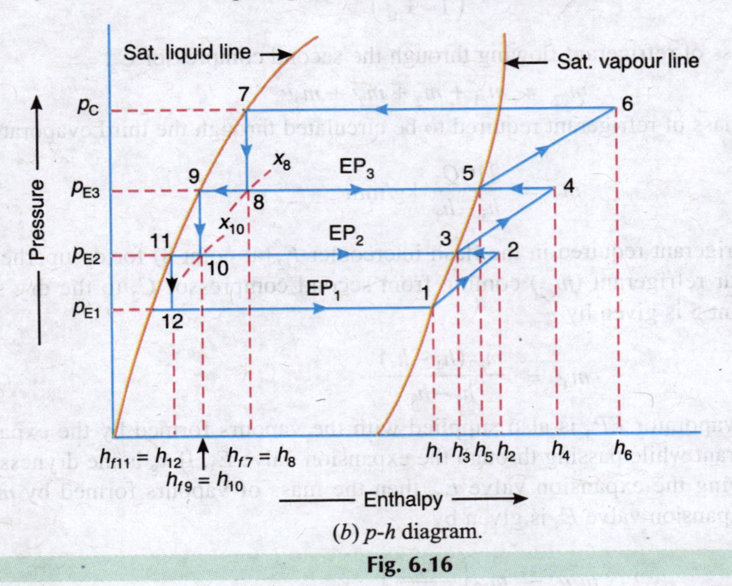

(*b*) *p-h* diagram.

Fig. 6.16

The arrangement, as shown in Fig. 6.16 (*a*), consists of three evaporators EP_1, EP_2 and EP_3 at different temperatures with compound compression (three-stage compression) and multiple expansion valves E_1, E_2 and E_3. The flash intercoolers F_1 and F_2 are provided in the system for cooling the superheated vapours leaving the compressors C_1 and C_2 respectively. The corresponding *p-h* diagram of the system is shown in Fig. 6.16 (*b*).

Let Q_1, Q_2 and Q_3 = Loads on the evaporators EP_1, EP_2 and EP_3 respectively in tonnes of refrigeration.

∴ Mass of refrigerant required to be circulated through the first evaporator EP_1 or passing through the first compressor C_1,

$$m_{c1} = m_1 = \frac{210 Q_1}{h_1 - h_{12}} \text{ kg/min}$$

Mass of refrigerant required to be circulated through the second evaporator EP_2,

$$m_2 = \frac{210 Q_2}{h_3 - h_{10}} \text{ kg/min}$$

Mass of refrigerant required in the flash intercooler F_1 (at point 10) for desuperheating the superheated vapour refrigerant (m_{c1}) coming from first compressor C_1 to the dry saturated condition as at point 3 is given by

$$m_2' = \frac{m_{c1}(h_2 - h_3)}{h_3 - h_{10}} \text{ kg/min}$$

It may be noted that the second evaporator EP_2 is also supplied with the vapours formed during expansion of m_{c1} kg/min refrigerant while passing through the expansion valve E_2. If x_{10} is the dryness fraction of the refrigerant leaving the expansion valve E_2 at point 10, then the mass of vapours formed by m_{c1} while passing through expansion valve E_2 is given by

$$m_2'' = m_{c1}\left(\frac{x_{10}}{1 - x_{10}}\right)$$

∴ Total mass of refrigerant flowing through the second compressor C_2,

$$m_{c2} = m_{c1} + m_2 + m_2' + m_2''$$

Similarly, mass of refrigerant required to be circulated through the third evaporator EP_3,

$$m_3 = \frac{210 Q_3}{h_5 - h_8} \text{ kg/min}$$

Mass of refrigerant required in the flash intercooler F_2 (at point 8) for desuperheating the superheated vapour refrigerant (m_{c2}) coming from second compressor C_2 to the dry saturated condition as at point 5 is given by

$$m_3' = \frac{m_{c2}(h_4 - h_5)}{h_5 - h_8}$$

The third evaporator EP_3 is also supplied with the vapours formed by the expansion of m_{c2} kg/min refrigerant while passing through the expansion valve E_3. If x_8 is the dryness fraction of refrigerant leaving the expansion valve E_3, then the mass of vapours formed by m_{c2} while passing through expansion valve E_3 is given by

$$m_3'' = m_{c2}\left(\frac{x_8}{1 - x_8}\right)$$

∴ Total mass of refrigerant passing through the third compressor C_3,

$$m_{c3} = m_{c2} + m_3 + m_3' + m_3''$$

We know that power required to drive the first compressor C_1,

$$P_1 = \frac{m_{c1}(h_2 - h_1)}{60} \text{ kW}$$

Power required to drive the second compressor C_2,

$$P_2 = \frac{m_{c2}(h_4 - h_3)}{60} \text{ kW}$$

and power required to drive the third compressor C_3,

$$P_3 = \frac{m_{c3}(h_6 - h_5)}{60} \text{ kW}$$

∴ Total power required to run the system,

$$P = P_1 + P_2 + P_3$$

We know that refrigerating effect of the system,

$$R_E = 210\,(Q_1 + Q_2 + Q_3) \text{ kJ/min}$$

Total work done in the three compressors,

$$W = P \times 60 \text{ kJ/min}$$

$$\therefore \text{ C.O.P. of the system } = \frac{R_E}{W} = \frac{210\,(Q_1 + Q_2 + Q_3)}{P \times 60}$$

Example 6.8. *A compound compression refrigerating system using R-12 as refrigerant consists of evaporators of capacities 30 TR at – 10°C, 20 TR at 5°C and 10 TR at 10°C. The vapours leaving the evaporators are dry and saturated. The system is provided with multiple expansion valves and flash intercoolers as shown in Fig. 6.16 (a). The condenser temperature is 40°C and the liquid refrigerant leaving the condenser is sub-cooled to 30°C. Assuming isentropic compression at each stage, find (a) the mass of refrigerant passing through each compressor ; (b) the power required to drive the system ; and (c) the C.O.P. of the system.*

Solution. Given : $Q_1 = 30$ TR ; $t_1 = t_{12} = -10$°C ; $Q_2 = 20$ TR ; $t_3 = t_{10} = t_{11} = 5$°C ; $Q_3 = 10$ TR ; $t_5 = t_8 = t_9 = 10$°C ; $t_c = 40$°C ; $t_7 = 30$°C

The *p-h* diagram of a refrigeration system consisting of three evaporators with compound compression, multiple expansion valve and flash intercoolers is shown in Fig. 6.17. The various values as read from the *p-h* diagram, are as follows :

Enthalpy of saturated vapour refrigerant leaving the first evaporator at –10°C or entering the first compressor at point 1,

$$h_1 = 183.19 \text{ kJ/kg}$$

Entropy of saturated vapour refrigerant at point 1,

$$s_1 = 0.702 \text{ kJ/kg K}$$

Enthalpy of saturated vapour refrigerant leaving the second evaporator at 5°C or entering the second compressor at point 3,

$$h_3 = 189.65 \text{ kJ/kg}$$

Entropy of saturated vapour refrigerant at point 3,

$$s_3 = 0.6943 \text{ kJ/kg}$$

Enthalpy of saturated vapour refrigerant leaving the third evaporator at 10°C or entering the third compressor at point 5,

$$h_5 = 191.74 \text{ kJ/kg}$$

Entropy of saturated vapour refrigerant at point 5,

$$s_5 = 0.6921 \text{ kJ/kg K}$$

In order to obtain the condition of refrigerant leaving the first compressor at point 2, draw a constant entropy line from point 1 having entropy $s_1 = 0.702$ kJ/kg K intersecting the second evaporator pressure line (corresponding to 5°C) at point 2. Similarly, the condition of refrigerant leaving the second compressor at point 4 and third compressor at point 6 may be obtained by drawing the constant entropy lines from points 3 and 5 with $s_3 = 0.6943$ kJ/kg K and $s_5 = 0.6921$ kJ/kg K, as shown in the *p-h* diagram. Now from the *p-h* diagram, we find that

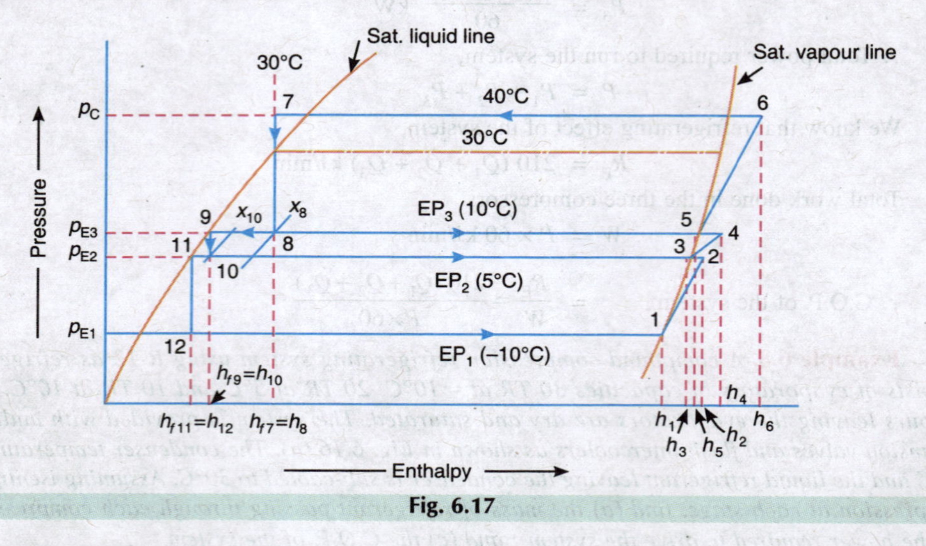

Fig. 6.17

Enthalpy of superheated vapour refrigerant leaving the first compressor at point 2,

$$h_2 = 194.1 \text{ kJ/kg}$$

Enthalpy of superheated vapour refrigerant leaving the second compressor at point 4,

$$h_4 = 195.3 \text{ kJ/kg}$$

Enthalpy of superheated vapour refrigerant leaving the third compressor at point 6,

$$h_6 = 207.18 \text{ kJ/kg}$$

Enthalpy of sub-cooled liquid refrigerant at 30°C at point 7,

$$h_{f7} = h_8 = 64.59 \text{ kJ/kg}$$

Condition of refrigerant (dryness fraction) leaving the expansion valve at point 8,

$$x_8 = 0.13$$

Enthalpy of saturated liquid refrigerant at point 9,

$$h_{f9} = h_{10} = 45.37 \text{ kJ/kg}$$

Condition of refrigerant (dryness fraction) leaving the expansion valve at point 10,

$$x_{10} = 0.04$$

Enthalpy of saturated liquid refrigerant at point 11,

$$h_{f11} = h_{12} = 40.69 \text{ kJ/kg}$$

(a) Mass of refrigerant passing through each compressor

We know that mass of refrigerant passing through the first evaporator or first compressor,

$$m_{c1} = m_1 = \frac{210\, Q_1}{h_1 - h_{12}} = \frac{210 \times 30}{183.19 - 40.69} = 44.2 \text{ kg/min} \textbf{ Ans.}$$

Mass of refrigerant required to be circulated through the second evaporator,

$$m_2 = \frac{210\,Q_2}{h_3 - h_{10}} = \frac{210 \times 20}{189.65 - 45.37} = 29.1 \text{ kg/min}$$

Mass of refrigerant required in the flash intercooler at point 10 for desuperheating the superheated vapour refrigerant m_{c1} = 44.2 kg/min coming from first compressor, to the dry saturated condition as at point 3 is

$$m_2' = \frac{m_{c1}(h_2 - h_3)}{h_3 - h_{10}} = \frac{44.2\,(194.1 - 189.65)}{189.65 - 45.37} = 1.36 \text{ kg/min}$$

The second evaporator is also supplied with vapours formed during expansion of m_{c1} = 44.2 kg/min refrigerant while passing through the second expansion valve.

∴ Mass of vapours formed by m_{c1} = 44.2 kg/min while passing through the second expansion valve,

$$m_2'' = m_{c1}\left(\frac{x_{10}}{1 - x_{10}}\right) = 44.2\left(\frac{0.04}{1 - 0.04}\right) = 1.84 \text{ kg/min}$$

∴ Total mass of refrigerant passing through the second compressor,

$$m_{c2} = m_{c1} + m_2 + m_2' + m_2''$$
$$= 44.2 + 29.1 + 1.36 + 1.84 = 76.5 \text{ kg/min} \quad \textbf{Ans.}$$

Now, mass of refrigerant required to be circulated through the third evaporator,

$$m_3 = \frac{210\,Q_3}{h_5 - h_8} = \frac{210 \times 10}{191.74 - 64.59} = 16.5 \text{ kg/min}$$

Mass of refrigerant required in the flash intercooler at point 8 for desuperheating the superheated vapour refrigerant m_{c2} = 76.5 kg/min coming from the second compressor to the dry saturated condition as at point 5 is

$$m_3' = \frac{m_{c2}\,(h_4 - h_5)}{h_5 - h_8} = \frac{76.5\,(195.3 - 191.74)}{191.74 - 64.59} = 2.14 \text{ kg/min}$$

The third evaporator is also supplied with the vapour formed by the expansion of m_{c2} = 76.5 kg/min refrigerant while passing through the third expansion valve.

∴ Mass of vapours formed by m_{c2} = 76.5 kg/min while passing through the third expansion valve,

$$m_3'' = m_{c2}\left(\frac{x_8}{1 - x_8}\right) = 76.5\left(\frac{0.13}{1 - 0.13}\right) = 11.47 \text{ kg/min}$$

∴ Total mass of refrigerant passing through the third compressor,

$$m_{c3} = m_{c2} + m_3 + m_3' + m_3''$$
$$= 76.5 + 16.5 + 2.14 + 11.47 = 106.61 \text{ kg/min} \quad \textbf{Ans.}$$

(b) Power required to drive the system

We know that power required to drive the first compressor,

$$P_1 = \frac{m_{c1}\,(h_2 - h_1)}{60} = \frac{44.2\,(194.1 - 183.19)}{60} = 8.04 \text{ kW}$$

Power required to drive the second compressor,

$$P_2 = \frac{m_{c2}(h_4 - h_3)}{60} = \frac{76.5(195.3 - 189.65)}{60} = 7.2 \text{ kW}$$

and power required to drive the third compressor,

$$P_3 = \frac{m_{c3}(h_6 - h_5)}{60} = \frac{106.61(201.18 - 191.74)}{60} = 27.4 \text{ kW}$$

∴ Total power required to drive the system,

$$P = P_1 + P_2 + P_3$$

$$= 8.04 + 7.2 + 27.4 = 42.64 \text{ kW} \textbf{ Ans.}$$

(c) C.O.P. of the system

We know that refrigerating effect of the system,

$$R_E = 210(Q_1 + Q_2 + Q_3)$$

$$= 210(30 + 20 + 10) = 12\,600 \text{ kJ/min}$$

and total work done in the three compressors,

$$W = P \times 60 = 42.64 \times 60 = 2558 \text{ kJ/min}$$

∴ C.O.P. of the system $= \frac{R_E}{W} = \frac{12600}{2558.4} = 4.9$ **Ans.**

EXERCISES

1. A single compressor, using R-12 as refrigerant, has three evaporators of capacities 10 TR, 30 TR and 20 TR. The temperatures in all the three evaporators is to be maintained at – 5°C. The vapours leaving the evaporators are dry and saturated. The condenser temperature is 40°C. The liquid refrigerant leaving the condenser is sub-cooled to 30°C. Assuming isentropic compression, find the power required to drive the compressor and C.O.P. of the system. **[Ans. 41.4 kW ; 5.07]**

2. A single compressor using R-12 as refrigerant has three evaporators of capacities 10 TR, 20 TR, and 10 TR as shown in Fig. 6.18.

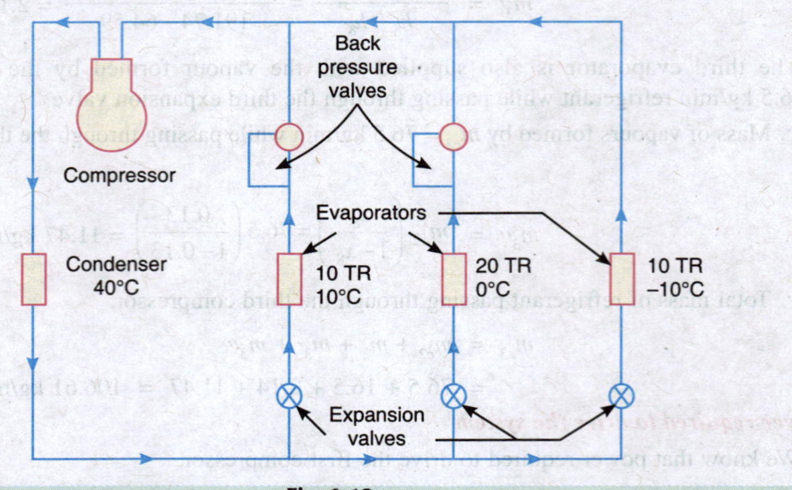

Fig. 6.18

The temperature in the three evaporators is to be maintained at 10°C, 0°C and – 10°C respectively. The condenser temperature is 40°C. The vapours leaving the evaporators are dry and saturated.

Assuming isentropic compression, find the power required to drive the compressor and the C.O.P. of the system. **[Ans.** 31.9 kW ; 4.38]

3. A single compressor system using R-12 as refrigerant has three evaporators of capacities 20 TR, 30 TR and 10 TR. The temperature in the three evaporators is to be maintained at – 5°C, 0°C and 7°C respectively. The system is provided with multiple expansion valves and back pressure valves. The condenser pressure is 9.609 bar. The liquid refrigerant leaving the condenser is sub-cooled to 30°C. The vapours leaving the evaporators are dry and saturated. Assuming isentropic compression, find (*a*) mass of refrigerant flowing through each evaporator, (*b*) power required to drive the compressor, and (*c*) C.O.P. of the system. **[Ans.** 28.13 kg/min, 44.82 kg/min, 29.55 kg/min ; 43.73 kW ; 4.8]

4. A refrigerating plant having R-12 as the working fluid, comprises three evaporators of 10 TR at – 10°C, 20 TR at 5°C and 30 TR at 10°C. Each of the evaporators is having the individual expansion valve arrangement. The refrigerant from the higher stage expansion valve is bled and mixed with delivery from the lower stage compressor and sent to the higher stage compressor suction thus effecting intercooling of the delivery gas from the lower stage and bringing the suction of the higher stage compressor to dry and saturated conditions. There is only one condenser operating at 40°C and sub-cooling the liquid refrigerant to 30°C. The compression during each stage of compression may be assumed isentropic. The evaporator discharge is dry saturated in each of the evaporators. Determine the power required by the entire system.

5. A refrigeration system using R-12 as refrigerant consists of three evaporators of capacities 20 TR at – 5°C, 30 TR at 0°C and 10 TR at 5°C. The vapours leaving the three evaporators are dry and saturated. The system is provided with individual compressors and multiple expansion valves. The condenser temperature is 40°C and the liquid refrigerant leaving the condenser is saturated. Assuming isentropic compression in each compressor, find (*a*) the mass of refrigerant flowing through each evaporator, (*b*) the power required to drive the system, and (*c*) the C.O.P. of the system.

   [Ans. 27.4 kg/min, 42.25 kg/min ; 34.12 kg/min ; 38.4 kW ; 5.56]

6. A refrigeration system using R-12 as refrigerant has three evaporators of capacities 30 TR at – 10°C, 20 TR at 5°C and 10 TR at 10°C. The refrigerant leaving the evaporators is dry and saturated. The system is provided with compound compression, individual expansion valves and flash intercoolers. The condenser temperature is 40°C. Assuming isentropic compression in each compressor, find the power required to run the system and C.O.P. of the system when the liquid refrigerant leaving the condenser is (*a*) saturated, and (*b*) sub-cooled to 30°C.

   [Ans. 45.46 kW, 4.62 ; 44 kW, 4.76]

QUESTIONS

1. Derive the expression for C.O.P. of a refrigerating system consisting of three evaporators at the same temperature with single compressor and expansion valve.
2. Draw a schematic diagram of a refrigerating system having three evaporators at different temperatures with single compressor, multiple expansion valves and back pressure valves. Explain the working of this system with the help of *p-h* diagram.
3. Derive the expression for C.O.P. of a refrigerating system consisting of two evaporators at different temperatures with individual compressors and individual expansion valves.
4. Explain with the help of a neat sketch, the working of a refrigerating system having three evaporators at different temperatures with individual compressors and multiple expansion valves.
5. Draw the schematic diagram of a refrigerating system having three evaporators at different temperatures with compound compression, individual expansion valves and flash intercoolers.

OBJECTIVE TYPE QUESTIONS

1. When different temperatures are to be maintained at different points, ... evaporator system may be used.

   (*a*) single (*b*) multiple

2. The fresh fruits, fresh vegetables, frozen products and dairy products .... at the same temperature and humidity.

   (*a*) can be maintained (*b*) can not be maintained

3. When food products or other hygroscopic materials kept in different compartments are to be maintained at the same temperature, then the evaporator

   (*a*) at the same temperature with single compressor is used

   (*b*) at different temperatures with single compressor is used

   (*c*) at different temperatures with individual compressor is used

   (*d*) none of the above

4. A system with multiple evaporators at different temperatures with compound compression will

   (*a*) increase the power requirements

   (*b*) reduce the power requirements

   (*c*) neither increase nor reduce the power requirements

5. In a system with multiple evaporators at different temperatures with compound compression and flash intercooler, the suction vapour to the high pressure stage is

   (*a*) dry saturated (*b*) superheated (*c*) saturated liquid

ANSWERS

1. (*b*) 2. (*b*) 3. (*a*) 4. (*b*) 5. (*a*)

CHAPTER

7 Vapour Absorption Refrigeration Systems

7.1 Introduction

The vapour absorption refrigeration system is one of the oldest methods of producing refrigerating effect. The principle of vapour absorption was first discovered by Michael Faraday in 1824 while performing a set of experiments to liquify certain gases. The first vapour absorption refrigeration machine was developed by a French scientist, Ferdinand Carre, in 1860. This system may be used in both the domestic and large industrial refrigerating plants. The refrigerant, commonly used in a vapour absorption system, is ammonia.

The vapour absorption system uses heat energy, instead of mechanical energy as in vapour compression systems, in order to change the conditions of the refrigerant required for the operation of the refrigeration cycle. We have discussed in the previous chapters that the function of a compressor, in a vapour compression system, is to withdraw the vapour refrigerant from the evaporator. It then raises its temperature and pressure higher than the cooling agent in the condenser so that the higher pressure vapours can reject heat in the condenser. The liquid refrigerant leaving the condenser is now ready to expand to the evaporator conditions again.

In the vapour absorption system, the compressor is replaced by an absorber, a pump, a generator and a pressure reducing valve. These components in vapour absorption system perform the same function as that of a compressor in vapour compression system. In this system, the vapour refrigerant from the evaporator is drawn into an absorber where it is absorbed by the weak solution of the refrigerant forming a strong solution. This strong solution is pumped to the generator where it is heated by some external source. During the heating process, the vapour refrigerant is driven off by the solution and enters into the condenser where it is liquified. The liquid refrigerant then flows into the evaporator and thus the cycle is completed.

7.2 Simple Vapour Absorption System

The simple vapour absorption system, as shown in Fig. 7.1, consists of an absorber, a pump, a generator and a pressure reducing valve to replace the compressor of vapour compression system. The other components of the system are condenser, receiver, expansion valve and evaporator as in the vapour compression system.

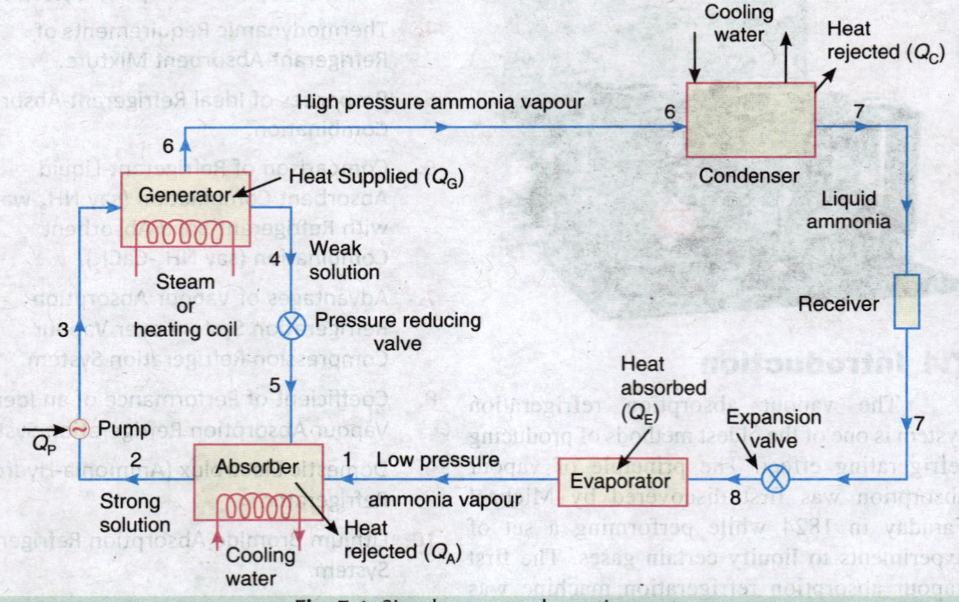

Fig. 7.1. Simple vapour absorption system.

In this system, the low pressure ammonia vapour leaving the evaporator enters the absorber where it is absorbed by the cold water in the absorber. The water has the ability to absorb very large quantities of ammonia vapour and the solution, thus formed, is known as *aqua-ammonia*. The absorption of ammonia vapour in water lowers the pressure in the absorber which in turn draws more ammonia vapour from the evaporator and thus raises the temperature of solution. Some form of cooling arrangement (usually water cooling) is employed in the absorber to remove the heat of solution evolved there. This is necessary in order to increase the absorption capacity of water, because at higher temperature water absorbs less ammonia vapour. The strong solution thus formed in the absorber is pumped to the generator by the liquid pump. The pump increases the pressure of the solution up to 10 bar.

Simple vapour absorption machine.

The *strong solution of ammonia in the generator is heated by some external source such as gas or steam. During the heating process, the ammonia vapour is driven off the solution at high pressure leaving behind the hot weak ammonia solution in the generator. This weak ammonia solution flows back to the absorber at low pressure after passing through the pressure reducing valve. The high pressure ammonia vapour from the generator is condensed in the condenser to a high pressure liquid ammonia. This liquid ammonia is passed to the expansion valve through the receiver and then to the evaporator. This completes the simple vapour absorption cycle.

7.3 Practical Vapour Absorption System

The simple absorption system as discussed in the previous article is not very economical. In order to make the system more practical, it is fitted with an analyser, a rectifier and two heat exchangers as shown in Fig. 7.2. These accessories help to improve the performance and working of the plant, as discussed below :

1. *Analyser.* When ammonia is vaporised in the generator, some water is also vaporised and will flow into the condenser along with the ammonia vapours in the simple system. If these unwanted water particles are not removed before entering into the condenser, they will enter into the expansion valve where they freeze and choke the pipeline. In order to remove these unwanted particles flowing to the condenser, an analyser is used. The analyser may be built as an integral part of the generator or made as a separate piece of equipment. It consists of a series of trays mounted above the generator. The strong solution from the absorber and the aqua from the rectifier are introduced at the top of the analyser and flow downward over the trays and into the generator. In this way, considerable liquid surface area is exposed to the vapour rising from the generator. The vapour is cooled and most of the water vapour condenses, so that mainly ammonia vapour (approximately 99%) leaves the top of the analyser. Since the aqua is heated by the vapour, less external heat is required in the generator.

2. *Rectifier.* In case the water vapours are not completely removed in the analyser, a closed type vapour cooler called rectifier (also known as dehydrator) is used. It is generally water cooled and may be of the double pipe, shell and coil or shell and tube type. Its function is to cool further the ammonia vapours leaving the analyser so that the remaining water vapours are condensed. Thus, only dry or anhydrous ammonia vapours flow to the condenser. The condensate from the rectifier is returned to the top of the analyser by a drip return pipe.

* A strong ammonia solution contains as much ammonia as possible whereas a weak ammonia solution contains considerably less ammonia.

3. *Heat exchangers.* The heat exchanger provided between the pump and the generator is used to cool the weak hot solution returning from the generator to the absorber. The heat removed from the weak solution raises the temperature of the strong solution leaving the pump and going to analyser and generator. This operation reduces the heat supplied to the generator and the amount of cooling required for the absorber. Thus the economy of the plant increases.

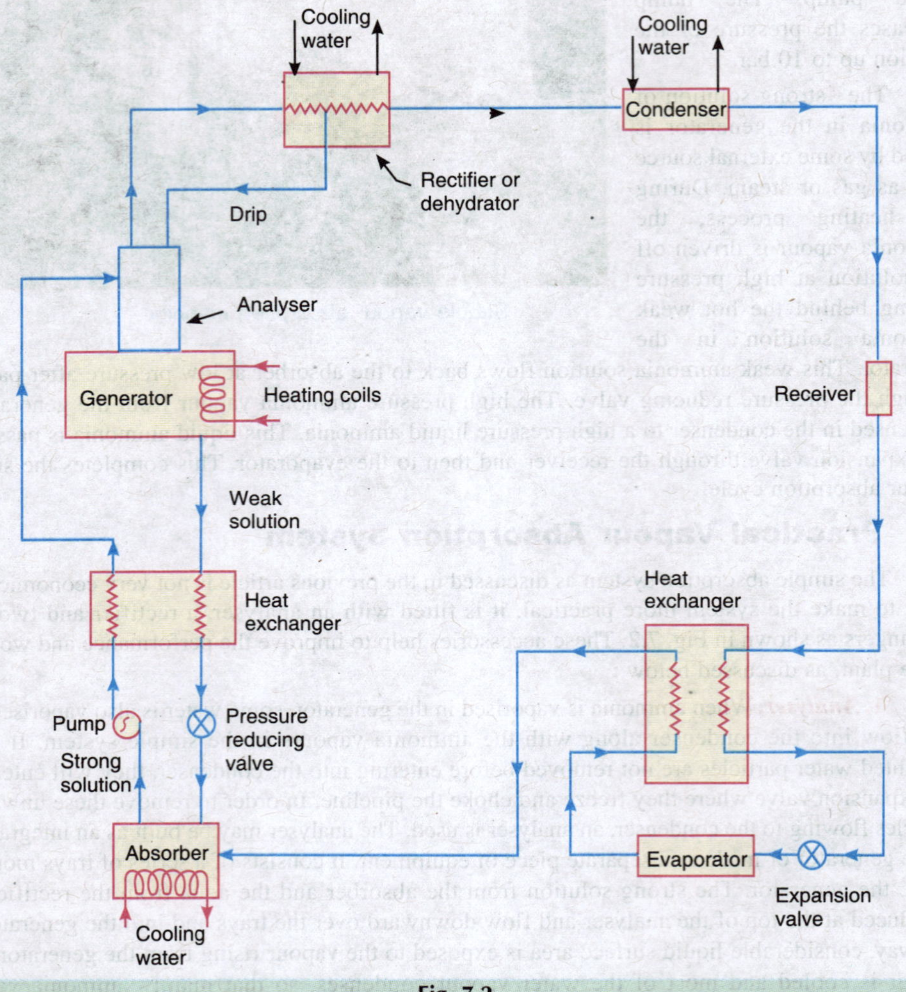

Fig. 7.2

The heat exchanger provided between the condenser and the evaporator may also be called liquid sub-cooler. In this heat exchanger, the liquid refrigerant leaving the condenser is sub-cooled by the low temperature ammonia vapour from the evaporator as shown in Fig. 7.2. This sub-cooled liquid is now passed to the expansion valve and then to the evaporator.

In this system, the net refrigerating effect is the heat absorbed by the refrigerant in the evaporator. The total energy supplied to the system is the sum of work done by the pump and the heat supplied in the generator. Therefore, the coefficient of performance of the system is given by

$$\text{C.O.P.} = \frac{\text{Heat absorbed in evaporator}}{\text{Work done by pump + Heat supplied in generator}}$$

7.4 Thermodynamic Requirements of Refrigerant-Absorbent Mixture

The two main thermodynamic requirements of the refrigerant-absorbent mixture are as follows :

1. *Solubility requirement*. The refrigerant should have more than Raoult's law solubility in the absorbent so that a strong solution, highly rich in the refrigerant, is formed in the absorber by the absorption of the refrigerant vapour.

2. *Boiling points requirement*. There should be a large difference in the normal boiling points of the two substances, at least 200°C, so that the absorbent exerts negligible vapour pressure at the generator temperature. Thus, almost absorbent-free refrigerant is boiled off from the generator and the absorbent alone returns to the absorber.

In addition, the refrigerant-absorbent mixture should possess the following desirable characteristics :

(*a*) It should have low viscosity to minimise pump work.
(*b*) It should have low freezing point.
(*c*) It should have good chemical and thermal stability.
(*d*) The irreversible chemical reactions of all kinds such as decomposition, polymerization, corrosion etc. are to be avoided.

7.5 Properties of Ideal Refrigerant-Absorbent Combination

The ideal refrigerant-absorbent combination should possess the following qualities :

1. The refrigerant should have high affinity for the absorber at low temperature and less affinity at high temperature.

2. The combination should have high degree of negative deviation from Raoult's law.

3. The mixture should have low specific heat and low viscosity.

4. The mixture (solution) should be non-corrosive.

5. The mixture should have a small heat.

6. The mixture should have low freezing point.

7. There should be a large difference in normal boiling points of the refrigerants and the absorbent.

Note : Out of the many combinations tried so far, the following two combinations are in common use in air conditioning applications :

(*a*) Ammonia-water combination ; and (*b*) Lithium-bromide water combination.

The ammonia-water absorption system finds a significant place in large tonnage industrial applications.

7.6 Comparison of Refrigerant-Liquid Absorbent Combination (say NH_3 - Water) with Refrigerant-Solid Absorbent Combination (say NH_3 – $CaCl_2$)

The solid absorbent system has the following two main advantages :

1. The amount of refrigerant cycled in relation to the amount of absorbent is larger for the solid absorbent than for the liquid absorbent. Therefore, in solid absorbent system, the thermal capacity of the salt contributes little, and hence heat required in the generator is less as compared to that for water in the aqua-ammonia system.

2. The solid absorbent system is extremely robust, not only in mechanical sense, but also with respect to adverse operating conditions. The performance is remarkably insensitive to changes in both condensing and evaporating temperatures. However the solid-absorption cycle has one major drawback. The heat of reaction is large compared with those found for liquid absorbent, approximately twice the latent heat of vaporisation.

However the C.O.P. for solid absorption cycle is still higher than that of liquid absorption cycle, particularly when the difference between the condensing temperature and the evaporating temperature is large.

The solid-absorbent combination is ideally suited for intermittent operation on solar energy.

7.7 Advantages of Vapour Absorption Refrigeration System over Vapour Compression Refrigeration System

Following are the advantages of vapour absorption system over vapour compression system :

1. In the vapour absorption system, the only moving part of the entire system is a pump which has a small motor. Thus, the operation of this system is essentially quiet and is subjected to little wear.

The vapour compression system of the same capacity has more wear, tear and noise due to moving parts of the compressor.

2. The vapour absorption system uses heat energy to change the condition of the refrigerant from the evaporator. The vapour compression system uses mechanical energy to change the condition of the refrigerant from the evaporator.

3. The vapour absorption systems are usually designed to use steam, either at high pressure or low pressure. The exhaust steam from furnaces and solar energy may also be used. Thus this system can be used where the electric power is difficult to obtain or is very expensive.

4. The vapour absorption systems can operate at reduced evaporator pressure and temperature by increasing the steam pressure to the generator, with little decrease in capacity. But the capacity of vapour compression system drops rapidly with lowered evaporator pressure.

5. The load variations do not affect the performance of a vapour absorption system. The load variations are met by controlling the quantity of aqua circulated and the quantity of steam supplied to the generator.

The performance of a vapour compression system at partial loads is poor.

6. In the vapour absorption system, the liquid refrigerant leaving the evaporator has no bad effect on the system except that of reducing the refrigerating effect. In the vapour compression system, it is essential to superheat the vapour refrigerant leaving the evaporator so that no liquid may enter the compressor.

7. The vapour absorption systems can be built in capacities well above 1000 tonnes of refrigeration each, which is the largest size for single compressor units.

8. The space requirements and automatic control requirements favour the absorption system more and more as the desired evaporator temperature drops.

7.8 Coefficient of Performance of an Ideal Vapour Absorption Refrigeration System

We have discussed earlier that in an ideal vapour absorption refrigeration system,

(*a*) the heat (Q_G) is given to the refrigerant in the generator,

(*b*) the heat (Q_C) is discharged to the atmosphere or cooling water from the condenser and absorber,

(*c*) the heat (Q_E) is absorbed by the refrigerant in the evaporator, and (*d*) the heat (Q_P) is added to the refrigerant due to pumpwork.

Neglecting the heat due to pumpwork (Q_P), we have according to First Law of Thermodynamics,

$$Q_C = Q_G + Q_E \quad \text{. . . (i)}$$

Let T_G = Temperature at which heat (Q_G) is given to the generator,

T_C = Temperature at which heat (Q_C) is discharged to atmosphere or cooling water from the condenser and absorber, and

T_E = Temperature at which heat (Q_E) is absorbed in the evaporator.

Since the vapour absorption system can be considered as a perfectly reversible system, therefore the initial entropy of the system must be equal to the entropy of the system after the change in its condition.

$$\therefore \quad \frac{Q_G}{T_G} + \frac{Q_E}{T_E} = \frac{Q_C}{T_C} \quad \text{. . . (ii)}$$

$$= \frac{Q_G + Q_E}{T_C} \quad \text{. . . [From equation (i)]}$$

or

$$\frac{Q_G}{T_G} - \frac{Q_G}{T_C} = \frac{Q_E}{T_C} - \frac{Q_E}{T_E}$$

$$Q_G\left(\frac{T_C - T_G}{T_G \times T_C}\right) = Q_E\left(\frac{T_E - T_C}{T_C \times T_E}\right)$$

$$\therefore \quad Q_G = Q_E\left[\frac{T_E - T_C}{T_C \times T_E}\right]\left[\frac{T_G \times T_C}{T_C - T_G}\right]$$

$$= Q_E\left[\frac{T_C - T_E}{T_C \times T_E}\right]\left[\frac{T_G \times T_C}{T_G - T_C}\right]$$

$$= Q_E\left(\frac{T_C - T_E}{T_E}\right)\left(\frac{T_G}{T_G - T_C}\right) \quad \text{. . . (iii)}$$

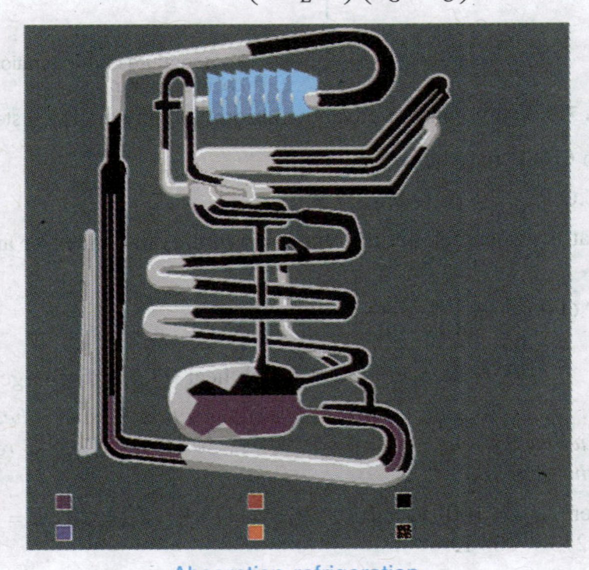

Absorption refrigeration.

Maximum coefficient of performance of the system is given by

$$(\text{C.O.P.})_{max} = \frac{Q_E}{Q_G} = \frac{Q_E}{Q_E\left(\frac{T_C - T_E}{T_E}\right)\left(\frac{T_G}{T_G - T_C}\right)}$$

$$= \left(\frac{T_E}{T_C - T_E}\right)\left(\frac{T_G - T_C}{T_G}\right) \quad \ldots (iv)$$

It may be noted that,

1. The expression $\frac{T_E}{T_C - T_E}$ is the C.O.P. of a Carnot refrigerator working between the temperature limits of T_E and T_C.

2. The expression $\frac{T_G - T_C}{T_G}$ is the efficiency of a Carnot engine working between the temperature limits of T_G and T_C.

Thus an ideal vapour absorption refrigeration system may be regarded as a combination of a Carnot engine and a Carnot refrigerator to produce the desired refrigeration effect as shown in Fig. 7.3.

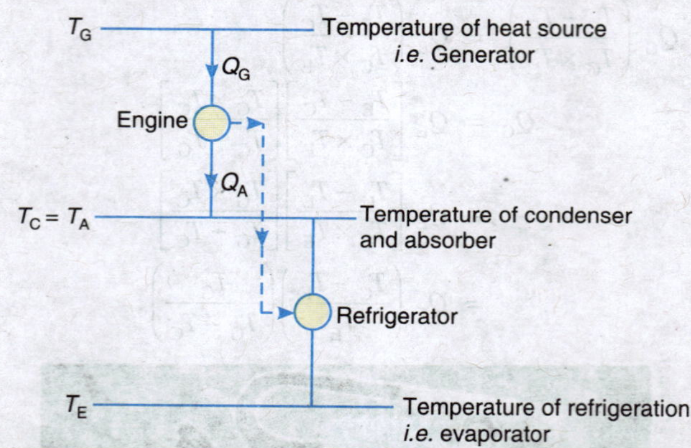

Fig. 7.3. Representation of vapour absorption refrigeration system.

The maximum C.O.P. may be written as

$$(\text{C.O.P.})_{max} = (\text{C.O.P.})_{carnot} \times \eta_{carnot}$$

In case the heat is discharged at different temperatures in condenser and absorber, then

$$(\text{C.O.P.})_{max} = \left[\frac{T_E}{T_C - T_E}\right]\left[\frac{T_G - T_A}{T_G}\right]$$

where T_A = Temperature at which heat (Q_A) is discharged in the absorber.

Example 7.1. *In a vapour absorption refrigeration system, heating, cooling and refrigeration take place at the temperatures of 100° C, 20° C and – 5° C respectively. Find the maximum C.O.P. of the system.*

Solution. Given : T_G = 100° C = 100 + 273 = 373 K ; T_C = 20° C = 20 + 273 = 293 K ; T_E = – 5° C = – 5 + 273 = 268 K

We know that maximum C.O.P. of the system

$$= \left(\frac{T_E}{T_C - T_E}\right)\left(\frac{T_G - T_C}{T_G}\right) = \left(\frac{268}{293 - 268}\right)\left(\frac{373 - 293}{373}\right) = 2.3 \text{ Ans.}$$

Example 7.2. *In an absorption type refrigerator, the heat is supplied to NH_3 generator by condensing steam at 2 bar and 90% dry. The temperature in the refrigerator is to be maintained at – 5° C. Find the maximum C.O.P. possible.*

If the refrigeration load is 20 tonnes and actual C.O.P. is 70% of the maximum C.O.P., find the mass of steam required per hour. Take temperature of the atmosphere as 30° C.

Solution. Given : $p = 2$ bar ; $x = 90\% = 0.9$; $T_E = -5°\text{ C} = -5 + 273 = 268$ K ; $Q = 20$ TR ; Actual C.O.P. = 70% of maximum C.O.P. ; $T_C = 30°\text{ C} = 30 + 273 = 303$ K

Maximum C.O.P.

From steam tables, we find that the saturation temperature of steam at a pressure of 2 bar is

$$T_G = 120.2°\text{ C} = 120.2 + 273 = 393.2 \text{ K}$$

We know that maximum C.O.P.

$$= \left[\frac{T_E}{T_C - T_E}\right]\left[\frac{T_G - T_C}{T_G}\right] = \left[\frac{268}{303 - 268}\right]\left[\frac{393.2 - 303}{393.2}\right] = 1.756 \text{ Ans.}$$

Mass of steam required per hour

We know that actual C.O.P.

$$= 70\% \text{ of maximum C.O.P.} = 0.7 \times 1.756 = 1.229$$

∴ Actual heat supplied

$$= \frac{\text{Refrigeration load}}{\text{Actual C.O.P.}} = \frac{20 \times 210}{1.229} = 3417.4 \text{ kJ/min}$$

Assuming that only latent heat of steam is used for heating purposes, therefore from steam tables, the latent heat of steam at 2 bar is

$$h_{fg} = 2201.6 \text{ kJ/kg}$$

∴ Mass of steam required per hour

$$= \frac{\text{Actual heat supplied}}{h_{fg}} = \frac{3417.4}{2201.6} = 1.552 \text{ kg/min}$$

$$= 1.552 \times 60 = 93.12 \text{ kg/h} \text{ Ans.}$$

Example 7.3. *In an aqua-ammonia absorption refrigeration system of 10TR capacity, the vapours leaving the generator are 100% pure NH_3 saturated at 40°C. The evaporator, absorber, condenser and generator temperatures are – 20°C, 30°C, 40°C and 70°C respectively. At absorber exit (strong solution), the concentration of ammonia in solution is x = 0.38 and enthalpy h = 22 kJ/kg. At generator exit (weak solution) x = 0.1 and h = 695 kJ/kg.*

1. *Determine mass flow rate of ammonia in the evaporator ;*
2. *Carry out overall mass conservation and mass conservation of ammonia in absorber to determine mass flow rates of weak and strong solutions ;*
3. *Determine the heat rejection in absorber and condenser ;*
4. *Heat added in generator ; and*
5. *C.O.P.*

Solution. Given : $Q = Q_E = 10$TR ; $T_6 = 40°C$; $T_E = -20°C$; $T_A = 30°C$; $T_C = 40°C$; $T_G = 70°C$; $x_2 = x_3 = 0.38$; $h_2 = h_3 = 22$ kJ/kg ; $x_4 = x_5 = 0.1$; $h_4 = h_5 = 695$ kJ/kg

The aqua-ammonia absorption refrigeration system is shown in Fig. 7.1. From the *p-h* diagram, we find that enthalpy of saturated ammonia vapour at 40° C,

$$h_6 = 1473 \text{ kJ / kg}$$

Enthalpy of saturated ammonia liquid at 40°C,

$$h_7 = h_8 = 372 \text{ kJ / kg}$$

and enthalpy of saturated ammonia vapour at –20°C,

$$h_1 = 1420 \text{ kJ / kg}$$

1. Mass flow rate of ammonia in the evaporator

We know that mass flow rate of ammonia in the evaporator,

$$m_1 = \frac{210\,Q_E}{h_1 - h_8} = \frac{210 \times 10}{1420 - 372} = 2 \text{ kg/min} \quad \textbf{Ans.}$$

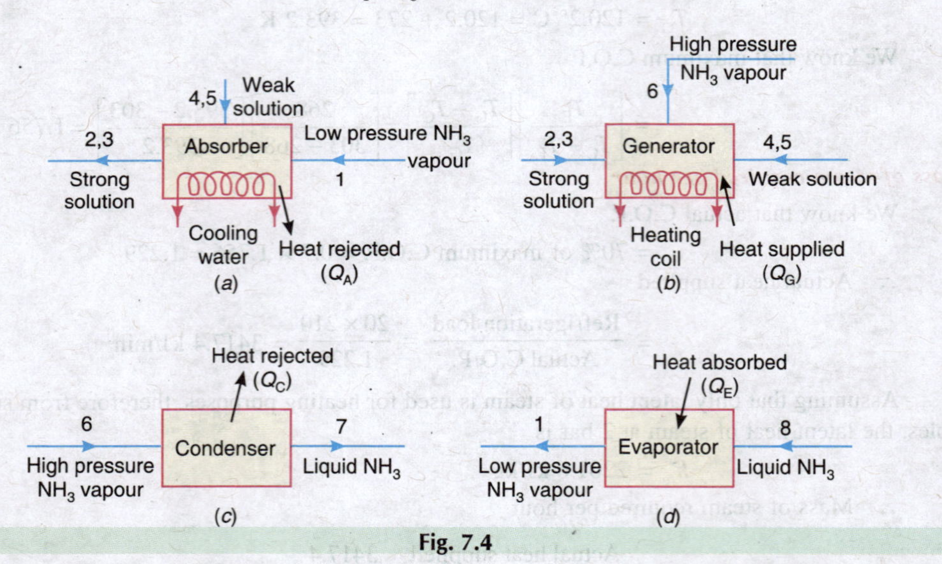

Fig. 7.4

2. Mass flow rate of weak and strong solutions

Let m_4 and m_2 = Mass flow rates of weak and strong solutions.

Considering the overall mass balance of NH_3 in the absorber [Refer Fig.7.4 (*a*)],

$$m_1 + m_4 = m_2 \quad \text{...}(i)$$

and considering the material balance of NH_3 in the absorber,

$$m_1 x_1 + m_4 x_4 = m_2 x_2 = (m_1 + m_4)\, x_2 \quad \text{...}(ii)$$

$$m_1(x_1 - x_2) = m_4(x_2 - x_4) \quad \text{...[From equation } (i)]$$

$$2\,(1 - 0.38) = m_4\,(0.32 - 0.1) = 0.22\, m_4$$

... [∵ For NH_3 vapour, $x_1 = 1$]

∴ $$m_4 = 5.636 \text{ kg / min} \quad \textbf{Ans.}$$

and $$m_2 = m_1 + m_4 = 2 + 5.636 = 7.636 \text{ kg / min} \quad \textbf{Ans.}$$

3. Heat rejection in absorber and condenser

Considering energy balance for absorber, the heat rejected to the atmosphere or cooling water,

$$Q_A = m_1 h_1 + m_4 h_4 - m_2 h_2$$
$$= 2 \times 1420 + 5.636 \times 695 - 7.636 \times 22$$
$$= 2840 + 3917 - 168 = 6589 \text{ kJ/min} \quad \textbf{Ans.}$$

Heat rejected from the condenser [Refer Fig. 7.4 (*c*)],

$$Q_C = m_6 (h_6 - h_7) = 2\,(1473 - 372) = 2202 \text{ kJ/min} \quad \textbf{Ans.}$$

... ($\because m_6 = m_1$)

4. Heat added in generator

Considering energy balance for generator, heat added in the generator [Refer Fig. 7.4 (*b*)],

$$Q_G = m_4 h_4 + m_6 h_6 - m_3 h_3$$
$$= 5.636 \times 695 + 2 \times 1473 - 7.636 \times 22$$

... ($\because m_3 = m_2$ and $h_3 = h_2$)

$$= 3917 + 2946 - 168 = 6695 \text{ kJ/min} \quad \textbf{Ans.}$$

5. C.O.P.

We know that $\text{C.O.P.} = \dfrac{Q_E}{Q_G} = \dfrac{10 \times 210}{6695} = 0.3137$ **Ans.**

... ($\because$ 1 TR = 210 kJ/min)

Example 7.4. *Simple absorption refrigeration cycle does not consider the rectification column and preheating exchanger. In such aqua-ammonia cycle, evaporator, absorber, condenser and generator temperatures are 233K, 303K, 313K and 373K respectively. The properties of aqua-ammonia are as follows:*

Particulars	*Concentration kg of NH_3 / kg solution*	*Enthalpy kJ/kg*
Strong solution leaving absorber	*0.421*	*30*
Weak solution leaving generator	*0.375*	*340*
Vapour leaving generator	*0.945*	*1870*
Liquid leaving condenser	*0.945*	*470*
Vapours leaving evaporator	*0.945*	*1388*

1. Draw schematic diagram of the system.

2. For 1 TR capacity, determine the mass flow rate of solution in evaporator.

3. Consider overall mass balance and material balance or partial mass balance of NH_3 in absorber. This will give two equations, solve them to determine mass flow rates of strong and weak solutions.

4. Consider energy balance for absorber and generator to find absorber heat rejection and generator heat transfer. Find condenser heat rejection also.

5. Check the overall energy balance neglecting pump work and then find C.O.P.

Solution. Given : $T_E = 233$ K ; $T_A = 303$ K ; $T_C = 313$ K ; $T_G = 373$ K ; $x_2 = 0.421$; $h_2 = 30$ kJ/kg ; $x_4 = x_5 = 0.375$; $h_4 = h_5 = 340$ kJ/kg ; $x_6 = 0.945$; $h_6 = 1870$ kJ/kg ; $x_7 = x_8 = 0.945$; $h_7 = h_8 = 470$ kJ/kg ; $x_1 = 0.945$; $h_1 = 1388$ kJ/kg ; $Q = Q_E = 1$ TR $= 210$ kJ/min

1. Schematic diagram of the system

The schematic diagram of the system is shown in Fig. 7.1.

2. Mass flow rate of solution in the evaporator

We know that mass flow rate in the evaporator,

$$m_1 = \frac{210\, Q_E}{h_1 - h_8} = \frac{210 \times 1}{1388 - 470} = 0.2288 \text{ kg / min} \quad \textbf{Ans.}$$

... (Here Q_E is in TR)

3. Mass flow rates of strong and weak solutions

Let m_2 and m_4 = Mass flow rates of strong and weak solutions respectively.

Considering the overall mass balance of NH_3 in the absorber,

$$m_1 + m_4 = m_2 \quad ...(i)$$

and considering material balance of NH_3 in the absorber,

$$m_1 x_1 + m_4 x_4 = m_2 x_2 = (m_1 + m_4)\, x_2 \quad ...[\text{From equation } (i)]$$

$$0.2288 \times 0.945 + m_4 \times 0.375 = (0.2288 + m_4)\, 0.421 = 0.0963 + 0.421\, m_4$$

$$0.2162 - 0.0963 = (0.421 - 0.375)\, m_4 = 0.046\, m_4$$

$$\therefore \quad m_4 = 0.1199/0.046 = 2.606 \text{ kg / min} \quad \textbf{Ans.}$$

and $$m_2 = m_1 + m_4 = 0.2288 + 2.606 = 2.8348 \text{ kg / min.} \quad \textbf{Ans.}$$

4. Absorber heat rejection, Generator heat transfer and Condenser heat rejection

Considering energy balance for absorber, the heat rejected to the atmosphere or cooling water,

$$Q_A = m_1 h_1 + m_4 h_4 - m_2 h_2$$

$$= 0.2288 \times 1388 + 2.606 \times 340 - 2.8348 \times 30$$

$$= 317.574 + 886.04 - 85.044 = 1118.57 \text{ kJ/min} \quad \textbf{Ans.}$$

Considering energy balance for generator, heat given to the refrigerant in the generator,

$$Q_G = m_4 h_4 + m_6 h_6 - m_3 h_3$$

$$= 2.606 \times 340 + 0.2288 \times 1870 - 2.8348 \times 30$$

...($\because$ $m_6 = m_1$, $m_3 = m_2$ and $h_3 = h_2$)

$$= 886.04 + 427.856 - 85.044 = 1228.852 \text{ kJ/min} \quad \textbf{Ans.}$$

and heat rejected from the condenser,

$$Q_C = m_6\,(h_6 - h_7) = 0.2288\,(1870 - 470) = 320.32 \text{ kJ/min} \quad \textbf{Ans.}$$

5. Check of overall energy balance and C.O.P.

Considering the overall energy balance, the heat absorbed by the generator and evaporator must be equal to the heat rejected from the absorber and condenser.

$\therefore$ Heat absorbed by the generator and evaporator

$$= Q_G + Q_E = 1228.852 + 210 = 1438.852 \text{ kJ/min}$$

and heat rejected from the absorber and condenser

$$= Q_A + Q_C = 1118.57 + 320.32 = 1438.89 \text{ kJ/min}$$

Since the heat absorbed and heat rejected are very close, therefore the overall energy balance is checked. **Ans.**

We know that $$\text{C.O.P.} = \frac{Q_E}{Q_G} = \frac{1 \times 210}{1228.852} = 0.171 \quad \textbf{Ans.}$$

...($\because$ $Q_E = Q = 1$ TR = 210 kJ/min)

7.9 Domestic Electrolux (Ammonia Hydrogen) Refrigerator

The domestic absorption type refrigerator was invented by two Swedish engineers, Carl Munters and Baltzer Von Platan, in 1925 while they were studying for their undergraduate course of Royal Institute of Technology in Stockholm. The idea was first developed by the 'Electrolux Company' of Luton, England.

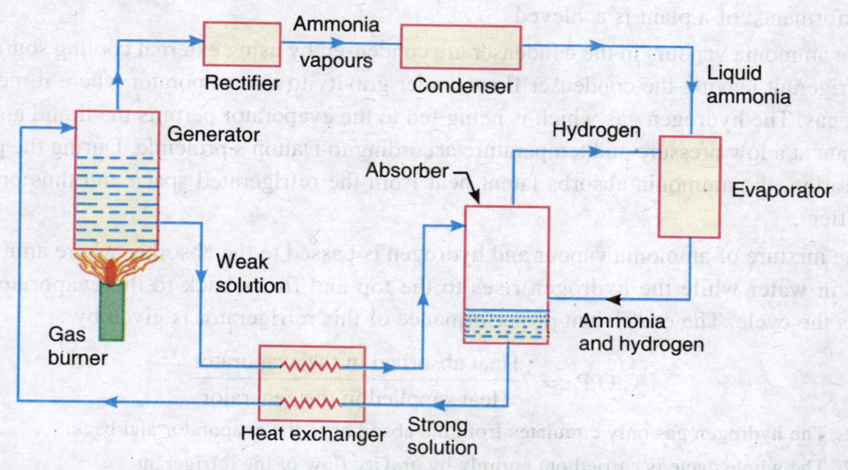

Fig. 7.5. Domestic electrolux type refrigerator.

This type of refrigerator is also called *three-fluid absorption system*. The main purpose of this system is to eliminate the pump so that in the absence of moving parts, the machine becomes noiseless. The three fluids used in this sytem are ammonia, hydrogen and water. The ammonia is used as a refrigerant because it possesses most of the desirable properties. It is toxic, but due to absence of moving parts, there is very little chance for the leakage and the total amount of refrigerant used is small. The hydrogen, being the lightest gas, is used to increase the rate of *evaporation of the liquid ammonia passing through the evaporator . The hydrogen is also non-corrosive and insoluble in water. This is used in the low-pressure side of the system. The water is used as a solvent because it has the ability to absorb ammonia readily. The principle of operation of a domestic electrolux type refrigerator, as shown in Fig. 7.5, is discussed below :

Absorption machine.

The strong ammonia solution from the absorber through heat exchanger is heated in the generator by applying heat from an external source, usually a gas burner. During this heating process, ammonia vapours are removed from the solution and passed to the condenser. A rectifier or a water separator fitted before the condenser removes

* It may be noted that lighter the gas, faster is the evaporation.

water vapour carried with the ammonia vapours, so that dry ammonia vapours are supplied to the condenser. These water vapours, if not removed, will enter into the evaporator causing freezing and choking of the machine. The hot weak solution left behind in the generator flows to the absorber through the heat exchanger. This hot weak solution while passing through the exchanger is cooled. The heat removed by the weak solution is utilised in raising the temperature of strong solution passing through the heat exchanger. In this way, the absorption is accelerated and the improvement in the performance of a plant is achieved.

The ammonia vapours in the condenser are condensed by using external cooling source. The liquid refrigerant leaving the condenser flows under gravity to the evaporator where it meets the hydrogen gas. The hydrogen gas which is being fed to the evaporator permits the liquid ammonia to evaporate at a low pressure and temperature according to Dalton's principle. During the process of evaporation, the ammonia absorbs latent heat from the refrigerated space and thus produces cooling effect.

The mixture of ammonia vapour and hydrogen is passed to the absorber where ammonia is absorbed in water while the hydrogen rises to the top and flows back to the evaporator. This completes the cycle. The coefficient of performance of this refrigerator is given by :

$$\text{C.O.P.} = \frac{\text{Heat absorbed in the evaporator}}{\text{Heat supplied in the generator}}$$

Notes : **1.** The hydrogen gas only circulates from the absorber to the evaporator and back.

2. The whole cycle is carried out entirely by gravity flow of the refrigerant.

3. It cannot be used for industrial purposes as the C.O.P. of the system is very low.

Example 7.5. *The total pressure maintained in an electrolux refrigerator is 14.71 bar. The temperature obtained in the evaporator is –15°C. The quantities of heat supplied in the generator are 418.7 kJ to dissociate 1kg of vapour and 1465.4 kJ/kg for increasing the total enthalpy of NH_3. The enthalpy of NH_3 entering the evaporator is 335 kJ/kg. Take the following properties of NH_3 at – 15°C :*

Pressure = 2.45 bar ; Enthalpy (h) of NH_3 vapour = 1666 kJ/kg ; Specific volume (v_s) = 0.5 m³/kg.

The hydrogen enters the evaporator at 25°C. Gas constant for H_2 = 4.218 kJ/kgK ; c_p for H_2 = 12.77 kJ/kg K.

Find the C.O.P. of the system. Assume NH_3 leaves the evaporator in saturated condition.

Solution. Given : $p_T = 14.71$ bar ; $T_E = -15°C = -15 + 273 = 258$ K ; $Q_{G1} = 418.7$ kJ/kg ; $Q_{G2} = 1465.4$ kJ/kg ; For NH_3, $h_1 = 335$ kJ/kg ; $p_2 = 2.45$ bar ; $h_2 = 1666$ kJ/kg ; $v = v_s = 0.5$ m³/kg ; For H_2, $T_3 = 25°C = 25 + 273 = 298$ K ; $R = 4.218$ kJ/kg K ; $c_p = 12.77$ kJ/kg K

The schematic diagram of an electrolux refrigerator is shown in Fig. 7.5. Consider the flow of 1 kg of NH_3, heat given to the generator,

$$Q_G = Q_{G1} + Q_{G2} = 418.7 + 1465.4 = 1884.1 \text{ kJ/kg}$$

Since in the evaporator, $p_T = 14.71$ bar and $p_2 = p_{NH3} = 2.45$ bar, therefore, pressure of hydrogen (H_2),

$$p_{H2} = p_T - p_2 = 14.71 - 2.45 = 12.26 \text{ bar} = 12.26 \times 10^5 \text{ N/m}^2 = 1226 \times 10^3 \text{ N/m}^2 = 1226 \text{ kN/m}^2$$

Refer Fig. 7.6. For NH_3, h_1 = 335 kJ/kg ; h_2 = 1666 kJ/kg.

Let m kg of H_2 flow through the evaporator per kg of NH_3. The ammonia (NH_3) occupies v = 0.5 m³/kg. The same is the volume of H_2 at T = – 15°C = – 15 + 273 = 258 K.

$$\therefore \quad m = \frac{p_{H2} v}{RT} = \frac{1226 \times 0.5}{4.218 \times 258} = 0.5633 \text{ kg}$$

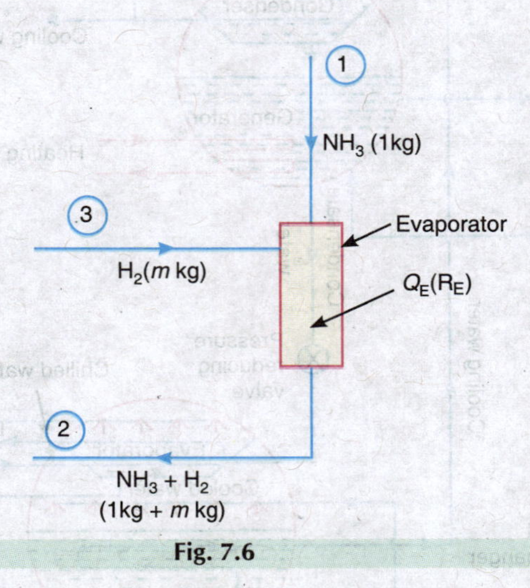

Fig. 7.6

Considering the energy balance for the evaporator (Refer Fig. 7.6),

$$1\,(h_2 - h_1) = c_p \times m\,(T_3 - T_2) + Q_E$$

$$1(1666 - 335) = 12.77 \times 0.5633\,(298 - 258) + Q_E$$

$$1331 = 288 + Q_E$$

$$\therefore \quad Q_E = 1331 - 288 = 1043 \text{ kJ/kg of } NH_3$$

We know that $\quad \text{C.O.P.} = \dfrac{Q_E}{Q_G} = \dfrac{1043}{1884.1} = 0.5536$ **Ans.**

7.10 Lithium Bromide Absorption Refrigeration System

The lithium bromide absorption refrigeration system uses a solution of lithium bromide in water. In this system, the *water is being used as a refrigerant whereas lithium bromide, which is a highly hydroscopic salt, as an absorbent. The lithium bromide solution has a strong affinity for water vapour because of its very low vapour pressure. Since lithium bromide solution is corrosive, therefore, inhibitors should be added in order to protect the metal parts of the system against corrosion. Lithium chromate is often used as a corrosion inhibitor. This system is very popular for air-conditioning in which low refrigeration temperatures (not below 0°C)** are required.

* In the domestic type absorption refrigerator (discussed in the previous article), the water is being used as absorbent and the ammonia as refrigerant.

** Since water is used as a refrigerant, therefore, the refrigeration temperature must be kept above the freezing point of water (0° C).

Fig. 7.7 shows a lithium bromide vapour absorption system. In this system, the absorber and the evaporator are placed in one shell which operates at the same low pressure of the system. The generator and condenser are placed in another shell which operates at the same high pressure of the system. The principle of operation of this system is discussed below :

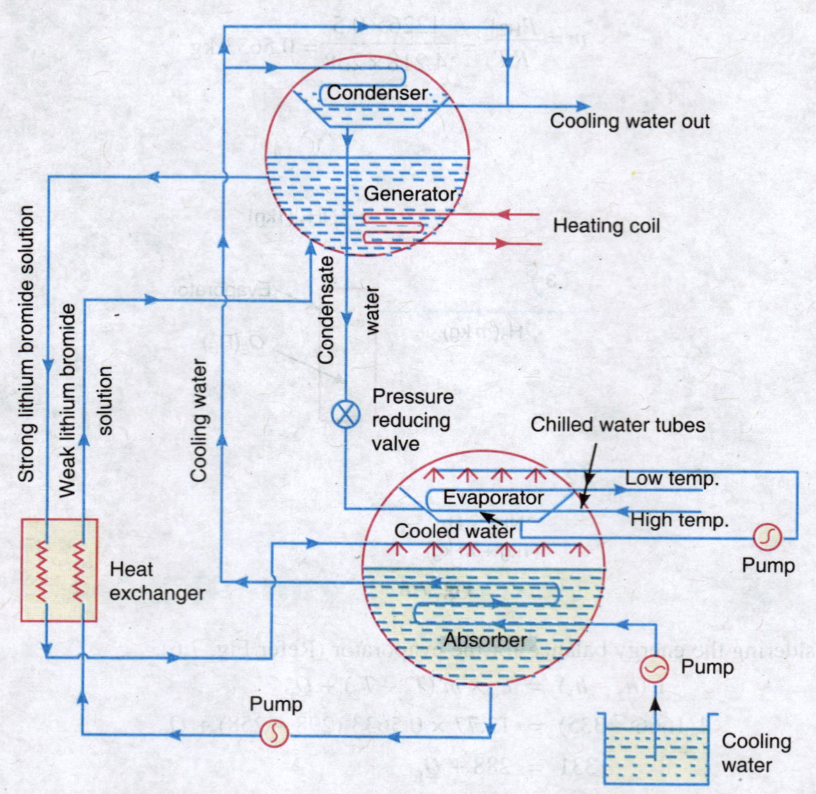

Fig. 7.7. Lithium-Bromide absorption refrigeration system.

The water for air-conditioning coils or process requirements is chilled as it is pumped through the chilled water tubes in the evaporator by giving up heat to the refrigerant water sprayed over the tubes. Since the pressure inside the evaporator is maintained very low, therefore, the refrigerant water evaporates. The water vapours thus formed will be absorbed by the strong lithium bromide solution which is sprayed in the absorber. In absorbing the water vapour, the lithium bromide solution helps in maintaining very low pressure (high vacuum) needed in the evaporator, and the solution becomes weak. This weak solution is pumped by a pump to the generator where it is heated up by using steam or hot water in the heating coils. A portion of water is evaporated by the heat and the solution now becomes more strong. This strong solution is passed through the heat exchanger and then sprayed in the absorber as discussed above. The weak solution of lithium bromide from the absorber to the generator is also passed through the heat exchanger. This weak solution gets heat from the strong solution in the heat exchanger, thus reducing the quantity of steam required to heat the weak solution in the generator.

The refrigerant water vapours formed in the generator due to heating of solution are passed to the condenser where they are cooled and condensed by the cooling water flowing through the condenser water tubes. The cooling water for condensing is pumped from the cooling water pond or tower. This cooling water first enters the absorber where it takes away the heat of condensation and dilution. The condensate from the condenser is supplied to the evaporator to compensate the water vapour formed in the evaporator. The pressure reducing valve reduces the pressure of condensate from the condenser pressure to the evaporator pressure. The cooled water from the evaporator is pumped and sprayed in the evaporator in order to cool the water for air-conditioning flowing through the chilled tubes. This completes the cycle.

Note : The pressure difference between the generator and the absorber and the gravity due to the height difference of the two shells is utilised to create the pressure for the spray.

Example 7.6. *The following data refer to a LiBr + H_2O absorption system :*

Generator temperature = 80°C ; Condenser temperature = Absorber temperature = 30°C ; Evaporator temperature = 10°C ; Condensate temperature = 25°C.

Steam enters the generator heating coil at 120° C (dry-saturated state steam) and leaves it at 100°C as condensate.

The concentration of liquid leaving the generator is 0.65 and its enthalpy is –75 kJ/kg. The concentration of liquid leaving the absorber is 0.51 and its enthalpy is –170 kJ/kg. The enthalpy of vapour leaving the generator is 2620 kJ/kg. The flow rate through the evaporator is 0.4 kg / s.

Find : 1. Pressure in generator, condenser, evaporator and absorber in mm of mercury head ; 2. Tonnage ; 3. Heat rejection to condenser and absorber ; 4. C.O.P ; and 5. Relative C.O.P.

Solution. Given : T_G = 80°C = 80 + 273 = 353 K ; $T_C = T_A$ = 30°C = 30 + 273 = 303 K ; T_E = 10°C = 10 + 273 = 283 K ; T_2 = 25°C ; T_H = 120°C ; T_L = 100°C ; x_5 = 0.65 ; h_5 = – 75 kJ/kg ; x_4 = 0.51 ; h_4 = – 170 kJ/kg ; h_1 = 2620 kJ/kg ; $m_1 = m_2 = m_3$ = 0.4 kg/s

The lithium bromide water absorption system may be represented as shown in Fig. 7.8.

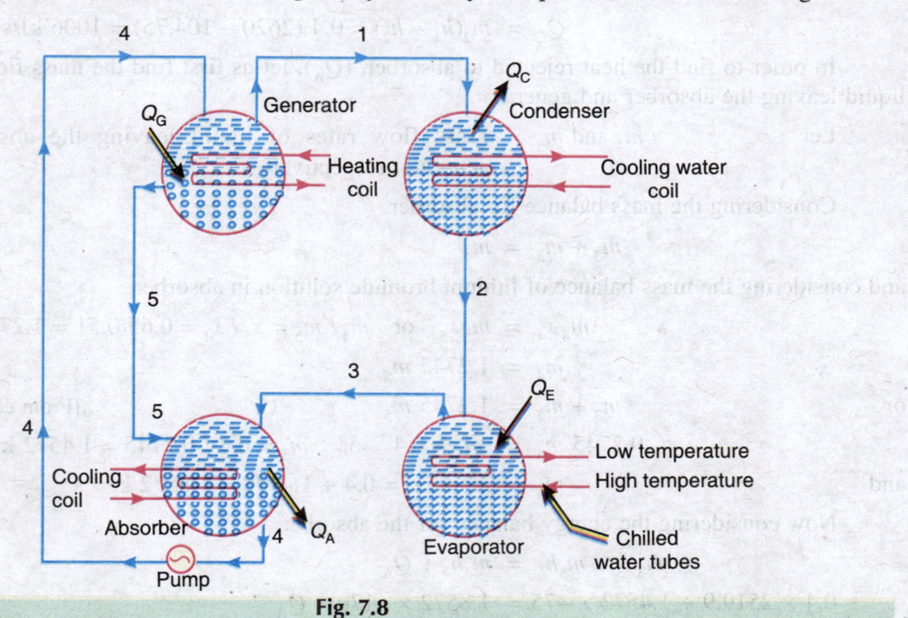

Fig. 7.8

1. Pressure in generator, condenser, evaporator and absorber in mm of Hg head

From steam tables of dry saturated steam corresponding to generator temperature of 80°C, we find that pressure in generator,

$$p_G = 0.4736 \text{ bar} = 0.4736 \times 760/1.013 = 355.3 \text{ mm of Hg} \quad \textbf{Ans.}$$

... (∵ 1.013 bar = 760 mm of Hg)

Corresponding to a condenser temperature of 30°C, pressure in condenser,

$$p_C = 0.042\,42 \text{ bar} = 0.042\,42 \times 760/1.013 = 31.82 \text{ mm of Hg} \quad \textbf{Ans.}$$

Corresponding to an evaporator temperature of 10°C, pressure in evaporator,

$$p_E = 0.012\,27 \text{ bar} = 0.012\,27 \times 760/1.013 = 9.2 \text{ mm of Hg} \quad \textbf{Ans.}$$

and corresponding to an absorber temperature of 30°C, pressure in absorber,

$$p_A = 0.042\,42 \text{ bar} = 0.042\,42 \times 760/1.013 = 31.82 \text{ mm of Hg} \quad \textbf{Ans.}$$

2. Tonnage or cooling capacity

From steam tables, corresponding to a condensate temperature of 25°C, we find that enthalpy of liquid water,

$$h_2 = 104.75 \text{ kJ/kg}$$

and enthalpy of vapour leaving the evaporator (assuming it as dry and saturated at 10°C),

$$h_3 = 2519.9 \text{ kJ/kg}$$

We know that cooling capacity or refrigerating effect (or heat absorbed in the evaporator),

$$Q_E = m_2(h_3 - h_2) = 0.4\,(2519.9 - 104.75) = 966 \text{ kJ/s}$$

$$= 966 / 3.5 = 276 \text{ TR} \quad \textbf{Ans.} \qquad ...(\because 1 \text{ TR} = 3.5 \text{ kJ/s})$$

3. Heat rejected to condenser and absorber

We know that heat rejected to condenser,

$$Q_C = m_2(h_1 - h_2) = 0.4\,(2620 - 104.75) = 1006 \text{ kJ/s} \quad \textbf{Ans.}$$

In order to find the heat rejected to absorber, (Q_A), let us first find the mass flow rates of liquid leaving the absorber and generator.

Let m_4 and m_5 = Mass flow rates of liquid leaving the absorber and generator respectively, in kg/s.

Considering the mass balance for absorber

$$m_3 + m_5 = m_4 \qquad ...(i)$$

and considering the mass balance of lithium bromide solution in absorber,

$$m_4 x_4 = m_5 x_5 \quad \text{or} \quad m_4 / m_5 = x_5 / x_4 = 0.65/0.51 = 1.2745$$

$$\therefore \quad m_4 = 1.2745\, m_5$$

or

$$m_3 + m_5 = 1.2745\, m_5 \qquad ...[\text{From equation } (i)]$$

$$0.2745\, m_5 = m_3 = 0.4 \quad \text{or} \quad m_5 = 0.4 / 0.2745 = 1.4572 \text{ kg / s}$$

and

$$m_4 = m_3 + m_5 = 0.4 + 1.4572 = 1.8572 \text{ kg / s}$$

Now considering the energy balance for the absorber,

$$m_3 h_3 + m_5 h_5 = m_4 h_4 + Q_A$$

$$0.4 \times 2519.9 + 1.4572 \times -75 = 1.8572 \times -170 + Q_A$$

$$1007.96 - 109.3 = -315.7 + Q_A$$

$$\therefore \quad Q_A = 1214.36 \text{ kJ/s} \quad \textbf{Ans.}$$

4. C.O.P.

First of all, let us find the heat transfer rate in the generator (Q_G).

Considering the energy balance for the generator,

$$m_1 h_1 + m_5 h_5 = m_4 h_4 + Q_G$$

$$0.4 \times 2620 + 1.4572 \times -75 = 1.8572 \times -170 + Q_G$$
$$1048 - 109.3 = -315.7 + Q_G$$
$$\therefore \quad Q_G = 1254.4 \text{ kJ/s}$$

We know that $\text{C.O.P.} = \dfrac{Q_E}{Q_G} = \dfrac{966}{1254.4} = 0.77$ **Ans.**

Note : The heat transfer rate in the generator (Q_G) is also given as

$$Q_G = Q_C + Q_A - Q_E = 1006 + 1214.36 - 966 = 1254.36 \text{ kJ/s}$$

5. Relative C.O.P.

We know that maximum C.O.P.,

$$(\text{C.O.P.})_{max} = \left[\frac{T_E}{T_C - T_E}\right]\left[\frac{T_G - T_A}{T_G}\right]$$

$$= \left[\frac{283}{303-283}\right]\left[\frac{353-303}{353}\right] = \frac{283}{20} \times \frac{50}{353} = 2.004$$

$$\therefore \quad \text{Relative C.O.P.} = \frac{\text{C.O.P.}}{(\text{C.O.P.})_{max}} = \frac{0.77}{2.004} = 0.384 \textbf{ Ans.}$$

EXERCISES

1. In a vapour absorption system, heat is supplied to the generator at a temperature of 90°C. The cooling in condenser and refrigeration evaporator takes place at 20°C and –10°C respectively. Find the maximum C.O.P. of the system. **[Ans. 1.69]**
2. In a vapour absorption system, the heat is supplied to the generator by condensing steam at 3 bar and 85% dry. The temperature in the evaporator is to be maintained at – 10°C. If the cooling water rejects heat at 30°C in the condenser, find the maximum C.O.P. of the system.

   When the refrigeration load is 10 tonnes and the actual C.O.P. is 40% of the maximum C.O.P., find the mass of steam required per hour. **[Ans. 1.674 ; 102.3 kg/h]**
3. In a 100TR aqua ammonia absorption plant, saturated liquid ammonia at 30°C leaves the condenser and enters the expansion valve. The evaporator pressure is 1.9 bar and the vapour temperature at evaporator exit is –10°C. The mass concentrations of ammonia in the weak and strong solutions are 0.25 and 0.325 respectively. Determine the mass flow rates (in kg / min) of the strong and weak solutions.

   [Ans. 185 kg / min ; 166.5 kg / min]
4. The following data refer to a lithium bromide water absorption refrigeration system :

   Capacity = 2 TR ; Concentration of LiBr and enthalpy values for weak solution leaving generator = 0.68 and 21 kJ/kg respectively; Concentration of LiBr and enthalpy values for strong solution leaving absorber = 0.58 and – 55 kJ/kg respectively. Temperature of water leaving condenser = 40°C ; Enthalpy of steam leaving evaporator = 2508 kJ/kg ; Specific heat of water = 4.2 kJ/kg K.

   Determine the mass flow rates of strong and weak solutions in kg/min and the heat transfer rates in the generator and the absorber in kJ/min.

   [Ans. 1.2195 kg/min ; 1.04 kg/min ; 573.56 kJ/min ; 539.09 kJ/min]
5. In a lithium bromide water absorption refrigeration system, absorber, condenser, evaporator and generator temperatures are 45°C, 45°C, 5°C and 105°C respectively. For strong solution leaving absorber, the enthalpy is –50 kJ/kg and concentration of LiBr is 0.60. For weak solution leaving generator, the enthalpy is 22 kJ/kg and concentration of LiBr is 0.67. The enthalpies of saturated steam and water at t°C may be determined from $h = 2501 + 1.88\,t$ kJ/kg and $h_f = 4.1867\,t$ kJ/kg.

Assuming simple absorption cycle, for a plant of 1TR capacity, determine : 1. mass flow rate of water by considering mass balance in absorber; 2. mass flow rate of strong and weak solutions by considering mass balance in absorber ; 3. heat transfer rates in the absorber, generator and condenser ; and 4. C.O.P.

[**Ans.** 0.09087 kg / min ; 0.8696 kg / min, 0.7787 kg / min ; 288.73 kJ/min, 305.96 kJ/min, 228.22 kJ/min ; 0.686]

QUESTIONS

1. What is the basic function of a compressor in vapour compression refrigeration system ? How this function is achieved in vapour absorption refrigeration system ?
2. Draw a neat sketch of a practical vapour absorption refrigeration cycle. Indicate thereon the phases of various fluids and the name of the equipments. Also indicate the direction of the external energy flow to or from the equipments.
3. What is the function of the following components in an absorption system :
   (*i*) Absorber (*ii*) Rectifier (*iii*) Analyser (*iv*) Heat exchangers.
4. Discuss the advantages of vapour absorption refrigeration system over vapour compression refrigeration system.
5. Derive an expression for the C.O.P. of an ideal vapour absorption system in terms of temperature T_G at which heat is supplied to the generator, the temperature T_E at which heat is absorbed in the evaporator and the temperature T_C at which heat is discharged from the condenser and absorber.
6. Draw a neat diagram of three-fluid system of refrigeration (electrolux refrigeration system) and explain its working.
7. Mention the function of each fluid in a three-fluid vapour absorption system.
8. Draw a neat diagram of lithium bromide water absorption system and explain its working. List the major field of applications of this system.

OBJECTIVE TYPE QUESTIONS

1. The refrigerant, commonly used in vapour absorption system, is
   (*a*) water (*b*) ammonia (*c*) freon (*d*) aqua-ammonia
2. A vapour absorption system
   (*a*) gives noisy operation (*b*) gives quiet operation
   (*c*) requires little power consumption (*d*) cools below 0°C
3. In aqua-ammonia absorption refrigeration system, incomplete rectification leads to accumulation of water in
   (*a*) condenser (*b*) evaporator (*c*) absorber (*d*) none of these
4. Comparing mechanical vapour compression refrigeration system with absorption refrigeration system, the compressor of the former is replaced in the latter by
   (*a*) an absorber and a liquid pump
   (*b*) an absorber, a generator, a liquid pump and a pressure reduction valve
   (*c*) an absorber, an evaporator, a liquid pump and an expansion valve
   (*d*) a generator, an evaporator, a liquid pump and an expansion valve
5. The C.O.P. of practical vapour compression system is ...... as compared to that for vapour absorption system.
   (*a*) more (*b*) less
6. If the C.O.P. of 1 TR ammonia-water absorption refrigeration plant is 0.5, then the heat supplied in the generator is
   (*a*) 1.5 kW (*b*) 3.5 kW (*c*) 7 kW (*d*) 10.5 kW

7. A rectifier is fitted in an ammonia absorption plant to
(*a*) superheat ammonia vapour
(*b*) remove the unwanted water vapour by heating the vapour mixture
(*c*) remove the unwanted water vapour by cooling the vapour mixture
(*d*) remove the unwanted water vapour by cooling the vapour mixture and condensing the water vapour

8. An electrolux refrigerator is a
(*a*) vapour compression refrigerator
(*b*) vapour absorption refrigerator with a liquid pump
(*c*) vapour absorption refrigerator without any pump
(*d*) none of the above

9. In electrolux refrigerator
(*a*) ammonia is absorbed in hydrogen
(*b*) ammonia is absorbed in water
(*c*) ammonia evaporates in hydrogen
(*d*) hydrogen evaporates in ammonia

10. An electrolux refrigerator is called a
(*a*) single-fluid absorption system
(*b*) two-fluid absorption system
(*c*) three-fluid absorption system
(*d*) none of these

11. The fluids used in the electrolux refrigerator are
(*a*) water and hydrogen
(*b*) ammonia and hydrogen
(*c*) ammonia, water and hydrogen
(*d*) none of these

12. Hydrogen is used in electrolux refrigeration system so as to .... the vapour pressure ammonia in evaporator.
(*a*) equalise
(*b*) reduce
(*c*) increase

13. An electrolux refrigerator has
(*a*) only one liquid pump
(*b*) only two liquid pumps
(*c*) no liquid pump
(*d*) none of these

14. In aqua-ammonia and Li-Br water absorption refrigeration systems, the refrigerant are respectively
(*a*) water and water
(*b*) water and Li-Br
(*c*) ammonia and Li-Br
(*d*) ammonia and water

15. In a lithium bromide absorption refrigeration system
(*a*) lithium bromide is used as a refrigerant and water as an absorbent
(*b*) water is used as a refrigerant and lithium bromide as an absorbent
(*c*) ammonia is used as a refrigerant and lithium bromide as an absorbent
(*d*) none of the above

ANSWERS

1. (*d*)	**2.** (*b*)	**3.** (*a*)	**4.** (*b*)	**5.** (*a*)
6. (*c*)	**7.** (*d*)	**8.** (*c*)	**9.** (*c*)	**10.** (*c*)
11. (*c*)	**12.** (*b*)	**13.** (*c*)	**14.** (*d*)	**15.** (*b*)

CHAPTER

8 Refrigerants

1. Introduction.
2. Desirable Properties of an Ideal Refrigerant.
3. Classification of Refrigerants.
4. Halo-carbon Refrigerants.
5. Azeotrope Refrigerants.
6. Inorganic Refrigerants.
7. Hydro-carbon Refrigerants.
8. Designation System for Refrigerants.
9. Substitutes for chloro-fluoro-carbon (CFC) Refrigerants.
10. Comparison of Refrigerants.
11. Thermodynamic Properties of Refrigerants.
12. Chemical Properties of Refrigerants.
13. Physical Properties of Refrigerants.
14. Secondary Refrigerants - Brines.

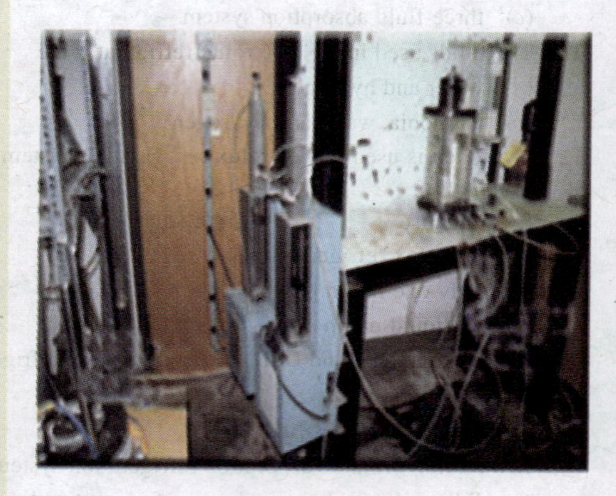

8.1 Introduction

The refrigerant is a heat carrying medium which during their cycle (*i.e.* compression, condensation, expansion and evaporation) in the refrigeration system absorbs heat from a low temperature system and discards the heat so absorbed to a higher temperature system.

The natural ice and a mixture of ice and salt were the first refrigerants. In 1834, ether, ammonia, sulphur dioxide, methyl chloride and carbon dioxide came into use as refrigerants in compression cycle refrigeration machines. Most of the early refrigerant materials have been discarded for safety reasons or for lack of chemical or thermal stability. In the present days, many new refrigerants including halo-carbon compounds, hydro-carbon compounds are used for air-conditioning and refrigeration applications.

The suitability of a refrigerant for a certain application is determined by its physical, thermodynamic, chemical properties and by various practical factors. There is no one refrigerant which can be used for all types of applications *i.e.* there is no ideal refrigerant. If one refrigerant has certain good advantages, it will have some disadvantages also. Hence, a refrigerant is chosen which has greater advantages and less disadvantages.

8.2 Desirable Properties of an Ideal Refrigerant

We have discussed above that there is no ideal refrigerant. A refrigerant is said to be ideal if it has all of the following properties :

1. Low boiling and freezing point,
2. High critical pressure and temperature,
3. High latent heat of vaporisation,
4. Low specific heat of liquid, and high specific heat of vapour,
5. Low specific volume of vapour,
6. High thermal conductivity,
7. Non-corrosive to metal,
8. Non-flammable and non-explosive,
9. Non-toxic,
10. Low cost,
11. Easily and regularly available,
12. Easy to liquify at moderate pressure and temperature,
13. Easy of locating leaks by odour or suitable indicator,
14. Mixes well with oil,
15. High coefficient of performance, and
16. Ozone friendly.

The standard comparison of refrigerants, as used in the refrigeration industry, is based on an evaporating temperature of – 15°C and a condensing temperature of + 30° C.

8.3 Classification of Refrigerants

The refrigerants may, broadly, be classified into the following two groups :

1. Primary refrigerants, and 2. Secondary refrigerants.

The refrigerants which directly take part in the refrigeration system are called *primary refrigerants* whereas the refrigerants which are first cooled by primary refrigerants and then used for cooling purposes are known as *secondary refrigerants.*

The primary refrigerants are further classified into the following four groups :

1. Halo-carbon or organic refrigerants,
2. Azeotrope refrigerants,
3. Inorganic refrigerants, and
4. Hydro-carbon refrigerants.

These above mentioned refrigerants are discussed, in detail, in the following pages.

8.4 Halo-carbon Refrigerants

The American Society of Heating, Refrigeration and Air-conditioning Engineers (ASHRAE) identifies 42 halo-carbon compounds as refrigerants, but only a few of them are commonly used. The following table gives some of the commonly used halo-carbon refrigerants :

Table 8.1. Commonly used halo-carbon refrigerants.

Refrigerant number	*Chemical name*	*Chemical formula*
R-11	Trichloro-monofluoro-methane	CCl_3F
R-12	Dichloro-difluoro-methane	CCl_2F_2
R-13	Monochloro-trifluoro-methane	$CClF_3$
R-14	Carbontetrafluoride	CF_4
R-21	Dichloro-monofluoro-methane	$CHCl_2F$
R-22	Monochloro-difluoro-methane	$CHClF_2$
R-30	Methylene chloride	CH_2Cl_2
R-40	Methyl chloride	CH_3Cl
R-100	Ethyl chloride	C_2H_5Cl
R-113	Trichloro-trifluoro-ethane	CCl_2FCClF_2 or $C_2Cl_3F_3$
R-114	Dichloro-tetrafluoro-ethane	$CClF_2CClF_2$ or $C_2Cl_2F_4$
R-115	Monochloro-pentafluoro-ethane	$CClF_2CF_3$ or C_2ClF_5
R-123	Dichloro-trifluoro-ethane	CF_3CHCl_2
R-124	Monochloro-tetrafluoro-ethane	CF_3CHClF
R-134 a	Tetrafluoro-ethane	CF_3CH_2F
R-152 a	Difluoro-ethane	CH_3CHF_2

The halo-carbon compounds are all synthetically produced and were developed as Freon family of refrigerants. Freon is a registered trade mark of E.I. Du Pont de Nemours and Co., America. Most of the halo-carbon refrigerants are now available from other manufacturers under various trade names such as Genetron, Isotron etc. The first of the halo-carbon refrigerants *i.e.* R-12 was developed in 1930 by Thomas Midgley. The various halo-carbon refrigerants mentioned above are now discussed, in detail, as below :

1. *R-11, Trichloro-monofluoro-methane (CCl_3F).* The R-11 is a synthetic chemical product which can be used as a refrigerant. It is stable, non-flammable and non-toxic. It is considered to be a low-pressure refrigerant. It has a low side pressure of 0.202 bar at – 15°C and high side pressure of 1.2606 bar at 30°C. The latent heat at – 15°C is 195 kJ/kg. The boiling point at atmospheric pressure is 23.77°C. Due to its low operating pressures, this refrigerant is exclusively used in large centrifugal compressor systems of 200 TR and above. The leaks may be detected by using a soap solution, a halide torch or by using an electronic detector.

R-11 is often used by service technicians as a flushing agent for cleaning the internal parts of a refrigerator compressor when overhauling systems. It is useful after a system had a motor burn out or after it has a great deal of moisture in the system. By flushing moisture from the system with R-11, evacuation time is shortened. R-11 is one of the safest cleaning solvents that can be used for this purpose. The cylinder colour code for R-11 is orange.

2. *R-12, Dichloro-difluoro-methane (CCl_2F_2).* The R-12 is a very popular refrigerant. It is a colourless, almost odourless liquid with boiling point of – 29°C at atmospheric pressure. It is non-toxic, non-corrosive, non-irritating and non-flammable. It has a relatively low latent heat value which is an advantage in small refrigerating machines. The large amount of refrigerant circulated will permit the use of less sensitive and more positive operating and regulating mechanisms. It operates at a low but positive head and back pressure and with a good volumetric efficiency. This refrigerant is used in many different types of industrial and commercial applications such as refrigerators, freezers, water coolers, room and window air-conditioning units etc. Its principal use is found in reciprocating and rotary compressors, but its use in centrifugal compressors for large commercial air-conditioning is increasing.

R-12 has a pressure of 0.82 bar at – 15°C and a pressure of 6.4 bar at 30°C. The latent heat of R-12 at – 15°C is 159 kJ/kg. The leak may be detected by soap solution, halide torch or an

Refrigerant cylinders.

electronic leak detector. Water is only slightly soluble in R- 12. At – 18°C, it will hold six parts per million by mass. The solution formed is very slightly corrosive to any of the common metals used in refrigerator construction. The addition of mineral oil to the refrigerant has no effect upon the corrosive action.

R-12 is more critical as to its moisture content when compared to R-22 and R-502. It is soluble in oil down to – 68°C. The oil will begin to separate at this temperature and due to its lightness than the refrigerant, it will collect on the surface of the liquid refrigerant. The refrigerant is available in a variety of cylinder sizes and the cylinder colour code is white.

3. *R-13, Monochloro-trifluoro-methane ($CClF_3$).* The R-13 has a boiling temperature of – 81.4°C at atmospheric pressure and a critical temperature of + 28.8°C. This refrigerant is used for the low temperature side of cascade systems. It is suitable with reciprocating compressors.

4. *R-14, Carbontetrafluoride (CF_4).* The R-14 has a boiling temperature of – 128°C at atmospheric pressure and critical temperature of – 45.5°C. It serves as an ultra-low temperature refrigerant for use in cascade systems.

5. *R-21, Dichloro-monofluoro-methane ($CHCl_2F$).* The R-21 has a boiling temperature of + 9°C at atmospheric pressure. It has found its principal use in centrifugal compressor systems for relatively high temperature refrigeration requirements.

6. *R-22, Monochloro-difluoro-methane ($CHClF_2$).* The R-22 is a man-made refrigerant developed for refrigeration installations that need a low evaporating temperature, as in fast freezing units which maintain a temperature of – 29°C to – 40°C. It has also been successfully used in air-conditioning units and in household refrigerators. It is used with reciprocating and centrifugal compressors. It is not necessary to use R-22 at below atmospheric pressures in order to obtain the low temperatures.

The boiling point of R-22 is – 41°C at atmospheric pressure. It has a latent heat of 216.5 kJ/kg at – 15°C. The normal head pressure at 30°C is 10.88 bar. This refrigerant is stable and is non-toxic, non-corrosive, non-irritating and non-flammable. The evaporator pressure of this refrigerant at – 15°C is 1.92 bar. Since water mixes better with R-22 than R- 12 by a ratio of 3 to 1, therefore driers (dessicants) should be used to remove most of the moisture to keep water to a minimum. This refrigerant has good solubility in oil down to – 9°C. However, the oil remains fluid enough to flow down the suction line at temperatures as low as – 40°C. The oil will begin to separate at this point. Since oil is lighter, therefore it will collect on the surface of the liquid refrigerant. The leaks may be detected with a soap solution, a halide torch or with an electronic leak detector. The cylinder colour code for R-22 is green.

7. R-30, Methylene chloride (CH_2Cl_2). The R-30 is a clear, water-white liquid with a sweet, non-irritating odour similar to that of chloroform. It has a boiling point of 39.8°C at atmospheric pressure. It is non-flammable, non-explosive and non-toxic. Due to its high boiling point, this refrigerant may be stored in closed cans instead of in compressed gas cylinders. The high and low sides of refrigeration system using R-30 operate under a vacuum. Since the volume of vapour at suction conditions is very high, therefore the use of R-30 is restricted to rotary or centrifugal compressors. This refrigerant was extensively used for air conditioning of theatres, auditoriums, and office buildings. Now-a-days, the refrigerant R-11 is used in place of R-30.

In order to detect leaks in a system using R-30, the pressure must be increased above atmosphere. A halide torch is used for detecting leaks.

8. R-40, Methyl-chloride (CH_3Cl). The R-40 is a colourless liquid with a faint, sweet, and non-irritating odour. Its boiling point at atmospheric pressure is – 23.7°C and the usual condenser pressure is 5 to 6.8 bar. The latent heat of vaporisation at – 15°C is 423.5 kJ/kg. It is flammable and explosive when mixed with air in concentrations from 8.1 to 17.2 per cent. This refrigerant is non-corrosive in its pure state, but it becomes corrosive in the presence of moisture. Aluminium, zinc and magnesium alloys should never be used with this refrigerant as they will corrode considerably and pollute the lubricating oil. Since the refrigerant R-40 is a solvent for many materials used in ordinary refrigeration compressors, therefore rubber and gaskets containing rubber should never be used. However, synthetic rubber is not affected by R-40. Thus metallic or asbestos-fibre gaskets containing insoluble binders should be used. The mineral oils are soluble in this refrigerant to a small extent.

This refrigerant has been used in domestic units with both reciprocating and rotary compressors and in commercial units with reciprocating compressors up to approximately 10 TR capacity. The leaks with R-40 may be detected by soap solution or electronic leak detector.

9. R-100, Ethyl chloride (C_2H_5Cl). The R-100 is a colourless liquid and in many respects it is similar to R-40 (Methyl chloride) but with low operating pressures. It has a boiling point of 13.1°C at atmospheric pressure. It is both toxic and flammable. Due to its low operating pressure, it is not used in refrigerating equipment.

10. R-113, Trichloro-trifluoro-ethane (CCl_2FCClF_2 or $C_2Cl_3F_3$). The R-113 has a boiling point of 47.6°C at atmospheric pressure. It is used in commercial and industrial air-conditioning with centrifugal compressor systems. Since this refrigerant has the advantage of remaining liquid at room temperatures and pressures, therefore it can be carried in sealed tins rather than cylinders.

11. R-114, Dichloro-tetrafluoro-ethane ($CClF_2CClF_2$ or $C_2Cl_2F_4$). The R-114 has a boiling point of 3.6°C at atmospheric pressure. At – 15°C, it evaporates at a pressure of 0.54 bar and at + 30°C it condenses at a pressure of 1.5 bar. Its latent heat of vaporisation at – 15°C is 143 kJ/kg. It is non-toxic, non-explosive and non corrosive even in the presence of water. It is used in fractional power household refrigerating systems and drinking water coolers employing rotary-vane type compressors.

12. R-123, Dichloro- trifluoro methane (CF_3CHCl_2). The R-123 is a potential substitute to R-11. It has about 4.3° C higher boiling point than R-11. It is, therefore, a lower pressure replacement for R-11, thus having larger specific volume of suction vapour. Hence, its use results in 10 to 15% reduction in capacity, if used in existing R-11 centrifugal compressors.

13. R-134 a, Tetrafluoro-ethane (CF_3CH_2F). The R-134a is considered to be the most preferred substitute for refrigerant R-12. Its boiling point is – 26.15° C which is quite close to the boiling point of R-12 which is – 29° C at atmospheric pressure. Since the refrigerent R-134a has no chlorine atom, therefore this refrigerant has zero *ozone depleting potential (ODP) and has 74% less global warming potential (GWP) as compared to R-12. It has lower suction pressure and large suction vapour volume. It is not soluble in mineral oil. Hence, for use in domestic refrigerators (with hermetic units), suitable synthetic oil (polyster based) is used. Care should be taken to prevent moisture from getting into the refrigeration system. For use in existing R-12 reciprocating compressors, it would require either an average increase in compressor speed of 5 to 8% or an equivalent increase in cylinder volume. Since the molecules of R-134a are smaller than R-12, therefore a very sensitive leak detector is used to detect leaks.

Note : The refrigerant R-134a is, now-a-days, widely used in car air-conditioners.

14. R-152a, Difluoro-ethane (CH_3CHF_2). The R-152a has similar characteristics as R-134a except that R-152a has slight vacuum in the evaporator at —25°C and the discharge temperature is higher because of its high value of the ratio of specific heats (γ).

8.5 Azeotrope Refrigerants

The term *'azeotrope'* refers to a stable mixture of refrigerants whose vapour and liquid phases retain identical compositions over a wide range of temperatures. However, these mixtures, usually, have properties that differ from either of their components. Some of the azeotropes are given in the following table :

Table 8.2. Azeotrope refrigerants.

Refrigerant number	*Azeotropic mixing refrigerants*	*Chemical formula*
R-500	73.8% R-12 and 26.2% R-152	CCl_2F_2 / CH_3CHF_2
R-502	48.8% R-22 and 51.2% R-115	$CHClF_2 / CClF_2CF_3$
R-503	40.1% R-23 and 59.9% R-13	$CHF_3 / CClF_3$
R-504	48.2% R-32 and 51.8% R-115	$CH_2F_2 / CClF_2CF_3$

These refrigerants are discussed, in detail, as below :

1. *R-500.* The R-500 is an azeotropic mixture of 73.8% R-12 (CCl_2F_2) and 26.2% of R-152 (CH_3CHF_2). It is non-flammable, low in toxicity and non-corrosive. It is used in both industrial and commercial applications but only in systems with reciprocating compressors. It has a fairly constant vapour pressure temperature curve which is different from the vaporizing curves for either R-152*a* or R-12.

This refrigerant offers about 20% greater refrigerating capacity than R-12 for the same size of motor when used for the same purpose. The evaporator pressure of this refrigerant is 1.37 bar at – 15°C and its condensing pressure is 7.78 bar at 30°C. It has a boiling point of – 33°C at atmospheric pressure. Its latent heat at – 15°C is 192 kJ/kg. It can be used whenever a higher capacity than that obtained with R-12 is needed. The solubility of water in R-500 is highly critical. It has fairly high solubility with oil. The leakage may be detected by using soap solution, a halide

* See Art.8.9.

torch, an electronic leak detector or a coloured tracing agent. The servicing refrigerators using this refrigerant does not present any unusual problem. Water is quite soluble in this refrigerant. It is necessary to keep moisture out of the system by careful dehydration and by using driers. The cylinder colour code for this refrigerant is yellow.

2. *R-502.* The R-502 is an azeotropic mixture of 48.8% R-22 ($CHClF_2$) and 51.2% of R-115 ($CClF_2CF_3$). It is a non-flammable, non-corrosive, practically non-toxic liquid. It is a good refrigerant for obtaining medium and low temperatures. It is suitable where temperatures from – 18°C to – 51°C are needed. It is often used in frozen food lockers, frozen food processing plants, frozen food display cases and in storage units for frozen foods and ice-cream. It is only used with reciprocating compressors. The boiling point of this refrigerant at atmospheric pressure is – 46°C. Its evaporating pressure at – 15°C is 2.48 bar and the condensing pressure at 30°C is 12.06 bar. Its latent heat at – 29°C is 168.6 kJ/kg.

The R-502 combines many of the good properties of R-12 and R-22. It gives a machine capacity equal to that of R-22 with just about the condensing temperature of a system using R-12. Since this refrigerant has a relatively low condensing pressure and temperature, therefore it increases the life of compressor valves and other parts. Better lubrication is possible because of the increased viscosity of the oil at low condensing temperature. It is possible to eliminate liquid injection to cool the compressor because of the low condensing pressure.

This refrigerant has all the qualities found in other halogenated (fluorocarbon) refrigerants. It is non-toxic, non-flammable, non-irritating, stable and non-corrosive. The leaks may be detected by soap solution, halide torch or electronic leak detector. It will hold 1.5 times more moisture at – 18°C than R-12. It has fair solubility in oil above 82°C. Below this temperature, the oil tries to separate and tends to collect on the surface of the liquid refrigerant. However, oil is carried back to the compressor at temperatures down to – 40°C. Special devices are sometimes used to return the oil to the compressor. The cylinder colour code for this refrigerant is orchid.

3. *R-503*. The R-503 is an azeotropic mixture of 40.1% R-23 (CHF_3) and 59.9% of R-13 ($CClF_3$). This is a non-flammable, non-corrosive, practically non-toxic liquid. Its boiling temperature at atmospheric pressure is – 88°C which is lower than either R-23 or R-13. Its evaporating pressure at – 15°C is 17.15 bar. Its critical temperature is 20°C and its critical pressure is 41.15 bar. This is a low temperature refrigerant and good for use in the low state of cascade systems which require temperatures in the range of – 73°C to – 87°C. The latent heat of vaporisation at atmospheric pressure is 173 kJ/kg.

The leaks in R-503 systems may be detected with the use of soap solution, a halide torch or an electronic leak detector. This refrigerant will hold more moisture than some other low temperature refrigerants. It may be noted that all low temperature applications must have extreme dryness, because any moisture not in solution with refrigerant is likely to form ice at the refrigerant control devices. The oil does not circulate well at low temperatures. The cascade and other low temperature equipments are normally fitted with oil separators and other devices for returning the oil to the compressor. The cylinder colour code for R-503 is aquamarine.

4. *R-504.* The R-504 is an azeotropic mixture of 48.2% R-32 (CH_2F_2) and 51.8 % R-115 ($CClF_2CF_3$). It is non-flammable, non-corrosive and non-toxic. The boiling temperature at atmospheric pressure is – 57°C. Its evaporating pressure at – 15°C is 5.88 bar and its critical pressure is 48 bar. As with all low temperature refrigerants, some difficulty may be experienced with the oil circulation. With the addition of 2 to 5% R-170 (ethane), the oil will be taken into the solution with the refrigerant and will circulate through the system with it.

The leaks in R-504 systems may be easily detected by using soap solution, a halide torch or an electronic leak detector. This refrigerant is used in industrial processes where a low temperature range of – 40°C to – 62°C is desired. The cylinder colour code for R-504 is tan.

8.6 Inorganic Refrigerants

The inorganic refrigerants were exclusively used before the introduction of halo-carbon refrigerants. These refrigerants are still in use due to their inherent thermodynamic and physical properties. The various inorganic refrigerants are given in the following table :

Table 8.3. Inorganic refrigerants.

Refrigerant number	*Chemical name*	*Chemical formula*
R-717	Ammonia	NH_3
R-729	Air	—
R-744	Carbon dioxide	CO_2
R-764	Sulphur dioxide	SO_2
R-118	Water	H_2O

These refrigerants are discussed, in detail, as below :

1. *R-717 (Ammonia).* The R-717, *i.e.* ammonia (NH_3), is one of the oldest and most widely used of all the refrigerants. Its greatest application is found in large and commercial reciprocating compression systems where high toxicity is secondary. It is also widely used in absorption systems. It is a chemical compound of nitrogen and hydrogen and under ordinary conditions, it is a colourless gas. Its boiling point at atmospheric pressure is – 33.3°C and its melting point from the solid is – 78°C. The low boiling point makes it possible to have refrigeration at temperatures considerably below 0°C without using pressures below atmospheric in the evaporator. Its latent heat of vaporisation at – 15°C is 1315 kJ/kg. Thus, large refrigerating effects are possible with relatively small sized machinery. The condenser pressure at 30°C is 10.78 bar. The condensers for R-717 are usually of water cooled type.

It is a poisonous gas if inhaled in large quantities. In lesser quantities, it is irritating to the eyes, nose and throat. This refrigerant is somewhat flammable and, when mixed with air in the ratio of 16% to 25% of gas by volume, will form an explosive mixture. The leaks of this refrigerant may be quickly and easily detected by the use of burning sulphur candle which in the presence of ammonia forms white fumes of ammonium sulphite. This refrigerant attacks copper and bronze in the presence of a little moisture but does not corrode iron or steel. It presents no special problems in connection with lubricant unless extreme temperatures are encountered. Since the refrigerant R-717 is lighter than oil, therefore, its separation does not create any problem. The excess oil in the evaporator may be removed by opening a valve in the bottom of the evaporator. This refrigerant is used in large compression

Ultra-low temp. refrigeration system using CO_2 and NH_3 for industrial applications.

machines using reciprocating compressors and in many absorption type systems. The use of this refrigerant is extensively found in cold storage, warehouse plants, ice cream manufacture, ice manufacture, beer manufacture, food freezing plants etc.

2. *R-729 (Air).* The dry air is used as a gaseous refrigerant in some compression systems, particularly in aircraft air-conditioning.

3. *R-744 (Carbon dioxide).* The principal refrigeration use of carbon dioxide is same as that of dry ice. It is non-toxic, non- irritating and non-flammable. The boiling point of this refrigerant is so extremely low (– 73.6°C) that at – 15°C, a pressure of well over 20.7 bar is required to prevent its evaporation. At a condenser temperature of +30°C, a pressure of approximately 70 bar is required to liquify the gas. Its critical temperature is 31°C and triple point is – 56.6°C. Due to its high operating pressure, the compressor of a carbon dioxide refrigerator unit is very small even for a comparatively large refrigerating capacity. However, because of its low efficiency as compared to other common refrigerants, it is seldom used in household units, but is used in some industrial applications and aboard ships.

4. *R-764 (Sulphur dioxide).* This refrigerant is produced by the combustion of sulphur in air. In the former years, it was widely used in household and small commercial units. The boiling point of sulphur dioxide is – 10°C at atmospheric pressure. The condensing pressure varies between 4.1 bar and 6.2 bar under normal operating conditions. The latent heat of sulphur dioxide at – 15°C is 396 kJ/kg. It is a very stable refrigerant with a high critical temperature and it is non-flammable and non-explosive. It has a very unpleasant and irritating odour. This refrigerant is not injurious to food and is used commercially as a ripener and preservative of foods. It is, however, extremely injurious to flowers, plants and shrubbery. The sulphur dioxide in its pure state is not corrosive, but when there is moisture present, the mixture forms sulphurous acid which is corrosive to steel. Thus it is very important that the moisture in the refrigerating system be held to a minimum.

The sulphur dioxide does not mix readily with oil. Therefore, an oil lighter than that used with other refrigerants may be used in the compressors. The refrigerant in the evaporator with oil floating on the top has a tendency to have a higher boiling point than that corresponding to its pressure. The modern evaporators overcome this by having the liquid introduced in such a way that the refrigerant is kept agitated while the unit is in operation. The leaks in the system with sulphur dioxide may be easily detected by means of soap solution or ammonia swab. A dense white smoke forms when sulphur dioxide and ammonia fumes come in contact.

5. *R-118 (Water).* The principal refrigeration use of water is as ice. The high freezing temperature of water limits its use in vapour compression systems. It is used as the refrigerant vapour in some absorption systems and in systems with steam jet compressors.

8.7 Hydro-carbon Refrigerants

Most of the hydro-carbon refrigerants are successfully used in industrial and commercial installations. They possess satisfactory thermodynamic properties but are highly flammable and explosive. The various hydro-carbon refrigerants are given in the following table :

Table 8.4. Hydro-carbon refrigerants.

Refrigerant number	*Chemical name*	*Chemical formula*
R-170	Ethane	C_2H_6
R-290	Propane	C_3H_3
R-600	Butane	C_4H_{10}
R-600*a*	Isobutane	C_4H_{10}
R-1120	Trichloroethylene	$C_2H_4Cl_3$
R-1130	Dichloroethylene	$C_2H_4Cl_2$
R-1150	Ethylene	C_2H_4
R-1270	Propylene	C_3H_6

Since the hydro-carbon refrigerants are not commonly used now-a-days, therefore, they are not discussed in detail.

Manufacture of Hydro-carbon refrigerants.

8.8 Designation System for Refrigerants

The refrigerants are internationally designated as 'R' followed by certain numbers such as R-11, R-12, R-114 etc. A refrigerant followed by a two-digit number indicates that a refrigerant is derived from methane base while three-digit number respresents ethane base. The numbers assigned to hydro-carbon and halo-carbon refrigerants have a special meaning. The first digit on the right is the number of fluorine (F) atoms in the refrigerant. The second digit from the right is one more than the number of hydrogen (H) atoms present. The third digit from the right is one less than the number of carbon (C) atoms, but when this digit is zero, it is omitted. The general chemical formula for the refrigerant, either for methane or ethane base, is given as $C_mH_nCl_pF_q$, in which $n + p + q = 2m + 2$

where m = Number of carbon atoms,

n = Number of hydrogen atoms,

p = Number of chlorine atoms, and

q = Number of fluorine atoms.

As discussed above, the number of the refrigerant is given by $R\ (m-1)(n+1)\ (q)$. Let us consider the following refrigerants to find its chemical formula and the number.

1. Dichloro-difluoro-methane

We see that in this refrigerant

Number of chlorine atoms, $p = 2$

Number of fluorine atoms, $q = 2$

and number of hydrogen atoms, $n = 0$

We know that $n + p + q = 2m + 2$

$0 + 2 + 2 = 2m + 2$ or $m = 1$

i.e. Number of carbon atoms = 1

Thus the chemical formula for dichloro-difluoro-methane becomes CCl_2F_2 and the number of refrigerant becomes R (1–1) (0+1)(2) or R-012 *i.e.* R-12.

2. Dichloro-tetrafluoro-ethane

We see that in this refrigerant

Number of chlorine atoms, $p = 2$

Number of fluorine atoms, $q = 4$

and number of hydrogen atoms, $n = 0$

We know that $n + p + q = 2m + 2$

$0 + 2 + 4 = 2m + 2$ or $m = 2$

i.e. Number of carbon atoms = 2

Thus the chemical formula for dichloro-tetrafluoro-ethane becomes $C_2Cl_2F_4$ and the number of refrigerant becomes *R* (2-1) (0+1) (4) or R-114.

3. Dichloro-trifluoro-ethane

We see that in this refrigerant

Number of chlorine atoms, $p = 2$

Number of fluorine atoms, $q = 3$

and number of hydrogen atoms, $n = 1$

We know that $n + p + q = 2m + 2$

$1 + 2 + 3 = 2m + 2$ or $m = 2$

i.e. Number of carbon atoms = 2

Thus the chemical formula for dichloro-trifluoro-ethane becomes $CHCl_2CF_3$ and the number of refrigerant becomes R(2 – 1) (1 + 1) (3) or R-123.

The inorganic refrigerants are designated by adding 700 to the molecular mass of the compound. For example, the molecular mass of ammonia is 17, therefore it is designated by R - (700 + 17) or R-717.

8.9 Substitutes for Chloro-fluoro-Carbon (CFC) Refrigerants

The most commonly used halo-carbon or organic refrigerants are the chloro-fluoro derivatives of methane (CH_4) and ethane (C_2H_6). The fully halogenated refrigerants with chlorine (Cl) atom in their molecules are referred to as chloro-fluoro-carbon (CFC) refrigerants. The refrigerants such as R-11, R-12, R-13, R-113, R-114 and R-115 are CFC refrigerants.

The refrigerants which contain hydrogen (H) atoms in their molecule along with chlorine (Cl) and fluorine (F) atoms are referred to as hydro-chloro-fluoro-carbon (HCFC) refrigerants. The refrigerants such as R-22, R-123 are HCFC refrigerants.

The refrigerants which contain no chlorine atom in their molecules are referred to as hydro-fluoro carbon (HFC) refrigerants. The refrigerants such as R-134a, R-152a are HFC refrigerants.

The refrigerants which contain no chlorine and fluorine atoms in their molecule are referred to as hydrocarbon (HC) refrigerants. The refrigerants such as R-290, R-600a are HC refrigerants.

It may be noted that the fluorine (F) atom in the molecule of the refrigerants makes them physiologically more favourable. The chlorine (Cl) atom in the molecule of the refrigerants is considered to be responsible for the depletion of ozone layer in the upper atmosphere which allows harmful ultra-violet rays from the sun to penetrate through the atmosphere and reach the earth's surface causing skin cancer. The chloro-fluoro-carbon (CFC) refrigerants have been linked to the depletion of this ozone layer. They have varying degree of *ozone depletion potential (ODP).

In addition to the ozone depletion effect on the environment, the halo-carbon refrigerants have a **global warming effect, which may cause serious changes in the environment.

According to an international agreement ***(Montreal Protocol, 1987), the use of halogenated CFC refrigerants that are considered to have high ODP (such as commonly used refrigerants R-11, R-12, R-113, R-114 and R-502) have been **phased out**. The refrigerant R-22 which is a hydro-chloro-fluoro carbon (HCFC) refrigerant is not covered under the original Montreal Protocol as its ODP is one-twentieth of R-11 and R-12. But because of its GWP, it has to be phased out. Nevertheless, R-22 is found to be of greater use these days as it is being employed not only on its existing R-22 applications but also as a substitute for R-11 in very large capacity air-conditioning applications with screw or centrifugal compressors.

The hydrocarbon (HC) and hydro-fluoro carbon (HFC) refrigerants provide an alternative to fully halogenated CFC refrigerants. Since they contain no chlorine atom at all, therefore they have zero ODP. Even hydro-chloro-fluoro carbon (HCFC) refrigerants which contain some chlorine (Cl) atoms, but in association with hydrogen (H) atoms, have much reduced ODP. However, the hydro-fluoro carbons (HFCs), because of their hydrogen (H) content, may be slightly flammable. The degree of flammability depends upon the number of H-atoms in the molecule. The pure hydrocarbons (HCs) are, of course, highly flammable.

At present, the following substitutes are available :

1. The HCFC refrigerant R-123 (CF_3CHCl_2) in place of R-11 (CCl_3F).
2. The HFC refrigerant R-134a (CF_3CH_2F) and R-152a (CH_3 CHF_2) in place of R-12.
3. The HFC refrigerant R-143a (CH_3CF_3) and R-125 (CHF_2CF_3) in place of R-502 (a mixture of R-22 and R-115).
4. The HC refrigerants, propane *i.e.* R-290 (C_3H_8) and isobutane *i.e.* R-600a (C_4H_{10}) may also be used in place of R-12.

8.10 Comparison of Refrigerants

There is no such refrigerant (*i.e.* ideal refrigerant) which can be used under all operating conditions. The characteristics of some refrigerants make them suitable for use with reciprocating compressors and other refrigerants are best suited to centrifugal or rotary compressors. Therefore in order to select a correct refrigerant, it is necessary that it should satisfy those properties which make it ideal to be used for the particular application. We shall now discuss the thermodynamic, chemical and physical properties of some important refrigerants.

* It may be noted that one chlorine atom can destroy 10^5 ozone molecules. The relative ability of a substance to deplete the ozone layer is called **ozone depletion potential** (ODP). The CFC refrigerants such as R-11 and R-12 have the highest (worst value) ODP = 1. The HCFC refrigerants have a relatively low ODP, *i.e.* R-22 has ODP = 0.05 and R-123 has ODP = 0.02. The HFC refrigerants do not cause any ozone depletion, *i.e.* R-134a has zero ODP.

** Global warming means the increase in average temperature of the earth. The causes of global warming are increase in CO_2 concentration, NO_2 emission and the use of CFC refrigerants. The ability of a substance to contribute to global warming is measured by the global warming potential (GWP). Some halocarbon refrigerants have a very high GWP. For example, GWP for R-22 = 100 and GWP for CO_2 = 1. Due to this reason, there is a concern about these refrigerants.

*** India became a party to Montreal Protocol in 1992.

8.11 Thermodynamic Properties of Refrigerants

The thermodynamic properties of refrigerants are discussed, in detail, as follows :

1. *Boiling temperature.* The boiling temperature of the refrigerant at atmospheric pressure should be low. If the boiling temperature of the refrigerant is high at atmospheric pressure, the compressor should be operated at high vacuum. The high boiling temperature reduces the capacity and operating cost of the system. The following table shows the boiling temperatures at atmospheric pressure of some commonly used refrigerants.

Table 8.5. Boiling temperatures.

Refrigerant	*Boiling temperature (° C) at atmospheric pressure*
R-11	+ 23.77
R-12	– 29
R-21	+ 9
R-22	– 41
R-30	+ 39.8
R-40	– 23.7
R-113	+ 47.6
R-114	+ 3.6
R-123	+ 27.85
R-134*a*	– 26.15
R-717	– 33.3
R-744	– 73.6
R-764	– 10

2. *Freezing temperature.* The freezing temperature of a refrigerant should be well below the operating evaporator temperature. Since the freezing temperature of most of the refrigerants are below – 35°C, therefore this property is taken into consideration only in low temperature operation. The following table shows the freezing temperatures of some common refrigerants.

Table 8.6. Freezing temperatures.

Refrigerant	*Freezing temperature (° C)*
R-11	– 111
R-12	– 157.5
R-21	– 135
R-22	– 160
R-30	– 96.6
R-40	– 97.5
R-113	– 35
R-114	– 94
R-134*a*	– 101
R-717	– 77.8
R-744	– 56.7
R-764	– 75.6

3. *Evaporator and condenser pressure.* Both the evaporating (low side) and condensing (high side) pressures should be positive (*i.e.* above atmospheric) and it should be as near to the atmospheric pressure as possible. The positive pressures are necessary in order to prevent leakage of air and moisture into the refrigerating system. It also permits easier detection of leaks. Too high evaporating and condensing pressures (above atmospheric) would require stronger refrigerating equipment (*i.e.* compressor, evaporator and condenser) resulting in higher initial cost. The following table shows the evaporating and condensing pressures, and compression ratio for various refrigerants when operating on the standard cycle of – 15°C evaporator temperature and + 30°C condenser temperature.

Table 8.7. Evaporator and condenser pressures.

Refrigerant	*Evaporator pressure (p_E) at – 15° C, in bar*	*Condenser pressure (p_C) at + 30° C, in bar*	*Compression ratio (p_C / p_E)*
R-11	0.2021	1.2607	6.24
R-12	1.8262	7.4510	4.08
R-21	0.3618	2.1540	5.95
R-22	2.9670	12.0340	4.05
R-30	0.0807	0.7310	9.06
R-40	1.4586	6.5310	4.47
R-113	0.0676	0.5421	8.02
R-114	0.4650	2.5161	5.41
R-123	0.0163	0.1095	4.69
R-134*a*	1.6397	7.7008	6.72
R-717	2.3634	11.67	4.94
R-744	22.90	71.930	3.14
R-764	0.8145	4.5830	5.63

The reciprocating compressors are used with refrigerants having low specific volumes, high operating pressures and high pressure ratios. The centrifugal compressors are used with refrigerants having high specific volumes, low operating pressures and low pressure ratios.

4. *Critical temperature and pressure.* The critical temperature of a refrigerant is the highest temperature at which it can be condensed to a liquid, regardless of a higher pressure. It should be above the highest condensing temperature that might be encountered. If the critical temperature of a refrigerant is too near the desired condensing temperature, the excessive power consumption results. The following table shows the critical temperature and pressures for the commonly used refrigerants. The critical temperature for most of the common refrigerants is well above the normal condensing temperature with the exception of carbon dioxide (R-744) whose critical temperature is 31°C.

Table 8.8. Critical temperature and pressures.

Refrigerant	*Critical temperature (° C)*	*Critical pressure (bar)*
R-11	198	43.8
R-12	112	41.2
R-21	178.5	51.65
R-22	96	49.38
R-30	216	44.14
R-40	143	66.83
R-113	214	34.14
R-114	145.8	32.61
R-123	183.8	36.74
R-134*a*	101.06	40.56
R-717	133	113.86
R-744	31	73.8
R-764	157	78.7

5. *Coefficient of performance and power requirements.* For an ideal refrigerant operating between – 15°C evaporator temperature and 30°C condenser temperature, the theoretical coefficient of performance for the reversed Carnot cycle is 5.74. The following table shows the values of theoretical coefficient of performance and power per tonne of refrigeration for some common refrigerants operating between – 15°C evaporator temperature and 30°C condenser temperature.

Table 8.9. Coefficient of performance and power per TR.

Refrigerant	*Coefficient of performance*	*kW / TR*
R-11	5.09	0.694
R-12	4.70	0.746
R-22	4.66	0.753
R-30	4.90	0.716
R-40	4.90	0.716
R-113	4.92	0.716
R-114	4.54	0.792
R-123	4.93	–
R-134*a*	4.61	0.762
R-717	4.76	0.738
R-729	5.74	0.619
R-744	2.56	1.372
R-764	4.87	0.724

From the above table, we see that R-11 has the coefficient of performance equal to 5.09 which is closest to the Carnot value of 5.74. The other refrigerants have also quite high values of coefficient of performance except R-744 (carbon dioxide) which has the value of coefficient of performance as 2.56 with a power requirement of 1.372 kW per tonne of refrigeration. This is due to its low critical point (31°C) and the condensing temperature is very close to it which is 30°C. Practically, all common refrigerants have approximately the same coefficient of performance and power requirement.

6. *Latent heat of vaporisation.* A refrigerant should have a high latent heat of vaporisation at the evaporator temperature. The high latent heat results in high refrigerating effect per kg of refrigerant circulated which reduces the mass of refrigerant to be circulated per tonne of refrigeration. Table 8.10 shows the refrigerating effect for the common refrigerants operating between – 15°C evaporator temperature and 30°C condenser temperature. It also shows the latent heat, mass of refrigerant circulated per tonne of refrigeration and the volume of the liquid refrigerant per tonne of refrigeration.

7. *Specific volume.* The specific volume of the refrigerant vapour at evaporator temperature (*i.e.* volume of suction vapour to the compressor) indicates the theoretical displacement of the compressor. The reciprocating compressors are used with refrigerants having high pressures and low volumes of the suction vapour. The centrifugal or turbo compressors are used with refrigerants having low pressures and high volumes of the suction vapour. The rotary compressors are used with refrigerants having intermediate pressures and volumes of the suction vapour. Table 8.11 shows the specific volume of the refrigerant vapour and theoretical piston displacements for various refrigerants.

Table 8.10. Refrigerating effect, latent heat of vaporisation, mass of refrigerant and volume of liquid refrigerant.

Refrigerant	*Refrigerating effect for standard cycle of – 15° C to + 30° C, in kJ/kg*	*Latent heat of vaporisation at – 15° C, in kJ/kg*	*Mass of refrigerant circulated per standard tonne, in kg/min*	*Volume of liquid refrigerant circulating per standard tonne at 30° C in litres per minute*
R-11	157.3	195.7	1.34	0.918
R-12	116.5	159.0	1.81	1.404
R-22	161.5	218.1	1.31	1.115
R-30	313.6	377.7	0.676	0.507
R-40	350.0	421.0	0.603	0.67
R-113	125.1	164.5	1.7	1.1
R-114	95.6	138.5	2.21	1.54
R-123	142.4	183.4	1.48	1.02
R-134*a*	148.1	209.5	1.42	1.19
R-717	1105.4	1316.5	0.19	0.32
R-744	129.3	274.0	1.63	2.73
R-764	329.5	394.7	0.64	0.474

Table 8.11. Specific volume and theoretical piston displacements.

Refrigerant	*Specific volume of the refrigerant vapour at – 15° C (m^3/kg)*	*Vapour displacement for standard tonne of refrigeration (m^3/min)*
R-11	0.77	1.016
R-12	0.091	0.163
R-22	0.078	0.101
R-30	3.12	2.08
R-40	0.28	0.167
R-113	1.7	2.08
R-114	0.263	0.58
R-123	0.902	1.33
R-134*a*	0.121	0.172
R-717	0.51	0.096
R-744	0.0167	0.027
R-764	0.401	0.254

8.12 Chemical Properties of Refrigerants

The chemical properties of refrigerants are discussed as follows :

1. *Flammability*. We have already discussed that hydro-carbon refrigerants such as ethane, propane etc. are highly flammable. Ammonia is also somewhat flammable and becomes explosive when mixed with air in the ratio of 16 to 25 per cent of gas by volume. The halo-carbon refrigerants are neither flammable nor explosive.

2. *Toxicity*. The toxicity of refrigerant may be of prime or secondary importance, depending upon the application. Some non-toxic refrigerants (*i.e.* all fluorocarbon refrigerants) when mixed with certain percentage of air become toxic.

The following table shows the relative toxicity of the common refrigerants, based upon the concentration and exposure time required to produce serious results.

Table 8.12. Toxicity based upon concentration and exposure time required.

Refrigerant	*Concentration in air to produce serious effects (% by volume)*	*Duration of exposure to produce serious effects (minutes)*	**Underwriters' Laboratories number group*
R-11	10	120	5
R-12	28.5 to 30.4	120	6
R-22	18 to 22.6	120	5A
R-30	5.1 to 5.3	30	4A
R-40	2 to 2.5	120	4
R-113	4.8 to 5.3	60	4
R-717	0.5 to 0.6	30	2
R-744	29 to 30	30 to 60	5A
R-764	0.7	5	1

From the above table, we see that R-717 (ammonia) and R-764 (sulphur dioxide) are highly toxic. These refrigerants are also strong irritants. Therefore these refrigerants are not used in domestic refrigeration and comfort air-conditioning. The use of toxic refrigerants is only limited to cold storages.

3. *Solubility of water.* Water is only slightly soluble in R-12. At – 18°C, it will hold six parts per million by weight. The solution formed is very slightly corrosive to any of the common metals. The solubility of water with R-22 is more than R-12 by a ratio of 3 to 1. If more water is present than can be dissolved by the refrigerant, the ice will be formed which chokes the expansion valve or capillary tube used for throttling in the system. This may be avoided by the proper dehydration of the refrigerating unit before charging and by the use of silica gel drier of the liquid line. Ammonia is highly soluble in water. Due to this reason, a wetted cloth is put at the point of leak to avoid harm to the persons working in ammonia refrigerating plants.

4. *Miscibility.* The ability of a refrigerant to mix with oil is called miscibility. This property of refrigerant is considered to be a secondary factor in the selection of a refrigerant. The degree of miscibility depends upon the temperature of the oil and pressure of the refrigerating vapour. The freon group of refrigerants are highly miscible refrigerants while ammonia, carbon dioxide, sulphur dioxide and methyl chloride are relatively non-miscible.

The non-miscible refrigerants require larger heat transfer surfaces due to poor heat conduction properties of oil. The miscible refrigerants are advantageous from the heat transfer point of view. They give better lubrication as the refrigerant acts as a carrier of oil to the moving parts. The miscible refrigerants also eliminate oil-separation problems and aid in the return of oil from the evaporator.

5. *Effect on perishable materials.* The refrigerants used in cold storage plant and in domestic refrigerators should be such that in case of leakage, it should have no effect on the perishable materials. The freon group of refrigerants have no effect upon dairy products, meats, vegetables, flowers and furs. There will be no change in colour, taste or texture of the material when exposed to freon.

Methyl chloride vapours have no effect upon furs, flowers, eating foods or drinking beverages. Sulphur dioxide destroys flowers, plants and furs, but it does not affect foods **. Ammonia dissolves easily in water and becomes alkaline in nature. Since most fruits and vegetables are acidic in nature, therefore ammonia reacts with these products and spoils the taste.

* The Underwriters' Laboratories, U.S.A., classifies refrigerants into six groups mainly on their degree of toxicity. The group one refrigerants are most toxic while group six refrigerants are the least toxic.

** The sulphur dioxide dissolves easily in water and becomes acidic in nature and most of the foods have acidic nature. So sulphur dioxide does not affect foods.

8.13 Physical Properties of Refrigerants

The physical properties of refrigerants are discussed as follows :

1. Stability and inertness. An ideal refrigerant should not decompose at any temperature normally encountered in the refrigerating system. It should not form higher boiling point liquids or solid substances through polymerization. Some refrigerants disintegrate forming non-condensable gases which causes high condensing pressure and vapour lock. The disintegration of refrigerant may be due to reaction with metals. In order to avoid this, a refrigerant should be inert with respect to all materials used in refrigerating system.

The freon group of refrigerants are stable up to a temperature of 535°C. Above this temperature, it decomposes and forms corrosive and poisonous products. The freon refrigerants are not used with rubber gaskets as it acts as a solvent with rubber. Since sulphur dioxide does not decompose below 1645°C, therefore it is one of the most stable refrigerants.

2. Corrosive property. The corrosive property of a refrigerant must be taken into consideration while selecting the refrigerant. The freon group of refrigerants are non-corrosive with practically all metals. Ammonia is used only with iron or steel. Sulphur dioxide is non-corrosive to all metals in the absence of water because sulphur dioxide reacts with water and forms sulphuric acid.

3. Viscosity. The refrigerant in the liquid and vapour states should have low viscosity. The low viscosity of the refrigerant is desirable because the pressure drops in passing through liquid and suction lines are small. The heat transfer through condenser and evaporator is improved at low viscosities. The following table shows the viscosities (in centipoises) at atmospheric pressure for the common refrigerants.

Table 8.13. Viscosities at atmospheric pressure.

Refrigerant	*Viscosity at liquid temperature*		*Viscosity at vapour temperature*	
	– 15° C	*37.5° C*	*– 15° C*	*60° C*
R-11	0.650	0.380	0.0096	0.0121
R-12	0.328	0.242	0.0114	0.0136
R-22	0.286	0.223	0.0114	0.0143
R-40	0.293	0.226	0.0095	0.0120
R-113	1.200	0.564	0.0093	0.0112
R-717	0.250	0.200	0.0085	0.0116
R-744	0.115	—	0.0132	0.0165
R-764	0.503	0.260	0.0111	0.0144

4. Thermal conductivity. The refrigerant in the liquid and vapour states should have high thermal conductivity. This property is required in finding the heat transfer coefficients in evaporators and condensers. Table 8.14 shows the thermal conductivities of common refrigerants.

5. Dielectric strength. The dielectric strength of a refrigerant is important in hermetically sealed units in which the electric motor is exposed to the refrigerant. The relative dielectric strength of the refrigerant is the ratio of the dielectric strength of nitrogen and the refrigerant vapour mixture to the dielectric strength of nitrogen at atmospheric pressure and room temperature. Table 8.15 shows the relative dielectric strengths of common refrigerants.

Table 8.14. Thermal conductivities.

Refrigerant	*Temperature* (° *C*)	*Thermal conductivity* (*W/mK*)
Liquid		
R-11	40	0.1022
R-12	40	0.0814
R-22	40	0.0970
R-30	30	0.1664
R-40	20	0.1612
R-113	40	0.0971
R-717	– 10 to + 20	0.5026
R-744	20	0.2080
R-764	20	0.3466
Vapour		
R-11	30	8.318×10^{-3}
R-12	30	9.705×10^{-3}
R-22	30	11.784×10^{-3}
R-30	0	6.759×10^{-3}
R-40	0	8.492×10^{-3}
R-113	30	7.798×10^{-3}
R-717	0	22.182×10^{-3}
R-744	0	14.037×10^{-3}
R-764	0	8.665×10^{-3}

Table 8.15. Dielectric strengths.

Refrigerant	*R-11*	*R-12*	*R-22*	*R-30*	*R-40*	*R-113*	*R-717*	*R-744*	*R-764*
Relative dielectric strength	3	2.4	1.31	1.11	1.06	2.6	0.82	0.88	1.9

6. *Leakage tendency.* The leakage tendency of a refrigerant should be low. If there is a leakage of refrigerant, it should be easily detectable. The leakage occurs due to opening in the joints or flaws in material used for construction. Since the fluorocarbon refrigerants are colourless, therefore, their leakage will increase the operating cost. The ammonia leakage is easily detected due to its pungent odour.

The leakage of fluorocarbon refrigerants may be detected by soap solution, a halide torch or an electronic leak detector. The latter is generally used in big refrigerating plants. The ammonia leakage is detected by using burning sulphur candle which in the presence of ammonia forms white fumes of ammonium sulphite.

7. *Cost.* The cost of refrigerant is not so important in small refrigerating units but it is very important in high capacity refrigerating systems like industrial and commercial. The ammonia, being the cheapest, is widely used in large industrial plants such as cold storages and ice plants. The refrigerant R-22 is costlier than refrigerant R-12. The cost of losses due to leakage is also important.

8.14 Secondary Refrigerants - Brines

Brines are secondary refrigerants and are generally used where temperatures are required to be maintained below the freezing point of water *i.e.* 0°C. In case the temperature involved is above the freezing point of water (0°C), then water is commonly used as a secondary refrigerant.

Brine is a solution of salt in water. It may be noted that when salt is mixed in water, then the freezing temperature of the solution becomes lower than that of the water. This is due to the fact that the salt while dissolving in water takes off its latent heat from the solution and cools it below the freezing point of water. The mass of the salt in the solution expressed as the percentage of the mass of the solution is known as *concentration* of the solution. As the concentration of the solution increases, its freezing point decreases. But if the concentration of the salt is increased beyond a certain point, the freezing point increases instead of decreasing. The point at which the freezing temperature is minimum, is known as *eutectic temperature* and the concentration at this point is known as *eutectic concentration.* The brine used in a particular application should have a concentration for which the freezing point of the brine is at least 5°C to 8°C lower than the brine temperature required.

Special secondary refrigerants.

The brines commonly used are calcium chloride ($CaCl_2$), sodium chloride *i.e.* common salt (NaCl) and glycols such as ethylene glycol, propylene glycol etc.

The calcium chloride brine has the eutectic temperature of – 55°C at salt concentration of 30% by mass. This brine is primarily used where temperatures below – 18°C are required. It is generally used in industrial process cooling and product freezing. The chief disadvantages of calcium chloride brine are its dehydrating effect and its tendency to impart a bitter taste to food products.

The sodium chloride brine has the eutectic temperature of – 21.1°C at salt concentration of 23% by mass. This brine is used in chilling and freezing of meat and fish.

Both of the above two brines are corrosive in nature for metallic containers which put limitation on their use. Also the thermal properties of the above two brines are less satisfactory.

Other water soluble compounds known as *antifreeze* are also used for decreasing the freezing point of water for certain refrigeration uses. Ethylene and propylene glycol have a number of good properties. Since they are non-corrosive and non-electrolytic even in the presence of water, therefore, these brines are most extensively used as antifreeze elements. The following table shows typical applications of various brines.

Table 8.16. Typical application of various brines.

Application	*Brine used*
1. Breweries	Propylene glycol
2. Chemical plants	Sodium chloride, Calcium chloride, Ethylene glycol
3. Dairies	Sodium chloride, Calcium chloride, Propylene glycol
4. Food process	Sodium chloride, Calcium chloride, Propylene glycol
5. Ice-creams	Calcium chloride, Propylene glycol
6. Ice-plant	Sodium chloride
7. Meat packing	Sodium chloride, Calcium chloride
8. Skating ring	Calcium chloride, Ethylene glycol
9. Special low temperature	Calcium chloride, Ethylene glycol

QUESTIONS

1. What are the desirable properties of an ideal refrigerant ?
2. Name the different refrigerants generally used.
3. Write the factors considered for the selection of refrigerant for a system.
4. Give the chemical formula and names of the refrigerants R-22 and R-114.
5. Write short notes on R-12 and R-22 as a refrigerant.
6. What type of compressor is preferred with refrigerant R-113 ?
7. What is an azeotrope ? Give some examples to indicate its importance.
8. Give azeotropic mixing refrigerants for the following refrigerants. Mention the chemical formula also.
   (*a*) R-500 (*b*) R-502 (*c*) R-503 (*d*) R-504
9. Describe the refrigerating properties of ammonia for use in domestic and commercial type of refrigerating appliances.
10. Discuss from the economical point of view whether sulphur dioxide or carbon dioxide is preferred as refrigerant.
11. How will you assign number to the refrigerants methyl chloride (CH_3Cl) and tetra-chloroethane ($C_2H_4Cl_4$) ?
12. Differentiate between physical and thermodynamic properties of a refrigerant. Explain which are more important giving specific examples.
13. Name the refrigerants and their operating pressures for the following applications :
    1. Domestic refrigerator, 2. Cold storage-100 TR, 3. Ice cream plant, 4. Room air-conditioner, and 5. 200-room hotel air-conditioning (centrifugal).
14. Compare the refrigerants R-11, R-12, R-22 and ammonia in regard to the following :
    (*a*) Normal boiling point,
    (*b*) Range of refrigeration temperatures, for which used,
    (*c*) Types of compressors used and their special features,
    (*d*) Maximum capacity of plants using these refrigerants, and
    (*e*) Types of heating, refrigerating and air-conditioning application.
15. Discuss, in detail, the secondary refrigerants.

OBJECTIVE TYPE QUESTIONS

1. The freon group of refrigerants are
   (*a*) halo-carbon refrigerants (*b*) azeotrope refrigerants
   (*c*) inorganic refrigerants (*d*) hydro-carbon refrigerants
2. Which of the following refrigerants has the lowest freezing point ?
   (*a*) R-11 (*b*) R-12 (*c*) R-22 (*d*) ammonia
3. A refrigerant with the highest critical pressure is
   (*a*) R-11 (*b*) R-12 (*c*) R-22 (*d*) ammonia
4. Which of the following is an azeotrope refrigerant ?
   (*a*) R-11 (*b*) R-40 (*c*) R-114 (*d*) R-502
5. The colour of the flame of halide torch, in case of leakage of freon refrigerant, will change to
   (*a*) bright green (*b*) yellow (*c*) red (*d*) orange
6. The boiling point of ammonia is
   (*a*) – 10.5° C (*b*) – 30° C (*c*) – 33.3° C (*d*) – 77.7° C
7. The refrigerant R-717 is
   (*a*) air (*b*) water (*c*) ammonia (*d*) carbon dioxide

8. The function of a halide torch is
(*a*) defrosting of the cooling coil
(*b*) superheating the vapour refrigerant
(*c*) detecting leakage of the refrigerant
(*d*) facilitating better lubrication in the refrigerator

9. The boiling point of carbon dioxide is
(*a*) – 20.5° C (*b*) – 50° C (*c*) – 73.6° C (*d*) – 78.3° C

10. The freezing point of R-12 is
(*a*) – 86.6° C (*b*) – 95.2° C (*c*) – 107.7° C (*d*) – 157.5° C

11. Which of the following refrigerants has the lowest boiling point ?
(*a*) ammonia (*b*) carbon dioxide (*c*) sulphur dioxide (*d*) R-12

12. Which of the following refrigerants has the highest freezing point ?
(*a*) ammonia (*b*) carbon dioxide (*c*) sulphur dioxide (*d*) R-12

13. Which of the following refrigerants is highly toxic and flammable ?
(*a*) ammonia (*b*) carbon dioxide (*c*) sulphur dioxide (*d*) R-12

14. The refrigerant widely used in domestic refrigerators is
(*a*) ammonia (*b*) carbon dioxide (*c*) sulphur dioxide (*d*) R-12

15. In larger industrial and commercial reciprocating compression systems, the refrigerant widely used is
(*a*) ammonia (*b*) carbon dioxide (*c*) sulphur dioxide (*d*) R-12

16. The ozone-friendly refrigerant R-134a contains
(*a*) one chlorine atom (*b*) two chlorine atoms
(*c*) four chlorine atoms (*d*) no chlorine atom

17. Which of the following method is not used for leakage detection of CFC refrigerants ?
(*a*) burning candle (*b*) soap solution
(*c*) halide torch (*d*) electronic leak detection device

18. For large commercial installations, ammonia is used as the refrigerant, because
(*a*) it has large latent heat (*b*) it is relatively cheap
(*c*) it has moderate working pressure (*d*) all of these

19. R-12 is generally preferred over R-22 in deep freezers since
(*a*) it has lower operating pressures
(*b*) it gives higher coefficient of performance
(*c*) it is miscible with oil over large range of temperatures
(*d*) it is immisicible with oil over large range of temperatures

20. Environmental protection agencies advise against the use of chloro-fluoro-carbon (CFC) refrigerants since
(*a*) these react with water vapour and cause acid rain
(*b*) these react with plants and cause greenhouse effect
(*c*) these react with oxygen and cause its depletion
(*d*) these react with ozone layer

ANSWERS

1. (*a*)	**2.** (*c*)	**3.** (*d*)	**4.** (*d*)	**5.** (*a*)
6. (*c*)	**7.** (*c*)	**8.** (*c*)	**9.** (*c*)	**10.** (*d*)
11. (*b*)	**12.** (*b*)	**13.** (*a*)	**14.** (*d*)	**15.** (*a*)
16. (*d*)	**17.** (*a*)	**18.** (*d*)	**19.** (*c*)	**20.** (*d*)

CHAPTER

Refrigerant Compressors

9

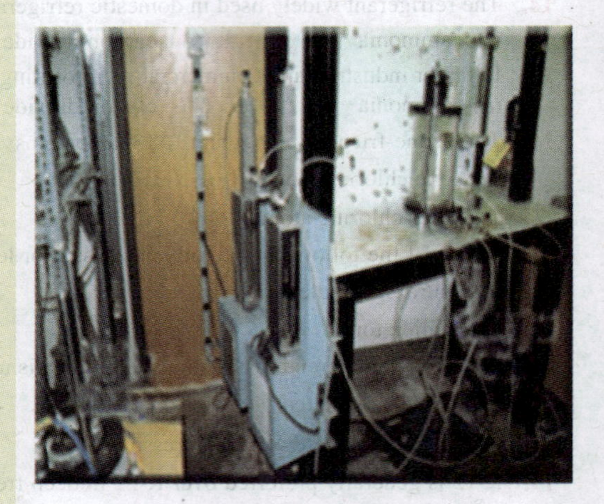

9.1 Introduction

A refrigerant compressor, as the name indicates, is a machine used to compress the vapour refrigerant from the evaporator and to raise its pressure so that the corresponding saturation temperature is higher than that of the cooling medium. It also continually circulates the refrigerant through the refrigerating system. Since the compression of refrigerant requires some work to be done on it, therefore, a compressor must be driven by some prime mover.

Note : Since the compressor virtually takes the heat at a low temperature from the evaporator and pumps it at the high temperature to the condenser, therefore it is often referred to as a heat pump.

9.2 Classification of Compressors

The compressors may be classified in many ways, but the following are important from the subject point of view :

1. *According to the method of compression*

 (*a*) Reciprocating compressors,

 (*b*) Rotary compressors, and

 (*c*) Centrifugal compressors.

2. *According to the number of working strokes*

 (*a*) Single acting compressors, and

 (*b*) Double acting compressors.

3. *According to the number of stages*

 (*a*) Single-stage (or single-cylinder) compressors, and

 (*b*) Multi-stage (or multi-cylinder) compressors.

4. *According to the method of drive employed*

 (*a*) Direct drive compressors, and

 (*b*) Belt drive compressors.

5. *According to the location of the prime mover*

 (*a*) Semi-hermetic compressors (direct drive, motor and compressor in separate housings), and

 (*b*) Hermetic compressors (direct drive, motor and compressor in same housings).

9.3 Important Terms

The following important terms, which will be frequently used in this chapter, should be clearly understood at this stage :

1. ***Suction pressure.*** It is the absolute pressure of refrigerant at the inlet of a compressor.

2. ***Discharge pressure.*** It is the absolute pressure of refrigerant at the outlet of a compressor.

3. ***Compression ratio (or pressure ratio).*** It is the ratio of absolute discharge pressure to the absolute suction pressure. Since the absolute discharge pressure is always more than the absolute suction pressure, therefore, the value of compression ratio is more than unity.

Note : The compression ratio may also be defined as the ratio of total cylinder volume to the clearance volume.

4. ***Suction volume.*** It is the volume of refrigerant sucked by the compressor during its suction stroke. It is usually denoted by v_s.

5. ***Piston displacement volume or stroke volume or swept volume.*** It is the volume swept by the piston when it moves from its top or inner dead position to bottom or outer dead centre position. Mathematically, piston displacement volume or stroke volume or swept volume,

$$v_p = \frac{\pi}{4} \times D^2 \times L$$

where D = Diameter of cylinder, and

L = Length of piston stroke.

6. ***Clearance factor.*** It is the ratio of clearance volume (v_c) to the piston displacement volume (v_p) . Mathematically, clearance factor,

$$C = \frac{v_c}{v_p}$$

7. *Compressor capacity*. It is the volume of the actual amount of refrigerant passing through the compressor in a unit time. It is equal to the suction volume (v_s). It is expressed in m^3/s.

8. *Volumetric efficiency*. It is the ratio of the compressor capacity or the suction volume (v_s) to the piston displacement volume (v_p). Mathematically, volumetric efficiency,

$$\eta_v = \frac{v_s}{v_p}$$

Note : A good compressor has a volumetric efficiency of 70 to 80 per cent.

9.4 Reciprocating Compressors

The compressors in which the vapour refrigerant is compressed by the reciprocating (*i.e.* back and forth) motion of the piston are called *reciprocating compressors*. These compressors are used for refrigerants which have comparatively low volume per kg and a large differential pressure, such as ammonia (R-717), R-12, R-22, and methyl chloride (R-40). The reciprocating compressors are available in sizes as small as 1/12 kW which are used in small domestic refrigerators and up to about 150 kW for large capacity installations.

The two types of reciprocating compressors in general use are single acting vertical compressors and double acting horizontal compressors. The single acting compressors usually have their cylinders arranged vertically, radially or in a *V* or *W* form. The double acting compressors usually have their cylinders arranged horizontally.

Fig. 9.1 shows a single stage, single acting reciprocating compressor in its simplest form. The principle of operation of the compression cycle is as discussed below :

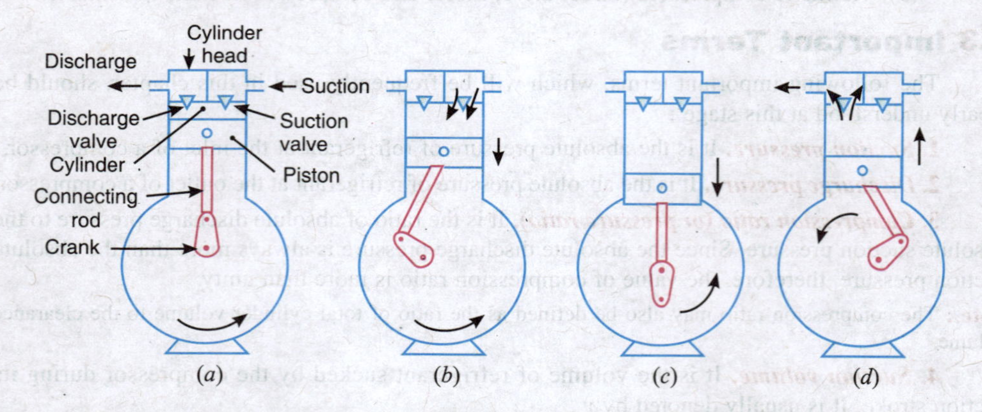

Fig. 9.1. Principle of operation of a single stage, single acting reciprocating compressor.

Let us consider that the piston is at the top of its stroke as shown in Fig. 9.1 (*a*). This is called top dead centre position of the piston. In this position, the suction valve is held closed because of the pressure in the clearance space between the top of the piston and the cylinder head. The discharge valve is also held closed because of the cylinder head pressure acting on the top of it.

When the piston moves downward (*i.e.* during suction stroke), as shown in Fig. 9.1 (*b*), the refrigerant left in the clearance space expands. Thus the volume of the cylinder (above the piston) increases and the pressure inside the cylinder decreases. When the pressure becomes slightly less than the suction pressure or atmospheric pressure, the suction valve gets opened and the vapour refrigerant flows into the cylinder. This flow continues until the piston reaches the bottom of its stroke (*i.e.* bottom dead centre). At the bottom of the stroke, as shown in Fig. 9.1 (*c*), the suction valve closes because of spring action. Now when the piston moves upward (*i.e.* during compression stroke), as shown in Fig. 9.1 (*d*), the volume of the cylinder decreases and the pressure inside the cylinder increases. When the pressure inside the cylinder becomes greater than that on the top of the discharge valve, the discharge valve gets opened and the vapour refrigerant is discharged into the condenser and the cycle is repeated.

It may be noted that in a single acting reciprocating compressor, the suction, compression and discharge of refrigerant takes place in two strokes of the piston or in one revolution of the crankshaft.

Notes : 1. In a double acting reciprocating compressor, the suction and compression takes place on both sides of the piston. It is thus obvious that such a compressor will supply double the volume of refrigerant than a single acting reciprocating compressor (neglecting volume of piston rod).

2. There must be a certain distance between the top of the piston and the cylinder head when the piston is on the top dead centre so that the piston does not strike the cylinder head. This distance is called *clearance space* and the volume therein is called the *clearance volume.* The refrigerant left in this space is at discharge pressure and its pressure must be reduced below that of suction pressure (atmospheric pressure) before any vapour refrigerant flows into the cylinder. The clearance space should be a minimum.

3. The low capacity compressors are air cooled. The cylinders of these compressors usually have fins to provide better air cooling. The high capacity compressors are cooled by providing water jackets around the cylinder.

9.5 Work Done by a Single-Stage Reciprocating Compressor

We have already discussed that in a reciprocating compressor, the vapour refrigerant is first sucked, compressed and then discharged. So there are three different operations of the compressor. Thus we see that work is done on the piston during the suction of refrigerant. Similarly, work is done by the piston during compression as well as discharge of the refrigerant. It may be noted that the work done by a reciprocating compressor is mathematically equal to the work done by the compressor during compression as well as discharge minus the work done on the compressor during suction.

Here we shall discuss the following two important cases of work done :

1. When there is no clearance volume in the cylinder, and
2. When there is some clearance volume.

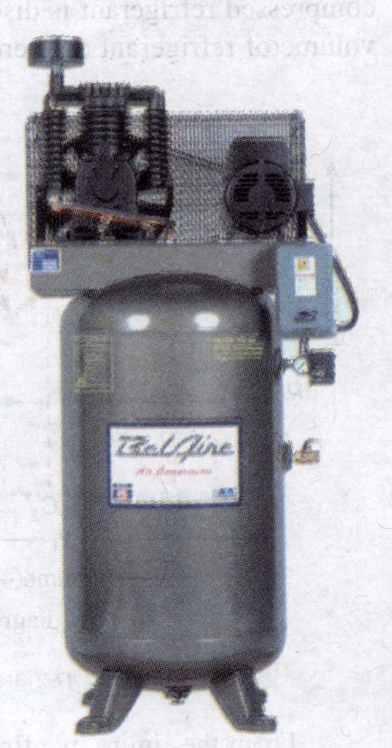

Reciprocating compressor.

9.6 Work Done by a Single-Stage, Single Acting Reciprocating Compressor without Clearance Volume

Consider a single-stage, single acting reciprocating compressor without clearance as shown by the *p-v* and *T-s* diagrams in Fig. 9.2.

Let p_1 = Suction pressure of the refrigerant (before compression),

v_1 = Suction volume of the refrigerant (before compression),

T_1 = Suction temperature of the refrigerant (before compression),

p_2, v_2 and T_2 = Corresponding values for the refrigerant at the discharge point *i.e.* after compression, and

r = Compression ratio or pressure ratio, p_2 / p_1.

As a matter of fact, the compression of refrigerant may be isothermal, polytropic, or isentropic (reversible adiabatic). Now we shall find out the amount of work done in compressing the refrigerant in all the abovementioned three cases.

1. *Work done during isothermal compression*

We have already discussed that when the piston moves from the top dead centre (point A), the refrigerant is admitted into the compressor cylinder and it continues till the piston reaches at its bottom dead centre (point B). Thus the line AB represents suction stroke and the area below this line (*i.e.* area $AB\,B'\,A'$) represents the work done during suction stroke. From the figure, we find that work done during suction stroke,

$$W_1 = \text{Area } A\,B\,B'\,A' = p_1 v_1$$

The refrigerant is compressed during the return stroke (or compression stroke BC_1) of the piston at constant temperature. The compression continues till the pressure (p_2) in the cylinder is sufficient to force open the discharge valve at C_1. After that no compression takes place with the inward movement of the piston. Now during the remaining part of the compression stroke, the compressed refrigerant is discharged to condenser till the piston reaches its top dead centre. The volume of refrigerant delivered (v_2) is represented by the line C_1D.

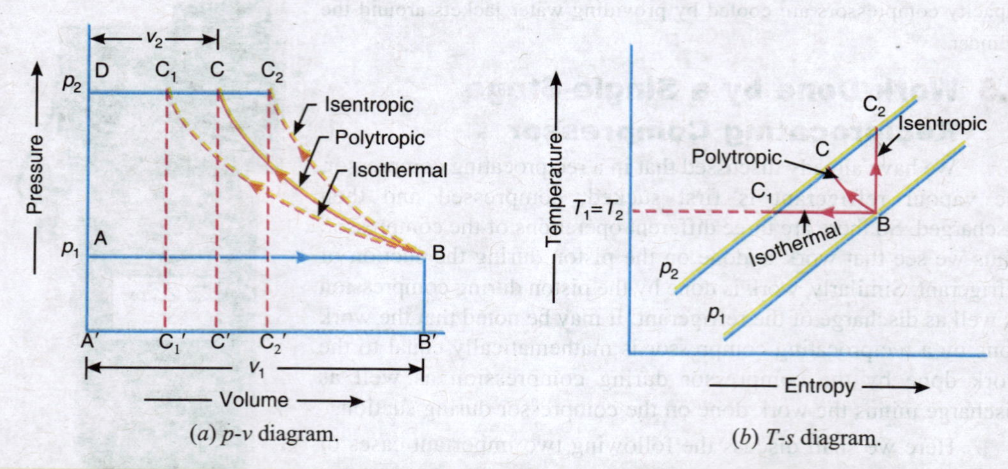

(*a*) *p-v* diagram.

(*b*) *T-s* diagram.

Fig. 9.2. *p-v* and *T-s* diagrams for a single-stage reciprocating compressor.

From the figure, we find that

Work done during compression,

$$W_2 = \text{Area } B\,C_1\,C_1'\,B' = p_1 v_1 \log_e \left(\frac{v_1}{v_2}\right)$$

and work done during discharge,

$$W_3 = \text{Area } C_1\,D\,A'\,C_1' = p_2 v_2$$

∴ Work done by the compressor per cycle,

$$W = \text{Area } ABC_1D$$

$$= \text{Area } C_1DA'C_1' + \text{Area } BC_1C_1'B' - \text{Area } ABB'A'$$

$$= W_3 + W_2 - W_1$$

$$= p_2v_2 + p_1v_1 \log_e\left(\frac{v_1}{v_2}\right) - p_1v_1 = p_1v_1 \log_e\left(\frac{v_1}{v_2}\right) \quad \ldots (\because p_1v_1 = p_2v_2)$$

$$= 2.3\, p_1v_1 \log\left(\frac{v_1}{v_2}\right) = 2.3\, p_1v_1 \log r \quad \ldots \left(\because \frac{v_1}{v_2} = \frac{p_2}{p_1} = r\right)$$

$$= 2.3\, m\, R\, T_1 \log r \quad \ldots (\because p_1v_1 = mRT_1)$$

2. Work done during polytropic compression (pv^n = Constant)

The polytropic compression is shown by the curve *BC* in Fig. 9.2. Now *CD* represents the volume of refrigerant delivered (v_2). We know that work done during suction,

$$W_1 = \text{Area } ABB'A' = p_1v_1$$

Work done during polytropic compression,

$$W_2 = \text{Area } BCC'B' = \frac{p_2v_2 - p_1v_1}{n-1}$$

and work done during discharge,

$$W_3 = \text{Area } CDA'C' = p_2v_2$$

∴ Work done by the compressor per cycle,

$$W = \text{Area } ABCD$$

$$= \text{Area } CDA'C' + \text{Area } BCC'B' - \text{Area } ABB'A'$$

$$= W_3 + W_2 - W_1$$

$$= p_2v_2 + \frac{p_2v_2 - p_1v_1}{n-1} - p_1v_1$$

$$= \frac{(n-1)p_2v_2 + p_2v_2 - p_1v_1 - (n-1)p_1v_1}{(n-1)}$$

$$= \frac{n}{n-1}(p_2v_2 - p_1v_1) \quad \ldots (i)$$

$$= \frac{n}{n-1} \times p_1v_1\left(\frac{p_2v_2}{p_1v_1} - 1\right) \quad \ldots (ii)$$

We know that for polytropic compression,

$$p_1v_1^{\,n} = p_2v_2^{\,n}$$

$$\therefore \quad \left(\frac{v_1}{v_2}\right) = \left(\frac{p_2}{p_1}\right)^{\frac{1}{n}} \text{ or } \left(\frac{v_2}{v_1}\right) = \left(\frac{p_1}{p_2}\right)^{\frac{1}{n}}$$

Substituting the value of v_2 / v_1 in equation (*ii*),

$$W = \frac{n}{n-1} \times p_1v_1\left[\frac{p_2}{p_1}\left(\frac{p_1}{p_2}\right)^{\frac{1}{n}} - 1\right]$$

$$= \frac{n}{n-1} \times p_1v_1\left[\frac{p_2}{p_1}\left(\frac{p_2}{p_1}\right)^{\frac{-1}{n}} - 1\right]$$

$$= \frac{n}{n-1} \times p_1 v_1 \left[\left(\frac{p_2}{p_1} \right)^{\frac{n-1}{n}} - 1 \right]$$

$$= \frac{n}{n-1} \times m\,R\,T_1 \left[\left(\frac{p_2}{p_1} \right)^{\frac{n-1}{n}} - 1 \right]$$

$$= \frac{n}{n-1} \times m\,R\,T_1 \left[\frac{T_2}{T_1} - 1 \right] \quad \ldots \left[\because \left(\frac{p_2}{p_1} \right)^{\frac{n-1}{n}} = \frac{T_2}{T_1} \right]$$

$$= \frac{n}{n-1} \times mR(T_2 - T_1)$$

3. Work done during isentropic compression

The isentropic compression is shown by the curve BC_2 in Fig. 9.2. In this case, the volume of refrigerant delivered (v_2) is represented by the line C_2D.

The work done by the compressor per cycle during compression may be worked out in the similar way as polytropic compression. The polytropic index n is changed to isentropic index γ in the previous results.

∴ Work done by the compressor per cycle,

$$W = \frac{\gamma}{\gamma-1} \times p_1 v_1 \left[\left(\frac{p_2}{p_1} \right)^{\frac{\gamma-1}{\gamma}} - 1 \right]$$

$$= \frac{\gamma}{\gamma-1} \times mR\,T_1 \left[\left(\frac{p_2}{p_1} \right)^{\frac{\gamma-1}{\gamma}} - 1 \right] = \frac{\gamma}{\gamma-1} \times mR\,(T_2 - T_1)$$

We know that ratio of specific heats,

$$\frac{c_p}{c_v} = \gamma \text{ ; and } c_p - c_v = R$$

or

$$R = c_p \left(1 - \frac{1}{\gamma} \right) = c_p \left(\frac{\gamma - 1}{\gamma} \right)$$

Substituting the value of R in the above expression, the work done by the compressor per cycle,

$$W = \frac{\gamma}{\gamma-1} \times m\,c_p \left(\frac{\gamma-1}{\gamma} \right) (T_2 - T_1) = m\,c_p (T_2 - T_1)$$

This shows that the work done by the compressor per cycle during isentropic compression is equal to the heat required to raise the temperature of refrigerant from T_1 to T_2 at constant pressure.

9.7 Power Required to Drive a Single-Stage Reciprocating Compressor

We have already obtained, in the previous article, the expressions for the work done (W) per cycle during isothermal, polytropic and isentropic compression. The power required to drive the compressor may be obtained from the usual relation,

Single stage reciprocating compressor.

$$P = \frac{W \text{ (in N-m/min)}}{60} \text{ watts}$$

$$= \frac{W \text{ (in N-m)} \times N_w}{60} \text{ watts}$$

If N is the speed of the compressor in r.p.m., then number of working strokes per minute,

$$N_w = N \quad \text{... (For single acting compressor)}$$

$$= 2N \quad \text{... (For double acting compressor)}$$

Note : Since the compression takes place in three different ways, therefore, the power obtained from different works done will be different. In general, following are the three values of power obtained :

1. Isothermal power $= \dfrac{W \text{ (in isothermal compression) } N_w}{60}$ watts

2 Indicated power $= \dfrac{W \text{ (in polytropic compression) } N_w}{60}$ watts

3. Isentropic power $= \dfrac{W \text{ (in isentropic compression) } N_w}{60}$ watts

Example 9.1. *A single-stage reciprocating compressor is required to compress 1.5 m³/min of vapour refrigerant from 1 bar to 8 bar. Find the power required to drive the compressor, if the compression of refrigerant is 1. isothermal ; 2. polytropic with polytropic index as 1.12 ; and 3. isentropic with isentropic index as 1.31.*

Solution. Given : $v_1 = 1.5$ m³/min ; $p_1 = 1$ bar $= 1 \times 10^5$ N/m² ; $p_2 = 8$ bar $= 8 \times 10^5$ N/m² ; $n = 1.12$; $\gamma = 1.31$

1. *Isothermal compression*

We know that work done by the compressor,

$$= 2.3\, p_1 v_1 \log\left(\frac{p_2}{p_1}\right) = 2.3 \times 1 \times 10^5 \times 1.5 \log\left(\frac{8}{1}\right) \text{ N-m/min}$$

$$= 3.45 \times 10^5 \times 0.9031 = 3.12 \times 10^5 \text{ N-m/min}$$

∴ Power required to drive the compressor

$$= 3.12 \times 10^5/60 = 5200 \text{ W} = 5.2 \text{ kW} \quad \textbf{Ans.}$$

2. ***Polytropic compression***

We know that work done by the compressor

$$= \frac{n}{n-1} \times p_1 v_1 \left[\left(\frac{p_2}{p_1} \right)^{\frac{n-1}{n}} - 1 \right]$$

$$= \frac{1.12}{1.12-1} \times 1 \times 10^5 \times 1.5 \left[\left(\frac{8}{1} \right)^{\frac{1.12-1}{1.12}} - 1 \right] \text{N-m/min}$$

$$= 14 \times 10^5 (1.25 - 1) = 3.5 \times 10^5 \text{ N-m/min}$$

∴ Power required to drive the compressor

$$= 3.5 \times 10^5/60 = 5833 \text{ W} = 5.833 \text{ kW} \quad \textbf{Ans.}$$

3. ***Isentropic compression***

We know that work done by the compressor,

$$= \frac{\gamma}{\gamma-1} \times p_1 v_1 \left[\left(\frac{p_2}{p_1} \right)^{\frac{\gamma-1}{\gamma}} - 1 \right]$$

$$= \frac{1.31}{1.31-1} \times 1 \times 10^5 \times 1.5 \left[\left(\frac{8}{1} \right)^{\frac{1.31-1}{1.31}} - 1 \right] \text{N-m/min}$$

$$= 6.34 \times 10^5 (1.63 - 1) = 4 \times 10^5 \text{ N-m/min}$$

∴ Power required to drive the compressor

$$= 4 \times 10^5/60 = 6667 \text{ W} = 6.667 \text{ kW} \quad \textbf{Ans.}$$

9.8 Work Done by Reciprocating Compressor with Clearance Volume

In the previous articles, we have assumed that there is no clearance volume in the compressor cylinder. In other words, the entire volume of the refrigerant, in the compressor cylinder, is compressed by the inward stroke of the piston. But in actual practice, it is not possible to reduce the clearance volume to zero, for mechanical reasons. Moreover, it is not desirable to allow the piston head to come in contact with the cylinder head. In addition to this, the passages leading to the suction and discharge valves always contribute to clearance volume. In general, the clearance volume is expressed as some percentage of the piston displacement.

Now consider a single stage, single acting horizontal reciprocating compressor with clearance volume as shown by the *p*-v diagram in Fig. 9.3.

Let p_1 = Suction pressure of refrigerant (before compression),

v_1 = Total volume of refrigerant in the compressor cylinder (before compression),

T_1 = Suction temperature of refrigerant (before compression),

p_2, v_2 and T_2 = Corresponding values at the discharge point 2 (*i.e.* after compression),

v_c = Clearance volume,

v_s = Actual volume of refrigerant sucked by the compressor *i.e.* suction volume = $v_1 - v_4$,

v_4 = Expanded clearance volume *i.e.* volume of refrigerant after expansion.

v_p = Stroke or effective swept volume or piston displacement of the compressor = $v_1 - v_3 = v_1 - v_c$,

n = Polytropic index for compression and expansion, and

r = Compression ratio or pressure ratio (*i.e.* p_2 / p_1).

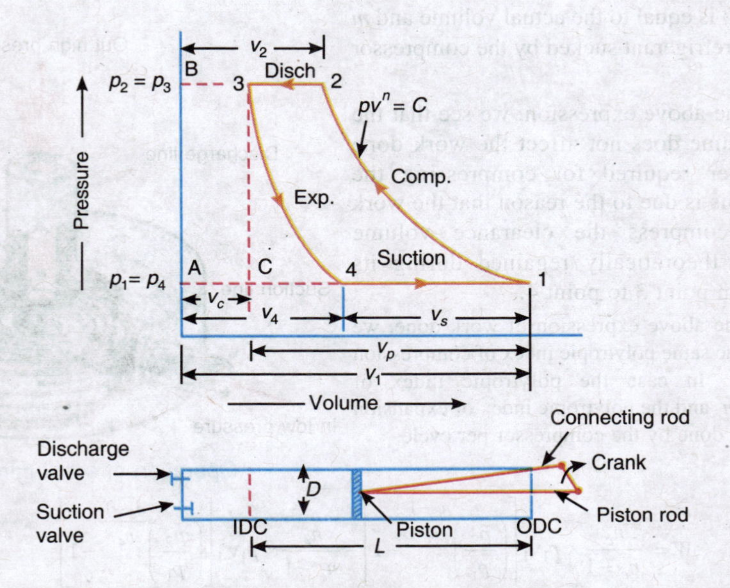

Fig. 9.3. *p-v* diagram with clearance volume.

We know that when the piston moves from outer dead centre (O.D.C.), during the return stroke (or inward stroke) of the piston, the vapour refrigerant is compressed as shown by the curve 1–2 in Fig. 9.3. This compression continues till the pressure p_2 in the cylinder is sufficient to force open the discharge valve at point 2. After that, no more compression takes place with the inward movement of the piston. Now, during the remaining part of the compression stroke, the compressed refrigerant is discharged to the condenser till the piston reaches at point 3. At this stage, there will be some refrigerant (equal to the clearance volume, v_c) left in the clearance space of the cylinder at the discharge pressure p_2. This entrapped refrigerant in the clearance space will now expand when piston moves from inner dead centre (I.D.C.) during some part of the outward stroke of the piston as shown by the curve 3–4. This expansion continues till the pressure p_1 is sufficient to force open the suction valve at point 4. Now the fresh charge of vapour refrigerant enters at point 4 during the suction stroke 4–1 at suction pressure p_1.

Though the compression and expansion may be isothermal, polytropic or isentropic, yet for all calculation purposes, it is assumed to be polytropic.

We know that work done by the compressor,

$$W = \text{Area 1-2–3–4} = \text{Area } A\text{-1-2-}B - A\text{-4-3-}B$$

$$= \frac{n}{n-1} \times p_1 v_1 \left[\left(\frac{p_2}{p_1}\right)^{\frac{n-1}{n}} - 1\right] - \frac{n}{n-1} \times p_4 v_4 \left[\left(\frac{p_2}{p_1}\right)^{\frac{n-1}{n}} - 1\right]$$

$$= \frac{n}{n-1} \times p_1 (v_1 - v_4) \left[\left(\frac{p_2}{p_1}\right)^{\frac{n-1}{n}} - 1\right] \quad \ldots (\because p_1 = p_4)$$

$$= \frac{n}{n-1} \times mRT_1 \left[\left(\frac{p_2}{p_1}\right)^{\frac{n-1}{n}} - 1\right]$$

where $(v_1 - v_4)$ is equal to the actual volume and m is the mass of refrigerant sucked by the compressor per cycle.

From the above expression, we see that the clearance volume does not affect the work done and the power required for compressing the refrigerant. This is due to the reason that the work required to compress the clearance volume refrigerant is theoretically regained during its expansion from point 3 to point 4.

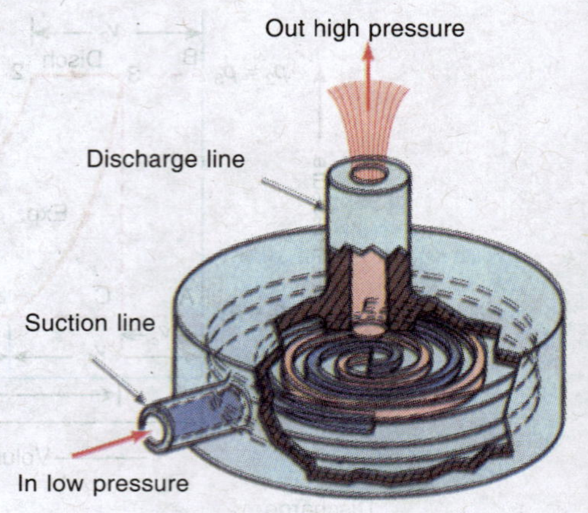

Operation of scroll compressor.

Notes : 1. In the above expression of work done, we have assumed the same polytropic index of compression and expansion. In case the polytropic index of compression is n_c and the polytropic index of expansion is n_e, then work done by the compressor per cycle,

$$W = \frac{n_c}{n_c - 1} \times p_1 v_1 \left[\left(\frac{p_2}{p_1}\right)^{\frac{n_c - 1}{n_c}} - 1\right] - \frac{n_e}{n_e - 1} \times p_1 v_4 \left[\left(\frac{p_2}{p_1}\right)^{\frac{n_e - 1}{n_e}} - 1\right]$$

2. The effect of clearance in reciprocating compressors is to reduce the volume of vapour refrigerant sucked during suction stroke.

9.9 Volumetric Efficiency of a Reciprocating Compressor

We have already discussed that the volumetric efficiency of a reciprocating compressor is the ratio of actual volume of refrigerant passing through the compressor per cycle (v_s) to the stroke volume of the compressor (v_p). Mathematically, volumetric efficiency,

$$\eta_v = \frac{v_s}{v_p} = \frac{v_1 - v_4}{v_1 - v_c} \quad \ldots (i)$$

The *p-v* diagram of a reciprocating compressor with clearance volume is shown in Fig. 9.4.

Since the refrigerant gases are expanded and compressed in every cycle, therefore, the curve 3–4 is called *re-expansion curve.* Applying $pv^n = C$ to points 3 and 4,

$$p_3 v_3^{\,n} = p_4 v_4^{\,n}$$

or

$$p_2 v_c^{\,n} = p_1 v_4^{\,n}$$

$$\therefore \quad v_4 = v_c \left(\frac{p_2}{p_1}\right)^{\frac{1}{n}}$$

Substituting the value of v_4 in equation (*i*),

$$\eta_v = \frac{v_1 - v_c\left(\frac{p_2}{p_1}\right)^{\frac{1}{n}}}{v_1 - v_c}$$

$$= \frac{v_1 + v_c - v_c - v_c\left(\frac{p_2}{p_1}\right)^{\frac{1}{n}}}{v_1 - v_c}$$

$$= \frac{v_1 - v_c}{v_1 - v_c} + \frac{v_c}{v_1 - v_c} - \frac{v_c\left(\frac{p_2}{p_1}\right)^{\frac{1}{n}}}{v_1 - v_c}$$

$$= 1 + C - C\left(\frac{p_2}{p_1}\right)^{\frac{1}{n}}$$

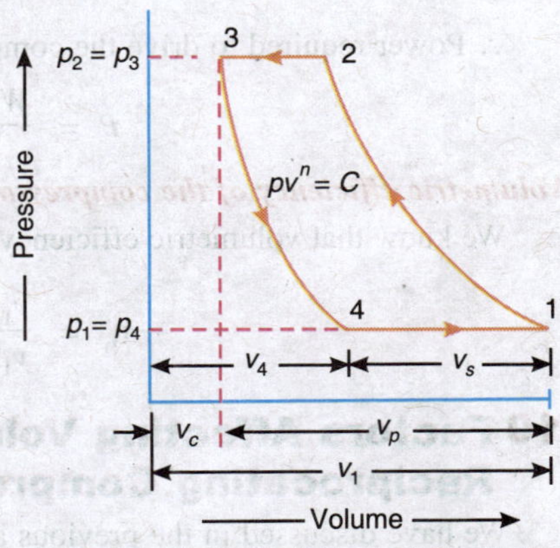

Fig. 9.4. *p-v* diagram with clearance volume.

where $\quad C$ = Clearance factor $= \dfrac{v_c}{v_1 - v_c}$

Example 9.2. *A single-stage, single acting reciprocating compressor has a bore of 200 mm and a stroke of 300 mm. It receives vapour refrigerant at 1 bar and delivers it at 5.5 bar. If the compression and expansion follows the law $pv^{1.3}$ = Constant and the clearance volume is 5 per cent of the stroke volume, determine : 1. The power required to drive the compressor, if it runs at 500 r.p.m. ; and 2. The volumetric efficiency of the compressor.*

Solution. Given : D = 200 mm = 0.2 m ; L = 300 mm = 0.3 m ; p_1 = 1 bar = 1 × 10⁵ N/m²; p_2 = 5.5 bar = 5.5 × 10⁵ N/m² ; n = 1.3 ; $v_c = 5\%\ v_p$; N = 500 r.p.m.

We know that the stroke volume,

$$v_p = \frac{\pi}{4} \times D^2 \times L = \frac{\pi}{4}(0.2)^2\, 0.3 = 0.0094 \text{ m}^3$$

Clearance volume, $\quad v_c = 5\%\ v_p = 0.05\ v_p = 0.05 \times 0.0094 = 0.000\,47 \text{ m}^3$

Total cylinder volume, $\quad v_1 = v_c + v_p = 0.000\,47 + 0.0094 = 0.009\,87 \text{ m}^3$

and expanded clearance volume,

$$v_4 = v_c\left(\frac{p_2}{p_1}\right)^{\frac{1}{n}} = 0.000\,47\left(\frac{5.5}{1}\right)^{\frac{1}{1.3}} = 0.001\,74 \text{ m}^3$$

1. ***Power required to drive the compressor***

We know that work done by the compressor,

$$W = \frac{n}{n-1} \times p_1(v_1 - v_4)\left[\left(\frac{p_2}{p_1}\right)^{\frac{n-1}{n}} - 1\right]$$

$$= \frac{1.3}{1.3-1} \times 1 \times 10^5\,(0.009\,87 - 0.001\,74)\left[\left(\frac{5.5}{1}\right)^{\frac{1.3-1}{1.3}} - 1\right] \text{ N-m}$$

$$= 3500\,(1.48 - 1) = 1695 \text{ N-m}$$

Since the compressor is single acting, therefore, number of working strokes per minute,

$$N_W = N = 500$$

∴ Power required to drive the compressor,

$$P = \frac{W \times N_W}{60} = \frac{1695 \times 500}{60} = 141\ 20 \text{ W} = 14.12 \text{ kW} \textbf{ Ans.}$$

2. *Volumetric efficiency of the compressor*

We know that volumetric efficiency of the compressor,

$$\eta_v = \frac{v_1 - v_4}{v_1 - v_c} = \frac{0.009\ 87 - 0.001\ 74}{0.00987 - 0.000\ 47} = 0.865 \text{ or } 86.5\% \textbf{ Ans.}$$

9.10 Factors Affecting Volumetric Efficiency of a Reciprocating Compressor

We have discussed in the previous article that the volumetric efficiency of a reciprocating compressor,

$$\eta_v = 1 + C - C\left(\frac{p_2}{p_1}\right)^{\frac{1}{n}} \quad \ldots (i)$$

where C = Clearance factor,

p_1 = Suction pressure,

p_2 = Discharge pressure, and

n = Polytropic index of expansion.

The various factors which affect the volumetric efficiency are discussed below:

1. *Effect of clearance factor*

The equation (*i*) may be written as

$$\eta_v = 1 + C\left[1 - \left(\frac{p_2}{p_1}\right)^{\frac{1}{n}}\right]$$

We know that the discharge pressure (p_2) is always more than the suction pressure (p_1).

Therefore, the ratio $\left(\frac{p_2}{p_1}\right)^{\frac{1}{n}}$ will be more than unity and the expression $\left[1 - \left(\frac{p_2}{p_1}\right)^{\frac{1}{n}}\right]$ will be negative. Thus, the increase in clearance factor (C) will increase the quantity to be subtracted from 1 to give volumetric efficiency. Hence the volumetric efficiency decreases as the clearance factor increases.

2. *Effect of valve pressure drops*

The *p-v* diagram of a reciprocating compressor considering the pressure drop at the suction and discharge valves is shown in Fig. 9.5. In actual practice, the compression will start from pressure p_s at point 1′ and expansion from pressure p_d at point 3′. The pressure $p_1 = p_4$ is the pressure at the flange on suction side and $p_2 = p_3$ is the pressure at the flange on discharge side.

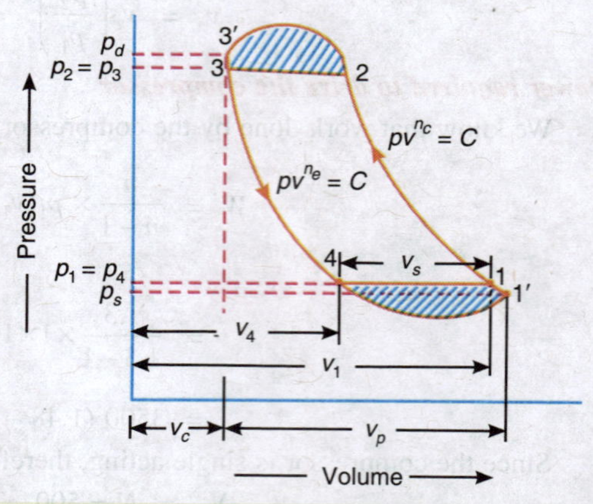

Fig. 9.5. *p-v* diagram with valve pressure drops.

Applying $pv^n = C$ to points 1 and 1′,

$$p_1(v_1)^{n_c} = p_s(v_c + v_p)^{n_c}$$

$$\therefore \quad v_1 = (v_c + v_p)\left(\frac{p_s}{p_1}\right)^{\frac{1}{n_c}} \quad \text{... (i)}$$

Similarly, applying $pv^{n_e} = C$ to points 3′ and 4,

$$p_d(v_c)^{n_e} = p_1(v_4)^{n_e}$$

$$\therefore \quad v_4 = v_c\left(\frac{p_d}{p_1}\right)^{\frac{1}{n_e}} \quad \text{... (ii)}$$

We know that volumetric efficiency,

$$\eta_v = \frac{v_1 - v_4}{v_p}$$

Substituting the values of v_1 and v_4 from equations (*i*) and (*ii*),

$$\eta_v = \frac{(v_c + v_p)\left(\frac{p_s}{p_1}\right)^{\frac{1}{n_c}} - v_c\left(\frac{p_d}{p_1}\right)^{\frac{1}{n_e}}}{v_p}$$

$$\therefore \quad = (C+1)\left(\frac{p_s}{p_1}\right)^{\frac{1}{n_c}} - C\left(\frac{p_d}{p_1}\right)^{\frac{1}{n_e}} \quad \text{... } \left(\because \frac{v_c}{v_p} = C\right)$$

$$= (1+C)\left(\frac{p_s}{p_1}\right)^{\frac{1}{n_c}} - C\left(\frac{p_d}{p_1}\right)^{\frac{1}{n_e}} \quad \text{... (iii)}$$

Notes : (*a*) When the pressure drop at suction valve is only considered, then $p_d = p_2 = p_3$. In this case,

$$\eta_v = (1+C)\left(\frac{p_s}{p_1}\right)^{\frac{1}{n_c}} - C\left(\frac{p_2}{p_1}\right)^{\frac{1}{n_e}}$$

(*b*) When the pressure drop at discharge valve is only considered, then $p_s = p_1 = p_4$. In this case,

$$\eta_v = 1 + C - C\left(\frac{p_d}{p_1}\right)^{\frac{1}{n_e}}$$

3. *Effect of heat gain from the cylinder walls and re-expansion index*

The effect of heat gain from the cylinder walls is to increase the expanded clearance volume (v_4) and to decrease the suction volume (v_s) which decreases the volumetric efficiency. In order to reduce the work of compression, the normal practice is to cool the cylinder either by water jackets provided around the cylinder as in the case of ammonia compressors or by surrounding air by natural convection through fins provided on the external surface of the cylinder as in the case of R-12 compressors.

The re-expansion index (n_e) is generally less than the isentropic index (γ), which gives a reduction in volumetric efficiency. Since the isentropic index for ammonia is high, therefore it will result in high discharge and cylinder wall temperatures. The cooling by water jacketing of ammonia compressors will result in lower discharge and cylinder wall temperatures and thus higher volumetric efficiency.

The isentropic index for the freon group of refrigerants is quite low. This gives large expanded clearance volume (v_4) and hence low volumetric efficiency. It is, therefore, necessary to provide minimum possible clearance in these compressors. It may be noted that the compressors used for freon group of refrigerants need to be air-cooled only, because of low isentropic index.

4. *Effect of valve and piston leakage*

The effect of gas leakage through the suction or discharge valves or past the piston rings is to decrease the volumetric efficiency of compressor. This factor is accounted for by allowing one per cent of the compression ratio (r), *i.e.*

$$\text{Leakage loss} = 0.01\, r = 0.01\left(\frac{p_d}{p_s}\right)$$

9.11 Overall or Total Volumetric Efficiency of a Reciprocating Compressor

The overall or total volumetric efficiency of a reciprocating compressor, considering all the factors discussed in the previous article, is given by

$$\text{Total } \eta_v = (1+C)\left(\frac{p_s}{p_1}\right)^{\frac{1}{n_c}} - C\left(\frac{p_d}{p_1}\right)^{\frac{1}{n_e}} - 0.01\left(\frac{p_d}{p_s}\right)$$

9.12 Multi-stage Compression

In the previous articles, we have taken into consideration the compression of vapour refrigerant in a single stage (*i.e.* single cylinder) compressor. In other words, the vapour refrigerant is compressed in a single cylinder and then delivered at a high pressure. But sometimes, the vapour refrigerant is required to be delivered at a very high pressure as in the case of low temperature refrigerating systems. In such cases, either we should compress the vapour refrigerant by employing a single stage compressor with a very high pressure ratio or compress it in two or more compressors placed in series. It has been experienced that if we employ a single stage compressor for producing very high pressure refrigerant, it suffers the following drawbacks :

1. The size of the cylinder will be too large.
2. Due to compression, there is a rise in temperature of the refrigerant. It is difficult to reject heat from the refrigerant in the small time available for compression.
3. Sometimes, the temperature of refrigerant at the end of compression is too high. It may heat up the cylinder head or burn the lubricating oil.
4. The friction losses and running cost are high.
5. The volumetric efficiency is low.

Multi-stage triaxial compression.

In order to overcome the abovementioned difficulties, the vapour refrigerant is compressed in two or more cylinders in series with intercooling arrangement between them. Such an arrangement is known as *compound* or *multi-stage compression.*

9.13 Advantages of Multi-stage Compression

Following are the main advantages of multi-stage compression over single stage compression :

1. The work done per kg of refrigerant is reduced in multi-stage compression with intercooler as compared to single stage compression for the same delivery pressure.
2. It improves the volumetric efficiency for the given pressure ratio.
3. The sizes of the two cylinders (*i.e.* high pressure and low pressure cylinders) may be adjusted to suit the volume and pressure of refrigerant.
4. It reduces the leakage loss considerably.
5. It gives more uniform torque, and hence a smaller size flywheel is required.
6. It provides effective lubrication because of lower temperature range.
7. It reduces the cost of compressor.

9.14 Two-Stage Reciprocating Compressor with Intercooler

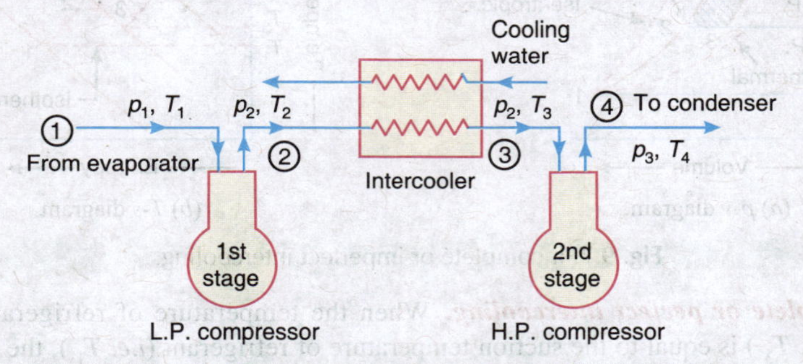

Fig. 9.6. Two-stage reciprocating compressor with intercooler.

The schematic arrangement for a two-stage reciprocating compressor with water-cooled intercooler is shown in Fig. 9.6.

First of all, the vapour refrigerant from the evaporator at pressure p_1 and temperature T_1 is sucked by the low pressure (L.P.) compressor at point 1 during its suction stroke. The refrigerant after compression in the L.P. compressor (first stage compressor) is delivered at point 2 to the intercooler at pressure p_2 and temperature T_2. In the intercooler, the compressed refrigerant is cooled from temperature T_2 to T_3 at constant pressure p_2. This cooled refrigerant is now sucked by the high pressure (H.P.) compressor at point 3 during its suction stroke. Finally, the refrigerant after compression in the H.P. compressor (second stage compressor) is delivered at point 4 to a condenser at pressure p_3 and temperature T_4.

9.15 Assumptions in Two-Stage Compression with Intercooler

The following simplifying assumptions are made in case of two stage-compression with intercooler :

1. The effect of clearance is neglected,
2. There is no pressure drop in the intercooler, and

3. The compression in both the low pressure (L.P.) and high pressure (H.P.) cylinders is polytropic *i.e.* $pv^n = C$.

9.16 Intercooling of Refrigerant in a Two-Stage Reciprocating Compressor

In Art. 9.14, we have discussed the working of a two-stage reciprocating compressor with an intercooler between the stages. As a matter of fact, the efficiency of an intercooler plays an important role in the working of a two-stage reciprocating compressor. Following two types of intercooling are important from the subject point of view :

1. *Incomplete or imperfect intercooling.* When the temperature of refrigerant leaving the intercooler (*i.e.* T_3) is more than the suction temperature of refrigerant (*i.e.* T_1), the intercooling is said to be *incomplete* or *imperfect intercooling*. In this case, the point 3 lies on the right side of the isothermal curve, as shown in Fig. 9.7 (*a*) and (*b*).

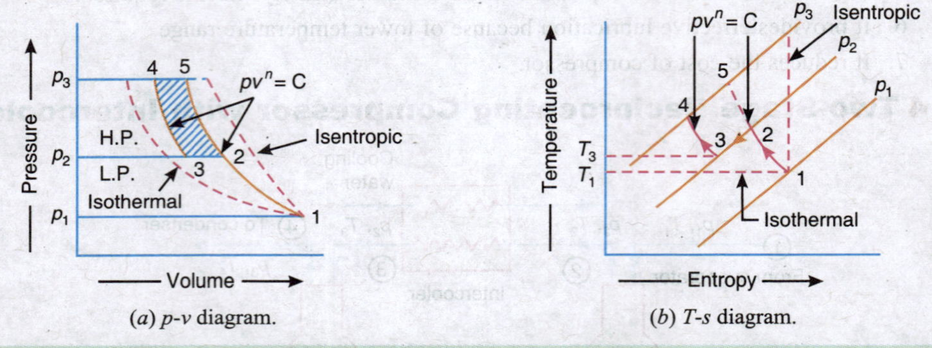

(*a*) *p-v* diagram. (*b*) *T-s* diagram.

Fig. 9.7. Incomplete or imperfect intercooling.

2. *Complete or perfect intercooling.* When the temperature of refrigerant leaving the intercooler (*i.e.* T_3) is equal to the suction temperature of refrigerant (*i.e.* T_1), the intercooling is said to be *complete* or *perfect intercooling*. In this case, the point 3 lies on the isothermal curve as shown in Fig. 9.8 (*a*) and (*b*).

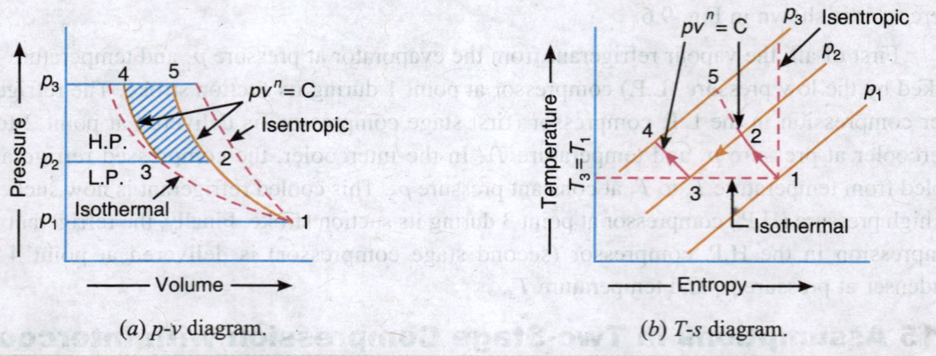

(*a*) *p-v* diagram. (*b*) *T-s* diagram.

Fig. 9.8. Complete or perfect intercooling.

Note : The amount of work saved due to intercooling is shown by the shaded area 2-3-4-5 in both the cases, to some scale.The amount of work saved with incomplete intercooling is less than that in case of complete intercooling.

9.17 Work Done by a Two-Stage Reciprocating Compressor with Intercooler

Consider a two-stage reciprocating compressor with intercooler compressing the refrigerant in its L.P. and H.P. cylinders.

Let p_1 = Evaporator pressure or suction pressure of L.P. compressor,

v_1 = Volume of L.P. cylinder,

T_1 = Suction temperature *i.e.* temperature of refrigerant entering the L.P. compressor,

p_2 = Discharge pressure of L.P. compressor or pressure at intercooler or suction pressure of H.P. compressor,

v_2 = Volume of H.P. cylinder,

T_2 = Temperature of vapour refrigerant leaving the L.P. compressor or entering the intercooler,

T_3 = Temperature of refrigerant leaving the intercooler,

p_3 = Discharge pressure of H.P. compressor or condenser pressure, and

n = Polytropic index for both the L.P. and H.P. compressors.

Now we shall consider both the cases of incomplete intercooling as well as complete intercooling one by one.

1. *When the intercooling is incomplete*

We know that work done per cycle in compressing the refrigerant in L.P. compressor,

$$W_L = \frac{n}{n-1} \times p_1 v_1 \left[\left(\frac{p_2}{p_1} \right)^{\frac{n-1}{n}} - 1 \right]$$

$$= \frac{n}{n-1} \times mRT_1 \left[\left(\frac{p_2}{p_1} \right)^{\frac{n-1}{n}} - 1 \right]$$

Similarly, work done per cycle in compressing the refrigerant in H.P. compressor,

$$W_H = \frac{n}{n-1} \times p_2 v_2 \left[\left(\frac{p_3}{p_2} \right)^{\frac{n-1}{n}} - 1 \right]$$

$$= \frac{n}{n-1} \times mRT_3 \left[\left(\frac{p_3}{p_2} \right)^{\frac{n-1}{n}} - 1 \right]$$

∴ Total work done per cycle,

$$W = W_L + W_H$$

$$= \frac{n}{n-1} \times p_1 v_1 \left[\left(\frac{p_2}{p_1} \right)^{\frac{n-1}{n}} - 1 \right] + \frac{n}{n-1} \times p_2 v_2 \left[\left(\frac{p_3}{p_2} \right)^{\frac{n-1}{n}} - 1 \right]$$

$$= \frac{n}{n-1} \times mR \left[T_1 \left(\frac{p_2}{p_1} \right)^{\frac{n-1}{n}} + T_3 \left(\frac{p_3}{p_2} \right)^{\frac{n-1}{n}} - T_1 - T_3 \right]$$

2. When the intercooling is complete

In case of complete intercooling, $p_1 v_1 = p_2 v_2$. Therefore, substituting $p_1 v_1 = p_2 v_2$ and $T_1 = T_3$ in the above expressions,

$$W = \frac{n}{n-1} \times p_1 v_1 \left[\left(\frac{p_2}{p_1} \right)^{\frac{n-1}{n}} + \left(\frac{p_3}{p_2} \right)^{\frac{n-1}{n}} - 2 \right]$$

$$= \frac{n}{n-1} \times mRT_1 \left[\left(\frac{p_2}{p_1} \right)^{\frac{n-1}{n}} + \left(\frac{p_3}{p_2} \right)^{\frac{n-1}{n}} - 2 \right]$$

9.18 Minimum Work Required for a Two-Stage Reciprocating Compressor

We have discussed in the previous article that the maximum work is saved in a two-stage reciprocating compressor with complete or perfect intercooling. But in actual practice, complete or perfect intercooling in refrigerant compressors cannot be achieved. We know that the total work done by the compressor per cycle, with incomplete or imperfect intercooling, is

$$W = \frac{n}{n-1} \times mR \left[T_1 \left(\frac{p_2}{p_1} \right)^{\frac{n-1}{n}} + T_3 \left(\frac{p_3}{p_2} \right)^{\frac{n-1}{n}} - T_1 - T_3 \right] \quad \ldots (i)$$

It is always necessary that the work done by the compressor should be minimum. In order to find the intermediate or intercooler pressure p_2 at which the work done is minimum, differentiate the above equation (*i*) with respect to p_2 and equate to zero. In other words, the work done by the compressor is minimum, when

$$\frac{dW}{dp_2} = 0$$

Let us now consider the following three cases to find p_2 for minimum work :

Two stage reciprocating compressor.

1. *When the temperature at the end of cooling in intercooler is fixed*

If ample cooling water supply is available, it is possible to cool the refrigerant to a fixed temperature T_3 for any temperature T_2 of the refrigerant entering the intercooler.

We know that for minimum work,

$$\frac{dW}{dp_2} = 0$$

$$\frac{d}{dp_2}\left[\frac{n}{n-1} \times mR\left\{T_1\left(\frac{p_2}{p_1}\right)^{\frac{n-1}{n}} + T_3\left(\frac{p_3}{p_2}\right)^{\frac{n-1}{n}} - T_1 - T_3\right\}\right] = 0$$

Now substituting $\frac{n-1}{n} = k$ (a constant) in the above equation, we have

$$\frac{d}{dp_2}\left[\frac{mR}{k}\left\{T_1\left(\frac{p_2}{p_1}\right)^{k} + T_3\left(\frac{p_3}{p_2}\right)^{k} - T_1 - T_3\right\}\right] = 0$$

$$T_1\, k\, p_2^{k-1} p_1^{-k} + T_3\, p_3^{\,k}(-k)(p_2)^{-k-1} - 0 - 0 = 0$$

$$\frac{T_1\, p_2^{k-1}}{p_1^{\,k}} - \frac{T_3\, p_3^{k}}{p_2^{\,k+1}} = 0$$

$$\frac{T_1\, p_2^{k-1}}{p_1^{\,k}} = \frac{T_3\, p_3^{k}}{p_2^{\,k+1}}$$

$$p_2^{\,k-1} \times p_2^{\,k+1} = \frac{T_3}{T_1}(p_1 p_3)^k$$

$$p_2^{\,2k} = \frac{T_3}{T_1}(p_1 p_3)^k$$

$$p_2 = \left[\frac{T_3}{T_1}(p_1 p_3)^k\right]^{\frac{1}{2k}} = \left(\frac{T_3}{T_1}\right)^{\frac{1}{2k}}(p_1 p_3)^{\frac{1}{2}}$$

$$= \left(\frac{T_3}{T_1}\right)^{\frac{n}{2(n-1)}}(p_1 p_3)^{\frac{1}{2}} \qquad \text{... (ii)}$$

$$= \sqrt{p_1 p_3\left(\frac{T_3}{T_1}\right)^{\frac{n}{n-1}}} \qquad \text{... (iii)}$$

Substituting the value of p_2 from equation (*ii*) in equation (*i*),

$$W = \frac{n}{n-1} \times mR \left[\frac{T_1}{(p_1)^{\frac{n-1}{n}}} \left\{ \left(\frac{T_3}{T_1}\right)^{\frac{n}{2(n-1)}} (p_1 p_3)^{\frac{1}{2}} \right\}^{\frac{n-1}{n}} + T_3 (p_3)^{\frac{n-1}{n}} \left\{ \left(\frac{T_3}{T_1}\right)^{\frac{n}{2(n-1)}} (p_1 p_3)^{\frac{1}{2}} \right\}^{\frac{1-n}{n}} - T_1 - T_3 \right]$$

$$= \frac{n}{n-1} \times mR \left[\frac{T_1}{(p_1)^{\frac{n-1}{n}}} \left\{ \left(\frac{T_3}{T_1}\right)^{\frac{1}{2}} (p_1 p_3)^{\frac{n-1}{2n}} \right\} + T_3 (p_3)^{\frac{n-1}{n}} \left\{ \left(\frac{T_3}{T_1}\right)^{-\frac{1}{2}} (p_1 p_3)^{\frac{1-n}{2n}} \right\} - T_1 - T_3 \right]$$

$$= \frac{n}{n-1} \times mR \left[\sqrt{T_1 T_3} \left(\frac{p_3}{p_1}\right)^{\frac{n-1}{2n}} + \sqrt{T_1 T_3} \left(\frac{p_3}{p_1}\right)^{\frac{n-1}{2n}} - T_1 - T_3 \right]$$

Note : For complete or perfect intercooling, $T_3 = T_1$. In this case

$$p_2 = \sqrt{p_1 p_3}$$

and

$$W = 2 \times \frac{n}{n-1} \times mRT_1 \left[\left(\frac{p_3}{p_1}\right)^{\frac{n-1}{2n}} - 1 \right]$$

2. When the cooling ratio is fixed

The cooling ratio is defined as the ratio of heat abstracted by the cooling system to the heat abstracted to bring the refrigerant to the initial temperature. Mathematically, cooling ratio,

$$q = \frac{T_2 - T_3}{T_2 - T_1}$$

If the cooling method used is such that the cooling ratio is constant, then

$$T_3 = T_2 - q\,(T_2 - T_1)$$

$$= T_1 \left[q + (1-q) \left(\frac{T_2}{T_1}\right) \right]$$

$$= T_1 \left[q + (1-q) \left(\frac{p_2}{p_1}\right)^{\frac{n-1}{n}} \right] \quad \text{... (iv)}$$

Differentiating this expression with respect to p_2,

$$\frac{dT_3}{dp_2} = \frac{T_1 \left[0 + (1-q) \left(\frac{n-1}{n}\right) (p_2)^{-\frac{1}{n}} \right]}{(p_1)^{\frac{n-1}{n}}} = \frac{T_1 (1-q) \left(\frac{n-1}{n}\right) (p_2)^{-\frac{1}{n}}}{(p_1)^{\frac{n-1}{n}}} \quad \text{... (v)}$$

We know that for minimum work,

$$\frac{dW}{dp_2} = 0$$

$$\frac{d}{dp_2}\left[\frac{n}{n-1}\times mR\left\{T_1\left(\frac{p_2}{p_1}\right)^{\frac{n-1}{n}} + T_3\left(\frac{p_3}{p_2}\right)^{\frac{n-1}{n}} - T_1 - T_3\right\}\right] = 0 \quad \text{... [From equation } (i)]$$

$$\frac{T_1\left(\frac{n-1}{n}\right)(p_2)^{-\frac{1}{n}}}{(p_1)^{\frac{n-1}{n}}} + \left[T_3(p_3)^{\frac{n-1}{n}}\left(-\frac{n-1}{n}\right)(p_2)^{\frac{1-2n}{n}} + \left(\frac{p_3}{p_2}\right)^{\frac{n-1}{n}}\frac{dT_3}{dp_2} - 0 - \frac{dT_3}{dp_2}\right] = 0 \quad \text{... } (vi)$$

Substituting the value of $\frac{dT_3}{dp_2}$ from equation (*iv*),

$$\frac{T_1\left(\frac{n-1}{n}\right)(p_2)^{-\frac{1}{n}}}{(p_1)^{\frac{n-1}{n}}} + \left[T_3(p_3)^{\frac{n-1}{n}}\left(\frac{1-n}{n}\right)(p_2)^{\frac{1-2n}{n}} + \left(\frac{p_3}{p_2}\right)^{\frac{n-1}{n}}\left\{T_1(1-q)\left(\frac{n-1}{n}\right)(p_2)^{-\frac{1}{n}}\right\} - \frac{T_1(1-q)\left(\frac{n-1}{n}\right)(p_2)^{-1}}{(p_1)^{\frac{n-1}{n}}}\right] = 0$$

Dividing throughout by $T_1\left(\frac{n-1}{n}\right)$, we get

$$\frac{(p_2)^{-\frac{1}{n}}}{(p_1)^{\frac{n-1}{n}}} + \frac{T_3}{T_1}(p_3)^{\frac{n-1}{n}}(-1)(p_2)^{\frac{1-2n}{n}} + \left(\frac{p_3}{p_2}\right)^{\frac{n-1}{n}}\frac{(1-q)(p_2)^{-\frac{1}{n}}}{(p_1)^{\frac{n-1}{n}}} - \frac{(1-q)(p_3)^{-\frac{1}{n}}}{(p_1)^{\frac{n-1}{n}}} = 0$$

$$(p_2)^{-\frac{1}{n}}(p_1)^{\frac{1-n}{n}} - \frac{T_3}{T_1}(p_3)^{\frac{n-1}{n}}(p_2)^{\frac{1-2n}{n}} + (p_3)^{\frac{n-1}{n}}(p_2)^{-1}(1-q)(p_1)^{\frac{1-n}{n}}$$

$$- (1-q)(p_2)^{-\frac{1}{n}}(p_1)^{\frac{1-n}{n}} = 0$$

Now substituting the value of $\frac{T_3}{T_1}$ from equation (*iv*),

$$(p_2)^{-\frac{1}{n}}(p_1)^{\frac{1-n}{n}} - \left\{q + (1-q)\left(\frac{p_2}{p_1}\right)^{\frac{n-1}{n}}\right\}\left\{(p_3)^{\frac{n-1}{n}}(p_3)^{\frac{1-2n}{n}}\right\}$$

$$+ (p_3)^{\frac{n-1}{n}}(p_2)^{-1}(1-q)(p_1)^{\frac{1-n}{n}} - (1-q)(p_2)^{-\frac{1}{n}}(p_1)^{\frac{1-n}{n}} = 0$$

$$(p_2)^{-\frac{1}{n}}(p_1)^{\frac{1-n}{n}} - q(p_3)^{\frac{n-1}{n}}(p_2)^{\frac{1-2n}{n}} - (1-q)(p_2)^{\frac{n-1}{n}}(p_1)^{\frac{1-n}{n}}(p_3)^{\frac{n-1}{n}}(p_2)^{\frac{1-2n}{n}}$$

$$+ (p_3)^{\frac{n-1}{n}}(p_2)^{-1}(1-q)(p_1)^{\frac{1-n}{n}} - (1-q)(p_2)^{-\frac{1}{n}}(p_1)^{\frac{1-n}{n}} = 0$$

$$(p_2)^{-\frac{1}{n}}(p_1)^{\frac{1-n}{n}} - q\,(p_3)^{\frac{n-1}{n}}\,(p_2)^{\frac{1-2n}{n}} - (p_2)^{-1}(p_1)^{\frac{1-n}{n}}\,(p_3)^{\frac{n-1}{n}}$$

$$+\,q(p_2)^{-1}(p_1)^{\frac{1-n}{n}}\,(p_3)^{\frac{n-1}{n}}\,(p_3)^{\frac{n-1}{n}}\,(p_2)^{-1}(p_1)^{\frac{1-n}{n}}$$

$$-\,q(p_3)^{\frac{n-1}{n}}\,(p_2)^{-1}(p_1)^{\frac{1-n}{n}} - (p_2)^{-\frac{1}{n}}(p_1)^{\frac{1-n}{n}} + q(p_2)^{-\frac{1}{n}}(p_1)^{\frac{1-n}{n}} = 0$$

$$(p_2)^{-\frac{1}{n}}(p_1)^{\frac{1-n}{n}} - q(p_3)^{\frac{n-1}{n}}\,(p_2)^{\frac{1-2n}{n}} - (p_2)^{-\frac{1}{n}}(p_1)^{\frac{1-n}{n}} + q(p_2)^{-\frac{1}{n}}(p_1)^{\frac{1-n}{n}} = 0$$

$$-\,q(p_3)^{\frac{n-1}{n}}\,(p_2)^{\frac{1-2n}{n}} + q\,(p_2)^{-\frac{1}{n}}(p_1)^{\frac{1-n}{n}} = 0$$

$$(p_3)^{\frac{n-1}{n}}\,(p_2)^{\frac{1-2n}{n}} = (p_2)^{-\frac{1}{n}}\,(p_1)^{\frac{1-n}{n}}$$

$$\frac{(p_2)^{\frac{1-2n}{n}}}{(p_2)^{-\frac{1}{n}}} = \frac{(p_1)^{\frac{1-n}{n}}}{(p_3)^{\frac{n-1}{n}}}$$

$$(p_2)^{\frac{1-2n}{n}}\,(p_2)^{\frac{1}{n}} = (p_1)^{\frac{1-n}{n}}\,(p_3)^{\frac{1-n}{n}}$$

$$(p_2)^{\frac{2(1-n)}{n}} = (p_1\,p_3)^{\frac{1-n}{n}}$$

$$p_2^2 = p_1 p_3$$

$$\therefore \quad p_2 = \sqrt{p_1 p_3}$$

Thus the intercooler or intermediate pressure is the geometric mean of the suction or evaporator pressure (p_1) and condenser pressure (p_3).

Substituting the value of p_2 in equation (*i*),

$$W = \frac{n}{n-1} \times mR\left[T_1\left(\frac{\sqrt{p_1 p_3}}{p_1}\right)^{\frac{n-1}{n}} + T_3\left(\frac{p_3}{\sqrt{p_1 p_3}}\right)^{\frac{n-1}{n}} - T_1 - T_3\right]$$

$$= \frac{n}{n-1} \times mR\left[T_1\left(\frac{p_3}{p_1}\right)^{\frac{n-1}{2n}} + T_3\left(\frac{p_3}{p_1}\right)^{\frac{n-1}{2n}} - T_1 - T_3\right]$$

where

$$T_3 = T_1\left[q + (1-q)\left(\frac{p_2}{p_1}\right)^{\frac{n-1}{n}}\right] \quad \text{... [Refer equation (iv)]}$$

Note : When the intercooling is perfect, then $T_3 = T_1$. In that case

$$W = 2 \times \frac{n}{n-1} \times mRT_1\left[\left(\frac{p_3}{p_1}\right)^{\frac{n-1}{2n}} - 1\right]$$

3. When the temperature after intercooling (T_3) is equal to the saturation temperature at intermediate or intercooler pressure

For the refrigeration compressors, it is always desirable to cool the refrigerant to a temperature T_3 which is also the saturation temperature at the intermediate pressure selected for minimum work. In such a case, we can write

$$T_3 = f(p_2)$$

where f is a function which corelates the saturation temperature of the refrigerant with the pressure. Hence to find the optimum pressure p_2, we have

$$\frac{dT_3}{dp_2} = f'(p_2) = \phi\,(p_2)$$

Substituting the values of T_3 and $\frac{dT_3}{dp_2}$ in equation (*vi*) and multiplying by $\left(\frac{n}{n-1} \times \frac{1}{T_1}\right)$, we get

$$(p_2)^{-\frac{1}{n}}(p_1)^{\frac{1-n}{n}} + \left[\frac{f(p_2)\,(p_3)^{\frac{n-1}{n}}\,(-1)\,(p_2)^{\frac{1-2n}{n}}}{T_1} + \frac{\left(\frac{p_3}{p_2}\right)^{\frac{n-1}{n}}\phi\,(p_2) \times n}{(n-1)T_1} - \phi\,(p_2)\left(\frac{n}{n-1}\right)\left(\frac{1}{T_1}\right)\right] = 0$$

$$(p_2)^{-\frac{1}{n}}(p_1)^{\frac{1-n}{n}} - \frac{f(p_2)}{T_1}(p_3)^{\frac{n-1}{n}}(p_2)^{\frac{1-2n}{n}} + \frac{n}{n-1} \times \frac{\phi\,(p_2)}{T_1}\left(\frac{p_3}{p_2}\right)^{\frac{n-1}{n}} - \frac{n}{n-1} \times \frac{\phi\,(p_2)}{T_1} = 0$$

The value of optimum pressure p_2 may be obtained by solving the above expression. The complexity depends upon the function f.

9.19 Performance Characteristics of Refrigerant Reciprocating Compressor

The performance of a refrigerant reciprocating compressor is measured in terms of its refrigerating capacity, brake power (B.P.) per tonne of refrigeration and the total brake power. The two important parameters in the design of a refrigerant reciprocating compressor are the evaporator (or suction) and condenser temperatures. The effect of these two parameters on the refrigerating capacity and B.P. are discussed below :

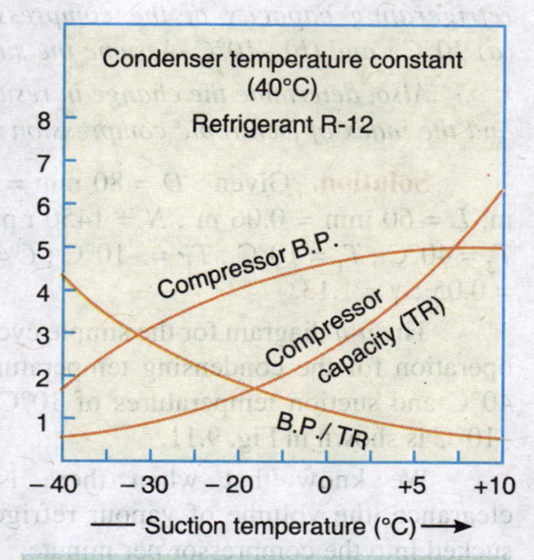

Fig. 9.9. Effect of suction temperature.

1. *Effect of suction temperature on compressor refrigerating capacity*

The effect of evaporator (or suction) temperature on the compressor refrigerating capacity (in tonnes) is shown in Fig. 9.9.

We see from the figure that the refrigerating capacity of a reciprocating compressor decreases with the decrease in evaporator or suction temperature. It is due to the fact that at low suction temperature, the vaporising pressure is low and therefore the density of suction vapour entering the compressor is low. Hence the mass of refrigerant circulated through the compressor per unit time decreases with the decrease in suction temperature for a given piston displacement.

2. *Effect of suction temperature on compressor B.P.*

The effect of suction temperature on the compressor B.P. per tonne of refrigeration and the total B.P. is shown in Fig. 9.9. We see from the figure that the compressor B.P. per tonne of refrigeration decreases with the increase in suction temperature, when the condenser temperature remains the same. But the total B.P. of the compressor is dependent on the following two factors :

(*a*) work of compression, and (*b*) mass of refrigerant circulated per minute.

A little consideration will show that with the decrease in suction temperature, the work of compression increases and the mass of refrigerant circulated per minute decreases. The decrease in the mass of refrigerant circulated outweighs the increase in work of compression per kg. Thus, the net effect is to decrease the total B.P. of the compressor with the decrease in suction temperature.

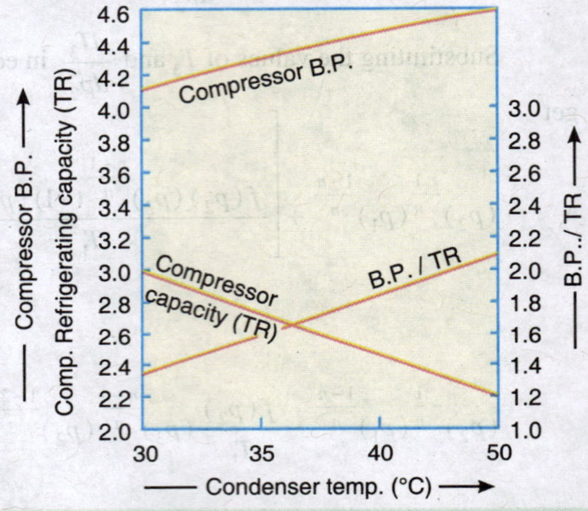

Fig. 9.10. Effect of condenser temperature.

3. *Effect of condenser temperature*

The effect of condenser temperature on the compressor refrigerating capacity, B.P. per tonne of refrigeration and the total B.P. is shown in Fig. 9.10. The suction temperature is kept constant in all the cases. We see from the figure that the increase in condenser temperature results in decrease of compressor refrigerating capacity, increase in B.P. per tonne of refrigeration and the total B.P.

Example 9.3. *A single-cylinder, single acting reciprocating compressor using R-12 as refrigerant has a bore 80 mm and stroke 60 mm. The compressor runs at 1450 r.p.m. If the condensing temperature is 40°C, find the mass of refrigerant circulated per minute and the refrigerating capacity of the compressor when the evaporating or suction temperature is (a) 10°C ; and (b) –10°C. Assume the simple cycle of operation and no clearance.*

Also, determine the change in results when the clearance factor of 5 per cent is considered and the index of isentropic compression is 1.13.

Solution. Given : D = 80 mm = 0.08 m; L = 60 mm = 0.06 m ; N = 1450 r.p.m. ; T_3 = 40°C ; T_1 = 10°C ; T_1' = –10°C ; C = 5% = 0.05 ; γ = 1.13

The *p-h* diagram for the simple cycle of operation for the condensing temperature of 40°C and suction temperatures of 10°C and –10°C is shown in Fig. 9.11.

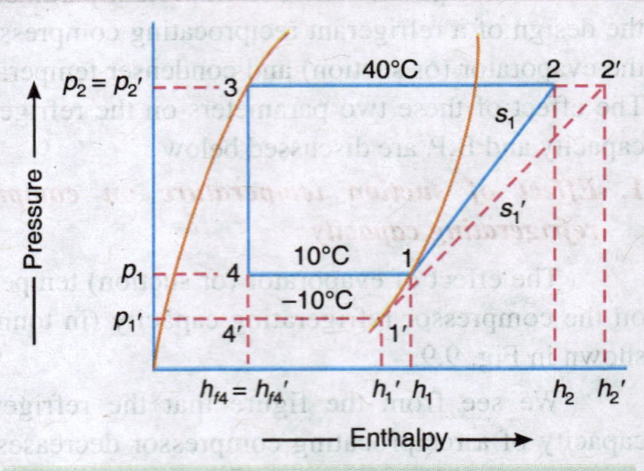

Fig. 9.11. *p-h* diagram.

We know that when there is no clearance, the volume of vapour refrigerant sucked into the compressor per minute,

$$v_1 = \text{Piston displacement per minute } (v_p)$$

$$= \frac{\pi}{4} \times D^2 \times L \times N = \frac{\pi}{4}(0.08)^2\, 0.06 \times 1450$$

$$= 0.4374 \text{ m}^3/\text{min}$$

(a) When the suction temperature (T_1) is 10°C

From *p-h* diagram, we find that the specific volume of R-12 saturated vapour refrigerant at point 1 (*i.e.* at 10°C),

$$v_{s1} = 0.041 \text{ m}^3/\text{kg}$$

Enthalpy of R-12 saturated vapour refrigerant at point 1 (*i.e.* at 10°C),

$$h_1 = 191.7 \text{ kJ/kg}$$

Enthalpy of R-12 saturated liquid refrigerant at point 3 (*i.e.* at 40°C),

$$h_{f3} = h_{f4} = h_{f4'} = 74.6 \text{ kJ/kg}$$

We know that mass of refrigerant circulated per minute,

$$m_1 = \frac{v_1}{v_{s1}} = \frac{0.4374}{0.041} = 10.67 \text{ kg/min } \textbf{Ans.}$$

and refrigerating capacity of the compressor,

$$R_1 = \frac{m_1(h_1 - h_{f4})}{210} = \frac{10.67(191.7 - 74.6)}{210} = 5.95 \text{ TR } \textbf{Ans.}$$

(b) When the suction temperature is –10°C

From *p-h* diagram, we find that the specific volume of R-12 saturated vapour refrigerant at point 1′ (*i.e.* at –10°C),

$$v_{s1'} = 0.077 \text{ m}^3/\text{kg}$$

and enthalpy of R-12 saturated vapour refrigerant at point 1′ (*i.e.* at –10°C),

$$h_1' = 183.2 \text{ kJ/kg}$$

We know that mass of refrigerant circulated per minute,

$$m_1' = \frac{v_1}{v_{s1}'} = \frac{0.4374}{0.077} = 5.68 \text{ kg/min } \textbf{Ans.}$$

and refrigerating capacity of the compressor,

$$R_1' = \frac{m_1'(h_1' - h_{f4}')}{210} = \frac{5.68(183.2 - 74.6)}{210} = 2.94 \text{ TR } \textbf{Ans.}$$

Change in results when clearance factor is 5 per cent

From *p-h* chart, we find that the pressure corresponding to 40°C saturation temperature,

$$p_2 = p_2' = 9.61 \text{ bar}$$

Pressure corresponding to 10°C saturation temperature,

$$p_1 = 4.23 \text{ bar}$$

and pressure corresponding to –10°C saturation temperature,

$$p_1' = 2.2 \text{ bar}$$

We know that the volumetric efficiency when suction temperature is 10°C,

$$\eta_v = 1 + C - C\left(\frac{p_2}{p_1}\right)^{\frac{1}{\gamma}} = 1 + 0.05 - 0.05\left(\frac{9.61}{4.23}\right)^{\frac{1}{1.13}}$$

$$= 1.05 - 0.103 = 0.947$$

and volumetric efficiency when suction temperature is –10°C,

$$\eta_v' = 1 + C - C\left(\frac{p_2'}{p_1'}\right)^{\frac{1}{\gamma}} = 1 + 0.05 - 0.05\left(\frac{9.61}{2.2}\right)^{\frac{1}{1.13}}$$

$$= 1.05 - 0.184 = 0.866$$

∴ Mass of refrigerant circulated per minute when evaporating temperature is 10°C

$$= m_1 \times \eta_v = 10.67 \times 0.947 = 10.1 \text{ kg/min}$$ **Ans.**

and mass of refrigerant circulated per minute when evaporating temperature is –10°C

$$= m_1' \times \eta_v' = 5.68 \times 0.866 = 4.92 \text{ kg/min}$$ **Ans.**

Refrigerating capacity of the compressor when evaporating temperature is 10°C

$$= R_1 \times \eta_v = 5.95 \times 0.947 = 5.63 \text{ TR}$$ **Ans.**

and refrigerating capacity of the compressor when evaporating temperature is –10°C

$$= R_1' \times \eta_v' = 2.94 \times 0.866 = 2.54 \text{ TR}$$ **Ans.**

Example 9.4. *A single cylinder, single acting reciprocating compressor using R-12 as refrigerant has a bore 80 mm and stroke 60 mm. The compressor runs at 1450 r.p.m. If the condensing temperature is 40°C, find the power per tonne of refrigeration and the total power required to drive the compressor when the suction temperature is (a) 10°C, and (b) – 10°C. Assume the simple cycle of operation and no clearance.*

Also determine the change in the results when the clearance factor of 5 per cent is considered and the index of isentropic compression is 1.13.

Solution. Given : D = 80 mm = 0.08 m ; L = 60 mm = 0.06 m ; N = 1450 r.p.m. ; T_3 = 40°C ; T_1 = 10°C ; T_1' = –10°C ; C = 5% = 0.05 ; γ = 1.13

Since the data given is same as that of previous example, therefore we shall use the same values as calculated already.

(a) When the suction temperature is 10°C

We have calculated in the previous example that mass of refrigerant circulated per min (at 10°C suction temperature),

$$m_1 = 10.67 \text{ kg/min}$$

and refrigerating capacity of the compressor,

$$R_1 = 5.95 \text{ TR}$$

From the *p-h* diagram as shown in Fig.9.11, we find that enthalpy of R-12 saturated vapour refrigerant at point 1 (*i.e.* at 10°C),

$$h_1 = 191.7 \text{ kJ/kg}$$

and enthalpy of superheated vapour refrigerant leaving the compressor at point 2,

$$h_2 = 206.2 \text{ kJ/kg}$$

We know that work done in compressing the refrigerant,

$$W_1 = m_1(h_2 - h_1) = 10.67(206.2 - 191.7) = 154.7 \text{ kJ/min}$$

∴ Power required per tonne of refrigeration,

$$P_1 = \frac{154.7}{60} \times \frac{1}{5.95} = 0.433 \text{ kW/TR}$$ **Ans.**

and total power required to drive the compressor,

$$P_T = 154.7/60 = 2.578 \text{ kW} \quad \textbf{Ans.}$$

(b) When the suction temperature is – 10°C

We have calculated in the previous example that the mass of refrigerant circulated per min. (at –10°C suction temperature),

$$m_1' = 5.68 \text{ kg/min}$$

and refrigerating capacity of the compressor,

$$R_1' = 2.94 \text{ TR}$$

From the *p-h* diagram as shown in Fig. 9.11, we find that enthalpy of R-12 saturated vapour refrigerant at point 1′ (*i.e.* at –10°C),

$$h_1' = 183.2 \text{ kJ/kg}$$

and enthalpy of R-12 superheated vapour refrigerant leaving the compressor at point 2′,

$$h_2' = 209.8 \text{ kJ/kg}$$

We know that work done in compressing the refrigerant,

$$W_1' = m_1'(h_2' - h_1') = 5.68\,(209.8 - 183.2) = 151 \text{ kJ/min}$$

∴ Power required per tonne of refrigeration,

$$P_1' = \frac{151}{60} \times \frac{1}{2.94} = 0.856 \text{ kW/TR} \quad \textbf{Ans.}$$

and total power required to drive the compressor,

$$P_T' = 151/60 = 2.517 \text{ kW} \quad \textbf{Ans.}$$

Change in results when clearance factor is 5 per cent

We have calculated in the previous example that the volumetric efficiency when suction temperature is 10°C,

$$\eta_v = 0.947$$

and volumetric efficiency when suction temperature is –10°C,

$$\eta_v' = 0.866$$

∴ Power required per TR when suction temperature is 10°C

$$= P_1 \times \eta_v = 0.433 \times 0.947 = 0.41 \text{ kW/TR} \quad \textbf{Ans.}$$

and total power required to drive the compressor when suction temperature is 10°C

$$= P_T \times \eta_v = 2.578 \times 0.947 = 2.44 \text{ kW} \quad \textbf{Ans.}$$

Similarly, power required per TR when suction temperature is –10°C

$$= P_1' \times \eta_v' = 0.856 \times 0.866 = 0.74 \text{ kW/TR} \quad \textbf{Ans.}$$

and total power required to drive the compressor when suction temperature is –10°C

$$= P_T' \times \eta_v' = 2.517 \times 0.866 = 2.18 \text{ kW} \quad \textbf{Ans.}$$

Example 9.5. *A single-cylinder, single acting reciprocating compressor using R-12 as refrigerant has a bore of 80 mm and a stroke of 60 mm. The compressor runs at 1450 r.p.m. Find the refrigerating capacity, power per tonne of refrigeration and the total power required to drive the compressor, when*

(a) the suction temperature is 10°C and condensing temperature is 50°C, and

(b) the suction temperature is –10°C and condensing temperature is 50°C

Compare the results obtained for (a) and (b) with those already obtained in Examples 9.3 and 9.4 for suction temperatures 10°C and –10°C and condensing temperature 40°C.

Assume simple cycle of operation and no clearance.

Solution. Given : D = 80 mm = 0.08 m ; L = 60 mm = 0.06 m ; N = 1450 r.p.m. ; T_1 = 10°C ; $T_1{}'$ = –10° C ; $T_3{}'$ = 50°C ; T_3 = 40°C

(a) When the suction temperature is 10°C and condensing temperature is 50°C

The *p-h* diagram for the simple cycle of operation for suction temperature of 10°C and condensing temperatures of 40°C and 50°C is shown in Fig. 9.12.

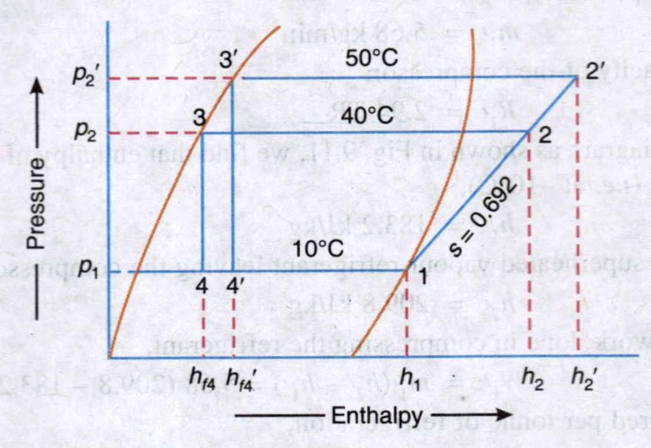

Fig. 9.12. *p-h* diagram.

We know that, when there is no clearance, the volume of vapour refrigerant sucked into the compressor per min,

$$v_1 = \text{Piston displacement per min}$$

$$= \frac{\pi}{4} \times D^2 \times L \times N = \frac{\pi}{4} (0.08)^2\, 0.06 \times 1450 = 0.4374 \text{ m}^3/\text{min}$$

From the *p-h* diagram, we find that specific volume of R-12 saturated vapour refrigerant at point 1 (*i.e.* at 10°C),

$$v_{s1} = 0.041 \text{ m}^3/\text{kg}$$

Enthalpy of R-12 saturated vapour refrigerant at point 1 (*i.e.* at 10°C),

$$h_1 = 191.7 \text{ kJ/kg}$$

Enthalpy of R-12 superheated vapour refrigerant at point 2′,

$$h_2{}' = 210.5 \text{ kJ/kg}$$

Enthalpy of R-12 liquid refrigerant at point 3′ (*i.e.* at 50°C),

$$h_{f3}{}' = h_{f4}{}' = 84.8 \text{ kJ/kg}$$

We know that mass of refrigerant circulated per min,

$$m_1 = \frac{v_1}{v_{s1}} = \frac{0.4374}{0.041} = 10.67 \text{ kg/min}$$

∴ Refrigerating capacity of the compressor,

$$R_1 = \frac{m_1 (h_1 - h_{f4'})}{210} = \frac{10.67(191.7 - 84.8)}{210} = 5.43 \text{ TR} \quad \textbf{Ans.}$$

Work done in compressing the refrigerant,

$$W_1 = m_1 (h_2{}' - h_1) = 10.67\,(210.5 - 191.7) = 200.6 \text{ kJ/min}$$

∴ Power per tonne of refrigeration,

$$P_1 = \frac{200.6}{60} \times \frac{1}{5.43} = 0.615 \text{ kW/TR} \quad \textbf{Ans.}$$

and total power required to drive the compressor,

$$P_T = 200.6/60 = 3.34 \text{ kW} \textbf{ Ans.}$$

(b) When the suction temperature is –10°C and condensing temperature is 50°C

The *p-h* diagram for the simple cycle of operation for suction temperature of –10°C and condensing temperatures of 40°C and 50°C is shown in Fig. 9.13.

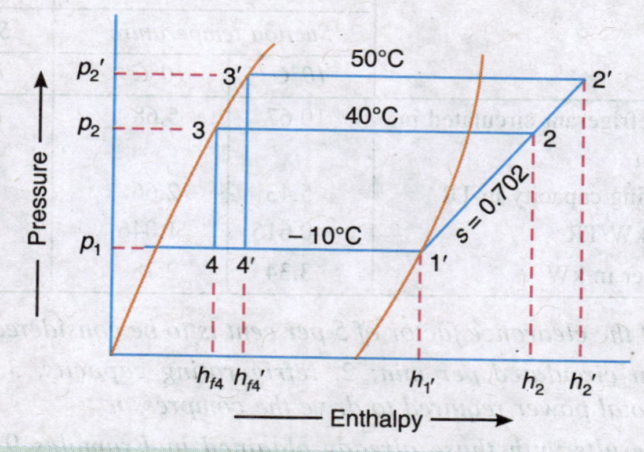

Fig. 9.13. *p-h* diagram.

From the *p-h* diagram, we find that specific volume of R-12 saturated vapour refrigerant at point 1′ (*i.e.* at –10°C),

$$v_{s1}' = 0.077 \text{ m}^3/\text{kg}$$

Enthalpy of R-12 saturated vapour refrigerant at point 1′ (*i.e.* at –10°C),

$$h_1' = 183.2 \text{ kJ/kg}$$

Enthalpy of R-12 superheated vapour refrigerant at point 2′,

$$h_2' = 212.6 \text{ kJ/kg}$$

and enthalpy of R-12 saturated liquid refrigerant at point 3′ (*i.e.* at 50°C),

$$h_{f3}' = h_{f4}' = 84.8 \text{ kJ/kg}$$

We know that mass of refrigerant circulated per min,

$$m_1' = \frac{v_1}{v_{s1}'} = \frac{0.4374}{0.077} = 5.68 \text{ kg/min}$$

∴ Refrigerating capacity of the compressor,

$$R_1' = \frac{m_1'(h_1' - h_{f4}')}{210} = \frac{5.68\,(183.2 - 84.8)}{210} = 2.66 \text{ TR} \textbf{ Ans.}$$

Work done in compressing the refrigerant,

$$W_1' = m_1'(h_2' - h_1') = 5.68\,(212.6 - 183.2) = 167 \text{ kJ/min}$$

∴ Power per tonne of refrigeration,

$$P_1' = \frac{167}{60} \times \frac{1}{2.66} = 1.046 \text{ kW/TR} \textbf{ Ans.}$$

and total power required to drive the compressor,

$$P_T' = 167/60 = 2.78 \text{ kW} \textbf{ Ans.}$$

The following table shows the comparison of results obtained for (*a*) and (*b*) with those already obtained in Examples 9.3 and 9.4, without clearance.

Table 9.1.

S.No.	Particulars	*Condensing temperature 50°C*		*Condensing temperature 40°C*	
		Suction temperature		*Suction temperature*	
		10°C	*–10°C*	*10°C*	*–10°C*
1.	Mass of refrigerant circulated in kg per min	10.67	5.68	10.67	5.68
2.	Refrigerating capacity in TR	5.43	2.66	5.95	2.94
3.	Power in kW/TR	0.615	1.046	0.433	0.856
4.	Total power in kW	3.34	2.78	2.578	2.517

Example 9.6. *If the clearance factor of 5 per cent is to be considered in Example 9.5, find 1. mass of refrigerant circulated per min; 2. refrigerating capacity, 3. power per tonne of refrigeration; and 4. total power required to drive the compressor.*

Compare the results with those already obtained in Examples 9.3 and 9.4, when the clearance factor is considered.

Solution. Given : $C = 5\% = 0.05$

From *p-h* diagram, we find that the pressure corresponding to 50°C saturation temperature,

$$p_{2'} = 12.2 \text{ bar}$$

Pressure corresponding to 10°C saturation temperature,

$$p_1 = 4.23 \text{ bar}$$

and pressure corresponding to –10°C saturation temperature,

$$p_{1'} = 2.2 \text{ bar}$$

We know that volumetric efficiency when suction temperature is 10°C,

$$\eta_v = 1 + C - C\left(\frac{p_{2'}}{p_1}\right)^{\frac{1}{\gamma}} = 1 + 0.05 - 0.05\left(\frac{12.2}{4.23}\right)^{\frac{1}{1.13}}$$

$$= 1.05 - 0.127 = 0.923$$

and volumetric efficiency when suction temperature is –10°C,

$$\eta_{v'} = 1 + C - C\left(\frac{p_{2'}}{p_{1'}}\right)^{\frac{1}{\gamma}} = 1 + 0.05 - 0.05\left(\frac{12.2}{2.2}\right)^{\frac{1}{1.13}}$$

$$= 1.05 - 0.228 = 0.822$$

(a) When suction temperature is 10°C and condensing temperature is 50°C

1. Mass of refrigerant circulated per min

$$= m_1 \times \eta_v = 10.67 \times 0.923 = 9.85 \text{ kg/min} \textbf{ Ans.}$$

2. Refrigerating capacity of the compressor

$$= R_1 \times \eta_v = 5.43 \times 0.923 = 5.01 \text{ TR} \textbf{ Ans.}$$

3. Power per tonne of refrigeration

$$= P_1 \times \eta_v = 0.615 \times 0.923 = 0.567 \text{ kW/TR}$$ **Ans.**

4. Total power required to drive the compressor

$$= P_T \times \eta_v = 3.34 \times 0.923 = 3.08 \text{ kW}$$ **Ans.**

(b) When suction temperature is –10°C and condensing temperature is 50°C

1. Mass of refrigerant circulated per min

$$= m_1' \times \eta_v' = 5.68 \times 0.822 = 4.67 \text{ kg/min}$$ **Ans.**

2. Refrigerating capacity of the compressor

$$= R_1' \times \eta_v' = 2.66 \times 0.822 = 2.186 \text{ TR}$$ **Ans.**

3. Power per tonne of refrigeration

$$= P_1' \times \eta_v' = 1.046 \times 0.822 = 0.86 \text{ kW/TR}$$ **Ans.**

4. Total power required to drive the compressor

$$= P_T' \times \eta_v' = 2.78 \times 0.822 = 2.28 \text{ kW}$$ **Ans.**

The following table shows the comparison of results obtained for (*a*) and (*b*) with those already obtained in Examples 9.3 and 9.4, considering 5 per cent clearance factor.

Table 9.2

S.No.	*Particulars*	*Condensing temperature 50°C*		*Condensing temperature 40°C*	
		Suction temperature		*Suction temperature*	
		10°C	*–10°C*	*10°C*	*–10°C*
1.	Mass of refrigerant circulated in kg/min	9.85	4.67	10.1	4.92
2.	Refrigerating capacity in TR	5.01	2.186	5.63	2.54
3.	Power in kW/TR	0.567	0.86	0.41	0.74
4.	Total power in kW	3.08	2.28	2.44	2.18

9.20 Hermetic Sealed Compressors

When the compressor and motor operate on the same shaft and are enclosed in a common casing, they are known as *hermetic sealed compressors.* These types of compressors eliminate the use of crankshaft seal which is necessary in ordinary compressors in order to prevent leakage of refrigerant. These compressors may operate on either reciprocating or rotary principle and may be mounted with the shaft in either the vertical or horizontal position. The hermetic units are widely used for small capacity refrigerating systems such as in domestic refrigerators, home freezers and window air conditioners.

The hermetic sealed compressors have the following advantages and disadvantages :

Advantages

1. The leakage of refrigerant is completely prevented.
2. It is less noisy.
3. It requires small space because of compactness.
4. The lubrication is simple as the motor and compressor operate in a sealed space with the lubricating oil.

Disadvantages

1. The maintenance is not easy because the moving parts are inaccessible.
2. A separate pump is required for evacuation and charging of refrigerant.

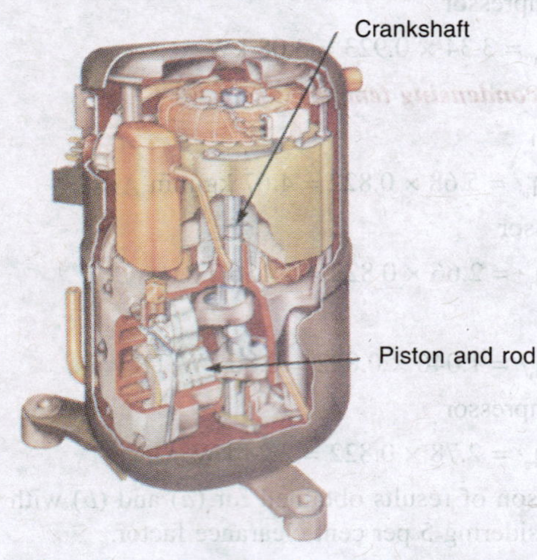

Internal working of a welded hermetic compressor.

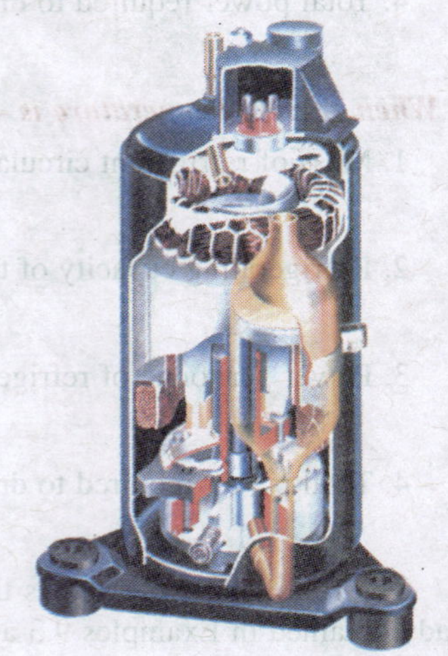

Rotary compressor.

9.21 Rotary Compressors

In rotary compressors, the vapour refrigerant from the evaporator is compressed due to the movement of blades. The rotary compressors are positive displacement type compressors. Since the clearance in rotary compressors is negligible, therefore they have high volumetric efficiency. These compressors may be used with refrigerants R-12, R-22, R-114 and ammonia. Following are the two basic types of rotary compressors :

1. *Single stationary blade type rotary compressor.* A single stationary blade type rotary compressor is shown in Fig. 9.14. It consists of a stationary cylinder, a roller (or impeller) and a shaft. The shaft has an eccentric on which the roller is mounted. A blade is set into the slot of a cylinder in such a manner that it always maintains contacts with the roller by means of a spring. The blade moves in and out of the slot to follow the rotor when it rotates. Since the blade separates the suction and discharge ports as shown in Fig. 9.14, therefore it is often called a sealing blade. When the shaft rotates, the roller also rotates so that it always touches the cylinder wall.

Fig. 9.14 (*a*) to (*d*) shows the various positions of roller as the vapour refrigerant is compressed. Fig. 9.14 (*a*) shows the completion of intake stroke (*i.e.* the cylinder is full of low pressure and temperature vapour refrigerant) and the beginning of compression stroke. When the roller rotates, the vapour refrigerant ahead of the roller is being compressed and the new intake from the evaporator is drawn into the cylinder, as shown in Fig. 9.14 (*b*). As the roller turns towards mid position as shown in Fig. 9.14 (*c*), more vapour refrigerant is drawn into the cylinder while the compressed refrigerant is discharged to the condenser. At the end of compression stroke, as shown in Fig. 9.14 (*d*), most of the compressed vapour refrigerant is passed through the discharge port to the condenser. A new charge of refrigerant is drawn into the cylinder. This, in turn, is compressed and discharged to the condenser. In this way, the low pressure and temperature vapour refrigerant is compressed gradually to a high pressure and temperature.

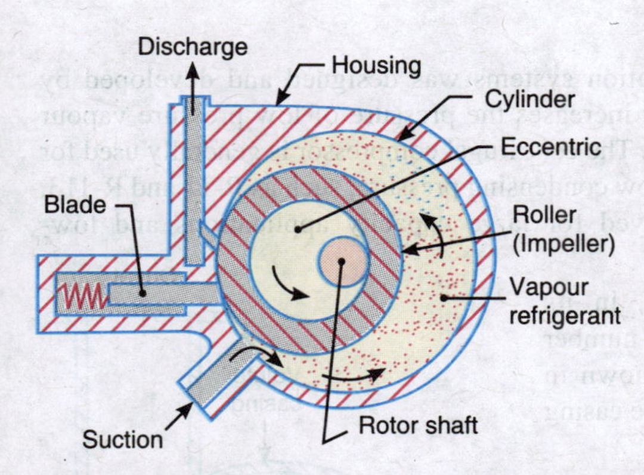

(*a*) Completion of intake stroke and beginning of compression.

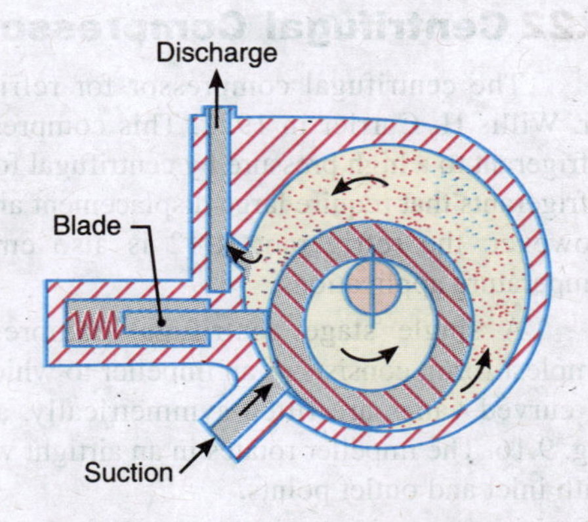

(*b*) Compression stroke continued and new intake stroke started.

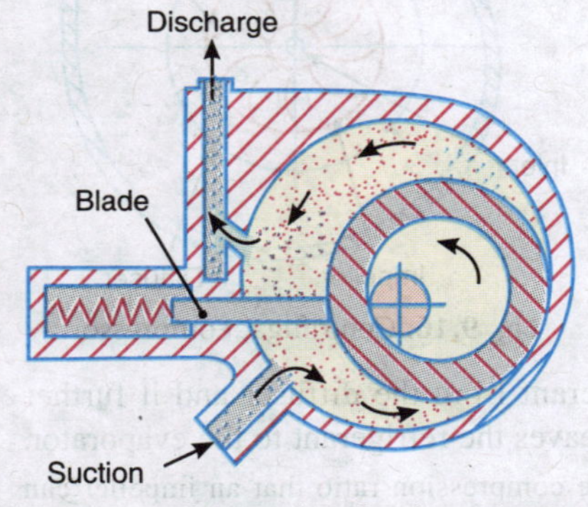

(*c*) Compression continued and new intake stroke continued.

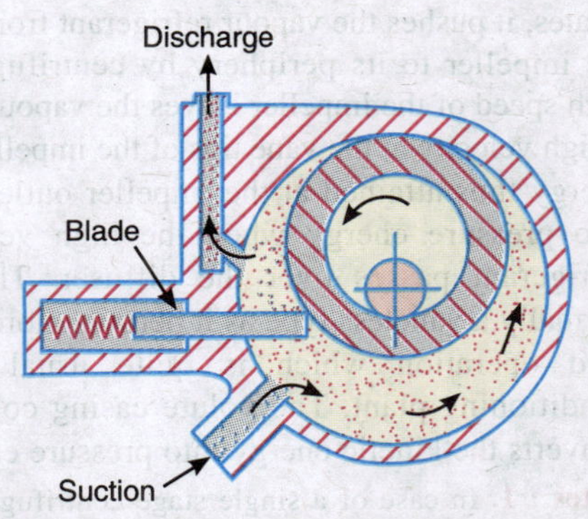

(*d*) Compressed vapour discharged to condenser and new intake stroke continued.

Fig. 9.14. Stationary single blade rotary compressor.

2. *Rotating blade type rotary compressor.* The rotating blade type rotary compressor is shown in Fig. 9.15. It consists of a cylinder and a slotted rotor containing a number of blades. The centre of the rotor is eccentric with the centre of the cylinder. The blades are forced against the cylinder wall by the centrifugal action during the rotation of the motor.

The low pressure and temperature vapour refrigerant from the evaporator is drawn through the suction port. As the rotor turns, the suction vapour refrigerant entrapped between the two adjacent blades is compressed. The compressed refrigerant at high pressure and temperature is discharged through the discharge port to the condenser.

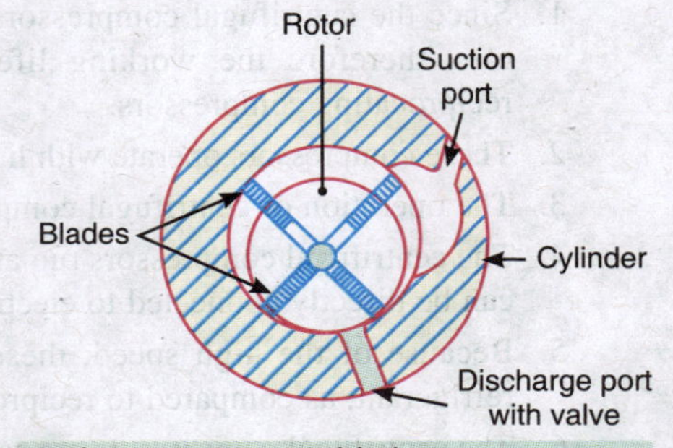

Fig. 9.15. Rotating blade type rotary compressor.

Note : The whole assembly of both the types of rotary compressors is enclosed in a housing which is filled with oil. When the compressor is working, an oil film forms the seal between the high pressure and low pressure side. But when the compressor stops, this seal is lost and therefore high pressure vapour refrigerant will flow into low pressure side. In order to avoid this, a check valve is usually provided in the suction line. This valve prevents the high pressure vapour refrigerant from flowing back to the evaporator.

9.22 Centrifugal Compressors

The centrifugal compressor for refrigeration systems was designed and developed by Dr. Willis H. Carrier in 1922. This compressor increases the pressure of low pressure vapour refrigerant to a high pressure by centrifugal force. The centrifugal compressor is generally used for refrigerants that require large displacement and low condensing pressure, such as R-11 and R-113. However, the refrigerant R-12 is also employed for large capacity applications and low-temperature applications.

A single stage centrifugal compressor, in its simplest form, consists of an impeller to which a number of curved vanes are fitted symmetrically, as shown in Fig. 9.16. The impeller rotates in an airtight volute casing with inlet and outlet points.

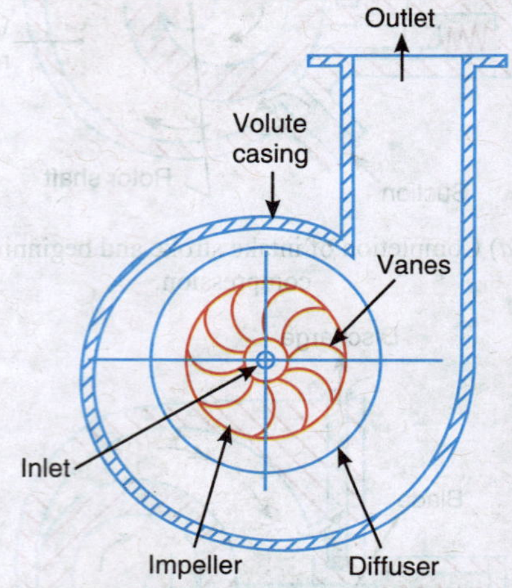

Fig. 9.16. Centrifugal compressor.

The impeller draws in low pressure vapour refrigerant from the evaporator. When the impeller rotates, it pushes the vapour refrigerant from the centre of the impeller to its periphery by centrifugal force. The high speed of the impeller leaves the vapour refrigerant at a high velocity at the vane tips of the impeller. The kinetic energy thus attained at the impeller outlet is converted into pressure energy when the high velocity vapour refrigerant passes over the diffuser. The diffuser is normally a vaneless type as it permits more efficient part load operation which is quite usual in any air-conditioning plant. The volute casing collects the refrigerant from the diffuser and it further converts the kinetic energy into pressure energy before it leaves the refrigerant to the evaporator.

Notes : 1. In case of a single stage centrifugal compressor, the compression ratio that an impeller can develop is limited to about 4.5. But when high compression ratio is desired, multi-stage centrifugal compressors with intercoolers are employed.

2. The centrifugal compressors have no valves, pistons and cylinders. The only wearing parts are the main bearings.

9.23 Advantages and Disadvantages of Centrifugal Compressors over Reciprocating Compressors

Following are the advantages and disadvantages of centrifugal compressors over reciprocating compressors :

Advantages

1. Since the centrifugal compressors have no valves, pistons, cylinders, connecting rod etc., therefore the working life of these compressors is more as compared to reciprocating compressors.
2. These compressors operate with little or no vibration as there are no unbalanced masses.
3. The operation of centrifugal compressors is quiet and calm.
4. The centrifugal compressors run at high speeds (3000 r.p.m. and above), therefore these can be directly connected to electric motors or steam turbines.
5. Because of the high speed, these compressors can handle large volume of vapour refrigerant, as compared to reciprocating compressors.
6. The centrifugal compressors are especially adapted for systems ranging from 50 to 5000 tonnes. They are also used for temperature ranges between – 90°C and + 10°C.

7. The efficiency of these compressors is considerably high.
8. The large size centrifugal compressors require less floor area as compared to reciprocating compressors.

Disadvantages

1. The main disadvantage in centrifugal compressors is *surging. It occurs when the refrigeration load decreases to below 35 per cent of the rated capacity and causes severe stress conditions in the compressor.
2. The increase in pressure per stage is less as compared to reciprocating compressors.
3. The centrifugal compressors are not practical below 50 tonnes capacity load.
4. The refrigerants used with these compressors should have high specific volume.

9.24 Capacity Control of Compressors

There are many refrigeration applications in which the refrigeration load is not constant. It is, therefore, necessary to provide some means to control the capacity of a compressor according to the load. It may be noted that the compressors operating under partial loads and low back pressure creates a condition where the coil may freeze or damage may result.

Here we shall discuss some of the control systems used for reciprocating and centrifugal compressors.

9.25 Capacity Control for Reciprocating Compressors

Following are the various methods used for controlling the capacity of reciprocating compressors :

1. By using multi-speed motor. It is a simple and satisfactory method of controlling capacity of a compressor. The capacity of a compressor is proportional to the speed of the driving motor. When the suction pressure of the refrigeration system is high, the motor speed and the compressor capacity must be increased. Due to this reason, electric motors with two or more speeds are used. The variable speed motors find limited application because the motors and their controls are expensive to use on large installations.

2. By suction valve lift control. In multi-cylinder compressors, the capacity may be controlled by forcing the suction valve to remain open in one or more cylinders and making them ineffective according to the load on the system.

3. By using multiple compressors. The multiple compressors of the same capacity can be used to provide capacity control. The operation of all units will provide the maximum desired capacity and the operation of the various combinations of the units will permit efficient capacity reduction. When this system of capacity control is used, the units are usually installed with common suction and discharge headers.

4. By cylinder by-pass system. The cylinder by-pass method of controlling capacity is activated by either temperature or pressure controls. In this method, used with multi-cylinder compressors, one or more cylinders may be made ineffective by by-passing the gas from the cylinder discharge to the intake port. A solenoid valve with thermostat is used to operate the by-pass. A check valve is provided in the system in order to isolate the inactive cylinders from the active cylinders. In this method, the power does not decrease in proportion to the load on the system.

5. By hot gas by-pass system. The capacity of a reciprocating compressor can be controlled by putting an artificial load on it. This is done by passing the hot gas from the discharge side to

* The reversal of flow of refrigerant from compressor to the evaporator when refrigeration load decreases, is called *surging*.

the suction side through a constant pressure expansion valve, as shown in Fig. 9.17. When the evaporator pressure tends to fall, the hot gas by-pass will maintain the constant suction pressure. The hot gas by-pass to the entrance and exit of the evaporator is shown in Fig. 9.17 (*a*) and (*b*) respectively. Since this is a method of loading and not unloading the compressor, therefore the brake power of the compressor remains constant irrespective of the load on the evaporator. Thus by employing the hot gas by-pass system, there is no saving of power. This method only prevents the evaporator surface from frosting up and avoids the working of compressor at excessive back pressure. When this system is used to supplement another control system discussed above, a saving in power can be obtained.

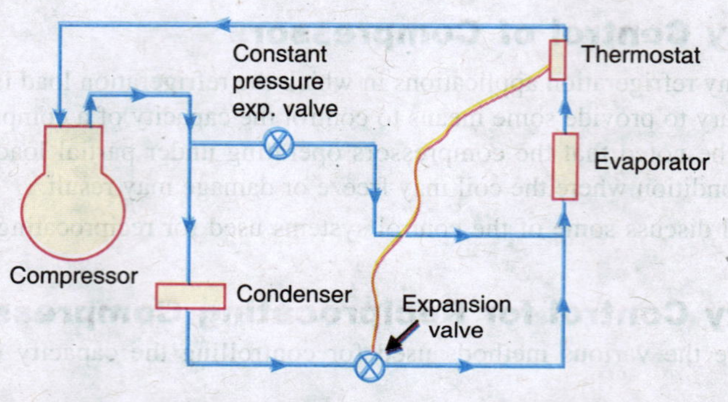

(*a*) Hot gas by-pass at the entrance of the evaporator.

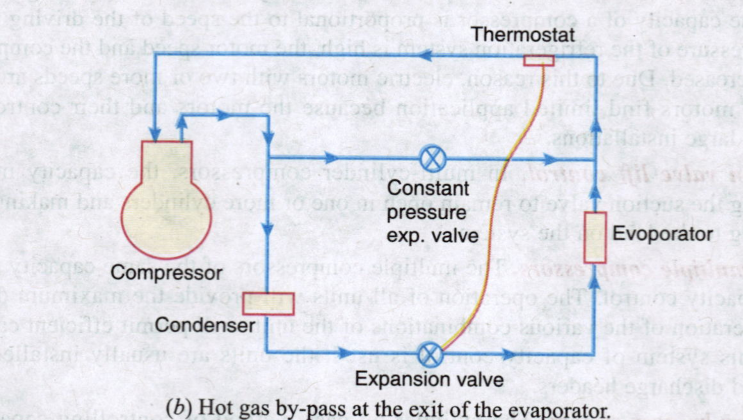

(*b*) Hot gas by-pass at the exit of the evaporator.

Fig. 9.17. Hot gas by-pass system.

9.26 Capacity Control of Centrifugal Compressors

The discharge pressure *versus* capacity relationship for a centrifugal compressor is very useful for capacity control. Generally, an impeller with backward curve blades is used because it gives a fairly flat pressure capacity characteristic. Fig. 9.18 shows the graph between discharge pressure and capacity in m^3/min, for an impeller with backward curve blades, running at two different constant speeds. The graph also shows the different system resistances. It may be noted that the capacity of a centrifugal compressor can be reduced by increasing the system resistance. This may be done by using the following methods :

1. *Condenser water control system.* The capacity of a centrifugal compressor may be controlled by increasing the condensing pressure and temperature. This is done by reducing the quantity of condenser cooling water. It may be noted that when the cooling water passing through the condenser is reduced, the rate of condensate also reduces. This gives rise in condenser pressure and temperature, thereby forcing the compressor to self-adjust to the new part load capacity as shown by point *a* in Fig. 9.18.

Curve 1 — Discharge pressure at constant speed.

Curve 2 — Discharge pressure, at same speed as that of curve 1, with inlet vane control.

Curve 3 — Discharge pressure at constant speed.

Fig. 9.18. Graph between discharge pressure and compressor capacity for a centrifugal compressor.

2. *Inlet vane control system.* The capacity of a centrifugal compressor may be controlled by the inlet vane which throttles the flow at inlet and reduces the inlet pressure. The discharge pressure at the same speed of compressor gets reduced. The curve 2 in Fig. 9.18 shows the performance with inlet vane control. Due to throttling at inlet, the system resistance increases at the same speed with the same condensing temperature as for the rated capacity shown by point *b*. The new operating point *c* will satisfy the part load requirement of the system.

3. *Speed control system.* The capacity of a centrifugal compressor may be controlled by varying the speed of the compressor as the discharge pressure is a function of the compressor speed. The variation in compressor speed changes system resistance due to the change in velocity of flow. The point *d*, as shown in Fig. 9.18, satisfies the part load requirements.

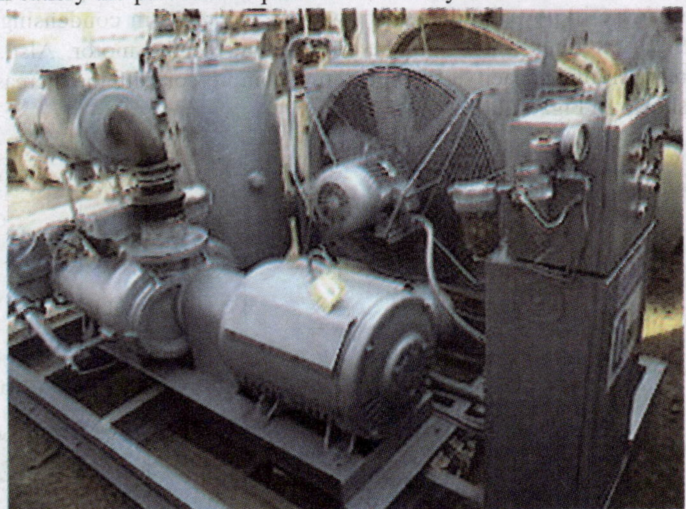

Centrifugal compressor.

Out of all the three methods discussed above, the speed control system is most efficient but it is expensive. The inlet vane control system is cheaper and less efficient than speed control system. The condenser water control system is the cheapest among all the systems, but it is least efficient.

9.27 Comparison of Performance of Reciprocating and Centrifugal Compressors

We have already discussed that the centrifugal compressors have high efficiency over a large range of load and a large volumetric displacement for its size, as compared to reciprocating compressors. In addition to this, the centrifugal compressor has many other desirable features. The most important feature is the flat-head capacity characteristic as compared to reciprocating compressor, as shown in Fig. 9.19. From the figure, we see that for a centrifugal compressor, evaporator temperature changes only from 2°C to 7.5°C when the load changes from 100 to 240 tonnes of refrigeration, whereas the evaporator temperature changes from –11°C to 6°C for the same load change for a reciprocating compressor.

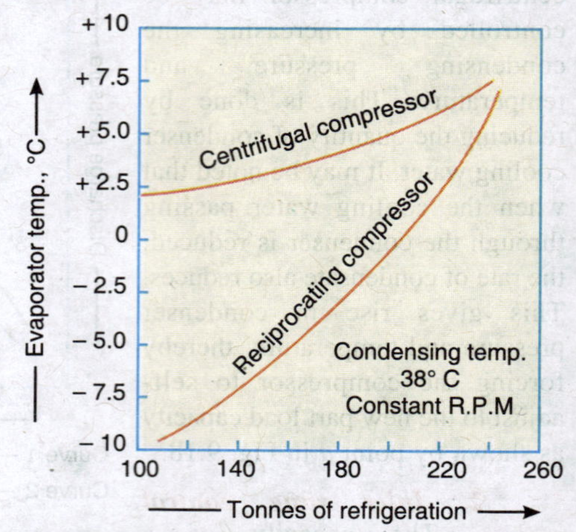

Fig. 9.19. Flat-head capacity characteristic variation with evaporator temperature.

Another advantageous feature of the centrifugal compressor is *non-overloading characteristic*, as shown in Fig. 9.20. We see from Fig. 9.20 (*a*) that for a centrifugal compressor, there is a reduction in power requirement with the increase in condensing temperature. Also, there is a reduction in refrigerating capacity of the centrifugal compressor with the increase in condensing temperature as shown in Fig. 9.20 (*b*). Hence there is no overloading of the motor as the condensing temperature increases. However, for reciprocating compressors, there is a small increase in power requirement with the increase in condensing temperature as shown in Fig. 9.20 (*a*). Thus, there will be overloading of the motor. Also, there is a small decrease in refrigerating capacity of reciprocating compressor with the increase in condensing temperature as shown in Fig. 9.20(*b*).

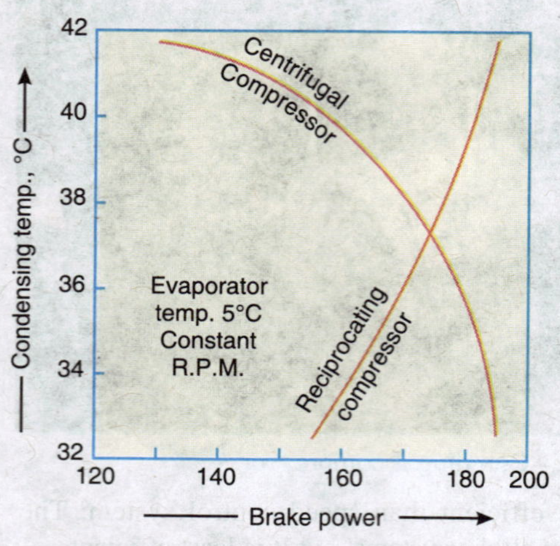

(*a*) B.P. variation with condensing temperature.

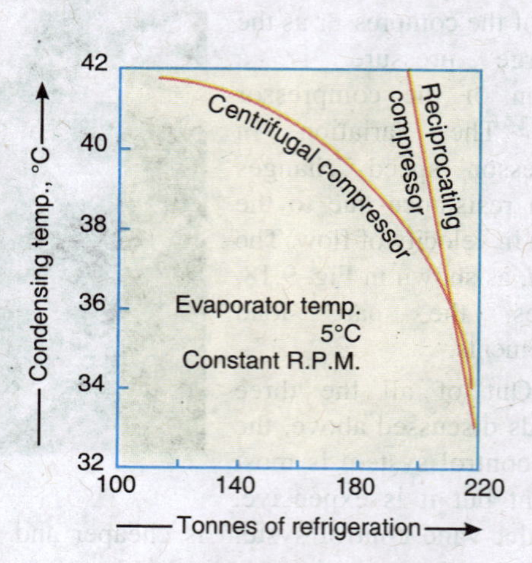

(*b*) Capacity variation with condensing temperature.

Fig. 9.20. Non-overloading characteristic.

The effect of speed on the refrigerating capacity and power requirements is shown in Fig. 9.21 (*a*) and (*b*) respectively for both the centrifugal and reciprocating compressors.

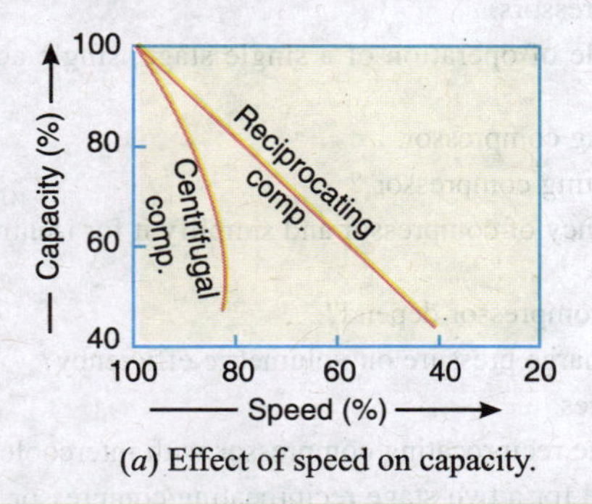

(*a*) Effect of speed on capacity.

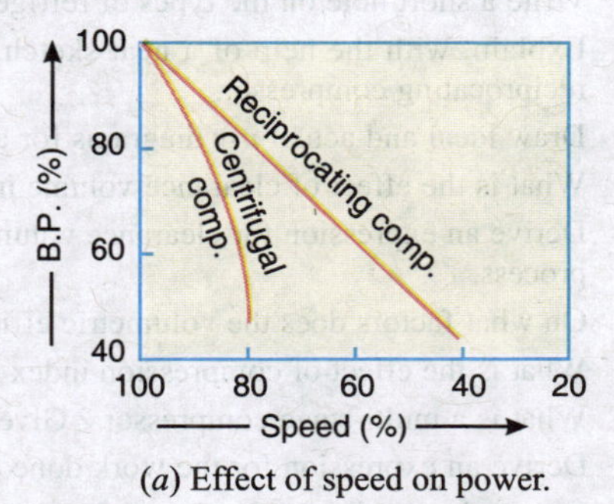

(*a*) Effect of speed on power.

Fig. 9.21. Effect of speed on the capacity and power.

EXERCISES

1. A reciprocating compressor compresses 1 m^3/min of gas from 1 bar to 5 bar. The clearance is 8% and the isentropic compression index is 1.4. Determine the percentage increase in compression work if the re-expansion index is 1.1 instead of 1.4. **[Ans.** 7.3 %]
2. The catalogue data of R-12 compressor shows that the compressor delivers 10 TR at – 5°C evaporator temperature and 40°C condensing temperature when the suction temperature of the compressor is 15°C. Determine the compressor capacity, if the suction temperature of the compressor is 0°C. Assume simple saturation cycle in both the cases. **[Ans.** 9.8 TR]
3. In an ammonia compressor, vapours enter at 2.908 bar saturated and leave at 15.55 bar and 112°C. Determine specific volumes at suction and discharge.

   The compressor is twin cylinder single acting with stroke to bore ratio of 1, speed = 1440 r.p.m. and rated for 10 TR at evaporator and condenser temperatures of –10°C and 40°C respectively. Find the diameters of the cylinders, assuming 4% clearance. Use the following data :

t (°C)	p(bar)	v_g (m^3/kg)	h_f (kJ/kg)	h_g (kJ/kg)	s_f(kJ/kg K)	s_g (kJ/kg K)
–10	2.908	0.418	134.95	1431.41	0.5435	5.4712
40	15.55	0.0833	371.47	1472.02	1.3574	4.8728

For superheated ammonia at 15.55 bar

Superheat (K)	v (m^3/kg)	h (kJ/kg)	s (kJ/kg K)
60	0.108	1647.9	5.3883
80	0.116	1700.3	5.5253

[Ans. 0.418 m^3/kg ; 0.1128 m^3/kg ; 74.3 mm]

4. A two cylinder R-12 reciprocating compressor has a bore 80 mm and stroke 60 mm. The compressor runs at 1450 r.p.m. The condensing temperature is 40°C. Find the mass of refrigerant circulated per minute and refrigerating capacity of the compressor when the evaporating temperature is (*a*) 5°C and (*b*) 0°C. Neglect clearance.

   [Ans. (*a*) 18.4 kg / min, 10.08 TR ; (*b*) 15.76 kg / min, 8.47 TR]
5. If the clearance of 5 per cent is to be considered in Question 4, find the mass of refrigerant circulated per minute, refrigerating capacity of the compressor and power required to drive the compressor.

   Assume the isentropic index of compression as 1.15.

   [Ans. (*a*) 17.17 kg / min, 9.4 TR ; (*b*) 14.43 kg / min, 7.76 TR]

QUESTIONS

1. Write a short note on the types of refrigeration compressors.
2. Explain, with the help of a neat sketch, the principle of operation of a single stage, single acting reciprocating compressor.
3. Draw ideal and actual *p-v* diagrams for a reciprocating compressor.
4. What is the effect of clearance volume in a reciprocating compressor ?
5. Derive an expression for clearance volumetric efficiency of compressor and simplify it for isentropic process.
6. On what factors does the volumetric efficiency of a compressor depend?
7. What is the effect of compression index and the discharge pressure on volumetric efficiency?
8. What is a multi-stage compressor ? Give its advantages.
9. Derive an expression for the work done of a two-stage reciprocating compressor with intercooler.
10. Obtain the conditions for the minimum work required for a two-stage reciprocating compressor.
11. Describe the effect of suction temperature on the refrigerating capacity and brake power of a reciprocating compressor.
12. What do you understand by hermetic sealed compressor ? Give its advantages.
13. Explain the working of a single stationary blade type rotary compressor.
14. Describe, with a sketch, a centrifugal compressor. Where are centrifugal compressors preferred over reciprocating compressors in refrigeration system ?
15. What are the advantages and disadvantages of centrifugal compressors over reciprocating compressors?
16. What type of compressor is preferred with the refrigerant R-113 ?
17. Discuss the capacity control systems for reciprocating and centrifugal compressors.
18. Compare the performance of reciprocating and centrifugal compressors.

OBJECTIVE TYPE QUESTIONS

1. A refrigerant compressor is used to
   (*a*) raise the pressure of the refrigerant
   (*b*) raise the temperature of the refrigerant
   (*c*) circulate the refrigerant through the refrigerating system
   (*d*) all of the above
2. The refrigerant supplied to a compressor must be
   (*a*) superheated vapour refrigerant
   (*b*) dry saturated liquid refrigerant
   (*c*) a mixture of liquid and vapour refrigerant
   (*d*) none of the above
3. The pressure at the inlet of a refrigerant compressor is called
   (*a*) suction pressure (*b*) discharge pressure
   (*c*) critical pressure (*d*) back pressure
4. The pressure at the outlet of a refrigerant compressor is called
   (*a*) suction pressure (*b*) discharge pressure
   (*c*) critical pressure (*d*) back pressure
5. The reciprocating refrigerant compressors are very suitable for
   (*a*) small displacements and low condensing pressures
   (*b*) large displacements and high condensing pressures
   (*c*) small displacements and high condensing pressures
   (*d*) large displacements and low condensing pressures

6. The work requirement for a reciprocating compressor is minimum when the compression process is
(*a*) isothermal (*b*) isentropic (*c*) polytropic (*d*) adiabatic

7. The clearance factor is the ratio of
(*a*) swept volume of the cylinder to the clearance volume
(*b*) total volume of the cylinder to the clearance volume
(*c*) clearance volume to the swept volume of the cylinder
(*d*) clearance volume to the total volume of the cylinder

8. The clearance volume in reciprocating refrigerant compressors ...... the work done and the power required for compressing the refrigerant.
(*a*) does not affect (*b*) increases (*c*) decreases

9. In compound compression refrigeration systems with intercooling, the optimum intercooler or intermediate pressure p_2, when the cooling ratio is fixed, is given by
(*a*) $p_2 = p_1 / p_3$ (*b*) $p_2 = p_3 / p_1$ (*c*) $p_2 = p_1 \times p_3$ (*d*) $p_2 = \sqrt{p_1 \times p_3}$
where p_1= Suction or evaporator pressure, and p_3 = Condenser pressure.

10. The centrifugal compressors are generally used for refrigerants that require
(*a*) small displacements and low condensing pressures
(*b*) large displacements and high condensing pressures
(*c*) small displacements and high condensing pressures
(*d*) large displacements and low condensing pressures

ANSWERS

1. (*d*)	**2.** (*a*)	**3.** (*a*)	**4.** (*b*)	**5.** (*c*)
6. (*a*)	**7.** (*c*)	**8.** (*a*)	**9.** (*d*)	**10.** (*d*)

CHAPTER 10

Condensers

1. Introduction.
2. Working of a Condenser.
3. Factors Affecting the Condenser Capacity.
4. Heat Rejection Factor.
5. Classification of Condensers.
6. Air-Cooled Condensers.
7. Types of Air-Cooled Condensers.
8. Water-Cooled Condensers.
9. Types of Water-Cooled Condensers.
10. Comparison of Air-Cooled and Water-Cooled Condensers.
11. Fouling Factor.
12. Heat Transfer in Condensers.
13. Condensing Heat Transfer Coefficient.
14. Air Side Coefficient.
15. Water Side Coefficient.
16. Finned Tubes.
17. Evaporative Condensers.
18. Cooling Towers and Spray Ponds.
19. Capacity of Cooling Towers and Spray Ponds.
20. Types of Cooling Towers.
21. Natural Draft Cooling Towers.
22. Mechanical Draft Cooling Towers.
23. Forced Draft Cooling Towers.
24. Induced Draft Cooling Towers.

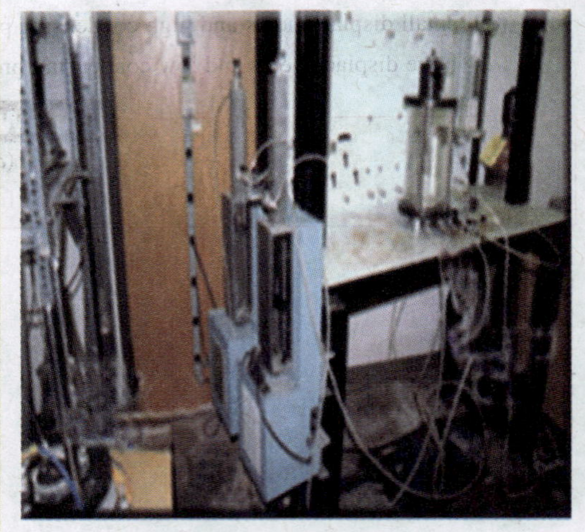

10.1 Introduction

The condenser is an important device used in the high pressure side of a refrigeration system. Its function is to remove heat of the hot vapour refrigerant discharged from the compressor. The hot vapour refrigerant consists of the heat absorbed by the evaporator and the heat of compression added by the mechanical energy of the compressor motor. The heat from the hot vapour refrigerant in a condenser is removed first by transferring it to the walls of the condenser tubes and then from the tubes to the condensing or cooling medium. The cooling medium may be air or water or a combination of the two.

The selection of a condenser depends upon the capacity of the refrigerating system, the type of refrigerant used and the type of cooling medium available.

10.2 Working of a Condenser

The working of a condenser may be best understood by considering a simple refrigerating system as shown in Fig. 10.1 (*a*). The corresponding *p-h* diagram showing three stages of a refrigerant cooling is shown in Fig. 10.1 (*b*). The compressor draws in the superheated vapour refrigerant that contains the heat it absorbed in the evaporator. The compressor adds more heat (*i.e.* the heat of compression) to the superheated vapour. This highly superheated vapour from the compressor is pumped to the condenser through the discharge line. The condenser cools the refrigerant in the following three stages :

1. First of all, the superheated vapour is cooled to saturation temperature (called desuperheating) corresponding to the pressure of the refrigerant. This is shown by the line 2–3 in Fig. 10.1 (*b*). The desuperheating occurs in the discharge line and in the first few coils of the condenser.

2. Now the saturated vapour refrigerant gives up its latent heat and is condensed to a saturated liquid refrigerant. This process, called condensation, is shown by the line 3-4.

3. The temperature of the liquid refrigerant is reduced below its saturation temperature (*i.e.* sub-cooled) in order to increase the refrigeration effect. This process is shown by the line 4-5.

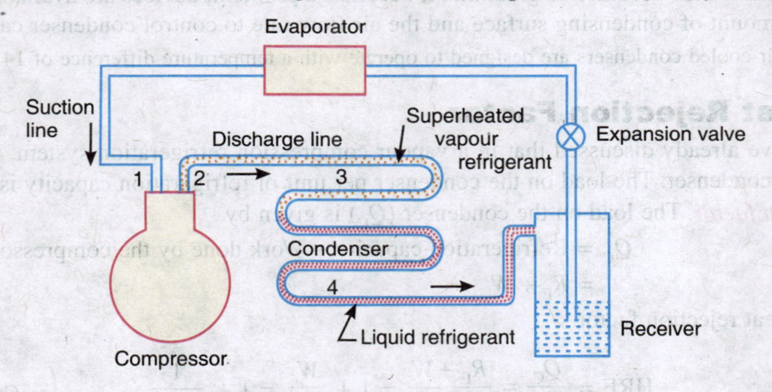

(*a*) Schematic diagram of a simple refrigerating system.

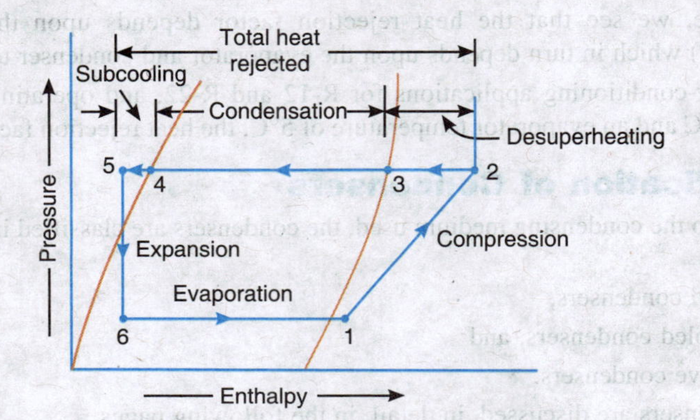

(*b*) *p-h* diagram of a simple refrigerating system.

Fig. 10.1

10.3 Factors Affecting the Condenser Capacity

The condenser capacity is the ability of the condenser to transfer heat from the hot vapour refrigerant to the condensing medium. The heat transfer capacity of a condenser depends upon the following factors :

1. ***Material.*** Since the different materials have different abilities of heat transfer, therefore the size of a condenser of a given capacity can be varied by selecting the right material. It may be noted that higher the ability of a material to transfer heat, the smaller will be the size of condenser.

2. ***Amount of contact.*** The condenser capacity may be varied by controlling the amount of contact between the condenser surface and the condensing medium. This can be done by varying the surface area of the condenser and the rate of flow of the condensing medium over the condenser surface. The amount of liquid refrigerant level in the condenser also affects the amount of contact between the vapour refrigerant and the condensing medium. The portion of the condenser used for liquid sub-cooling cannot condense any vapour refrigerant.

3. ***Temperature difference.*** The heat transfer capacity of a condenser greatly depends upon the temperature difference between the condensing medium and the vapour refrigerant. As the temperature difference increases, the heat transfer rate increases and therefore the condenser capacity increases. Generally, this temperature difference cannot be controlled. But when the temperature difference becomes so great that it becomes a problem, devices are available that will change the amount of condensing surface and the air flow rate to control condenser capacity.

Note : Most air-cooled condensers are designed to operate with a temperature difference of 14° C.

10.4 Heat Rejection Factor

We have already discussed that in a vapour compression refrigeration system, the heat is rejected in a condenser. The load on the condenser per unit of refrigeration capacity is known as *heat rejection factor.* The load on the condenser (Q_C) is given by

$$Q_C = \text{Refrigeration capacity} + \text{Work done by the compressor}$$

$$= R_E + W$$

∴ Heat rejection factor,

$$\text{HRF} = \frac{Q_C}{R_E} = \frac{R_E + W}{R_E} = 1 + \frac{W}{R_E} = 1 + \frac{1}{\text{COP}} \quad \ldots \left(\because \text{COP} = \frac{R_E}{W} \right)$$

From above, we see that the heat rejection factor depends upon the coefficient of performance (COP) which in turn depends upon the evaporator and condenser temperatures.

In actual air-conditioning applications for R-12 and R-22, and operating at a condenser temperature of 40°C and an evaporator temperature of 5°C, the heat rejection factor is about 1.25.

10.5 Classification of Condensers

According to the condensing medium used, the condensers are classified into the following three groups :

1. Air-cooled condensers,
2. Water-cooled condensers, and
3. Evaporative condensers.

These condensers are discussed, in detail, in the following pages.

10.6 Air-Cooled Condensers

An air-cooled condenser is one in which the removal of heat is done by air. It consists of steel or copper tubing through which the refrigerant flows. The size of tube usually ranges from 6 mm to 18 mm outside diameter, depending upon the size of condenser. Generally copper tubes are used because of its excellent heat transfer ability. The condensers with steel tubes are used in ammonia refrigerating systems. The tubes are usually provided with plate type fins to increase the surface area for heat transfer, as shown in Fig. 10.2. The fins are usually made from aluminium because of its light weight. The fin spacing is quite wide to reduce dust clogging.

Air-cooled condenser.

The condensers with single row of tubing provides the most efficient heat transfer. This is because the air temperature rises as it passes through each row of tubing. The temperature difference between the air and the vapour refrigerant decreases in each row of tubing and therefore each row becomes less effective. However, single-row condensers require more space than multi-row condensers. The single-row condensers are usually used in small capacity refrigeration systems such as domestic refrigerators, freezers, water coolers and room air conditioners.

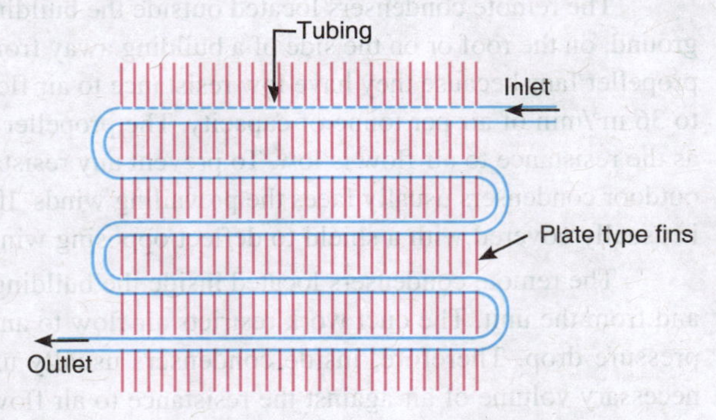

Fig. 10.2. Air-cooled condenser.

The air-cooled condensers may have two or more rows of tubing, but the condensers with up to six rows of tubing are common. Some condensers have seven or eight rows. However more than eight rows of tubing are usually not efficient. This is because the air temperature will be too close to the condenser temperature to absorb any more heat after passing through eight rows of tubing.

Note : The main disadvantage of an air-cooled condenser is that it operates at a higher condensing temperature than a water-cooled condenser. The higher condensing temperature causes the compressor to work more.

10.7 Types of Air-Cooled Condensers

Following are the two types of air-cooled condensers :

1. ***Natural convection air-cooled condensers.*** In natural convection air-cooled condenser, the heat transfer from the condenser coils to the air is by natural convection. As the air comes in contact with the warm condenser tubes, it absorbs heat from the refrigerant and thus the temperature of air increases. The warm air, being lighter, rises up and the cold air from below rises to take away the heat from the condenser. This cycle continues in natural convection air-cooled condensers. Since the rate of heat transfer in natural convection condenser is slower, therefore they require a larger surface area as compared to forced convection condensers. The natural convection air-cooled condensers are used only in small-capacity applications such as domestic refrigerators, freezers, water coolers and room air-conditioners.

2. ***Forced convection air-cooled condensers.*** In forced convection air-cooled condensers, the fan (either propeller or centrifugal) is used to force the air over the condenser coils to increase its heat transfer capacity. The forced convection condensers may be divided into the following two groups :

(*a*) Base mounted air-cooled condensers, and (*b*) Remote air-cooled condensers.

The base mounted air-cooled condensers, using propeller fans, are mounted on the same base of compressor, motor, receiver and other controls. The entire arrangement is called a *condensing unit.* In small units, the compressor is belt-driven from the motor and the fan required to force the air through the condenser is mounted on the shaft of the motor. The use of this type of compressor for indoor units is limited up to 3 kW capacity motor only. These condensing units are used on packaged refrigeration systems of 10 tonnes or less.

The remote air-cooled condensers are used on systems above 10 tonnes and are available up to 125 tonnes. The systems above 125 tonnes usually have two or more condensers. These condensers may be horizontal or vertical. They can be located either outside or inside the building.

The remote condensers located outside the building can be mounted on a foundation on the ground, on the roof or on the side of a building away from the walls. These condensers usually use propeller fans because they have low resistance to air flow and free air discharge. They require 18 to 36 m^3/min of air per tonne of capacity. The propeller fans can move this volume of air as long as the resistance to air flow is low. To prevent any resistance to air flow, the fan intake on vertical outdoor condensers usually faces the prevailing winds. If this is not possible, the air discharge side is usually covered with a shield to deflect opposing winds.

The remote condensers located inside the building usually require duct work to carry air to and from the unit. The duct work restricts air flow to and from the condenser and causes high air pressure drop. Therefore, inside condensers usually use centrifugal fans which can move the necessary volume of air against the resistance to air flow.

10.8 Water-Cooled Condensers

A water-cooled condenser is one in which water is used as the condensing medium. They are always preferred where an adequate supply of clear inexpensive water and means of water disposal are available. These condensers are commonly used in commercial and industrial refrigerating units. The water-cooled condensers may use either of the following two water systems :

1. Waste water system, or 2. Recirculated water system.

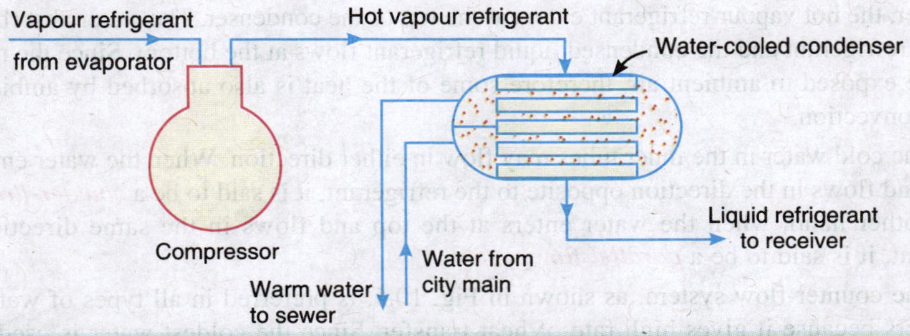

Fig. 10.3. Water-cooled condenser using waste water system.

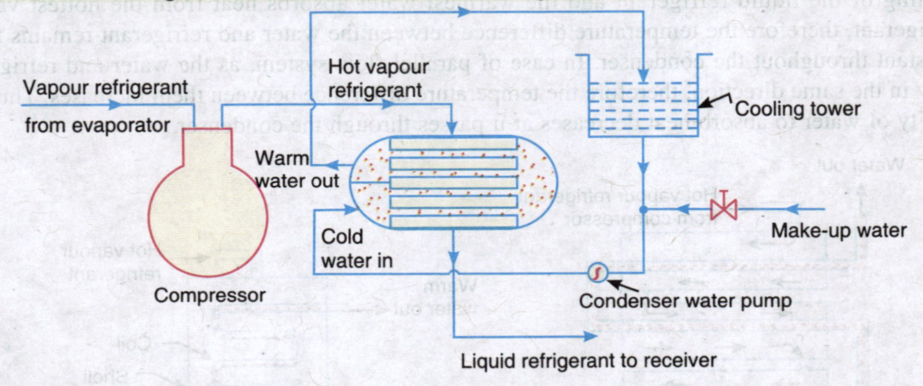

Fig. 10.4. Water-cooled condenser with recirculating water system.

In a *waste water system,* the water after circulating in the condenser is discharged to a sewer, as shown in Fig. 10.3. This system is used on small units and in locations where large quantities of fresh inexpensive water and a sewer system large enough to handle the waste water are available. The most common source of fresh water supply is the city main.

In a *recirculated water system,* as shown in Fig. 10.4, the same water circulating in the condenser is cooled and used again and again. Thus this system requires some type of water-cooling device. The cooling water towers and spray ponds are the most common cooling devices used in a recirculated water system. The warm water from the condenser is led to the cooling tower where it is cooled by self evaporation into a stream of air. The water pumps are used to circulate the water through the system and then to the cooling tower which is usually located on the roof. Once a recirculated water system is filled with water, the only additional water required is make-up water. The make-up water simply replaces the water that evaporates from the cooling tower or spray pond.

Note : The water-cooled condensers operate at a lower condensing temperature than an air-cooled condenser. This is because the supply water temperature is normally lower than the ambient air temperature, but the difference between the condensing and cooling medium temperatures is normally the same (*i.e.* 14°C). Thus, the compressor for a water-cooled condenser requires less power for the same capacity.

10.9 Types of Water-Cooled Condensers

The water-cooled condensers are classified, according to their construction, into the following three groups :

1. ***Tube-in-tube or double-tube condensers.*** The tube-in-tube or double-tube condenser, as shown in Fig. 10.5, consists of a water tube inside a large refrigerant tube. In this type of condenser, the hot vapour refrigerant enters at the top of the condenser. The water absorbs the heat from the refrigerant and the condensed liquid refrigerant flows at the bottom. Since the refrigerant tubes are exposed to ambient air, therefore some of the heat is also absorbed by ambient air by natural convection.

The cold water in the inner tubes may flow in either direction. When the water enters at the bottom and flows in the direction opposite to the refrigerant, it is said to be a *counter-flow system.* On the other hand, when the water enters at the top and flows in the same direction as the refrigerant, it is said to be a *parallel-flow system.*

The counter-flow system, as shown in Fig. 10.5, is preferred in all types of water-cooled condensers because it gives high rate of heat transfer. Since the coldest water is used for final cooling of the liquid refrigerant and the warmest water absorbs heat from the hottest vapour refrigerant, therefore the temperature difference between the water and refrigerant remains fairly constant throughout the condenser. In case of parallel-flow system, as the water and refrigerant flow in the same direction, therefore the temperature difference between them increases. Thus the ability of water to absorb heat decreases at it passes through the condenser.

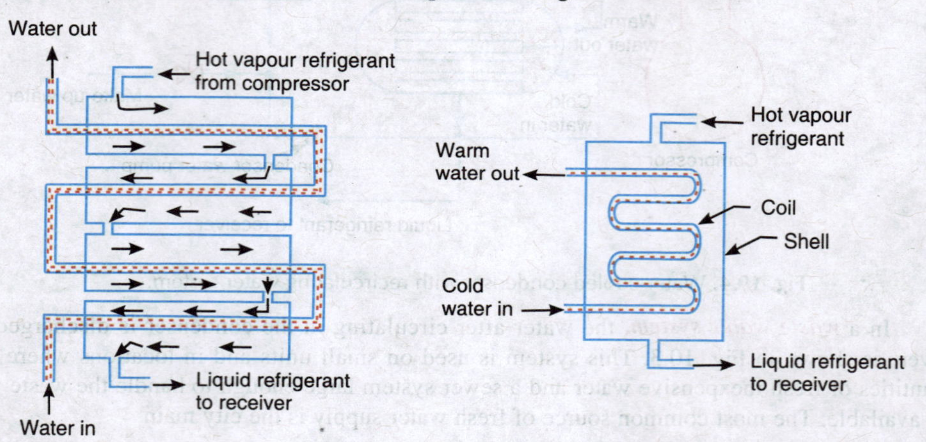

Fig. 10.5. Tube-in-tube condenser.

Fig. 10.6. Shell and coil condenser.

2. Shell and coil condensers. A shell and coil condenser, as shown in Fig. 10.6, consists of one or more water coils enclosed in a welded steel shell. Both the finned and bare coil types are available.

The shell and coil condenser may be either vertical (as shown in the figure) or horizontal. In this type of condenser, the hot vapour refrigerant enters at the top of the shell and surrounds the water coils. As the vapour condenses, it drops to the bottom of the shell which often serves as a receiver. Most vertical type shell and coil condensers use counter-flow water system as it is more efficient than parallel-flow water system. In the shell and coil condensers, coiled tubing is free to expand and

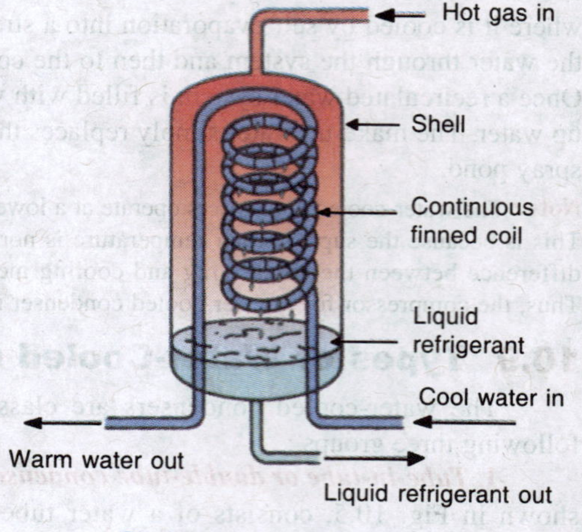

Shell and coil condenser.

contract with temperature changes because of its spring action and can withstand any strain caused by temperature changes. Since the water coils are enclosed in a welded steel shell, therefore the mechanical cleaning of these coils is not possible. The coils are cleaned with chemicals. The shell and coil condensers are used for units up to 50 tonnes capacity.

3. *Shell and tube condensers.* The shell and tube condenser, as shown in Fig. 10.7, consists of a cylindrical steel shell containing a number of straight water tubes. The tubes are expanded into grooves in the tube sheet holes to form a vapour-tight fit. The tube sheets are welded to the shell at both the ends. The removable water boxes are bolted to the tube sheet at each end to facilitate cleaning of the condenser. The intermediate supports are provided in the shell to avoid sagging of the tubes.

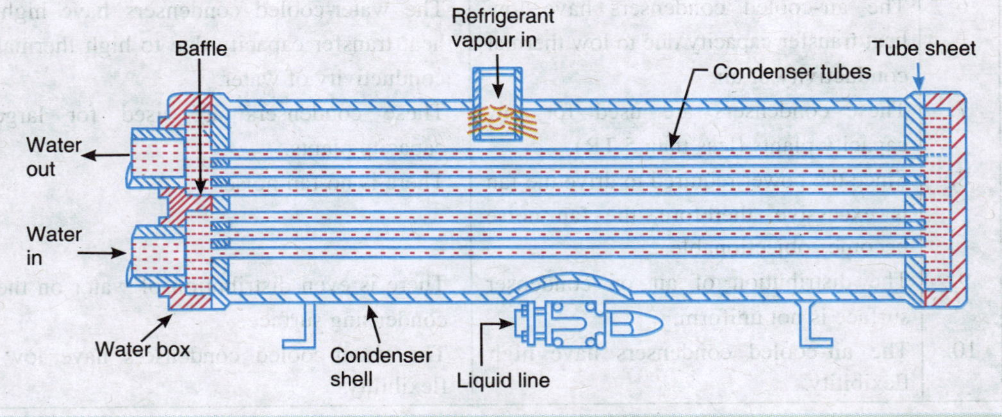

Fig. 10.7. Shell and tube condenser.

The condenser tubes are made either from steel or copper, with or without fins. The steel tubes without fins are usually used in ammonia refrigerating systems because ammonia corrodes copper tubing.

In this type of condenser, the hot vapour refrigerant enters at the top of the shell and condenses as it comes in contact with water tubes. The condensed liquid refrigerant drops to the bottom of the shell which often serves as a receiver. However, if the maximum storage capacity for the liquid refrigerant is less than the total charge of the system, then a receiver of adequate capacity has to be added in case the pump down facility is to be provided as in ice plants, cold storages etc. In some condensers, extra rows of water tubes are provided at the lower end of the condenser for sub-cooling of the liquid refrigerant below the condensing temperature.

10.10 Comparison of Air-Cooled and Water-Cooled Condensers

Following are the comparison between air-cooled and water-cooled condensers :

S.No.	*Air-cooled condenser*	*Water-cooled condenser*
1.	Since the construction of air-cooled condenser is very simple, therefore the initial cost is less. The maintenance cost is also low.	Since the construction of water-cooled condenser is complicated, therefore the initial cost is high. The maintenance cost is also high.
2.	There is no handling problem with air-cooled condensers.	The water-cooled condensers are difficult to handle.

S.No.	Air-cooled condenser	Water-cooled condenser
3.	The air-cooled condensers do not require piping arrangement for carrying the air.	The pipes are required to take water to and from the condenser.
4.	There is no problem in disposing of used air.	There is a problem of disposing the used water unless a recirculation system is provided.
5.	Since there is no corrosion, therefore fouling effect is low.	Since corrosion occurs inside the tubes carrying the water, therefore fouling effects are high.
6.	The air-cooled condensers have low heat transfer capacity due to low thermal conductivity of air.	The water-cooled condensers have high heat transfer capacity due to high thermal conductivity of water.
7.	These condensers are used for low capacity plants (less than 5 TR).	These condensers are used for large capacity plants.
8.	Since the power required to drive the fan is excessive, therefore, the fan noise becomes objectionable.	There is no fan noise.
9.	The distribution of air on condenser surface is not uniform.	There is even distribution of water on the condensing surface.
10.	The air-cooled condensers have high flexibility.	The water-cooled condensers have low flexibility.

10.11 Fouling Factor

The water used in water-cooled condensers always contains a certain amount of minerals and other foreign materials, depending upon its source. These materials form deposits inside the condenser water tubes. This is called *water fouling.* The deposits insulate the tubes, reduce their heat transfer rate and restrict the water flow.

The fouling factor is the reciprocal of heat transfer coefficient for the material of scale. The following are the recommended fouling factors :

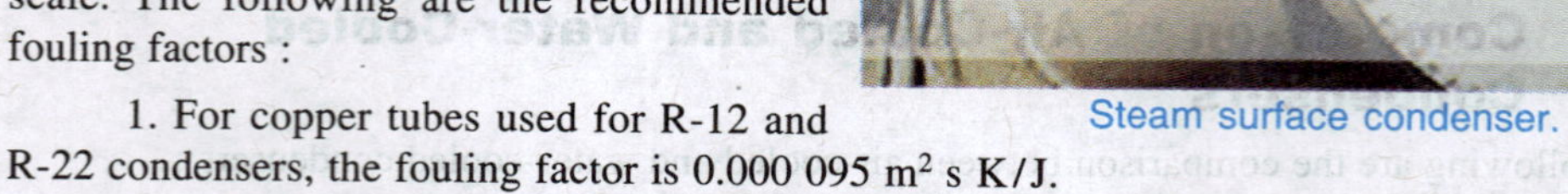

Steam surface condenser.

1. For copper tubes used for R-12 and R-22 condensers, the fouling factor is 0.000 095 m^2 s K/J.

2. For steel tubes used in ammonia condensers, the fouling factor is 0.000 18 m^2 s K/J.

10.12 Heat Transfer in Condensers

The heat transfer (Q) in water-cooled condensers is given by

$$Q = UA\,\Delta T = \frac{\Delta T}{R}$$

where U = Overall heat transfer coefficient,

A = Surface area of the condenser,

ΔT = Overall temperature difference, and

R = Overall thermal resistance of the condenser = $1/U\,A$

In order to find the overall thermal resistance, let us consider that the water is flowing inside the tube and the refrigerant outside the tube in a shell of a condenser, as shown in Fig. 10.8. When steady state is reached, there is a film of water inside the tube over the scale formed due to hardness of water. Another layer of film over the tube is formed by the refrigerant. The heat transfer from the vapour refrigerant to the water in tubes takes place in the following manner.

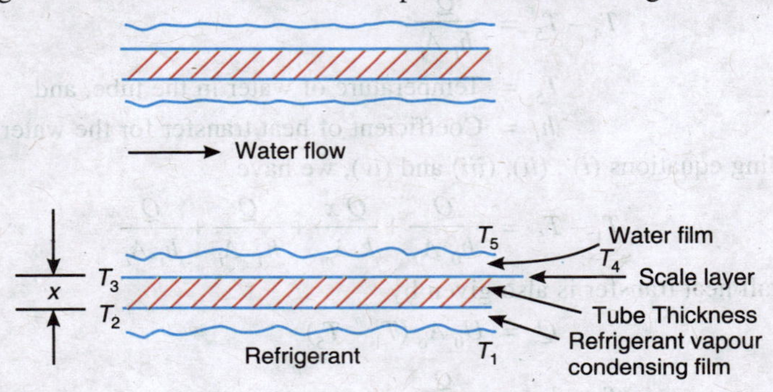

Fig. 10.8. Heat transfer in condensers.

1. The heat transfer takes place from the vapour refrigerant to the outside of the tube through the condensing film. The value of this heat transfer is given by :

$$Q = h_0\,A_0\,(T_1 - T_2)$$

or
$$T_1 - T_2 = \frac{Q}{h_0\,A_0} \quad \dots (i)$$

where T_1 = Temperature of the refrigerant vapour condensing film,

T_2 = Temperature at the outside surface of the tube,

h_0 = Coefficient of heat transfer for the refrigerant vapour condensing film, and

A_0 = Condensing area.

2. The heat transfer takes place from the outside surface to the inside surface of the tube. The value of this heat transfer is given by

$$Q = \frac{k\,A_m\,(T_2 - T_3)}{x}$$

or
$$T_2 - T_3 = \frac{Q\,x}{k\,A_m} \quad \dots (ii)$$

where T_3 = Temperature at the inside surface of the tube,

x = Thickness of the tube,

k = Thermal conductivity of the tube material, and

A_m = Mean surface area of the tube.

3. The heat transfer takes place through the layer of scale. The value of this heat transfer is given by

$$Q = h_f\,A_i\,(T_3 - T_4)$$

or
$$T_3 - T_4 = \frac{Q}{h_f\,A_i} \quad \dots (iii)$$

where T_4 = Temperature at the boundary layer of the scale,

h_f = Coefficient of heat transfer through the scale, and

A_i = Inside area of the tube.

4. The heat transfer takes place from the boundary layer film to the water inside the tube. The value of this heat transfer is given by

$$Q = h_i A_i (T_4 - T_5)$$

or

$$T_4 - T_5 = \frac{Q}{h_i A_i} \quad \dots (iv)$$

where T_5 = Temperature of water in the tube, and

h_i = Coefficient of heat transfer for the water side.

Now adding equations (*i*), (*ii*), (*iii*) and (*iv*), we have

$$T_1 - T_5 = \frac{Q}{h_0 A_0} + \frac{Q x}{k A_m} + \frac{Q}{h_f A_i} + \frac{Q}{h_i A_i} \quad \dots (v)$$

The overall heat transfer is also given by

$$Q = U_0 A_o (T_1 - T_5)$$

or

$$T_1 - T_5 = \frac{Q}{U_0 A_0} \quad \dots (vi)$$

From equations (*v*) and (*vi*),

$$\frac{1}{U_0 A_0} = \frac{1}{h_0 A_0} + \frac{x}{k A_m} + \frac{1}{h_f A_i} + \frac{1}{h_i A_i}$$

10.13 Condensing Heat Transfer Coefficient

Since the refrigerants are usually good wetting agents, therefore in condensation, they follow a film-wise pattern. The following equations have been suggested by Nusselt using the laminar liquid film theory.

1. Overall coefficient of heat transfer for condensation on a vertical surface of height x is given by

$$h = 0.943 \left[\frac{(k_f)^3 (\rho_f - \rho_g)\, g \times h_{fg}}{\mu \times x \times \Delta t} \right]^{\frac{1}{4}} \quad \dots (i)$$

2. Average coefficient of heat transfer for vapour condensing outside of horizontal tubes of diameter D is given by

$$h_o = 0.725 \left[\frac{(k_f)^3 (\rho_f)^2\, g \times h_{fg}}{N \times D \times \mu_f \times \Delta t} \right]^{\frac{1}{4}} \quad \dots (ii)$$

where k_f = Thermal conductivity of liquid condensate,

ρ_f = Density of liquid condensate,

ρ_g = Density of vapour and is very small as compared to ρ_f,

g = Acceleration due to gravity,

h_{fg} = Latent heat of vaporisation,

μ_f = Viscosity of condensed refrigerant film,

Δt = Difference of temperatures between condensing vapour and outside surface,

N = Number of tubes in a vertical row, and

D = Diameter of tube.

10.14 Air-side Coefficient

The average heat transfer coefficient for the forced convection of air across a tube may be determined by using the Grimson's equation, which is given as

$$\frac{h\,D}{k} = C\left(\frac{D\,u_{\infty}\,\rho}{\mu}\right)^{n} P_r^{1/3}$$

The values of constants C and n as functions of Reynold's number may be taken from the following table.

Table 10.1. Values of Constants C and *n*.

Reynolds number (R_e)	*C*	*n*
0.4 – 4	0.989	0.33
4 – 40	0.911	0.385
40 – 4000	0.683	0.466
4000 – 40 000	0.193	0.618
40 000 – 400 000	0.0266	0.805

The Reynold number is based on velocity u_{∞} at a distance from the tube. In case of bank of tubes, u_{∞} is given by approximately

$$u_{\infty} = \frac{\text{Volume flow rate}}{\text{Face area}} = \frac{Q_V}{A_F}$$

However, the Reynold number is based on the maximum velocity (u_{max}). The value of maximum velocity in case of bank of tubes is given by

$$u_{max} = u_{\infty} \times \frac{S_n}{S_n - D}$$

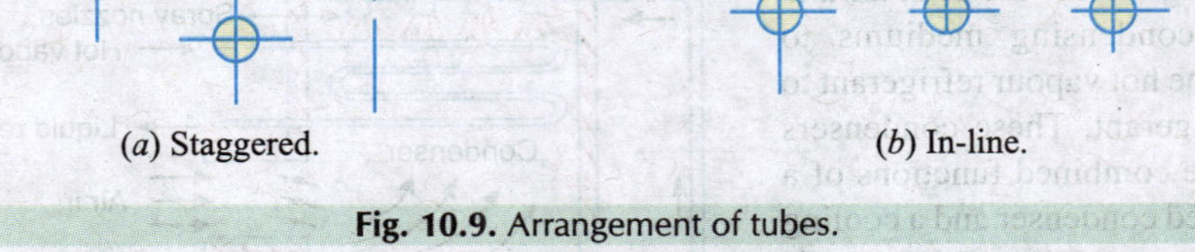

Sn

(*a*) Staggered. (*b*) In-line.

Fig. 10.9. Arrangement of tubes.

where S_n is the vertical pitch of tubes which may be staggered or in-line type as shown in Fig. 10.9.

Note : In case of natural convection air condensers,

$$h = 1.32\ (\Delta T/D)^{0.25}$$

This expression is used for laminar flows where $G_r \cdot P_r$ is more than 10^4 and less than 10^9. In case of turbulent flow where $G_r \cdot P_r$ is more than 10^9, then the value of h is given by

$$h = 1.24\ (\Delta T)^{1/3}$$

10.15 Water-side Coefficient

The water-side coefficient for forced convection turbulent flow inside the tubes is given by Dittus-Boelter equation as follows :

$$h_i = 0.023 \times \frac{k}{D}\left(\frac{D\,V\,\rho}{\mu}\right)^{0.8}\left(\frac{c\,\mu}{k}\right)^{0.4}$$

where D = Inside diameter of tube,

k = Thermal conductivity of water,

V = Average velocity of water,

ρ = Density of water,

μ = Viscosity of water, and

c = Specific heat of water.

Note : The expression $DV\rho/\mu$ is called Reynolds number and $c\mu/k$ is called Prandtl number.

10.16 Finned Tubes

We have discussed in Art.10.12 that

$$\frac{1}{U_o\,A_o} = \frac{1}{h_o\,A_o} + \frac{x}{k\,A_m} + \frac{1}{h_f\,A_i} + \frac{1}{h_i\,A_i}$$

From this expression, it is very clear that the overall heat transfer coefficient U_o has a lower value than the lower of the two film coefficients h_o and h_i. It is, therefore, advantageous to reduce the major heat transfer resistance k. The best way of doing this is to provide extended surface or fins on the side of lower heat transfer coefficient. The following are the typical values of heat transfer coefficients for shell and tube condensers :

Water-side coefficient = 6000 W/m^2 K

Condensing coefficient = 8000 W/m^2 K for ammonia

= 1500 W/m^2 K for R-12

Thus, it is desirable to use fins on refrigerant side for R-12 and other halocarbon refrigerants. Since in air-cooled condensers, the air-side coefficient is much lower than that of the refrigerant side, therefore, fins must be provided on the air side.

10.17 Evaporative Condensers

The evaporative condensers, as shown in Fig. 10.10, use both air and water as condensing mediums to condense the hot vapour refrigerant to liquid refrigerant. These condensers perform the combined functions of a water-cooled condenser and a cooling tower. In its operation, the water is pumped from the sump to a spray header and sprayed through nozzles over the condenser coils through which the hot vapour refrigerant from the compressor is passing. The heat

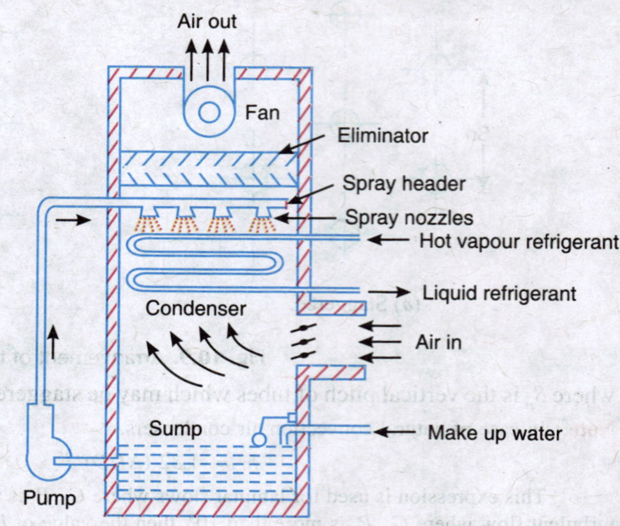

Fig. 10.10. Evaporative condenser.

transfers from the refrigerant through the condensing tube walls and into the water that is wetting the outside surface of the tubes. At the same time, a fan draws air from the bottom side of the condenser and discharged out at the top of the condenser. The air causes the water from the surface of the condenser coils to evaporate and absorb the latent heat of evaporation from the remaining water to cool it. Though most of the cooling takes place by evaporation, the air can also absorb some sensible heat from water. Since the heat for vaporising the water is taken from the refrigerant, therefore the vapour refrigerant condenses into a liquid refrigerant.

The cold water that drops down into a sump is recirculated. In order to make up the deficiency caused by the evaporated water, additional water is supplied to the sump. A float valve in the sump controls the make-up supply. The eliminator is provided above the spray header to stop particles of water escaping along with the discharge air.

10.18 Cooling Towers and Spray Ponds

A *cooling tower* is an enclosed tower-like structure through which atmospheric air circulates to cool large quantities of warm water by direct contact. A *spray pond* consists of a piping and spray nozzle arrangement suspended over an outdoor open reservoir or pond. It can also cool large quantities of warm water. The cooling towers and spray ponds, used for refrigeration and air conditioning systems, cool the warm water pumped from the water-cooled condensers. Then the same water can be used again and again to cool the condenser.

The principle of cooling the water in cooling towers and spray ponds is similar to that of evaporative condensers, *i.e.* the warm water is cooled by means of evaporation. The air surrounding the falling water droplets from the spray nozzles causes some of the water droplets to evaporate. The evaporating water absorbs latent heat of evaporation from the remaining water and thus cools it. The air also absorbs a small amount of sensible heat from the the remaining water. The cooled water collects in the pond or in a sump at the cooling tower which is recirculated through the condenser.

10.19 Capacity of Cooling Towers and Spray Ponds

The capacity of cooling towers and spray ponds depends upon the amount of evaporation of water that takes place. The amount of evaporation of water, in turn, depends upon the following factors :

1. The amount of water surface exposed to the air,
2. The length of the exposure time,
3. The velocity of air passing over the water droplets formed in cooling towers, and
4. The wet bulb temperature of the atmospheric air.

It may be noted that when the wet bulb temperature of the air decreases, the air can absorb more water vapour and therefore evaporate more water. Thus the capacity of the cooling tower and spray ponds increases as the wet bulb temperature of air decreases. The dry bulb temperature of air has less influence on their capacities, because air absorbs little sensible heat from water. However, the amount of sensible heat that the air can absorb increases as the dry bulb temperature difference between the water and air increases, such as during cold weather operation.

10.20 Types of Cooling Towers

The cooling towers are mainly divided, according to their method of air circulation, into the following two groups :

1. Natural draft cooling towers, and 2. Mechanical draft cooling towers.

In *natural draft cooling towers,* the air circulates through the tower by natural convection whereas in *mechanical draft cooling towers,* the air is forced through the tower by means of fans

Natural draft cooling towers.

Mechanical draft cooling towers.

or blowers. The natural draft and mechanical draft cooling towers are discussed, in detail, in the following pages.

10.21 Natural Draft Cooling Towers

Since the air circulating through the natural draft cooling towers is atmospheric air, therefore these cooling towers are also known as *atmospheric natural draft cooling towers* or simply *atmospheric cooling towers.* The two types of atmospheric natural draft cooling towers are spray type and splash deck type as discussed below :

1. ***Atmospheric natural draft (spray type) cooling towers.*** The atmospheric natural draft (spray type) cooling tower, as shown in Fig. 10.11, consists of a box-shaped structure with louvers. The louvers allow the atmospheric air to pass through the tower, but slant down towards the inside of the tower to retain water in it. The framework and louvers are usually made of steel. The size of the cooling tower depends upon the capacity of the unit. The atmospheric natural draft (spray type) cooling towers should be located in the open space or on the roof of a building where the air can blow freely through them.

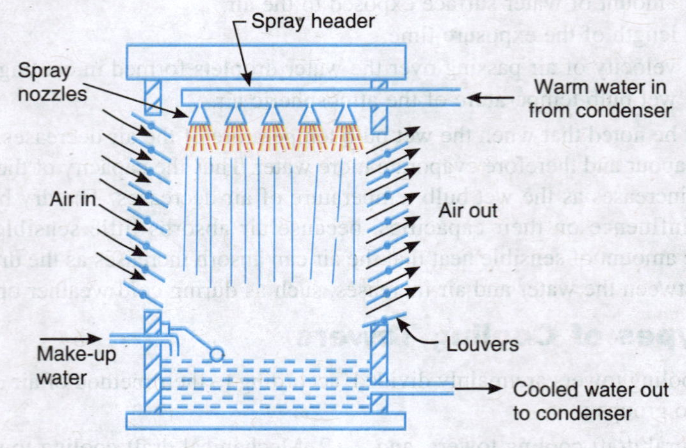

Fig. 10.11. Atmospheric natural draft (spray type) cooling tower.

In this type of cooling tower, warm water from the condenser is pumped to a spray header provided at the top of a tower. It is sprayed down into the tower through the nozzles. Since the heat transfer from water to air is dependent upon the surface of water exposed to the air stream, therefore a spray nozzle having finer spray pattern is essential for good performance of the cooling tower. It may be noted that the finer spray exposes more water surface to air. However, if the spray is too fine, too much water is blown away. The water spray blown away by the air is called *drift*. The drift increases water loss in the tower, but does not affect the cooling action.

2. Atmospheric natural draft (splash deck type) cooling tower. The atmospheric natural draft (splash deck type) cooling tower, as shown in Fig. 10.12, is similar to a spray type cooling tower except that it contains decking (also called *fill*) made from redwood hollow tiles, ceramic, metal sheets or plastic. Many splash deck towers do not employ nozzles. The water splashes on the decking from the holes in the bottom of a water box on the top of a tower, as shown in Fig. 10.12. The object of decking is to increase the rate of heat transfer, by exposing a large amount of wetted surface to the air. The decking also helps to break up the water into small droplets and slows down the fall of water to the bottom of tower.

This type of cooling tower is 20 to 30 per cent more efficient than the spray type for the same size and same quantity of water flow.

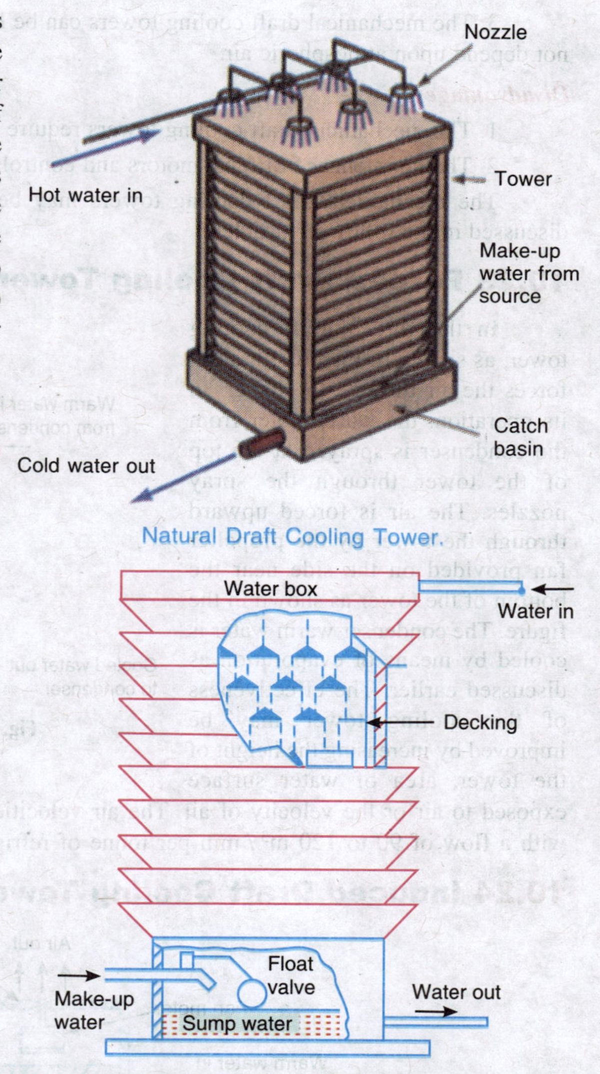

Fig. 10.12. Atmospheric natural draft (splash deck type) cooling tower.

10.22 Mechanical Draft Cooling Towers

The mechanical draft cooling towers are similar to atmospheric natural draft cooling towers except that the fans are used to force the air through them. These towers may use either propeller or centrifugal fans. The mechanical draft cooling towers have the following main advantages and disadvantages over the atmospheric natural draft cooling towers.

Advantages

1. The mechanical draft cooling towers are smaller in size than natural draft cooling towers of the same capacity, because the large volume of forced air increases the cooling capacity.

2. The cooling capacity of mechanical draft cooling towers can be controlled by controlling the amount of forced air.

3. The mechanical draft cooling towers can be located inside the building because they do not depend upon atmospheric air.

Disadvantages

1. The mechanical draft cooling towers require additional power to operate the fans.

2. The maintenance of fans, motors and controls increases the operating cost.

The mechanical draft cooling towers may be either forced draft or induced draft, as discussed in the following articles.

10.23 Forced Draft Cooling Towers

In the forced draft cooling tower, as shown in Fig. 10.13, a fan forces the air through the tower. In its operation, the warm water from the condenser is sprayed at the top of the tower through the spray nozzles. The air is forced upward through the tower by the propeller fan provided on the side near the bottom of the tower as shown in the figure. The condenser warm water is cooled by means of evaporation as discussed earlier. The effectiveness of the cooling tower may be improved by increasing the height of the tower, area of water surface exposed to air or the velocity of air. The air velocities from 75 to 120 m/min is recommended with a flow of 90 to 120 m^3 / min per tonne of refrigeration capacity.

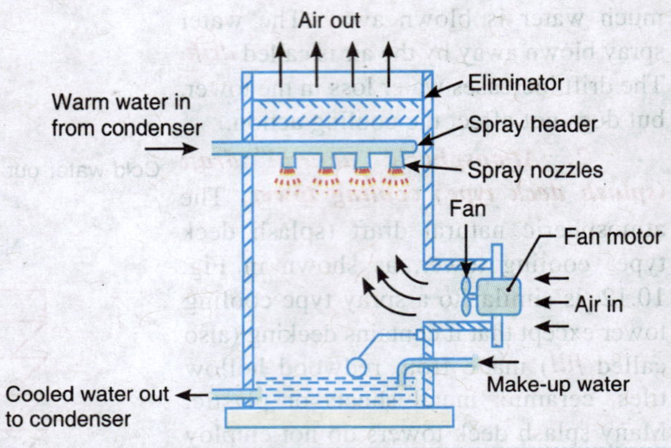

Fig. 10.13. Forced draft cooling tower.

10.24 Induced Draft Cooling Towers

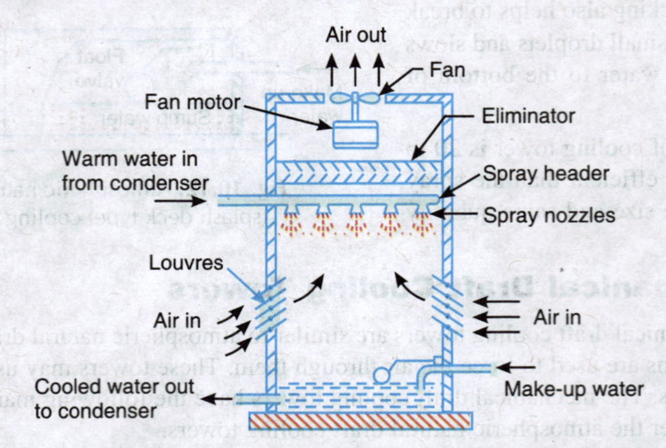

Fig. 10.14. Induced draft cooling tower.

In the induced draft cooling tower, as shown in Fig. 10.14, the fan sucks the air through the tower. The induced draft cooling towers are similar to forced draft cooling towers except that the fans are located at the top instead of at the bottom and draw the air upward through the tower.

QUESTIONS

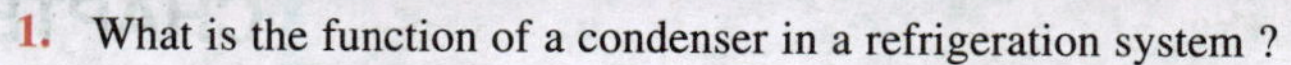

1. What is the function of a condenser in a refrigeration system ?
2. What are the points to be considered for selecting a condenser for a refrigeration system ?
3. Explain the term heat rejection ratio. On what factor does it depend?
4. Give the main types of condensers in use with specific application of each type.
5. Discuss the natural convection and forced convection types of air-cooled condensers.
6. Explain a recirculated water system used in water-cooled condensers.
7. Describe, with neat sketches, the working of (*a*) shell and coil condenser; and (*b*) shell and tube condenser.
8. Give the comparison of air-cooled condenser and water-cooled condenser.
9. Describe the working of an evaporative condenser.
10. Write a short note on cooling towers.

OBJECTIVE TYPE QUESTIONS

1. A condenser is used in the .... pressure side of a refrigerating system.
   (*a*) low (*b*) high
2. The heat rejection factor (HRF) is given by
   (*a*) 1 + C.O.P. (*b*) 1 – C.O.P. (*c*) $1 + \dfrac{1}{\text{C.O.P.}}$ (*d*) $1 - \dfrac{1}{\text{C.O.P.}}$
3. In actual air conditioning applications for R-12 and R-22 and operating at a condenser temperature of 40°C and an evaporator temperature of 5°C, the heat rejection factor is about
   (*a*) 1 (*b*) 1.25 (*c*) 2.15 (*d*) 5.12
4. Most air-cooled condensers are designed to operate with a temperature difference of
   (*a*) 5°C (*b*) 8°C (*c*) 14°C (*d*) 22°C
5. The natural convection air-cooled condensers are used in
   (*a*) domestic refrigerators (*b*) water coolers
   (*c*) room air conditioners (*d*) all of these
6. A water-cooled condenser operates at a .... condensing temperature than an air-cooled condenser.
   (*a*) higher (*b*) lower
7. In a shell and coil condenser,
   (*a*) water flows in the shell and the refrigerant in the coil
   (*b*) water flows in the coil and the refrigerant in the shell
   (*c*) only water flows through the shell as well as coil
   (*d*) only refrigerant flows through the shell as well as coil
8. For ammonia refrigerating systems, the tubes of a shell and tube condenser are made of
   (*a*) copper (*b*) aluminium (*c*) steel (*d*) brass
9. The condensing medium used in evaporative condensers is
   (*a*) air only (*b*) water only (*c*) both air and water
10. The mechanical draft cooling towers are .... in size than the natural draft cooling towers of the same capacity.
    (*a*) smaller (*b*) larger

ANSWERS

1. (*b*)	2. (*c*)	3. (*b*)	4. (*c*)	5. (*a*)
6. (*b*)	7. (*b*)	8. (*c*)	9. (*c*)	10. (*a*)

CHAPTER

Evaporators 11

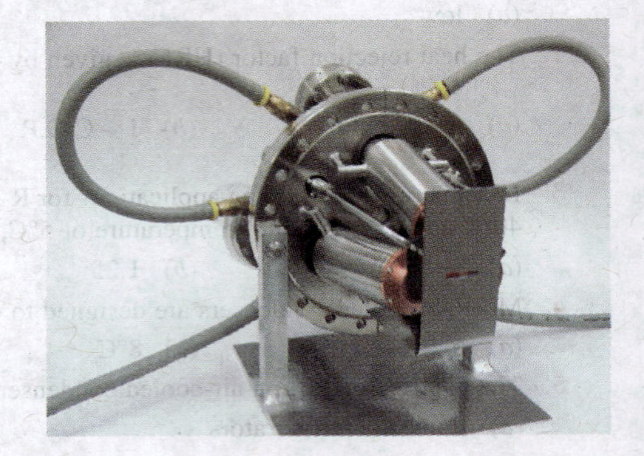

11.1 Introduction

The evaporator is an important device used in the low pressure side of a refrigeration system. The liquid refrigerant from the expansion valve enters into the evaporator where it boils and changes into vapour. The function of an evaporator is to absorb heat from the surrounding location or medium which is to be cooled, by means of a refrigerant. The temperature of the boiling refrigerant in the evaporator must always be less than that of the surrounding medium so that the heat flows to the refrigerant. The evaporator becomes cold and remains cold due to the following two reasons :

1. The temperature of the evaporator coil is low due to the low temperature of the refrigerant inside the coil.

2. The low temperature of the refrigerant remains unchanged because any heat it absorbs is converted to latent heat as boiling proceeds.

Note : The evaporator is also known as a *cooling coil*, a *chilling coil* or a *freezing coil*. The evaporator cools by using the refrigerant's latent heat of vaporisation to absorb heat from the medium being cooled.

11.2 Working of an Evaporator

The working of an evaporator may be best understood by considering the simple refrigerating system, as shown in Fig. 11.1(*a*). The corresponding *p-h* diagram is shown in Fig. 11.1 (*b*). The point 5 in the figure represents the entry of liquid refrigerant into the expansion valve. Under proper operating conditions, the liquid refrigerant is sub-cooled (*i.e.* cooled below its saturation

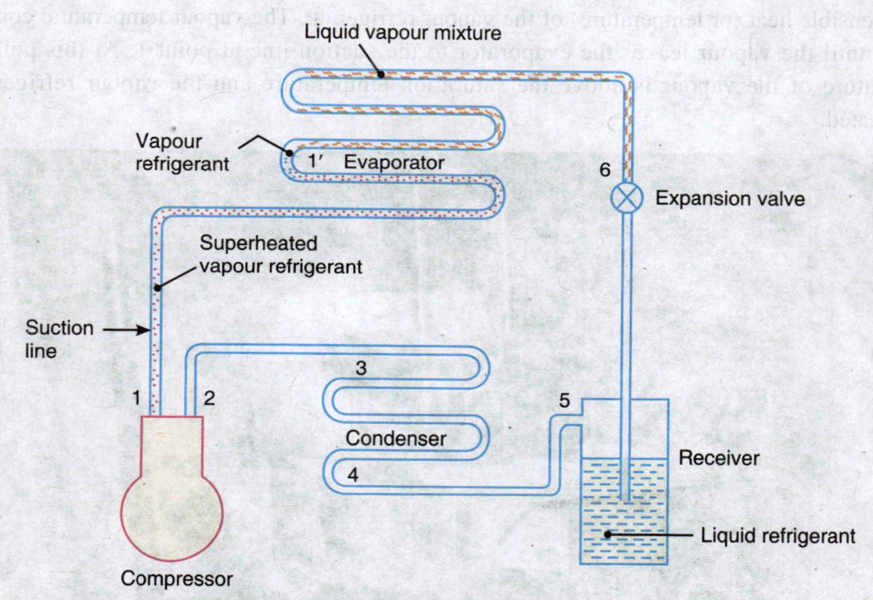

(*a*) Schematic diagram of a simple refrigerating system.

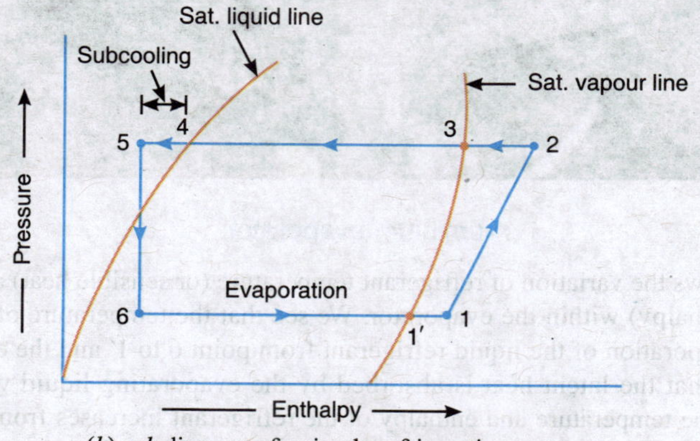

(*b*) *p-h* diagram of a simple refrigerating system.

Fig. 11.1

temperature). The sub-cooling ensures that the expansion valve receives pure liquid refrigerant with no vapour present to restrict the flow of refrigerant through the expansion valve.

The liquid refrigerant at low pressure enters the evaporator at point 6, as shown in Fig. 11.1. As the liquid refrigerant passes through the evaporator coil, it continually absorbs heat through the coil walls, from the medium being cooled. During this, the refrigerant continues to boil and evaporate. Finally at point 1′, all the liquid refrigerant has evaporated and only vapour refrigerant remains in the evaporator coil. The liquid refrigerant's ability to convert absorbed heat to latent heat is now used up.

Since the vapour refrigerant at point 1′ is still colder than the medium being cooled, therefore the vapour refrigerant continues to absorb heat. This heat absorption causes an increase in the sensible heat (or temperature) of the vapour refrigerant. The vapour temperature continues to rise until the vapour leaves the evaporator to the suction line at point 1. At this point, the temperature of the vapour is above the saturation temperature and the vapour refrigerant is superheated.

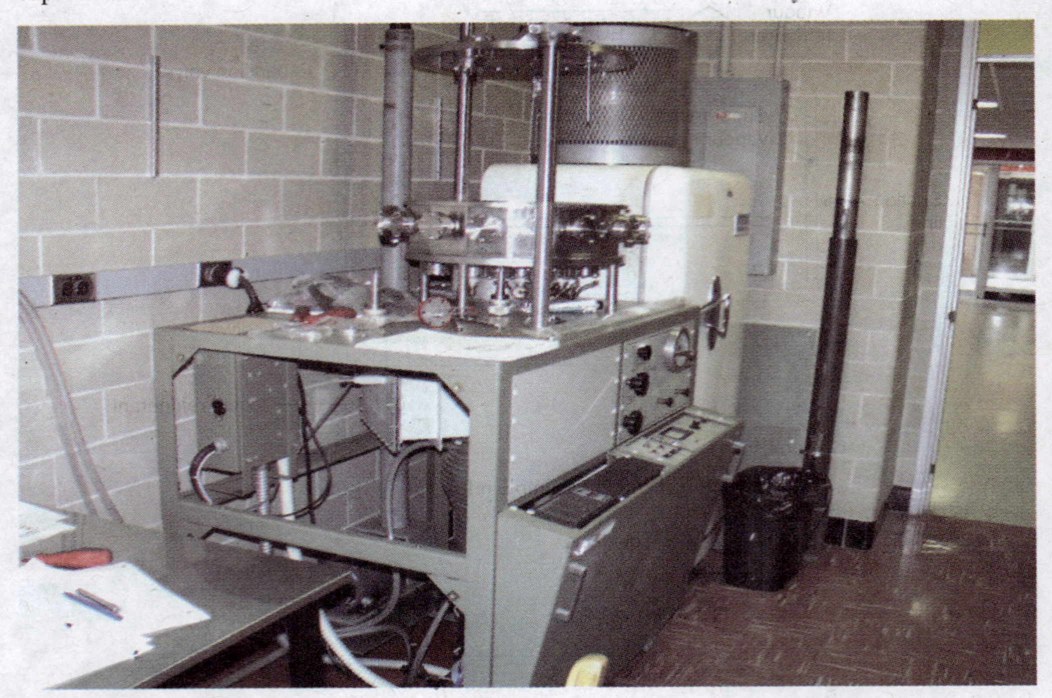

Aluminium evaporator.

Fig. 11.2 shows the variation of refrigerant temperature (or sensible heat) and the refrigerant heat content (or enthalpy) within the evaporator. We see that the temperature of the refrigerant is constant during evaporation of the liquid refrigerant from point 6 to 1′ and the enthalpy increases steadily. It shows that the latent heat is absorbed by the evaporating liquid with no change in temperature. Both the temperature and enthalpy of the refrigerant increases from point 1′ to 1. At point 1, all the liquid refrigerant has evaporated. The line 1′-1 shows the increase in sensible heat of the vapour refrigerant.

Note : The heat absorbed by the refrigerant, while it is evaporating, accounts for most of the cooling accomplished by the evaporator. In other words, the cooling is due to the evaporation of the refrigerant, rather than to increase in the refrigerant's sensible heat.

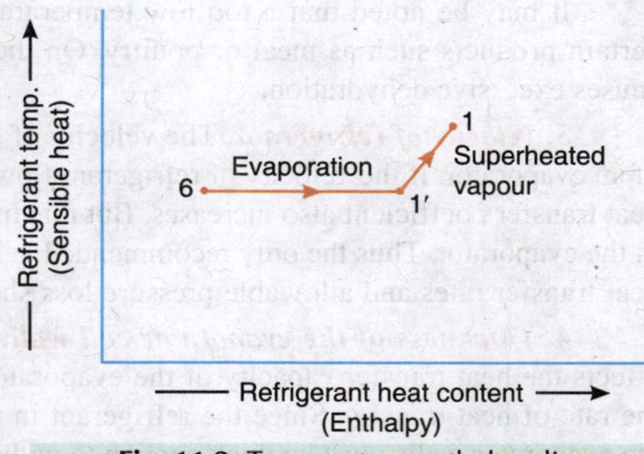

Fig. 11.2. Temperature-enthalpy diagram for an evaporator.

11.3 Capacity of an Evaporator

The capacity of an evaporator is defined as the amount of heat absorbed by it over a given period of time. The heat absorbed or heat transfer capacity of an evaporator is given by

$$Q = UA\,(T_2 - T_1)\ \text{W} \quad \text{or} \quad \text{J/s}$$

where U = Overall heat transfer coefficient in W/m^2 °C,

A = Area of evaporator surface in m^2,

T_2 = Temperature of medium to be cooled (or temperature outside the evaporator) in °C, and

T_1 = Saturation temperature of refrigerant at evaporator pressure (or temperature inside the evaporator) in °C.

11.4 Factors Affecting the Heat Transfer Capacity of an Evaporator

Though there are many factors upon which the heat transfer capacity of an evaporator depends, yet the following are important from the subject point of view :

1. *Material.* In order to have rapid heat transfer in an evaporator, the material used for the construction of an evaporator coil should be a good conductor of heat. The material which is not affected by the refrigerant must also be selected. Since metals are best conductors of heat, therefore they are always used for evaporators. Iron and steel can be used with all common refrigerants. Brass and copper are used with all refrigerants except ammonia. Aluminium should not be used with freon.

2. *Temperature difference.* The temperature difference between the refrigerant within the evaporator and the product to be cooled plays an important role in the heat transfer capacity of an evaporator. The following table shows the suggested temperature difference for some of the products to be cooled.

Table 11.1. Suggested temperature difference for some of the products

Product	*Cooling temperature (°C)*	*Refrigerant temperature (°C)*	*Temperature difference (°C)*
Cheese, eggs and butter	0 to – 2	– 10	8
Poultry and meats (frozen)	– 20 to – 15	– 30	10
Vegetables and fruits	+7	– 5	12
Cream and milk	+3	– 8	11
Beer	+4.5	– 7	11.5

It may be noted that a too low temperature difference (below 8°C) may cause slime on certain products such as meat or poultry. On the other hand, too high a temperature difference causes excessive dehydration.

3. *Velocity of refrigerant.* The velocity of refrigerant also affects the heat transfer capacity of an evaporator. If the velocity of refrigerant flowing through the evaporator increases, the overall heat transfer coefficient also increases. But this increased velocity will cause greater pressure loss in the evaporator. Thus the only recommended velocities for different refrigerants which give high heat transfer rates and allowable pressure loss should be used.

4. *Thickness of the evaporator coil wall.* The thickness of the evaporator coil wall also affects the heat transfer capacity of the evaporator. In general, the thicker the wall, the slower is the rate of heat transfer. Since the refrigerant in the evaporator coils is under pressure, therefore the evaporator walls must be thick enough to withstand the effects of that pressure. It may be noted that the thickness has only a slight effect on total heat transfer capacity because the evaporators are usually made from highly conductive materials.

5. *Contact surface area.* An important factor affecting the evaporator capacity is the contact surface available between the walls of evaporator coil and the medium being cooled. The amount of contact surface, in turn, depends basically on the physical size and shape of the evaporator coil.

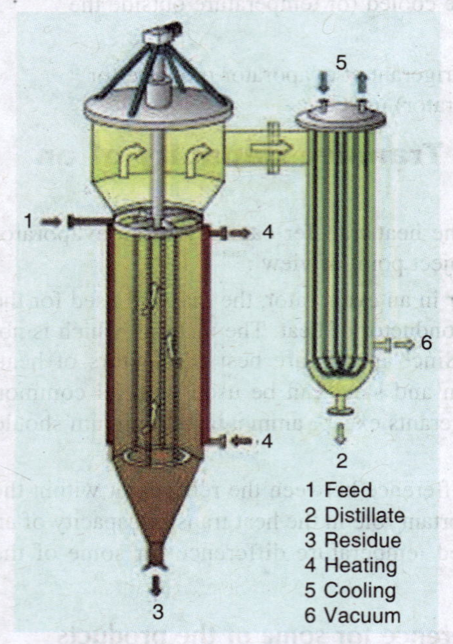

Wiped film evaporator with external condenser.

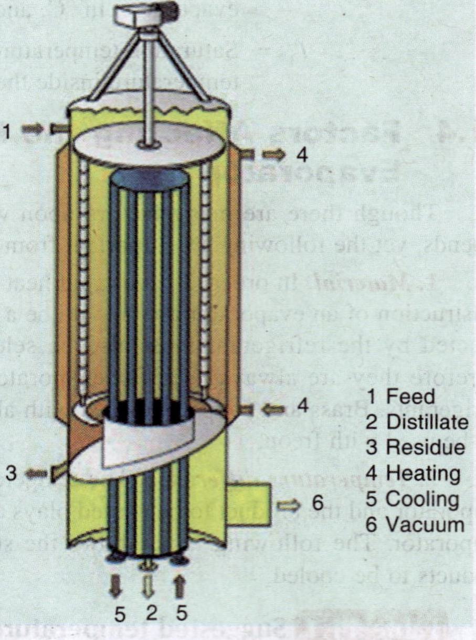

Short path evaporator with internal condenser.

11.5 Heat Transfer in Evaporators

The heat transfer in evaporators has the following three resistances in its path :

1. The resistance of medium being cooled. This may be air, water, brine or any other fluid or a wetted surface of a cooling and dehumidifying coil.

2. The resistance of metallic wall of tube liquid.

3. The resistance of cooling medium *i.e.* refrigerant film which gets heat from solid metallic walls.

11.6 Heat Transfer During Boiling

The mechanism of boiling is so complex that it is difficult to predict the heat transfer coefficient. It is due to the factors such as latent heat effects, surface tension, saturation temperature and nature of the solid surface. The boiling occurs in the following two ways :

1. Pool boiling as it occurs in flooded evaporators ; and
2. Flow boiling or forced convection boiling as it occurs in direct expansion evaporators.

When heat is added to a liquid from a submerged solid surface, the boiling process is called *pool boiling.* A necessary condition for the occurrence of pool boiling is that the temperature of the heating surface exceeds the saturation temperature of the liquid. In this process, the vapour produced may form bubbles, which grow and subsequently detach themselves from the surface, rising to the free surface due to buoyancy effects. On the other hand, the *flow boiling* or *forced convection boiling* occurs in a flowing stream and the boiling surface may itself be a portion of the flow passage. This phenomenon is generally associated with two phase flows through confined passages.

The process of pool boiling is shown in Fig. 11.3. In this figure, the heat flux (Q/A in W/m^2) is plotted against the excess temperature ($\Delta t = t_w - t_s$, in °C). The pool boiling experiments have shown that the boiling process of liquid at its saturation temperature has the following three distinct regimes :

(a) Interface evaporation. This occurs at the free surface without the formation of bubbles when the solid wall temperature t_w is only few degrees (about 5° C) above the saturation temperature of evaporating substance (t_s).

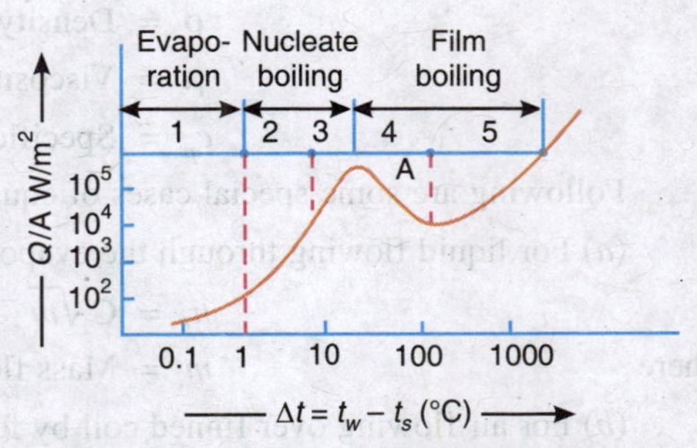

Fig. 11.3. Characteristics of boiling process.

(b) Nucleate boiling. When the excess temperature (Δt) increases, vapour bubbles are formed. These bubbles rise above the metal surface but condense before reaching the liquid surface. When the excess temperature (Δt) is further increased, the bubbles rise and collapse through the free surface by limit surface tension. This process is called nucleate boiling.

(c) Film boiling. When the maximum heat flux limit is reached at point A as shown in Fig. 11.3, all the heating surface gets covered with vapour bubbles causing the film boiling process.

The heat transfer in flooded evaporators is corresponding to second regime of nucleate boiling.

11.7 Heat Transfer Coefficient for Nucleate Pool Boiling

The experimental data for nucleate pool boiling as corelated by Rohsenow is given as

$$\frac{c_f\,(T_w - T_s)}{h_{fg}} = c_{sf}\left[\frac{Q/A}{\mu_f / h_{fg}}\sqrt{\frac{\sigma}{g\,(\rho_f - \rho_g)}}\right]^{1/3}\left(\frac{c\,\mu}{K}\right)_f^s$$

where c_f = Specific heat of saturated liquid, in J/kg K.

$T_w - T_s$ = Excess temperature in K,

h_{fg} = Enthalpy of vaporisation in J/kg.

c_{sf} = An empirical constant. Its value is 0.013 for combination of copper tube with R-12, R-22 etc.

Q/A = Heat flux in W/m^2,

μ_f = Viscosity of saturated liquid in kg/m-s.

σ = Surface tension of liquid vapour interface in N/m,

g = Acceleration due to gravity in m/s^2,

ρ_f = Density of saturated liquid in kg/m^3,

ρ_g = Density of saturated vapour in kg/m^3.

11.8 Fluid Side Heat Transfer Coefficient

The fluid side heat transfer coefficient (h_f) is given by the following relation :

$$h_f = \frac{CK}{D}\left(\frac{V D \rho}{\mu}\right)^{0.8}\left(\frac{c_p\,\mu}{K}\right)^{0.4} \quad \ldots (i)$$

where h_f = Fluid side heat transfer coefficient, in W/m^2 K,

C = Constant,

K = Conductivity in W/m K,

D = Inside diameter of tube in metres,

V = Velocity in m/s,

ρ = Density of fluid in kg/m^3,

μ = Viscosity in kg/m-s,

c_p = Specific heat at constant pressure in kJ/kg K,

Following are some special cases of equation (*i*) :

(*a*) For liquid flowing through the evaporator shell,

$$h_f = C\sqrt{m}$$

where m = Mass flow rate of the liquid.

(*b*) For air flowing over finned coil by forced convection,

$$h_f = C\,(m)^{0.4}$$

(*c*) For air flowing over cold pipes by natural convection,

$$h_f = 0.2\left(\frac{T_a - T_m}{D}\right)^{1/4}$$

where T_a = Dry bulb temperature of air, and

T_m = Mean temperature of pipe surface.

11.9 Types of Evaporators

Though there are many types of evaporators, yet the following are important from the subject point of view :

1. *According to the type of construction*

(*a*) Bare tube coil evaporator,

(*b*) Finned tube evaporator,

(*c*) Plate evaporator,

(*d*) Shell and tube evaporator,

(*e*) Shell and coil evaporator, and

(*f*) Tube-in-tube evaporator,

2. *According to the manner in which liquid refrigerant is fed*

(*a*) Flooded evaporator, and

(*b*) Dry expansion evaporator

3. *According to the mode of heat transfer*

(*a*) Natural convection evaporator, and

(*b*) Forced convection evaporator

4. *According to operating conditions*

(*a*) Frosting evaporator,

(*b*) Non-frosting evaporator, and

(*c*) Defrosting evaporator

We shall now discuss the above mentioned evaporators, in detail, in the following pages.

11.10 Bare Tube Coil Evaporators

The simplest type of evaporator is the bare tube coil evaporator, as shown in Fig. 11.4.

The bare tube coil evaporators are also known as *prime-surface evaporators.* Because of its simple construction, the bare tube coil is easy to clean and defrost. A little consideration will show that this type of evaporator offers relatively little surface contact area as compared to other types of coils. The amount of surface area may be increased by simply extending the length of the tube, but there are disadvantages of excessive tube length. The effective length of the tube is limited by the capacity of expansion valve. If the tube is too long for the valve's capacity, the liquid refrigerant will tend to completely vaporise early in its progress through the tube, thus leading to excessive superheating at the outlet. The long tubes will also cause considerably greater pressure drop between the inlet and outlet of the evaporator. This results in a reduced suction line pressure.

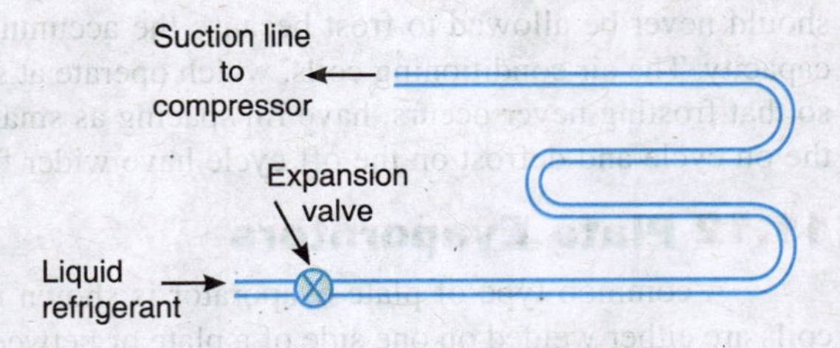

Fig. 11.4. Bare tube coil evaporator.

The diameter of the tube in relation to tube length may also be critical. If the tube diameter is too large, the refrigerant velocity will be too low and the volume of refrigerant will be too great in relation to the surface area of the tube to allow complete vaporisation. This, in turn, may allow liquid refrigerant to enter the suction line with possible damage to the compressor (*i.e.* slugging). On the other hand, if the diameter is too small, the pressure drop due to friction may be too high and will reduce the system efficiency.

The bare tube coil evaporators may be used for any type of refrigeration requirement. Its use is, however, limited to applications where the box temperatures are under 0°C and in liquid cooling, because the accumulation of ice or frost on these evaporators has less effect on the heat transfer than on those equipped with fins. The bare tube coil evaporators are also extensively used in household refrigerators because they are easier to keep clean.

11.11 Finned Evaporators

The finned evaporator, as shown in Fig. 11.5, consists of bare tubes or coils over which the metal plates or fins are fastened.

The metal fins are constructed of thin sheets of metal having good thermal conductivity. The shape, size or spacing of the fins can be adapted to provide best rate of heat transfer for a given application. Since the fins greatly increase the contact surfaces for heat transfer, therefore the finned evaporators are also called *extended surface evaporators.*

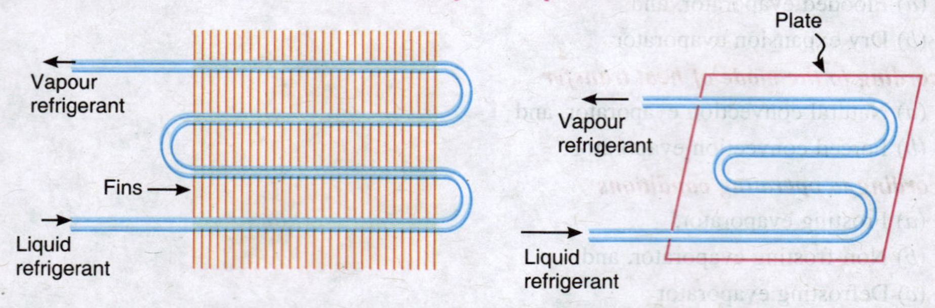

Fig. 11.5. Finned evaporator. **Fig. 11.6.** Plate evaporator.

The finned evaporators are primarily designed for air conditioning applications where the refrigerator temperature is above 0°C. Because of the rapid heat transfer of the finned evaporator, it will defrost itself on the off cycle when the temperature of the coil is near 0°C. A finned coil should never be allowed to frost because the accumulation of frost between the fins reduces the capacity. The air conditioning coils, which operate at suction temperatures which are high enough so that frosting never occurs, have fin spacing as small as 3 mm. The finned coils which frost on the on cycle and defrost on the off cycle have wider fin spacing.

11.12 Plate Evaporators

A common type of plate evaporator is shown in Fig. 11.6. In this type of evaporator, the coils are either welded on one side of a plate or between the two plates which are welded together at the edges. The plate evaporators are generally used in household refrigerators, home freezers, beverage coolers, ice cream cabinets, locker plants etc.

11.13 Shell and Tube Evaporators

The shell and tube evaporator, as shown in Fig. 11.7, is similar to a shell and tube condenser. It consists of a number of horizontal tubes enclosed in a cylindrical shell. The inlet and outlet headers with perforated metal tube sheets are connected at each end of the tubes.

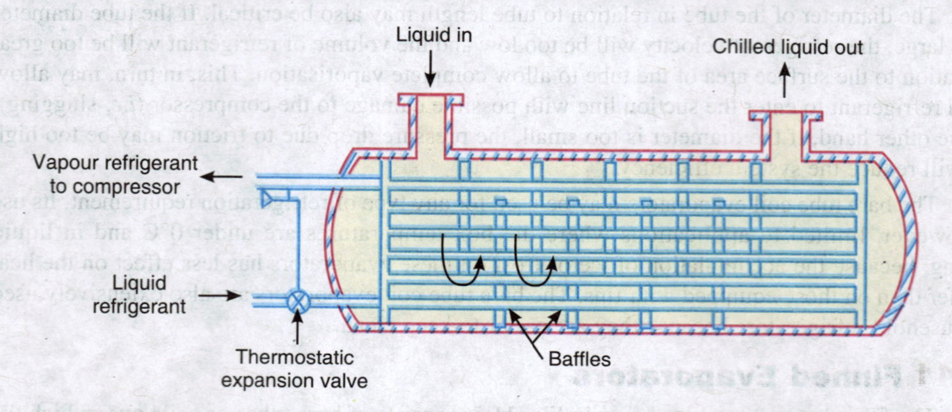

Fig. 11.7. Shell and tube evaporator.

These evaporators are generally used to chill water or brine solutions. When it is operated as a dry expansion evaporator, the refrigerant circulates through the tubes and the liquid to be cooled fills the space around the tubes within the shell. The dry expansion shell and tube evaporators are used for refrigerating units of 2 to 250 TR capacity. When it is operated as a flooded evaporator, the water or brine flows through the tubes and the refrigerant circulates around the tubes. The flooded shell and tube evaporators are used for refrigerating units of 10 to 5000 TR capacity.

11.14 Shell and Coil Evaporators

The shell and coil evaporators, as shown in Fig. 11.8, are generally dry expansion evaporators to chill water. The cooling coil is a continuous tube that can be in the form of a single or double spiral. The shell may be sealed or open. The sealed shells are usually found in shell and coil evaporators used to cool drinking water. The evaporators having flanged shells are often used to chill water in secondary refrigeration systems.

Another type of shell and coil evaporator is shown in Fig. 11.9. Both types of evaporators are usually used where small capacity (2 to 10 TR) liquid cooling is required. It may be noted that the shell and coil evaporator is restricted to operation above 5°C in order to prevent the freezing problems.

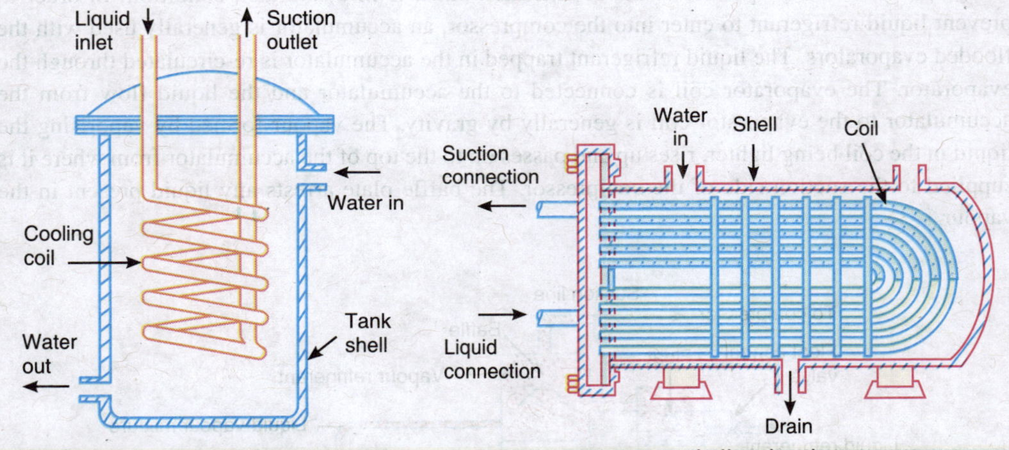

Fig. 11.8. Shell and coil evaporator.

Fig. 11.9. Shell and coil evaporator.

11.15 Tube-in-Tube or Double-Tube Evaporators

The tube-in-tube evaporator (or double-tube evaporator), as shown in Fig. 11.10, consists of one tube inside another tube. The liquid to be cooled flows through the inner tube while the primary refrigerant or secondary refrigerant (*i.e.* water, air or brine) circulates in the space between the two tubes. The tube-in-tube evaporator provides high heat transfer rates. However, they require more space than shell and tube evaporators of the same capacity. These evaporators are used for wine cooling and in petroleum industry for chilling of oil.

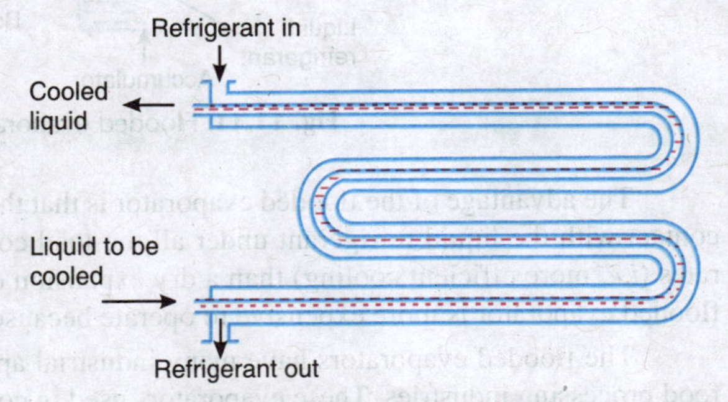

Fig. 11.10. Tube-in-tube or double-tube evaporator.

11.16 Flooded Evaporators

In a flooded evaporator, as shown in Fig. 11.11, a constant liquid refrigerant level is always maintained. A float control valve is used as an expansion device which maintains constant liquid level in the evaporator. The liquid refrigerant from the receiver passes through a low side float control valve and accumulator before entering the evaporator coil. The accumulator (also called a *surge drum* or *surge tank*) serves as a storage tank for the liquid refrigerant. It maintains a constant liquid level in the evaporator and helps to separate the liquid refrigerant from the vapour returning to the compressor. Due to the heat supplied by the substance to be cooled, the liquid refrigerant in the evaporator coil vaporises and thus the liquid level falls down. The accumulator supplies more liquid to the evaporator in order to keep the liquid refrigerant in the evaporator at proper level. In this way, the level of liquid refrigerant in the accumulator also falls down. Since the float within the float chamber rests on liquid refrigerant at the same level as that in the accumulator, therefore the float also falls down and open the float valve. Now the liquid refrigerant from the receiver is admitted into the accumulator. As the liquid level in the accumulator rises and reaches to the constant level, the float also rises with it until the float control valve closes.

Since the evaporator is almost completely filled with liquid refrigerant, therefore the vapour refrigerant from the evaporator is not superheated but it is in a saturated condition. In order to prevent liquid refrigerant to enter into the compressor, an accumulator is generally used with the flooded evaporators. The liquid refrigerant trapped in the accumulator is re-circulated through the evaporator. The evaporator coil is connected to the accumulator and the liquid flow from the accumulator to the evaporator coil is generally by gravity. The vapour formed by vaporising the liquid in the coil being lighter, rises up and passes on to the top of the accumulator from where it is supplied to the suction side of the compressor. The baffle plate arrests any liquid present in the vapour.

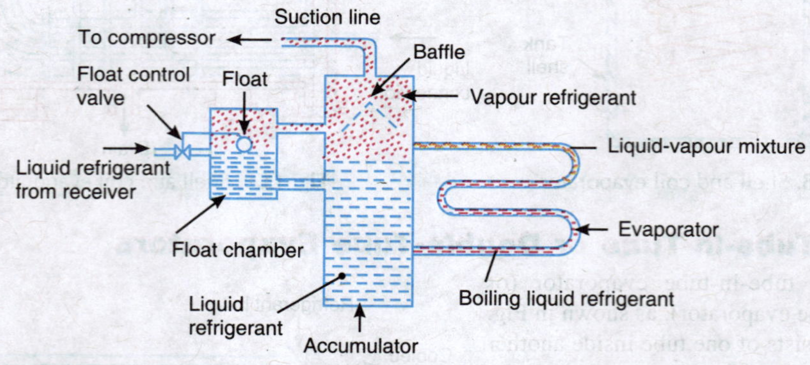

Fig. 11.11. Flooded evaporator.

The advantage of the flooded evaporator is that the whole surface of the evaporator coil is in contact with the liquid refrigerant under all the load conditions. Thus, it gives high heat transfer rates (*i.e.* more efficient cooling) than a dry expansion evaporator of the same size. However, the flooded evaporator is more expensive to operate because it requires more refrigerant charge.

The flooded evaporators have many industrial applications, especially in the chemical and food processing industries. These evaporators used in comfort and process air cooling installations may be of the bare coil type or finned type. Another type of the flooded evaporator is the plate evaporator which is found in cold storage boxes and freezers.

11.17 Dry Expansion Evaporators

The dry expansion evaporators are not really dry at all. They simply use relatively little refrigerant as compared to flooded evaporators having the same coil volume. The dry expansion evaporators are usually only one-fourth to one-third filled with liquid refrigerant. A simple bare-tube dry expansion evaporator is shown in Fig. 11.12. The finned coil dry expansion evaporators are also available.

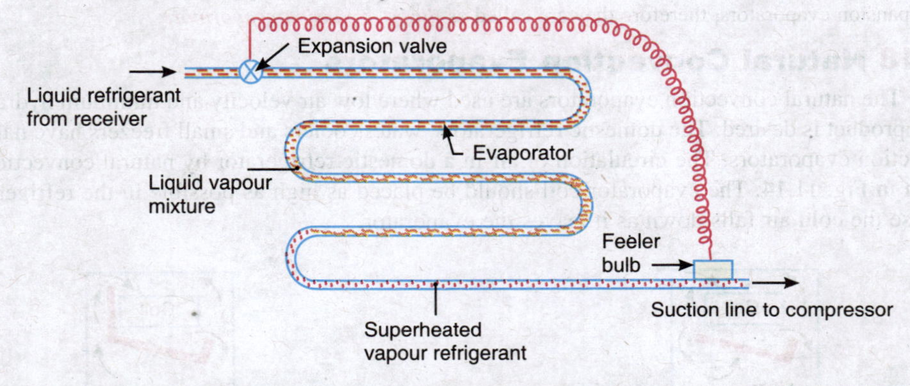

Fig. 11.12. Dry expansion evaporator.

In dry expansion evaporators, the liquid refrigerant from the receiver is fed by the expansion valve to the evaporator coil. The expansion valve controls the rate of flow of liquid refrigerant in such a way that all the liquid refrigerant is vaporised by the time it reaches at the end of the evaporator coils or the suction line to the compressor. The vapour is also superheated to a limited extent. It may be noted that in a dry expansion system, the refrigerant does not recirculate within the evaporator as in flooded type evaporator.

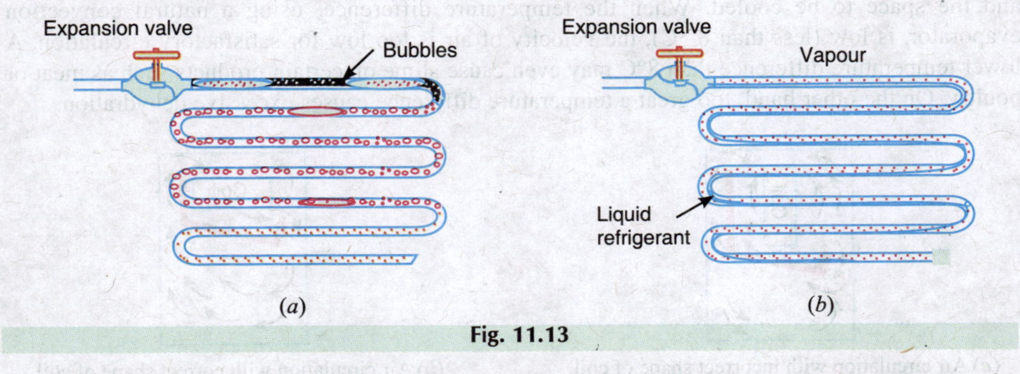

Fig. 11.13

The rate at which the liquid refrigerant is fed to the evaporator generally depends upon the rate of vaporisation and increases or decreases as the load on the evaporator increases or decreases. When the cooling load on the evaporator is light, the quantity of liquid refrigerant in the evaporator is small. We know that when liquid refrigerant passes through the expansion valve, some vapour (or flash gas) is formed. The flash gas causes bubbles in the evaporator. As more refrigerant vaporises, the bubbles get larger. If the coil diameter is small, the bubbles can cause dry areas on the interior walls of the coil, as shown in Fig. 11.13 (*a*). These dry areas reduce the rate of heat transfer. Thus, the evaporator efficiency decreases as dry areas increase, *i.e.* when the load on the evaporator is light. If the cooling load on the evaporator is heavy, the expansion valve allows a larger quantity of liquid refrigerant into the evaporator coil in order to accommodate the heavy load. In this case, the liquid and vapour separate. The liquid refrigerant flows along the bottom of the coil and vapour rises towards the top as shown in Fig. 11.13 (*b*). Thus, when the evaporator operates in this way, its efficiency is greatest. However, this efficiency depends upon the diameter of evaporator tubes, quantity of refrigerant in the evaporator and the velocity of the liquid refrigerant within the evaporator coil.

Note : Since the medium to be cooled comes in direct contact with the evaporator surfaces, in flooded and dry expansion evaporators, therefore they are called as *direct expansion evaporators*.

11.18 Natural Convection Evaporators

The natural convection evaporators are used where low air velocity and minimum hydration of the product is desired. The domestic refrigerators, water coolers and small freezers have natural convection evaporators. The circulation of air in a domestic refrigerator by natural convection is shown in Fig. 11.14. The evaporator coil should be placed as high as possible in the refrigerator because the cold air falls down as it leaves the evaporator.

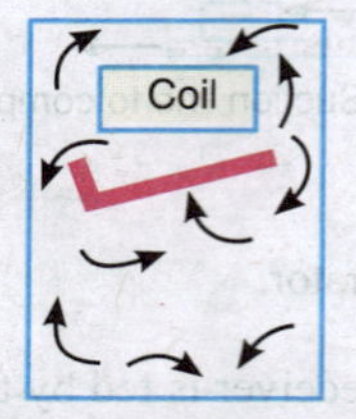

(*a*) Air circulation without baffles.

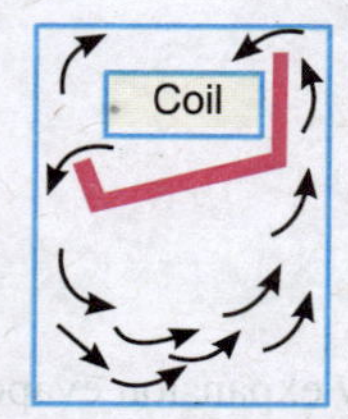

(*b*) Air circulation with baffles.

Fig. 11.14. Circulation of air in natural convection evaporator.

The velocity of air over the evaporator coil considerably affects the capacity. In natural convection, the velocity of air depends upon the temperature difference between the evaporator and the space to be cooled. When the temperature difference, using a natural convection evaporator, is low (less than 8° C), the velocity of air is too low for satisfactory circulation. A lower temperature difference than 8°C may even cause slime on certain products such as meat or poultry. On the other hand, too great a temperature difference causes excessive dehydration.

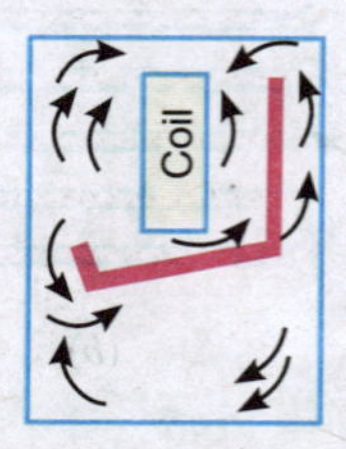

(*a*) Air circulation with incorrect shape of coil.

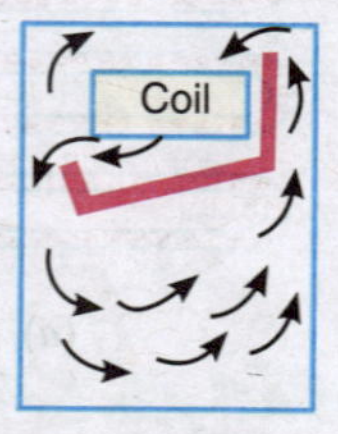

(*b*) Air circulation with correct shape of coil.

Fig. 11.15

The circulation of air around the coil depends upon its size, shape and location. A small, compact coil in a large box will cause only a small portion of air in the box to come in contact with the coil. The remainder of the air in the box will then be cooled by induced currents instead of by direct contact. This will result in large variations of temperature in various parts of the box. The effect on air distribution with incorrect shape of coil is shown in Fig. 11.15. The coil should occupy at least 2/3rd of the width of the path of the air circulation and 3/4th the length of box. The natural convection can be improved by the use of baffles as shown in Fig. 11.14 and 11.15. These are simply sheet metal plates which guide the ascending and descending air currents in their proper channels.

11.19 Forced Convection Evaporators

In forced convection evaporators, the air is forced over the refrigerant cooled coils and fins. This is done by a fan driven by an electric motor. The fins are provided to increase the heat transfer rate.

Citrus evaporator.

The forced convection evaporators are more efficient than natural convection evaporators because they require less cooling surface and high evaporator pressures can be used which save considerable power input to the compressor. These types of evaporators are suited for air cooling units as well as for refrigerator cabinets used to store bottled beverages or foods in sealed containers.

The forced circulation air cooling units may be divided into the following three groups according to the velocity of air leaving the unit.

1. *Low velocity cooling units.* These units have a discharge air rate from 60 m/min to 90 m/min. The low velocity cooling units are used in comfort air conditioning where low noise and low air velocity rates are needed. Both centrifugal and propeller type fans are used with low velocity cooling units.

2. *Medium velocity cooling units.* These units have an exit velocity of air from 150 m/min to 240 m/min. They are frequently used in refrigerators and freezers where drafts and noise are not a problem. The propeller fans are usually the source of air circulation in these units.

3. *High velocity cooling units.* These units have a discharge air rate above 240 m/min. They are used principally in blast freezers in special product refrigerators requiring quick pull-down of temperature. The high velocity cooling units usually use centrifugal fans as the source of air circulation.

11.20 Frosting Evaporators

The frosting type evaporators always operate at temperatures below 0°C. This means that the coil frosts continually when in use and it must be removed at regular intervals either manually or automatically for most efficient operation. The frost which forms on the evaporator comes from the moisture in the air. Some evaporators run at extremely low temperature in order to keep the refrigeration fixture cool. This allows frost and ice to build up. It may be noted that as the frost grows in thickness, the coil or cooling efficiency decreases until the ice and frost is removed. The evaporators in household refrigerators, bare pipe coils in storage boxes and low temperature evaporators fall under the frosting evaporators.

11.21 Non-Frosting Evaporators

The non-frosting evaporators operate at a temperature above 0°C at all times. Therefore, the frost does not form on the evaporator. Occasionally, the evaporator may build up a light coat of frost just before the compressor shuts off. This frost immediately melts on the off cycle.

The big advantage of a non-frosting evaporator is its operation at a temperature close to freezing (0.6°C to 1°C). At this temperature, it does not draw the moisture out of the air rapidly. This permits to maintain a *relative humidity from 75 to 85 per cent in the cabinet. This helps to keep the food fresh and stops shrinking in weight. The use of non-frosting evaporators are limited to high temperature work such as air conditioning, bakery refrigeration and candy storage.

Note : Since the non-frosting evaporators have greater area, therefore they are heavy than the frosting evaporators. They also require baffles to direct the air flow in one direction past the coils. The coil efficiency is increased as the baffles speed up the air cooling process.

11.22 Defrosting Evaporators

A defrosting evaporator is one in which frost accumulates on the coils when the compressor is running and melts after the compressor shuts off. During the running of the compressor, the evaporator remains at a temperature of about – 7° C to – 6° C. At these temperatures, the frost forms on the coil surface. But after the compressor stops, the coil warms upto a temperature slightly above 0°C and the frost melts before the compressor starts again.

A defrosting evaporator is usually a finned coil because of the necessity of rapid heat transfer during the off cycle. It also keeps a high relative humidity of about 90 to 95 per cent. However, this sacrifices the temperature difference between the evaporator refrigerant and the air in the cabinet. It requires greater evaporator area to make up this loss.

Sometimes, the defrosting evaporator presents a problem. In some installations, the top of the evaporator may defrost and moisture flows down the evaporator surface. The moisture may not have sufficient time to escape before the compressor lowers the temperature of the evaporator to – 7° C to – 6° C range. When this happens, the ice accumulates at the lower parts of the evaporator. This ice accumulation of the evaporator fins may block the free circulation of air around the coils, thus reducing the effectiveness of the evaporator.

11.23 Methods of Defrosting an Evaporator

When air is cooled below its **dew point temperature, it gives up the moisture and deposits on the nearest colder surface. Since the colder surface in a refrigerated area is the evaporator, therefore the moisture (in the form of dew) deposits on the surface of the evaporator. When the temperature of the evaporator falls below 0°C, the moisture deposited on the surface of the evaporator freezes and forms a coating of frost. If this frost is not removed periodically, it acts as an insulator and retards the heat transfer rate between the air and the evaporator which causes the compressor to run at a lower suction pressure. Under such conditions, the capacity and efficiency of the refrigerating system decreases. Thus when an evaporator operates below 0°C, it must be defrosted periodically for the efficient operation of the system. The various methods used for defrosting the evaporators are as follows :

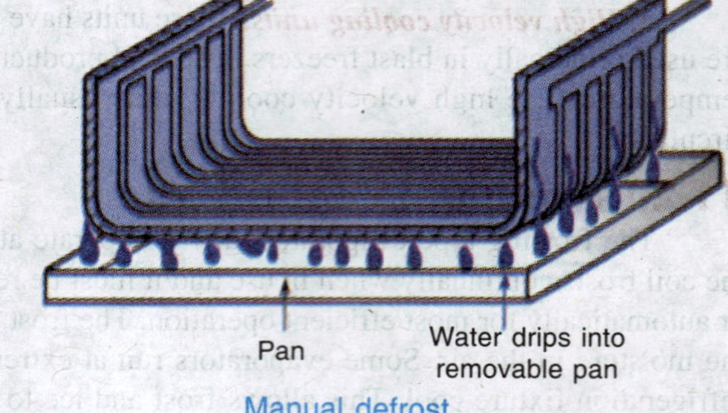

Manual defrost.

* The relative humidity is the ratio of the amount of water in air space to the amount of water that the air space can hold at a given temperature. The humidity is important in refrigeration because fresh foods and dairy products require high humidity conditions to reduce moisture loss.

** When warm air cools, its capacity to hold moisture decreases and relative humidity increases. When the air temperature reaches the point where the relative humidity is 100%, the air is saturated because it cannot hold any more moisture. This temperature is called *dew point temperature.*

1. Manual defrosting method,
2. Pressure control defrosting method,
3. Temperature control defrosting method,
4. Water defrosting method,
5. Reverse cycle defrosting method,
6. Simple hot gas defrosting method,
7. Automatic hot gas defrosting method,
8. Thermobank defrosting method, and
9. Electric defrosting method,

All the above defrosting methods are discussed, in detail, in the following pages.

11.24 Manual Defrosting Method

It is the earliest and simplest method of defrosting. In this method, either the compressor is stopped or the refrigerant to the evaporator (in large installations) is closed until the accumulated frost or ice is melted. At that point, the compressor is started or the refrigerant valve is re-opened. This method is still employed on large installations. The major drawback of this system is that it requires long period for defrosting.

11.25 Pressure Control Defrosting Method

We have already discussed that the accumulation of the frost on the surface of the evaporator reduces the heat transfer rate between the air and the evaporator which causes the compressor to run at a lower suction pressure. When the pressure falls below the predetermined value due to accumulation of frost, the pressure operated control, as shown in Fig. 11.16, comes into action and stops the compressor. The control is set in such a way as to permit the coil to warm up to 0° C, and defrost before it starts again. The stopping and starting of the system is done automatically according to the change in pressure.

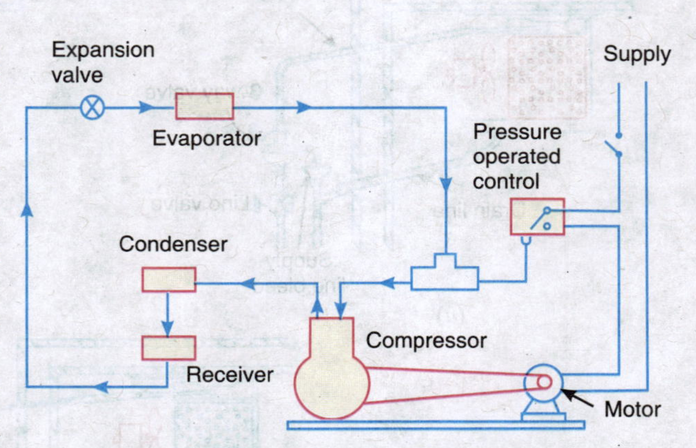

Fig. 11.16. Pressure defrosting method.

11.26 Temperature Control Defrosting Method

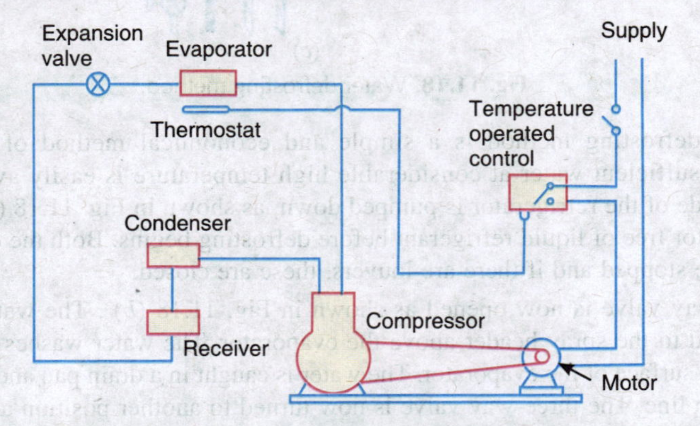

Fig. 11.17. Temperature control defrosting method.

The temperature control defrosting method, as shown in Fig. 11.17, is similar to pressure control defrosting method. The accumulation of frost on the surface of the evaporator causes reduction in heat transfer rate between the air and the evaporator. Due to this, the temperature of air passing over the coils increases. When the temperature increases above the requirements of the system, a temperature operated control comes into action and stops the compressor. The stopping and starting of the system is done automatically according to the change in temperature.

The temperature control defrost method makes it possible to maintain temperatures regardless of the load variations. This method is preferred to the pressure control defrost method for cold locations because the operation of the compressor is controlled directly by the temperature. The disadvantage of this method is that it does not assure complete defrost especially in hot weather or other high load conditions. This disadvantage may be overcome by an occasional manual defrost method.

11.27 Water Defrosting Method

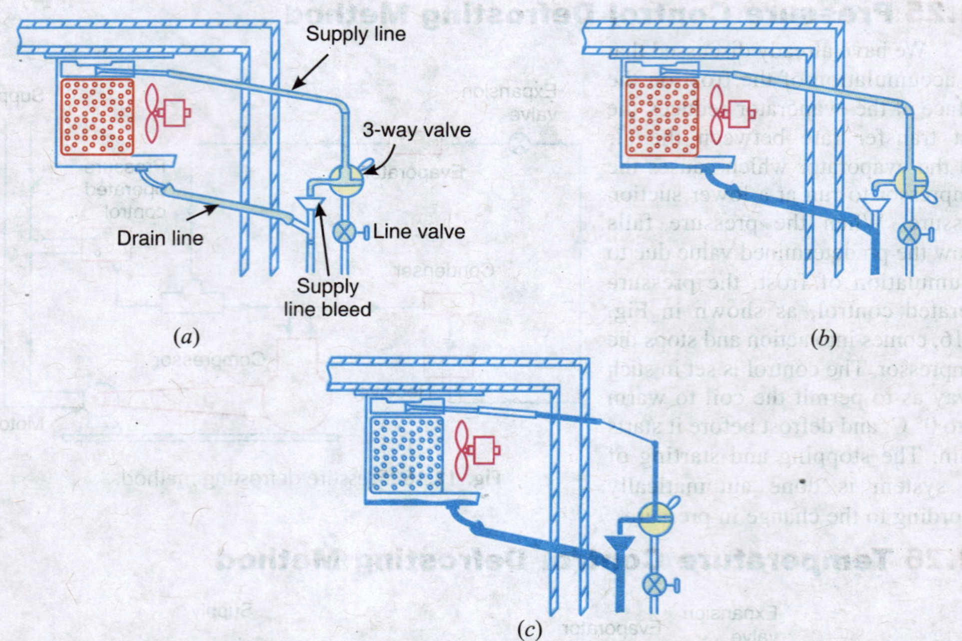

Fig. 11.18. Water defrosting method.

The water defrosting method is a simple and economical method of defrosting the evaporator where sufficient water at considerable high temperature is easily available. In this method, the low side of the refrigerator is pumped down, as shown in Fig. 11.18 (*a*) . This action leaves the evaporator free of liquid refrigerant before defrosting begins. Both the compressor and evaporator fans are stopped and if there are louvers, these are closed.

The three-way valve is now opened as shown in Fig. 11.18 (*b*) . The water in sufficient quantity is supplied to the spray header above the evaporator. The water washes out the ice and frost formed on the surface of the evaporator. The water is caught in a drain pan and must be drawn out through a drain line. The three-way valve is now turned to another position as shown in Fig. 11.18 (*c*). This permits the removal of water from the supply line and other drain lines during the defrosting cycle.

11.28 Reverse Cycle Defrosting Method

The normal cycle of operation in a refrigerating system is shown in Fig. 11.19. The evaporator may be defrosted by reversing the cycle of operation in the system as shown in Fig. 11.20. When the cycle is reversed, the evaporator becomes the condenser and the condenser becomes an evaporator. When the evaporator functions as a condenser, it melts the accumulated frost.

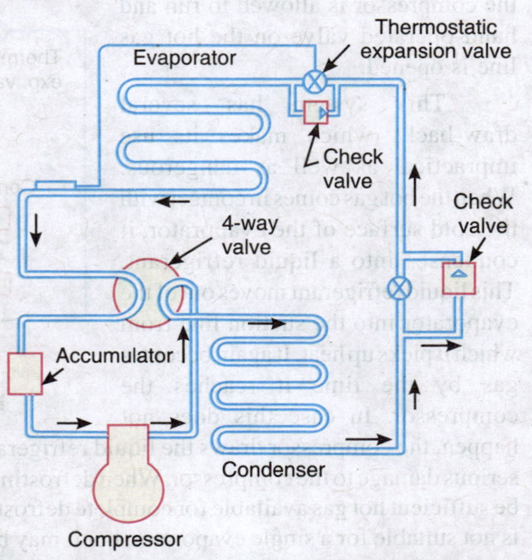

Fig. 11.19. Four-way valve position for normal refrigerating system.

The reversing of cycle is handled by installing a four-way valve. During the reverse cycle defrosting system, the four-way valve is turned manually or automatically and the hot gas from the compressor is passed to the evaporator. The hot gas condenses and heats the evaporator coil. Thus the frost accumulated on the evaporator surface is melted out. The condensed gas from the evaporator bypasses the thermostatic expansion valve by means of a check valve. The refrigerant evaporates in the condenser and is returned to the compressor in a vapour state.

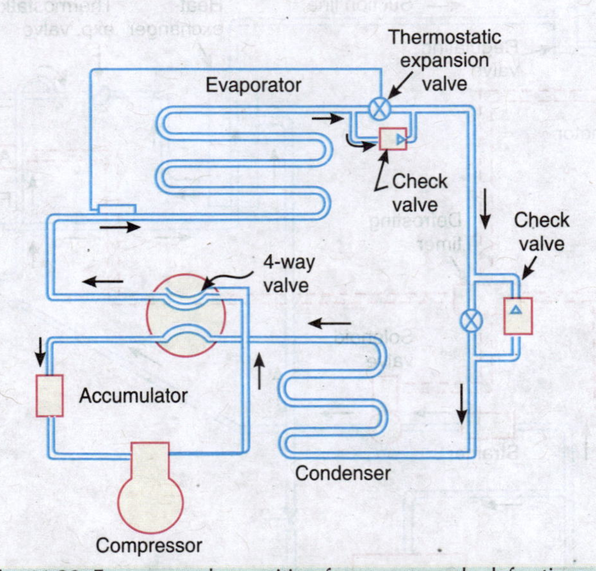

Fig. 11.20. Four-way valve position for reverse cycle defrosting system.

11.29 Simple Hot Gas Defrosting Method

The very simplest form of hot gas defrosting method is shown in Fig. 11.21. In this method, the hot gas (vapour refrigerant) discharged from the compressor is run through a bypass or hot gas line to the evaporator at a point between the evaporator and the thermostatic expansion valve. As the hot gas gives up heat to the cold evaporator, the frost accumulated on the surface of the evaporator melts. When defrosting, the compressor is allowed to run and hand-operated valve on the hot gas line is opened.

This system has several draw-backs which makes its use impractical as well as dangerous. When the hot gas comes in contact with the cold surface of the evaporator, it condenses into a liquid refrigerant. This liquid refrigerant moves out of the evaporator into the suction line from which it picks up heat. It again becomes gas by the time it reaches the compressor. In case this does not happen, the compressor draws the liquid refrigerant. This may cause oil pumping and in some cases serious damage to the compressor. When defrosting, the coil is absorbing heat, therefore there may not be sufficient hot gas available for complete defrosting. This simple form of hot gas defrosting method is not suitable for a single evaporator, but it may be used if there are two or more evaporators on one compressor.

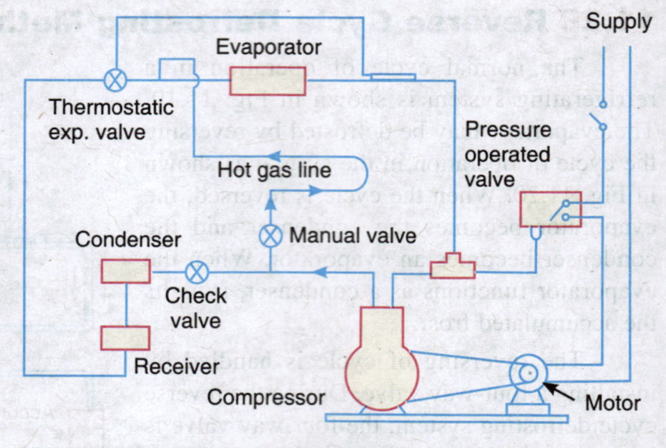

Fig. 11.21. Simple hot gas defrosting method.

11.30 Automatic Hot Gas Defrosting Method

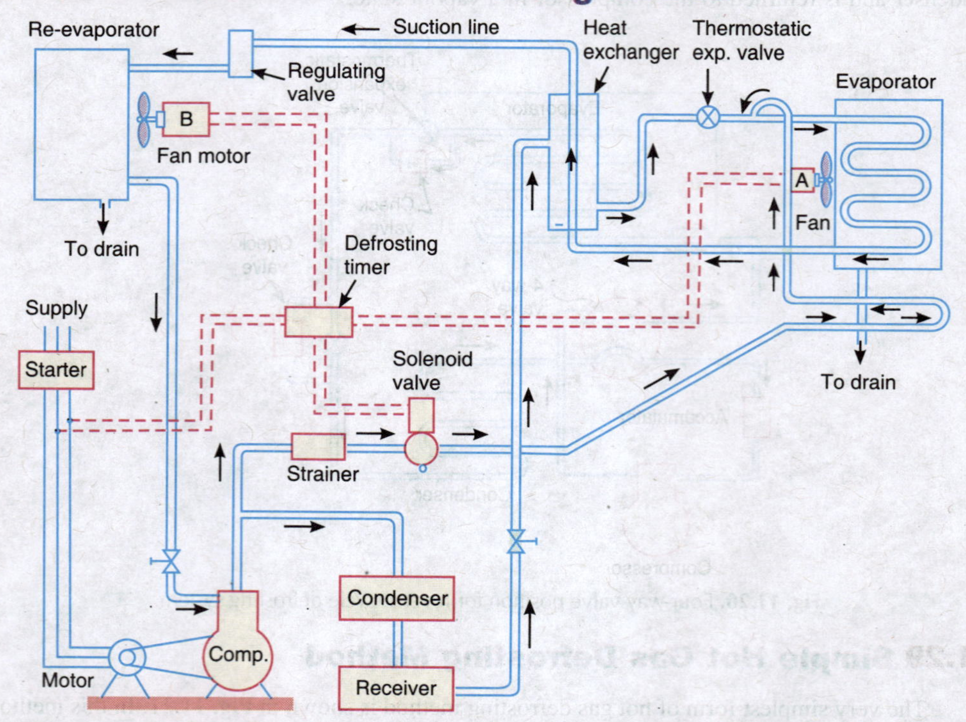

Fig. 11.22. Automatic hot gas defrosting method with re-evaporator.

The automatic hot gas defrosting method with re-evaporator is shown in Fig. 11.22. In this method, the defrosting timer starts the defrosting cycle by opening the solenoid valve in the hot gas line. During this operation, the hot gas (vapour refrigerant) discharged from the compressor is run through the hot gas line to the evaporator at a point between the evaporator and the thermostatic expansion valve. As the hot gas gives up heat to the cold evaporator, the frost accumulated on the surface of the evaporator melts.

When the hot gas comes in contact with the cold surface of the evaporator, it condenses into a liquid refrigerant. This liquid refrigerant is now passed to a re-evaporator where it is again evaporated before being fed to the compressor. During the defrosting cycle, the fan motor *A* is in non-working condition while the fan motor *B* remains in working condition. The opening of solenoid valve, stopping of fan *A* and starting of fan *B* is done simultaneously by the automatic defrosting timer.

When the defrosting is completed, the defrosting timer shuts off the solenoid valve in order to stop the flow of hot gas to the evaporator and puts the system in normal cycle of operation. During this normal cycle, the solenoid valve is closed and the hot gas discharged from the compressor is passed through the condenser, receiver, heat exchanger, thermostatic expansion valve, evaporator and back to the compressor. The fan motor *A* remains in working condition during this normal cycle while the fan motor *B* remains in non-working condition.

11.31 Thermobank Defrosting Method

The thermobank defrosting method, as shown in Fig. 11.23, is developed by Kramer Trenton and Company. This method affords completely automatic defrosting using hot gas and a heat accumulator (called thermobank) in the suction line. As the name suggests, the thermobank is a bank for storing heat. The heat from the compressor discharge is stored in a small tank of liquid during the normal refrigerating cycle. During the defrosting cycle, this stored heat goes to the suction line instead of directly to the hot gas line. This ensures that the liquid from the evaporator is vaporised early in the defrost cycle. This prevents the liquid slugging of the compressor.

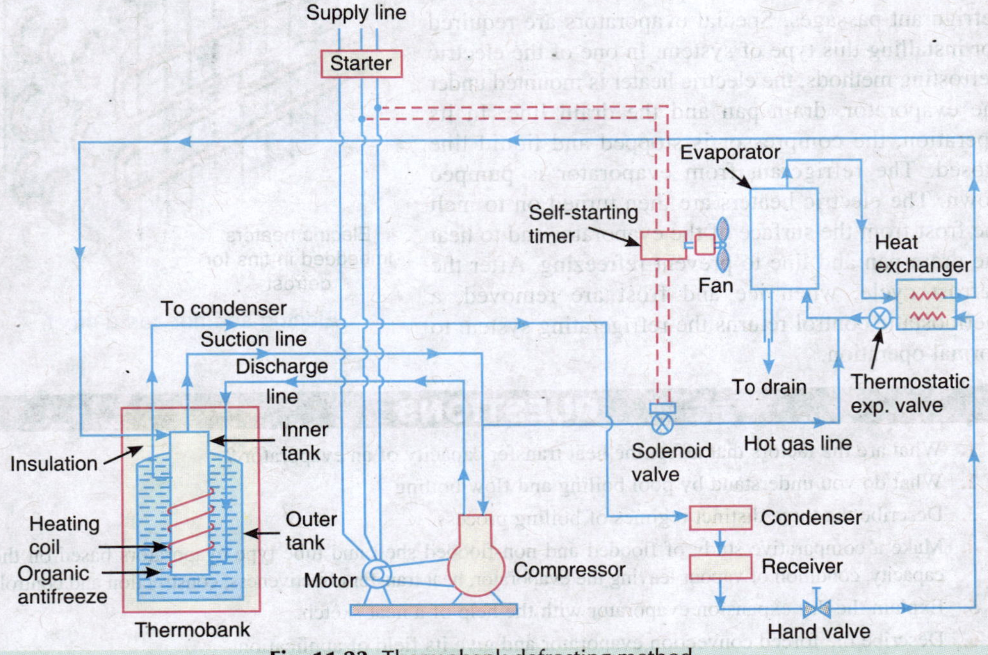

Fig. 11.23. Thermobank defrosting method.

The thermobank, as shown in Fig. 11.23, consists of a tank within another tank. The inner tank is longer than the outer tank. The suction connection is made at the top of the inner tank. The outer tank contains a quantity of antifreeze liquid in which there is a submerged coil. The coil carries the hot gas discharged from the compressor. Both the tanks are insulated to prevent the loss of heat. A solenoid valve placed in the hot gas line is electrically connected to a self-starting timer.

During the normal cycle of operation, the solenoid valve is closed. The hot gas discharged from the compressor passes through the submerged coil in the outer tank, heats the antifreeze and then condensed in a condenser. The liquid refrigerant from the receiver flows through the heat exchanger, thermostatic expansion valve and then to the evaporator. The vapour refrigerant from the evaporator enters the upper part of the thermobank . Since the vapour refrigerant (*i.e.* suction gas) enters and leaves the tank at the upper end, therefore it does not absorb any heat.

When a predetermined lapse of time has passed and a sufficient quantity of frost has accumulated on the surface of the evaporator, the timer automatically opens the solenoid valve and shuts off the fan motor. During the defrosting cycle, the compressor continues to operate and the hot gas goes through the solenoid valve to the evaporator. The hot gas gives up its heat to the evaporator and the accumulated frost is melted. When the hot gas comes in contact with the cold surface of the evaporator, it condenses into a liquid refrigerant. This liquid refrigerant from the evaporator then flows to the suction line to the thermobank. Due to the stored heat in the thermobank, it re-evaporates and goes to the compressor. This defrosting cycle continues for approximately six minutes, after which the timer closes the solenoid valve and starts the fan motor. The system then returns to normal cycle of operation.

11.32 Electric Defrosting Method

The electric defrosting method is popularly used for low temperature evaporators. In this system, the heating coils for heating may be installed within the evaporator, around the evaporator, or within the refrigerant passages. Special evaporators are required for installing this type of system. In one of the electric defrosting methods, the electric heater is mounted under the evaporator, drain pan and the drain line. In its operation, the compressor is stopped and liquid line closed. The refrigerant from evaporator is pumped down. The electric heaters are then turned on to melt the frost from the surface of the evaporator and to heat the drain pan and line to prevent refreezing. After the defrost cycle, when ice and frost are removed, a thermostatic control returns the refrigerating system to normal operation.

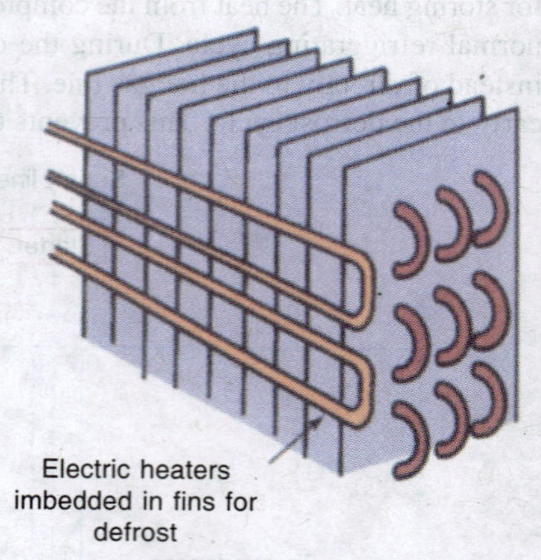

Electric and hot gas defrost.

QUESTIONS

1. What are the factors that affect the heat transfer capacity of an evaporator ?
2. What do you understand by pool boiling and flow boiling ?
3. Describe the three distinct regimes of boiling process.
4. Make a comparative study of flooded and non-flooded shell and tube type evaporators based on the capacity, condition of vapour leaving the evaporator, heat transfer effectiveness, construction and control.
5. Explain the dry expansion evaporator with the help of a neat sketch.
6. Describe the forced convection evaporator and give its field of application.
7. Discuss the frosting and defrosting evaporators.
8. Write a short note on methods of defrosting.

OBJECTIVE TYPE QUESTIONS

1. An evaporator is used in the .... of a refrigeration system.
   (*a*) low pressure side (*b*) high pressure side
2. An evaporator is also known as
   (*a*) freezing coil (*b*) cooling coil (*c*) chilling coil (*d*) all of these
3. The temperature of the boiling refrigerant in the evaporator must always be .... than that of the surrounding medium.
   (*a*) lower (*b*) higher
4. The evaporator changes the low pressure liquid refrigerant from the expansion valve into
   (*a*) high pressure liquid refrigerant
   (*b*) low pressure liquid and vapour refrigerant
   (*c*) low pressure vapour refrigerant
   (*d*) none of the above
5. The fluid side heat transfer coefficient (h_f) when liquid flows through the evaporator shell is given by
   (*a*) $C\sqrt{m}$ (*b*) $C.m$ (*c*) $C\,(m)^2$ (*d*) $C\,(m)^4$
6. The fluid side heat transfer coefficient when air flows over finned coil by forced convection is given by
   (*a*) $C\,(m)^{0.1}$ (*b*) $C\,(m)^{0.2}$ (*c*) $C\,(m)^{0.3}$ (*d*) $C\,(m)^{0.4}$
7. The bare tube evaporators are also known as .... evaporators.
   (*a*) prime surface (*b*) extended surface
8. The evaporator generally used in home freezers, ice cream cabinets etc. is
   (*a*) plate evaporator (*b*) finned evaporator
   (*c*) shell and tube evaporator (*d*) shell and coil evaporator
9. The evaporator generally used for wine cooling and in petroleum industry for chilling oil is
   (*a*) plate evaporator (*b*) finned evaporator
   (*c*) tube-in-tube evaporator (*d*) shell and tube evaporator
10. The evaporator used in household refrigerators is
    (*a*) frosting evaporator (*b*) non-frosting evaporator
    (*c*) defrosting evaporator (*d*) none of these

ANSWERS

1. (*a*)	2. (*d*)	3. (*b*)	4. (*c*)	5. (*a*)
6. (*d*)	7. (*a*)	8. (*a*)	9. (*c*)	10. (*a*)

CHAPTER

12

Expansion Devices

1. Introduction.
2. Types of Expansion Devices.
3. Capillary Tube.
4. Hand-operated Expansion Valve.
5. Automatic (or Constant Pressure) Expansion Valve.
6. Thermostatic Expansion Valve.
7. Low-side Float Valve.
8. High-side Float Valve.

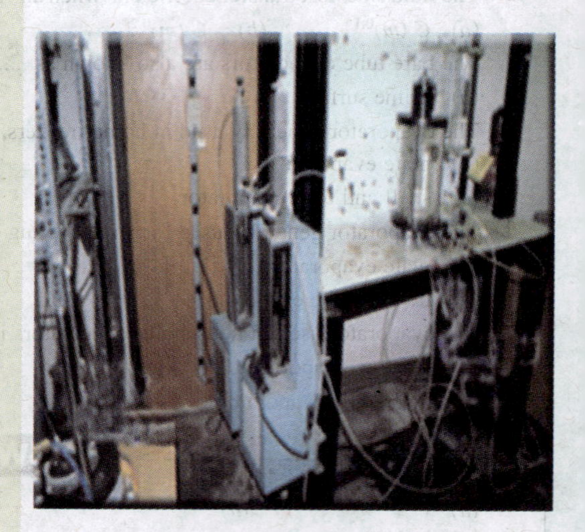

12.1 Introduction

The expansion device (also known as *metering device* or *throttling device*) is an important device that divides the high pressure side and the low pressure side of a refrigerating system. It is connected between the receiver (containing liquid refrigerant at high pressure) and the evaporator (containing liquid refrigerant at low pressure). The expansion device performs the following functions :

1. It reduces the high pressure liquid refrigerant to low pressure liquid refrigerant before being fed to the evaporator.

2. It maintains the desired pressure difference between the high and low pressure sides of the system, so that the liquid refrigerant vaporises at the designed pressure in the evaporator.
3. It controls the flow of refrigerant according to the load on the evaporator.

Note: The expansion devices used with dry expansion evaporators are usually called *expansion valves* whereas the expansion devices used with flooded evaporators are known as *float valves*.

12.2 Types of Expansion Devices

Following are the main types of expansion devices used in industrial and commercial refrigeration and air conditioning system.

1. Capillary tube,
2. Hand-operated expansion valve,
3. Automatic or constant pressure expansion valve,
4. Thermostatic expansion valve,
5. Low-side float valve, and
6. High-side float valve.

The above types of expansion devices are discussed, in detail, in the following pages.

12.3 Capillary Tube

The capillary tube, as shown in Fig. 12.1, is used as an expansion device in small capacity hermetic sealed refrigeration units such as in domestic refrigerators, water coolers, room air-conditioners and freezers. It is a copper tube of small internal diameter and of varying length depending upon the application. The inside diameter of the tube used in refrigeration work is generally about 0.5 mm to 2.25 mm and the length varies from 0.5 m to 5 m. It is installed in the liquid line between the condenser and the evaporator as shown in Fig. 12.1. A fine mesh screen is provided at the inlet of the tube in order to protect it from contaminants. A small filter drier is used on some systems to provide additional freeze-up application.

In its operation, the liquid refrigerant from the condenser enters the capillary tube. Due to the frictional resistance offered by a small diameter tube, the pressure drops. Since the frictional resistance is directly proportional to the length and inversely proportional to the diameter, therefore longer the capillary tube and smaller its inside diameter, greater is the pressure drop created in the refrigerant flow. In other words, greater pressure difference between the condenser and evaporator

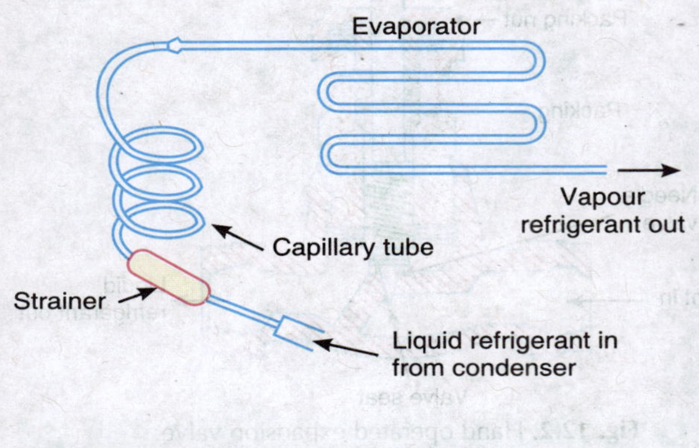

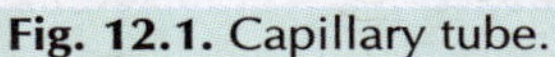
Fig. 12.1. Capillary tube.

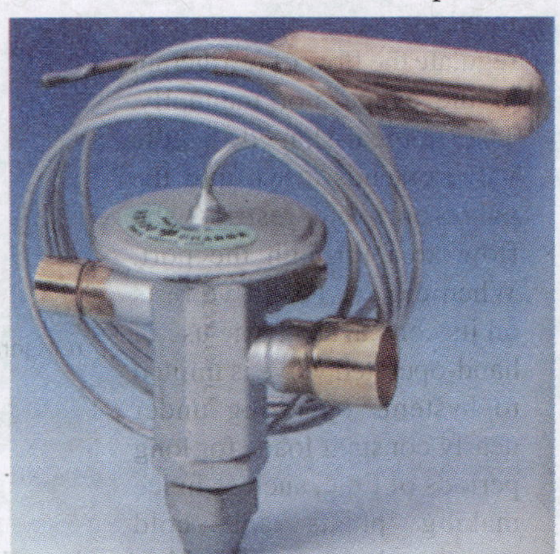
Flare type valve.

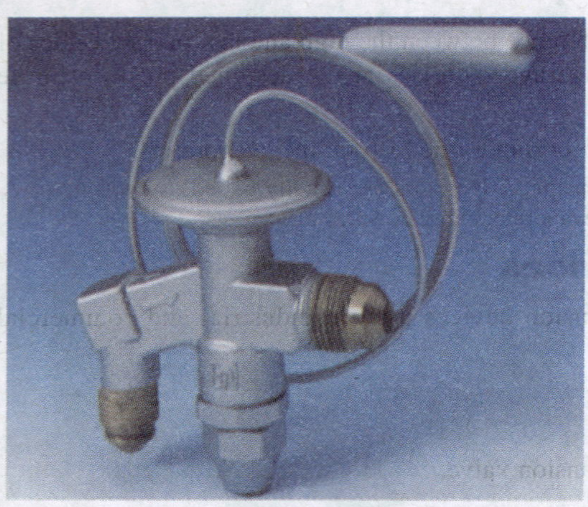

TVXs have remote sensing elements.

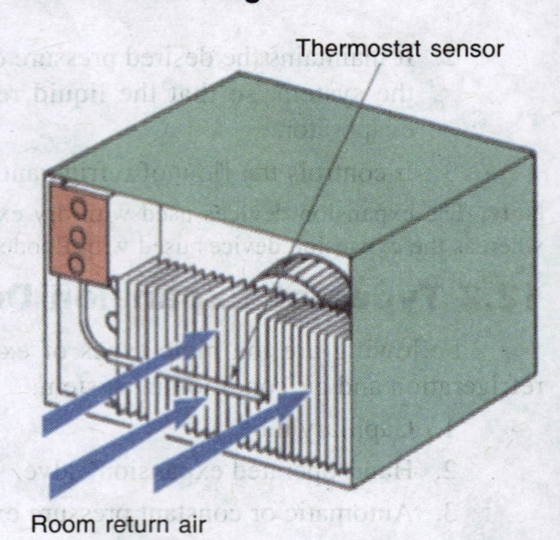

The thermostat is located in the return airstream.

(*i.e.* high side and low side) is needed for a given flow rate of the refrigerant. The diameter and length of the capillary tube once selected for a given set of conditions and load cannot operate efficiently at other conditions.

The refrigeration system using capillary tube have the following advantages :

1. The cost of capillary tube is less than all other forms of expansion devices.
2. When the compressor stops, the refrigerant continues to flow into the evaporator and equalises the pressure between the high side and low side of the system. This considerably decreases the starting load on the compressor. Thus a low starting torque motor (low-cost motor) can be used to drive the compressor, which is a great advantage.
3. Since the refrigerant charge in a capillary tube system is critical, therefore no receiver is necessary.

12.4 Hand-operated Expansion Valve

The hand-operated expansion valve, as shown in Fig. 12.2, is the most simple type of expansion valve but it requires an operator to regulate the flow of refrigerant to the evaporator manually. The conical-shaped needle valve extends down into the valve port and restricts the flow area through the port. When closed, the valve rests on its conical seat. The use of hand-operated valve is limited to systems operating under nearly constant loads for long periods of time, such as in ice making plants and cold storages. It is not suitable for installations where the load varies and the compressor runs intermittently to maintain a constant temperature.

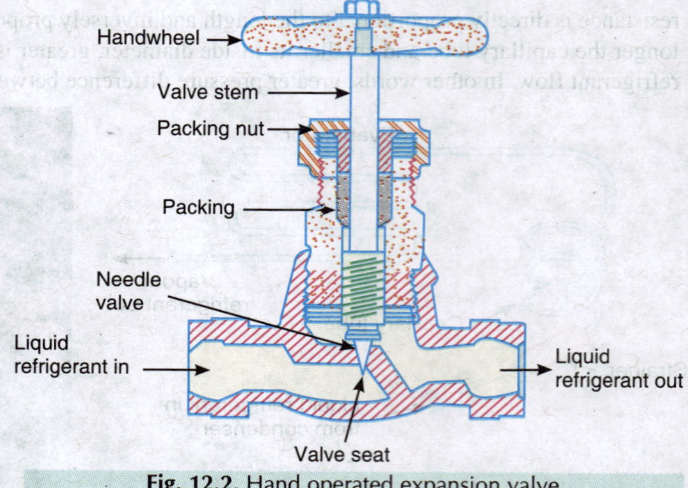

Fig. 12.2. Hand operated expansion valve.

12.5 Automatic (or Constant Pressure) Expansion Valve

The automatic expansion valve is also known as *constant pressure expansion valve*, because it maintains constant evaporator pressure regardless of the load on the evaporator. Its main moving force is the evaporator pressure. It is used with dry expansion evaporators where the load is relatively constant.

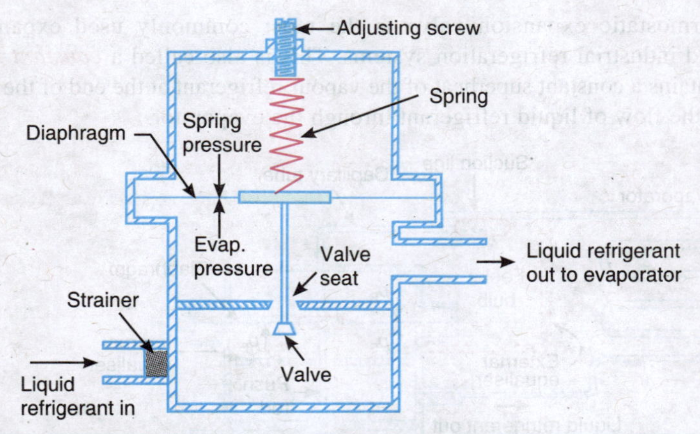

Fig. 12.3. Automatic (or constant pressure) expansion valve.

The automatic expansion valve, as shown in Fig. 12.3, consists of a needle valve and a seat (which forms an orifice), a metallic diaphragm or bellows, spring and an adjusting screw. The opening and closing of the valve with respect to the seat depends upon the following two opposing forces acting on the diaphragm :

1. The spring pressure and atmospheric pressure acting on the top of the diaphragm, and
2. The evaporator pressure acting below the diaphragm.

When the compressor is running, the valve maintains an evaporator pressure in equilibrium with the spring pressure and the atmospheric pressure. The spring pressure can be varied by adjusting the tension of the spring with the help of spring adjusting screw. Once the spring is adjusted for a desired evaporator pressure, then the valve operates automatically to maintain constant evaporator pressure by controlling the flow of refrigerant to the evaporator.

When the evaporator pressure falls down, the diaphragm moves downwards to open the valve. This allows more liquid refrigerant to enter into the evaporator and thus increasing the evaporator pressure till the desired evaporator pressure is reached. On the other hand, when the evaporator pressure rises, the diaphragm moves upwards to reduce the opening of the valve. This decreases the flow of liquid refrigerant to the evaporator which, in turn, lowers the evaporator pressure till the desired evaporator pressure is reached.

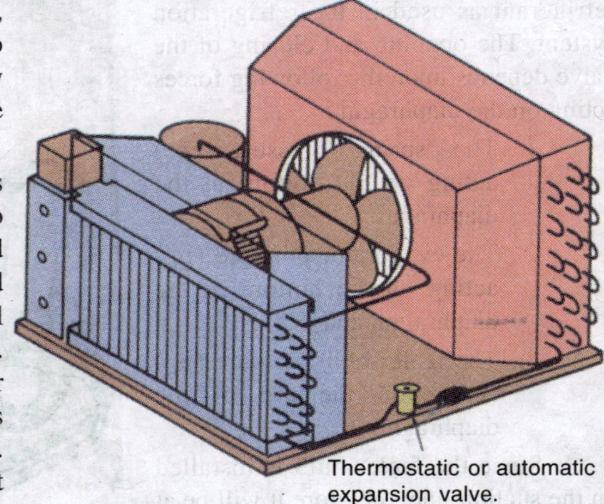

A thermostatic or automatic expansion valve may sometimes be installed instead of replacing a capillary tube.

When the compressor stops, the liquid refrigerant continues to flow into the evaporator and increases the pressure in the evaporator. This increase in evaporator pressure causes the diaphragm to move upwards and the valve is closed. It remains closed until the compressor starts again and reduces the pressure in the evaporator.

12.6 Thermostatic Expansion Valve

The thermostatic expansion valve is the most commonly used expansion device in commercial and industrial refrigeration systems. This is also called a *constant superheat valve* because it maintains a constant superheat of the vapour refrigerant at the end of the evaporator coil, by controlling the flow of liquid refrigerant through the evaporator.

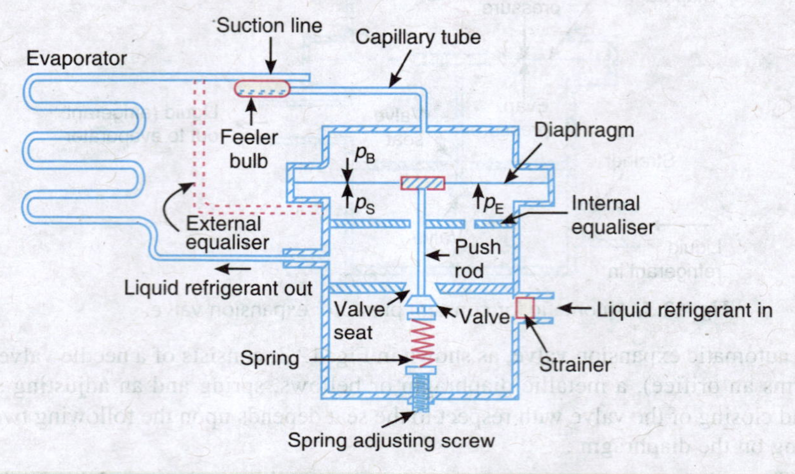

Fig. 12.4. Thermostatic expansion valve.

The thermostatic expansion valve, as shown in Fig. 12.4, consists of a needle valve and a seat, a metallic diaphragm, spring and an adjusting screw. In addition to this, it has a feeler or thermal bulb which is mounted on the suction line near the outlet of the evaporator coil. The feeler bulb is partly filled with the same liquid refrigerant as used in the refrigeration system. The opening and closing of the valve depends upon the following forces acting on the diaphragm :

1. The spring pressure (p_S) acting on the bottom of the diaphragm,
2. The evaporator pressure (p_E) acting on the bottom of the diaphragm, and
3. The feeler bulb pressure (p_B) acting on the top of the diaphragm.

Thermostatic expansion valve.

Since the feeler bulb is installed on the suction line, therefore it will be at the same temperature as the refrigerant at that point. Any change in the temperature of the refrigerant will cause a change in pressure in the feeler bulb which will be transmitted to the top of the diaphragm. Under normal operating conditions, the feeler bulb pressure acting at the top of the diaphragm is balanced by the spring pressure and the evaporator pressure acting at the bottom of the diaphragm. The force tending to close the valve is dependent upon the spring pressure and the evaporator pressure which, in turn, depends upon the saturation temperature of the refrigerant in the evaporator coil. The force tending to open the valve depends upon the feeler bulb pressure which, in turn, depends upon the temperature of refrigerant in the bulb. Thus the operation of valve is controlled by the difference between the two temperatures (*i.e.* saturation temperature and feeler bulb temperature) which is the superheat. The degree of superheat of the vapour refrigerant leaving the evaporator depends upon the initial setting of the spring tension, which can be changed with the help of spring adjusting screw. When the valve is set for a certain superheat, then it maintains that setting under all load conditions on the evaporator.

If the load on the evaporator increases, it causes the liquid refrigerant to boil faster in the evaporator coil. The temperature of the feeler bulb increases due to early vaporisation of the liquid refrigerant. Thus the feeler bulb pressure increases and this pressure is transmitted through the capillary tube to the diaphragm. The diaphragm moves downwards and opens the valve to admit more quantity of liquid refrigerant to the evaporator. This continues till the pressure equilibrium on the diaphragm is reached. On the other hand, when the load on the evaporator decreases, less liquid refrigerant evaporates in the evaporator coil. The excess liquid refrigerant flows towards the evaporator outlet which cools the feeler bulb with the result the feeler bulb pressure decreases due to decrease in its temperature. The low feeler bulb pressure is transmitted through the capillary tube to the diaphragm and moves it upward. This reduces the opening of the valve and thus the flow of liquid refrigerant to the evaporator. The evaporator pressure decreases due to reduced quantity of liquid refrigerant flowing to the evaporator. This continues till the evaporator pressure and the spring pressure maintain equilibrium with the feeler bulb pressure.

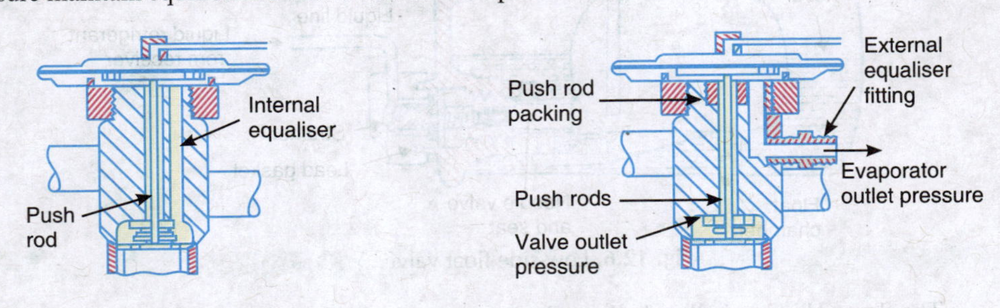

(*a*) Valve with internal equaliser. (*b*) Valve with external equaliser.

Fig. 12.5. Thermostatic expansion valve with equaliser.

The thermostatic expansion valve may be either *internally equalised* or *externally equalised* as shown in Fig. 12.5. The standard thermostatic expansion valves are internally equalised.

In a thermostatic expansion valve with internal equaliser, as shown in Fig. 12.5 (*a*), the pressure acting on the bottom of the diaphragm is equal to the evaporator inlet pressure. A hole drilled in the valve body transmits this pressure. The standard thermostatic expansion valves work well on evaporators having low pressure drops below 0.14 bar. If the pressure drop in evaporator is high (above 0.14 bar), then the pressure at the outlet of the evaporator (or at the feeler bulb location) will be less by the amount equal to the pressure drop. In such a case, the feeler bulb pressure should rise to maintain equilibrium with the inlet evaporator pressure and the spring pressure. The rise in feeler bulb pressure will raise its temperature and thus the degree of superheat. This means that an internally equalised thermostatic expansion valve will operate with excessive superheat. Thus the flow of refrigerant to the evaporator and hence the net refrigerating effect reduces.

In order to overcome this effect due to pressure drop in evaporators, a thermostatic expansion valve with external equaliser, as shown in Fig. 12.5 (*b*), is used. In this type of valve, the pressure at the bottom of the diaphragm is equal to the outlet evaporator pressure. Thus the ill effect of the evaporator pressure drop is overcome. In an externally equalised thermostatic expansion valve, a small diameter equaliser tube connects the diaphragm with the evaporator outlet as shown in Fig. 12.4. This connection is made immediately downstream of the feeler bulb location. In this position, the bulb temperature will not be affected by the occasional slugs of oil or small leaks of the refrigerant past the gland packing of the push rod. The externally equalised thermostatic expansion valve operates at the desired superheat regardless of the evaporator pressure drop.

Notes: 1. The thermostatic expansion valves are usually rated in tonnes of refrigeration.

2. Most thermostatic expansion valves are set for 5°C of superheat.

12.7 Low-side Float Valve

As the name indicates, the low-side float valve, as shown in Fig. 12.6, is located in the low pressure side (*i.e.* between the evaporator and compressor suction line) of the refrigeration system. It maintains a constant level of liquid refrigerant in the evaporator and float chamber by opening and closing a needle valve. A refrigeration system with low-side float valve is shown in Fig. 12.7.

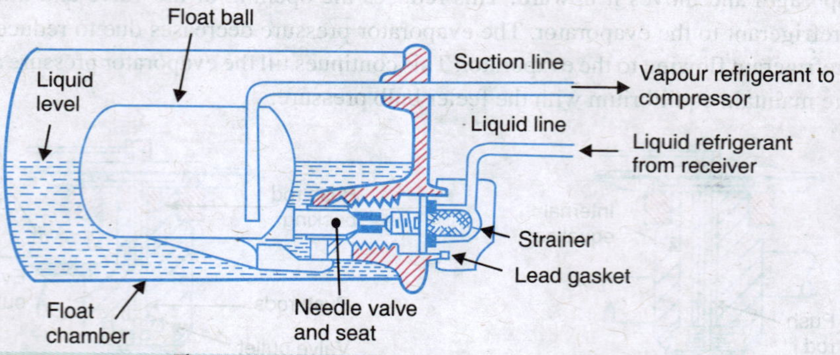

Fig. 12.6. Low-side float valve.

The float valve is a hollow ball attached to one end of a float arm as shown in Fig. 12.6. The other end of the arm is connected to a needle valve. The movement of the float ball (rise or fall) is transmitted to the needle valve by the float arm which closes or opens the flow of liquid refrigerant. Since the float valve is hollow, therefore it floats on the liquid refrigerant, in the float chamber.

When the liquid refrigerant in the evaporator vaporises, its level falls down. This causes the float to drop and thus opens the needle valve, thereby allowing liquid refrigerant from the liquid line to the float chamber and then to the evaporator to make up for the amount of vaporisation. When the desired liquid level is reached, the float rises and closes the needle valve. The major advantage of the low-side float valve is that it maintains a constant liquid level in the evaporator under all loading conditions regardless of the evaporator pressure and temperature.

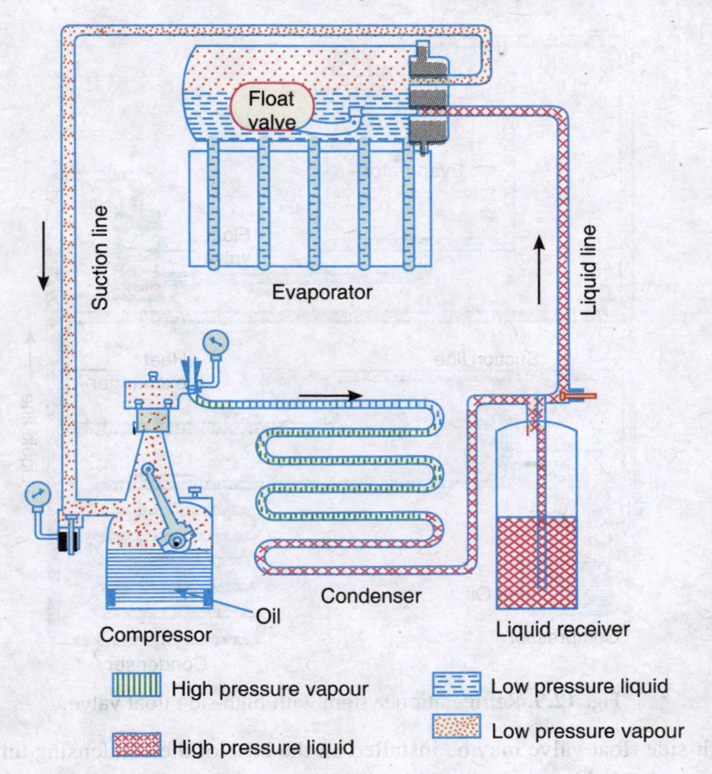

Fig. 12.7. Refrigeration system with low-side float valve.

12.8 High-side Float Valve

As the name indicates, the high-side float valve, as shown in Fig. 12.8, is located on the high pressure side (*i.e.* between the condenser and evaporator) of the refrigeration system. It controls the flow of liquid refrigerant to the evaporator according to the load and maintains a constant liquid level in the evaporator and float chamber by opening or closing of a needle valve. A refrigeration system with high-side float valve is shown in Fig. 12.9.

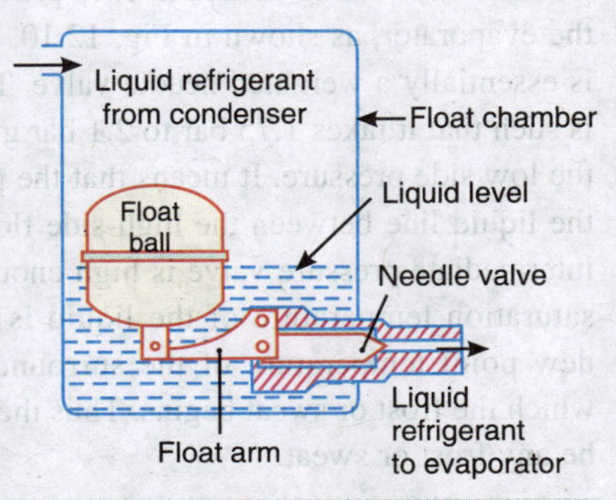

Fig. 12.8. High-side float valve.

The liquid refrigerant from the condenser flows to the float chamber. As the level of liquid refrigerant in the float chamber rises, the float ball also rises, thereby opening the needle valve. This allows the liquid refrigerant to flow into the evaporator. When the liquid level in the float chamber falls down, the float ball also drops, thereby closing the needle valve. It may be noted that the condenser supplies the liquid refrigerant at the same rate as it evaporates in the evaporator. Since the rate of vaporisation depends upon the load on the evaporator, therefore the high-side float valve functions according to the load. The high-side float valve may be used with dry expansion evaporators.

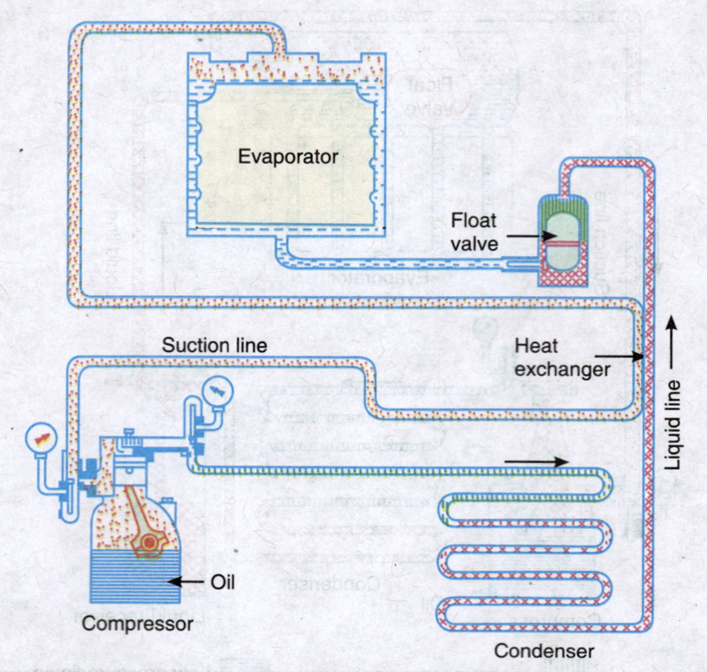

Fig. 12.9. Refrigeration system with high-side float valve.

The high-side float valve may be installed on the base of the condensing unit or near to the evaporator, as it is dependent on the liquid level in the evaporator. When it is installed on the base of the condensing unit, the liquid line from the float to the evaporator will frost or sweat. In order to overcome this undesirable condition, the liquid line is insulated. Another way of preventing the frosting or sweating condition is to provide an intermediate pressure valve in the liquid line near the evaporator, as shown in Fig. 12.10. This valve is essentially a weighted needle valve. The weight is such that it takes 1.75 bar to 2.1 bar greater than the low side pressure. It means that the pressure in the liquid line between the high-side float and the intermediate pressure valve is high enough so that saturation temperature of the liquid is above the dew point temperature of the surrounding air at which the frost or sweat begins. Thus there will not be any frost or sweat.

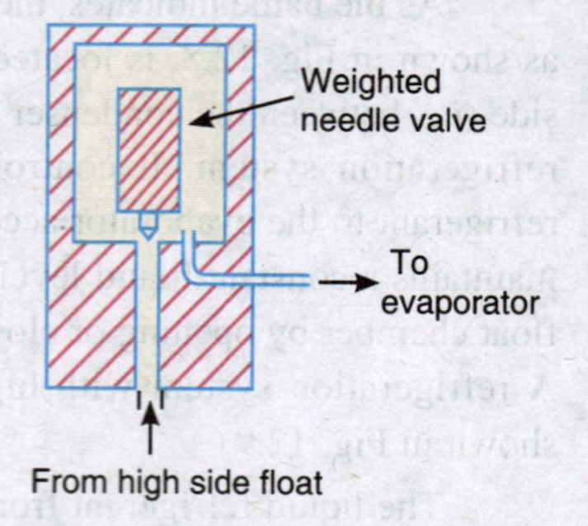

Fig. 12.10. Intermediate pressure valve.

In those applications where the high-side float valve is installed on the top of the refrigerator near to the evaporator, no intermediate pressure valve is needed because the liquid line is inside the cabinet.

The refrigerant liquid level in a high-side float system is critical. Too much liquid refrigerant causes flooding of the suction line and too little liquid refrigerant causes a low evaporator level resulting in low capacity of the refrigerating system.

QUESTIONS

1. Discuss the operation of a capillary tube in a refrigeration system.
2. Explain in brief as to why capillary tube is preferred to other throttling devices in household refrigerator.
3. Draw a neat sketch of a hand-operated expansion valve and explain its working.
4. Explain the working of an automatic expansion valve. Why it is called constant pressure expansion valve ?
5. Explain the working principle of thermostatic expansion valve with the help of a neat diagram.
6. Differentiate between low-side float valve and high-side float valve.

OBJECTIVE TYPE QUESTIONS

1. In a refrigeration system, the expansion device is connected between the
   (*a*) compressor and condenser (*b*) condenser and receiver
   (*c*) receiver and evaporator (*d*) evaporator and compressor
2. The expansion devices used with flooded evaporators are called
   (*a*) float valves (*b*) expansion valves
3. The capillary tube, as an expansion device, is used in
   (*a*) domestic refrigerators (*b*) water coolers
   (*c*) room air-conditioners (*d*) all of these
4. The pressure in a capillary tube decreases due to
   (*a*) frictional resistance offered by the tube wall (*b*) acceleration of refrigerant in the tube
   (*c*) heat transfer from the tube (*d*) both (*a*) and (*b*)
5. Capillary tube is not used in large capacity refrigeration systems because
   (*a*) cost is too high (*b*) capacity control is not possible
   (*c*) it is made of copper (*d*) required pressure drop cannot be achieved
6. The capillary tube used as expansion device in vapour compression system works on the principle of
   (*a*) isothermal expansion causing pressure drop (*b*) adiabatic expansion causing pressure drop
   (*c*) flow through pipe with friction causing pressure drop
   (*d*) throttle expansion causing pressure drop
7. Thermostatic expansion valve is used in .... type of evaporators.
   (*a*) flooded (*b*) DX coil (*c*) dry
8. The thermostatic expansion valve operates on the changes in the
   (*a*) degree of superheat at exit from the evaporator (*b*) temperature of the evaporator
   (*c*) pressure in the evaporator (*d*) none of these
9. The thermostatic expansion valve is also called
   (*a*) constant pressure valve (*b*) constant temperature valve
   (*c*) constant superheat valve (*d*) none of these
10. Most thermostatic expansion valves are set for a superheat of
   (*a*) 5°C (*b*) 10°C (*c*) 15°C (*d*) 20°C

ANSWERS

1. (*c*)	**2.** (*a*)	**3.** (*d*)	**4.** (*d*)	**5.** (*b*)
6. (*c*)	**7.** (*c*)	**8.** (*a*)	**9.** (*c*)	**10.** (*a*)

CHAPTER

Food Preservation 13

13.1 Introduction

The food, like air and water, is necessary for the human being to live. All the foods used by human being may be obtained from the plant or animal kingdom. All the foods are not produced during the whole year. The different foods are produced at different places and in a particular season. Since some of the foods are required all round the year in various parts of the country, therefore it is very much essential to preserve them during transportation and subsequent storage until they are finally consumed.

The food preservation may be defined as the state in which the food may be retained over a period of time without being contaminated by pathogenic organisms or chemicals and without losing its colour, texture, flavour and nutrition value.

13.2 Advantages of Food Preservation

Following are the advantages of food preservation:

1. It adds variety to the food.
2. It increases the shelf life of food.
3. It increases the food supply.
4. It decreases the wastage of food.

13.3 Causes of Food Spoilage

Food preservation equipment.

All types of foods contain proteins, carbohydrates, fats (lipids), vitamins and minerals. The proteins and minerals like iron, calcium and phosphorus help for tissue building and growth of the body. The carbohydrates and fats provide energy to the body. The vitamins and minerals are essential to safeguard the body against diseases.

The destruction of any one of the above-mentioned components causes the spoilage of food. The spoilage period depends upon the type of food. The perishable foods like meat, fish, milk and many fruits and vegetables begin to deteriorate immediately unless properly preserved. The semi-perishable foods like eggs, onions and potatoes can be kept for several weeks in a cool dry place. The non-perishable foods like cereals, pulses and nuts can be stored for long periods of time. The spoilage of food comes in the form of bad odour, uncommon colour, bad taste and physical appearance.

The spoilage of food is due to the physical and chemical changes taking place in it, as discussed below :

1. *Spoilage due to physical changes*

One of the causes of the spoilage of unpackaged fresh foods, such as meat, poultry, fish, fruit, vegetables, cheese, eggs etc. is the loss of moisture from the surface of the product by evaporation into the surrounding air. This process is known as *desiccation* or *dehydration.* In fruits and vegetables, desiccation is accompanied by a considerable loss in both weight and vitamin content. The loss of weight also affects the taste of food. In meat, cheese, etc., desiccation causes discoloration, shrinkage and heavy trim losses. Eggs lose moisture through the porous shell, with a resulting loss of weight and general downgrading of the eggs.

The spoilage of food, particularly fruits and vegetables, due to impact, bruising and squeezing is very common in packing.

2. *Spoilage due to chemical changes*

The spoilage of food is caused by a series of complex chemical changes in the food. These chemical changes are brought about by both internal and external spoiling agents. The former are the natural enzymes, whereas the latter are the micro-organisms. These two types of spoiling agents responsible for the spoilage of food are discussed, in detail, as below :

(a) Enzymes. The enzymes are inherent in all organic substances such as fruits, vegetables and animals. They are organic catalysts produced by cells. The life of every cell of plant or animal tissue depends upon the chemical reactions activated by these organic catalysts. Chemically, enzymes are proteins in nature and hence may be denatured by heat. There are various types of enzymes and specific enzymes act on some specific foods starting the chemical action which is responsible for the spoilage of food. The following table shows the different types of enzymes and their chemical action with the substance.

Table 13.1. Different types of enzymes and their chemical action with the substance.

S.No.	Type of enzymes	Chemical action of enzymes
1.	Ptyalin	Converts cooked starches into soluble sugars-maltose.
2.	Pepsin	Converts proteins to peptones.
3.	Rennin	Converts caseinogen to casein.
4.	Amylase	Converts sugars and starches to maltose.
5.	Lipase	Reduces fats to glycerin and fatty acids.
6.	Trypsin	Reduces proteins and peptones to polypeptides and amino acids.
7.	Erepsin	Reduces all protein substances to amino acids.

The moisture (water) is necessary for the proper activity of the enzymes. The optimum temperature at which most enzymes act rapidly is about 37°C, but they are destroyed by heating the foods or their activity can be reduced by freezing the foods.

(b) Micro-organisms. The micro-organisms are undetectable living organisms present in the surroundings. They grow in and on the surface of the food. The various micro-organisms responsible for the spoilage of food are discussed, in detail, as below :

(i) Bacteria. The bacteria are single-celled organisms found in soil, water, dust and air. Many bacteria are useful in preserving the food and their presence is necessary in some specific foods such as those which ferment apple juice to produce cider. Some bacteria cause the spoilage of foods.

Though the size, shape and structure of bacteria vary according to the environment in which they grow, yet the following three main shapes of bacteria exist :

1. Spheres called cocci,
2. Rods called bacilli, and
3. Twisted rods called spirilla

Since bacteria are capable of withstanding extreme temperature, therefore they may be classified according to their temperature ranges into three general groups, as shown in the following table :

Table 13.2. Types of bacteria.

S.No.	Type of bacteria	Temperature in °C		
		Minimum	Optimum	Maximum
1.	Psychrophillic	0	15–20	30
2.	Mesophillic	15–30	25–40	50
3.	Thermophillic	25–45	45–55	55–85

The psychrophillic bacteria are those organisms which play an important part in the spoilage of food in the refrigerator and in cold storages. The kneaded dough left in the refrigerator shows grey or black specks due to the activity of psychrophillic bacteria. The thermophillic bacteria are those organisms which are capable of withstanding high temperatures. The food and canning industry and milk processing plants are generally affected by thermophillic bacteria. Since bacteria may be *aerobic* or *anaerobic*, therefore they are likely to flourish anywhere and everywhere. Some of them may cause spoilage of food while others may cause food poisoning and diseases borne through food.

The bacteria may be destroyed by sunlight, ultra-violet rays, extreme heat and by the use of certain chemical substances.

(ii) Yeast. It is another micro-organism which is responsible for food spoilage. They require water and a source of energy (usually sugar) for their growth. The growth is most rapid at temperature between 25°C and 30°C. Since sugar serves as a source of energy to yeast, therefore they are generally found in places where sugar is available. The yeasts find their way into the ground when they are washed or blown from the surface of fruits, particularly grapes. The yeast cells which are always present in the atmosphere may contaminate food and cause its spoilage. They produce pigments and undesirable chemical products during their metabolism. The yeast may cause spoilage of fruit juices, syrups, molasses, honey, jellies and other foods, converting their sugar into alcohol and carbon dioxide.

Generally all types of yeast will be destroyed when subjected to a temperature of 100°C and their activity will be stopped under low temperatures, but they remain in food for a considerably long period as their cells are hard.

(iii) Moulds (Fungus). The moulds are multi-cellular, filamentous fungi that contain sporangium. The spores in the sporangium spread through the air and start new mould plants. When these spores find a favourable environment, they germinate and produce a fluffy growth. They are found in different colours such as white, grey, blue, green, red, orange or some other colour depending upon the variety of the mould. Most moulds grow between 25°C and 30°C in warm damp places. Some moulds can grow even at refrigerator temperature.

The growth of mould is rapid on acidic foods such as lemon and on foods having high sugar content such as jams and jellies. They also grow on neutral foods such as bread and other starchy foods which are spoiled by the rhizopus (commonly known as black mould). The green fuzz seen on decaying fruits is usually a member of the penicillium genus. Most moulds are not harmful. A small proportion of moulds found on foodstuffs is capable of producing toxic materials known as *mycotoxins*. The best known of these are *aflatoxins* produced by moulds growing on peanuts, ragi, wheat and millet which have not been dried as soon as they are harvested.

13.4 Methods of Food Preservation

All the methods of food preservation must provide such an environment in and around the preserved food so that it produces one or more unfavourable conditions to the continued activity of the spoilage agents. When the product is to be preserved for a long time, the unfavourable conditions must be of sufficient severity to eliminate the spoilage agents entirely or at least make them ineffective. Following are the various methods used for food preservation :

1. ***Heat processing.*** All types of spoilage agents (*i.e.* enzymes, bacteria and moulds) are destroyed when subjected to high temperatures over a period of time. The temperature of the product is raised to a level fatal to all spoilage agents and is maintained at this level until they are all destroyed. The product is then sealed in sterilized, airtight containers. A product so processed will remain in a preserved state for a long time.

2. ***Dehydration.*** The process of removing the moisture from the product is called dehydration (*i.e.* drying). It is one of the oldest methods of preserving foods and still it is widely used. Since both enzymes and micro-organisms require moisture for their growth, therefore it is necessary to stop completely their growth by dehydrating the foods. A variety of dehydrated products are available in the market. They include dried milk, dehydrated soups, instant coffee, pre-cooked peas and cereals. A very common method used for dehydration is sun-drying. This method cannot preserve the taste of foods.

3. *Chemical preservation.* This method may employ high concentrations of salt, sugar and acids.

Salt as a preservative is used for preserving vegetables and fruits like tamarind, raw mango, amla, fish and meat. It may be used in dry or brine form. The presence of a high concentration of salt prevents the water from being available for bacterial growth. This is because the concentration of salt in the water is higher than that in the bacterial cells. Thus the water cannot be absorbed by the cellular membrane of the bacteria.

The principle of sugar as preservative is same as that of salt. However, preservation with salt is a cold process while in case of sugar, the mixture is heated. Sugar acts as a preservative because the high concentration of sugar solution withdraws water from the micro-organisms, thereby preventing their growth. The moulds will, however, grow on the surface of jams, jellies, if proper sterility is not maintained.

When the medium in which food is preserved is strongly acidic, then most of the micro-organisms cannot survive. The use of vinegar (acetic acid) and lemon juice (citric acid) is common in home methods of pickling. The benzoic acid as a preservative is used up to a concentration of 0.1% for all coloured fruits and vegetables.

The pesticides are sprayed over fruits and vegetables and foodgrains to prevent spoilage. If these pesticides are used beyond safe levels, they are highly toxic.

4. *Oils and spices.* The oils and spices along with salt and sugar provide a medium that resists the activity of the micro-organisms in food. Moreover, they improve the flavour of the food being preserved. Spices such as chillis, fenugreek, mustard and pepper are used in pickling. When oil is used in pickling, the top layer of oil prevents the micro-organisms in the air from coming into contact with the food.

5. *Canning.* The canning is the preservation of food in sealed containers, usually after the application of heat, through steam under pressure. During this process, some of the micro-organisms are destroyed and the rest are rendered inactive. The enzymes are also inactivated. The containers are then sealed to prevent recontamination of foods. During aseptic canning, high temperature is applied for a very short time. The product is first sterilized at 135°C to 172°C in a few seconds, then cooked and filled in sterilized containers. This method is used for fluid products such as fruit juices, syrups and sauces.

Refrigeration system.

6. *Pasteurization.* This method is generally used on large scale to protect milk against bacterial infection. The milk used for the preparation of milk products like cheese, butter and ice cream is pasteurized. The pasteurization may be brought about by the *holding process* or *high temperature short time method.* In the holding process, the milk is heated to at least 62°C and kept at that temperature for at least 30 minutes. In the high temperature short time method, the milk is heated to 70°C and kept at that temperature for at least 15 seconds.

The milk may be sterilized either by boiling for a period of time or by the application of heat as in the preparation of evaporated milk. The sterilization deepens the colour of milk and gives it a slightly caramelized flavour, while pasteurization does not change the colour or flavour of milk.

13.5 Food Preservation by Refrigeration

The refrigeration is the only means of preserving food in its original freshness. It may be noted that when food is to be preserved by refrigeration, the refrigerating process must begin very soon after harvesting or killing of animal and must be continuous until the food is finally consumed.

The preservation of perishable foods by refrigeration involves the use of low temperature as a means of eliminating or retarding the activity of spoilage agents. The low temperatures are not as effective as high temperatures in bringing about the destruction of spoilage agents. The storage of perishable foods at low temperatures provides a practical means of preserving perishable foods in their original fresh state for longer periods of time. The degree of low temperature required for adequate preservation varies with the type of product stored and the length of time the product is to be kept in storage.

The application of refrigeration for preserving foods is common in domestic refrigerators, commercial refrigerators and cold storages, as discussed in the following pages.

13.6 Domestic Refrigerators for Food Preservation

The main purpose of a domestic refrigerator is to provide a low temperature for storage and distribution of foods and drinks. It has been found that the growth of food destroying micro-organisms is much faster (about 1000 times) at a temperature of 10°C than at 4°C. This is enough to emphasize the use of refrigerators for preserving fruits, vegetables, fish, meat, milk etc., which would otherwise be spoiled at higher temperature (generally in summer season) in a short time.

The storage in domestic refrigerators is generally short-term or temporary storage. These are small in size and vary in capacity from 0.1 m^3 to 0.25 m^3 (100 litres to 250 litres). The domestic refrigerators have generally hermetically sealed units, in which the motor and compressor are enclosed in one chamber. Due to short-term storage, the load on the domestic refrigerator is intermittent. The refrigerant used should preferably be non-irritant and non-toxic. Generally, methyl chloride, R-12 and R-11 are used as refrigerants.

13.7 Commercial Refrigerators for Food Preservation

The commercial refrigerators are usually small ready built, refrigerated fixtures used for processing, storing, displaying and dispensing of perishable commodities. These are generally used by retail stores, markets, hotels, restaurants and establishments. The commercial refrigerators are classified into the following three main groups :

1. *Reach-in refrigerators.* These refrigerators are the most versatile and the most widely used of all commercial fixtures. These refrigerators may be used for both storage and display of the product. When used for storage only, they have solid doors, while for display of product, glazed doors are provided. These refrigerators are commonly used in grocery stores, meat markets, bakeries, lunch counters, restaurants and hotels.

Display cases.

2. *Walk-in cooler.* The walk-in coolers are mainly used as storage fixtures. They are available in a wide variety of sizes. These are used by nearly all retail stores, markets, hotels, restaurants etc. for the storage of perishable foods. Some walk-in coolers are provided with glazed reach-in doors for storing, displaying and dispensing of dairy products, eggs and beverages. Such type of coolers are widely used in grocery stores, particularly drive-in groceries.

3. *Display cases.* The principal function of display cases is to display the product or commodity in a most attractive manner in order to stimulate sales. Thus, in the design of a refrigerated display fixture, the first consideration is given to the product being displayed. The storage life of a product in a display fixture is very limited. It ranges from a few hours in certain cases to a week or more in others, depending upon the type of product and the type of fixture. The display fixtures are generally of the following two types :

(*a*) Self-service case, and (*b*) Service case.

In a *self-service case*, the customer serves himself directly while in the *service case*, the customer is served by the attendant.

The self-service case is very popular in super-markets, retail stores and other self-service establishments. The service case is mainly used in small groceries, markets, bakeries etc. The self-service cases may be of open type or closed type. The open type display cases are more popular than closed type in the super-markets and retail stores.

13.8 Cold Storages for Food Preservation

Many food products may be stored at some temperature above the freezing point. The storage may be short-term storage or long-term storage. The storages which are used for short-term storage purposes are known as *cold storages*. The short-term storage is usually meant for retail establishments where rapid turnover of the product is normally expected. The period for short-term storage ranges from one to two days or to a week or more in some cases, but not more than fifteen days under any circumstances. The long-term storage is usually carried out by wholesalers and commercial storage warehouses. The storage period depends on the type of product stored and its condition on entering the storage. The maximum storage period for long-term storage ranges from seven to ten days for some sensitive products like ripe tomatoes and up to six or eight months for more durable products such as onions and smoked meat. When perishable foods are to be stored

* See Art. 13.9.

for a longer period, they should be frozen and stored in frozen storages. However, some fresh foods like tomatoes are damaged by the freezing process and therefore cannot be successfully frozen.

In general, the conditions required for short-term storage are more flexible than those required for long-term storage and higher storage temperatures are permissible for short-term storage.

The following points should be kept in mind while storing the foods in cold storages:

1. ***Storage temperature.*** Most of the foods for short-term storage are stored at a temperature slightly above their freezing point. The effect of incorrect storage temperature is to lower the quality of the stored product. Table 13.3 (Page 418) shows the recommended storage conditions for various food products.

2. ***Relative humidity and air motion.*** The relative humidity and air motion are important factors which must be controlled in the storage of all perishable foods in their natural state. The control of these factors is necessary in order to prevent excessive loss of moisture from the product (dehydration). It may be noted that low relative humidity and high air velocity causes excessive dehydration in the stored product. We have already discussed that the loss of moisture from the fruits and vegetables causes reduction in weight and vitamins. Its effect is more aggravated on meat causing shrinkage and discoloration.

3. ***Mixed storage.*** When different types of food products are stored in a common storage, then it is called mixed storage. Since the storage temperatures are different for different foods, therefore the mixed storage causes a problem. The storage conditions in such spaces represent a compromise among the storage conditions required for various food products separately. The higher temperatures are used in mixed storage to minimise the chances of damaging the more sensitive products against cold storage diseases when stored at temperature below their critical temperature.

Though the higher storage temperature tends to shorten the life of certain products held in mixed storage, yet it is not a serious problem for short-term or temporary storage. For long-term storage, most of the large wholesale and commercial storage warehouses have a number of separate storage spaces. The general practice in such cases is to group the various products for storage and only those products requiring approximately the same storage conditions are placed together in common storage.

Another important problem associated with the mixed storage is the absorption of odour and flavour. Some products either absorb or give off odours while in storage. Therefore care should be taken not to store such products together even for a short period. The dairy products are highly sensitive in absorbing odours from other products held in mixed storage. The potatoes are worst among all which give off very bad flavours to other products in storage and should never be stored with fruits, eggs, dairy products and nuts.

4. ***Condition of products at the time of entering storage.*** The condition of the products at the time of entering the storage is one of the important factors for determining the storage life of a refrigerated product. It may be noted that refrigeration only arrests or retards the natural processes of deterioration. But the already deteriorated products cannot be restored to good condition. Hence, only fruits and vegetables in good condition should be accepted for storage. The fruits and vegetables

Self-contained refrigerated box has its condensing unit built in.

intended for storage should be harvested before they fully mature because maturation and ripening of these products start after harvesting.

The storage life of fully matured or damaged fruits and vegetables is extremely short even under the best storage conditions. In order to assure maximum storage life with minimum loss of quality, the product should be chilled to the storage temperature immediately after harvesting or killing the product. When products are to be shipped over long distances, they should be pre-cooled and shipped by refrigerated transport.

5. *Product chilling.* The product chilling is different from product storage. In chilling rooms, the product enters at a high temperature and it is chilled quickly to the storage temperature. The product is removed from the chilling room and placed in a holding cooler for storage. The handling of the product during the chilling period has a great influence on the ultimate quality and storage life of the product.

In chilling rooms, the relative humidity depends upon the product being chilled. But when the product is chilled in vapour-proof containers, the humidity is unimportant in chilling rooms. If the products are chilled in unpackaged state, the product temperature and vapour pressure are high during the early stages of chilling, and they lose moisture rapidly and produce fog in the chilling room. In order to avoid excessive moisture loss and shrinkage, rapid chilling and high air velocity are desirable for lowering the temperature and vapour pressure of the product as quickly as possible during the early stages of chilling. The high air velocity is also desirable to carry away the vapour and thus to prevent condensation of moisture on the surface of the product.

When the products subjected to dehydration are to be chilled, then the humidity should be kept at a high level. Some extremely sensitive products, such as poultry and fish, are frequently chilled in ice to reduce moisture losses during chilling. When products packed in ice are placed in refrigerated storage, the slowly melting ice keeps the surface of the product moist and prevents excessive dehydration. The eggs are sometimes dipped in a light mineral oil before chilling and storage only due to this reason.

13.9 Frozen Storages for Food Preservation

When the food products are to be preserved in its original fresh state for longer periods, then they are usually frozen and stored approximately at – 15°C or below. Such storages are known as *frozen storages.* They differ from the cold storages in their size and the temperature range. The size of the frozen storages is considerably smaller than cold storages and the temperatures used for preserving the foods are also low as compared with the temperatures used for cold storages. The food products which are commonly preserved in frozen storages include fruits, vegetables, fruit juices, meat, poultry, breads, pastries, ice cream and a wide variety of pre-cooked foods.

The high quality products in good condition should only be frozen. The vegetables and fruits to be frozen should be harvested at the peak of maturity and frozen as quickly as possible after harvesting. Both the vegetables and fruits require considerable processing before freezing. The vegetables after harvesting are cleaned and washed to remove foreign material like leaves, dirt, insects, juices, etc., and then blanched into hot water or steam at 100°C. The blanching destroys natural enzymes and increases the storage life of frozen vegetables. The blanching time depends upon the type and variety of the vegetables. It ranges from 1 to 1½ minute for green beans to 11 minutes for large ears of corn. Many bacteria may survive even after blanching. Therefore to prevent spoilage by these bacteria, the vegetables should be chilled to – 15°C immediately after blanching. The vegetables are packaged before freezing at – 20°C or below.

The fruits after harvesting are also cleaned and washed for removing foreign material and to reduce microbial contamination. They are then freezed quickly at – 20°C or below. It may be noted that fruits are never blanched because it affects the state of fruits. In order to reduce oxidation and rapid browning of the fresh fruit, they are covered with a light sugar syrup or in some cases with ascorbic acid, citric acid or sulphur dioxide.

The meat products do not require any special processing before freezing. They are directly freezed taking them into frozen storages after washing.

13.10 Methods of Food Freezing

Following are the two methods mainly used for food freezing :

1. *Slow or sharp freezing.* In the slow freezing method, the food products are placed in a low temperature room and allowed to freeze slowly in still air by natural convection. The temperature maintained in sharp freezers ranges from – 17°C to – 40°C. Since the circulation of air in the stored room is by natural convection, therefore the time required for freezing the food is considerably large. It may be from three hours to three days depending upon the bulk of product and the conditions in the sharp freezer. This method is commonly used for beef and pork, boxed poultry, fish, fruit in barrels and other large containers, and eggs in cans.

2. *Quick freezing.* In the quick freezing method for freezing the food products in storage, the forced circulation of cold air is necessary. The difference between quick and slow freezing is only in size of ice crystals formed within the food during freezing.

The quick freezing is generally done in any one of the following ways :

(a) Immersion freezing. In this method of quick freezing, the food products are immersed into low temperature liquids. Since the liquids are good conductors of heat and in good thermal contact with all the products, therefore the heat transfer is rapid and the product is completely frozen in a very short time. The liquid mediums used for freezing the foods should be non-toxic and should not produce any bad effect on the immersed foods. The liquids used for this purpose are sodium chloride, brine, sugar brine and propylene glycol.

The fish and shrimp are the two animal products which are most frequently frozen by immersion. This freezing method produces a thin coating of ice on the surface of the product resulting in prevention of dehydration of unpackaged products during the storage period. The only disadvantage of this system is the extraction of the juices from the products by osmosis resulting into contamination and weakening of the freezing solution. This defect can be avoided by freezing the products in canned or packaged forms.

(b) Indirect contact freezing. In this method of quick freezing, the food product is freezed by the direct contact with metal surface cooled by any of the refrigerants such as ammonia, R-12, R-22 or cold brine. For this purpose, various types of plate freezers consisting of metal plates with food products placed on them and a refrigerant circulating through them, are used. They are available in different sizes ranging from 100 to 1200 kg loading capacity. The freezers are commonly used for freezing the fruits, vegetables, meat and fish.

Since the food products preserved in plate freezers do not come in direct contact with the cooling medium, therefore the foods maintain their taste and colour. The only disadvantage of this method is that the ice crystals formed on the surface of the food product are of little bigger size than immersion freezing method.

(c) Air blast freezing. In this method of quick freezing, the food products are freezed by the contact with cold air. The air blast freezing is widely used because it provides excellent quality of the food among all other types. In this method, a very low temperature air is circulated with a very high velocity around the various parts of the product kept in insulated tunnel type storages. The temperatures of – 20°C to – 40°C are commonly used for this method of freezing. The velocity of air varies from 30 m/min to 120 m/min according to the type of food to be freezed. It may be noted that dehydration of the product may occur in freezing unpacked whole or dressed fish in blast freezer unless the velocity of air is kept to about 160 m/min and the period of exposure of air is controlled.

The air blast freezing is mainly used for freezing fishery products like shrimp, fish fillets, steaks, scallops or pre-cooled products packed in small packages.

Table 13.3 Recommended storage conditions of perishable products.

Product	*Range of storage temperature in °C*		*Relative humidity in percentage*		*Freezing point in °C*	*Composition in % water*	*Maximum storage period*
	Short-term storage	*Long-term storage*	*Gravity air circulation*	*Forced air circulation*			
(1)	*(2)*	*(3)*	*(4)*	*(5)*	*(6)*	*(7)*	*(8)*
Fruits							
Apples	1.7 to 4.4	–1.1 to 0	85	88	–1.94	85	8 months
Bananas	12.8 to 13.3	12.8 to 13.3	80	80	–3.3 to –1.1	75	10 days
Berries	1.7 to 4.4	– 0.6 to 0.6	80	85	–3.3 to –1.7	84	7-20 days
Grapes	1.7 to 4.4	– 1.1 to 0	80	85	–2.2	77	1-6 months
Grapefruit	4.4 to 7.2	0 to 1.1	85	90	–1.94	77	3 months
Lemons	12.8 to 15.6	10 to 12.8	80	85	–2.2	89	3 months
Oranges	4.4 to 7.2	0 to 1.1	80	85	–2.2	86	2 months
Peaches	1.7 to 4.4	– 0.6 to 0.6	80	85	–1.4	88	1 month
Pears	1.7 to 4.4	–1.1 to 0	85	90	–1.94	84	1-7 months
Pineapples	4.4 to 7.2	3.3 to 4.4	75	85	–2.2	88	1 month
Vegetables							
Artichokes	4.4 to 7.2	– 0.6 to 0.6	90	90	–2.5	92	2-5 months
Avocados	4.4 to 7.2	0.6 to 1.7	85	85	–0.6	94	10 days
Beans (green)	4.4 to 7.2	0 to 1.1	85	90	–1.1	68.5	1 month
Beans (dried)	10 to 15.6	0 to 1.7	60	60	—	12.5	12 months

(1)	*(2)*	*(3)*	*(4)*	*(5)*	*(6)*	*(7)*	*(8)*
Beats	4.4 to 7.2	0 to 1.1	85	90	– 2.8	85.5	7–90 days
Cabbage	1.7 to 4.4	0 to 1.1	90	95	– 0.6	91.5	4 months
Carrots	1.7 to 4.4	0 to 1.1	90	95	–1.4	88	2–4 months
Cauliflower	1.7 to 4.4	0 to 1.1	85	90	–1.1	92.5	10 days
Celery	1.7 to 4.4	–0.6 to 0.6	90	90	–1.1	94.5	2–4 months
Corn (green)	1.7 to 4.4	–0.6 to 0.6	85	90	–1.7	75.5	10 days
Corn (dried)	10 to 15.6	1.7 to 4.4	69	60	—	10.5	12 months
Cucumbers	10 to 15.6	10 to 15.6	80	85	– 0.83	95.5	10 days
Eggplant	10 to 15.6	10 to 15.6	85	90	– 0.83	92	10 days
Mangoes	—	0.6 to 1.7	85	85	0	93	10 days
Onions	10 to 15.6	0 to 1.1	70	75	– 1.1	87.5	5–6 months
Potatoes	2.2 to 10	3.3 to 5.6	85	90	– 1.7	78.5	6 months
Tomatoes (green)	12.8 to 15.6	11.1 to 12.8	80	85	– 0.83	94	6 weeks
Tomatoes (ripe)	18.3 to 21.1	18.3 to 21.1	80	85	– 0.83	94.5	—
Vegetables (mixed)	4.4 to 7.2	1.7 to 4.4	85	87	– 1.1	90	—
Meats							
Beef (fresh)	1.7 to 4.4	–1.1 to 0	84	87	– 2.8	68	3 weeks
Fish (fresh)	1.1 to 3.3	–1.1 to 0	85	85	– 2.2	70	15 days
Fish (frozen)	–9.4 to –6.7	–15 to –12.2	80	80	—	70	6 months
Lamb	1.1 to 3.3	–2.2 to –1.1	85	90	– 1.7	58	2 weeks
Livers	–2.2 to –1.1	–6.7 to –5.6	80	80	– 1.7	65.5	6 months

(1)	(2)	(3)	(4)	(5)	(6)	(7)	(8)
Oysters (shell)	1.7 to 4.4	0 to 3.3	90	90	–2.8	80.4	15 days
Oysters (tub)	1.7 to 4.4	0 to 3.3	70	70	–2.8	87	10 days
Pork (fresh)	1.1 to 3.3	– 1.1 to 0	80	85	–2.2	60	15 days
Pork (smoked)	4.4 to 7.2	– 2.2 to –1.1	80	85	—	57	15 days
Poultry (fresh)	– 2.2 to –1.1	– 2.2 to –1.1	84	87	–2.8	74	10 days
Poultry (frozen)	– 9.4 to –6.7	– 17.8 to –15	85	85	–2.8	74	10 months
Sausage (fresh)	1.7 to 4.4	– 6.1 to –2.8	80	85	–3.3	65	15 days
Sausage (smoked)	4.4 to 7.2	0 to 44	75	80	–3.9	60	6 months
Miscellaneous							
Beer	1.7 to 4.4	1.1 to 3.3	70	70	–2.2	92	6 months
Butter	1.7 to 4.4	– 0.4	80	80	–17.8 to –1.1	15	6 months
Cheese (curing)	10 to 15.6	10 to 15.6	80	85	—	—	—
Beam (40%)	1.7 to 4.4	– 15 to –12.2	80	80	–2.2	55	4 months
Eggs (crated)	4.4 to 7.2	– 1.1 to – 0.6	85	85	–2.8	73	12 months
Eggs (frozen)	– 9.4 to –6.7	– 17.8 to –15	60	60	–2.8	—	18 months
Flowers	4.4	1.7	—	90	0	—	1 week
Honey	4.4 to 7.2	– 0.6 to 0.6	70	70	—	18	12 months
Ice cream	– 17.8 to –12.2	– 28.9 to –17.8	85	85	–2.8 to –17.8	60	2 weeks
Lard	7.2 to 10	0 to 1.1	80	80	—	—	6 months
Milk	1.7 to 4.4	1.7 to 4.4	70	70	–0.6	87.5	5 days
Nuts (dried)	1.7 to 4.4	– 1.1 to 0	75	75	—	3 to 10	8–12 months

QUESTIONS

1. Define food preservation. What are its advantages ?
2. What are the major causes of food spoilage ?
3. How each of the following spoiling agents acts to spoil the foods ?
   (*a*) Enzymes (*b*) Yeast (*c*) Bacteria (*d*) Mould
4. Explain the applications of refrigeration for food preservation. Explain how the refrigeration controls the spoilage of food.
5. What do you understand by short-term and long-term storages ?
6. Discuss the factors to be kept in mind while storing the foods in cold storages.
7. In what respect the frozen food storage differs from cold storage ?
8. Explain, in brief, the processing of vegetables and fruits before preserving them in frozen storages.
9. Why the foods are freezed ?
10. Describe the three basic methods of freezing food products.

OBJECTIVE TYPE QUESTIONS

1. The spoilage of food is due to
   (*a*) desiccation (*b*) enzymes (*c*) micro-organisms (*d*) all of these
2. The optimum temperature at which most enzymes act rapidly is about
   (*a*) 14°C (*b*) 28°C (*c*) 37°C (*d*) 53°C
3. The bacteria may be destroyed by
   (*a*) sunlight (*b*) ultra-violet rays (*c*) extreme heat (*d*) all of these
4. All types of yeast will be destroyed when subjected to a temperature of
   (*a*) 40°C (*b*) 60°C (*c*) 80°C (*d*) 100°C
5. The process of removing the moisture from the food product is called
   (*a*) heat processing (*b*) dehydration (*c*) canning (*d*) pasteurization
6. The method generally used on large scale to protect milk against bacterial infection, is
   (*a*) heat processing (*b*) dehydration (*c*) canning (*d*) pasteurization
7. The maximum storage period for long-term storage ranges from .... for ripe tomatoes.
   (*a*) seven to ten days (*b*) six to eight months
8. The quick freezing of food products is done by
   (*a*) immersion freezing (*b*) indirect contact freezing
   (*c*) air blast freezing (*d*) all of these
9. In air blast freezing method,
   (*a*) a very low temperature air is circulated with a very low velocity
   (*b*) a very low temperature air is circulated with a very high velocity
   (*c*) a very high temperature air is circulated with a very high velocity
   (*d*) a very high temperature air is circulated with a very low velocity
10. In air blast freezing, the velocity of air varies from
    (*a*) 5 to 10 m/min (*b*) 10 to 20 m/min (*c*) 20 to 30 m/min (*d*) 30 to 120 m/min

ANSWERS

1. (*d*)	2. (*c*)	3. (*d*)	4. (*d*)	5. (*b*)
6. (*d*)	7. (*a*)	8. (*d*)	9. (*b*)	10. (*d*)

CHAPTER

14

Low Temperature Refrigeration (Cryogenics)

1. Introduction.
2. Limitations of Vapour Compression Refrigeration Systems for Production of Low Temperature.
3. Cascade Refrigeration System.
4. Coefficient of Performance of a Two-Stage Cascade System.
5. Solid Carbon Dioxide (Dry Ice).
6. Manufacture of Solid Carbon Dioxide or Dry Ice.
7. Liquefaction of Gases.
8. Linde System for Liquefaction of Air.
9. Claude System for Liquefaction of Air.
10. Advantages of Claude System over Linde System.
11. Liquefaction of Hydrogen.
12. Liquefaction of Helium.
13. Production of Low Temperature by Adiabatic Demagnetisation of Paramagnetic Salt.

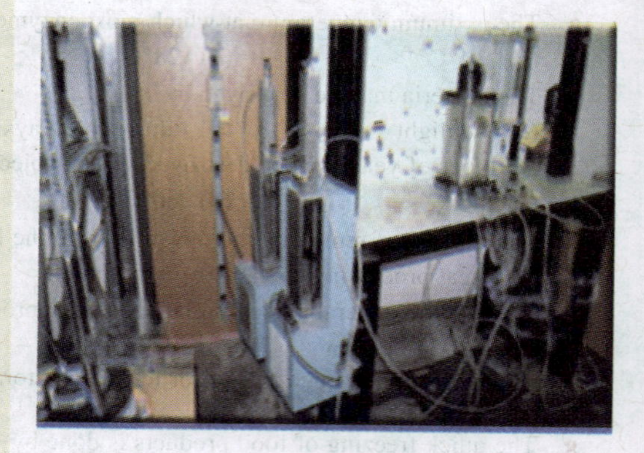

14.1 Introduction

The term 'cryogenic' is derived from the Greek word Kryos which means cold or frost. It is frequently applied to very low temperature refrigeration applications such as in the liquefaction of gases and in the study of physical phenomenon at temperatures approaching absolute zero.

The first low temperature refrigeration system was primarily developed for the solidification of carbon dioxide and the liquefaction and subsequent fractional distillation of gases such as air, oxygen, nitrogen, hydrogen and helium. Oxygen was liquefied in 1877 by Coilletet and Pietet. Hydrogen was liquefied in 1898 by Dewar using Joule-Thomson expansion of gases. The liquid oxygen boils at 90.2 K (–182.8°C) and the liquid hydrogen at 20.4 K (–252.6°C). The liquefaction of helium was accomplished in 1908 by H. Kamerlingh Onnes in the famous cryogenic laboratories of the University of Leiden. Initially, by evaporation of liquid helium under high vacuum, the temperature as low as 1.1 K (–271.9°C) was obtained. But by making improvements in the apparatus, a temperature of 0.7 K was reached by the year 1928. In 1933, Giauque and Debye proposed the adiabatic demagnetisation of paramagnetic salts for attaining the lower temperature and 0.1 K was reached through their methods. Now-a-days, the lowest temperature successfully achieved by this method is 0.001 K (*i.e.* very close to absolute zero).

Note : In refrigeration, the temperatures from –100°C to –273°C (or absolute zero) are treated as *low temperatures.*

14.2 Limitations of Vapour Compression Refrigeration Systems for Production of Low Temperature

The single-stage vapour compression refrigeration systems for different refrigerants are limited to an evaporator temperature of – 40°C. Below this temperature, the use of vapour compression systems has many drawbacks as discussed below :

1. The use of vapour compression refrigeration system for the production of low temperatures is limited inherently by the solidification temperature of the refrigerants. The following table shows the freezing temperatures for the commonly used refrigerants in vapour compression refrigeration systems.

Table 14.1. Freezing temperatures for the commonly used refrigerants.

Refrigerant	*R-11*	*R-12*	*R-21*	*R-22*	*R-30*	*R-40*	*R-113*	*R-717* (NH_3)	*R-744* (CO_2)	*R-764* (SO_2)
Freezing temperature (°C) at atmospheric pressure	–111	–157.5	–135	–160	–96.6	–97.5	–35	–77.8	–56.7	–75.6

The refrigerant used must have a freezing temperature well below the required temperature to be attained. Thus, the refrigerants R-113, R-717, R-744 and R-764 cannot be used for low temperature refrigeration systems.

2. The pressure in the evaporator is extremely low (below atmospheric) and the suction volume is very large when a refrigerant with high boiling temperature is used.

3. The pressure in the condenser is extremely high when a refrigerant with low boiling temperature is used.

4. The coefficient of performance is low because of very high pressure ratios.

5. The difficulties encountered in the operation of any mechanical equipment at very low temperatures.

We have discussed in chapter 5 (Art. 5.1) that multistage compression system is used when low evaporator temperature is required and the pressure ratio is high. A two-stage compression system using R-12 is employed for evaporator temperatures up to – 60°C and a three-stage compression system up to – 68°C. The compression

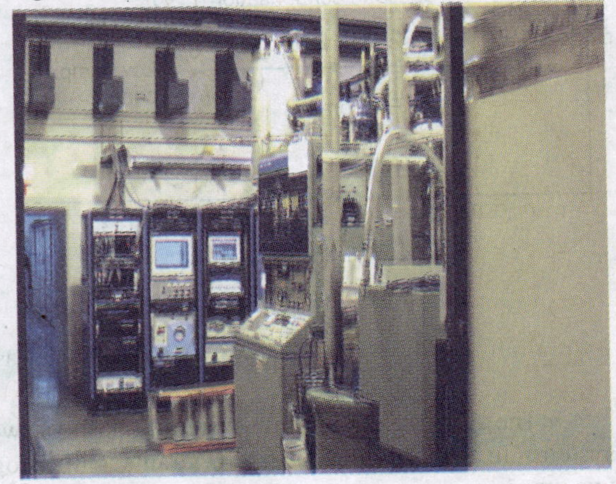

Cryogenics.

systems with R-22 operate satisfactorily at 5°C to 10°C lower temperatures. R-13 can be used to temperatures well below – 100°C, but must be cascaded to prevent extremely high condenser pressure as the critical temperature is 28.8°C.

When the vapour compression system is to be used for the production of low temperature, the common alternative to stage compression is the *cascade system.* In this system, a series of refrigerants with progressively lower boiling temperatures are used in a series of single-stage units.

14.3 Cascade Refrigeration System

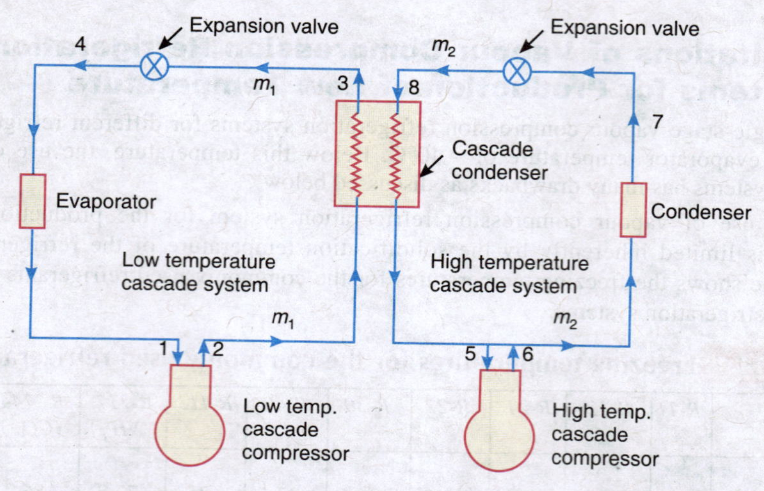

(*a*) Schematic diagram of a two-stage cascade system.

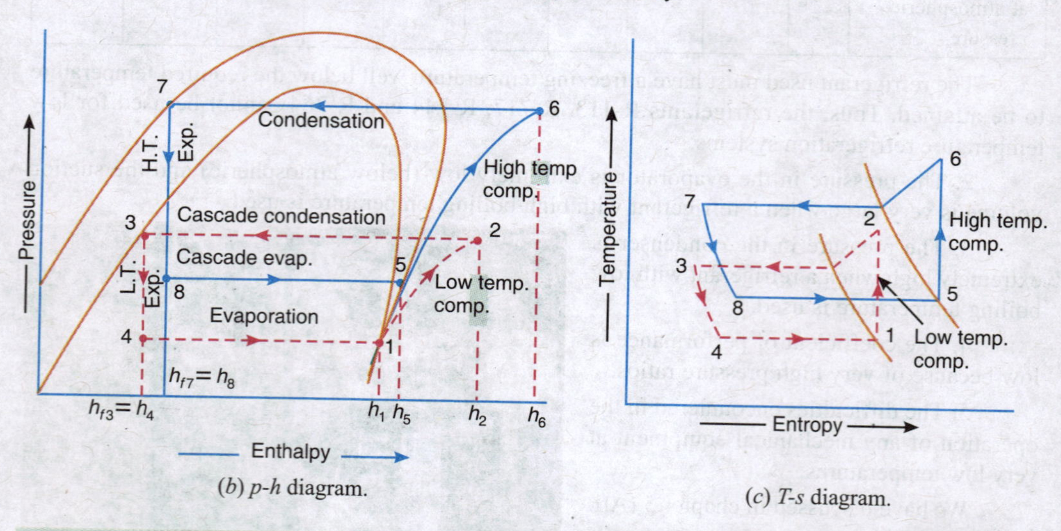

(*b*) *p-h* diagram.

(*c*) *T-s* diagram.

Fig. 14.1. Two-stage cascade system.

The cascade refrigeration system consists of two or more vapour compression refrigeration systems in series which use refrigerants with progressively lower boiling temperatures. A two-stage cascade system using two refrigerants is shown in Fig. 14.1(*a*) and its corresponding *p-h* and *T-s* diagrams are shown in Fig. 14.1 (*b*) and (*c*) respectively.

In this system, a cascade condenser serves as an evaporator for the high temperature cascade system and a condenser for the low temperature cascade system. The only useful refrigerating effect is produced in the evaporator of the low temperature cascade system. The principal advantage of the cascade system is that it permits the use of two different refrigerants. The high temperature cascade system uses a refrigerant with high boiling temperature such as R-12 or R-22. The low temperature cascade system uses a refrigerant with low boiling temperature such as R-13 or R-13 BI. These low boiling temperature refrigerants have extremely high pressure which ensures a smaller compressor displacement in the low temperature cascade system and a higher coefficient of performance.

The cascade system was first used by Pietet in 1877 for liquefaction of oxygen, employing sulphur dioxide (SO_2) and carbon dioxide (CO_2) as intermediate refrigerants. Another set of refrigerants commonly used for liquefaction of gases in a three-stage cascade system is *ammonia (NH_3), ethylene (C_2H_4) and methane (CH_4). A three-stage cascade system is shown in Fig. 14.2 (*a*) and its corresponding *p-h* and *T-s* diagrams are shown in Fig. 14.2 (*b*) and (*c*) respectively.

The additional advantage of a cascade system over the multistage system is that the lubricating oil from one compressor cannot wander to other compressors.

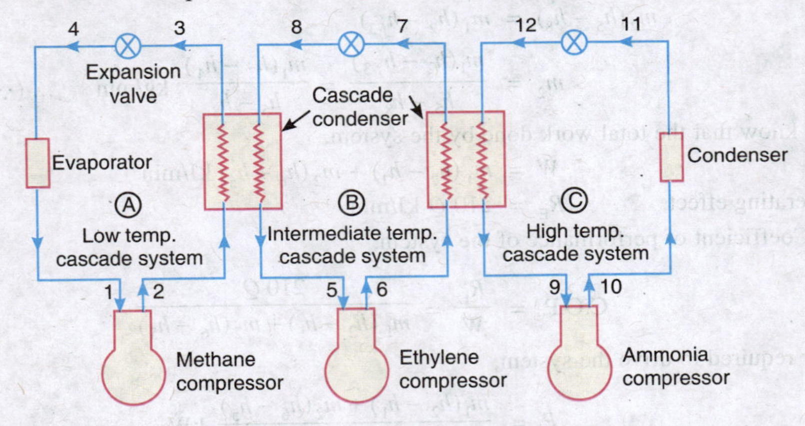

(*a*) Schematic diagram of a three-stage cascade system.

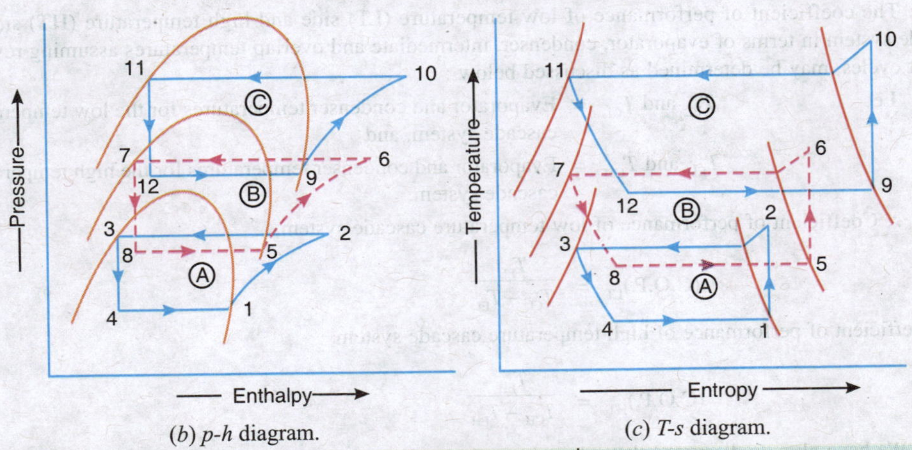

(*b*) *p-h* diagram.

(*c*) *T-s* diagram.

Fig. 14.2. Three-stage cascade system.

* Ammonia is also used in the high temperature cascade system for the manufacture of solid carbon dioxide.

Note : The difference in low temperature cascade condenser temperature and high temperature cascade evaporator temperature is called *temperature overlap.* This is necessary for heat transfer. If these temperatures are equal, then it is known as *intermediate temperature.*

14.4 Coefficient of Performance of a Two-Stage Cascade System

The schematic and *p-h* diagram of a two-stage cascade system is shown in Fig. 14.1. If Q tonnes of refrigeration is the load on the low temperature cascade system, then the mass of refrigerant flowing through the low temperature cascade system is given by

$$m_1 = \frac{210\,Q}{h_1 - h_4} \text{ kg/min}$$

The mass of refrigerant m_2 required in the high temperature cascade system in order to liquefy the refrigerant of low temperature cascade system in the cascade condenser may be obtained by balancing the heat of both the systems. In other words, the heat absorbed in the high temperature cascade system must be equal to the heat rejected in the low temperature cascade system. Mathematically,

$$m_2(h_5 - h_8) = m_1(h_2 - h_{f3})$$

$$\therefore \quad m_2 = \frac{m_1(h_2 - h_{f3})}{h_5 - h_8} = \frac{m_1(h_2 - h_4)}{h_5 - h_8} \text{ kg/min} \qquad \ldots (\because h_{f3} = h_4)$$

We know that the total work done by the system,

$$W = m_1(h_2 - h_1) + m_2(h_6 - h_5) \text{ kJ/min}$$

and refrigerating effect,

$$R_E = 210\,Q \text{ kJ/min}$$

$\therefore$ Coefficient of performance of the system,

$$\text{C.O.P.} = \frac{R_E}{W} = \frac{210\,Q}{m_1(h_2 - h_1) + m_2(h_6 - h_5)}$$

and power required to drive the system,

$$P = \frac{m_1(h_2 - h_1) + m_2(h_6 - h_5)}{60} \text{ kW}$$

Note : The coefficient of performance of low temperature (LT) side and high temperature (HT) side of cascade system in terms of evaporator, condenser, intermediate and overlap temperatures assuming reverse Carnot cycles, may be determined as discussed below :

Let T_{EL} and T_{CL} = Evaporator and condenser temperatures for the low temperature cascade system, and

T_{EH} and T_{CH} = Evaporator and condenser temperatures for the high temperature cascade system.

$\therefore$ Coefficient of performance of low temperature cascade system,

$$(\text{C.O.P.})_{LT} = \frac{T_{EL}}{T_{CL} - T_{EL}}$$

and coefficient of performance of high temperature cascade system,

$$(\text{C.O.P.})_{HT} = \frac{T_{EH}}{T_{CH} - T_{EH}}$$

We have already discussed that when the low temperature cascade condenser temperature (T_{CL}) is equal to the high temperature cascade evaporator temperature (T_{EH}), it is then known as intermediate temperature (T_I). In other words,

$$T_{CL} = T_{EH} = T_I$$

Thus, the above expressions for C.O.P. in terms of intermediate temperature (T_I) may be written as

$$(\text{C.O.P.})_{LT} = \frac{T_{EL}}{T_I - T_{EL}}$$

and

$$(\text{C.O.P.})_{HT} = \frac{T_I}{T_{CL} - T_I}$$

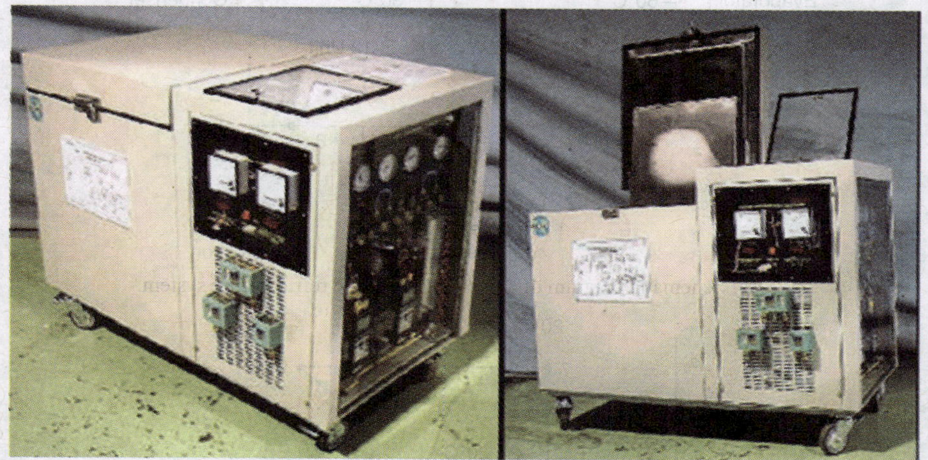

Cascade refrigeration system.

Example 14.1. *A cascade refrigeration system is designed to supply 10 tonnes of refrigeration at an evaporator temperature of – 60°C and a condenser temperature of 25°C. The load at – 60°C is absorbed by a unit using R-22 as the refrigerant and is rejected to a cascade condenser at – 20°C. The cascade condenser is cooled by a unit using R-12 as the refrigerant and operating between – 30°C evaporating temperature and 25°C condenser temperature. The refrigerant leaving the R-12 condenser is subcooled to 20°C but there is no subcooling of R-22 refrigerant. The gas leaving both the evaporators is dry and saturated and the compressions are isentropic. Neglecting losses, determine : 1. Compression ratio for each unit ; 2. Quantity of refrigerant circulated per minute for each unit; 3. C.O.P. for each unit; 4. C.O.P. of the whole system; and 5. Theoretical power required to run the system.*

Solution. Given : Q = 10 TR ; t_E (R–22) = – 60°C ; t_c (R–12) = 25°C ; t_c (R–22) = –20°C; t_E (R–12) = –30°C ; t_7 = 20°C

The schematic diagram of a two-stage cascade refrigerating system using R-22 and R-12 is shown in Fig. 14.3 (*a*). The corresponding *p-h* diagram of the system is shown in Fig. 14.3 (*b*). The cycle 1-2-3-4 is for R-22 unit whereas the cycle 5-6-7-8 is for R-12 unit. Both these cycles are superimposed in the *p-h* diagram.

From the *p-h* diagram for R-22, we find that the pressure at point 1 corresponding to – 60°C,

$$p_1 = 0.3745 \text{ bar}$$

Enthalpy of saturated vapour refrigerant at point 1,

$$h_1 = 223.7 \text{ kJ/kg}$$

Entropy of saturated vapour refrigerant at point 1,

$$s_1 = 1.054 \text{ kJ/kg K}$$

Enthalpy of superheated vapour refrigerant at point 2,

$$h_2 = 275 \text{ kJ/kg}$$

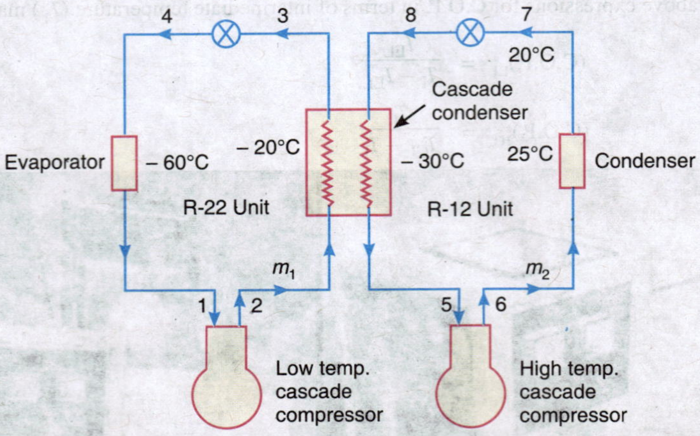

(*a*) Schematic diagram of a two-stage cascade refrigerating system.

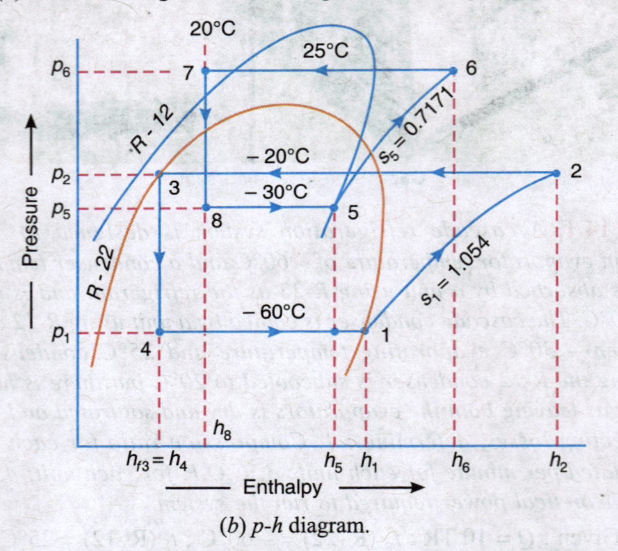

(*b*) *p-h* diagram.

Fig. 14.3

Pressure at point 2 corresponding to – 20°C,

$$p_2 = 2.458 \text{ bar}$$

and enthalpy of saturated liquid refrigerant at point 3 (corresponding to – 20°C)

$$h_{f3} = h_4 = 22.2 \text{ kJ/kg}$$

Now from the *p-h* diagram for R-12, we find that the pressure at point 5 corresponding to – 30°C,

$$p_5 = 1.044 \text{ bar}$$

Enthalpy of saturated vapour refrigerant at point 5,

$$h_5 = 174.2 \text{ kJ/kg}$$

Entropy of saturated vapour refrigerant at point 5,

$$s_5 = 0.7171 \text{ kJ/kg K}$$

Enthalpy of superheated vapour refrigerant at point 6,

$$h_6 = 207 \text{ kJ/kg}$$

Pressure at point 6 corresponding to 25°C,

$$p_6 = 6.518 \text{ bar}$$

and enthalpy of liquid refrigerant at point 7 corresponding to 20°C,

$$h_{f7} = h_8 = 54.9 \text{ kJ/kg}$$

1. *Compression ratio for each unit*

We know that the compression ratio for R-22 unit

$$= \frac{p_2}{p_1} = \frac{2.458}{0.3745} = 6.56 \text{ Ans.}$$

Similarly, compression ratio for R-12 unit

$$= \frac{p_6}{p_5} = \frac{6.518}{1.044} = 6.24 \text{ Ans.}$$

2. *Quantity of refrigerant circulated per minute for each unit*

We know that mass of refrigerant circulated for R-22 unit,

$$m_1 = \frac{210\,Q}{h_1 - h_4} = \frac{210 \times 10}{223.7 - 22.2} = 10.4 \text{ kg/min Ans.}$$

and mass of refrigerant circulated for R-12 unit,

$$m_2 = \frac{m_1(h_2 - h_4)}{h_5 - h_8} = \frac{10.4(275 - 22.2)}{174.2 - 54.9} = 22.04 \text{ kg/min Ans.}$$

3. *C.O.P. for each unit*

We know that refrigerating effect for R-22 unit,

$$R_{E1} = m_1\,(h_1 - h_4) = 210\,Q = 210 \times 10 = 2100 \text{ kJ/min}$$

Refrigerating effect for R-12 unit,

$$R_{E2} = m_2\,(h_5 - h_8) = 22.04\,(174.2 - 54.9) = 2629.4 \text{ kJ/min}$$

Work done in the compressor for R-22 unit,

$$W_1 = m_1\,(h_2 - h_1) = 10.4\,(275 - 223.7) = 533.5 \text{ kJ/min}$$

and work done in the compressor for R-12 unit,

$$W_2 = m_2\,(h_6 - h_5) = 22.04\,(207 - 174.2) = 723 \text{ kJ/min}$$

$\therefore$ C.O.P. for R-22 unit,

$$= \frac{R_{E1}}{W_1} = \frac{2100}{533.5} = 3.93 \text{ Ans.}$$

and C.O.P. for R-12 unit

$$= \frac{R_{E2}}{W_2} = \frac{2629.4}{723} = 3.64 \text{ Ans.}$$

4. *C.O.P. of the whole system*

We know that C.O.P. of the whole system

$$= \frac{R_E}{W} = \frac{210\,Q}{W_1 + W_2} = \frac{210 \times 10}{533.5 + 723} = 1.67 \text{ Ans.}$$

5. *Theoretical power required to run the system*

We know that the theoretical power required to run the system,

$$P = \frac{W_1 + W_2}{60} = \frac{533.5 + 723}{60} = 20.94 \text{ kW Ans.}$$

14.5 Solid Carbon Dioxide or Dry Ice

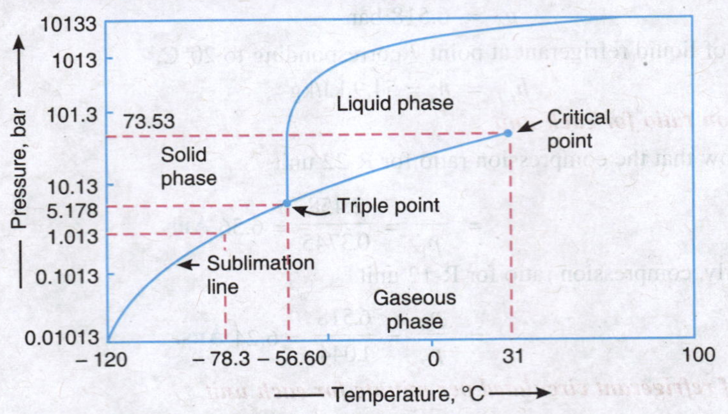

Fig. 14.4. Pressure-temperature diagram for carbon dioxide.

The solid carbon dioxide (also known as dry ice), despite its high manufacturing cost, is used as a refrigerant due to the following reasons :

1. It is non-toxic and non-corrosive.
2. It can be easily handled and cut.
3. It has the advantage over water ice of lower temperature, the absence of objectionable liquid and smaller space requirement for a required cooling capacity.
4. The *triple point temperature for carbon dioxide is – 56.6°C at 5.178 bar. From the pressure-temperature diagram for carbon dioxide as shown in Fig. 14.4, we see that when the pressure is above the triple point pressure, the solid phase passes first into the liquid phase and then into the gaseous phase. But when the pressure is below the triple point pressure, the solid phase sublimates directly into the gaseous phase. At atmospheric pressure (1.013 bar), the solid carbon dioxide sublimates to gas at a temperature of – 78.3°C. Thus solid carbon dioxide may be used at atmospheric pressure for preservation of perishables which require low temperature during transportation.

Note : Below the triple point temperature, the heat required to change the substance from a solid directly to a gas is termed as *latent heat of sublimation.* At the triple point temperature, the heat required to change from a solid to a liquid is called *latent heat of fusion.* Above the triple point temperature, the heat required to change from a liquid to a vapour is termed as *latent heat of vaporisation.*

14.6 Manufacture of Solid Carbon Dioxide or Dry Ice

The carbon dioxide gas is, first of all, compressed to a pressure of 60 to 70 bar and then condensed by using cooling water. The liquid carbon dioxide thus obtained is solidified in order to produce solid carbon dioxide or dry ice.

One of the most widely used methods of solidification of the liquid carbon dioxide is by the expansion of liquid carbon dioxide to a pressure below that of its triple point pressure. The simplest form of equipment used for the production of solid carbon dioxide

Dry ice manufacture and cleaning.

* The triple point is that point at which the solid, liquid and vapour phases are in equilibrium.

or dry ice by the expansion of liquid carbon dioxide is shown in Fig. 14.5. The equipment is similar to that of vapour compression system with the exceptions that means must be provided for supplying the make-up carbon dioxide gas and for the removal of solid carbon dioxide. The corresponding *p-h* diagram for the production of solid carbon dioxide is shown in Fig. 14.6. First of all, the carbon dioxide gas is compressed isentropically in a compressor from point 1 to point 2 to a pressure sufficiently high as represented by the curve 1-2 in Fig. 14.6. It is then condensed in a condenser by cooling water from point 2 to point 3. The high pressure liquid carbon dioxide thus obtained is now expanded in an expansion valve from point 3 to point 4, to a pressure below that of the triple point pressure. The carbon dioxide at point 4 is a mixture of solid carbon dioxide and gaseous carbon dioxide. The solid carbon dioxide or dry ice can be removed at point 5 and pressed mechanically into blocks. The carbon dioxide gas removed at point 6 is mixed with the make-up carbon dioxide gas (equal to the weight of solid carbon dioxide removed) at point 7 before being fed to the compressor at point 1.

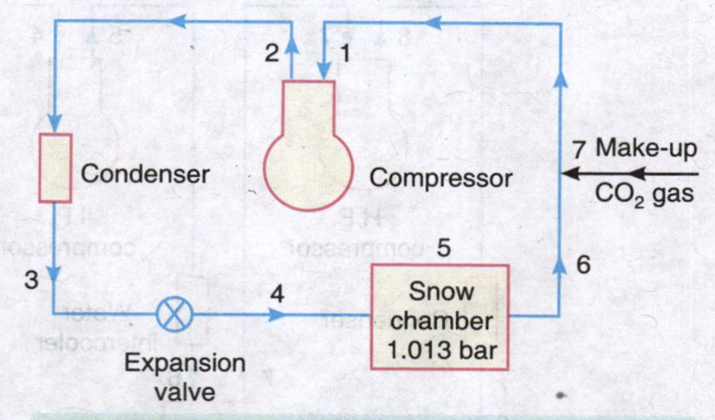

Fig. 14.5. Schematic diagram of equipment for production of solid carbon dioxide or dry ice.

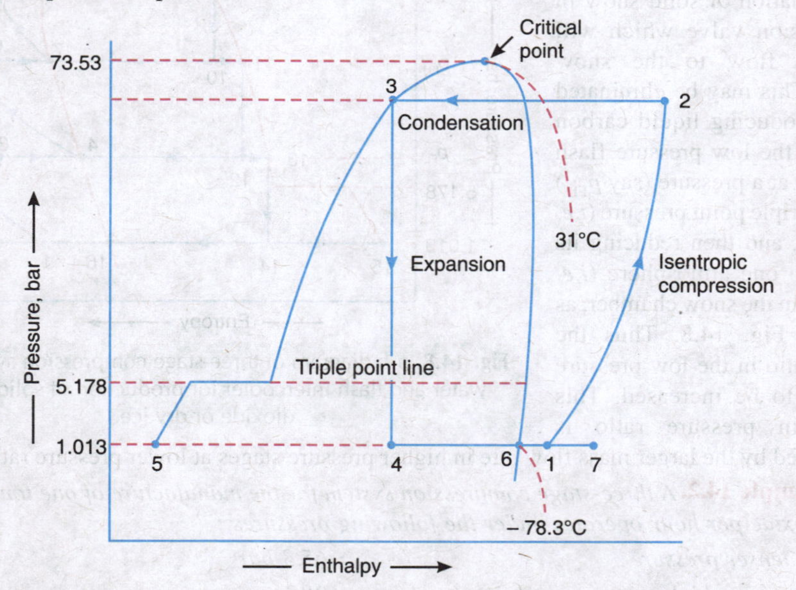

Fig. 14.6. *p-h* diagram for production of solid carbon dioxide or dry ice.

The power consumption for the satisfactory operation of this single-stage compression system is approximately 330 kW hour per tonne of solid carbon dioxide produced. One of the main reasons for high power consumption is the high pressure ratio which is about 70. Thus, it is very uneconomical for the production of solid carbon dioxide. In order to reduce the power consumption, multi-stage compression (usually three-stage compression) with water intercoolers and flash intercoolers, as shown in Fig. 14.7, may be used. Its corresponding *p-h* diagram is shown in Fig. 14.8.

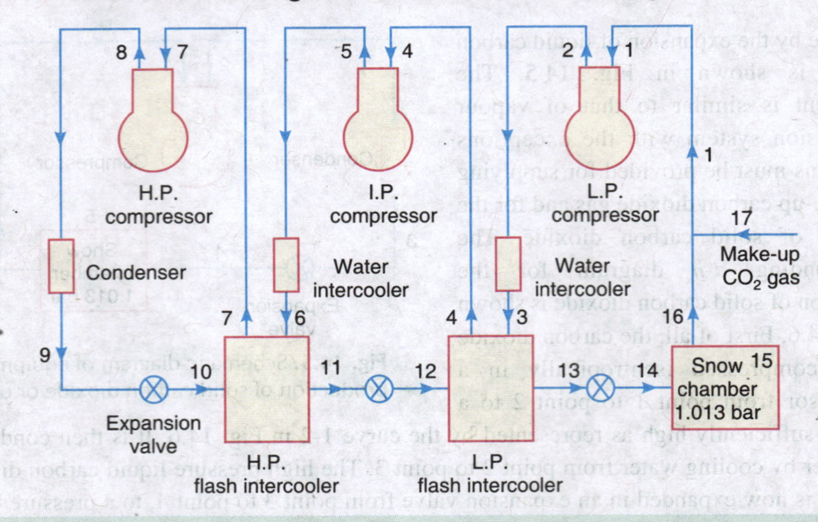

Fig. 14.7. Three-stage compression system with water and flash intercooler for production of solid carbon dioxide or dry ice.

Another problem in the process of manufacture of dry ice is the formation of solid snow in the expansion valve which will block the flow to the snow chamber. This may be eliminated by first producing liquid carbon dioxide in the low pressure flash intercooler at a pressure (say p_{FL}) above the triple point pressure (*i.e.* 5.178 bar), and then reducing its pressure to one atmosphere (*i.e.* 1.013 bar) in the snow chamber, as shown in Fig. 14.8. Thus the pressure ratio in the low pressure stage has to be increased. This increase in pressure ratio is compensated by the larger mass flow rate in higher pressure stages at lower pressure ratios.

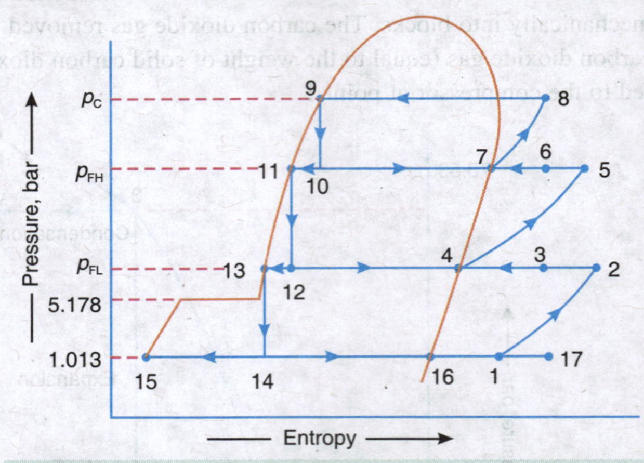

Fig. 14.8. *p-h* diagram of three-stage compression system with water and flash intercooler for production of solid carbon dioxide or dry ice.

Example 14.2. *A three-stage compression system for the manufacture of one tonne of solid carbon dioxide per hour operates under the following pressures :*

Condenser pressure	*= 65.3 bar*
Pressure in high pressure flash intercooler	*= 28 bar*
Pressure in low pressure flash intercooler	*= 6 bar*
Pressure in snow chamber	*= 1.013 bar*

Both the flash and water intercoolers are used between the stages as shown in Fig. 14.7. The make-up carbon dioxide gas enters at 1.013 bar and 21°C. The gas leaving both the water intercoolers is at 21°C. Calculate the theoretical power required to run the system.

Solution. Given : m_{15} = 1 t / h = 1000 kg / h ; p_C = 65.3 bar ; p_{FH} = 28 bar ; p_{FL} = 6 bar ; p_{SC} = 1.013 bar ; $t_{17} = t_3 = t_6$ = 21°C

The *p-h* diagram of a three-stage compression system with flash and water intercoolers is shown in Fig. 14.9. We know that at point 14, the carbon dioxide is a mixture of solid carbon dioxide and gaseous carbon dioxide. The solid carbon dioxide is removed at point 15 and the gaseous carbon dioxide at point 16.

Let m_{14} = Mass of solid carbon dioxide and gaseous carbon dioxide = $m_{15} + m_{16}$

m_{15} = Mass of solid carbon dioxide removed at point 15

= 1000 kg/h ...(Given)

m_{16} = Mass of carbon dioxide gas removed at point 16.

Considering the heat balance of the snow chamber, we have

$$m_{14}\,h_{14} = m_{15}\,h_{15} + m_{16}\,h_{16} \qquad ... (i)$$

From the *p-h* diagram, we find that enthalpy at point 14,

$$h_{14} = -24.5 \text{ kJ/kg}$$

Enthalpy at point 15, $h_{15} = -82.5$ kJ/kg

and enthalpy at point 16, $h_{16} = 310$ kJ/kg

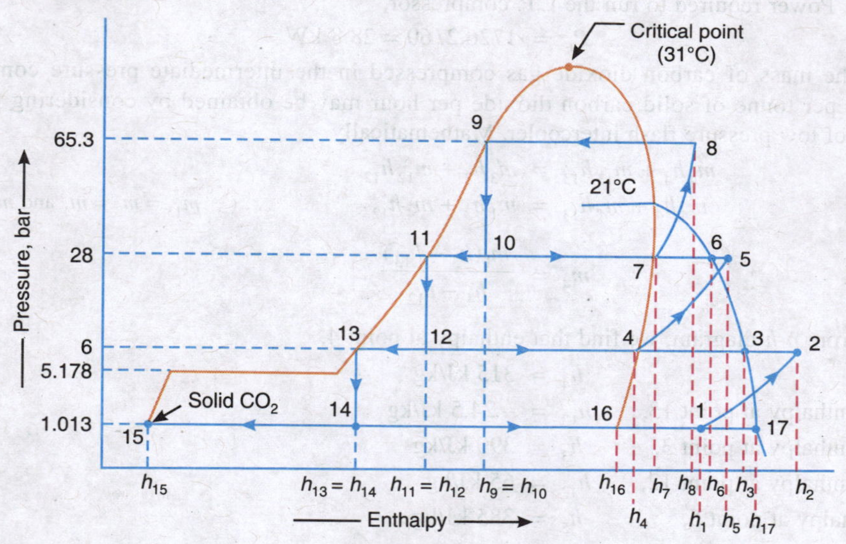

Fig. 14.9

Substituting the values in equation (*i*), we have

$$m_{14} \times -24.5 = 1000 \times -82.5 + (m_{14} - 1000)\,310$$

$$\ldots (\because m_{16} = m_{14} - m_{15})$$

$$-24.5\,m_{14} = -82\,500 + 310\,m_{14} - 310\,000$$

$$-24.5\,m_{14} - 310\,m_{14} = -392\,500$$

$$\therefore \quad m_{14} = 1173 \text{ kg/h}$$

The enthalpy at point 1 (h_1) may be obtained by considering the heat balance between points 1, 16 and 17. Mathematically,

$$m_1\,h_1 = m_{16}\,h_{16} + m_{17}\,h_{17} \qquad ... (ii)$$

The make-up carbon dioxide gas supplied at a temperature of 21°C at point 17 is equal to the mass of solid carbon dioxide removed at point 15, *i.e.*

$$m_{17} = m_{15} = 1000 \text{ kg/h}$$

and $$m_1 = m_{14} = 1173 \text{ kg/h}$$

From p-h diagram, we find that enthalpy at point 17,

$$h_{17} = 400 \text{ kJ/kg}$$

Substituting the values in equation (*ii*), we have

$$1173 \times h_1 = (1173 - 1000)\, 310 + 1000 \times 400 = 453\,630$$

$$\therefore \quad h_1 = 386.7 \text{ kJ/kg}$$

From point 1, draw a constant entropy line 1-2 to intersect the constant pressure line of 6 bar at point 2. From p-h diagram, we find that enthalpy at point 2,

$$h_2 = 475 \text{ kJ/kg}$$

We know that work done in the L.P. compressor,

$$W_L = m_1 (h_2 - h_1)$$

$$= 1173(475 - 386.7) = 103\,576 \text{ kJ/h} \qquad \ldots (\because m_1 = m_{14})$$

$$= 1726.2 \text{ kJ/min}$$

∴ Power required to run the L.P. compressor,

$$P_L = 1726.2/60 = 28.8 \text{ kW}$$

The mass of carbon dioxide gas compressed in the intermediate pressure compressor (*i.e.* m_4) per tonne of solid carbon dioxide per hour may be obtained by considering the heat balance of low pressure flash intercooler. Mathematically

$$m_4 h_4 + m_{13} h_{13} = m_3 h_3 + m_{12} h_{12}$$

or $$m_4 h_4 + m_1 h_{13} = m_1 h_3 + m_4 h_{12} \qquad \ldots (\because m_{13} = m_3 = m_1 \text{ and } m_{12} = m_4)$$

$$\therefore \quad m_4 = \frac{m_1 (h_3 - h_{13})}{h_4 - h_{12}} \qquad \ldots (iii)$$

From p-h diagram, we find that enthalpy at point 4,

$$h_4 = 315 \text{ kJ/kg}$$

Enthalpy at point 13, $h_{13} = -24.5$ kJ/kg

Enthalpy at point 3, $h_3 = 390$ kJ/kg

Enthalpy at point 12, $h_{12} = 65$ kJ/kg

and enthalpy at point 5, $h_5 = 385$ kJ/kg

Substituting the values in equation (*iii*), we have

$$m_4 = \frac{1173\,[390 - (-24.5)]}{315 - 65} = 1945 \text{ kg/h}$$

We know that work done in the I.P. compressor,

$$W_I = m_4 (h_5 - h_4) = 1945\,(385 - 315) = 136\,150 \text{ kJ/h}$$

$$= 2269.2 \text{ kJ/min}$$

∴ Power required to run the I.P. compressor,

$$P_I = 2269.2 / 60 = 37.8 \text{ kW}$$

Now to find the mass of carbon dioxide gas compressed in the high pressure compressor (*i.e.* m_7) per tonne of solid carbon dioxide per hour may be obtained by considering the heat balance of high pressure flash intercooler. Mathematically,

$$m_7 h_7 + m_{11} h_{11} = m_6 h_6 + m_{10} h_{10}$$

or

$$m_7 h_7 + m_4 h_{11} = m_4 h_6 + m_7 h_{10} \quad \ldots (\because m_{11} = m_6 = m_4 \text{ and } m_{10} = m_7)$$

$$\therefore \quad m_7 = \frac{m_4 (h_6 - h_{11})}{h_7 - h_{10}} \quad \ldots (iv)$$

From *p-h* diagram, we find that enthalpy at point 6,

$$h_6 = 362 \text{ kJ/kg}$$

Enthalpy at point 11, $h_{11} = 65$ kJ/kg

Enthalpy at point 7, $h_7 = 320$ kJ/kg

Enthalpy at point 10, $h_{10} = 160$ kJ/kg

and enthalpy at point 8, $h_8 = 356$ kJ/kg

Substituting the values in equation (*iv*), we have

$$m_7 = \frac{1945\,(362 - 65)}{320 - 160} = 3610 \text{ kg/h}$$

We know that work done in the H.P. compressor,

$$W_H = m_7 (h_8 - h_7) = 3610\,(356 - 320) = 129\,960 \text{ kJ/h}$$
$$= 2166 \text{ kJ/min}$$

$\therefore$ Power required to run the H.P. compressor,

$$P_H = 2166/60 = 36.1 \text{ kW}$$

and total power required to run the system,

$$P = P_L + P_I + P_H = 28.8 + 37.8 + 36.1 = 102.7 \text{ kW}$$ **Ans.**

14.7 Liquefaction of Gases

The gas must be liquefied in order to produce low temperature for refrigeration purposes. The following two methods of expansion of gases may be used for producing the low temperature of gas or to produce refrigeration.

1. *Isentropic expansion.* The gas may be expanded isentropically to produce the low temperature. This method of expanding the gas is employed in air refrigeration cycle and at present it is utilised in aircraft refrigeration systems. Since very low temperature of the gas may be obtained by isentropic expansion, therefore its use is limited to comparatively high temperatures.

2. *Free, irreversible expansion.* The production of low temperature by the free, irreversible expansion (or throttling) of the gas from a comparatively high pressure through an orifice or other restriction to a lower pressure, is of great importance. We know that the free, irreversible expansion of a perfect gas results in no change of temperature, but no real gas is really perfect. Thus the actual drop in temperature experienced by most gases upon free expansion between two pressures may be used as a measure of the degree of imperfection of that gas and also as a means of refrigeration.

The change in temperature with drop in pressure at constant enthalpy is termed as *Joule-Thomson coefficient* (μ). Mathematically,

$$\mu = \left(\frac{dT}{dp}\right)_H$$

which varies with both the temperature and pressure of the gas.

The magnitude of the Joule-Thomson coefficient is a measure of the imperfection of a gas or its deviation from perfect gas behaviour. For real gases, μ may be either positive or negative depending upon the thermodynamic state of the gas. When μ is zero, the temperature of the gas remains constant with throttling. The temperature at which $\mu = 0$, is called the *inversion temperature* for a given pressure. If μ is greater than zero, the temperature of the gas decreases with throttling and when μ is less than zero, the temperature of the gas increases with throttling. Thus, in cooling of a gas by throttling, we require that the gas shows a large positive value of μ.

Liquefaction of gases.

The Joule-Thomson coefficient is not a constant but is a function of both pressure and temperature. We shall now derive the functional relationship for the coefficient. According to First Law of Thermodynamics,

$$\delta q = du + \delta w = du + p\,dv \quad \text{... } (\because \delta w = p\,dv)$$

or

$$du = \delta q - p\,dv = T\,ds - p\,dv \quad \text{... } (\because \delta q = T\,ds) \quad \text{... (i)}$$

We know that

$$d\,(pv) = p\,dv + v\,dp$$

or

$$p\,dv = d\,(pv) - v\,dp$$

Substituting the value of $p\,dv$ in equation *(i)*, we have

$$du = T\,ds - [\,d\,(pv) - v\,dp] = T\,ds - d\,(pv) + v\,dp$$

$$d\,(u + pv) = T\,ds + v\,dp$$

$$dh = T\,ds + v\,dp \quad \text{... } (\because u + pv = \text{Enthalpy, } h) \quad \text{... (ii)}$$

Since entropy (s) is a function of temperature (T) and pressure (p), therefore

$$ds = \left(\frac{\partial s}{\partial T}\right)_p dT + \left(\frac{\partial s}{\partial p}\right)_T dp$$

Substituting the value of ds in equation *(ii)*,

$$dh = T\left[\left(\frac{\partial s}{\partial T}\right)_p dT + \left(\frac{\partial s}{\partial p}\right)_T dp\right] + v\,dp \quad \text{... (iii)}$$

The partial derivative of equation *(iii)* with respect to p with T constant gives the equation

$$\left(\frac{\partial h}{\partial p}\right)_T = T\left(\frac{\partial s}{\partial p}\right)_p + v \quad \text{... (iv)}$$

and the partial derivative of equation (*iii*) with respect to p with h constant gives the equation

$$\left(\frac{\partial h}{\partial p}\right)_h = T\left(\frac{\partial s}{\partial T}\right)_p\left(\frac{\partial T}{\partial p}\right)_h + T\left(\frac{\partial s}{\partial p}\right)_p + v = 0$$

or
$$T\left(\frac{\partial s}{\partial T}\right)_p\left(\frac{\partial T}{\partial p}\right)_h = -\left[T\left(\frac{\partial s}{\partial p}\right)_p + v\right] = -\left(\frac{\partial h}{\partial p}\right)_T \quad \text{... [From equation } (iv)]$$

$$= -\left[\frac{\partial}{\partial p}(u + pv)\right]_T = -\left(\frac{\partial u}{\partial p}\right)_T - \left[\frac{\partial}{\partial p}(pv)\right]_T \quad \text{... } (v)$$

Since the specific heat at constant pressure (c_p) is the quantity of heat required to raise the temperature through 1°C, therefore

$$c_p = \left(\frac{dh}{dT}\right)_p = \left(\frac{Tds}{dT}\right)_p = T\left(\frac{ds}{dT}\right)_p$$

Now the equation (v) may be written as

$$c_p\left(\frac{\partial T}{\partial p}\right)_h = -\left(\frac{\partial u}{\partial p}\right)_T - \left[\frac{\partial}{\partial p}(pv)\right]_T \quad \text{... } (vi)$$

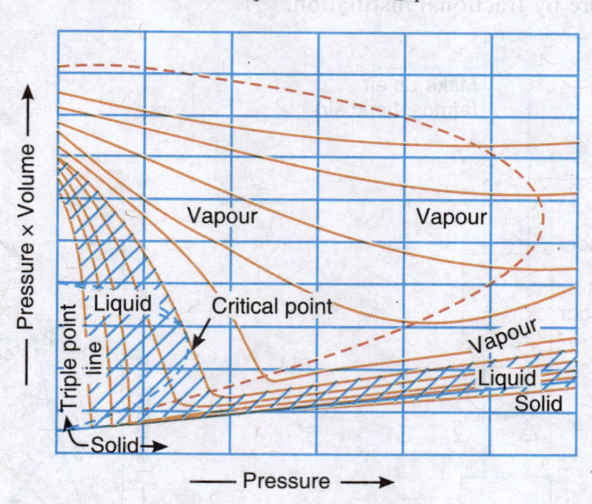

Fig. 14.10. Isotherms on p and pv coordinates.

The term $\left(\frac{\partial T}{\partial p}\right)_h$ is the Joule-Thomson coefficient. From equation (*vi*), we see that the Joule-Thomson coefficient is dependent upon two quantities, *i.e.* $\left(\frac{\partial u}{\partial p}\right)_T$ and $\left[\frac{\partial}{\partial p}(pv)\right]_T$. The term $\left(\frac{\partial u}{\partial p}\right)_T$ is the change in internal energy per unit change in pressure which is usually negative. The term $\left[\frac{\partial}{\partial p}(pv)\right]_T$ represents the deviation of the gas from the perfect gas laws and may be either positive or negative. The variation that may be expected in the term $\left[\frac{\partial}{\partial p}(pv)\right]_T$ is shown in Fig. 14.10, in which pv is plotted against p.

When the right-hand side of equation (*vi*) is zero, this defines the conditions for the inversion temperature or the temperature and pressure conditions under which expansion results in neither heating nor cooling of the gas. Under these conditions,

$$\left(\frac{\partial u}{\partial p}\right)_T = \left[\frac{\partial}{\partial p}(pv)\right]_T \qquad \ldots (vii)$$

Above these conditions, expansion will result in heating and below these conditions, expansion will result in cooling.

The value of the Joule-Thomson coefficients is negative for all gases under high temperatures and pressures. Under these conditions, the throttling results in a warming of the gas. The inversion temperatures for most common gases are above those of temperatures normally encountered. Therefore, if throttling occurs at ordinary temperatures and from high pressures that are not too high, then a cooling effect results. However, hydrogen and helium behave at ordinary temperatures as most gases do at high temperatures, since their temperatures are –77.8°C and –250°C respectively. Therefore, if throttling of these gases is to be used to accomplish refrigeration, it is first necessary to cool them below their inversion temperatures by some other means.

14.8 Linde System for Liquefaction of Air

The liquefaction of air is an important industrial process not only for the production of liquid air but also in the separation of oxygen, nitrogen, hydrogen, helium and many other gases from the atmosphere by fractional distillation.

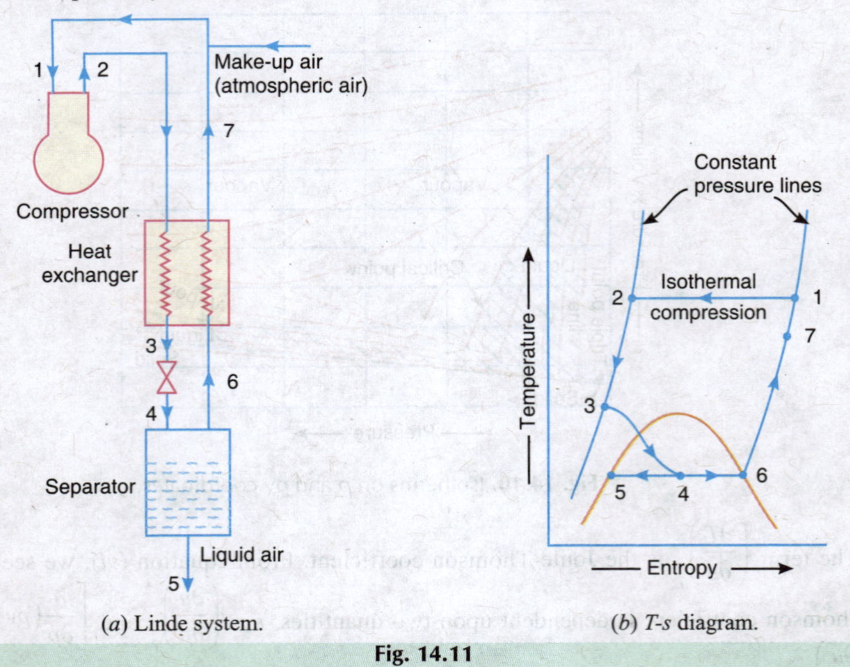

(*a*) Linde system. (*b*) *T-s* diagram.

Fig. 14.11

The simplest method of air liquefaction is the *Linde* or *Hampson system*, as shown in Fig. 14.11 (*a*). The equipment includes a compressor, a heat exchanger and a separator. In this system, the atmospheric air is compressed isothermally in a compressor between points 1 and 2 to pressure of 100 to 200 atmosphere. This high pressure air is cooled to about –106.7°C in the heat exchanger between points 2 and 3. The cooled air from the heat exchanger is throttled between points 3 and 4 to atmospheric pressure and a temperature of –190°C. A portion of the air is liquefied and removed from the separator at point 5. The remainder of the cold air leaves the separator at point 6 and returns to the compressor through the heat exchanger where it cools the incoming high pressure air. The make-up air from the atmosphere (equal to the amount of air liquefied) is also supplied to the compressor along with the air removed from the heat exchanger at point 7. The Linde process is represented on temperature-entropy diagram as shown in Fig. 14.11 (*b*).

Liquefaction of air.

Let m_2 = Mass of air compressed in the compressor or mass of high pressure air passing through the heat exchanger between points 2 and 3,

m_5 = Mass of air liquefied in the separator, and

m_6 = Mass of low pressure air in passing through the heat exchanger between points 6 and 1 = $m_2 - m_5$

Now for the heat balance of the heat exchanger,

$$m_2(h_2 - h_3) = m_6(h_1 - h_6)$$

$$= (m_2 - m_5)(h_1 - h_6) \quad \dots (\because m_6 = m_2 - m_5) \qquad \dots (i)$$

and for the heat balance of the separator,

$$m_2 h_4 = m_5 h_5 + m_6 h_6 = m_5 h_5 + (m_2 - m_5)\, h_6 \qquad \dots (ii)$$

Since the process 3-4 is a throttling process, therefore

$$h_3 = h_4 \qquad \dots (iii)$$

From equations (*i*) and (*ii*), the mass of air liquefied by the Linde process may be obtained.

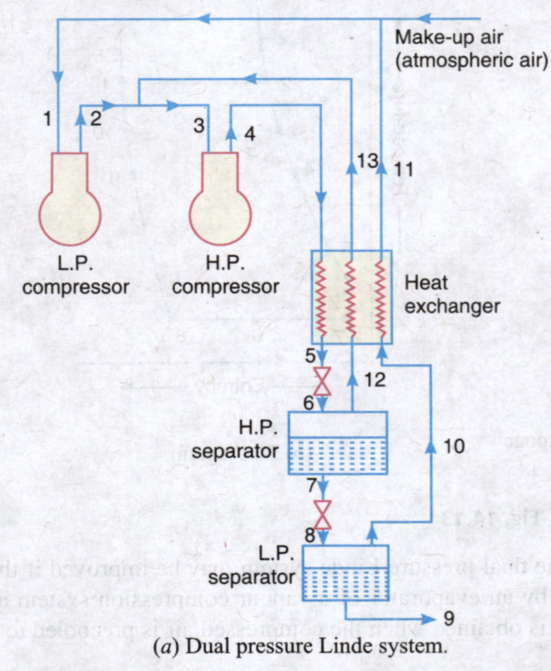

(*a*) Dual pressure Linde system.

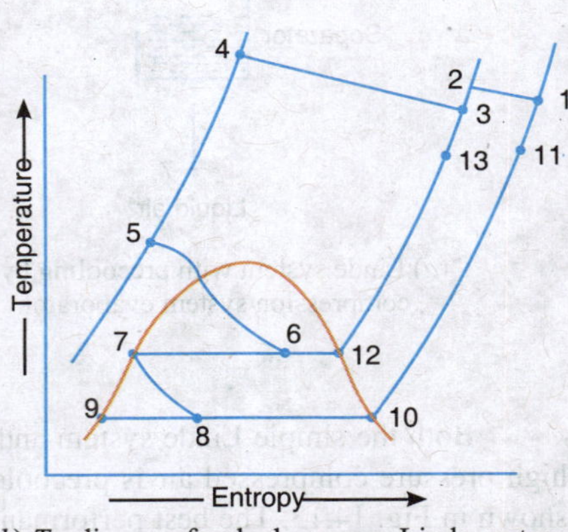

(*b*) *T-s* diagram for dual pressure Linde system.

Fig. 14.12

The simple Linde or Hampson system for liquefying air is comparatively inefficient and is used only when small quantity of liquid air is needed. In order to increase the efficiency, a dual pressure Linde system as shown in Fig. 14.12 (*a*) is used. Its corresponding *T-s* diagram is shown in Fig. 14.12 (*b*). In this system, a portion of the gas after the first throttling process from point 5 to point 6 is bled back through a heat exchanger into the discharge of the first (low pressure) stage of compression. The increased economy results from this system because the majority of the gas undergoes only one expansion to the intermediate pressure and one compression from this intermediate pressure to the high pressure.

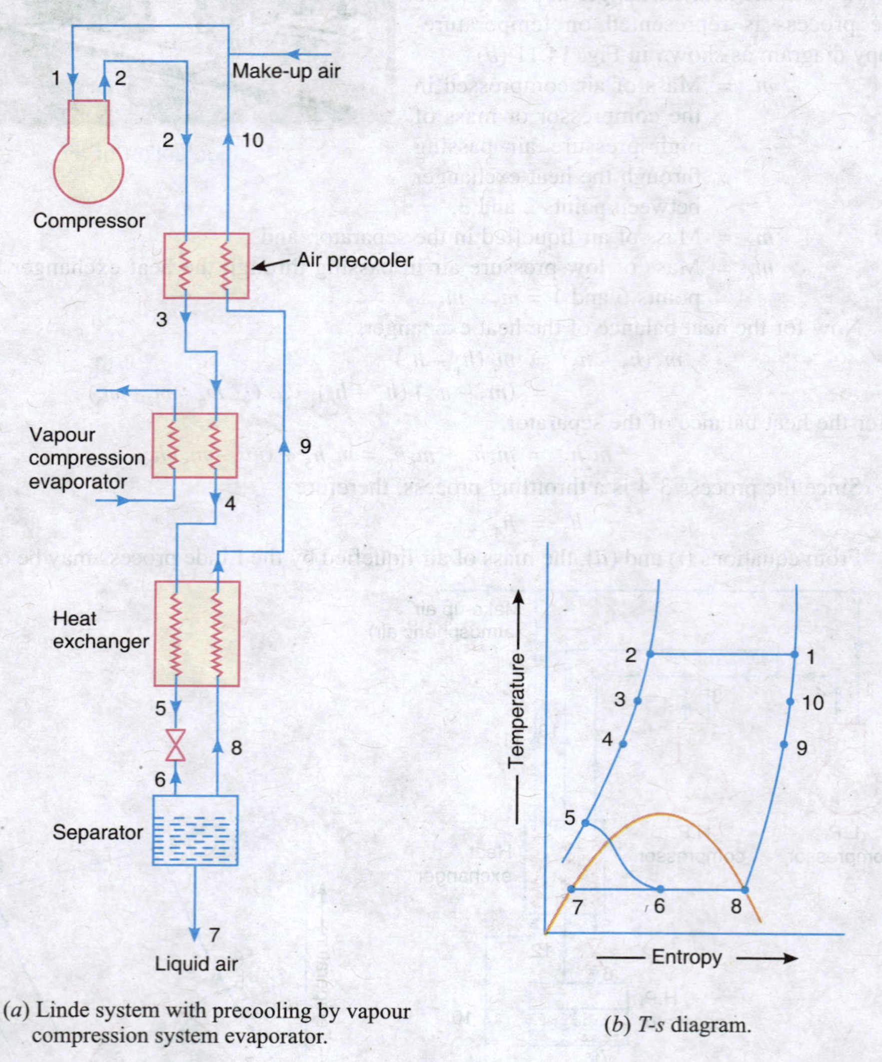

(*a*) Linde system with precooling by vapour compression system evaporator.

(*b*) *T-s* diagram.

Fig. 14.13

Both the simple Linde system and the dual pressure Linde system may be improved if the high pressure compressed air is precooled by an evaporator of a vapour compression system as shown in Fig. 14.13. The best performance is obtained when the compressed air is precooled to a relatively low temperature.

Example 14.3. *Dry air at 20°C and 1 bar is to be liquefied by the simple Linde method. The air is isothermally compressed at 20°C to 170 bar. The make-up air is supplied to the system at 20°C and 1 bar. Find the yield of liquid air in kg per kg of air compressed and the temperature of air before throttling.*

Solution. Given : $t_1 = t_2 = 20°C$; $p_1 = p_6 = 1$ bar ; $p_2 = 170$ bar

The *T-s* diagram of air for simple Linde system is shown in Fig. 14.14. From the diagram, we find that enthalpy at point 1,

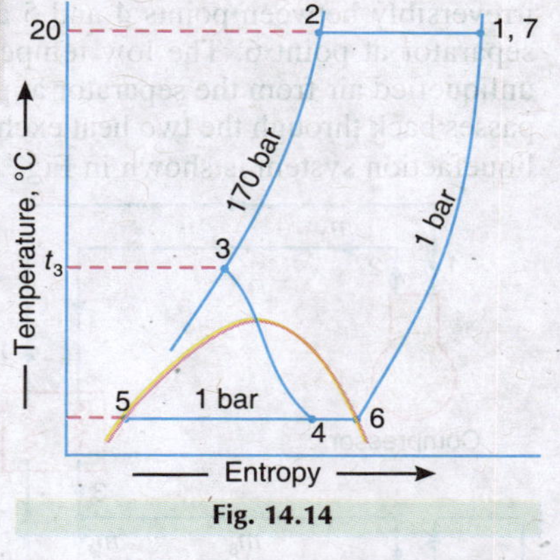

Fig. 14.14

$$h_1 = 506 \text{ kJ/kg}$$

Enthalpy at point 2, $h_2 = 473$ kJ/kg

Enthalpy at point 5, $h_5 = 92$ kJ/kg

and enthalpy at point 6, $h_6 = 292$ kJ/kg

Yield of liquid air per kg of air compressed

Let m_5 = Yield of liquid air in kg per kg of air compressed, and

m_2 = Mass of air compressed = 1 kg ... (Given)

We know that for the heat balance of the heat exchanger,

$$m_2 (h_2 - h_3) = (m_2 - m_5)(h_1 - h_6)$$

$$1(473 - h_3) = (1 - m_5)(506 - 292)$$

$$473 - h_3 = 214 - 214\, m_5$$

$$\therefore \quad h_3 = 259 + 214\, m_5 \quad \text{... (i)}$$

Now for the heat balance of the separator,

$$m_2 h_4 = m_5 h_5 + (m_2 - m_5) h_6$$

$$1 \times h_3 = m_5 \times 92 + (1 - m_5)\, 292 \quad \text{... } (\because h_4 = h_3)$$

$$\therefore \quad h_3 = 92\, m_5 + 292 - 292\, m_5$$

$$= 292 - 200\, m_5 \quad \text{... (ii)}$$

Equating equations (*i*) and (*ii*), we have

$$259 + 214\, m_5 = 292 - 200\, m_5$$

$$214\, m_5 + 200\, m_5 = 292 - 259 = 33$$

$$\therefore \quad m_5 = 0.08 \text{ kg} \quad \textbf{Ans.}$$

Temperature of air before throttling

Let t_3 = Temperature of air before throttling.

Substituting the value of m_5 in equation (*i*), we have

$$h_3 = 259 + 214 \times 0.08 = 276.1 \text{ kJ/kg}$$

From the *T-s* diagram for air, we find that the temperature of air corresponding to 276.1 kJ/kg is

$$t_3 = 175 \text{ K} = -98°C \quad \textbf{Ans.}$$

14.9 Claude System for Liquefaction of Air

The Claude air liquefaction system differs from the simple Linde system by the addition of an expander and a second heat exchanger as shown in Fig. 14.15 (*a*). In this system, the air is compressed isothermally in a compressor to approximately 40 atmospheres between points 1 and 2. This high pressure air is partially cooled by passing through the first heat exchanger between points 2 and 3.

A portion of air (about 80 per cent) at point 3 is bled and cooled by expansion in an expander between points 3 and 8. The remaining portion of air (*i.e.* 20 per cent) passes through the second heat exchanger between points 3 and 4. The air from the second heat exchanger is throttled irreversibly between points 4 and 5 at atmospheric pressure. The liquid air is removed from the separator at point 6. The low temperature air from the expander at point 8 is mixed with the unliquefied air from the separator at point 7, giving increased mass flow of air at point 9. This air passes back through the two heat exchangers to the compressor. The *T-s* diagram for the Claude air liquefaction system is shown in Fig. 14.15 (*b*).

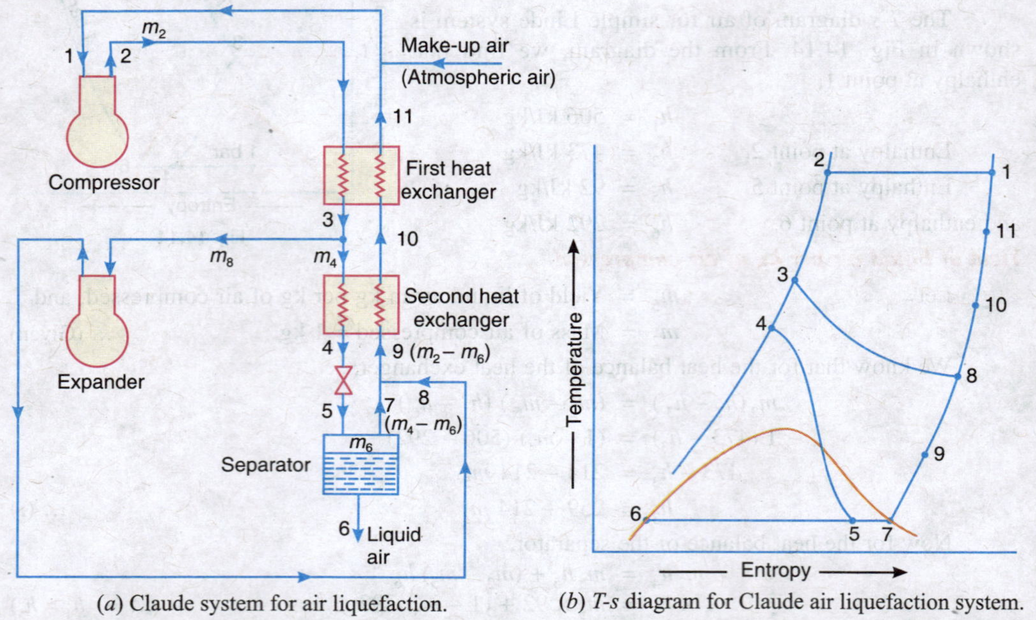

(*a*) Claude system for air liquefaction. (*b*) *T-s* diagram for Claude air liquefaction system.

Fig. 14.15

The Claude system is more efficient than the Linde system because the expansion of air in the expander results in lower temperature than the Linde system.

Let m_2 = Mass of compressed air,

m_8 = Mass of air bypassed to expander,

m_4 = Mass of air passing through the second heat exchanger $= m_2 - m_8$, and

m_6 = Mass of air liquefied in the separator,

Now for the heat balance of the first heat exchanger,

$$m_2 (h_2 - h_3) = (m_2 - m_6)(h_1 - h_{10}) \quad \dots (i)$$

For the heat balance of the second heat exchanger

$$m_4 (h_3 - h_4) = (m_2 - m_6)(h_{10} - h_9) \quad \dots (ii)$$

and for the heat balance of the separator,

$$m_4 h_5 = m_6 h_6 + (m_4 - m_6) h_7 \quad \dots (iii)$$

The air from the expander at point 8 is mixed with the unliquefied air from the separator at point 7. The combined condition of air is represented by point 9 on the *T-s* diagram as shown in Fig. 14.15 (*b*). The enthalpy at point 9 is given by

$$(m_4 - m_6 + m_8) h_9 = m_8 h_8 + (m_4 - m_6) h_7$$

$$\therefore \quad h_9 = \frac{m_8 h_8 + (m_4 - m_6) h_7}{m_4 - m_6 + m_8} = \frac{m_8 h_8 + (m_4 - m_6) h_7}{m_2 - m_6} \quad \dots (iv)$$

$$\dots (\because m_4 = m_2 - m_8)$$

From the above expressions, we may find out the mass of air liquefied by the Claude system.

14.10 Advantages of Claude System over Linde System

The following are the advantages of Claude system over Linde system for the liquefaction of gases :

1. In Claude system, the air is to be compressed only up to 40 atmospheres, as compared to 100-200 atmospheres in Linde system.
2. About 80 per cent of the air is expanded reversibly in the expander and the remaining 20 per cent of the compressed air is subjected to irreversible throttling. In Linde system, all the air is throttled irreversibly.
3. Claude system gives an enhanced liquefaction.
4. The specific work of Claude system is less than that of simple Linde system.

Example 14.4. *Dry air at 20°C and 1 bar is to be liquefied by the Claude method. The air is compressed isothermally at 20°C to 170 bar. Assume that 80 per cent of the total mass of air compressed passes through the expander. The temperature of air entering the expander is –80°C while the temperature of air leaving the expander is –140°C. The make-up air is supplied also at 20°C and 1 bar. Determine the yield of liquid air in kg per kg of air compressed and the temperature of air before throttling.*

Solution. Given : $t_1 = t_2 = 20°C$; $p_1 =$ 1 bar ; $p_2 = 170$ bar ; $m_8 = 0.8\, m_2$; $t_3 = -80°C$; $t_8 = -140°C$

The *T-s* diagram of air for Claude system is shown in Fig. 14.16. From the diagram, we find that enthalpy at point 1,

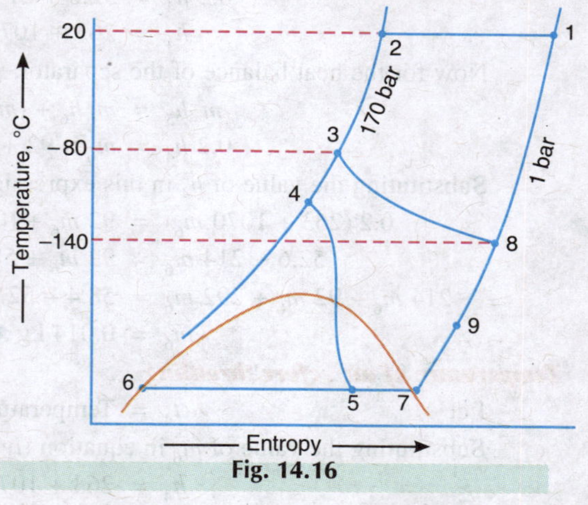

Fig. 14.16

$$h_1 = 506 \text{ kJ/kg}$$

Enthalpy at point 2, $h_2 = 473$ kJ/kg

Enthalpy at point 3, $h_3 = 309$ kJ/kg

Enthalpy at point 6, $h_6 = 92$ kJ/kg

Enthalpy at point 7, $h_7 = 292$ kJ/kg

and enthalpy at point 8, $h_8 = 343$ kJ/kg

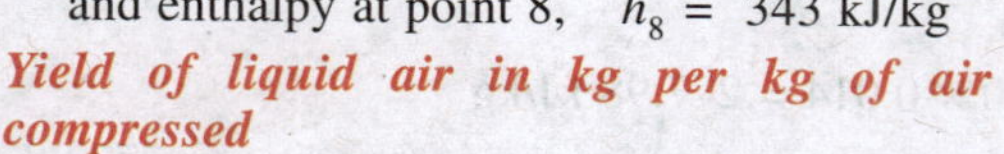

Yield of liquid air in kg per kg of air compressed

Let

m_6 = Yield of liquid air in kg per kg of air compressed,

m_2 = Mass of air compressed = 1 kg, ... (Given)

m_8 = Mass of air bypassed to expander, ... (Given)

$= 0.8\, m_2 = 0.8$ kg,

m_4 = Mass of air passing through the second heat exchanger.

$= m_2 - m_8 = 1 - 0.8 = 0.2$ kg

The air from the expander at point 8 is mixed with unliquefied air from the separator at point 7. The condition of air after mixing is represented by point 9 on the *T-s* diagram. The enthaply at point 9 is given by

$$h_9 = \frac{m_8 h_8 + (m_4 - m_6)\, h_7}{m_2 - m_6} = \frac{0.8 \times 343 + (0.2 - m_6)292}{1 - m_6}$$

$$= \frac{274.4 + 58.4 - 292\, m_6}{1 - m_6} = \frac{332.8 - 292\, m_6}{1 - m_6} \qquad \text{... } (i)$$

We know that for the heat balance of the first heat exchanger,

$$m_2(h_2 - h_3) = (m_2 - m_6)(h_1 - h_{10})$$
$$1(473 - 309) = (1 - m_6)(506 - h_{10})$$
$$164 = 506 - 506\, m_6 - h_{10}(1 - m_6)$$
$$h_{10}(1 - m_6) = 342 - 506\, m_6 \qquad \text{... } (ii)$$

For the heat balance of the second heat exchanger,

$$m_4(h_3 - h_4) = (m_2 - m_6)(h_{10} - h_9)$$
$$0.2(309 - h_4) = (1 - m_6)\left(h_{10} - \frac{332.8 - 292\, m_6}{1 - m_6}\right)$$

... [Substituting the value of h_9 from equation (*i*)]

$$61.8 - 0.2\, h_4 = (1 - m_6)\, h_{10} - 332.8 + 292\, m_6$$
$$h_{10}(1 - m_6) = 394.6 - 292\, m_6 - 0.2\, h_4 \qquad \text{... } (iii)$$

Equating equations (*ii*) and (*iii*), we have

$$342 - 596\, m_6 = 394.6 - 292\, m_6 - 0.2\, h_4$$
$$0.2\, h_4 = 52.6 + 214\, m_6$$
$$\therefore \quad h_4 = 263 + 1070\, m_6 \qquad \text{... } (iv)$$

Now for the heat balance of the separator,

$$m_4 h_5 = m_6 h_6 + (m_4 - m_6) h_7$$
$$0.2\, h_4 = m_6 \times 92 + (0.2 - m_6)\, 292 \qquad \text{... } (\because h_5 = h_4)$$

Substituting the value of h_4 in this expression, we have

$$0.2(263 + 1070\, m_6) = 92\, m_6 + (0.2 - m_6)\, 292$$
$$52.6 + 214\, m_6 = 92\, m_6 + 58.4 - 292\, m_6$$
$$214\, m_6 - 92\, m_6 + 292\, m_6 = 58.4 - 52.6 = 5.8$$
$$\therefore \quad m_6 = 0.014 \text{ kg } \textbf{Ans.}$$

Temperature of air before throttling

Let t_4 = Temperature of air before throttling.

Substituting the value of m_6 in equation (*iv*),

$$h_4 = 263 + 1070 \times 0.014 = 277.98 \text{ kJ/kg}$$

From the *T-s* diagram for air, we find that the temperature of air corresponding to 277.98 kJ/kg is

$$t_4 = -100°\text{C } \textbf{Ans.}$$

14.11 Liquefaction of Hydrogen

The hydrogen is the most difficult gas to liquefy because of its extremely low liquefaction temperature. The schematic arrangement of equipment for liquefying hydrogen is shown in Fig. 14.17. In this system, pure hydrogen gas at a pressure about 100 atmospheres and 27°C from the compressor is precooled in two heat exchangers *A* and *B*. In heat exchanger *A*, the incoming high pressure hydrogen is cooled by the outgoing low pressure hydrogen while in heat exchanger *B*, it is cooled by nitrogen. The high pressure hydrogen gas from both the heat exchangers is passed through a third heat exchanger *C* where the hydrogen gas is further cooled to about –207°C by nitrogen boiling under reduced pressure.

Liquefaction of hydrogen.

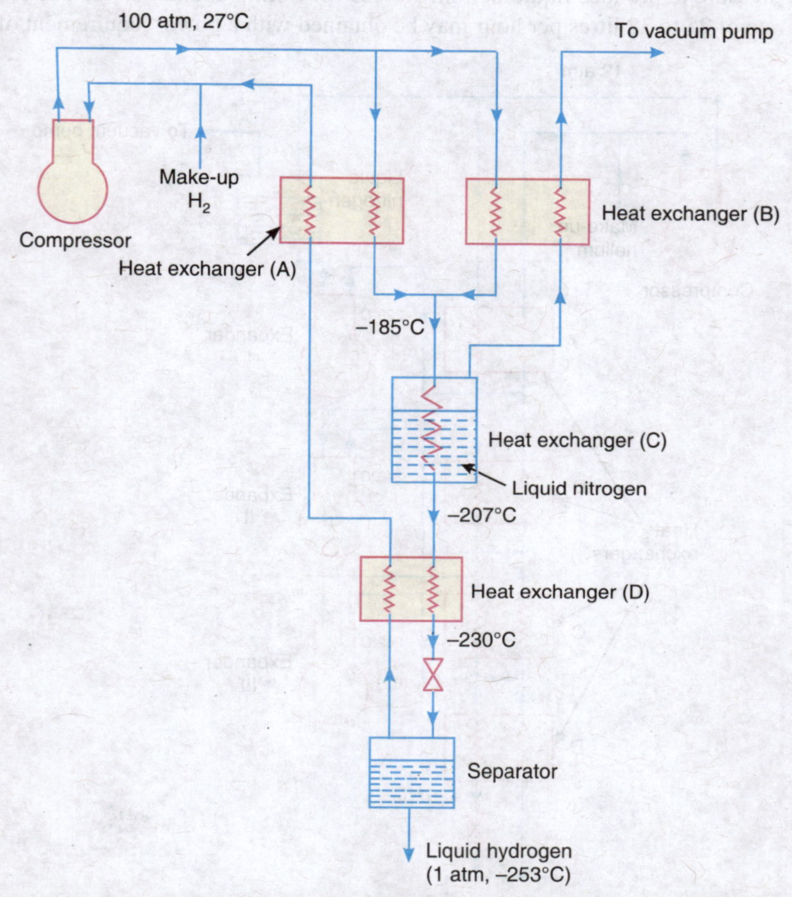

Fig. 14.17. Schematic diagram for hydrogen liquefaction.

This hydrogen gas is further cooled to about –230°C in the fourth heat exchanger *D* by the low pressure hydrogen gas returning from the separator. The liquid hydrogen is produced by throttling the hydrogen gas from the heat exchanger *D* to atmospheric pressure.

14.12 Liquefaction of Helium

The helium is the most difficult of all gases to liquefy. At atmospheric pressure, it boils at approximately –269°C. Its maximum inversion temperature is about –234°C. The helium was liquefied by H.K. Onnes of the University of Leiden in 1908. It may be liquefied by an arrangement similar to hydrogen liquefier (Fig. 14.17) where both liquid nitrogen and liquid hydrogen are used for precooling. The disadvantages of this system include the high cost and hazardous nature of liquid hydrogen.

The helium may also be liquefied by using Claude principle, where expanders are used for producing refrigeration. Fig. 14.18 shows a system developed by S.C. Collins for liquefaction of helium. In this system, helium at a pressure of approximately 12 atmospheres is supplied to the liquefier by a four-stage compressor. The part of the helium is precooled by liquid nitrogen. By using the combination of heat exchangers and expanders, as shown in Fig. 14.18, the high pressure helium gas may be cooled to about –257°C. The helium gas thus obtained is throttled to atmospheric pressure to produce liquid helium at –269°C. Collins found that for one such plant, a liquefaction rate of 25 to 32 litres per hour may be obtained with a power requirement of 45 kW.

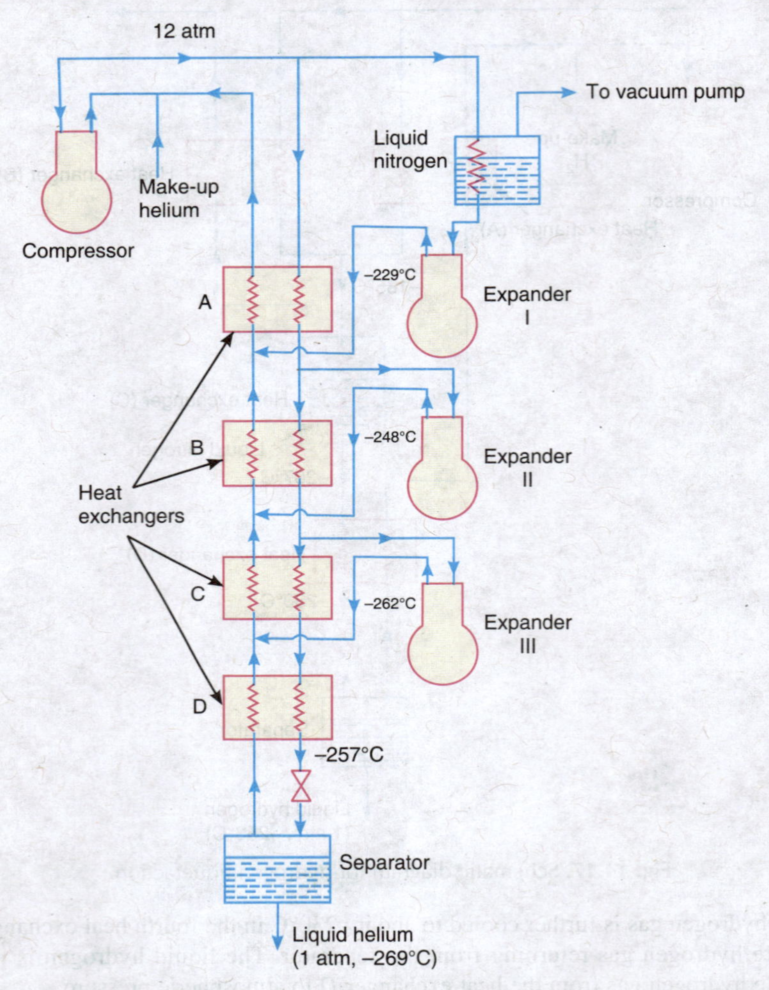

Fig. 14.18. Schematic diagram for helium liquefaction.

14.13 Production of Low Temperature by Adiabatic Demagnetisation of a Paramagnetic Salt

It is possible to attain a temperature of about 0.8 K (–272.2°C) through the lowering of pressure over liquid helium. In 1933, Giauque and Debye proposed the adiabatic demagnetisation of magnetic salts for attaining the lower temperatures. The lowest recorded temperature as low as 0.001 K (*i.e.* approaching to absolute zero) may be obtained by adiabatic demagnetisation of certain paramagnetic salts previously cooled by liquid helium and subjected to a strong magnetic field.

All substances may be divided into two classes with respect to their magnetic properties. Those substances which are repelled by a magnetic pole are called *diamagnetic* while those attracted by a magnetic pole (such as iron) are called *paramagnetic*. Some of the paramagnetic salts such as gadolinium sulphate are found to be best suited for obtaining low temperatures by their adiabatic demagnetisation. If such salts are precooled to a very low temperature so that any thermal motion of the molecules will be at a minimum, the molecules may be considered as tiny magnets. When the salt is not magnetised, the molecules are oriented in a random manner such that the magnetic forces are in balance. When such a substance is exposed to a strong magnetic field, the molecules attempt to align themselves in the direction of the magnetic field.

The realignment (or demagnetisation) of molecules requires work. This work is converted into heat and causes a temperature rise unless the heat is removed by some form of cooling. When the magnetic field is removed, the molecules readjust their positions to the original random arrangement. Such readjustment requires that the salt perform work. In the absence of external heat exchange, the internal energy of the salt decreases. Consequently the salt must cool itself.

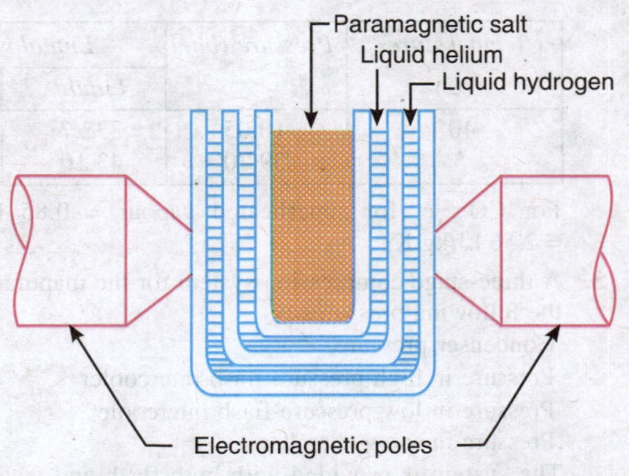

Fig. 14.19. Adiabatic demagnetisation apparatus.

Fig. 14.19 shows a schematic arrangement for adiabatic demagnetisation of a paramagnetic salt. The inner chamber containing the salt specimen is initially filled with gaseous helium. This chamber is surrounded by a bath of liquid helium, which, in turn, is surrounded by a bath of liquid hydrogen. In actual practice, the procedure of cooling magnetically is accomplished in the following four steps:

1. First of all, the paramagnetic salt is cooled to about 0.8 K (–272.2°C) by surrounding it with liquid helium boiling under reduced pressure.

2. The salt is exposed to a strong magnetic field of about 25 000 gauss. The heat produced by magnetisation of the salt is transferred to the liquid helium without change in salt temperature.

3. The inner chamber containing the salt is evacuated of gaseous helium and the substance is thermally isolated at a temperature below 1 K under the stress of a strong magnetic field.

4. When the magnetic field is released, the salt temperature decreases in an almost perfectly isentropic way. The temperature of the salt as low as 0.001 K may be attained.

Such temperatures (*i.e.* close to absoulte zero) cannot be measured by ordinary means but can be calculated by measuring the magnetic susceptibility. According to Curie's law, the magnetic susceptibility (x) is inversely proportional to the absolute temperature (T). Mathematically,

$$x \propto \frac{1}{T} \quad \text{or} \quad x = \frac{C}{T}$$

where C is a constant.

EXERCISES

1. A cascade refrigeration system is required to supply 20 tonnes of refrigeration at an evaporator temperature of –50°C and a condenser temperature of 40°C. The load at –50°C is absorbed by a unit using R-22 as the refrigerant and is rejected to a cascade condenser at –10°C. This cascade condenser is cooled by a unit using R-12 as the refrigerant and operating between –25°C evaporating temperature and 40°C condensing temperature. The refrigerant leaving the R-12 condenser is subcooled by 5°C and there is no subcooling of R-22 refrigerant. The gas leaving both the evaporators is dry and saturated and the compressions are isentropic. Neglecting losses, find (*a*) compression ratio for each unit ; (*b*) quantity of refrigerant circulated per minute for each unit ; (*c*) C.O.P. of the whole system ; and (*d*) theoretical power required to drive the system.

[Ans. 5.5, 7.75 ; 21.7 kg/min, 47.7 kg/min ; 1.6 ; 43.1 kW]

2. The following data refer to a cascade refrigeration system using NH_3 and CO_2. The evaporating and condensing temperatures of CO_2 are – 40°C and 5°C respectively. The evaporating and condensing temperatures of NH_3 are – 10°C and 35°C respectively. In both the circuits, the liquid leaving the condenser is saturated, the vapour leaving the evaporator is dry and saturated and the compression is isentropic. Calculate the mass flow rates of NH_3 and CO_2 and the C.O.P. For CO_2, the following data may be used.

Temperature (°C)	*Pressure (bar)*	*Enthalpy (kJ/kg)*		*Entropy (kJ/kg K)*	
		Liquid	*Vapour*	*Liquid*	*Vapour*
– 40	10.05	332.7	652.8	3.8531	5.2262
5	39.70	43.10	649.8	4.2231	5.0097

For CO_2, c_p for superheated vapour = 0.85 kJ/kg K. For NH_3, c_p for superheated vapour = 2.56 kJ/kg K. **[Ans.** 0.252 kg / min ; 0.95 kg/min ; 1.776]

3. A three-stage compression system for the manufacture of 500 kg per hour of dry ice operates under the following pressures :

Condenser pressure = 64.42 bar
Pressure in high pressure flash intercooler = 28 bar
Pressure in low pressure flash intercooler = 6.86 bar
Pressure in snow chamber = 1.013 bar

The system is provided with both flash and water intercoolers between the stages. The make-up carbon dioxide gas is supplied at 1.013 bar and 25°C. The gas leaving both the water intercoolers is at 30°C. Find the theoretical power required to run the system. **[Ans.** 79.84 kW]

4. A three-stage compression system for the manufacture of solid CO_2 operates with the following pressures :

Condenser pressure = 68 bar
Intermediate stage compressor discharge pressure = 30 bar
Low stage compressor discharge pressure = 6 bar
Snow or solid CO_2 chamber pressure = 1.013 bar

Both flash and water intercoolers are used between the stages. The make-up CO_2 enters at 25°C and 1.013 bar and the gas leaving the water intercoolers is at 20°C. Find (*a*) theoretical power per tonne of solid CO_2 per hour ; and (*b*) theoretical piston displacement in m³/min for each tonne of solid CO_2 per hour. **[Ans.** 162.8 kW, 18.523 m³/min]

5. Dry air at 30°C and 1 bar is to be liquefied by simple Linde system. The air is compressed isothermally at 30°C and 200 bar. If the make-up air is supplied at 30°C and 1 bar, find the mass of air liquefied per kg of air compressed. **[Ans.** 0.08 kg]

QUESTIONS

1. What do you understand by 'cryogenics' ?
2. Discuss the limitations of vapour compression refrigeration system for production of low temperature.
3. Sketch and explain a cascade refrigeration system. Draw cascade refrigeration cycle on temperature - entropy and pressure - enthalpy diagrams.
4. What is meant by overlap temperature and intermediate temperature ?
5. Write down expressions for C.O.P. of low temperature (LT) side and high temperature (HT) side of cascade in terms of evaporator, condenser, intermediate and overlap temperatures assuming reverse Carnot cycles. Optimize the product of $(C.O.P.)_{LT}$ and $(C.O.P.)_{HT}$ to determine the optimum intermediate temperature.
6. Explain the working of a system used for the production of dry ice.
7. Explain the importance of Joule-Thomson coefficient and inversion temperature when operating a system for liquefaction of gases.
8. Write a short note on 'air liquefaction system'.
9. Describe the method adopted for liquefaction of hydrogen and helium.
10. Discuss the arrangement used for producing low temperatures by adiabatic demagnetisation of a paramagnetic salt.

OBJECTIVE TYPE QUESTIONS

1. The liquid oxygen boils at
   (*a*) 20.1 K (*b*) 51.1 K (*c*) 70.4 K (*d*) 90.2 K
2. The triple point for carbon dioxide at 5.18 bar is
   (*a*) –20.4°C (*b*) – 40.6°C (*c*) – 56.6°C (*d*) –76.4°C
3. The solidification of liquid carbon dioxide is done by expanding it to a pressure .... that of its triple point pressure.
   (*a*) above (*b*) below
4. In the simple Linde system, the high pressure air from the compressor is cooled in a heat exchanger to a temperature of
   (*a*) – 53.6°C (*b*) – 72.8°C (*c*) –106.7°C (*d*) –138.5°C
5. The Claude air liquefaction system is .... efficient than the Linde system.
   (*a*) more (*b*) less

ANSWERS

1. (*d*) 2. (*c*) 3. (*b*) 4. (*c*) 5. (*a*)

CHAPTER

15

Steam Jet Refrigeration System

1. Introduction.
2. Principle of Steam Jet Refrigeration System.
3. Water as a Refrigerant.
4. Working of Steam Jet Refrigeration System.
5. Steam Ejector.
6. Analysis of Steam Jet Refrigeration System.
7. Efficiencies used in Steam Jet Refrigeration System.
8. Mass of Motive Steam Required.
9. Advantages and Disadvantages of Steam Jet Refrigeration System.

15.1 Introduction

The steam jet refrigeration system (also known as ejector refrigeration system) is one of the oldest methods of producing refrigerating effect. The basic components of this system are an evaporator, a compression device, a condenser, and a refrigerant control device. This system employs a steam ejector or booster (instead of mechanical compressor) to compress the refrigerant to the required condenser pressure level. In this system, water is used as the refrigerant. Since the freezing point of water is 0°C, therefore, it cannot be used for applications below 0°C. The steam jet refrigeration system is widely used in food processing plants for precooling of vegetables and concentrating fruit juices, gas plants, paper mills, breweries etc.

15.2 Principle of Steam Jet Refrigeration System

The boiling point of a liquid changes with change in external pressure. In normal conditions, pressure exerted on the surface of any liquid is the atmospheric pressure. If this atmospheric pressure is reduced on the surface of a liquid, by some means, then the liquid will start boiling at lower temperature, because of reduced pressure. This basic principle of boiling of liquid at lower temperature by reducing the pressure on its surface is used in steam jet refrigeration system.

The boiling point of pure water at standard atmospheric pressure of 760 mm of Hg (1.013 bar) is 100°C. It may be noted that water boils at 12°C if the pressure on the surface of water is kept at 0.014 bar and at 7°C if the pressure on the surface of water is 0.01 bar. The reduced pressure on the surface of water is maintained by throttling the steam through the jets or nozzles.

15.3 Water as a Refrigerant

We have already discussed that water is used as a refrigerant in steam jet refrigeration system, and the cooling effect is produced by the continuous vaporisation of a part of water in the evaporator at reduced pressure. When water is to be chilled from 10°C to 5°C, at least one per cent of water flowing through the evaporator must be vaporised.

Let m = Mass of water in the evaporator in kg,
s = Specific heat of water = 4.2 kJ/kg °C,
h_{fg} = Latent heat of vaporisation of water at some reduced pressure in kJ/kg, and
q_R = Heat removed from the water in kJ/kg.

Consider that one per cent of m kg of water is evaporated by throttling the steam through the nozzle at some reduced pressure (say at a pressure of 0.085 bar). Thus, the total heat removed by this one per cent of evaporated water

$$= \frac{m}{100} \times h_{fg}$$

∴ Fall in temperature of the remaining water will be

$$\Delta T_F = \frac{q_R}{\left(m - \frac{m}{100}\right)s} = \frac{h_{fg}}{\left(m - \frac{m}{100}\right)s} \quad \ldots (\because q_R = h_{fg})$$

Now for a mass, m = 100 kg and h_{fg} = 2400.5 kJ/kg at some reduced pressure (at a pressure of 0.085 bar, from steam tables),

$$\Delta T_F = \frac{2400.5}{\left(100 - \frac{100}{100}\right)4.2} = 5.77°C$$

It means that if one kg of water is removed by boiling on reducing some pressure, 2400.5 kJ/kg of heat is removed from the water which is required for its evaporation and water becomes colder by 5.77°C. This removal of heat is a continuous process. By evaporating one more kg of water, the remaining water will become colder by another 5.77°C. In this way, the temperature of water at some point will become 0°C or less than 0°C due to which the remaining water will freeze. Due to this limitation, the steam jet refrigeration is used only where the lowest required temperature for air-conditioning is above the freezing point of water *i.e.* 0°C. Hence it is confined to high temperature water cooling and air-conditioning.

15.4 Working of Steam Jet Refrigeration System

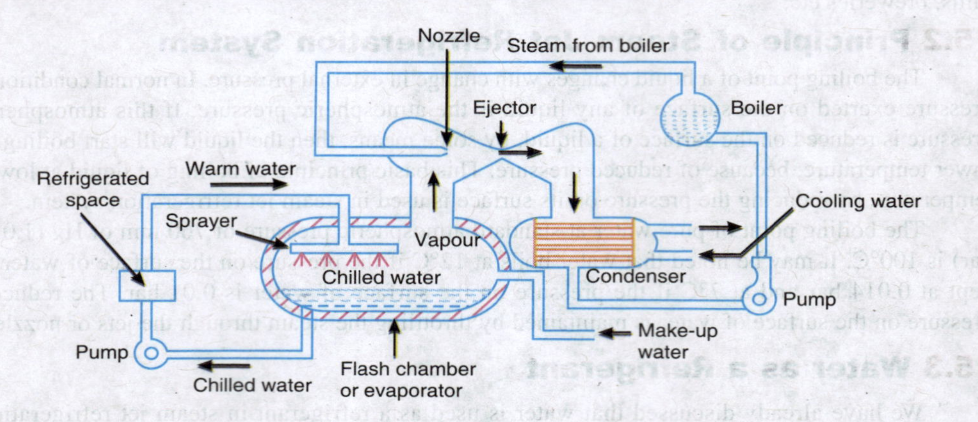

Fig. 15.1. Steam jet refrigeration system.

The main components of the steam jet refrigeration system, as shown in Fig. 15.1, are the flash chamber or evaporator, steam nozzles, ejector and condenser.

The flash chamber or evaporator is a large vessel and is heavily insulated to avoid the rise in temperature of water due to high ambient temperature. It is fitted with perforated pipes for spraying water. The warm water coming out of the refrigerated space is sprayed into the flash water chamber where some of which is converted into vapours after absorbing the latent heat, thereby cooling the rest of water.

The high pressure steam from the boiler is passed through the steam nozzles thereby increasing its velocity. This high velocity steam in the ejector would entrain the water vapours from the flash chamber which would result in further formation of vapours. The mixture of steam and water vapour passes through the venturi-tube of the ejector and gets compressed. The temperature and pressure of the mixture rises considerably and fed to the water cooled condenser where it gets condensed. The condensate is again fed to the boiler as feed water. A constant water level is maintained in the flash chamber and any loss of water due to evaporation is made up from the make-up water line.

Notes : 1. We have already discussed that the steam jet refrigeration system is based on the reduced boiling point at reduced pressure theory. In order to maintain the required reduced pressure in the flash chamber, the water vapours produced should be removed as early as possible. The removed vapour should then be compressed to the point where it condenses at a temperature above the temperature of the medium available for such purpose.

2. When the chilled water from the flash chamber is directly used in the form of spray for cooling the air, then it is known as *open cooling system.*

3. When the chilled water is passed through the coils and it does not come in contact with the air to be cooled, then it is known as *closed cooling system.*

15.5 Steam Ejector

The steam ejector is one of the important components of a steam jet refrigeration system. It is used to compress the water vapours coming out of the flash chamber. It uses the energy of fast moving jet of steam to entrain the vapours from the flash chamber and then compress it. The essential components of a steam ejector are shown in Fig. 15.2.

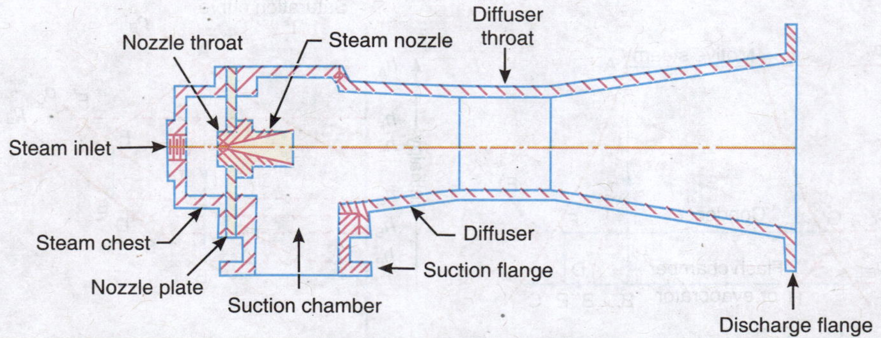

Fig. 15.2. Steam ejector.

The high pressure steam from the boiler (generally called *primary fluid* or *motive steam*) expands while flowing through the convergent divergent nozzle. The expansion causes a very low pressure and increases steam velocity. The steam attains very high velocities in the range of 1000 m/s to 1350 m/s. The nozzles are designed for lowest operating pressure ratio between nozzle throat and exit. The nozzle pressure ratio of less than 200 are undesirable because of poor ejector efficiency when operating at low steam pressure.

Multi-stage steam ejector.

The water vapours from the flash chamber are entrained by the high velocity steam and both are mixed in the mixing section at constant pressure. The mean velocity of the mixture will be supersonic, after the mixing is complete. This supersonic steam gets a normal shock, in the constant area throat of the diffuser. This results in the rise of pressure and subsonic flow. The function of the diverging portion of the diffuser is to recover the velocity head as pressure head by gradually reducing the velocity.

15.6 Analysis of Steam Jet Refrigeration System

The temperature-entropy (*T-s*) and enthalpy-entropy (*h-s*) diagrams for a steam jet refrigeration system are shown in Fig. 15.3 (*a*) and (*b*) respectively.

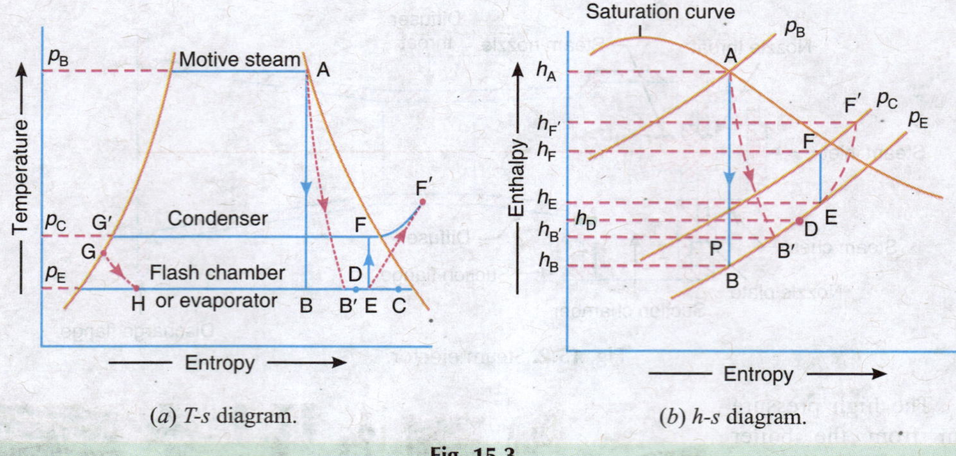

(*a*) *T-s* diagram. (*b*) *h-s* diagram.

Fig. 15.3

The point *A* represents the initial condition of the motive steam before passing through the nozzle and the point *B* is the final condition of the steam, assuming isentropic expansion. The point *C* represents the initial condition of the water vapour in the flash chamber or evaporator and the point *E* is the condition of the mixture of high velocity steam from the nozzle and the entrained water vapour before compression. Assuming isentropic compression, the final condition of the mixture discharged to the condenser is represented by point *F*. The condition of motive steam just before mixing with the water vapour is shown at point *D*. The make-up water is supplied at point *G* whose temperature is slightly lower than the condenser temperature and is throttled to point *H* in the flash chamber.

Note : Due to certain unavoidable losses in the expansion and compression, the actual expansion of motive steam is represented by *AB* ′ and the actual compression of the mixture of motive steam and water vapour is represented by *EF* ′.

15.7 Efficiencies used in Steam Jet Refrigeration System

The various efficiencies used in steam jet refrigeration system are discussed below :

1. *Nozzle efficiency.* It is defined as the ratio of actual enthalpy drop to the isentropic enthalpy drop of the motive steam passing through the nozzle. Mathematically, nozzle efficiency,

$$\eta_N = \frac{\text{Actual enthalpy drop}}{\text{Isentropic enthalpy drop}} = \frac{AP}{AB} = \frac{h_A - h_{B'}}{h_A - h_B}$$

... (Refer Fig. 15.3)

The nozzle efficiency may vary from 85 to 90 per cent.

2. *Entrainment efficiency.* The water vapours formed in the flash chamber or evaporator comes out with a very low velocity as compared to the velocity of the steam (*V*) coming out of the nozzle which is given by

$$^{*}V = \sqrt{2000\,(h_A - h_{B'})} = 44.72\sqrt{h_A - h_{B'}}$$

The expression $(h_A - h_B')$ represents the kinetic energy of the motive steam. This kinetic energy gives the required momentum to the water vapours coming out of the flash chamber or evaporator. The process of giving the momentum to the water vapour formed in the flash chamber by the high velocity steam is called *entrainment of vapour*. During the entrainment of water vapour

* For further details, refer Authors' popular book on 'A Textbook of Thermal Engineering'.

from the flash chamber, the motive steam loses some of its kinetic energy. This process of entrainment is inefficient and part of the original motive force available for compression is reduced. This is taken into consideration by a factor known as entrainment efficiency. Mathematically, entrainment efficiency,

$$\eta_E = \frac{h_A - h_D}{h_A - h_{B'}}$$

The entrainment efficiency may be taken as 65 per cent.

3. *Compression efficiency.* It is defined as the ratio of the isentropic enthalpy increase to the actual enthalpy increase required for the compression of the mixture of motive steam and the water vapours, in the diffuser. Mathematically, compression efficiency,

$$\eta_C = \frac{\text{Isentropic enthalpy increase}}{\text{Actual enthalpy increase}} = \frac{h_F - h_E}{h_{F'} - h_E}$$

The compression efficiency may be taken as 75 to 80 per cent.

Note : The compression efficiency is also known as *diffuser efficiency*.

Corrosion resistant filter.

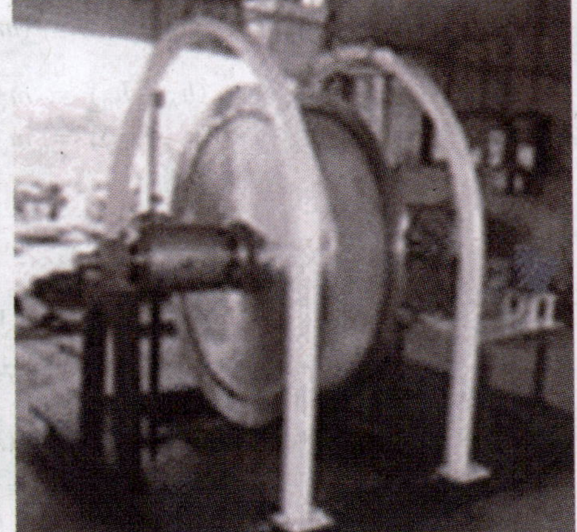

Rotary vacuum filter.

15.8 Mass of Motive Steam Required

According to the law of conservation of energy, the available energy for compression must be equal to the energy required for compression.

Let m_s = Mass of motive steam supplied in kg/min,

m_v = Mass of water vapours formed from the flash chamber or evaporator in kg/min,

m = Mass of the mixture for compression in kg/min = $m_s + m_v$

We know that available energy for compression

$$= m_s (h_A - h_D) \quad \text{... (i)}$$

and energy required for compression

$$= m (h_{F'} - h_E) = (m_s + m_v) (h_{F'} - h_E) \quad \text{... (ii)}$$

Now according to law of conservation of energy,

$$m_s (h_A - h_D) = (m_s + m_v) (h_{F'} - h_E) \quad \text{... (iii)}$$

We have already discussed that the nozzle efficiency,

$$\eta_N = \frac{h_A - h_{B'}}{h_A - h_B} \quad \text{or} \quad h_A - h_{B'} = \eta_N (h_A - h_B) \qquad \ldots (iv)$$

Entrainment efficiency,

$$\eta_E = \frac{h_A - h_D}{h_A - h_{B'}} \quad \text{or} \quad h_A - h_D = \eta_E (h_A - h_{B'}) \qquad \ldots (v)$$

and compression efficiency,

$$\eta_C = \frac{h_F - h_E}{h_{F'} - h_E} \quad \text{or} \quad h_{F'} - h_E = \frac{h_F - h_E}{\eta_C} \qquad \ldots (vi)$$

Substituting the value of $(h_A - h_D)$ and $(h_{F'} - h_E)$ from equations (v) and (vi) in equation (iii), we have

$$m_s \times \eta_E (h_A - h_{B'}) = (m_s + m_v)\left(\frac{h_F - h_E}{\eta_C}\right)$$

$$m_s \times \eta_E \times \eta_N (h_A - h_B) = (m_s + m_v)\left(\frac{h_F - h_E}{\eta_C}\right) \qquad \ldots \text{[From equation } (iv)]$$

$$m_s \times \eta_E \times \eta_N \times \eta_C (h_A - h_B) = m_s (h_F - h_E) + m_v (h_F - h_E)$$

$$m_s [(h_A - h_B)\, \eta_N\, \eta_E\, \eta_C - (h_F - h_E)] = m_v (h_F - h_E)$$

or

$$\frac{m_s}{m_v} = \frac{(h_F - h_E)}{(h_A - h_B)\eta_N \eta_E \eta_C - (h_F - h_E)}$$

where $\frac{m_s}{m_v}$ = Mass of motive steam required per kg of water vapour produced in the flash chamber.

We have discussed in Art.15.6 that the make-up water is supplied at point G whose temperature is slightly lower than the condenser temperature and is throttled to point H in the flash chamber and leaves it corresponding to the condition at point C. Since the enthalpy of water at point G is equal to the enthalpy of water at point H, therefore for each kg of water vapour formed, heat absorbed is $(h_C - h_{fg})$ kJ/kg. In other words, net refrigerating effect,

$$R_E = m_v (h_C - h_{fg}) \text{ kJ/min}$$

If Q tonnes of refrigeration is the refrigerating load, then the heat absorbed or net refrigerating effect,

$$R_E = 210\, Q \text{ kJ/min}$$

From the above expressions, we find that the mass of water vapour formed,

$$m_v = \frac{210\, Q}{h_C - h_{fg}} \text{ kg/min}$$

Since one kg of water vapour requires m_s kg of motive steam, therefore,

Mass of motive steam required per Q tonne of refrigerating load

= Mass of water vapour per minute × Motive steam required per kg of vapour

$$= m_v \times m_s$$

$$= \frac{210\, Q}{h_C - h_{fg}} \times m_s$$

Note : The coefficient of performance of the system is given by

$$\text{C.O.P.} = \frac{m_v (h_C - h_{fg})}{m_s (h_A - h_{fg})}$$

15.9 Advantages and Disadvantages of Steam Jet Refrigeration System

Following are the advantages and disadvantages of a steam jet refrigeration system :

Advantages

1. It is simple in construction and rigidly designed.
2. It is a vibration-free system as pumps are the only moving parts.
3. It has low maintenance cost, low production cost and high reliability.
4. It has relatively less plant mass (kg / TR). Hence, there are now a number of air-conditioning applications ranging up to 300 TR in capacity as well as many industrial applications of even larger size.
5. It uses water as a refrigerant. Water is very safe to use as it is non-poisonous and non-inflammable.
6. This system has an ability to adjust quickly to load variations.
7. The running cost of this system is quite low.

Disadvantages

1. The system is not suitable for water temperature below 4°C.
2. For proper functioning of this system, maintenance of high vacuum in the evaporator is necessary. This is done by direct vaporisation to produce chilled water which is usually limited as tremendous volume of vapour is to be handled.

Example 15.1. *A steam ejector refrigeration system is supplied with motive steam at 7 bar saturated with the water in the flash chamber at 4.5°C. The make-up water is supplied to the cooling system at 18°C and the condenser is operated at 0.058 bar.The nozzle efficiency is 88%, the entrainment efficiency is 65% and the compression efficiency is 80%. The quality of steam and flash vapour at the beginning of compression is 92%.*

Determine : 1. mass of motive steam required per kg of flash vapour ; 2. quality of vapour flashed from the flash chamber ; 3. refrigerating effect per kg of flash vapour ; 4. mass of motive steam required per hour per tonne of refrigeration ; 5. volume of vapour removed from the flash chamber per hour per tonne of refrigeration ; and 6. coefficient of performance of the system.

Solution. Given : $p_B = 7$ bar ; $t_w = 4.5°C$; $t_{mw} = 18°C$; $p_C = 0.058$ bar ; $\eta_N = 88\% = 0.88$; $\eta_E = 65\% = 0.65$; $\eta_C = 80\% = 0.8$; $x_E = 92\% = 0.92$

The *T-s* and *h-s* diagrams for the steam ejector refrigeration system is shown in Fig. 15.4.

From steam tables of dry saturated steam, corresponding to a pressure of 7 bar, we find that

$h_A = 2762$ kJ/kg ; $s_A = 6.705$ kJ/kg K ; $t_A = 165°C$

and corresponding to a temperature of *4.5°C, we find that

$h_{fB} = 18.9$ kJ/kg ; $h_{fgB} = 2490.9$ kJ/kg ; $s_{fB} = 0.0685$ kJ/kg K ; $s_{fgB} = 8.9715$ kJ/kg K

First of all, let us find the dryness fraction of the steam at point *B* (*i.e.* x_B). We know that for isentropic expansion *AB*,

* From steam tables, pressure corresponding to 4.5°C is 0.008 43 bar.

Entropy before expansion (s_A)

$$= \text{Entropy after expansion } (s_B)$$

or

$$6.705 = s_{fB} + x_B \times s_{fgB} = 0.0685 + x_B \times 8.9715$$

$$\therefore \quad x_B = \frac{6.705 - 0.0685}{8.9715} = 0.74$$

and enthalpy at B,

$$h_B = h_{fB} + x_B \times h_{fgB} = 18.9 + 0.74 \times 2490.9 = 1862.16 \text{ kJ/kg}$$

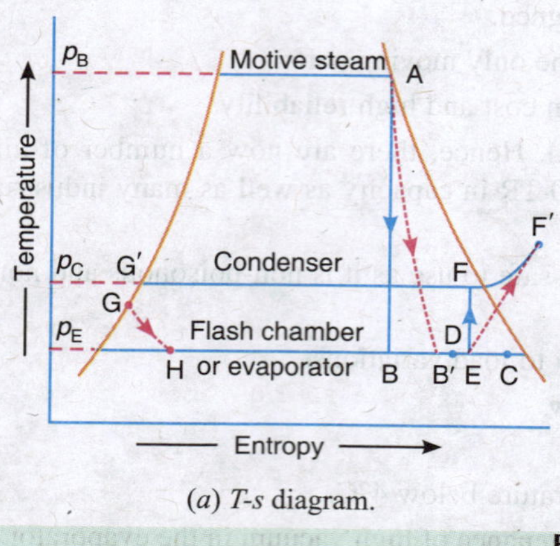

(a) T-s diagram.

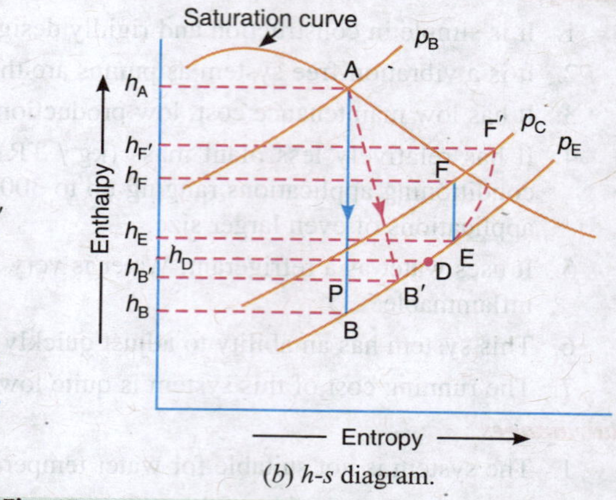

(b) h-s diagram.

Fig. 15.4

We know that nozzle efficiency (η_N),

$$0.88 = \frac{h_A - h_{B'}}{h_A - h_B} = \frac{2762 - h_{B'}}{2762 - 1862.16}$$

$$\therefore \quad h_{B'} = 2762 - 0.88\,(2762 - 1862.16) = 1970.14 \text{ kJ/kg}$$

Now let us find the dryness fraction of steam at point B' (*i.e.* $x_{B'}$). Since the points B, B', D and E lie on the same pressure line (corresponding to 4.5°C), therefore

$$h_{fB} = h_{fB'} = h_{fD} = h_{fE} = 18.9 \text{ kJ/kg}$$

and

$$h_{fgB} = h_{fgB'} = h_{fgD} = h_{fgE} = 2490.9 \text{ kJ/kg}$$

We know that enthalpy at B',

$$h_{B'} = h_{fB'} + x_{B'} \times h_{fgB'}$$

$$1970.14 = 18.9 + x_{B'} \times 2490.9$$

$$\therefore \quad x_{B'} = \frac{1970.14 - 18.9}{2490.9} = 0.78$$

Let h_D = Enthalpy of steam at D, and

x_D = Dryness fraction of steam at D.

We know that entrainment efficiency (η_E),

$$0.65 = \frac{h_A - h_D}{h_A - h_{B'}} = \frac{2762 - h_D}{2762 - 1970.14}$$

Multiple-effect evaporator.

$$\therefore \quad h_D = 2762 - 0.65\,(2762 - 1970.14) = 2247.3 \text{ kJ/kg}$$

We also know that enthalpy at point D (h_D),

$$2247.3 = h_{fD} + x_D \times h_{fgD} = 18.9 + x_D \times 2490.9$$

$$\therefore \quad x_D = \frac{2247.3 - 18.9}{2490.9} = 0.894$$

Enthalpy at point E,

$$h_E = h_{fE} + x_E \times h_{fgE} = 18.9 + 0.92 \times 2490.9$$

$$= 2310.5 \text{ kJ/kg} \qquad \ldots (\because \text{ It is given that } x_E = 0.92)$$

Now let us find the dryness fraction of the mixture of the motive steam and water vapour after isentropic compression at point F.

Let x_F = Dryness fraction at point F.

We know that entropy at point E,

$$s_E = s_{fE} + x_E \times s_{fgE} = 0.0685 + 0.92 \times 8.9715$$

$$= 8.3223 \text{ kJ/kg K} \qquad \ldots (\because s_{fE} = s_{fB} \text{ and } s_{fgE} = s_{fgB})$$

From steam tables, corresponding to a condenser pressure of 0.058 bar, we find that

$h_{fF} = 148.86$ kJ/kg ; $h_{fgF} = 2417.5$ kJ/kg

$s_{fF} = 0.512$ kJ/kg K ; $s_{fgF} = 7.831$ kJ/kg K

Since the compression of the mixture is isentropic, therefore

Entropy before compression (s_E)

= Entropy after compression (s_F)

$$8.3223 = s_{fF} + x_F \times s_{fgF} = 0.512 + x_F \times 7.831$$

$$\therefore \quad x_F = \frac{8.3223 - 0.512}{7.831} = 0.997$$

We know that enthalpy at point F,

$$h_F = h_{fF} + x_F \times h_{fgF} = 148.86 + 0.997 \times 2417.5 = 2559.1 \text{ kJ/kg}$$

We also know that compression efficiency (η_C),

$$0.8 = \frac{h_F - h_E}{h_{F'} - h_E} = \frac{2559.1 - 2310.5}{h_{F'} - 2310.5}$$

$$\therefore \quad h_{F'} = \frac{2559.1 - 2310.5}{0.8} + 2310.5 = 2621.2 \text{ kJ/kg}$$

1. *Mass of motive steam required per kg of the flash vapour*

We know that mass of motive steam required per kg of the flash vapour,

$$\frac{m_s}{m_v} = \frac{h_F - h_E}{(h_A - h_B)\eta_N \eta_E \eta_C - (h_F - h_E)}$$

$$= \frac{2559.1 - 2310.5}{(2762 - 1862.16)0.88 \times 0.65 \times 0.8 - (2559.1 - 2310.5)}$$

$$= \frac{248.6}{411.8 - 248.6} = 1.523 \text{ kg/kg of flash vapour} \quad \textbf{Ans.}$$

2. *Quality of vapour flashed from the flash chamber*

Let x_C = Dryness fraction of the vapour flashed from the flash chamber.

First of all, let us find the enthalpy at point C. We know that

$$m_v\,h_C + m_s\,h_D = (m_s + m_v)\,h_E$$

$$h_C + \frac{m_s}{m_v} \times h_D = \left(\frac{m_s}{m_v} + 1\right) h_E$$

$$h_C + 1.523 \times 2247.3 = (1.523 + 1)\,2310.5$$

$$h_C + 3422.6 = 5829.4$$

$$\therefore \quad h_C = 2406.8 \text{ kJ/kg}$$

We also know that enthalpy at point C (h_C),

$$2406.8 = h_{fC} + x_C \times h_{fB} = 18.9 + x_C \times 2490.9$$

... ($\because$ $h_{fC} = h_{fB}$ and $h_{fgC} = h_{fgB}$)

$$\therefore \quad x_C = \frac{2406.8 - 18.9}{2490.9} = 0.96 \text{ Ans.}$$

3. ***Refrigerating effect per kg of flash vapour***

We know that refrigerating effect per kg of flash vapour,

$$R_E = h_C - h_{fG} = 2406.8 - 75.5 = 2331.3 \text{ kJ/kg}$$

... ($\because$ From steam tables, h_{fG} at 18°C = 75.5 kJ/kg)

4. ***Mass of motive steam required per hour per tonne of refrigeration***

We know that mass of motive steam required per hour per tonne of refrigeration

$$= \frac{210\,Q}{h_C - h_{fG}} \times \frac{m_s}{m_v} = \frac{210 \times 1}{2406.8 - 75.5} \times 1.523$$

$$= 0.133 \text{ kg/min/TR}$$ ... ($\because$ Q = 1 TR)

$$= 0.133 \times 60 = 7.98 \text{ kg/h/TR Ans.}$$

5. ***Volume of vapour removed from the flash chamber per hour per tonne of refrigeration***

We know that volume of vapour (per kg) removed from the flash chamber,

$$v_C = \text{Volume of liquid at } C + x_C \text{ (Volume of saturated vapour} - \text{Volume of liquid)}$$

$$= 1 + 0.95\,(152.22 - 1) = 144.66 \text{ m}^3\text{/kg}$$

. . . (From steam tables, volume of saturated vapour corresponding to 4.5°C = 152.22 m³/kg)

$\therefore$ Volume of vapour removed from the flash chamber per hour per tonne of refrigeration

$$= v_C \times \frac{210\,Q}{h_C - h_{fG}} \times 60$$

$$= 144.66 \times \frac{210 \times 1}{2406.8 - 75.5} \times 60 = 782 \text{ m}^3\text{/h/TR Ans.}$$

6. ***Coefficient of performance of the system***

From steam tables, corresponding to a condenser pressure of 0.058 bar, we find that enthalpy of liquid at point G',

$$h_{fG'} = 148.8 \text{ kJ/kg}$$

We know that coefficient of performance of the system,

$$\text{C.O.P.} = \frac{m_v\,(h_C - h_{fG})}{m_s\,(h_A - h_{fG'})} = \frac{1\,(2406.8 - 75.5)}{1.523\,(2762 - 148.8)} = 0.586 \text{ Ans.}$$

Example 15.2. *A thermal power plant of 2200 kW capacity is supplied steam from the boiler at 30 bar and 400°C. A refrigeration load of 150 tonnes is taken by a steam jet refrigeration system and to operate this system, steam is bled off from the steam turbine at 5 bar. The vacuum maintained in the condenser is 730 mm of Hg. The steam enters in thermo-compressor at 0.01 bar and with dryness fraction of 0.94. The make-up water enters into the flash chamber at 18°C. The various efficiencies may be taken as follows :*

Isentropic efficiency of the steam turbine = 90% ; nozzle efficiency = 90%; entrainment efficiency = 65% ; thermo-compressor efficiency = 65%.

Using the Mollier diagram, find the following :

1. Dryness fraction of steam leaving the flash chamber ;

2. Mass of motive steam bled off from the steam turbine ;

3. Mass of steam generated in the boiler ;

4. Mass of additional steam to be generated by the boiler due to refrigeration load ; and

5. Coefficient of performance of the system.

The same condenser is used for power turbine and refrigeration system.

Solution. Given : $P = 2200$ kW ; $p_1 = 30$ bar ; $t_1 = 400$°C ; $Q = 150$ TR ; $p_2 = p_2' = 5$ bar; $p_7 = p_7' = p_8 = p_8' = 760 - 730 =$ *30 mm of Hg = 0.04 bar ; $p_3 = p_3' = p_4 = p_5 = p_6 = 0.01$ bar ; $x_4 = 0.94$; $\eta_T = 90\% = 0.9$; $\eta_N = 90\% = 0.9$; $\eta_E = 65\% = 0.65$; $\eta_C = 65\% = 0.65$

The schematic diagram of a thermal power plant is shown in Fig. 15.5 and the various processes on the Mollier diagram are shown in Fig. 15.6. The point 1 represents the entry of steam at a pressure of 30 bar and 400°C from the boiler to the steam turbine. The process 1-2 represents the isentropic expansion whereas the process 1-2′ represents the actual expansion of steam in the turbine. From the Mollier diagram, we find that enthalpy at point 1 corresponding to 30 bar and 400°C is

$$h_1 = 3232 \text{ kJ/kg}$$

and enthalpy at point 2, $$h_2 = 2792 \text{ kJ/kg}$$

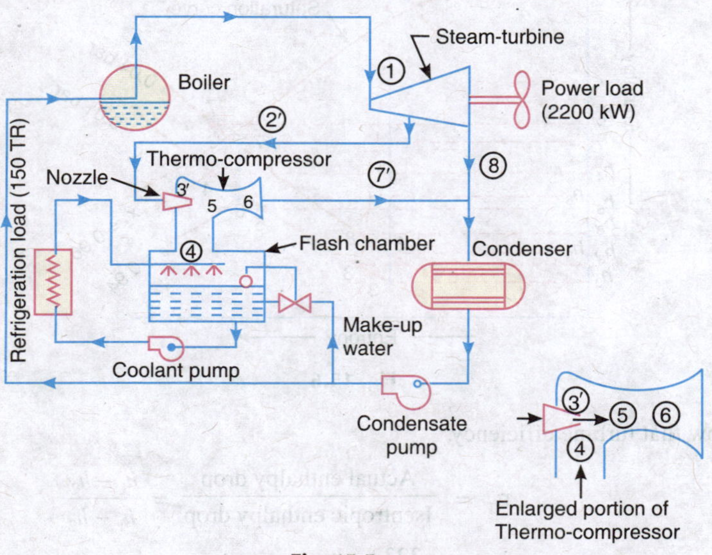

Fig. 15.5

* We know that condenser pressure = Atmospheric pressure – Vacuum pressure
= 760 – 730 = 30 mm of Hg
= 30 × 0.001 33 = 0.0399 or 0.04 bar

Thermal power plant.

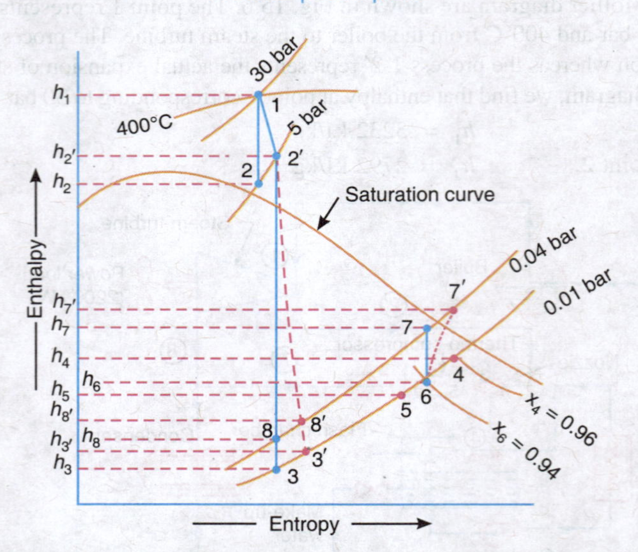

Fig. 15.6

We know that turbine efficiency,

$$\eta_T = \frac{\text{Actual enthalpy drop}}{\text{Isentropic enthalpy drop}} = \frac{h_1 - h_{2'}}{h_1 - h_2}$$

$$0.9 = \frac{3232 - h_{2'}}{3232 - 2792}$$

$$\therefore \quad h_{2'} = 3232 - 0.9\,(3232 - 2792) = 2836 \text{ kJ/kg}$$

The steam required for the steam jet refrigeration system is bled off from the turbine at point 2′, *i.e.* at 5 bar and enthalpy $h_{2'} = 2836$ kJ/kg. The process 2′-3 represents the isentropic expansion of steam whereas the process 2′-3′ represents the actual expansion of steam through the nozzle up to the flash chamber pressure of 0.01 bar. From the Mollier diagram, we find that enthalpy at point 3,

$$h_3 = 1968 \text{ kJ/kg}$$

We know that nozzle efficiency

$$\eta_N = \frac{\text{Actual enthalpy drop}}{\text{Isentropic enthalpy drop}} = \frac{h_{2'} - h_{3'}}{h_{2'} - h_3}$$

$$0.9 = \frac{2836 - h_{3'}}{2836 - 1968}$$

$$\therefore \quad h_{3'} = 2836 - 0.9\,(2836 - 1968) = 2055 \text{ kJ/kg}$$

The remaining steam expands in the turbine from point 2′ to a condenser pressure of 0.04 bar. The process 2′- 8 represents the isentropic expansion whereas the process 2′- 8′ represents the actual expansion of the remaining steam in the turbine. From the Mollier diagram, we find that enthalpy at point 8,

$$h_8 = 2115 \text{ kJ/kg}$$

We know that turbine efficiency,

$$\eta_T = \frac{\text{Actual enthalpy drop}}{\text{Isentropic enthalpy drop}} = \frac{h_{2'} - h_{8'}}{h_{2'} - h_8}$$

$$0.9 = \frac{2836 - h_{8'}}{2836 - 2115}$$

$$\therefore \quad h_{8'} = 2836 - 0.9\,(2836 - 2115) = 2187 \text{ kJ/kg}$$

The steam coming out from the nozzle at a very high velocity at point 5 entrains the water vapours from the flash chamber at a low velocity.

We know that entrainment efficiency,

$$\eta_E = \frac{h_{2'} - h_5}{h_{2'} - h_{3'}}$$

$$0.65 = \frac{2836 - h_5}{2836 - 2055}$$

$$\therefore \quad h_5 = 2836 - 0.65\,(2836 - 2055) = 2328 \text{ kJ/kg}$$

The steam at point 5 mixes with water vapours from the flash chamber at point 4 (which is unknown). The resulting condition of the mixture is represented by point 6 and its condition is given, *i.e.* pressure 0.01 bar and dryness fraction (x_6) is 0.94. From the Mollier diagram, we find that enthalpy at point 6,

$$h_6 = 2365 \text{ kJ/kg}$$

The steam at point 6 is now compressed in the thermo-compressor. The process 6-7 represents the isentropic compression whereas the process 6-7′ represents the actual compression. From the Mollier diagram, we find that enthalpy at point 7,

$$h_7 = 2545 \text{ kJ/kg}$$

We know that thermo-compressor or compression efficiency,

$$\eta_C = \frac{\text{Isentropic enthalpy increase}}{\text{Actual enthalpy increase}} = \frac{h_7 - h_6}{h_{7'} - h_6}$$

$$0.65 = \frac{2545 - 2365}{h_{7'} - 2365}$$

$$\therefore \quad h_{7'} = \frac{2545 - 2365}{0.65} + 2365 = 2642 \text{ kJ/kg}$$

1. *Dryness fraction of steam leaving the flash chamber*

We know that the mass of motive steam required per kg of flash vapour produced in the flash chamber,

$$\frac{m_s}{m_v} = \frac{h_7 - h_6}{(h_{2'} - h_3)\eta_N \eta_E \eta_C - (h_7 - h_6)}$$

$$= \frac{2545 - 2365}{(2836 - 1968)0.9 \times 0.65 \times 0.65 - (2545 - 2365)}$$

$$= \frac{180}{330 - 180} = 1.2 \text{ kg/kg of flash vapour}$$

For enthalpy balance,

$$m_v\, h_4 + m_s\, h_5 = (m_s + m_v)\, h_6$$

$$h_4 + \frac{m_s}{m_v} \times h_5 = \left(\frac{m_s}{m_v} + 1\right) h_6$$

$$h_4 + 1.2 \times 2328 = (1.2 + 1)\, 2365$$

$$\therefore \quad h_4 = 2409 \text{ kJ/kg}$$

Now mark point 4 on the Mollier diagram on a pressure line of 0.01 bar and enthalpy of $h_4 = 2409$ kJ/kg. From the Mollier diagram, we find that dryness fraction of steam leaving the flash chamber (*i.e.* at point 4),

$$x_4 = 0.96 \textbf{ Ans.}$$

2. *Mass of motive steam bled off from the steam turbine*

We know that refrigerating effect per kg of flash vapour,

$$R_E = h_4 - \text{Enthalpy of make-up water}$$

$$= 2409 - 4.2 \times 18 = 2333.4 \text{ kJ/kg of flash vapour}$$

and amount of flashed vapour per hour for 150 TR

$$= \frac{150 \times 210 \times 60}{2333.4} = 810 \text{ kg/h}$$

∴ Mass of motive steam bled off from the steam turbine

$$= \frac{m_s}{m_v} \times 810 = 1.2 \times 810 = 972 \text{ kg/h} \textbf{ Ans.}$$

3. *Mass of steam generated in the boiler*

Let m_B = Mass of steam generated in the boiler or mass of steam supplied to the power turbine.

$\therefore\ m_B(h_1 - h_{2'}) + (m_B - 972)(h_{2'} - h_{8'})$

$$= \text{Power in kW} \times 3600 \quad \text{... } (\because 1 \text{ kW} = 1 \text{kJ/s} = 1 \times 3600 \text{ kJ/h})$$

$$m_B(3232 - 2836) + (m_B - 972)(2836 - 2187) = 2200 \times 3600$$

$$396\, m_B + 649\, m_B - 630\,828 = 7920 \times 10^3$$

$$\therefore \quad m_B = 8183 \text{ kg / h} \textbf{ Ans.}$$

4. Mass of additional steam to be generated by the boiler due to refrigeration load

If the refrigeration load is not there, the mass of steam required for 2200 kW is given by

$$m_{B'} = \frac{2200 \times 3600}{h_1 - h_{8'}} = \frac{2200 \times 3600}{3232 - 2187} = 7579 \text{ kg/h}$$

$\therefore$ Mass of additional steam to be generated by the boiler due to refrigeration load

$$= m_B - m_{B'} = 8183 - 7579 = 604 \text{ kg/h} \textbf{ Ans.}$$

5. Coefficient of performance of the system

We know that enthalpy of steam supplied to the refrigeration system at point 2′,

$$h_{2'} = 2836 \text{ kJ/kg}$$

The recoverable heat from the steam is the enthalpy of water at 0.04 bar which is equal to 121.4 kJ/kg (from steam tables).

$\therefore$ Heat to be actually supplied per kg of steam for refrigeration purposes

$$= h_{2'} - \text{Enthalpy of water at 0.04 bar}$$

$$= 2836 - 121.4 = 2714.6 \text{ kJ/kg}$$

We know that coefficient of performance of the system,

$$\text{C.O.P.} = \frac{\text{Refrigerating load in kJ / h}}{\text{Mass of motive steam bled / h} \times \text{Heat supplied / kg of steam}}$$

$$= \frac{150 \times 210 \times 60}{972 \times 2714.6} = 0.716 \textbf{ Ans.}$$

EXERCISES

1. In an industrial plant, it is required to supply cooled water at 10°C, the temperature of the make-up and recirculated water being 25°C. For this purpose, a motive steam at 8.5 bar and a temperature of 190°C is supplied to a steam jet ejector. The quality of vapour leaving the flash chamber and entering the ejector is 0.95. The condenser pressure is 60 mm of Hg. It may be assumed that nozzle efficiency = 0.93; entrainment efficiency = 0.65 and compression efficiency = 0.75. The quality of steam and flash vapour at the beginning of compression is 0.918.

   Find : 1. mass of motive steam required to produce unit mass of flash vapour at 10°C ; 2. total motive steam required to produce 1 tonne of refrigeration per hour ; and 3. coefficient of performance of the system. **[Ans.** 1.5 kg ; 8.25 kg / h / TR ; 0.58]

2. A steam ejector water vapour refrigeration system is supplied with motive dry saturated steam at 6 bar. It expands through a nozzle down to flash chamber pressure meant to chill water at 5°C.

   Taking nozzle efficiency = 0.92 ; entrainment efficiency = 0.6 ; and compression efficiency = 0.75, obtain 1. amount of water to be evaporated, and motive steam per tonne of cooling ; 2. tonnage of the plant for 2 kg / s of evaporation of water ; and 3. C.O.P.

   Assume the condensing temperature to be 36°C and make-up water temperature as 30°C.

QUESTIONS

1. What is the principle of a steam jet refrigeration system ?
2. Explain, with the help of a neat sketch, the working of a steam jet refrigeration system.
3. Draw the temperature-entropy and enthalpy-entropy diagram of a steam jet refrigeration system and write the expressions for the following efficiencies:

   (*a*) Nozzle efficiency ;

   (*b*) Entrainment efficiency ; and

   (*c*) Compression efficiency.
4. Derive an expression for finding out the mass of motive steam required per kg of water vapour produced.
5. What are the advantages and disadvantages of steam jet refrigeration system over other types of refrigeration system ?

OBJECTIVE TYPE QUESTIONS

1. In a steam jet refrigeration system, the motive steam expands in

   (*a*) convergent nozzle (*b*) divergent nozzle

   (*c*) convergent - divergent nozzle (*d*) any nozzle
2. The velocity of steam at the exit of the nozzle is

   (*a*) supersonic (*b*) sonic (*c*) sub-sonic (*d*) none of these
3. The compression device used in a steam jet refrigeration system is a

   (*a*) vapour compressor (*b*) steam ejector

   (*c*) diffuser (*d*) liquid pump
4. The ratio of isentropic enthalpy increase to the actual enthalpy increase required for the compression of the motive steam and the water vapours is known as

   (*a*) nozzle efficiency (*b*) boiler efficiency

   (*c*) entrainment efficiency (*d*) compression efficiency
5. The coefficient of performance of the steam jet refrigeration system varies from

   (*a*) 0.5 to 0.8 (*b*) 2 to 4 (*c*) 5 to 10 (*d*) none of these

ANSWERS

1. (*c*)	2. (*a*)	3. (*b*)	4. (*d*)	5. (*a*)

CHAPTER

16 Psychrometry

16.1 Introduction

The psychrometry is that branch of engineering science which deals with the study of moist air *i.e.* dry air mixed with water vapour or humidity. It also includes the study of behaviour of dry air and water vapour mixture under various sets of conditions. Though the earth's atmosphere is a mixture of gases including nitrogen (N_2), oxygen (O_2), argon (Ar) and carbon dioxide (CO_2), yet for the purpose of psychrometry, it is considered to be a mixture of dry air and water vapour only.

16.2 Psychrometric Terms

Though there are many psychrometric terms, yet the following are important from the subject point of view :

1. *Dry air.* The pure dry air is a mixture of a number of gases such as nitrogen, oxygen, carbon dioxide, hydrogen, argon, neon, helium etc. But the nitrogen and oxygen have the major portion of the combination.

The dry air is considered to have the composition as given in the following table :

Table 16.1. Composition of dry air.

S.No.	*Constituent*	*By volume*	*By mass*	*Molecular mass*
1.	Nitrogen (N_2)	78.03%	75.47%	28
2.	Oxygen (O_2)	20.99%	23.19%	32
3.	Argon (Ar)	0.94%	1.29%	40
4.	Carbon-dioxide (CO_2)	0.03%	0.05%	44
5.	Hydrogen (H_2)	0.01%	—	2

The molecular mass of dry air is taken as 28.966 and the gas constant of air (R_a) is equal to 0.287 kJ/kg K or 287 J/kg K.

The molecular mass of water vapour is taken as 18.016 and the gas constant for water vapour (R_v) is equal to 0.461 kJ/kg K or 461 J/kg K.

Notes : (*a*) The pure dry air does not ordinarily exist in nature because it always contains some water vapour.

(*b*) The term air, wherever used in this text, means dry air containing moisture in the vapour form.

(*c*) Both dry air and water vapour can be considered as perfect gases because both exist in the atmosphere at low pressure. Thus all the perfect gas terms can be applied to them individually.

(*d*) The density of dry air is taken as 1.293 kg/m^3 at pressure 1.0135 bar or 101.35 kN/m^2 and at temperature 0°C (273 K).

2. *Moist air.* It is a mixture of dry air and water vapour. The amount of water vapour present in the air depends upon the absolute pressure and temperature of the mixture.

3. *Saturated air.* It is a mixture of dry air and water vapour, when the air has diffused the maximum amount of water vapour into it. The water vapours, usually, occur in the form of superheated steam as an invisible gas. However, when the saturated air is cooled, the water vapour in the air starts condensing, and the same may be visible in the form of moist, fog or condensation on cold surfaces.

4. *Degree of saturation.* It is the ratio of actual mass of water vapour in a unit mass of dry air to the mass of water vapour in the same mass of dry air when it is saturated at the same temperature.

5. *Humidity.* It is the mass of water vapour present in 1 kg of dry air, and is generally expressed in terms of gram per kg of dry air (g / kg of dry air). It is also called *specific humidity* or *humidity ratio*.

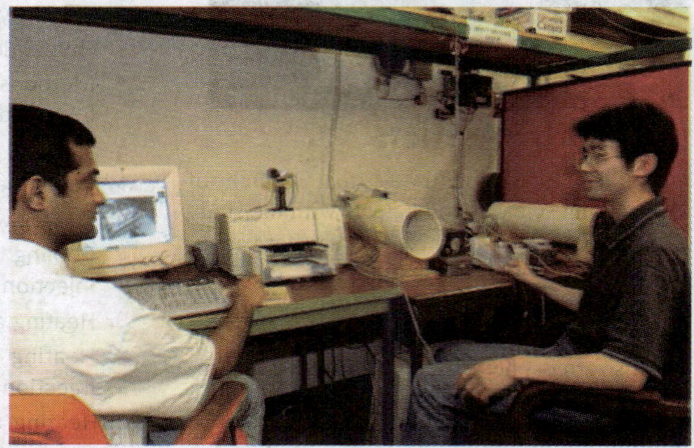

Psychrometric properties of air.

6. *Absolute humidity.* It is the mass of water vapour present in 1 m^3 of dry air, and is generally expressed in terms of gram per cubic metre of dry air (g/m^3 of dry air). It is also expressed in terms of grains per cubic metre of dry air. Mathematically, one kg of water vapour is equal to 15 430 grains.

7. *Relative humidity.* It is the ratio of actual mass of water vapour in a given volume of moist air to the mass of water vapour in the same volume of saturated air at the same temperature and pressure. It is briefly written as RH.

8. *Dry bulb temperature.* It is the temperature of air recorded by a thermometer, when it is not affected by the moisture present in the air. The dry bulb temperature (briefly written as DBT) is generally denoted by t_d or t_{db}.

9. *Wet bulb temperature.* It is the temperature of air recorded by a thermometer, when its bulb is surrounded by a wet cloth exposed to the air. Such a thermometer is called *wet bulb thermometer. The wet bulb temperature (briefly written as WBT) is generally denoted by t_w or t_{wb}.

10. *Wet bulb depression.* It is the difference between dry bulb temperature and wet bulb temperature at any point. The wet bulb depression indicates relative humidity of the air.

11. *Dew point temperature.* It is the temperature of air recorded by a thermometer, when the moisture (water vapour) present in it begins to condense. In other words, the dew point temperature is the saturation temperature (t_{sat}) corresponding to the partial pressure of water vapour (p_v). It is, usually, denoted by t_{dp}. Since p_v is very small, therefore the saturation temperature by water vapour at p_v is also low (less than the atmospheric or dry bulb temperature). Thus the water vapour in air exists in the superheated state and the moist air containing moisture in such a form (*i.e.* superheated state) is said to be *unsaturated air.* This condition is shown by point *A* on temperature-entropy (*T-s*) diagram as shown in Fig. 16.1. When the partial pressure of water vapour (p_v) is equal to the saturation pressure (p_s), the water vapour is in dry condition and the air will be *saturated air.*

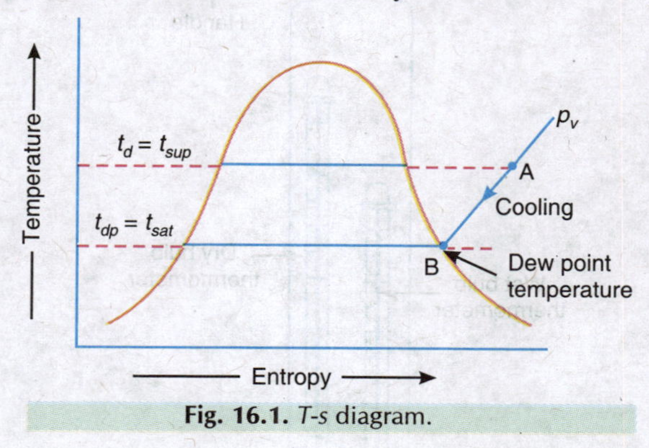

Fig. 16.1. *T-s* diagram.

If a sample of unsaturated air, containing superheated water vapour, is cooled at constant pressure, the partial pressure (p_v) of each constituent remains constant until the water vapour reaches the saturated state as shown by point *B* in Fig. 16.1. At this point *B*, the first drop of dew will be formed and hence the temperature at point *B* is called *dew point temperature.* Further cooling will cause condensation of water vapour.

From the above we see that the dew point temperature is the temperature at which the water vapour begins to condense.

Note : For saturated air, the dry bulb temperature, wet bulb temperature and dew point temperature is same.

* A wet bulb thermometer has its bulb covered with a piece of soft cloth (or silk wick) which is exposed to the air. The lower part of this cloth is dipped in a small basin of water. The water from the basin rises up in the cloth by the capillary action, and then gets evaporated. It may be noted that if relative humidity of air is high (*i.e.* the air contains more water vapour), there will be little evaporation and thus there will be a small cooling effect. On the other hand, if relative humidity of air is low (*i.e.* the air contains less water vapour), there will be more evaporation, and thus there will be more cooling effect.

12. ***Dew point depression.*** It is the difference between the dry bulb temperature and dew point temperature of air.

13. ***Psychrometer.*** There are many types of psychrometers, but the sling psychrometer, as shown in Fig. 16.2, is widely used. It consists of a dry bulb thermometer and a wet bulb thermometer mounted side by side in a protective case that is attached to a handle by a swivel connection so that the case can be easily rotated. The dry bulb thermometer is directly exposed to air and measures the actual temperature of the air. The bulb of the wet bulb thermometer is covered by a wick thoroughly wetted by distilled water. The temperature measured by this wick covered bulb of a thermometer is the temperature of liquid water in the wick and is called wet bulb temperature.

The sling psychrometer is rotated in the air for approximately one minute after which the readings from both the thermometers are taken. This process is repeated several times to assure that the lowest possible wet bulb temperature is recorded.

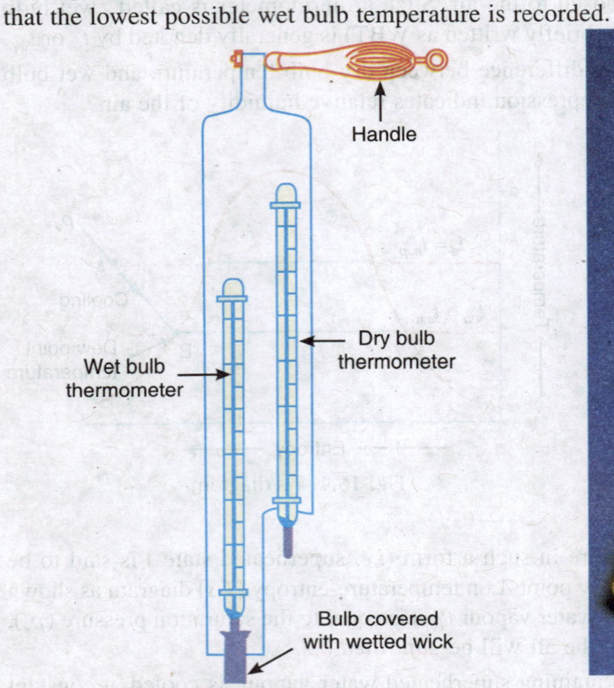

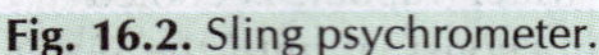
Fig. 16.2. Sling psychrometer.

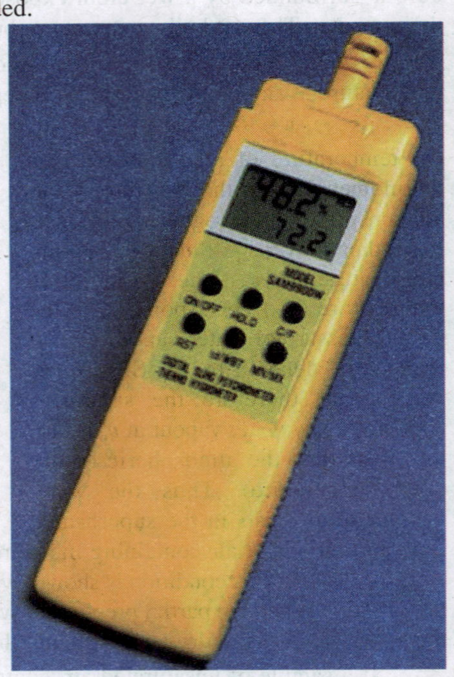

Digital psychrometer.

16.3 Dalton's Law of Partial Pressures

It states, *"The total pressure exerted by the mixture of air and water vapour is equal to the sum of the pressures, which each constituent would exert, if it occupied the same space by itself."* In other words, the total pressure exerted by air and water vapour mixture is equal to the barometric pressure. Mathematically, barometric pressure of the mixture,

$$p_b = p_a + p_v$$

where p_a = Partial pressure of dry air, and

p_v = Partial pressure of water vapour.

16.4 Psychrometric Relations

We have already discussed some psychrometric terms in Art. 16.2. These terms have some relations between one another. The following psychrometric relations are important from the subject point of view :

1. *Specific humidity, humidity ratio or moisture content.* It is the mass of water vapour present in 1 kg of dry air (in the air-vapour mixture) and is generally expressed in g/kg of dry air. It may also be defined as the ratio of mass of water vapour to the mass of dry air in a given volume of the air-vapour mixture.

Let p_a, v_a, T_a, m_a and R_a = Pressure, volume, absolute temperature, mass and gas constant respectively for dry air, and

p_v, v_v, T_v, m_v and R_v = Corresponding values for the water vapour.

Assuming that the dry air and water vapour behave as perfect gases, we have for dry air,

$$p_a v_a = m_a R_a T_a \qquad \text{... (i)}$$

and for water vapour,

$$p_v v_v = m_v R_v T_v \qquad \text{... (ii)}$$

Also

$$v_a = v_v$$

and

$$T_a = T_v = T_d \qquad \text{... (where } T_d \text{ is dry bulb temperature)}$$

From equations (*i*) and (*ii*), we have

$$\frac{p_v}{p_a} = \frac{m_v R_v}{m_a R_a}$$

∴ Humidity ratio,

$$W = \frac{m_v}{m_a} = \frac{R_a p_v}{R_v p_a}$$

Substituting R_a = 0.287 kJ/kg K for dry air and R_v = 0.461 kJ/kg K for water vapour in the above equation, we have

$$W = \frac{0.287 \times p_v}{0.461 \times p_a} = 0.622 \times \frac{p_v}{p_a} = 0.622 \times \frac{p_v}{p_b - p_v}$$

$$\text{... } (\because \; p_b = p_a + p_v)$$

Consider unsaturated air containing superheated vapour at dry bulb temperature t_d and partial pressure p_v as shown by point *A* on the *T-s* diagram in Fig. 16.3. If water is added into this unsaturated air, the water will evaporate which will increase the moisture content (specific humidity) of the air and the partial pressure p_v increases. This will continue until the water vapour becomes saturated at that temperature, as shown by point *C* in Fig. 16.3, and there will be more evaporation of water. The partial pressure p_v increases to the saturation pressure p_s and it is maximum partial pressure of water vapour at temperature t_d. The air containing moisture in such a state (point *C*) is called *saturated air.*

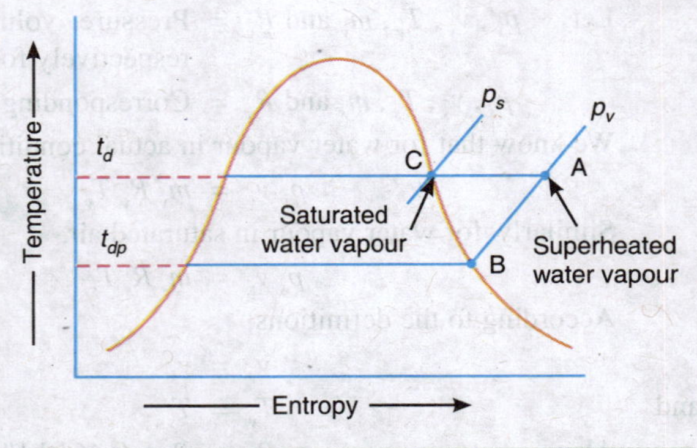

Fig. 16.3. *T-s* diagram.

For saturated air (*i.e.* when the air is holding maximum amount of water vapour), the humidity ratio or maximum specific humidity,

$$W_s = W_{max} = 0.622 \times \frac{p_s}{p_b - p_s}$$

where p_s = Partial pressure of air corresponding to saturation temperature (*i.e.* dry bulb temperature t_d).

2. *Degree of saturation or percentage humidity.* We have already discussed that the degree of saturation is the ratio of actual mass of water vapour in a unit mass of dry air to the mass of water vapour in the same mass of dry air when it is saturated at the same temperature (dry bulb temperature). In other words, it may be defined as the ratio of actual specific humidity to the specific humidity of saturated air at the same dry bulb temperature. It is, usually, denoted by μ. Mathematically, degree of saturation,

$$\mu = \frac{W}{W_s} = \frac{\dfrac{0.622\, p_v}{p_b - p_v}}{\dfrac{0.622\, p_s}{p_b - p_s}} = \frac{p_v}{p_s}\left(\frac{p_b - p_s}{p_b - p_v}\right) = \frac{p_v}{p_s}\left[\frac{1 - \dfrac{p_s}{p_b}}{1 - \dfrac{p_v}{p_b}}\right]$$

Notes : ***(a)*** The partial pressure of saturated air (p_s) is obtained from the steam tables corresponding to dry bulb temperature t_d.

(b) If the relative humidity, $\phi = p_v / p_s$ is equal to zero, then the humidity ratio, $W = 0$, *i.e.* for dry air, $\mu = 0$.

(c) If the relative humidity, $\phi = p_v / p_s$ is equal to 1, then $W = W_s$ and $\mu = 1$. Thus μ varies between 0 and 1.

3. *Relative humidity.* We have already discussed that the relative humidity is the ratio of actual mass of water vapour (m_v) in a given volume of moist air to the mass of water vapour (m_s) in the same volume of saturated air at the same temperature and pressure. It is usually denoted by φ. Mathematically, relative humidity,

$$\phi = \frac{m_v}{m_s}$$

Let p_v, v_v, T_v, m_v and R_v = Pressure, volume, temperature, mass and gas constant respectively for water vapour in actual conditions, and

p_s, v_s, T_s, m_s and R_s = Corresponding values for water vapour in saturated air.

We know that for water vapour in actual conditions,

$$p_v\, v_v = m_v\, R_v\, T_v \quad \text{... (i)}$$

Similarly, for water vapour in saturated air,

$$p_s\, v_s = m_s\, R_s\, T_s \quad \text{... (ii)}$$

According to the definitions,

$$v_v = v_s$$

and $$T_v = T_s$$

Also $$R_v = R_s = 0.461 \text{ kJ/kg K}$$

∴ From equations (*i*) and (*ii*), relative humidity,

$$\phi = \frac{m_v}{m_s} = \frac{p_v}{p_s}$$

Thus, the relative humidity may also be defined as the ratio of actual partial pressure of water vapour in moist air at a given temperature (dry bulb temperature) to the saturation pressure of water vapour (or partial pressure of water vapour in saturated air) at the same temperature.

The relative humidity may also be obtained as discussed below :

We know that degree of saturation,

$$\mu = \frac{p_v}{p_s}\left[\frac{1-\dfrac{p_s}{p_b}}{1-\dfrac{p_v}{p_b}}\right] = \phi\left[\frac{1-\dfrac{p_s}{p_b}}{1-\phi\times\dfrac{p_s}{p_b}}\right] \qquad \ldots \left(\because \phi = \frac{p_v}{p_s}\right)$$

$$\therefore \quad \phi = \frac{\mu}{1-(1-\mu)\dfrac{p_s}{p_b}}$$

Note : For saturated air, the relative humidity is 100%.

4. Pressure of water vapour. According to Carrier's equation, the partial pressure of water vapour,

$$p_v = p_w - \frac{(p_b - p_w)(t_d - t_w)}{1544 - 1.44\, t_w}$$

where p_w = Saturation pressure corresponding to wet bulb temperature (from steam tables),

p_b = Barometric pressure,

t_d = Dry bulb temperature, and

t_w = Wet bulb temperature.

5. Vapour density or absolute humidity. We have already discussed that the vapour density or absolute humidity is the mass of water vapour present in 1 m^3 of dry air.

Let v_v = Volume of water vapour in m^3/kg of dry air at its partial pressure,

v_a = Volume of dry air in m^3/kg of dry air at its partial pressure,

ρ_v = Density of water vapour in kg/m^3 corresponding to its partial pressure and dry bulb temperature t_d, and

ρ_a = Density of dry air in kg/m^3 of dry air.

We know that mass of water vapour,

$$m_v = v_v\,\rho_v \qquad \ldots (i)$$

and mass of dry air,

$$m_a = v_a\,\rho_a \qquad \ldots (ii)$$

Dividing equation (*i*) by equation (*ii*),

$$\frac{m_v}{m_a} = \frac{v_v\,\rho_v}{v_a\,\rho_a}$$

Since $v_a = v_v$, therefore humidity ratio,

$$W = \frac{m_v}{m_a} = \frac{\rho_v}{\rho_a} \quad \text{or} \quad \rho_v = W\rho_a \qquad \ldots (iii)$$

We know that $\quad p_a\,v_a = m_a\,R_a\,T_d$

Since $v_a = \dfrac{1}{\rho_a}$ and m_a = 1 kg, therefore substituting these values in the above expression, we get

$$p_a \times \frac{1}{\rho_a} = R_a\, T_d \quad \text{or} \quad \rho_a = \frac{p_a}{R_a\, T_d}$$

Substituting the value of ρ_a in equation (*iii*), we have

$$\rho_v = \frac{W\, p_a}{R_a\, T_d} = \frac{W\,(p_b - p_v)}{R_a\, T_d} \qquad \ldots (\because\ p_b = p_a + p_v)$$

where p_a = Pressure of air in kN/m^2,

R_a = Gas constant for air = 0.287 kJ/ kg K, and

T_d = Dry bulb temperature in K.

16.5 Enthalpy (Total Heat) of Moist Air

The enthalpy of moist air is numerically equal to the enthalpy of dry air plus the enthalpy of water vapour associated with dry air. Let us consider one kg of dry air. We know that enthalpy of 1 kg of dry air,

$$h_a = c_{pa}\, t_d \qquad \ldots (i)$$

where c_{pa} = Specific heat of dry air which is normally taken as 1.005 kJ / kg K, and

t_d = Dry bulb temperature.

Enthalpy of water vapour associated with 1 kg of dry air,

$$h_v = W\, h_s \qquad \ldots (ii)$$

where W = Mass of water vapour in 1 kg of dry air (*i.e.* specific humidity), and

h_s = Enthalpy of water vapour per kg of dry air at dew point temperature (t_{dp}).

If the moist air is superheated, then the enthalpy of water vapour

$$= W\, c_{ps}\,(t_d - t_{dp}) \qquad \ldots (iii)$$

where c_{ps} = Specific heat of superheated water vapour which is normally taken as 1.9 kJ/kg K, and

$t_d - t_{dp}$ = Degree of superheat of the water vapour.

$\therefore$ Total enthalpy of superheated water vapour,

$$h = c_{pa}\, t_d + W\, h_s + W\, c_{ps}\,(t_d - t_{dp})$$

$$= c_{pa}\, t_d + W\,[h_{fdp} + h_{fgdp} + c_{ps}\,(t_d - t_{dp})] \qquad \ldots (\because\ h_s = h_{fdp} + h_{fgdp})$$

$$= c_{pa}\, t_d + W\,[4.2\, t_{dp} + h_{fgdp} + c_{ps}\,(t_d - t_{dp})] \qquad \ldots (\because\ h_{fdp} + 4.2\, t_{dp})$$

$$= c_{pa}\, t_d + 4.2\, W\, t_{dp} + W\, h_{fgdp} + W\, c_{ps}\, t_d - W\, c_{ps}\, t_{dp}$$

$$= (c_{pa} + W\, c_{ps})\, t_d + W\,[h_{fgdp} + t_{dp}\,(4.2 - c_{ps})]$$

$$= (c_{pa} + W\, c_{ps})\, t_d + W\,[h_{fgdp} + t_{dp}\,(4.2 - 1.9)]$$

$$= (c_{pa} + W\, c_{ps})\, t_d + W\,[h_{fgdp} + 2.3\, t_{dp}]$$

The term $(c_{pa} + W\, c_{ps})$ is called *humid specific heat* (c_{pm}). It is the specific heat or heat capacity of moist air, *i.e.* (1 + *W*) kg / kg of dry air. At low temperature of air conditioning range, the value of W is very small. The general value of humid specific heat in air conditioning range is taken as 1.022 kJ/kg K.

$$\therefore \quad h = 1.022\, t_d + W\,(h_{fgdp} + 2.3\, t_{dp}) \text{ kJ/kg}$$

where h_{fgdp} = Latent heat of vaporisation of water corresponding to dew point temperature (from steam tables).

An approximate result may be obtained by the following relation:

$$h = 1.005\, t_d + W\,[2500 + 1.9\, t_d] \text{ kJ/kg}$$

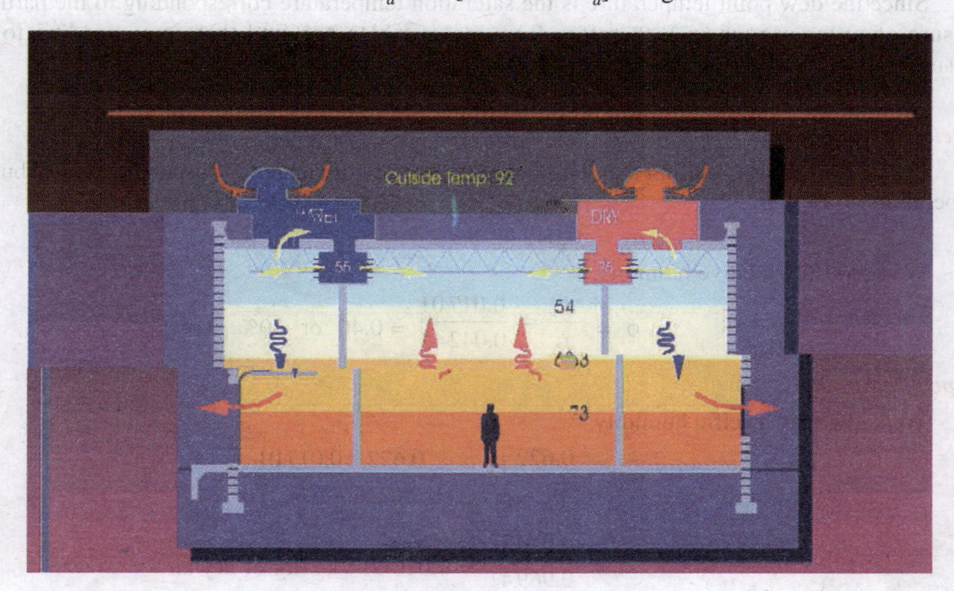

Psychrometric control

Example 16.1. *The readings from a sling psychrometer are as follows :*

Dry bulb temperature = 30° C ; Wet bulb temperature = 20° C ; Barometer reading = 740 mm of Hg.

Using steam tables, determine : 1. Dew point temperature ; 2. Relative humidity ; 3. Specific humidity ; 4. Degree of saturation ; 5. Vapour density ; and 6. Enthalpy of mixture per kg of dry air.

Solution. Given : $t_d = 30°\text{C}$; $t_w = 20°\text{C}$; $p_b = 740$ mm of Hg

1. *Dew point temperature*

First of all, let us find the partial pressure of water vapour (p_v).

From steam tables, we find that the saturation pressure corresponding to wet bulb temperature of 20° C is

$$p_w = 0.023\ 37 \text{ bar}$$

We know that barometric pressure,

$$p_b = 740 \text{ mm of Hg} \qquad \text{... (Given)}$$

$$= 740 \times 133.3 = 98\ 642 \text{ N/m}^2 \qquad \text{... } (\because 1 \text{ mm of Hg} = 133.3 \text{ N/m}^2)$$

$$= 0.986\ 42 \text{ bar} \qquad \text{... } (\because 1 \text{ bar} = 10^5 \text{ N/m}^2)$$

∴ Partial pressure of water vapour,

$$p_v = p_w - \frac{(p_b - p_w)(t_d - t_w)}{1544 - 1.44\, t_w}$$

$$= 0.023\,37 - \frac{(0.986\,42 - 0.02337)(30 - 20)}{1544 - 1.44 \times 20}$$

$$= 0.023\,37 - 0.006\,36 = 0.017\,01 \text{ bar}$$

Since the dew point temperature is the saturation temperature corresponding to the partial pressure of water vapour (p_v), therefore from steam tables, we find that corresponding to a pressure of 0.017 01 bar, the dew point temperature is

$$t_{dp} = 15° \text{ C} \quad \textbf{Ans.}$$

2. *Relative humidity*

From steam tables, we find that the saturation pressure of vapour corresponding to dry bulb temperature of 30°C is

$$p_s = 0.042\,42 \text{ bar}$$

We know that relative humidity,

$$\phi = \frac{p_v}{p_s} = \frac{0.01701}{0.04242} = 0.40 \text{ or } 40\% \quad \textbf{Ans.}$$

3. *Specific humidity*

We know that specific humidity,

$$W = \frac{0.622\, p_v}{p_b - p_v} = \frac{0.622 \times 0.01701}{0.986\,42 - 0.01701}$$

$$= \frac{0.01058}{0.96941} = 0.010\,914 \text{ kg/kg of dry air}$$

$$= 10.914 \text{ g/kg of dry air} \quad \textbf{Ans.}$$

4. *Degree of saturation*

We know that specific humidity of saturated air,

$$W_s = \frac{0.622\, p_s}{p_b - p_s} = \frac{0.622 \times 0.04242}{0.98642 - 0.04242}$$

$$= \frac{0.02638}{0.944} = 0.027\,945 \text{ kg/kg of dry air}$$

We know that degree of saturation,

$$\mu = \frac{W}{W_s} = \frac{0.010\,914}{0.027\,945} = 0.391 \text{ or } 39.1\% \quad \textbf{Ans.}$$

Note : The degree of saturation (μ) may also be calculated from the following relation :

$$\mu = \frac{p_v}{p_s}\left(\frac{p_b - p_s}{p_b - p_v}\right)$$

$$= \frac{0.01701}{0.04242}\left[\frac{0.98642 - 0.04242}{0.98642 - 0.01701}\right]$$

$$= 0.391 \text{ or } 39.1\% \quad \textbf{Ans.}$$

5. *Vapour density*

We know that vapour density,

$$\rho_v = \frac{W(p_b - p_v)}{R_a T_d} = \frac{0.010\,914\,(0.986\,42 - 0.017\,01)\,10^5}{287\,(273 + 30)}$$

$$= 0.012\,16 \text{ kg/m}^3 \text{ of dry air} \quad \textbf{Ans.}$$

6. *Enthalpy of mixture per kg of dry air*

From steam tables, we find that the latent heat of vaporisation of water at dew point temperature of 15°C is

$$h_{fgdp} = 2466.1 \text{ kJ/kg}$$

∴ Enthalpy of mixture per kg of dry air,

$$h = 1.022\, t_d + W\,[h_{fgdp} + 2.3\, t_{dp}]$$

$$= 1.022 \times 30 + 0.010\,914\,[2466.1 + 2.3 \times 15]$$

$$= 30.66 + 27.29 = 57.95 \text{ kJ/kg of dry air} \quad \textbf{Ans.}$$

Example 16.2. *On a particular day, the atmospheric air was found to have a dry bulb temperature of 30° C and a wet bulb temperature of 18° C. The barometric pressure was observed to be 756 mm of Hg. Using the tables of psychrometric properties of air, determine the relative humidity, the specific humidity, the dew point temperature, the enthalpy of air per kg of dry air and the volume of mixture per kg of dry air.*

Solution. Given : $t_d = 30$°C ; $t_w = 18$°C ; $p_b = 756$ mm of Hg

Relative humidity

First of all, let us find the partial pressure of water vapour (p_v). From steam tables, we find that the saturation pressure corresponding to wet bulb temperature of 18°C is,

$$p_w = 0.020\,62 \text{ bar} = 0.020\,62 \times 10^5 = 2062 \text{ N/m}^2$$

$$= \frac{2062}{133.3} = 15.47 \text{ mm of Hg} \qquad \ldots (\because 1 \text{ mm of Hg} = 133.3 \text{ N/m}^2)$$

We know that
$$p_v = p_w - \frac{(p_b - p_w)(t_d - t_w)}{1544 - 1.44\, t_w}$$

$$= 15.47 - \frac{(756 - 15.47)(30 - 18)}{1544 - 1.44 \times 18} \text{ mm of Hg}$$

$$= 15.47 - 5.85 = 9.62 \text{ mm of Hg}$$

From steam tables, we find that the saturation pressure of vapour corresponding to dry bulb temperature of 30°C is

$$p_s = 0.042\,42 \text{ bar} = 0.042\,42 \times 10^5 = 4242 \text{ N/m}^2$$

$$= \frac{4242}{133.3} = 31.8 \text{ mm of Hg}$$

We know that the relative humidity,

$$\phi = \frac{p_v}{p_s} = \frac{9.62}{31.8} = 0.3022 \text{ or } 30.22\% \quad \textbf{Ans.}$$

Specific humidity

We know that specific humidity,

$$W = \frac{0.622\, p_v}{p_b - p_v} = \frac{0.622 \times 9.62}{756 - 9.62} = 0.008 \text{ kg/kg of dry air} \quad \textbf{Ans.}$$

Dew point temperature

Since the dew point temperature is the saturation temperature corresponding to the partial pressure of water vapour (p_v), therefore from steam tables, we find that corresponding to 9.62 mm of Hg or 9.62 × 133.3 = 1282.3 N/m^2 = 0.012 823 bar, the dew point temperature is,

$$t_{dp} = 10.6° \text{ C} \quad \textbf{Ans.}$$

Enthalpy of air per kg of dry air

From steam tables, we also find that latent heat of vaporisation of water at dew point temperature of 10.6°C,

$$h_{fgdp} = 2476.5 \text{ kJ/kg}$$

We know that enthalpy of air per kg of dry air,

$$h = 1.022\, t_d + W\,(h_{fgdp} + 2.3\, t_{dp})$$

$$= 1.022 \times 30 + 0.008\,(2476.5 + 2.3 \times 10.6)$$

$$= 30.66 + 20 = 50.66 \text{ kJ/kg of dry air} \quad \textbf{Ans.}$$

Volume of the mixture per kg of dry air

From psychrometric tables, we find that specific volume of the dry air at 760 mm of Hg and 30°C dry bulb temperature is 0.8585 m^3/kg of dry air. We know that one kg of dry air at a partial pressure of (756 – 9.62) mm of Hg occupies the same volume as W = 0.008 kg of vapour at its partial pressure of 9.62 mm of Hg. Moreover, the mixture occupies the same volume but at a total pressure of 756 mm of Hg.

∴ Volume of the mixture (v) at a dry bulb temperature of 30°C and a pressure of 9.62 mm of Hg

= Volume of 1 kg of dry air (v_a) at a pressure of (756 – 9.62) or 746.38 mm of Hg

$$= 0.8585 \times \frac{760}{746.38} = 0.8741 \text{ kg/kg of dry air} \quad \textbf{Ans.}$$

Note : The volume of mixture per kg of dry air may be calculated as discussed below :

We know that $$v = v_a = \frac{R_a\, T_d}{p_a}$$

where R_a = Gas constant for air = 287 J/kg K

T_d = Dry bulb temperature in K

= 30 + 273 = 303 K, and

p_a = Pressure of air in N/m^2

= $p_b - p_v$ = 756 – 9.62 = 746.38 mm of Hg

= 746.38 × 133.3 = 994 92 N/m^2

Substituting the values in the above equation,

$$v = \frac{287 \times 303}{994\,92} = 0.8741 \text{ m}^3\text{/kg of dry air} \quad \textbf{Ans.}$$

Example 16.3. *The humidity ratio of atmospheric air at 28°C dry bulb temperature and 760 mm of mercury is 0.016 kg / kg of dry air. Determine: 1. partial pressure of water vapour ; 2. relative humidity ; 3. dew point temperature ; 4. specific enthalpy; and 5. vapour density.*

Solution. Given : $t_d = 28°C$; $p_b = 760$ mm of Hg ; $W = 0.016$ kg/kg of dry air

1. *Partial pressure of water vapour*

Let p_v = Partial pressure of water vapour.

We know that humidity ratio (W),

$$0.016 = \frac{0.622\, p_v}{p_b - p_v} = \frac{0.622\, p_v}{760 - p_v}$$

$$12.16 - 0.016\, p_v = 0.622\, p_v \quad \text{or} \quad 0.638\, p_v = 12.16$$

$$\therefore \quad p_v = 12.16 / 0.638 = 19.06 \text{ mm of Hg}$$

$$= 19.06 \times 133.3 = 2540.6 \text{ N/m}^2 \quad \textbf{Ans.}$$

2. *Relative humidity*

From steam tables, we find that the saturation pressure of vapour corresponding to dry bulb temperature of 28°C is

$$p_s = 0.03778 \text{ bar} = 3778 \text{ N/m}^2$$

∴ Relative humidity,

$$\phi = \frac{p_v}{p_s} = \frac{2540.6}{3778} = 0.672 \text{ or } 67.2\% \quad \textbf{Ans.}$$

3. *Dew point temperature*

Since the dew point temperature is the saturation temperature corresponding to the partial pressure of water vapour (p_v), therefore from steam tables, we find that corresponding to a pressure of 2540.6 N/m^2 (0.025406 bar), the dew point temperature is,

$$t_{dp} = 21.1° \text{ C} \quad \textbf{Ans.}$$

4. *Specific enthalpy*

From steam tables, latent heat of vaporisation of water corresponding to a dew point temperature of 21.1° C,

$$h_{fgdp} = 2451.76 \text{ kJ/kg}$$

We know that specific enthalpy,

$$h = 1.022\, t_d + W\,(h_{fgdp} + 2.3\, t_{dp})$$

$$= 1.022 \times 28 + 0.016\,(2451.76 + 2.3 \times 21.1)$$

$$= 28.62 + 40 = 68.62 \text{ kJ / kg of dry air} \quad \textbf{Ans.}$$

5. *Vapour density*

We know that vapour density,

$$\rho_v = \frac{W\,(p_b - p_v)}{R_a\, T_d} = \frac{0.016\,(760 - 19.06)\,133.3}{287\,(273 + 28)}$$

$$= 0.0183 \text{ kg / m}^3 \text{ of dry air} \quad \textbf{Ans.}$$

Example 16.4. *A room 7 m × 4 m × 4 m is occupied by an air-water vapour mixture at 38°C. The atmospheric pressure is 1 bar and the relative humidity is 70%. Determine the humidity ratio, dew point, mass of dry air and mass of water vapour. If the mixture of air-water vapour is further cooled at constant pressure until the temperature is 10°C, find the amount of water vapour condensed.*

Solution. Given : $v = 7 \times 4 \times 4 = 112\ m^3$; $T_d = 38°C = 38 + 273 = 311$ K ; $p_b = 1$ bar ; $\phi = 70\% = 0.7$

Humidity ratio

First of all, let us find the partial pressure of water vapour (p_v). From steam tables, we find that saturation pressure of vapour corresponding to a temperature of 38° C is,

$$p_s = 0.066\ 24 \text{ bar}$$

We know that relative humidity (ϕ),

$$0.7 = \frac{p_v}{p_s}$$

$$\therefore \quad p_v = 0.7\ p_s = 0.7 \times 0.066\ 24 = 0.046\ 368 \text{ bar}$$

We also know that humidity ratio,

$$W = \frac{0.622\ p_v}{p_b - p_v} = \frac{0.622 \times 0.046\ 368}{1 - 0.046\ 368}$$

$$= \frac{0.028\ 841}{0.953\ 632} = 0.0302 \text{ kg/kg of dry air}$$

$$= 30.2 \text{ g/kg of dry air} \quad \textbf{Ans.}$$

Dew point temperature

Since the dew point temperature (t_{dp}) is the saturation temperature corresponding to the partial pressure of water vapour (p_v), therefore, from steam tables, we find that corresponding to a pressure of 0.046 368 bar, the dew point temperature is

$$t_{dp} = 31.56°\text{ C} \quad \textbf{Ans.}$$

Mass of dry air

Let $\quad m_a$ = Mass of dry air, and

p_a = Pressure of dry air = $p_b - p_v$

$= 1 - 0.046\ 368 = 0.953\ 632$ bar

$= 0.953\ 632 \times 10^5 = 95363.2\ N/m^2$ ... ($\because$ 1 bar = $10^5\ N/m^2$)

We know that $\quad p_a\, v = m_a\, R_a\, T_d$

$$\therefore \quad m_a = \frac{p_a\, v}{R_a\, T_d} = \frac{953\,63.2 \times 112}{287 \times 311} = 119.7 \text{ kg} \quad \textbf{Ans.}$$

... (Taking R_a = 287 J/kg K)

Mass of water vapour

Let $\quad m_v$ = Mass of water vapour.

We know that humidity ratio (W),

$$0.0302 = \frac{m_v}{m_a} = \frac{m_v}{119.7} \quad \text{or} \quad m_v = 0.0302 \times 119.7 = 3.61 \text{ kg} \quad \textbf{Ans.}$$

Amount of water vapour condensed

If the temperature is 10° C, the air will be saturated before some water condenses out. From steam tables, we find that saturation pressure of vapour corresponding to 10° C is

$$p_s = p_v = 0.012\,27 \text{ bar} \qquad \ldots (\because \text{ Pressure is constant})$$

We know that humidity ratio,

$$W = \frac{0.622\, p_v}{p_b - p_v} = \frac{0.622 \times 0.012\,27}{1 - 0.012\,27} = \frac{0.007\,632}{0.987\,73}$$

$$= 0.007\,73 \text{ kg/kg of dry air} = 7.73 \text{ g/kg of dry air}$$

We know that pressure of dry air,

$$p_a = p_b - p_v = 1 - 0.012\,27 = 0.987\,73 \text{ bar}$$

$$= 0.987\,73 \times 10^5 = 987\,73 \text{ N/m}^2$$

$\therefore$ Mass of dry air, $m_a = \dfrac{p_a\, v}{R_a\, T} = \dfrac{987\,73 \times 112}{287\,(10 + 273)} = 136.2$ kg

and mass of water vapour, $m_v = W \times m_a = 0.007\,73 \times 136.2 = 1.053$ kg

$\therefore$ Amount of water vapour condensed

$$= 3.61 - 1.053 = 2.557 \text{ kg} \quad \textbf{Ans.}$$

16.6 Thermodynamic Wet Bulb Temperature or Adiabatic Saturation Temperature

The thermodynamic wet bulb temperature or adiabatic saturation temperature is the temperature at which the air can be brought to saturation state, adiabatically, by the evaporation of water into the flowing air.

The equipment used for the adiabatic saturation of air, in its simplest form, consists of an insulated chamber containing adequate quantity of water. There is also an arrangement for extra water (known as make-up water) to flow into the chamber from its top, as shown in Fig. 16.4.

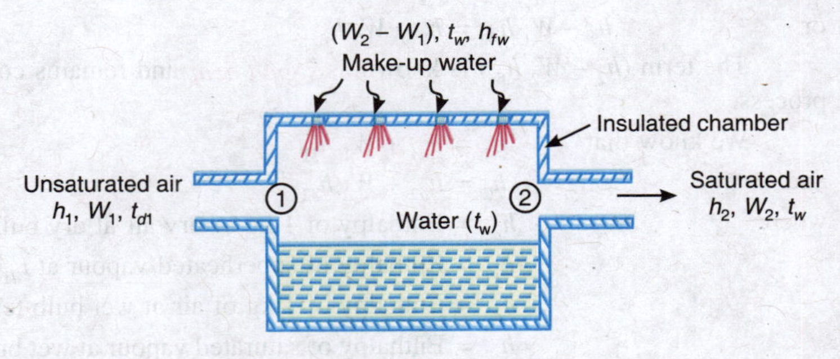

Fig. 16.4. Adiabatic saturation of air.

Let the unsaturated air enters the chamber at section 1. As the air passes through the chamber over a long sheet of water, the water evaporates which is carried with the flowing stream of air, and the specific humidity of the air increases. The make-up water is added to the chamber at this temperature to make the water level constant. Both the air and water are cooled as the evaporation takes place. This process continues until the energy transferred from the air to the water is equal to the energy required to vaporise the water. When steady conditions are reached, the air flowing at section 2 is saturated with water vapour. The temperature of the saturated air at section 2 is known as *thermodynamic wet bulb temperature* or *adiabatic saturation temperature.*

The adiabatic saturation process can be represented on *T-s* diagram as shown by the curve 1-2 in Fig. 16.5.

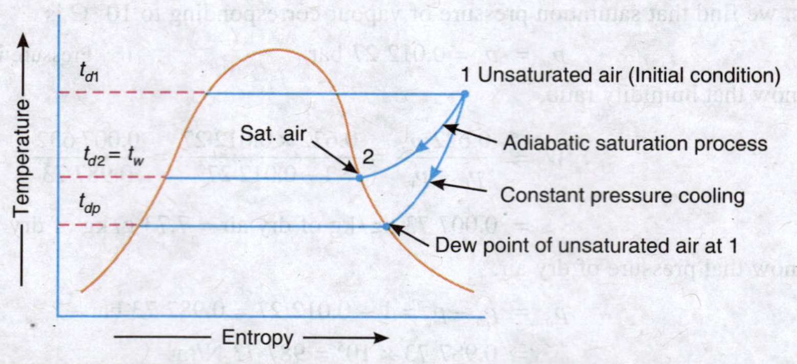

Fig. 16.5. *T-s* diagram for adiabatic saturation process.

During the adiabatic saturation process, the partial pressure of vapour increases, although the total pressure of the air-vapour mixture remains constant. The unsaturated air initially at dry bulb temperature t_{d1} is cooled adiabatically to dry bulb temperature t_{d2} which is equal to the adiabatic saturation temperature t_w. It may be noted that the adiabatic saturation temperature is taken equal to the wet bulb temperature for all practical purposes.

Let h_1 = Enthalpy of unsaturated air at section 1,

W_1 = Specific humidity of air at section 1,

h_2, W_2 = Corresponding values of saturated air at section 2, and

h_{fw} = Sensible heat of water at adiabatic saturation temperature.

Balancing the enthalpies of air at inlet and outlet (*i.e.* at sections 1 and 2),

$$h_1 + (W_2 - W_1)\, h_{fw} = h_2 \quad \ldots (i)$$

or

$$h_1 - W_1\, h_{fw} = h_2 - W_2\, h_{fw} \quad \ldots (ii)$$

The term $(h_2 - W_2\, h_{fw})$ is known as *sigma heat* and remains constant during the adiabatic process.

We know that $h_1 = h_{a1} + W_1\, h_{s1}$

and $h_2 = h_{a2} + W_2\, h_{s2}$

where h_{a1} = Enthalpy of 1 kg of dry air at dry bulb temperature t_{d1},

*h_{s1} = Enthalpy of superheated vapour at t_{d1} per kg of vapour,

h_{a2} = Enthalpy of 1 kg of air at wet bulb temperature t_w, and

h_{s2} = Enthalpy of saturated vapour at wet bulb temperature t_w per kg of vapour.

Now the equation (*ii*) may be written as :

$$(h_{a1} + W_1\, h_{s1}) - W_1\, h_{fw} = (h_{a2} + W_2\, h_{s2}) - W_2\, h_{fw}$$

$$W_1\,(h_{s1} - h_{fw}) = W_2\,(h_{s2} - h_{fw}) + h_{a2} - h_{a1}$$

$$\therefore \quad W_1 = \frac{W_2\,(h_{s2} - h_{fw}) + h_{a2} - h_{a1}}{h_{s1} - h_{fw}}$$

* In psychrometry, the enthalpy of superheated vapour at dry bulb temperature t_{d1} is taken equal to the enthalpy of saturated vapour corresponding to dry bulb temperature t_{d1}.

Example 16.5. *Atmospheric air at 0.965 bar enters the adiabatic saturator. The wet bulb temperature is 20° C and dry bulb temperature is 31° C during adiabatic saturation process. Determine : 1. humidity ratio of the entering air ; 2. vapour pressure and relative humidity at 31° C ; and 3. dew point temperature.*

Solution. Given : $p_b = 0.965$ bar ; $t_w = 20°$ C ; $t_d = 31°$ C

1. Humidity ratio of the entering air

Let W_1 = Humidity ratio of the entering air, and

W_2 = Humidity ratio of the saturated air.

First of all, let us find the value of W_2. From psychrometric or steam tables, we find that saturation pressure of vapour at 20° C,

$$p_{v2} = 0.023\ 37 \text{ bar}$$

Enthalpy of saturated vapour at 20° C,

$$h_{s2} = h_{g2} = 2538.2 \text{ kJ/kg}$$

Sensible heat of water at 20° C,

$$h_{fw} = 83.9 \text{ kJ/kg}$$

and enthalpy of saturated vapour at 31° C,

$$h_{s1} = h_{g1} = 2558.2 \text{ kJ/kg}$$

We know that enthalpy of unsaturated air corresponding to dry bulb temperature of 31° C,

$$h_{a1} = m\, c_p\, t_d = 1 \times 1.005 \times 31 = 31.155 \text{ kJ/kg}$$

... (Taking c_p for air = 1.005 kJ/kg°C)

Enthalpy of 1 kg of saturated air corresponding to wet bulb temperature of 20° C,

$$h_{a2} = m\, c_p\, t_w = 1 \times 1.005 \times 20 = 20.1 \text{ kJ/kg}$$

We know that
$$W_2 = \frac{0.622\, p_{v2}}{p_b - p_{v2}} = \frac{0.622 \times 0.023\,37}{0.965 - 0.023\,37} = 0.0154 \text{ kg/kg of dry air}$$

$$\therefore \quad W_1 = \frac{W_2\,(h_{s2} - h_{fw}) + h_{a2} - h_{a1}}{h_{s1} - h_{fw}}$$

$$= \frac{0.0154\,(2538.2 - 83.9) + 20.1 - 31.155}{2558.2 - 83.9}$$

$$= 0.0108 \text{ kg/kg of dry air} \quad \textbf{Ans.}$$

2. Vapour pressure and relative humidity at 31° C

Let p_{v1} = Vapour pressure at 31° C.

We know that humidity ratio of the entering air (W_1),

$$0.0108 = \frac{0.622\, p_{v1}}{p_b - p_{v1}} = \frac{0.622\, p_{v1}}{0.965 - p_{v1}}$$

or $$0.0104 - 0.0108\, p_{v1} = 0.622\, p_{v1}$$

$$\therefore \quad 0.6328\, p_{v1} = 0.0104 \quad \text{or} \quad p_{v1} = 0.0164 \text{ bar} \quad \textbf{Ans.}$$

From psychrometric or steam tables, we find that the saturation pressure corresponding to 31° C is

$$p_s = 0.044\,91 \text{ bar}$$

∴ Relative humidity,

$$\phi = \frac{p_{v1}}{p_s} = \frac{0.0164}{0.044\,91} = 0.365 \text{ or } 36.5\% \quad \textbf{Ans.}$$

3. Dew point temperature

Since the dew point temperature (t_{dp}) is the saturation temperature corresponding to the partial pressure of water vapour (p_{v1}), therefore from psychrometric or steam tables, we find that corresponding to a pressure of 0.0164 bar, the dew point temperature is

$$t_{dp} = 14.5° \text{ C} \quad \textbf{Ans.}$$

16.7 Psychrometric Chart

It is a graphical representation of the various thermodynamic properties of moist air. The psychrometric chart is very useful for finding out the properties of air (which are required in the field of air conditioning) and eliminate lot of calculations. There is a slight variation in the charts prepared by different air-conditioning manufactures but basically they are all alike. The psychrometric chart is normally drawn for standard atmospheric pressure of 760 mm of Hg (or 1.01325 bar).

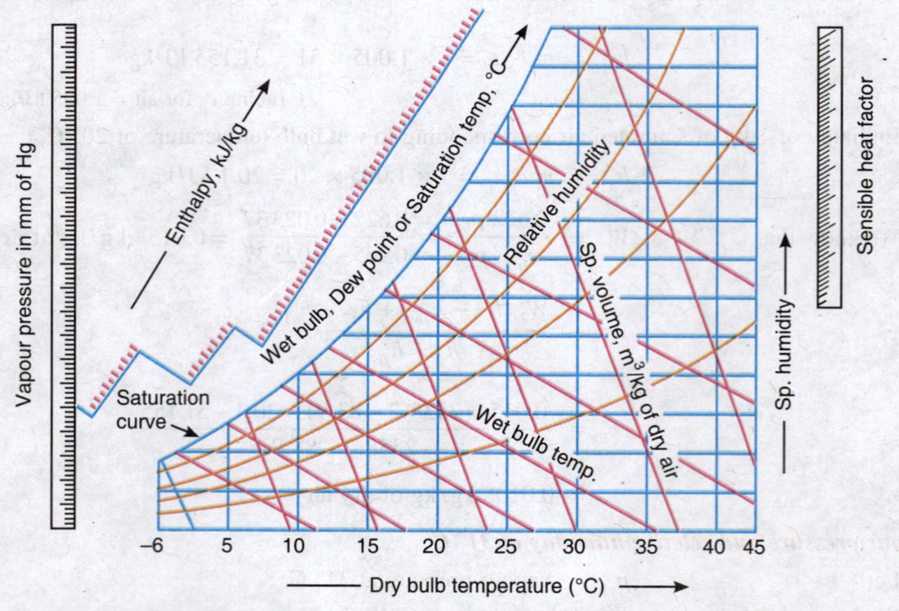

Fig. 16.6. Psychrometric chart.

In a psychrometric chart, dry bulb temperature is taken as abscissa and specific humidity *i.e.* moisture contents as ordinate, as shown in Fig. 16.6. Now the saturation curve is drawn by plotting the various saturation points at corresponding dry bulb temperatures. The saturation curve represents 100% relative humidity at various dry bulb temperatures. It also represents the wet bulb and dew point temperatures.

Though the psychrometric chart has a number of details, yet the following lines are important from the subject point of view :

1. *Dry bulb temperature lines.* The dry bulb temperature lines are vertical *i.e.* parallel to the ordinate and uniformly spaced as shown in Fig. 16.7. Generally the temperature range of these lines on psychrometric chart is from – 6° C to 45° C. The dry bulb temperature lines are drawn with difference of every 5°C and up to the saturation curve as shown in the figure. The values of dry bulb temperatures are also shown on the saturation curve.

2. Specific humidity or moisture content lines. The specific humidity (moisture content) lines are horizontal *i.e.* parallel to the abscissa and are also uniformly spaced as shown in Fig. 16.8. Generally, moisture content range of these lines on psychrometric chart is from 0 to 30 g / kg of dry air (or from 0 to 0.030 kg / kg of dry air). The moisture content lines are drawn with a difference of every 1 g (or 0.001 kg) and up to the saturation curve as shown in the figure.

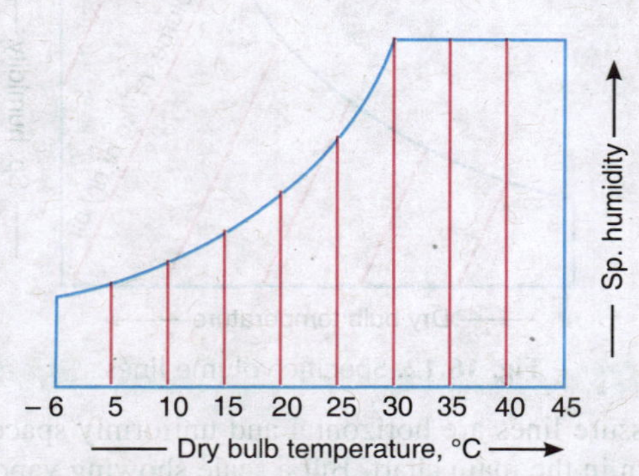

Fig. 16.7. Dry bulb temperature lines.

30
25
20
15
10
5
0
Sp. humidity
Dry bulb temperature

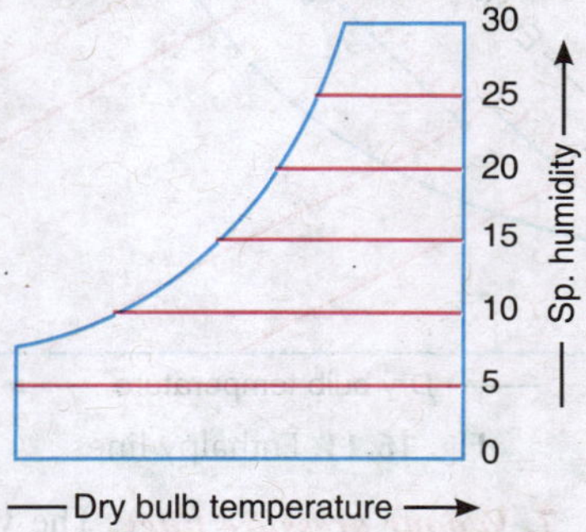

Fig. 16.8. Specific humidity lines.

3. *Dew point temperature lines.* The dew point temperature lines are horizontal *i.e.* parallel to the abscissa and non-uniformly spaced as shown in Fig. 16.9. At any point on the saturation curve, the dry bulb and dew point temperatures are equal.

The values of dew point temperatures are generally given along the saturation curve of the chart as shown in the figure.

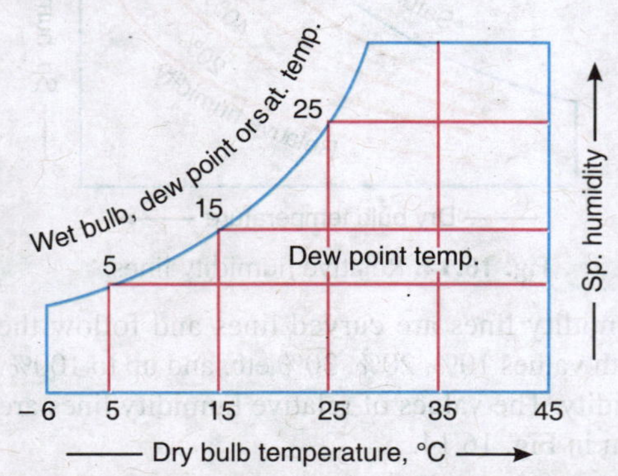

Fig. 16.9. Dew point temperature lines.

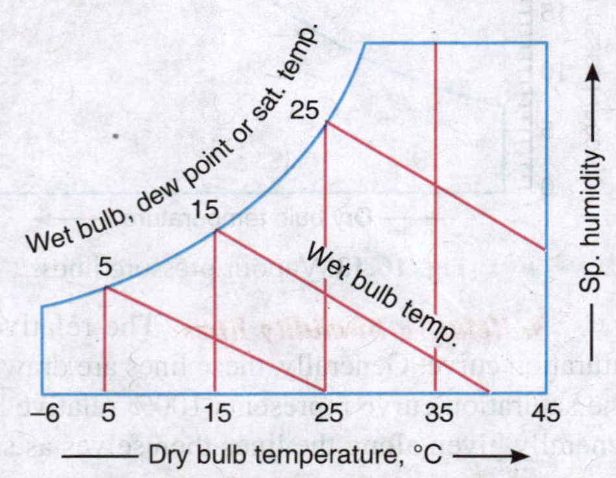

Fig. 16.10. Wet bulb temperature lines.

4. *Wet bulb temperature lines.* The wet bulb temperature lines are inclined straight lines and non-uniformly spaced as shown in Fig. 16.10. At any point on the saturation curve, the dry bulb and wet bulb temperatures are equal.

The values of wet bulb temperatures are generally given along the saturation curve of the chart as shown in the figure.

5. *Enthalpy (total heat) lines.* The enthalpy (or total heat) lines are inclined straight lines and uniformly spaced as shown in Fig. 16.11. These lines are parallel to the wet bulb temperature lines, and are drawn up to the saturation curve. Some of these lines coincide with the wet bulb temperature lines also.

The values of total enthalpy are given on a scale above the saturation curve as shown in the figure.

6. *Specific volume lines.* The specific volume lines are obliquely inclined straight lines and uniformly spaced as shown in Fig. 16.12. These lines are drawn up to the saturation curve.

The values of volume lines are generally given at the base of the chart.

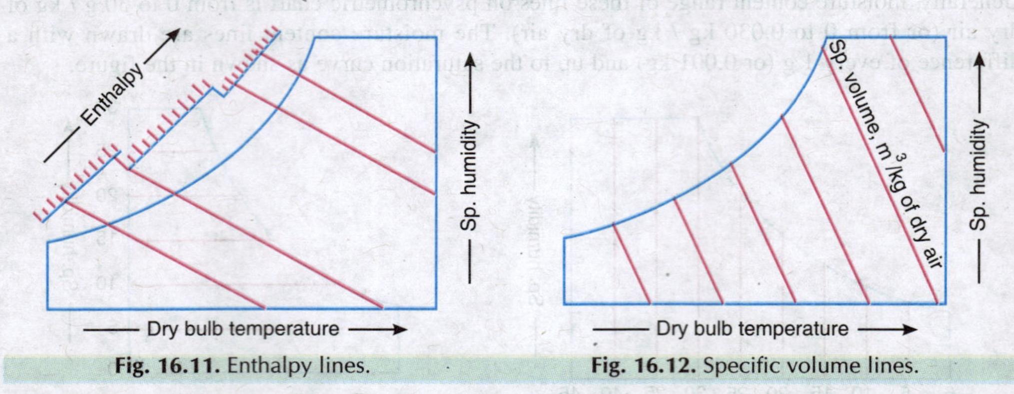

Fig. 16.11. Enthalpy lines. **Fig. 16.12.** Specific volume lines.

7. *Vapour pressure lines.* The vapour pressure lines are horizontal and uniformly spaced. Generally, the vapour pressure lines are not drawn in the main chart. But a scale showing vapour pressure in mm of Hg is given on the extreme left side of the chart as shown in Fig. 16.13.

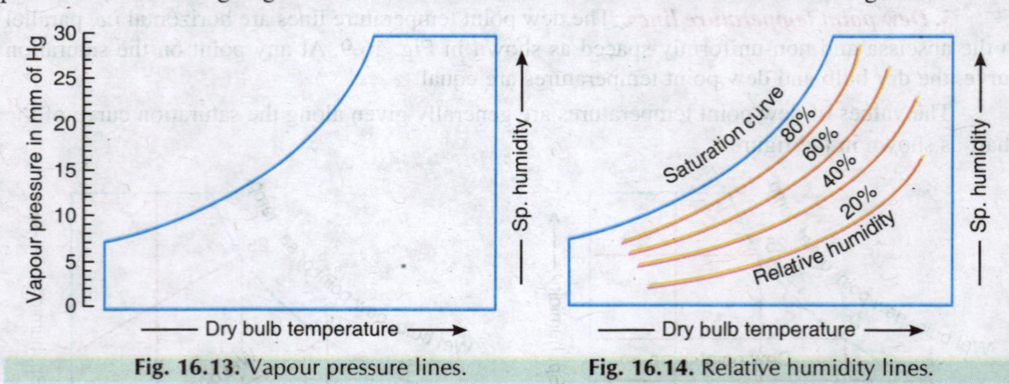

Fig. 16.13. Vapour pressure lines. **Fig. 16.14.** Relative humidity lines.

8. *Relative humidity lines.* The relative humidity lines are curved lines and follow the saturation curve. Generally, these lines are drawn with values 10%, 20%, 30% etc. and up to 100%. The saturation curve represents 100% relative humidity. The values of relative humidity lines are generally given along the lines themselves as shown in Fig. 16.14.

Example 16.6. *For a sample of air having 22°C DBT, relative humidity 30 per cent at barometric pressure of 760 mm of Hg, calculate : 1. Vapour pressure, 2. Humidity ratio, 3. Vapour density, and 4. Enthalpy.*

Verify your results by psychrometric chart.

Solution. Given : $t_d = 22°$ C ; $\phi = 30\% = 0.3$; $p_b = 760$ mm of Hg $= 760 \times 133.3$ $= 101\ 308$ N/m² $= 1.01308$ bar

1. *Vapour pressure*

Let p_v = Vapour pressure.

From steam tables, we find that the saturation pressure of vapour corresponding to dry bulb temperature of 22° C is

$$p_s = 0.026\,42 \text{ bar}$$

We know that relative humidity (ϕ),

$$0.3 = \frac{p_v}{p_s} = \frac{p_v}{0.026\,42}$$

$$\therefore \quad p_v = 0.3 \times 0.026\,42 = 0.007\,926 \text{ bar} \quad \textbf{Ans.}$$

2. *Humidity ratio*

We know that humidity ratio,

$$W = \frac{0.622\, p_v}{p_b - p_v} = \frac{0.622 \times 0.007\,926}{1.013\,08 - 0.007\,926}$$

$$= 0.0049 \text{ kg/kg of dry air} \quad \textbf{Ans.}$$

3. *Vapour density*

We know that vapour density,

$$\rho_v = \frac{W\,(p_b - p_v)}{R_a\, T_d} = \frac{0.0049\,(1.013\,08 - 0.007\,926)\,10^5}{287\,(273 + 22)}$$

$$= 0.005\,82 \text{ kg/m}^3 \text{ of dry air} \quad \textbf{Ans.}$$

4. *Enthalpy*

From steam tables, we find that saturation temperature or dew point temperature corresponding to a pressure of p_v = 0.007 926 bar is

$$t_{dp} = 3.8° \text{ C}$$

and latent heat of vaporisation of water at dew point temperature of 3.8° C is

$$h_{fgdp} = 2492.6 \text{ kJ/kg}$$

We know that enthalpy,

$$h = 1.022\, t_d + W\,(h_{fgdp} + 2.3\, t_{dp})$$

$$= 1.022 \times 22 + 0.0049\,(2492.6 + 2.3 \times 3.8)$$

$$= 22.484 + 12.256 = 34.74 \text{ kJ/kg of dry air} \quad \textbf{Ans.}$$

Verification from psychrometric chart

The initial condition of air *i.e.* 22° C dry bulb temperature and 30% relative humidity is marked on the psychrometric chart at point *A* as shown in Fig. 16.15.

From point *A*, draw a horizontal line meeting the vapour pressure line at point *B* and humidity ratio line at *C*. From the psychrometric chart, we find that vapour pressure at point *B*,

$$p_v = 5.94 \text{ mm of Hg}$$

$$= 5.94 \times 133.3 = 791.8 \text{ N/m}^2 = 0.007\,918 \text{ bar} \quad \textbf{Ans.}$$

and humidity ratio at point *C*,

$$W = 5 \text{ g/kg of dry air} = 0.005 \text{ kg/kg of dry air} \quad \textbf{Ans.}$$

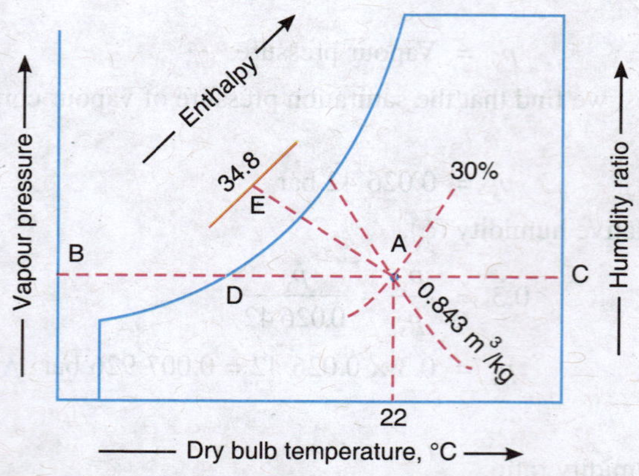

Fig. 16.15

We also find from the psychrometric chart that the specific volume at point A is 0.843 m^3/kg of dry air.

$\therefore$ Vapour density, $\rho_v = W/\rho_a = 0.005/0.843 = 0.0058$ kg/m^3 of dry air **Ans.**

Now from point A, draw a line parallel to the wet bulb temperature line meeting the enthalpy line at point E. Now the enthalpy of air as read from the chart is 34.8 kJ/kg of dry air. **Ans.**

16.8 Psychrometric Processes

The various psychrometric processes involved in air conditioning to vary the psychrometric properties of air according to the requirement are as follows :

1. Sensible heating, 2. Sensible cooling, 3. Humidification and dehumidification, 4. Cooling and adiabatic humidification, 5. Cooling and humidification by water injection, 6. Heating and humidification, 7. Humidification by steam injection, 8. Adiabatic chemical dehumidification, 9. Adiabatic mixing of air streams.

We shall now discuss these psychrometric processes, in detail, in the following pages.

16.9 Sensible Heating

The heating of air, without any change in its specific humidity, is known as *sensible heating*. Let air at temperature t_{d1} passes over a heating coil of temperature t_{d3}, as shown in Fig. 16.16 (*a*). It may be noted that the temperature of air leaving the heating coil (t_{d2}) will be less than t_{d3}. The process of sensible heating, on the psychrometric chart, is shown by a horizontal line 1-2 extending from left to right as shown in Fig. 16.16 (*b*). The point 3 represents the surface temperature of the heating coil.

The heat absorbed by the air during sensible heating may be obtained from the psychrometric chart by the enthalpy difference $(h_2 - h_1)$ as shown in Fig. 16.16 (*b*). It may be noted that the specific humidity during the sensible heating remains constant (*i.e.* $W_1 = W_2$). The dry bulb temperature increases from t_{d1} to t_{d2} and relative humidity reduces from ϕ_1 to ϕ_2 as shown in Fig. 16.16 (*b*). The amount of heat added during sensible heating may also be obtained from the relation :

Heat added,
$$q = h_2 - h_1$$
$$= c_{pa}(t_{d2} - t_{d1}) + W\,c_{ps}(t_{d2} - t_{d1})$$
$$= (c_{pa} + W.c_{ps})(t_{d2} - t_{d1}) = c_{pm}(t_{d2} - t_{d1})$$

The term $(c_{pa} + W\, c_{ps})$ is called *humid specific heat* (c_{pm}) and its value is taken as 1.022 kJ /kg K.

$\therefore$ Heat added, $q = 1.022\,(t_{d2} - t_{d1})$ kJ/kg

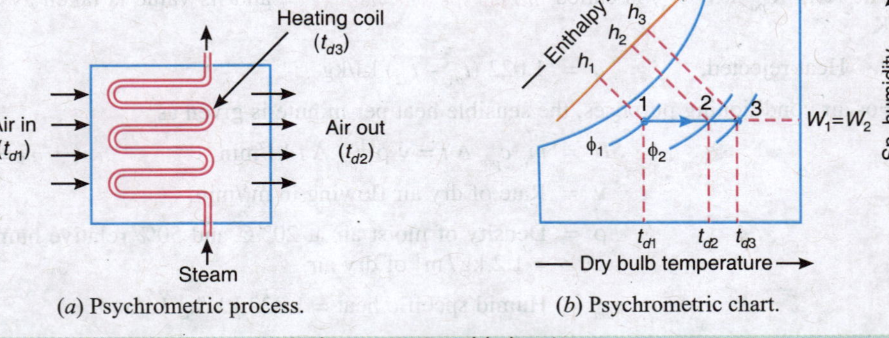

(*a*) Psychrometric process. (*b*) Psychrometric chart.

Fig. 16.16. Sensible heating.

Notes : 1. For sensible heating, steam or hot water is passed through the heating coil. The heating coil may be electric resistance coil.

2. The sensible heating of moist air can be done to any desired temperature.

16.10 Sensible Cooling

The cooling of air, without any change in its specific humidity, is known as *sensible cooling.* Let air at temperature t_{d1} passes over a cooling coil of temperature t_{d3} as shown in Fig. 16.17 (*a*). It may be noted that the temperature of air leaving the cooling coil (t_{d2}) will be more than t_{d3}. The process of sensible cooling, on the psychrometric chart, is shown by a horizontal line 1-2 extending from right to left as shown in Fig. 16.17 (*b*). The point 3 represents the surface temperature of the cooling coil.

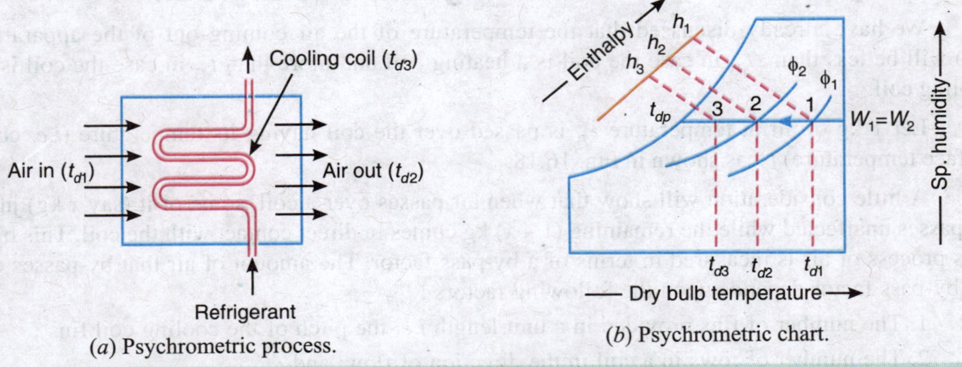

(*a*) Psychrometric process. (*b*) Psychrometric chart.

Fig. 16.17. Sensible cooling.

The heat rejected by air during sensible cooling may be obtained from the psychrometric chart by the enthalpy difference $(h_1 - h_2)$ as shown in Fig. 16.17 (*b*).

It may be noted that the specific humidity during the sensible cooling remains constant (*i.e.* $W_1 = W_2$). The dry bulb temperature reduces from t_{d1} to t_{d2} and relative humidity increases from ϕ_1 to ϕ_2, as shown in Fig. 16.17 (*b*). The amount of heat rejected during sensible cooling may also be obtained from the relation :

Heat rejected, $q = h_1 - h_2$

$$= c_{pa}\,(t_{d1} - t_{d2}) + W\,c_{ps}\,(t_{d1} - t_{d2})$$

$$= (c_{pa} + W\,c_{ps})\,(t_{d1} - t_{d2}) = c_{pm}\,(t_{d1} - t_{d2})$$

The term $(c_{pa} + W\,c_{ps})$ is called *humid specific heat* (c_{pm}) and its value is taken as 1.022 kJ /kg K.

$\therefore$ Heat rejected, $q = 1.022\,(t_{d1} - t_{d2})$ kJ/kg

For air conditioning purposes, the sensible heat per minute is given as

$$SH = m_a\,c_{pm}\,\Delta t = v\,\rho\,c_{pm}\,\Delta t \text{ kJ/min} \qquad \ldots (\because\ m = v\,\rho)$$

where v = Rate of dry air flowing in m³/min,

ρ = Density of moist air at 20° C and 50% relative humidity = 1.2 kg / m³ of dry air,

c_{pm} = Humid specific heat = 1.022 kJ /kg K, and

$\Delta t = t_{d1} - t_{d2}$ = Difference of dry bulb temperatures between the entering and leaving conditions of air in ° C.

Substituting the values of ρ and c_{pm} in the above expression, we get

$$SH = v \times 1.2 \times 1.022 \times \Delta t = 1.2264\,v \times \Delta t \text{ kJ/min}$$

$$= \frac{1.2264\,v \times \Delta t}{60} = 0.020\,44\,v \times \Delta t \text{ kJ/s or kW}$$

$$\ldots (\because\ 1 \text{ kJ/s} = 1 \text{ kW})$$

Notes : 1. For sensible cooling, the cooling coil may have refrigerant, cooling water or cool gas flowing through it.

2. The sensible cooling can be done only up to the dew point temperature (t_{dp}) as shown in Fig. 16.17 (*b*). The cooling below this temperature will result in the condensation of moisture.

16.11 By-pass Factor of Heating and Cooling Coil

We have already discussed that the temperature of the air coming out of the apparatus (t_{d2}) will be less than *t_{d3} in case the coil is a heating coil and more than t_{d3} in case the coil is a cooling coil.

Let 1 kg of air at temperature t_{d1} is passed over the coil having its temperature (*i.e.* coil surface temperature) t_{d3} as shown in Fig. 16.18.

A little consideration will show that when air passes over a coil, some of it (say x kg) just by-passes unaffected while the remaining $(1 - x)$ kg comes in direct contact with the coil. This by-pass process of air is measured in terms of a by-pass factor. The amount of air that by-passes or the by-pass factor depends upon the following factors :

1. The number of fins provided in a unit length *i.e.* the pitch of the cooling coil fins ;
2. The number of rows in a coil in the direction of flow; and
3. The velocity of flow of air.

It may be noted that the by-pass factor of a cooling coil decreases with decrease in fin spacing and increase in number of rows.

* Under ideal conditions, the dry bulb temperature of the air leaving the apparatus (t_{d2}) should be equal to that of the coil (t_{d3}). But it is not so, because of the inefficiency of the coil. This phenomenon is known as *by-pass factor*.

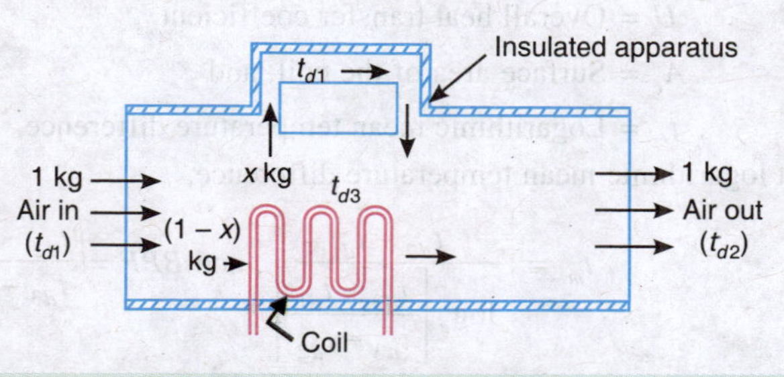

Fig. 16.18. By-pass factor.

Balancing the enthalpies, we get

$$x\, c_{pm}\, t_{d1} + (1 - x)\, c_{pm}\, t_{d3} = 1 \times c_{pm}\, t_{d2} \qquad \text{... (where } c_{pm} = \text{Specific humid heat)}$$

or
$$x\,(t_{d3} - t_{d1}) = t_{d3} - t_{d2}$$

$$\therefore \qquad x = \frac{t_{d3} - t_{d2}}{t_{d3} - t_{d1}}$$

where x is called the *by-pass factor* of the coil and is generally written as *BPF*. Therefore, by-pass factor for heating coil,

$$BPF = \frac{t_{d3} - t_{d2}}{t_{d3} - t_{d1}}$$

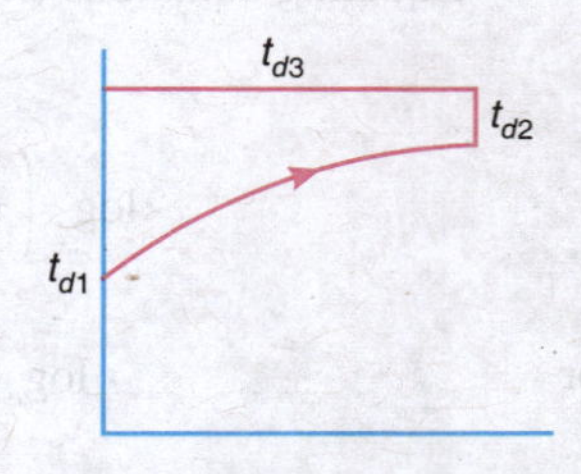

Fig. 16.19

Similarly, *by-pass factor for cooling coil,

$$BPF = \frac{t_{d2} - t_{d3}}{t_{d1} - t_{d3}}$$

The by-pass factor for heating or cooling coil may also be obtained as discussed below :

Let the air passes over a heating coil. Since the temperature distribution of air passing through the heating coil is as shown in Fig. 16.19, therefore sensible heat given out by the coil,

$$Q_s = U\, A_c\, t_m \qquad \ldots (i)$$

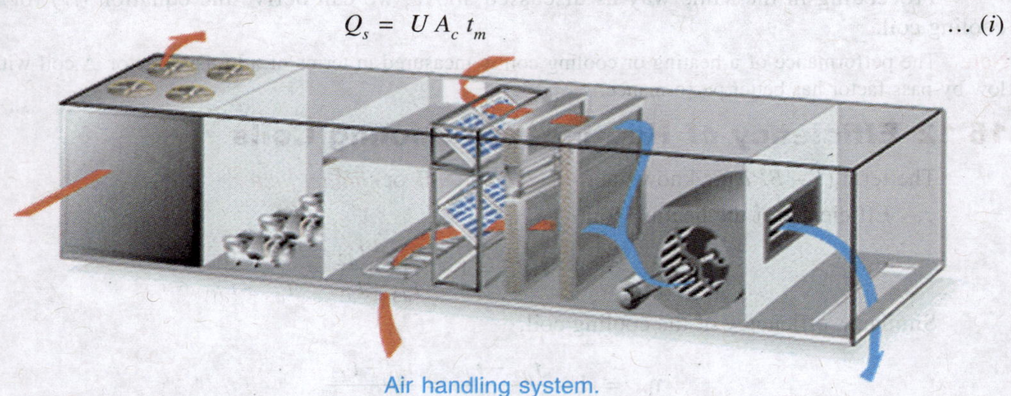

Air handling system.

* If *BPF* of one row of the coil is x, then *BPF* of n rows of similar coil will be $(x)^n$.

where U = Overall heat transfer coefficient,

A_c = Surface area of the coil, and

t_m = Logarithmic mean temperature difference.

We know that logarithmic mean temperature difference,

$$t_m = \frac{t_{d2} - t_{d1}}{\log_e \left[\frac{t_{d3} - t_{d1}}{t_{d3} - t_{d2}}\right]}, \text{ and } BPF = \frac{t_{d3} - t_{d2}}{t_{d3} - t_{d1}}$$

$$\therefore \quad t_m = \frac{t_{d2} - t_{d1}}{\log_e (1/BPF)}$$

Now the equation (i) may be written as

$$Q_s = U \times A_c \times \frac{t_{d2} - t_{d1}}{\log_e (1/BPF)} \quad \ldots (ii)$$

We have already discussed that the heat added during sensible heating,

$$Q_s = m_a\, c_{pm}\, (t_{d2} - t_{d1}) \quad \ldots (iii)$$

where c_{pm} = Humid specific heat = 1.022 kJ/kg K, and

m_a = Mass of air passing over the coil.

Equating equations (ii) and (iii), we have

$$UA_c = m_a\, c_{pm} \log_e (1/BPF)$$

$$\log_e \left(\frac{1}{BPF}\right) = \frac{UA_c}{m_a\, c_{pm}}$$

or

$$\log_e (BPF) = -\frac{UA_c}{m_a\, c_{pm}}$$

$$\therefore \quad BPF = e^{-\left(\frac{UA_c}{m_a\, c_{pm}}\right)} = e^{-\left(\frac{UA_c}{1.022\, m_a}\right)} \quad \ldots (iv)$$

Proceeding in the same way as discussed above, we can derive the equation (iv) for a cooling coil.

Note : The performance of a heating or cooling coil is measured in terms of a by-pass factor. A coil with low by-pass factor has better performance.

16.12 Efficiency of Heating and Cooling Coils

The term $(1 - BPF)$ is known as *efficiency of coil* or *contact factor.*

∴ Efficiency of the heating coil,

$$\eta_H = 1 - BPF = 1 - \frac{t_{d3} - t_{d2}}{t_{d3} - t_{d1}} = \frac{t_{d2} - t_{d1}}{t_{d3} - t_{d1}}$$

Similarly, efficiency of the cooling coil,

$$\eta_C = 1 - \frac{t_{d2} - t_{d3}}{t_{d1} - t_{d3}} = \frac{t_{d1} - t_{d2}}{t_{d1} - t_{d3}}$$

Example 16.7. *In a heating application, moist air enters a steam heating coil at 10° C, 50% RH and leaves at 30° C. Determine the sensible heat transfer, if mass flow rate of air is 100 kg of dry air per second. Also determine the steam mass flow rate if steam enters saturated at 100°C and condensate leaves at 80° C.*

Solution. Given : t_{d1} = 10° C ; ϕ_1 = 50% ; t_{d2} = 30° C ; m_a = 100 kg/s ; t_s = 100° C ; t_C = 80° C

Sensible heat transfer

First of all, mark the initial condition of air, *i.e.* 10° C dry bulb temperature and 50% relative humidity on the psychrometric chart at point 1, as shown in Fig. 16.20. Draw a constant specific humidity line from point 1 to intersect the vertical line drawn through 30° C dry bulb temperature at point 2. The line 1-2 represents sensible heating of air.

Enthalpy
h_1
h_2
50%
1
2
10
30
Dry bulb temperature, °C
Sp. humidity

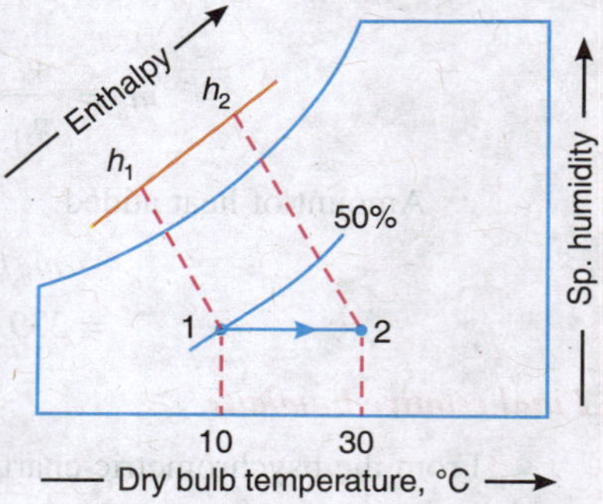

Fig. 16.20

From the psychrometric chart, we find that enthalpy at point 1,

$$h_1 = 19.3 \text{ kJ/kg of dry air}$$

and enthalpy at point 2,

$$h_2 = 39.8 \text{ kJ/kg of dry air}$$

We know that sensible heat transfer,

$$Q = m_a (h_2 - h_1) = 100 (39.8 - 19.3) = 2050 \text{ kJ/s} \quad \textbf{Ans.}$$

Steam mass flow rate

From steam tables, corresponding to a temperature of 100° C , we find that the enthalpy of saturated steam,

$$h_g = 2676 \text{ kJ/kg}$$

and enthalpy of condensate, corresponding to 80° C,

$$h_f = 334.9 \text{ kJ/kg}$$

∴ Steam mass flow rate

$$= \frac{Q}{h_g - h_f} = \frac{2050}{2676 - 334.9} = 0.8756 \text{ kg/s}$$

$$= 0.8756 \times 3600 = 3152 \text{ kg/h} \quad \textbf{Ans.}$$

Example 16.8. *A quantity of air having a volume of 300 m³ at 30° C dry bulb temperature and 25° C wet bulb temperature is heated to 40° C dry bulb temperature. Estimate the amount of heat added, final relative humidity and wet bulb temperature. The air pressure is 1.013 25 bar.*

Solution. Given : v_1 = 300 m³ ; t_{d1} = 30° C ; t_{w1} = 25° C ; t_{d2} = 40° C ; p_b = 1.013 25 bar

First of all, mark the initial condition of air *i.e.* at 30° C dry bulb temperature and 25° C wet bulb temperature on the psychrometric chart at point 1, as shown in Fig. 16.21. Draw a constant specific humidity line from point 1 to intersect the vertical line drawn through 40° C dry bulb temperature at point 2. The line 1-2 represents sensible heating of air.

Amount of heat added

From the psychrometric chart, we find that specific volume of air at point 1,

$$v_{s1} = 0.883 \text{ m}^3\text{/kg of dry air}$$

Enthalpy at point 1,

$$h_1 = 76 \text{ kJ/kg of dry air}$$

and enthalpy at point 2,

$$h_2 = 86.4 \text{ kJ/kg of dry air}$$

We know that amount of air supplied,

$$m_a = \frac{v_1}{v_{s1}} = \frac{300}{0.883} = 339.75 \text{ kg}$$

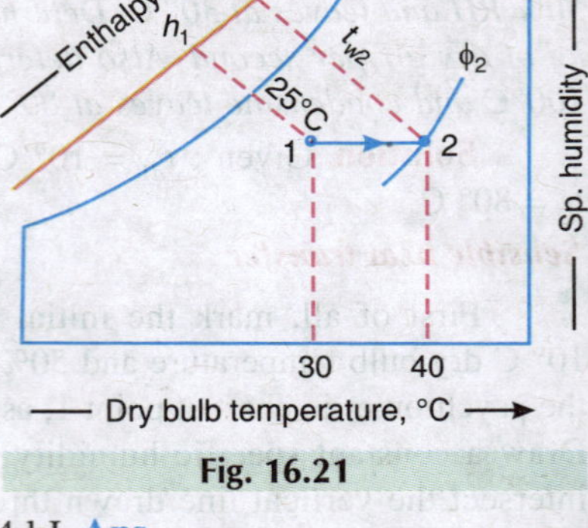

Fig. 16.21

$\therefore$ Amount of heat added

$$= m_a (h_2 - h_1)$$
$$= 339.75 (86.4 - 76) = 3533.4 \text{ kJ} \quad \textbf{Ans.}$$

Final relative humidity

From the psychrometric chart, we find that the relative humidity at point 2 is

$$\phi_2 = 39\% \quad \textbf{Ans.}$$

Wet bulb temperature

From the psychrometric chart, we find that the wet bulb temperature at point 2 is

$$t_{w2} = 27.5° \text{ C} \quad \textbf{Ans.}$$

Example 16.9. *The air enters a duct at 10° C and 80% RH at the rate of 150 m³/min and is heated to 30° C without adding or removing any moisture. The pressure remains constant at 1 atmosphere. Determine the relative humidity of air at exit from the duct and the rate of heat transfer.*

Solution. Given : $t_{d1} = 10°$ C ; $\phi_1 = 80\%$; $v_1 = 150$ m³/min ; $t_{d2} = 30°$ C ; $p = p_b = 1$ atm = 1.013 bar

Relative humidity of air at exit

First of all, mark the initial condition of air *i.e.* at 10° C dry bulb temperature and 80% relative humidity, on the psychrometric chart at point 1, as shown in Fig. 16.22. Since air is heated to 30° C without adding or removing any moisture, therefore it is a case of sensible heating. Draw a constant specific humidity line from point 1 to intersect the vertical line drawn through 30° C dry bulb temperature, at point 2. The line 1-2 represents sensible heating of air.

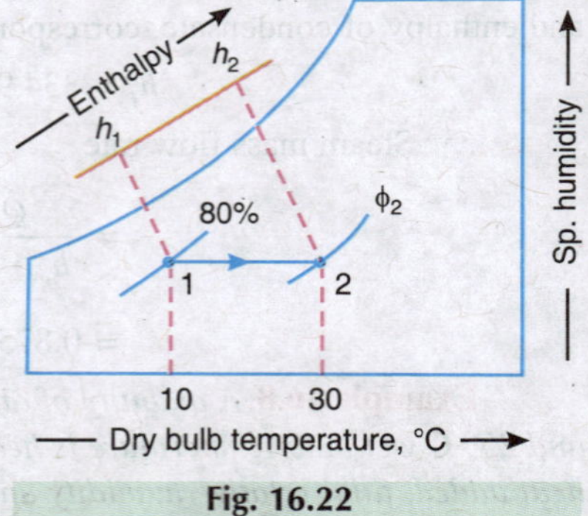

Fig. 16.22

From the psychrometric chart, we find that the relative humidity of air at exit *i.e.* at point 2,

$$\phi_2 = 23.5\% \quad \textbf{Ans.}$$

Rate of heat transfer

From the psychrometric chart, we also find that the specific volume of air at point 1,

$$v_{s1} = 0.81 \text{ m}^3\text{/kg of dry air}$$

Enthalpy of air at point 1,

$$h_1 = 26 \text{ kJ/kg of dry air}$$

and enthalpy of air at point 2,

$$h_2 = 46 \text{ kJ/kg of dry air}$$

We know that amount of air supplied,

$$^*m_a = \frac{v_1}{v_{s1}} = \frac{150}{0.81} = 185.2 \text{ m}^3/\text{min}$$

∴ Rate of heat transfer

$$= m_a (h_2 - h_1) = 185.2 (46 - 26) = 3704 \text{ kJ/min} \quad \textbf{Ans.}$$

Example 16.10. *Atmospheric air with dry bulb temperature of 28° C and a wet bulb temperature of 17° C is cooled to 15° C without changing its moisture content. Find : 1. Original relative humidity ; 2. Final relative humidity ; and 3. Final wet bulb temperature.*

Solution. Given : $t_{d1} = 28°$ C ; $t_{w1} = 17°$ C ; $t_{d2} = 15°$ C

The initial condition of air, *i.e.* 28° C dry bulb temperature and 17° C wet bulb temperature is marked on the psychrometric chart at point 1, as shown in Fig. 16.23. Now mark the final condition of air by drawing a horizontal line through point 1 (because there is no change in moisture content of the air) to meet the 15° C dry bulb temperature line at point 2, as shown in Fig. 16.23.

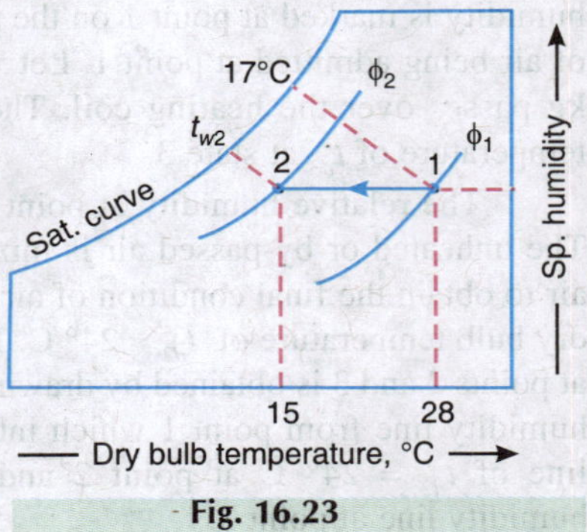

Fig. 16.23

1. *Original relative humidity*

From the psychrometric chart, we find that the original relative humidity at point 1,

$$\phi_1 = 34\% \quad \textbf{Ans.}$$

2. *Final relative humidity*

From psychrometric chart, we find that the final relative humidity at point 2,

$$\phi_2 = 73\% \quad \textbf{Ans.}$$

3. *Final wet bulb temperature*

From the psychrometric chart, we find that the final wet bulb temperature at point 2,

$$t_{w2} = 12.2° \text{ C} \quad \textbf{Ans.}$$

Example 16.11. *The moist air is heated by steam condensing inside the tubes of a heating coil as shown in Fig. 16.24. The part of the air passes through the coil and part is by-passed around the coil. The barometric pressure is 1 bar. Determine : 1. The air per minute (in 3′) which by-pass the coil ; and 2. The heat added by the coil.*

---

* The amount of air supplied (m_a) may also be obtained as discussed below :

From steam tables, we find that saturation pressure of vapour corresponding to dry bulb temperature of 10° C is

$$p_{s1} = 0.012\ 27 \text{ bar}$$

We know that partial pressure of vapour,

$$p_{v1} = \phi_1 \times p_{s1} = 0.8 \times 0.012\ 27 = 0.009\ 82 \text{ bar} \qquad \ldots (\because \ \phi_1 = p_{v1}/p_{s1})$$

$$\therefore \quad m_a = \frac{(p_b - p_{v1})\, v_1}{R_a\, T_{d1}} = \frac{(1.013 - 0.009\ 82)\, 10^5 \times 150}{287\,(273 + 10)} = 185.2 \text{ m}^3/\text{min}$$

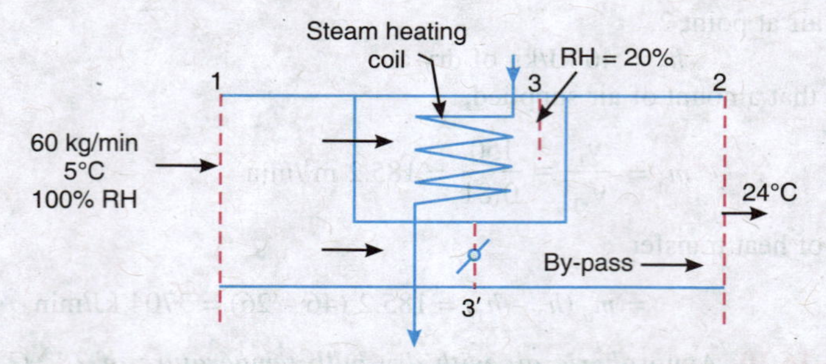

Fig. 16.24

Solution. Given : $m_a = 60$ kg/min ; $t_{d1} = 5°$ C ; $\phi_1 = 100\%$; $\phi_3 = 20\%$; $t_{d2} = 24°$ C

First of all, the initial condition of air at a dry bulb temperature of 5° C and 100% relative humidity is marked at point 1 on the psychrometric chart , as shown in Fig. 16.25. Consider 1 kg of air being admitted at point 1. Let x kg of air by-passes unaffected while the remaining $(1 - x)$ kg passes over the heating coil. The heating coil raises the temperature of air to a dry bulb temperature of t_{d3} at state 3.

The relative humidity at point 3 is given as 20%. The unheated or by-passed air is mixed with the heated air to obtain the final condition of air at point 2 having a dry bulb temperature of $t_{d2} = 24°$ C. The condition of air at points 2 and 3 is obtained by drawing constant specific humidity line from point 1 which intersects the vertical line of $t_{d2} = 24°$ C at point 2 and the 20% relative humidity line at point 3.

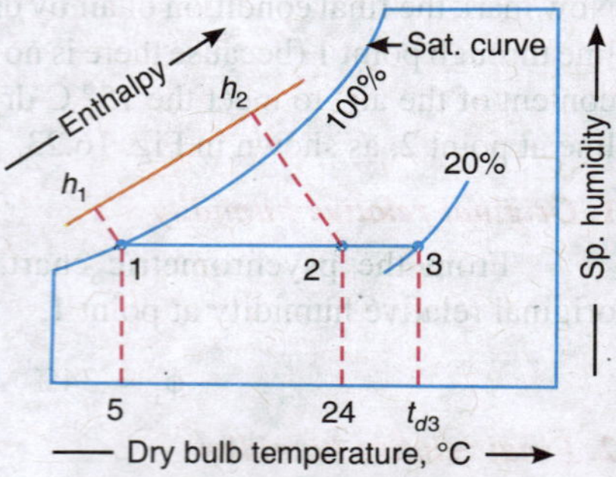

Fig. 16.25

From the psychrometric chart, we find that dry bulb temperature of air leaving the heating coil, at point 3,

$$t_{d3} = 30.8° \text{ C}$$

Enthalpy of air at point 1,

$$h_1 = 19 \text{ kJ/kg of dry air}$$

Enthalpy of air at point 2,

$$h_2 = 38 \text{ kJ/kg of dry air}$$

and enthalpy of air at point 3,

$$h_3 = 44.8 \text{ kJ/kg of dry air}$$

1. ***Air per minute which by-pass the coil***

We know that

$$*x = \frac{t_{d3} - t_{d2}}{t_{d3} - t_{d1}} = \frac{30.8 - 24}{30.8 - 5} = \frac{6.8}{25.8} = 0.264$$

Air which by-pass the coil

$$= x \times m_a = 0.264 \times 60 = 15.84 \text{ kg/min}$$ **Ans.**

* The value of x may also be obtained by balancing the enthalpies as follows :

$$h_2 = x\,h_1 + (1 - x)\,h_3$$

or

$$x = \frac{h_3 - h_2}{h_3 - h_1} = \frac{44.8 - 38}{44.8 - 19} = \frac{6.8}{25.8} = 0.264$$

2. Heat added by the coil

We know that heat added by the coil

$$= m_a (h_2 - h_1) = 60 (38 - 19) = 1140 \text{ kJ/min} \quad \textbf{Ans.}$$

Example 16.12. *The atmospheric air at 760 mm of Hg, dry bulb temperature 15° C and wet bulb temperature 11° C enters a heating coil whose temperature is 41° C . Assuming by-pass factor of heating coil as 0.5, determine dry bulb temperature; wet bulb temperature and relative humidity of the air leaving the coil. Also determine the sensible heat added to the air per kg of dry air.*

Solution. Given : p_b = 760 mm of Hg ; t_{d1} = 15° C ; t_{w1} = 11° C ; t_{d3} = 41° C ; BPF = 0.5

The initial condition of air entering the coil at dry bulb temperature of 15° C and wet bulb temperature of 11° C is shown by point 1 on the psychrometric chart as shown in Fig. 16.26. Now draw a constant specific humidity line from point 1 to intersect the vertical line drawn through 41° C at point 3. The point 2 lies on the line 1-3.

Dry bulb temperature of the air leaving the coil

Let t_{d2} = Dry bulb temperature of the air leaving the coil.

We know that by-pass factor (BPF),

$$0.5 = \frac{t_{d3} - t_{d2}}{t_{d3} - t_{d1}} = \frac{41 - t_{d2}}{41 - 15} = \frac{41 - t_{d2}}{26}$$

$$\therefore \quad t_{d2} = 41 - 0.5 \times 26 = 28° \text{ C} \quad \textbf{Ans.}$$

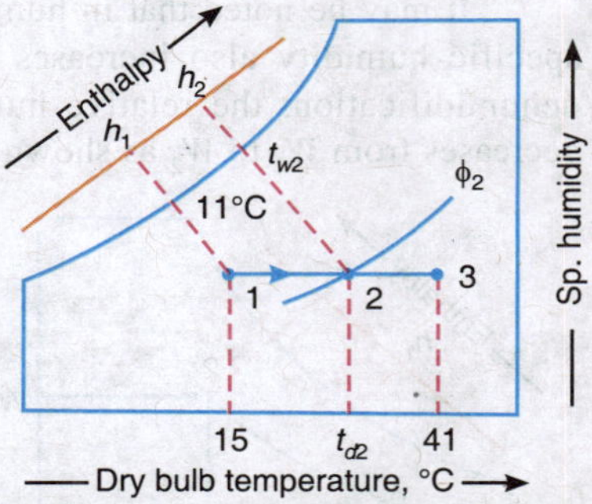

Enthalpy
h_2
h_1
t_{w2}
ϕ_2
11°C
1
2
3
Sp. humidity
15
t_{d2}
41
Dry bulb temperature, °C

Fig. 16.26

Wet bulb temperature of the air leaving the coil

From the psychrometric chart, we find that the wet bulb temperature of the air leaving the coil at point 2 is

$$t_{w2} = 16.1° \text{ C} \quad \textbf{Ans.}$$

Relative humidity of the air leaving the coil

From the psychrometric chart, we find that the relative humidity of the air leaving the coil at point 2 is

$$\phi_2 = 29\% \quad \textbf{Ans.}$$

Sensible heat added to the air per kg of dry air

From the psychrometric chart, we find that enthalpy of air at point 2,

$$h_2 = 46 \text{ kJ/kg of dry air}$$

and enthalpy of air at point 1,

$$h_1 = 31.8 \text{ kJ/kg of dry air}$$

∴ Sensible heat added to the air per kg of dry air,

$$= h_2 - h_1 = 46 - 31.8 = 14.2 \text{ kJ/kg of dry air} \quad \textbf{Ans.}$$

16.13 Humidification and Dehumidification

The addition of moisture to the air, without change in its dry bulb temperature, is known as *humidification*. Similarly, removal of moisture from the air, without change in its dry bulb temperature, is known as *dehumidification*. The heat added during humidification process and heat removed during dehumidification process is shown on the psychrometric chart in Fig. 16.27 (*a*) and (*b*) respectively.

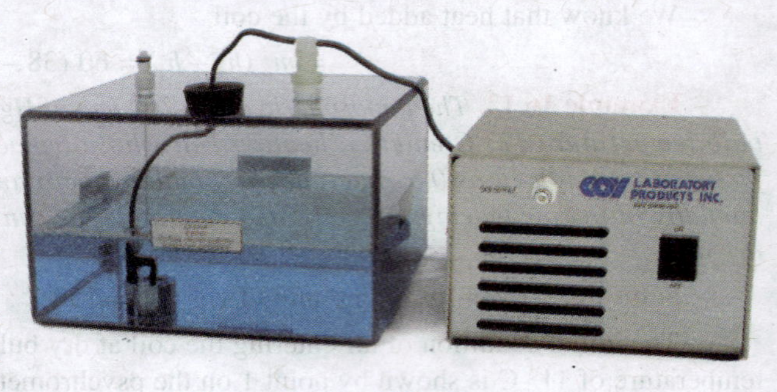

Ultrasonic humidification system.

It may be noted that in humidification, the relative humidity increases from ϕ_1 to ϕ_2 and specific humidity also increases from W_1 to W_2 as shown in Fig. 16.27 (*a*). Similarly, in dehumidification, the relative humidity decreases from ϕ_1 to ϕ_2 and specific humidity also decreases from W_1 to W_2 as shown in Fig. 16.27 (*b*).

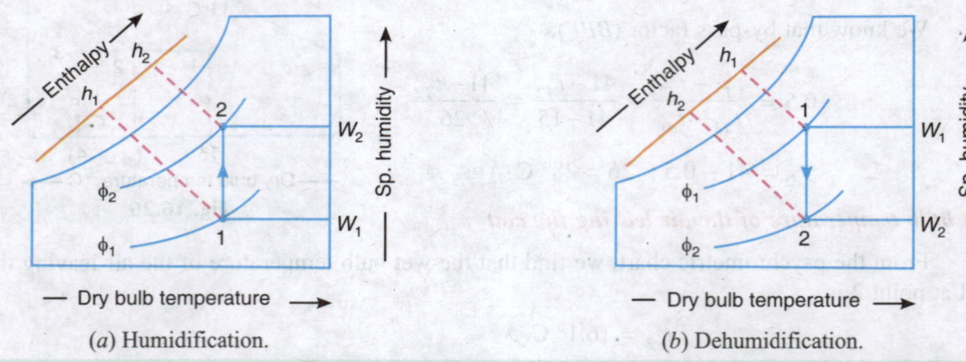

(*a*) Humidification. (*b*) Dehumidification.

Fig. 16.27. Humidification and dehumidification.

It may be noted that in humidification, change in enthalpy is shown by the intercept $(h_2 - h_1)$ on the psychrometric chart. Since the dry bulb temperature of air during the humidification remains constant, therefore its sensible heat also remains constant. It is thus obvious that the change in enthalpy per kg of dry air due to the increased moisture content equal to $(W_2 - W_1)$ kg per kg of dry air is considered to cause a latent heat transfer (*LH*). Mathematically,

$$LH = (h_2 - h_1) = h_{fg}\ (W_2 - W_1)$$

where h_{fg} is the latent heat of vaporisation at dry bulb temperature (t_{d1}).

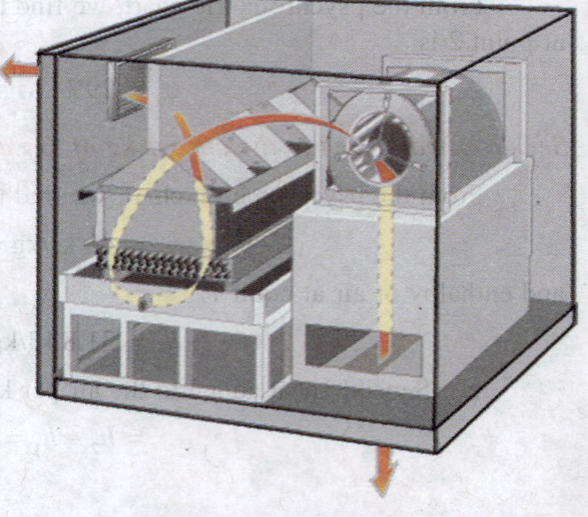

Multiple small plate dehumidification system

Notes : 1. For dehumidification, the above equation may be written as :

$$LH = (h_1 - h_2) = h_{fg}\ (W_1 - W_2)$$

2. Absolute humidification and dehumidification processes are rarely found in practice. These are always accompanied by heating or cooling processes.

3. In air conditioning, the latent heat load per minute is given as

$$LH = m_a\ \Delta h = m_a\ h_{fg}\ \Delta W = v\ \rho\ h_{fg}\ \Delta W \qquad \ldots (\because\ m_a = v\ \rho)$$

where v = Rate of dry air flowing in m^3/min,

ρ = Density of moist air = 1.2 kg/m^3 of dry air,

h_{fg} = Latent heat of vaporisation = 2500 kJ /kg, and

ΔW = Difference of specific humidity between the entering and leaving conditions of air = $(W_2 - W_1)$ for humidification and $(W_1 - W_2)$ for dehumidification.

Substituting these values in the above expression, we get

$$LH = v \times 1.2 \times 2500 \times \Delta W = 3000\ v \times \Delta W \text{ kJ/min}$$

$$= \frac{3000\ v \times \Delta W}{60} = 50\ v \times \Delta W \text{ kJ/s or kW}$$

16.14 Methods of Obtaining Humidification and Dehumidification

The humidification is achieved either by supplying or spraying steam or hot water or cold water into the air. The humidification may be obtained by the following two methods :

1. ***Direct method.*** In this method, the water is sprayed in a highly atomised state into the room to be air-conditioned. This method of obtaining humidification is not very effective.

2. ***Indirect method.*** In this method, the water is introduced into the air in the air-conditioning plant, with the help of an air-washer, as shown in Fig. 16.28. This conditioned air is then supplied to the room to be air-conditioned. The air-washer humidification may be accomplished in the following three ways :

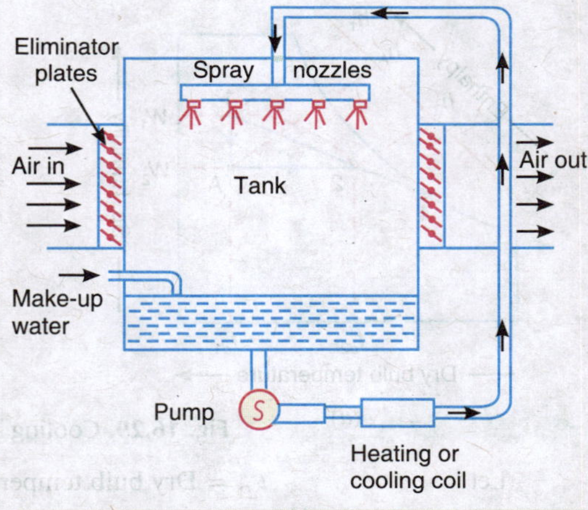

Fig. 16.28. Air-washer.

(*a*) by using re-circulated spray water without prior heating of air,

(*b*) by pre-heating the air and then washing it with re-circulated water, and

(*c*) by using heated spray water.

The dehumidification may be accomplished with the help of an air-washer or by using chemicals. In the air-washer system, the outside or entering air is cooled below its dew point temperature so that it loses moisture by condensation. The moisture removal is also accomplished when the spray water is chilled water and its temperature is lower than the dew point temperature of the entering air. Since the air leaving the air-washer has its dry bulb temperature much below the desired temperature in the room, therefore a heating coil is placed after the air-washer. The dehumidification may also be achieved by using chemicals which have the capacity to absorb moisture in them. Two types of chemicals known as absorbents (such as calcium chloride) and adsorbents (such as silica gel and activated alumina) are commonly used for this purpose.

16.15 Sensible Heat Factor

As a matter of fact, the heat added during a psychrometric process may be split up into sensible heat and latent heat. The ratio of the *sensible heat to the total heat is known as *sensible heat factor* (briefly written as *SHF*) or *sensible heat ratio* (briefly written as *SHR*). Mathematically,

$$SHF = \frac{\text{Sensible heat}}{\text{Total heat}} = \frac{SH}{SH + LH}$$

where SH = Sensible heat, and

LH = Latent heat.

The sensible heat factor scale is shown on the right hand side of the psychrometric chart.

16.16 Cooling and Dehumidification

This process is generally used in summer air conditioning to cool and dehumidify the air. The air is passed over a cooling coil or through a cold water spray. In this process, the dry bulb temperature as well as the specific humidity of air decreases. The final relative humidity of the air is generally higher than that of the entering air. The dehumidification of air is only possible when the effective surface temperature of the cooling coil (*i.e.* t_{d4}) is less than the dew point temperature of the air entering the coil (*i.e.* t_{dp1}). The effective surface temperature of the coil is known as *apparatus dew point* (briefly written as *ADP*). The cooling and dehumidification process is shown in Fig. 16.29.

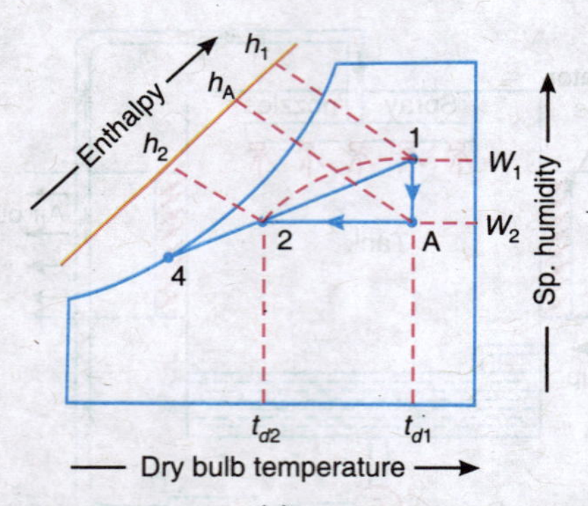

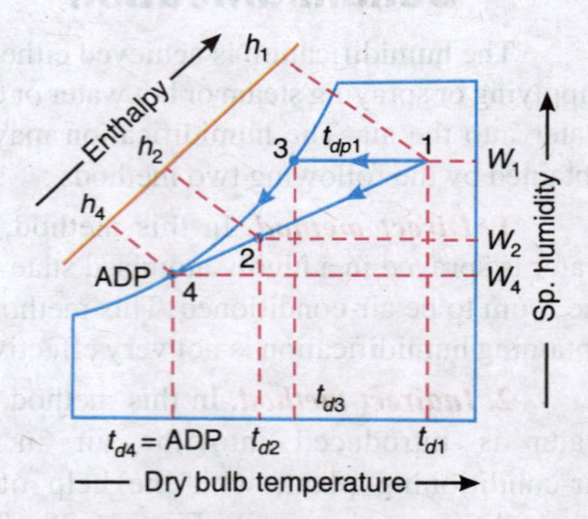

(*a*) **Fig. 16.29.** Cooling and dehumidification. (*b*)

Let t_{d1} = Dry bulb temperature of air entering the coil,

t_{dp1} = Dew point temperature of the entering air = t_{d3}, and

t_{d4} = Effective surface temperature or *ADP* of the coil.

Under ideal conditions, the dry bulb temperature of the air leaving the cooling coil (*i.e.* t_{d4}) should be equal to the surface temperature of the cooling coil (*i.e. ADP*), but it is never possible due to inefficiency of the cooling coil. Therefore, the resulting condition of air coming out of the coil is shown by a point 2 on the straight line joining the points 1 and 4. The by-pass factor in this case is given by

$$BPF = \frac{t_{d2} - t_{d4}}{t_{d1} - t_{d4}} = \frac{t_{d2} - ADP}{t_{d1} - ADP}$$

Also
$$BPF = \frac{W_2 - W_4}{W_1 - W_4} = \frac{h_2 - h_4}{h_1 - h_4}$$

* Refer also Chapter 18, Art 18.13.

Actually, the cooling and dehumidification process follows the path as shown by a dotted curve in Fig. 16.29 (*a*), but for the calculation of psychrometric properties, only end points are important. Thus the cooling and dehumidification process shown by a line 1-2 may be assumed to have followed a path 1-*A* (*i.e.* dehumidification) and *A*-2 (*i.e.* cooling) as shown in Fig. 16.29 (*a*). We see that the total heat removed from the air during the cooling and dehumidification process is

$$q = h_1 - h_2 = (h_1 - h_A) + (h_A - h_2) = LH + SH$$

where $LH = h_1 - h_A$ = Latent heat removed due to condensation of vapour of the reduced moisture content $(W_1 - W_2)$, and

$SH = h_A - h_2$ = Sensible heat removed.

We know that sensible heat factor,

$$SHF = \frac{\text{Sensible heat}}{\text{Total heat}} = \frac{SH}{LH + SH} = \frac{h_A - h_2}{h_1 - h_2}$$

Note : The line 1-4 (*i.e.* the line joining the point of entering air and the apparatus dew point) in Fig. 16.29 (*b*) is known as *sensible heat factor line.*

Example 16.13. *In a cooling application, moist air enters a refrigeration coil at the rate of 100 kg of dry air per minute at 35° C and 50% RH. The apparatus dew point of coil is 5° C and by-pass factor is 0.15. Determine the outlet state of moist air and cooling capacity of coil in TR.*

Solution. Given : $m_a = 100$ kg/min ; $t_{d1} = 35°$ C ; $\phi_1 = 50\%$; $ADP = 5°$ C ; $BPF = 0.15$

Outlet state of moist air

Let t_{d2} and ϕ_2 = Temperature and relative humidity of air leaving the cooling coil.

First of all, mark the initial condition of air, *i.e.* 35° C dry bulb temperature and 50% relative humidity on the psychrometric chart at point 1, as shown in Fig. 16.30. From the psychrometric chart, we find that the dew point temperature of the entering air at point 1,

$$t_{dp1} = 23° \text{ C}$$

Since the coil or apparatus dew point (*ADP*) is less than the dew point temperature of entering air, therefore it is a process of cooling and dehumidification.

We know that by-pass factor,

$$BPF = \frac{t_{d2} - t_{d4}}{t_{d1} - t_{d4}} = \frac{t_{d2} - ADP}{t_{d1} - ADP}$$

or

$$0.15 = \frac{t_{d2} - 5}{35 - 5} = \frac{t_{d2} - 5}{30}$$

$$\therefore \quad t_{d2} = 0.15 \times 30 + 5 = 9.5° \text{ C} \quad \textbf{Ans.}$$

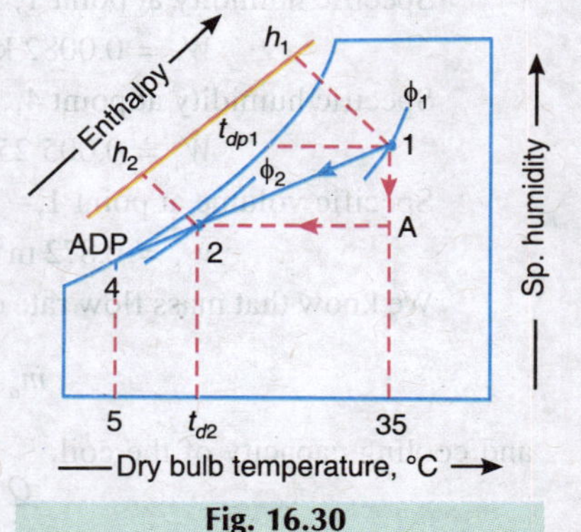

Fig. 16.30

From the psychrometric chart, we find that the relative humidity corresponding to a dry bulb temperature (t_{d2}) of 9.5° C on the line 1-4 is $\phi_2 = 99\%$. **Ans.**

Cooling capacity of the coil

The resulting condition of the air coming out of the coil is shown by point 2, on the line joining the points 1 and 4, as shown in Fig. 16.30. The line 1-2 represents the cooling and dehumidification process which may be assumed to have followed the path 1-*A* (*i.e.*

dehumidification) and A-2 (*i.e.* cooling). Now from the psychrometric chart, we find that enthalpy of entering air at point 1,

$$h_1 = 81 \text{ kJ/kg of dry air}$$

and enthalpy of air at point 2,

$$h_2 = 28 \text{ kJ/kg of dry air}$$

We know that cooling capacity of the coil

$$= m_a (h_1 - h_2) = 100\,(81 - 28) = 5300 \text{ kJ/min}$$

$$= 5300/210 = 25.24 \text{ TR} \quad \textbf{Ans.} \qquad \ldots (\because \; 1 \text{ TR} = 210 \text{ kJ/min})$$

Example 16.14. *39.6 m³/min of a mixture of recirculated room air and outdoor air enters a cooling coil at 31°C dry bulb temperature and 18.5°C wet bulb temperature. The effective surface temperature of the coil is 4.4°C. The surface area of the coil is such as would give 12.5 kW of refrigeration with the given entering air state. Determine the dry and wet bulb temperatures of the air leaving the coil and the by-pass factor.*

Solution. Given : $v_1 = 39.6 \text{ m}^3/\text{min}$; $t_{d1} = 31°\text{C}$; $t_{w1} = 18.5°\text{C}$; $ADP = t_{d4} = 4.4°\text{C}$; $Q = 12.5$ kW $= 12.5$ kJ/s $= 12.5 \times 60$ kJ/min

Dry and wet bulb temperature of the air leaving the coil

Let t_{d2} and t_{w2} = Dry and wet bulb temperature of the air leaving the coil.

First of all, mark the initial condition of air, *i.e.* 31°C dry bulb temperature and 18.5°C wet bulb temperature on the psychrometric chart at point 1, as shown in Fig. 16.31. Now mark the effective surface temperature (*ADP*) of the coil at 4.4°C at point 4.

From the psychrometric chart, we find that enthalpy at point 1,

$$h_1 = 52.5 \text{ kJ / kg of dry air}$$

Enthalpy at point 4,

$$h_4 = 17.7 \text{ kJ / kg of dry air}$$

Specific humidity at point 1,

$$W_1 = 0.0082 \text{ kg / kg of dry air}$$

Specific humidity at point 4,

$$W_4 = 0.005\,25 \text{ kg / kg of dry air}$$

Specific volume at point 1,

$$v_{s1} = 0.872 \text{ m}^3/\text{kg}$$

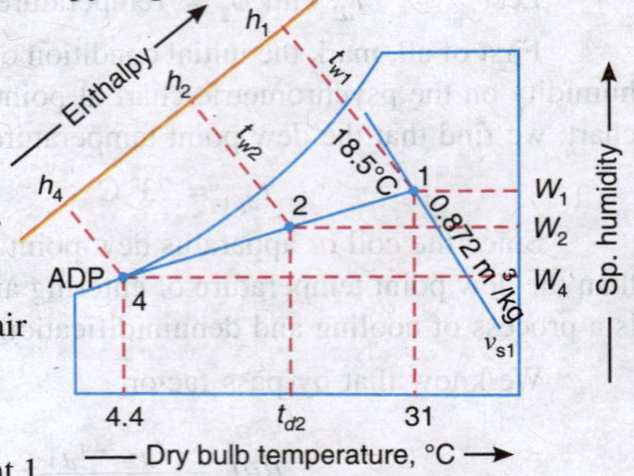

Fig. 16.31

We know that mass flow rate of dry air at point 1,

$$m_a = \frac{v_1}{v_{s1}} = \frac{39.6}{0.872} = 44.41 \text{ kg/min}$$

and cooling capacity of the coil,

$$Q = m_a (h_1 - h_2)$$

or
$$h_1 - h_2 = \frac{Q}{m_a} = \frac{12.5 \times 60}{44.41} = 16.89 \text{ kJ / kg of dry air}$$

$$\therefore \quad h_2 = h_1 - 16.89 = 52.5 - 16.89 = 35.61 \text{ kJ / kg of dry air}$$

The equation for the condition line 1–2–4 is given as

$$\frac{W_2 - W_4}{W_1 - W_4} = \frac{h_2 - h_4}{h_1 - h_4}$$

$$\frac{W_2 - 0.005\,25}{0.0082 - 0.005\,25} = \frac{35.61 - 17.7}{52.5 - 17.7}$$

$$\therefore \quad W_2 = 0.006\,77 \text{ kg / kg of dry air}$$

Now plot point 2 on the psychrometric chart such as enthalpy, h_2 = 35.61 kJ/kg of dry air and specific humidity, W_2 = 0.00677 kg/kg of dry air. At point 2, we find that

$$t_{d2} = 18.5°\text{C} \; ; \text{ and } t_{w2} = 12.5°\text{C} \textbf{ Ans.}$$

By-pass factor

We know that by-pass factor,

$$BPF = \frac{h_2 - h_4}{h_1 - h_4} = \frac{35.61 - 17.7}{52.5 - 17.7} = 0.5146 \textbf{ Ans.}$$

Example 16.15. *The atmospheric air at 30° C dry bulb temperature and 75% relative humidity enters a cooling coil at the rate of 200 m³/min. The coil dew point temperature is 14° C and the by-pass factor of the coil is 0.1. Determine : 1. the temperature of air leaving the cooling coil; 2. the capacity of the cooling coil in tonnes of refrigeration and in kilowatt; 3. the amount of water vapour removed per minute; and 4. the sensible heat factor for the process.*

Solution. Given : t_{d1} = 30° C ; ϕ_1 = 75% ; v_1 = 200 m³/min ; $ADP = t_{d4}$ = 14° C ; BPF = 0.1

1. Temperature of air leaving the cooling coil

Let t_{d2} = Temperature of air leaving the cooling coil.

First of all, mark the initial condition of the air, *i.e.* 30° C dry bulb temperature and 75% relative humidity on the psychrometric chart at point 1, as shown in Fig. 16.32. From the psychrometric chart, the dew point temperature of the entering air at point 1.

$$t_{dp1} = 25.2°\text{C}$$

Since the coil dew point temperature (or ADP) is less than the dew point temperature of entering air, therefore it is a process of cooling and dehumidification.

We know that by-pass factor,

$$BPF = \frac{t_{d2} - t_{d4}}{t_{d1} - t_{d4}} = \frac{t_{d2} - ADP}{t_{d1} - ADP}$$

$$0.1 = \frac{t_{d2} - 14}{30 - 14}$$

$$\therefore \quad t_{d2} = 15.6° \text{ C } \textbf{Ans.}$$

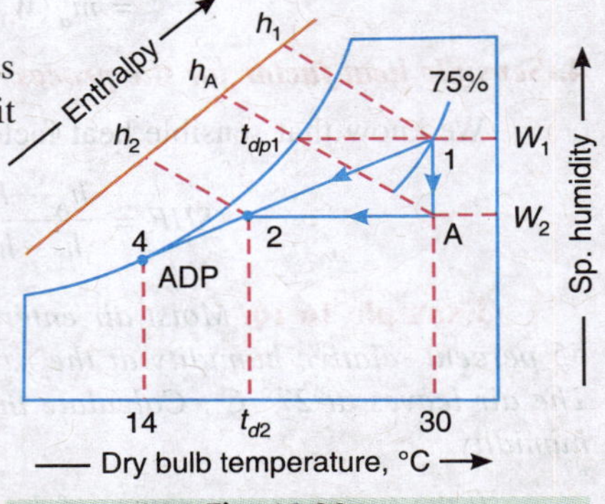

Fig. 16.32

2. Capacity of the cooling coil

The resulting condition of the air coming out of the coil is shown by point 2, on the line joining the points 1 and 4, as shown in Fig. 16.32. The line 1-2 represents the cooling and dehumidification process which may be assumed to have followed the path 1-*A* (*i.e.* dehumidification) and *A*-2 (*i.e.* cooling). Now from the psychrometric chart, we find that

Water vapour in the entering air or the specific humidity of entering air at point 1,

$$W_1 = 0.0202 \text{ kg/kg of dry air}$$

Water vapour in the leaving air or the specific humidity of leaving air at point 2,

$$W_2 = 0.011 \text{ kg/kg of dry air}$$

Specific volume of entering air at point 1,

$$v_{s1} = 0.884 \text{ m}^3\text{/kg of dry air}$$

Enthalpy of entering air at point 1,

$$h_1 = 82 \text{ kJ/kg of dry air}$$

Enthalpy of air at point *A*,

$$h_A = 58 \text{ kJ/kg of dry air}$$

and enthalpy of leaving air at point 2,

$$h_2 = 43.5 \text{ kJ/kg of dry air}$$

We know that mass of air flowing through the cooling coil,

$$m_a = \frac{v_1}{v_{s1}} = \frac{200}{0.884} = 226.2 \text{ kg/min}$$

∴ Capacity of the cooling coil in tonnes of refrigeration

$$= m_a (h_1 - h_2) = 226.2\,(82 - 43.5) = 8709 \text{ kJ/min}$$

$$= 8709/210 = 41.5 \text{ TR} \quad \textbf{Ans.} \qquad \ldots (\because 1 \text{ TR} = 210 \text{ kJ/min})$$

and capacity of the cooling coil in kilowatt

$$= 8709/60 = 145.15 \text{ kW} \quad \textbf{Ans.}$$

3. *Amount of water vapour removed per minute*

We know that amount of water vapour removed

$$= m_a (W_1 - W_2) = 226.2\,(0.0202 - 0.011) = 2.08 \text{ kg/min} \quad \textbf{Ans.}$$

4. *Sensible heat factor for the process*

We know that sensible heat factor,

$$SHF = \frac{h_A - h_2}{h_1 - h_2} = \frac{58 - 43.5}{82 - 43.5} = 0.377 \quad \textbf{Ans.}$$

Example 16.16. *Moist air enters a refrigeration coil at 35° C dry bulb temperature and 55 percent relative humidity at the rate of 100 m³/min. The barometric pressure is 1.013 bar. The air leaves at 27° C. Calculate the tonnes of refrigeration required and the final relative humidity.*

If the surface temperature of the cooling coil is 10° C and by-pass factor 0.1, calculate the tonnes of refrigeration required and the condensate flow.

Solution. Given : $t_{d1} = 35°$ C ; $\phi_1 = 55\%$; $v_1 = 100$ m³/min ; *$p_b = 1.013$ bar ; $t_{d2} = 27°$ C

Tonnes of refrigeration required

First of all, mark the initial condition of the air, *i.e.* 35° C dry bulb temperature and 55% relative humidity on the psychrometric chart at point 1, as shown in Fig. 16.33. From the psychrometric chart, we find that the dew point temperature at point 1,

$$t_{dp1} = 24.5° \text{ C}$$

* The values as read from the psychrometric chart are at standard atmospheric pressure of 1.013 bar.

Since the temperature of air leaving the coil (*i.e.* 27° C) or the effective temperature of the coil is above the dew point temperature of entering air (*i.e.* 24.5° C), therefore no dehumidification occurs. Thus, it is a sensible cooling process from 35° C dry bulb temperature and 55% relative humidity to 27° C dry bulb temperatures as shown by the horizontal line 1-2 on the psychrometric chart. The point 2 represents the condition of air leaving the coil.

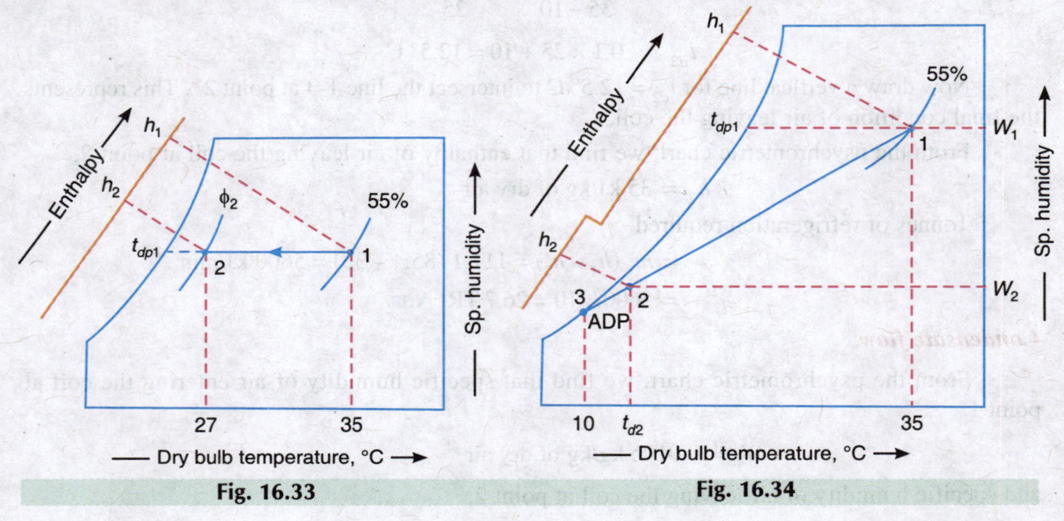

Fig. 16.33 **Fig. 16.34**

From the psychrometric chart, we find that enthalpy of air entering the coil at point 1,

$$h_1 = 85.4 \text{ kJ/kg of dry air}$$

Specific volume of air entering the coil at point 1,

$$v_{s1} = 0.9 \text{ m}^3\text{/kg of dry air}$$

and enthalpy of air leaving the coil at point 2,

$$h_2 = 77 \text{ kJ/kg of dry air}$$

We know that mass of air entering the coil,

$$m_1 = \frac{v_1}{v_{s1}} = \frac{100}{0.9} = 111.1 \text{ kg/min}$$

∴ Tonnes of refrigeration required

$$= m_1 (h_1 - h_2) = 111.1 (85.4 - 77) = 933.24 \text{ kJ/min}$$

$$= 933.24/210 = 4.44 \text{ TR} \quad \textbf{Ans.}$$

Final relative humidity

From the psychrometric chart, we find that the relative humidity of air leaving the coil at point 2,

$$\phi_2 = 86\% \quad \textbf{Ans.}$$

Tonnes of refrigeration required when surface temperature of cooling coil is 10° C

Given : $BPF = 0.1$

Since the surface temperature or apparatus dew point (*i.e.* 10° C) of cooling coil is less than the dew point temperature of entering air (24.5° C), therefore, it is a cooling and dehumidification process, as shown in Fig. 16.34. Join point 1 (*i.e.* 35° C dry bulb temperature and 55% relative humidity) to point 3 on the saturation curve (*i.e.* ADP = 10° C). Mark point 2 (final condition of air) on the line 1-3 as discussed below:

Let t_{d2} = Dry bulb temperature of air at point 2.

We know that by-pass factor of the coil,

$$BPF = \frac{t_{d2} - ADP}{t_{d1} - ADP}$$

$$0.1 = \frac{t_{d2} - 10}{35 - 10} = \frac{t_{d2} - 10}{25}$$

$$\therefore \quad t_{d2} = 0.1 \times 25 + 10 = 12.5° \text{ C}$$

Now draw a vertical line for t_{d2} = 12.5° C to intersect the line 1-3 at point 2*. This represents the final condition of air leaving the coil.

From the psychrometric chart, we find that enthalpy of air leaving the coil at point 2,

$$h_2 = 35 \text{ kJ/kg of dry air}$$

Tonnes of refrigeration required

$$= m_1 (h_1 - h_2) = 111.1 (85.4 - 35) = 5600 \text{ kJ/min}$$

$$= 5600/210 = 26.7 \text{ TR} \quad \textbf{Ans.}$$

Condensate flow

From the psychrometric chart, we find that specific humidity of air entering the coil at point 1,

$$W_1 = 0.0196 \text{ kg/kg of dry air}$$

and specific humidity of air leaving the coil at point 2,

$$W_2 = 0.0088 \text{ kg/kg of dry air}$$

∴ Condensate flow *i.e.* water condensed at the coil

$$= m_1 (W_1 - W_2) = 111.1 (0.0196 - 0.0088) = 1.2 \text{ kg / min} \quad \textbf{Ans.}$$

16.17 Cooling with Adiabatic Humidification

When the air is passed through an insulated chamber, as shown in Fig. 16.35 (*a*), having sprays of water (known as air washer) maintained at a temperature (t_I) higher than the dew point temperature of entering air (t_{dp1}), but lower than its dry bulb temperature (t_{d1}) of entering air or equal to the wet bulb temperature of the entering air (t_{w1}), then the air is said to be cooled and humidified. Since no heat is supplied or rejected from the spray water as the same water is re-circulated again and again, therefore, in this case, a condition of adiabatic saturation will be reached. The temperature of spray water will reach the thermodynamic wet bulb temperature of the air entering the spray water. This process is shown by line 1-3 on the psychrometric chart as shown in Fig. 16.35 (*b*), and follows the path along the constant wet bulb temperature line or constant enthalpy line.

In an ideal case *i.e.* when the humidification is perfect (or the humidifying efficiency of the spray chamber is 100%), the final condition of the air will be at point 3 (*i.e.* at temperature t_{d3} and relative humidity 100%). In actual practice, perfect humidification is never achieved. Therefore, the final condition of air at outlet is represented by point 2 on the line 1-3, as shown in Fig. 16.35 (*b*).

* The point 2 may also be obtained by dividing the line 1-3 in such a way that

$$\frac{\text{Length } 2\text{-}3}{\text{Length } 1\text{-}3} = BPF = 0.1$$

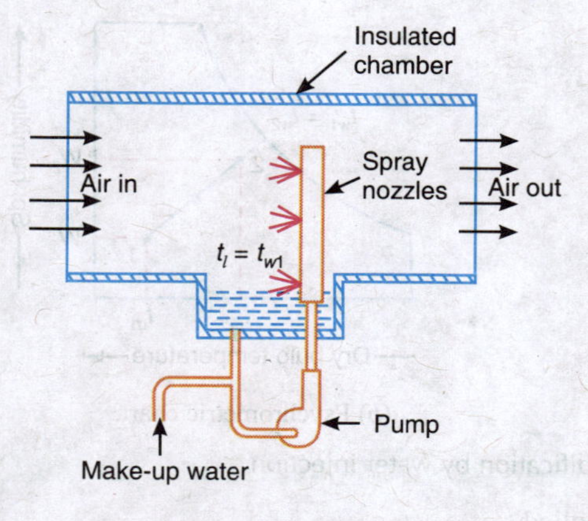

(a) Psychrometric process.

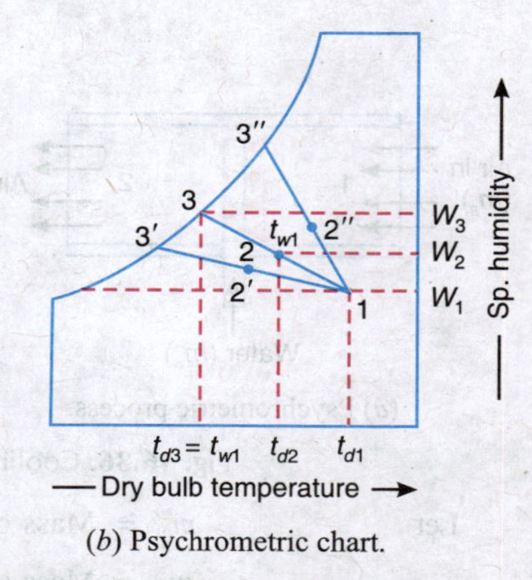

(b) Psychrometric chart.

Fig. 16.35. Cooling with adiabatic humidification.

The effectiveness or the humidifying efficiency of the spray chamber is given by

$$\eta_H = \frac{\text{Actual drop in DBT}}{\text{Ideal drop in DBT}} = \frac{\text{Actual drop in sp. humidity}}{\text{Ideal drop in sp. humidity}}$$

$$= \frac{t_{d1} - t_{d2}}{t_{d1} - t_{d3}} = \frac{W_2 - W_1}{W_3 - W_1}$$

Notes : 1. When the sprays of water are maintained at a temperature lower than the wet bulb temperature of the entering air (*i.e.* t_l is less than t_{w1}) by cooling the spray water by coolers before it is pumped to the spray nozzles, then for the ideal condition, the process follows the path 1-3′, as shown in Fig. 16.35 (*b*). In such cases, the effectiveness or the humidifying efficiency of the spray chamber is given by

$$\eta_H = \frac{t_{d1} - t_{d2'}}{t_{d1} - t_{d3}'} = \frac{W_2' - W_1}{W_3' - W_1}$$

where t_{d3}' is less than t_{w1}.

2. When the sprays of water are maintained at a temperature higher than the wet bulb temperature of the entering air (*i.e.* t_l is greater than t_{w1}) but lower than the dry bulb temperature of the entering air (t_{d1}) by heating the spray water by heaters before it is pumped to the spray nozzles, then for the ideal condition, the process follows the path 1-3″ as shown in Fig. 16.35 (*b*). In such cases, the effectiveness or the humidifying efficiency of the spray chamber is given by

$$\eta_H = \frac{t_{d1} - t_{d2''}}{t_{d1} - t_{d3}''} = \frac{W_2'' - W_1}{W_3'' - W_1}$$

where t_{d3}'' is greater than t_{w1}.

16.18 Cooling and Humidification by Water Injection (Evaporative Cooling)

Let water at a temperature t_l is injected into the flowing stream of dry air as shown in Fig. 16.36 (*a*). The final condition of air depends upon the amount of water evaporation. When the water is injected at a temperature equal to the wet bulb temperature of the entering air (t_{w1}), then the process follows the path of constant wet bulb temperature line, as shown by the line 1-2 in Fig. 16.36 (*b*).

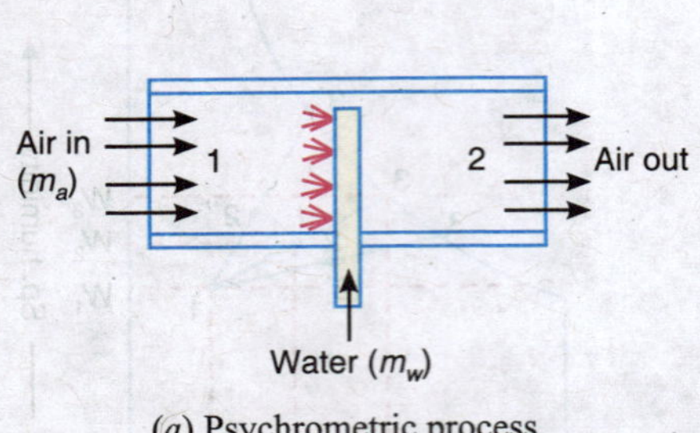

(*a*) Psychrometric process.

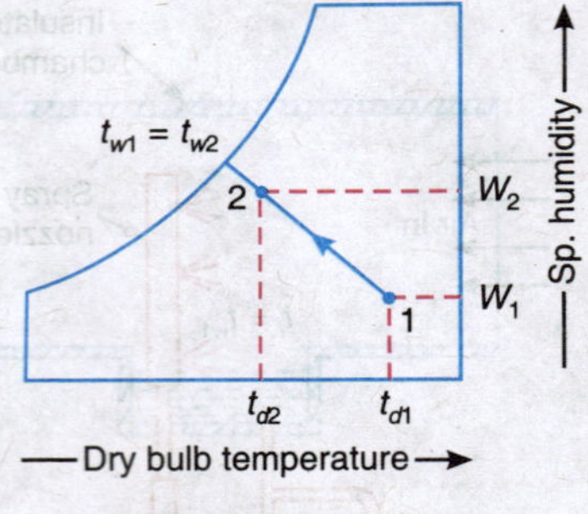

(*b*) Psychrometric chart.

Fig. 16.36. Cooling and humidification by water injection.

Let m_w = Mass of water supplied,

m_a = Mass of dry air,

W_1 = Specific humidity of entering air,

W_2 = Specific humidity of leaving air, and

h_w = Enthalpy of water injected into the air.

Now for the mass balance,

$$W_2 = W_1 + \frac{m_w}{m_a} \quad \ldots (i)$$

and for heat balance,

$$h_2 = h_1 + \frac{m_w}{m_a} \times h_{fw}$$

$$= h_1 + (W_2 - W_1)\, h_{fw} \quad \ldots \text{[From equation } (i)]$$

Since $(W_2 - W_1)\, h_{fw}$ is very small as compared to h_1 and h_2, therefore it may be neglected. Thus the water injection process is a constant enthalpy process, irrespective of the temperature of water injected (*i.e.* whether the temperature $t_l < t_w$ or $t_l > t_w$).

Example 16.17. *200 m³ of air per min. is passed through the adiabatic humidifier. The condition of air at inlet is 40° C dry bulb temperature and 15% relative humidity and the outlet condition is 25° C dry bulb temperature and 20° C wet bulb temperature. Find the dew point temperature and the amount of water vapour added to the air per minute.*

Solution. Given : $v_1 = 200$ m³/min ; $t_{d1} = 40°$ C ; $\phi_1 = 15\%$; $t_{d2} = 25°$ C ; $t_{w2} = 20°$ C

First of all, mark the inlet condition of air at 40°C dry bulb temperature and 15% relative humidity on the psychrometric chart at point 1, as shown in Fig. 16.37. Now mark the outlet condition of air at 25° C dry bulb temperature and 20° C wet bulb temperature, as point 2. The line 1-2 represents the adiabatic humidification.

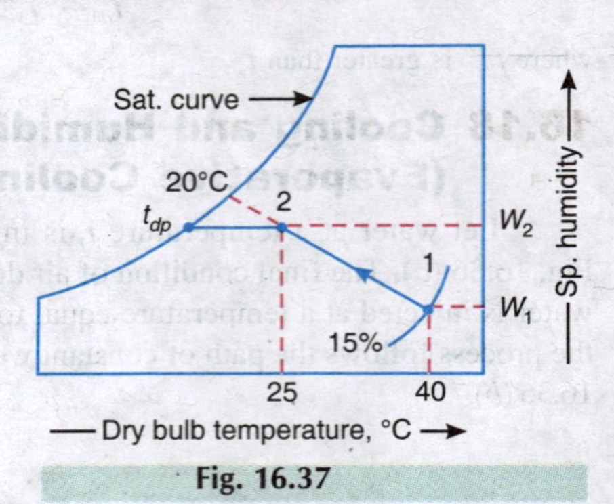

Fig. 16.37

Dew point temperature

On the psychrometric chart, draw a horizontal line through point 2 upto the saturation curve. From the chart, we find that dew point temperature,

$$t_{dp} = 17.6° \text{ C} \quad \textbf{Ans.}$$

Amount of water vapour added to the air per minute

From the psychrometric chart, we find that specific volume of air at point 1,

$$v_{s1} = 0.896 \text{ m}^3\text{/kg of dry air}$$

Specific humidity at point 1,

$$W_1 = 0.007 \text{ kg/kg of dry air}$$

and specific humidity at point 2,

$$W_2 = 0.0126 \text{ kg/kg of dry air}$$

We know that mass of air supplied,

$$m_a = \frac{v_1}{v_{s1}} = \frac{200}{0.896} = 223.2 \text{ kg/min}$$

∴ Amount of water vapour added to the air

$$= m_a (W_2 - W_1)$$

$$= 223.2 (0.0126 - 0.007) = 1.25 \text{ kg/min} \quad \textbf{Ans.}$$

Evaporative cooling system.

Example 16.18. *A drying room is to be maintained at 32° C and 30% RH. The sensible heat gain to the room is 150 000 kJ/h. The moisture to be evaporated from the objects during drying is 18 kg / h. If there is no direct heat source to provide for evaporation in the room, calculate the state and rate of supply air at 15° C dry bulb temperature.*

Solution. Given : $t_{d2} = 32°$ C ; $\phi_2 = 30\%$; $RSH = 150\ 000$ kJ / h ; $m_w = 18$ kg / h ; $t_{d1} = 15°$ C

First of all, let us find the mass of supply air (m_a). We know that sensible heat gain to the room (RSH),

$$150\,000 = m_a\, c_p\, (t_{d2} - t_{d1}) = m_a \times 1.005\ (32 - 15) = 17.085\ m_a$$

$$\therefore \quad m_a = 150\,000/17.085 = 8780 \text{ kg/h} \quad \textbf{Ans.} \quad (\text{Taking } c_p = 1.005 \text{ kJ/kg°C})$$

The initial and final conditions of the air are marked on the psychrometric chart as points 1 and 2 respectively, as shown in Fig. 16.38.

Let ϕ_1 = Relative humidity of air at point 1,

and

W_1 = Specific humidity of air at point 1.

From the psychrometric chart, we find that specific humidity of air at point 2,

$$W_2 = 0.0088 \text{ kg/kg of dry air}$$

We also know that the specific humidity of leaving air at point 2,

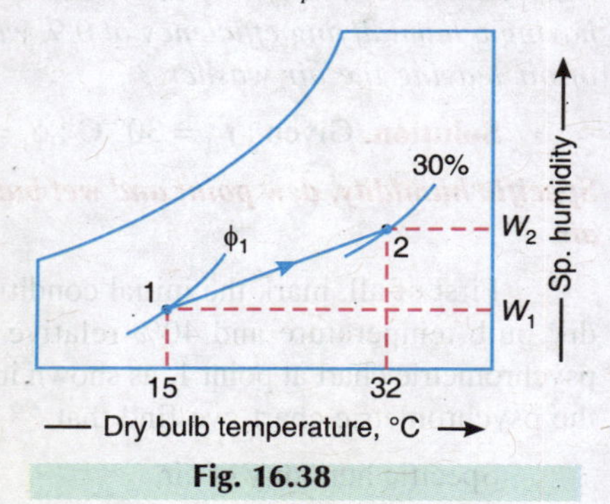

Fig. 16.38

$$W_2 = W_1 + \frac{m_w}{m_a} = W_1 + \frac{18}{m_a}$$

or

$$W_1 = W_2 - \frac{18}{m_a} = 0.0088 - \frac{18}{8780} = 0.006\ 75 \text{ kg/kg of dry air}$$

Now from the psychrometric chart, we find that corresponding to 15°C dry bulb temperature and 0.006 75 kg/kg of dry air of specific humidity, the relative humidity at point 1, is

$$\phi_1 = 65\% \quad \textbf{Ans.}$$

Example 16.19. *Determine the final dry bulb temperature and relative humidity of air washed with recirculated spray water if the air is initially at dry bulb temperature 35° C and 50% relative humidity as it enters an air washer which has humidifying efficiency of 85 per cent.*

Solution. Given : $t_{d1} = 35°$ C ; $\phi_1 = 50\%$; $\eta_H = 85\% = 0.85$

First of all, mark the initial condition of air at 35° C dry bulb temperature and 50% relative humidity on the psychrometric chart at point 1, as shown in Fig. 16.39. The wet bulb temperature of the entering air as read from the psychrometric chart is

$$t_{w1} = 26.1° \text{ C} = t_{d3}$$

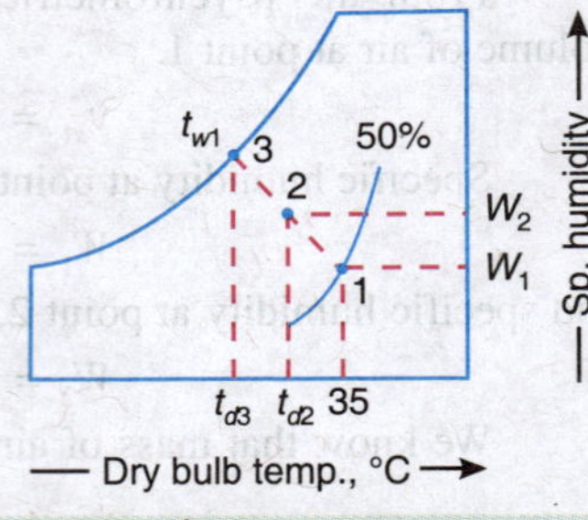

Fig. 16.39

Final dry bulb temperature

Let t_{d2} = Final dry bulb temperature of the air leaving the air washer.

We know that humidifying efficiency of an air washer (η_H),

$$0.85 = \frac{t_{d1} - t_{d2}}{t_{d1} - t_{d3}} = \frac{35 - t_{d2}}{35 - 26.1} = \frac{35 - t_{d2}}{8.9}$$

$$\therefore \quad t_{d2} = 35 - 0.85 \times 8.9 = 27.435° \text{ C} \quad \textbf{Ans.}$$

Final relative humidity

On the constant wet bulb temperature line 1-3, mark point 2 such that $t_{d2} = 27.435°$C. Now the relative humidity of the air leaving the air washer (corresponding to point 2) as read from the psychrometric chart is

$$\phi_2 = 90\% \quad \textbf{Ans.}$$

Example 16.20. *At a certain locality, the dry bulb temperature of air is 30° C and the relative humidity is 40%. Determine the specific humidity and the dew point and wet bulb temperatures of air. If this air is cooled in an air washer using recirculated spray water and having a humidifying efficiency of 0.9, what are dry bulb temperature and dew point temperature of air leaving the air washer ?*

Solution. Given : $t_{d1} = 30°$ C ; $\phi_1 = 40\%$; $\eta_H = 0.9$

Specific humidity, dew point and wet bulb temperature of air.

First of all, mark the initial condition of air at 30° C dry bulb temperature and 40% relative humidity, on the psychrometric chart at point 1, as shown in Fig. 16.40. From the psychrometric chart, we find that

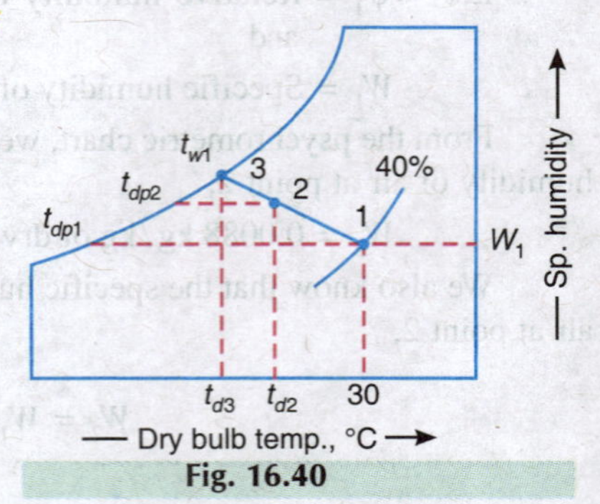

Fig. 16.40

Specific humidity of air,

$$W_1 = 0.0106 \text{ kg/kg of dry air} \quad \textbf{Ans.}$$

Dew point temperature of air,

$$t_{dp1} = 15° \text{ C} \quad \textbf{Ans.}$$

and wet bulb temperature of air,

$$t_{w1} = 19.8° \text{ C} \quad \textbf{Ans.}$$

Dry bulb temperature and dew point temperature of air leaving the air washer

Let t_{d2} = Dry bulb temperature of air leaving the air washer.

We know that humidifying efficiency of an air washer (η_H),

$$0.9 = \frac{t_{d1} - t_{d2}}{t_{d1} - t_{d3}} = \frac{30 - t_{d2}}{30 - 19.8} = \frac{30 - t_{d2}}{10.2} \quad \ldots (\because t_{d3} = t_{w1})$$

$$\therefore \quad t_{ap2} = 30 - 0.9 \times 10.2 = 20.82° \text{ C} \quad \textbf{Ans.}$$

Now mark point 2 on the constant wet bulb temperature line 1-3 such that t_{d2} = 20.82° C. The dew point temperature at point 2 is read as

$$\therefore \quad t_{dp2} = 19.4° \text{ C} \quad \textbf{Ans.}$$

Example 16.21. *The atmospheric air at 40° C dry bulb temperature and 18° C wet bulb temperature is flowing at the rate of 100 m³/min through the space. Water at 18° C is injected into the air stream at the rate of 48 kg / h. Determine the specific humidity and enthalpy of the leaving air. Also determine the dry bulb temperature, wet bulb temperature and relative humidity of the leaving air.*

Solution. Given : t_{d1} = 40° C ; t_{w1} = 18° C ; v_1 = 100 m³/min ; t_l = 18° C ; m = 48 kg/h = 0.8 kg / min

Specific humidity of the leaving air

Let W_2 = Specific humidity of the leaving air.

First of all, mark the initial condition of air *i.e.* at 40° C dry bulb temperature and 18° C wet bulb temperature, on the psychrometric chart at point 1, as shown in Fig. 16.41. Now from the psychrometric chart, we find that specific volume of air at point 1,

$$v_{s1} = 0.89 \text{ m}^3/\text{kg of dry air}$$

Specific humidity of air at point 1,

$$W_1 = 0.004 \text{ kg/kg of dry air}$$

Enthalpy of air at point 1,

$$h_1 = 51 \text{ kJ / kg of dry air}$$

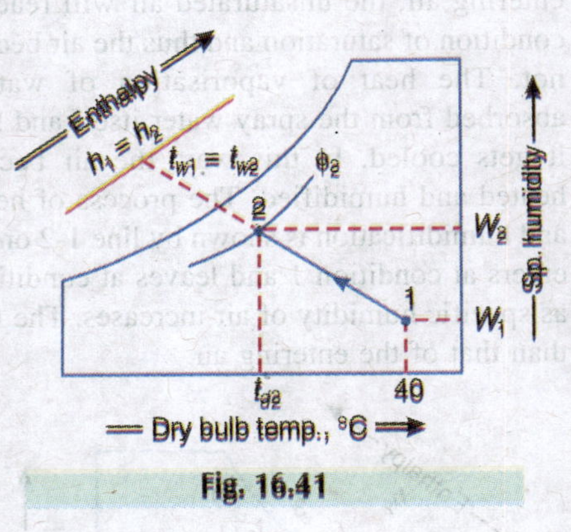

Fig. 16.41

We know that mass of air flowing,

$$m_a = \frac{v_1}{v_{s1}} = \frac{100}{0.89} = 112.4 \text{ kg/min}$$

$$\therefore \quad W_2 = W_1 + \frac{m_w}{m_a} = 0.004 + \frac{0.8}{112.4} = 0.0111 \text{ kg/kg of dry air} \quad \textbf{Ans.}$$

Enthalpy of the leaving air

Since the water is injected at a temperature (t_l = 18° C) equal to the wet bulb temperature of the entering air (t_{w1} = 18° C), therefore the process follows the path of constant wet bulb temperature line or constant enthalpy line, as shown in Fig. 16.41.

∴ Enthalpy of leaving air,

$$h_2 = \text{Enthalpy of entering air} = 51 \text{ kJ/kg of dry air} \quad \textbf{Ans.}$$

Dry bulb temperature, wet bulb temperature and relative humidity of the leaving air

Mark the condition of the leaving air on the psychrometric chart as point 2 corresponding to W_2 = 0.0111 kg / kg of dry air and h_2 = 51 kJ / kg of dry air. Now from the psychrometric chart, corresponding to point 2, we find that dry bulb temperature of the leaving air,

$$t_{d2} = 22.4° \text{ C} \quad \textbf{Ans.}$$

Wet bulb temperature of the leaving air,

$$t_{w2} = t_{w1} = 18° \text{ C} \quad \textbf{Ans.}$$

and relative humidity of the leaving air,

$$\phi_2 = 65\% \quad \textbf{Ans.}$$

16.19 Heating and Humidification

This process is generally used in winter air conditioning to warm and humidify the air. It is the reverse process of cooling and dehumidification. When air is passed through a humidifier having spray water temperature higher than the dry bulb temperature of the entering air, the unsaturated air will reach the condition of saturation and thus the air becomes hot. The heat of vaporisation of water is absorbed from the spray water itself and hence it gets cooled. In this way, the air becomes heated and humidified. The process of heating and humidification is shown by line 1-2 on the psychrometric chart as shown in Fig. 16.42. The air enters at condition 1 and leaves at condition 2. In this process, the dry bulb temperature as well as specific humidity of air increases. The final relative humidity of the air can be lower or higher than that of the entering air.

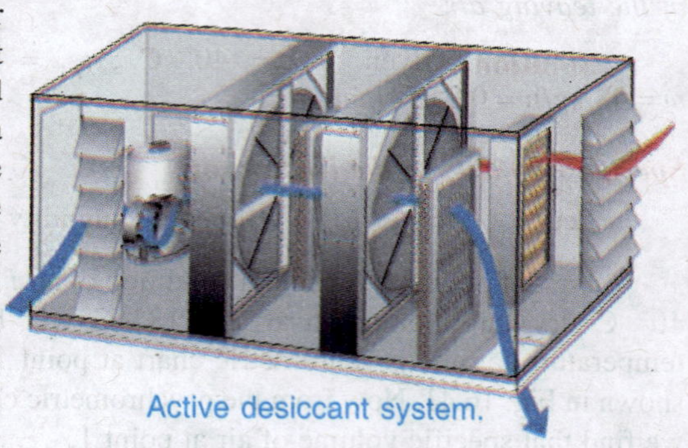

Active desiccant system.

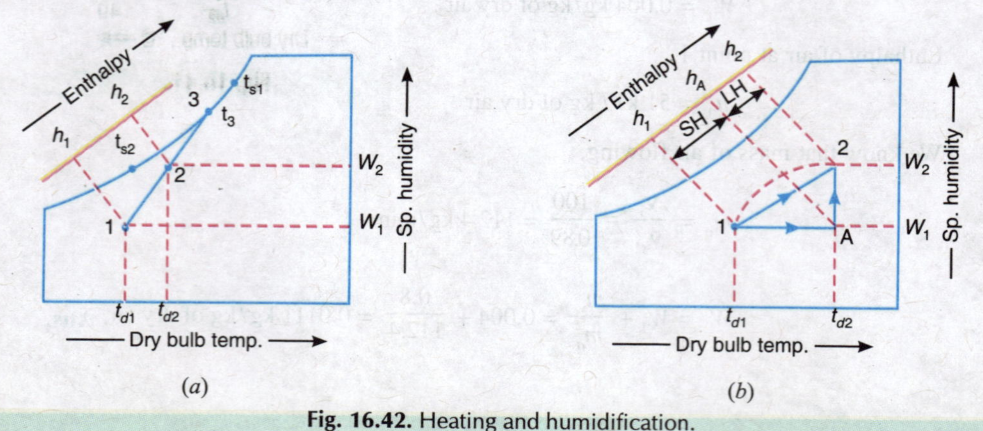

Fig. 16.42. Heating and humidification.

Let m_{w1} and m_{w2} = Mass of spray water entering and leaving the humidifier in kg,

h_{fw1} and h_{fw2} = Enthalpy of spray water entering and leaving the humidifier in kJ/kg,

W_1 and W_2 = Specific humidity of the entering and leaving air in kg/kg of dry air,

h_1 and h_2 = Enthalpy of entering and leaving air in kJ/kg of dry air, and

m_a = Mass of dry air entering in kg.

For mass balance of spray water,

$$(m_{w1} - m_{w2}) = m_a (W_2 - W_1)$$

or

$$m_{w2} = m_{w1} - m_a (W_2 - W_1) \qquad \ldots (i)$$

and for enthalpy balance,

$$m_{w1}\, h_{fw1} - m_{w2}\, h_{fw2} = m_a (h_2 - h_1) \qquad \ldots (ii)$$

Substituting the value of m_{w2} from equation (*i*), we have

$$m_{w1}\, h_{fw1} - [m_{w1} - m_a (W_2 - W_1)]\, h_{fw2} = m_a (h_2 - h_1)$$

$$\therefore \quad h_2 - h_1 = \frac{m_{w1}}{m_a} (h_{fw1} - h_{fw2}) + (W_2 - W_1)\, h_{fw2}$$

The temperatures t_{s1} and t_{s2} shown in Fig. 16.42 (*a*) denote the temperatures of entering and leaving spray water respectively. The temperature t_3 is the mean temperature of the spray water which the entering air may be assumed to approach.

Actually, the heating and humidification process follows the path as shown by dotted curve in Fig. 16.42 (*b*), but for the calculation of psychrometric properties, only the end points are important. Thus, the heating and humidification process shown by a line 1-2 on the psychrometric chart may be assumed to have followed the path 1-*A* (*i.e.* heating) and *A*-2 (*i.e.* humidification), as shown in Fig. 16.42 (*b*). We see that the total heat added to the air during heating and humidification is

$$q = h_2 - h_1 = (h_2 - h_A) + (h_A - h_1) = q_L + q_S$$

where $q_L = (h_2 - h_A)$ = Latent heat of vaporisation of the increased moisture content $(W_2 - W_1)$, and

$q_S = (h_A - h_1)$ = Sensible heat added

We know that sensible heat factor,

$$SHF = \frac{\text{Sensible heat}}{\text{Total heat}} = \frac{q_S}{q} = \frac{q_S}{q_S + q_L} = \frac{h_A - h_1}{h_2 - h_1}$$

Note : The line 1-2 in Fig. 16.42 (*b*) is called *sensible heat factor line.*

16.20 Heating and Humidification by Steam Injection

The steam is normally injected into the air in order to increase its specific humidity as shown in Fig. 16.43 (*a*). This process is used for the air conditioning of textile mills where high humidity is to be maintained. The dry bulb temperature of air changes very little during this process, as shown on the psychrometric chart in Fig. 16.43 (*b*).

Let m_s = Mass of steam supplied,

m_a = Mass of dry air entering,

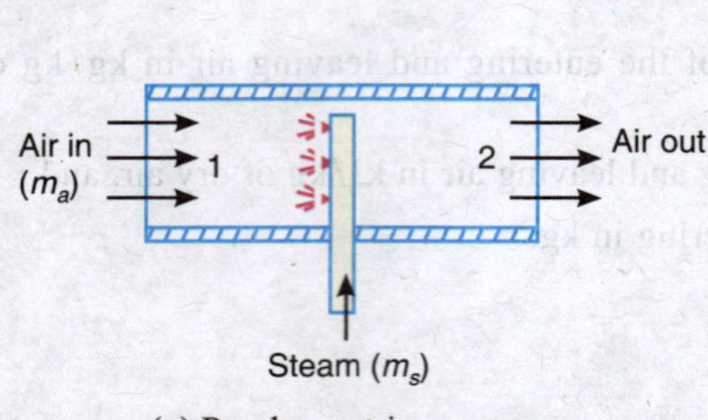

(a) Psychrometric process.

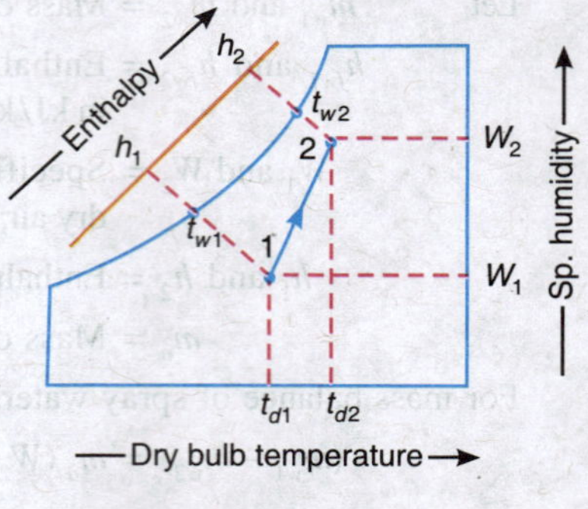

(b) Psychrometric chart.

Fig. 16.43. Heating and humidification by steam injection.

W_1 = Specific humidity of air entering,

W_2 = Specific humidity of air leaving,

h_1 = Enthalpy of air entering,

h_2 = Enthalpy of air leaving, and

h_s = Enthalpy of steam injected into the air.

Now for the mass balance,

$$W_2 = W_1 + \frac{m_s}{m_a} \quad \ldots (i)$$

and for the heat balance,

$$h_2 = h_1 + \frac{m_s}{m_a} \times h_s = h_1 + (W_2 - W_1)\, h_s \quad \ldots \text{[From equation (i)]}$$

Example 16.22. *The atmospheric air at 25° C dry bulb temperature and 12° C wet bulb temperature is flowing at the rate of 100 m³/min through the duct. The dry saturated steam at 100° C is injected into the air steam at the rate of 72 kg per hour. Calculate the specific humidity and enthalpy of the leaving air. Also determine the dry bulb temperature, wet bulb temperature and relative humidity of the leaving air.*

Solution. Given : t_{d1} = 25° C ; t_{w1} = 12° C ; v_1 = 100 m³/min ; t_s = 100° C ; m_s = 72 kg / h = 1.2 kg / min

Specific humidity of the leaving air

Let W_2 = Specific humidity of the leaving air.

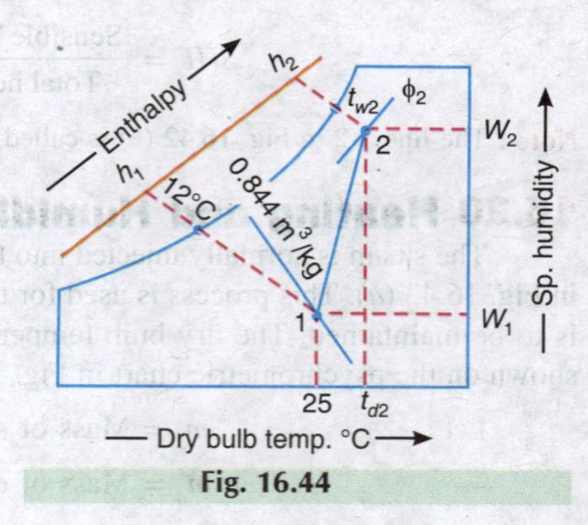

Fig. 16.44

First of all, mark the initial condition of air *i.e.* at 25° C dry bulb temperature and 12° C wet bulb temperature on the psychrometric chart at point 1, as shown in Fig. 16.44. Now from the psychrometric chart, we find that the specific volume of air at point 1,

$$v_{s1} = 0.844 \text{ m}^3/\text{kg of dry air}$$

Specific humidity of air at point 1,

$$W_1 = 0.0034 \text{ kg/kg of dry air}$$

Enthalpy of air at point 1,

$$h_1 = 34.2 \text{ kJ/kg of dry air}$$

We know that mass of air flowing,

$$m_a = \frac{v_1}{v_{s1}} = \frac{100}{0.844} = 118.5 \text{ kg/min}$$

We know that $W_2 = W_1 + \dfrac{m_s}{m_a} = 0.0034 + \dfrac{1.2}{118.5} = 0.0135$ kg/kg of dry air **Ans.**

Enthalpy of leaving air

Let h_2 = Enthalpy of leaving air.

From steam tables, we find that enthalpy of dry saturated steam corresponding to 100° C is

$$h_s = 2676 \text{ kJ/kg}$$

We know that $h_2 = h_1 + \dfrac{m_s}{m_a} \times h_s = 34.2 + \dfrac{1.2}{118.5} \times 2676$

$$= 61.3 \text{ kJ/kg of dry air } \textbf{Ans.}$$

Dry bulb temperature, wet bulb temperature and relative humidity of the leaving air

Mark the condition of leaving air on the psychrometric chart as point 2 corresponding to $W_2 = 0.0135$ kg / kg of dry air and $h_2 = 61.3$ kJ / kg of dry air. Now from the psychrometric chart corresponding to point 2,

Dry bulb temperature of the leaving air,

$$t_{d2} = 26.1° \text{ C } \textbf{Ans.}$$

Wet bulb temperature of the leaving air,

$$t_{w2} = 21.1° \text{ C } \textbf{Ans.}$$

and relative humidity of the leaving air,

$$\phi_2 = 62\% \textbf{ Ans.}$$

Example 16.23. *The moist air enters a chamber at 5°C dry bulb temperature and 2.5°C wet bulb temperature at the rate of 90 m³/min. The barometric pressure is 1.01325 bar. While passing through the chamber, the air absorbs sensible heat at the rate of 40.7 kW and picks up 40 kg/h of saturated steam at 110°C. Determine the dry bulb and wet bulb temperatures of the leaving air.*

Solution. Given : $t_{d1} = 5°C$; $t_{w1} = 2.5°C$; $v_1 = 90 \text{ m}^3/\text{min}$; $p_b = 1.01325$ bar ; $q_s = 40.7$ kW $= 40.7 \times 3600$ kJ/h ; $m_s = 40$ kg/h ; $t_s = 110°C$

Dry bulb and wet bulb temperatures of the leaving air

Let t_{d3} and t_{w3} = Dry bulb and wet bulb temperatures of the air.

First of all, mark the initial condition of air, *i.e.* at 5°C dry bulb temperature and 2.5°C wet bulb temperature on the psychrometric chart at point 1, as shown in Fig. 16.45. The line 1-2 shows the sensible heating of air and the line 2-3 represents the injection of steam.

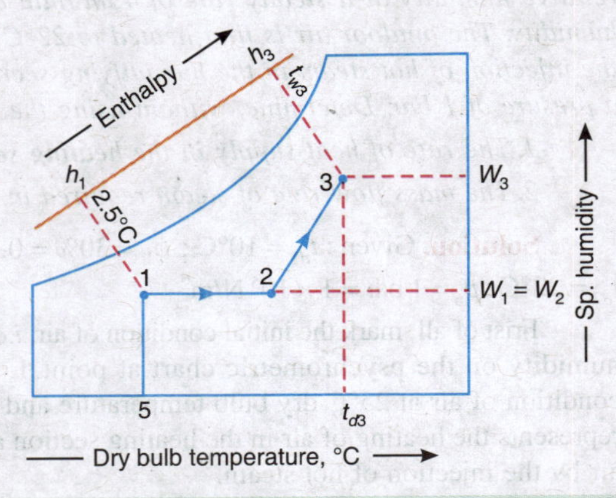

Fig. 16.45

Now from the psychrometric chart, we find that specific volume of air at point 1,

$$v_{s1} = 0.792 \text{ m}^3/\text{kg}$$

Specific humidity of air at point 1,

$$W_1 = 0.0035 \text{ kJ / kg of air}$$

and enthalpy of air at point 1,

$$h_1 = 13.8 \text{ kJ / kg}$$

We know that mass of air flowing,

$$m_a = \frac{v_1}{v_{s1}} = \frac{90}{0.792} = 113.636 \text{ kg / min} = 6818 \text{ kg / h}$$

and specific humidity at point 3, $W_3 = W_1 + \dfrac{m_s}{m_a} = 0.0035 + \dfrac{40}{6818} = 0.0094$ kg / kg of dry air

From steam tables, we find that enthalpy of dry saturated steam corresponding to 110°C is,

$$h_s = 2691.3 \text{ kJ / kg}$$

Now for the heat balance,

$$h_3 = h_1 + \frac{m_s}{m_a} \times h_s + \frac{q_s}{m_a}$$

$$= 13.8 + \frac{40}{6818} \times 2691.3 + \frac{40.7 \times 3600}{6818}$$

$$= 13.8 + 15.8 + 21.5 = 51.1 \text{ kJ/kg}$$

Mark the condition of leaving air on the psychrometric chart as point 3 corresponding to $W_3 = 0.0094$ kg / kg of dry air, and $h_3 = 51.1$ kJ / kg of dry air. Now from the psychrometric chart, corresponding to point 3,

Dry bulb temperature of the leaving air,

$$t_{d3} = 26°\text{C}$$ **Ans.**

and wet bulb temperature of the leaving air,

$$t_{w3} = 18.5°\text{C}$$ **Ans.**

Example 16.24. *An air conditioning system is to take in outdoor air at 10°C and 30% relative humidity at a steady rate of 45m³/min and to condition it to 25°C and 60% relative humidity. The outdoor air is first heated to 22°C in the heating section and then humidified by the injection of hot steam in the humidifying section. Asuming the entire process takes place at a presure of 1 bar. Determine, without using the pyschrometric chart:*

1. The rate of heat supply in the heating section ; and

2. The mass flow rate of steam required in the humidifying section.

Solution. Given : $t_{d1} = 10°\text{C}$; $\phi_1 = 30\% = 0.3$; $v_1 = 45 \text{ m}^3/\text{min}$; $t_{d3} = 25°\text{C}$; $\phi_3 = 60\% = 0.6$; $t_{d2} = 22°\text{C}$; $p_b = 1 \text{ bar} = 1 \times 10^5 \text{ N/m}^2$

Frist of all, mark the initial condition of air, *i.e.*, at 10°C dry bulb temperature and 30% relative humidity on the psychrometric chart at point 1, as shown in Fig. 16.46. Then mark the final condition of air at 25°C dry bulb temperature and 60% relative humidity at point 3. The line 1-2 represents the heating of air in the heating section and the line 2-3 represents the humdification of air by the injection of hot steam.

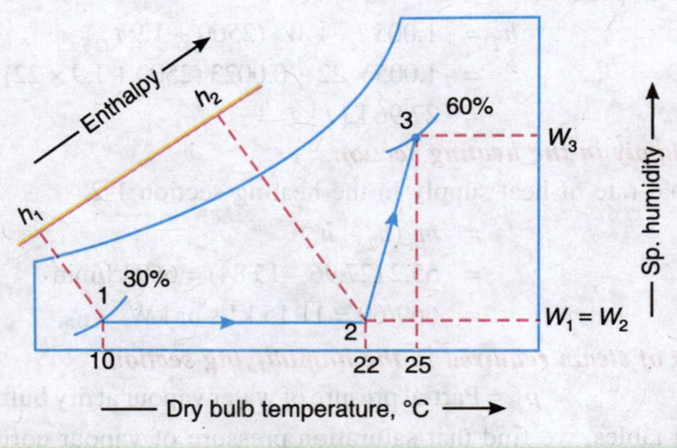

Fig. 16.46

Let p_{v1} = Partial pressure of water vapour at dry bulb temperature of 10°C.

From steam tables, we find that saturation pressure of vapour corresponding to dry bulb temperature of 10°C is,

$$p_{s1} = 0.012\,27 \text{ bar} = 1227 \text{ N/m}^2 \qquad \ldots (\because 1 \text{ bar} = 10^5 \text{ N/m}^2)$$

We know that relative humidity (ϕ_1),

$$0.3 = \frac{p_{v1}}{p_{s1}}$$

$$\therefore \quad p_{v1} = 0.3 \times p_{s1} = 0.3 \times 0.012\,27 = 0.003\,681 \text{ bar} = 368.1 \text{ N/m}^2$$

and specific humidity at point 1,

$$W_1 = \frac{0.622\,p_{v1}}{p_b - p_{v1}} = \frac{0.622 \times 0.003\,681}{1 - 0.003\,681} = 0.0023 \text{ kg / kg of dry air}$$

We know that pressure of dry air,

$$p_a = p_b - p_{v1} = 1 - 0.003\,681 = 0.996\,319 \text{ bar} = 99631.9 \text{ N/m}^2$$

$\therefore$ Specific volume of dry air,

$$v_{s1} = \frac{R_a T_{d1}}{p_a} = \frac{287(10+273)}{99631.9} = 0.815 \text{ m}^3\text{/kg}$$

$$\ldots (\because R_a = 287 \text{ J/kgK})$$

and mass of air flowing, $m_a = \dfrac{v_1}{v_{s1}} = \dfrac{45}{0.815} = 55.2$ kg / min

We know that enthalpy of air at point 1,

$$h_1 = 1.005\, t_{d1} + W_1\,(2500 + 1.9\, t_{d1})$$

$$= 1.005 \times 10 + 0.0023\,(2500 + 1.9 \times 10) = 15.84 \text{ kJ/kg}$$

and enthalpy of air at point 2,

$$h_2 = 1.005\, t_{d2} + W_2\,(2500 + 1.9\, t_{d2})$$
$$= 1.005 \times 22 + 0.0023\,(2500 + 1.9 \times 22) \qquad \ldots(\because W_2 = W_1)$$
$$= 27.96 \text{ kJ / kg}$$

1. *Rate of heat supply in the heating section*

We know that rate of heat supply in the heating section 1-2

$$= m_a\,(h_2 - h_1)$$
$$= 55.2\,(27.96 - 15.84) = 669 \text{ kJ/min}$$
$$= 669/60 = 11.15 \text{ kJ/s or kW} \quad \textbf{Ans.}$$

2. *Mass flow rate of steam required in the humidifying section*

Let p_{v3} = Partial presure of water vapour at dry bulb temperature of 25°C.

From steam tables, we find that saturation pressure of vapour corresponding to dry bulb temperature of 25°C is

$$p_{s3} = 0.03166 \text{ bar} = 3166 \text{ N/m}^2$$

We know that relative humidity (ϕ_3),

$$0.6 = \frac{p_{v3}}{p_{s3}}$$

$$\therefore \quad p_{v3} = 0.6 \times p_{s3} = 0.6 \times 0.03166 = 0.019 \text{ bar}$$

and specific humidity at point 3,

$$W_3 = \frac{0.622\, p_{v3}}{p_b - p_{v3}} = \frac{0.622 \times 0.019}{1 - 0.019} = 0.012 \text{ kg / kg of dry air}$$

∴ Mass flow of steam required

$$= \text{Increase in moisture contents from point 2 to point 3}$$
$$= m_a\,(W_3 - W_2) = 55.2\,(0.012 - 0.0023) = 0.535 \text{ kg / min} \quad \textbf{Ans.}$$

Example 16.25. *Atmospheric air at a dry bulb temperature of 16° C and 25% relative humidity passes through a furnace and then through a humidifier, in such a way that the final dry bulb temperature is 30° C and 50% relative humidity. Find the heat and moisture added to the air. Also determine the sensible heat factor of the process.*

Solution. Given : t_{d1} = 16° C ; ϕ_1 = 25% ; t_{d2} = 30° C ; ϕ_2 = 50%

Heat added to the air

First of all, mark the initial condition of air *i.e.* at 16° C dry bulb temperature and 25% relative humidity on the psychrometric chart at point 1, as shown in Fig. 16.47. Then mark the final condition of air at 30° C dry bulb temperature and 50% relative humidity on the psychrometric chart at point 2. Now locate the point *A* by drawing horizontal line through point 1 and vertical line through point 2. From the psychrometric chart, we find that enthalpy of air at point 1,

$$h_1 = 23 \text{ kJ/kg of dry air}$$

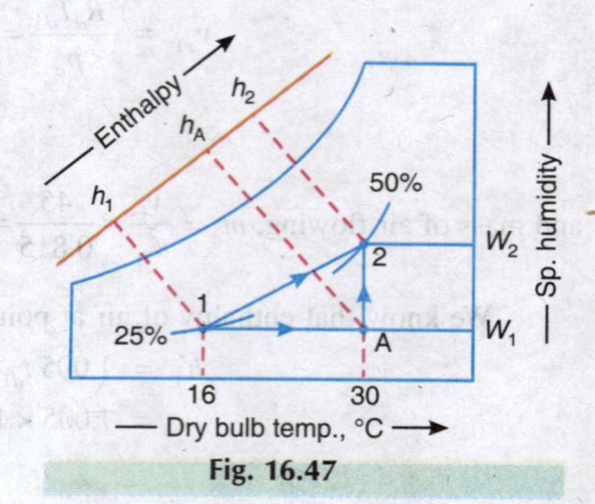

Fig. 16.47

Enthalpy of air at point A,

$$h_A = 38 \text{ kJ/kg of dry air}$$

and enthalpy of air at point 2,

$$h_2 = 64 \text{ kJ/kg of dry air}$$

∴ Heat added to the air

$$= h_2 - h_1 = 64 - 23 = 41 \text{ kJ/kg of dry air} \quad \textbf{Ans.}$$

Moisture added to the air

From the psychrometric chart, we find that the specific humidity in the air at point 1,

$$W_1 = 0.0026 \text{ kg/kg of dry air}$$

and specific humidity in the air at point 2,

$$W_2 = 0.0132 \text{ kg/kg of dry air}$$

∴ Moisture added to the air

$$= W_2 - W_1 = 0.0132 - 0.0026 = 0.0106 \text{ kg/kg of dry air} \quad \textbf{Ans.}$$

Sensible heat factor of the process

We know that sensible heat factor of the process,

$$SHF = \frac{h_A - h_1}{h_2 - h_1} = \frac{38 - 23}{64 - 23} = 0.366 \quad \textbf{Ans.}$$

Example 16.26. *Air at 10° C dry bulb temperature and 90% relative humidity is to be heated and humidified to 35° C dry bulb temperature and 22.5° C wet bulb temperature. The air is pre-heated sensibly before passing to the air washer in which water is recirculated. The relative humidity of the air coming out of the air washer is 90%. This air is again reheated sensibly to obtain the final desired condition. Find : 1. the temperature to which the air should be preheated. 2. the total heating required ; 3. the make up water required in the air washer ; and 4. the humidifying efficiency of the air washer.*

Solution. Given : $t_{d1} = 10°$ C ; $\phi_1 = 90\%$; $t_{d2} = 35°$ C ; $t_{w2} = 22.5°$ C

First of all, mark the initial condition of air *i.e.* at 10° C dry bulb temperature and 90% relative humidity, on the psychrometric chart at point 1, as shown in Fig. 16.48. Now mark the final condition of air *i.e.* at 35° C dry bulb temperature and 22.5° C wet bulb temperature at point 2.

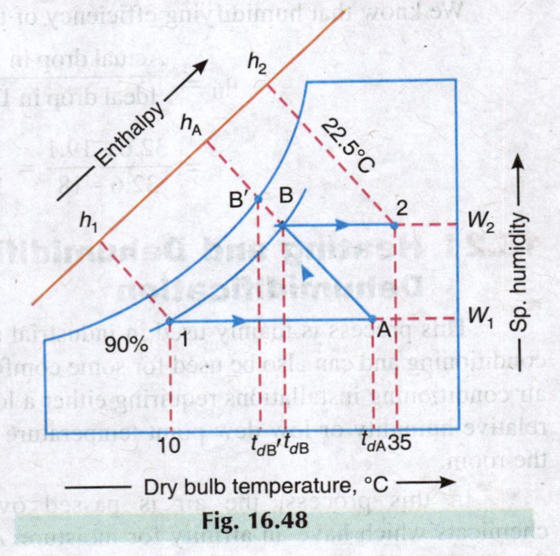

Fig. 16.48

From point 1, draw a horizontal line to represent sensible heating and from point 2 draw horizontal line to intersect 90% relative humidity curve at point B. Now from point B, draw a constant wet bulb temperature line which intersects the horizontal line drawn through point 1 at point A. The line 1-A represents preheating of air, line AB represents humidification and line B-2 represents reheating to final condition.

1. Temperature to which the air should be preheated

From the psychrometric chart, the temperature to which the air should be preheated (corresponding to point A) is

$$t_{dA} = 32.6° \text{ C} \quad \textbf{Ans.}$$

2. Total heating required

From the psychrometric chart, we find that enthalpy of air at point 1,

$$h_1 = 27.2 \text{ kJ/kg of dry air}$$

Enthalpy of air at point *A*,

$$h_A = 51 \text{ kJ/kg of dry air}$$

and enthalpy of air at point 2,

$$h_2 = 68 \text{ kJ/kg of dry air}$$

We know that heat required for preheating of air

$$= h_A - h_1 = 51 - 27.2 = 23.8 \text{ kJ/kg of dry air}$$

and heat required for reheating of air

$$= h_2 - h_B = 68 - 51 = 17 \text{ kJ/kg of dry air} \quad \ldots (\because \; h_B = h_A)$$

$\therefore$ Total heat required

$$= 23.8 + 17 = 40.8 \text{ kJ/kg of dry air}$$ **Ans.**

3. Make up water required in the air washer

From the psychrometric chart, we find that specific humidity of entering air,

$$W_1 = 0.0068 \text{ kg/kg of dry air}$$

and specific humidity of leaving air,

$$W_2 = 0.0122 \text{ kg/kg of dry air}$$

$\therefore$ Make up water required in the air washer

$$= W_B - W_A = W_2 - W_1$$

$$= 0.0122 - 0.0068 = 0.0054 \text{ kg/kg of dry air}$$ **Ans.**

4. Humidifying efficiency of the air washer

From the psychrometric chart, we find that

$$t_{dB} = 19.1° \text{ C} \quad \text{and} \quad t_{dB'} = 18° \text{ C}$$

We know that humidifying efficiency of the air washer,

$$\eta_H = \frac{\text{Actual drop in DBT}}{\text{Ideal drop in DBT}} = \frac{t_{dA} - t_{dB}}{t_{dA} - t_{dB'}}$$

$$= \frac{32.6 - 19.1}{32.6 - 18} = \frac{13.5}{14.6} = 0.924 \text{ or } 92.4\%$$ **Ans.**

16.21 Heating and Dehumidification -Adiabatic Chemical Dehumidification

This process is mainly used in industrial air conditioning and can also be used for some comfort air conditioning installations requiring either a low relative humidity or low dew point temperature in the room.

In this process, the air is passed over chemicals which have an affinity for moisture. As the air comes in contact with these chemicals, the moisture gets condensed out of the air and gives up its latent heat. Due to the condensation, the specific humidity decreases and the heat of condensation supplies sensible heat for heating the air and thus increasing its dry bulb temperature.

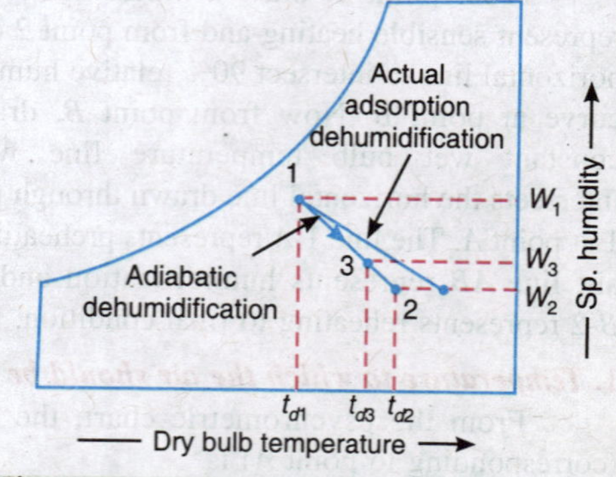

Fig. 16.49. Heating and dehumidification.

The process, which is the reverse of adiabatic saturation process, is shown by the line 1-2 on the psychrometric chart as shown in Fig. 16.49. The path followed during the process is along the constant wet bulb temperature line or constant enthalpy line.

The effectiveness or efficiency of the dehumidifier is given as

$$\eta_H = \frac{\text{Actual increase in dry bulb temperature}}{\text{Ideal increase in dry bulb temperature}} = \frac{t_{d3} - t_{d1}}{t_{d2} - t_{d1}}$$

Notes : 1. In actual practice, the process is accompanied with a release of heat called *heat of adsorption,* which is very large. Thus the sensible heat gain of air exceeds the loss of latent heat and the process is shown above the constant wet bulb temperature line in Fig. 16.49.

2. Two types of chemicals used for dehumidification are *absorbents* and *adsorbents.* The absorbents are substances which can take up moisture from air and during this process change it chemically, physically or in both respects. These includes water solutions or brines of calcium chloride, lithium chloride, lithium bromide and ethylene glycol. These are used as air dehydrators by spraying or otherwise exposing a large surface of the solution in the air stream.

The adsorbents are substances in the solid state which can take up moisture from the air and during this process do not change it chemically or physically. These include silca gel (which is a form of silicon dioxide prepared by mixing fused sodium silicate and sulphuric acid) and activated alumina (which is a porous amorphous form of aluminium oxide).

Example 16.27. *Saturated air at 21° C is passed through a drier so that its final relative humidity is 20%. The drier uses silica gel adsorbent. The air is then passed through a cooler until its final temperature is 21° C without a change in specific humidity. Determine : 1. the temperature of air at the end of the drying process; 2. the heat rejected during the cooling process ; 3. the relative humidity at the end of cooling process; 4. the dew point temperature at the end of the drying process ; and 5. the moisture removed during the drying process.*

Solution. Given : $t_{d1} = t_{d3} = 21°$ C ; $\phi_2 = 20\%$

1. ***Temperature of air at the end of drying process***

First of all, mark the initial condition of air *i.e.* at 21° C dry bulb temperature upto the saturation curve (because the air is saturated) on the psychrometric chart at point 1, as shown in Fig. 16.50. Since the drying process is a chemical dehumidification process, therefore, it follows a path along the constant wet bulb temperature or the constant enthalpy line as shown by the line 1- 2 in Fig. 16.50. Now mark the point 2 at relative humidity of 20%.

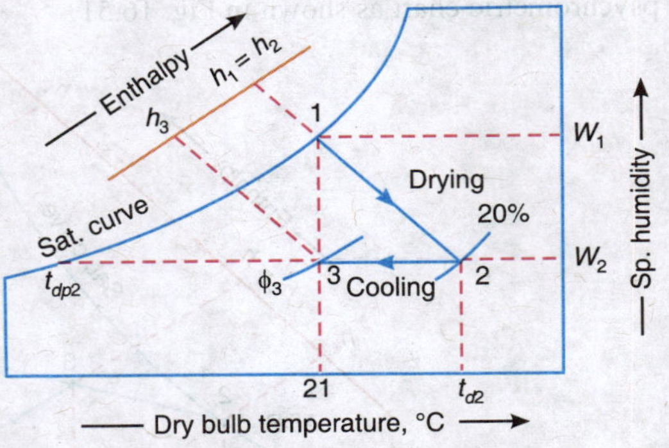

Fig. 16.50

From the psychrometric chart, the temperature at the end of drying process at point 2,

$$t_{d2} = 38.5° \text{ C } \textbf{Ans.}$$

2. ***Heat rejected during the cooling process***

The cooling process is shown by the line 2-3 on the psychrometric chart as shown in Fig. 16.50. From the psychrometric chart, we find that enthalpy of air at point 2,

$$h_2 = 61 \text{ kJ/kg of dry air}$$

and enthalpy of air at point 3,

$$h_3 = 43 \text{ kJ/kg of dry air}$$

∴ Heat rejected during the cooling process

$$= h_2 - h_3 = 61 - 43 = 18 \text{ kJ/kg of dry air}$$ **Ans.**

3. Relative humidity at the end of cooling process

From the psychrometric chart, we find that relative humidity at the end of cooling process (*i.e.* at point 3),

$$\phi_3 = 55\%$$ **Ans.**

4. Dew point temperature at the end of drying process

From the psychrometric chart, we find that the dew point temperature at the end of the drying process,

$$t_{dp2} = 11.6° \text{ C}$$ **Ans.**

5. Moisture removed during the drying process

From the psychrometric chart, we find that moisture in air before the drying process at point 1,

$$W_1 = 0.0157 \text{ kg/kg of dry air}$$

and moisture in air after the drying process at point 2,

$$W_2 = 0.0084 \text{ kg/kg of dry air}$$

∴ Moisture removed during the drying process

$$= W_1 - W_2 = 0.0157 - 0.0084 = 0.0073 \text{ kg/kg of dry air}$$ **Ans.**

Example 16.28. *300 m³ of air is supplied per minute from outdoor conditions of 40° C dry bulb temperature and 26° C wet bulb temperature to an air- conditioned room. The air is dehumidified first by a cooling coil having by-pass factor 0.32 and dew point temperature 15° C and then by a chemical dehumidifier. Air leaves the chemical dehumidifier at 30° C dry bulb temperature. Air is then passed over a cooling coil whose surface temperature is 15° C and by-pass factor 0.26. Calculate the capacities of the two cooling coils and dehumidifier.*

Solution. Given : $v_1 = 300 \text{ m}^3/\text{min}$; $t_{d1} = 40°$ C ; $t_{w1} = 26°$ C ; $ADP = 15°$ C ; $t_{d4} = 30°$ C

The various processes involved in an air-conditioning of a room are shown on the psychrometric chart as shown in Fig. 16.51.

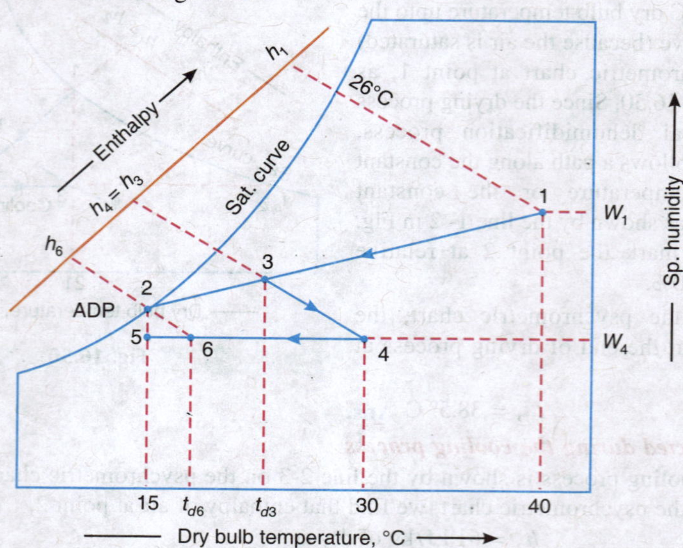

Fig. 16.51

First of all, mark the outdoor conditions of air at 40° C dry bulb temperature and 26°C wet bulb temperature on the psychrometric chart, at point 1, as shown in Fig. 16.51. Now mark point 2 on the saturation curve representing the apparatus dew point (*i.e.* ADP = 15° C) of the cooling coil. Since the by-pass factor of the cooling coil is 0.32, therefore mark point 3 on the line 1-2 such that

Length of line 2-3 = 0.32 × Length of line 2-1

The point 3 represents the condition of air leaving the cooling coil. This air now passes through a chemical dehumidifier. From point 3, draw a constant enthalpy line to intersect the vertical line drawn through 30° C dry bulb temperature, at point 4. The air after leaving the chemical dehumidifier, passes over another cooling coil having surface temperature of 15° C. Therefore through point 4, draw a horizontal line intersecting the vertical line drawn through 15° C *DBT*, at point 5. Since the by-pass factor of this coil is 0.26, therefore mark point 6 on the line 4-5 such that

Length of line 5-6 = 0.26 × Length of line 5-4.

The point 6 represents the condition of air leaving the second cooling coil.

Capacities of the two cooling coils

From the psychrometric chart, we find that specific humidity of air at point 1,

$$W_1 = 0.0156 \text{ kg/kg of dry air}$$

Enthalpy of air at point 1,

$$h_1 = 80.6 \text{ kJ/kg of dry air}$$

Specific volume of air at point 1,

$$v_{s1} = 0.94 \text{ m}^3\text{/kg of dry air}$$

Enthalpy of air at point 3,

$$h_3 = 54.5 \text{ kJ/kg of dry air}$$

Enthalpy of air at point 4,

$$h_4 = h_3 = 54.5 \text{ kJ/kg of dry air}$$

Specific humidity of air at point 4,

$$W_4 = 0.0094 \text{ kJ/kg of dry air}$$

Enthalpy of air at point 6,

$$h_6 = 43 \text{ kJ/kg of dry air}$$

We know that mass of air supplied,

$$m_a = \frac{v_1}{v_{s1}} = \frac{300}{0.94} = 319 \text{ kg/min}$$

We know that the capacity of the first cooling coil

$$= m_a (h_1 - h_3) = 319 (80.6 - 54.5) = 8326 \text{ kJ/min} \quad \textbf{Ans.}$$

and capacity of the second cooling coil

$$= m_a (h_4 - h_6) = 319 (54.5 - 43) = 3668 \text{ kJ/min} \quad \textbf{Ans.}$$

Capacity of the dehumidifier

We know that the capacity of the dehumidifier

$$= m_a (W_1 - W_4) = 319 (0.0156 - 0.0094) = 1.98 \text{ kg/min} \quad \textbf{Ans.}$$

16.22 Adiabatic Mixing of Two Air Streams

When two quantities of air having different enthalpies and different specific humidities are mixed, the final condition of the air mixture depends upon the masses involved, and on the enthalpy and specific humidity of each of the constituent masses which enter the mixture.

Now consider two air streams 1 and 2 mixing adiabatically as shown in Fig. 16.52 (*a*).

Let m_1 = Mass of air entering at 1,

h_1 = Enthalpy of air entering at 1,

W_1 = Specific humidity of air entering at 1,

m_2, h_2, W_2 = Corresponding values of air entering at 2, and

m_3, h_3, W_3 = Corresponding values of the mixture leaving at 3.

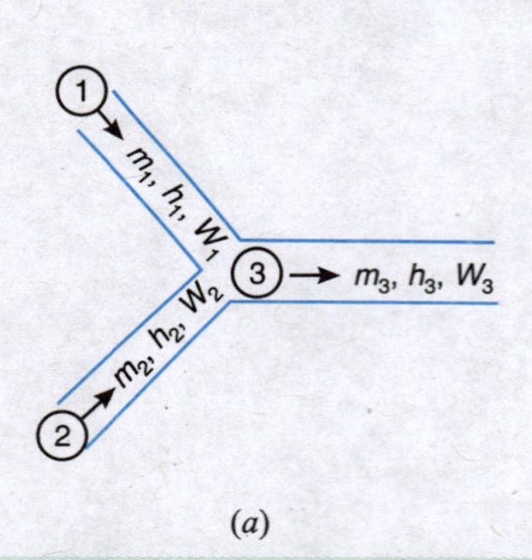

(*a*)

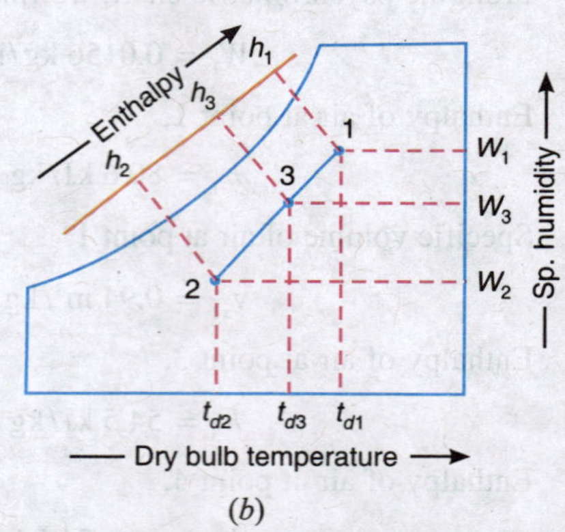

(*b*)

Fig. 16.52. Adiabatic mixing of two air streams.

Assuming no loss of enthalpy and specific humidity during the air mixing process, we have for the mass balance,

$$m_1 + m_2 = m_3 \quad \ldots (i)$$

For the energy balance,

$$m_1 h_1 + m_2 h_2 = m_3 h_3 \quad \ldots (ii)$$

and for the mass balance of water vapour,

$$m_1 W_1 + m_2 W_2 = m_3 W_3 \quad \ldots (iii)$$

Substituting the value of m_3 from equation (*i*) in equation (*ii*),

$$m_1 h_1 + m_2 h_2 = (m_1 + m_2) h_3 = m_1 h_3 + m_2 h_3$$

or

$$m_1 h_1 - m_1 h_3 = m_2 h_3 - m_2 h_2$$

$$m_1 (h_1 - h_3) = m_2 (h_3 - h_2)$$

$$\therefore \quad \frac{m_1}{m_2} = \frac{h_3 - h_2}{h_1 - h_3} \quad \ldots (iv)$$

Similarly, substituting the value of m_3 from equation (*i*) in equation (*iii*), we have

$$\frac{m_1}{m_2} = \frac{W_3 - W_2}{W_1 - W_3} \qquad \dots (v)$$

Now from equations (*iv*) and (*v*),

$$\frac{m_1}{m_2} = \frac{h_3 - h_2}{h_1 - h_3} = \frac{W_3 - W_2}{W_1 - W_3} \qquad \dots (vi)$$

The adiabatic mixing process is represented on the psychrometric chart as shown in Fig. 16.52 (*b*). The final condition of the mixture (point 3) lies on the straight line 1-2. The point 3 divides the line 1-2 in the inverse ratio of the mixing masses. By calculating the value of W_3 from equation (*vi*), the point 3 is plotted on the line 1-2.

It may be noted that when warm and high humidity air is mixed with cold air, the resulting mixture will be a fog and the final condition (point 3) on the psychrometric chart will lie to the left or above the saturation curve which represents the fog region, as shown in Fig. 16.53. The temperature of the fog is that of the extended wet bulb line passing through point 3.

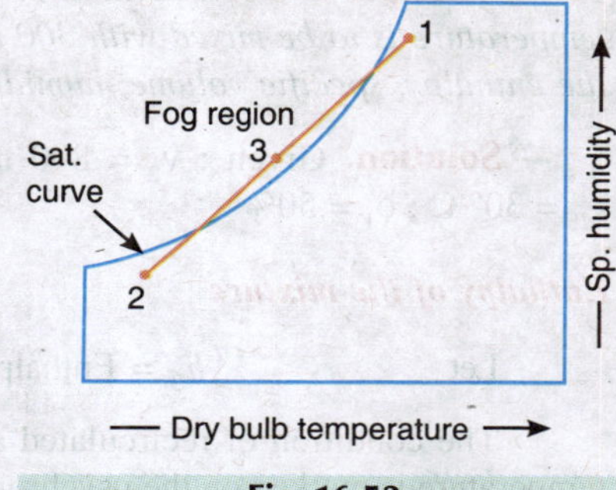

Fig. 16.53

The fog may also result when steam or a very fine water spray is injected into air in a greater quantity than required to saturate the air. Even lesser quantity of steam, if not mixed properly, may result fog.

The fog can be cleared by heating the fog, mixing the fog with warmer unsaturated air or mechanically separating the water droplets from the air.

Example 16.29. *One kg of air at 40° C dry bulb temperature and 50% relative humidity is mixed with 2 kg of air at 20° C dry bulb temperature and 20°C dew point temperature. Calculate temperature and specific humidity of the mixture.*

Solution. Given : $m_1 = 1$ kg ; $t_{d1} = 40°$ C ; $\phi_1 = 50\%$; $m_2 = 2$ kg ; $t_{d2} = 20°$ C ; $t_{dp} = 20°$ C

Specific humidity of the mixture

Let W_3 = Specific humidity of the mixture.

The condition of first mass of air at 40° C dry bulb temperature and 50% relative humidity is marked on the psychrometric chart at point 1, as shown in Fig. 16.54. Now mark the condition of second mass of air at 20° C dry bulb temperature and 20° C dew point temperature at point 2, as shown in the figure. This point lies on the saturation curve. Join the points 1 and 2. From the psychrometric chart, we find that specific humidity of the first mass of air,

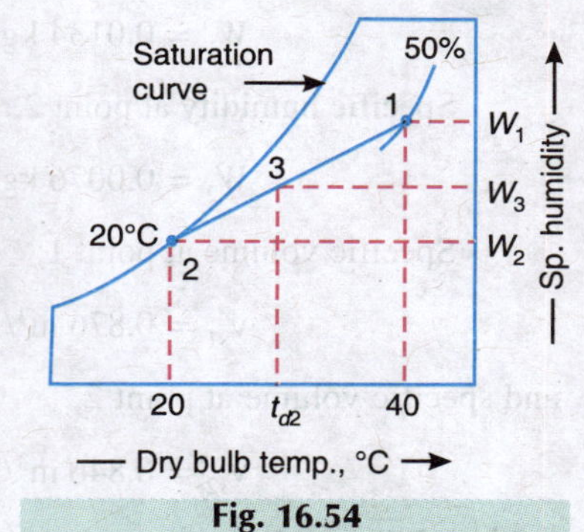

Fig. 16.54

$$W_1 = 0.0238 \text{ kg/kg of dry air}$$

and specific humidity of the second mass of air,

$$W_2 = 0.0148 \text{ kg/kg of dry air}$$

We know that

$$\frac{m_1}{m_2} = \frac{W_3 - W_2}{W_1 - W_3}$$

$$\frac{1}{2} = \frac{W_3 - 0.0148}{0.0238 - W_3}$$

or $$0.0238 - W_3 = 2W_3 - 0.0296$$

$$\therefore \quad W_3 = 0.0178 \text{ kg/kg of dry air} \quad \textbf{Ans.}$$

Temperature of the mixture

Now plot point 3 on the line joining the points 1 and 2 corresponding to specific humidity $W_3 = 0.0178$ kg / kg of dry air, as shown in Fig. 16.51. We find that at point 3, the dry bulb temperature of the mixture is

$$t_{d3} = 26.8° \text{ C} \quad \textbf{Ans.}$$

Example 16.30. *800 m³/min of recirculated air at 22° C DBT and 10° C dew point temperature is to be mixed with 300 m³/min of fresh air at 30° C DBT and 50% RH. Determine the enthalpy, specific volume, humidity ratio and dew point temperature of the mixture.*

Solution. Given : $v_2 = 800$ m³/min ; $t_{d2} = 22°$ C ; $t_{dp2} = 10°$ C ; $v_1 = 300$ m³/min ; $t_{d1} = 30°$ C ; $\phi_1 = 50\%$

Enthalpy of the mixture

Let h_3 = Enthalpy of the mixture.

The condition of recirculated air at 22° C dry air bulb temperature and 10° C dew point temperature is marked on the psychrometric chart at point 2 as shown in Fig. 16.55. Now mark the condition of fresh air at 30° C dry bulb temperature and 50% relative humidity at point 1 as shown in the figure. Join points 1 and 2.

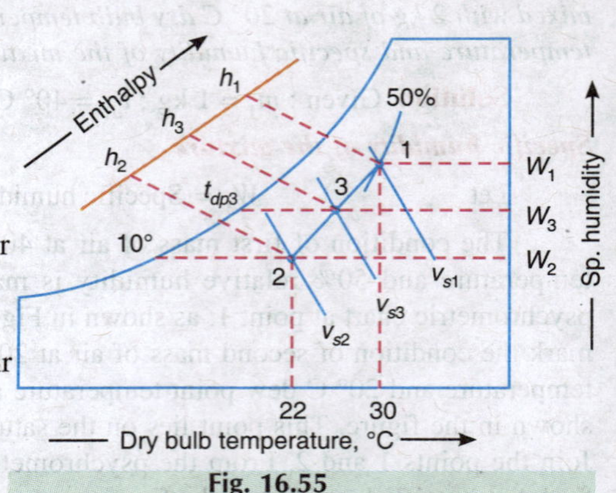

Fig. 16.55

From the psychrometric chart, we find that enthalpy at point 1,

$$h_1 = 64.6 \text{ kJ/kg of dry air}$$

Enthalpy at point 2,

$$h_2 = 41.8 \text{ kJ/kg of dry air}$$

Specific humidity at point 1,

$$W_1 = 0.0134 \text{ kg / kg of dry air}$$

Specific humidity at point 2,

$$W_2 = 0.0076 \text{ kg / kg of dry air}$$

Specific volume at point 1,

$$v_{s1} = 0.876 \text{ m}^3\text{/kg of dry air}$$

and specific volume at point 2,

$$v_{s2} = 0.846 \text{ m}^3\text{/kg of dry air}$$

We know that mass of fresh air at point 1,

$$m_1 = \frac{v_1}{v_{s1}} = \frac{300}{0.876} = 342.5 \text{ kg/min}$$

and mass of recirculated air at point 2,

$$m_2 = \frac{v_2}{v_{s2}} = \frac{800}{0.846} = 945.6 \text{ kg/min}$$

We know that

$$\frac{m_1}{m_2} = \frac{h_3 - h_2}{h_1 - h_3} \quad \text{or} \quad \frac{342.5}{945.6} = \frac{h_3 - 41.8}{64.6 - h_3}$$

$$\therefore \qquad 0.362\,(64.6 - h_3) = h_3 - 41.8$$

or

$$h_3 = 47.86 \text{ kJ/kg of dry air} \quad \textbf{Ans.}$$

Specific volume, humidity ratio and dew point temperature of the mixture

Plot point 3 on line joining the points 1 and 2 corresponding to enthalpy h_3 = 47.86 kJ / kg of dry air, as shown in Fig. 16.52.

From point 3 on the psychrometric chart, we find that specific volume of the mixture at point 3,

$$v_{s3} = 0.855 \text{ m}^3\text{/kg of dry air} \quad \textbf{Ans.}$$

Humidity ratio of the mixture at point 3,

$$W_3 = 0.0092 \text{ kg/kg of dry air} \quad \textbf{Ans.}$$

and dew point temperature of the mixture at point 3,

$$t_{dp3} = 13° \text{ C} \quad \textbf{Ans.}$$

Example 16.31. *The saturated air leaving the cooling section of an air conditioning system at 14°C at the rate of 50 m³/min is mixed adiabatically with the outside air at 32°C and 60% relative humidity at a rate of 20 m³/min. Assuming that the mixing process occurs at a pressure of 1 atmosphere, determine the specific humidity, relative humidity, dry bulb temperature and the volume flow rate of the mixture.*

Solution. Given : t_{d2} = 14°C ; v_2 = 50 m³/min ; t_{d1} = 32°C ; ϕ_1 = 60 % ; v_1 = 20 m³/min ; p = 1 atm

The condition of saturated air at dry bulb temperature of 14°C is marked on the psychrometric chart at point 2 (on the saturation curve) as shown in Fig 16.56. Now mark the conditon of outside air at 32°C dry bulb temperature and 60% relative humidity at point 1. Join the points 1 and 2.

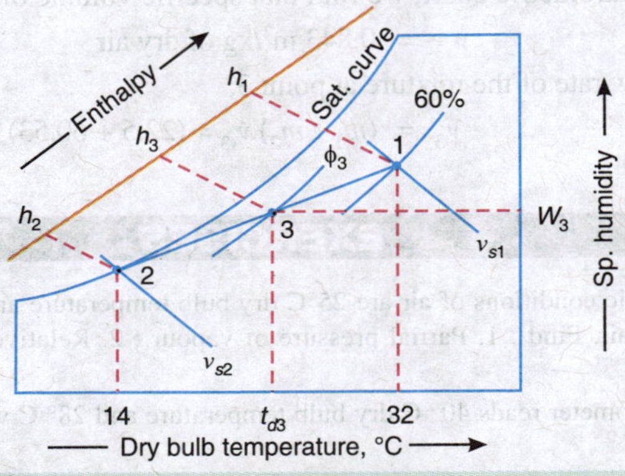

Fig. 16.56

From the psychrometric chart, we find that enthalpy at point 1,

$$h_1 = 78 \text{ kJ/kg of dry air}$$

Enthalpy at point 2, $h_2 = 39.4$ kJ/kg of dry air

Specific volume at point 1,

$$v_{s1} = 0.889 \text{ m}^3\text{/kg of dry air}$$

Specific volume at point 2,

$$v_{s2} = 0.826 \text{ m}^3\text{/kg of dry air}$$

Let h_3 = Enthalpy of air after mixing at point 3.

We know that mass of outside air at point 1,

$$m_1 = \frac{v_1}{v_{s1}} = \frac{20}{0.889} = 22.5 \text{ kg / min}$$

and mass of saturated air leaving the cooling section,

$$m_2 = \frac{v_2}{v_{s2}} = \frac{50}{0.826} = 60.53 \text{ kg / min}$$

We know that $\dfrac{m_1}{m_2} = \dfrac{h_3 - h_2}{h_1 - h_3}$ or $\dfrac{22.5}{60.53} = \dfrac{h_3 - 39.4}{78 - h_3}$

$$\therefore \quad 0.3717\,(78 - h_3) = h_3 - 39.4$$

or

$$h_3 = 49.86 \text{ kJ / kg of dry air}$$

Specific humidity, relative humidity, dry bulb temperature

Plot point 3 on the line joining the points 1 and 2 corresponding to enthalpy $h_3 = 49.86$ kJ/kg of dry air, as shown in Fig. 16.56.

From point 3 on the psychrometric chart, we find that specific humidity of the mixture at point 3,

$$W_3 = 0.0122 \text{ kg / kg of dry air} \quad \textbf{Ans.}$$

Relative humidity of the mixture at point 3,

$$\phi_3 = 90\% \quad \textbf{Ans.}$$

and dry bulb temperature of the mixture at point 3,

$$t_{d3} = 19°\text{C} \quad \textbf{Ans.}$$

Volume flow rate of the mixture

From the psychrometric chart, we find that specific volume of the mixture at point 3,

$$v_{s3} = 0.843 \text{ m}^3\text{/kg of dry air}$$

∴ Volume flow rate of the mixture at point 3,

$$v_3 = (m_1 + m_2)\, v_{s3} = (22.5 + 60.53)\, 0.843 = 70 \text{ m}^3\text{/min} \quad \textbf{Ans.}$$

EXERCISES

1. The atmospheric conditions of air are 25°C dry bulb temperature and specific humidity of 0.01 kg per kg of dry air. Find : 1. Partial pressure of vapour ; 2. Relative humidity ; and 3. Dew point temperature. **[Ans.** 0.016 bar ; 50.6% ; 14.1° C**]**
2. A sling psychrometer reads 40° C dry bulb temperature and 28° C wet bulb temperature. Calculate the following :

1. Specific humidity ; 2. Relative humidity ; 3. Vapour density in air ; 4. Dew point temperature ; 5. Enthalpy of mixture per kg of dry air.

[Ans. 0.019 kg / kg of dry air ; 40.7% ; 0.0208 kg / m^3 of dry air ; 24° C ; 88.38 kJ / kg of dry air]

3. The pressure and temperature of a mixture of dry air and water vapour are 736 mm of Hg and 21°C. The dew point temperature of the mixture is 15°C. Determine the following using steam tables :
1. Partial pressure of water vapour ; 2. Relative humidity ;
3. Specific humidity ; 4. Enthalpy of mixture per kg of dry air ;
5. Specific volume of the mixture per kg of dry air.

[Ans. 12.78 mm of Hg ; 68.53% ; 0.011 kg / kg of dry air ; 48.97 kJ/kg of dry air ; 0.875 m^3/kg of dry air]

4. A sample of air is having dry bulb temperature 21°C and relative humidity 30% at barometric pressure of 760 mm of Hg. Find : 1. Partial pressure of vapour ; 2. Specific humidity ; 3. Wet bulb temperature and corresponding saturation pressure ; 4. Precentage humidity or degree of saturation ; 5. Specific volume of dry air ; 6. Dew point temperature ; and 7. Enthalpy of moist air per kg of dry air.

Given : R = 0.287 kJ/kg K ; c_p(dry air) = 1.005 kJ/kg K ; specific heat of superheated vapour = 1.884 kJ/kg K and latent heat of vaporisation at dew point temperature = 2493 kJ/kg. Do not use psychrometric chart. Psychrometric tables can be used.

[Ans. 5.595 mm of Hg ; 0.00461 kg / kg of dry air ; 11.5°C, 10.258 mm of Hg ; 0.2945 ; 0.839 m^3/kg of dry air ; 3°C ; 32.812 kJ/kg of dry air]

5. A sample of moist air has a dry bulb temperature of 25° C and a relative humidity of 50 per cent. The barometric pressure is 740 mm of Hg. Calculate : 1. partial pressure of water vapour and dry air ; 2. dew point temperature and specific humidity of air ; 3. enthalpy of air per kg of dry air.

[Ans. 0.01583 bar ; 14° C ; 0.0101 kg / kg of dry air ; 50.81 kJ / kg of dry air]

6. The moist air exists at a total pressure of 1.01325 bar and 25° C dry bulb temperature. If the degree of saturation is 50%, determine the following using steam tables :
1. Specific humidity ; 2. Dew point temperature ; and 3. Specific volume of moist air.

[Ans. 10.03 g / kg of dry air ; 14° C ; 0.857 m^3 / kg]

7. The atmospheric conditions of air are 35° C dry bulb temperature, 60% relative humidity and 1.01325 bar pressure. If 0.005 kg of moisture per kg of dry air is removed, the temperature becomes 25° C. Determine the final relative humidity and dew point temperature. [Ans. 88.6% ; 23° C]

8. The atmospheric air enters the adiabatic saturator at 33° C dry bulb temperature and 23° C wet bulb temperature. The barometric pressure is 740 mm of Hg. Determine the specific humidity and vapour pressure at 33° C. [Ans. 0.012 kg / kg of dry air ; 13 mm of Hg]

9. The atmospheric air has 35° C dry bulb temperature and 50% relative humidity. Using psychrometric chart, find (*i*) wet bulb temperature, (*ii*) humidity ratio, (*iii*) dew point temperature, and (*iv*) enthalpy of air per kg of dry air. [Ans. 26.2° C ; 0.0178 kg / kg of dry air ; 23° C ; 81 kJ / kg of dry air]

10. The atmospheric air at 750 mm of Hg has 34° C dry bulb temperature and 19° C wet bulb temperature. Using psychrometric chart, find (*a*) partial pressure of vapour, (*b*) saturation pressure corresponding to 34° C, and (*c*) volume of air per kg of dry air.

[Ans. 8.5 mm of Hg ; 39.9 mm of Hg ; 0.088 m^3/kg of dry air]

11. The atmospheric air at 15° C dry bulb temperature and 80% relative humidity is supplied to the heating chamber at the rate of 100 m^3/min. The leaving air has a temperature of 22° C without change in its moisture contents. Determine the heat added to the air per min. and final relative humidity of the air. [Ans. 843 kJ /min ; 52%]

12. 100 m^3 of air at 35°C and 70% relative humidity is cooled to 20° C and 55% relative humidity by passing it through a cooling coil. Using psychrometric tables only, find : 1. the amount of water vapour removed in kg / h; and 2. the cooling capacity of the coil in tonnes of refrigeration. Assume specific heat of superheated vapour in air as 1.88 kJ/kg K and R for air = 287.14 J/kg K.

[Ans. 144 kg / h ; 37 TR]

13. The air at 35° C dry bulb temperature and 25° C wet bulb temperature is passed through a cooling coil at the rate of 280 m^3/min. The air leaves the cooling coil at 26.5° C dry bulb temperature and 50% relative humidity. Find : 1. Capacity of the cooling coil in tonnes of refrigeration ; 2. Wet bulb temperature of the leaving air; 3. Water vapour removed per minute ; and 4. Sensible heat factor.

[Ans. 31.98 TR ; 19.2° C ; 1.56 kg / kg of dry air ; 0.39]

14. Air at 30° C dry bulb temperature and 60% relative humidity is passed through a cooling coil at the rate of 250 m^3/min. The air leaves the cooling coil at 14° C dry bulb temperature. If the by-pass factor of the cooling coil is 0.1, find : (*a*) surface temperature of the cooling coil or ADP, (*b*) relative humidity of the air leaving the cooling coil, (*c*) capacity of the cooling coil in kW, and (*d*) sensible heat factor. **[Ans.** 12.2°C ; 95% ; 157.15 kW ; 0.5]

15. The outside air at 31°C dry bulb temperature and 18.5° C wet bulb temperature enters a cooling coil at the rate of 40 m^3/min. The effective surface temperature of the cooling coil is 4.5° C and its cooling capacity is 12.5 kW of refrigeration. Find : (*a*) dry bulb and wet bulb temperatures of the air leaving the coil, (*b*) enthalpy of air leaving the coil, and (*c*) by-pass factor of the coil.

[Ans. 18.8° C ; 12.7°C ; 35.6 kJ / kg of dry air ; 0.52]

16. The moist air at 30°C and 50% RH enters a steady state dehumidifier at the rate of 280 m^3/min. The air passes over a cooling coil and water vapour condenses. The saturated moist air exists at 10° C. The condensate also leaves the dehumidifier at 10° C. The pressure is atmospheric.

Determine : 1. the mass flow rate of dry air ; 2. the rate at which water is condensed ; and 3. the required refrigerating capacity. **[Ans.** 319.4 kg /min ; 1.854 kg /min ; 53.6 TR]

17. Air at 40° C and 30% relative humidity is passed through an adiabatic air washer at the rate of 28 m^3/ min. Find the state of air leaving the air washer, if the effectiveness of the air washer is 80%.

[Ans. 32.5° C DBT, 55% RH]

18. Air at 40° C dry bulb temperature and 15% relative humidity is passed through the adiabatic humidifier at the rate of 200 m^3/min. The outlet conditions of air are 25° C dry bulb temperature and 20° C wet bulb temperature. Find. 1. dew point temperature ; 2. relative humidity of exit air ; and 3. amount of water vapour added to the air per minute. **[Ans.** 17.8° C ; 65% ; 1.26 kg /min]

19. The atmospheric air at 20° C and 60% relative humidity is heated and humidified in such a way that the final dry bulb temperature is 30° C and relative humidity is 50%. Determine the heat and moisture added to the air per minute, if the volume of entering air is 100 m^3/min.

[Ans. 2636.5 kJ/min ; 0.57 kg / min]

20. Air at 0° C and 95% relative humidity has to be heated and humidified to 25° C and 40% relative humidity by the following three processes :

(*a*) preheating ; (*b*) adiabatic saturation in a recirculated water air-washer ; and (*c*) reheating to final state.

Calculate: 1. The heating required in two heaters ; 2. The makeup water required in washer and temperature of washer. Assume effectiveness of washer as 80 per cent.

[Ans. 27.3 kJ / kg of dry air ; 9 kJ / kg of dry air ; 4.1 g / kg of dry air ; 12.6° C]

21. In an air washer installation, 100 m^3 of air per minute at 5° C dry bulb temperature and 80% relative humidity has to be heated and humidified to 25° C and 45% of relative humidity by the following methods :-

(*i*) by blowing the air through a heated water spray air washer, (*ii*) by preheating, adiabatic saturation in a re-circulated air washer upto 95% relative humidity and then re-heating to final state.

Calculate for (*i*) and (*ii*) : (*a*) total heat added, (*b*) make up water required in washer, and (*c*) humidifying efficiency of the re-circulated water air washer.

[Ans. 4045 kJ / min ; 0.554 kg / min ; 4045 kJ / min ; 0.554 kJ / min ; 96.4 %]

22. A small auditorium is required to be maintained at 22° C dry bulb temperature and 70% relative humidity. The ambient conditions are at 30° C dry bulb temperature and 75% relative humidity. The amount of free air circulated is 200 m^3/min. The required conditions are achieved by first cooling and dehumidifying through a cooling coil having apparatus dew point of 14° C and then by heating. With the help of psychrometric chart, find :

1.The capacity of the cooling coil in tonnes of refrigeration and its by-pass factor ; 2. The amount of water vapour removed by the cooling coil in kg / h ; and 3. The capacity of the heating coil in kW and its surface temperature. Assume the by-pass factor as 0.2.

[Ans. 39.6 TR , 0.138 ; 116.6 kg /h ; 15.7 kW , 23.4° C]

23. Air at 10° C dry bulb temperature and 8° C wet bulb temperature is supplied at the rate of 15 m³/min. It is brought to 20° C dry bulb temperature and 60% relative humidity by heating and then by adiabatic humidification. Find : (*a*) capacity of the heating coil in kW, (*b*) surface temperature of the coil, if the by-pass factor is 0.32, and (*c*) capacity of the humidifier.

[Ans. 3.6 kW ; 26.77° C ; 0.0555 kg / min]

24. 1 kg of air at 20° C dry bulb temperature and 40% relative humidity is mixed adiabatically with 2 kg of air at 40° C dry bulb temperature and 40% relative humidity. Find specific humidity and enthalpy of the final condition of air. **[Ans.** 0.0144 kg / kg of air ; 71.67 kJ / kg of dry air]

25. 30 m³/min of moist air at 15° C dry bulb temperature and 13° C wet bulb temperature is mixed with 12 m³/min at 25° C dry bulb temperature and 50% relative humidity. Determine the dry bulb temperature and wet bulb temperature of the resulting mixture. **[Ans.** 16° C ; 14.5° C]

QUESTIONS

1. What do you understand by the term 'psychrometry'?

2. Define the following :

1. Specific humidity ; 2. Absolute humidity ; 3. Relative humidity ; 4. Dew point temperature.

3. What is a sling psychrometer ? Make a neat sketch and explain its use.

4. Establish the following expression for air-vapour mixture :

$$\text{Specific humidity,} \quad W = 0.622 \times \frac{p_v}{p_b - p_v}$$

where p_v = Partial pressure of water vapour, and

p_b = Barometric pressure.

5. Derive the relationship between relative humidity (ϕ) and degree of saturation (μ) in the form

$$\phi = \frac{\mu}{1-(1-\mu)\left(\dfrac{p_s}{p}\right)}$$

where p is the atmospheric pressure and p_s is the saturation pressure at temperature of the mixture.

6. What is the difference between 'wet bulb temperature' and 'thermodynamic wet bulb temperature' ?

7. Prove that the partial pressure of water vapour in the atmospheric air remains constant as long as the specific humidity remains constant.

8. An unsaturated air stream is undergoing adiabatic saturation process and the outgoing air is saturated. Draw the system schematically. Represent the process on *T-s* and psychrometric charts. Find, for the incoming stream of air, humidity ratio from the adiabatic saturation equation obtained from heat balance.

9. Prove that the enthalpy of the humid air remains constant along a wet bulb temperature line on the psychrometric chart.

10. When is dehumidification of air necessary and how it is achieved.

11. Write a short note on by-pass factor for cooling coils.

12. Define sensible heat factor.

13. With the help of psychrometric chart, explain the following processes :

1. Sensible heating and sensible cooling process ;

2. Heating and dehumidification process ;

3. Cooling and humidification process ; and
4. Cooling and dehumidification process.

14. Show the following processes on the skeleton psychrometric chart :

(*a*) Dehumidification of moist air by cooling ; and

(*b*) Adiabatic mixing of two air streams.

15. What is fog ? Show on the psychrometric chart when two air streams yield fogged state of air.

OBJECTIVE TYPE QUESTIONS

1. A mixture of dry air and water vapour, when the air has diffused the maximum amount of water vapour into it, is called

(*a*) dry air (*b*) moist air (*c*) saturated air (*d*) specific humidity

2. The temperature of air recorded by a thermometer, when it is not effected by the moisture present in it, is called

(*a*) wet bulb temperature (*b*) dry bulb temperature
(*c*) dew point temperature (*d*) none of these

3. For unsaturated air, the dew point temperature is .... wet bulb temperature.

(*a*) equal to (*b*) less than (*c*) more than

4. The difference between dry bulb temperature and wet bulb temperature, is called

(*a*) dry bulb depression (*b*) wet bulb depression
(*c*) dew point depression (*d*) degree of saturation

5. The wet bulb depression is zero, when relative humidity is equal to

(*a*) zero (*b*) 0.5 (*c*) 0.75 (*d*) 1.0

6. The relative humidity of air is defined as the ratio of

(*a*) mass of water vapour in a given volume to the total mass of the mixture of air and water vapour

(*b*) mass of water vapour in a given volume to the mass of water vapour, if air is saturated at the same temperature

(*c*) mass of water vapour in a given volume to the mass of air

(*d*) mass of air to the mass of water vapour in the mixture of air and water vapour

7. On a psychrometric chart, sensible cooling is represented by

(*a*) horizontal line (*b*) inclined line
(*c*) vertical line (*d*) none of these

8. If relative humidity is 100%, then

(*a*) dry bulb temperature is greater than wet bulb temperature

(*b*) wet bulb temperature is greater than dry bulb temperature

(*c*) dry bulb temperature is equal to wet bulb temperature

(*d*) dry bulb temperature is one-half of wet bulb temperature

9. The vertical and uniformly spaced lines on a psychrometric chart indicates

(*a*) dry bulb temperature (*b*) wet bulb temperature
(*c*) dew point temperature (*d*) specific humidity

10. The curved lines on a psychrometric chart indicates

(*a*) dry bulb temperature (*b*) wet bulb temperature
(*c*) specific humidity (*d*) relative humidity

11. During sensible cooling of air, the specific humidity

(*a*) increases (*b*) decreases (*c*) remains constant

12. During sensible cooling of air, the dry bulb temperature

(*a*) increases (*b*) decreases (*c*) remains constant

13. During sensible cooling of air, the wet bulb temperature

(*a*) increases (*b*) decreases (*c*) remains constant

14. During sensible cooling of air, the coil efficiency is given by

(*a*) $1 - BPF$ (*b*) $1 + BPF$ (*c*) $\sqrt{1 - BPF}$ (*d*) $1/BPF$

where BPF = By-pass factor.

15. The by-pass factor of two rows of similar coil is 0.01. The by pass factor of one row of the coil will be

(*a*) 0.005 (*b*) 0.05 (*c*) 0.1 (*d*) 0.2

16. The minimum temperature to which moist air can be cooled under ideal conditions in a spray washer is

(*a*) dew point temperature of inlet air (*b*) wet bulb temperature of inlet air

(*c*) water inlet temperature (*d*) water outlet temperature

17. The by-pass factor of a cooling coil decreases with

(*a*) decrease in fin spacing and increase in number of rows

(*b*) increase in fin spacing and increase in number of rows

(*c*) increase in fin spacing and decrease in number of rows

(*d*) decrease in fin spacing and decrease in number of rows

18. The ratio of sensible heat to total heat is known as

(*a*) specific humidity (*b*) relative humidity

(*c*) apparatus dew point (*d*) sensible heat factor

19. The process generally used in winter air conditioning to warm and humidify the air, is called

(*a*) humidificaton (*b*) dehumidification

(*c*) heating and humidification (*d*) cooling and dehumidification

20. In order to cool and dehumidify a stream of moist air, it must be passed over the coil at a temperature

(*a*) which lies between the dry bulb and wet bulb temperature of the incoming stream

(*b*) which lies between the wet bulb and dew point temperature of the incoming stream

(*c*) which is lower than the dew point temperature of the incoming stream

(*d*) of adiabatic saturation of incoming stream

ANSWERS

1. (*c*)	**2.** (*b*)	**3.** (*b*)	**4.** (*b*)	**5.** (*d*)
6. (*b*)	**7.** (*a*)	**8.** (*c*)	**9.** (*a*)	**10.** (*d*)
11. (*c*)	**12.** (*b*)	**13.** (*b*)	**14.** (*a*)	**15.** (*b*)
16. (*b*)	**17.** (*a*)	**18.** (*d*)	**19.** (*c*)	**20.** (*c*)

CHAPTER

Comfort Conditions 17

17.1 Introduction

Strictly speaking, the human comfort depends upon physiological and psychological conditions. Thus it is difficult to define the term 'human comfort'. There are many definitions given for this term by different bodies. But the most accepted definition, from the subject point of view, is given by the American Society of Heating, Refrigeration and Air Conditioning Engineers (ASHRAE) which states : *human comfort is that condition of mind, which expresses satisfaction with the thermal environment.*

17.2 Thermal Exchanges of Body with Environment

The human body works best at a certain temperature, like any other machine, but it cannot tolerate wide range of variations in their environmental temperatures like machines. The human body maintains its thermal equilibrium with the environment by means of three modes of heat transfer *i.e.* evaporation, radiation and convection. The way in which the individual's body maintains itself in comfortable equilibrium will be by its automatic use of one or more of the three modes of heat transfer. A human body feels comfortable when the heat produced by metabolism of human body is equal to the sum of the heat dissipated to the surroundings and the heat stored in human body by raising the temperature of body tissues. This phenomenon may be represented by the following equation :

$$Q_M - W = Q_E \pm Q_R \pm Q_C \pm Q_S$$

where Q_M = Metabolic heat produced within the body,

W = Useful rate of working,

$Q_M - W$ = Heat to be dissipated to the atmosphere,

Q_E = Heat lost by evaporation,

*Q_R = Heat lost or gained by radiation,

*Q_C = Heat lost or gained by convection, and

**Q_S = Heat stored in the body.

It may be noted that

1. The metabolic heat produced (Q_M) depends upon the rate of food energy consumption in the body. A fasting, weak or sick man, will have less metabolic heat production.

2. The heat loss by evaporation is always positive. It depends upon the vapour pressure difference between the skin surface and the surrounding air. The heat loss of evaporation (Q_E) is given by

$$Q_E = C_d A (p_s - p_v) h_{fg} C_c$$

where C_d = Diffusion coefficient in kg of water evaporated per unit surface area and pressure difference per hour,

A = Skin surface area = 1.8 m² for normal man,

p_s = Saturation vapour pressure corresponding to skin temperature,

p_v = Vapour pressure of surrounding air,

h_{fg} = Latent heat of vaporisation = 2450 kJ / kg,

C_c = Factor which accounts for clothing worn.

The value of Q_E becomes zero when $p_s = p_v$, *i.e.* when the surrounding air temperature is equal to the skin temperature and air is saturated or when it is higher than the skin temperature and the air is nearly saturated.

---

* The *plus* sign is used when heat is lost to the surroundings and *negative* sign is used when heat is gained from surroundings.

** The *plus* sign is used when the temperature of the body rises and *negative* sign is used when the temperature of the body falls.

The value of Q_E is never negative as when p_s is less than p_v, the skin will not absorb moisture from the surrounding air as it is in saturated state. The only way for equalising the pressure difference is by increasing p_s to p_v by rise of skin temperature from the sensible heat flow from air to skin.

3. The heat loss or gain by radiation (Q_R) from the body to the surroundings depends upon the *mean radiant temperature*. It is the average surface temperature of the surrounding objects when properly weighted, and varies from place to place inside the room. When the mean radiant temperature is lower than the dry bulb temperature of air in the room, Q_R is *positive i.e.* the body will undergo a radiant heat loss. On the other hand, if the mean radiant temperature is higher than the dry bulb temperature of air in the room, Q_R is *negative i.e.* the body will undergo a radiant heat gain.

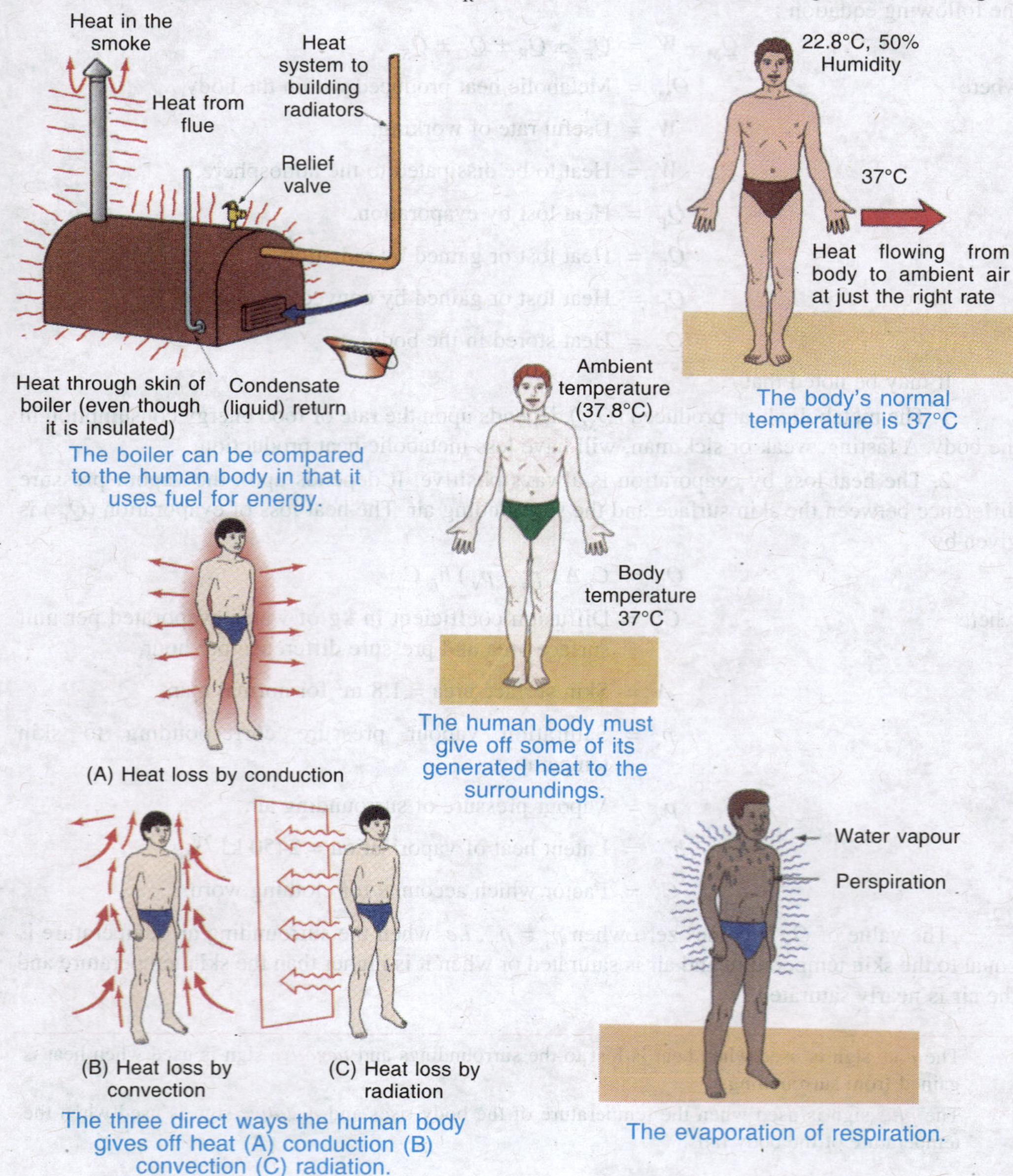

The boiler can be compared to the human body in that it uses fuel for energy.

The body's normal temperature is 37°C

The human body must give off some of its generated heat to the surroundings.

(A) Heat loss by conduction

(B) Heat loss by convection

(C) Heat loss by radiation

The three direct ways the human body gives off heat (A) conduction (B) convection (C) radiation.

The evaporation of respiration.

4. The heat loss by convection (Q_C) from the body to the surroundings is given by

$$Q_C = UA\,(t_B - t_S)$$

where U = Body film coefficient of heat transfer,

A = Body surface area = 1.8 m^2 for normal man,

t_B = Temperature of the body, and

t_S = Temperature of the surroundings.

When the temperature of the surroundings (t_S) is higher than the temperature of the body (t_B), then Q_C will be *negative*, *i.e.* the heat will be gained by the body. On the other hand, if the temperature of the surroundings (t_S) is lower than the temperature of the body (t_B), then Q_C will be the *positive*, *i.e.* the heat will be lost by the body. Since the body film coefficient of heat transfer increases with the increase in air velocity, therefore higher air velocities will produce uncomfort when t_S is higher than t_B. The higher air velocities are recommended when t_S is lower than t_B.

5. When Q_E, Q_R and Q_C are high and positive and ($Q_E + Q_R + Q_C$) is greater than ($Q_M - W$), then the heat stored in the body (Q_S) will be *negative* *i.e.* the body temperature falls down . Thus the sick, weak, old or a fasting man feels more cold. On the other hand, a man gets fever when high internal body activities increases Q_M to such an extent so that Q_S becomes *positive* for the given Q_E, Q_R and Q_C.

The heat stored in the body has maximum and minimum limits which when exceeded brings death. The usual body temperature, for a normal man (when Q_S = 0) is 37°C (98.6°F). The temperature of the body when falls below 36.5°C (98°F) and exceeds 40.5°C (105°F) is dangerous. There is some kind of thermostatic control called *vasomotor control mechanism* in the human body which maintains the temperature of body at the normal level of 37°C, by regulating the blood supply to the skin. When the temperature of the body falls (*i.e.* the heat stored Q_S in the body is *negative*), then the vasomotor control decreases the circulation of blood which decreases conductivity of nerve cells and other tissues between the skin and the inner body cells. This allows skin temperature to fall but allows higher inner temperature of body cells beneath. When the temperature of the body rises (*i.e.* the heat stored Q_S in the body is *positive*) , then the vasomotor control increases blood circulation which increases conductivity of tissues and hence allows less temperature drop between the skin and inner body cells.

The human body feels comfortable when there is no change in the body temperature, *i.e.* when the heat stored in the body Q_S is zero. Any variation in the body temperature acts as a stress to the brain which ultimately results in either perspiration or shivering.

17.3 Physiological Hazards Resulting from Heat

In summer, the temperature of the surroundings is always higher than the temperature of the body. Thus the body will gain heat from the surroundings by means of radiation and convection processes. The body can dissipate heat only through evaporation of sweat. When the heat loss by evaporation is unable to cope with the heat gain, there will be storage of heat in the body and the temperature of body rises. Several physiological hazards exist, the severity of which depends upon the extent and time duration of body temperature rise. Following are some of the physiological hazards which may result due to the rise in body temperature.

1. ***Heat exhaustion.*** It is due to the failure of normal blood circulation. The symptoms of heat exhaustion include fatigue, headache, dizziness, vomiting and abnormal mental reactions such as irritability. Severe heat exhaustion may cause fainting . It does not cause permanent injury to the body and recovery is usually rapid when the person is removed to a cool place.

2. *Heat cramp.* It results from loss of salt due to an excessive rate of body perspiration. It causes severe pain in the calf and thigh muscles. The heat cramp may be largely avoided by using salt tablets.

3. *Heat stroke.* It is the most serious hazard. When a man is exposed to excessive heat and work, the body temperature may rise rapidly to 40.5°C (105°F) or higher. At such elevated temperatures, sweating ceases and the man may enter a coma, with death imminent. A person experiencing a heat stroke may have permanent damage to the brain. The heat stroke may be avoided by taking sufficient water at frequent intervals. It has been found that a man doing hard work in the sun requires about one litre of water per hour.

17.4 Factors Affecting Human Comfort

In designing winter or summer air conditioning system, the designer should be well conversant with a number of factors which physiologically affect human comfort. The important factors are as follows :

1. Effective temperature, 2. Heat production and regulation in human body, 3. Heat and moisture losses from the human body, 4. Moisture content of air, 5. Quality and quantity of air. 6. Air motion, 7. Hot and cold surfaces, and 8. Air stratification.

These factors are discussed, in detail, in the following articles :

17.5 Effective Temperature

The degree of warmth or cold felt by a human body depends mainly on the following three factors :

1. Dry bulb temperature, 2. Relative humidity, and 3. Air velocity.

In order to evaluate the combined effect of these factors, the term *effective temperature* is employed. It is defined as that index which corelates the combined effects of air temperature, relative humidity and air velocity on the human body. The numerical value of effective temperature is made equal to the temperature of still (*i.e.* 5 to 8 m/min air velocity) saturated air, which produces the same sensation of warmth or coolness as produced under the given conditions.

The practical application of the concept of effective temperature is presented by the *comfort chart*, as shown in Fig. 17.1. This chart is the result of research made on different kinds of people subjected to wide range of environmental temperature, relative humidity and air movement by the American Society of Heating, Refrigeration and Air conditioning Engineers (ASHRAE). It is applicable to reasonably still air (5 to 8 m/min air velocity) to situations where the occupants are seated at rest or doing light work and to spaces whose enclosing surfaces are at a mean temperature equal to the air dry bulb temperature.

In the comfort chart, as shown in Fig. 17.1, the dry bulb temperature is taken as abscissa and the wet bulb temperature as ordinates. The relative humidity lines are replotted from the psychrometric chart. The statistically prepared graphs corresponding to summer and winter season are also superimposed. These graphs have effective temperature scale as abscissa and % of people feeling comfortable as ordinate.

A close study of the chart reveals that the several combinations of wet and dry bulb temperatures with different relative humidities will produce the same *effective temperature.

* From the comfort chart, we see that for a point corresponding to dry bulb temperature of 17.5°C, wet bulb temperature of 12.5°C and relative humidity of 60%, the effective temperature is 16°C. Now for the same feeling of comfort and warmth, there is another point on 100% relative humidity line at which dry bulb temperature and wet bulb temperature are both equal to 16°C. Thus both have an effective temperature of 16°C.

However, all points located on a given effective temperature line do not indicate conditions of equal comfort or discomfort. The extremely high or low relative humidities may produce conditions of discomfort regardless of the existent effective temperature. The moist desirable relative humidity range lies between 30 and 70 per cent. When the relative humidity is much below 30 per cent, the mucous membranes and the skin surface become too dry for comfort and health. On the other hand, if the relative himidity is above 70 per cent, there is a tendency for a clammy or sticky sensation to develop. The curves at the top and bottom, as shown in Fig. 17.1, indicate the percentages of person participating in tests, who found various effective temperatures satisfactory for comfort.

The comfort chart shows the range for both summer and winter condition within which a condition of comfort exists for most people. For summer conditions, the chart indicates that a maximum of 98 percent people felt comfortable for an effective temperature of 21.6°C . For winter conditions, chart indicates that an effective temperature of 20°C was desired by 97.7 percent people. It has been found that for comfort, women require 0.5°C higher effective temperature than men. All men and women above 40 years of age prefer 0.5°C higher effective temperature than the persons below 40 years of age.

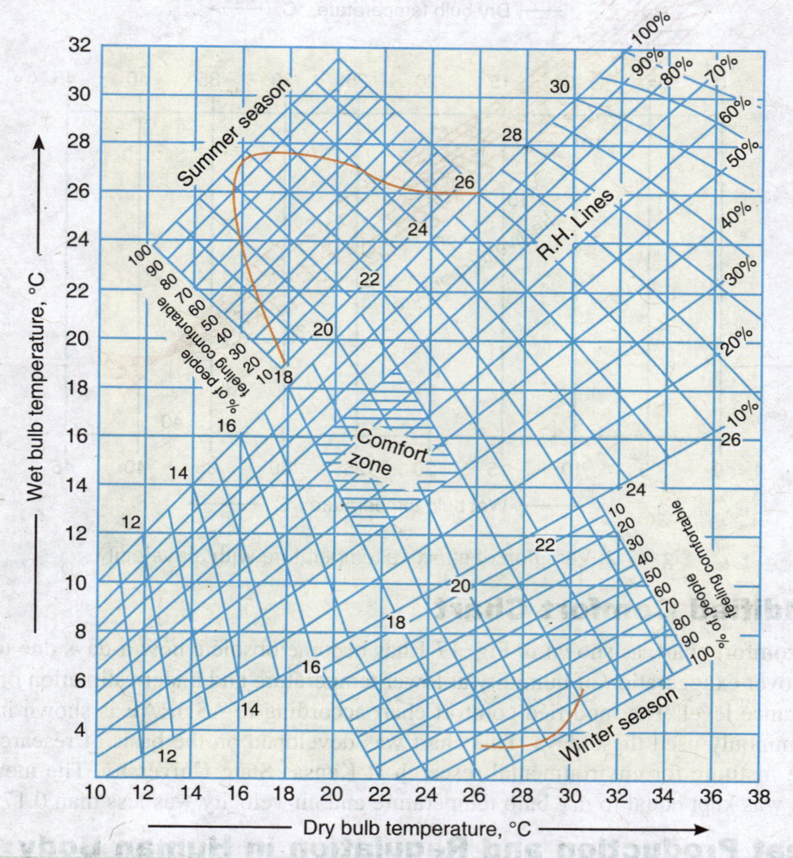

Fig. 17.1. Comfort chart for still air (air velocities from 5 to 8 m/min)

It may be noted that the comfort chart, as shown in Fig. 17.1, does not take into account the variations in comfort conditions when there are wide variations in the mean radiant temperature (MRT). In the range of 26.5°C, a rise of 0.5°C in mean radiant temperature above the room dry bulb temperature raises the effective temperature by 0.5°C. The effect of mean radiant temperature on comfort is less pronounced at high temperatures than at low temperatures.

The comfort conditions for persons at work vary with the rate of work and the amount of clothing worn. In general, the greater the degree of activity, the lower the effective temperature necessary for comfort.

Fig. 17.2 shows the variation in effective temperature with different air velocities. We see that for the atmospheric conditions of 24°C dry bulb temperature and 16°C wet bulb temperature correspond to about 21°C with nominally still air (velocity 6 m / min) and it is about 17°C at an air velocity of 210 m / min. The same effective temperature is observed at higher dry bulb and wet bulb temperatures with higher velocities. The case is reversed after 37.8°C as in that case higher velocities will increase sensible heat flow from air to body and will decrease comfort. The same effective temperature means same feeling of warmth, but it does not mean same comfort.

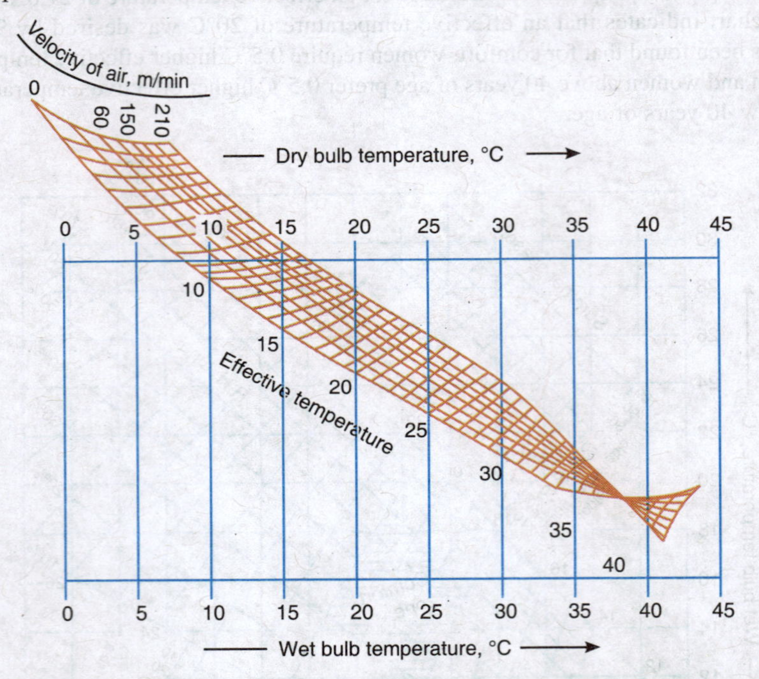

Fig. 17.2. Variation of effective temperature with air velocity.

17.6 Modified Comfort Chart

The comfort chart, as shown in Fig. 17.1,has become obsolete now-a-days due to its short comings of over exaggeration of humidity at lower temperature and under estimation of humidity at heat tolerance level. The modified comfort chart according to ASHRAE is shown in Fig.17.3 and it is commonly used these days. This chart was developed on the basis of research done in 1963 by the institute for environmental research at Kansas State University. The mean radiant temperature was kept equal to dry bulb temperature and air velocity was less than 0.17 m / s.

17.7 Heat Production and Regulation in Human Body

The human body acts like a heat engine which gets its energy from the combustion of food within the body. The process of combustion (called metabolism) produces heat and energy due to the oxidation of products in the body by oxygen obtained from inhaled air. The rate of heat production depends upon the individual's health, his physical activity and his environment. The rate at which the body produces heat is termed as *metabolic rate.* The heat production from a normal healthy person when asleep (called *basal metabolic rate*) is about 60 watts and it is about ten times more for a person carrying out sustained very hard work.

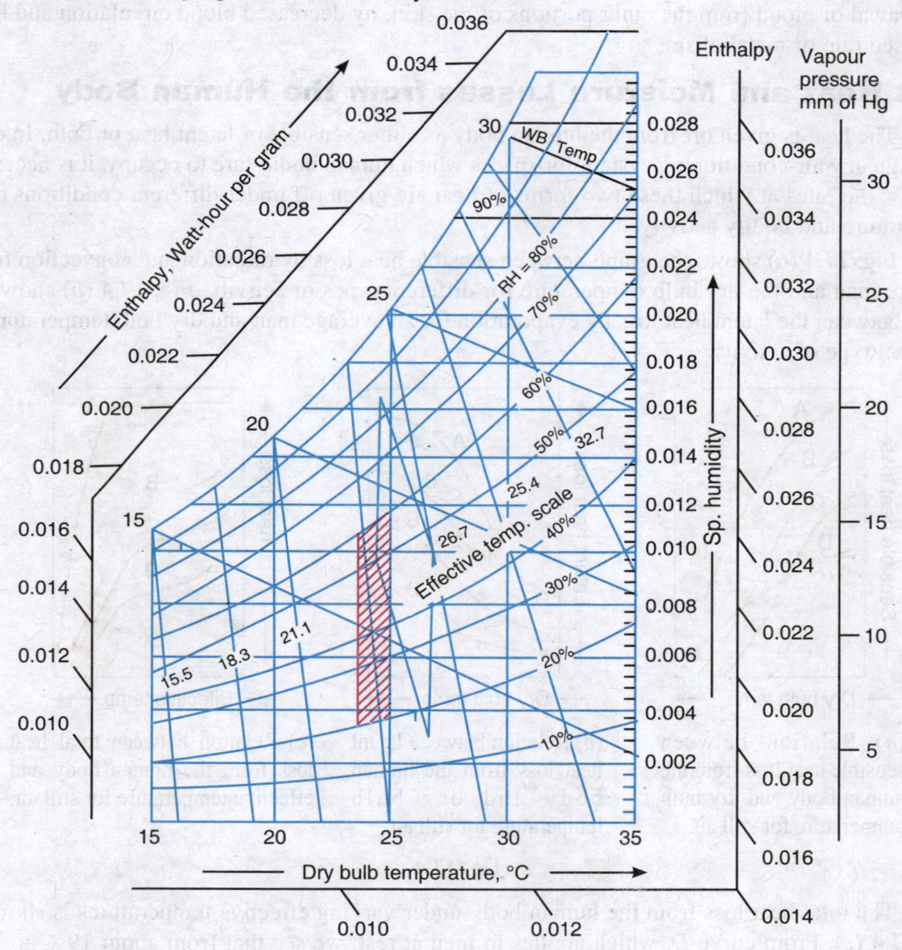

Fig. 17.3. Modified comfort chart.

Since the body has a thermal efficiency of 20 per cent, therefore the remaining 80 per cent of the heat must be rejected to the surrounding environment, otherwise accumulation of heat results which causes discomfort. The rate and the manner of rejection of heat is controlled by the automatic regulation system of a human body.

In order to effect the loss of heat from the body to produce cold, the body may react to bring more blood to the capillaries in the skin. The heat losses from the skin, now, may take place by radiation, convection and by evaporation. When the process of radiation or convection or both fails to produce necessary loss of heat, the sweat glands become more active and more moisture is deposited on the skin, carrying heat away as it evaporates . It may be noted that when the temperature of surrounding air and objects is below the blood temperature, the heat is removed by radiation and convection. On the other hand, when the temperature of surrounding air is above the blood temperature, the heat is removed by evaporation only. In case the body fails to throw off the requisite amount of heat, the blood temperature rises. This results in the accumulation of heat which will cause discomfort.

The human body attempts to maintain its temperature when exposed to cold by the withdrawal of blood from the outer portions of the skin, by decreased blood circulation and by an increased rate of metabolism.

17.8 Heat and Moisture Losses from the Human Body

The heat is given off from the human body as either sensible or latent heat or both. In order to design any air-conditioning system for spaces which human bodies are to occupy, it is necessary to know the rates at which these two forms of heat are given off under different conditions of air temperature and bodily activity.

Fig. 17.4 (*a*) shows the graph between sensible heat loss by radiation and convection for an average man and the dry bulb temperature for different types of activity. Fig. 17.4 (*b*) shows the graph between the latent heat loss by evaporation for an average man and dry bulb temperature for different type of activity.

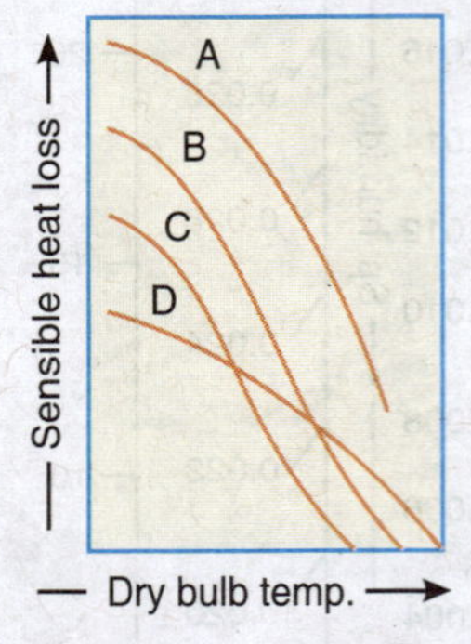

(*a*) Relation between sensible heat loss from the human body and dry bulb temperature for still air.

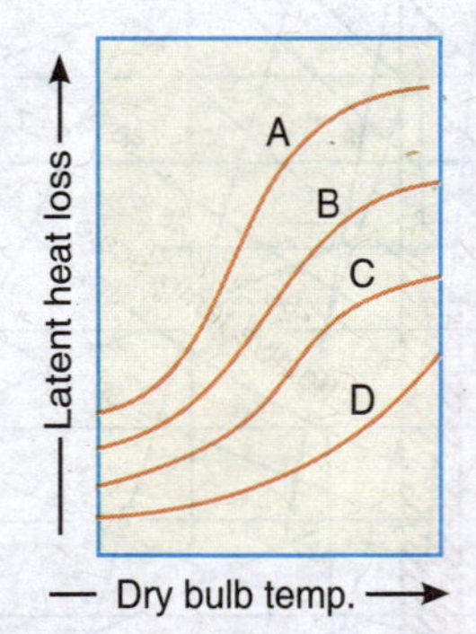

(*b*) Relation between latent heat loss from the human body and dry bulb temperature for still air.

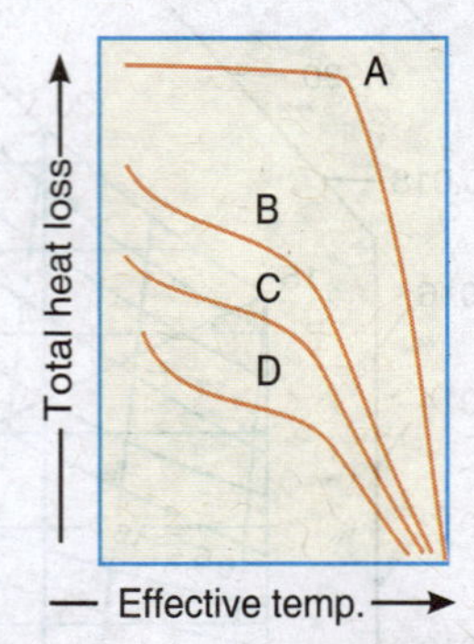

(*c*) Relation between total heat loss from the human body and effective temperature for still air.

Fig. 17.4

The total heat loss from the human body under varying effective temperatures is shown in Fig. 17.4 (*c*). From curve *D*, which applies to men at rest, we see that from about 19°C to 30°C effective temperature, the heat loss is constant. At the lower effective temperature, the heat dissipation increases which results in a feeling of coolness. At higher effective temperature, the ability to lose heat rapidly decreases resulting in severe discomfort. The curves *A, B, C* and *D* shown in Fig. 17.4 represents as follows :

Curve *A* — Men working at the rate of 90 kN-m / h

Curve *B* — Men working at the rate of 45 kN-m / h

Curve *C* — Men working at the rate of 22.5 kN-m / h

Curve *D* — Men at rest.

17.9 Moisture Content of Air

We have seen in Art. 17.5 that the dry bulb temperature, relative humidity and air motion are inter-related. The moisture content of outside air during winter is generally low and it is above the average during summer, because the capacity of the air to carry moisture is dependent upon its dry bulb temperature. This means that in winter, if the cold outside air having a low moisture content leaks into the conditioned space, it will cause a low relative humidity unless moisture is added to the air by the process of humidification. In summer, the reverse will take place unless moisture is removed from the inside air by the dehumidification process. Thus, while designing an air- conditioning system, the proper dry bulb temperature for either summer or winter must be selected in accordance with the practical consideration of relative humidities which are feasible. In general, for winter conditions in the average residence, relative humidities above 35 to 40 per cent are not practical. In summer comfort cooling, the air of the occupied space should not have a relative humidity above 60 per cent. With these limitations, the necessary dry bulb temperature for the air may be determined from the comfort chart.

17.10 Quality and Quantity of Air

The air in an occupied space should, at all times, be free from toxic, unhealthful or disagreeable fumes such as *carbon dioxide. It should also be free from dust and odour. In order to obtain these conditions, enough clean outside air must always be supplied to an occupied space to counteract or adequately dilute the sources of contamination.

The concentration of odour in a room depends upon many factors such as dietary and hygienic habits of occupants, type and amount of outdoor air supplied, room volume per occupant and types of odour sources. In general, when there is no smoking in a room, 1 m^3/min per person of outside air will take care of all the conditions. But when smoking takes place in a room, 1.5 m^3/min per person of outside air is necessary. In most air-conditioning systems, a large amount of air is recirculated over and above the required amount of outside air to satisfy the minimum ventilation conditions in regard to odour and purity. For general application, a minimum of 0.3 m^3/min of outside air per person, mixed with 0.6 m^3/min of recirculated air is good. The recommended and minimum values for the outside air required per person are given in Chapter 19 on cooling load estimation (Table 19.10).

17.11 Air Motion

The air motion which includes the distribution of air is very important to maintain uniform temperature in the conditioned space. No air conditioning system is satisfactory unless the air handled is properly circulated and distributed. Ordinarily, the air velocity in the occupied zone should not exceed 8 to 12 m / min. The air velocities in the space above the occupied zone should be very high in order to produce good distribution of air in the occupied zone, provided that the air in motion does not produce any objectionable noise. The flow of air should be preferably towards the faces of the individuals rather than from the rear in the occupied zone. Also for the proper and perfect distribution of air in the air-conditioned space, down flow should be preferred instead of up flow.

The air motion without proper air distribution produces local cooling sensation known as *draft*.

17.12 Cold and Hot Surfaces

The cold or hot objects in a conditioned space may cause discomfort to the occupants. A single glass of large area when exposed to the outdoor air during winter will produce discomfort

* The atmospheric air contains 0.03% to 0.04% by volume of carbon dioxide and it should not increase 0.6% which is necessary for proper functioning of respiratory system. The carbon dioxide, in excess of 2% dilutes oxygen contents and makes breathing difficult. When the carbon dioxide exceeds 6%, breathing is very difficult and 10% carbon dioxide causes loss of consciousness. A normal man at rest in breathing, exhales about 0.015 to 0.018 m^3/h of carbon dioxide.

to the occupants of a room by absorbing heat from them by radiation. On the other hand, a ceiling that is warmer than the room air during summer causes discomfort. Thus, in the designing of an air conditioning system, the temperature of the surfaces to which the body may be exposed must be given considerable importance.

A jump duct allows for balanced pressure between rooms and improved comfort conditions.

17.13 Air Stratification

When air is heated, its density decreases and thus it rises to the upper part of the confined space. This results in a considerable variation in the temperatures between the floor and ceilng levels. The movement of the air to produce the temperature gradient from floor to ceiling is termed as *air stratification.* In order to achieve comfortable conditions in the occupied space, the air conditioning system must be designed to reduce the air stratification to a minimum.

17.14 Factors Affecting Optimum Effective Temperature

The important factors which affect the optimum effective temperature are as follows :

1. *Climatic and seasonal differences.* It is a known fact that the people living in colder climates feel comfortable at a lower effective temperatures than those living in warmer regions. There is a relationship between the optimum indoor effective temperature and the optimum outdoor temperature, which changes with seasons. We see from the comfort chart (Fig. 17.1) that in winter, the optimum effective temperature is 19°C whereas in summer this temperature is 22°C.

2. *Clothing.* It is another important factor which affects the optimum effective temperature. It may be noted that the person with light clothings need less optimum temperature than a person with heavy clothings.

3. *Age and Sex.* We have already discussed that the women of all ages require higher effective temperature (about 0.5°C) than men. Similar is the case with young and old people. The children also need higher effective temperature than adults. Thus, the maternity halls are always kept at an effective temperature of 2 to 3°C higher than the effective temperature used for adults.

4. *Duration of stay.* It has been established that if the stay in a room is shorter (as in the case of persons going to banks), then higher effective temperature is required than that needed for long stay (as in the case of persons working in an office).

5. *Kind of activity.* When the activity of the person is heavy such as people working in a factory, dancing hall, then low effective temperature is needed than for the people sitting in cinema hall or auditorium.

6. *Density of occupants.* The effect of body radiant heat from person to person particularly in a densely occupied space like auditorium is large enough which require a slight lower effective temperature.

17.15 Inside Summer Design Conditions

The following table shows the recommended inside design conditions for summer comfort cooling :

Table 17.1. Recommended inside design conditions.

Outside DBT(°C)	*Occupancy over 40 minutes*				*Occupancy below 40 minutes*			
	DBT (°C)	*WBT (°C)*	*RH (%)*	*ET (°C)*	*DBT (°C)*	*WBT (°C)*	*RH (%)*	*ET (°C)*
	23.9	18.3	60	21.7	24.4	18.9	61	22.2
26.5	25	17.2	47	21.7	25.6	17.8	47	22.2
	26.1	16.1	35	21.7	26.7	16.7	36	22.2
	24.4	18.9	61	22.2	25	19.4	61	22.8
29.5	25.6	17.8	47	22.2	26.1	18.3	48	22.8
	26.7	16.7	36	22.2	27.2	17.2	36	22.8
	25	19.4	61	22.8	25.6	20.6	64	23.3
32.5	26.1	18.3	48	22.8	26.7	19.4	52	23.3
	27.2	17.2	36	22.8	27.8	18.3	40	23.3
	25.6	20.6	64	23.3	26.1	21.1	65	23.9
35.5	26.7	19.4	52	23.3	27.2	20	52	23.9
	27.8	18.3	40	23.3	28.3	18.9	41	23.9
	26.1	21.1	65	23.9	27.2	21.7	63	24.4
37.5	27.2	20	52	23.9	28.3	20.6	50	24.4
	28.3	18.9	41	23.9	29.5	19.4	38	24.4
	26.6	21.7	65	24.2	27.2	22.2	65	24.7
40.5	27.8	20.6	52	24.2	28.3	21.1	54	24.7
	28.9	19.4	42	24.2	29.5	20	41	24.7

17.16 Outside Summer Design Conditions

The following table shows the outside summer design conditions for important cities in India.

Table 17.2. Outside summer design conditions.

City	DBT (°C)	WBT (°C)	RH (%)	ET (°C)
Agra	41.5	22	17	29.7
Ahmedabad	43.2	26.4	37	32.5
Ahmednagar	42.2	31.1	45	29.5
Alibagh	31.6	25.8	63	28.4
Aligarh	42.2	28.3	35	29.7
Allahabad	41.7	25	26.5	30.8
Ambala	43.3	26.7	29	29.5
Amritsar	40.1	28.1	—	29.9
Banaras	40.8	26.1	31	31.9
Bangalore	32.9	24.7	52	28.1
Baroda	40.3	29.1	45	31.7
Bhopal	40.2	23.2	24	29.7
Chandigarh	40.1	23.9	27	29.9
Chennai	38.5	28.6	47	25.6
Cochin	35	27.8	59	28.1
Coimbatore	34.7	27.5	58	30
Cuttack	40.6	30.6	48	32
Darjeeling	17.2	14.5	77	16.8
Dehradun	40.6	26.7	35	27.8
Delhi	40.4	23.9	26	29.9
Guwahati	30.9	25.4	62	27.6
Hyderabad	39.5	26.6	37	30.8
Indore	39.4	25.1	32	29.7
Jaipur	40.8	21.7	18	28.6
Jamnagar	37.8	27.2	44	30.6
Jamshedpur	39.4	28.4	45	31
Jodhpur	43.3	26.7	29	30.6
Kanpur	41.2	21.5	17	28.6
Kathmandu	29.4	23.9	65	24.9
Kolkata	35.3	27.9	60	30.3
Lucknow	42.8	28.3	34	30.3
Ludhiana	40.1	27.9	41	29.9
Madurai	38.3	25.6	38	30.6
Mahabaleshwar	28.8	19.2	40	24.2
Mangalore	35.6	27.8	60	29
Mumbai	32.8	26.7	64	29

City	DBT (°C)	WBT (°C)	RH (%)	ET (°C)
Mysore	33.3	25.6	52	29.2
Nagpur	42.6	24.6	27	30.6
Patna	37.9	26.7	42	30.5
Pune	37.1	25.6	40	30.5
Puri	32	27.8	75	29.5
Rajpur	43.3	28.3	34	30.5
Rajkot	40.5	28.9	43	31.2
Roorkee	39	26.7	40	29.5
Shillong	29.4	21.1	48	26.6
Shrinagar	25	18.4	55	22.2
Simla	22.9	12.5	28	19.2
Surat	36.3	21.6	26	30.5
Trivandrum	30.7	29.4	90	29.7
Tiruchirappalli	38.6	25.6	32	30.2
Visakhapatnam	33.3	26.9	39	29.3

QUESTIONS

1. Explain in brief as to how the human body reacts to changes in temperature of environment. Also explain the effect of activities on the heat load calculation for comfort application.
2. Distinguish clearly between heat stroke, heat exhaustion and heat cramp.
3. State the factors that determine human comfort.
4. Define the term 'effective temperature' and explain its significance in the design of air conditioning systems.
5. What is 'effective temperature' ? What factors affect effective temperature ?
6. Sketch 'comfort chart' and show on it the 'comfort zone' ?
7. Explain clearly the different stages of human body defence against variations of weather conditions during summer and winter.
8. Discuss, briefly, the factors which govern the optimum effective temperature for comfort.

OBJECTIVE TYPE QUESTIONS

1. A fasting, weak or sick man will have .... metabolic heat production.
   (*a*) less (*b*) more
2. When the temperature of the surroundings is higher than the temperature of the body, then the heat loss by convection from the body to the surroundings will be
   (*a*) positive (*b*) negative (*c*) zero (*d*) none of these
3. The human body feels comfortable when the heat stored in the body is
   (*a*) positive (*b*) negative (*c*) zero (*d*) none of these
4. The degree of warmth or cold felt by a human body depends mainly on
   (*a*) dry bulb temperature (*b*) relative humidity
   (*c*) air velocity (*d*) all of these
5. The index which corelates the combined effects of air temperature, relative humidity and air velocity on the human body is known as
   (*a*) mean radiant temperature (*b*) effective temperature
   (*c*) dew point temperature (*d*) none of these

6. The effective temperature .... with decrease in relative humidity at the same dry bulb temperature.

(*a*) decreases (*b*) increases

7. For comfort, all men and women above 40 years of age prefer .... effective temperature than the persons below 40 years of age.

(*a*) higher (*b*) lower

8. The heat production from a normal healthy man when asleep is about

(*a*) 20 watts (*b*) 40 watts (*c*) 60 watts (*d*) 80 watts

9. In summer comfort cooling, the air of the occupied space should not have a relative humidity above

(*a*) 30% (*b*) 40% (*c*) 50% (*d*) 60%

10. The optimum effective temperature for human comfort is

(*a*) higher in winter than in summer (*b*) lower in winter than in summer

(*c*) same in winter and summer (*d*) not dependent on season

ANSWERS

1. (*a*)	**2.** (*b*)	**3.** (*c*)	**4.** (*d*)	**5.** (*b*)
6. (*a*)	**7.** (*a*)	**8.** (*c*)	**9.** (*d*)	**10.** (*b*)

CHAPTER

18 Air Conditioning Systems

18.1 Introduction

The air conditioning is that branch of engineering science which deals with the study of conditioning of air *i.e.* supplying and maintaining desirable internal atmospheric conditions for human comfort, irrespective of external conditions. This subject, in its broad sense, also deals with the conditioning of air for industrial purposes, food processing, storage of food and other materials.

18.2 Factors Affecting Comfort Air Conditioning

The four important factors for comfort air conditioning are discussed as below :

1. *Temperature of air.* In air conditioning, the control of temperature means the maintenance of any desired temperature within an enclosed space even though the temperature of the outside air is above or below the desired room temperature. This is accomplished either by the addition or removal of heat from the enclosed space as and when demanded. It may be noted that a human being feels comfortable when the air is at 21°C with 56% relative humidity.

2. *Humidity of air.* The control of humidity of air means the decreasing or increasing of moisture contents of air during summer or winter respectively in order to produce comfortable and healthy conditions. The control of humidity is not only necessary for human comfort but it also increases the efficiency of the workers. In general, for summer air conditioning, the relative humidity should not be less than 60% whereas for winter air conditioning it should not be more than 40%.

3. *Purity of air.* It is an important factor for the comfort of a human body. It has been noticed that people do not feel comfortable when breathing contaminated air, even if it is within acceptable temperature and humidity ranges. It is thus obvious that proper filtration, cleaning and purification of air is essential to keep it free from dust and other impurities.

4. *Motion of air.* The motion or circulation of air is another important factor which should be controlled, in order to keep constant temperature throughout the conditioned space. It is, therefore, necessary that there should be equi-distribution of air throughout the space to be air conditioned.

18.3 Air Conditioning System

We have already discussed in Art. 18.2, the four important factors which affect the human comfort. The system which effectively controls these conditions to produce the desired effects upon the occupants of the space is known as an *air conditioning system.*

18.4 Equipments Used in an Air Conditioning System

Following are the main equipments or parts used in an air conditioning system :

1. *Circulation fan.* The main function of this fan is to move air to and from the room.

2. *Air conditioning unit.* It is a unit which consists of cooling and dehumidifying processes for summer air conditioning or heating and humidification processes for winter air conditioning.

3. *Supply duct.* It directs the conditioned air from the circulating fan to the space to be air conditioned at proper point.

4. *Supply outlets.* These are grills which distribute the conditioned air evenly in the room.

5. *Return outlets.* These are the openings in a room surface which allow the room air to enter the return duct.

6. *Filters.* The main function of the filters is to remove dust, dirt and other harmful bacteria from the air.

18.5 Classification of Air Conditioning Systems

The air conditioning systems may be broadly classified as follows :

1. *According to the purpose*

(*a*) Comfort air conditioning system, and

(*b*) Industrial air conditioning system.

2. *According to season of the year*

(*a*) Winter air conditioning system,

(*b*) Summer air conditioning system, and

(*c*) Year-round air conditioning system.

3. According to the arrangement of equipment

(*a*) Unitary air conditioning system, and

(*b*) Central air conditioning system.

In this chapter, we shall discuss all the above mentioned air conditioning systems one by one.

18.6 Comfort Air Conditioning System

In comfort air conditioning, the air is brought to the required dry bulb temperature and relative humidity for the human health, comfort and efficiency. If sufficient data of the required condition is not given, then it is assumed to be 21°C dry bulb temperature and 50% relative humidity. The sensible heat factor is, generally, kept as following :

For residence or private office = 0.9

For restaurant or busy office = 0.8

Auditorium or cinema hall = 0.7

Ball room dance hall etc. = 0.6

The comfort air conditioning may be adopted for homes, offices, shops, restaurants, theatres, hospitals, schools etc.

Example 18.1. *An air conditioning plant is required to supply 60 m³ of air per minute at a DBT of 21°C and 55% RH. The outside air is at DBT of 28°C and 60% RH. Determine the mass of water drained and capacity of the cooling coil. Assume the air conditioning plant first to dehumidify and then to cool the air.*

Solution. Given : $v_2 = 60$ m³/min ; $t_{d2} = 21$°C ; $\phi_2 = 55\%$; $t_{d1} = 28$°C ; $\phi_1 = 60\%$

Mass of water drained

First of all, mark the initial condition of air at 28°C dry bulb temperature and 60% relative humidity on the psychrometric chart as point 1, as shown in Fig. 18.1. Now mark the final condition of air at 21°C dry bulb temperature and 55% relative humidity as point 2. From the psychrometric chart, we find that

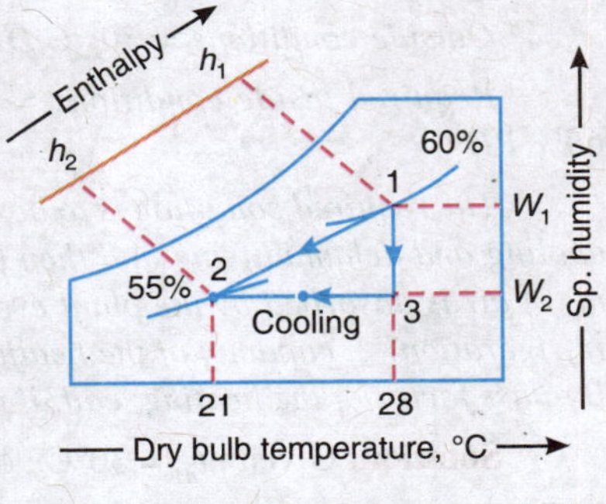

Fig. 18.1

Specific humidity of air at point 1,

$$W_1 = 0.0142 \text{ kg / kg of dry air}$$

Specific humidity of air at point 2,

$$W_2 = 0.0084 \text{ kg / kg of dry air}$$

and specific volume of air at point 2,

$$v_{s2} = 0.845 \text{ m}^3\text{ / kg of dry air}$$

We know that mass of air circulated,

$$m_a = \frac{v_2}{v_{s2}} = \frac{60}{0.845} = 71 \text{ kg / min}$$

∴ Mass of water drained

$$= m_a(W_1 - W_2) = 71(0.0142 - 0.0084) = 0.412 \text{ kg / min}$$

$$= 0.412 \times 60 = 24.72 \text{ kg / h}$$ **Ans.**

Capacity of the cooling coil

From the psychrometric chart, we find that

Enthalpy of air at point 1,

$$h_1 = 64.8 \text{ kJ / kg of dry air}$$

and enthalpy of air at point 2,

$$h_2 = 42.4 \text{ kJ / kg of dry air}$$

∴ Capacity of the cooling coil

$$= m_a (h_1 - h_2) = 71(64.8 - 42.4) = 1590.4 \text{ kJ / min}$$

$$= 1590.4 / 210 = 7.57 \text{ TR}$$ **Ans.**

18.7 Industrial Air Conditioning System

It is an important system of air conditioning these days in which the inside dry bulb temperature and relative humidity of the air is kept constant for proper working of the machines and for the proper research and manufacturing processes. Some of the sophisticated electronic and other machines need a particular dry bulb temperature and relative humidity. Sometimes, these machines also require a particular method of psychrometric processes. This type of air conditioning system is used in textile mills, paper mills, machine-parts manufacturing plants, tool rooms, photo-processing plants etc.

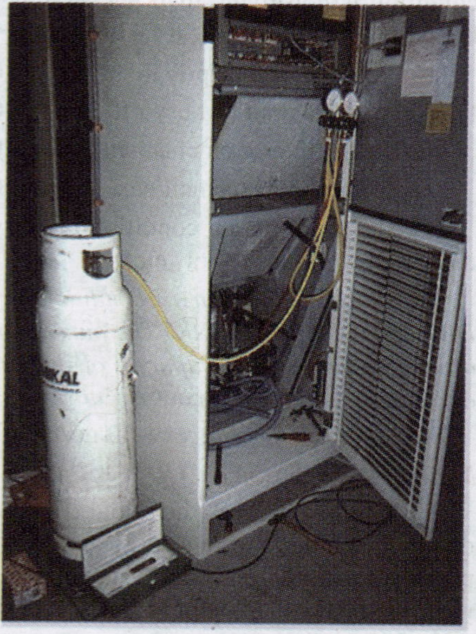

Industrial air-conditioning system.

Example 18.2. *Following data refers to an air conditioning system to be designed for an industrial process for hot and wet climate :*

Outside conditions = 30° C DBT and 75% RH

Required inside conditions = 20° C DBT and 60% RH

The required condition is to be achieved first by cooling and dehumidifying and then by heating. If 20 m³ of air is absorbed by the plant every minute, find : 1. capacity of the cooling coil in tonnes of refrigeration; 2. capacity of the heating coil in kW ; 3. amount of water removed per hour; and 4. By-pass factor of the heating coil, if its surface temperature is 35°C.

Solution. Given : $t_{d1} = 30°C$; $\phi_1 = 75\%$; $t_{d3} = 20°C$; $\phi_3 = 60\%$; $v_1 = 20 \text{ m}^3/\text{min}$; $t_{d4} = 35°C$

1. *Capacity of the cooling coil in tonnes of refrigeration*

First of all, mark the initial condition of air at 30°C dry bulb temperature and 75% relative humidity on the psychrometric chart as point 1, as shown in Fig. 18.2. Then mark the final condition of air at 20°C dry bulb temperature and 60% relative humidity on the chart as point 3.

Now locate the points 2′ and 2 on the saturation curve by drawing horizontal lines through points 1 and 3 as shown in Fig. 18.2. On the chart, the process 1-2′ represents the sensible cooling, 2′-2 represents dehumidifying process, and 2-3 represents the sensible heating process. From the psychrometric chart, we find that the specific volume of air at point 1,

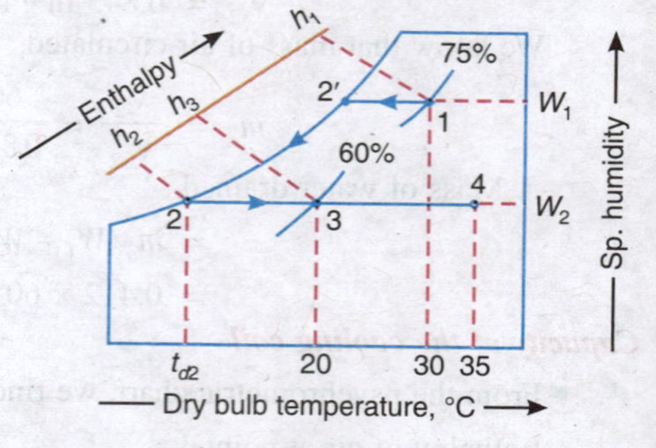

Fig. 18.2

$$v_{s1} = 0.886 \text{ m}^3 / \text{kg of dry air}$$

Enthalpy of air at point 1,

$$h_1 = 81.8 \text{ kJ / kg of dry air}$$

and enthalpy of air at point 2,

$$h_2 = 34.2 \text{ kJ / kg of dry air}$$

We know that mass of air absorbed by the plant,

$$m_a = \frac{v_1}{v_{s1}} = \frac{20}{0.866} = 22.6 \text{ kg / min}$$

∴ Capacity of the cooling coil

$$= m_a (h_1 - h_2) = 22.6 (81.8 - 34.2) = 1075.76 \text{ kJ / min}$$

$$= 1075.76/210 = 5.1 \text{ TR}$$ **Ans.**

2. Capacity of the heating coil in kW

From the psychrometric chart, we find that enthalpy of air at point 3,

$$h_3 = 42.6 \text{ kJ / kg of dry air}$$

∴ Capacity of the heating coil

$$= m_a (h_3 - h_2) = 22.6 (42.6 - 34.2) = 189.84 \text{ kJ / min}$$

$$= 189.84/60 = 3.16 \text{ kW}$$ **Ans.**

3. Amount of water removed per hour

From the psychrometric chart, we find that specific humidity of air at point 1,

$$W_1 = 0.0202 \text{ kg / kg of dry air}$$

and specific humidity of air at point 2,

$$W_2 = 0.0088 \text{ kg / kg of dry air}$$

∴ Amount of water removed per hour

$$= m_a (W_1 - W_2) = 22.6 (0.0202 - 0.0088) = 0.258 \text{ kg / min}$$

$$= 0.258 \times 60 = 15.48 \text{ kg / h}$$ **Ans.**

4. By-pass factor of the heating coil

We know that by-pass factor,

$$BPF = \frac{t_{d4} - t_{d3}}{t_{d4} - t_{d2}} = \frac{35 - 20}{35 - 12.2} = 0.658$$ **Ans.**

... [From psychrometric chart, $t_{d2} = 12.2°C$]

18.8 Winter Air Conditioning System

In winter air conditioning, the air is heated, which is generally accompanied by humidification. The schematic arrangement of the system is shown in Fig. 18.3.

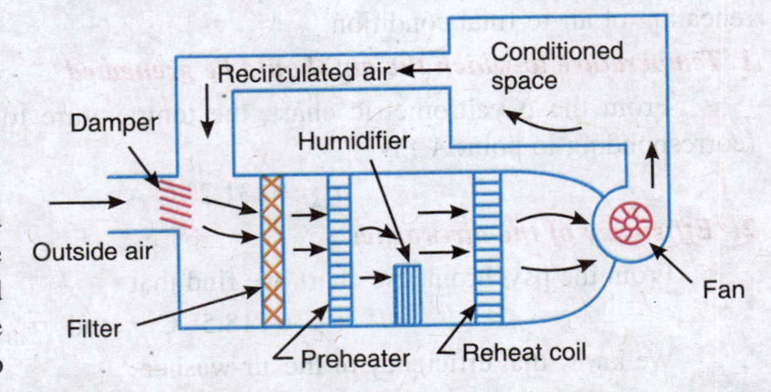

Fig. 18.3. Winter air conditioning system.

The outside air flows through a damper and mixes up with the recirculated air (which is obtained from the conditioned space). The mixed air passes through a filter to remove dirt, dust and other impurities. The air now passes through a preheat coil in order to prevent the possible freezing of water and to control the evaporation of water in the humidifier. After that, the air is made to pass through a reheat coil to bring the air to the designed dry bulb temperature. Now, the conditioned air is supplied to the conditioned space by a fan. From the conditioned space, a part of the used air is exhausted to the atmosphere by the exhaust fans or ventilators. The remaining part of the used air (known as recirculated air) is again conditioned as shown in Fig. 18.3.

The outside air is sucked and made to mix with recirculated air, in order to make up for the loss of conditioned (or used) air through exhaust fans or ventilation from the conditioned space.

Example 18.3. *Air at 10°C DBT and 90% RH is to be brought to 35°C DBT and 22.5°C WBT with the help of winter air conditioner. If the humidified air comes out of the humidifier at 90% RH, draw the various processes involved on a skeleton psychrometric chart and find : 1. the temperature to which the air should be preheated, and 2. the efficiency of the air-washer.*

Solution. Given : $t_{d1} = 10°\text{ C}$; $\phi_1 = 90\%$; $t_{d2} = 35°\text{C}$; $t_{w2} = 22.5°\text{C}$

First of all, mark the initial condition of air at 10°C dry bulb temperature and 90% relative humidity on the psychrometric chart as point 1, as shown in Fig. 18.4. Now mark the final condition of air at 35°C dry bulb temperature and 22.5°C wet bulb temperature, as point 2. Since the final condition of air is obtained with the help of a winter air conditioner, therefore the processes involved are as follows :

1. Preheating of entering air in a preheater,
2. Humidification of preheated air in a humidifier or air-washer, and
3. Reheating of humidified air in a reheater.

These processes are shown by 1-*A*, *A*-*B*, and *B*-2 respectively on the psychrometric chart as shown in Fig. 18.4. In order to obtain these processes on the psychrometric chart, draw a horizontal line through point 1 to represent sensible heating of air and from point 2 draw a horizontal line to intersect 90% relative humidity line at *B*. Now from point *B*, draw a constant wet bulb temperature line which intersects the horizontal line drawn through point 1 at point *A*. Now line 1-*A* represents preheating of air, line *A*-*B* represents humidification and line *B*-2 represents reheating of air to final condition.

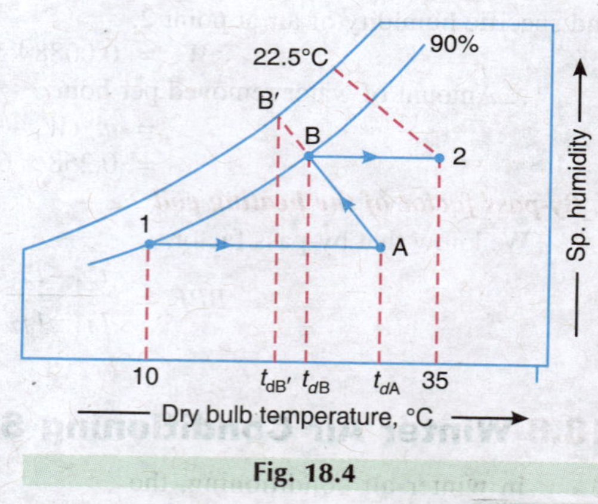

Fig. 18.4

1. *Temperature to which the air should be preheated*

From the psychrometric chart, the temperature to which the air should be preheated (corresponding to point *A*) is

$$t_{dA} = 31.2°\text{ C}\ \textbf{Ans.}$$

2. *Efficiency of the air-washer*

From the psychrometric chart, we find that

$$t_{dB} = 18.5°\text{ C}; \quad \text{and} \quad t_{dB'} = 17.5°\text{ C}$$

We know that efficiency of the air-washer

$$= \frac{\text{Actual drop in DBT}}{\text{Ideal drop in DBT}} = \frac{t_{dA} - t_{dB}}{t_{dA} - t_{dB'}}$$

$$= \frac{31.2 - 18.5}{31.2 - 17.5} = 0.927 \text{ or } 92.7\%\ \textbf{Ans.}$$

18.9 Summer Air Conditioning System

It is the most important type of air conditioning, in which the air is cooled and generally dehumidified. The schematic arrangement of a typical summer air conditioning system is shown in Fig. 18.5.

Summer air-conditioning system.

The outside air flows through the damper, and mixes up with recirculated air (which is obtained from the conditioned space). The mixed air passes through a filter to remove dirt, dust and other impurities. The air now passes through a cooling coil. The coil has a temperature much below the required dry bulb temperature of the air in the conditioned space. The cooled air passes through a perforated membrane and loses its moisture in the condensed form which is collected in a sump. After that, the air is made to pass through a heating coil which heats up the air slightly. This is done to bring the air to the designed dry bulb temperature and relative humidity.

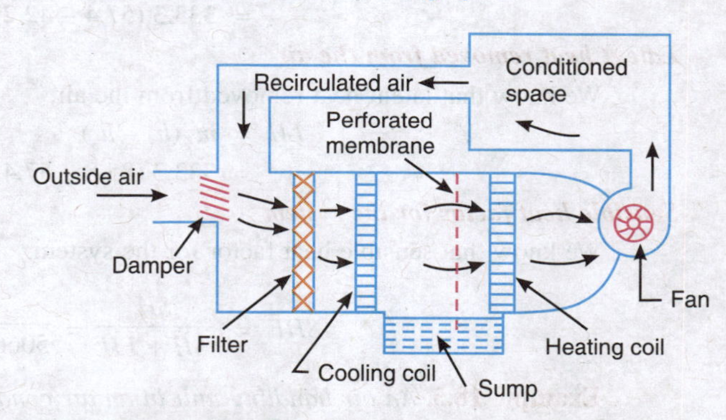

Fig. 18.5. Summer air conditioning system.

Now the conditioned air is supplied to the conditioned space by a fan. From the conditioned space, a part of the used air is exhausted to the atmosphere by the exhaust fans or ventilators. The remaining part of the used air (known as recirculated air) is again conditioned as shown in Fig. 18.5. The outside air is sucked and made to mix with the recirculated air in order to make up for the loss of conditioned (or used) air through exhaust fans or ventilation from the conditioned space.

Example 18.4. *The amount of air supplied to an air conditioned hall is 300 m³/min. The atmospheric conditions are 35°C DBT and 55% RH. The required conditions are 20°C DBT and 60% RH. Find out the sensible heat and latent heat removed from the air per minute. Also find sensible heat factor for the system.*

Solution. Given : $v_1 = 300\ m^3/min$; $t_{d1} = 35°C$; $\phi_1 = 55\%$; $t_{d2} = 20°C$; $\phi_2 = 60\%$

First of all, mark the initial condition of air at 35°C dry bulb temperature and 55% relative humidity on the psychrometric chart at point 1, as shown in Fig. 18.6. Now mark the final condition of air at 20°C dry bulb temperature and 60% relative humidity on the chart as point 2. Locate point 3 on the chart by drawing horizontal line through point 2 and vertical line through point 1. From the psychrometric chart, we find that specific volume of air at point 1,

$$v_{s1} = 0.9\ m^3/kg \text{ of dry air}$$

∴ Mass of air supplied,

$$m_a = \frac{v_1}{v_{s1}} = \frac{300}{0.9} = 333.3 \text{ kg / min}$$

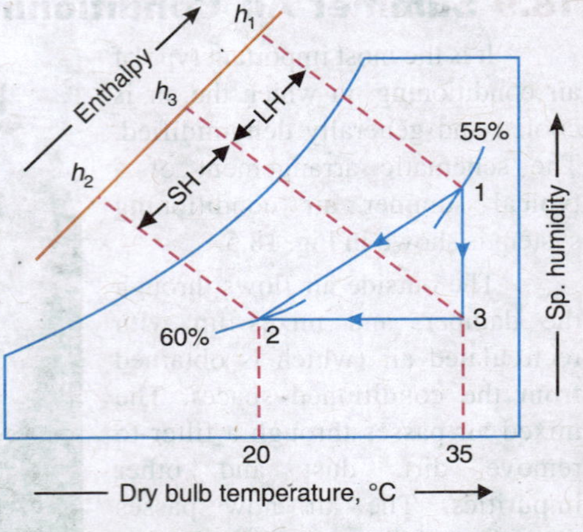

Fig. 18.6

Sensible heat removed from the air

From the psychrometric chart, we find that enthalpy of air at point 1,

$$h_1 = 85.8 \text{ kJ/kg of dry air}$$

Enthalpy of air at point 2,

$$h_2 = 42.2 \text{ kJ/kg of dry air}$$

and enthalpy of air at point 3,

$$h_3 = 57.4 \text{ kJ/kg of dry air}$$

We know that sensible heat removed from the air,

$$SH = m_a (h_3 - h_2)$$
$$= 333.3 (57.4 - 42.2) = 5066.2 \text{ kJ/min} \quad \textbf{Ans.}$$

Latent heat removed from the air

We know that latent heat removed from the air,

$$LH = m_a (h_1 - h_3)$$
$$= 333.3 (85.8 - 57.4) = 9465.7 \text{ kJ/min} \quad \textbf{Ans.}$$

Sensible heat factor for the system

We know that sensible heat factor for the system,

$$SHF = \frac{SH}{SH + LH} = \frac{5066.2}{5066.2 + 9465.7} = 0.348 \quad \textbf{Ans.}$$

Example 18.5. *An air handling unit in an air conditioning plant supplies a total of 4500 m³/min of dry air which comprises by mass 20% of fresh air at 40°C DBT and 27°C WBT and 80% recirculated air at 25°C DBT and 50% RH. The air leaves the cooling coil at 13°C saturated. Calculate the total cooling load and room heat gain. The following data can be used :*

Condition	*DBT* °C	*WBT* °C	*RH* %	*Sp. humidity* g of water vapour / kg of dry air	*Enthalpy* kJ / kg of dry air
Outside	*40*	*27*	—	*17.2*	*85*
Inside	*25*	—	*50*	*10.0*	*51*
ADP	*13*	—	*100*	*9.4*	*36.8*

Specific volume of air entering the cooling coil is 0.869 m³ / kg of dry air.

Solution. Given : $v_3 = 4500$ m³/min ; $t_{d1} = 40$°C ; $t_{w1} = 27$°C ; $t_{d2} = 25$°C ; $\phi_2 = 50$% ; $t_{d4} = ADP = 13$°C ; $W_1 = 17.2$ g / kg of dry air = 0.0172 kg / kg of dry air ; $W_2 = 10$ g / kg of dry air = 0.01 kg / kg of dry air ; $W_4 = 9.4$ g / kg of dry air = 0.0094 kg / kg of dry air ; $h_1 = 85$ kJ/kg of dry air ; $h_2 = 51$ kJ/kg of dry air ; $h_4 = 36.8$ kJ/kg of dry air ; $v_{s3} = 0.869$ m³/kg of dry air

First of all, mark the condition of fresh air at 40°C dry bulb temperature and 27°C wet bulb temperature on the psychrometric chart as point 1, as shown in Fig. 18.7. Now mark the condition of recirculated air at 25°C dry bulb temperature and 50% relative humidity as point 2. The condition of air entering the cooling coil *(point 3) is marked on the line 1-2, such that the specific volume of air at this point is 0.869 m³/kg of dry air. The point 4 represents the condition of air leaving the cooling coil at 13°C on the saturation curve.

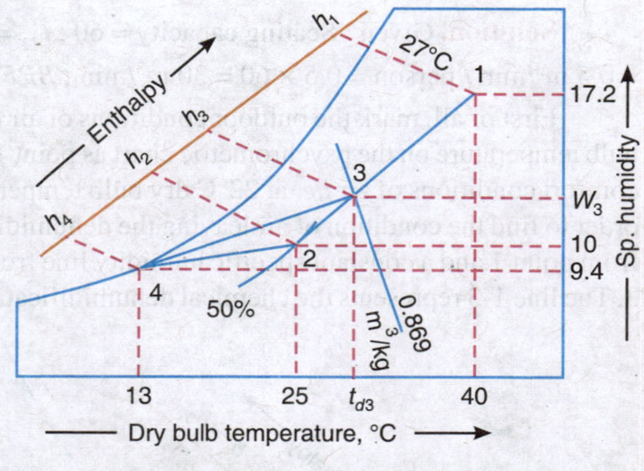

Fig. 18.7

From the psychrometric chart, we find that enthalpy of air entering the cooling coil at point 3,

$$h_3 = 57.8 \text{ kJ / kg of dry air}$$

Specific humidity of air entering the cooling coil at point 3,

$$W_3 = 0.0116 \text{ kg / kg of dry air}$$

and dry bulb temperature of air entering the cooling coil at point 3,

$$t_{d3} = 28.3°\text{C}$$

Total cooling load

We know that mass of air entering the cooling coil,

$$m_{a3} = \frac{v_3}{v_{s3}} = \frac{4500}{0.869} = 5178 \text{ kg / min}$$

∴ Total cooling load $= m_{a3}(h_3 - h_4) = 5178\ (57.8 - 36.8) = 108\ 738$ kJ / min

$= 108\ 738 / 210 = 517.8$ TR **Ans.**

Room heat gain

Since the total mass of air (m_{a3} = 5178 kg / min) comprises 20% of fresh air, therefore mass of fresh air supplied at point 1,

$$m_{a1} = 0.2 \times 5178 = 1035.6 \text{ kg / min}$$

and fresh air load $= m_{a1}(h_1 - h_2) = 1035.6\ (85 - 51) = 35\ 210$ kJ / min

$= 35\ 210 / 210 = 168$ TR **Ans.**

∴ Room heat gain = Total cooling load – Fresh air load

$= 517.8 - 168 = 349.8$ TR **Ans.**

Example 18.6. *A conference room of 60 seating capacity is to be air conditioned for comfort conditions of 22°C dry bulb temperature and 55% relative humidity. The outdoor conditions are 32°C dry bulb temperature and 22°C wet bulb temperature. The quantity of air supplied is 0.5 m³/min/person. The comfort conditions are achieved first by chemical dehumidification and by cooling coil. Determine : 1. Dry bulb temperature of air at exit of dehumidifier; 2. Capacity of dehumidifier; 3. Capacity and surface temperature of cooling coil, if the by-pass factor is 0.30.*

* The point 3 also represents the mixing of 20% of fresh air and 80% of recirculated air. The point 3 may also be located on the line 1-2, such that

Length 2-3 = Length 2-1 × 0.2

Solution. Given : Seating capacity = 60 ; t_{d2} = 22°C ; ϕ_2 = 55% ; t_{d1} = 32°C ; t_{w1} = 22°C ; v_1 = 0.5 m³/min / person = 0.5 × 60 = 30 m³/min ; *BPF* = 0.3

First of all, mark the outdoor conditions of air *i.e.* at 32°C dry bulb temperature and 22°C wet bulb temperature on the psychrometric chart as point 1, as shown in Fig. 18.8. Now mark the required comfort conditions of air *i.e.* at 22°C dry bulb temperature and 55% relative humidity, as point 2. In order to find the condition of air leaving the dehumidifier, draw a constant wet bulb temperature line from point 1 and a constant specific humidity line from point 2. Let these two lines intersect at point 3. The line 1-3 represents the chemical dehumidification and the line 3-2 represents sensible cooling.

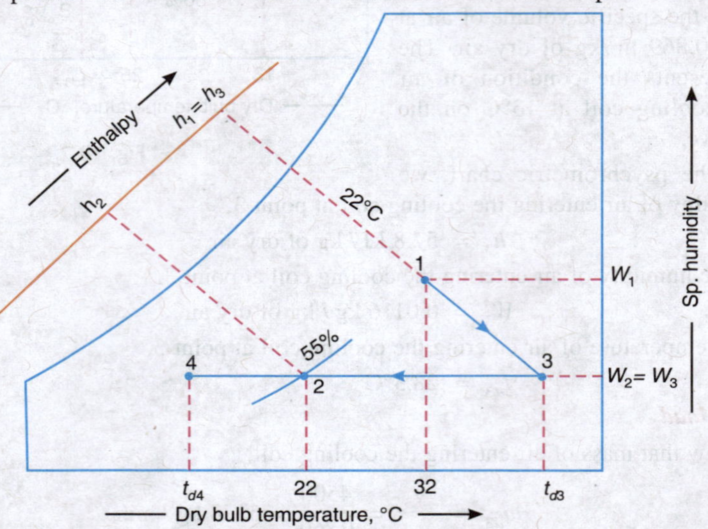

Fig. 18.8

1. Dry bulb temperature of air at exit of dehumidifier

From the psychrometric chart, we find that dry bulb temperature of air at exit of dehumidifier *i.e.* at point 3,

$$t_{d3} = 41\text{°C} \quad \textbf{Ans.}$$

2. Capacity of dehumidifier

From the psychrometric chart, we find that enthalpy of air at point 1,

$$h_1 = h_3 = 64.5 \text{ kJ / kg of dry air}$$

Enthalpy of air at point 2,

$$h_2 = 45 \text{ kJ / kg of dry air}$$

Specific humidity of air at point 1,

$$W_1 = 0.0123 \text{ kg / kg of dry air}$$

Specific humidity of air at point 3,

$$W_3 = W_2 = 0.0084 \text{ kg / kg of dry air}$$

and specific volume of air at point 1,

$$v_{s1} = 0.881 \text{ m}^3\text{ / kg of dry air}$$

We know that mass of air supplied,

$$m_a = \frac{v_1}{v_{s1}} = \frac{30}{0.881} = 34.05 \text{ kg / min}$$

∴ Capacity of the dehumidifier

$$= m_a (W_1 - W_3)$$
$$= 34.05 (0.0123 - 0.0084) = 0.1328 \text{ kg / min}$$
$$= 0.1328 \times 60 = 7.968 \text{ kg / h}$$ **Ans.**

3. Capacity and surface temperature of cooling coil

We know that capacity of the cooling coil

$$= m_a (h_3 - h_2) = 34.05 (64.5 - 45) = 664 \text{ kJ/min}$$
$$= 664 / 210 = 3.16 \text{ TR}$$ **Ans.** ...(∵ 1TR = 210 kJ/min)

Let t_{d4} = Surface temperature of the cooling coil.

We know that by-pass factor (*BPF*),

$$0.3 = \frac{t_{d2} - t_{d4}}{t_{d3} - t_{d4}} = \frac{22 - t_{d4}}{41 - t_{d4}}$$

$$0.3 (41 - t_{d4}) = 22 - t_{d4} \quad \text{or} \quad 12.3 - 0.3\, t_{d4} = 22 - t_{d4}$$

$$\therefore \quad t_{d4} = \frac{22 - 12.3}{0.7} = 13.86\text{°C}$$ **Ans.**

Example 18.7. *The following data refer to air conditioning of a public hall :*

Outdoor conditions = 40° C DBT , 20° C WBT

Required comfort conditions = 20° C DBT , 50% RH

Seating capacity of hall = 1000

Amount of outdoor air supplied = 0.3 m³/min/person

If the required condition is achieved first by adiabatic humidifying and then cooling, find :

1. The capacity of the cooling coil and surface temperature of the coil if the by-pass factor is 0.25 ; and 2. The capacity of the humidifier and its efficiency.

Solution. Given : $t_{d1} = 40\text{°C}$; $t_{w1} = 20\text{°C}$; $t_{d2} = 20\text{°C}$; $\phi_2 = 50\%$; Seating capacity = 1000 ; $v_1 = 0.3 \text{ m}^3\text{/min/person} = 0.3 \times 1000 = 300 \text{ m}^3\text{/min}$; $BPF = 0.25$

First of all, mark the outdoor conditions of air *i.e.* at 40°C dry bulb temperature and 20°C wet bulb temperature on the psychrometric chart as point 1, as shown in Fig. 18.9. Now mark the required comfort conditions of air *i.e.* at 20°C dry bulb temperature and 50% relative humidity, as point 2. From point 1, draw a constant wet bulb temperature line and from point 2 draw a constant specific humidity line. Let these two lines intersect at point 3. The line 1-3 represents adiabatic humidification and the line 3-2 represents sensible cooling.

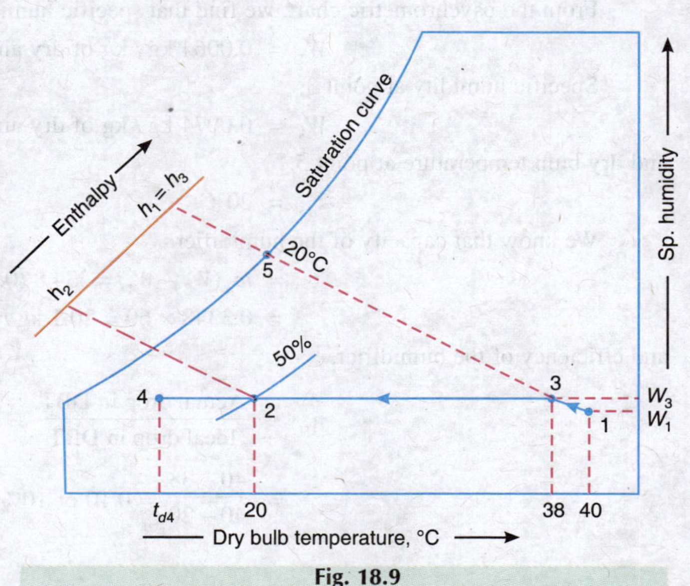

Fig. 18.9

From the psychrometric chart, we find that specific volume of air at point 1,

$$v_{s1} = 0.896 \text{ m}^3/\text{kg of dry air}$$

∴ Mass of air supplied,

$$m_a = \frac{v_1}{v_{s1}} = \frac{300}{0.896} = 334.8 \text{ kg/min}$$

1. *Capacity of the cooling coil and surface temperature of the coil*

From the psychrometric chart, we find that enthalpy of air at point 3,

$$h_3 = 57.6 \text{ kJ/kg of dry air}$$

Enthalpy of air at point 2,

$$h_2 = 39 \text{ kJ/kg of dry air}$$

Dry bulb temperature of air after humidification *i.e.* at point 3,

$$t_{d3} = 38°\text{C}$$

We know that capacity of the cooling coil

$$= m_a(h_3 - h_2) = 334.8\,(57.6 - 39) = 6227 \text{ kJ/min}$$

$$= 6227/210 = 29.6 \text{ TR } \textbf{Ans.}$$

Let t_{d4} = Surface temperature of the coil

We know that by-pass factor (BPF),

$$0.25 = \frac{t_{d2} - t_{d4}}{t_{d3} - t_{d4}} = \frac{20 - t_{d4}}{38 - t_{d4}}$$

$$0.25\,(38 - t_{d4}) = 20 - t_{d4} \quad \text{or} \quad 9.5 - 0.25\,t_{d4} = 20 - t_{d4}$$

$$\therefore \quad t_{d4} = \frac{20 - 9.5}{0.75} = 14°\text{C } \textbf{Ans.}$$

2. *Capacity of the humidifier and its efficiency*

From the psychrometric chart, we find that specific humidity at point 1,

$$W_1 = 0.0064 \text{ kg/kg of dry air}$$

Specific humidity at point 3,

$$W_3 = 0.0074 \text{ kg/kg of dry air}$$

and dry bulb temperature at point 5,

$$t_{d5} = 20°\text{C}$$

We know that capacity of the humidifier

$$= m_a(W_3 - W_1) = 334.8\,(0.0074 - 0.0064) = 0.3348 \text{ kg/min}$$

$$= 0.3348 \times 60 = 20.1 \text{ kg/h } \textbf{Ans.}$$

and efficiency of the humidifier,

$$\eta_H = \frac{\text{Actual drop in DBT}}{\text{Ideal drop in DBT}} = \frac{t_{d1} - t_{d3}}{t_{d1} - t_{d5}}$$

$$= \frac{40 - 38}{40 - 20} = 0.10 \text{ or } 10\% \textbf{ Ans.}$$

18.10 Year-Round Air Conditioning System

The year-round air conditioning system should have equipment for both the summer and winter air conditioning. The schematic arrangement of a modern summer year-round air conditioning system is shown in Fig.18.10.

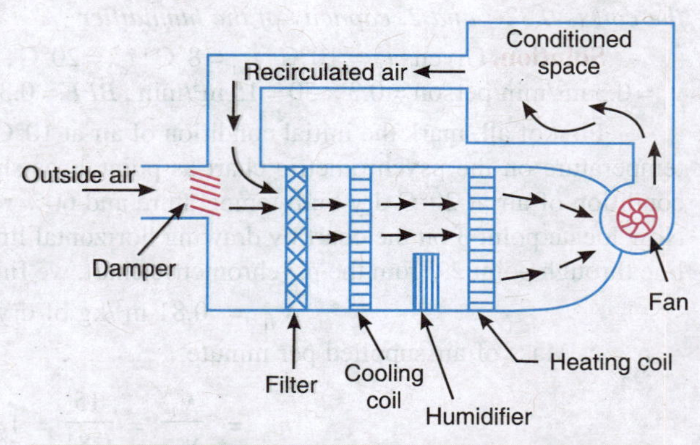

Fig. 18.10. Year-round air conditioning system.

The outside air flows through the damper and mixes up with the recirculated air (which is obtained from the conditioned space). The mixed air passes through a filter to remove dirt, dust and other impurities. In summer air conditioning, the cooling coil operates to cool the air to the desired value. The dehumidification is obtained by operating the cooling coil at a temperature lower than the dew point temperature (apparatus dew point). In winter, the cooling coil is made inoperative and the heating coil operates to heat the air. The spray type humidifier is also made use of in the dry season to humidify the air.

18.11 Unitary Air Conditioning System

In this system, factory assembled air conditioners are installed in or adjacent to the space to be conditioned. The unitary air conditioning systems are of the following two types :

1. *Window units*. These are self-contained units of small capacity of 1 TR to 3 TR, and are mounted in a window or through the wall. They are employed to condition the air of one room only. If the room is bigger in size, then two or more units are installed.

2. *Vertical packed units*. These are also self-contained units of bigger capacity of 5 to 20 TR and are installed adjacent to the space to be conditioned. This is very useful for conditioning the air of a restaurant, bank or small office.

The unitary air conditioning system may be adopted for winter, summer or year-round air conditioning.

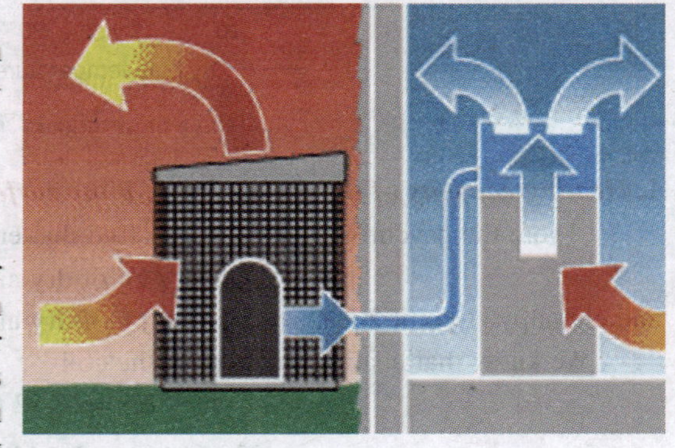

Central air-conditioning system.

18.12 Central Air Conditioning System

This is the most important type of air conditioning system, which is adopted, when the cooling capacity required is 25 TR or more. The central air conditioning system is also adopted when the air flow is more than 300 m^3/min or different zones in a building are to be air conditioned.

Example 18.8. *An air conditioning plant is to be designed for a small office for winter conditions with the following data:*

Outdoor conditions	*= 10°C DBT and 8°C WBT*
Required indoor conditions	*= 20°C DBT and 60% RH*
Amount of air circulation	*= 0.3 m^3/min/person*
Seating capacity of the office	*= 50 persons*

The required condition is achieved first by heating and then by adiabatic humidifying. Find: 1. Heating capacity of the coil in kW and the surface temperature, if the by-pass factor of the coil is 0.32 ; and 2. capacity of the humidifier.

Solution. Given : $t_{d1} = 10°C$; $t_{w1} = 8°C$; $t_{d2} = 20°C$; $\phi_2 = 60\%$; seating capacity = 50 persons; $v_1 = 0.3$ m³/min/person $= 0.3 \times 50 = 15$ m³/min ; $BPF = 0.32$

First of all, mark the initial condition of air at 10°C dry bulb temperature and 8°C wet bulb temperature on the psychrometric chart as point 1, as shown in Fig. 18.11. Now mark the final condition of air at 20°C dry bulb temperature and 60% relative humidity on the chart as point 2. Now locate point 3 on the chart by drawing horizontal line through point 1 and constant enthalpy line through point 2, From the psychrometric chart, we find that the specific volume at point 1,

$$v_{s1} = 0.81 \text{ m}^3/\text{kg of dry air}$$

∴ Mass of air supplied per minute,

$$m_a = \frac{v_1}{v_{s1}} = \frac{15}{0.81} = 18.52 \text{ kg / min}$$

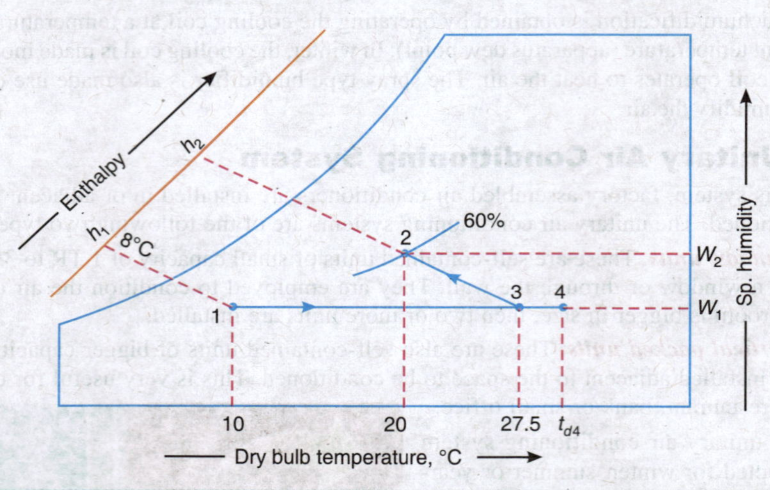

Fig. 18.11

1. *Heating capacity of the coil in kW and the surface temperature*

From the psychrometric chart, we find that enthalpy at point 1,

$$h_1 = 24.8 \text{ kJ / kg of dry air}$$

and enthalpy at point 2, $h_2 = 42.6$ kJ / kg of dry air

We know that heating capacity of the coil

$$= m_a (h_2 - h_1) = 18.52 (42.6 - 24.8) = 329.66 \text{ kJ/min}$$

$$= 329.66/60 = 5.5 \text{ kW } \textbf{Ans.}$$

Let t_{d4} = Surface temperature of the coil.

We know that by-pass factor (BPF),

$$0.32 = \frac{t_{d4} - t_{d3}}{t_{d4} - t_{d1}} = \frac{t_{d4} - 27.5}{t_{d4} - 10}$$ ...[From psychrometric chart, $t_{d3} = 27.5°C$]

or $0.32 (t_{d4} - 10) = t_{d4} - 27.5$ or $0.32\, t_{d4} - 3.2 = t_{d4} - 27.5$

∴ $t_{d4} = 24.3 / 0.68 = 35.7°C$ **Ans.**

2. *Capacity of the humidifier*

From the psychrometric chart, we find that specific humidity at point 1,

$$W_1 = 0.0058 \text{ kg / kg of dry air}$$

and specific humidity at point 2,

$$W_2 = 0.0088 \text{ kg / kg of dry air}$$

We know that capacity of the humidifier,

$$= m_a(W_2 - W_1) = 18.52\ (0.0088 - 0.0058) = 0.055 \text{ kg / min}$$

$$= 0.055 \times 60 = 3.3 \text{ kg / h } \textbf{Ans.}$$

Example 18.9. *A small office hall of 25 persons capacity is provided with summer air conditioning system with the following data :*

Outside conditions	*= 34°C DBT and 28°C WBT*
Inside conditions	*= 24°C DBT and 50% RH*
Volume of air supplied	*= 0.4 m³ / min / person*
Sensible heat load in room	*= 125 600 kJ / h*
Latent heat load in the room	*= 42 000 kJ / h*

Find the sensible heat factor of the plant.

Solution. Given : Seating capacity = 25 persons ; t_{d1} = 34°C ; t_{w1} = 28°C ; t_{d2} = 24°C ; ϕ_2 = 50% ; v_1 = 0.4 m³/min/person = 0.4 × 25 = 10 m³/min ; S.H. load = 125 600 kJ / h ; L.H. load = 42 000 kJ / h

First of all, mark the initial condition of air at 34°C dry bulb temperature and 28°C wet bulb temperature on the psychrometric chart as point 1, as shown in Fig. 18.12. Now mark the final condition of air at 24°C dry bulb temperature and 50% relative humidity on the chart as point 2. Now locate point 3 on the chart by drawing horizontal line through point 2 and vertical line through point 1. From the psychrometric chart, we find that specific volume at point 1,

$$v_{s1} = 0.9 \text{ m}^3 \text{ / kg of dry air}$$

Enthalpy of air at point 1,

$$h_1 = 90 \text{ kJ / kg of dry air}$$

Enthalpy of air at point 2,

$$h_2 = 48 \text{ kJ / kg of dry air}$$

and enthalpy of air at point 3,

$$h_3 = 58 \text{ kJ / kg of dry air}$$

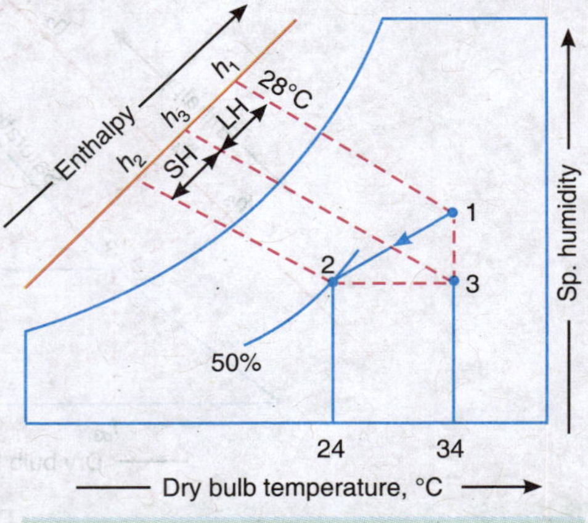

We know that mass of air supplied per min,

$$m_a = \frac{v_1}{v_{s1}} = \frac{10}{0.9} = 11.1 \text{ kg / min}$$

Fig. 18.12

and sensible heat removed from the air

$$= m_a(h_3 - h_2) = 11.1(58 - 48) = 111 \text{ kJ / min}$$

$$= 111 \times 60 = 6660 \text{ kJ / h}$$

∴ Total sensible heat of the room,

$$SH = 6660 + 125\ 600 = 132\ 260 \text{ kJ / h}$$

We know that latent heat removed from the air

$$= m_a (h_1 - h_3) = 11.1(90 - 58) = 355 \text{ kJ / min}$$
$$= 355 \times 60 = 21\ 300 \text{ kJ / h}$$

∴ Total latent heat of the room,

$$LH = 21\ 300 + 42\ 000 = 63\ 300 \text{ kJ / h}$$

We know that sensible heat factor,

$$SHF = \frac{SH}{SH + LH} = \frac{132\ 260}{132\ 260 + 63\ 300} = 0.676 \text{ Ans.}$$

Example 18.10. *A restaurant with a capacity of 100 persons is to be air-conditioned with the following conditions :*

Outside conditions	*: 30°C DBT and 70% RH*
Desired inside conditions	*: 23°C DBT and 55% RH*
Quantity of air supplied	*: 0.5 m³ / min / person*

The desired conditions are achieved by cooling, dehumidifying and then heating . Determine : 1. Capacity of cooling coil in tonnes of refrigeration ; 2. Capacity of heating coil ; 3. Amount of water removed by dehumidifier ; and 4. By-pass factor of the heating coil if its surface temperature is 35°C.

Solution. Given : Number of persons = 100 ; t_{d1} = 30°C ; ϕ_1 = 70% ; t_{d4} = 23°C ; ϕ_4 = 55% ; v_1 = 0.5 m³ / min / person = 0.5 × 100 = 50 m³/min

First of all, mark the outside conditions of air at 30°C dry bulb temperature and 70% relative humidity on the psychrometric chart as point 1, as shown in Fig. 18.13. Now mark the desired inside conditions of air at 23°C dry bulb temperature and 55% relative humidity on the chart as point 4. The process 1-2 represents the sensible cooling, process 2-3 represents dehumidification and the process 3-4 represents the sensible heating.

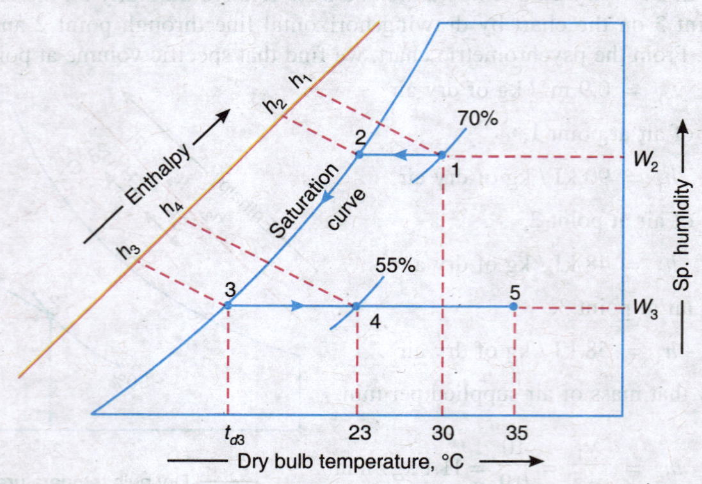

Fig. 18.13

From the psychrometric chart, we find that the specific volume at point 1,

$$v_{s1} = 0.885 \text{ m}^3\text{/kg of dry air}$$

∴ Mass of air supplied,

$$m_a = \frac{v_1}{v_{s1}} = \frac{50}{0.885} = 56.5 \text{ kg / min}$$

1. *Capacity of cooling coil in tonnes of refrigeration*

From the psychrometric chart, we find that enthalpy of air at point 1,

$$h_1 = 78.5 \text{ kJ/kg of dry air}$$

Enthalpy of air at point 3,

$$h_3 = 37.8 \text{ kJ/kg of dry air}$$

∴ Capacity of the cooling coil

$$= m_a (h_1 - h_3) = 56.5\ (78.5 - 37.8) = 2300 \text{ kJ/min}$$

$$= 2300 / 210 = 10.95 \text{ TR}$$ **Ans.**

2. *Capacity of heating coil*

From the psychrometric chart, we find that enthalpy of air at point 4,

$$h_4 = 47.6 \text{ kJ / kg of dry air}$$

∴ Capacity of the heating coil

$$= m_a (h_4 - h_3) = 56.5\ (47.6 - 37.8) = 554 \text{ kJ/min}$$

$$= 554 / 60 = 9.23 \text{ kW}$$ **Ans.**

3. *Amount of water removed by dehumidifier*

From the psychrometric chart, we find that specific humidity at point 2,

$$W_2 = 0.0188 \text{ kg / kg of dry air}$$

and specific humidity at point 3,

$$W_3 = 0.0095 \text{ kg / kg of dry air}$$

∴ Amount of water removed by dehumidifier

$$= m_a (W_2 - W_3) = 56.5\ (0.0188 - 0.0095) = 0.525 \text{ kg / min}$$

$$= 0.525 \times 60 = 31.5 \text{ kg / h}$$ **Ans.**

4. *By-pass factor of the heating coil*

Let t_{d5} = Surface temperature of the heating coil = 35°C ... (Given)

From the psychrometric chart, we find that dry bulb temperature at point 3,

$$t_{d3} = 13.5°\text{C}$$

We know that by-pass factor of the heating coil,

$$BPF = \frac{t_{d5} - t_{d4}}{t_{d5} - t_{d3}} = \frac{35 - 23}{35 - 13.5} = 0.558$$ **Ans.**

Air-conditioning system.

18.13 Room Sensible Heat Factor

It is defined as the ratio of the room sensible heat to the room total heat. Mathematically, room sensible heat factor,

$$RSHF = \frac{RSH}{RTH} = \frac{RSH}{RSH + RLH}$$

where RSH = Room sensible heat,
RLH = Room latent heat, and
RTH = Room total heat.

The conditioned air supplied to the room must have the capacity to take up simultaneously both the room sensible heat and room latent heat loads. The point S on the psychrometric chart, as shown in Fig. 18.14, represents the supply air condition and the point R represents the required final condition in the room (*i.e.* room design condition). The line SR is called the *room sensible heat factor line* ($RSHF$ line). The slope of this line gives the ratio of the room sensible heat (RSH) to the room latent heat (RLH). Thus the supply air having its conditions given by any point on this line will satisfy the requirements of the room with adequate supply of such air. In other words, the supply air having conditions marked by points S_1, S_2, S_3, S_4 etc., will satisfy the requirement but the quantity of air supplied will be different for different supply air points. The supply condition at S requires minimum air and at point S_4, it is maximum of all the four points.

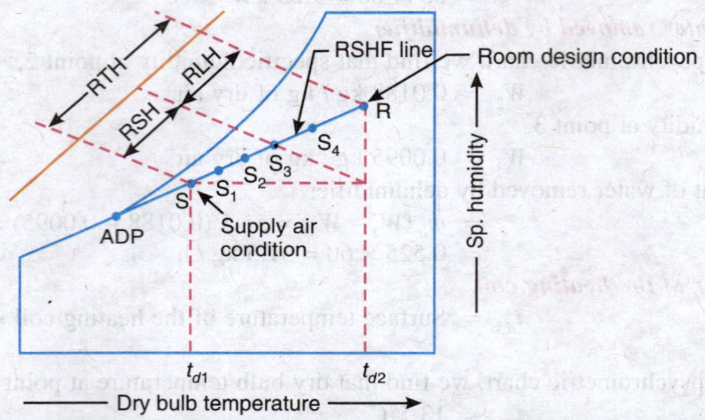

Fig. 18.14. Representation of supply air condition and room design condition.

When the supply air conditions are not known, which in fact is generally required to be found out, the room sensible heat factor line may be drawn from the calculated value of room sensible heat factor ($RSHF$), as discussed below :

1. Mark point a on the sensible heat factor scale given on the right hand corner of the psychrometric chart as shown in Fig. 18.15. The point a represents the calculated value of $RSHF$.

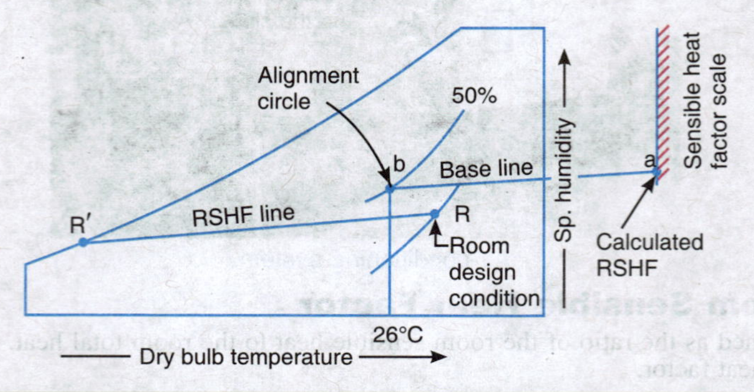

Fig. 18.15. Room sensible heat factor ($RSHF$) line.

2. Join point a with the *alignment circle or the reference point b. The line ab is called base line.
3. Mark point R on the psychrometric chart to represent the room design conditions.
4. Through point R draw a line RR' parallel to the base line ab. This line is the required room sensible heat factor line.

Note : In a cooling and dehumidification process, the temperature at which the room sensible heat factor line intersects the saturation curve is called *room apparatus dew point (ADP)*.

18.14 Grand Sensible Heat Factor

It is defined as the ratio of the total sensible heat to the grand total heat which the cooling coil or the conditioning apparatus is required to handle. Mathematically, grand sensible heat factor,

$$GSHF = \frac{TSH}{GTH} = \frac{TSH}{TSH + TLH} = \frac{RSH + OASH}{(RSH + OASH) + (RLH + OALH)}$$

where TSH = Total sensible heat = $RSH + OASH$

TLH = Total latent heat = $RLH + OALH$

GTH = Grand total heat = $TSH + TLH = RSH + RLH + OATH$

$= RSH + RLH + (OASH + OALH)$

Let v_1 = Volume of outside air or ventilation in m^3/min,

t_{d1} = Dry bulb temperature of outside air in °C,

W_1 = Specific humidity of outside air in kg / kg of dry air,

h_1 = Enthalpy of outside air in kJ / kg of dry air,

t_{d2} = Dry bulb temperature of room air in °C,

W_2 = Specific humidity of room air in kg / kg of dry air, and

h_2 = Enthalpy of room air in kJ / kg of dry air.

∴ Outside air sensible heat,

$$OASH = 0.020\ 44\ v_1 (t_{d1} - t_{d2}) \text{ kW}$$ ... (Refer Art. 16.10)

Outside air latent heat,

$$OALH = 50\ v_1 (W_1 - W_2) \text{ kW}$$ ... (Refer Art. 16.13)

and outside air total heat,

$$OATH = OASH + OALH$$

The outside air total heat may also be calculated from the following relation :

$$OATH = 0.02\ v_1 (h_1 - h_2) \text{ kW}$$

Generally, the air supplied to the air conditioning plant is a mixture of fresh air (or outside air or ventilation) and the recirculated air having the properties of room air. On the psychrometric chart, as shown in Fig. 18.16, the point 1 represents the outside condition of air, the point 2 represents the room air condition and the point 3 represents the mixture condition of air entering the cooling coil. When the mixture condition enters the cooling coil or conditioning apparatus, it is cooled and dehumidified. The point 4 shows the supply air or leaving condition of air from the cooling coil or conditioning apparatus. When the point

* The alignment circle is marked on the psychrometric chart at 26°C dry bulb temperature and 50% relative humidity.

3 is joined with the point 4, it gives a *grand sensible heat factor line* (*GSHF* line) as shown in Fig. 18.16. This line, when produced up to the saturation curve, gives apparatus dew point (*ADP*).

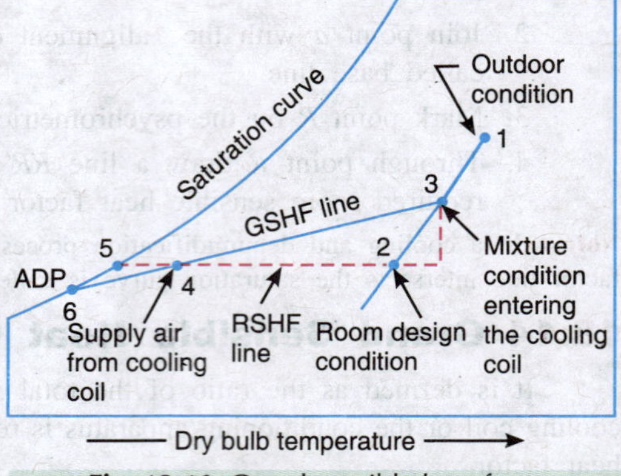

Fig. 18.16. Grand sensible heat factor.

If the mixture condition entering the cooling coil or conditioning apparatus and the grand sensible heat factor (*GSHF*) are known, then the *GSHF* line may be drawn on the psychrometric chart in the similar way as discussed for *RSHF* line. The point 4, as shown in Fig. 18.16, is the intersection of *GSHF* line and *RSHF* line. This point gives the ideal conditions for supply air to the room.

18.15 Effective Room Sensible Heat Factor

It is defined as the ratio of the effective room sensible heat to the effective room total heat. Mathematically, effective room sensible heat factor,

$$ERSHF = \frac{ERSH}{ERTH} = \frac{ERSH}{ERSH + ERLH}$$

where
$$ERSH = \text{Effective room sensible heat} = RSH + OASH \times BPF$$
$$= RSH + 0.020\,44\, v_1 (t_{d1} - t_{d2})\, BPF$$
$$ERLH = \text{Effective room latent heat} = RLH + OALH \times BPF$$
$$= RLH + 50\, v_1 (W_1 - W_2)\, BPF$$
$$ERTH = \text{Effective room total heat} = ERSH + ERLH$$
$$BPF = \text{By-pass factor}$$

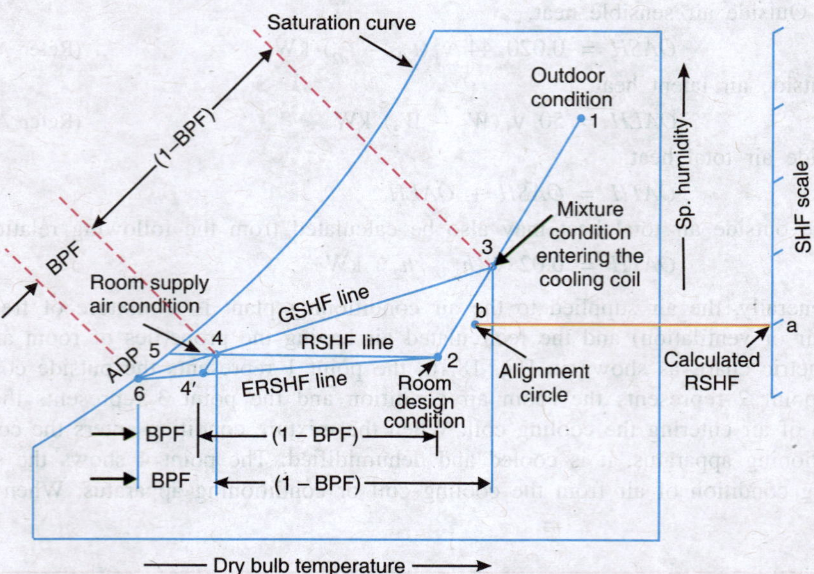

Fig. 18.17. Effective room sensible heat factor.

The line joining the point 2 and point 6 *i.e.* *ADP*, as shown in Fig. 18.17, gives the effective room sensible heat factor line (*ERSHF* line). From point 4, draw 4-4′ parallel to 3-2. Therefore from similar triangles 6-4-4′ and 6-3-2,

$$BPF = \frac{\text{Length 4-6}}{\text{Length 3-6}} = \frac{\text{Length 4}'\text{-6}}{\text{Length 2-6}}$$

The by-pass factor is also given by,

$$BPF = \frac{t_{d4} - ADP}{t_{d3} - ADP} = \frac{t_{d4'} - ADP}{t_{d2} - ADP}$$

Notes : 1. The effective room sensible heat (*ERSH*), effective room latent heat (*ERLH*) and effective room total heat (*ERTH*) may also be obtained from the following relations :

$$ERSH = 0.020\,44\ v_d\,(t_{d2} - ADP)\,(1 - BPF)\ \text{kW}$$

$$ERLH = 50\ v_d\,(W_2 - W_{ADP})\,(1 - BPF)\ \text{kW}$$

and

$$ERTH = 0.02\ v_d\,(h_2 - h_{ADP})\,(1 - BPF)\ \text{kW}$$

where v_d = Volume of dehumidified air to room or space in m³/min,

ADP = Apparatus dew point in °C,

W_{ADP} = Specific humidity at apparatus dew point in kg / kg of dry air, and

h_{ADP} = Enthalpy at apparatus dew point in kJ / kg of dry air.

2. The mass of dehumidified air is given by

$$m_d = \frac{\text{Room total heat}}{h_2 - h_4}$$

where h_2 = Enthalpy of air at room condition, and

h_4 = Enthalpy of supply air to room from the cooling coil.

Example 18.11. *A room has a sensible heat gain of 24 kW and a latent heat gain of 5.2 kW and it has to be maintained at 26°C DBT and 50% RH. 180 m³/min of air is delivered to the room. Determine the state of supply air.*

Solution. Given : $RSH = 24$ kW ; $RLH = 5.2$ kW ; $t_{d1} = 26$°C ; $\phi_1 = 50\%$; $v = 180$ m³/min

Let t_{d2} = Dry bulb temperature of the supply air.

We know that room sensible heat load,

$$RSH = 0.020\,44\ v\,(t_{d1} - t_{d2})$$

$$24 = 0.020\,44 \times 180\,(26 - t_{d2})$$

$$= 3.68\,(26 - t_{d2})$$

$$\therefore \quad t_{d2} = 19.5\text{°C}$$

We also know that room sensible heat factor,

$$RSHF = \frac{RSH}{RSH + RLH} = \frac{24}{24 + 5.2} = 0.822$$

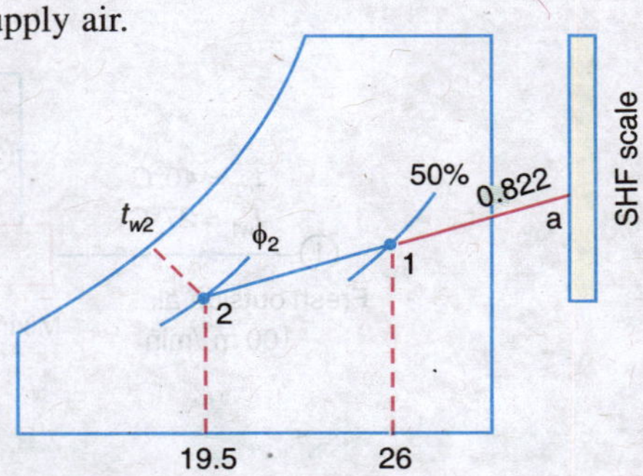

Fig. 18.18

First of all, mark room condition of the air *i.e.* 26°C dry bulb temperature and 50% relative humidity as point 1 on the psychrometric chart, as shown in Fig. 18.18. The point 1 also represents the alignment circle (*i.e.* 26°C *DBT* and 50% *RH*). Now mark the calculated value of *RSHF* = 0.822 on the sensible heat factor scale as point *a* and join with the alignment circle *i.e.* point 1.

Produce this line up to point 2 such that t_{d2} = 19.5°C. The point 2 represents the supply condition of air. From the psychrometric chart, we find that wet bulb temperature of supply air,

$$t_{w2} = 16° \text{ C } \textbf{Ans.}$$

and relative humidity, $\phi_2 = 71\%$ **Ans.**

Example 18.12. *In an airconditioning system, the inside and outside conditions are dry bulb temperature 25°C, relative humidity 50% and dry bulb temperature 40°C, wet bulb temperature 27°C respectively. The room sensible heat factor is 0.8. 50% of the room air is rejected to atmosphere and an equal quantity of fresh air added before air enters the air conditioning apparatus. If the fresh air added is 100 m³/min, determine :*

1. Room sensible and latent heat load ;

2. Sensible and latent heat load due to fresh air ;

3. Apparatus dew point ;

4. Humidity ratio and dry bulb temperature of air entering air conditioning apparatus.

Assume by-pass factor as zero, density of air as 1.2 kg / m³ at a total pressure of 1.01325 bar.

Solution. Given : t_{d1} = 40°C ; t_{w1} = 27°C ; t_{d2} = 25°C ; ϕ_2 = 50% ; RSHF = 0.8 ; v_1 = 100 m³/min; ρ_a = 1.2 kg/m³

The flow diagram for the air conditioning system is shown in Fig. 18.19, and it is represented on the psychrometric chart as discussed below :

First of all, mark the outside condition of air at 40°C dry bulb temperature and 27°C wet bulb temperature on the psychrometric chart as point 1, as shown in Fig. 18.20. Now mark the inside condition of air at 25°C dry bulb temperature and 50% relative humidity as point 2. Since 50% of the room air and 50% of fresh air is added before entering the air conditioning apparatus, therefore mark point 3 on the line 1-2 such that

$$\text{Length 2-3} = \frac{\text{Length 1-2}}{2}$$

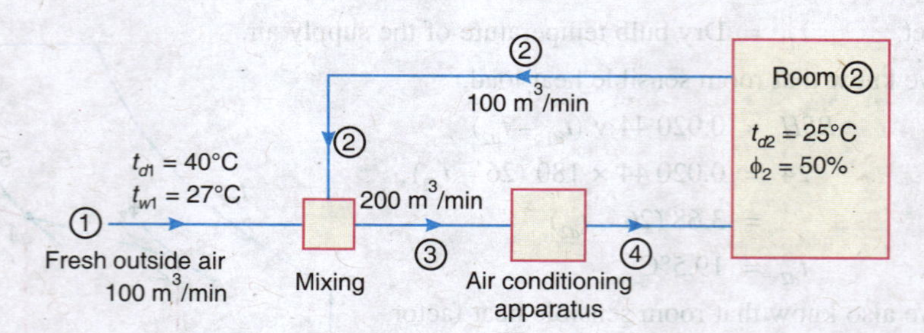

Fig. 18.19

Now mark the given value of RSHF (*i.e.* 0.8) on the room sensible heat factor scale and join this with the alignment circle (*i.e.* 26°C DBT and 50% RH). From point 2, draw a line 2-4 parallel to this line. This line is called RSHF line. The point 4 represents the apparatus dew point (*ADP*). From the psychrometric chart, we find the enthalpy of air at point 1,

$$h_1 = 85.2 \text{ kJ / kg of dry air}$$

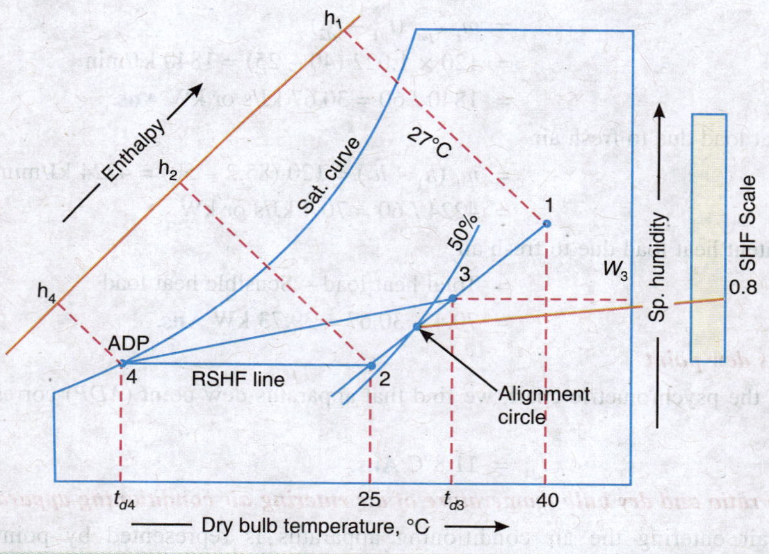

Fig. 18.20

Enthalpy of air at point 2,

$$h_2 = 50 \text{ kJ / kg of dry air}$$

and enthalpy of air at point 4,

$$h_4 = 33 \text{ kJ / kg of dry air}$$

1. Room sensible and latent heat load

We know that mass of air supplied to the room,

$$m_a = v_3 \times \rho_a = (100 + 100)1.2 = 240 \text{ kg / min}$$

∴ Room sensible heat load,

$$RSH = m_a\, c_{pm}\,(t_{d2} - t_{d4})$$

$$= 240 \times 1.022\,(25 - 11.8) = 3238 \text{ kJ / min}$$

$$= 3238/60 = 53.96 \text{ kJ/s or kW} \quad \textbf{Ans.}$$

... [∵ From psychrometric chart, $t_{d4} = 11.8$°C]

and room total heat load,

$$RTH = m_a\,(h_2 - h_4) = 240\,(50 - 33) = 4080 \text{ kJ/min}$$

$$= 4080/60 = 68 \text{ kJ/s or kW}$$

∴ Room latent heat load,

$$RLH = RTH - RSH$$

$$= 68 - 53.96 = 14.04 \text{ kW} \quad \textbf{Ans.}$$

2. Sensible and latent heat load due to fresh air

We know that mass of fresh air supplied,

$$m_F = v_1 \times \rho_a = 100 \times 1.2 = 120 \text{ kg / min}$$

∴ Sensible heat load due to fresh air

$$= m_F \, c_{pm} \, (t_{d1} - t_{d2})$$
$$= 120 \times 1.022 \, (40 - 25) = 1840 \text{ kJ/min}$$
$$= 1840 / 60 = 30.67 \text{ kJ/s or kW}$$ **Ans.**

and total heat load due to fresh air

$$= m_F \, (h_1 - h_2) = 120 \, (85.2 - 50) = 4224 \text{ kJ/min}$$
$$= 4224 / 60 = 70.4 \text{ kJ/s or kW}$$

∴ Latent heat load due to fresh air

$$= \text{Total heat load} - \text{Sensible heat load}$$
$$= 70.4 - 30.67 = 39.73 \text{ kW}$$ **Ans.**

3. Apparatus dew point

From the psychrometric chart, we find that apparatus dew point (*ADP*) corresponding to point 4 is

$$t_{d4} = 11.8°\text{C}$$ **Ans.**

4. Humidity ratio and dry bulb temperature of air entering air conditioning apparatus

The air entering the air conditioning apparatus is represented by point 3 on the psychrometric chart as shown in Fig. 18.20. From the psychrometric chart, we find that humidity ratio corresponding to point 3,

$$W_3 = 0.0138 \text{ kg / kg of dry air}$$ **Ans.**

and dry bulb temperature corresponding to point 3,

$$t_{d3} = 32.5°\text{C}$$ **Ans.**

Example 18.13. *An air conditioned space is maintained at 27°C dry bulb temperature and 50% relative humidity. The ambient conditions are 40°C dry bulb temperature and 27°C wet bulb temperature. The space has a sensible heat gain of 14kW. The air is supplied to the space at 7°C saturated. Calculate :*

1. Mass of moist air supplied to the space in kg / h ; 2. Latent heat gain of space in kW ; and 3. Cooling load of air-washer in kW if 30 per cent of air supplied to the space is fresh, the remainder being recirculated.

Solution. Given : $t_{d2} = 27°\text{C}$; $\phi_2 = 50\%$; $t_{d1} = 40°\text{C}$; $t_{w1} = 27°\text{C}$; $Qs = 14$ kW $= 14$ kJ/s $= 14 \times 3600$ kJ / h ; $t_{d4} = 7°\text{C}$

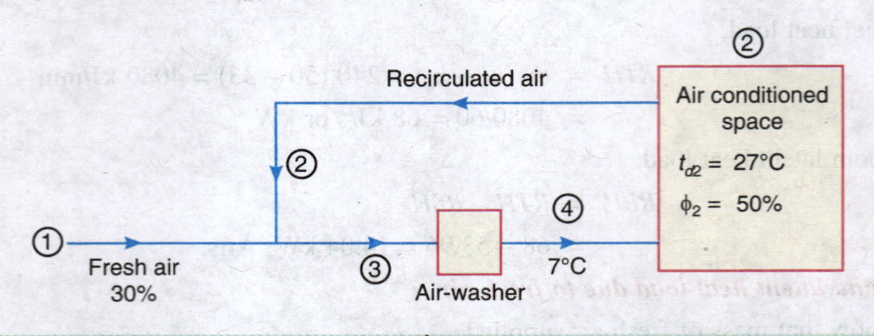

Fig. 18.21

The line diagram for the air conditioned space is shown in Fig. 18.21 and it is represented on the psychrometric chart as discussed below:

First of all, mark the ambient (outside) conditions of air at 40°C dry bulb temperature and 27°C wet bulb temperature on the psychrometric chart as point1, as shown in Fig. 18.22. Now mark the inside conditions of the space at 27°C dry bulb temperature and 50% relative humidity. Since the air is supplied to the space at 7°C saturated, therefore mark point 4 on the saturation curve at 7°C. Also 30 per cent of air supplied to the space (*i.e.* at point 2) is fresh, therefore mark point 3 on the line 2-1, such that

Length 2 - 3 = 0.3 × Length 2 - 1

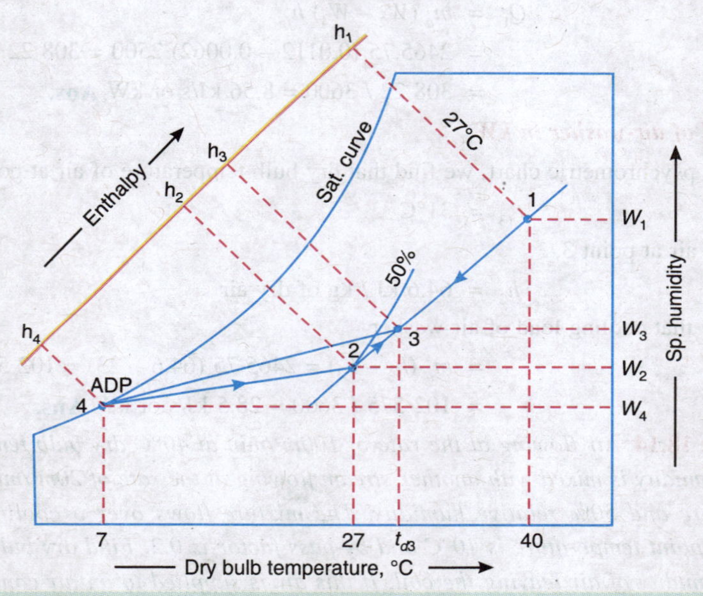

Fig. 18.22

Now from the psychrometric chart, we find that enthalpy of air at point 1,

$$h_1 = 85 \text{ kJ / kg of dry air}$$

Specific humidity of air at point 1,

$$W_1 = 0.0172 \text{ kg / kg of dry air}$$

Enthalpy of air at point 2,

$$h_2 = 56 \text{ kJ / kg of dry air}$$

Specific humidity of air at point 2,

$$W_2 = 0.0112 \text{ kg / kg of dry air}$$

Enthalpy of air at point 4,

$$h_4 = 23 \text{ kJ / kg of dry air}$$

Specific humidity of air at point 4,

$$W_4 = 0.0062 \text{ kg / kg of dry air}$$

1. *Mass of moist air supplied to the space in kg / h*

We know that mass of dry air supplied to the space,

$$m_a = \frac{Q_S}{c_{pm}(t_{d2} - t_{d4})} = \frac{14 \times 3600}{1.022\,(27 - 7)} = 2465.75 \text{ kg / h}$$

... [$\because$ c_{pm} = Humid specific heat = 1.022 kJ / kg K]

∴ Mass of moist air supplied to the space

$$= m_a (1 + W_4) = 2465.75 (1 + 0.0062)$$
$$= 2481 \text{ kg / h}$$ **Ans.**

2. Latent heat gain of space in kW

We know that latent heat gain of space,

$$Q_L = m_a (W_2 - W_4) h_{fg}$$
$$= 2465.75 (0.0112 - 0.0062) 2500 = 308\ 22 \text{ kJ/h}$$
$$= 308\ 22 / 3600 = 8.56 \text{ kJ/s or kW}$$ **Ans.**

3. Cooling load of air-washer in kW

From the psychrometric chart, we find that dry bulb temperature of air at point 3,

$$*t_{d3} = 31°\text{C}$$

and enthalpy of air at point 3,

$$h_3 = 64.6 \text{ kJ / kg of dry air}$$

We know that cooling load of air-washer

$$= m_a (h_3 - h_4) = 2465.75 (64.6 - 23) = 102\ 575 \text{ kJ/h}$$
$$= 102\ 575 / 3600 = 28.5 \text{ kJ/s or kW}$$ **Ans.**

Example 18.14. *Air flowing at the rate of 100m³/min at 40°C dry bulb temperature and 50% relative humidity is mixed with another stream flowing at the rate of 20m³/min at 26°C dry bulb temperature and 50% relative humidity. The mixture flows over a cooling coil whose apparatus dew point temperature is 10°C and by-pass factor is 0.2. Find dry bulb temperature and relative humidity of air leaving the coil. If this air is supplied to an air-conditioned room where dry bulb temperature of 26°C and relative humidity of 50% are maintained, estimate : 1. Room sensible heat factor ; and 2. Cooling load capacity of the coil in tonnes of refrigeration.*

Solution. Given : $v_1 = 100$ m³/min ; $t_{d1} = 40$°C ; $\phi_1 = 50\%$; $v_2 = 20$ m³/min ; $t_{d2} = 26$°C ; $\phi_2 = 50\%$; $ADP = 10$°C ; $BPF = 0.2$

The flow diagram for an air-conditioned room is shown in Fig 18.23 and it is represented on the psychrometric chart as discussed below :

First of all, mark the initial condition of air at 40°C dry bulb temperature and 50% relative humidity on the psychrometric chart as point 1, as shown in Fig. 18.24. Now mark the room condition of air at 26°C dry bulb temperature and 50% relative humidity as point 2. From the psychrometric chart, we find that enthalpy of air at point 1,

$$h_1 = 99.8 \text{ kJ / kg of dry air}$$

Enthalpy of air at point 2,

$$h_2 = 53.5 \text{ kJ / kg of dry air}$$

Specific volume of air at point 1,

$$v_{s1} = 0.92 \text{ m}^3 \text{ / kg of dry air}$$

---

* The value of t_{d3} may be obtained as follows :

$$t_{d3} = 0.7\ t_{d2} + 0.3\ t_{d1} = 0.7 \times 27 + 0.3 \times 40 = 30.9°\text{C}$$

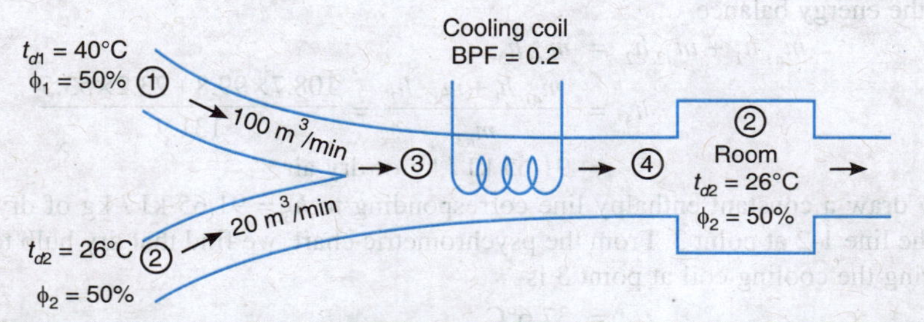

Fig. 18.23

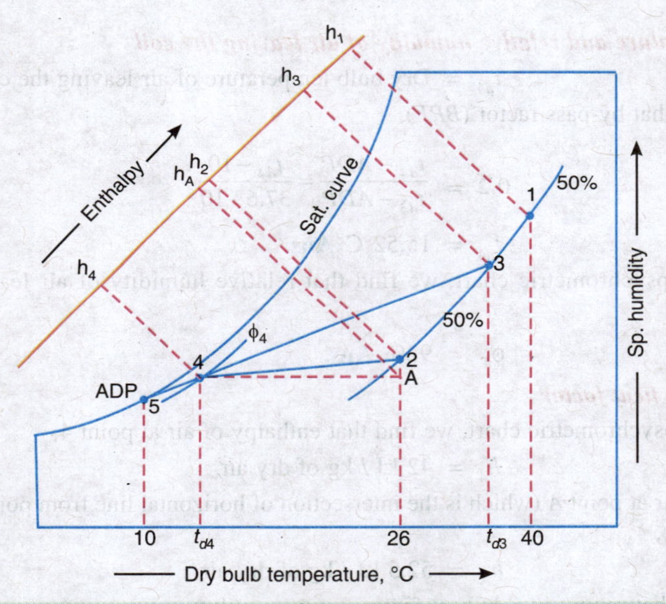

Fig. 18.24

and specific volume of air at point 2,

$$v_{s2} = 0.862 \text{ m}^3 / \text{kg of dry air}$$

We know that mass of air supplied at point 1,

$$m_{a1} = \frac{v_1}{v_{s1}} = \frac{100}{0.92} = 108.7 \text{ kg / min}$$

and mass of air supplied at point 2,

$$m_{a2} = \frac{v_2}{v_{s2}} = \frac{20}{0.862} = 23.2 \text{ kg / min}$$

∴ Mass of air flowing through the cooling coil at point 3,

$$m_{a3} = m_{a1} + m_{a2} = 108.7 + 23.2 = 131.9 \text{ kg / min}$$

For the energy balance,

$$m_{a1}\, h_1 + m_{a2}\, h_2 = m_{a3}\, h_3$$

$$\therefore \quad h_3 = \frac{m_{a1}\, h_1 + m_{a2}\, h_2}{m_{a3}} = \frac{108.7 \times 99.8 + 23.2 \times 53.5}{131.9}$$

$$= 91.65 \text{ kJ / kg of dry air}$$

Now draw a constant enthalpy line corresponding to h_3 = 91.65 kJ / kg of dry air which intersects the line 1-2 at point 3. From the psychrometric chart, we find that dry bulb temperature of air entering the cooling coil at point 3 is

$$t_{d3} = 37.6°\text{C}$$

Mark point 5 on the saturation curve such that *ADP* = 10°C, and draw a line 3-5. The point 4 lies on this line.

Dry bulb temperature and relative humidity of air leaving the coil

Let t_{d4} = Dry bulb temperature of air leaving the coil.

We know that by-pass factor (*BPF*),

$$0.2 = \frac{t_{d4} - ADP}{t_{d3} - ADP} = \frac{t_{d4} - 10}{37.6 - 10}$$

$$\therefore \quad t_{d4} = 15.52°\text{C} \quad \textbf{Ans.}$$

From the psychrometric chart, we find that relative humidity of air leaving the coil at point 4 is

$$\phi_4 = 92\% \quad \textbf{Ans.}$$

1. *Room sensible heat factor*

From the psychrometric chart, we find that enthalpy of air at point 4,

$$h_4 = 42 \text{ kJ / kg of dry air.}$$

and enthalpy of air at point *A* (which is the intersection of horizontal line from point 4 and vertical line from point 2),

$$h_A = 52.5 \text{ kJ / kg of dry air}$$

We know that room sensible heat factor,

$$RSHF = \frac{h_A - h_4}{h_2 - h_4} = \frac{52.5 - 42}{53.5 - 42} = 0.913 \quad \textbf{Ans.}$$

2. *Cooling load capacity of the coil*

We know that cooling load capacity of the coil

$$= m_{a3}\,(h_3 - h_4) = 131.9\,(91.65 - 42) = 6548.8 \text{ kJ/min}$$

$$= 6548.8 / 210 = 31.185 \text{ TR} \quad \textbf{Ans.}$$

...($\because$ 1 TR = 210 kJ/min)

Example 18.15. *An air conditioned auditorium is to be maintained at 27°C dry bulb temperature and 60% relative humidity. The ambient condition is 40°C dry bulb temperature and 30°C wet bulb temperature. The total sensible heat load is 100 000 kJ/h and the total latent heat load is 40 000 kJ/h. 60% of the return air is recirculated and mixed with 40% of make-up air after the cooling coil. The condition of air leaving the cooling coil is at 18°C.*

Determine : 1. Room sensible heat factor ; 2. The condition of air entering the auditorium ; 3. The amount of make-up air ; 4. Apparatus dew point ; and 5. By-pass factor of the cooling coil.

Show the processes on the psychrometric chart.

Solution. Given : $t_{d4} = 27°C$; $\phi_4 = 60\%$; $t_{d1} = 40°C$; $t_{w1} = 30°C$; $RSH = 100\,000$ kJ / h ; $RLH = 40\,000$ kJ / h ; $t_{d2} = 18°C$

1. ***Room sensible heat factor***

We know that room sensible heat factor,

$$RSHF = \frac{RSH}{RSH + RLH} = \frac{100\,000}{100\,000 + 40\,000} = 0.714 \text{ Ans.}$$

2. ***Condition of air entering the auditorium***

The line diagram for processes involved in the air conditioning of an auditorium is shown in Fig. 18.25. These processes are shown on the psychrometric chart as discussed below :

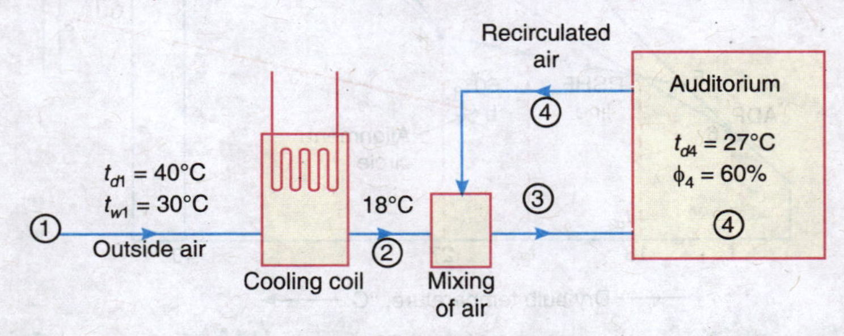

Fig. 18.25

First of all, mark the ambient condition of air (outside air) *i.e.* at 40°C dry bulb temperature and 30°C wet bulb temperature on the psychrometric chart as point 1, as shown in Fig. 18.26. Now mark the condition of air in the auditorium, *i.e.* at 27°C dry bulb temperature and 60% relative humidity, as point 4.

Mark the calculated value of *RSHF* = 0.714 on the sensible heat factor scale as point *a* and join with point *b* which is the alignment circle (*i.e.* 26°C *DBT* and 50% *RH*) as shown in Fig. 18.26. Now from point 4, draw a line 4-5 (known as *RSHF* line) parallel to the line *ab*. Since the condition of air leaving the cooling coil is at 18°C, therefore, mark point 2 such that $t_{d2} = 18°C$. Join points 1 and 2 and produce up to point 6 on the saturation curve. The line 1-2-6 is the *GSHF* line. It is given that 60% of the air from the auditorium is recirculated and mixed with 40% of the make-up air after the cooling coil. The mixing condition of air is shown at point 3 such that

$$\frac{\text{Length } 2-3}{\text{Length } 2-4} = 0.6$$

The condition of air entering the auditorium is given by point 3. From the psychrometric chart, we find that at point 3,

Dry bulb temperature, $t_{d3} = 23°$ C **Ans.**

Wet bulb temperature, $t_{w3} = 19.5°$ C **Ans.**

and relative humidity, $\phi_3 = 72\%$ **Ans.**

3. ***Amount of make-up air***

From the psychrometric chart, we find that enthalpy of air at point 4,

$$h_4 = 61 \text{ kJ/kg of dry air}$$

and enthalpy of air at point 3, $h_3 = 56$ kJ/kg of dry air

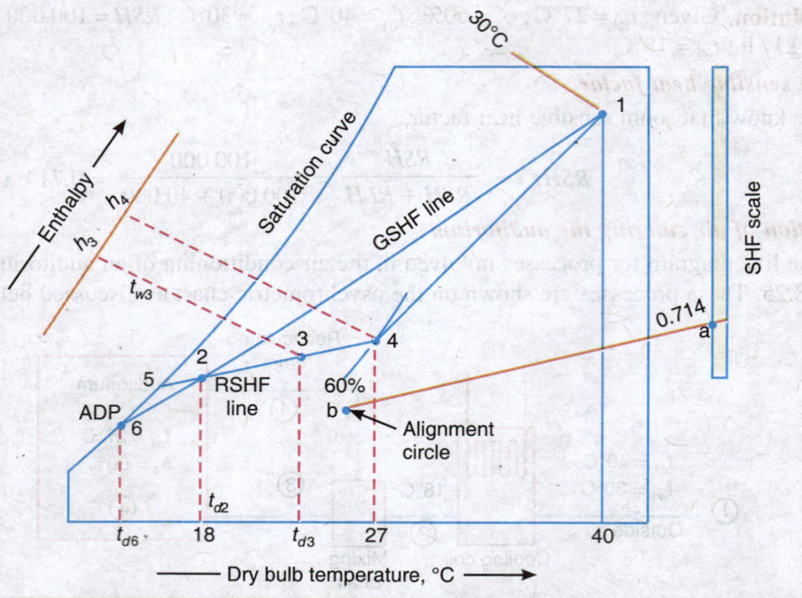

Fig. 18.26

We know that mass of supply air to the auditorium,

$$m_S = \frac{\text{Room total heat}}{h_4 - h_3} = \frac{RSH + RLH}{h_4 - h_3}$$

$$= \frac{100\,000 + 40\,000}{61 - 56} = 28\,000 \text{ kg / h}$$

Since the make-up air is 40% of supply air, therefore mass of make-up air

$$= 0.4 \times 28\,000 = 11\,200 \text{ kg / h} \quad \textbf{Ans.}$$

4. *Apparatus dew point*

From the psychrometric chart, we find that the apparatus dew point of the cooling coil at point 6 is

$$ADP = t_{d6} = 13° \text{ C} \quad \textbf{Ans.}$$

5. *By-pass factor of the cooling coil*

We know that by-pass factor of the cooling coil,

$$BPF = \frac{t_{d2} - ADP}{t_{d1} - ADP} = \frac{18 - 13}{40 - 13} = \frac{5}{27} = 0.185 \quad \textbf{Ans.}$$

Example 18.16. *An air conditioned hall is to be maintained at 27°C dry bulb temperature and 21°C wet bulb temperature. It has a sensible heat load of 46.5 kW and latent heat load of 17.5 kW. The air supplied from outside atmosphere at 38°C dry bulb temperature and 27°C wet bulb temperature is 25 m³/min, directly into the room through ventilation and infiltration. Outside air to be conditioned is passed through the cooling coil whose apparatus dew point is 15°C. The quantity of recirculated air from the hall is 60%. This quantity is mixed with the conditioned air after the cooling coil. Determine : 1. condition of air after the coil and before the recirculated air mixes with it ; 2. condition of air entering the hall, i.e. after mixing with recirculated air ; 3. mass of fresh air entering the cooler ; 4. by-pass factor of the cooling coil; and 5. refrigerating load on the cooling coil.*

Solution. Given : $t_{d4} = 27°C$; $t_{w4} = 21°C$; $Q_{S4} = 46.5$ kW ; $Q_{L4} = 17.5$ kW ; $t_{d1} = 38°C$; $t_{w1} = 27°C$; $v_1 = 25$ m³/min ; $ADP = 15°C$

The line diagram for the processes involved in the air conditioning of a hall is shown in Fig. 18.27. These processes are shown on the psychrometric chart as discussed below :

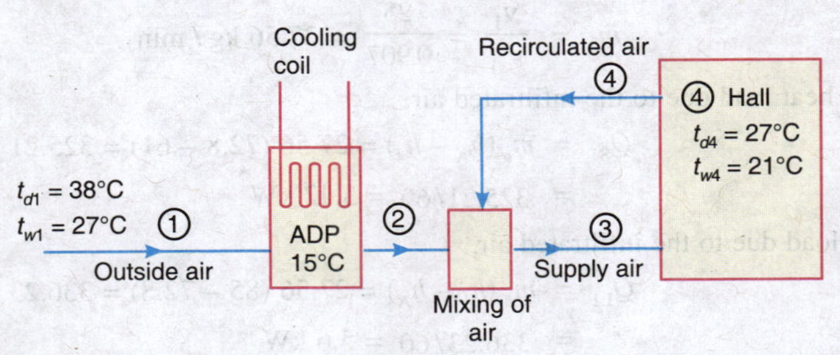

Fig. 18.27

First of all, mark the condition of outside air *i.e.* at 38°C dry bulb temperature and 27°C wet bulb temperature on the psychrometric chart as point 1, as shown in Fig. 18.28. Now mark the condition of air in the hall, *i.e.* at 27°C dry bulb temperature and 21°C wet bulb temperature, at point 4. Mark point *A* by drawing vertical and horizontal lines from points 1 and 4 respectively. Since 25 m³/min of outside air at $t_{d1} = 38°C$ and $t_{w1} = 27°C$ is supplied directly into the room through ventilation and infiltration, therefore the sensible heat and latent heat of 25 m³/min infiltrated air are added to the hall in addition to the sensible heat load of 46.5 kW and latent heat load of 17.5 kW.

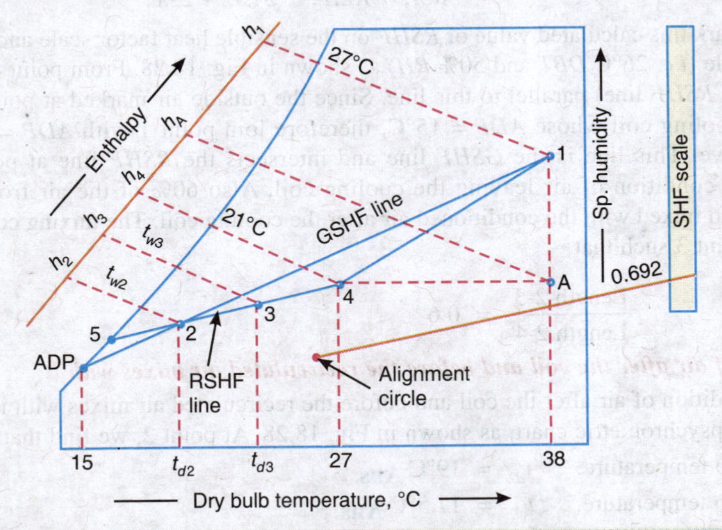

Fig. 18.28

From the psychrometric chart, we find that enthalpy of air at point 1,

$$h_1 = 85 \text{ kJ/kg of dry air}$$

Enthalpy of air at point 4,

$$h_4 = 61 \text{ kJ/kg of dry air}$$

and enthalpy of air at point *A* ,

$$h_A = 72.8 \text{ kJ / kg of dry air}$$

Also specific volume of air at point 1,

$$v_{s1} = 0.907 \text{ m}^3/\text{kg of dry air}$$

∴ Mass of air infiltrated into the hall,

$$m_a = \frac{v_1}{v_{s1}} = \frac{25}{0.907} = 27.56 \text{ kg / min}$$

Sensible heat load due to the infiltrated air,

$$Q_{S1} = m_a (h_A - h_4) = 27.56\,(72.8 - 61) = 325.21 \text{ kJ/min}$$

$$= 325.21/60 = 5.42 \text{ kW}$$

and latent heat load due to the infiltrated air,

$$Q_{L1} = m_a (h_1 - h_A) = 27.56\,(85 - 72.8) = 336.23 \text{ kJ/min}$$

$$= 336.23/60 = 5.6 \text{ kW}$$

∴ Total room sensible heat load,

$$RSH = Q_{S4} + Q_{S1} = 46.5 + 5.42 = 51.92 \text{ kW}$$

and total room latent heat load

$$RLH = Q_{L4} + Q_{L1} = 17.5 + 5.6 = 23.1 \text{ kW}$$

We know that room sensible heat factor,

$$RSHF = \frac{RSH}{RSH + RLH} = \frac{51.92}{51.92 + 23.1} = 0.692$$

Now mark this calculated value of *RSHF* on the sensible heat factor scale and join with the alignment circle (*i.e.* 26°C *DBT* and 50% *RH*) as shown in Fig. 18.28. From point 4, draw a line 4-5 (known as *RSHF* line) parallel to this line. Since the outside air marked at point 1 is passed through the cooling coil whose *ADP* = 15°C, therefore join point 1 with *ADP* = 15°C on the saturation curve. This line is the *GSHF* line and intersects the *RSHF* line at point 2, which represents the condition of air leaving the cooling coil. Also 60% of the air from the hall is recirculated and mixed with the conditioned air after the cooling coil. The mixing condition of air is shown at point 3 such that

$$\frac{\text{Length 2-3}}{\text{Length 2-4}} = 0.6$$

1. *Condition of air after the coil and before the recirculated air mixes with it*

The condition of air after the coil and before the recirculated air mixes with it is shown by point 2 on the psychrometric chart, as shown in Fig. 18.28. At point 2, we find that

Dry bulb temperature, $t_{d2} = 19°\text{C}$ **Ans.**

Wet bulb temperature, $t_{w2} = 17.5°\text{C}$ **Ans.**

2. *Condition of air entering the hall, i.e. after mixing with recirculated air*

The condition of air entering the hall, *i.e.* after mixing with recirculated air, is shown by point 3 on the psychrometric chart, as shown in Fig. 18.28. At point 3, we find that

Dry bulb temperature, $t_{d3} = 24°\text{C}$ **Ans.**

Wet bulb temperature, $t_{w3} = 19.8°\text{C}$ **Ans.**

3. *Mass of fresh air entering the cooler*

The mass of fresh air passing through the cooling coil to take up the sensible and latent heat of the hall is given by

$$m_F = \frac{\text{Total heat removed}}{h_4 - h_2} = \frac{RSH + RLH}{h_4 - h_2}$$

$$= \frac{51.92 + 23.1}{61 - 49} = 6.25 \text{ kg/s} = 6.25 \times 60 = 375 \text{ kg / min} \textbf{ Ans.}$$

... (From psychrometric chart, h_2 = 49 kJ / kg of dry air)

4. *By-pass factor of the cooling coil*

We know that by-pass factor of the cooling coil,

$$BPF = \frac{t_{d2} - ADP}{t_{d1} - ADP} = \frac{19 - 15}{38 - 15} = 0.174 \textbf{ Ans.}$$

5. *Refrigerating load on the cooling coil*

We know that the refrigerating load on the cooling coil

$$= m_F (h_1 - h_2) = 375(85 - 49) = 13\ 500 \text{ kJ/min}$$

$$= 13\ 500/210 = 64.3 \text{ TR} \textbf{ Ans.}$$

Example 18.17. *The room sensible and latent heat loads for an air conditioned space are 25 kW and 5 kW respectively. The room condition is 25°C dry bulb temperature and 50% relative humidity. The outdoor condition is 40°C dry bulb temperature and 50% relative humidity. The ventilation requirement is such that on mass flow rate basis 20% of fresh air is introduced and 80% of supply air is recirculated. The by-pass factor of the cooling coil is 0.15.*

Determine : 1. supply air flow rate ; 2. outside air sensible heat ; 3. outside air latent heat ; 4. grand total heat ; and 5. effective room sensible heat factor.

Solution. Given : RSH = 25 kW ; RLH = 5 kW ; t_{d2} = 25°C ; ϕ_2 = 50% ; t_{d1} = 40°C ; ϕ_1 = 50% ; BPF = 0.15

The flow diagram for the air conditioned space is shown in Fig. 18.29 and it is represented on the psychrometric chart as discussed below :

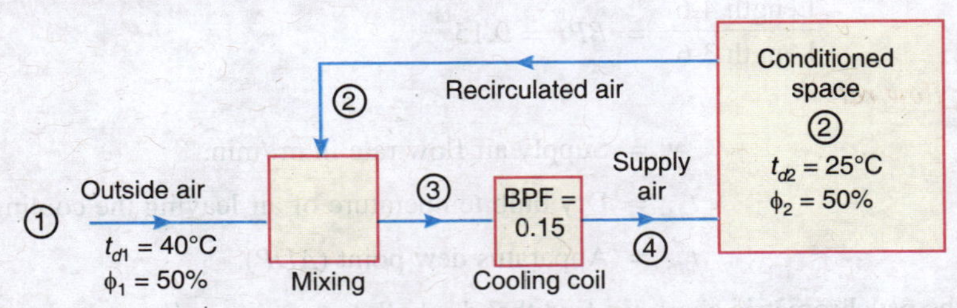

Fig. 18.29

First of all, mark the initial condition of air at 40°C dry bulb temperature and 50% relative humidity on the psychrometric chart as point 1, as shown in Fig. 18.30. Now mark the room condition of air at 25°C dry bulb temperature and 50% relative humidity as point 2. We know that room sensible heat factor,

$$RSHF = \frac{RSH}{RSH + RLH} = \frac{25}{25 + 5} = 0.833$$

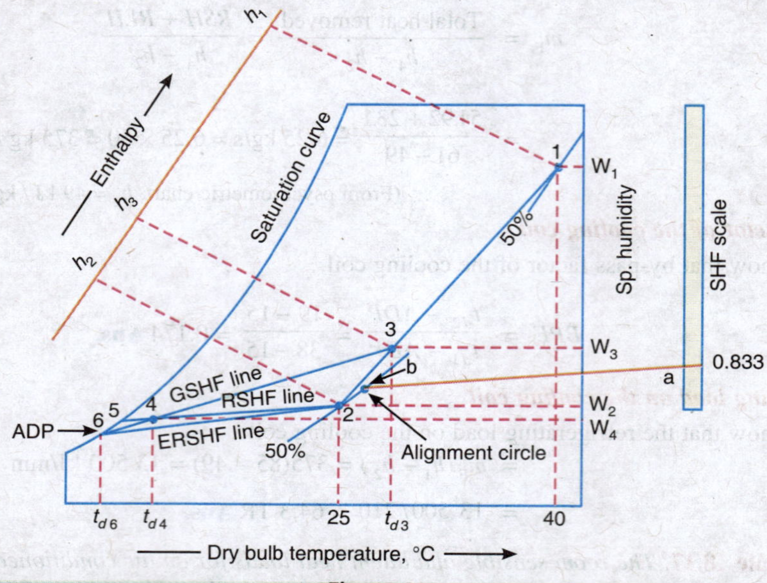

Fig. 18.30

Now mark this calculated value of $RSHF = 0.833$ on the sensible heat factor scale as point a and join with point b which is the alignment circle (*i.e.* 26°C dry bulb temperature and 50% relative humidity). From point 2, draw a line 2-5 parallel to the line ab. The line 2-5 is called *RSHF* line. Since 20% of fresh or outside air is mixed with 80% of supply air, therefore the condition of air entering the cooling coil after mixing process is marked on the line 1-2 by point 3, such that

$$\text{Length 2-3} = \text{Length 1-2} \times 0.2$$

Through point 3, draw a line 3-6 (known as *GSHF* line) intersecting the *RSHF* line at point 4 and the saturation curve at point 6, such that

$$\frac{\text{Length 4-6}}{\text{Length 3-6}} = BPF = 0.15$$

1. *Supply air flow rate*

Let v = Supply air flow rate in m³/min,

t_{d4} = Dry bulb temperature of air leaving the cooling coil, and

t_{d6} = Apparatus dew point (*ADP*).

From the psychrometric chart, we find that dry bulb temperature of air entering the cooling coil at point 3,

$$t_{d3} = 28° \text{ C}$$

We know that by-pass factor (*BPF*),

$$0.15 = \frac{t_{d4} - ADP}{t_{d3} - ADP} = \frac{t_{d4} - t_{d6}}{28 - t_{d6}}$$

By trial and error, we find that

$$t_{d4} = 13.72° \text{ C and } t_{d6} = 11.2° \text{ C}$$

We know that room sensible heat load,

$$RSH = 0.020\ 44\ v\ (t_{d2} - t_{d4})$$

$$25 = 0.020\ 44\ v\ (25 - 13.72) = 0.23\ v$$

$$\therefore \quad v = 25 / 0.23 = 108.7\ m^3/min$$ **Ans.**

2. *Outside air sensible heat*

Since the outside air is 20% of the supply air, therefore outside air flow rate,

$$v_0 = 0.2\ v = 0.2 \times 108.7 = 21.74\ m^3/min$$

We know that outside air sensible heat,

$$OASH = 0.020\ 44\ v_0\ (t_{d1} - t_{d2})$$

$$= 0.020\ 44 \times 21.74\ (40 - 25) = 6.66\ kW$$ **Ans.**

3. *Outside air latent heat*

From the psychrometric chart, we find that specific humidity of outside air at point 1,

$$W_1 = 0.0236\ kg / kg\ of\ dry\ air$$

and specific humidity of room air at point 2,

$$W_2 = 0.0098\ kg / kg\ of\ dry\ air$$

We know that outside air latent heat,

$$OALH = 50\ v_0\ (W_1 - W_2)$$

$$= 50 \times 21.74\ (0.0236 - 0.0098) = 15\ kW$$ **Ans.**

4. *Grand total heat*

We know that total sensible heat,

$$TSH = RSH + OASH = 25 + 6.66 = 31.66\ kW$$

and total latent heat,

$$TLH = RLH + OALH = 5 + 15 = 20\ kW$$

∴ Grand total heat,

$$GTH = TSH + TLH = 31.66 + 20 = 51.66\ kW$$ **Ans.**

Note : The total sensible heat (*TSH*) and total latent heat (*TLH*) may also be calculated as follows :

From psychrometric chart, we find that specific humidity at point 3,

$$W_3 = 0.0127\ kg / kg\ of\ dry\ air$$

and specific humidity at point 4,

$$W_4 = 0.009\ kg / kg\ of\ dry\ air$$

We know that total sensible heat,

$$TSH = 0.020\ 44\ v\ (t_{d3} - t_{d4})$$

$$= 0.020\ 44 \times 108.7\ (28 - 13.72) = 31.7\ kW$$

and total latent heat,

$$TLH = 50\ v\ (W_3 - W_4)$$

$$= 50 \times 108.7\ (0.0127 - 0.009) = 20.1\ kW$$

5. *Effective room sensible heat factor*

We know that effective room sensible heat,

$$ERSH = RSH + OASH \times BPF = 25 + 6.66 \times 0.15 = 26\ kW$$

and effective room latent heat,

$$ERLH = RLH + OALH \times BPF = 5 + 15 \times 0.15 = 7.25 \text{ kW}$$

∴ Effective room sensible heat factor,

$$ERSHF = \frac{ERSH}{ERSH + ERLH} = \frac{26}{26 + 7.25} = 0.782 \textbf{ Ans.}$$

The line 2-6 represents the effective room sensible heat factor (*ERSHF*) line.

Example 18.18. *A retail shop located in a city at 30°N latitude has the following loads :*

Room sensible heat = 58.15 kW

Room latent heat = 14.54 kW

The summer outside and inside design conditions are

Outside : 40°C DBT, 27°C WBT

Inside : 25°C DBT, 50% RH

70 m³/min of ventilation air is used. Determine the following, if the by-pass factor of the cooling coil is 0.15 :

1. Ventilation load ; 2. Grand total heat ; 3. Effective room sensible heat factor ; 4. Apparatus dew point ; 5. Dehumidified air quantity ; 6. Condition of air entering and leaving the apparatus.

Solution. Given : RSH = 58.15 kW ; RLH = 14.54 kW ; t_{d1} = 40°C ; t_{w1} = 27°C ; t_{d2} = 25°C ; ϕ_2 = 50% ; v_1 = 70 m³/min ; BPF = 0.15

The flow diagram is shown in Fig 18.31 and it is represented on the psychrometric chart as discussed below :

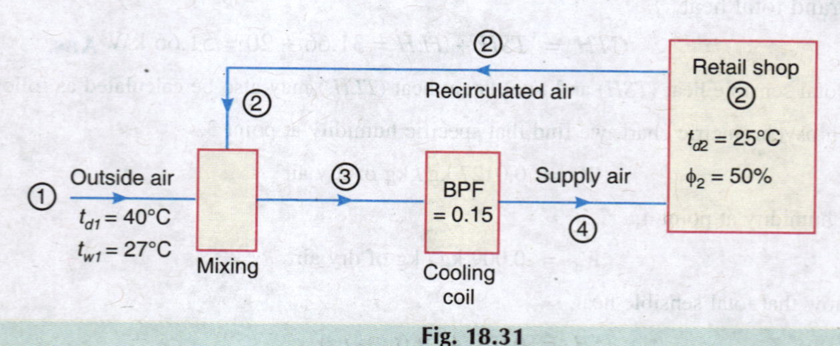

Fig. 18.31

First of all, mark the outside condition of air at 40°C dry bulb temperature and 27°C wet bulb temperature on the psychrometric chart as point 1, as shown in Fig. 18.32. Now mark the inside condition of air at 25°C dry bulb temperature and 50% relative humidity as point 2. From the psychrometric chart, we find that enthalpy of air at point 1,

$$h_1 = 85.2 \text{ kJ / kg of dry air}$$

and enthalpy of air at point 2,

$$h_2 = 50 \text{ kJ / kg of dry air}$$

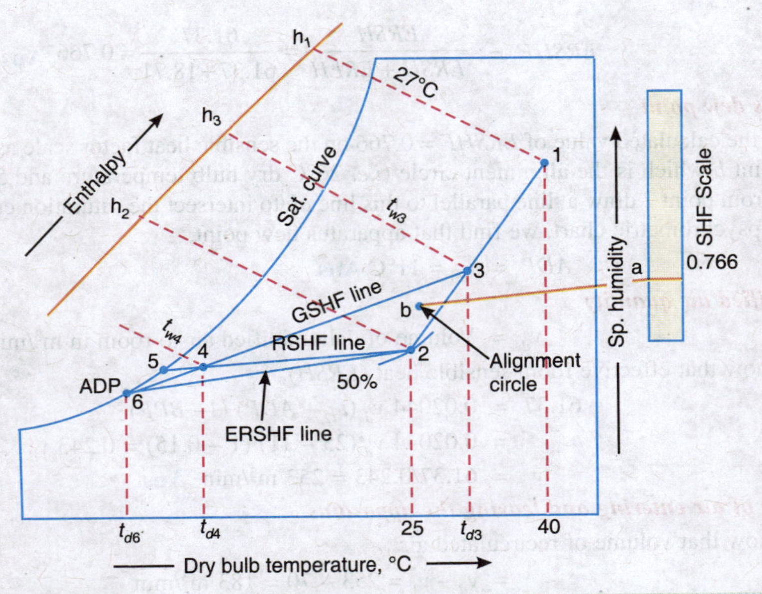

Fig. 18.32

1. *Ventilation load*

We know that outside air sensible heat,

$$OASH = 0.02044\ v_1(t_{d1} - t_{d2})$$
$$= 0.02044 \times 70\ (40 - 25) = 21.46 \text{ kW}$$

and outside air total heat,

$$OATH = 0.02\ v_1\ (h_1 - h_2) = 0.02 \times 70\ (85.2 - 50) = 49.28 \text{ kW}$$

$\therefore$ Ventilation load $= OATH = 49.28$ kW **Ans.**

2. *Grand total heat*

We know that outside air latent heat,

$$OALH = OATH - OASH = 49.28 - 21.46 = 27.82 \text{ kW}$$

Total sensible heat,

$$TSH = RSH + OASH = 58.15 + 21.46 = 79.61 \text{ kW}$$

Total latent heat,

$$TLH = RLH + OALH = 14.54 + 27.82 = 42.36 \text{ kW}$$

We know that grand total heat,

$$GTH = TSH + TLH = 79.61 + 42.36 = 121.97 \text{ kW}$$ **Ans.**

3. *Effective room sensible heat factor*

We know that effective room sensible heat,

$$ERSH = RSH + OASH \times BPF$$
$$= 58.15 + 21.46 \times 0.15 = 61.37 \text{ kW}$$

and effective room latent heat,

$$ERLH = RLH + OALH \times BPF$$
$$= 14.54 + 27.82 \times 0.15 = 18.71 \text{ kW}$$

∴ Effective room sensible heat factor,

$$ERSHF = \frac{ERSH}{ERSH + ERLH} = \frac{61.37}{61.37 + 18.71} = 0.766 \text{ Ans.}$$

4. Apparatus dew point

Mark the calculated value of $ERSHF = 0.766$ on the sensible heat factor scale as point a and join with point b which is the alignment circle (*i.e.* 26°C dry bulb temperature and 50% relative humidity). From point 2 draw a line parallel to this line ab to intersect the saturation curve at point 6. From the psychrometric chart, we find that apparatus dew point,

$$ADP = t_{d6} = 11°\text{C} \text{ Ans.}$$

5. Dehumidified air quantity

Let v_d = Volume of dehumidified air to room in m³/min.

We know that effective room sensible heat ($ERSH$),

$$61.37 = 0.02044\, v_d\, (t_{d2} - ADP)\, (1 - BPF)$$

$$= 0.02044\, v_d\, (25 - 11)\, (1 - 0.15) = 0.243\, v_d$$

$$\therefore \quad v_d = 61.37/0.243 = 253 \text{ m}^3/\text{min} \text{ Ans.}$$

6. Condition of air entering and leaving the apparatus

We know that volume of recirculated air

$$= v_d - v_1 = 253 - 70 = 183 \text{ m}^3/\text{min}$$

Thus 183 m³/min of recirculated air is mixed with 70 m³/min of ventilation air. The mixing condition is shown at point 3, such that

$$\text{Length 2-3} = \text{Length 1-2} \times \frac{70}{253}$$

From the psychrometric chart, we find that dry bulb temperature and wet bulb temperature of air entering the apparatus at point 3,

$$*t_{d3} = 29°\text{C, and } t_{w3} = 20.7\ °\text{C} \text{ Ans.}$$

Through point 3, draw a line 3-6 (known as GSHF line) and mark point 4 on this line such that

$$\frac{\text{Length 4-6}}{\text{Length 3-6}} = BPF = 0.15$$

From the psychrometric chart, we find that dry bulb temperature and wet bulb temperature of air leaving the apparatus at point 4,

$$t_{d4} = 13.7°\text{ C, and } t_{w4} = 12.7\ °\text{C} \text{ Ans.}$$

Notes : 1. The dry bulb temperature at point 4 may also be obtained as follows :

$$RSH = 0.02044\, v_d\, (t_{d2} - t_{d4})$$

$$58.15 = 0.02044 \times 253\, (25 - t_{d4})$$

$$\therefore \quad t_{d4} = 13.76°\text{ C}$$

2. The point 4, as shown in Fig. 18.32, is the intersection of $GSHF$ line 3-6 and RSHF line 2-5. This point gives the leaving condition of air from the cooling coil or conditioning apparatus and it is the ideal condition for supply air to the room.

* The temperature of air entering the coil at point 3 may also be obtained as follows :

$$t_{d3} = \frac{183 \times 25 + 70 \times 40}{253} = 29.15°\text{C}$$

Example 18.19. *The following data refer to summer air conditioning of a building :*

Outside design conditions	*= 43°C DBT, 27°C WBT*
Inside design conditions	*= 25°C DBT, 50% RH*
Room sensible heat gain	*= 84 000 kJ / h*
Room latent heat gain	*= 21 000 kJ / h*
By-pass factor of the cooling coil used	*= 0.2*

The return air from the room is mixed with the outside air before entry to cooling coil in the ratio of 4 : 1 by mass. Determine : (a) Apparatus dew point of the cooling coil ; (b) Entry and exit conditions of air for cooling coil ; (c) Fresh air mass flow rate; and (d) Refrigeration load on the cooling coil.

Solution. Given : t_{d1} = 43°C ; t_{w1} = 27°C ; t_{d2} = 25°C ; ϕ_2 = 50% ; $R\dot{S}H$ = 84 000 kJ / h ; RLH = 21 000 kJ / h ; BPF = 0.2

The flow diagram for the conditioned space is shown in Fig. 18.33 and it is represented on the psychrometric chart as discussed below :

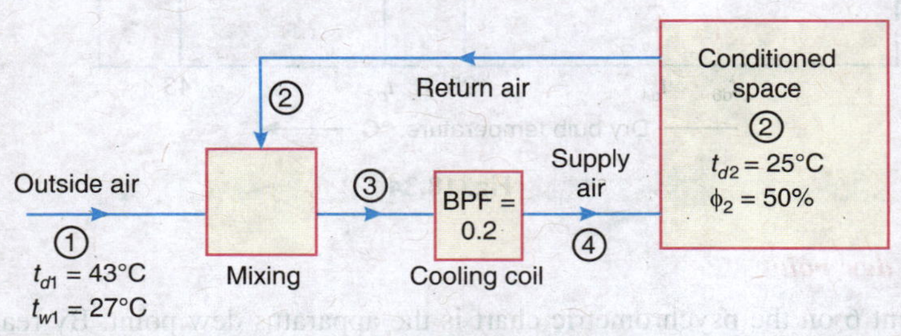

Fig. 18.33

First of all, mark the outside condition of air at 43°C dry bulb temperature and 27°C wet bulb temperature on the psychrometric chart as point 1, as shown in Fig. 18.34. Now mark the inside conditions of air at 25°C dry bulb temperature and 50% relative humidity as point 2. We know that room sensible heat factor,

$$RSHF = \frac{RSH}{RSH + RLH} = \frac{84\,000}{84\,000 + 21\,000} = 0.8$$

Now mark this calculated value of *RSHF* on the sensible heat factor scale and join with the alignment circle (*i.e.* 26°C *DBT* and 50% *RH*). From point 2, draw a line 2-5 parallel to this line. This line 2-5 is called *RSHF* line. Since the return air from the conditioned space is mixed with outside air before entry to the cooling coil in the ratio of 4 : 1, therefore the condition of air entering the cooling coil after mixing process is marked on the line 1-2 by point 3, such that

$$\text{Length 2-3} = \frac{\text{Length 1-2}}{5}$$

Through point 3, draw a line 3-6 (known as *GSHF* line) intersecting the *RSHF* line at point 4 and the saturation curve at point 6, such that

$$\frac{\text{Length 4-6}}{\text{Length 3-6}} = BPF = 0.2$$

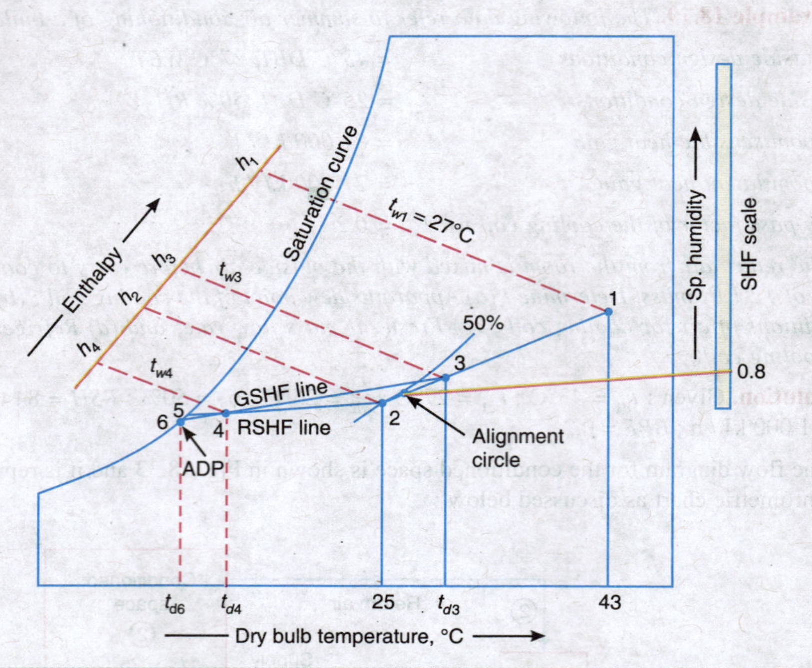

Fig. 18.34

(a) Apparatus dew point

The point 6 on the psychrometric chart is the apparatus dew point. By reading the value from the chart, we find that

Apparatus dew point (*ADP*)

$$= t_{d6} = 11°\text{C} \textbf{ Ans.}$$

Note : The apparatus dew point (*ADP* or t_{d6}) may be obtained by using any one of the following relations :

$$BPF = \frac{t_{d4} - t_{d6}}{t_{d3} - t_{d6}} = \frac{h_4 - h_6}{h_3 - h_6} = \frac{W_4 - W_6}{W_3 - W_6}$$

The value of t_{d3} as read from the psychrometric chart is 28.8° C. Using first relation, we have

$$0.2 = \frac{t_{d4} - t_{d6}}{28.8 - t_{d6}}$$

By trial and error, we find that

$$t_{d4} = 14.56° \text{ C and } t_{d6} = 11° \text{ C}$$

(b) Entry and exit conditions of air for cooling coil

The point 3 and point 4 represent the entry and exit condition of air for cooling coil as shown in Fig. 18.34. From the psychrometric chart, we find that

Dry bulb temperature of entering air,

$$t_{d3} = 28.8° \text{ C} \textbf{ Ans.}$$

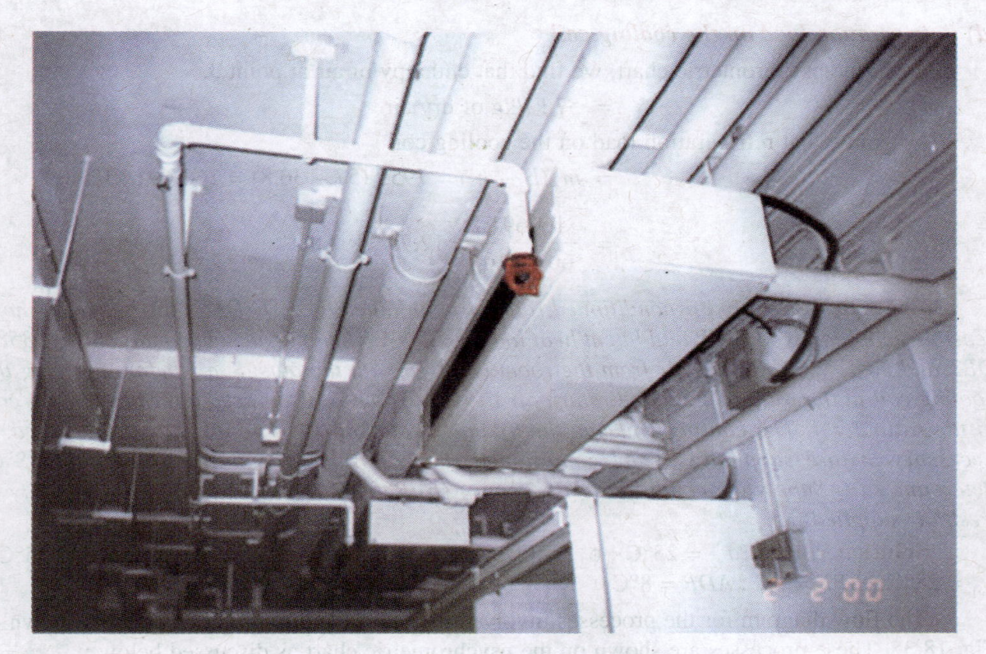

Central chilled water air-conditioning system.

Wet bulb temperature of entering air,

$$t_{w3} = 19.9° \text{ C } \textbf{Ans.}$$

Dry bulb temperature of exit air,

$$t_{d4} = 14.5° \text{ C } \textbf{Ans.}$$

and wet bulb temperature of exit air,

$$t_{w4} = 13° \text{ C } \textbf{Ans.}$$

(c) Fresh air mass flow rate

From the psychrometric chart, we find that enthalpy of air at point 2 ,

$$h_2 = 50 \text{ kJ/kg of dry air}$$

and enthalpy of air at point 4,

$$h_4 = 36.8 \text{ kJ/kg of dry air}$$

We know that mass of dehumidified air or the total mass of air flowing,

$$m_a = \frac{\text{Room total heat}}{h_2 - h_4} = \frac{RSH + RLH}{h_2 - h_4}$$

$$= \frac{84\,000 + 21\,000}{50 - 36.8} = 7955 \text{ kg / h}$$

Since this mass of air contains return air and fresh air in the ratio 4 : 1, therefore fresh air mass flow rate,

$$m_F = 7955 \times \frac{1}{5} = 1591 \text{ kg / h } \textbf{Ans.}$$

(d) Refrigeration load on the cooling coil

From the psychrometric chart, we find that enthalpy of air at point 3,

$$h_3 = 57 \text{ kJ/kg of dry air}$$

We know that refrigeration load on the cooling coil

$$= m_a (h_3 - h_4) = 7955\ (57 - 36.8) = 160.691 \text{ kJ/h}$$

$$= \frac{160691}{60 \times 210} = 12.75 \text{ TR } \textbf{Ans.}$$

Example 18.20. *An air conditioned room is maintained at 25°C DBT and 50% RH whose sensible heat load is 11.5 kW and latent heat load is 7.5 kW when the outside conditions are 35°C DBT and 28°C WBT. Return air from the room is mixed with the outside air before entering the cooling coil in the ratio of 4 : 1 and return air from the room is also mixed after the cooling coil in the ratio 1 : 4. The cooling coil has the by-pass factor of 0.1. The air may be reheated, if necessary, before supplying to the conditioned room. Assuming apparatus dew point 8°C, determine : (a) Supply air condition to the room ; (b) Refrigeration load ; and (c) Quantity of fresh air supplied.*

Solution. Given : t_{d2} = 25°C ; ϕ_2 = 50% ; RSH = 11.5 kW ; RLH = 7.5 kW ; t_{d1} = 35°C ; t_{w1} = 28°C ; BPF = 0.1 ; ADP = 8°C

The flow diagram for the processes involved in the air conditioning of a room is shown in Fig. 18.35. These processes are shown on the psychrometric chart as discussed below :

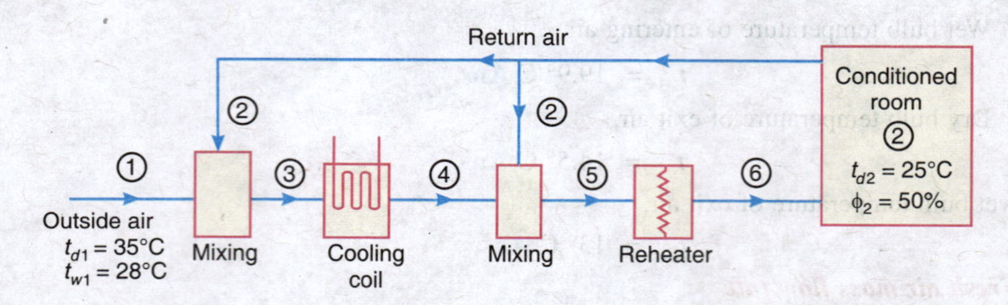

Fig. 18.35

First of all, mark the condition of outside air at 35°C dry bulb temperature and 28°C wet bulb temperature on the psychrometric chart as point 1, as shown in Fig. 18.36. Now mark the condition of air inside the room at 25°C dry bulb temperature and 50% relative humidity as point 2. Since the return air from the room and the outside air is mixed before entering the cooling coil in the ratio 4 : 1, therefore mark the condition of air after mixing and before entering to the cooling coil (*i.e.* point 3) on the line 1-2, such that

$$\text{Length 2-3} = \frac{\text{Length 2-1}}{5}$$

From the psychrometric chart, we find that dry bulb temperature at point 3,

$$t_{d3} = 27°\text{C}$$

and enthalpy at point 3, $\quad h_3 = 58.4 \text{ kJ/kg of dry air}$

Let t_{d4} be the dry bulb temperature of air leaving the coil. We know that the by-pass factor of the cooling coil (BPF),

$$0.1 = \frac{t_{d4} - ADP}{t_{d3} - ADP} = \frac{t_{d4} - 8}{27 - 8}$$

$$\therefore \quad t_{d4} = 0.1\ (27 - 8) + 8 = 9.9°C$$

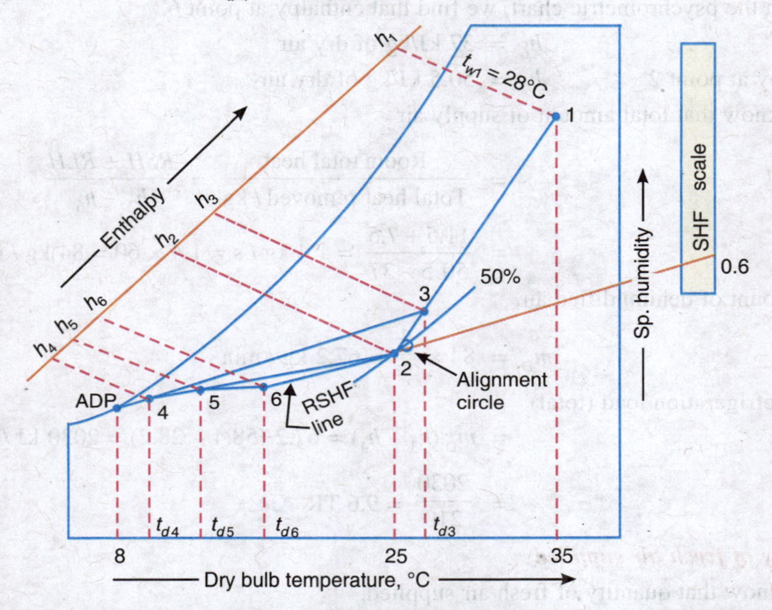

Fig. 18.36

Mark the point 4 (*i.e.* condition of air leaving the cooling coil) on the line joining point 3 and ADP = 8°C on the saturation curve, such that its dry bulb temperature is equal to 9.9°C. From the psychrometric chart, we find that enthalpy at point 4,

$$h_4 = 28.2 \text{ kJ/kg of dry air}$$

Since the return air from the room is also mixed with the air leaving the cooling coil in the ratio 1 : 4, therefore mark the condition of air after mixing (*i.e.* point 5) on the line 4-2, such that

$$\text{Length 4-5} = \frac{\text{Length 4-2}}{5}$$

From the psychrometric chart, we find that dry bulb temperature at point 5,

$$t_{d5} = 13°C$$

and enthalpy at point 5, $\quad h_5 = 33 \text{ kJ/kg of dry air}$

We know that room sensible heat factor,

$$RSHF = \frac{RSH}{RSH + RLH} = \frac{11.5}{11.5 + 7.5} = 0.6$$

It may be noted that the reheating of air is necessary to take care of the room sensible heat factor which is equal to 0.6. Now draw a horizontal line from point 5 to intersect the $RSHF$ line at point 6, which represents the condition of supply air to the room.

(a) Supply air condition to the room

From the psychrometric chart, we find that

Dry bulb temperature at point 6,

$$t_{d6} = 17° \text{ C } \textbf{Ans.}$$

Wet bulb temperature at point 6,

$$t_{w6} = 13.2° \text{ C } \textbf{Ans.}$$

(b) Refrigeration load

From the psychrometric chart, we find that enthalpy at point 6,

$$h_6 = 37 \text{ kJ/kg of dry air}$$

and enthalpy at point 2, $h_2 = 50.5$ kJ/kg of dry air

We know that total amount of supply air

$$= \frac{\text{Room total heat}}{\text{Total heat removed / kg}} = \frac{RSH + RLH}{h_2 - h_6}$$

$$= \frac{11.5 + 7.5}{50.5 - 37} = 1.4 \text{ kg / s} = 1.4 \times 60 = 84 \text{ kg / min}$$

and the amount of dehumidified air,

$$m_D = 84 \times \frac{4}{5} = 67.2 \text{ kg / min}$$

∴ Refrigeration load (total)

$$= m_D (h_3 - h_4) = 67.2 \ (58.4 - 28.2) = 2030 \text{ kJ / min}$$

$$= \frac{2030}{210} = 9.6 \text{ TR } \textbf{Ans.}$$

(c) Quantity of fresh air supplied

We know that quantity of fresh air supplied,

$$m_F = \frac{m_D}{5} = \frac{67.2}{5} = 13.44 \text{ kg / min } \textbf{Ans.}$$

EXERCISES

1. An air conditioning is handling 30 m^3 of air per minute at 32°C dry bulb temperature and 22°C wet bulb temperature. If the final conditions of air are 22°C dry bulb temperature and 50% relative humidity, find the heating capacity of the dehumidifier. **[Ans.** 7.3 kg / h]

2. It is desired to maintain a hall at 23°C dry bulb temperature and 60% relative humidity. The outdoor conditions of air are 42°C dry bulb temperature and 21% relative humidity. Suggest an air conditioning system and explain its working. Represent the process on the psychrometric chart and express equation for capacity of equipment used. **[Ans.** 18.8 *m* kJ / min]

3. A restaurant with a capacity of 100 persons is to be air conditioned with the following conditions :

Outside conditions : 30°C *DBT* and 70% *RH* ;

Desired inside conditions : 23°C *DBT* and 55% *RH* ; Quantity of air supplied : 0.5 m^3/min/person

The desired conditions are achieved by cooling, dehumidifying and then heating. Determine :

1. Capacity of cooling coil in tonnes of refrigeration ;
2. Capacity of heating coil ;
3. Amount of water removed by dehumidifier ; and
4. By-pass factor of heating coil if its surface temperature is 35° C.

[Ans. 10.95TR ; 9.23 kW ; 31.5 kg/h ; 0.558]

4. The following data refer to an air conditioning system for industrial process for hot and wet summer conditions :

Outdoor conditions	= 30°C *DBT* and 75% *RH*
Required conditions	= 22°C *DBT* and 70% *RH*
Amount of out door air supplied	= 200 m^3/min
Coil dew point temperature	= 14°C

If the required condition is achieved by first cooling and dehumidifying and then by heating, find :

1. The capacity of the cooling coil and its by-pass factor.
2. The capacity of the heating coil and surface temperature of the heating coil if the by-pass factor is 0.2. **[Ans.** 37.3 kW, 0.188 ; 4.5 kW, 23.25°C **]**

5. A conference room of 60 seating capacity is to be air conditioned for comfort conditions of 22°C *DBT* and 55% *RH*. The outdoor conditions are 32° C *DBT* and 22° C *WBT*. The quantity of air supplied is 0.5 m^3/min/person. The comfort conditions are achieved by chemical dehumidification and by cooling coil. Determine : 1. *DBT* of air at exit of dehumidifier; 2. Capacity of dehumidifier; 3. Capacity and surface temperature of cooling coil if *BPF* is 0.30.

[Ans. 41°C ; 6.95 kg/h ; 3.1TR ; 14.63°C**]**

6. A cinema hall of seating capacity 1500 persons has been provided with an air conditioned plant with the following data :

Outdoor conditions	= 40°C *DBT* and 20°C *WBT*
Required indoor conditions	= 20°C *DBT* and 60% *RH*
Amount of outdoor air supplied	= 0.3 m^3/min/person

If the required condition is achieved first by adiabatic humidifying and then by cooling, find : 1. the capacity of cooling coil and surface temperature of the coil if by-pass factor is 0.25 ; and 2. the capacity of the humidifier and its efficiency. **[Ans.** 118.8 kW ; 15.27°C ; 72 kg / h, 29% **]**

7. The following data refer to an air-conditioning system of a cinema hall for winter conditions :

Outdoor conditions	= 10°C *DBT*, 60% *RH*
Required comfort conditions	= 22°C *DBT*, 60% *RH*
Seating capacity	= 2000
Amount of outdoor air supplied	= 0.25 m^3/min/person

The required condition is achieved by heating, humidifying and then again by heating. The air coming out of the humidifier is having 75% relative humidity. Find : (*a*) the heating capacity of the first heater in kW and the surface temperature of the coil if its by-pass factor is 0.3; (*b*) the capacity of the humidifier in kg / h; and (*c*) the heating capacity of the second heater and its by-pass factor, if the surface temperature of the coil is 25°C. **[Ans.** 220 kW , 40°C ; 200 kg / h ; 47.3 kW , 0.44 **]**

8. The following data refer for a space to be air conditioned :

Inside design conditions	= 25°C *DBT*, 50% *RH*
Outdoor air conditions	= 43°C *DBT*, 27.5°C *WBT*
Room sensible heat gain	= 20 kW
Room latent heat gain	= 5 kW
By-pass factor of the cooling coil	= 0.1

The return air from the space is mixed with the outside air before entering the cooling coil in the ratio of 4 : 1 by mass. Determine (*a*) apparatus dew point ; (*b*) condition of air entering and leaving the cooling coil ; (*c*) dehumidified air quantity ; (*d*) fresh air mass flow and volume flow rate; and (*e*) total refrigeration load on the air-conditioning plant.

[Ans. 11.8°C ; 28.6°C , 13.5°C ; 6265 kg / h ; 1253 kg / h , 19.2 m^3/min ; 40.7 kW **]**

9. The following data refer to summer air conditioning of a restaurant :

Inside design conditions	= 27°C *DBT* and 21°C *WBT*
Outside design conditions	= 38°C *DBT* and 27°C *WBT*
Sensible heat load	= 126 000 kJ / h
Latent heat load	= 50 400 kJ /h

The outside air is supplied at the rate of 20 m^3 /min directly into the room through ventilators and by infiltration. The outside air to be conditioned is passed through a cooling coil which has an apparatus dew point of 12°C and 60% of the total air is recirculated from the conditioned space and mixed with conditioned air after the cooling coil. Find : (*a*) condition of air after the cooling coil before mixing with recirculated air ; (*b*) condition of air entering the restaurant ; (*c*) mass of fresh air entering the cooling coil ; (*d*) by-pass factor of the cooling coil ; and (*e*) total refrigeration load of the cooling coil.

[Ans. 18.5°C , 16.5°C ; 22.5°C, 8.5°C ; 267 kg/min ; 0.25 ; 47.5 TR **]**

10. The following data apply to an air-conditioning system :

Room sensible heat	= 5.8 kW
Room latent heat	= 5.8 kW
Inside design conditions	= 25°C *DBT* and 50% *RH*
Outside design conditions	= 35°C *DBT* and 28°C *WBT*

The room air is mixed with the outside air before entering the cooling coil in the ratio of 4 : 1. The coil by-pass factor is 0.1 and apparatus dew point is 10°C. The room air is again mixed with the air leaving the cooling coil in the ratio of 1 : 4 and the mixture is then allowed to enter the reheater before being supplied into the room. Determine : (*a*) supply air condition to the room ; (*b*) reheater capacity; (*c*) refrigeration capacity of the cooling coil ; and (*d*) quantity of fresh air (outdoor) supplied.

[Ans. 21.5°C ; 15.3°C , 11.6 kW ; 33.8 kW ; 4695 kg / h **]**

11. The outdoor summer design condition for a bank for 100 persons at a place is 35°C dry bulb temperature and 24°C wet bulb temperature. The required inside conditions are 24°C dry bulb temperature and 50% relative humidity. The room sensible heat is 58 kW and the room latent heat is 15 kW. The ventilation requirement per person is 0.54 m^3/min. The by-pass factor is 0.15.

Determine : 1. Grand total heat ; 2. Effective sensible heat factor ; 3. Apparatus dew point ; and 4. Volume flow rate of dehumidified air. **[Ans.** 99.18 kW ; 0.778 ; 10°C ; 246.2 m^3 / min **]**

12. The following data are given for the space to be air conditioned :

Outside air conditions	= 43°C *DBT* and 27°C *WBT*
Inside design conditions	= 25°C *DBT* and 50% *RH*
Room sensible heat load	= 40 kW
Room latent heat load	= 10 kW
By-pass factor of the cooling coil	= 0.2

The return air from the room is mixed with the outside air before entry to the cooling coil in the ratio of 3 : 1 by mass.

Determine : 1. supply air flow rate ; 2. outside air sensible heat ; 3. outside air latent heat ; 4. grand total heat ; and 5. effective room sensible heat factor.

[Ans. 191 m^3/min ; 17.57 kW ; 14.75 kW ; 82.32 kW ; 0.771 **]**

QUESTIONS

1. Write a short note on the factors affecting comfort air conditioning.

2. Explain the difference between winter air conditioning and summer air conditioning.

3. Draw a neat diagram of air-conditioning system required for winter season. Explain the working of different components in the circuit.

4. Draw a neat labelled diagram of a year-round air conditioning system.

5. Describe unitary and central air conditioning system.
6. Define room sensible heat factor. How room sensible heat factor line is drawn on the psychrometric chart ?
7. Explain the procedure to draw a grand sensible heat factor line on a psychrometric chart.
8. What do you understand by effective room sensible heat factor ?

OBJECTIVE TYPE QUESTIONS

1. In summer air conditioning, the air is
   (*a*) cooled and humidified (*b*) cooled and dehumidified
   (*c*) heated and humidified (*d*) heated and dehumidified
2. In winter air conditioning, the air is
   (*a*) cooled and humidified (*b*) cooled and dehumidified
   (*c*) heated and humidified (*d*) heated and dehumidified
3. For summer air conditioning, the relative humidity should not be less than
   (*a*) 40% (*b*) 60% (*c*) 75% (*d*) 90%
4. For winter air conditioning, the relative humidity should not be more than
   (*a*) 40% (*b*) 60% (*c*) 75% (*d*) 90%
5. The sensible heat factor for auditorium or cinema hall is generally kept as
   (*a*) 0.6 (*b*) 0.7 (*c*) 0.8 (*d*) 0.9
6. The conditioned air supplied to the room must have the capacity to take up
   (*a*) room sensible heat load only (*b*) room latent heat load only
   (*c*) both room sensible heat and latent heat loads
7. The alignment circle is marked on the psychrometric chart at
   (*a*) 20°C *DBT* and 50% *RH* (*b*) 26°C *DBT* and 50% *RH*
   (*c*) 20°C *DBT* and 60% *RH* (*d*) 26°C *DBT* and 60% *RH*
8. The supply air state of cooling coil with a by-pass factor (*BPF*) lies at
   (*a*) intersection of *RSHF* line with saturation curve
   (*b*) intersection of *GSHF* line with saturation curve
   (*c*) point dividing *RSHF* line in proportions of *BPF* and (1-*BPF*)
   (*d*) intersection of *RSHF* line and *GSHF* line
9. The effective room sensible heat factor (*ERSHF*) is given by

   (*a*) $\dfrac{RSH}{RLH}$ (*b*) $\dfrac{RSH + OASH}{RLH + OALH}$

   (*c*) $\dfrac{RSH + OASH \times BPF}{(RSH + RLH) + (OASH + OALH)\, BPF}$ (*d*) none of these

   where RSH = Room sensible heat,
   RLH = Room latent heat,
   $OASH$ = Outside air sensible heat,
   $OALH$ = Outside air latent heat,
   BPF = By-pass factor.

10. In Fig. 18.37, line 2-6 is the effective room sensible heat factor (*ERSHF*) line, the line 2-5 is the room sensible heat factor (*RSHF*) line and the line 3-6 is the grand sensible heat factor (*GSHF*) line. Which of the following statements is correct ?

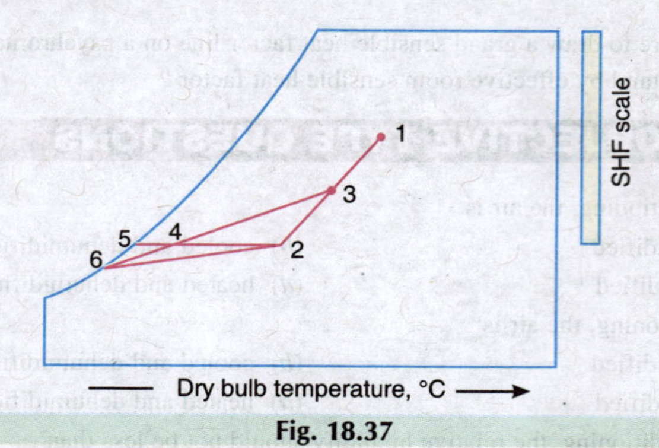

Fig. 18.37

(*a*) *GSHF* is greater than *RSHF* and *ERSHF* both

(*b*) *GSHF* is less than *RSHF* and *ERSHF* both

(*c*) *GSHF* is greater than *RSHF* but less than *ERSHF*

(*d*) *GSHF* is less than *RSHF* but greater than *ERSHF*

ANSWERS

1. (*b*)	**2.** (*c*)	**3.** (*b*)	**4.** (*a*)	**5.** (*b*)
6. (*c*)	**7.** (*b*)	**8.** (*d*)	**9.** (*c*)	**10.** (*d*)

CHAPTER

19 Cooling Load Estimation

19.1 Introduction

The total heat required to be removed from the space in order to bring it at the desired temperature by the air conditioning and refrigeration equipment is known as *cooling load.*

The purpose of a load estimation is to determine the size of the air conditioning and refrigeration equipment that is required to maintain inside design conditions during periods of maximum outside temperatures. The design load is based on inside and outside design conditions and it is air conditioning and refrigeration equipment capacity to produce and maintain satisfactory inside conditions.

19.2 Components of a Cooling Load

The two main components of a cooling load imposed on an air conditioning plant operating during hot weather are as follows :

1. *Sensible heat gain.* When there is a direct addition of heat to the enclosed space, a gain in the sensible heat is said to occur. This sensible heat is to be removed during the process of summer air conditioning. The sensible heat gain may occur due to any one or all of the following sources of heat transfer :

(*a*) The heat flowing into the building by conduction through exterior walls, floors, ceilings, doors and windows due to the temperature difference on their two sides.

(*b*) The heat received from solar radiation. It consists of

(*i*) The heat transmitted directly through glass of windows, ventilators or doors, and

(*ii*) The heat absorbed by walls and roofs exposed to solar radiation and later on transferred to the room by conduction.

Carrier infinity system.

(*c*) The heat conducted through interior partition from rooms in the same building which are not conditioned.

(*d*) The heat given off by lights, motors, machinery, cooking operations, industrial processes etc.

(*e*) The heat liberated by the occupants.

(*f*) The heat carried by the outside air which leaks in (infiltrating air) through the cracks in doors, windows and through their frequent openings.

(*g*) The heat gain through the walls of ducts carrying conditioned air through unconditioned space in the building.

(*h*) The heat gain from the fan work.

2. *Latent heat gain.* When there is an addition of water vapour to the air of enclosed space, a gain in latent heat is said to occur. This latent heat is to be removed during the process of summer air-conditioning. The latent heat gain may occur due to any one or all of the following sources :

(*a*) The heat gain due to moisture in the outside air entering by infiltration.

(*b*) The heat gain due to condensation of moisture from occupants.

(*c*) The heat gain due to condensation of moisture from any process such as cooking foods which takes place within the conditioned space.

(*d*) The heat gain due to moisture passing directly into the conditioned space through permeable walls or partitions from the outside or from adjoining regions where the water vapour pressure is higher.

The total heat load to be removed by the air-conditioning and refrigeration equipment is the sum of sensible and latent heat loads as discussed above.

Note : When the outside air is introduced for ventilation purposes, there is a sensible heat gain as well as latent heat gain. The sensible heat gain is due to the temperature difference between the fresh air and the air in space whereas the latent heat gain is due to the difference of humidity.

19.3 Sensible Heat Gain through Building Structure by Conduction

The heat gain through a building structure such as walls, floors, ceilings, doors and windows constitutes the major portion of sensible heat load.

Consider a building wall composed of a single homogeneous material as shown in Fig. 19.1. A little consideration will show that the heat passing through a wall is first received at the wall surface exposed to the region of higher air temperature by radiation, convection and conduction. It then flows through the material of the wall to the surface exposed at the region of lower air temperature. Now the heat is dispersed through the processes of radiation, convection and conduction. Thus, the heat transferred or gained through a wall under steady state condition is

$$Q = f_o (t_o - t_1) A + \frac{k}{x}(t_1 - t_2) A + f_i (t_2 - t_i) A$$

$$= UA\,(t_o - t_i)$$

where f_o = Outside film or surface conductance,

f_i = Inside film or surface conductance,

A = Outside area of wall,

t_o = Outside air temperature,

t_i = Inside air temperature,

x = Thickness of wall,

k = Thermal conductivity for the material of the wall, and

U = Overall coefficient of heat transmission of the wall.

$$= \frac{1}{\dfrac{1}{f_o} + \dfrac{x}{k} + \dfrac{1}{f_i}}$$

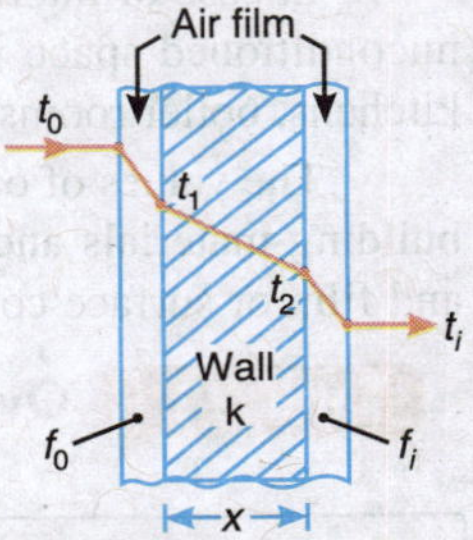

Fig. 19.1. Heat transfer through a single wall.

When a wall is made up of layers of different materials as shown in Fig. 19.2, then the overall coefficient of heat transmission is given by

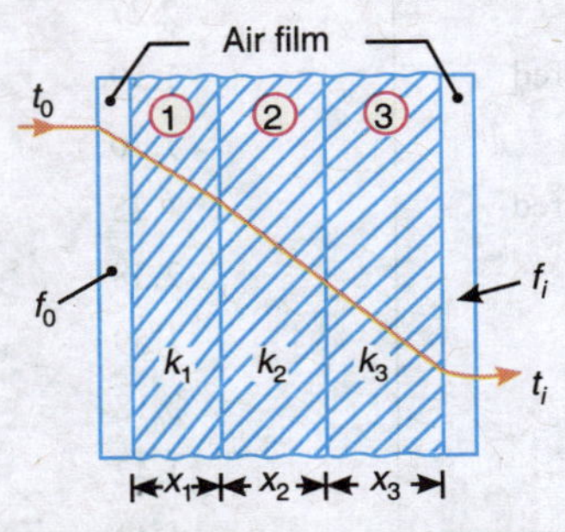

Fig. 19.2. Heat transfer through a composite wall.

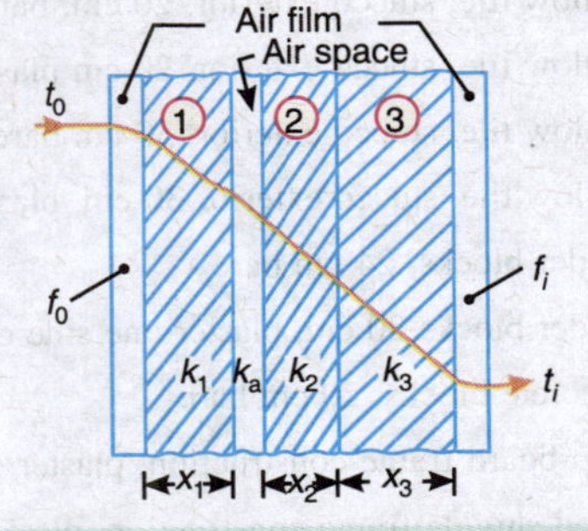

Fig. 19.3. Heat transfer through a composite wall with air space.

$$U = \frac{1}{\frac{1}{f_o} + \frac{x_1}{k_1} + \frac{x_2}{k_2} + \frac{x_3}{k_3} + \ldots + \frac{1}{f_i}}$$

These types of walls are made for a cold storage with a maximum low temperature insulation of 10 cm.

When air space is provided between the materials as shown in Fig. 19.3, then the overall coefficient of heat transmission is given by

$$U = \frac{1}{\frac{1}{f_o} + \frac{x_1}{k_1} + \frac{1}{k_a} + \frac{x_2}{k_2} + \frac{x_3}{k_3} + \ldots + \frac{1}{f_i}}$$

where k_a = Thermal conductance of air space.

In case of interior walls or partitions, floors, and ceilings, the temperature in the adjoining unconditioned space is usually 5 to 10 degree below the outdoor design temperatures. But for kitchens, boiler rooms, attices and show windows, it is often higher than the outdoor temperature.

The values of overall coefficient of heat transmission for structures, thermal conductivity of building materials and insulating materials, conductance of building materials and construction, and film or surface conductance for air film and air spaces, are given in the following tables.

Table 19.1. Overall coefficient of heat transmission (*U*) for structures with a wind velocity of 24 km/h outside.

Structure	*Overall coefficient of heat transmission (U) in W/m² K*
Brick wall, 20 cm, bare	2.84
Brick wall, 20 cm, plaster one side on brick	2.61
Brick wall, 20 cm, plaster one side on metal lath-furred	1.82
Brick wall, 40 cm, bare	1.59
Brick wall, 40 cm, plaster one side on brick	1.53
Brick wall, 40 cm, plaster one side on metal lath-furred	1.20
Hollow tile, stucco exterior, 20 cm, bare	2.27
Hollow tile, stucco exterior, 20 cm plaster on metal lath-furred	1.59
Hollow tile, stucco exterior, 30 cm bare	1.70
Hollow tile, stucco exterior 30 cm, plaster on metal lath-furred	1.25
Cinder blocks, 24 cm bare	2.38
Cinder blocks 20 cm, plaster one side on metal lath-furred	1.59
Concrete blocks, 20 cm bare	3.18
Clap-board frame construction, plaster on wood lath	1.42
Wood shingle frame construction, plaster on wood lath	1.42
Stucco frame construction, plaster on wood lath	1.70
Brick veneer frame construction, plaster on wood lath	1.53

Table 19.2. Thermal conductivity (*k*) of building materials.

Material	*Thermal conductivity (k) in W / m K*
Brick, low density	8.66
Brick, high density	15.93
Cement mortar	20.82
Cement plaster, typical	20.82
Concrete, typical	20.82
Concrete, typical fibre gypsum, 87.5% gypsum, 12.5% wood chips	1.61
Sand and gravel	21.86
Lime stone	18.72
Cinder	8.50
Cinder, boiler (12.5 mm to 19 mm)	2.13
Title or terazzo, typical	20.82
Asbestos building board	4.71
Gypsum, between layers of heavy paper	2.44
Gypsum, plaster, typical	5.72
Wood across grain, typical	1.73
Balsa	0.657
Maple, hard, 16% moisture	2.0
Pine, yellow short leaf, 16% moisture	1.45

Table 19.3. Thermal conductivity (*k*) of insulating materials.

Material	*Thermal conductivity (k) in W /m K*
Asbestos packed	2.812
Asbestos loose	1.852
Asbestos, paper, thin layers organic binder	0.85
Asbestos mill board	1.454
Asbestos wood	4.85
Asphalt roofing (felt)	1.212
Balsa	0.607
Cement wood (saw dust and Portland cement)	1.683
Charcoal (hard woods) coarse	0.617
Cork board, typical	0.468 to 0.486
Cork regranulated, coarse	0.536
Cork regranulated	0.468
Cotton	0.676
Diatomaceous earth	0.460
Eel grass	0.412
Glass wool, high grade	0.388 to 0.460
Glass wool, commercial grade	0.444
Hair felt, not compressed	0.444
Insulation boards, various fibres	0.553 to 0.657
Kapok, loosely packed	0.416
Planer shavings, various woods	0.692
Rock wool	0.468 to 0.486
Rubber, expanded	0.364
Saw dust, various woods	0.710

Table 19.4. Conductance of building materials and construction for thickness indicated.

Material	*Conductance in W/m² K*
Plaster board 9 mm thick	21.17
Plaster board 12.5 mm thick	16.14
Roofs :	
Asphalt, composition or prepared roofing	36.85
Shingles, asbestos	34.05
Shingles, asphalt	36.65
Shingles, slate	58.96
Metal lath and plaster, total thickness 12.5 mm	24.96
Wood lath and plaster, total thickness 18 mm	14.20
For shealthing 25 mm and building paper	4.88
For shealthing 25 mm building paper and yellow pine lap siding	2.83
For shealthing 25 mm building paper and stucco	4.65
Air space over 18 mm faced ordinary building materials	6.15
Typical hollow clay tile, 10 cm	5.67
Typical hollow clay tile, 15 cm	3.64
Typical hollow clay tile, 20 cm	3.40
Typical hollow clay tile, 25 cm	3.29
Typical hollow clay tile, 30 cm	2.277
Typical hollow clay tile, 40 cm	1.762
Concrete block, sand and gravel, typical, 20 cm	5.675
Concrete block, sand and gravel, typical, 30 cm	4.547
Concrete block, cinder aggregate, typical, 20 cm	3.404
Concrete block, cinder aggregate, typical, 30 cm	3.006
Surface coefficient, outside for 24 km. p.h. wind	34.05

Table 19.5. Film or surface conductance for air film and air spaces.

Material	*Surface position*	*Thermal conductance (f) in W/m²K*
Air film (Surface)		
Still air (f_i) heat flowing up	Horizontal	9.25
	Sloping, 45°	9.1
Still air, heat flowing down	Horizontal	6.13
	Sloping 45°	7.5
Still air, heat flowing horizontal	Vertical	8.3
Wind, 24 km / h (f_o)	Any position	34.0
Wind, 12 km / h (f_o)	Any position	22.7

Material	*Surface position*	*Thermal conductance (f) in W/m²K*
Air spaces bounded by structural material		
Heat flowing up	Horizontal, 19 to 100 mm	6.7
	Sloping 45°, 19 to 100 mm	6.3
Heat flowing horizontal	Vertical, 19 to 100 mm	5.85
Heat flowing down	Sloping 45°, 19 to 100 mm	5.5
	Horizontal, 19 mm	5.56
	Horizontal, 38 mm	4.94
	Horizontal, 100 mm	4.6

Note : The above values of thermal conductance are for a mean temperature of 10°C to 32°C.

19.4 Heat Gain from Solar Radiation

The solar radiation striking the outside surfaces of a building may contribute appreciably to the peak load on the air conditioning and refrigeration equipment and must, therefore, be considered. The amount of heat that flows towards the interior of a building due to solar radiation depends upon the following factors :

1. Altitude angle of the sun,
2. Clearness of the sky,
3. Position of the surface with respect to the direction of the sun's rays,
4. Absorptivity of the surface,
5. Ratio of the overall coefficient of heat transfer of the wall to the coefficient of heat transfer of the outside air film, and
6. Temperature of the ground and surrounding objects with which the heated surface may interchange radiant heat.

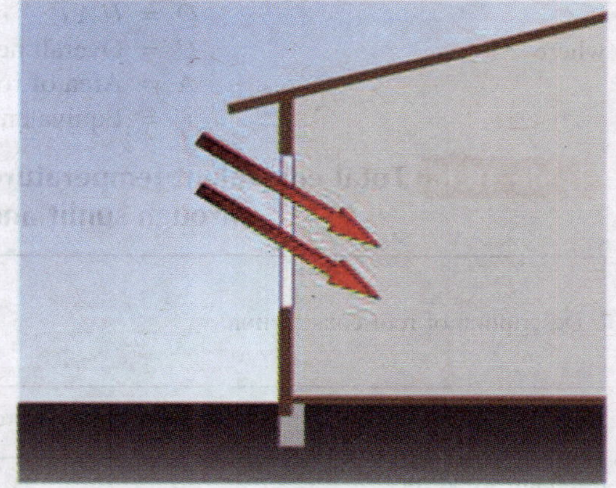

Solar heat gain coefficient is the fraction of incident solar radiation admitted through a window.

The altitude angle of the sun, in turn, depends upon the altitude of the locality, season of year and hour of day.

The heat from solar radiation is received by building surfaces in two forms *i.e.* *direct radiation* and *sky* or *diffuse radiation*. The direct radiation is the impingement of the sun's rays upon the surface. The sky or diffuse radiation is received from moisture and dust particles in atmosphere which absorbs part of the energy of the sun's rays thereby becoming heated to a temperature above that of the air. The sky radiation is received by surfaces which do not face the sun.

19.5 Solar Heat Gain (Sensible) through Outside Walls and Roofs

The transmission of heat through the walls exposed to the outdoors and roof is not steady (*i.e.* the flow of heat is periodic) due to variation in the outside air temperature and the solar radiation intensity over a period of 24 hours. A little consideration will show that the temperature of wall rises with the rise in outside air temperature and the heat is stored in the wall which has a considerable storage capacity. Thus the heat transfered to the room is reduced. The stored heat in the wall is given off to the room when the outside air temperature falls. Since the outside air temperature changes continuously over a cycle of 24 hours, therefore instantaneous heat gain from outside is not equal to the instantaneous heat gain inside the room, the difference being stored or rejected by the wall. The heat stored by the wall is given off later in the evening. Thus peak of incoming heat rate is delayed by the storage effect of the walls and it is also reduced. Fig. 19.4 shows the curves of instantaneous load coming from outside and the actual load felt inside. The area under the two curves is equal. The shaded area above the actual load shows the heat stored and below the actual load shows the heat released by the walls and other structures.

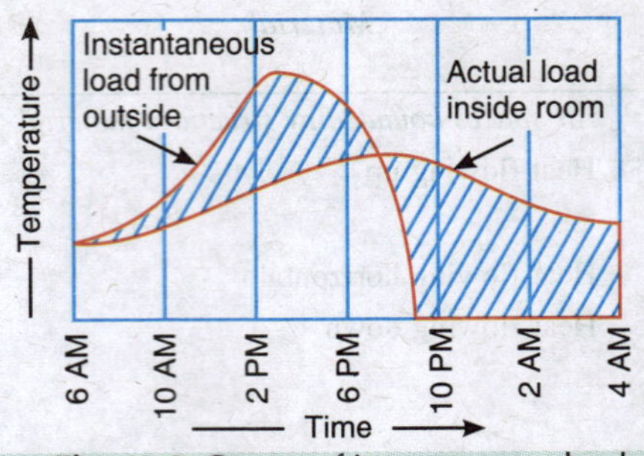

Fig. 19.4. Curves of instantaneous load.

A convenient method of taking into account this lagging effect of storage and the solar radiation is to use an *equivalent temperature differential*. The equivalent temperature differentials for roofs are given in Table 19.6, and those applying to walls is given in Table 19.7. The heat gain through outside walls and roofs is given by

$$Q = U\,A\,t_e$$

where
U = Overall heat transmission coefficient of roof or wall,
A = Area of roof or wall, and
t_e = Equivalent temperature differential.

Table 19.6. Total equivalent temperature differentials for calculating heat gain through sunlit and shaded roofs.

Description of roof construction[1]	Sun-time								
	A.M.			P.M.					
	8	10	12	2	4	6	8	10	12
Light construction roofs-exposed to sun									
2.5 cm wood[2] or 2.5 cm wood + 2.5 cm or 5 cm insulation or 5 cm wood	12	38	54	62	50	26	10	4	0
Medium construction roofs-exposed to sun									
5 cm concrete or 5 cm concrete + 2.5 cm or 5 cm insulation or 5 cm wood	6	30	48	58	50	32	14	6	2
5 cm gypsum or 5 cm gypsum + 2.5 cm insulation 2.5 cm wood[2] / 5 cm wood[2] or / 5 cm concrete or / 5 cm gypsum + 10 cm rock wool in furred ceiling	0	20	40	52	54	42	20	10	6

1. Includes 9 mm felt, roofing with or without slag. May also be used for single roof.
2. Nominal thickness of wood.

10 cm concrete or 10 cm concrete + 5 cm insulation	0	20	38	50	52	40	22	12	6

Heavy construction roofs-exposed to sun

15 cm concrete	4	6	24	38	46	44	32	18	12
15 cm concrete + 5 cm insulation	6	6	20	34	42	44	34	20	14

Roofs covered with water-exposed to sun

Light construction roof with 2.5 cm water	0	4	16	22	18	14	10	2	0
Heavy construction roof with 2.5 cm water	– 2	– 2	– 4	10	14	16	14	10	6
Any roof with 15 cm water	– 2	0	0	6	10	10	8	4	0

Roofs with roof sprays-exposed to sun

Light construction	0	4	12	18	16	14	10	2	0
Heavy construction	– 2	– 2	2	8	12	14	12	10	6

Roofs in shade

Light construction	– 4	0	5	12	14	12	8	2	0
Medium construction	– 4	– 2	2	8	12	12	10	6	2
Heavy construction	– 2	– 2	7	4	8	10	10	8	4

Table 19.7. Total equivalent temperature differentials for calculating heat gain through sun lit and shaded walls.

North Latitude Wall Facing	Sun-Time																		South Latitude Wall Facing
	A.M.						P.M.												
	8		10		12		2		4		6		8		10		12		
	Exterior colour of wall ; D-dark, L-light																		
	D	L	D	L	D	L	D	L	D	L	D	L	D	L	D	L	D	L	
Frame																			
NE	22	13	24	12	14	10	12	10	14	14	14	14	10	10	6	4	2	2	SE
E	30	14	36	18	32	16	12	12	14	14	14	14	10	10	6	6	2	2	E
SE	13	6	26	16	28	18	24	16	16	14	14	14	10	10	6	4	2	2	NE
S	– 4	– 4	4	6	22	12	30	20	26	20	16	14	10	10	6	6	2	2	N
SW	– 4	– 4	0	–2	6	4	26	22	40	28	42	28	24	20	6	4	2	2	NW
W	– 4	– 4	0	6	6	6	20	12	40	28	48	34	22	22	8	8	2	2	W
NW	– 4	– 4	0	–2	6	4	12	10	24	20	40	26	34	24	6	4	2	2	SW
N (shade)	– 4	– 4	– 2	–2	4	4	10	10	14	14	2	12	8	8	4	4	0	2	S (shade)

10 cm brick or stone veneer frame

NE	–2	– 4	24	12	20	10	10	6	12	10	14	14	12	12	10	10	5	4	SE
E	2	0	30	14	31	17	14	14	12	12	14	14	12	12	10	8	6	6	E
SE	2	–2	20	10	28	16	26	16	18	14	14	14	12	12	10	8	6	6	NE
S	– 4	– 4	–2	–2	12	6	24	16	26	18	20	16	12	12	8	8	4	4	N
SW	0	–2	0	–2	2	2	12	8	32	22	36	26	34	24	10	8	6	6	NW
W	0	–2	0	0	4	2	10	8	26	18	40	28	42	28	16	14	6	6	W
NW	– 4	– 4	–2	–2	2	2	8	6	12	12	30	22	34	24	12	10	6	6	SW
N (shade)	– 4	– 4	–2	–2	0	0	6	6	10	10	12	12	12	12	8	8	4	4	S (shade)

20 cm hollow tile or 20 cm cinder block

NE	0	0	0	0	20	10	16	10	10	6	12	10	14	12	12	10	8	8	SE
E	4	2	12	4	24	12	26	14	20	12	12	10	14	12	14	10	10	8	E
SE	2	0	2	0	16	8	20	12	20	14	14	12	14	12	12	10	8	6	NE
S	0	0	0	0	2	0	12	6	24	14	26	16	20	14	12	10	8	9	N
SW	2	0	2	0	2	0	6	4	12	10	26	18	30	20	26	18	8	6	NW
W	4	2	4	2	4	2	6	4	10	8	18	14	30	22	32	22	18	11	W
NW	0	0	0	0	2	0	4	2	8	6	12	10	22	18	30	22	10	14	SW
N (shade)	–2	–2	–2	–2	–2	–2	0	0	6	6	10	11	10	10	10	10	6	6	S (shade)

20 cm brick or 30 cm hollow tile or 30 cm cinder block

NE	2	2	2	2	10	2	16	8	14	8	10	6	10	8	10	10	10	8	SE
E	8	6	8	6	14	8	18	10	18	10	14	8	14	10	14	10	12	10	E
SE	8	4	6	4	6	4	14	10	18	12	16	12	12	10	12	10	12	10	NE
S	4	2	4	2	4	2	4	2	10	6	16	10	16	12	12	10	12	8	N
SW	8	4	6	4	6	14	8	4	10	6	12	8	20	12	24	16	20	14	NW
W	8	4	6	4	6	6	8	6	10	6	14	8	20	16	24	16	24	16	W
NW	2	2	2	2	2	2	4	2	6	4	8	6	10	8	16	14	18	14	SW
N (shade)	0	0	0	0	0	0	0	0	2	2	6	6	8	8	8	8	6	6	S (shade)

30 cm brick

NE	8	6	8	6	8	4	8	4	10	4	12	6	12	6	10	6	10	6	SE
E	12	8	12	8	12	8	10	6	12	8	14	10	14	10	14	8	14	8	E
SE	10	6	10	6	10	6	10	6	10	6	12	8	14	10	14	10	12	8	NE
S	8	6	8	6	6	4	6	4	6	4	8	4	10	6	12	8	12	8	N
SW	10	6	10	6	10	6	10	10	10	6	10	6	10	8	12	8	14	10	NW
W	12	8	12	8	12	8	10	10	10	6	10	6	10	6	12	8	10	10	W
NW	8	6	8	6	8	4	8	4	8	4	8	4	8	6	10	6	10	6	SW
N (shade)	4	4	2	2	2	2	2	2	2	2	2	2	2	2	4	4	4	6	S (shade)

20 cm concrete or Stone or 15 cm or 20 cm concrete block

NE	4	2	4	0	16	8	14	8	10	6	12	8	12	10	10	8	8	6	SE
E	6	4	14	8	24	12	24	12	18	10	14	10	14	10	12	10	10	8	E
SE	6	2	6	4	16	10	18	12	18	12	14	12	12	10	12	10	10	8	NE
S	2	1	2	1	4	1	12	6	16	12	18	12	14	12	10	8	1	6	E
SW	6	2	4	2	6	2	8	4	14	10	22	16	24	16	22	16	10	3	NW
W	6	4	6	4	6	4	8	6	12	8	20	14	28	18	26	13	14	10	W
NW	4	2	4	0	4	2	4	4	6	6	12	10	20	14	22	16	8	6	SW
N (shade)	0	0	0	0	0	0	2	2	4	4	6	6	8	8	6	6	4	4	S (shade)

100 cm concrete or stone

NE	6	4	6	2	6	2	14	8	14	8	10	8	10	8	12	10	10	8	SE
E	10	6	8	6	10	6	18	10	18	12	16	10	12	10	14	10	14	10	E
SE	8	4	8	4	6	4	14	8	16	10	16	10	14	10	12	10	12	10	NE
S	6	4	4	2	4	2	4	2	10	6	14	10	16	12	14	10	10	8	N
SW	8	4	8	4	6	4	6	4	8	6	10	8	8	14	20	14	18	16	NW
W	10	6	8	6	8	6	10	6	10	6	12	8	16	10	24	14	22	16	W
NW	6	4	6	2	6	2	6	4	6	4	8	6	16	8	18	12	20	14	SW
N (shade)	0	0	0	0	0	0	0	0	2	2	4	4	0	6	8	8	6	8	S (shade)

19.6 Sol Air Temperature

It is a hypothetical temperature used to calculate the heat received by the outside surface of a building wall by the combined effect of convection and radiation. The heat received by the outside surface of the wall by convection is given by

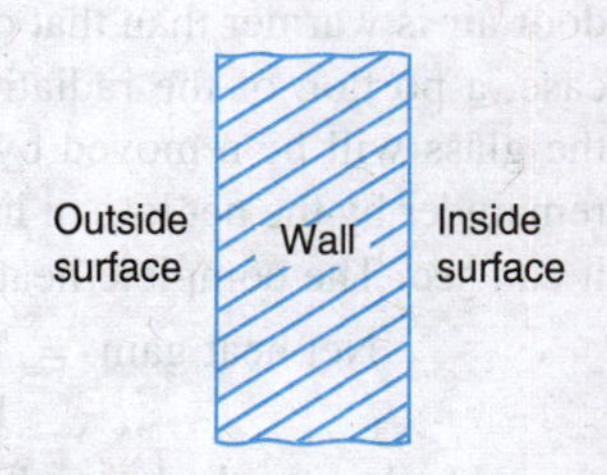

Fig. 19.5. Sol air temperature.

$$q_c = f_o (t_o - t_{os})$$

where f_o = Outside film coefficient,

t_o = Temperature of outside air, and

t_{os} = Temperature of the outside surface of the wall.

The heat received by the outside surface of the wall by radiation is given by

$$q_r = I\,\alpha$$

where I = Total radiation intensity, and

α = Absorptivity of the surface.

∴ Total heat received by the outside surface of the wall,

$$q_{os} = q_c + q_r = f_o (t_o - t_{os}) + I\,\alpha = f_o \times t_o - f_o \times t_{os} + I\,\alpha$$

$$= f_o \left(t_o + \frac{I\,\alpha}{f_o} \right) - f_o \times t_{os}$$

$$= f_o \times t_e - f_o \times t_{os} = f_o (t_e - t_{os})$$

where $t_e = t_o + \dfrac{I\,\alpha}{f_o}$

This temperature t_e is known as *sol air temperature.*

19.7 Solar Heat Gain through Glass Areas

A new air-conditioning system keeps the room dry and at a constant cool temperature.

The heat gain through the glass areas constitutes a major portion of the load on the cooling apparatus. When a sheet of glass is subjected to solar radiation (direct and diffuse), a part of it is absorbed, a part is reflected and the remainder is transmitted directly to the interior of the building.

In the absorption process, the temperature of the glass increases until it is in a position to lose heat at the same rate in an interchange of energy with surfaces inside and outside of the building. A simultaneous interchange of radiant energy takes place between the exterior surface of the glass and the heated particles in the atmosphere and between the interior surface and the various objects in the room. In addition to the radiant effects, the net heat gain into the interior of a building through a sheet of glass is affected by convection air currents on both sides. If the temperature of the glass is higher than that of the out-door air, there will be no gain by conduction in spite of the fact that the out-door air is warmer than that of inside. In that case, a portion of the radiation absorbed by the glass will be removed by inside air currents, another portion by outside air currents and the remainder by the net loss in interchanges of radiant energy between the glass and all of the surfaces it can see. The complete heat balance can be written as follows :

Net heat gain = Transmitted solar radiation + Heat flow by convection and radiation heat exchanges between glass and indoor surfaces

Neglecting the heat which may be stored in the glass or which may be given up by the glass under changing conditions, the second term of the above expression may be written as follows :

Heat flow by convection and radiation heat exchanges between glass and indoor surfaces

= Absorbed solar radiation ± Radiation and convection heat exchanges between glass and outdoor surfaces

Notes : 1. The solar heat gain through glass areas varies from hour to hour, from day to day and from latitude to latitude. For complete details of the solar radiation with respect to time of day and situation of glass areas, the guide published by American Society of Heating and Ventilating Engineers (ASHVE) may be referred.

2. The values of solar heat gain are usually given in examination problems.

19.8 Heat Gain due to Infiltration

The infiltration air is the air that enters a conditioned space through window cracks and opening of doors. This is caused by pressure difference on the two sides of the windows and doors and it depends upon the wind velocity and its direction and difference in densities due to the temperature difference between the inside and outside air.

There are two methods of estimating the infiltrated air :

1. Crack length method, and 2. Air change method.

The crack length method is usually used where greater accuracy is required. In most cases, the air change method is used for calculating the quantity of infiltrated air. According to this method, the amount of infiltrated air through windows and walls is

$$= \frac{L \times W \times H \times A_C}{60} \text{ m}^3/\text{min}$$

where L = Room length is metres,

W = Room width in metres,

H = Room height in metres, and

A_C = Air changes per hour.

The total room infiltration air for an entire building is taken one-half of the above calculated value because infiltration takes place on the windward side of a building.

The following table shows the number of air changes per hour for a variety of room types and exposures.

Table 19.8. Number of air changes per hour.

Kind of room or building	*Number of air changes per hour (A_C)*
Rooms with no windows or outside doors	0.5 to 0.75
Rooms, one wall exposed	1
Rooms, two walls exposed	1.5
Rooms, three walls exposed	2
Rooms, four walls exposed	2
Entrance halls	2 to 3
Reception halls	2
Bath rooms	2

For each person passing through a door leading to the outside or to an unconditioned space, the values given in the following table for the door infiltration should be added to the infiltration air through windows and walls, in order to find the total building infiltration.

Table 19.9. Infiltrated air.

Usage of door	*Infiltrated air for 1.8 metres revolving door in m³ per person per passage*	
	Freely revolving door	*Door equipped with brake*
Infrequent	2.5	2
Average	2	1.75
Heavy	1.5	1.25

Notes : 1. For rooms with weather stripped windows or storm sash, use 1/2 of the values given in Table 19.8 but it should never be less than 0.5.

2. The figures in Table 19.9 are based on the assumption that there is no wind pressure and the swinging doors are in use in one wall only. Any swinging door in other walls should be kept closed to ensure air conditioning in accordance with the recommended standards.

3. The door infiltration for a 0.9 m swinging door may be taken as 3 m³/min.

19.9 Heat Gain due to Ventilation

The ventilation (*i.e.* supply of outside air) is provided to the conditioned space in order to minimise odour, concentration of smoke, carbon dioxide and other undesirable gases so that freshness of air could be maintained. The quantity of outside air used for ventilation should provide at least one-half air change per hour in buildings with normal ceiling heights. Also, if the infiltration air quantity is larger than the ventilation quantity, then the latter should be increased to at least equal to the infiltration air. The outside air adds sensible as well as latent heat.

The following table shows the recommended and minimum values for the outside air required per person in the building.

Table 19.10. Outside air required per person.

Application	*Smoking*	*Outside air in m^3/min/person*	
		Recommended	*Minimum*
(1)	(2)	(3)	(4)
Apartment	Some	0.6	0.45
Banking space	Occasional	0.3	0.23
Barber shops	Considerable	0.45	0.3
Beauty Parlour	Occasional	0.3	0.23
Cocktail bars	Heavy	0.9	0.75
Departmental stores	None	0.23	0.15
Drug stores	Considerable	0.3	0.23
Factories	None	0.3	0.23
Funeral parlours	None	0.3	0.23
Hospitals, private rooms	None	0.9	0.75
Hospitals, wards	None	0.6	0.45
Hotel rooms	Heavy	0.9	0.75
Meeting rooms	Very heavy	1.5	0.9
Offices, general	Some	0.45	0.3
Offices, private	None	0.75	0.45
Restaurants	Considerable	0.45	0.36
Cafeterias	Considerable	0.36	0.3
Theatres	None	0.23	0.15

19.10 Heat Gain from Occupants

The human body in a cooled space constitutes cooling load of sensible heat and latent heat. The heat gain from occupants is based on the average number of people that are expected to be present in the conditioned space. The heat load produced by each person depends upon the acitivity of the person. Table 19.11 shows a wide range of activities from the rest position to one of heavy work.

Notes : 1. The values given in Table 19.11 are based on 27°C dry bulb temperature. For 25°C room dry bulb, the total heat remains the same, but the sensible heat values should be increased approximately 10 per cent and the latent heat values decreased accordingly.

2. The adjusted total heat gain is based on normal percentage of men, women and children for the application listed, with the postulate that the gain from an adult female is 85 per cent of that for an adult male and that the gain from a child is 75 per cent of that for an adult male.

3. The adjusted total heat value for sedentary work, restaurant includes 17.5 W for food per individual (8.75 W sensible and 8.75 W latent).

4. For bowling, figure one person per alley actually bowling and all other as sitting (117 W) and standing (161 W).

Table 19.11. Heat gain from occupants (in watts).

Degree of activity (1)	*Typical application* (2)	*Total heat Adults-Male* (3)	*Total heat adjusted* (4)	*Sensible heat* (5)	*Latent heat* (6)
1. Seated-at rest	Theatre-Matinee	114	97	53	44
	Theatre-Evening	114	102	57	45
2. Seated- very light work	Offices, Hotels, Apartments	132	117	57	67
3. Moderately active office work	Offices, Hotels, Apartments	139	132	59	73
4. Standing, light work, walking slowly	Department store, Retail store	161	132	59	73
5. Walking, seated, standing, walking slowly	Drug store, Bank	161	146	58	88
6. Sedentary work	Restaurant	144	161	64	97
7. Light bench work	Factory	234	220	64	155
8. Moderate dancing	Dance hall	264	249	72	177
9. Walking 4.8 km / h, moderately heavy work	Factory	293	293	88	205
10. Bowling	Bowling alley	440	425	136	288
11. Heavy work	Factory	440	425	136	288

19.11 Heat Gain from Appliances

The appliances frequently used in air conditioned spaces may be electrical, gas-fired or steam heated. Following table gives most of the commonly used appliances together with approximate values of sensible heat and latent heat.

Table 19.12. Heat gain from appliances without hoods (in watts).

Appliances	*Electrical*		*Gas*	
	Sensible (2)	*Latent* (3)	*Sensible* (4)	*Latent* (5)
(1)				
1. Coffee brewer, $2\frac{1}{4}$ litres	264	64	396	103
2. Coffee warmer, $2\frac{1}{4}$ litres	68	18	117	29
3. Coffee brewer with tank, 20 litres	1406	352	2110	528
4. Coffee urn, $13\frac{1}{2}$ litres	645	436	733	733
5. Coffee urn, $22\frac{1}{2}$ litres	996	674	1143	1143

(1)	(2)	(3)	(4)	(5)
6. Egg boiler, 2 cups	352	234	—	—
7. Food warmer with plate warmer, per sq m of top surface	1140	1140	—	—
8. Food warmer only per sq m of top surface	650	1140	2756	1401
9. Fry kettle per sq m of fry area	1026	1465	1758	1172
10. Griddle, per sq m of fry area	880	470	—	—
11. Grill, meat, per sq m of fry area	1378	733	2930	733
12. Grill, sand witch, per sq m of fry area	790	205	—	—
13. Hair dryer, blower type	674	117	—	—
14. Hair dryer, helmet type	547	98	—	—
15. Toaster, 360 slices per hour	1495	380	2257	967
16. Stoves, short order				
open top, per sq m top	—	—	1230	1230
closed top, per sq m top	—	—	967	967
Fry top, per sq m top	—	—	1055	1055
17. Permanent wave machine	250	44	—	—
18. Sterilizer instrument for physicians	190	352	—	—

19.12 Heat Gain from Products

The heat emitted from the products to be stored is very important in case of cold storages. The loads to be considered in the cold storages are divided into the following groups :

1. *Chilling load above freezing*. The product chilling load above freezing depends upon the mass of product (m), mean specific heat of the product (c_{pm}), entering product temperature (T_1), final product temperature desired (T_2) and the chilling time (t_{ch}). Mathematically, chilling load above freezing,

$$Q_{ch} = \frac{m\, c_{pm}(T_1 - T_2)}{t_{ch}}$$

2. *Freezing load*. The freezing load depends upon the mass of the product (m), its latent heat of freezing (h_{fg}) and the freezing time (t_F). Mathematically, freezing load,

$$Q_F = \frac{m \times h_{fg}}{t_F}$$

3. *Cooling load below freezing*. The cooling load below freezing depends upon the mass of the product (m), mean specific heat of the freezing product (c'_{pm}), actual storage temperature of the product (T_1'), desired freezing temperature of the product (T_2') and the cooling time (t_c). Mathematically, cooling load below freezing,

$$Q_c = \frac{m\, c'_{pm}(T_1' - T_2')}{t_c}$$

4. *Product reaction or respiration heat*. During the maturing of some food products, even in cold storage, reaction or respiration heat is evolved. This heat gain is given by

$$Q_R = m \times \text{Evolution of heat per kg of food per hour}$$

The rate of heat production for different types of food is given in the following table.

Table 19.13. Rate of heat production for different type of foods.

Commodity (1)	*Temperature in °C* (2)	*Heat evolved per tonne per 24 hours in kJ* (3)
1. Apples	0	690 – 1045
	4.5	1170 – 1865
	15.5	4650 – 6950
	30	6950 – 16 250
2. Bananas, green and ripe	20	8835
3. Beets	0	2800
	4.5	4270
	15.5	7640
4. Cabbage	0	1240
	4.5	1770
	15.5	4260
5. Carrots	0	2240
	4.5	3660
	15.5	8520
6. Cauliflower	4.5	4650
	15.5	10 500
7. Cherries (sour)	0	1380 – 1840
	15.5	11 600 – 13 920
8. Cucumbers	15.5	2250 – 6850
9. Grape fruit	0	480
	4.5	1130
	15.5	2930
10. Lemons	0	610
	4.5	860
	15.5	2180
11. Melons	0	1350
	4.5	2050
	15.5	8840
12. Mushrooms (cultivated)	0	6490
	10	23 215
13. Onions	0	690 – 1170
	10	1860 – 2090
14. Oranges	0	730 – 965
	4.5	1465
	15.5	5275
15. Peaches	0	900 – 1450
	4.5	1450 – 2050
	15.5	7560 – 9820
16. Pears	0	690 – 920

(1)	(2)	(3)
17. Peppers	0	2870
	4.5	4960
	15.5	8940
18. Peas	0	8500 – 8700
	4.5	13 500 – 16 600
	15.5	40 800 – 46 200
19. Potatoes	0	460 – 920
	4.5	1170 – 1860
	20	2320 – 3725
20. Raspberries	2.2	4650 – 6970
	15.5	16 245 – 18 570
21. Straw berries	0	2890 – 4020
	4.5	5400 – 6970
	15.5	16 500 – 20 200
22. Tomatoes (green)	0	610
	4.5	1130
	15.5	6570
23. Tomatoes (ripe)	0	1090
	4.5	1340
	15.5	5950
24. Turnips	0	1960
	4.5	2290
	15.5	5500

The latent heat load consists of product dehydration or condensation of the moisture from the products in storage. The following table shows the product moisture loss in storage for various products.

Table 19.14. Moisture loss for various products.

Commodity (1)	*Temperature (°C)* (2)	*Humidity (%)* (3)	*Period for loss* (4)	*Loss of moisture (%)* (5)
Apples	0	90	6 – 8 months	3 – 4
	2.8	95	1 – 50 days	0.03 – 1.1
	2.8	75	1 – 50 days	0.08 – 3.5
	15	95	1 – 50 days	0.06 – 2.5
	15	75	1 – 50 days	0.18 – 8.2
Beef : 0.9 kg steak	—	90 and 80	1 day	0.9 and 1.8
18 kg roast	—	90 and 80	1 day	0.4 and 8
Beef	Normal storage	Normal storage	1 month	3
Beets	0	98	1 month	2.5
Cabbage : curled	– 0.5	85	2 – 7 months	10 – 15
red	– 0.5	85	6 – 7 months	6 – 8
white	– 0.5	85	6 – 7 months	8 – 10
Carrots	0	98	1 month	1.5
Cauliflower	0	98	1 month	11
Cheese	15	75	—	10

(1)	(2)	(3)	(4)	(5)
Cucumber	0	98	1 month	6
Currants	– 0.5	85	2 months	6
Eggs	0	99	3 months	2
	0	88	3 months	7
Goose berries	–1.1	85	2 months	6
Onions	– 0.5	85	6 – 8 months	6 – 8
Peaches	–1.1	90	1 – 2 months	8 – 12
Pears	0	90	4 – 6 months	3 – 6
Peppers	0	98	1 month	3
Plums	–1.1	85	2 months	4 – 6
Spinach	–1.1	90	6 – 7 months	4 – 5
Squash	0	98	1 month	1
Straw berries	1.1	90	1 month	4

19.13 Heat Gain from Lighting Equipments

The heat gained from electric lights depends upon the rating of lights in watts, use factor and allowance factor. Mathematically, the heat gained from electric lights is given by

$$Q = \text{Total wattage of lights} \times \text{Use factor} \times \text{Allowance factor}$$

The use factor is the ratio of actual wattage in use to the installed wattage. Its value depends upon the type of use to which the room is put. In case of residences, commercial stores and shops, its value is usually taken as unity, whereas for industrial workshops it is taken below 0.5.

The allowance factor is generally used in case of flourescent tubes to allow for the power used by the ballast. Its value is usually taken as 1.25.

19.14 Heat Gain from Power Equipments

The power equipments such as fan, motor or any other equipment of this type also adds heat in the air-conditioned space. The power consumed by the air-conditioning fan is converted into heat energy and imparted to the air. If the fan is located before the air conditioner, the heat energy

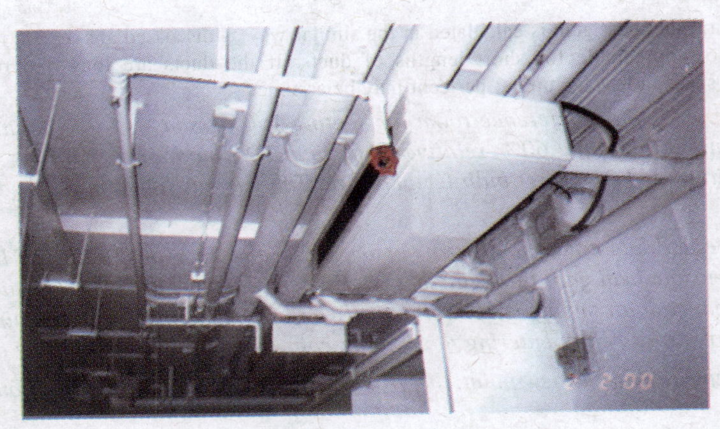

Air Conditioning system.

must be added to the total load. If the fan is located after the air-conditioner, the heat energy is added to the room sensible heat load.

The electric motors used to operate the conditioning equipment within the conditioned space adds at least a part of the heat equivalent of the power consumed to the cooling load. The heat gain from the electric motor is given by

$$Q_{em} = \frac{\text{Power rating of motor in kW}}{\text{Motor efficiency}} \times \text{Load factor}$$

The load factor is the fraction of the total load at which the motor is working.

19.15 Heat Gain through Ducts

The heat gain due to supply-duct depends upon the temperature of air in the duct and the temperature of the space surrounding the duct. The heat gain through the duct (Q_D) is calculated by using the relation,

$$Q_D = UA_D (t_a - t_s)$$

where U = Overall heat-transfer coefficient,

A_D = Surface area of the duct,

t_a = Temperature of ambient air, and

t_s = Temperature of supply air.

When all of the duct is in the air conditioned space, then the heat gain is zero. If the duct is located in an unconditioned space, there is a gain in heat and condensation takes place. In order to prevent condensation and to reduce duct heat gain, the duct must be insulated. The heat gain through supply duct is roughly taken as 5% of the room sensible heat.

The loss due to supply air leakage is not easy to estimate, because it depends upon the workmanship of duct construction and the length of run. It has been found that the duct leakages are of the order of 5 to 30 per cent. The air leakage from supply ducts results in the loss of cooling capacity unless the leakages take place within the conditioned space. The following are the minimum recommended values of air leakages through ducts :

Long runs 10%

Medium runs 5%

Short runs . . . . Neglect

Note: The return-duct heat gain is calculated in the similar way as discussed above. But the air-leakage in this case is not appreciable for short lengths of ducts. If the ducts are long and runs through the unconditioned space, a leakage upto 3 per cent may be assumed.

Example 19.1. *A conference room for seating 100 persons is to be maintained at 22°C dry bulb temperature and 60% relative humidity. The outdoor conditions are 40°C dry bulb temperature and 27°C wet bulb temperature. The various loads in the auditorium are as follows:*

Sensible and latent heat loads per person, 80 W and 50 W respectively ; lights and fans, 15 000 W ; sensible heat gain through glass, walls, ceiling etc., 15 000 W. The air infiltration is 20 m³/min and fresh air supply is 100 m³/min. Two-third of recirculated room air and one-third of fresh air are mixed before entering the cooling coil. The by-pass factor of the coil is 0.1.

Determine apparatus dew point, the grand total heat load and effective room sensible heat factor.

Solution. Given : No. of persons = 100 ; t_{d2} = 22°C ; ϕ_2 = 60% ; t_{d1} = 40°C ; t_{w1} = 27°C ; Q_S per person = 80 W ; Q_L per person = 50 W ; Q_{SL} = 15 000 W = 15 kW ; Q_{SG} = 15 000 W = 15 kW ; v_1 = 20 m³/min ; v_F = 100 m³/min ; BPF = 0.1

The flow diagram for the given air conditioning system is shown in Fig. 19.6. The various points on the psychrometric chart, as shown in Fig. 19.7, are marked as discussed below :

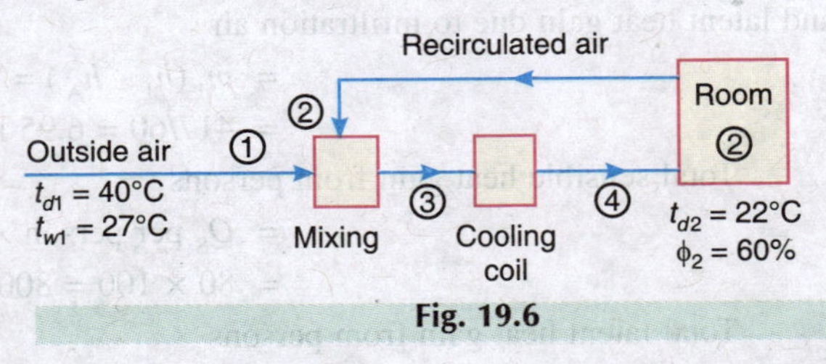

Fig. 19.6

First of all, mark the outside condition of air *i.e.* 40°C dry bulb temperature and 27°C wet bulb temperature on the psychrometric chart as point 1, as shown in Fig. 19.7. Now mark the inside condition of air in the room *i.e.* 22°C dry bulb temperature and 60% relative humidity as point 2. Locate point *A* by drawing vertical and horizontal lines through points 1 and 2 respectively.

From the psychrometric chart, we find that specific volume of air at point 1,

$$v_{s1} = 0.911 \text{ m}^3\text{/kg of dry air}$$

Enthalpy of air at point 1,

$$h_1 = 85 \text{ kJ/kg of dry air}$$

Enthalpy of air at point 2,

$$h_2 = 47.5 \text{ kJ/kg of dry air}$$

and enthalpy of air at point *A*,

$$h_A = 66 \text{ kJ/kg of dry air}$$

We know that mass of infiltrated air at point 1,

$$m_a = \frac{v_1}{v_{s1}} = \frac{20}{0.911} = 21.95 \text{ kg/min}$$

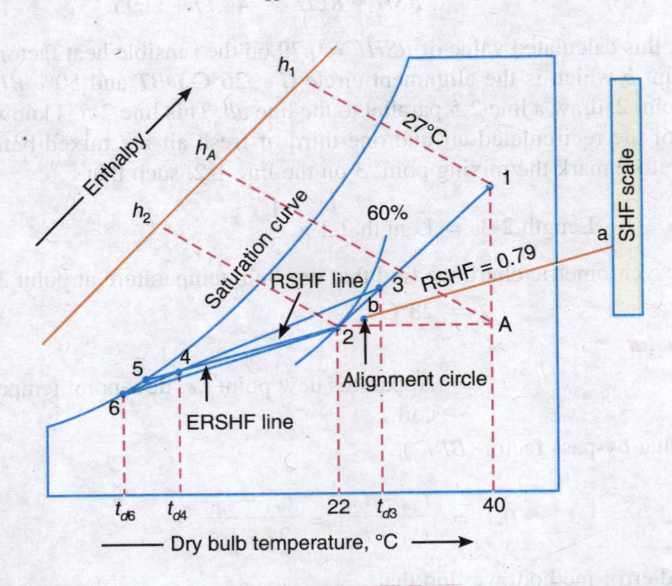

Fig. 19.7

∴ Sensible heat gain due to infiltration air

$$= m_1 (h_A - h_2) = 21.95 (66 - 47.5) = 406 \text{ kJ/min}$$
$$= 406/60 = 6.77 \text{ kW}$$

and latent heat gain due to infiltration air

$$= m_1 (h_1 - h_A) = 21.95 (85 - 66) = 417 \text{ kJ/min}$$
$$= 417/60 = 6.95 \text{ kW}$$

Total sensible heat gain from persons

$$= Q_S \text{ per person} \times \text{No. of persons}$$
$$= 80 \times 100 = 8000 \text{ W} = 8 \text{ kW}$$

Total latent heat gain from persons

$$= Q_L \text{ per person} \times \text{No. of persons}$$
$$= 50 \times 100 = 5000 \text{ W} = 5 \text{ kW}$$

∴ Total sensible heat gain in the room,

RSH = Sensible heat gain due to infiltration air + Sensible heat gain from persons + Sensible heat gain due to lights and fans (Q_{SL}) + Sensible heat gain through glass, walls and ceiling etc. (Q_{SG})

$$= 6.77 + 8 + 15 + 15 = 44.77 \text{ kW}$$

and total latent heat gain in the room,

RLH = Latent heat gain due to infiltration air + Latent heat gain from persons

$$= 6.95 + 5 = 11.95 \text{ kW}$$

We know that room sensible heat factor,

$$RSHF = \frac{RSH}{RSH + RLH} = \frac{44.77}{44.77 + 11.95} = 0.79$$

Now mark this calculated value of $RSHF = 0.79$ on the sensible heat factor scale as point a and join with point b which is the alignment circle (*i.e.* 26°C *DBT* and 50% *RH*), as shown in Fig.19.7. From point 2, draw a line 2-5 parallel to the line ab. This line 2-5 is known as *RSHF* line. Since two-third of the recirculated air and one-third of fresh air are mixed before entering the cooling coil, therefore mark the mixing point 3 on the line 1-2, such that

$$\text{Length 2-3} = \text{Length 2-1} \times \frac{1}{3}$$

From the psychrometric chart, we find that dry bulb temperature at point 3,

$$t_{d3} = 28°\text{C}$$

Apparatus dew point

Let t_{d6} = Apparatus dew point *i.e.* dew point temperature of the coil.

We know that by-pass factor (BPF),

$$0.1 = \frac{t_{d4} - t_{d6}}{t_{d3} - t_{d6}} = \frac{t_{d4} - t_{d6}}{28 - t_{d6}}$$

By trial and error method, we find that

$$t_{d4} = 13.15°\text{C and } t_{d6} = 11.5°\text{C}$$ **Ans.**

Grand total heat load

We know that outside air total heat,

$$OATH = 0.02\ v_1 (h_1 - h_2)$$
$$= 0.02 \times 20 (85 - 47.5) = 15 \text{ kW}$$

∴ Grand total heat load,

$$GTH = RSH + RLH + OATH$$
$$= 44.77 + 11.95 + 15 = 71.72 \text{ kW}$$ **Ans.**

Effective room sensible heat factor

Join point 6 with point 2 and produce upto the sensible heat factor scale on the psychrometric chart. The line 6-2 is known as effective room sensible heat factor (*ERSHF*) line and its value is given as

$$ERSHF = 0.77$$ **Ans.**

Example 19.2. *The following data relates to the office air conditioning plant having maximum seating capacity of 25 occupants :*

Outside design conditions	= 34°C DBT, 28°C WBT
Inside design conditions	= 24°C DBT, 50% RH
Solar heat gain	= 9120 W
Latent heat gain per occupant	= 105 W
Sensible heat gain per occupant	= 90 W
Lightening load	= 2300 W
Sensible heat load from other sources	= 11 630 W
Infiltration load	= 14 m³/min

Assuming 40% fresh air and 60% of recirculated air passing through the evaporator coil and the by-pass factor of 0.15, find the dew point temperature of the coil and capacity of the plant.

Solution. Given : No. of occupants = 25 ; $t_{d1} = 34°$ C ; $t_{w1} = 28°$ C ; $t_{d2} = 24°$ C ; $\phi_2 = 50\%$; $Q_{SS} = 9120$ W ; Q_L per occupant = 105 W ; Q_S per occupant = 90 W ; $Q_{SL} = 2300$ W ; Q_S from other sources = 11 630 W ; $v_1 = 14\ m^3/min$; $BPF = 0.15$

The flow diagram for the given air conditioning system is shown in Fig. 19.8. The various points on the psychrometric chart, as shown in Fig. 19.9, are marked as discussed below :

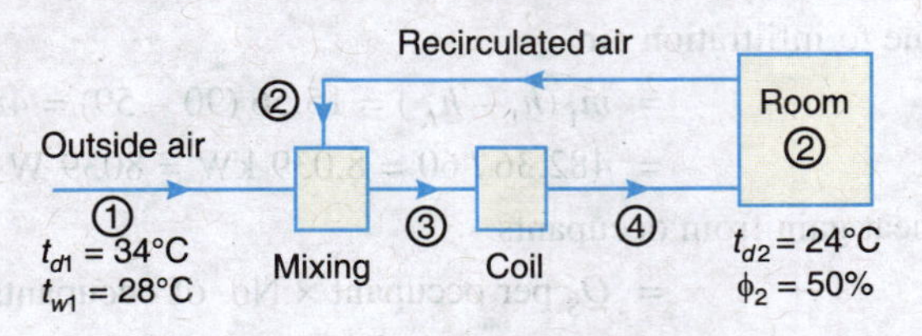

Fig. 19.8

First of all, mark the outside condition of air *i.e.* 34°C dry bulb temperature and 28°C wet bulb temperature on the psychrometric chart as point 1, as shown in Fig. 19.9. Now mark the inside condition of air *i.e.* 24°C dry bulb temperature and 50% relative humidity as point 2. Locate point *A* by drawing vertical and horizontal lines through points 1 and 2, respectively.

From the psychrometric chart, we find that specific volume of air at point 1,

$$v_{s1} = 0.9 \text{ m}^3/\text{kg of dry air}$$

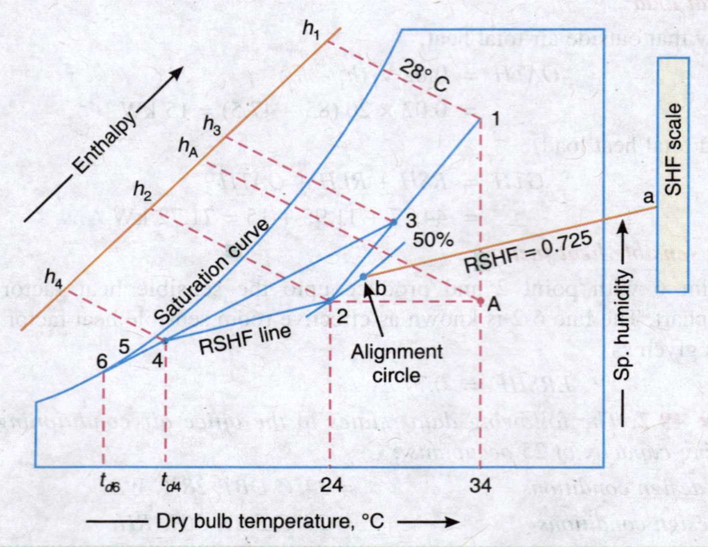

Fig. 19.9

Enthalpy of air at point 1,

$$h_1 = 90 \text{ kJ/kg of dry air}$$

Enthalpy of air at point 2,

$$h_2 = 48 \text{ kJ/kg of dry air}$$

and enthalpy of air at point A,

$$h_A = 59 \text{ kJ/kg of dry air}$$

We know that mass of infiltrated air at point 1,

$$m_1 = \frac{v_1}{v_{s1}} = \frac{14}{0.9} = 15.56 \text{ kg / min}$$

∴ Sensible heat gain due to infiltration air

$$= m_1 (h_A - h_2) = 15.56 (59 - 48) = 171.16 \text{ kJ/min}$$

$$= 171.16 / 60 = 2.853 \text{ kW} = 2853 \text{ W}$$

and latent heat gain due to infiltration air

$$= m_1 (h_1 - h_A) = 15.56 (90 - 59) = 482.36 \text{ kJ/min}$$

$$= 482.36 / 60 = 8.039 \text{ kW} = 8039 \text{ W}$$

Total sensible heat gain from occupants

$$= Q_S \text{ per occupant} \times \text{No. of occupants}$$

$$= 90 \times 25 = 2250 \text{ W}$$

Total latent heat gain from occupants

$$= Q_L \text{ per occupant} \times \text{No. of occupants}$$

$$= 105 \times 25 = 2625 \text{ W}$$

∴ Total sensible heat gain in the room,

RSH = Solar heat gain + Sensible heat gain due to infiltration air + Sensible heat gain from occupants + Sensible heat gain due to lightening + Sensible heat gain from other sources

$$= 9120 + 2853 + 2250 + 2300 + 11\,630 = 28\,153 \text{ W}$$

$$= 28.153 \text{ kW}$$

and total latent heat gain in the room,

RLH = Latent heat gain due to infiltration air + Latent heat gain from occupants

$$= 8039 + 2625 = 10\,664 \text{ W} = 10.664 \text{ kW}$$

We know that room sensible heat factor,

$$RSHF = \frac{RSH}{RSH + RLH} = \frac{28.153}{28.153 + 10.664} = 0.725$$

Now mark this calculated value of $RSHF = 0.725$ on the sensible heat factor scale as point a and join with point b which is the alignment circle (*i.e.* 26°C *DBT* and 50% *RH*), as shown in Fig.19.9. From point 2, draw a line 2-5 parallel to the line *ab*. The line 2-5 is known as *RSHF* line. Since 40% of fresh air (or outside air) and 60% of recirculated air is passed through the evaporator coil, therefore mark the mixing point 3 on the line 1-2, such that

$$\text{Length 2-3} = \text{Length 2-1} \times 0.4$$

Dew point temperature of the coil

Let t_{d6} = Dew point temperature of the coil.

We know that by-pass factor of the coil (*BPF*),

$$0.15 = \frac{t_{d4} - t_{d6}}{t_{d3} - t_{d6}} = \frac{t_{d4} - t_{d6}}{28 - t_{d6}}$$

By trial and error method, we find that

$$t_{d4} = 10.4\text{°C} \quad \text{and} \quad t_{d6} = 7.3\text{°C} \textbf{ Ans.}$$

Capacity of the plant

From the psychrometric chart, we find that enthalpy of air at point 4,

$$h_4 = 29 \text{ kJ/kg of dry air}$$

and enthalpy of air at point 3,

$$h_3 = 64.8 \text{ kJ/kg of dry air}$$

We know that mass of air entering the room,

$$m_a = \frac{\text{Total room heat}}{\text{Total heat removed}} = \frac{RSH + RLH}{h_2 - h_4}$$

$$= \frac{28.153 + 10.664}{48 - 29} = 2.043 \text{ kg / s} = 122.58 \text{ kg / min}$$

∴ Capacity of the plant

$$= m_a (h_3 - h_4) = 122.58 (64.8 - 29) = 4388.4 \text{ kJ/min}$$

$$= 4388.4 / 210 = 20.9 \text{ TR } \textbf{Ans.}$$

Example 19.3. *An air conditioning system is designed for a restaurant when the following data is available :*

Total heat flow through the walls, roof and floor = 6.2 kW

Solar heat gain through glass = 2 kW

Equipment sensible heat gain = 2.9 kW

Main air conditioning plant room with all original fittings.

Equipment latent heat gain	= 0.7 kW
Total infiltration air	= 400 m³/h
Outdoor conditions	= 35°C DBT ; 26°C WBT
Inside designed conditions	= 27°C DBT, 55% RH
Minimum temperature of air supplied to room	= 17°C DBT
Total amount of fresh air supplied	= 1600 m³/h
Seating chairs for dining	= 50
Employees serving the meals	= 5
Sensible heat gain per person	= 58 W
Latent heat gain per sitting person	= 44 W
Latent heat gain per employee	= 76 W
Sensible heat added from meals	= 0.17 kW
Latent heat added from meals	= 0.3 kW
Motor power connected to fan	= 7.6 kW

If the fan is situated before the conditioner, then find the following :

(a) Amount of air delivered to the room in m^3/h;

(b) Percentage of recirculated air ;

(c) Refrigeration load on the coil in tonnes of refrigeration ; and

(d) Dew point temperature of the cooling coil and by-pass factor.

Solution. Given : $Q_{SW} = 6.2$ kW ; $Q_{SG} = 2$ kW ; $Q_{SE} = 2.9$ kW ; $Q_{LE} = 0.7$ kW ; $v_1 = 400 \text{ m}^3/\text{h}$; $t_{d1} = 35°$ C ; $t_{w1} = 26°$ C ; $t_{d2} = 27°$ C ; $\phi_2 = 55\%$; $t_{d4} = 17°$C ; $v_F = 1600 \text{ m}^3/\text{h}$; Dinning chairs = 50 ; Employees = 5 ; Q_S per person = 58 W ; Q_L per person = 44 W ; Q_L per employee = 76 W ; $Q_{SM} = 0.17$ kW ; $Q_{LM} = 0.3$ kW ; $Q_M = 7.6$ kW

The flow diagram for the given air conditioning system is shown in Fig. 19.10.

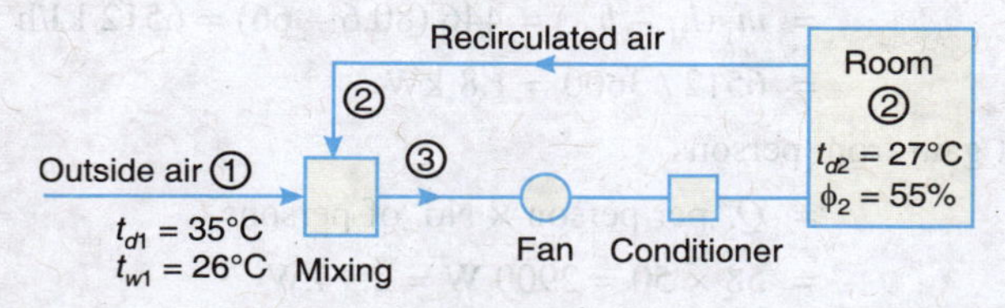

Fig. 19.10

First of all, mark the outdoor conditions of air *i.e.* 35°C *DBT* and 26°C *WBT* on the psychrometric chart as point 1, as shown in Fig. 19.11. Now mark the inside designed conditions of air *i.e.* 27°C *DBT* and 55% *RH* as point 2. Locate point *A* by drawing vertical and horizontal lines from points 1 and 2 respectively.

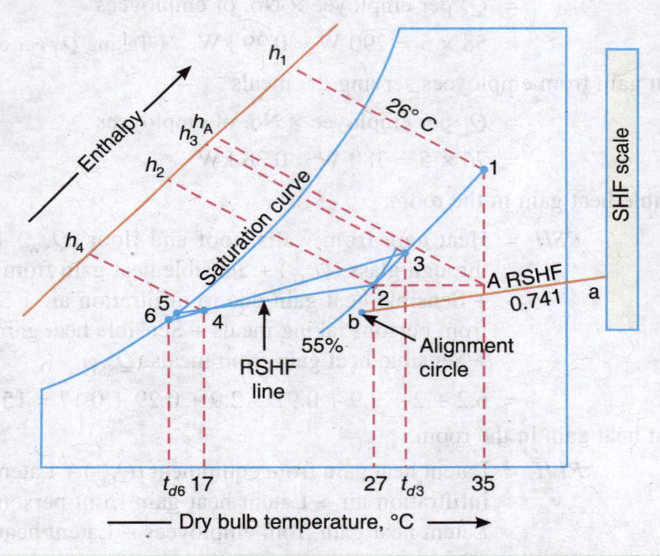

Fig. 19.11

From the psychrometric chart, we find that specific volume of air at point 1,

$$v_{s1} = 0.897 \text{ m}^3\text{/kg of dry air}$$

Enthalpy of air at point 1,

$$h_1 = 80.6 \text{ kJ/kg of dry air}$$

Enthalpy of air at point 2,

$$h_2 = 58.2 \text{ kJ/kg of dry air}$$

and enthalpy of air at point *A*,

$$h_A = 66 \text{ kJ/kg of dry air}$$

We know that mass of infiltration air,

$$m_1 = \frac{v_1}{v_{s1}} = \frac{400}{0.897} = 446 \text{ kg / h}$$

∴ Sensible heat gain due to infiltration air

$$= m_1 (h_A - h_2) = 446 (66 - 58.2) = 3480 \text{ kJ/h}$$

$$= 3480 / 3600 = 0.97 \text{ kW} \quad \ldots (\because \ 1 \text{ kW} = 1 \text{ kJ/s} = 3600 \text{ kJ/h})$$

and latent heat gain due to infiltration air

$= m_1 (h_1 - h_A) = 446 (80.6 - 66) = 6512$ kJ/h

$= 6512 / 3600 = 1.8$ kW

Sensible heat gain from persons

$= Q_S$ per person × No. of persons

$= 58 \times 50 = 2900$ W $= 2.9$ kW

Latent heat gain from persons

$= Q_L$ per person × No. of persons

$= 44 \times 50 = 2200$ W $= 2.2$ kW

Sensible heat gain from employees serving the meals

$= Q_S$ per employee × No. of employees

$= 58 \times 5 = 290$ W $= 0.29$ kW... (Taking Q_S per employee = 58 W)

Latent heat gain from employees serving the meals

$= Q_L$ per employee × No. of employees

$= 76 \times 5 = 380$ W $= 0.38$ kW

Total sensible heat gain in the room,

RSH = Heat gain from walls, roof and floor (Q_{SW}) + Solar heat gain through glass (Q_{SG}) + Sensible heat gain from equipment (Q_{SE}) + Sensible heat gain due to infiltration air + Sensible heat gain from persons taking meals + Sensible heat gain from employees + Sensible heat gain from meals (Q_{SM})

$= 6.2 + 2 + 2.9 + 0.97 + 2.9 + 0.29 + 0.17 = 15.43$ kW

Total latent heat gain in the room,

RLH = Latent heat gain from equipment (Q_{LE}) + Latent heat gain due to infiltration air + Latent heat gain from persons taking meals + Latent heat gain from employees + Latent heat gain from meals (Q_{LM})

$= 0.7 + 1.8 + 2.2 + 0.38 + 0.3 = 5.38$ kW

∴ Room sensible heat factor,

$$RSHF = \frac{RSH}{RSH + RLH} = \frac{15.43}{15.43 + 5.38} = 0.741$$

Now mark this calculated value of *RSHF* = 0.741 on the sensible heat factor scale as point *a* and join with point *b* which is the alignment circle (*i.e.* 26°C *DBT* and 50% *RH*), as shown in Fig.19.11. From point 2, draw a line 2-5 parallel to line *ab*. This line 2-5 is known as *RSHF* line. Now draw a vertical line through 17°C *DBT* (minimum temperature of air supplied to the room) which cuts *RSHF* line at point 4.

From the psychrometric chart, we find that specific volume of air supplied to the room at point 4,

$v_{s4} = 0.836$ m³ / kg of dry air

and enthalpy of air at point 4,

$h_4 = 45$ kJ / kg of dry air

(a) Amount of air delivered to the room in m³/h

We know that the amount of air delivered to the room,

$$m_a = \frac{\text{Total room heat}}{\text{Total heat removed}} = \frac{RSH + RLH}{h_2 - h_4}$$

$$= \frac{15.43 + 5.38}{58.2 - 45} = 1.576 \text{ kg / s} = 5675 \text{ kg / h}$$

∴ Amount of air delivered to the room in m³/h,

$$v_a = m_a \times v_{s4} = 5675 \times 0.836 = 4745 \text{ m}^3\text{/ h} \textbf{ Ans.}$$

(b) Percentage of recirculated air

We know that mass of fresh air supplied,

$$m_F = \frac{v_F}{v_{s1}} = \frac{1600}{0.897} = 1784 \text{ kg / h}$$

∴ Mass of recirculated air

$$= m_a - m_F = 5675 - 1784 = 3891 \text{ kg / h}$$

and percentage of recirculated air

$$= \frac{3891}{5675} = 0.686 = 68.6\% \textbf{ Ans.}$$

(b) Refrigeration load on the coil

Since the recirculated air at point 2 is 68.6%, therefore the fresh air supplied to the room at point 1 is 31.4%. The mixing of recirculated air and fresh air before entering to the cooling coil is marked at point 3 on the psychrometric chart, such that

$$\text{Length 2-3} = \text{Length 1-2} \times 0.314$$

From the psychrometic chart, we find that enthalpy of air at point 3,

$$h_3 = 64.6 \text{ kJ/kg of dry air}$$

∴ Refrigeration load on the coil

$$= m_a (h_3 - h_4) + \text{Heat added by fan motor}$$

$$= 1.576\,(64.6 - 45) + 7.6 = 38.5 \text{ kW}$$

$$= 38.5 / 3.5 = 11 \text{ TR} \textbf{ Ans.} \quad \ldots (\because 1 \text{ TR} = 3.5 \text{ kW})$$

(d) Dew point temperature of the cooling coil and by-pass factor

Join point 3 with point 4 and produce the line to intersect the saturation curve at point 6. From the psychrometric chart, we find that dew point temperature of the cooling coil,

$$t_{dp} = t_{d6} = 14.6°\text{C} \textbf{ Ans.}$$

We know that by-pass factor,

$$BPF = \frac{t_{d4} - t_{d6}}{t_{d3} - t_{d6}} = \frac{17 - 14.6}{29.5 - 14.6} = 0.161 \textbf{ Ans.}$$

Example 19.4. *A hall is to be maintained at 24°C dry bulb temperature and 60% relative humidity under the following conditions :*

Outdoor conditions	*= 38°C DBT and 28°C WBT*
Sensible heat load in the room	*= 46.4 kW*
Latent heat load in the room	*= 11.6 kW*
Total infiltration air	*= 1200 m³/h*
Apparatus dew point temperature	*= 10°C*
Quantity of recirculated air from the hall	*= 60%*

If the quantity of recirculated air is mixed with the conditioned air after the cooling coil, find the following :

(a) The condition of air leaving the conditioner coil and before mixing with the recirculated air;

(b) The condition of air before entering the hall ;

(c) The mass of air entering the cooler ;

(d) The mass of total air passing through the hall ;

(e) The by-pass factor of the cooling coil ; and

(f) The refrigeration load on the cooling coil in tonnes of refrigeration.

Solution. Given : t_{d2} = 24°C ; ϕ_2 = 60% ; t_{d1} = 38°C ; t_{w1} = 28°C ; Q_S = 46.4 kW ; Q_L = 11.6 kW ; v_1 = 1200 m³/h ; ADP = 10°C

The flow diagram for the given air conditioning system is shown in Fig. 19.12.

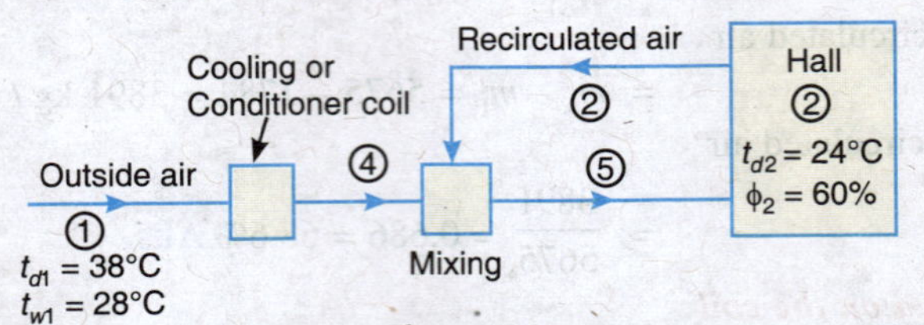

Fig. 19.12

(a) Condition of air leaving the conditioner coil and before mixing with the recirculated air

First of all, mark the outdoor conditions of air *i.e.* 38°C *DBT* and 28°C *WBT* on the psychrometric chart as point 1, as shown in Fig. 19.13. Mark the inside conditions of air *i.e.* 24°C *DBT* and 60% *RH* as point 2. Locate point *A* by drawing vertical and horizontal lines from points 1 and 2 respectively. From the psychrometric chart, we find that specific volume of infiltrated air at point 1,

$$v_{s1} = 0.907 \text{ m}^3/\text{kg of dry air}$$

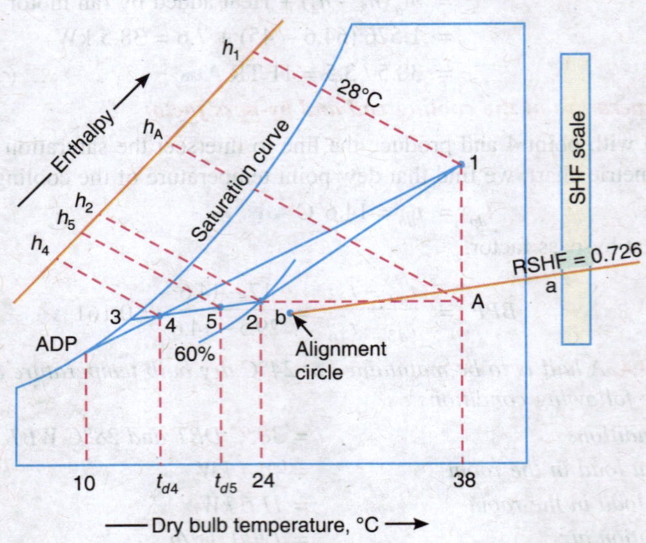

Fig. 19.13

Enthalpy of air at point 1,

$$h_1 = 89.6 \text{ kJ/kg of dry air}$$

Enthalpy of air at point 2,

$$h_2 = 53 \text{ kJ/kg of dry air}$$

and enthalpy of air at point *A*,

$$h_A = 68 \text{ kJ/kg of dry air}$$

We know that mass of infiltrated air,

$$m_1 = \frac{v_1}{v_{s1}} = \frac{1200}{0.907} = 1323 \text{ kg / h}$$

∴ Sensible heat gain due to infiltrated air

$$= m_1 (h_A - h_2) = 1323 (68 - 53) = 19\,845 \text{ kJ/h}$$

$$= 19\,845 / 3600 = 5.5 \text{ kW} \quad \ldots (\because 1 \text{ kW} = 1 \text{ kJ/s} = 3600 \text{ kJ/h})$$

and latent heat gain due to infiltrated air

$$= m_1 (h_1 - h_A) = 1323 (89.6 - 68) = 28\,577 \text{ kJ/h}$$

$$= 28\,577 / 3600 = 7.94 \text{ kW}$$

∴ Total sensible heat gain in the room,

$$RSH = 46.4 + 5.5 = 51.9 \text{ kW}$$

and total latent heat gain in the room,

$$RLH = 11.6 + 7.94 = 19.54 \text{ kW}$$

We know that room sensible heat factor,

$$RSHF = \frac{RSH}{RSH + RLH} = \frac{51.9}{51.9 + 19.54} = 0.726$$

Now mark this calculated value of *RSHF* = 0.726 on the sensible heat factor scale as point *a* and join with point *b* which is the alignment circle (*i.e.*26°C *DBT* and 50% *RH*), as shown in Fig. 19.13. From point 2, draw a line 2-3 parallel to the line *ab*. This line 2-3 is known as *RSHF* line. Draw a line joining the points 1 and *ADP* = 10°C which intersects the line 2-3 at point 4. This point 4 represents the condition of air leaving the conditioner coil (or coiling coil) and before mixing with the recirculated air.

From the psychrometric chart, we find that dry bulb temperature at point 4,

$$t_{d4} = 16°\text{C} \quad \textbf{Ans.}$$

and relative humidity at point 4,

$$\phi_4 = 90\% \quad \textbf{Ans.}$$

(b) Condition of air before entering the hall

Since 60% of the air from the room at point 2 is recirculated and mixed with 40% of air leaving the cooling or conditioner coil at point 4, therefore mark the mixing point 5, such that

$$\text{Length 2-5} = \text{Length 2-4} \times 0.4$$

The point 5 represents the condition of air before entering the hall.

From the psychrometric chart, we find that dry bulb temperature at point 5,

$$t_{d5} = 20.8°\text{C} \quad \textbf{Ans.}$$

and relative humidity at point 5,

$$\phi_5 = 70\% \quad \textbf{Ans.}$$

(c) *Mass of air entering the cooler*

We know that mass of air entering the cooler,

$$m_C = \frac{\text{Total room heat}}{\text{Total heat removed}} = \frac{RSH + RLH}{h_2 - h_4}$$

$$= \frac{51.9 + 19.54}{53 - 42} = 6.5 \text{ kg / s} = 6.5 \times 60 = 390 \text{ kg / min}$$ **Ans.**

(d) *Mass of total air passing through the hall*

Let m_H = Mass of total air passing through the hall.

Since the total air (m_H) passing through the hall consists of air passing through the cooler and recirculated air from the hall (*i.e.* 60% m_H), therefore

$$390 + 0.6\, m_H = m_H$$

$$\therefore \quad m_H = 390 / 0.4 = 975 \text{ kg / min}$$ **Ans.**

(e) *By-pass factor of the cooling coil*

We know that by-pass factor of the cooling coil,

$$BPF = \frac{t_{d4} - ADP}{t_{d1} - ADP} = \frac{16 - 10}{38 - 10} = 0.214$$ **Ans.**

(f) *Refrigeration load on the cooling coil*

We know that refrigeration load on the cooling coil

$$= m_C (h_1 - h_4) = 390\,(89.6 - 42) = 18\,564 \text{ kJ/min}$$

$$= 18\,564 / 210 = 88 \text{ TR}$$ **Ans.**

Example 19.5. *An air conditioning system is to be designed for a restaurant with the following data :*

Outside design conditions	= *40°C DBT, 28°C WBT*
Inside design conditions	= *25°C DBT, 50% RH*
Solar heat gain through walls, roof and floor	= *5.87 kW*
Solar heat gain through glass	= *5.52 kW*
Occupants	= *25*
Sensible heat gain per person	= *58 W*
Latent heat gain per person	= *58 W*
Internal lighting load	= *15 lamps of 100 W*
	10 fluorescent tubes of 80 W
Sensible heat gain from other sources	= *11.63 kW*
Infiltrated air	= *15 m³/min*

If 25% fresh air and 75% recirculated air is mixed and passed through the conditioner coil, find :

(a) The amount of total air required in m^3/h *;*

(b) The dew point temperature of the coil ;

(c) The condition of supply air to the room ; and

(d) The capacity of the conditioning plant.

Assume the by-pass factor equal to 0.2.

Draw the schematic diagram of the system and show the system on skeleton psychrometric chart and insert the temperature and enthalpy values at salient points.

Solution. Given : t_{d1} = 40°C ; t_{w1} = 28°C ; t_{d2} = 25°C ; ϕ_2 = 50% ; Q_{SW} = 5.87 kW ; Q_{SG} = 5.52 kW ; Occupants = 25 ; Q_S per person = 58 W ; Q_L per person = 58 W ; Q_{SL} = 15 × 100 + 10 × 80 = 2300 W = 2.3 kW ; Q_S from other sources = 11.63 kW ; v_1 = 15 m³/min ; *BPF* = 0.2

The schematic diagram of the given air conditioning system is shown in Fig. 19.14. The various points on the psychrometric chart, as shown in Fig. 19.15, are marked as discussed below :

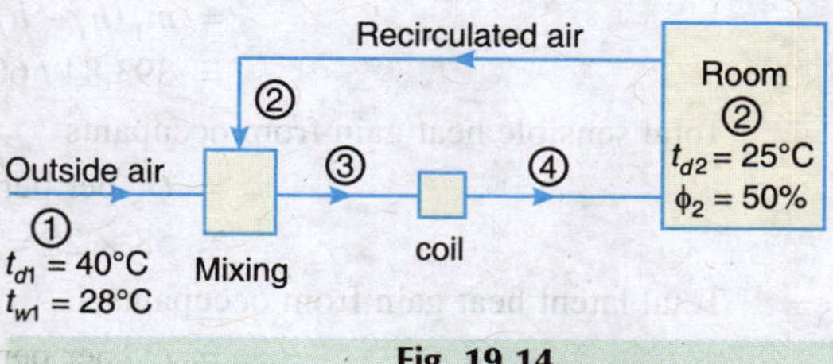

Fig. 19.14

First of all, mark point 1 on the psychrometric chart representing the outside conditions of air *i.e.* 40°C *DBT* and 28°C *WBT*. Now mark point 2 representing the inside conditions of air *i.e.* 25°C *DBT* and 50% *RH*. Locate point *A* by drawing vertical and horizontal lines from points 1 and 2 respectively.

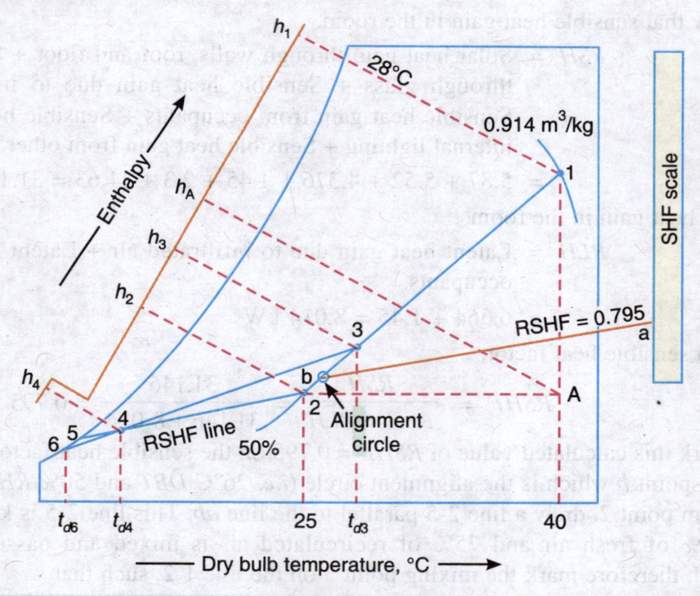

Fig. 19.15

From the psychrometric chart, we find that the specific volume of air at point 1,

$$v_{s1} = 0.914 \text{ m}^3\text{/kg of dry air}$$

Enthalpy of air at point 1,

$$h_1 = 90 \text{ kJ/kg of dry air}$$

Enthalpy of air at point 2,

$$h_2 = 50 \text{ kJ/kg of dry air}$$

and enthalpy of air at point *A*,

$$h_A = 66 \text{ kJ/kg of dry air}$$

We know that mass of infiltrated air at point 1,

$$m_1 = \frac{v_1}{v_{s1}} = \frac{15}{0.914} = 16.41 \text{ kg / min}$$

∴ Sensible heat gain due to infiltrated air

$= m_1 (h_A - h_2) = 16.41(66 - 50) = 262.56$ kJ/min

$= 262.56 / 60 = 4.376$ kW

and latent heat gain due to infiltrated air

$= m_1 (h_1 - h_A) = 16.41(90 - 66) = 393.84$ kJ/min

$= 393.84 / 60 = 6.564$ kW

Total sensible heat gain from occupants

$= Q_S$ per person × No. of occupants

$= 58 \times 25 = 1450$ W $= 1.45$ kW

Total latent heat gain from occupants,

$= Q_L$ per person × No. of occupants

$= 58 \times 25 = 1450$ W $= 1.45$ kW

Sensible heat gain due to internal lighting load,

$= 15 \times 100 + 10 \times 80 = 2300$ W $= 2.3$ kW

We know that sensible heat gain in the room,

RSH = Solar heat gain through walls, roof and floor + Solar heat gain through glass + Sensible heat gain due to infiltrated air + Sensible heat gain from occupants + Sensible heat gain due to internal lighting + Sensible heat gain from other sources

$= 5.87 + 5.52 + 4.376 + 1.45 + 2.3 + 11.63 = 31.146$ kW

and total latent heat gain in the room,

RLH = Latent heat gain due to infiltrated air + Latent heat gain from occupants

$= 6.564 + 1.45 = 8.014$ kW

∴ Room sensible heat factor,

$$RSHF = \frac{RSH}{RSH + RLH} = \frac{31.146}{31.146 + 8.014} = 0.795$$

Now mark this calculated value of *RSHF* = 0.795 on the sensible heat factor scale as point *a* and join with point *b* which is the alignment circle (*i.e.* 26°C *DBT* and 50% *RH*), as shown in Fig. 19.15. From point 2, draw a line 2-5 parallel to the line *ab*. This line 2-5 is known as *RSHF* line. Since 25% of fresh air and 75% of recirculated air is mixed and passed through the conditioner coil, therefore mark the mixing point 3 on the line 1-2, such that

Length 2-3 = Length 2-1 × 0.25

From the psychrometric chart, we find that dry bulb temperature of air at point 3,

$t_{d3} = 28.8$°C

and enthalpy of air at point 3,

$h_3 = 60$ kJ/kg of dry air

The point 4 on the line 2-5 and point 6 on the saturation curve (line 3-6) is marked as discussed below :

We know that by-pass factor of the coil (*BPF*),

$$0.2 = \frac{t_{d4} - t_{d6}}{t_{d3} - t_{d6}} = \frac{t_{d4} - t_{d6}}{28.8 - t_{d6}}$$

By trial and error method, we find that

$t_{d4} = 14.4$°C and $t_{d6} = 10.8$°C

The point 4 represents the condition of air leaving the coil and entering the room. The point 6 represents the dew point temperature of the coil. From the psychrometric chart, we find that enthalpy of air at point 4,

$$h_4 = 36.5 \text{ kJ/kg of dry air}$$

and specific volume of air at point 4,

$$v_{s4} = 0.826 \text{ m}^3\text{/kg of dry air}$$

(a) Amount of total air required in m³/h

Let v_a = Amount of total air required in m³/h.

We know that the amount of total air required,

$$m_a = \frac{\text{Total room heat}}{\text{Total heat removed}} = \frac{RSH + RLH}{h_2 - h_4}$$

$$= \frac{31.146 + 8.014}{50 - 36.5} = 2.9 \text{ kg/s} = 2.9 \times 3600 = 10\,440 \text{ kg/h}$$

$$\therefore \quad v_a = m_a \times v_{s4}$$

$$= 10\,440 \times 0.826 = 8623.4 \text{ m}^3\text{/h} \textbf{ Ans.}$$

(b) Dew point temperature of the coil

We have obtained above that the dew point temperature of the coil,

$$t_{dp} = t_{d6} = 10.8°\text{C} \textbf{ Ans.}$$

(c) Condition of supply air to the room

The point 4, as shown in Fig. 19.15, represents the condition of supply air to the room. At this point 4, dry bulb temperature,

$$t_{d4} = 14.4°\text{C} \textbf{ Ans.}$$

and relative humidity at point 4,

$$\phi_4 = 88\% \textbf{ Ans.}$$

(d) Capacity of the conditioning plant

We know that capacity of the conditioning plant

$$= m_a (h_3 - h_4) = 2.9 (60 - 36.5) = 39.15 \text{ kW}$$

$$= 39.15 / 3.5 = 11.2 \text{ TR} \textbf{ Ans.} \quad \ldots (\because 1 \text{ TR} = 3.5 \text{ kW})$$

Example 19.6. *A cold storage is to be designed to store 500 tonnes of vegetables when the following data are available :*

Outdoor conditions	*= 35°C DBT and 28°C WBT*
Required indoor conditions	*= 20°C DBT and 60% RH*
Water contents of the vegetables	*= 76%*
Loss of water content	*= 0.01% per hour*
People working in the cold storage	*= 25*
Fresh air supplied from outside	*= 4400 m³ / h*
Sensible heat gain through glass	*= 5.8 kW*
Sensible heat gain through walls and ceilings	*= 11.6 kW*
Heat from equipment and reaction heat of vegetables	*= 3.5 kW*
Infiltrated air	*= 200 m³ / h*

If the air conditioning is achieved by first cooling and dehumidifying and then heating, find the following :

(a) Amount of recirculated air, if the recirculated air is mixed with fresh air before entering the cooling coil,

(b) Capacity of the cooling coil in tonnes of refrigeration and its by-pass factor if dew point temperature of the coil is 5°C, and

(c) Capacity of the heating coil in kW.

The temperature of the air entering the room is not to exceed 15° C.

Solution. Given : m_v = 500 t = 500 × 10^3 kg ; t_{d1} = 35°C ; t_{w1} = 28°C ; t_{d2} = 20°C ; ϕ_2 = 60% ; W_v = 76% = 0.76 ; W_{Loss} = 0.01% per hour ; No. of people = 25 ; v_F = 4400 m³/h ; Q_{SG} = 5.8 kW ; Q_{SW} = 11.6 kW ; Q_{SR} = 3.5 kW ; v_1 = 200 m³/h

The flow diagram for the given air conditioning system is shown in Fig. 19.16.

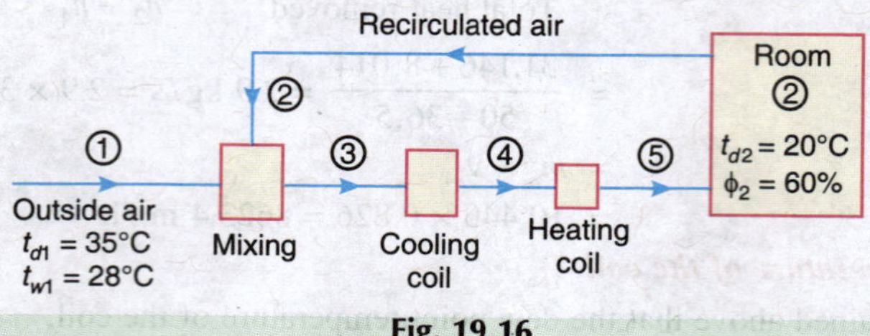

Fig. 19.16

First of all, mark the outdoor conditions of air at 35°C *DBT* and 28°C *WBT* on the psychrometric chart as point 1, as shown in Fig. 19.17. Now mark the required indoor conditions of air at 20°C *DBT* and 60% *RH* as point 2. Locate point *A* by drawing vertical and horizontal lines from points 1 and 2 respectively.

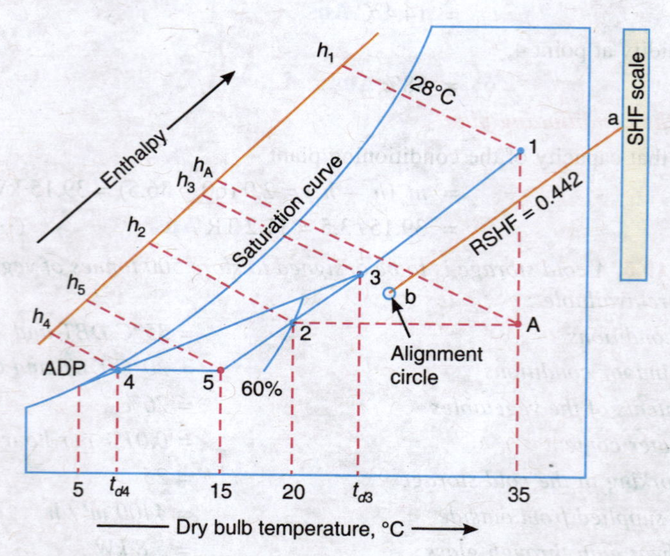

Fig. 19.17

From the psychrometric chart, we find that the specific volume of air at point 1,

$$v_{s1} = 0.903 \text{ m}^3\text{/kg of dry air}$$

Enthalpy of air at point 1,

$$h_1 = 90 \text{ kJ/kg of dry air}$$

Enthalpy of air at point 2,

$$h_2 = 42.3 \text{ kJ/kg of dry air}$$

and enthalpy of air at point A,

$$h_A = 58.2 \text{ kJ/kg of dry air}$$

We know that mass of infiltrated air supplied at point 1,

$$m_1 = \frac{v_1}{v_{s1}} = \frac{200}{0.903} = 221.5 \text{ kg / h}$$

∴ Sensible heat gain due to infiltrated air

$$= m_1 (h_A - h_2) = 221.5 (58.2 - 42.3) = 3522 \text{ kJ / h}$$

$$= 3522 / 3600 = 0.98 \text{ kW} \quad \text{... } (\because 1 \text{ kW} = 3600 \text{ kJ/h})$$

and latent heat gain due to infiltrated air

$$= m_1 (h_1 - h_A) = 221.5 (90 - 58.2) = 7044 \text{ kJ / h}$$

$$= 7044 / 3600 = 1.96 \text{ kW}$$

Since the water content of the vegetables is 76% and the loss of water content is 0.01% per hour, therefore total water loss from the vegetables which is carried by the air

$$= 500 \times 10^3 \times 0.76 \times \frac{0.01}{100} = 38 \text{ kg / h}$$

and latent heat gain due to evaporation of water from vegetables

$$= 38 \times 2500 = 95\,000 \text{ kJ / h} = 95\,000 / 3600 = 26.4 \text{ kW}$$

... (Taking latent heat of evaporation of water as 2500 kJ/kg)

Assuming the sensible and latent heat gain per person working in the cold storage as 250 kJ/h and 210 kJ/h respectively, we have

Total sensible heat gain from persons

$$= 25 \times 250 = 6250 \text{ kJ / h} = 6250 / 3600 = 1.74 \text{ kW}$$

and total latent heat gain from persons

$$= 25 \times 210 = 5250 \text{ kJ / h} = 5250 / 3600 = 1.46 \text{ kW}$$

Total sensible heat gain in cold storage,

RSH = Sensible heat gain through walls and ceiling (Q_{SW}) + Sensible heat gain through glass (Q_{SG}) + Heat from equipment and reaction heat of vegetables (Q_{SR}) + Sensible heat gain due to infiltrated air + Sensible heat gain from persons

$$= 11.6 + 5.8 + 3.5 + 0.98 + 1.74 = 23.62 \text{ kW}$$

Total latent heat gain in cold storage,

RLH = Latent heat gain due to evaporation of water from vegetables + Latent heat gain due to infiltrated air + Latent heat gain from persons

$$= 26.4 + 1.96 + 1.46 = 29.82 \text{ kW}$$

We know that room sensible heat factor,

$$RSHF = \frac{RSH}{RSH + RLH} = \frac{23.62}{23.62 + 29.82} = 0.442$$

Now mark this calculated value of *RSHF* = 0.442 on the sensible heat factor scale as point *a* and join with point *b* which is the alignment circle (*i.e.* 26°C *DBT* and 50% *RH*) as shown in Fig. 19.17. From point 2, draw a line parallel to the line *ab*, intersecting the vertical line drawn through 15°C *DBT* (*i.e.* temperature of air entering the room) at point 5. This line 2-5 is known as *RSHF* line. The point 5 represents the condition of air entering the room.

From the psychrometric chart, we find that enthalpy of air at point 5,

$$h_5 = 31.6 \text{ kJ/kg of dry air}$$

(a) Amount of recirculated air

We know that total mass of air passing through the room,

$$m_a = \frac{\text{Total room heat}}{\text{Total heat removed}} = \frac{RSH + RLH}{h_2 - h_5} = \frac{23.62 + 29.82}{42.3 - 31.6}$$

$$= 5 \text{ kg / s} = 5 \times 3600 = 18\,000 \text{ kg / h}$$

and mass of fresh air supplied from outside,

$$m_F = \frac{v_F}{v_{s1}} = \frac{4400}{0.903} = 4873 \text{ kg / h}$$

$\therefore$ Mass of recirculated air,

$$m_R = m_a - m_F = 18\,000 - 4873 = 13\,127 \text{ kg / h}$$

We know that percentage of recirculated air

$$= \frac{m_R}{m_a} \times 100 = \frac{13\,127}{18\,000} \times 100 = 73\% \textbf{ Ans.}$$

(b) Capacity of the cooling coil and its by-pass factor

Since 73% of the total air passing through the room at point 2 is recirculated and mixed with 27% of air at point 1, therefore mark the mixing point 3 on the chart, such that

$$\text{Length 2-3} = \text{Length 2-1} \times 0.27$$

Now draw a line joining points 3 and $ADP = 5°C$ intersecting the constant specific humidity line drawn through point 5, at point 4. The point 4 represents the condition of air leaving the cooling coil. The line 4-5 represents the heating of air. From the psychrometric chart, we find that dry bulb temperature of air at point 3,

$$t_{d3} = 24.3°C$$

Dry bulb temperature of air at point 4,

$$t_{d4} = 7.8°C$$

Enthalpy of air at point 3,

$$h_3 = 56 \text{ kJ/kg of dry air}$$

Enthalpy of air at point 4,

$$h_4 = 24 \text{ kJ/kg of dry air}$$

We know that capacity of the cooling coil

$$= m_a (h_3 - h_4) = 5 (56 - 24) = 160 \text{ kW}$$

$$= 160 / 3.5 = 45.7 \text{ TR } \textbf{Ans.}$$

and by-pass factor of the cooling coil

$$BPF = \frac{t_{d4} - ADP}{t_{d3} - ADP} = \frac{7.8 - 5}{24.3 - 5} = 0.145 \textbf{ Ans.}$$

(c) Capacity of the heating coil

From the psychrometric chart, we find that enthalpy of air at point 5,

$$h_5 = 31.6 \text{ kJ/kg of dry air}$$

∴ Capacity of the heating coil

$$= m_a (h_5 - h_4) = 5 (31.6 - 24) = 38 \text{ kW} \textbf{ Ans.}$$

Example 19.7. *A laboratory 30 m × 20 m × 4 m high is to be air conditioned. The 30 m wall faces north. The north wall has two doors of 2.5 m × 3 m each. The south wall has four glass windows of 2 m × 1.5 m each. The east and west walls also have four windows of the same size. The lighting load is 15 W fluorescent per m² floor area. The infiltration is one air change. The solar heat gain factors (SHGF) for south, east and west glass are 150, 50 and 350 W/m² respectively. The overall heat transfer coefficients for walls, roof, floor, door and windows are 2.5, 2 , 3, 1.5 and 6 W/m² K respectively. The corrected equivalent temperature differences for north, south, east, west walls, roof and floor are 12, 15, 12, 17, 20 and 2.5°C respectively. There are 100 persons with sensible and latent heat loads of 75 W and 55 W each respectively. The ventilation requirement is 0.3 m³/min per person. The outdoor condition is 43°C dry bulb temperature and 0.0277 kg / kg of dry air of humidity ratio. The indoor condition is 25°C dry bulb temperature and 0.01 kg / kg of dry air of humidity ratio. Use a factor of 1.25 for fluorescent light.*

Determine room sensible heat load and room latent heat load with 5% safety factor, 5% for fan power, 1% leakage of supply air and 0.5% heat leakage to supply air duct.

Solution. Given : $L = 30$ m ; $B = 20$ m ; $H = 4$ m ; Door size = 2.5 m × 3 m ; Window size = 2.5 m × 3 m ; $Q_{SL} = 15$ W/m² floor area ; No. of air changes (A_C) = 1 ; *SHGF* for south glass = 150 W/m² ; *SHGF* for east glass = 50 W/m² ; *SHGF* for west glass = 350 W/m² ; $U_{walls} = 2.5$ W/m² K ; $U_{roof} = 2$ W/m² K ; $U_{floor} = 3$ W/m² K ; $U_{door} = 1.5$ W/m² K ; $U_{windows} = 6$ W/m² K ; t_e for north wall = 12°C ; t_e for south wall = 15°C ; t_e for east wall = 12°C ; t_e for west wall = 17°C ; t_e for roof = 20°C ; t_e for floor = 2.5°C ; No. of persons = 100 ; Q_S per person = 75 W ; Q_L per person = 55 W ; $v_1 = 0.3$ m³ / min per person ; $t_{d1} = 43$°C ; $W_1 = 0.0277$ kg/kg of dry air ; $t_{d2} = 25$°C ; $W_2 = 0.01$ kg /kg of dry air ; Allowance factor for fluorescent light = 1.25 ; Safety factor = 5% ; Factor for fan power = 5% ; Factor for leakage of supply air = 1% ; Factor for heat leakage to supply air duct = 0.5%

The plan of a laboratory room is shown in Fig. 19.18. The sensible heat gain from various sources is given in the following table.

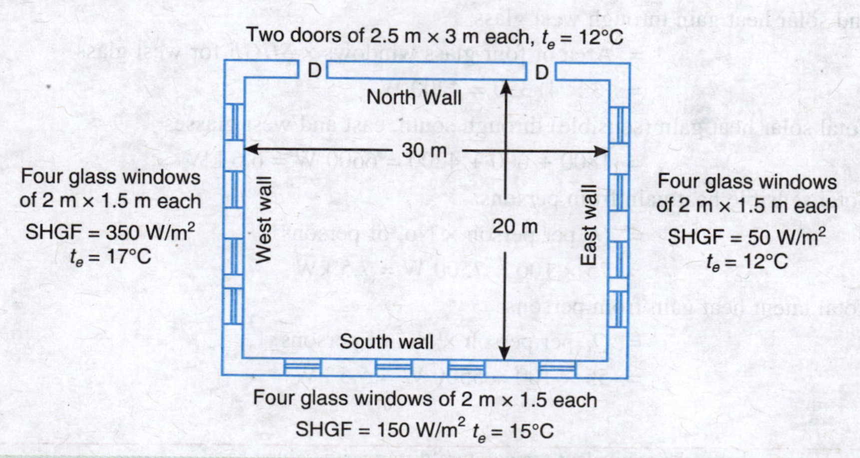

Fig. 19.18

Table 19.15. Estimation of sensible heat gain.

Area of one door = 2.5 × 3 = 7.5 m^2

Area of one window = 2 × 1.5 = 3 m^2

Source	Overall heat transfer coefficient (U) W/m^2 K	Area (A) m^2	Equivalent temperature difference (t_e)°C	Sensible heat gain = U.A. t_e W
1. North wall	2.5	30 × 4 – 2 × 7.5 = 105	12	2.5 × 105 × 12 = 3150
2. South wall	2.5	30 × 4 – 4 × 3 = 108	15	2.5 × 108 × 15 = 4050
3. East wall	2.5	20 × 4 – 4 × 3 = 68	12	2.5 × 68 × 12 = 2040
4. West wall	2.5	20 × 4 – 4 × 3 = 68	17	2.5 × 68 × 17 = 2890
5. Roof	2	30 × 20 = 600	20	2 × 600 × 20 = 24 000
6. Floor	3	30 × 20 = 600	2.5	3 × 600 × 2.5 = 4500
7. Doors (2 Nos.) in north wall	1.5	2 × 7.5 = 15	12 (Same as north wall)	1.5 × 15 × 12 = 270
8. Windows				
(a) South wall (4 Nos)	6	3 × 4 = 12	*18	6 × 12 × 18 = 1296
(b) East wall (4 Nos)	6	3 × 4 = 12	18	6 × 12 × 18 = 1296
(c) West wall (4 Nos)	6	3 × 4 = 12	18	6 × 12 × 18 = 1296
				Total = 44 788 W = 44.788 kW

Solar heat gain through south glass

$= $ Area of four glass windows × *SHGF* for south glass

$= (3 \times 4)\ 150 = 1800$ W

Similarly, solar heat gain through east glass

$= $ Area of four glass windows × *SHGF* for east glass

$= (3 \times 4)\ 50 = 600$ W

and solar heat gain through west glass

$= $ Area of four glass windows × *SHGF* for west glass

$= (3 \times 4)\ 350 = 4200$ W

Total solar heat gain (sensible) through south, east and west glasses

$= 1800 + 600 + 4200 = 6600$ W = 6.6 kW

Total sensible heat gain from persons

$= Q_S$ per person × No. of persons

$= 75 \times 100 = 7500$ W = 7.5 kW

Total latent heat gain from persons

$= Q_L$ per person × No. of persons

$= 55 \times 100 = 5500$ W = 5.5 kW

* For glass windows, the equivalent temperature difference is given as

$$t_e = t_{d1} - t_{d2} = 43 - 25 = 18°\text{C}$$

We know that the amount of infiltrated air,

$$v_1 = \frac{L \times W \times H \times A_C}{60} = \frac{30 \times 20 \times 4 \times 1}{60} = 40 \text{ m}^3/\text{min} \quad (\because A_C = 1)$$

∴ *Sensible heat gain due to infiltration air

$$= 0.020\,44\, v_1\,(t_{d1} - t_{d2}) = 0.020\,44 \times 40\,(43 - 25) = 14.7 \text{ kW}$$

and *latent heat gain due to infiltration air

$$= 50\, v_1\,(W_1 - W_2) = 50 \times 40\,(0.0277 - 0.01) = 35.4 \text{ kW}$$

We know that volume of ventilation or outside air,

$$v = 0.3 \text{ m}^3/\text{min/person} = 0.3 \times 100 = 30 \text{ m}^3/\text{min}$$

∴ **Outside air sensible heat,

$$OASH = 0.020\,44\, v\,(t_{d1} - t_{d2}) = 0.020\,44 \times 30\,(43 - 25) = 11.04 \text{ kW}$$

and **outside air latent heat,

$$OALH = 50\, v\,(W_1 - W_2) = 50 \times 30\,(0.0277 - 0.01) = 26.6 \text{ kW}$$

Sensible heat gain due to lighting

= Total wattage of lights × Use factor × Allowance factor

$$= 15\,(30 \times 20) \times 1 \times 1.25 = 11\,250 \text{ W} = 11.25 \text{ kW}$$

... ($\because$ Use factor = 1)

Assuming that the fan is placed before the conditioner (*i.e.* neglecting factor of 5% for fan power), the total room sensible heat (*RSH*) is to be increased by 6.5% (*i.e.* 5% for safety factor ; 1% for leakage of supply air and 0.5% for heat leakage to duct).

∴ Total room sensible heat,

RSH = 1.065 [Heat gain from walls, roof, floor and windows (From Table 19.15) + Solar heat gain through glasses + Sensible heat gain from persons + Sensible heat gain due to infiltration air + Sensible heat gain due to ventilation (*OASH*) + Sensible heat gain due to lighting]

= 1.065 [44.788 + 6.6 + 7.5 + 14.7 + 11.04 + 11.25]

= 102 kW **Ans.**

The total room latent heat (*RLH*) is to be increased by 6% (*i.e.* 5% for safety factor and 1% for leakage of supply air).

∴ Total room latent heat,

RLH = 1.06 [Latent heat gain from persons + Latent heat gain due to infiltration air + Latent heat due to ventilation (*OALH*)]

= 1.06 [5.5 + 35.4 + 26.6] = 71.6 kW **Ans.**

If the fan is located after the conditioner, then the 5% for fan power should be added to *RSH*. The *RLH* will remain unaffected.

∴ *RSH* = (1.065 + 0.05) (44.788 + 6.6 + 7.5 + 14.7 + 11.04 + 11.25) kW

= 107 kW **Ans.**

---

* Refer Chapter 16, Art. 16.10 and Art. 16.13.

** Refer Chapter 18, Art. 18.14.

EXERCISES

1. The following data is available for the design of air conditioning of a small theatre :

Outdoor design conditions	= 34°C *DBT* and 70% *RH*
Comfort conditions required	= 22°C *DBT* and 50% *RH*
Total seating capacity	= 350 persons
Sensible heat gain per person	= 90 W
Latent heat gain per person	= 30 W
Sensible heat due to solar gain and infiltrated air	= 46.6 kW
Latent heat gain due to infiltrated air	= 23.3 kW
Fresh air supplied	= 0.4 m^3/min/person
Desirable temperature rise in the theatre	= 8°C

Assume that the recirculated air is mixed with the fresh air after leaving the conditioner. Find : (*a*) Room sensible heat factor ; (*b*) The percentage of total air circulated ; (*c*) The refrigeration capacity of the conditioner coil.

Assume that the air leaves the conditioner coil with 100% *RH*. **[Ans.** 0.72 ; 76% ; 50 TR]

2. The design data for an air-conditioning plant of a restaurant is given below :

Outdoor design conditions	= 35°C *DBT* and 24°C *WBT*
Indoor design conditions	= 27°C *DBT* and 55% *RH*
Seating capacity of the restaurant	= 50
Latent heat gain per person	= 44 W
Latent heat gain from meals per person	= 6 W
Sensible heat gain per person	= 58 W
Sensible heat gain from meals per person	= 3.5 W
Number of service employees	= 5
Latent heat gain per employee	= 75 W
Sensible heat gain per employee	= 58 W
Sensible heat gain from outside	= 8.14 kW
Sensible heat gain from inside equipment	= 2.9 kW
Latent heat gain from inside equipment	= 0.7 kW
Rate of infiltrated air	= 400 m^3/h
Rate of fresh air supply	= 1600 m^3/h
Minimum temperature of air supplied to room	= 17°C *DBT*

The fan is situated before the conditioner and has a motor of 11 kW. Calculate : (*a*) volume of air passing through the room in m^3/h ; (*b*) percentage of recirculated air ; (*c*) apparatus dew point temperature and by-pass factor ; and (*d*) cooling capacity in tonnes of refrigeration.

[**Ans.** 4745 m^3/h ; 68.6 % ; 14.6°C, 0.161 ; 11.9 TR]

3. The following are the design data for an air conditioning system proposed for a restaurant :

Outdoor conditions	= 34°C *DBT* and 28°C *WBT*
Indoor conditions	= 24°C *DBT* and 50% *RH*
Solar heat gain through walls, roof and floor	= 4.7 kW
Solar heat gain through glass areas	= 4.4 kW
Occupants	= 25

Sensible heat gain per person = 85 W
Latent heat gain per person = 105 W
Internal lighting load = 15 lamps of 100 watts capacity each plus 10 fluorescent fixtures of 80 watts each
Sensible heat gain from other sources = 11.6 kW
Infiltration air = 14 m^3/min
Coil by-pass factor = 0.15

If the return and out-door air are adiabatically mixed in the ratio of 3 : 2 (by mass) and then passed through the conditioner, determine :

(*a*) dry bulb and wet bulb temperatures of supply air ; (*b*) apparatus dew point; and (*c*) capacity of the air-conditioning plant. **[Ans.** 10.4°C ; 10°C ; 7.3°C ; 72 kW]

4. A space to be air conditioned for comfort at 25°C , 50% *RH* has the following peak load data :

Room size : 20 m wide, 20 m long and 4 m high.

North glass : 4 windows of 2 m × 1.5 m each with solar heat gain factor (*SHGF*) = 50 W/m^2.

East glass : 4 windows of 1 m × 0.75 m each with *SHGF* = 60 W/m^2.

West glass : 4 windows of 2 m × 1.5 m each with *SHGF* = 350 W/m^2.

East wall and North wall have two doors each 2.5 m × 3 m.

Overall heat transfer coefficient (U_o) in W/m^2 K are : – Roof = 2.0 ; Walls = 2.5 ; Glass = 6.0 ; Door = 1.5 ; and Floor = 3.0.

Corrected equivalent temperature difference (ΔT_{eq}) : East wall = 10 K : West wall = 15 K ; North wall = 10 K ; South wall = 13 K ; Roof = 18 K and floor = 2.5 K.

Infiltration : One air change
Occupancy : 100 persons dissipating 75 W sensible heat and 65 W latent heat
Lighting : Fluorescent 10 W/m^2 of floor area
Ventilation : 0.3 m^3/min per person
Out-door conditions : 43°C and 30% *RH*
Indoor conditions : 25°C and 50% *RH*

Determine room sensible heat (*RSH*) load ; room latent heat (*RLH*) load ; outside air sensible heat load (*OASH*) : outside air latent heat load (*OALH*).

Given apparatus dew point (*ADP*) = 7°C, find effective sensible heat factor (*ESHF*) and supply air flow rate. Humidity ratio of outdoor and indoor air may be taken as 0.017 and 0.01 kg / kg of dry air

[Ans. 69.3 kW ; 14.95 kW ; 79.2 kW ; 24.2 kW ; 0.8 ; 404.6 kg / min]

QUESTIONS

1. What do you understand by the term cooling load ?
2. What are the different factors considered in load estimation sheet for comfort application ?
3. Discuss briefly the different types of heat loads which have to be taken into account in order to estimate the total heat load of a large restaurant for summer air conditioning.
4. Define sol-air temperature and equivalent temperature difference.
5. Explain the methods of estimating heat gain due to infiltrated air.
6. Write the expression for calculating the heat gain through the ducts.

OBJECTIVE TYPE QUESTIONS

1. When the outside air is introduced for ventilation purposes, there is a
   (*a*) sensible heat gain (*b*) latent heat gain
   (*c*) sensible heat gain as well as latent heat gain (*d*) none of these
2. The recommended outside air required per person for theatres is
   (*a*) 0.23 m^3/min (*b*) 0.36 m^3/min (*c*) 0.45 m^3/min (*d*) 1.5 m^3/min
3. The human body in a cooled space constitutes cooling load of
   (*a*) sensible heat only (*b*) latent heat only
   (*c*) sensible heat and latent heat (*d*) none of these
4. The sensible heat produced by a person working in a bank is about
   (*a*) 53 W (*b*) 58 W (*c*) 88 W (*d*) 136 W
5. The latent heat gain from an electrical egg boiler is about
   (*a*) 234 W (*b*) 352 W (*c*) 440 W (*d*) 730 W

ANSWERS

1. (*c*) **2.** (*a*) **3.** (*c*) **4.** (*b*) **5.** (*a*)

CHAPTER 20

Ducts

1. Introduction.
2. Classification of Ducts.
3. Duct Material.
4. Duct Construction.
5. Duct Shape.
6. Pressure in Ducts.
7. Continuity Equation for Ducts.
8. Bernoulli's Equation for Ducts.
9. Pressure Losses in Ducts.
10. Pressure Loss due to Friction in Ducts.
11. Friction Factor for Ducts.
12. Equivalent Diameter of a Circular Duct for a Rectangular Duct.
13. Friction Chart for Circular Ducts.
14. Dynamic Losses in Ducts.
15. Pressure Loss due to Enlargement in Area and Static Regain.
16. Pressure Loss due to Contraction in Area.
17. Pressure Loss at Suction and Discharge of a Duct.
18. Pressure Loss due to Obstruction in a Duct.
19. Duct Design.
20. Methods of Determination of Duct Size.
21. System Resistance.
22. Systems in Series.
23. Systems in Parallel.

20.1 Introduction

The conditioned air (cooled or heated) from the air conditioning equipment must be properly distributed to rooms or spaces to be conditioned in order to provide comfort conditions. When the conditioned air cannot be supplied directly from the air conditioning equipment to the spaces to be conditioned, then the ducts are installed. The duct systems convey the conditioned air from the air conditioning equipment to the proper air distribution points or air supply outlets in the room and carry the return air from the room back to the air conditioning equipment for reconditioning and recirculation.

It may be noted that the duct system for proper distribution of conditioned air cost nearly 20 to 30 per cent of the total cost of the equipments required and the power required by fans forms the substantial part of the running cost. Thus, it is necessary to design the air duct system in such a way that the capital cost of ducts and the cost of running the fans is lowest.

20.2 Classification of Ducts

The ducts may be classified as follows :

1. *Supply air duct.* The duct which supplies the conditioned air from the air conditioning equipment to the space to be conditioned is called supply air duct.

2. *Return air duct.* The duct which carries the recirculating air from the conditioned space back to the air conditioning equipment is called return air duct.

3. *Fresh air duct.* The duct which carries the outside air is called fresh air duct.

4. *Low pressure duct.* When the static pressure in the duct is less than 50 mm of water gauge, the duct is said to be a low pressure duct.

5. *Medium pressure duct.* When the static pressure in the duct is up to 150 mm of water gauge, the duct is said to be a medium pressure duct.

6. *High pressure duct.* When the static pressure in the duct is from 150 to 250 mm of water gauge, the duct is said to be a high pressure duct.

7. *Low velocity duct.* When the velocity of air in the duct is up to 600 m/min, the duct is said to be a low velocity duct.

8. *High velocity duct.* When the velocity of air in the duct is more than 600 m/min, the duct is said to be a high velocity duct.

20.3 Duct Material

The ducts are usually made from galvanised iron sheet metal, aluminium sheet metal or black steel. The most commonly used duct material in air conditioning systems is galvanised sheet metal, because the zinc coating of this metal prevents rusting and avoids the cost of painting. The sheet thickness of galvanised iron (*G.I.*) duct varies from 26 gauge (0.55 mm) to 16 gauge (1.6 mm). The aluminium is used because of its lighter weight and resistance to moisture. The black sheet metal is always painted unless they withstand high temperature.

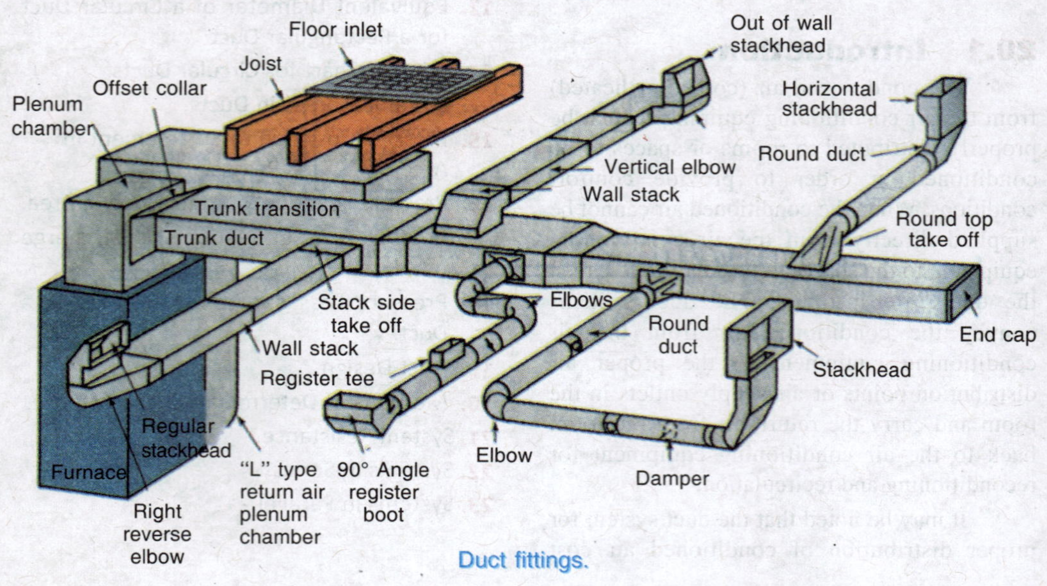

Duct fittings.

Now-a-days, the use of non-metal ducts has increased. The resin bonded glass fibre ducts are used because they are quite strong and easy to manufacture according to the desired shape and size. They are used in low velocity applications less than 600 metres/min and for static pressures below 5 mm of water gauge. The cement asbestos ducts may be used for underground air distribution and for exhausting corrosive materials. The wooden ducts may be used in places where moisture content in the air is not very large.

20.4 Duct Construction

The sheet metal ducts expand and contract as they heat and cool. The fabric joints are often used to absorb this movement. In order to prevent most of the fan and furnace noise from travelling along the duct metal, the fabric joints should also be used where ducts fasten to a furnace or an air conditioner. But, in fact, most duct joints are made of sheet metal. The various types of sheet metal joints used in the construction of ducts are shown in Fig. 20.1.

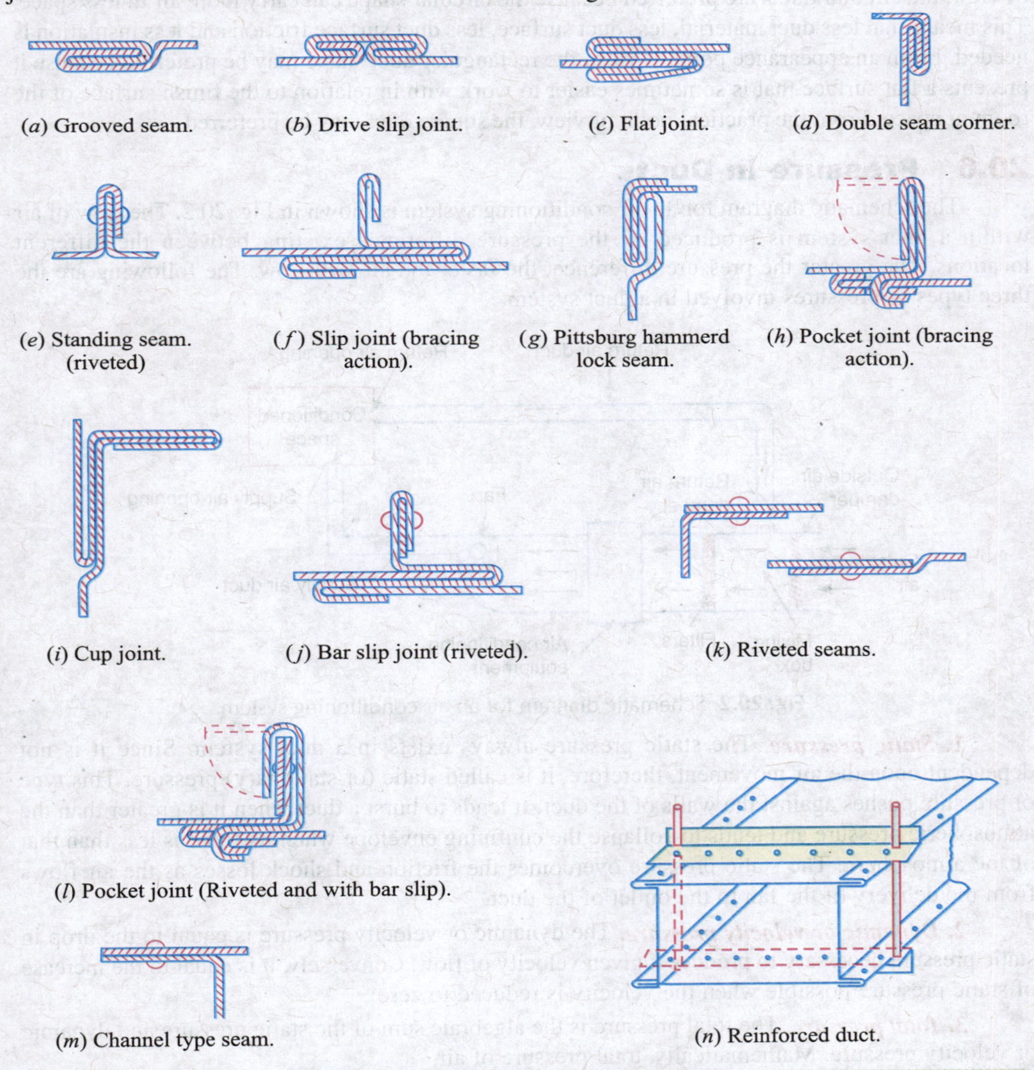

(*a*) Grooved seam. (*b*) Drive slip joint. (*c*) Flat joint. (*d*) Double seam corner.

(*e*) Standing seam. (riveted) (*f*) Slip joint (bracing action). (*g*) Pittsburg hammerd lock seam. (*h*) Pocket joint (bracing action).

(*i*) Cup joint. (*j*) Bar slip joint (riveted). (*k*) Riveted seams.

(*l*) Pocket joint (Riveted and with bar slip).

(*m*) Channel type seam. (*n*) Reinforced duct.

Fig. 20.1. Sheet metal duct joints.

The joint should be airtight and strong. Many joints are riveted for added strength and tightness. Many of these joints are also sealed with special duct tape to make them leakproof. The sealants are put in the duct seam for the same purpose.

When ducts travel through unconditioned space, they are often insulated to reduce noise as well as to decrease the rate of heat loss or gain through their walls. The insulation is fastened to the duct with adhesives. In some cases, metal clips are used to hold the insulation in place. For larger ducts and ducts under high pressures, the duct reinforcement as shown in Fig.20.1 is recommended. The duct reinforcement prevents bulging or collapsing of the duct and keeps the seams from separating. They also help to reduce noise associated with the duct vibration.

20.5 Duct Shape

The ducts may be made in circular, rectangular or square shapes. From an economical point of view, the circular ducts are preferred because the circular shape can carry more air in less space. This means that less duct material, less duct surface, less duct surface friction and less insulation is needed. From an appearance point of view, the rectangular duct shape may be preferred because it presents a flat surface that is sometimes easier to work with in relation to the finish surface of the room or space. From the practical point of view, the square duct may be preferred.

20.6 Pressure in Ducts

The schematic diagram for an air conditioning system is shown in Fig. 20.2. The flow of air within a duct system is produced by the pressure difference existing between the different locations. The greater the pressure difference, the faster the air will flow. The following are the three types of pressures involved in a duct system.

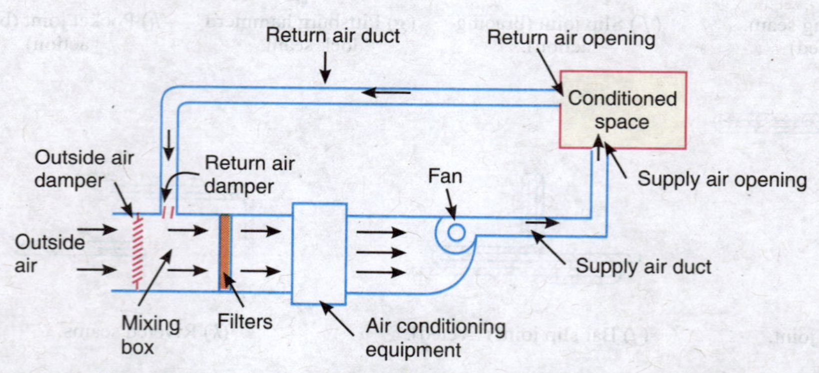

Fig. 20.2. Schematic diagram for an air conditioning system.

1. *Static pressure.* The static pressure always exists in a duct system. Since it is not dependent upon the air movement, therefore, it is called static (or stationary) pressure. This type of pressure pushes against the walls of the duct. It tends to burst a duct when it is greater than the atmospheric pressure and tends to collapse the confining envelope when its force is less than that of the atmosphere. The static pressure overcomes the friction and shock losses as the air flows from the delivery of the fan to the outlet of the duct.

2. *Dynamic or velocity pressure.* The dynamic or velocity pressure is equal to the drop in static pressure necessary to produce a given velocity of flow. Conversely, it is equal to the increase of static pressure possible when the velocity is reduced to zero.

3. *Total pressure.* The total pressure is the algebraic sum of the static pressure and dynamic or velocity pressure. Mathematically, total pressure of air,

$$p_T = p_s + p_v$$

where $\qquad p_s$ = Static pressure of air, and

p_v = Dynamic or velocity pressure of air.

The total pressure overcomes the pressure losses caused by the various obstacles on the way from the fan to the conditioned space.

Note : The static and total pressure may either be positive or negative. The dynamic or velocity pressure is always positive.

20.7 Continuity Equation for Ducts

Consider the flow of air through a duct between the two sections 1-1 and 2-2, as shown in Fig. 20.3 (*a*).

Let $\qquad Q_1$ = Quantity of air passing through section 1-1,

m_1 = Mass flow rate of air through section 1-1,

A_1 = Cross-sectional area of duct at section 1-1,

V_1 = Velocity of air at section 1-1,

ρ_1 = Density of air at section 1-1,

Q_2, m_2, A_2, V_2 and ρ_2 = Corresponding values at section 2-2.

We know that mass flow rate of air through section 1-1,

$$m_1 = \rho_1 Q_1 = \rho_1 A_1 V_1 \qquad \ldots (\because Q_1 = A_1 V_1) \qquad \ldots (i)$$

and mass flow rate of air through section 2-2,

$$m_2 = \rho_2 Q_2 = \rho_2 A_2 V_2 \qquad \ldots (\because Q_2 = A_2 V_2) \qquad \ldots (ii)$$

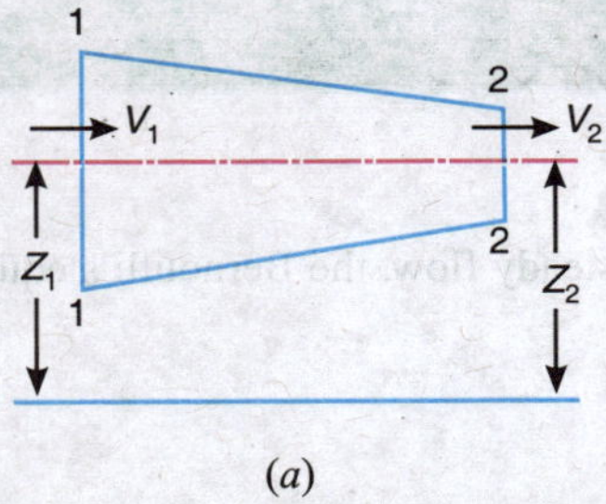

(*a*) (*b*)

Fig. 20.3. Flow of air through a duct.

Since the mass flow rate of air through sections 1-1 and 2-2 is same, therefore equating equations (*i*) and (*ii*), we get

$$\rho_1 A_1 V_1 = \rho_2 A_2 V_2$$

Normally, the density of air is assumed constant (1.2 kg / m^3) for air conditioning purposes, therefore

$$A_1 V_1 = A_2 V_2 \quad \text{or} \quad Q_1 = Q_2$$

This is called the *continuity equation* for flowing air through ducts.

When the duct 1 branches into two ducts 2 and 3 as shown in Fig. 20.3 (*b*), then

$$m_1 = m_2 + m_3 \quad \text{or} \quad \rho_1 Q_1 = \rho_2 Q_2 + \rho_3 Q_3$$

Since the density of air is assumed constant in a duct system, therefore

$$Q_1 = Q_2 + Q_3$$

Coaxial ducts.

20.8 Bernoulli's Equation for Ducts

We know that, for frictionless, incompressible and steady flow, the Bernoulli's equation is

$$\frac{p}{\rho} + \frac{V^2}{2} + g\,Z = \text{constant}$$

or

$$p + \frac{\rho V^2}{2} + \rho\,g\,Z = \text{constant}$$

Applying this equation to the two cross-sections 1-1 and 2-2 of a duct,

$$p_{s1} + \frac{\rho_1 (V_1)^2}{2} + \rho_1\,g\,Z_1 = p_{s2} + \frac{\rho_2 (V_2)^2}{2} + \rho_2\,g\,Z_2$$

Since $\rho_1 = \rho_2$ and $Z_1 = Z_2$, therefore the above expression may be written as

$$p_{s1} + \frac{\rho_1 (V_1)^2}{2} = p_{s2} + \frac{\rho_2 (V_2)^2}{2}$$

or

$$p_{s1} + p_{v1} = p_{s2} + p_{v2}$$

where p_{s1} and p_{s2} = Static gauge pressure, and

p_{v1} and p_{v2} = Velocity pressure.

From the above expression, we see that when the flow is frictionless and there is no pressure drop between the two sections, then the total pressure at the two sections will be equal. In other words, the total pressure,

$$p_T = p_{s1} + p_{v1} = p_{s2} + p_{v2} \quad \dots (i)$$

In actual practice, there is always a pressure drop in a duct due to friction and other causes such as sudden changes in the cross-section and direction. If p_L is the total pressure drop or loss between the two sections 1-1 and 2-2, then equation (*i*) is written as

$$p_{s1} + p_{v1} = p_{s2} + p_{v2} + p_L \quad \text{... (ii)}$$

In case a fan or blower is introduced between the two sections of a duct, then equation (*ii*) may be written as

$$p_{s1} + p_{v1} + p_{TF} = p_{s2} + p_{v2} + p_L \quad \text{... (iii)}$$

where p_{TF} = Rise in pressure due to fan work and is known as fan total pressure.

Notes : 1. The pressures in the duct are usually expressed in mm of water.

2. The air flowing through the duct is taken as standard air unless it is stated. The standard air is the air which corresponds to 20°C, an atmospheric pressure of 1.013 bar (or 101.3 kN/m²) and a relative humidity of 62 per cent.

3. For the standard air, the mass density (ρ_a) is equal to 1.2 kg / m³ (or 11.72 N/m³).

4. If the velocity of air (V) flowing through the duct is in m/s, then velocity pressure in the duct,

$$p_v = \frac{\rho_a V^2}{2} = \frac{1.2V^2}{2} = 0.6\, V^2 \text{ N/m}^2$$

$$= \frac{0.6V^2}{9.81} = \frac{V^2}{16.35} = \left(\frac{V}{4.04}\right)^2 \text{ mm of water}$$

$$\text{... } \left(\because 1 \text{ N/m}^2 = \frac{1}{9.81} \text{ mm of water}\right)$$

When the velocity of air is in m / min, then velocity pressure in the duct,

$$p_v = \left(\frac{V}{60 \times 4.04}\right)^2 = \left(\frac{V}{242.4}\right)^2 \text{ mm of water}$$

Example 20.1. *The main air supply duct of an air conditioning system is 800 mm × 600 mm in cross-section, and carries 300 m³/min of standard air. It branches into two ducts of cross-section 600 mm × 500 mm and 600 mm × 400 mm. If the mean velocity in the larger branch is 480 m / min, find : 1. mean velocity in the main duct and the smaller branch, and 2. mean velocity pressure in each duct.*

Solution. Given : a_1 = 800 mm = 0.8 m ; b_1 = 600 mm = 0.6 m ; Q_1 = 300 m³/min = 5 m³/s ; a_2 = 600 mm = 0.6 m ; b_2 = 500 mm = 0.5 m ; a_3 = 600 mm = 0.6 m ; b_3 = 400 mm = 0.4 m ; V_2 = 480 m / min = 8 m / s

The cross-section of the duct is shown in Fig. 20.4.

Cross-sectional area of the main duct,

$$A_1 = a_1 b_1 = 0.8 \times 0.6 = 0.48 \text{ m}^2$$

Cross-sectional area of the larger branch,

$$A_2 = a_2 b_2 = 0.6 \times 0.5 = 0.3 \text{ m}^2$$

and cross-sectional area of the smaller branch,

$$A_3 = a_3 b_3 = 0.6 \times 0.4 = 0.24 \text{ m}^2$$

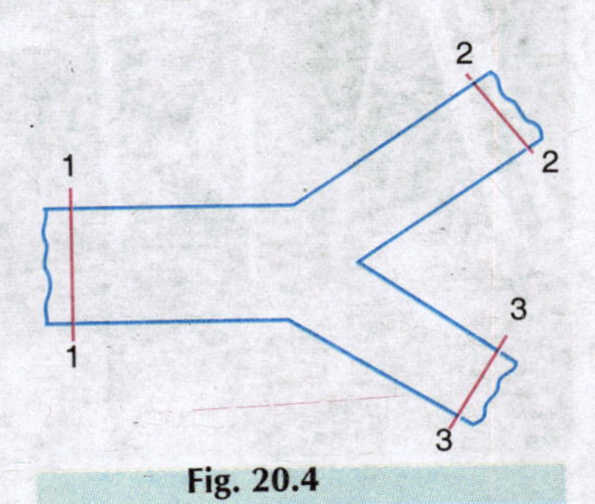

Fig. 20.4

1. *Mean velocity in the main duct and the smaller branch*

We know that the mean velocity in the main duct,

$$V_1 = \frac{Q_1}{A_1} = \frac{5}{0.4} = 10.4 \text{ m/s} \quad \textbf{Ans.}$$

Quantity of air passing through the larger branch,

$$Q_2 = A_2 V_2 = 0.3 \times 8 = 2.4 \text{ m}^3/\text{s}$$

∴ Quantity of air passing through the smaller branch,

$$Q_3 = Q_1 - Q_2 = 5 - 2.4 = 2.6 \text{ m}^3/\text{s}$$

We know that mean velocity in the smaller branch,

$$V_3 = \frac{Q_3}{A_3} = \frac{2.6}{0.24} = 10.8 \text{ m/s} \textbf{ Ans.}$$

2. ***Mean velocity pressure in each duct***

We know that velocity pressure in the main duct,

$$p_{v1} = \left(\frac{V_1}{4.04}\right)^2 = \left(\frac{10.4}{4.04}\right)^2 = 6.62 \text{ mm of water } \textbf{Ans.}$$

Mean velocity pressure in the larger branch,

$$p_{v2} = \left(\frac{V_2}{4.04}\right)^2 = \left(\frac{8}{4.04}\right)^2 = 3.92 \text{ mm of water } \textbf{Ans.}$$

and mean velocity pressure in the smaller branch,

$$p_{v3} = \left(\frac{V_3}{4.04}\right)^2 = \left(\frac{10.8}{4.04}\right)^2 = 7.14 \text{ mm of water } \textbf{Ans.}$$

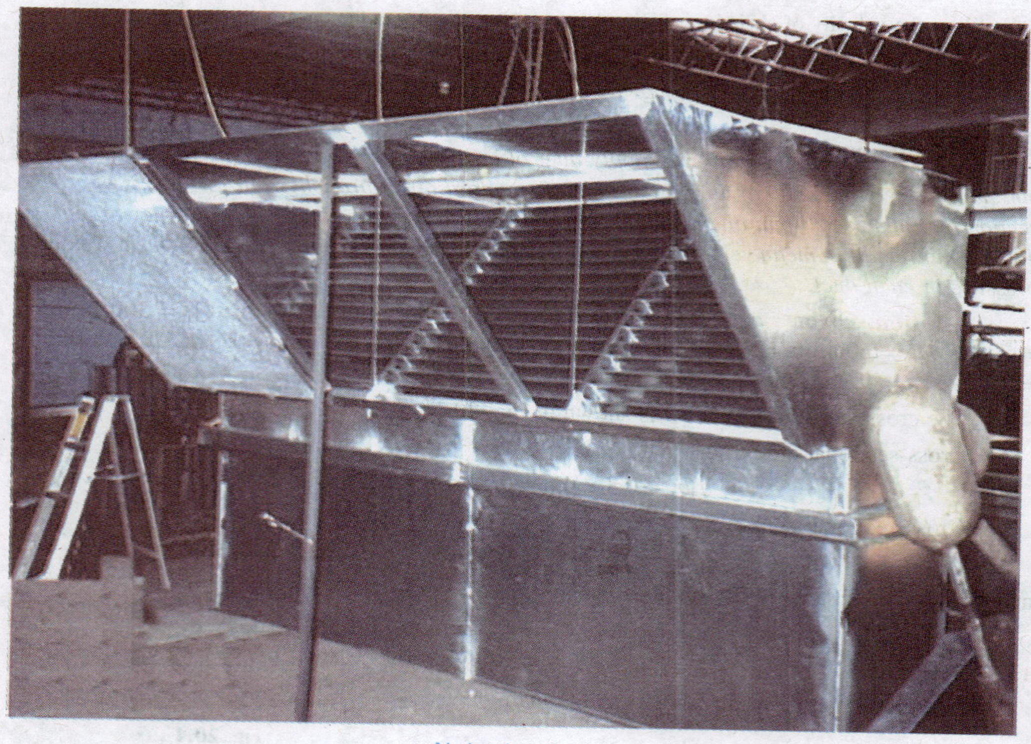

Air intake ducts.

20.9 Pressure Losses in Ducts

A little consideration will show that the pressure is lost due to friction between the moving particles of the fluid (*i.e.* air) and the interior surfaces of a duct. When the pressure loss occurs in a straight duct, it is usually termed as *friction loss*. The pressure is also lost dynamically at the changes of direction such as in bends, elbows etc. and at the changes of cross-section of the duct. This type of pressure loss is usually termed as *dynamic loss*.

We shall now discuss these pressure losses, in detail, in the following pages.

20.10 Pressure Loss due to Friction in Ducts

The pressure loss due to friction in ducts may be obtained by using the D'Arcy's formula or the Fanning's equation, *i.e.*,

$$p_f = \frac{f L \rho_a V^2}{2m} \qquad \ldots (i)$$

where p_f = Pressure loss due to friction in N/m²,

f = Friction factor depending upon the surface of the duct (dimensionless),

L = Length of the duct in metres,

V = Mean velocity of the air flowing through the duct in m/s, and

m = Hydraulic mean depth in metres

$$= \frac{\text{Cross-sectional area of the duct } (A)}{\text{Wetted perimeter of the duct } (P)}$$

In air conditioning, the pressure loss due to friction in ducts is generally expressed in mm of water. From equation (*i*),

$$p_f = \frac{f L}{m}\left(\frac{\rho_a V^2}{2}\right) = \frac{f L}{m} \times p_v \text{ mm of water} \qquad \ldots (ii)$$

We have seen in Art. 20.8 that the velocity pressure in the duct for standard air is given by

$$p_v = \frac{\rho_a V^2}{2} = \left(\frac{V}{4.04}\right)^2 \text{ mm of water}$$

Thus equation (*ii*) may be written as

$$p_f = \frac{f L}{m}\left(\frac{V}{4.04}\right)^2 \text{ mm of water} \qquad \ldots (iii)$$

Notes : 1. When air is at a temperature $t°$ C, then pressure loss due to friction in the duct is given by

$$p_f = \frac{f L}{m}\left(\frac{V}{4.04}\right)^2 \left(\frac{273 + 20}{273 + t}\right)$$

$$= \frac{f L}{m}\left(\frac{V}{4.04}\right)^2 \left(\frac{293}{273 + t}\right) \text{ mm of water}$$

2. For a circular duct of diameter D, the hydraulic mean depth,

$$m = \frac{A}{P} = \frac{\frac{\pi}{4} \times D^2}{\pi D} = \frac{D}{4}$$

3. For a rectangular duct of sides a and b, the hydraulic mean depth,

$$m = \frac{A}{P} = \frac{ab}{2(a+b)}$$

Example 20.2. *A duct of 15 m length passes air at the rate of 90 m³/min. Assuming the friction factor as 0.005, calculate the pressure drop in the duct in mm of water when (a) the duct is circular of diameter 0.3 m ; and (b) the duct is of 0.3 m square cross-section.*

Solution. Given : $L = 15$ m ; $Q = 90$ m³/min ; $f = 0.005$

(a) Pressure drop in a circular duct

Let D = Diameter of the circular duct = 0.3 m ... (Given)

Cross-sectional area of the duct,

$$A = \frac{\pi}{4} \times D^2 = \frac{\pi}{4}(0.3)^2 = 0.07 \text{ m}^2$$

∴ Velocity of air passing through the duct,

$$V = \frac{Q}{A} = \frac{90}{0.07} = 1285.7 \text{ m/min} = 21.4 \text{ m/s}$$

Wetted perimeter of the duct,

$$P = \pi D = \pi \times 0.3 = 0.94 \text{ m}$$

∴ Hydraulic mean depth of the duct,

$$m = \frac{A}{P} = \frac{0.07}{0.94} = 0.074 \text{ m}$$

We know that pressure drop in a duct,

$$p_f = \frac{fL}{m}\left(\frac{V}{4.04}\right)^2 = \frac{0.005 \times 15}{0.074}\left(\frac{21.4}{4.04}\right)^2 \text{ mm of water}$$

$$= 28.47 \text{ mm of water } \textbf{Ans.}$$

(b) Pressure drop in a square duct

Let $a = b$ = Sides of the square duct = 0.3 m ... (Given)

Cross-sectional area of the duct,

$$A = a \times b = 0.3 \times 0.3 = 0.09 \text{ m}^2$$

∴ Velocity of air passing through the duct,

$$V = \frac{Q}{A} = \frac{90}{0.09} = 1000 \text{ m/min} = 16.67 \text{ m/s}$$

Wetted perimeter of the duct,

$$P = 2(a + b) = 2(0.3 + 0.3) = 1.2 \text{ m}$$

∴ Hydraulic mean depth of the duct,

$$m = \frac{A}{P} = \frac{0.09}{1.2} = 0.075$$

We know that pressure drop in a duct,

$$p_f = \frac{fL}{m}\left(\frac{V}{4.04}\right)^2 = \frac{0.005 \times 15}{0.075}\left(\frac{16.67}{4.04}\right)^2 \text{ mm of water}$$

$$= 17.02 \text{ mm of water } \textbf{Ans.}$$

20.11 Friction Factor for Ducts

The value of friction factor, for smooth ducts, may be obtained by using the following expressions :

1. For laminar flow, the friction factor,

$$f = \frac{64}{R_N}$$

2. For turbulent flow, the friction factor,

$$f = \frac{0.3164}{(R_N)^{0.25}}$$

where R_N is the Reynold number. It is defined as the ratio of the inertia forces to the viscous forces. It may be noted that the Reynold number is a dimensionless quantity and gives us the information about the type of flow (*i.e.* *laminar or turbulent). Mathematically, Reynold number,

$$R_N = \frac{\text{Inertia forces}}{\text{Viscous forces}} = \frac{\rho_a V^2}{\mu V / D} = \frac{\rho_a DV}{\mu} = \frac{DV}{\mu / \rho_a} = \frac{DV}{K}$$

where
V = Velocity of air in m/s ,
D = Diameter of the duct in metres,
ρ_a = Mass density of air in kg / m^3,
μ = Absolute or dynamic viscosity in N-s/m^2,
K = Kinematic viscosity in m^2/s = μ / ρ_a.

The following table shows the values of absolute viscosity and mass density of air at different temperatures and at atmospheric pressure of 1.013 bar.

Table 20.1. Values of absolute viscosity and density of air at different temperatures.

Temperature in °C	*Absolute or dynamic viscosity (μ) in N-s / m²*	*Density in kg / m³*
0	17.24×10^{-6}	1.29
10	17.71×10^{-6}	1.25
20	18.18×10^{-6}	1:20
30	18.65×10^{-6}	1.16
40	19.12×10^{-6}	1.13
50	19.60×10^{-6}	1.09

In case of rough pipes or ducts, the friction factor (f) depends upon the roughness factor e / D, where e is the absolute roughness of the surface and D is the diameter of the duct. Table 20.2 shows the values of e for different types of commercial pipes or duct work.

The friction factor (f) for rough pipes or ducts may be obtained from the following equation :

$$f = \frac{1}{\left[1.74 - 2\log\left(\frac{2e}{D}\right)\right]^2}$$

The values of friction factor (f) for different Reynold numbers (R_N) and different roughness factors (e/D) may be read directly from the Moody chart as shown in Fig. 20.5.

* When the Reynold number (R_N) is less than 2000, the flow is said to be *laminar flow.* But if the Reynold number is between 2000 and 4000, the flow is neither laminar nor turbulent flow. When the Reynold number exceeds 4000, the flow is said to be *turbulent flow.*

It may be noted that the values given in Fig. 20.5 are independent of the fluid and apply equally well to water, air or other gases when they are evaluated in terms of Reynold number.

Note : When the friction factor 'f' is obtained from the Moody chart as shown in Fig. 20.5, then the pressure loss due to friction in a circular duct of diameter 'D' is given by

$$p_f = \frac{fL}{D} \times p_v = \frac{fL}{D}\left(\frac{V}{4.04}\right)^2 \text{ mm of water}$$

Table 20.2. Recommended surface roughness (e) for different material pipes or ducts.

Types of pipe or duct	*Absolute surface roughness (e) in mm*
Smooth-drawn tubes (glass, brass and lead)	0.0015
Commercial steel or wrought iron pipe	0.045
Galvanised iron or steel air ducts	0.15
Cast iron	0.255
Riveted steel (light weight, small rivets)	0.9
Riveted steel (heavy weight, large rivets)	9.0
Smooth concrete	0.3
Average concrete	1.2
Very rough concrete	3.0
Brick conduits	9.0
Wood-stave conduits	0.9

Example 20.3. *A galvanised steel duct of 0.4 m diameter and 20 m long carries air at 20°C and 1.013 bar. If the flow rate of air through the duct is 60 m³/min, determine the pressure loss due to friction in (a) N/m² ; and (b) in mm of water.*

Solution. Given : $D = 0.4$ m ; $L = 20$ m ; $t = 20°C$; $p = 1.013$ bar ; $Q = 60$ m³/min $= 1$ m³/s

We know that velocity of air in the duct,

$$V = \frac{Q}{A} = \frac{Q}{\pi D^2 / 4} = \frac{1 \times 4}{\pi (0.4)^2} = 7.96 \text{ m/s}$$

From Table 20.1, we find that for air at 20°C and atmospheric pressure of 1.013 bar, mass density of air,

$$\rho_a = 1.2 \text{ kg/m}^3$$

and absolute or dynamic viscosity

$$\mu = 18.18 \times 10^{-6} \text{ N-s/m}^2$$

∴ Reynold number,

$$R_N = \frac{\rho_a D V}{\mu} = \frac{1.2 \times 0.4 \times 7.96}{18.18 \times 10^{-6}} = 210\,165$$

Now from Table 20.2, surface roughness for a galvanised steel duct,

$$e = 0.15 \text{ mm} = 0.000\,15 \text{ m}$$

∴ Roughness factor,

$$\frac{e}{D} = \frac{0.000\,15}{0.4} = 0.000\,375$$

From the Moody chart as shown in Fig. 20.5, we find that corresponding to $R_N = 210\,165$ and $e/D = 0.000\,375$, the friction factor,

$$f = 0.016$$

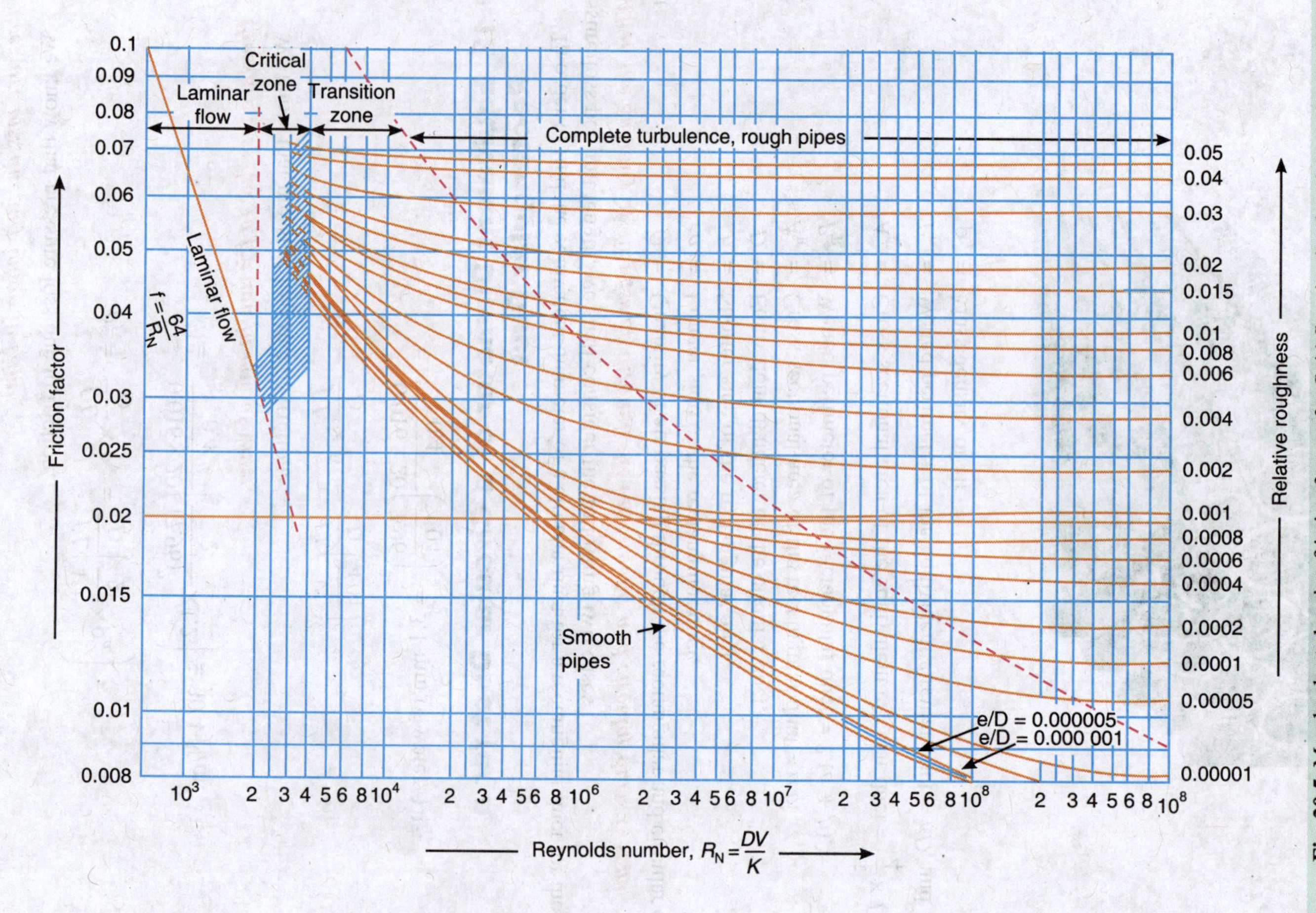

Fig. 20.5. Moody chart showing friction factors for fluid flow in circular pipes or ducts.

(a) Pressure loss due to friction in N/m²

We know that pressure loss due to friction,

$$p_f = \frac{f\,L}{D} \times p_v = \frac{f\,L}{D}\left(\frac{V^2}{2} \times \rho_a\right)$$

$$= \frac{0.016 \times 20}{0.4}\left(\frac{(7.96)^2}{2} \times 1.2\right) = 30.4 \text{ N/m}^2 \textbf{ Ans.}$$

(b) Pressure loss due to friction in mm of water

We know that pressure loss due to friction,

$$p_f = \frac{f\,L}{D} \times p_v = \frac{f\,L}{D}\left(\frac{V}{4.04}\right)^2$$

$$= \frac{0.016 \times 20}{0.4}\left(\frac{7.96}{4.04}\right)^2 = 3.1 \text{ mm of water } \textbf{Ans.}$$

20.12 Equivalent Diameter of a Circular Duct for a Rectangular Duct

In order to find the equivalent diameter of a circular duct for a rectangular duct for the same pressure loss per unit length, we shall consider the following two cases :

1. *When the quantity of air passing through the rectangular and circular ducts is same*

Let Q = Quantity of air passing through the rectangular and circular ducts,

a = Longer side of the rectangular duct,

b = Shorter side of the rectangular duct,

D = Equivalent diameter of the circular duct,

A_R = Cross-sectional area of the rectangular duct = $a.b$,

P_R = Wetted perimeter of the rectangular duct = $2\,(a + b)$,

A_C = Cross-sectional area of the equivalent circular duct = $\frac{\pi}{4} \times D^2$,

P_C = Wetted perimeter of the equivalent circular duct = πD, and

ρ_a = Mass density of air.

Spiral ducts.

Velocity of air passing through the circular duct,

$$V_C = \frac{Q}{A_C}$$

and velocity of air passing through the rectangular duct,

$$V_R = \frac{Q}{A_R}$$

We know that pressure loss due to friction,

$$p_f = \frac{f L \rho_a V^2}{2\,m} = \frac{f L \rho_a}{2\,m}\left(\frac{Q}{A}\right)^2 \qquad \ldots \left(\because V = \frac{Q}{A}\right)$$

and hydraulic mean depth,

$$m = \frac{\text{Cross-sectional area of duct } (A)}{\text{Wetted perimeter of duct } (P)}$$

∴ Pressure loss due to friction for the circular duct,

$$p_{fC} = \frac{f L \rho_a}{2}\left(\frac{P_C}{A_C}\right)\left(\frac{Q}{A_C}\right)^2 = \frac{f L \rho_a Q^2}{2}\left(\frac{P_C}{(A_C)^3}\right) \qquad \ldots (i)$$

and pressure loss due to friction for the rectangular duct,

$$p_{fR} = \frac{f L \rho_a}{2}\left(\frac{P_R}{A_R}\right)\left(\frac{Q}{A_R}\right)^2 = \frac{f L \rho_a Q^2}{2}\left(\frac{P_R}{(A_R)^3}\right) \qquad \ldots (ii)$$

Since the pressure loss, friction factor, length, density and quantity of air for the circular and rectangular ducts is same, therefore from equations (*i*) and (*ii*),

$$\frac{P_C}{(A_C)^3} = \frac{P_R}{(A_R)^3} \quad \text{or} \quad \frac{\pi D}{\left(\frac{\pi}{4} \times D^2\right)^3} = \frac{2(a+b)}{(ab)^3}$$

$$\therefore \quad \frac{32}{\pi^2 D^5} = \frac{a+b}{a^3 b^3} \quad \text{or} \quad D^5 = \frac{32\,a^3 b^3}{\pi^2 (a+b)}$$

or

$$D = \left[\frac{32\,a^3 b^3}{\pi^2 (a+b)}\right]^{1/5} = 1.265\left(\frac{a^3 b^3}{a+b}\right)^{1/5} \qquad \ldots (iii)$$

2. *When the velocity of air passing through the rectangular and circular ducts is same*

Let V = Velocity of air passing through the rectangular and circular ducts.

We know that the pressure loss due to friction for a circular duct,

$$p_{fC} = \frac{f L \rho_a V^2}{2}\left(\frac{P_C}{A_C}\right) \qquad \ldots (iv)$$

and pressure loss due to friction for a rectangular duct,

$$p_{fR} = \frac{f L \rho_a V^2}{2}\left(\frac{P_R}{A_R}\right) \qquad \ldots (v)$$

Since the pressure loss, velocity of air, friction factor, density and length for the circular and rectangular ducts is same, therefore from equations (*iv*) and (*v*),

$$\frac{P_C}{A_C} = \frac{P_R}{A_R} \quad \text{or} \quad \frac{\pi D}{\frac{\pi}{4} \times D^2} = \frac{2(a+b)}{ab}$$

$$\therefore \qquad D = \frac{2ab}{a+b} = \frac{2a}{a/b+1} \qquad \text{... (vi)}$$

where a / b is known as *aspect ratio*.

Note : The aspect ratio, for rectangular ducts, should not be greater than 8 in any case.

Example 20.4. *A rectangular duct section of 500 mm × 350 mm size carries 75 m³/min of air having density of 1.15 kg / m³. Determine the equivalent diameter of a circular duct if (a) the quantity of air carried in both the cases is same, and (b) the velocity of air in both the cases is same.*

If f = 0.01 for sheet metal, find the pressure loss per 100 m length of duct.

Solution. Given : a = 500 mm = 0.5 m ; b = 350 mm = 0.35 m ; Q = 75 m³/min ; ρ_a = 1.15 kg / m³ ; f = 0.01 ; L = 100 m

(a) Equivalent diameter of a circular duct if the quantity of air carried in both the cases is same

When the quantity of air carried by rectangular and circular ducts is same, then equivalent diameter of a circular duct,

$$D = 1.265\left(\frac{a^3 b^3}{a+b}\right)^{1/5} = 1.265\left[\frac{(0.5)^3(0.35)^3}{0.5+0.35}\right]^{1/5}$$

$$= 1.265\left(\frac{0.005\,36}{0.85}\right)^{0.2} = 1.265 \times 0.363 = 0.46 \text{ m} \quad \textbf{Ans.}$$

(b) Equivalent diameter of a circular duct if the velocity of air in both the cases is same

When the velocity of air passing through the rectangular and circular ducts is same, then equivalent diameter of a circular duct,

$$D = \frac{2ab}{a+b} = \frac{2 \times 0.5 \times 0.35}{0.5+0.35} = 0.41 \text{ m} \quad \textbf{Ans.}$$

Pressure loss

We know that velocity of air passing through the duct,

$$V = \frac{Q}{A} = \frac{Q}{ab} = \frac{75}{0.5 \times 0.35} = 428.6 \text{ m/min} = 7.143 \text{ m/s}$$

and mean hydraulic depth of the duct,

$$m = \frac{A}{P} = \frac{ab}{2(a+b)} = \frac{0.5 \times 0.35}{2\,(0.5+0.35)} = 0.103 \text{ m}$$

$\therefore$ Pressure loss,

$$p_f = \frac{f L \rho_a V^2}{2m} = \frac{0.01 \times 100 \times 1.15\,(7.143)^2}{2 \times 0.103} = 284.8 \text{ N/m}^2$$

$$= 284.8 / 9.81 = 29 \text{ mm of water} \quad \textbf{Ans.}$$

$$\text{... } \left(\because\ 1 \text{ N/m}^2 = \frac{1}{9.81} \text{ mm of water}\right)$$

Example 20.5. *A duct 2m by 1m in size carrying conditioned air runs in a straight line for 50 m from the supply fan. It divides into two parts each of 80 m long and 2 m by 1 m in cross-section as shown in Fig. 20.6.*

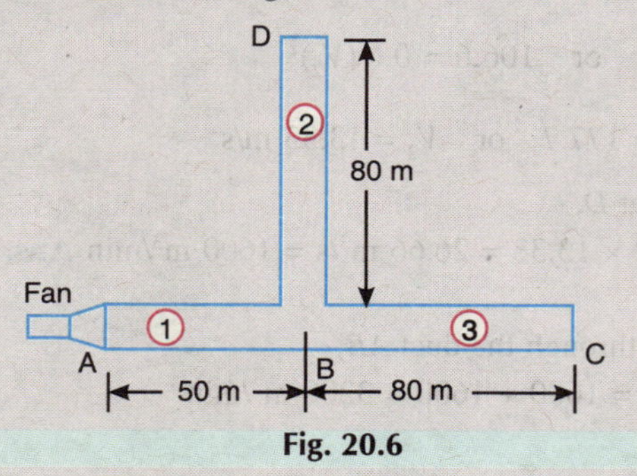

Fig. 20.6

Insulated air ducts.

If the quantity of air discharged at C is 1600 m³/min, calculate the quantity discharged at D and the static pressure at the fan outlet A. Calculate the duct friction loss in N/m² taking the value of friction factor as 0.005.

Solution. Given : $a = 2$ m ; $b = 1$ m ; $L_1 = 50$ m ; $L_2 = 80$ m ; $L_3 = 80$ m ; $Q_3 = 1600$ m³/min ; $f = 0.005$

Quantity of air discharged at D

Let Q_2 = Quantity of air discharged at *D*.

Cross-sectional area of ducts *AB*, *DB* and *BC*,

$$A_1 = A_2 = A_3 = a \times b = 2 \times 1 = 2 \text{ m}^2$$

Velocity of air in duct *BC*,

$$V_3 = \frac{Q_3}{A_3} = \frac{1600}{2} = 800 \text{ m / min} = 13.33 \text{ m/s}$$

We know that velocity pressure,

$$p_v = \frac{V^2}{2} \times \rho_a = \frac{V^2}{2} \times 1.2 = 0.6\, V^2 \text{ N/m}^2$$

... (Taking $\rho_a = 1.2$ kg/m³)

∴ Velocity pressure in duct *BC*,

$$p_{v3} = 0.6\,(V_3)^2 = 0.6\,(13.33)^2 = 106.6 \text{ N/m}^2$$

Since each duct is 2 m by 1 m, therefore hydraulic mean depth for ducts *AB*, *BD*, and *BC* is

$$m_1 = m_2 = m_3 = \frac{ab}{2(a+b)} = \frac{2 \times 1}{2(2+1)} = \frac{1}{3}$$

∴ Pressure loss due to friction in duct *BC*,

$$p_{f3} = \frac{f\,L_3}{m_3}(p_{v3}) = \frac{0.005 \times 80}{1/3} \times 106.6 = 128 \text{ N/m}^2$$

and total pressure at *B*,

$$p_{TB} = p_{f3} + p_{v3} = 128 + 106.6 = 234.6 \text{ N/m}^2 \quad \text{... (i)}$$

Let p_{v2} = Velocity pressure in the duct *BD*, and

V_2 = Velocity in the duct *BD*.

We know that the pressure loss due to friction in the duct *BD*,

$$p_{f2} = \frac{f\,L_2}{m_2}(p_{v2}) = \frac{0.005 \times 80}{1/3}(p_{v2}) = 1.2\,(p_{v2}) \text{ N/m}^2 \quad \text{... (ii)}$$

From equations (*i*) and (*ii*), we get

$$p_{v2} = \frac{234.6}{1.2} = 106.6 \text{ N/m}^2$$

We know that $p_{v2} = 0.6\,(V_2)^2$ or $106.6 = 0.6\,(V_2)^2$

$$\therefore \quad (V_2)^2 = \frac{106.6}{0.6} = 177.7 \quad \text{or} \quad V_2 = 13.33 \text{ m/s}$$

We know that quantity of air discharged at *D*,

$$Q_2 = A_2 V_2 = 2 \times 13.33 = 26.66 \text{ m}^3\text{/s} = 1600 \text{ m}^3\text{/min} \textbf{ Ans.}$$

Static pressure at the fan outlet

We know that the quantity of air flowing through the duct *AB*,

$$Q_1 = Q_2 + Q_3 = 1600 + 1600 = 3200 \text{ m}^3\text{/min}$$

∴ Velocity of air in the duct *AB*,

$$V_1 = \frac{Q_1}{A_1} = \frac{3200}{2} = 1600 \text{ m / min} = 26.67 \text{ m / s}$$

Velocity pressure in the duct *AB*,

$$p_{v1} = 0.6\,(V_1)^2 = 0.6\,(26.67)^2 = 427 \text{ N/m}^2$$

Pressure loss due to friction in the duct *AB*,

$$p_{f1} = \frac{f\,L_1}{m_1} \times p_{v1} = \frac{0.005 \times 50}{1/3} \times 427 = 320 \text{ N/m}^2$$

Total pressure at the fan outlet (*i.e.* at *A*),

$$p_{TA} = p_{TB} + p_{f1} = 234.6 + 320 = 554.6 \text{ N/m}^2$$

∴ Static pressure at the fan outlet (*i.e.* at *A*),

$$p_{SA} = p_{TA} - p_{v1} = 554.6 - 427 = 127.6 \text{ N/m}^2 \textbf{ Ans.}$$

Example 20.6. *Fig. 20.7 shows a duct system containing a dehumidifying coil. The air enters at a dry bulb temperature of 26° C and relative humidity of 60% and leaves at a dry bulb temperature of 14°C and relative humidity of 90%. The chilled water enters the cooling coil at the rate of 15 kg / s and its temperature rise is 4.03°C. The velocity in the main duct is 10 m/s. The aspect ratio for each duct is 2 : 1. The main duct line is A-B-C with pressure drop per unit length being same and there is no damper in the system. The pressure drop due to 90° bend is equivalent to 3 times the diameter of the duct at the point considered.*

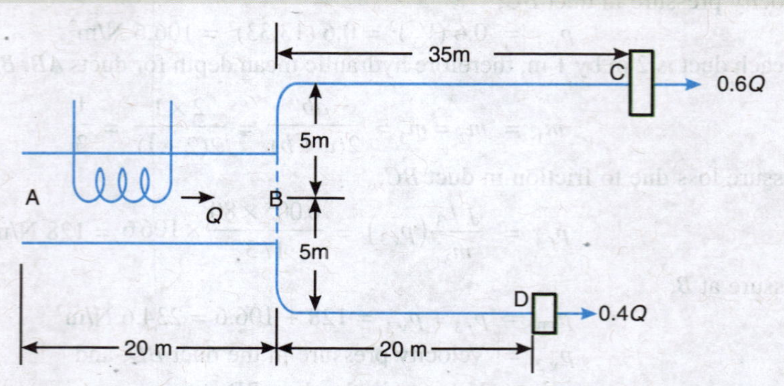

Fig. 20.7

Determine : 1. Air flow rate (m³/s) ; 2. Main duct dimensions ; 3. Size of duct BC ; 4. Total pressure ; and 5. Dimensions of duct BD.

Solution. Given : $t_{dA} = 26°C$; $\phi_A = 60\%$; $t_{dB} = 14°$ C ; $\phi_B = 90\%$; $m_w = 15$ kg / s ; $\Delta t_w = 4.03°C$; $V_{AB} = 10$ m / s ; Aspect ratio = $a / b = 2 : 1$; $p_b = 3D_d$; $L_{AB} = 20$ m ; $L_{BC} = 5 + 35 = 40$ m ; $L_{BD} = 5 + 20 = 25$ m

1. *Air flow rate*

Let Q_{AB} = Air flow rate through the duct *AB* in m³/s.

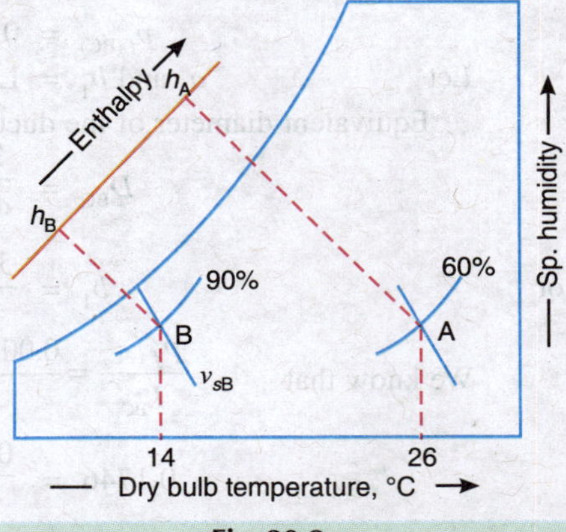

Fig. 20.8

The conditions of entering and leaving air are marked as points *A* and *B* respectively on the psychrometric chart, as shown in Fig. 20.8. From the psychrometric chart, we find that enthalpy of entering air at point *A*,

$$h_A = 58.4 \text{ kJ / kg of dry air}$$

Enthalpy of leaving air at point *B*,

$$h_B = 36.5 \text{ kJ / kg of dry air}$$

Specific volume of leaving air at point 2,

$$v_{sB} = 0.825 \text{ m}^3 \text{ / kg of dry air}$$

Let *m* be the mass rate of air (in kg / s) flowing through the duct *AB*, then the decrease in enthalpy of air is equal to the heat carried away by chilled water.

or
$$m\,(h_A - h_B) = m_w \times c_w \times \Delta t_w$$
$$m\,(58.4 - 36.5) = 15 \times 4.187 \times 4.03 = 253$$

... (Taking specific heat of water, c_w = 4.187 kJ / kg K)

and air flow rate,
$$Q_{AB} = m / v_{sB} = 11.55 / 0.825 = 14 \text{ m}^3\text{/s} \quad \textbf{Ans.}$$

2. *Main duct dimensions*

Let a and b = Longer and shorter side of the main duct *AB*.

We know that cross-sectional area of the main duct *AB*,

$$A_{AB} = a \times b = 2b \times b = 2b^2 \qquad ...(\because a/b = 2)$$

and quantity of air flowing through the main duct *AB*,

$$Q_{AB} = A_{AB} \times V_{AB} = 2b^2 \times 10 = 20\,b^2$$

$$\therefore \quad b^2 = Q_{AB} / 20 = 14 / 20 = 0.7 \quad \text{or} \quad b = 0.836 \text{ m} \quad \textbf{Ans.}$$

and
$$a = 2b = 2 \times 0.836 = 1.672 \text{ m} \quad \textbf{Ans.}$$

3. *Size of duct BC*

We know that equivalent diameter of the duct *AB*,

$$D_{AB} = \frac{2ab}{a+b} = \frac{2 \times 1.672 \times 0.836}{1.672 + 0.836} = 1.115$$

and pressure loss due to friction in duct *AB* per metre length,

$$\frac{p_{f(AB)}}{L_{AB}} = \frac{0.002\,268\,(Q_{AB})^{1.852}}{(D_{AB})^{4.973}} = \frac{0.002\,268\,(14)^{1.852}}{(1.115)^{4.973}}$$

$$= \frac{0.3}{1.718} = 0.1746 \text{ mm of water / m length}$$

This is constant for all sections, *i.e.*

$$\frac{p_{f(AB)}}{L_{AB}} = \frac{p_{f(BC)}}{L_{BC}} = \frac{p_{f(BD)}}{L_{BD}} = 0.1746 \text{ mm of water / m length}$$

∴ Pressure loss due to friction in duct AB, ...(i)

$$p_{f(AB)} = 0.1746 \times L_{AB} = 0.1746 \times 20 = 3.492 \text{ mm of water}$$

Pressure loss due to friction in duct BC,

$$p_{f(BC)} = 0.1746 \times L_{BC} = 0.1746 \times 40 = 6.984 \text{ mm of water}$$

Let a_1 and b_1 = Longer and shorter side of the duct BC.

∴ Equivalent diameter of the duct BC,

$$D_{BC} = \frac{2a_1 b_1}{a_1 + b_1} = \frac{2 \times 2b_1 \times b_1}{2b_1 + b_1} = \frac{4b_1}{3} \quad \text{...}(\because a_1 = 2b_1)$$

or

$$b_1 = \frac{3D_{BC}}{4} = 0.75\, D_{BC} \quad \text{...(ii)}$$

We know that

$$\frac{p_{f(BC)}}{L_{BC}} = \frac{0.002\ 268\ (Q_{BC})^{1.852}}{(D_{BC})^{4.973}}$$

$$0.1746 = \frac{0.002\ 268\ (0.6 \times 14)^{1.852}}{(D_{BC})^{4.973}} = \frac{0.1168}{(D_{BC})^{4.973}}$$

...($\because Q_{BC} = 0.6\ Q_{AB}$)

or

$$D_{BC} = \left(\frac{0.1168}{0.1746}\right)^{1/4.973} = (0.669)^{0.2011} = 0.922 \text{ m}$$

We know that $b_1 = 0.75\ D_{BC} = 0.75 \times 0.922 = 0.6915$ m **Ans.**

...[From equation (ii)]

and $a_1 = 2b_1 = 2 \times 0.6915 = 1.383$ m **Ans.**

4. *Total pressure*

We know that velocity of air flowing through the duct AB,

$$V_{AB} = \frac{Q_{AB}}{A_{AB}} = \frac{Q_{AB}}{a \times b} = \frac{14}{1.672 \times 0.836} = 10 \text{ m / s}$$

and velocity of air flowing through the duct BC,

$$V_{BC} = \frac{Q_{BC}}{A_{BC}} = \frac{0.6\ Q_{AB}}{a_1 \times b_1} = \frac{0.6 \times 14}{1.383 \times 0.6915} = 8.783 \text{ m / s}$$

∴ Velocity pressure in duct AB,

$$p_{v(AB)} = \left(\frac{V_{AB}}{4.04}\right)^2 = \left(\frac{10}{4.04}\right)^2 = 6.127 \text{ mm of water}$$

and velocity pressure in duct BC,

$$p_{v(BC)} = \left(\frac{V_{BC}}{4.04}\right)^2 = \left(\frac{8.783}{4.04}\right)^2 = 4.726 \text{ mm of water}$$

The dynamic losses between A and C are as follows :

(a) Pressure loss at discharge opening at C

$$= p_{v(BC)} = 4.726 \text{ mm of water}$$

(b) Elbow loss at B $= 0.25\ p_{v(BC)}$

$$= 0.25 \times 4.726 = 1.1815 \text{ mm of water}$$

(c) Fitting loss at B $= 0.25\ [\,p_{v(AB)} - p_{v(BC)}\,]$

$$= 0.25\ [6.127 - 4.726] = 0.35 \text{ mm of water}$$

(*d*) Presure loss in dehumidifier coil

$= $ Usually 2 to 10 mm of water. Let us take it as 5 mm of water

∴ Total dynamic loss between *A* and *C*,

$$p_d = (a) + (b) + (c) + (d)$$

$$= 4.726 + 1.1815 + 0.35 + 5 = 11.2575 \text{ mm of water}$$

and total pressure (at fan exit)

$$= p_{f(\text{AB})} + p_{f(\text{AB})} + p_d$$

$$= 3.492 + 6.984 + 11.2575 = 21.7335 \text{ mm of water} \quad \textbf{Ans.}$$

5. *Dimensions of duct BD*

Let a_2 and b_2 = Longer and shorter sides of the duct *BD*.

We know that equivalent diameter of the duct *BD*,

$$D_{\text{BD}} = \frac{2a_2b_2}{a_2+b_2} = \frac{2\times 2b_2 \times b_2}{2b_2+b_2} = \frac{4b_2^2}{3b_2} = \frac{4b_2}{3} \quad \ldots (\because a_2 = 2b_2)$$

or $$b_2 = 0.75\, D_{BD}$$

We know that

$$\frac{p_{f(\text{BD})}}{L_{\text{BD}}} = \frac{0.002\ 268\ (Q_{\text{BD}})^{1.852}}{(D_{\text{BD}})^{4.973}}$$

$$0.1746 = \frac{0.002\ 268\ (0.4\times 14)^{1.852}}{(D_{\text{BD}})^{4.973}} = \frac{0.055}{(D_{\text{BD}})^{4.973}} \quad \ldots (\because Q_{\text{BD}} = 0.4\, Q_{\text{AB}})$$

∴ $$D_{\text{BD}} = \left(\frac{0.055}{0.1746}\right)^{1/4.973} = (0.3157)^{0.2011} = 0.793 \text{ m}$$

We know that $b_2 = 0.75\, D_{\text{BD}} = 0.75 \times 0.793 = 0.595$ m **Ans.**

and $a_2 = 2b_2 = 2 \times 0.595 = 1.19$ m **Ans.**

Example 20.7. *A ventilation duct 1.8 m by 1.2 m runs in a straight line for 45 m from the supply fan, then bifurcates into parts each 60 m long and each 1.8 m by 1.2 m, as shown in Fig. 20.9 (a). In order to reduce the air noise to a minimum, the branch BD is divided into three parts and lined with absorbent material of negligible thickness as shown in Fig. 20.9 (b). If the quantity of air discharged at C is 1360 m³/min, calculate the quantity of air discharged at D and the static pressure at the fan outlet. The friction factor 'f' is 0.0055 for sheet metal and 0.007 for the absorbent material.*

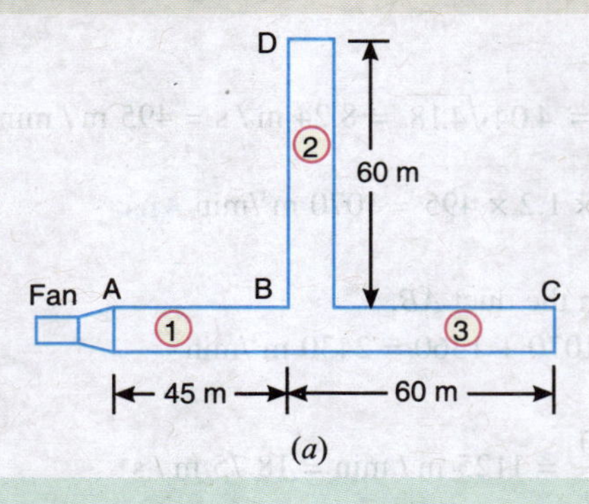

(*a*)

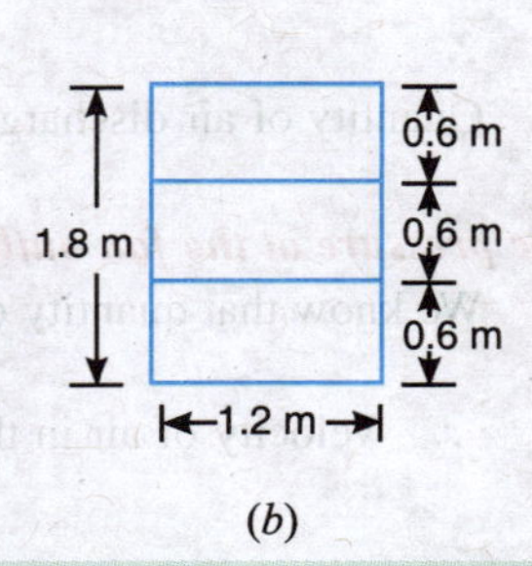

(*b*)

Fig. 20.9

Solution. Given : $a = 1.8$ m ; $b = 1.2$ m ; $L_1 = 45$ m ; $L_2 = L_3 = 60$ m ; $Q_3 = 1360$ m³/min ; $f_1 = f_3 = 0.0055$; $f_2 = 0.007$

Quantity of air discharged at D

We know that cross-sectional area of the duct *AB* or *BC*,

$$A_1 = A_3 = a \times b = 1.8 \times 1.2 = 2.16 \text{ m}^2$$

∴ Velocity of air in the duct *BC*,

$$V_3 = \frac{Q_3}{A_3} = \frac{1360}{2.16} = 630 \text{ m/min} = 10.5 \text{ m/s}$$

and velocity pressure in the duct *BC*,

$$p_{v3} = \left(\frac{V_3}{4.04}\right)^2 = \left(\frac{10.5}{4.04}\right)^2 = 6.76 \text{ mm of water}$$

Hydraulic mean depth of the duct *AB* or *BC*,

$$m_1 = m_3 = \frac{ab}{2(a+b)} = \frac{1.8 \times 1.2}{2\,(1.8 + 1.2)} = 0.36 \text{ m}$$

Pressure loss due to friction in the duct *BC*,

$$p_{f3} = \frac{f_3 L_3}{m_3} \times p_{v3} = \frac{0.0055 \times 60}{0.36} \times 6.76 = 6.2 \text{ mm of water}$$

∴ Total pressure at *B*,

$$p_{TB} = p_{f3} + p_{v3} = 6.2 + 6.76 = 12.96 \text{ mm of water} \quad \ldots (i)$$

Let p_{v2} = Velocity pressure in the duct *BD*, and

V_2 = Velocity in the duct *BD*.

Since the duct *BD* is divided into three parts, each being 1.2 m by 0.6 m, therefore hydraulic mean depth of the duct *BD*,

$$m_2 = \frac{3 \times 1.2 \times 0.6}{3 \times 2\,(1.2 + 0.6)} = 0.2 \text{ m}$$

∴ Pressure loss due to friction in the duct *BD*,

$$p_{f2} = \frac{f_2 L_2}{m_2} \times p_{v2} = \frac{0.007 \times 60}{0.2} \times p_{v2} = 2.1\, p_{v2} \text{ mm of water}$$

and total pressure at *B*, $p_{TB} = p_{f2} + p_{v2} = 2.1\, p_{v2} + p_{v2} = 3.1\, p_{v2}$ mm of water ... (*ii*)

From equations (*i*) and (*ii*), we get

$$p_{v2} = \frac{12.96}{3.1} = 4.18 \text{ mm of water}$$

We know that

$$p_{v2} = \left(\frac{V_2}{4.04}\right)^2$$

$$\therefore \quad V_2 = 4.04\sqrt{p_{v2}} = 4.04\sqrt{4.18} = 8.24 \text{ m/s} = 495 \text{ m/min}$$

Quantity of air discharged at *D*,

$$Q_2 = A_2 V_2 = 1.8 \times 1.2 \times 495 = 1070 \text{ m}^3/\text{min}$$ **Ans.**

Static pressure at the fan outlet

We know that quantity of air flowing through the duct *AB*,

$$Q_1 = Q_2 + Q_3 = 1070 + 1360 = 2430 \text{ m}^3/\text{min}$$

∴ Velocity of air in the duct *AB*,

$$V_1 = \frac{Q_1}{A_1} = \frac{2430}{2.16} = 1125 \text{ m/min} = 18.75 \text{ m/s}$$

Velocity pressure in the duct AB,

$$p_{v1} = \left(\frac{V_1}{4.04}\right)^2 = \left(\frac{18.75}{4.04}\right)^2 = 21.5 \text{ mm of water}$$

Pressure loss due to friction in the duct AB,

$$p_{f1} = \frac{f_1 L_1}{m_1} \times p_{v1} = \frac{0.0055 \times 45}{0.36} \times 21.5 = 14.8 \text{ mm of water}$$

Total pressure at the fan outlet (*i.e.* at A),

$$p_{TA} = p_{TB} + p_{f1} = 12.96 + 14.8 = 27.76 \text{ mm of water}$$

∴ Static pressure at the fan outlet (*i.e.* at A),

$$p_{SA} = p_{TA} - p_{vA} = 27.76 - 21.15 = 6.26 \text{ mm of water} \quad \textbf{Ans.}$$

... ($\because p_{vA} = p_{v1}$)

Example 20.8. *A rectangular duct 0.15 m by 0.12 m is 20 m long and carries standard air at the rate of 0.3 m³/s. Calculate the total pressure required at the inlet to the duct in order to maintain this flow and the air power. Assume that for the duct, the friction factor f = 0.005.*

Solution. Given : $a = 0.15$ m ; $b = 0.12$ m ; $L = 20$ m ; $Q = 0.3$ m³/s ; $f = 0.005$

Total pressure required at the inlet to the duct

Let p_T = Total pressure required at the inlet to the duct.

We know that cross-sectional area of the duct,

$$A = a \times b = 0.15 \times 0.12 = 0.018 \text{ m}^2$$

∴ Velocity of air in the duct,

$$V = \frac{Q}{A} = \frac{0.3}{0.018} = 16.67 \text{ m/s}$$

and velocity pressure,

$$p_v = \frac{\rho_a V^2}{2} = \frac{1.2 V^2}{2} = 0.6\, V^2 = 0.6\,(16.67)^2 = 166.7 \text{ N/m}^2$$

... ($\because \rho_a$ for standard air = 1.2 kg / m³)

Wetted perimeter of the duct,

$$P = 2(a + b) = 2(0.15 + 0.12) = 0.54 \text{ m}$$

Hydraulic mean depth of the duct,

$$m = \frac{A}{P} = \frac{0.018}{0.54} = 0.033 \text{ m}$$

We know that pressure loss due to friction in the duct,

$$p_f = \frac{f\,L}{m} \times p_v = \frac{0.005 \times 20}{0.033} \times 166.7 = 505.15 \text{ N/m}^2$$

∴ Total pressure required at the inlet to the duct,

$$p_T = p_f + p_v = 505.15 + 166.7 = 671.85 \text{ N/m}^2 \quad \textbf{Ans.}$$

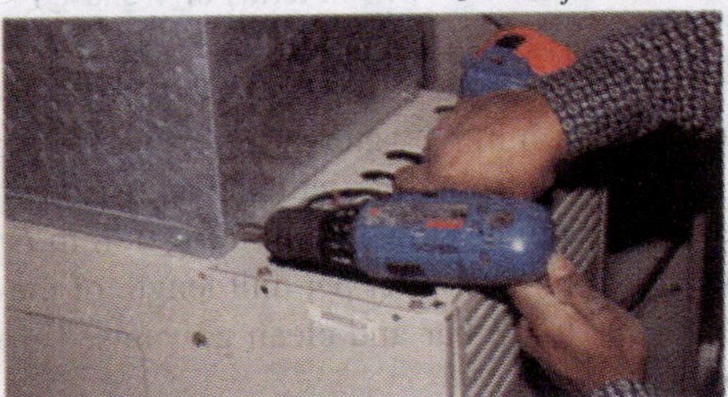

Fastening duct using a portable electric drill and sheet metal screws.

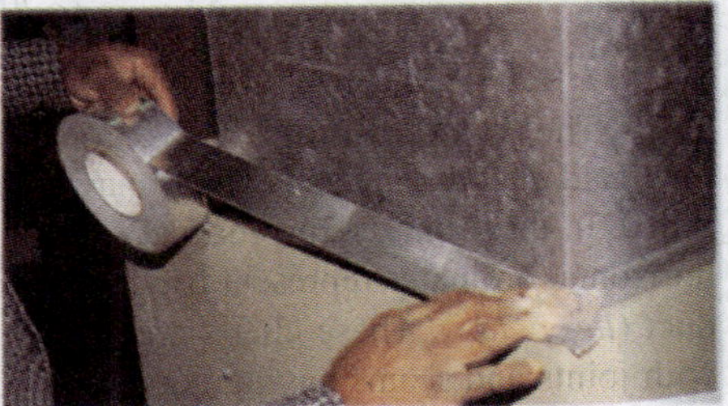

Taped duct connection.

Air power

We know that the air power

$$= Q \times p_T = 0.3 \times 671.85 = 201.6 \text{ N-m/s} = 201.6 \text{ W} \quad \textbf{Ans.}$$

... ($\because$ 1 N–m/s = 1 W)

20.13 Friction Chart for Circular Ducts

We have already discussed in Art. 20.10, that the frictional pressure loss in ducts according to D' Arcy's formula or Fanning's equation is given by

$$p_f = \frac{f L \rho_a V^2}{2\,m}$$

and hydraulic mean depth for circular ducts,

$$m = \frac{A}{P} = \frac{\frac{\pi}{4} \times D^2}{\pi D} = \frac{D}{4}$$

$\therefore$ Frictional pressure loss for circular ducts,

$$p_f = \frac{f L \rho_a V^2}{2}\left(\frac{4}{D}\right) = \frac{4 f L}{D}\left(\frac{\rho_a V^2}{2}\right)$$

$$= \frac{4 f L}{D} \times p_v \text{ (in N/m}^2) \qquad \text{... } (i)$$

According to Fritzsche, the frictional pressure loss (p_f) in circular ducts may be obtained from the following relation :

$$p_f = \frac{0.014\,22 L\,(V)^{1.852}}{(D)^{1.269}} \text{ (in N/m}^2) \qquad \text{... } (ii)$$

where V is in m/s and L and D are in metres.

We know that mean velocity of air flowing through the duct,

$$V = \frac{\text{Volume flow rate}}{\text{Cross-sectional area}} = \frac{Q}{A} = \frac{4Q}{\pi D^2}$$

Substituting this value of V in equation (*ii*), we have

$$p_f = \frac{0.222\,43 L\,(Q)^{1.852}}{(D)^{4.973}} \text{ N/m}^2 \qquad \text{... } (iii)$$

$$= \frac{0.002\,268 L\,(Q)^{1.852}}{(D)^{4.973}} \text{ mm of water} \qquad \text{... } (iv)$$

$$= \frac{0.012\,199 L\,(V)^{2.4865}}{(Q)^{0.6343}} \text{ N/m}^2 \qquad \text{... (in terms of } V \text{ and } Q) \qquad \text{... } (v)$$

The different equations as obtained above from the Fritzsche formula give quite accurate results.

The frictional pressure loss for circular ducts (in mm of water) for various velocities (in m/s) and duct diameters (in mm) may be obtained directly from the friction chart as shown in Fig. 20.10. In these charts, the vertical ordinates represent volume flow rate of air (Q) in m^3/s and the horizontal ordinates represent frictional pressure loss in mm of water per unit length of the circular duct (*i.e.* p_f / L). These charts are valid for 20°C and 1.013 bar and clean galvanised iron ducts with joints and seams having good commercial practice.

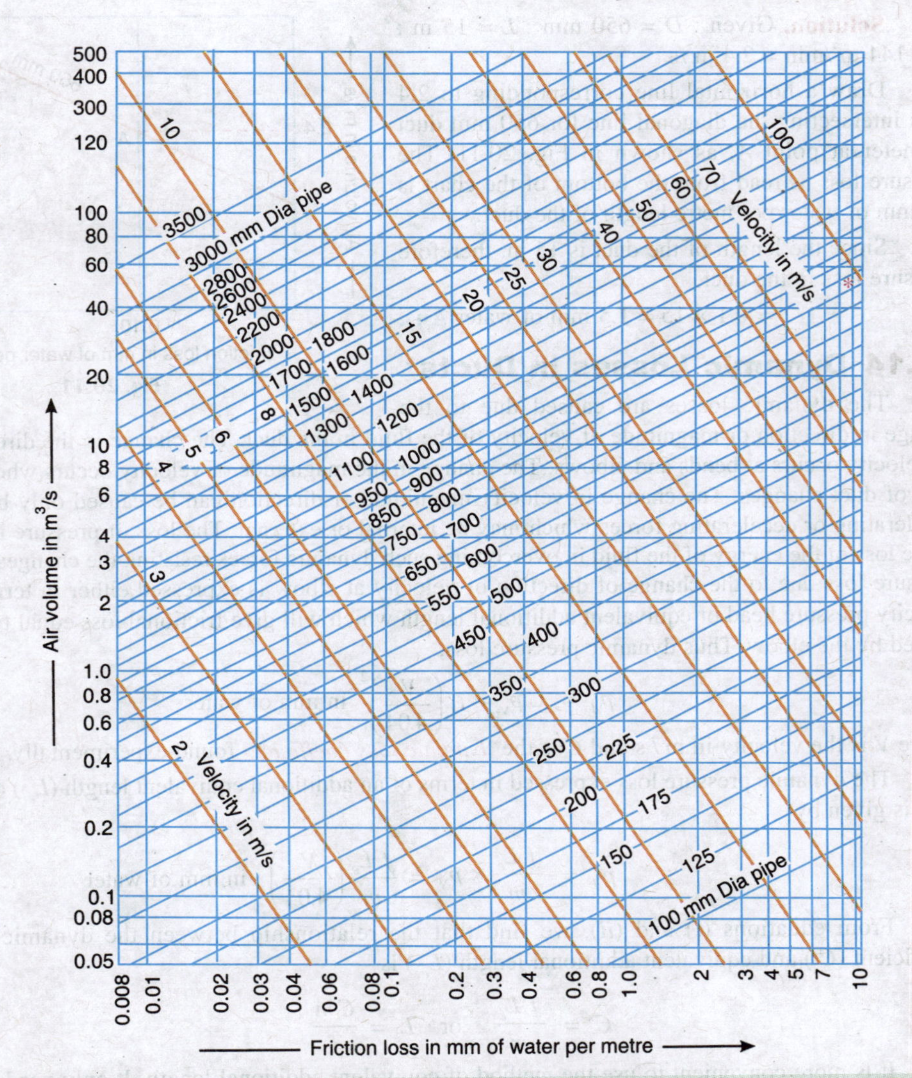

Fig. 20.10. Friction chart for circular ducts.

Notes : 1. If the duct is made of other material, such as plastics, concrete, wood, fibreglass etc., then correction factor should be applied. For small differences in density of air, the correction shall be made according to

$$p_f \propto \rho_a$$

2. If the air is at other temperatures, then the correction shall be made according to

$$p_f \propto \frac{1}{(T)^{0.857}}$$

3. The friction chart, as shown in Fig. 20.10, may be used to determine the pressure loss in rectangular ducts if the equivalent diameter of a circular duct for the rectangular duct is obtained first as discussed in Art. 20.12.

Example 20.9. *A duct of 650 mm diameter and 15 m long carries 144 m³/mm of air at 20°C. Find the pressure loss in the duct by using the friction chart.*

Solution. Given : D = 650 mm ; L = 15 m ; Q = 144 m³/min = 2.4 m³/s

Draw a horizontal line corresponding to 2.4 m³/s intersecting the diagonal line for 650 mm duct diameter at point A, as shown in Fig. 20.11. The pressure loss as read from the bottom of the chart is 0.1 mm of water per metre length of the duct.

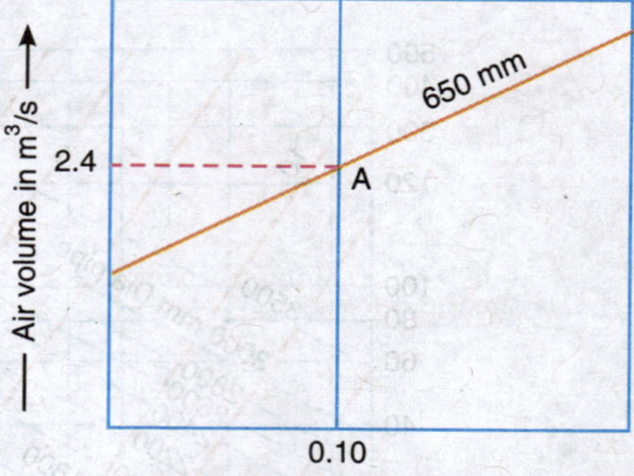

Fig. 20.11

Since the length of the duct is 15 m, therefore pressure loss in the duct,

$$= 0.1 \times 15 = 1.5 \text{ mm of water} \quad \textbf{Ans.}$$

20.14 Dynamic Losses in Ducts

The dynamic losses are caused due to the change in direction or magnitude of velocity of the fluid in the duct. The change in the direction of velocity occurs at bends and elbows. The change in the magnitude of velocity occurs when the area of duct changes. The change in velocity magnitude or direction can be caused only by the accelerating or decelerating forces which may be internal or external. The loss of pressure is due to the loss of the energy of the fluid in overcoming such dynamic forces resisting the changes. The pressure loss due to the change of direction of velocity at elbow is expressed either in terms of velocity pressure head or equivalent additional length which will give frictional loss equal to that caused by the elbow. Thus dynamic pressure loss,

$$p_d = C\,p_v = C\left(\frac{V}{4.04}\right)^2 \text{ in mm of water} \qquad \ldots (i)$$

where V is the velocity in m / s and C is the *dynamic loss coefficient* found experimentally.

The dynamic pressure loss expressed in terms of an additional equivalent length (L_e) of the duct is given by

$$p_d = \frac{f\,L_e}{m} \times p_v = \frac{f\,L_e}{m}\left(\frac{V}{4.04}\right)^2 \text{ in mm of water} \qquad \ldots (ii)$$

From equations (i) and (ii), we find that the relationship between the dynamic loss coefficient (C) and equivalent additional length (L_e) is

$$C = \frac{f\,L_e}{m} \quad \text{or} \quad L_e = \frac{C\,m}{f}$$

It is more convenient to use the method of equivalent additional length. It only needs this additional length to be added to the actual total length and the total pressure loss can then be found as frictional loss only. In considering the actual total length, the length of bends and elbows should also be taken in addition to the straight lengths of the ducts, as the equivalent additional length accounts for only the dynamic losses and not the actual frictional losses in length of bends or elbows.

20.15 Pressure Loss due to Enlargement in Area and Static Regain

When the area of a duct changes, the velocity of air flowing through the duct changes. A little consideration will show that when the area increases, the velocity decreases with a rise in pressure and the conversion of velocity head (or velocity pressure) into pressure head (or static pressure) takes place. The increase in static pressure as a result of the conversion from velocity pressure is termed as *static regain.*

Consider a duct A B C through which air is flowing and having a sudden or abrupt enlargement at B. As a result of this enlargement eddies will be formed in the corner of the duct at B as shown in Fig. 20.12. The loss of pressure takes place due to the eddies formed at the suddenly enlarged section.

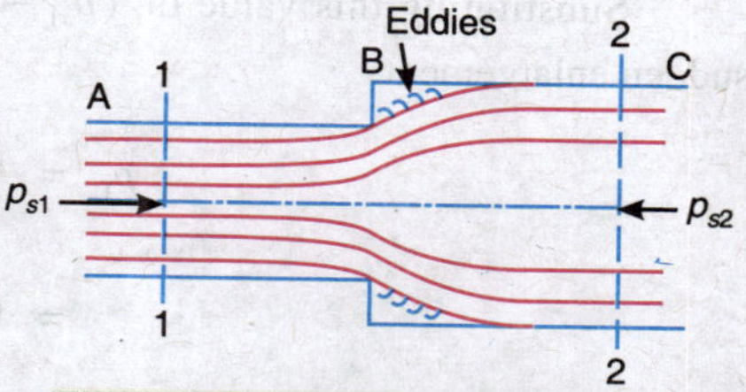

Fig. 20.12. Sudden enlargement.

Let p_{s1} = Static pressure of air at section 1-1,

A_1 = Cross-sectional area of the duct at section 1-1,

V_1 = Velocity of air at section 1-1,

p_{s2}, A_2, V_2 = Corresponding values at section 2-2, and

p_e = Pressure loss due to sudden enlargement.

Applying Bernoulli's equation to sections 1-1 and 2-2, we have

$$p_{s1} + p_{v1} = p_{s2} + p_{v2} + p_L$$

$$p_{s1} + \frac{\rho_a (V_1)^2}{2} = p_{s2} + \frac{\rho_a (V_2)^2}{2} + p_L$$

$$\therefore \quad p_L = [p_{s1} - p_{s2}] + \left[\frac{\rho_a (V_1)^2}{2} - \frac{\rho_a (V_2)^2}{2}\right] \quad \text{... (i)}$$

We know that momentum of air per second at section 1-1

$$= \text{Mass} \times \text{Velocity} = (\text{Volume} \times \text{Density}) \times \text{Velocity}$$

$$= (A_1 V_1 \rho_a) V_1 = \rho_a A_1 (V_1)^2 \quad \text{... } (\because \text{Volume} = \text{Area} \times \text{velocity})$$

Similarly, momentum of air per second at sections 2-2,

$$= \rho_a A_2 (V_2)^2$$

$\therefore$ Change of momentum per second

$$= \rho_a A_1 (V_1)^2 - \rho_a A_2 (V_2)^2 \quad \text{... (ii)}$$

Since the flow between sections 1-1 and 2-2 is continuous, therefore

$$A_1 V_1 = A_2 V_2 \quad \text{or} \quad A_1 = \frac{A_2 V_2}{V_1}$$

Substituting this value of A_1 in equation (*ii*), we have change of momentum per second,

$$= \frac{\rho_a A_2 V_2 (V_1)^2}{V_1} - \rho_a A_2 (V_2)^2$$

$$= \rho_a A_2 V_2 V_1 - \rho_a A_2 (V_2)^2 \quad \text{... (iii)}$$

The force responsible for this change of momentum

$$= (p_{s2} - p_{s1}) A_2 \quad \text{... (iv)}$$

Equating equations (*iii*) and (*iv*), we have

$$(p_{s2} - p_{s1}) A_2 = \rho_a A_2 V_2 V_1 - \rho_a A_2 (V_2)^2$$

or

$$p_{s2} - p_{s1} = \rho_a [V_1 V_2 - (V_2)^2]$$

and

$$p_{s1} - p_{s2} = \rho_a [(V_2)^2 - V_1 V_2]$$

Substituting this value of $(p_{s1} - p_{s2})$ in equation (*i*), we find that pressure loss due to sudden enlargement,

$$p_L = \rho_a\left[(V_2)^2 - V_1V_2\right] + \left[\frac{\rho_a (V_1)^2}{2} - \frac{\rho_a (V_2)^2}{2}\right]$$

$$= \rho_a\left[\frac{2(V_2)^2 - 2V_1 V_2 + (V_1)^2 - (V_2)^2}{2}\right]$$

$$= \rho_a\left[\frac{(V_1)^2 + (V_2)^2 - 2V_1 V_2}{2}\right] = \rho_a\left[\frac{(V_1 - V_2)^2}{2}\right]^* \text{ N/m}^2$$

$$= 1.2\left[\frac{(V_1 - V_2)^2}{2}\right] = 0.6\,(V_1 - V_2)^2 \text{ N/m}^2$$

... (Taking $\rho_a = 1.2$ kg / m^3)

$$= \frac{0.6}{9.81}(V_1 - V_2)^2 = \left(\frac{V_1 - V_2}{4.04}\right)^2 \text{ mm of water} \quad \text{... (v)}$$

... ($\because$ 1 N/m^2 = $\frac{1}{9.81}$ mm of water)

We know that $A_1 V_1 = A_2 V_2$

$$\therefore \quad V_2 = \frac{A_1 V_1}{A_2} \quad \text{or} \quad V_1 = \frac{A_2 V_2}{A_1}$$

Substituting the value of V_2 in equation (*v*), we have

$$p_L = \left[\frac{V_1 - \dfrac{A_1 V_1}{A_2}}{4.04}\right]^2 = \left(1 - \frac{A_1}{A_2}\right)^2\left(\frac{V_1}{4.04}\right)^2$$

$$= C_1\left(\frac{V_1}{4.04}\right)^2 = C_1 p_{v1} \quad \text{... (vi)}$$

where $C_1 = \left(1 - \dfrac{A_1}{A_2}\right)^2$, known as loss coefficient, and

$$p_{v1} = \left(\frac{V_1}{4.04}\right)^2 \text{ in mm of water}$$

Now substituting the value of V_1 in equation (v),

$$p_L = \left[\frac{\left(\dfrac{A_2 V_2}{A_1} - V_2\right)}{4.04}\right]^2 = \left(\frac{A_2}{A_1} - 1\right)^2\left(\frac{V}{4.04}\right)^2$$

$$= C_2\left(\frac{V_2}{4.04}\right)^2 = C_2 p_{v2} \quad \text{... (vii)}$$

---

* In case the pressure loss due to sudden enlargement (p_L) is required in metres of air, then

$$p_L = \frac{(V_1 - V_2)^2}{2g} \text{ m of air}$$

where
$$C_2 = \left(\frac{A_2}{A_1} - 1\right)^2, \text{ known as loss coefficient, and}$$

$$p_{v2} = \left(\frac{V_2}{4.04}\right)^2 \text{ in mm of water}$$

When there is a gradual enlargement in area of the duct, as shown in Fig. 20.13, then the pressure loss due to gradual enlargement is given by

$$p_L = C_r C_1 \left(\frac{V_1}{4.04}\right)^2 = C_r C_2 \left(\frac{V_2}{4.04}\right)^2$$

where C_r is the loss coefficient giving the ratio of the actual loss to the loss for sudden enlargement. The following table shows the values of C_r as a function of included angle θ of the sides.

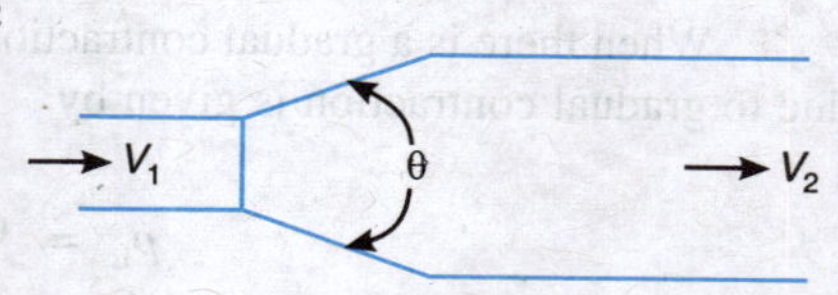

Fig. 20.13. Gradual enlargement.

Table 20.3. Values of loss coefficient.

Condition (θ°)	5	7	10	20	30	40
Loss coefficient (C_r)	0.17	0.22	0.28	0.45	0.59	0.73

Notes : 1. When the enlargement is not accompanied with pressure loss, then there will be full conversion of the velocity pressure into static pressure. In such a case, static pressure regain.

$$SPR = p_{s2} - p_{s1} = p_{v1} - p_{v2}$$

2. When the enlargement is accompanied with pressure loss, the increase in static pressure or static regain is reduced by the amount of the pressure loss. In such a case, static pressure regain,

$$SPR = p_{s2} - p_{s1} = (p_{v1} - p_{v2}) - p_L = R\,(p_{v1} - p_{v2})$$

where R is the static regain factor.

20.16 Pressure Loss due to Contraction in Area

Consider a duct *ABC* through which air is flowing and having a sudden or abrupt contraction at *B*, as shown in Fig. 20.14. It may be noted that when air is flowing through a duct of such a section, the eddies are formed at two places, *i.e.* at the shoulders of the large section and beyond the entry at the smaller section forming a vena contracta at section 1-1. Strictly speaking, the loss of pressure due to sudden contraction is not due to contraction itself, but it is due to the sudden enlargement of flow area from vena contracta (*i.e.* section 1-1) to the section of the smaller duct (*i.e.* section 2-2).

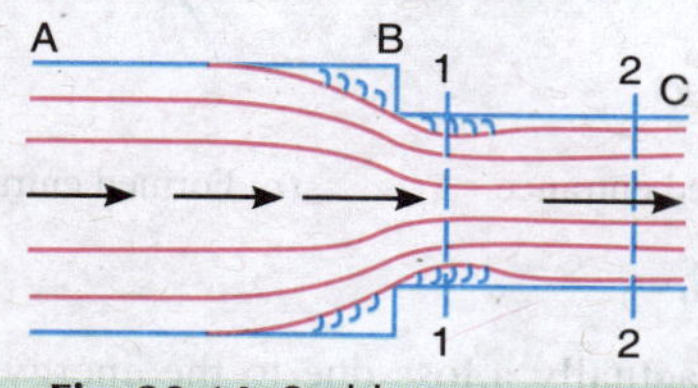

Fig. 20.14. Sudden contraction.

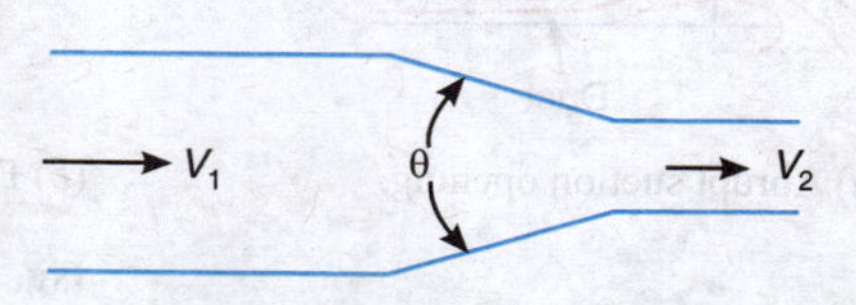

Fig. 20.15. Gradual contraction.

Let A_1 = Area of duct at section 1-1,

V_1 = Velocity of air at section 1-1,

A_2, V_2 = Corresponding values at section 2-2, and

p_L = Pressure loss due to sudden contraction.

Since the pressure loss due to sudden contraction is equal to the pressure loss due to sudden enlargement from section 1-1 to section 2-2, therefore pressure loss due to sudden contraction,

$$p_L = \left(1 - \frac{A_1}{A_2}\right)^2 \left(\frac{V_1}{4.04}\right)^2 = C_1 \left(\frac{V_1}{4.04}\right)^2 = C_2 \left(\frac{V_2}{4.04}\right)^2$$

When there is a gradual contraction in area of the duct as shown in Fig. 20.15, then the loss due to gradual contraction is given by

$$p_L = C_r C_1 \left(\frac{V_1}{4.04}\right)^2 = C_r C_2 \left(\frac{V_2}{4.04}\right)^2$$

where C_r is the loss coefficient. The following table shows the values of C_r as a function of included angle θ of the sides.

Table 20.4. Values of loss coefficient.

Condition (θ°)	30°	45°	60°
Loss coefficient (C_r)	0.02	0.04	0.07

20.17 Pressure Loss at Suction and Discharge of a Duct

The pressure loss at suction to the duct is given by

$$p_L = \frac{C V^2}{2g} \text{ m of air} = \frac{C V^2}{2} \times \rho_a \text{ N/m}^2$$

$$= C \left(\frac{V}{4.04}\right)^2 \text{ mm of water}$$

where V is the velocity of air in the duct in m/s and C is the loss coefficient.

In case of an abrupt suction opening, as shown in Fig. 20.16 (*a*), the air is accelerated as it approaches to the opening, forming a vena contracta inside the duct. The area changes from infinity to the duct area. In such a case, the loss coefficient C is taken as 0.85. By making a flanged entrance as shown in Fig. 20.16 (*b*), the loss coefficient is reduced to 0.34. The loss coefficient can be further reduced to 0.03 by making formed entrance of bell-mouth shape as shown in Fig. 20.16 (*c*).

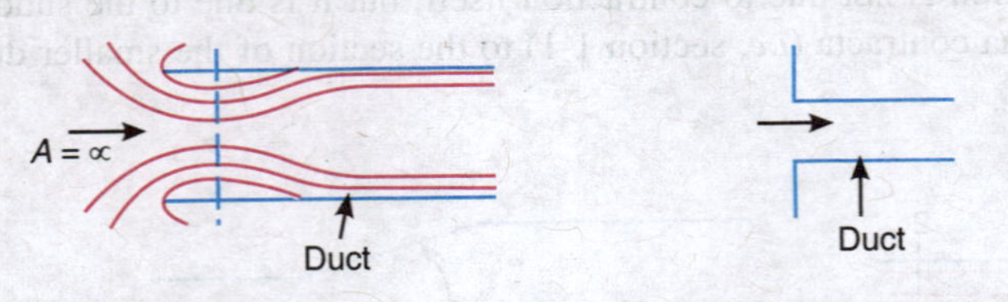

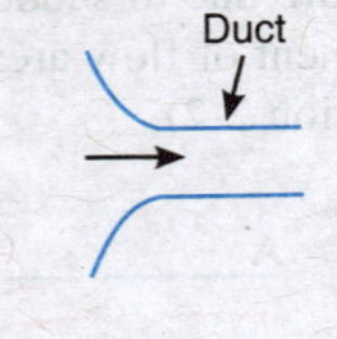

(*a*) Abrupt suction opening. (*b*) Flanged entrance. (*c*) Formed entrance.

Fig. 20.16

The pressure loss at the discharge of a duct is actually a loss due to the energy of head, which the flowing air has, by virtue of its motion. The value of pressure loss at the discharge or exit of a duct is given by

$$p_L = \frac{V^2}{2g} \text{ m of air}$$

$$= \frac{V^2}{2} \times \rho_a \text{ N/m}^2 = \left(\frac{V}{4.04}\right)^2 \text{ mm of water}$$

20.18 Pressure Loss due to an Obstruction in a Duct

In the previous articles, we have discussed the frictional and dynamic losses in ducts. In addition to these losses, the pressure losses also occur due to various obstructions in the path of air flow from the fan to the outlet (*i.e.* air conditioned room). The possible losses for different obstructions are given in the following table.

Table 20.5. Possible pressure loss for different obstructions.

Type of obstruction	*Possible pressure loss in mm of water*
Air heaters or cooler with several rows	5 to 10
Air washers	6.25 to 10
Air filters	5 to 10
Screen grills	2.5 to 5

Note : The actual pressure losses in the above equipments must be obtained from the data of the manufacturer whose equipment is to be used in the air conditioning system under consideration.

Example 20.10. *A length of main circular duct has three branch ducts taking equal air volumes at equal intervals. Each interval duct has a friction loss of 1.3 mm of water and a static pressure of 5 mm of water is necessary at each branch to cope with its friction loss. If the initial velocity in the main duct of 1.2 m diameter is 600 m/min, calculate the velocities and diameters of the second and third lengths, whereby the static pressure regain is sufficient to overcome the friction loss in the succeeding length of main duct up to the next branch. The static pressure regain factor is 0.6.*

Draw a simple sketch of the duct system and identify total, static and velocity pressures at the appropriate points of change.

Solution. Given : $p_{f1} = p_{f2} = p_{f3} = 1.3$ mm of water ; $p_{SB} = p_{SD} = p_{SF} = 5$ mm of water ; $D_1 = 1.2$ m ; $V_1 = 600$ m/min $= 10$ m/s ; $R = 0.6$

The duct system is shown in Fig. 20.17.

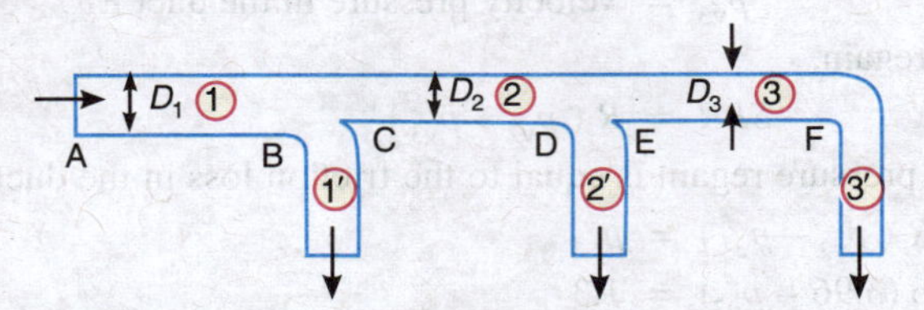

Fig. 20.17

Velocity and diameter of CD

Let V_2 = Velocity of air in the duct *CD*,

D_2 = Diameter of the duct *CD*, and

p_{v2} = Velocity pressure in the duct *CD*.

We know that the quantity of air passing through the duct AB,

$$Q_1 = A_1 V_1 = \frac{\pi}{4}(D_1)^2 V_1 = \frac{\pi}{4}(1.2)^2\, 600 = 678.6 \text{ m}^3/\text{min}$$

Quantity of air passing through each branch,

$$Q_1' = Q_2' = Q_3' = \frac{678.6}{3} = 226.2 \text{ m}^3/\text{min}$$

Velocity pressure in the duct AB,

$$p_{v1} = \left(\frac{V_1}{4.04}\right)^2 = \left(\frac{10}{4.04}\right)^2 = 6.13 \text{ mm of water}$$

We know that static pressure regain,

$$SPR = R\,(p_{v1} - p_{v2})$$

Since the static pressure regain is equal to the friction loss in the duct CD, therefore

$$R\,(p_{v1} - p_{v2}) = p_{f2}$$

or $$0.6\,(6.13 - p_{v2}) = 1.3$$

$\therefore$ $$p_{v2} = 6.13 - 1.3 / 0.6 = 3.96 \text{ mm of water}$$

Velocity pressure in the duct CD,

$$p_{v2} = \left(\frac{V_2}{4.04}\right)^2$$

or $$V_2 = 4.04\sqrt{p_{v2}} = 4.04\sqrt{3.96} = 8.04 \text{ m/s} = 482.4 \text{ m/min} \quad \textbf{Ans.}$$

Quantity of air passing through the duct CD,

$$Q_2 = Q_1 - Q_1' = 678.6 - 226.2 = 452.4 \text{ m}^3/\text{min}$$

We know that $$Q_2 = A_2 V_2 = \frac{\pi}{4}(D_2)^2 V_2$$

$\therefore$ $$(D_2)^2 = \frac{4Q_2}{\pi V_2} = \frac{4 \times 452.4}{\pi \times 482.4} = 1.194 \text{ m}^2 \quad \text{or} \quad D_2 = 1.09 \text{ m} \quad \textbf{Ans.}$$

Velocity and diameter of duct EF

Let V_3 = Velocity in the duct EF,

D_3 = Diameter of the duct EF, and

p_{v3} = Velocity pressure in the duct EF.

Static pressure regain,

$$SPR = R\,(p_{v2} - p_{v3})$$

Since the static pressure regain is equal to the friction loss in the duct EF, therefore

$$R\,(p_{v2} - p_{v3}) = p_{f3}$$

or $$0.6\,(3.96 - p_{v3}) = 1.3$$

$\therefore$ $$p_{v3} = 3.96 - 1.3 / 0.6 = 1.79 \text{ mm of water}$$

Velocity pressure in the duct EF,

$$p_{v3} = \left(\frac{V_3}{4.04}\right)^2$$

$\therefore$ $$V_3 = 4.04\sqrt{p_{v3}} = 4.04\sqrt{1.79} = 5.4 \text{ m/s}$$

$$= 324.3 \text{ m/min} \quad \textbf{Ans.}$$

Quantity of air passing through the duct *EF*,

$$Q_3 = Q_2 - Q_2' = 452.4 - 226.2 = 226.2 \text{ m}^3/\text{min}$$

We know that
$$Q_3 = A_3 V_3 = \frac{\pi}{4}(D_3)^2 V_3$$

$$\therefore \quad (D_3)^2 = \frac{4Q_3}{\pi V_3} = \frac{4 \times 226.2}{\pi \times 324.3} = 0.888 \text{ m}^2 \quad \text{or} \quad D_3 = 0.94 \text{ m} \quad \textbf{Ans.}$$

Total, static and velocity pressures at the appropriate points of change

The total, static and velocity pressures at the appropriate points of change are tabulated in the following table :

Table 20.6

Pressure in mm of water	*Appropriate points of change*				
	B	*C*	*D*	*E*	*F*
Total pressure (p_T)	11.13	10.26	8.96	8.09	6.79
Static pressure (p_S)	5	6.3	5	6.3	5
Velocity pressure (p_V)	6.13	3.96	3.96	1.79	1.79

Fabricated ducts.

Example 20.11. *A circular duct of 100 mm diameter converges gradually to a 75 mm duct. The static pressure just upstream of the reducer is 30 mm of water and the velocity is 450 m/min. The loss of pressure in the reducer is 0.1 of the velocity head in the duct downstream of the reducer. Calculate the total pressures upstream and downstream of the reducer. Also determine the pressure indicated by a U-tube water manometer connected differentially to pressure tappings upstream and downstream of the reducer.*

Solution. Given : $D_1 = 100$ mm ; $D_2 = 75$ mm ; $p_{s1} = 30$ mm of water ; $V_1 = 450$ m / min = 7.5 m /s ; $p_L = 0.1\, p_{v2}$

The cross-section of the duct is shown in Fig. 20.18.

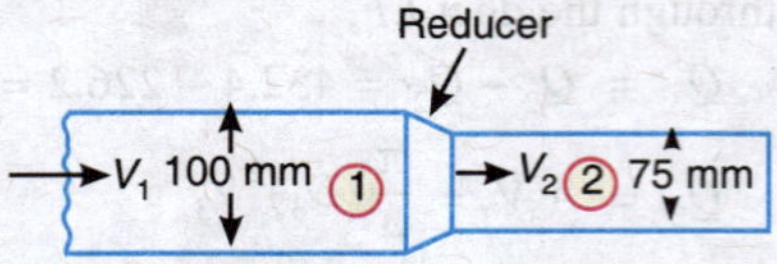

Fig. 20.18

Total pressure upstream and downstream of the reducer

Let p_{T1} = Total pressure upstream of the reducer,

p_{T2} = Total pressure downstream of the reducer, and

V_2 = Velocity downstream of the reducer.

We know that velocity pressure upstream of the reducer,

$$p_{v1} = \left(\frac{V_1}{4.04}\right)^2 = \left(\frac{7.5}{4.04}\right)^2 = 3.45 \text{ mm of water}$$

$$\therefore \quad p_{T1} = p_{s1} + p_{v1} = 30 + 3.45 = 33.45 \text{ mm of water} \quad \textbf{Ans.}$$

We know that $A_1 V_1 = A_2 V_2$ or $\frac{\pi}{4}(D_1)^2 V_1 = \frac{\pi}{4}(D_2)^2 V_2$

$$\therefore \quad V_2 = V_1\left(\frac{D_1}{D_2}\right)^2 = 7.5\left(\frac{100}{75}\right)^2 = 13.35 \text{ m/s}$$

Velocity pressure downstream of the reducer,

$$p_{v2} = \left(\frac{V_2}{4.04}\right)^2 = \left(\frac{13.35}{4.04}\right)^2 = 10.9 \text{ mm of water}$$

and pressure loss in the reducer,

$$p_L = 0.1\, p_{v2} = 0.1 \times 10.9 = 1.09 \text{ mm of water}$$

We know that $p_{T1} = p_{T2} + p_L$

$$\therefore \quad p_{T2} = p_{T1} - p_L = 33.45 - 1.09 = 32.36 \text{ mm of water} \quad \textbf{Ans.}$$

Pressure indicated by a U-tube water manometer

We know that static pressure downstream of the reducer,

$$p_{s2} = p_{T2} - p_{v2} \quad \ldots (\because p_{T2} = p_{s2} + p_{v2})$$

$$= 32.36 - 10.9 = 21.46 \text{ mm of water}$$

The pressure indicated by a U-tube water manometer is the difference between the static pressures at the tapping sections.

$$\therefore \quad p_{s1} - p_{s2} = 30 - 21.46 = 8.54 \text{ mm of water} \quad \textbf{Ans.}$$

Example 20.12. *The duct, as shown in Fig. 20.19 , is supplied with air at A and discharges freely to atmosphere at F. The duct AB is 0.6 m in diameter and 45 m long, the duct CD is 1.2 m in diameter and 150 m long and the duct EF is 0.45 m in diameter and 15 m long. The velocity of air in duct AB is 600 m/min.*

The loss in the expander BC is 0.5 times the velocity pressure in duct AB and the loss in the reducer DE is 0.2 times the velocity pressure in the duct EF.

Using the expression

$$p_f = \frac{0.263 V^{1.35}}{D^{1.27}}$$

where p_f is the pressure loss due to friction in mm of water per 100 m length of duct, V is the duct velocity in m/s and D is the diameter of duct in m, calculate the static pressure at point A.

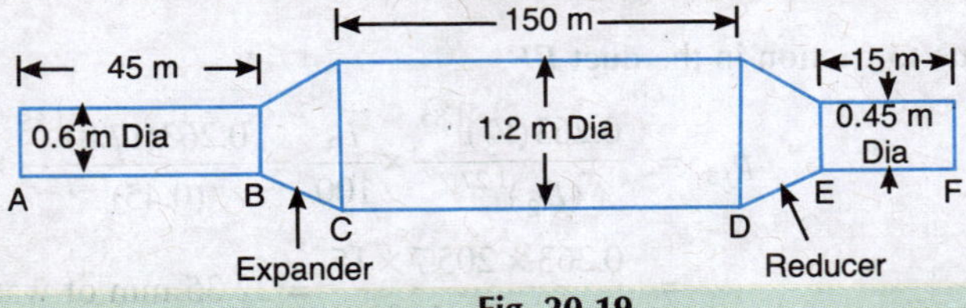

Fig. 20.19

Solution. Given : $D_1 = 0.6$ m ; $L_1 = 45$ m ; $D_2 = 1.2$ m ; $L_2 = 150$ m ; $D_3 = 0.45$ m ; $L_3 = 15$ m ; $V_1 = 600$ m / min = 10 m / s

Let V_2 = Velocity of air in duct *CD*, and

V_3 = Velocity of air in duct *EF*.

We know that $A_1V_1 = A_2V_2 = A_3V_3$

$$\therefore \quad V_2 = \frac{A_1V_1}{A_2} = V_1\left(\frac{D_1}{D_2}\right)^2 = 10\left(\frac{0.6}{1.2}\right)^2 = 2.5 \text{ m/s}$$

and

$$V_3 = \frac{A_1V_1}{A_3} = V_1\left(\frac{D_1}{D_3}\right)^2 = 10\left(\frac{0.6}{0.45}\right)^2 = 17.8 \text{ m/s}$$

Assuming standard air, velocity pressure in the duct *AB*,

$$p_{v1} = \left(\frac{V_1}{4.04}\right)^2 = \left(\frac{10}{4.04}\right)^2 = 6.13 \text{ mm of water}$$

Velocity pressure in the duct *CD*,

$$p_{v2} = \left(\frac{V_2}{4.04}\right)^2 = \left(\frac{2.5}{4.04}\right)^2 = 0.383 \text{ mm of water}$$

and velocity pressure in the duct *EF*,

$$p_{v3} = \left(\frac{V_3}{4.04}\right)^2 = \left(\frac{17.8}{4.04}\right)^2 = 19.4 \text{ mm of water}$$

Pressure loss in the expander *BC*,

$$p_{L1} = 0.5 \times p_{v1} = 0.5 \times 6.13 = 3.065 \text{ mm of water}$$

Pressure loss in the reducer *DE*,

$$p_{L2} = 0.2 \times p_{v3} = 0.2 \times 19.4 = 3.88 \text{ mm of water}$$

We know that the pressure loss due to friction in the duct *AB*,

$$p_{f1} = \frac{0.263(V_1)^{1.85}}{(D_1)^{1.27}} \times \frac{L_1}{100} = \frac{0.263(10)^{1.85}}{(0.6)^{1.27}} \times \frac{45}{100} \text{ mm of water}$$

$$= \frac{0.263 \times 70.8 \times 45}{0.523 \times 100} = 16.02 \text{ mm of water}$$

Pressure loss due to friction in the duct *CD*,

$$p_{f2} = \frac{0.263(V_2)^{1.85}}{(D_2)^{1.27}} \times \frac{L_2}{100} = \frac{0.263(2.5)^{1.85}}{(1.2)^{1.27}} \times \frac{150}{100} \text{ mm of water}$$

$$= \frac{0.263 \times 5.45 \times 150}{1.26 \times 100} = 1.7 \text{ mm of water}$$

and pressure loss due to friction in the duct *EF*,

$$p_{f3} = \frac{0.263 (V_3)^{1.85}}{(D_2)^{1.27}} \times \frac{L_3}{100} = \frac{0.263 (17.8)^{1.85}}{(0.45)^{1.27}} \times \frac{15}{100} \text{ mm of water}$$

$$= \frac{0.263 \times 205.7 \times 15}{0.363 \times 100} = 22.35 \text{ mm of water}$$

∴ Total pressure loss in the duct between *A* and *F*,

$$\Sigma p_L = p_{f1} + p_{L1} + p_{f2} + p_{L2} + p_{f3}$$

$$= 16.02 + 3.065 + 1.7 + 3.88 + 22.35 = 47.015 \text{ mm of water}$$

Let p_{s1} = Static pressure at point *A*.

Now applying Bernoulli's equation between *A* and *F*,

$$p_{s1} + p_{v1} = p_{s3} + p_{v3} + \Sigma p_L$$

Since the discharge at *F* is to atmosphere, therefore $p_{s3} = 0$. Substituting the values in the above equation, we have

$$p_{s1} + 6.13 = 0 + 19.4 + 47.015$$

$$\therefore \quad p_{s1} = 60.285 \text{ mm of water } \textbf{Ans.}$$

Example 20.13. *A centrifugal fan of outlet 0.9 m by 0.7 m is moving standard air at a rate of 680 m³/min through a system which consists of straight inlet and outlet ducts. The inlet duct is 0.9 m diameter and 15 m long and the discharge duct is 1 m diameter and 60 m long. There is a diffuser between the fan discharge and the 1 m diameter duct for which the loss of pressure may be taken as 0.3 times the difference between velocity pressures. The loss at entry to the inlet duct is 0.5 times of the velocity pressure there and the friction factor 'f' for the inlet duct is 0.004 and for the outlet duct is 0.0035. Determine : 1. fan total pressure; 2. static pressures at the fan inlet and outlet; and 3. plot the variation of the total pressure and static pressure along the system.*

Assume that the air is sucked in by the inlet duct and delivered by the outlet duct is at atmospheric pressure.

Solution. Given : $a = 0.9$ m ; $b = 0.7$ m ; $Q = 680$ m³/min ; $D_1 = 0.9$ m ; $L_1 = 15$ m ; $D_4 = 1$ m ; $L_4 = 60$ m ; $f_1 = 0.004$; $f_4 = 0.0035$

The duct system is shown in Fig. 20.20 (*a*),

1. *Fan total pressure*

Cross-sectional area of the inlet duct 1-2,

$$A_1 = A_2 = \frac{\pi}{4}(D_1)^2 = \frac{\pi}{4}(0.9)^2 = 0.636 \text{ m}^2$$

∴ Velocity of air in the inlet duct 1-2,

$$V_1 = V_2 = \frac{Q}{A_1} = \frac{680}{0.636} = 1069.2 \text{ m / min} = 17.82 \text{ m / s}$$

and velocity pressure in the inlet duct 1-2,

$$p_{v1} = p_{v2} = \left(\frac{V_1}{4.04}\right)^2 = \left(\frac{17.82}{4.04}\right)^2 = 19.45 \text{ mm of water}$$

Cross-sectional area at the fan outlet,

$$A_3 = a \times b = 0.9 \times 0.7 = 0.63 \text{ m}^2$$

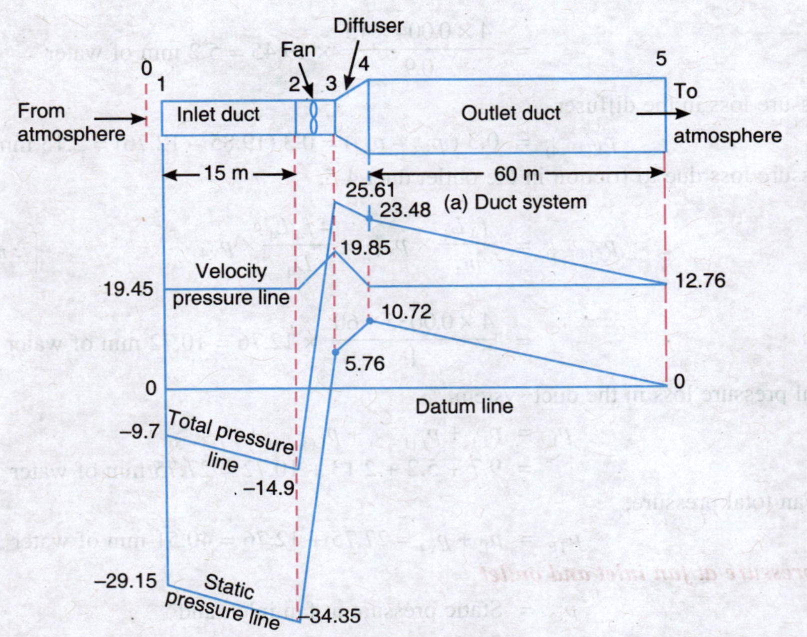

(*b*) Variation of total pressure and static pressure along the duct system.

Fig. 20.20

∴ Velocity of air at the fan outlet,

$$V_3 = \frac{Q}{A_3} = \frac{680}{0.63} = 1079.4 \text{ m / min} = 18 \text{ m / s}$$

and velocity pressure at the fan outlet,

$$p_{v3} = \left(\frac{V_3}{4.04}\right)^2 = \left(\frac{18}{4.04}\right)^2 = 19.85 \text{ mm of water}$$

Cross-sectional area of the outlet duct 4-5,

$$A_4 = A_5 = \frac{\pi}{4}(D_4)^2 = \frac{\pi}{4} \times 1^2 = 0.7855 \text{ m}^2$$

∴ Velocity of air in the outlet duct 4-5,

$$V_4 = V_5 = \frac{Q}{A_4} = \frac{680}{0.7855} = 865.7 \text{ m / min} = 14.43 \text{ m / s}$$

and velocity pressure in the outlet duct 4-5,

$$p_{v4} = p_{v5} = \left(\frac{V_4}{4.04}\right)^2 = \left(\frac{14.43}{4.04}\right)^2 = 12.76 \text{ mm of water}$$

Pressure loss at entry to the inlet duct at point 1,

$$p_{L1} = 0.5\, p_{v1} = 0.5 \times 19.45 = 9.7 \text{ mm of water}$$

Pressure loss due to friction in the inlet duct 1-2,

$$p_{f(1-2)} = \frac{f_1 L_1}{m_1} \times p_{v1} = \frac{4 f_1 L_1}{D_1} \times p_{v1} \qquad \ldots \left(\because m_1 = \frac{D_1}{4}\right)$$

$$= \frac{4 \times 0.004 \times 15}{0.9} \times 19.45 = 5.2 \text{ mm of water}$$

Pressure loss in the diffuser,

$$p_{L(3-4)} = 0.3\,(p_{v3} - p_{v4}) = 0.3\,(19.85 - 12.76) = 2.13 \text{ mm of water}$$

Pressure loss due to friction in the outlet duct 4-5,

$$p_{f(4-5)} = \frac{f_4 L_4}{m_4} \times p_{v4} = \frac{4 f_4 L_4}{D_4} \times p_{v4} \qquad \ldots \left(\because m_4 = \frac{D_4}{4}\right)$$

$$= \frac{4 \times 0.0035 \times 60}{1} \times 12.76 = 10.72 \text{ mm of water}$$

Total pressure loss in the duct system,

$$p_L = p_{L1} + p_{f(1-2)} + p_{L(3-4)} + p_{f(4-5)}$$

$$= 9.7 + 5.2 + 2.13 + 10.72 = 27.75 \text{ mm of water}$$

∴ Fan total pressure,

$$p_{TF} = p_L + p_{v4} = 27.75 + 12.76 = 40.51 \text{ mm of water} \quad \textbf{Ans.}$$

2. *Static pressure at fan inlet and outlet*

Let p_{s2} = Static pressure at fan inlet, and

p_{s3} = Static pressure at fan outlet.

Applying Bernoulli's equation to a section at O in the atmosphere upstream of the entrance to inlet duct and to a point at fan inlet (*i.e.* at section 2), we have

Total pressure at O = Total pressure at 2

$$p_{TO} = p_{T2} = p_{s2} + p_{v2} + p_{L1} + p_{f(1-2)}$$

Since the total pressure in the atmosphere (p_{TO}) is zero, therefore

$$0 = p_{s2} + 19.45 + 9.7 + 5.2 = p_{s2} + 34.35$$

$$\therefore \quad p_{s2} = -34.35 \text{ mm of water} \quad \textbf{Ans.}$$

Again applying Bernoulli's equation to fan inlet (*i.e.* section 2) and fan outlet (*i.e.* section 3), we have

$$p_{T2} = p_{T3}$$

$$p_{s2} + p_{v2} + p_{TF} = p_{s3} + p_{v3}$$

$$-34.35 + 19.45 + 40.51 = p_{s3} + 19.85$$

or $$p_{s3} = 5.76 \text{ mm of water} \quad \textbf{Ans.}$$

3. *Variation of total pressure and static pressure along the system*

We know that total pressure at entrance to inlet duct (*i.e.* at section 1),

$$p_{T1} = p_{TO} - p_{L1} = 0 - 9.7 = -9.7 \text{ mm of water}$$

Static pressure at entrance to inlet duct,

$$p_{s1} = p_{T1} - p_{v1} = -9.7 - 19.45 = -29.15 \text{ mm of water}$$

Total pressure at fan inlet (*i.e.* at section 2),

$$p_{T2} = p_{T1} - p_{f(1-2)} = -9.7 - 5.2 = -14.9 \text{ mm of water}$$

Static pressure at fan inlet,

$$p_{s2} = p_{T2} - p_{v2} = -14.9 - 19.45 = -34.35 \text{ mm of water}$$

Total pressure at fan outlet or inlet to diffuser (*i.e.* at section 3),

$$p_{T3} = p_{T2} + p_{TF} = -14.9 + 40.51 = 25.61 \text{ mm of water}$$

Static pressure at fan outlet,

$$p_{s3} = p_{T3} - p_{v3} = 25.61 - 19.85 = 5.76 \text{ mm of water}$$

... (same as before)

Total pressure at outlet of diffuser or entrance to outlet duct (*i.e.* at section 4),

$$p_{T4} = p_{T3} - p_{L(3-4)} = 25.61 - 2.13 = 23.48 \text{ mm of water}$$

Static pressure at outlet of diffuser or entrance to outlet duct,

$$p_{s4} = p_{T4} - p_{v4} = 23.48 - 12.76 = 10.72 \text{ mm of water}$$

Total pressure at exit of outlet duct (*i.e.* at section 5),

$$p_{T5} = p_{T4} - p_{f(4-5)} = 23.48 - 10.72 = 12.76 \text{ mm of water}$$

Static pressure at exit of outlet duct,

$$p_{s5} = p_{T5} - p_{v5} = 12.76 - 12.76 = 0$$

The variation of total pressure and static pressure along the system is shown in Fig. 20.20 (*b*). **Ans.**

20.19 Duct Design

The object of duct design is to determine the dimensions of all ducts in the given system. The ducts should carry the necessary volume of conditioned air from the fan outlet to the conditioned space with minimum frictional and dynamic losses. The duct layout must be made so as to reach the outlet without least number of bends, obstructions and area changes. The area changes must be gradual where possible and limited to not more than 20° for diverging area and 60° for converging area. For rectangular ducts, the aspect ratio of 4 and less is desirable but it should not be greater than 8 in any case. The minimum sheet metal is required with square cross-section for given cross-sectional area.

The velocities in the ducts must be high enough to reduce the size of the ducts but it should be low enough to reduce the noise and pressure losses to economise power requirement. The velocities recommended for various applications are given in the following table :

Table 20.7. Recommended velocities for various applications.

Designation	*Recommended velocities in m/min*		
	Residences	*Schools, theatres and public buildings*	*Industrial buildings*
Outdoor air intakes	150	150	150
Filters	75	90	105
Heating coils	135	150	180
Air washers	150	150	150
Fan outlets	300 – 480	400 – 600	480 – 720
Main ducts	200 – 300	300 – 400	350 – 550
Branch duct	180	180 – 270	240 – 300
Branch risers	150	180 – 210	240

After the layout of the duct is decided and the requirements of air quantities at various outlets are known, then the size of the ducts may be obtained as discussed in the following pages.

20.20 Methods for Determination of Duct Size

The following three methods for determination of duct size are important from the subject point of view :

1. ***Velocity reduction method.*** In this method, the velocities in the ducts are assumed such that they progressively decrease as the flow proceeds. The pressure drops are calculated for these velocities for respective branches and the main duct. The duct sizes are determined for assumed velocities and known quantities of air to be supplied through the respective ducts. The pressure at the outlet is adjusted by dampers in the respective ducts. The fan is designed to overcome the pressure losses along any single run including losses of the main duct, branch duct, elbows, enlargements and contractions of areas etc. In case the fan is already selected, the velocities in reducing order are adjusted to consume pressure available in the longest run or the run in which the maximum pressure loss is expected. The pressure in the remaining branches is adjusted by dampers.

Heating and air - conditioning ducts.

This method is the easiest in sizing ducts and the velocities can be adjusted to avoid noise. The major disadvantage of this system is that considerable experience and judgement is required in selecting velocities so as to make the system optimum in economy and power.

The velocity reduction method of designing ducts is usually adopted for very simple systems.

2. ***Equal pressure drop (or friction loss) method.*** In this method, the size of duct is decided to give equal pressure drop (or friction loss) per metre length in all ducts. If the layout of the ducts is symmetrical giving the same length of the various runs, this method gives equal pressure loss in various branches and no dampering is required to balance them. In case the runs are of different lengths, then the shortest run will have minimum loss and consequently high pressure at the outlet. It is, therefore, necessary to reduce this high pressure by heavy dampering or modifying this method to provide higher velocities in shorter runs. But the high velocities in short run to reduce high pressure may create objectionable noise. Thus noise absorbing outlets and fittings must be provided. The dampers if provided near the main duct will help in reducing the noise as the branch duct will dissipate some noise.

The velocities, in this method , are automatically reduced in the branch ducts as the flow is decreased. This method does not however balance the pressures at the outlets if the branches are of different lengths and hence dampers are required for balancing the pressure drops in various branches.

A modification of this method is to design the main duct for equal friction and branch ducts for consuming the pressure available at the take-off from the main duct. In such a design, the pressures at the outlet will be same and no dampering is required for balancing the pressure drops in various runs. However, dampers are provided for small adjustment.

3. *Static regain method.* In this method, the size of the duct is decided to give equal pressure at all outlets, for perfect balancing of the air duct layout system. This may be done by equalising the pressure losses in various branches. This is possible if the friction loss in each branch is made equal to the gain in pressure due to reduction in velocity. The gain in pressure (or static pressure regain) due to change in velocity is given by

$$SPR = R\rho_a\left[\frac{(V_1)^2-(V_2)^2}{2}\right] = R\,(p_{v1}-p_{v2}) \quad \dots \left(\because p_v = \frac{\rho_a V^2}{2}\right)$$

where R is the static regain factor.

It may not be possible to design economically very long branches and the branches very near to the fan for complete regain. In such cases, it is sufficient to design the main duct for complete regain and provide same pressure at all outlets from the main duct for branches. The partial regain may be considered a good practice for a few outlets from the main duct, so that same pressure loss is allowed in the beginning.

This method allows for balancing but reducing velocity increases duct size and it should not be taken beyond the economic limit.

Example 20.14. *In the duct system, as shown in Fig. 20.21, the cross-section of the main duct is such that the width is always twice the depth. The total quantity of air entering at A is 690 m³/min and the static pressure at A is 18 mm of water.*

A branch 0.9 m broad and 0.6 m deep is led off at B for a length of 45 m. This discharges at atmospheric pressure and there is a loss of 0.5 velocity head at the conversion piece. Another branch at C is so designed that it discharges 120 m³/min.

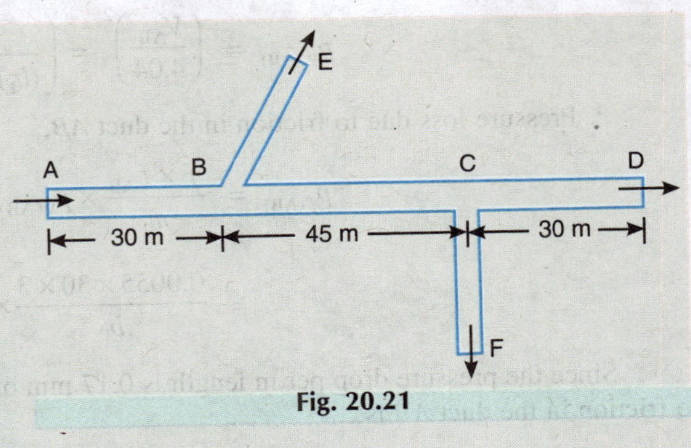

Fig. 20.21

Calculate the dimensions of the three sections of the main duct, designing on a basis of uniform pressure drop. Neglect velocity changes in the main duct and assume that the discharge pressure at D is atmospheric. The value of friction factor may be taken as 0.0055.

Solution. Given : Q_{AB} = 690 m³/min ; p_{SA} = 18 mm of water ; a = 0.9 m ; b = 0.6 m ; L_{BE} = 45 m ; $p_{L(BE)} = 0.5\, p_{v\,(BE)}$; Q_{CF} = 120 m³/min ; f = 0.0055 ; L_{AB} = 30 m ; L_{BC} = 45 m ; L_{CD} = 30 m

We know that when the velocity changes in the main duct AD are neglected, then the static pressure at A must be equal to the pressure loss due to friction between A and D.

$$\therefore \quad p_{f\,(AB)} + p_{f\,(BC)} + p_{f\,(CD)} = p_{SA} = 18 \text{ mm of water}$$

Total length of the duct AD,

$$L = L_{AB} + L_{BC} + L_{CD} = 30 + 45 + 30 = 105 \text{ m}$$

∴ Pressure drop per m length of the duct AD

$$= \frac{18}{105} = 0.17 \text{ mm of water}$$

Dimensions of the duct AB

Let b_1 = Depth of the duct AB, and

a_1 = Width of the duct $AB = 2\,b_1$ ... (Given)

Cross-sectional area of the duct AB,

$$A_1 = a_1 b_1 = 2b_1 \times b_1 = 2(b_1)^2$$

Wetted perimeter of the duct AB,

$$P_1 = 2\,(a_1 + b_1) = 2\,(2b_1 + b_1) = 6\,b_1$$

∴ Hydraulic mean depth of the duct AB,

$$m_1 = \frac{A_1}{P_1} = \frac{2(b_1)^2}{6\,b_1} = \frac{b_1}{3}$$

Velocity of air in the duct AB,

$$V_{AB} = \frac{Q_{AB}}{A_1} = \frac{690}{2(b_1)^2} = \frac{345}{(b_1)^2} \text{ m / min} = \frac{5.75}{(b_1)^2} \text{ m / s}$$

Velocity pressure in the duct AB,

$$p_{v\,(AB)} = \left(\frac{V_{AB}}{4.04}\right)^2 = \left(\frac{5.75}{(b_1)^2 \times 4.04}\right)^2 = \frac{2}{(b_1)^4}$$

∴ Pressure loss due to friction in the duct AB,

$$p_{f\,(AB)} = \frac{f \times L_{AB}}{m_1} \times p_{v(AB)}$$

$$= \frac{0.0055 \times 30 \times 3}{b_1} \times \frac{2}{(b_1)^4} = \frac{0.99}{(b_1)^5} \quad \text{... (i)}$$

Since the pressure drop per m length is 0.17 mm of water, therefore total pressure loss due to friction in the duct AB is

$$p_{f\,(AB)} = 0.17 \times 30 = 5.1 \text{ mm of water} \quad \text{... (ii)}$$

Equating equations (*i*) and (*ii*), we have

$$\frac{0.99}{(b_1)^5} = 5.1 \quad \text{or} \quad (b_1)^5 = \frac{0.99}{5.1} = 0.194$$

∴ $b_1 = 0.72$ m **Ans.**

and $a_1 = 2b_1 = 2 \times 0.72 = 1.44$ m **Ans.**

Dimensions of the duct BC

Let b_2 = Depth of the duct BC, and

a_2 = Width of the duct $BC = 2b_2$ ... (Given)

First of all, let us find out the quantity of air flowing in the duct BE. We know that cross-sectional area of the duct BE,

$$A = ab = 0.9 \times 0.6 = 0.54 \text{ m}^2$$

Wetted perimeter of the duct BE,

$$P = 2\,(a + b) = 2\,(0.9 + 0.6) = 3 \text{ m}$$

and hydraulic mean depth of the duct *BE*,

$$m = \frac{A}{P} = \frac{0.54}{3} = 0.18 \text{ m}$$

∴ Pressure loss due to friction in the duct *BE*,

$$p_{f\,(BE)} = \frac{f \times L_{BE}}{m} \times p_{v\,(BE)} = \frac{0.0055 \times 45}{0.18} \times p_{v\,(BE)} = 1.375\, p_{v\,(BE)}$$

Pressure loss at the conversion piece

$$= 0.5\, p_{v\,(BE)} \quad \text{... (Given)}$$

∴ Pressure at $B = 1.375\, p_{v\,(BE)} + 0.5\, p_{v\,(BE)} = 1.875\, p_{v\,(BE)}$ ... (*iii*)

Also Pressure at $B = p_{SA} - p_{f\,(AB)} = 18 - 5.1 = 12.9$ mm of water ... (*iv*)

Equating equations (*iii*) and (*iv*),

$$1.875\, p_{v\,(BE)} = 12.9 \quad \text{or} \quad p_{v\,(BE)} = \frac{12.9}{1.875} = 6.88 \text{ mm of water}$$

We know that velocity of air in the duct *BE*,

$$V_{BE} = 4.04\sqrt{p_{v\,(BE)}} = 4.04\sqrt{6.88} = 10.5 \text{ m/s} = 630 \text{ m/min}$$

∴ Quantity of air flowing through the duct *BE*,

$$Q_{BE} = A \times V_{BE} = 0.54 \times 630 = 340 \text{ m}^3\text{/min}$$

and quantity of air flowing through the duct *BC*,

$$Q_{BC} = Q_{AB} - Q_{BE} = 690 - 340 = 350 \text{ m}^3\text{/min}$$

Cross-sectional area of the duct *BC*,

$$A_2 = a_2 b_2 = 2b_2 \times b_2 = 2(b_2)^2$$

Wetted perimeter of the duct *BC*,

$$P_2 = 2\,(a_2 + b_2) = 2\,(2b_2 + b_2) = 6\,b_2$$

∴ Hydraulic mean depth of the duct *BC*,

$$m_2 = \frac{A_2}{P_2} = \frac{2(b_2)^2}{6\,b_2} = \frac{b_2}{3}$$

Velocity of air in the duct *BC*,

$$V_{BC} = \frac{Q_{BC}}{A_2} = \frac{350}{2(b_2)^2} = \frac{175}{(b_2)^2} \text{ m/min} = \frac{2.92}{(b_2)^2} \text{ m/s}$$

Velocity pressure in the duct *BC*,

$$p_{v\,(BC)} = \left[\frac{V_{BC}}{4.04}\right]^2 = \left[\frac{2.92}{(b_2)^2 \times 4.04}\right]^2 = \frac{0.52}{(b_2)^4}$$

∴ Pressure loss due to friction in the duct *BC*,

$$p_{f\,(BC)} = \frac{f \times L_{BC}}{m_2} \times p_{v\,(BC)} = \frac{0.0055 \times 45 \times 3}{b_2} \times \frac{0.52}{(b_2)^4} = \frac{0.386}{(b_2)^5} \quad \text{... (v)}$$

Since the pressure drop per m length is 0.17 mm of water, therefore total pressure loss due to friction in the duct *BC* is

$$p_{f\,(BC)} = 0.17 \times 45 = 7.65 \text{ mm of water} \quad \text{... (vi)}$$

Equating equations (*v*) and (*vi*),

$$\frac{0.386}{(b_2)^5} = 7.65 \quad \text{or} \quad (b_2)^5 = \frac{0.386}{7.65} = 0.05$$

$\therefore \quad b_2 = 0.55$ m **Ans.**

and $\quad a_2 = 2b_2 = 2 \times 0.55 = 1.1$ m **Ans.**

Dimensions of the duct CD

Let $\quad b_3$ = Depth of the duct *CD*, and

a_3 = Width of the duct $CD = 2\,b_3$ ... (Given)

Cross-sectional area of the duct *CD*,

$$A_3 = a_3 b_3 = 2b_3 \times b_3 = 2\,(b_3)^2$$

Wetted perimeter of the duct *CD*,

$$P_3 = 2\,(a_3 + b_3) = 2\,(2b_3 + b_3) = 6\,b_3$$

$\therefore$ Hydraulic mean depth of the duct *CD*,

$$m_3 = \frac{A_3}{P_3} = \frac{2(b_3)^2}{6\,b_3} = \frac{b_3}{3}$$

We know that the quantity of the air flowing through the duct *CD*,

$$Q_{CD} = Q_{BC} - Q_{CF} = 350 - 120 = 230 \text{ m}^3\text{/min}$$

Velocity of air in the duct *CD*,

$$V_{CD} = \frac{Q_{CD}}{A_3} = \frac{230}{2(b_3)^2} = \frac{115}{(b_3)^2} \text{ m / min} = \frac{1.92}{(b_3)^2} \text{ m / s}$$

Velocity pressure in the duct *CD*,

$$p_{v\,(CD)} = \left(\frac{V_{CD}}{4.04}\right)^2 = \left[\frac{1.92}{(b_3)^2 \times 4.04}\right]^2 = \frac{0.226}{(b_3)^4} \text{ mm of water}$$

$\therefore$ Pressure loss due to friction in the duct *CD*,

$$p_{f\,(CD)} = \frac{f \times L_{CD}}{m_3} \times p_{v\,(CD)} = \frac{0.0055 \times 30 \times 3}{b_3} \times \frac{0.226}{(b_3)^4}$$

$$= \frac{0.112}{(b_3)^5} \text{ mm of water} \quad \text{... }(vii)$$

Also pressure loss due to friction in the duct *CD*,

$$p_{f\,(CD)} = 0.17 \times 30 = 5.1 \text{ mm of water} \quad \text{... }(viii)$$

Equating equations (*vii*) and (*viii*), we have

$$\frac{0.112}{(b_3)^5} = 5.1 \quad \text{or} \quad (b_3)^5 = \frac{0.112}{5.1} = 0.022$$

$\therefore \quad b_3 = 0.466$ m **Ans.**

and $\quad a_3 = 2b_3 = 2 \times 0.466 = 0.932$ m **Ans.**

Carbon fibre air duct system.

Example 20.15. *An air conditioning system has volume flow rate of 7.5 m³/s and fan outlet velocity is 10 m/s. The duct has four branches with 90° elbows. The first branch is 10 m from fan. The distance between branches is 10 m and the main duct has 90° elbow 10m after the fourth branch. The volume flow rate in each branch is 1.5 m³/s. The main duct runs 10 m after the 90° bend.*

1. Using equal friction method, determine the equivalent diameter of duct and dimensions of rectangular duct if one side of the duct is 0.5 m.

2. Determine the total pressure drop. Given:

$$\frac{p_f}{L} = \frac{0.002\ 268\ (Q)^{1.852}}{(D)^{4.973}} \text{ mm of water ; } p_v = \left(\frac{V}{4.04}\right)^2 \text{ mm of water ;}$$

$$D = \frac{1.3(ab)^{0.625}}{(a+b)^{0.25}} \text{ ; Elbow loss} = 0.25\ p_v$$

Fitting losses where changes in area occur = 0.25 × Difference of velocity pressures

Dynamic loss in branch $= 0.2\ p_v$ + *Elbow loss*

Solution. Given : $Q_{AB} = 7.5$ m³/s ; $V_{AB} = 10$ m/s; $L_{AB} = L_{BC} = L_{CD} = L_{DE} = L_{EF} = 10$ m

The duct system is shown in Fig. 20.22.

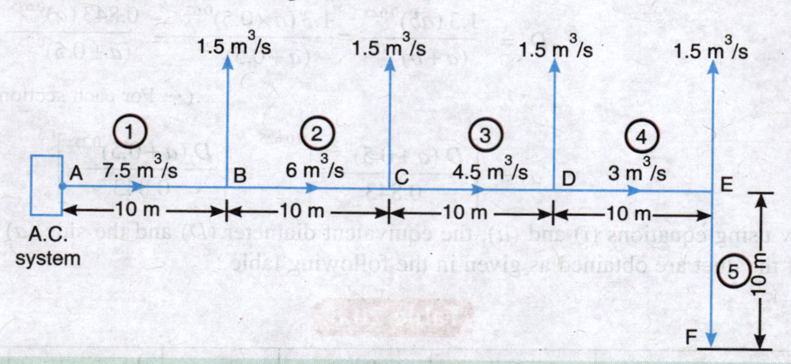

Fig. 20.22

Since the volume flow rate in each branch at *B*, *C*, *D* and *E* is 1.5 m³/s, therefore $Q_{BC} = 7.5 - 1.5 = 6$ m³/s ; $Q_{CD} = 6 - 1.5 = 4.5$ m³/s ; $Q_{DE} = 4.5 - 1.5 = 3$ m³/s

(a) Equivalent diameter of the duct and dimensions of the rectangular duct

Let a = Longer side of the duct *AB* in metres, and

b = Shorter side of the duct *AB* = 0.5 m ...(Given)

∴ Cross-sectional area of the duct B,

$$A_{AB} = a \times b = a \times 0.5 = 0.5\ a \text{ m}^2$$

We know that quantity of air passing through the duct *AB*,

$$Q_{AB} = A_{AB} \times V_{AB} = 0.5a \times 10 = 5a$$

$$\therefore \quad a = Q_{AB} / 5 = 7.5 / 5 = 1.5 \text{ m} \quad \textbf{Ans.}$$

and equivalent diameter for the duct *AB*,

$$D_{AB} = \frac{1.3\,(ab)^{0.625}}{(a+b)^{0.25}} = \frac{1.3\,(1.5\times0.5)^{0.625}}{(1.5+0.5)^{0.25}} = \frac{1.3\times0.8354}{1.1892}$$

$$= 0.9132 \text{ m} \textbf{ Ans.}$$

We know that pressure loss due to friction in the duct *AB* per metre length,

$$\frac{p_{f(AB)}}{L_{AB}} = \frac{0.002\ 268\ (Q_{AB})^{1.852}}{(D_{AB})^{4.973}} = \frac{0.002\ 268(7.5)^{1.852}}{(0.9132)^{4.973}} = \frac{0.094\ 68}{0.636\ 64}$$

$$= 0.1487 \text{ mm of water}$$

This is constant for all sections.

∴ Equivalent diameter for each section,

$$D = \left[\frac{0.002\ 268(Q)^{1.852}}{0.1487}\right]^{1/4.973} = 0.4312\ (Q)^{0.3724} \qquad \ldots(i)$$

It is given that equivalent diameter,

$$D = \frac{1.3\,(ab)^{0.625}}{(a+b)^{0.25}} = \frac{1.3\,(a\times0.5)^{0.625}}{(a+0.5)^{0.25}} = \frac{0.843\,(a)^{0.625}}{(a+0.5)^{0.25}}$$

...(∵ For each section, b = 0.5 m)

$$\therefore \quad a = \left[\frac{D\,(a+0.5)^{0.25}}{0.843}\right]^{1/0.625} = \left[\frac{D\,(a+0.5)^{0.25}}{0.843}\right]^{1.6} \qquad \ldots(ii)$$

Now using equations (*i*) and (*ii*), the equivalent diameter (*D*) and the side (*a*) for various sections of the duct are obtained as given in the following table :

Table 20.8.

Section	*Length (L) m*	*Flow rate (Q) m³/s*	*Equivalent diameter (D) m from equaton (i)*	**Side (a) m from equation (ii)*	*Cross-sectional area (A) m² A = a × b = a × 0.5*	*Velocity (V)m/s V = Q/A*	*Velocity pressure (p_v) mm of water $p_v = \left[\frac{V}{4.04}\right]^2$*
AB	10	7.5	0.9132	1.5	0.75	10	6.1268
BC	10	6	0.8404	1.2424	0.6212	9.6587	5.7158
CD	10	4.5	0.7550	0.980	0.49	9.1837	5.1674
DE	10	3	0.6492	0.710	0.355	8.4507	4.3754
EF	10	1.5	0.5015	0.4216	0.2108	7.1157	3.1022

From the above table, we find that equivalent diameter for various sections are : D_{AB} = 0.9132 m ; D_{BC} = 0.8404 m ; D_{CD} = 0.7550 m ; D_{DE} = 0.6492 m ; and D_{EF} = 0.5015 m. **Ans.**

* The side *a* is obtained by iteration from equation (*ii*)

and the dimensions of rectangular duct for various sections are:

Section AB = 1.5 m × 0.5 m ; Section BC = 1.2424 m × 0.5 m ;

Section CD = 0.980 m × 0.5 m ; Section DE = 0.7110 m × 0.5 m ; and

Section EF = 0.4216 m × 0.5 m **Ans.**

2. Total pressure drop

We know that total pressure loss due to friction between AF,

$$p_{f(AF)} = 0.1487 \times L_{AF} = 0.1487 \times 50 = 7.435 \text{ mm of water}$$

... ($\because$ p_f per m length = 0.1487 mm of water)

Now let us find the dynamic losses between A and F as discussed below :

(*a*) Loss in discharge opening

$$= p_{v(EF)} = 3.1022 \text{ mm of water}$$

(*b*) Elbow loss $= 0.25\, p_{v(EF)} = 0.25 \times 3.1022 = 0.7755$ mm of water

(*c*) Fitting loss = 0.25 × Difference of velocity pressures

$\therefore$ Fitting loss at B $= 0.25\,[\,p_{v(AB)} - p_{v(BC)}\,]$

$= 0.25\,(6.1268 - 5.7158) = 0.1027$ mm of water

Fitting loss at C $= 0.25\,[\,p_{v(BC)} - p_{v(CD)}\,]$

$= 0.25\,[\,5.7158 - 5.1674\,] = 0.1371$ mm of water

Fitting loss at D $= 0.25\,[\,p_{v(CD)} - p_{v(DE)}\,]$

$= 0.25\,[\,5.1674 - 4.3754\,] = 0.198$ mm of water

Fitting loss at E $= 0.25\,[\,p_{v(DE)} - p_{v(EF)}\,]$

$= 0.25\,[\,4.3754 - 3.1022\,] = 0.3183$ mm of water

$\therefore$ Total fitting loss $= 0.1027 + 0.1371 + 0.198 + 0.3183 = 0.7561$ mm of water

and total dynamic losses between A and F,

$$p_{d(AF)} = 3.1022 + 0.7755 + 0.7561 = 4.6338 \text{ mm of water}$$

$\therefore$ Total pressure drop = Total frictional and dynamic losses between A and F

$$p_T = p_{f(AF)} + p_{d(AF)}$$

$$= 7.435 + 4.6338 = 12.0688 \text{ mm of water}$$ **Ans.**

Example 20.16. *In the air duct system, as shown in Fig. 20.23, air enters at A with a static pressure of 7.5 mm of water. The branch at B is 15 m long and delivers 120 m³/min. The branch at C is 22.5 m long and delivers 140 m³/min. At the end D of the main duct, the air delivered is 200 m³/min.*

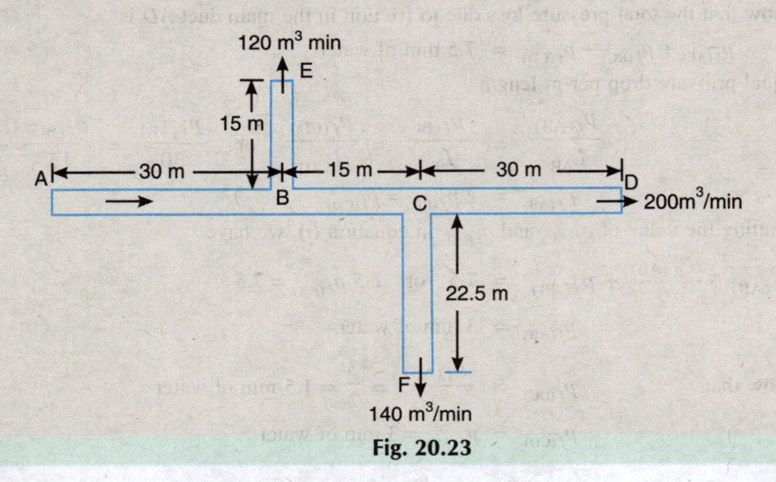

Fig. 20.23

Using friction chart and equal pressure drop method, determine the correct diameter and velocity pressures in lengths, AB, BC, CD, branch BE and branch CF. Consider friction losses only.

Solution. Given : p_{SA} = 7.5 mm of water ; L_{BE} = 15 m ; Q_{BE} = 120 m³/min ; L_{CF} = 22.5 m ; Q_{CF} = 140 m³/min ; Q_{CD} = 200 m³/min ; L_{AB} = 30 m ; L_{BC} = 15 m ; L_{CD} = 30 m

Diameter of lengths AB, BC, CD, branch BE and branch CF

According to Bernoulli's equation, the pressure losses between the main duct *AD* plus velocity pressure at *D* must be equal to the total pressure at *A*. Thus

$$p_{f(AB)} + p_{f(BC)} + p_{f(CD)} + p_{v(D)} = p_{SA} + p_{v(AB)}$$

Neglecting velocity pressure in the main duct, total pressure losses between the main duct *AD*,

$$p_f = p_{f(AB)} + p_{f(BC)} + p_{f(CD)} = 7.5 \text{ mm of water}$$

Total length of the main duct *AD*,

$$L = L_{AB} + L_{BC} + L_{CD} = 30 + 15 + 30 = 75 \text{ m}$$

Pressure drop per metre length of the duct

$$= 7.5 / 75 = 0.1 \text{ mm of water}$$

∴ Total pressure loss due to friction in the duct *AB*,

$$^{*}p_{f(AB)} = 0.1 \times 30 = 3 \text{ mm of water}$$

Total pressure loss due to friction in the duct *BC*,

$$p_{f(BC)} = 0.1 \times 15 = 1.5 \text{ mm of water}$$

and total pressure loss due to friction in the duct *CD*,

$$p_{f(CD)} = 0.1 \times 30 = 3 \text{ mm of water}$$

We know that the quantity of air passing through the duct *AB*,

$$Q_{AB} = Q_{BE} + Q_{CF} + Q_{CD}$$

$$= 120 + 140 + 200 = 460 \text{ m}^3/\text{min} = 7.67 \text{ m}^3/\text{s}$$

* The total pressure loss due to friction in the ducts *AB*, *BC* and *CD* may be calculated as discussed below :

We know that the total pressure loss due to friction in the main duct *AD* is

$$p_{f(AB)} + p_{f(BC)} + p_{f(CD)} = 7.5 \text{ mm of water} \quad \ldots (i)$$

For equal pressure drop per m length

$$\frac{p_{f(AB)}}{L_{AB}} = \frac{p_{f(BC)}}{L_{BC}} = \frac{p_{f(CD)}}{L_{CD}} \quad \text{or} \quad \frac{p_{f(AB)}}{30} = \frac{p_{f(BC)}}{15} = \frac{p_{f(CD)}}{30}$$

$$\therefore \quad p_{f(AB)} = 2\,p_{f(BC)} = p_{f(CD)}$$

Substituting the value of $p_{f(BC)}$ and $p_{f(CD)}$ in equation (*i*), we have

$$p_{f(AB)} + \frac{p_{f(AB)}}{2} + p_{f(AB)} = 7.5 \quad \text{or} \quad 2.5\,p_{f(AB)} = 7.5$$

$$\therefore \quad p_{f(AB)} = 3 \text{ mm of water}$$

We know that

$$p_{f(BC)} = \frac{p_{f(AB)}}{2} = \frac{3}{2} = 1.5 \text{ mm of water}$$

and

$$p_{f(CD)} = p_{f(AB)} = 3 \text{ mm of water}$$

Quantity of air passing through the duct *BC*,

$$Q_{BC} = Q_{CF} + Q_{CD} = 140 + 200 = 340 \text{ m}^3/\text{min} = 5.67 \text{ m}^3/\text{s}$$

and quantity of air passing through the duct *CD*,

$$Q_{CD} = 200 \text{ m}^3/\text{min} = 3.33 \text{ m}^3/\text{s} \quad \text{... (Given)}$$

From the friction chart, as shown in Fig. 20.10, we find that for 7.67 m^3/s and 0.1 mm of water, the diameter of duct *AB*,

$$D_{AB} = 0.95 \text{ m} \quad \textbf{Ans.}$$

Similarly, for 5.67 m^3/s and 0.1 mm of water, the diameter of duct *BC*,

$$D_{BC} = 0.85 \text{ m} \quad \textbf{Ans.}$$

and for 3.33 m^3/s and 0.1 mm of water, the diameter of duct *CD*,

$$D_{CD} = 0.72 \text{ m} \quad \textbf{Ans.}$$

We know that static pressure at *B*,

$$p_{SB} = p_{SA} - p_{f(AB)}$$
$$= 7.5 - 3 = 4.5 \text{ mm of water}$$

Since the branch *BE* is 15 m long, therefore pressure drop required per metre length

$$^*p_{f(BE)} = \frac{4.5}{15} = 0.3 \text{ mm of water}$$

Now from the friction chart, we find that for Q_{BE} = 120 m^3/min (or 2 m^3/s) and 0.3 mm of water, the diameter of branch *BE*,

$$D_{BE} = 0.46 \text{ m} \quad \textbf{Ans.}$$

---

* The pressure drop in the branches *BE* and *CF* may be obtained as discussed below:

We know that pressure drop in the main duct per m length

$$= 0.1 \text{ mm of water}$$

Since the pressure drops are equal, therefore

Pressure drop in the branch *BE* for 15 m length

$$= \text{Pressure drop in the main duct } BD \text{ of 45 m length}$$
$$= 0.1 \times 45 = 4.5 \text{ mm of water}$$

∴ Pressure drop in the branch *BE* per m length

$$p_{f(BE)} = \frac{4.5}{15} = 0.3 \text{ mm of water}$$

Similarly, pressure drop in the branch *CF* for 22.5 m length

$$= \text{Pressure drop in the main duct } CD \text{ for 30 m length}$$
$$= 0.1 \times 30 = 3 \text{ mm of water}$$

∴ Pressure drop in the branch *CF* per m length,

$$p_{f(CF)} = \frac{3}{22.5} = 0.133 \text{ mm of water}$$

We know that static pressure at *C*,

$$p_{SC} = p_{SB} - p_{f\,(BC)}$$
$$= 4.5 - 1.5 = 3 \text{ mm of water}$$

Since the branch *CF* is 22.5 m, therefore pressure drop required per metre length

$$p_{f\,(CF)} = \frac{3}{22.5} = 0.133 \text{ mm of water}$$

From the friction chart, we find that for $Q_{CD} = 140$ m³/min (or 2.33 m³/s) and 0.133 mm of water, the diameter of branch *CF*,

$$D_{CF} = 0.6 \text{ m}$$ **Ans.**

Velocity pressure in lengths AB, BC, CD, branch BE and branch CF

The velocity pressure in lengths *AB*, *BC*, *CD* and branch *BE* and branch *CF* are calculated in the following table :

Table 20.9

Length	*Diameter (D) m*	*Cross-sectional area* $\left(A = \frac{\pi}{4} D^2\right)$ *m²*	*Quantity of air (Q) m³/min*	*Velocity* $\left(V = \frac{Q}{A}\right)$ *m/min*	*Velocity pressure* $*p_v = \left(\frac{V}{242.4}\right)^2$ *mm of water*
AB	0.95	0.709	460	648.8	7.16
BC	0.85	0.567	340	599.6	6.12
CD	0.72	0.407	200	491.4	4.11
BE	0.46	0.166	120	722.9	8.89
CF	0.6	0.283	140	494.7	4.16

Example 20.17. *An air duct system is provided as shown in Fig 20.24.*

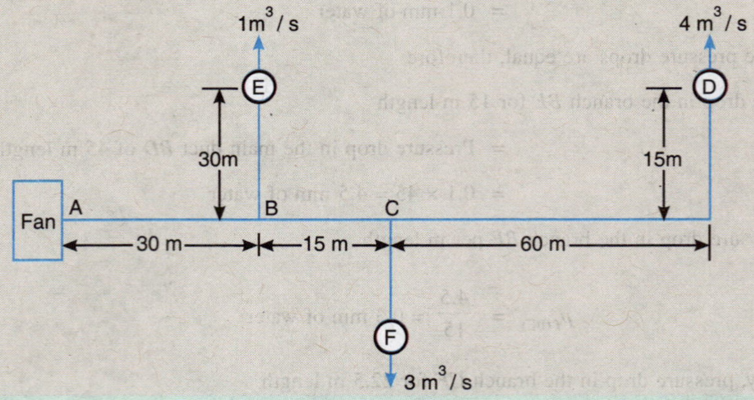

Fig. 20.24

1. Determine the dimensions of AB, BC and CD using the equal friction method. Choose a friction rate of 0.08 mm of water per metre length of duct. Use the following formula for friction rate :

---

* When velocity (*V*) is in m / min, then velocity pressure,

$$p_v = \left(\frac{V}{60 \times 4.04}\right)^2 = \left(\frac{V}{242.4}\right)^2 \text{ mm of water}$$

$$\frac{p_f}{L} = \frac{2.268 \times 10^{-3} (Q)^{1.852}}{(D)^{4.973}} \text{ mm of water/m}$$

where Q is m³/s and D and L are in metres.

2. *Determine the total and static pressures at point A. Assume free exit at each outlet. Losses are given by :*

   For elbow: 0.25 p_v ; Branch : 0.2 p_v + Elbow loss

   For straight-through section : 0.25 ($p_{v1} - p_{v2}$)

3. *Find the diameter of BE so that no dampering is required in the section.*

Solution. Given : L_{AB} = 30 m ; L_{BC} = 15 m ; L_{CD} = 60 + 15 = 75 m ; Q_{BE} = 1 m³/s; Q_{CF} = 3 m³/s ; Q_{CD} = 4 m³/s ; p_f / L = 0.08 mm of water / m

1. *Dimensions of AB, BC and CD*

We know that quantity of air passing through the duct *AB*,

$$Q_{AB} = Q_{BE} + Q_{CF} + Q_{CD} = 1 + 3 + 4 = 8 \text{ m}^3/\text{s}$$

Quantity of air passing through the duct *BC*,

$$Q_{BC} = Q_{CF} + Q_{CD} = 3 + 4 = 7 \text{ m}^3/\text{s}$$

and quantity of air passing through the duct *CD*,

$$Q_{CD} = 4 \text{ m}^3/\text{s}$$

We know that pressure loss due to friction per metre length of duct,

$$\frac{p_f}{L} = \frac{2.268 \times 10^{-3} (Q)^{1.852}}{(D)^{4.973}} = 0.08 \text{ mm of water} \quad \text{...(Given)}$$

$$\therefore \quad D = \left[\frac{2.268 \times 10^{-3} (Q)^{1.852}}{0.08} \right]^{1/4.973} = 0.488\ 46\ (Q)^{0.3724}$$

The diameter and velocity pressure in lengths *AB*, *BC* and *CD* are calculated in the following table :

Table 20.10

Section	*Length (L)* m	*Quantity of air (Q)* m³/s	*Diameter (D)* $0.488\ 46(Q)^{0.3724}$ m	*Cross-sectional area (A)* $\left[\frac{\pi}{4}D^2\right]$ m²	*Velocity* $V = \frac{Q}{A}$ m / s	*Velocity pressure* $p_v = \left(\frac{V}{4.04}\right)^2$ mm of water
AB	30	8	1.0596	0.8819	9.071	5.041
BC	15	7	1.0082	0.7984	8.767	4.709
CD	75	4	0.8185	0.5262	7.602	3.541

From the above table, we find that diameter of duct *AB*,

$$D_{AB} = 1.0596 \text{ m} \quad \textbf{Ans.}$$

Diameter of duct *BC*, D_{BC} = 1.0082 m **Ans.**

and diameter of duct *CD*, D_{CD} = 0.8185 m **Ans.**

2. Total and static pressure at point A

We know that pressure loss due to friction per metre length of duct,

$$\frac{p_f}{L} = 0.08 \text{ mm of water}$$

∴ Total pressure loss due to friction in the duct *AB*,

$$p_{f\,(AB)} = 0.08 \times L_{AB} = 0.08 \times 30 = 2.4 \text{ mm of water}$$

Total pressure loss due to friction in the duct *BC*,

$$p_{f(BC)} = 0.08 \times L_{BC} = 0.08 \times 15 = 1.2 \text{ mm of water}$$

Total pressure loss due to friction in the duct *CD*,

$$p_{f(CD)} = 0.08 \times L_{CD} = 0.08 \times 75 = 6 \text{ mm of water}$$

and total pressure loss due to friction from *A* to *D*,

$$p_f = p_{f\,(AB)} + p_{f(BC)} + p_{f\,(CD)}$$

$$= 2.4 + 1.2 + 6 = 9.6 \text{ mm of water}$$

The dynamic losses between *A* and *D* are as follows :

(*a*) Loss in discharge opening $= p_{v\,(CD)} = 3.541$ mm of water

(*b*) Elbow loss $= 0.25\, p_{v\,(CD)} = 0.25 \times 3.541 = 0.885$ mm of water

(*c*) Fitting loss at *B* $= 0.25\, [\, p_{v(AB)} - p_{v(BC)}\,]$

$= 0.25\,(5.041 - 4.709) = 0.083$ mm of water

and fitting loss at *C* $= 0.25\, [\, p_{v(BC)} - p_{v(CD)}\,]$

$= 0.25\,[4.709 - 3.541] = 0.292$ mm of water

∴ Total dynamic loss, $p_d = 3.541 + 0.885 + 0.083 + 0.292 = 4.801$ mm of water

We know that total friction and dynamic losses between *A* and *D*

= Total pressure at fan exit at *A*

$$p_T = p_{TA} = p_f + p_d = 9.6 + 4.801 = 14.401 \text{ mm of water} \textbf{ Ans.}$$

and static pressure at *A*, $p_{sA} = p_{TA} - p_{v(AB)} = 14.401 - 5.041 = 9.36$ mm of water **Ans.**

3. Diameter of BE so that no dampering is required

Let D_{BE} = Diameter of branch *BE*.

We know that total pressure at *B*,

$$p_{TB} = p_{TA} - p_{f(AB)} = 14.401 - 2.4 = 12.001 \text{ mm of water}$$

and the total pressure loss in the branch *BE*,

$$p_{T(BE)} = 0.25\, p_v + 0.2\, p_v + p_v + p_{f(BE)}$$

$$= 1.45\, p_v + p_{f(BE)}$$

The first term on the right hand side represents the sum of dynamic losses in the elbow at *B*, the area change in the branch and the velocity pressure at discharge. Equating it to the total pressure at *B* (p_{TB}) for complete balancing, we have

$$1.45\, p_v + p_{f\,(BE)} = p_{TB} = 12.001$$

This equation can only be solved by trial and error.

Asuming $D_{BE} = 0.4$ m

Pressure loss due to friction in the branch *BE*,

$$p_{f(\mathrm{BE})} = \frac{2.268\times10^{-3}(Q_{\mathrm{BE}})^{1.852}}{(D_{\mathrm{BE}})^{4.973}}\times L_{\mathrm{BE}}$$

$$= \frac{2.268\times10^{-3}(1)^{1.852}}{(0.4)^{4.973}}\times30 = 6.48 \text{ mm of water}$$

and velocity of air in the branch *BE*,

$$V_{\mathrm{BE}} = \frac{Q_{\mathrm{BE}}}{A_{\mathrm{BE}}} = \frac{4Q_{\mathrm{BE}}}{\pi\,(D_{\mathrm{BE}})^2} = \frac{4\times1}{\pi\,(0.4)^2} = 7.957 \text{ m/s}$$

∴ Velocity pressure in the branch *BE*,

$$p_{v(\mathrm{BE})} = \left(\frac{V_{\mathrm{BE}}}{4.04}\right)^2 = \left(\frac{7.957}{4.04}\right)^2 = 3.88 \text{ mm of water}$$

We know that total pressure loss in the branch *BE*,

$$p_{\mathrm{T(BE)}} = 1.45\,p_v + p_{f(\mathrm{BE})}$$

$$= 1.45 \times 3.88 + 6.48 = 12.106 \text{ mm of water}$$

$$\ldots [\because p_v = p_{v(\mathrm{BE})}]$$

This is approximately equal to the total pressure at B (p_{TB}). Hence an assumed diameter for branch *BE* (*i.e.* D_{BE}) of 0.4 m is satisfactory. **Ans.**

20.21 System Resistance

A single duct-line having lengths of different cross-sections, bends, expanders, reducers, dampers and registers etc. constitutes a system of resistances in series, in which all the pressure losses occurring at a given volumetric flow rate are added together to give the system resistance or total pressure loss at the given volumetric flow rate.

We know that the pressure loss (frictional or dynamic) is directly proportional to the velocity pressure, therefore total pressure loss in a duct system,

$$p_{\mathrm{L}} = C_1 p_{v1} + C_2 p_{v2} + \ldots \qquad \ldots (i)$$

We also know that $p_{v1} \propto (V_1)^2$, $p_{v2} \propto (V_2)^2$, and so on. Thus, the equation (*i*) may be written as

$$p_{\mathrm{L}} = C_1 (V_1)^2 + C_2 (V_2)^2 + \ldots \qquad \ldots (ii)$$

But for a system in series, the volumetric flow rate (Q) is same. Therefore from the continuity equation,

$$Q = A_1 V_1 = A_2 V_2 = \ldots$$

i.e. $V_1 \propto Q$, $V_2 \propto Q$ and so on.

∴ Equation (*ii*) may be written as

$$p_{\mathrm{L}} = K_1 Q^2 + K_2 Q^2 + \ldots$$

$$= (K_1 + K_2 + \ldots)\,Q^2 = KQ^2$$

This expression represents the relationship between the pressure loss (or resistance) in the system and the volume flow rate. Thus, system resistance,

$$R = KQ^2$$

where K is a constant of the system.

20.22 Systems in Series

If a number of systems R_1, R_2, R_3 etc. having constants K_1, K_2, K_3 etc. are connected in series, as shown in Fig. 20.25(*a*), it can be reduced to a single equivalent system as shown in Fig. 20.25(*b*). The resistance of a single equivalent system or the overall system resistance for the given flow rate is obtained by adding the individual system resistances, *i.e.*

$$R_e = R_1 + R_2 + R_3$$

or

$$K\,Q^2 = K_1\,Q^2 + K_2\,Q^2 + K_3\,Q^2$$

and

$$K = K_1 + K_2 + K_3$$

(*a*) Systems in series. (*b*) Single equivalent system.

Fig. 20.25

20.23 Systems in Parallel

If a number of systems R_1, R_2, R_3 etc. having constants K_1, K_2, K_3 etc. are connected in parallel, as shown in Fig. 20.26 (*a*), it can be reduced to a single equivalent system as shown in Fig. 20.26 (*b*). The resistance of a single equivalent system or the overall system resistance for the given flow rate may be obtained by calculating the constant K for the equivalent system from the constants K_1, K_2, K_3 etc. of the parallel system.

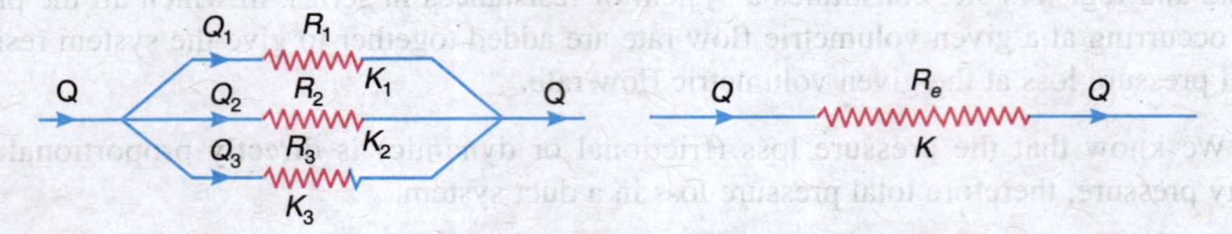

(*a*) Systems in parallel. (*b*) Single equivalent system.

Fig. 20.26

We know that
$$Q = Q_1 + Q_2 + Q_3$$

$$\sqrt{\frac{R_e}{K}} = \sqrt{\frac{R_1}{K_1}} + \sqrt{\frac{R_2}{K_2}} + \sqrt{\frac{R_3}{K_3}}$$

For the systems in parallel, the pressure loss is same, *i.e.*

$$R_e = R_1 = R_2 = R_3$$

$$\therefore \quad \sqrt{\frac{1}{K}} = \sqrt{\frac{1}{K_1}} + \sqrt{\frac{1}{K_2}} + \sqrt{\frac{1}{K_3}}$$

Example 20.18. *A duct system is represented diagramatically as shown in Fig. 20.27, where*

R_1 = 60 mm of water at 180 m³/min,

R_2 = 18 mm of water at 60 m³/min, and

R_3 = 30 mm of water at 75 m³/min.

Calculate the constant of an equivalent resistance of the above system and hence the pressure loss if the volume flow rate through R_1 is 120 m³/min.

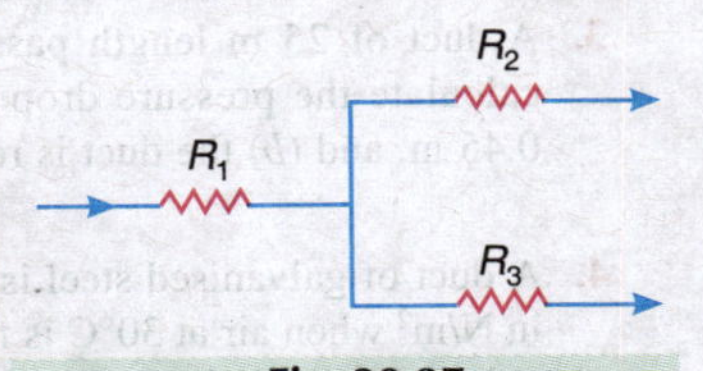

Fig. 20.27

Solution. Given : R_1 = 60 mm of water ; Q_1 = 180 m³/min; R_2 = 18 mm of water ; Q_2 = 60 m³/min ; R_3 = 30 mm of water ; Q_3 = 75 m³/min ; Q_1' = 120 m³/min

We know that the constant for the resistance R_1,

$$K_1 = \frac{R_1}{(Q_1)^2} = \frac{60}{(180)^2} = 1.85 \times 10^{-3}$$

Now, let us find the constant for the single equivalent resistance (R_e) for the parallel resistances R_2 and R_3. We know that the constant for the resistance R_2,

$$K_2 = \frac{R_2}{(Q_2)^2} = \frac{18}{(60)^2} = 5 \times 10^{-3}$$

Similarly, constant for the resistance R_3,

$$K_3 = \frac{R_3}{(Q_3)^2} = \frac{30}{(75)^2} = 5.3 \times 10^{-3}$$

Let K = Constant for the single equivalent resistance R_e.

$$\therefore \quad \frac{1}{\sqrt{K}} = \frac{1}{\sqrt{K_2}} + \frac{1}{\sqrt{K_3}} = \frac{1}{\sqrt{5 \times 10^{-3}}} + \frac{1}{\sqrt{5.3 \times 10^{-3}}}$$

$$= 14.14 + 13.7 = 27.84$$

or $$K = \frac{1}{(27.84)^2} = 1.3 \times 10^{-3}$$

Since the resistance R_1 of constant K_1 is connected with single equivalent resistance R_e of constant K in series, therefore constant for the whole system,

$$K' = K_1 + K$$

$$= 1.85 \times 10^{-3} + 1.3 \times 10^{-3} = 3.15 \times 10^{-3}$$

and the equivalent system resistance,

$$R_{e'} = K'\,(Q_{1'})^2 = 3.15 \times 10^{-3}\,(Q_{1'})^2$$ **Ans.**

Since the system resistance means the pressure loss, therefore pressure loss for a volume flow rate of Q_1' = 120 m³/min

$$= 3.15 \times 10^{-3}\,(120)^2 = 45.36 \text{ mm of water}$$ **Ans.**

EXERCISES

1. A main duct 0.6 m by 0.6 m carries air at the rate of 280 m³/min. It is divided into two ducts, one being 0.6 m by 0.3 m and the other is 0.6 m by 0.45 m. If the mean velocity in the larger branch is 9 m/s, calculate the mean velocity in the main duct and the other branch. Also find the mean velocity pressure in each duct assuming standard air density.
[Ans. 12.96 m/s ; 12.4 m/s ; 10.3 mm of water ; 9.4 mm of water ; 4.96 mm of water]

2. A grill has 15 m³/min of air passing through it and a duct connecting to it has 0.1 m² cross-sectional area. The static pressure behind the grill is 2.5 mm of water. Find the effective area of the grill in square metres.
[Ans. 0.036 m²]

3. A duct of 25 m length passes air at the rate of 120 m^3/min. Assuming friction factor as 0.006, calculate the pressure drop in the duct in mm of water when (*a*) the duct is circular of diameter 0.45 m, and (*b*) the duct is rectangular of sides 0.6 m and 0.45 m.

[**Ans.** 13 mm of water ; 3.93 mm of water]

4. A duct of galvanised steel is of 300 mm diameter and 15 m long. Find the pressure drop in the duct in N/m^2 when air at 30°C is flowing with a mean velocity of 480 m/min. Use Moody chart to find the value of friction factor '*f*'.

[**Ans.** 33.4 N/m^2]

5. A duct of 0.45 m diameter and 90 m long leads from a fan discharge chamber where the pressure is 15 mm of water to a plenum chamber where the pressure is 10 mm of water.

   In order to increase the flow, two alternatives are considered. One is to lay a duct of 0.3 m diameter and 90 m long in parallel with the duct of 0.45 m diameter. The other is to increase the diameter of 0.45 m diameter duct for the last 60 m length. Calculate the increased diameter so that this method gives the same flow as the 0.45 m and 0.3 m ducts in parallel. Assume that the pressures in the fan chamber and plenum chamber are unaffected by changes in the flow and consider duct friction losses only. The friction factor may be taken as 0.0055.

[**Ans.** 0.567 m]

6. A 0.3 m diameter circular duct carries standard air at a velocity of 360 m/min. It is replaced by a rectangular duct having the same pressure loss per unit length due to friction. Determine the dimensions of the rectangular duct if the aspect ratio is to be 1.5 for (*a*) the same volume flow through the two ducts, and (*b*) the same velocity in the two ducts.

[**Ans.** 0.336 m , 0.224 m ; 0.375 m, 0.25 m]

7. A fan delivers air at 8 m/s to the main line of air conditioning duct. After a straight run of 10 m, a branch of 4 m length delivers air to a room with 2 m^3/s. Another branch after a 5 m further run is taken from the main branch. This branch is 5 m in length and delivers 3 m^3/s. The main branch runs straight for another 5 m and then turns by 90° and runs for another 5 m length and delivers 3 m^3/s of air. The losses are given as

   For elbow : 0.25 p_v where $p_v = \left(\dfrac{V}{4.04}\right)^2$ mm of water.

   For branch : 0.2 p_v + Elbow loss

   For straight through section : 0.25 × Difference of velocity pressure

   Fitting losses where change in area occurs.

   Find the size of all the ducts and determine the static pressure requirement. Use the following formula for friction rate :

$$\frac{p_f}{L} = \frac{0.002\,268\,(Q)^{1.852}}{(D)^{4.973}}$$

   where Q is in m^3/s and D is in m.

   Find the friction rate in main line and then assume same friction rate for whole of duct work.

[**Ans.** 1.128 m , 1.01 m ; 0.783 m ; 0.905 mm of water ; 0.0586 mm of water / m length]

8. A main air duct of constant rectangular cross-section 2 m by 0.6 m has several branch pieces leading off it. One of these branches is shown in Fig. 20.28. The branch leaves the main at right angles and has a 90° bend in it of radius 0.6 m, as well as a damper for flow regulation and an exit grill. The branch is 0.6 m by 0.3 m and 3 m long. The flow through the main duct before this off take is 750 m^3/min and the flow required in the branch is 90 m^3/min. Calculate the

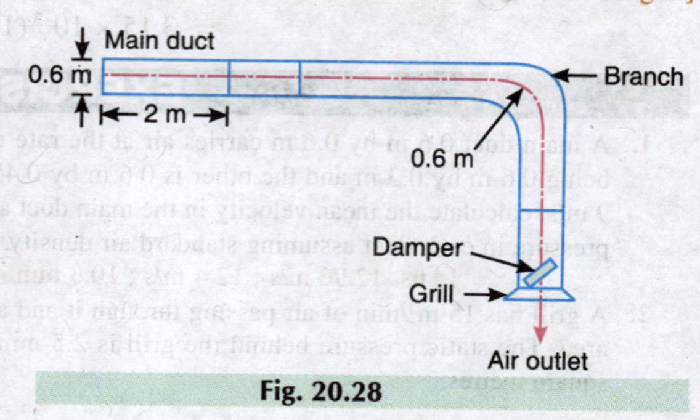

Fig. 20.28

damper resistance necessary to give the required flow when the static pressure in the main duct before off-take is 10 mm of water.

If the branch pieces are placed 12 m apart along the main duct, calculate the required percentage static regain in the main duct across the off take shown in order that the static pressure in the next off take will be 10 mm of water also. The dynamic loss coefficient for the branch, 90° bend and exit grill may be taken as 0.65, 0.3 and 0.8 respectively. **[Ans.** 9.94 mm of water ; 66.2%**]**

9. An air duct system is provided as shown in Fig. 20.29. The standard air enters at point *A* with a static pressure of 12 mm of water.

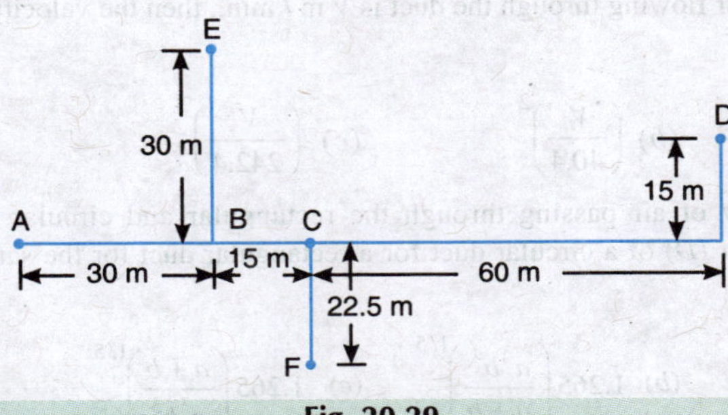

Fig. 20.29

The branch *BE* delivers 60 m³/min at *E*, branch *CF* delivers 180 m³/min at *F* and the main duct delivers 240 m³/min at *D*.

Using equal pressure drop method, find the duct dimensions. Assume free exit at each outlet.

[Ans. D_{AB} = 1 m ; D_{BC} = 0.9 m ; D_{CD} = 0.75 m ; D_{BE} = 0.375 m ; D_{CF} = 0.55 m **]**

10. An air conditioning duct runs straight from fan over 60 m length. It has four equally spaced outlet diffusers mounted on duct, the last one being at the end of duct. The volume flow rate through each diffuser is 1 m³/s. The velocity at duct inlet is 15 m/s. Carry out the duct design by static regain method if static regain factor is 0.75 at each transition and frictional pressure drop is given by

$$\frac{p_f}{L} = \frac{0.002\,268\,(Q)^{1.852}}{(D)^{4.973}} \text{ mm of water}$$

and $$p_v = \left(\frac{V}{4.04}\right)^2 \text{ mm of water}$$

where Q is in m³/s and V is in m /s.

[Ans. D_1 = 0.583 m ; V_2 = 9.15 m / s, D_2 = 0.646 m ; V_3 = 5.82 m /s, D_3 = 0.661 m ; V_4 = 3.6 m / s, D_4 = 0.595 m **]**

[Hint : Solve this question by using the procedure given in Example 20.10]

QUESTIONS

1. Why the ducts are used in an air conditioning system ?
2. Which material is commonly used for making ducts in the air conditioning systems?
3. What do you understand by static and velocity pressure in a duct ?
4. Derive an expression for the equivalent diameter of circular duct corresponding to a rectangular duct of sides *a* and *b*, for the same pressure loss per unit length, when (*i*) the quantity of air passing through both the ducts is same, and (*ii*) the velocity of air flowing through both the ducts is same. The friction factor remains the same for both the ducts.
5. Describe the different methods of air conditioning duct design. Why are dampers required in some systems ?

OBJECTIVE TYPE QUESTIONS

1. A duct is said to be a low velocity duct if the velocity of air in the duct is up to
   (a) 600 m / min (b) 800 m / min (c) 1200 m / min (d) 1600 m / min
2. The duct is made of
   (a) galvanised iron (b) aluminium (c) fibreglass (d) any one of these
3. The fibreglass ducts may be used in .... velocity applications.
   (a) high (b) low
4. If the velocity of air flowing through the duct is V m / min., then the velocity pressure (p_v) in mm of water is given by
   (a) $\left(\frac{V}{4.04}\right)^2$ (b) $\left(\frac{V}{40.4}\right)^2$ (c) $\left(\frac{V}{242.4}\right)^2$ (d) $\left(\frac{V}{2424}\right)^2$
5. When the quantity of air passing through the rectangular and circular ducts is same, then the equivalent diameter (D) of a circular duct for a rectangular duct for the same pressure loss per unit length is given by
   (a) $1.265\left(\frac{a^3b^3}{a+b}\right)^{1/5}$ (b) $1.265\left(\frac{a^4b^4}{a+b}\right)^{1/5}$ (c) $1.265\left(\frac{a+b}{a^3b^3}\right)^{1/5}$ (d) $1.265\left(\frac{a+b}{a^4b^4}\right)^{1/5}$
   where a and b = Longer and shorter sides of the rectangular duct respectively.
6. When the velocity of air passing through the rectangular and circular ducts is same, then the equivalent diameter (D) of a circular duct for a rectangular duct for the same pressure loss per unit length is given by
   (a) $\frac{a+b}{ab}$ (b) $\frac{2ab}{a+b}$ (c) $\frac{2a}{a-b}$ (d) $\frac{2b}{a+b}$
7. For rectangular ducts, the aspect ratio is equal to
   (a) sum of longer and shorter sides (b) difference of longer and shorter sides
   (c) product of longer and shorter sides (d) ratio of longer and shorter sides
8. The aspect ratio, for rectangular ducts, should not be greater than .... in any case.
   (a) 8 (b) 10 (c) 12 (d) 16
9. In designing ducts, the equal friction method is ideal
   (a) only for return ducts (b) when the system is balanced
   (c) when the system is not balanced (d) none of these
10. The static regain method of designing the ducts as compared to equal friction method
    (a) increases balancing problems
    (b) increases the cost of sheet metal for the duct
    (c) decreases the cost of sheet metal for the duct
    (d) none of the above

ANSWERS

1. (a) 2. (d) 3. (b) 4. (c) 5. (a)
6. (b) 7. (d) 8. (a) 9. (b) 10. (b)

CHAPTER

21 Fans

1. Introduction.
2. Types of Fans.
3. Centrifugal Fans.
4. Axial Flow Fans.
5. Total Pressure Developed by a Fan.
6. Fan Air Power.
7. Fan Efficiencies.
8. Fan performance Curves.
9. Velocity Triangles for Moving Blades of a Centrifugal Fan for Radial Entry of Air.
10. Work done and Theoretical Total Head Developed by a Centrifugal Fan for Radial Entry of Air.
11. Specific Speed of a Centrifugal Fan.
12. Fan Similarity Laws.
13. Fan and System Characteristic.
14. Fans in Series.
15. Fans in Parallel.

21.1 Introduction

A fan is a kind of pump which is used for pumping or circulating the air through the entire duct system and the conditioned space. It is usually located at the inlet of the air conditioner. A fan, essentially, consists of a rotating wheel (called impeller) which is surrounded by a stationary member known as housing. The energy is transmited to the air by the power driven wheel and a pressure difference is created to provide flow of air. The air may be moved by either creating an above-atmospheric pressure (*i.e.* positive pressure) or a below-atmospheric pressure (*i.e.* negative pressure). All fans produce both the conditions. The air at inlet to the fan is below atmospheric pressure while at the exhaust or outlet of the fan is above atmospheric pressure. The air feed into a fan is called *induced draft* while the air exhaust from a fan is called *forced draft.*

The fans, irrespective of their type of construction, may function as either blowers or exhausters. The blowers discharge air against a pressure at their outlet whereas exhausters remove gases from a space by suction.

21.2 Types of Fans

The following two types of fans may be used for the transmission of air :

1. Centrifugal or radial flow fans, and 2. Axial flow fans.

When the air enters the impeller axially and is discharged radially from the impeller, it is called a *centrifugal* or *radial flow fan.*

When the air flows parallel to the axis of impeller, it is called an *axial flow fan.*

21.3 Centrifugal Fans

The centrifugal fans are widely used for duct air conditioning system, because they can efficiently move large or small quantities of air over a greater range of operating pressures. All centrifugal fans have an impeller or wheel mounted in a scroll type of housing, as shown in Fig. 21.1. The impeller is turned either by the direct drive or more frequently by an electric motor employing pulleys and belt. The centrifugal force created by the rotating impeller moves the air outward along the blade channels. The outward moving air streams are combined by the scroll into a single large air stream. This air stream leaves the fan through the discharge outlet.

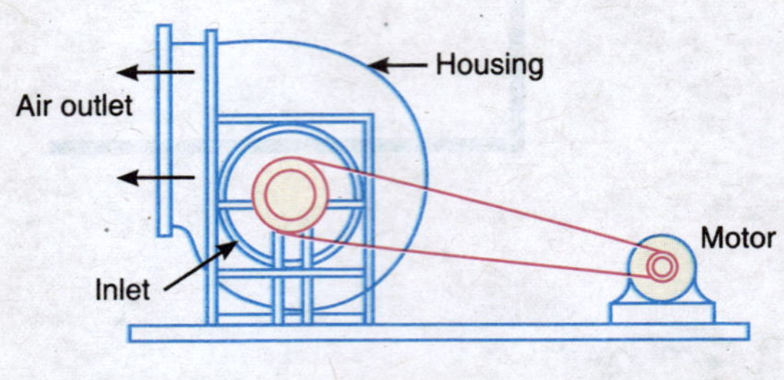

Fig. 21.1. Centrifugal fan.

The fan impeller may have the following three types of blades :

1. Radial or straight blades, 2. Forward curved blades, and 3. Backward curved blades.

The centrifugal fans with *radial blades,* as shown in Fig. 21.2 (*a*), have simple impeller construction. The blades run straight out from a central hub. Some fans of this type have heavy steel blades with high structural strength. These fans provide very high pressure at high speeds.

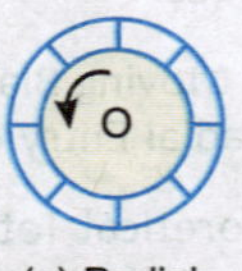

(*a*) Radial.

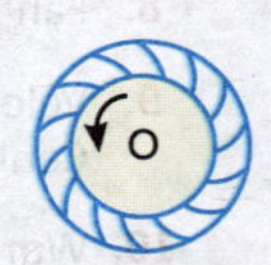

(*b*) Forward curved.

(*c*) Backward curved.

Fig. 21.2. Three types of fan impeller blades.

A large number of centrifugal fans installed in air conditioning systems have impellers with *forward curved blades,* as shown in Fig. 21.2 (*b*). Since the blades are very shallow in depth, therefore the diameter of the housing air-inlet opening more nearly approaches to that of the impeller. The ample inlet opening, together with stream-lined hub of the wheel, promotes a smooth flow of air into the rotating blades. This increases the efficiency of the fan and reduces its noise. The forward curved blades are more capable of overcoming the attached duct system resistance when their operation is at low speeds.

The centrifugal fan impeller may have *backward curved blades,* as shown in Fig. 21.2 (*c*). The backward curved blades must be operated at a much higher speed of rotation than the forward curved blades, if the same static pressure is to be produced in each case. In some cases, the higher speed may be an advantage because of a possible direct connection to the driving motor. The fan impellers having backward curved blades operate at high efficiency and have no overloading power characteristic. They also offer the advantage of wide ranges of capacity at constant speed with small changes in the power requirements.

Radial centrifugal fans.

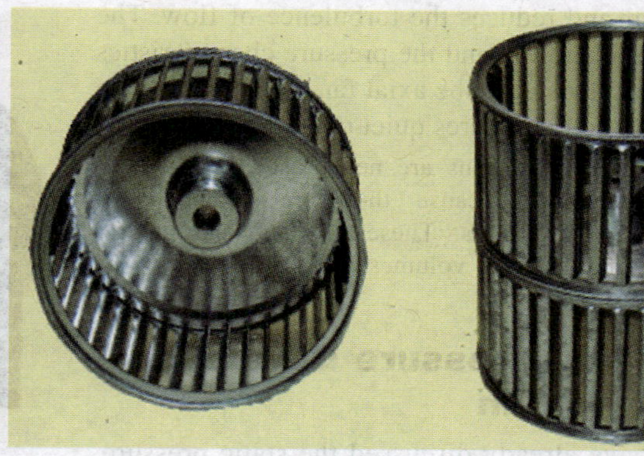

Centrifugal blowers.

Note : The number of impeller blades varies in centrifugal fans. The radial blade impellers seldom have more than 8 or 10 blades. The forward curved impellers usually have 24 to 64 blades whereas the backward curved impeller usually have 10 to 16 blades.

21.4 Axial Flow Fans

The axial flow fans are divided into the following three groups :

1. *Propeller fan.* A propeller type of axial flow fan consists of a propeller or disc type wheel which operates within a mounting ring as shown in Fig. 21.3 (*a*). The design of the ring surrounding the wheel is important because it prevents the air discharged from being drawn backward into the wheel around its periphery. The propeller fans are used only when the resistance to air movement is small. They are useful for the ventilation of attic spaces, lavatories and bathrooms, removal of cooking odours from kitchens and many other applications where little or no duct work is involved.

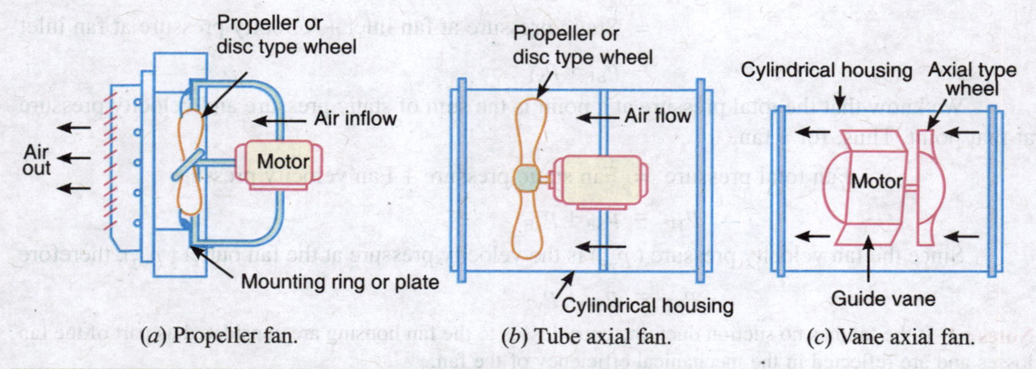

Fig. 21.3. Types of axial flow fans.

2. *Tube axial fan.* A tube axial fan, consists of a propeller wheel housed in a simple cylinder as shown in Fig. 21.3 (*b*). The wheel may be driven either from an electric motor within the cylinder directly connected to its shaft or may be driven through a belt arrangement from a motor mounted outside the housing. These fans are easily installed in round ducts. They are more efficient than propeller fans. The air discharge from tube axial fan follows a spiral path as it leaves the cylindrical housing.

3. *Vane axial fan.* A vane axial fan combines a tube axial fan wheel mounted in a cylinder with a set of air guide vanes, as shown in Fig. 21.3 (*c*). This fan eliminates spiral flow of the discharge air and reduces the turbulence of flow. The efficiency of operation and the pressure characteristics are better than those of tube axial fan. The straight line flow leaving the fan assures quiet operation.

Note : The axial flow fans are never used for duct air conditioning system because they are incapable of developing high pressures. These fans are particularly suitable for handling large volumes of air at relatively low pressures.

Axial flow fans.

21.5 Total Pressure Developed by a Fan

We have already discussed the static pressure, velocity pressure and total pressure of air in ducts. In case of a fan, the *fan static pressure* (p_{SF}) is the pressure increase produced by a fan. The *fan velocity pressure* (p_{vF}) is the velocity pressure corresponding to the mean velocity of air at the fan outlet based on the total outlet area without any deductions for motors, fairings, or other bodies. The total pressure created by a fan or the *fan total pressure* (p_{TF}) is the algebraic difference between the total pressure at the fan outlet and the total pressure at the fan inlet. Mathematically, fan total pressure,

$$p_{TF} = p_{T2} - p_{T1}$$

where p_{T2} = Total pressure at fan outlet

= Static pressure at fan outlet + Velocity pressure at fan outlet

$= p_{S2} + p_{v2}$, and

p_{T1} = Total pressure at fan inlet

= Static pressure at fan inlet + Velocity pressure at fan inlet

$= p_{S1} + p_{v1}$

We know that the total pressure at a point is the sum of static pressure and velocity pressure at that point. Thus, for a fan,

Fan total pressure = Fan static pressure + Fan velocity pressure

i.e.

$$p_{TF} = p_{SF} + p_{vF}$$

Since the fan velocity pressure (p_{vF}) is the velocity pressure at the fan outlet (p_{v2}), therefore

$$p_{TF} = p_{SF} + p_{v2}$$

Notes : 1. If the fan has no suction duct, the entry losses to the fan housing are considered as part of the fan losses and are reflected in the mechanical efficiency of the fan.

In an actual system, the fan has a suction duct and apparatus such as filters and coils. In such a system, the total pressure at the fan inlet is always equal to the total frictional resistance in that part of the system. Also, the total pressure at the fan inlet in such a system is always *negative* and it is numerically less than the static pressure at the fan inlet.

2. If the fan has no discharge duct (*i.e.* the fan delivers air directly into a free open space), the discharge static pressure is zero (*i.e.* $p_{s2} = 0$). Thus the total pressure at the fan outlet is equal to the velocity pressure (*i.e.* $p_{T2} = p_{v2}$). In an actual system, the fan has a discharge duct. In such a system, the total pressure at the fan outlet is equal to the velocity pressure at the point of discharge plus all pressure losses in the path taken by air to reach that point.

21.6 Fan Air Power

The power output of a fan is expressed in terms of air power and represents the work done by the fan. Mathematically, total fan air power (based on fan total pressure, p_{TF}),

$$P_{at} = \frac{9.81\, Q \times p_{TF} \times K_P}{60} \text{ (in watts)}$$

where Q = Total quantity of air flowing at the fan inlet in m^3/min,

p_{TF} = Fan total pressure in mm of water, and

K_P = Compressibility coefficient.

Similarly, static fan air power based on the fan static pressure (p_{SF}) ,

$$P_{as} = \frac{9.81\, Q \times p_{SF} \times K_P}{60} \text{ (in watts)}$$

Note : If Q is expressed in m^3/s and p_{TF} and p_{SF} are in N/m^2, then total fan air power (in watts),

$$P_{at} = Q \times p_{TF} \times K_P$$

and static fan air power, $P_{as} = Q \times p_{SF} \times K_P$

21.7 Fan Efficiencies

The ratio of the total fan air power to the driving power (or brake power) required at the fan shaft is known as *total fan efficiency*. It is also called *mechanical efficiency* of the fan. Mathematically, total fan efficiency,

$$\eta_{TF} = \frac{\text{Total fan air power } (P_{at})}{\text{Input or brake power } (B.P.)}$$

Similarly, static fan efficiency,

$$\eta_{SF} = \frac{\text{Static fan air power } (P_{as})}{\text{Input or brake power } (B.P.)}$$

Example 21.1. *A centrifugal fan has a circular inlet duct of 0.45 m diameter and a rectangular outlet duct of 0.45 m by 0.375 m. The static pressure at the fan inlet is –12.5 mm of water and the static pressure at the fan outlet is 25 mm of water when the fan delivers 115 m³/min and absorbs 1 kW.*

Assuming standard air density in both ducts and compressibility factor as 1, determine (a) total pressure at fan inlet and outlet, (b) fan total pressure and fan static pressure, and (c) fan efficiency and fan static efficiency.

Solution. Given : D = 0.45 m ; a = 0.45 m ; b = 0.375 m ; p_{S1} = – 12.5 mm of water ; p_{S2} = 25 mm of water ; Q = 115 m^3/min ; $B.P.$ = 1 kW ; K_P = 1

(a) Total pressure at fan inlet and outlet

Let p_{T1} = Total pressure at fan inlet, and

p_{T2} = Total pressure at fan outlet.

Cross-sectional area of circular inlet duct,

$$A_1 = \frac{\pi}{4} \times D^2 = \frac{\pi}{4}(0.45)^2 = 0.16 \text{ m}^2$$

∴ Velocity of air in the inlet duct,

$$V_1 = \frac{Q}{A_1} = \frac{115}{0.16} = 718.7 \text{ m}^3/\text{min} = 11.98 \text{ m}^3/\text{s}$$

We know that velocity pressure in the inlet duct,

$$p_{v1} = \left(\frac{V_1}{4.04}\right)^2 = \left(\frac{11.98}{4.04}\right)^2 = 8.8 \text{ mm of water}$$

∴ Total pressure at fan inlet,

$$p_{T1} = p_{S1} + p_{v1} = -12.5 + 8.8 = -3.7 \text{ mm of water} \textbf{ Ans.}$$

Cross-sectional area of rectangular outlet duct,

$$A_2 = a \times b = 0.45 \times 0.375 = 0.17 \text{ m}^2$$

∴ Velocity of air in the outlet duct,

$$V_2 = \frac{Q}{A_2} = \frac{115}{0.17} = 676.5 \text{ m}^3/\text{min} = 11.3 \text{ m}^3/\text{s}$$

We know that velocity pressure in the outlet duct,

$$p_{v2} = \left(\frac{V_2}{4.04}\right)^2 = \left(\frac{11.3}{4.04}\right)^2 = 7.8 \text{ mm of water}$$

∴ Total pressure at fan outlet,

$$p_{T2} = p_{S2} + p_{v2} = 25 + 7.8 = 32.8 \text{ mm of water} \textbf{ Ans.}$$

(b) Fan total pressure and fan static pressure

We know that fan total pressure,

$$p_{TF} = p_{T2} - p_{T1} = 32.8 - (-3.7) = 36.5 \text{ mm of water} \textbf{ Ans.}$$

and fan static pressure,

$$p_{SF} = p_{TF} - p_{v2} = 36.5 - 7.8 = 28.7 \text{ mm of water} \textbf{ Ans.}$$

(c) Fan total efficiency and fan static efficiency

We know that fan air power,

$$P_{at} = \frac{9.81\, Q \times p_{TF} \times K_P}{60} = \frac{9.81 \times 115 \times 36.5 \times 1}{60} \text{ W}$$

$$= 686.3 \text{ W} = 0.6863 \text{ kW}$$

and static fan air power,

$$P_{as} = \frac{9.81\, Q \times p_{SF} \times K_P}{60} = \frac{9.81 \times 115 \times 28.7 \times 1}{60} \text{ W}$$

$$= 539.6 \text{ W} = 0.5396 \text{ kW}$$

∴ Fan total effciency,

$$\eta_{TF} = \frac{P_{at}}{B.P.} = \frac{0.6863}{1} = 0.6863 \text{ or } 68.63\% \textbf{ Ans.}$$

and fan static efficiency,

$$\eta_{SF} = \frac{P_{as}}{B.P.} = \frac{0.5396}{1} = 0.5396 \text{ or } 53.96\% \textbf{ Ans.}$$

21.8 Fan Performance Curves

A fan performance curve is a graph of a fan's volume rate plotted against pressure, power, or efficiency. The performance curves for the various types of fans are shown in Fig. 21.4 to Fig. 21.7.

In all the figures, the abscissas represent the range of air flow capacity expressed as a percentage of the amount of air delivered when the fan is discharging freely into an open space. The ordinates represent the percentages of efficiency, power at free delivery and static pressure with outlet closed.

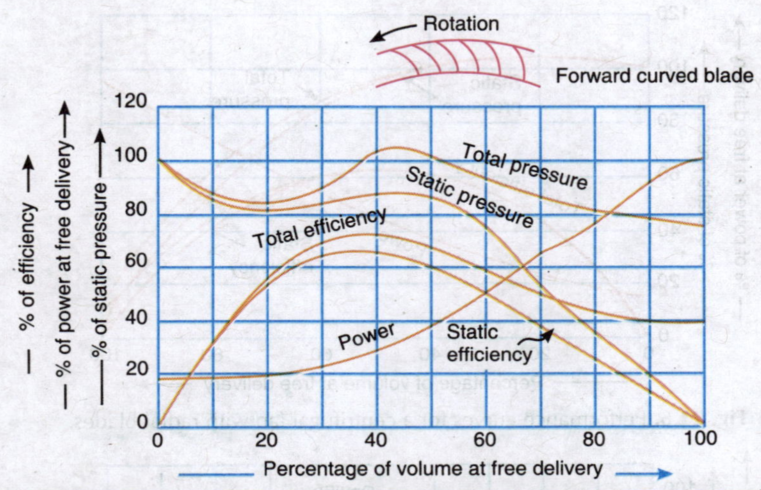

Fig. 21.4. Performance curves for a centrifugal fan with forward curved blades.

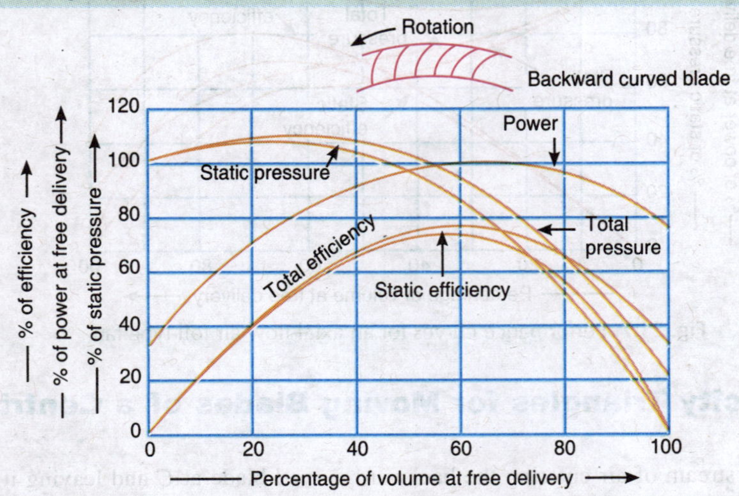

Fig. 21.5. Performance curves for a centrifugal fan with backward curved blades.

From Fig. 21.4, we see that the centrifugal fans with forward curved blades require an ever-increasing amount of power as the air volume is increased. However, this type of centrifugal fan provides greater static pressure for a given blade-tip velocity than the other types and it is commonly used in air conditioning systems inspite of this disadvantage.

From Fig. 21.5, we see that centrifugal fans with backward curved blades require maximum power. The operating condition which requires the maximum power is close to the combination of volume and static pressure under which the fan operates most efficiently. The fans of this type are said to have non- overloading power characteristic which means that the driving motor cannot be overloaded if the fan and motor are properly selected.

The axial flow fans, as shown in Fig. 21.7, also have non over-loading power characteristics.

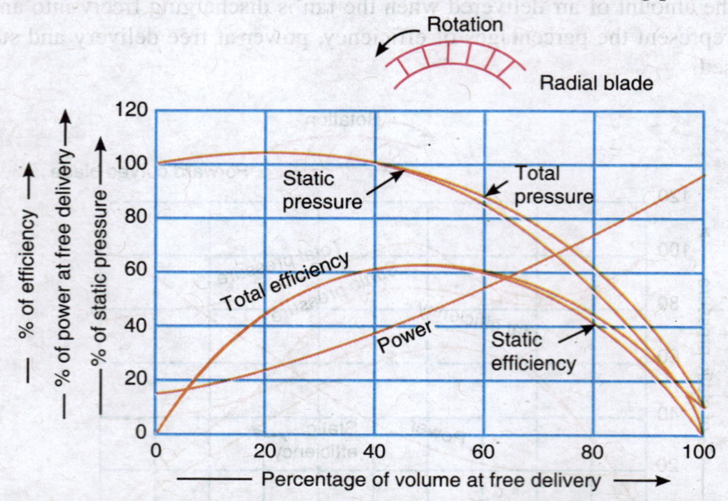

Fig. 21.6. Performance curves for a centrifugal fan with radial blades.

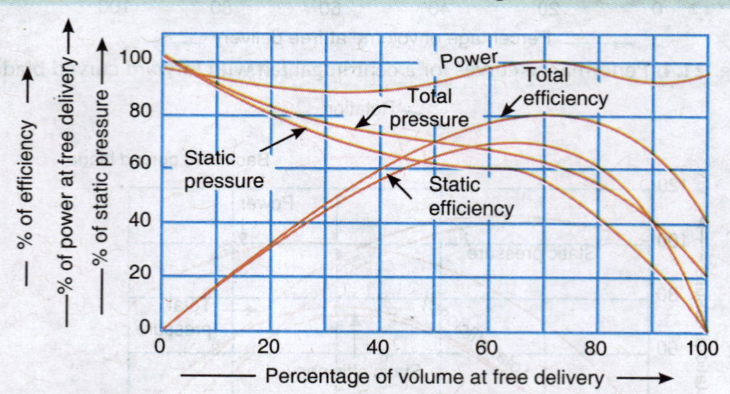

Fig. 21.7. Performance curves for an axial flow air foil-type fan .

21.9 Velocity Triangles for Moving Blades of a Centrifugal Fan

Consider a stream of air entering the backward curved blade at C and leaving it at D, as shown in Fig. 21.8 (*a*). The velocity triangles at the inlet and outlet tips of the blade are shown in Fig. 21.8 (*b*).

Let V_{b1} = Linear or tangential velocity of the moving blade at inlet (BA).

V_1 = Absolute velocity of air entering the blade ($A'C$).

V_{f1} = Velocity of flow at inlet (AC). It is the radial component of V_1.

V_{r1} = Relative velocity of air to the moving blade at inlet (BC). It is vectorial difference between V_{b1} and V_1.

V_{w1} = Velocity of whirl at inlet. It is the tangential component of V_1.

β_1 = Blade angle at inlet. It is the angle which the relative velocity (V_{r1}) makes with the tangent at the blade inlet. It is equal to the angle between V_{r1} and V_{b1} (*i.e.* angle CBA).

$V_{b2}, V_2, V_{f2}, V_{r2}, V_{w2}, \beta_2$ = Corresponding values at outlet of the blade tip.

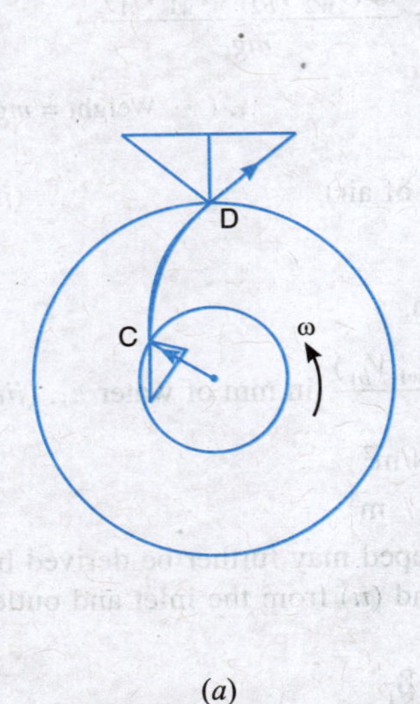

(*a*)

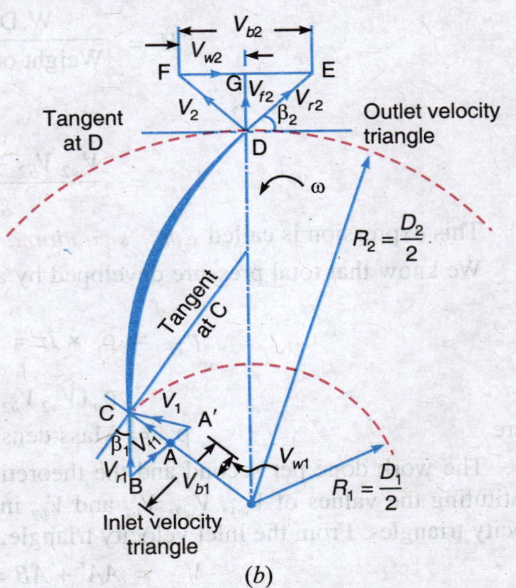

(*b*)

Fig. 21.8. Velocity triangles for a centrifugal fan.

It may be seen from the above that the suffix 1 stands for the blade inlet and the suffix 2 stands for blade outlet. A little consideration will show that as the air enters and leaves the blades without any shock (or in other words tangentially), therefore the shape of the blades will be such that V_{r1} and V_{r2} are along the tangents to the blades at inlet and outlet respectively.

Let m = Mass of air flowing through the impeller in kg per second,

R_1 = Internal radius of the impeller = $D_1/2$

R_2 = External radius of the impeller = $D_2/2$, and

ω = Angular velocity of the impeller in radians per second

$$= \frac{V_{b1}}{R_1} \text{ or } \frac{V_{b2}}{R_2}$$

We know that the angular momentum entering the impeller per second

= Mass of air flowing per second × Velocity of whirl × Radius of impeller

$= m\, V_{w1}\, R_1$

Similarly, angular momentum leaving the impeller per second

$= m\, V_{w2}\, R_2$

According to Newton's second law of angular motion, the torque in the direction of motion of blades is equal to the rate of change of angular momentum.

∴ Torque in the direction of motion of blades

$$= mV_{w2} R_2 - mV_{w1} R_1 = m (V_{w2} R_2 - V_{w1} R_1)$$

and work done per second in the direction of motion of blades,

$$\text{W.D./second} = \text{Torque} \times \text{Angular velocity} = m (V_{w2} R_2 - V_{w1} R_1)\, \omega$$

Since $V_{b1} = \omega R_1$ and $V_{b2} = \omega R_2$, therefore

$$\text{W.D./second} = m (V_{w2} V_{b2} - V_{w1} V_{b1}) \quad \text{... (i)}$$

∴ Theoretical total head developed by a centrifugal fan,

$$H = \frac{\text{W.D./second}}{\text{Weight of air / second}} = \frac{m (V_{w2} V_{b2} - V_{w1} V_{b1})}{mg}$$

$$\text{... (} \because \text{Weight} = mg \text{)}$$

$$= \frac{V_{w2} V_{b2} - V_{w1} V_{b1}}{g} \text{ (in m of air)} \quad \text{... (ii)}$$

This expression is called *Euler's equation.*

We know that total pressure developed by a centrifugal fan,

$$p_{TF} = \rho_a \times H = \frac{\rho_a (V_{w2} V_{b2} - V_{w1} V_{b1})}{g} \text{ in mm of water} \quad \text{... (iii)}$$

$$= \rho_a (V_{w2} V_{b2} - V_{w1} V_{b1}) \text{ in N/m}^2$$

where ρ_a = Mass density of air in kg / m³.

The work done per second and the theoretical head developed may further be derived by substituting the values of V_{w1}, V_{w2}, V_{b1} and V_{b2} in equation (*i*) and (*ii*) from the inlet and outlet velocity triangles. From the inlet velocity triangle, we find that

$$V_{b1} = AA' + AB = V_{w1} + V_{r1} \cos \beta_1$$

From the outlet velocity triangle,

$$V_{b2} = FG + GE = V_{w2} + V_{r2} \cos \beta_2$$

or

$$V_{w2} = V_{b2} - V_{r2} \cos \beta_2 \quad \text{... (iv)}$$

We know that

$$(V_2)^2 = (V_{b2})^2 + (V_{r2})^2 - 2 (V_{b2})(V_{r2}) \cos \beta_2$$

$$\therefore \quad \cos \beta_2 = \frac{(V_{b2})^2 + (V_{r2})^2 - (V_2)^2}{2 V_{b2} V_{r2}}$$

Substituting the value of cos β_2 in equation (*iv*), we get

$$V_{w2} = V_{b2} - V_{r2} \left[\frac{(V_{b2})^2 + (V_{r2})^2 - (V_2)^2}{2 V_{b2} V_{r2}} \right]$$

$$= V_{b2} - \left[\frac{(V_{b2})^2 + (V_{r2})^2 - (V_2)^2}{2 V_{b2}} \right]$$

$$= \frac{2 (V_{b2})^2 - (V_{b2})^2 - (V_{r2})^2 + (V_2)^2}{2 V_{b2}} = \frac{(V_{b2})^2 - (V_{r2})^2 + (V_2)^2}{2 V_{b2}}$$

Similarly for the inlet velocity triangle,

$$V_{w1} = \frac{(V_{b1})^2 - (V_{r1})^2 + (V_1)^2}{2\, V_{b1}}$$

Substituting the value of V_{w1} and V_{w2} in equation (*i*), we get

$$\text{W.D./second} = m\left[\frac{(V_{b2})^2 - (V_{r2})^2 + (V_2)^2}{2\, V_{b2}} \times V_{b2} - \frac{(V_{b1})^2 - (V_{r1})^2 + (V_1)^2}{2\, V_{b1}} \times V_{b1}\right]$$

$$= m\left[\frac{(V_{b2})^2 - (V_{b1})^2}{2} + \frac{(V_{r1})^2 - (V_{r2})^2}{2} + \frac{(V_2)^2 - (V_1)^2}{2}\right] \text{ in N-m or J}$$

and

$$H = \frac{\text{W.D / second}}{mg}$$

$$= \frac{(V_{b2})^2 - (V_{b1})^2}{2g} + \frac{(V_{r1})^2 - (V_{r2})^2}{2g} + \frac{(V_2)^2 - (V_1)^2}{2g} \text{ in m of air}$$

The *first term* of this expression represents the increase in static head which is due to the forced vortex formed by the rotating impeller. The *second term* represents that part of the static head increase which, according to Bernoulli's equation, results from the change in relative velocity of air passing through the impeller. The *third term* is the change in velocity head of the air flowing through the impeller. Since the first two terms represents the change in static head and the third term represents the velocity head, therefore H is the fan total head developed.

Note : If the speed of impeller is N r.p.m., then the blade velocity at inlet or outlet (V_{b1} or V_{b2}) may be obtained by the relations :

$$V_{b1} = \frac{\pi\, D_1 N}{60}\,, \quad \text{and} \quad V_{b2} = \frac{\pi\, D_2 N}{60}$$

where D_1 and D_2 are the internal and external diameters of the impeller respectively.

21.10 Work Done and Theoretical Total Head Developed by a Centrifugal Fan for Radial Entry of Air

We have seen in the previous article that the work done per second by a centrifugal fan,

$$\text{W.D./second} = m\,(V_{w2} \times V_{b2} - V_{w1} \times V_{b1}) \text{ in N-m or J}$$

and theoretical total head developed by a centrifugal fan,

$$H = \frac{1}{g}(V_{w2} \times V_{b2} - V_{w1} \times V_{b1}) \text{ in m of air}$$

When the air enters the blades at right angles (*i.e.* radially) to the direction of motion of the blade, then

$$V_1 = V_{f1} \quad \text{and} \quad V_{w1} = 0$$

∴ For radial entry of air, workdone per second,

$$\text{W.D./second} = m\,(V_{w2} \times V_{b2}) \text{ in N-m or J} \quad \text{... }(i)$$

and theoretical total head developed,

$$H = \frac{1}{g}(V_{w2} \times V_{b2}) \text{ in m of air} \quad \text{... }(ii)$$

Let Q = Quantity of air flowing through the fan in m^3/s,

D_1 = Internal diameter of impeller in metres,
D_2 = External diameter of impeller in metres,
b_1 = Width of impeller at inlet in metres, and
b_2 = Width of impeller at outlet in metres.

Since the quantity of air flowing through the impeller is constant, therefore

$$Q = \pi D_1 b_1 V_{f1} = \pi D_2 b_2 V_{f2}$$

or

$$V_{f1} = \frac{Q}{\pi D_1 b_1}\text{ ; and } V_{f2} = \frac{Q}{\pi D_2 b_2}$$

Now from the outlet velocity triangle as shown in Fig. 21.8,

$$\cot \beta_2 = \frac{GE}{GD} = \frac{V_{b2} - V_{w2}}{V_{f2}} \quad \text{... } (iii)$$

$$\therefore \quad V_{b2} - V_{w2} = V_{f2} \cot \beta_2 = \frac{Q \cot \beta_2}{\pi D_2 b_2}$$

and

$$V_{w2} = V_{b2} - \frac{Q \cot \beta_2}{\pi D_2 b_2}$$

Substituting the value of V_{w2} in equations (*i*) and (*ii*), we get

$$\text{W.D./second} = m \left[V_{b2} - \frac{Q \cot \beta_2}{\pi D_2 b_2} \right] V_{b2}$$

$$= m \left[(V_{b2})^2 - \frac{Q \cot \beta_2 \times V_{b2}}{\pi D_2 b_2} \right] \text{ in N-m or J} \quad \text{... } (iv)$$

and theoretical total head developed,

$$H = \frac{1}{g} \left[(V_{b2})^2 - \frac{Q \cot \beta_2 \times V_{b2}}{\pi D_2 b_2} \right] \text{ in m of air} \quad \text{... } (v)$$

Notes : 1. We know that total pressure developed by a centrifugal fan,

$$p_{TF} = \rho_a \times H = \frac{\rho_a \times V_{w2} \times V_{b2}}{g} \text{ in mm of water}$$

... [From equation (*ii*)]

$$= \rho_a \times V_{w2} \times V_{b2} \text{ in N/m}^2$$

$$= \frac{\rho_a}{g} \left[(V_{b2})^2 - \frac{Q \cot \beta_2 \times V_{b2}}{\pi D_2 b_2} \right] \text{ in mm of water}$$

... [From equation (*v*)]

$$= \rho_a \left[(V_{b2})^2 - \frac{Q \cot \beta_2 \times V_{b2}}{\pi D_2 b_2} \right] \text{ in N/m}^2$$

where ρ_a = Mass density of air in kg / m^3.

2. From equation (*iii*),

$$\cot \beta_2 = \frac{V_{b2} - V_{w2}}{V_{f2}} \quad \text{or} \quad V_{w2} = V_{b2} - V_{f2} \cot \beta_2$$

It may be noted that

(*a*) For radial blades, $\beta_2 = 90°$. Thus $V_{w2} = V_{b2}$.

(*b*) For forward curved blades, β_2 is greater than 90°. Thus V_{w2} is greater than V_{b2}.

(*c*) For backward curved blades, β_2 is less than 90°. Thus V_{w2} is less than V_{b2}.

We know that the total pressure developed by a centrifugal fan is

$$p_{TF} = \rho_a \times V_{w2} \times V_{b2} \text{ in N/m}^2$$

Thus, for a given blade velocity (V_{b2}), the total pressure developed by a centrifugal fan having forward curved blades is greatest.

Example 21.2. *A centrifugal fan delivers 120 m³/min when running at 960 r.p.m. The impeller diameter is 0.7 m and the diameter at the blade inlet is 0.48 m. The air enters the impeller with a small whirl component in the direction of impeller rotation, but the relative velocity meets the blade tangentially. The impeller width at inlet is 160 mm and at outlet is 110 mm. The blades are backward curved making angles of 22.5° and 50° with the tangents at inlet and outlet respectively. Draw the inlet and outlet velocity triangles and determine the theoretical total head developed by the impeller.*

Assuming that the losses at inlet, in the impeller and in the casing amount to 70 per cent of the velocity head at impeller outlet and the velocity head at the fan discharge is 10 per cent of the velocity head at impeller outlet, calculate the fan static pressure in mm of water. Take the mass density of air to be 1.2 kg / m³ and neglect the effect of blade thickness and interblade circulation.

Solution. Given : $Q = 120$ m³/min = 2 m³/s ; $N = 960$ r.p.m. ; $D_2 = 0.7$ m ; $D_1 = 0.48$ m ; $b_1 = 160$ mm = 0.16 m ; $b_2 = 110$ mm = 0.11 m ; $\beta_1 = 22.5°$; $\beta_2 = 50°$; $\rho_a = 1.2$ kg / m³

Theoretical total head developed by the impeller

Let H = Theoretical total head developed by the impeller.

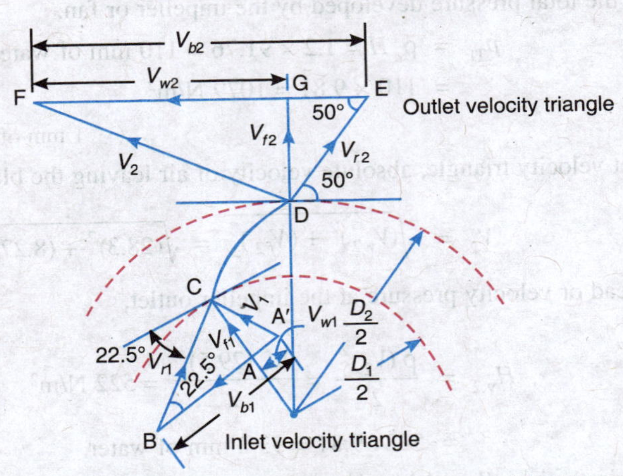

Fig. 21.9

The inlet and outlet velocity triangles are shown in Fig. 21.9. We know that blade velocity at inlet,

$$V_{b1} = \frac{\pi D_1 N}{60} = \frac{\pi \times 0.48 \times 960}{60} = 24 \text{ m / s}$$

Blade velocity at outlet,

$$V_{b2} = \frac{\pi D_2 N}{60} = \frac{\pi \times 0.7 \times 960}{60} = 35.2 \text{ m / s}$$

Quantity of air delivered

$$Q = \pi D_1 b_1 V_{f1} = \pi D_2 b_2 V_{f2}$$

∴ Velocity of flow at inlet,

$$V_{f1} = \frac{Q}{\pi D_1 b_1} = \frac{2}{\pi \times 0.48 \times 0.16} = 8.3 \text{ m/s}$$

and velocity of flow at outlet,

$$V_{f2} = \frac{Q}{\pi D_2 b_2} = \frac{2}{\pi \times 0.7 \times 0.11} = 8.27 \text{ m/s}$$

From the inlet velocity triangle,

$$AB = AC \cot 22.5° \quad \text{or} \quad V_{b1} - V_{w1} = V_{f1} \cot 22.5°$$

$$\therefore \quad 24 - V_{w1} = 8.3 \times 2.4142 = 20 \quad \text{or} \quad V_{w1} = 24 - 20 = 4 \text{ m/s}$$

From the outlet velocity triangle,

$$GE = GD \cot 50° \quad \text{or} \quad V_{b2} - V_{w2} = V_{f2} \cot 50°$$

$$\therefore \quad 35.2 - V_{w2} = 8.27 \times 0.8391 = 6.9 \quad \text{or} \quad V_{w2} = 35.2 - 6.9 = 28.3 \text{ m/s}$$

We know that the theoretical total head developed by the impeller,

$$H = \frac{1}{g}(V_{w2} \times V_{b2} - V_{w1} \times V_{b1})$$

$$= \frac{1}{9.81}(28.3 \times 35.2 - 4 \times 24) = 91.76 \text{ m of air} \quad \textbf{Ans.}$$

Fan static pressure

We know that the total pressure developed by the impeller or fan,

$$p_{TF} = \rho_a H = 1.2 \times 91.76 = 110 \text{ mm of water}$$

$$= 110 \times 9.81 = 1079 \text{ N/m}^2$$

... ($\because$ 1 mm of water = 9.81 N/m^2)

From the outlet velocity triangle, absolute velocity of air leaving the blade,

$$V_2 = \sqrt{(V_{w2})^2 + (V_{f2})^2} = \sqrt{(28.3)^2 + (8.27)^2} = 29.5 \text{ m/s}$$

∴ *Velocity head or velocity pressure at the impeller outlet,

$$H_{v2} = \frac{\rho (V_2)^2}{2} = \frac{1.2\,(29.5)^2}{2} = 522 \text{ N/m}^2$$

$$= 522 / 9.81 = 53.2 \text{ mm of water}$$

Losses at inlet, in the impeller and in the casing

= 70% of velocity head at outlet ...(Given)

$= 0.7\, H_{v2} = 0.7 \times 53.2 = 37.24$ mm of water

* The velocity head (in mm of water) may be calculated by using the relation as discussed in Chapter 20 on 'Ducts'. We know that the velocity head or velocity pressure,

$$H_{v2} = \left(\frac{V_2}{4.04}\right)^2 = \left(\frac{29.5}{4.04}\right)^2 = 53.2 \text{ mm of water}$$

and velocity head or velocity pressure at the fan discharge

$$= 10\% \text{ of velocity head at outlet}$$

$$= 0.1\, H_{v2} = 0.1 \times 53.2 = 5.32 \text{ mm of water}$$

Now applying Bernoulli's equation to fan inlet and outlet, we have

Total pressure at fan inlet + Total pressure developed by impeller

= Losses + Total pressure at fan outlet

Since the total pressure at the fan inlet is zero and the total pressure at the fan outlet (p_{TF}) is the sum of static pressure at fan outlet or fan static pressure (p_{S2}) and the velocity pressure at fan outlet, therefore

$$0 + 110 = 37.24 + p_{S2} + 5.32$$

$$\therefore \quad p_{S2} = 110 - 37.24 - 5.32 = 67.44 \text{ mm of water}$$ **Ans.**

21.11 Specific Speed of a Centrifugal Fan

The specific speed of a centrifugal fan is defined as the speed of a geometrically similar fan which would deliver 1 m³ of air per second against a head of 1 m of air. It is usually denoted by N_S.

Let
Q = Total quantity of air flowing through the fan,
D = Diameter of impeller,
b = Width of impeller,
V_f = Velocity of flow,
N = Speed of impeller, and
H = Head developed by the fan.

We know that
$$Q = \pi D b V_f \quad \text{or} \quad Q \propto D b V_f \qquad \text{... (i)}$$

Also
$$D \propto b \qquad \text{... (ii)}$$

$\therefore$ From equations (*i*) and (*ii*),

$$Q \propto D^2 V_f \qquad \text{... (iii)}$$

We also know that blade velocity or tangential velocity of the impeller,

$$V_b = \frac{\pi D N}{60} \quad \text{or} \quad V_b \propto DN \qquad \text{... (iv)}$$

Also
$$V_b \propto V_f \propto \sqrt{H} \qquad \text{... (v)}$$

From equations (*iv*) and (*v*),

$$DN \propto \sqrt{H} \quad \text{or} \quad D \propto \frac{\sqrt{H}}{N} \qquad \text{... (vi)}$$

Substituting the value of D in equation (*iii*), we have

$$Q \propto \frac{H}{N^2} \times V_f \quad \text{or} \quad Q \propto \frac{H}{N^2} \times \sqrt{H} \propto \frac{(H)^{3/2}}{N^2}$$

$$\text{... } (\because V_f \propto \sqrt{H})$$

$$\therefore \quad Q = \frac{K(H)^{3/2}}{N^2} \qquad \text{... (vii)}$$

where K is constant of proportionality. According to the definition, if Q = 1 m³/s and H = 1 m, then $N = N_S$. Substituting these values in equation (*vii*), we have

$$1 = \frac{K \times 1^{3/2}}{(N_S)^2} \quad \text{or} \quad K = (N_S)^2$$

Now from equation (*vii*),

$$Q = \frac{(N_S)^2 (H)^{3/2}}{N^2} \quad \text{or} \quad N_S = \frac{N\sqrt{Q}}{H^{3/4}}$$

21.12 Fan Similarity Laws

The two fans are said to be geometrically similar when all of their wheel dimensions have the same proportionate ratios.

For any series of geometrically similar fans and for any point on their characteristic curves, the following fan laws hold :

1. The volume flow rate or capacity (Q) of a fan is directly proportional to the fan speed (N) and cube of the impeller diameter (D). In other words

$$Q \propto N \propto D^3$$

or
$$\frac{Q_1}{N_1} = \frac{Q_2}{N_2} \quad \text{and} \quad \frac{Q_1}{(D_1)^3} = \frac{Q_2}{(D_2)^3}$$

It may also be written as

$$\frac{Q_1}{N_1 (D_1)^3} = \frac{Q_2}{N_2 (D_2)^3} = \text{Constant}$$

where suffix 1 represents the actual fan and suffix 2 represents the geometrically similar fan.

2. The total pressure developed by a fan (p_{TF}) is directly proportional to the square of the fan speed (N^2), square of the impeller diameter (D^2), and density of the air (ρ_a). In other words,

$$p_{TF} \propto N^2 \propto D^2 \propto \rho_a$$

or
$$\frac{p_{TF1}}{(N_1)^2} = \frac{p_{TF2}}{(N_2)^2} \quad \text{and} \quad \frac{p_{TF1}}{(D_1)^2} = \frac{p_{TF2}}{(D_2)^2}$$

Also
$$\frac{p_{TF1}}{\rho_{a1}} = \frac{p_{TF2}}{\rho_{a2}}$$

It may also be written as

$$\frac{p_{TF1}}{(N_1)^2 (D_1)^2 \rho_{a1}} = \frac{p_{TF2}}{(N_2)^2 (D_2)^2 \rho_{a2}} = \text{Constant}$$

This is also applicable to fan static pressure (p_{SF}) and fan velocity pressure (p_{vF}).

We know that the total pressure developed by the fan,

$$p_{TF} = \rho_a H$$

where H is the total head developed by the fan,

$$\therefore \quad \frac{\rho_{a1} H_1}{(N_1)^2 (D_1)^2 \rho_{a1}} = \frac{\rho_{a2} H_2}{(N_2)^2 (D_2)^2 \rho_{a2}} \quad \text{or} \quad \frac{H_1}{(N_1)^2 (D_1)^2} = \frac{H_2}{(N_2)^2 (D_2)^2} = \text{Constant}$$

3. The power (P) of a fan is directly proportional to the cube of the fan speed (N^3), fifth power of the impeller diameter (D^5) and density of the air (ρ_a). In other words,

$$P \propto N^3 \propto D^5 \propto \rho_a$$

or
$$\frac{P_1}{(N_1)^3} = \frac{P_2}{(N_2)^3} \quad \text{and} \quad \frac{P_1}{(D_1)^5} = \frac{P_2}{(D_2)^5}$$

Also $$\frac{P_1}{\rho_{a1}} = \frac{P_2}{\rho_{a2}}$$

It may also be written as

$$\frac{P_1}{(N_1)^3 (D_1)^5 \rho_{a1}} = \frac{P_2}{(N_2)^3 (D_2)^5 \rho_{a2}} = \text{Constant}$$

4. The efficiency (η) is constant. In other words,

$$\eta_1 = \eta_2 = \text{Constant}$$

Example. 21.3. *A fan for the ventilation plant is to be exported to an area where the air density is 0.96 kg / m³ and is scheduled to deliver 6 m³/s against a static pressure of 50 mm of water, with a static efficiency of 65 per cent.*

If it is driven by a constant speed motor, calculate the static pressure and shaft power in the maker's works where the air density is 1.2 kg / m³.

Solution. Given : $\rho_1 = 0.96$ kg / m³ ; $Q_1 = 6$ m³/s ; $p_{SF1} = 50$ mm of water ; $\eta_S = 65\% = 0.65$; $\rho_2 = 1.2$ kg / m³

Static pressure in the maker's works

Let p_{SF2} = Static pressure in the maker's works.

We know that $$\frac{p_{SF1}}{\rho_1} = \frac{p_{SF2}}{\rho_2}$$

$$\therefore \quad p_{SF2} = p_{SF1} \times \frac{\rho_2}{\rho_1} = 50 \times \frac{1.2}{0.96} = 62.5 \text{ mm of water}$$

$$= 62.5 \times 9.81 = 613 \text{ N/m}^2 \textbf{ Ans.}$$

... ($\because$ 1 mm of water = 9.81 N/m²)

Shaft power in the maker's works

Since the speed (N) and diameter of impeller (D) is constant, therefore from the relation

$$\frac{Q_1}{N_1 (D_1)^3} = \frac{Q_2}{N_2 (D_2)^3}, \text{ we get } Q_1 = Q_2 = 6 \text{ m}^3/\text{s}$$

We know that static fan air power,

$$P_{aS2} = Q \times p_{SF2} = 6 \times 613 = 3678 \text{ N-m / s or W}$$

... ($\because$ 1 N-m/s = 1 W)

$\therefore$ Shaft power in the maker's works,

$$P_{S2} = \frac{P_{aS2}}{\eta_S} = \frac{3678}{0.65} = 5660 \text{ W} = 5.66 \text{ kW } \textbf{Ans.}$$

Example 21.4. *A fan of diameter 0.7 m running at 1500 r.p.m. delivers 140 m³/min of air at 15°C against 75 mm of water of total pressure when its total efficiency is 86 per cent. Determine the volume of air delivered, total pressure developed and power consumed, if*

(a) the air temperature is 50°C,

(b) the air temperature is 50°C and the fan speed is increased to 1700 r.p.m., and

(c) the conditions are same as in (b) but a 0.6 m diameter, geometrically similar fan is used.

Solution. Given : $D_1 = 0.7$ m ; $N_1 = 1500$ r.p.m ; $Q_1 = 140$ m³/min ; $t_1 = 15$°C ; $p_{TF1} = 75$ mm of water ; $\eta_T = 86\% = 0.86$; $t_2 = 50$°C

First of all, let us find the ratio of the densities at temperatures 15°C and 50°C.

Let ρ_{a1} = Density of air at 15°C, and

ρ_{a2} = Density of air at 50°C.

We know that for the constant barometric pressure, the density of air at 50°C,

$$\rho_{a2} = \rho_{a1}\left(\frac{273 + t_1}{273 + t_2}\right)$$

$$\therefore \quad \frac{\rho_{a1}}{\rho_{a2}} = \frac{273 + t_2}{273 + t_1} = \frac{273 + 50}{273 + 15} = 1.12$$

(a) Volume of air delivered, total pressure developed and power consumed when the air temperature is 50°C

Let Q_2 = Volume of air delivered,

p_{TF2} = Total pressure developed, and

P_2 = Power consumed.

We know that

$$\frac{Q_1}{N_1 (D_1)^3} = \frac{Q_2}{N_2 (D_2)^3}$$

Since there is no change in speed or diameter, therefore

$$Q_2 = Q_1 = 140 \text{ m}^3\text{/min} \textbf{ Ans.}$$

Inline centrifugal fans.

Now

$$\frac{p_{TF1}}{p_{TF2}} = \frac{\rho_{a1}}{\rho_{a2}}$$

$$\therefore \quad p_{TF2} = p_{TF1} \times \frac{\rho_{a2}}{\rho_{a1}}$$

$$= 75 \times \frac{1}{1.12} = 67 \text{ mm of water}$$

$$= 67 \times 9.81 = 657.3 \text{ N/m}^2 \textbf{ Ans.}$$

We know that power consumed,

$$P_2 = Q_2 \times p_{TF2} \times \frac{1}{\eta} = 140 \times 657.3 \times \frac{1}{0.86}$$

$$= 107\ 002 \text{ N-m/min} = 107\ 002 / 60 = 1783.4 \text{ W} \textbf{ Ans.}$$

(b) Volume of air delivered, total pressure developed and power consumed when air temperature is 50°C and speed is increased to 1700 r.p.m.

Let Q_3 = Volume of air delivered,

p_{TF3} = Total pressure developed,

P_3 = Power consumed, and

N_3 = Increased speed = 1700 r.p.m. ... (Given)

We know that

$$\frac{Q_3}{N_3} = \frac{Q_2}{N_2}$$

$$\therefore \quad Q_3 = Q_2 \times \frac{N_3}{N_2} = 140 \times \frac{1700}{1500} = 158.7 \text{ m}^3\text{/min} \textbf{ Ans.}$$

... (Here $N_2 = N_1$)

Now

$$\frac{p_{TF3}}{(N_3)^2} = \frac{p_{TF2}}{(N_2)^2}$$

$$\therefore \quad p_{TF3} = p_{TF2}\left(\frac{N_3}{N_2}\right)^2 = 67\left(\frac{1700}{1500}\right)^2 = 86 \text{ mm of water} \textbf{ Ans.}$$

and $$\frac{P_3}{(N_3)^3} = \frac{P_2}{(N_2)^3}$$

$$\therefore \quad P_3 = P_2\left(\frac{N_3}{N_2}\right)^3 = 1783.4\left(\frac{1700}{1500}\right)^3 = 2596 \text{ W } \textbf{Ans.}$$

(c) Volume of air delivered, total pressure developed and power consumed for the conditions as in (b) and when diameter is 0.6 m

Let Q_4 = Volume of air delivered,

p_{TF4} = Total pressure developed,

P_4 = Power consumed, and

D_4 = New impeller diameter = 0.6 m . . . (Given)

We know that $$\frac{Q_4}{(D_4)^3} = \frac{Q_3}{(D_3)^3}$$

$$\therefore \quad Q_4 = Q_3\left(\frac{D_4}{D_3}\right)^3 = 158.7\left(\frac{0.6}{0.7}\right)^3 = 99.94 \text{ m}^3/\text{min } \textbf{Ans.}$$

... (Here $D_3 = D_1$)

Now $$\frac{p_{TF4}}{(D_4)^2} = \frac{p_{TF3}}{(D_3)^2}$$

$$\therefore \quad p_{TF4} = p_{TF3}\left(\frac{D_4}{D_3}\right)^2 = 86\left(\frac{0.6}{0.7}\right)^2 = 63.2 \text{ mm of water } \textbf{Ans.}$$

and $$\frac{P_4}{(D_4)^5} = \frac{P_3}{(D_3)^5}$$

$$\therefore \quad P_4 = P_3\left(\frac{D_4}{D_3}\right)^5 = 2596\left(\frac{0.6}{0.7}\right)^5 = 1201 \text{ W } \textbf{Ans.}$$

21.13 Fan and System Characteristic

We have already discussed that all of the duct work elements such as elbows, tees, registers, dampers etc., offer resistance to the flow of air and cause loss in pressure. The change in pressure loss or resistance with the change in flow rate is called *system characteristic.* Any air-conditioning or ventilating system that has a duct work, heating and cooling coils, dampers, registers etc. has a definite system characteristic. The system characteristic is independent of the fan used in that system.

We have seen in the previous chapter that the system resistance or pressure loss of any fixed system varies as the square of the flow rate, *i.e.*

$$R \text{ or } p_L = KQ^2$$

If the resistance of a system is plotted against the varying amounts of flow rates, a

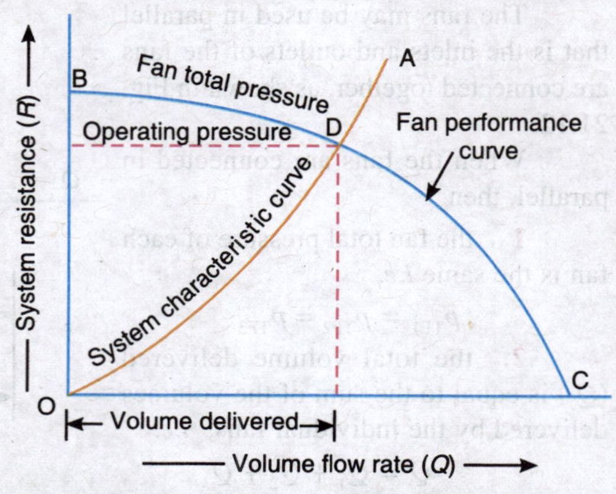

Fig. 21.10. Fan and system characteristic.

curve such as *OA* is obtained as shown in Fig. 21.10. This curve is a parabola and is usually known as a *system characteristic curve.*

When a fan operates in conjunction with a particular system, then the loss of total pressure in the system at a given volume flow must be equal to the total pressure developed by the fan (*i.e.* fan total pressure) at the same volume. This condition is satisfied by the point of intersection (point *D*) of system characteristic curve *OA* and the fan performance curve *BC*. This point of intersection is called the *operating point* of that particular fan in that particular system, as it indicates the volume and pressure at which the fan operates.

21.14 Fans in Series

Some times it is necessary to use more than one fan in conjunction with a given system. The fans may be used in series that is the outlet of the first fan is connected to the inlet of second fan and outlet of the second fan is connected to the inlet of the third fan and so on, as shown in Fig. 21.11.

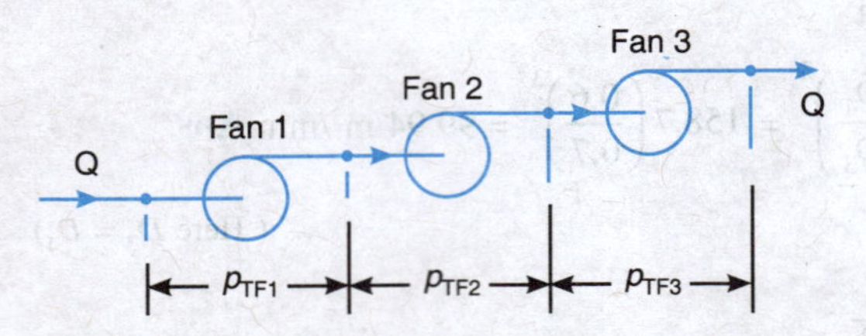

Fig. 21.11. Fans in series.

Series centrifugal fans.

When the fans are connected in series, then

1. the volume flow rate (Q) through each fan is same, *i.e.*

$$Q = Q_1 = Q_2 = Q_3$$

2. the overall fan total pressure (p_{TF}) is equal to the sum of the fan total pressures developed by the individual fans, *i.e.*

$$p_{TF} = p_{TF1} + p_{TF2} + p_{TF3}$$

21.15 Fans in Parallel

The fans may be used in parallel that is the inlets and outlets of the fans are connected together, as shown in Fig. 21.12.

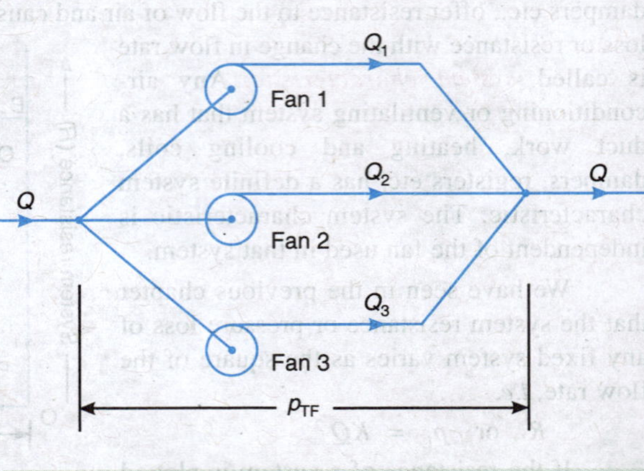

Fig. 21.12. Fans in parallel.

When the fans are connected in parallel, then

1. the fan total pressure of each fan is the same *i.e.*

$$p_{TF1} = p_{TF2} = p_{TF3}$$

2. the total volume delivered (Q) is equal to the sum of the volumes delivered by the individual fans, *i.e.*

$$Q = Q_1 + Q_2 + Q_3$$

Example 21.5. *A fan delivers air to a system as shown in Fig. 21.13, where R_1 = 50 mm of water for a volume flow of 180 m³/min and $R_2 = R_3$ = 17.5 mm of water for a volume flow of 60 m³/min. The fan performance is as follows :*

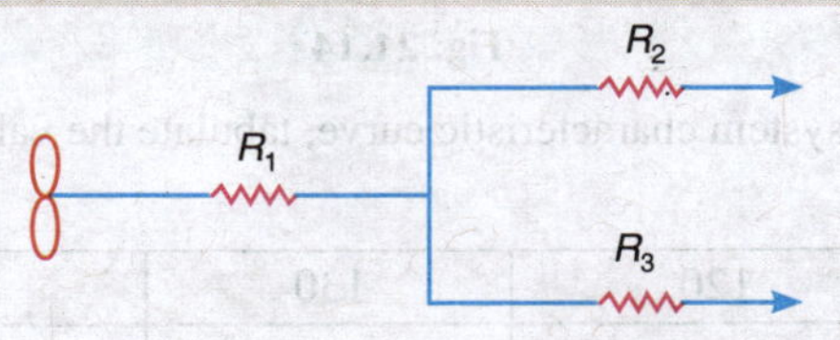

Fig. 21.13

Fan volume flow, m³/min	*120*	*150*	*180*	*210*
Fan total pressure, mm of water	*70*	*55*	*40*	*18*

Find the volume of air handled by the fan. By what percentage would it increase if a third branch of resistance $R_4 = R_2$ is connected in parallel with the existing two branches.

Solution. Given : R_1 = 50 mm of water ; Q_1 = 180 m³/min ; $R_2 = R_3$ = 17.5 mm of water ; $Q_2 = Q_3$ = 60 m³/min

Volume of air handled by the fan

First of all, let us find the resistance of the single equivalent system for the parallel resistances R_2 and R_3. Let R_e be the resistance of the equivalent system and K_e the equivalent constant.

We know that constant for R_1,

$$K_1 = \frac{R_1}{(Q_1)^2} = \frac{50}{(180)^2} = 1.54 \times 10^{-3}$$

Similarly, constant for R_2,

$$K_2 = \frac{R_2}{(Q_2)^2} = \frac{17.5}{(60)^2} = 4.86 \times 10^{-3}$$

and constant for R_3,

$$K_3 = \frac{R_3}{(Q_3)^2} = \frac{17.5}{(60)^2} = 4.86 \times 10^{-3}$$

We know that

$$\sqrt{\frac{1}{K_e}} = \sqrt{\frac{1}{K_2}} + \sqrt{\frac{1}{K_3}}$$

or

$$\sqrt{\frac{1}{K_e}} = \sqrt{\frac{1}{4.86 \times 10^{-3}}} + \sqrt{\frac{1}{4.86 \times 10^{-3}}} = 14.34 + 14.34 = 28.68$$

Squaring both sides, we have

$$\frac{1}{K_e} = (28.68)^2 = 822.5$$

or

$$K_e = \frac{1}{822.5} = 1.2 \times 10^{-3}$$

The whole system now reduces to a system in series as shown in Fig. 21.14. If Q is the volume flow rate of air in m³/min, then the system resistance,

$$R' = R_1 + R_e = K_1 Q^2 + K_e Q^2$$
$$= 1.54 \times 10^{-3} Q^2 + 1.2 \times 10^{-2} Q^2 = 2.74 \times 10^{-3} Q^2$$

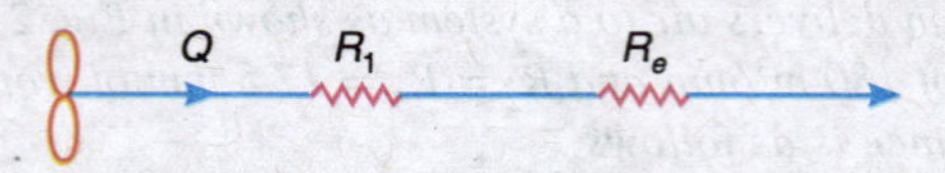

Fig. 21.14

In order to draw the system characteristic curve, tabulate the values of R' for different values of Q as given below :

Q (m^3/min)	120	130	140	150
R' (mm of water)	39.45	46.3	53.7	61.65

From the above values of Q and R', draw the system characteristic curve as shown in Fig. 21.15. This curve intersects the fan performance curve plotted for the given values, at point P as shown in Fig. 21.15. The point P is the operating point of the fan.

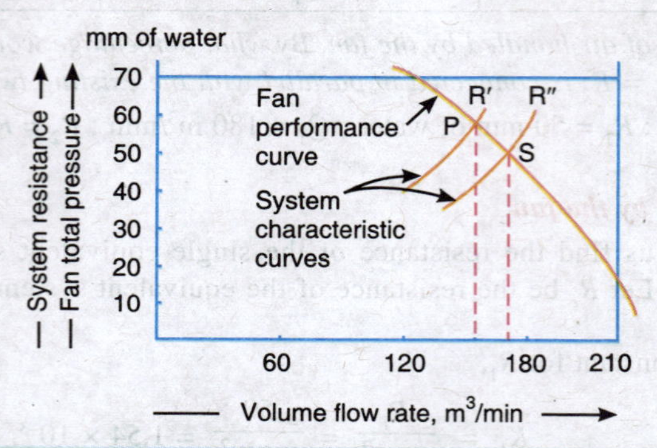

Fig. 21.15

∴ Volume of air handled by the fan

$$= \text{Volume of air at } P = 145 \text{ m}^3/\text{min} \quad \textbf{Ans.}$$

Percentage increase in volume

When a third branch of resistance $R_4 = R_2$ is connected in parallel, with the existing two branches as shown in Fig. 21.16, then the constant for the equivalent resistance R'_e is given by

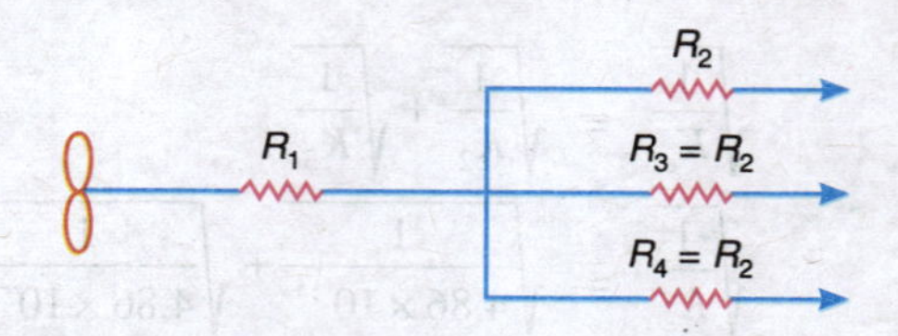

Fig. 21.16

$$\sqrt{\frac{1}{K'_e}} = \sqrt{\frac{1}{K_2}} + \sqrt{\frac{1}{K_3}} + \sqrt{\frac{1}{K_4}}$$

Since $R_2 = R_3 = R_4$, therefore $K_2 = K_3 = K_4 = 4.86 \times 10^{-3}$

$$\therefore \quad \sqrt{\frac{1}{K'_e}} = 3 \times \sqrt{\frac{1}{K_2}} = 3 \times \sqrt{\frac{1}{4.86 \times 10^3}} = 43.02$$

Squaring both sides, we have

$$\frac{1}{K_e'} = (43.02)^2 = 1850.7$$

or

$$K_e' = \frac{1}{1850.7} = 0.54 \times 10^{-3}$$

The whole system now reduces to a system in series as shown in Fig. 21.17. If Q is the volume flow rate of air in m³/min, then the system resistance,

$$R'' = R_1 + R_e' = K_1 Q^2 + K_e' Q^2$$

$$= 1.54 \times 10^{-3} Q^2 + 0.54 \times 10^{-3} Q^2 = 2.04 \times 10^{-3} Q^2$$

Fig. 21.17

Now tabulate the values of R'' for different values of Q as given below :

Q (m³/min)	140	150	160	170	180
R″ (mm of water)	40	45.9	52.2	58.9	66.1

From these values, draw the system characteristic curve as shown in Fig. 21.15. This curve intersects the fan performance curve at point S. At this point, volume of air handled by the fan

$$= 158 \text{ m}^3/\text{min}$$

∴ Percentage increase in volume

$$= \frac{158 - 145}{145} \times 100 = 8.96\% \textbf{ Ans.}$$

Example 21.6. *The fans A and B supply equal volumes of air to a system as shown in Fig. 21.18.*

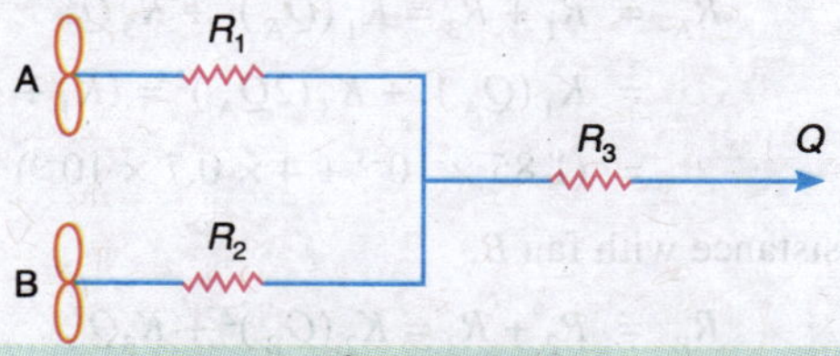

Fig. 21.18

The performance of the two fans is given below :

Volume flow (m³/min)	*Fan total pressure (mm of water)*	
	Fan A	*Fan B*
60	*70*	*76.5*
90	*69.5*	*65.5*
120	*67*	*48*
150	*60*	*26*
180	*48*	—

The resistances of the system are

$$R_1 = 15 \text{ mm of water at } 90 \text{ m}^3/\text{min} ;$$
$$R_2 = 7.5 \text{ mm of water at } 120 \text{ m}^3/\text{min, and}$$
$$R_3 = 2.5 \text{ mm of water at } 60 \text{ m}^3/\text{min.}$$

Determine the operating points of each of the two fans and the volume flowing through the common branch.

Solution. Given : $R_1 = 15$ mm of water ; $Q_1 = 90$ m³/min ; $R_2 = 7.5$ mm of water ; $Q_2 = 120$ m³/min ; $R_3 = 2.5$ mm of water $Q_3 = 60$ m³/min

Operating points of each of the two fans

We know that constant for R_1,

$$K_1 = \frac{R_1}{(Q_1)^2} = \frac{15}{(90)^2} = 1.85 \times 10^{-3}$$

Similarly, constant for R_2,

$$K_2 = \frac{R_2}{(Q_2)^2} = \frac{7.5}{(120)^2} = 0.52 \times 10^{-3}$$

and constant for R_3, $$K_3 = \frac{R_3}{(Q_3)^2} = \frac{2.5}{(60)^2} = 0.7 \times 10^{-3}$$

Tube axial flow fans impeller.

Let Q_A and Q_B be the volume of air supplied by the fans A and B. Since the volume of air supplied by the fans A and B is equal, therefore volume of air flowing through the common duct,

$$Q = 2Q_A = 2Q_B \quad \text{... } (i)$$

We know that system resistance with fan A,

$$R_A = R_1 + R_3 = K_1 (Q_A)^2 + K_3 Q^2$$
$$= K_1 (Q_A)^2 + K_3 (2Q_A)^2 = (K_1 + 4K_3)(Q_A)^2 \quad \text{... } [\because Q = Q_A]$$
$$= (1.85 \times 10^{-3} + 4 \times 0.7 \times 10^{-3})(Q_A)^2 = 4.65 \times 10^{-3} (Q_A)^2$$

Similarly, system resistance with fan B,

$$R_B = R_2 + R_3 = K_2 (Q_B)^2 + K_3 Q^2$$
$$= K_2 (Q_A)^2 + K_3 (2Q_A)^2$$
$$= (K_2 + 4K_3)(Q_A)^2 \quad \text{... [From equation } (i)]$$
$$= (0.52 \times 10^{-3} + 4 \times 0.7 \times 10^{-3})(Q_A)^2$$
$$= 3.32 \times 10^{-3} (Q_A)^2$$

Now tabulate the values of R_A and R_B for different values of Q_A, as given below :

Q_A (*m³/min*)	100	110	120	130
R_A (*mm of water*)	46.5	56.25	67	78.6
R_B (*mm of water*)	33.2	40.17	48	56.1

From these values, draw the system characteristic curves as shown in Fig. 21.19. The curve for R_A intersects the fan performance curve for fan A, at point P and the curve for R_B intersects the fan performance curve for fan B, at point S. The points P and S are the operating points of fans A and B respectively. From Fig. 21.19, we find that the operating point P for fan A lies at 120 m³/min and 67 mm of water. The operating point S for fan B lies at 120 m³/min and 48 mm of water. **Ans.**

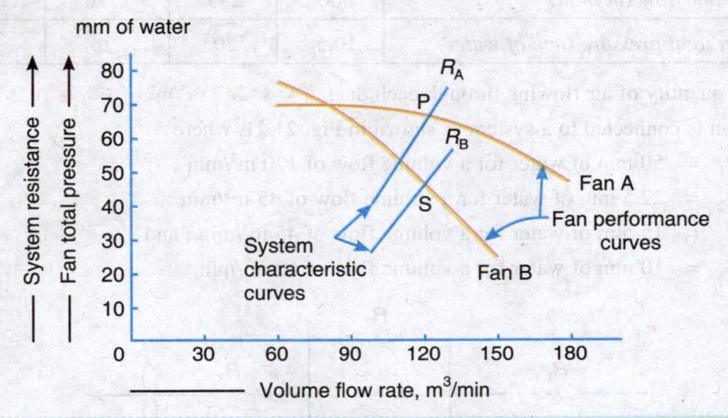

Fig. 21.19

Volume flowing through the common branch

Volume flowing through the common branch,

$$Q = \text{Volume at } P + \text{Volume at } Q$$
$$= 120 + 120 = 240 \text{ m}^3/\text{min} \quad \textbf{Ans.}$$

EXERCISES

1. A fan draws in air freely and discharges through a test duct of cross-section 0.07 m² in which the static pressure is 20 per cent of the velocity pressure. If the total efficiency of the fan is 65 percent and the input power is 1 kW, find the quantity of air being delivered in m³/min. **[Ans. 99 m³/min]**

2. A centrifugal fan having 0.75 m diameter impeller and rotating at 960 r.p.m. delivers 150 m³/min of air at 75 mm of total water column. If the air enters the impeller at outlet is 120 mm, determine the required blade angle at outlet. Assume that 45 per cent of the theoretical head is dissipated as impeller and casing losses and take the mass density of air as 1.2 kg / m³. **[Ans. 23.5°]**

3. A fan is to deliver 500 m³/min at a static pressure of 25 mm of water when running at 250 r.p.m. and requiring 5 kW. If the fan speed is changed to 300 r.p.m., find the capacity, static pressure and the power required. **[Ans. 600 m³/min ; 36 mm of water ; 8.64 kW]**

4. A fan delivers air to a system as shown in Fig. 21.20, where

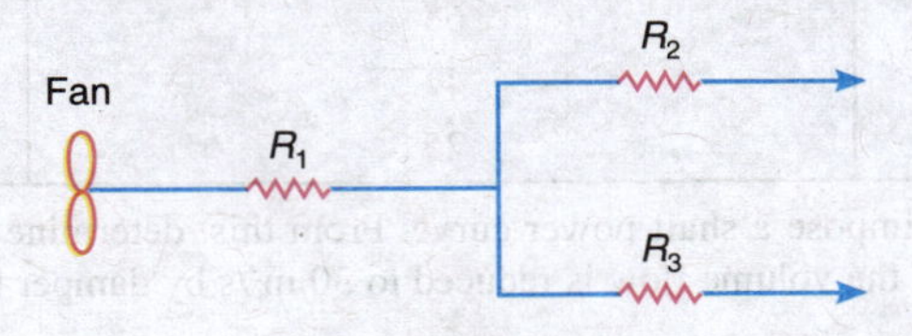

Fig. 21.20

R_1 = 12.5 mm of water for a volume flow of 300 m³/min ;

R_2 = 12.5 mm of water for a volume flow of 180 m³/min ; and

R_3 = 17.5 mm of water for a volume flow of 120 m³/min.

The fan performance is as follows :

Volume flow (m³/min)	300	285	270	255	240
Fan total pressure (mm of water)	10.5	20	26	30	35.5

Find the quantity of air flowing through each duct. [**Ans.** 280 m³/min ; 178 m³/min ; 102 m³/min]

5. A fan is connected to a system as shown in Fig. 21.21, where

R_1 = 50 mm of water for a volume flow of 120 m³/min ;

R_2 = 22.5 mm of water for a volume flow of 45 m³/min ;

R_3 = 15 mm of water for a volume flow of 45 m³/min ; and

R_4 = 10 mm of water for a volume flow of 60 m³/min.

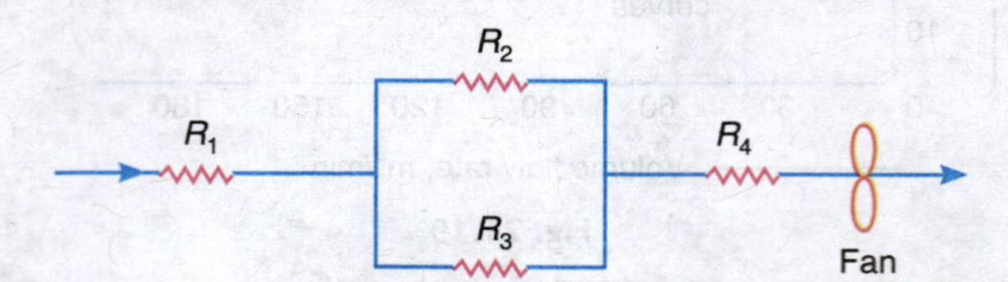

Fig. 21.21

The fan performance is as follows :

Volume flow (m³/min)	60	90	120	150
Fan total pressure (mm of water)	94.5	92.5	87.5	80

Determine the fan operating point and the volumes of air flowing in R_2 and R_3.

[**Ans.** 103.5 m³/min at 90.5 mm of water ; 46.7 m³/min ; 56.8 m³/min]

6. The performance for a centrifugal fan driven by a constant speed motor is given below :

Volume flow (m³/s)	*Static pressure (mm of water)*	*Efficiency (%)*
0	85	0
10	92.5	46
20	95	66
30	90	70
40	80	67
50	65	60
60	47.5	48
70	25	32

Plot these and superimpose a shaft power curve. From this, determine the shaft power at 50 m³/s. Also determine the power if the volume flow is reduced to 30 m³/s by damper regulation.

If instead of using damper regulation, the fan speed is reduced approximately by a hydraulic coupling of constant torque and zero slip, calculate the reduction in power input to the fan shaft.

[**Ans.** 53 kW ; 37.8 kW ; 18.7 kW]

QUESTIONS

1. What is the function of a fan in an air-conditioning system ?
2. Describe a centrifugal fan with the help of a neat sketch.
3. Explain the various types of axial flow fans.
4. Define the following :
   (*a*) Fan total pressure, (*b*) Fan air power, and (*c*) Fan total efficiency.
5. Write in brief explanatory note on the comparative study of the characteristics of backward and forward curved blade fans.
6. Define specific speed for a centrifugal fan. Derive its expression.
7. What do you understand by a geometrically similar fan ? Discuss the various fan similarity laws.

OBJECTIVE TYPE QUESTIONS

1. A fan may be considered as a pump, because it
   (*a*) looks like most other kind of pumps (*b*) circulate fluids, like other pumps
   (*c*) rotates (*d*) all of these
2. The air at inlet of a fan is .... atmospheric pressure.
   (*a*) above (*b*) below
3. In axial flow fans,
   (*a*) the air flows parallel to the axis of impeller
   (*b*) the air flows perpendicular to the axis of impeller
   (*c*) the air may flow either parallel or perpendicular to the axis of impeller
   (*d*) none of the above
4. The air guide vanes are sometimes installed in axial flow fans in order to
   (*a*) increase the static pressure (*b*) eliminate spiral flow of discharge air
   (*c*) reduce high frequency sound generation (*d*) all of these
5. The axial flow fans are particularly suitable for handling
   (*a*) large volumes of air at relatively low pressures (*b*) small volumes of air at relatively low pressures
   (*c*) large volumes of air at relatively high pressures (*d*) small volumes of air at relatively high pressures
6. The fan total pressure is the algebraic .... between the total pressure at the fan outlet and the total pressure at the fan inlet.
   (*a*) sum (*b*) difference
7. Two fans that are of different sizes, but have the same basic shape to their performance curves, are called
   (*a*) base fans (*b*) tubeaxial fans
   (*c*) geometrically similar fans (*d*) multi-stage fans
8. The capacity of a fan is .... cube of the impeller diameter.
   (*a*) directly proportional to (*b*) inversely proportional to
9. If N is the fan speed, then power of a fan is directly proportional to
   (*a*) N (*b*) N^2 (*c*) N^3 (*d*) N^4
10. If D is the impeller diameter, then power of a fan is directly proportional to
    (*a*) D^2 (*b*) D^3 (*c*) D^4 (*d*) D^5

ANSWERS

1. (*a*)	2. (*b*)	3. (*a*)	4. (*b*)	5. (*a*)
6. (*b*)	7. (*c*)	8. (*a*)	9. (*c*)	10. (*d*)

CHAPTER

22

Applications of Refrigeration and Air Conditioning

22.1 Introduction

Over the span of last few decades, refrigeration industry has grown into full-fledged industry in developed or northern countries. The refrigeration has become as essential feature rather than a luxury. The refrigeration has brought much more laurels and comforts to human beings than any other devices of human comfort. We can see the use of refrigeration and air conditioning practically in all spheres and walks of life. The application of refrigeration can be classified in the following six categories :

1. Domestic,
2. Commercial,
3. Industrial,
4. Marine,
5. Air conditioning, and
6. *Food preservation.

22.2 Domestic Refrigerator and Freezer

Now-a-days, the refrigerator has become an essential part of a household rather than a luxury. It is used for preserving food and thereby reducing waste. The primary function of a refrigerator or freezer is to provide food storage space maintained at low temperature for the preservation of food. Its essential secondary function is the formation of ice cubes for domestic consumption. They are usually specified by the internal gross volume and the deep freezer's volume. A storage temperature of 0ºC to 4ºC (273 K to 277 K) is satisfactory for the preservation of most of the fresh foods. For the short term storage of frozen foods (such as in a domestic refrigerator), temperatures much below the freezing point are required. The freezers are generally provided at the top portion of the refrigerator space. In some refrigerators, freezers are provided at bottom. This arrangement seems to be based on the heat transfer considerations but it may be noted that the time taken to cool products kept at upper portion would be more.

Domestic Refrigerator.

The refrigerators may be single-door, double-door top freezer, double-door bottom freezer, and side-by-side door freezer. The double-door refrigerators are very commonly used now-a-days because of the need for larger storage space and better preservation of frozen foods. These refrigerators are divided into two separate compartments, one for the fresh food or general items and the other for the storage of frozen foods. Since the requirement of food or general items in a day is quite high, therefore the frequency of opening the door is also high, but the daily use of frozen items is very much limited so with double-door system, the freezer space is not subjected to wide temperature variations. This helps in maintaining a stable temperature for the preservation of the frozen foods.

The mechanical vapour compression cycle as well as absorption cycle may be adopted for domestic refrigerators and freezers, but the mechanical vapour compression system is actually used over absorption system, because of its compactness and more efficient use of electrical energy, as shown in Fig. 22.1. The refrigerants used are generally R-12 or R-22. The compressor is mounted at the bottom of the refrigerator frame. The power of compressor can vary according to size of the refrigerator (*i.e.* 75 W, 92 W, 125 W, 180 W and 370 W etc.). The condenser is put at the back about 40 to 60 mm away from the cabinet. The condenser may be either chassis type or tube and wire type. In the former, the condenser tube is mounted on a metal sheet which acts as fins. The tube and wire type condensers are quite simple in which few tubes are held tightly under wire frame from both sides. These wires act as cylindrical fins increasing the rate of heat transfer. The capillary tube is kept in contact with the evaporator inlet pipe. A drier is connected between the receiver and the evaporator to eliminate traces of moisture if any. In some cases, the temperature of two cabinets of the refrigerator have to be controlled independently. Under such circumstances, independent compressors and cooling coils are used.

The evaporator coil is wrapped around the freezer in a suitable manner to give efficient heat transfer. Sometimes, the freezer chamber is made from a pair of sheet joined together in such a way that the passage between the sheets act as an evaporator coil. The cooling of lower space is

* Food preservation has been discussed, in detail, in Chapter 13.

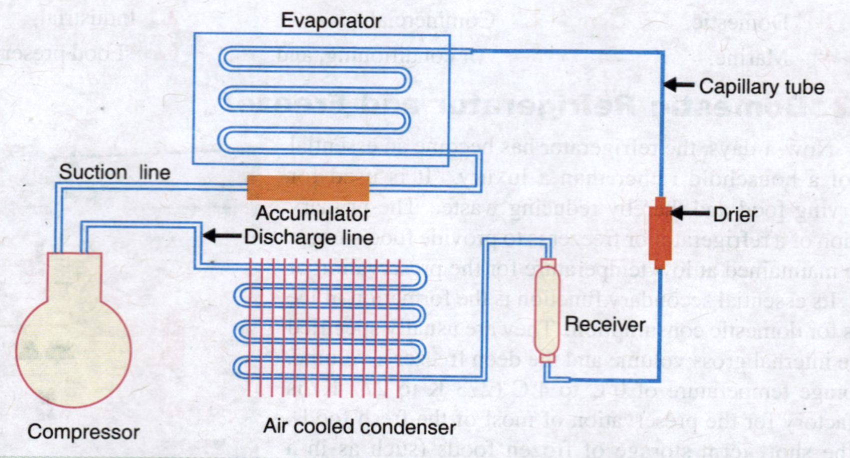

Fig. 22.1. Refrigeration system for a domestic refrigerator.

accomplished by free convection (due to density gradient). The thermostatic sensing element is provided to the evaporator coil which can control temperature in the freezer upto −15ºC in steps or continuously depending upon the type of controlling switch employed.

The refrigerator body is provided with good quality insulation in order to prevent heat transfer into the system. Usually 60 to 100 mm thick glass loose-fill fibre or glass rolls or thermocole is used since the conductivity of these insulating materials are quite low. In freezers where temperature has to be maintained quite low (– 40ºC), the insulation thickness may be about 200 mm.

22.3 Defrosting in Refrigerators

Since the evaporator in a refrigerator operate at temperatures below 0ºC (*i.e.* below the freezing point of water), therefore it is subjected to the accumulation of frost or ice. The lower the evaporator temperature, the thicker will be the frost formation. The frost acts as an insulation that impedes the heat transfer to the evaporator. This leads to further thickening of frost as the temperature tends to go down because of decreased heat transfer rate. This ultimately leads to substantial reduction in the evaporator capacity and the system efficiency. Thus, the removal of frost or defrosting the evaporator at regular intervals is an absolute necessity.

One of the simplest method of defrosting is by manually putting 'Off' the refrigerator and restarting only after complete defrosting of the evaporator. Now-a-days, the refrigerators are provided with push button defrost thermostats. A push button is provided in the centre of the thermostat knob. The defrosting can be initiated by simply pressing the push button which causes the leverage in the thermostat to break and keeps the electrical contact of the thermostat open until the evaporator temperature rises above freezing point and defrosting takes place. The refrigerator returns to normal functioning automatically, once the defrosting is complete.

In a double-door refrigerator, the evaporator in the fresh food or general storage compartment is generally designed for natural cycle defrost. The defrosting takes place every time the compressor switches off the thermostat, as the storage temperature is above the freezing point. However, the defrosting may not be always complete. Over a period of continuous use, some amount of residual frost after the defrost cycle can get collected, particularly on the lower portion of the evaporator. Such accumulation of frost due to the residual frost of the natural defrost cycle has to be removed periodically by manually stopping the machine. The freezer compartment is not provided with automatic defrost system, as the temperature of the frozen food should not be allowed to go above the freezing point. Therefore, manual defrosting has to be carried out periodically. Obviously, for defrosting, the frozen foods should be removed or defrosting should be carried out when the freezer

Freezer plus three fridge compartments including drinks fridge.

is empty. The defrost water from the evaporator flows to a condensate pan provided below the evaporator of the fresh food compartment. From there it drains into a tray in the compressor compartment. The water collected in the tray will evaporate due to the hot environment in the compressor compartment. It may be necessary to empty the tray manually once in a while, as the water accumulates in the tray.

22.4 Controls in Refrigerator

The controls are very essential for satisfactory and economical working of any refrigerator. The electrical connection diagram of a domestic refrigerator is shown in Fig. 22.2. The refrigerator is fitted with the following controls :

(a) ***Starting relay.*** The starting relay is used to provide the necessary starting torque required to start the motor. It also disconnects the starting winding of the motor when the motor speed increases. When the compressor motor is to be started, the thermostat is in the closed position. When the electric supply is given, an electric current passes through the running winding of the motor and the starting relay. Due to the flow of electric current through relay coil and due to electromagnetism, its armature is pulled thereby closing the starting winding contacts. The current through starting winding provides the starting torque and the motor starts. As the motor speed increases, the running winding current decreases. The current in the starting relay is no longer able to hold the relay and it gets released thereby opening the starting winding contacts. Thus, the starting winding gets disconnected.

(b) ***Overload protector.*** The basic function of an overload protector is to protect the compressor motor winding from damage due to excessive current, in the event of overloading or due to some fault in the electric circuit. It consists of a bimetallic strip. During the normal working of the compressor, the contacts are closed. Whenever there is any abnormal behaviour (*i.e.* overheating, overcurrent due to fault or overload), the bimetallic strip gets heated and bends, thereby opening the motor contacts, and de-energising it. The overload protector is fitted on the body of the compressor and operates due to the combined action of heat produced when current passes through the bimetallic strip and a heater element, and heat transferred from the compressor body. It may be noted that the abnormal behaviour of the compressor may be due to low voltage, high voltage, high load, low suction pressure, high suction and discharge pressure.

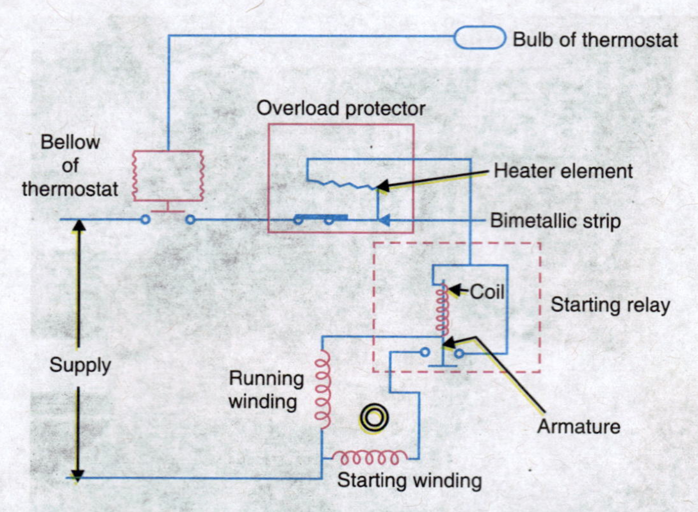

Fig. 22.2. Electrical connection diagram of a domestic refrigerator.

(*c*) ***Thermostat.*** A thermostat is used to control the temperature in the refrigerator. The bulb of the thermostat is clamped to the evaporator or freezer. The thermostat bulb is charged with few drops of refrigerant. The thermostat can be set to maintain different temperatures at a time. When the desired temperature is obtained, the bulb of the thermostat senses it, the liquid in it compresses and operate bellows of the thermostat and opens the compressor motor contacts. The temperature at which compressor motor stops is called *cut-out temperature.* When the temperature increases, the liquid in the bulb expands thereby closing the bellows contact of the compressor motor. The temperature at which the compressor motor starts, is called *cut-in temperature.* A thermostat is very crucial in the operation of a refrigerator as the running time of the compressor is reduced considerably thereby cutting the operation cost as well as enhancing the compressor life due to non-continuous working.

22.5 Room Air Conditioner

A room air conditioner is a compact, self contained air-conditioning unit which is normally installed in a window or wall opening of the room and is widely known as window type air conditioner. It works on vapour compression cycle. A complete unit of a room air conditioner consists of the refrigeration system, the control system (thermostat and selector switch), electrical protection system (motor overload switches and winding protection thermostat on the compressor motor), air circulation system (fan motor, centrifugal evaporator blower), ventilation (fresh air damper) and exhaust system.

The refrigeration system consists of a hermetic type compressor, forced air-cooled finned condenser coil, finned cooling coil, capillary tube as the throttling device and a refrigerant drier. The refrigerant used is R-12 or R-22. In hermetic compressors, a winding thermostat is embedded in the compressor motor windings. It puts off the compressor if the winding temperature exceeds the safe limit, thus protecting the winding against high temperature.

The condenser is a continuous coil made of copper tubing with aluminium fins attached to it to increase the heat transfer rate (rejecting heat to atmosphere). A propeller type fan provides the necessary air to cool the refrigerant in the condenser and also exhausts air from the air-conditioned space when the exhaust damper is opened. The evaporator is a cooling coil also made of copper with aluminium fins attached to it to increase the heat transfer rate (taking in heat from the room air).

The room air-conditioner is installed in such a way that the evaporator faces the room. A centrifugal blower is installed behind the cooling coil which sends cool air in the room. A filter is installed on the fresh air entering side of the evaporator to remove any dirt from the air. A damper inside the cabinet regulates the fresh air intake of the room air-conditioner. The quality of fresh air may be varied by adjusting the dampers. If all the air in the room is to be exhausted, the fan control of the unit is set to 'Exhaust' position. The condenser fan or blower exhausts all the air to the atmosphere. Thus, smoke and odour are removed by the condenser fan which draws air through the dampers and exhausts it through the louvers in the rear of the unit.

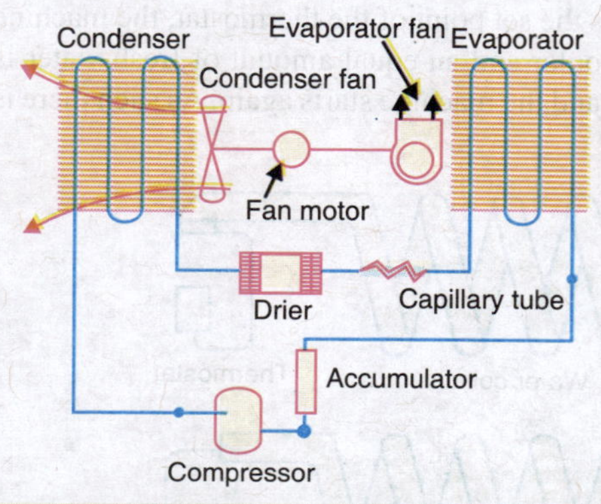

Fig. 22.3. Layout of a room air conditioner.

Room air conditioner.

A thermostat element is located in the return air passage of the unit. It controls the opertion of the compressor based on the return air temperature, which indicates the room temperature. It may be noted that when the required temperature is obtained, the compressor is stopped.

A selector switch often known as master control, controls the compressor motor, condenser fan motor, and evaporator fan motor. When the control switch is in 'Ventilate' position, only evaporator blower motor operates and outside fresh air is supplied in the room which is not cool as the compressor is not working. In the 'Exhaust' position, the condenser fan motor operates and all the room air is exhausted to the atmosphere. In the 'cool' position, all the motors *i.e.* compressor motor, condenser motor, and evaporator motor are in working state and cool air is supplied to the room.

Notes : 1. By installing a reversing valve, the air-conditioner unit can be used for heating the room during winter. The reversing valve is a two position valve with four ports. The discharge and suction lines of the compressor are connected to two ports. The other two ports are connected to the inlet side of the air-cooled condenser and the suction outlet of the evaporator.

2. The advantage of using a reversing valve for heating is that the energy required for heating the room will be much less than that required for heating with electrical strip heaters.

22.6 Water Coolers

The purpose of a water cooler is to make water available at a constant temperature irrespective of ambient temperature. They are meant to produce cold water at about 7ºC to 13ºC (280 K to 286 K) for quenching the thirst of the people working in hot environment. The warm or normal water can serve the physical requirement of our system for the proper functioning of the body organs but it does not quench the thirst especially in hot summers. The temperature of cold water is controlled with the help of a thermostatic switch set within 7ºC to 13ºC range.

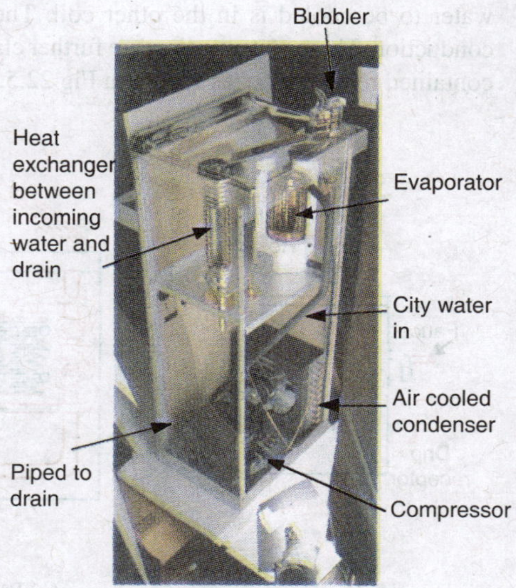

Water cooler.

There are two types of unitary water coolers *i.e.* the storage type and the instantaneous type. In the *storage type water coolers*, the evaporator coil is soldered on to the walls of the storage tank of the cooler, generally on the outside surface of the walls. The tank may be of galvanized steel or stainless steel sheets. The water level in the tank is maintained by a float valve. In this type of water cooler, the machine will have to run for a long time to bring down the temperature of the mass of water in the storage tank. Once the temperature touches the set point of the thermostat, the machine cycle is stopped. When the water is drawn from the cooler and an equal amount of fresh water is allowed in the tank, the temperature will rise up slowly and the machine starts again. As such there is always a reservoir of cold water all the time.

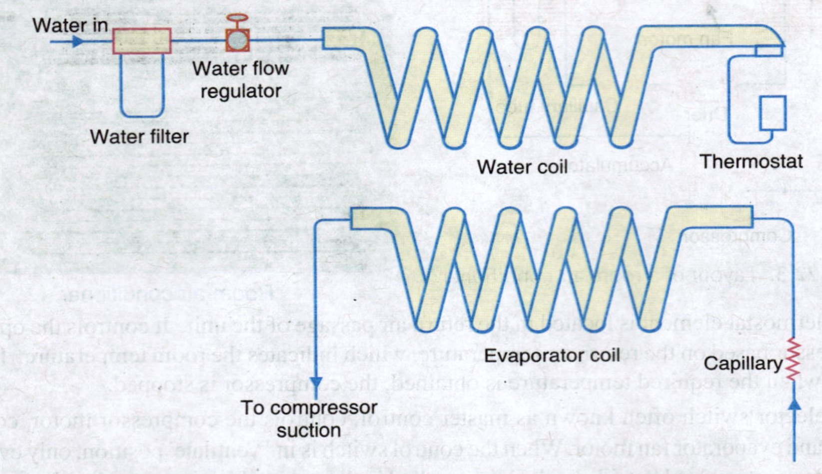

Fig. 22.4. Cooling coil of an instantaneous type cooler.

In case of *instantaneous type water coolers*, the evaporator, as shown in Fig 22.4, consists of two separate cylindrically wound coils made of copper or stainless steel tube (In the figure, the cooling coil and the water coil are shown separately for clarity in explanation, otherwise, the coils are entwined and bonded together by soldering). The evaporating refrigerant is in one of the coils and the water to be cooled is in the other coil. The water is cooled by the refrigerant in evaporator by conduction. These water coolers are further classified as (*a*) bottle type, (*b*) pressure type, and (*c*) self-contained remote type, as shown in Fig 22.5. These are discussed, in detail, as follows :

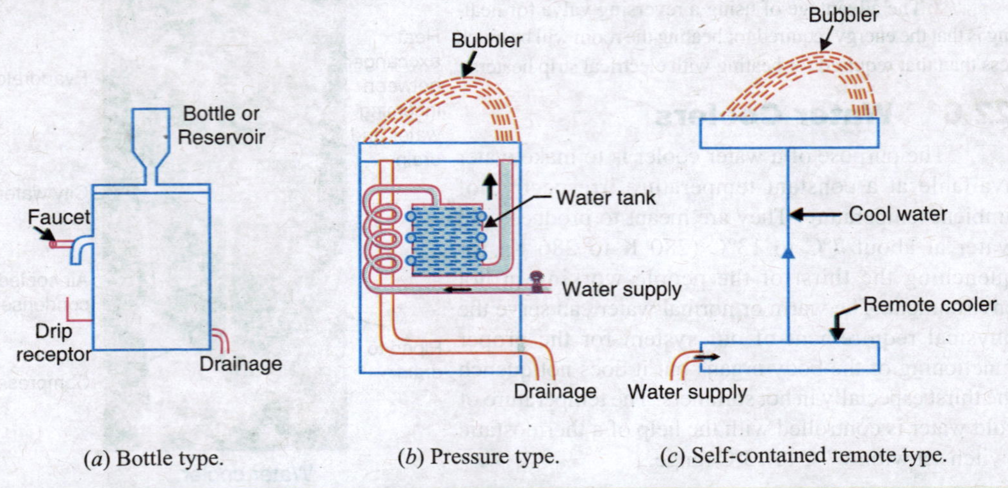

(*a*) Bottle type. (*b*) Pressure type. (*c*) Self-contained remote type.

Fig. 22.5. Instantaneous type of water coolers.

(a) ***Bottle type.*** As the name suggests, this type of instantaneous water cooler employs a bottle or reservoir for storing water to be cooled. No city main inlet connection is required as it is normally used to cool water supplied in 25 litre glass bottles, which are placed on top of the unit, as shown in Fig. 22.5(*a*).

(b) ***Pressure type.*** In this type of instantaneous water cooler, as shown in Fig. 22.5(*b*), water is supplied under pressure. The city main water enters the cooler through the inlet connection at the rear of the cooler. It then passes through a pre-cooler. The pre-cooler is cooled by the waste water of the cooler. As the waste water temperature is low, it is made use of cooling the supply water by passing through a pipe coil wrapped around the drainage line (a counter-flow heat exchanger). This arrangement helps in reducing the cooling load for the cooler. The amount of cooling depends upon the quantity of waste water and the length of the pipe coil comprising of pre-cooler.

The pre-cooled water then enters the storage chamber and loses its heat to the refrigerant. The outlet water pipe is connected at the bottom of the storage tank, which is fitted with a self-closing valve or bubbler. A thermostat controls the temperature of the water in the pipe to a set point.

(c) ***Self contained remote type cooler.*** This type of cooler employs a mechanical refrigeration system. The water cooled from the remote cooler is supplied to desired drinking place, away from the system. This type of arrangement does not require extra space near the place of work and is quite useful.

Notes : 1. The faucet or push type water taps are generally provided for drawing cold water in both the types in order to minimize the wastage of refrigerated water.

2. The thermostat controls the operation of the refrigeration compressor to maintain the water temperature within the set limits. In case of instantaneous cooler, the feeler bulb of the thermostat is clamped on to the water pipe at its outlet end whereas in case of storage type cooler, the bulb is kept immersed in water in the tank or clamped to the wall of the storage tank on the outside, at a lower level, much below the lower most evaporator refrigerant tube, soldered on the tank.

3. In the case of the instantaneous type cooler, it is very important that the flow rate of water is adjusted to match its capacity. If the rate of flow is higher, the cooler will not be able to bring down the temperature of water to the set level. It may be noted that with a very high flow rate, the refrigeration system will work at a very high evaporator temperature (so at a higher suction pressure) which may adversely affect the compressor motor of the cooling unit.

22.7 Capacity of Water Coolers

The cooling load for the water cooler (Q) may be obtained from the following relation :

$$Q = m_w \, c_p \, (T_i - T_o)$$

where m_w = Rate of water consumption,

c_p = Specific heat of water = 4.18 kJ/kg K,

T_i = Inlet temperature of water, and

T_o = Outlet temperature of water.

The amount of cold water requirements under various conditions is given in Table 22.1. These figures are based on extensive statistical survey. The refrigerants such as ammonia, sulphur dioxide etc. are now-a-days not used because of safety reasons. Generally R-12 is the most common refrigerant up to one tonne refrigeration (1TR) capacity and R-22 for two tonne refrigeration (2TR) capacity and appropriate combinations for larger size units.

The amount of wastage of cold water should be included while estimating the amount of water consumption. Usually heavy insulation around 40 mm to 60 mm thick glass wool or thermocole is provided rendering insignificant heat transfer through insulation.

Table 22.1. Amount of drinking water requirements under various conditions.

Applications	*Temperature in °C*	*Amount of water used including drainage in litres per hour*
Restaurants	7 – 10	0.5 per person
Schools	7 – 10	0.56 per student
Offices	7 – 10	0.56 per person
Stores	7 – 10	4.5 per 100 customers
Hotels	7 – 10	0.4 per room
Hospitals	7 – 10	0.4 per bed and per attendant
Theatres	7 – 10	4.5 per 100 seats
Light duty manufacturing shop	10 – 13	0.9 per person
Heavy duty manufacturing shop	10 – 13	1.1 per person

Example 22.1. *Calculate the refrigeration tonnage of a water cooler meant for drinking water at 10°C for 500 workers of an industrial organisation of 6 hours duty time. The water is available at 30°C. The heat transfer through insulation is 5% of the total heat load. Recommend the tonnage of water cooler, assuming heavy duty. Also calculate the total water consumption per day.*

Solution. Given : $T_o = 10°C$; Workers = 500 ; Time = 6 h ; $T_i = 30°C$; Heat transfer = 5% of total heat load

Refrigeration tonnage of water cooler

Let Q_T = Total refrigeration tonnage of water cooler.

From Table 22.1, we find that the drinking water requirement per person for heavy duty manufacturing shop is 1.1litres per hour.

∴ Cold water supply per hour,

$$m_w = 1.1 \times 500 = 550 \ l/\text{h or } 550 \text{ kg/h} \quad \text{...(∵ Sp. gr. of water = 1)}$$

We know that cooling load for the water cooler,

$$Q = m_w \, c_p \, (T_i - T_o) = 550 \times 4.18 \, (30 - 10) = 46\,090 \text{ kJ/h}$$

$$\text{...(∵ } c_p \text{ for water = 4.18 kJ/kg K)}$$

Since heat transfer through insulation is 5% of the total load, therefore total cooling load,

$$Q_T = 0.05 \, Q_T + Q \quad \text{or} \quad 0.95 \, Q_T = Q$$

$$\therefore \quad Q_T = Q / 0.95 = 46\,090 / 0.95 = 48\,515 \text{ kJ/h} = 808.6 \text{ kJ/min}$$

or refrigeration tonnage of water cooler,

$$Q_T = 808.6/210 = 3.85 \text{ TR } \textbf{Ans.} \quad \text{...(∵ 1 TR = 210 kJ/min)}$$

Water consumption per day

We know that water consumption per day (for 6 hours duty time)

$$= 550 \times 6 = 3300 \text{ litres / day } \textbf{Ans.}$$

22.8 Applications of Air Conditioning in Industry

The air conditioning has numerous applications in industry. Generally, in an industry air conditioning is required to control the temperature, air flow, dust content and humidity to get a quality product. In addition, air conditioning is required for the following purposes:

1. To provide comforts to the workers.

2. To provide necessary low temperature conditions required for the manufacture of certain products in industries such as textile, printing and refineries.

3. To provide a clean room for the precision work, laboratories and quality control rooms.

4. To preserve food during storage and transportation.
5. For drying of products.

The application of air-conditioning in different industries is discussed briefly as follows :

(a) Textile industry. The air-conditioning plays an important role in textile industry and is essential for the production of quality textile products in addition to reduction in wastage. The industries producing cotton, silk, rayon, wool and nylon require maintenance of specific relative humidity during manufacturing process. The finished fabrics are soft and durable when manufactured under proper humidity conditions. If these are manufactured under dry humidity conditions, the finished product becomes brittle and weak.

The different manufacturing processes require different temperatures as well as humidity conditions. For example, low humidity (55 - 60%) is required in spinning department to avoid sticky action of cotton to leather aprons. However, high humidity (70 - 85%) in weaving department is maintained to reduce the static electricity effects and to increase the strength of the cotton threads. Table 22.2 shows some of the recommended environmental conditions for textile industry.

Table 22.2. Recommended environmental conditions for textile industry.

S. No.	*Type of thread*	*Operation*	*Temperature (K)*	*Relative humidity (%)*
1.	Cotton	Combing	297	55 – 60
		Drawing or roving	300	60
		Frame spinning	300 – 303	55 – 60
			298 – 300	70 – 85
2.	Silk	Preparatory	300	60 – 65
		Weaving	300	60 – 70
		Spinning	300	65 – 70
		Throwing	300	60
3.	Rayon	Spinning	300 – 305	50 – 60
		Weaving	300	50 - 65
		Knitting	300 – 303	60 – 65
4.	Woollen	Carding	300 – 303	65 – 70
		Spinning	300 – 303	50 – 60
		Weaving	300 – 303	60 – 65
		Drawing	297	50 – 60

(b) Photographic industry. In photographic industry, the manufacturing processes require accurate control of temperature and humidity. The photographic film is made of cellulose ester, coated with silver salt emulsion. The photographic paper used for films is manufactured by applying a coating of emulsion on a particular, highly pure wood pulp paper. All these processes are carried out under controlled conditions, as these are sensitive to variations in temperature and humidity. The dust particles are undesirable during manufacturing processes in the photographic industry. Special type of filters are employed to supply clean cool air at the required humidity. The absolute filters are normally employed as the filtering media along with pre-filters.

(c) Printing industry. In printing industry, the required temperature and humidity are in the range of 23°C to 27°C (296 K to 300 K) and 20% to 50% relative humidity. In textile printing factory, it is very essential to maintain correct humidity conditions as low humidity may result in light prints and high humidity may result in blurring or blots.

In paper printing industries, in low humidity conditions paper becomes dry and under high humidity conditions, the printing ink dries slowly which may result in non-uniform printing.

(d) ***Food industry.*** In food industry, air-conditioning is universally employed for the preservation of perishable foods such as cheese, butter, fruits, milk, vegetables, meats etc. Also, during the manufacturing processes, air-conditioning is used for cooling the materials, *e.g.* during milk and butter processing, cooling is provided by refrigeration equipment.

In cold storage, the temperature, humidity and dust-particles are controlled to prolong their life. It may be noted that in high humidity conditions, biscuits, pastries and cakes absorb moisture and product deterioration is inevitable. In high temperatures, fungus and bacteria growth multiply in perishable foods, thus controlled conditions of temperature and humidity are very essential.

Air-conditioning plays an equally important role in the transportation of food from one place to another. For this, refrigerated trucks are employed. Dry ice is commonly used for the transportation of food with an advantage that it does not spoil the food as it directly gets converted from solid to vapour.

22.9 Refrigerated Trucks

The large quantities of fresh and frozen perishable products are transported in refrigerated trucks and trailers. Thus, the trucks, trailers or containers are equipped with refrigeration systems to maintain the required temperature in the cargo space for the preservation of these food articles and substances during transportation. Generally, truck bodies are designed for temperatures upto 4ºC to carry cold fruits, vegetables, meats, dairy products and upto –20ºC for carrying frozen foods like ice-cream.

The simplest form of such a transport is an insulated vehicle without a refrigeration system. The body of truck is constructed of metal or wood. The metal construction is preferred as it has longer life and is sturdy. The body is very well insulated with a material of high insulating value. The following methods of cooling are adopted in insulated trucks:

1. ***Using water ice.*** The top of the product is suitably iced after loading it into the truck. However, this can be adopted only for short distance haulage and that too only for products that will not get affected by water (formed by melting of ice on absorption of heat of transmission). The refrigerating effect produced by the melting of ice is 335 kJ/kg.

2. ***Water ice in bunkers with forced air circulation.*** The ice bunkers are fitted in front of the insulated vehicles as shown in Fig. 22.6 . These can be either removable type or permanently installed type. The fans are used to circulate cool air in the body of the truck. The air is usually drawn from the floor of the body, cooled through ice and circulated over the space in which the products are stacked. The fan may be driven either by the truck engine or electrically operated by the battery of the truck.

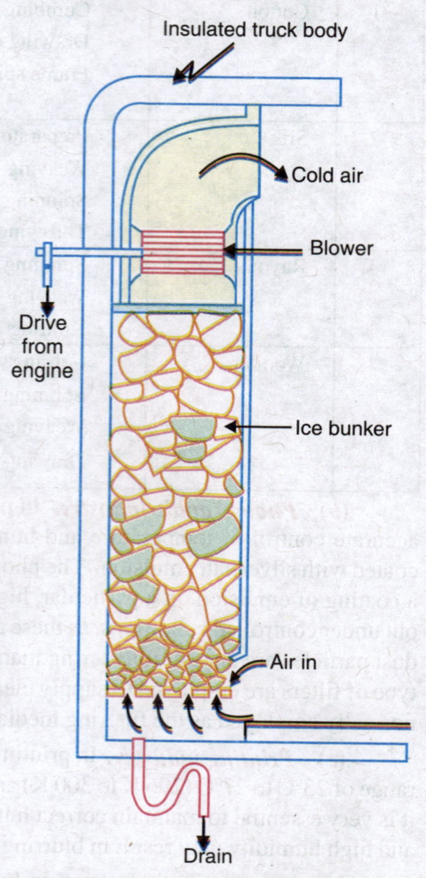

Fig. 22.6. Ice bunker used in transport refrigeration.

In another system, a tank containing a mixture of ice and salt in brine solution is circulated in coil fitted with fins. A fan fitted behind the coil and powered by the truck engine forces cool air inside the body where products are stored.

3. ***Using dry ice.*** The dry ice is used in many small retail trucks for the delivery of frozen foods and ice cream. The dry ice blocks are kept in the ceiling and the cooling is by natural convection. If forced circulation is employed, the dry ice may be placed in bunkers as like water ice.

The dry ice (or solid CO_2) is preferably used where it is easily available than any other system of refrigeration because of its trouble free operation and being very light in weight. It may be noted that dry ice sublimates (*i.e.* on heating, it changes directly into the vapour state without passing through the liquid state), thus the products will not get wet as in case of ice. The refrigerating effect produced by the sublimation of dry ice (which takes place at a temperature of –78.3° C) is 605.5 kJ/kg.

4. ***Eutectic plates.*** The eutectic plates forming passages are fixed on side walls and ceiling of the insulated cargo space of the truck. The eutectic solutions are special liquids with low freezing points. The heat of transmission is taken care of by the latent heat of fusion of the melting frozen solution. The plates have inbuilt refrigerant passages, which are connected to a refrigeration plant provided at the terminals of the transport operator to refreeze the eutectic solution.

Note : For long distance transportation, the vehicle is equipped with a mechanical refrigeration system driven by an independent oil engine. In some cases, an electric motor is incorporated to run the compressor at terminals by external electric supply.

22.10 Marine Air-conditioning

The air-conditioning of the passenger ships as well as cargo ships is recently developed and is used by many companies of the world. It may be noted that the general principles of air-conditioning which apply to land installations are also applicable to marine, provided all the factors affecting the construction and operation of a ship are given proper considerations. The system should be flexible to work under different climates, ranging from extreme hot to extreme cold as the ship may pass through regions of different climates. This should be taken into account for selecting the equipment for outside design conditions.

In case of cargo ships, the refrigeration system should be capable of providing any temperature between – 23.5° C and 12.5° C. The reciprocating compressors with R-12 as refrigerant are universally employed. It may be noted that R-22 is not recommended because of its critical oil miscibility. Also, compound compression system is generally employed with R-12 instead of parallel operation as with R-12 there is a tendency for oil to migrate and flood one compressor while starving another. Thus, this increases the number of compressors as each evaporator has an independent compressor. The condensers are of the shell and tube type using sea water for cooling. The corrosion resistant cupronickel material is, therefore, used for tubes and end covers. A receiver capable of holding 20 per cent more charge in addition to whole charge is essential. The liquid line should emerge from both ends of the condenser, and later joining into one is required as it ensures continuous draining of the liquid during roll or pitch of the ship.

However, the various factors that are also considered while designing air-conditioning system for ships are as follows :

1. The working of the system should be free from objectionable noise and vibration. The vibration of ducts , water piping and refrigerant piping may cause fatigue and failure. In order to avoid this, it is necessary to provide proper vibration dampers and canvas duct joints from the vibration machinery to the conditioned air carrying ducts.

2. The equipments should be designed to occupy minimum space because of limited space on ship board.

3. The design of air-circulation system needs special attention, as the heat load in every compartment is likely to be different. The load in each room also differs in day and night. Thus, duct layout is more difficult due to lack of space for carrying the ducts.

4. The low ceiling height of rooms in the ship needs special attention for layout of supply and return air ducts. The attention should be given to the room dimensions, ceiling heights, volume of air handled and air temperature difference between the supply and room air.

5. The certain quantity of spare parts of essential items and extra refrigerant charge should be carried as the period of voyage may be long.

Note : The refrigeration system can be made economical by the use of parallel brine circuits in cooling coils and fewer condensing units.

22.11 Ice Manufacture

The commercial ice is produced by freezing potable water in standard cans placed in rectangular tanks. The tanks are filled with chilled brine. For increasing the heat transfer from the water in the can to the chilled brine, the brine solution is kept in constant motion by agitators. The agitators can be either horizontal or vertical and are operated by means of electric motors. The brine temperature is maintained by the refrigeration plant at −10°C to −11°C.

The ammonia gas is used as the refrigerant because of its excellent thermal properties. It also produces very high refrigerating effect per kg of refrigerant and low specific volume of the refrigerant in vapour state.

The high temperature, high pressure ammonia vapours are condensed in a condenser which may be of shell and tube type or evaporative type. The condensed liquid ammonia is collected in the receiver and then expanded through the expansion valve. Due to the expansion, the pressure of the liquid ammonia is considerably reduced. It then passes through the evaporator coils surrounding a brine tank in which brine solution is filled. The low pressure liquid ammonia absorbs heat from the brine solution, equivalent to its latent heat of vaporisation, gets converted to vapour state and is once again fed to compressor to complete the cycle. The layout of an ice manufacturing system is shown in Fig. 22.7.

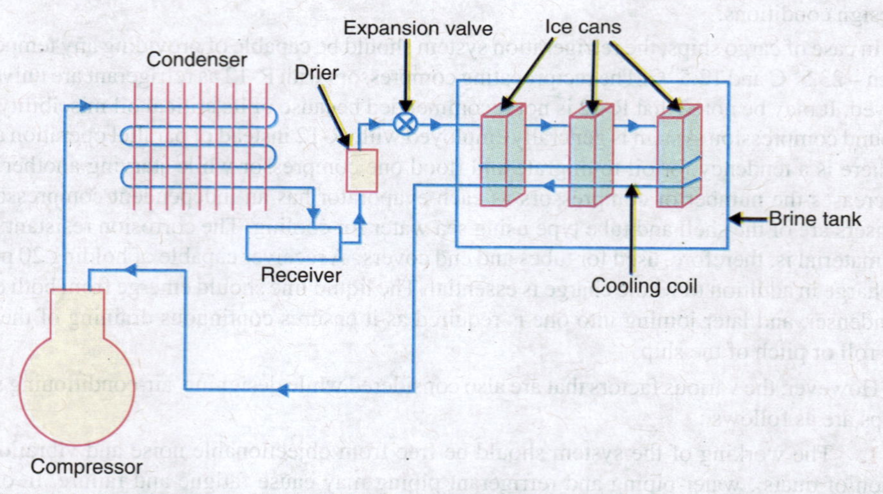

Fig. 22.7. Ice plant layout.

The brine tanks are usually fabricated of 6 mm thick mild steel plates with tie rods welded end to end. The depth of tank is such that the brine level is around 25 mm higher than the water level in the cans. The tank is insulated on all the four sides and from the bottom. The insulated wooden lids are provided to cover the top insegments, to facilitate the removal or replacing of ice-cans. The ice-cans are fabricated from galvanized steel sheets and are given chromium treatment to prevent corrosion due to chemical reactions. It also prevents corrosion from reaction with ammonia (NH_3) in case it leaks from the system.

In order to get clear transparent ice, water in the can is agitated by the use of low pressure air through the tubes suspended from the top. Due to agitation, the dissolved impurities such as salts, even colours get collected in the unfrozen water core. It is desirable that it should be taken out and replaced with fresh water.

It may be noted that the ice of potable water (treated or untreated) frozen at a temperature lower than –12°C can crack. Therefore, brine temperature is kept at a higher level, say –11° C to –10° C. Water in the ice cans placed in the brine cools rapidly up to a temperature of about 3° C to 4° C. Thereafter, it takes more time for water to touch 0°C. The rate of freezing decreases substantially as the thickness of the ice layer increases. This is because of the reason that the ice layer offers more thermal resistance to the heat flow from the water to the brine.

Notes : 1. Generally sodium chloride (NaCl) or calcium chloride ($CaCl_2$) is used as brine. Due to its low cost and being less injurious, sodium chloride brine is preferably used.

2. The freezing time is dependent upon the brine temperature and the extent of brine and water agitation.

22.12 Cooling of Milk (Milk Processing)

The milk is known to be one of the most perishable foods and if it is not maintained at sufficient low temperature, it gets spoiled due to the growth of bacteria and other organisms. Some kind of bacteria rapidly multiplies in number at temperatures of 21°C to 38°C (294 K to 311 K) and the milk gets spoilt and sour. As the temperature of the milk is reduced, the bacterial growth decreases and practically ceases at 0°C to 5°C (273 K to 278 K), though the bacilli are not killed even at very low temperatures. The bacterial content can be eliminated to a great extent by heating the milk to 62°C (335 K) and holding it at that temperature for about 30 minutes. Thereafter to minimize the bacterial growth and preservation, the milk is cooled to 4°C to 5°C (277 K to 278 K). This process of heating and immediately cooling the milk for controlling the bacterial growth is known as *pasteurization.*

The pasteurization is generally done in a batch type process. In this process, the raw milk is heated by hot water or steam to 62°C. In case the heating is done by hot water, it is sprayed around the outside lining of the vat by a distributor, which gets collected in a sump at the bottom of the vat, reheated and once again sprayed. The steam heating is carried out by flowing steam in the space provided between lining and the casing of the vat. The heated milk is then cooled, first by the cooling tower water and then by the chilled water or brine to 4°C to 5°C, which is the desired temperature for filling the milk in bottles. The heating and cooling is done by passing the milk (by milk pumps) through the heat exchange plates. The milk flows between the two plates and the hot water, cooling tower water or chilled water (brine) is circulated through alternate pairs of plates. The direction of flow of heating or cooling fluids is opposite to that of milk to obtain better heat transfer.

In order to control the fat content of milk, it is desired to churn the milk. Such milk is retailed as 'toned milk'. The fat thus removed is processed as butter and stored at 4°C to 5°C. The cheese is another product from milk and is stored at about 4°C.

The ice-cream is another product, which is manufactured by using milk fat along with sugar and other ingredients of ice-cream. The mixture is pasteurized to a temperature of 70°C to 80°C (343 K to 353 K) and homogenized. The homogenization is the process of breaking down the fat and solid content of the mixture to a very finely divided state to avoid any lump formation while freezing the mixture. The plate type heat exchangers are used for heating and cooling the liquid mix to 5°C. Then the mixture is frozen in ice-cream freezers. During freezing, the ice-cream is whipped to mix certain amount of air. Generally the ice-cream is frozen to about – 2.5°C to – 5°C (270.5 K to 268 K), depending upon the type of mixture used. Thereafter, it is hardened in the hardening rooms to a temperature of –18°C (255 K). The hardening rooms are maintained at about –30°C to –35°C (243 K to 238 K) for rapid freezing of the balance water content in the ice-cream. The rapid freezing ensures small size of the ice crystals.

22.13 Cold Storages

The *cold storage is a building designed to store certain goods like food stuffs, fruits, vegetables, and dairy products within well defined temperature range and relative humidity (RH).The cold storage is also an application of air conditioning, in a way that the air is cooled by passing it over a cooling coil of refrigeration plant and supplied back to the room.

The temperature and humidity conditions maintained inside a cold storage depend upon the type of product stored inside it, *e.g.* vegetables require the maintenance of a temperature of around 0°C to 5°C (273 K to 278 K) with high RH of 80 to 90%, 4°C to 5°C (277 K to 278 K) for milk processing, −25°C to −30°C (248 K to 243 K) for quick freezing of fish, −20°C to −45°C (253 K to 228 K) for chlorine liquifier, etc. Thus, the conditions required for storage can be divided into the following two categories :

(*a*) Cold storages for products which are to be maintained at temperatures of 0°C and above, and

(*b*) Cold storages for products which are to be maintained at temperatures below 0°C.

It may be noted that the refrigeration does not improve the quality of food products, it only slows down its deterioration. The product must be under refrigeration for the entire course of the passage from the producer to the consumer and this continuity is known as the cold chain. During storage, the fresh vegetables and fruits produce *'heat of respiration'*. Thus, the refrigeration plant must be designed to take care of this load in addition to the usual heat loads, load due to heat released on defrosting etc.

Cold storage.

22.14 Quick Freezing

The rapidly frozen foods have much longer life than foods stored in cold storages. The products that are generally preserved by quick freezing are meat, fish, some vegetables such as peas, spinach, cauliflower, carrots, beans etc. During the process of quick freezing, the product temperature drops down fast to the freezing temperature of 0°C. Thereafter, the latent heat of water content in the product is removed. During this process, the temperature of the product does not fall . For complete details on quick freezing refer Chapter 13 (Art.13.10).

* For complete details on Cold Storages, refer Chapter 13 (Art. 13.8).

22.15 Cooling and Heating of Foods

The process of sudden cooling and heating of food products is related with transient state of the process because they are to be preserved for very short duration. Therefore, food technologists are concerned with the heat transfer during transient period which involves study of unsteady state heat transfer situation. The analytical solution of the differential equation defining such situation have been plotted in a series of graphs called *Gurnie – Lurie chart.*

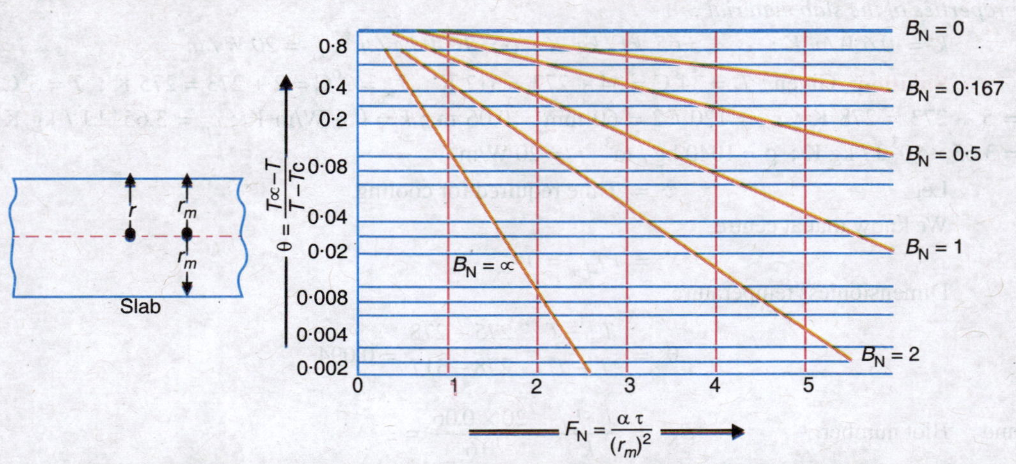

Fig. 22.8. Gurnie – Lurie chart for cooling or heating of a slab of food.

Fig. 22.8 represents the solution of transient conditions involved in cooling or heating of an infinite slab of food item. The figure gives a plot of dimensionless temperature defined by

$$\theta = \frac{T_\infty - T}{T - T_i}$$

where T_∞ = Ambient temperature,

T = Temperature after elapse of a time period τ, and

T_i = Initial uniform temperature of the slab against dimensionless time defined by the Fourier number F_N.

Mathematically, Fourier number,

$$F_N = \frac{k\,\tau}{\rho\, c_p\,(r_m)^2} = \frac{\alpha\,\tau}{(r_m)^2}$$

where α = Thermal diffusivity of the material = $k / \rho\, c_p$,

k = Thermal conductivity of the slab material,

ρ = Density of the slab,

τ = Time of heating or cooling, and

r_m = Half of the thickness of infinite slab.

The parameter X defines the dimensionless distance,

$$X = \frac{r}{r_m}$$

where r = Distance from the centre line to the point under consideration.

and the parameter B_N (Biot number) defines, the ratio of the convective resistance to the conductive resistance. Mathematically,

$$B_N = \frac{f_g \cdot r_m}{k}$$

where f_g = Heat transfer coefficient at slab surface.

Example 22.2. *A thin slab of meat is to be cooled from 34°C to a temperature of 5°C with the help of a cooling medium at a temperature of 2°C. Determine the time required at the centre of a thin slab when the slab is 120 mm thick and to be cooled from both the sides. Following are the heat properties of the slab material :*

$$k = 0.6 \text{ W/m K}; \; c_p = 3.65 \text{ kJ/kg K}; \; \rho = 1040 \text{ kg/m}^3; \; f_g = 20 \text{ W/m}^2$$

Solution. Given : $T_i = 34°C = 34 + 273 = 317$ K ; $T_\infty = 2°C = 2 + 273 = 275$ K ; $T = 5°C = 5 + 273 = 278$ K ; $r_m = 120 / 2 = 60$ mm $= 0.06$ m ; $k = 0.6$ W/m K ; $c_p = 3.65$ kJ / kg K $= 3.65 \times 10^3$ J / kg K ; $\rho = 1040$ kg / m^3 ; $f_g = 20$ W/m^2

Let τ = Time required for cooling.

We know that at centre,

$$X = r / r_m = 0$$

Dimensionless temperature,

$$\theta = \frac{T_\infty - T}{T - T_i} = \frac{275 - 278}{278 - 317} = 0.094$$

and Biot number,

$$B_N = \frac{f_g \cdot r_m}{k} = \frac{20 \times 0.06}{0.6} = 2$$

From the Gurnie-Lurie chart (shown in Fig. 22.8), we find that corresponding to $\theta = 0.094$ and $B_N = 2$, the value of Fourier number (F_N) is 2.2.

We know that

$$F_N = \frac{k \tau}{\rho \, c_p \, (r_m)^2} = \frac{0.6 \, \tau}{1040 \, (3.65 \times 10^3) \, (0.06)^2} = 0.044 \times 10^{-3} \, \tau$$

$$\therefore \quad \tau = \frac{F_N}{0.044 \times 10^{-3}} = \frac{2.2}{0.044 \times 10^{-3}} = 50 \times 10^3 \text{ s}$$

$$= 50 \times 10^3 / 3600 = 13.9 \text{ h} \quad \textbf{Ans.}$$

22.16 Freeze Drying

The freeze drying is the process of liquid separation from a product in a frozen state, achieved by sublimation under very low pressures near to vacuum. The sublimation is the process by virtue of which a substance in a solid state gets directly converted into a vapour state. This will happen when pressure of surroundings is below the triple point pressure.

The sublimation serves to obtain a product that retains even its volatile components and initial quality and the vacuum is used to maintain the physical state as frozen and to direct the vapour flow. At present, the main problems to the applications of the process, in general, are as follows:

1. The relatively high cost of the freeze-dried product by sublimation-dehydration due to high vacuum,
2. The very low temperature refrigeration to prefreeze the product and then to condense the sublimated vapour,
3. The complicated operational control,
4. The long duration of the freeze drying time.

However, the cost is not the determining factor for the manufacture of certain products including blood plasma, life saving drugs such as gamma-globulin, high value food products such as mushrooms, shrimp, prawns etc.

Two models of the freeze drying process, one with radiant heating and the other with microwave heating are shown in Fig. 22. 9 and Fig. 22.10. The thick line shows the temperature distribution. The dehydration takes place in the vacuum chamber. The pressure in the chamber must be maintained below the triple-point pressure, normally below 5 mm of Hg, otherwise the product will begin to thaw.

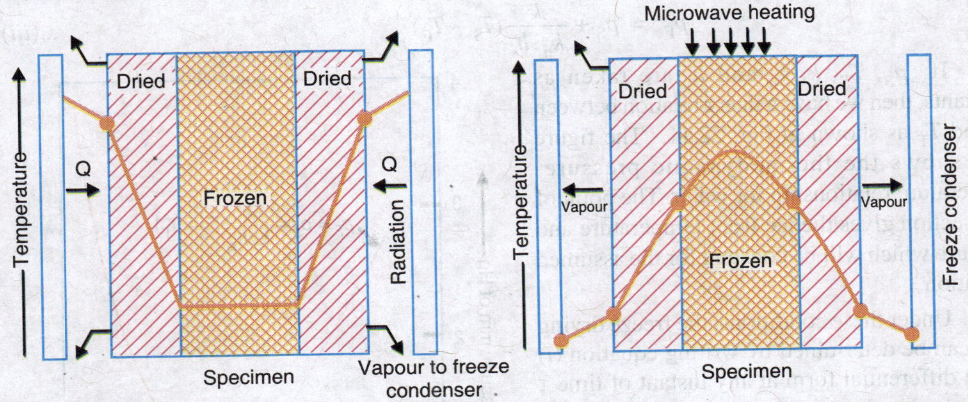

Fig. 22.9. Temperature distribution during freeze drying of a specimen with radiant heating.

Fig. 22.10. Temperature distribution during freeze drying of a specimen with microwave heating.

The rate of dehydration is governed by either one of the following factors :

(*a*) The rate of heat transfer from the heat source to the ice front, or

(*b*) The rate of vapour diffusion from the ice front to the freeze condenser.

Normally, the rate of dehydration is controlled by the transfer of heat which may be supplied by either a radiant-heat source, or heating platens sandwiching the product, but having pores to permit the sublimated vapours to escape. In such a case, there is an obvious similarity between freezing and freeze drying. Neuman and Plank's solutions for freezing can, therefore, be used to calculate freeze-drying time as well. A simple heat and mass transfer model is, however, presented below. Numerical methods can also be successfully used. Efforts are also being made to use microwave heating to improve heat transfer as well as vapour diffusion as is evident from Fig. 22.10.

It must be mentioned that the heat input is limited by the maximum allowable temperature of the dried material at the surface so that the product is not damaged by overheating.

22.17 Heat and Mass Transfer through the Dried Material

Let T_S = Maximum allowable surface temperature of the material,

T_F = Ice front temperature,

p_S = Constant partial pressure of water vapour in the drying chamber,

p_F = Ice front pressure,

A = Cross-sectional area,

x = Thickness of the dried material at any instant of time,

k = Thermal conductivity of the dried material, and

k_d = Diffusion coefficient or permeability of the dried material.

We know that rate of heat transfer,

$$Q = \frac{k\,A}{x}(T_S - T_F) \qquad \ldots(i)$$

and the rate of mass transfer,

$$m = \frac{k_d A}{x}(p_F - p_S) \qquad ...(ii)$$

Assuming that the rate of heat transfer is equal to the rate of mass transfer multiplied by the latent heat of sublimation h_s, we obtain the energy balance from equations (*i*) and (*ii*) as follows :

$$p_F = p_S + \frac{k}{k_d . h_s}(T_S - T_F) \qquad ...(iii)$$

If p_S, T_S, k, k_d and h_s are taken as constants, then we have a linear relation between p_F and T_F as shown in Fig. 22.11. The figure also shows the thermodynamic pressure-temperature relationship for water. The point of intersection gives the ice front temperature and pressure which will be constant for the assumed condition.

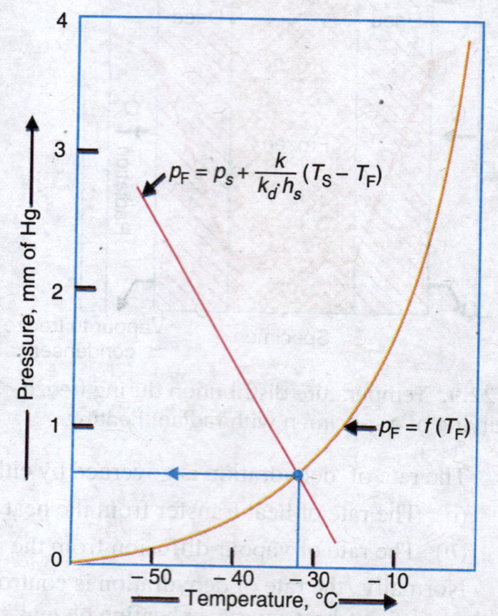

Fig. 22.11. Graphical representation of ice front temperature and vapour pressure during freeze drying.

Under these conditions, the freeze drying time can be determined by writing equation (*i*) in the differential form at any instant of time τ when the ice front is at a distance *x* from the surface, *i.e.*

$$Q\,d\tau = k A (T_S - T_F)\frac{d\tau}{dx}$$

$$= \rho A x (w_i - w_f)\, h_s \qquad ...(iv)$$

where ρ = Density of solids in the dried material,

w_i = Initial moisture content per unit mass of solids, and

w_f = Final moisture content per unit mass of solids.

The integration of equation (*iv*) with ($T_S - T_F$) constant, yields for the drying time,

$$\tau = \frac{x^2\, e(w_i - w_f)\, h_s}{8k\,(T_S - T_F)}$$

It may be noted that the largest source of error in the above method is in generalizing the surface temperature and assuming it to be constant. In actual practice, as the dried layer thickness grows, the surface temperature has to be increased to overcome the increased thermal resistance. Similarly, the ice front temperature is also affected by the pressure drop through the dried layer. Numerical techniques can be used to obtain a more accurate result for the drying time. As an approximate method, the Neuman solution for freezing can also be applied to freeze drying.

QUESTIONS

1. Describe the functioning of a domestic refrigerator. Explain the various controls in a refrigerator.
2. Describe the different types of water coolers.
3. Explain the various industrial applications of air-conditioning system.
4. Write short notes on :

   (*a*) Room air conditioner ; (*b*) Refrigerated trucks; and (*c*) Marine air-conditioning
5. Discuss heat and mass transfer through dried material.

OBJECTIVE TYPE QUESTIONS

1. The commonly used refrigerant in a domestic refrigerator is

(*a*) R - 11 (*b*) R - 12 (*c*) R - 22 (*d*) NH_3

2. The expansion of the refrigerant in domestic refrigerator is carried out in

(*a*) accumulator (*b*) drier (*c*) capillary tube (*d*) none of these

3. The room air conditioner controls the

(*a*) temperature of the air (*b*) temperature and humidity of the air

(*c*) temperature and dust of air (*d*) none of these

4. The commonly used refrigerant in ice plant is

(*a*) NH_3 (*b*) CO_2 (*c*) R–12 (*d*) none of these

5. The temperature maintained in the brine tank of ice plant is

(*a*) –10°C (*b*) –15°C (*c*) –18°C (*d*) –20°C

6. The milky white ice is obtained if

(*a*) air is present in it (*b*) impurities are present in water

(*c*) CO_2 is present in it (*d*) none of these

7. The dry ice is produced by

(*a*) drying the ice (*b*) keeping ice in an insulated chamber

(*c*) by solidifying liquid CO_2 (*d*) none of these

8. The milk is stored at a temperature of

(*a*) – 5°C (*b*) 4°C (*c*) 10°C (*d*) 12°C

9. The process of heating and immediately cooling the milk for controlling the bacterial growth is known as

(*a*) pasteurization (*b*) regeneration (*c*) blending (*d*) none of these

10. Ice of potable water can crack if frozen at a temperature lower than

(*a*) –12°C (*b*) –15°C (*c*) –20°C (*d*) –25°C

ANSWERS

1. (*b*)	**2.** (*c*)	**3.** (*b*)	**4.** (*a*)	**5.** (*a*)
6. (*a*)	**7.** (*c*)	**8.** (*b*)	**9.** (*a*)	**10.** (*a*)

INDEX